Some Important Organic Functional Groups

	Functional Group*	Example	IUPAC Name
Acid anhydride	$\overset{:O:}{\underset{}{-C}}-\overset{:O:}{\underset{}{O}}-\overset{:O:}{\underset{}{C}}-$	$\underset{O}{CH_3C}\underset{O}{OCCH_3}$	Ethanoic anhydride (Acetic anhydride)
Acid chloride	$\overset{:O:}{\underset{}{-C}}-\ddot{C}\ddot{l}:$	$\underset{O}{CH_3C}Cl$	Ethanoyl chloride (Acetyl chloride)
Alcohol	$-\ddot{O}H$	CH_3CH_2OH	Ethanol (Ethyl alcohol)
Aldehyde	$\overset{:O:}{\underset{}{-C}}-H$	$\underset{O}{CH_3C}H$	Ethanal (Acetaldehyde)
Alkane	-----	CH_3CH_3	Ethane
Alkene	$\underset{}{C}=\underset{}{C}$	$CH_2=CH_2$	Ethene (Ethylene)
Alkyne	$-C\equiv C-$	$HC\equiv CH$	Ethyne (Acetylene)
Amide	$\overset{:O:}{\underset{}{-C}}-\ddot{N}\underset{}{-}$	$\underset{O}{CH_3C}NH_2$	Ethanamide (Acetamide)
Amine, primary	$-\ddot{N}H_2$	$CH_3CH_2NH_2$	Ethylamine
Amine, secondary	$-\ddot{N}H-$	$(CH_3CH_2)_2NH$	Diethylamine
Amine, tertiary	$-\ddot{N}\underset{}{-}$	$(CH_3CH_2)_3N$	Triethylamine

	Functional Group*	Example	IUPAC Name
Carboxylic acid	$\overset{:O:}{\underset{}{-C}}-\ddot{O}-H$	$\underset{O}{CH_3C}OH$	Ethanoic acid (Acetic acid)
Disulfide	$-\ddot{S}-\ddot{S}-$	CH_3SSCH_3	Dimethyl disulfide
Epoxide	$\overset{:O:}{\underset{}{C}}\underset{}{-C}$	H_2C-CH_2 (epoxide ring)	Oxirane (Ethylene oxide)
Ester	$\overset{:O:}{\underset{}{-C}}-\ddot{O}-C-$	$\underset{O}{CH_3C}OCH_3$	Methyl ethanoate (Methyl acetate)
Ether	$-\ddot{O}-$	CH_3OCH_3	Dimethyl ether
Haloalkane	$-\ddot{X}:$ $X = F, Cl, Br, I$	CH_3CH_2Cl	Chloroethane (Ethyl chloride)
Ketone	$\overset{:O:}{\underset{}{-C}}-$	$\underset{O}{CH_3C}CH_3$	Propanone (Acetone)
Nitrile	$-C\equiv N:$	$CH_3-C\equiv N$	Ethanenitrile (Acetonitrile)
Nitro	$-\overset{:\ddot{O}:^-}{\underset{:\ddot{O}:}{N^+}}$	CH_3NO_2	Nitromethane
Phenol	(benzene ring with $-\ddot{O}H$)	(benzene ring with $-OH$)	Phenol
Sulfide	$-\ddot{S}-$	CH_3SCH_3	Dime...
Thiol	$-\ddot{S}-H$	CH_3CH_2SH	Etha... (Eth...

* Where bonds to an atom are not specified, the atom is assumed to be bonded to one or more carbon or hydrogen atoms in the rest of the molecule.

Online Web Learning
UMassAmherst

Get a Better Grade in Chemistry!

Log in now to the leading online learning system for chemistry. Score better on exams, get homework help, and more!

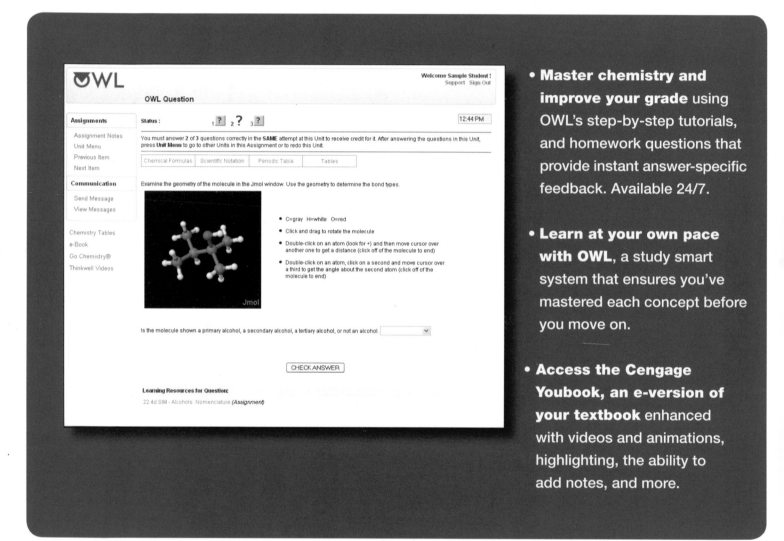

- **Master chemistry and improve your grade** using OWL's step-by-step tutorials, and homework questions that provide instant answer-specific feedback. Available 24/7.

- **Learn at your own pace with OWL**, a study smart system that ensures you've mastered each concept before you move on.

- **Access the Cengage Youbook, an e-version of your textbook** enhanced with videos and animations, highlighting, the ability to add notes, and more.

To get started, use the access code that may have been packaged with your text or purchase access online. Check with your instructor to verify that OWL is required for your course before purchasing.

www.cengage.com/OWL

Organic Chemistry Hybrid

SIXTH EDITION

William H. Brown
Beloit College

Christopher S. Foote
University of California, Los Angeles

Brent L. Iverson
University of Texas, Austin

Eric V. Anslyn
University of Texas, Austin

Chapter 29 was originally contributed by
Bruce M. Novak
North Carolina State University

BROOKS/COLE
CENGAGE Learning

Australia • Brazil • Japan • Korea • Mexico • Singapore • Spain • United Kingdom • United States

BROOKS/COLE
CENGAGE Learning

Organic Chemistry Hybrid, **Sixth Edition**
William H. Brown, Christopher
S. Foote, Brent L. Iverson, Eric V. Anslyn

Executive Editor: Lisa Lockwood

Senior Developmental Editor: Sandra Kiselica

Assistant Editor: Elizabeth Woods

Editorial Assistant: Laura Bowen

Senior Media Editor: Lisa Weber

Media Editor: Stephanie Van Camp

Senior Marketing Manager: Barb Bartoszek

Marketing Assistant: Julie Stefani

Marketing Communications Manager:
Linda Yip

Content Project Manager: Teresa L. Trego

Design Director: Rob Hugel

Art Director: John Walker

Print Buyer: Judy Inouye

Rights Acquisitions Specialist:
Tom McDonough

Production Service: PreMediaGlobal

Text Designer: Ellen Pettengell

Photo Researcher: Bill Smith Group

Copy Editor: PreMediaGlobal

OWL producers: Stephen Battisti, Cindy
Stein, David Hart (Center for Educational
Software Development, University of
Massachusetts, Amherst)

Illustrator: Greg Gambino, PreMediaGlobal

Cover Designer: RHDG | Riezebos Holzbaur

Cover Image: © Corbis Images/Tobias
Bernhard

Compositor: PreMediaGlobal

© 2012, 2009 Brooks/Cole, Cengage Learning

ALL RIGHTS RESERVED. No part of this work covered by the copyright
herein may be reproduced, transmitted, stored, or used in any form
or by any means graphic, electronic, or mechanical, including but
not limited to photocopying, recording, scanning, digitizing, taping,
Web distribution, information networks, or information storage
and retrieval systems, except as permitted under Section 107 or 108
of the 1976 United States Copyright Act, without the prior written
permission of the publisher.

For product information and technology assistance, contact us at
Cengage Learning Customer & Sales Support, 1-800-354-9706.

For permission to use material from this text or product,
submit all requests online at **www.cengage.com/permissions.**
Further permissions questions can be e-mailed to
permissionrequest@cengage.com.

Library of Congress Control Number: 2010939137

ISBN-13: 978-1-111-98776-3
ISBN-10: 1-111-98776-9

Brooks/Cole
20 Davis Drive
Belmont, CA 94002-3098
USA

Cengage Learning is a leading provider of customized learning
solutions with office locations around the globe, including Singapore,
the United Kingdom, Australia, Mexico, Brazil, and Japan. Locate your
local office at **www.cengage.com/global.**

Cengage Learning products are represented in Canada by Nelson
Education, Ltd.

To learn more about Brooks/Cole, visit **www.cengage.com/brookscole**

Purchase any of our products at your local college store or at our
preferred online store **www.cengagebrain.com.**

Printed in China by China Translation & Printing Services Limited
2 3 4 5 6 7 14 13 12 11

Dedication

This Sixth Edition is dedicated to the memory of our dear friend and colleague, Christopher Foote. Chris' insights, encouragement, and dedication to this project can never be replaced. His kind and nurturing spirit lives on in all who are lucky enough to have known him.

About the Authors

WILLIAM H. BROWN is an Emeritus Professor of Chemistry at Beloit College, where he has twice been named Teacher of the Year. His teaching responsibilities included organic chemistry, advanced organic chemistry, and special topics in pharmacology and drug synthesis. He received his Ph.D. from Columbia University under the direction of Gilbert Stork and did postdoctoral work at California Institute of Technology and the University of Arizona.

CHRISTOPHER S. FOOTE received his B.S. from Yale University and his Ph.D. from Harvard University. His scholarly credits include Sloan Fellow; Guggenheim Fellow; ACS Baekland Award; ACS Cope Scholar; Southern California Section ACS Tolman Medal; President, American Society for Photobiology; and Senior Editor, Accounts of Chemical Research. He was a Professor of Chemistry at UCLA.

BRENT L. IVERSON received his B.S. from Stanford University and his Ph.D. from the California Institute of Technology. He is a University Distinguished Teaching Professor at The University of Texas, Austin as well as a respected researcher. Brent's research spans the interface of organic chemistry and molecular biology. His group has developed several patented technologies, including an effective treatment for anthrax.

ERIC V. ANSLYN is a University Distinguished Teaching Professor at The University of Texas at Austin. He earned his bachelor's degree from California State University, Northridge, his Ph.D. from the California Institute of Technology and did postdoctoral work at Columbia University under the direction of Ronald Breslow. Eric has won numerous teaching awards and his research focuses on the physical and bioorganic chemistry of synthetic and natural receptors and catalysts.

Contents in Brief

Contents

Chapter 6 Reactions of Alkenes   175

Chapter 7 Alkynes   227

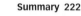

Chapter 8 Haloalkanes, Halogenation, and Radical Reactions   250

Chapter 20 Dienes, Conjugated Systems, and Pericyclic Reactions 657

Chapter 21 Benzene and the Concept of Aromaticity 692

Chapter 22 Reactions of Benzene and Its Derivatives 728

Chapter 23 Amines 755

Appendices:

List of Mechanisms

Chapter 6 Reactions of Alkenes

Electrophilic Addition of HBr to 2-Butene (**Section 6.3A**)

Acid-Catalyzed Hydration of Propene (**Section 6.3B**)

Carbocation Rearrangement in the Addition of HCl to an Alkene (**Section 6.3C**)

Addition of Bromine with Anti Stereoselectivity (**Section 6.3D**)

Halohydrin Formation and Its Anti Stereoselectivity (**Section 6.3E**)

Oxymercuration-Reduction of an Alkene (**Section 6.3F**)

Hydroboration (**Section 6.4**)

Oxidation of a Trialkylborane by Alkaline Hydrogen Peroxide (**Section 6.4**)

Formation of an Ozonide (**Section 6.5B**)

Chapter 7 Alkynes

Addition of HBr to an Alkyne (**Section 7.6B**)

$HgSO_4/H_2SO_4$ Catalyzed Hydration of an Alkyne (**Section 7.7B**)

Reduction of an Alkyne by Sodium in Liquid Ammonia (**Section 7.8C**)

Chapter 8 Haloalkanes, Halogenation, and Radical Reactions

Radical Chlorination of Ethane (**Section 8.5B**)

Allylic Bromination of Propene Using NBS (**Section 8.6A**)

Radical-Initiated Non-Markovnikov Addition of HBr to Alkenes (**Section 8.8**)

Chapter 9 Nucleophilic Substitution and β-Elimination

An S_N2 Reaction (**Section 9.2A**)

An S_N1 Reaction (**Section 9.2B**)

Rearrangement During Solvolysis of 2-Chloro-3-phenylbutane (**Section 9.3F**)

E1 Reaction of 2-Bromo-2-methylpropane (**Section 9.6A**)

E2 Reaction of 2-Bromopropane (**Section 9.6B**)

E2 Reaction of meso-1,2-Dibromo-1,2-diphenylethane (**Section 9.7C**)

E2 Reaction of the Enantiomers of 1,2-Dibromo-1,2-diphenylethane (**Section 9.7C**)

E2 Reaction of cis-1-Chloro-2-isopropylcyclohexane (**Section 9.7C**)

Hydrolysis of a Sulfur Mustard—Participation by a Neighboring Group (**Section 9.10**)

Chapter 10 Alcohols

Reaction of a 3° Alcohol with HBr—An S_N1 Reaction (**Section 10.5A**)

Reaction of a 1° Alcohol with HBr—An S_N2 Reaction (**Section 10.5A**)

Rearrangement upon Treatment of Neopentyl Alcohol with HCl (**Section 10.5A**)

Chapter 17 Carboxylic Acids

Formation of a Methyl Ester Using Diazomethane (Section 17.7B)

Decarboxylation of a β-Ketocarboxylic Acid (Section 17.9A)

Decarboxylation of a β-Dicarboxylic Acid (Section 17.9B)

Chapter 18 Functional Derivatives of Carboxylic Acids

Fischer Esterification (Section 18.3E)

Hydrolysis of an Acid Chloride (Section 18.4A)

Hydrolysis of an Ester in Aqueous Base (Saponification) (Section 18.4C)

Hydrolysis of an Amide in Aqueous Acid (Section 18.4D)

Hydrolysis of an Amide in Aqueous Base (Section 18.4D)

Hydrolysis of a Cyano Group to an Amide in Aqueous Base (Section 18.4E)

Reaction of an Acid Chloride and Ammonia (Section 18.6A)

Reaction of an Ester with a Grignard Reagent (Section 18.9A)

Reduction of an Ester by Lithium Aluminum Hydride (Section 18.10A)

Reduction of an Amide by Lithium Aluminum Hydride (Section 18.10B)

Chapter 19 Enolate Anions and Enamines

Base-Catalyzed Aldol Reaction (Section 19.2A)

Acid-Catalyzed Aldol Reaction (Section 19.2A)

Acid-Catalyzed Dehydration of an Aldol Product (Section 19.2A)

Claisen Condensation (Section 19.3A)

Alkylation of an Enamine (Section 19.5A)

Michael Reaction—Conjugate Addition of Enolate Anions (Section 19.8A)

Chapter 20 Dienes, Conjugated Systems, and Pericylic Reactions

1,2- and 1,4-Addition to a Conjugated Diene (Section 20.2A)

The Claisen Rearrangement (Section 20.6A)

The Cope Rearrangement (Section 20.6B)

Chapter 21 Benzene and the Concept of Aromaticity

Kolbe Carboxylation of Phenol (Section 21.4E)

Chapter 22 Reactions of Benzene and its Derivatives

Electrophilic Aromatic Substitution—Chlorination (Section 22.1A)

Formation of the Nitronium Ion (Section 22.1B)

Friedel-Crafts Alkylation (Section 22.1C)

Friedel-Crafts Acylation—Generation of an Acylium Ion (Section 22.1C)

Nucleophilic Aromatic Substitution via a Benzyne Intermediate (Section 22.3A)

Nucleophilic Aromatic Substitution by Addition-Elimination (Section 22.3B)

Preface

Thank you for purchasing the hybrid edition of "Organic Chemistry." This trimmer version of the book includes no end-of-chapter questions. Instead, because you are using OWL (Online Web Learning), all end-of-chapter questions are assignable online in OWL, along with tutorials, mastery questions, visualizations, and the new, enhanced, and customizable Cengage YouBook. Your students can print questions from OWL with correct answers and valuable problem-solving feedback to help guide them to superior understanding and mastery of concepts.

The reduced page count in this book allows us to offer it at a lower price to students as well as lessen the environmental impact of this edition, and still provide a useful print product.

Introduction

This Sixth Edition of *Organic Chemistry* significantly extends the transformation started in the Fifth Edition. Students taking an organic chemistry course have two objectives; the first is to learn organic chemistry, and the second is to establish the intellectual foundation for other molecular science courses. Most often, these other courses involve biochemistry or specialized topics such as materials science. This textbook addresses these two objectives head-on by first presenting mechanistic and synthetic organic chemistry geared toward giving students a fundamental understanding of organic molecules and reactions, as well as their mechanisms and uses in organic synthesis. The text then builds upon the fundamentals by providing an emphasis on bridging concepts that will prepare students for subsequent science courses. Several unique elements including comprehensive end of chapter summaries, a new paradigm for learning mechanisms and a new learning tool we call Organic Chemistry Road Maps have been included to increase the efficiency of student studying and learning.

Making Connections

All the important reaction mechanistic and synthetic details are still found throughout the text, but we have increased the important connections between different reaction mechanisms. The intent is to make the study of organic chemistry a process involving the learning and application of fundamental principles and not an exercise in memorization. Throughout this edition, the uniting concept of nucleophiles reacting with electrophiles is highlighted. Especially helpful is the use of electrostatic potential surface models of reacting molecules. These maps emphasize, in an easily interpreted, color-coded fashion, how the majority of reactions involve areas of higher electron density on one reactant (a nucleophile) interacting with areas of lower electron density on the other reactant (an electrophile).

A Fresh Look at Mechanisms

Starting in Chapter 6, this edition introduces a revolutionary new paradigm for learning organic chemistry mechanisms. Students are introduced to a small set of individual mechanism elements in Chapter 6. The mechanism elements are explained in detail, including when they are appropriate to use. Reaction mechanisms throughout the book are then described as combinations of these individual mechanism elements, which are written in stepwise fashion. This new approach not only simplifies the learning of mechanisms for students, it makes it easier for them to recognize similarities and differences among related reactions. Most important, it makes the *prediction* of reaction mechanisms simpler, analogous to a multiple choice situation in which the correct mechanism element is chosen from a small menu of choices. Also, to give students more confidence with writing mechanisms, Appendix 10 on *Common Mistakes in Arrow Pushing* has been added. In addition, many mechanisms (particularly in Chapter 18) have been broken down into more descriptive steps to make it easier for students to understand the concepts involved.

A Fresh Look at Orbitals

An organic chemist's theoretical framework for understanding electron density within molecules is based on atomic and molecular orbitals. Paradoxically, organic chemistry texts generally provide only passing coverage of orbitals, never revealing their true shapes or full significance. The Sixth Edition is the first organic text to paint a detailed picture of the orbital nature of electron density in Chapter 1 by focusing on the interplay between the two complementary approaches to orbital descriptions, valence bond theory and molecular orbital theory. Chapter 1 provides a comprehensive description of how electronic theory is used by organic chemists to understand structure, bonding, and reactivity. Significantly, students are given easy-to-use guidelines that detail *when* and *how* to use electronic theory, even in complex situations, such as molecules described by multiple resonance contributing structures. The inclusion of calculated orbital diagrams side-by-side with the familiar orbital cartoons gives students a greater appreciation for orbital sizes and shapes that are reinforced throughout the book. The intent is to provide students with a strong theoretical foundation that will give them unprecedented insight and intuition into molecular structure and reactivity.

Mastering Skills

Mastering organic chemistry requires the development of certain intellectual skills. To this end, eleven *How To* boxes highlight "survival skills" for organic chemistry students. Topics include, *How To Draw Alternative Chair Conformations of Cyclohexanes* (Section 2.5), *How To Draw Curved Arrows and Push Electrons* (Section 1.8), and *How to Write Mechanisms for Interconversions of Carboxylic Acid Derivatives* (Section 18.4).

Helping Students Prepare More Efficiently

A key feature of the Sixth Edition is the **end-of-chapter summaries**, which are mini study guides designed to help students prepare for class exams and later for standardized tests such as the MCAT. When preparing for exams, students will benefit from the bulleted lists of important concepts with highlighted keywords. These mini study guides make it easier for students to identify hard-to-grasp material by referring them to the sections of the text for a full explanation and then providing them with end-of-chapter problems that test and reinforce their comprehension.

As a companion to the summary outlines, newly expanded **end-of-chapter summaries of reactions** systematically list the reactions covered in each chapter. These include prose descriptions of mechanisms as well as important information such as observed stereochemistry or regiochemistry. Students will find these reaction summaries particularly efficient when preparing for exam questions requiring application of reactions in the context of new molecules or even multi-step syntheses.

The appendix reference material has been enhanced with two unique items to provide students with a quickly accessible source of important information. The first is a thorough list of stereochemical definitions (Appendix 8). **Stereochemical**

terms are subtle and difficult to master, so having them compiled in one location allows students to compare and contrast any new terms with those learned in earlier chapters, as well as prepare for exams. In addition, Appendix 9, *Summary of the Rules of Nomenclature* provides a practical listing of the nomenclature rules described throughout the text. In response to student requests, this appendix provides a single location for the rules students need when naming complex molecules that contain multiple functional groups.

Organic Chemistry Applied to the Synthesis of Important Molecules

Organic chemistry enables the synthesis of thousands of useful molecules. Synthetic applications of the reactions covered in this text are emphasized throughout, partly through the many new challenging *Synthesis* problems, the goal of which is to demonstrate to students how synthetic organic chemistry is used in pharmaceutical research and in the production of useful pharmaceuticals. The text provides applications of the reactions to the synthesis of important molecules such as Valium, fluoxetine (Prozac), meperidine (Demerol), albuterol (Proventil), tamoxifene, and sildefanil (Viagra). *Multi-Step Synthesis* problems challenge students to develop their own multi-step synthetic plan for converting a relatively simple starting material into a more complex target molecule. Multi-step synthesis is supported by an expanded description of retrosynthetic analysis in multiple chapters, including tips on recognizing when to use certain reactions, such as those involving enolates in the construction of complex structures.

Road Maps for Organic Chemistry, an Innovative and Powerful Way to Visualize Organic Reactions

In this Sixth Edition, we introduce an innovation in organic chemistry learning that we refer to as the **Organic Chemistry Road Map**. It is a graphical representation of the different organic reactions taught in the context of the important functional groups. The functional groups of an organic chemistry road map are analogous to cities on a real road map, and the reactions are like the roads between those cities. Arrows are used to represent routes that are known between functional groups, and the reagents required to bring about each reaction are written next to the corresponding arrow. Multi-step synthesis questions are often the most challenging for organic chemistry students even though synthesis is at the core of organic chemistry as a discipline. The power of an Organic Chemistry Road Map is that it helps students visualize the reactions needed to interconvert key functional groups in multi-step synthesis problems. The construction and use of Organic Chemistry Road Maps are introduced in the end-of-chapter problems beginning in Chapter 6 and is presented in complete form in a new Appendix 11.

Organic Chemistry Applied to Biology

The application of organic chemistry principles to important biological molecules is integrated where appropriate to establish a bridge with biochemistry courses. In particular, *Connections to Biological Chemistry* gives special attention to those aspects of organic chemistry that are essential to an understanding of the chemistry of living systems. For example, the organic chemistry of amino acids is highlighted beginning in Section 3.9, along with the importance of alkene geometry to both membrane fluidity and nutrition (Section 5.4). How hydrogen bonding is involved with drug-receptor interactions (Section 10.2) is discussed. Importantly, these *Connections to Biological Chemistry* features have been added throughout the book, not just at the end, in recognition of the fact that not all instructors make it through the biological chemistry chapters at the end of the text. Relevance to practical application is also emphasized in an expanded array of essays titled **Chemical Connections**. Topics include: medicines like penicillins and cephalosporins (Section 18.1), food supplements like antioxidants (Section 8.7) and materials science concepts such as spider silk (Section 27.6). These sections provide a bridge between the theory of Organic Chemistry and well-known, current, practical applications. A list of the *Chemical Connections* as well as *Connections to Biological Chemistry* essays can be found on the inside back cover of this text.

Unique Organizational Elements

❯ Together, Chapter 1 (comprehensive description of electronic theory) and Chapter 3 (detailed description of acids and bases in organic chemistry) provide a fundamental grasp of molecular structure and properties, giving students the basis to *understand all* aspects of the mechanistic discussions that follow. Equipping students with the proper tools from the beginning gives them a predictive command of reactivity and foster chemical intuition, while discouraging superficial memorization.

❯ Because of the increased use of NMR spectroscopy in chemical and biochemical research, as well as the growing dependence on MRI for medical diagnosis, Chapter 13, Nuclear Magnetic Resonance Spectroscopy, is detailed and up-to-date. The practical and theoretical aspects concerning NMR spectra and signal splitting patterns are highlighted and a complete description of FT-NMR provides a stronger technical connection to MRI.

❯ Carbonyl chemistry (Chapters 16–19) is placed earlier than most texts so professors have the time to teach this material to the majority of students in an organic chemistry class, who are geared toward a life-science degree and/or career in the health professions. Carbonyl chemistry is fundamental to the chemistry of living systems and connections between carbonyl chemistry and the chemistry of carbohydrates is highlighted earlier in the book. This latter change mirrors the increasing importance of carbohydrate chemistry on the MCAT.

❯ Chapter 24, Carbon–Carbon Bonding Forming Reactions, combines knowledge from previous chapters and challenges students to devise syntheses. The intent to is expose students to the excitement and challenge of modern synthetic chemistry.

New to the Sixth Edition

In this edition, we introduce revolutionary new approaches to teaching organic chemistry that are designed to promote an unprecedented level of student learning and comprehension.

❯ Chapter 1 was extensively rewritten. The two different approaches to electronic theory and bonding are described in a comprehensive fashion. Importantly, students are shown how to use these theories, even in complex situations, such as those involving resonance contributing structures.

❯ Calculated orbitals were added throughout the text to provide students with important insights concerning electron density location extending far beyond that obtained using only the cursory description and cartoon depiction of orbitals.

❯ Energy diagrams are introduced in the acid-base chapter (Chapter 4) to give students an appreciation for the energetic aspects of chemistry in the context of the mechanistically most straightforward reaction they will encounter, namely proton transfer. Placing energy diagrams in Chapter 4 also introduces important thermodynamic and kinetic concepts early in the course to help lay the foundation before discussing specific functional group reactions.

❯ Chapter 6 has new material on the correct use of arrows to indicate electron movement and simplifies the learning of mechanisms. We introduce a small number of mechanism elements and then explain how these can be assembled in predicable ways to construct in systematic fashion the vast majority of organic reaction mechanisms. This paradigm-shifting approach to teaching mechanism gives students the tools they need to understand and predict mechanisms eliminating the need for superficial memorization.

❯ Organic Chemistry Road Maps are introduced in Chapter 6. This innovation in organic chemistry learning gives students a visual representation of the different reactions and shows how these road maps can be used in specific sequences for the multi-step synthesis of complex molecules.

❯ Chapter 9 was extensively rewritten to provide a clear description of the interplay of parameters that determine mechanism among substitution and elimination reactions. Students are shown how to analyze structures of the nucleophile and electrophile and the reaction conditions that enable the accurate prediction of reaction outcome.

❯ In Chapter 16, there is an increased emphasis on mechanism as the new format was used for all mechanisms, and several new mechanism boxes were added.

❯ In Chapter 18, Section 18.3 on Characteristic Reactions was expanded to make it easier to understand and an entirely new section was added in order to underscore the key concept of microscopic reversibility.

❯ Chapter 20 was reorganized and renamed to reflect an expanded description of pericyclic reaction theory with application to the Diels-Alder reaction and sigmatropic shifts. The advanced concept of frontier molecular orbital (FMO) interactions is now introduced and used to explain the various reactions in the chapter.

❯ Chapter 24 has been expanded to include the Stille and Sonagashira couplings.

❯ **OWL (Online Web Learning) for Organic Chemistry** This fully integrated online system features more than 6,000 practice and homework problems. OWL for Organic Chemistry provides students with instant analysis and feedback to homework problems, modeling questions, end-of-chapter questions, molecular-structure building exercises, and animations created specifically for *Organic Chemistry*, Sixth Edition.

Online Web Learning

Special Features

❯ **New A revolutionary new paradigm** for learning organic chemistry mechanisms is introduced in Chapter 6 and then used throughout the book.

❯ **New Organic Chemistry Road Maps** are introduced as an innovation in organic chemistry learning. Organic chemistry road maps are presented in end-of-chapter problems and a new Appendix 11.

❯ **New Accurate Orbital Diagrams** have been added throughout the text to provide students with a more realistic understanding of electronic theory as applied to organic chemistry.

❯ **New Two new appendices** Appendix 10 on Common Mistakes in Arrow Pushing and Appendix 11 on Organic Chemistry Road Maps.

❯ **Updated Chemical Connections** These essays illustrate applications of organic chemistry to everyday settings. Topics range from Chiral Drugs to Drugs That Lower Plasma Levels of Cholesterol and The Chemistry of Superglue. A complete list can be found on the inside of the back cover.

❯ **Updated Connections to Biological Chemistry** Application of organic chemistry to biology is emphasized throughout the text, in the *Connections to Biological Chemistry* essays and in end-of-chapter problems. See the inside of the back cover for a complete list. New essays include pyridoxine (Vitamin B_6) and electron transfer agents in biological oxidation-reduction reactions.

❯ **Updated** Eleven *How To* features are included in the first part of the book. These describe "survival skills" for the organic chemistry student. Interactive versions of these boxes are assignable in OWL.

❯ **Updated In-Chapter Examples** There are an abundance of in-chapter examples, each with a detailed solution, so students can *immediately* see how the concepts just discussed relate to specific questions and their answers. Following each in-chapter example is a comparable in-chapter problem designed to give students the opportunity to solve a related problem on their own.

❱ Updated **End-of-Chapter Summaries** highlight in outline form all the important ideas of the chapter. Each concept is keyed to the section in the chapter containing a full explanation, as well as to the problems that reinforce understanding.

❱ Updated **End-of-Chapter Summaries of Key Reactions** list the reactions described in the chapter, complete with a prose description of the mechanism and important considerations such as stereochemistry or regiochemistry.

❱ Updated **End-of-Chapter Problems** There are plentiful end-of-chapter problems, with the majority categorized by topic. A red problem number indicates an applied, real-world problem. *Multi-Step Synthesis* problems, many dealing with the synthesis of important pharmaceuticals and *Reactions in Context* problems dealing with functional group transformations of more complex molecules are included.

❱ Updated **Glossary of Key Terms** Throughout the book definitions for new terms are placed in the margin for easy reference. In addition, all definitions are collected in a handy glossary at the end of the text and keyed to the section where the term is introduced.

❱ Updated **A Unique Appendix on Precise Stereochemical Definitions** A comprehensive listing of stereochemical terms in a single collection provides students with a resource that can be referred to often as new terms are encountered.

❱ Updated **A Unique Nomenclature Appendix** Appendix 9 provides a comprehensive listing of all the rules introduced in the text governing nomenclature of complex molecules.

❱ Updated **A Unique Arrow Pushing Appendix** The correct use of arrow pushing is emphasized and students are encouraged to avoid common mistakes.

❱ Updated **Full-Color Art Program** One of the most distinctive features of this text is its visual impact. The text's extensive full-color art program includes a large number of molecular models generated with a three-dimensional look as well as applied photos. In addition, special colors are used to highlight parts of molecules and to follow the course of reactions.

❱ Updated **Electrostatic Potential Maps** are provided at appropriate places throughout the text to illustrate the important concepts of resonance, electrophilicity and nucleophilicity.

Support Package

For the Student and Instructor
❱ **OWL for Organic Chemistry**

Instant Access OWL with eBook for Text (6 months) ISBN 1-111-47204-1

Instant Access OWL with eBook for Text (24 months) ISBN 1-111-47206-8

By Steve Hixson and Peter Lillya of the University of Massachusetts, Amherst, and William Vining of the State University of New York at Oneonta. End-of chapter questions by David W. Brown, Florida Gulf Coast University. **OWL** Online Web Learning offers more assignable, gradable content (including end-of chapter questions specific to this textbook) and more reliability and flexibility than any other system. OWL's powerful course management tools allow instructors to control due dates, number of attempts, and whether students see answers or receive feedback on how to solve problems. OWL includes the **Cengage YouBook**, a Flash-based eBook that is interactive and customizable. It features a text edit tool that allows instructors to modify the textbook narrative as needed. With the Cengage YouBook, instructors can quickly re-order entire sections and chapters or hide any content they don't teach to create an eBook that perfectly matches their syllabus. Instructors can further customize the Cengage YouBook by publishing web links. It includes animated figures, video clips, highlighting, notes, and more.

ᗠWL

Online Web Learning

Developed by chemistry instructors for teaching chemistry, OWL is the only system specifically designed to support **mastery learning**, where students work as long as they need to master each chemical concept and skill. OWL has already helped hundreds of thousands of students master chemistry through a wide range of assignment types, including tutorials and algorithmically generated homework questions that provide instant, answer-specific feedback.

OWL is continually enhanced with online learning tools to address the various learning styles of today's students such as:

❭ **Quick Prep** review courses that help students learn essential skills to succeed in General and Organic Chemistry

❭ **Jmol** molecular visualization program for rotating molecules and measuring bond lengths and angles

In addition, when you become an OWL user, you can expect service that goes far beyond the ordinary. To learn more or to see a demo, please contact your Cengage Learning representative or visit us at **www.cengage.com/owl**.

Animated Mechanisms

A major advance in the 6th edition of the book is the inclusion of animated mechanisms. These interactive animations guide the students through each step in a reaction, along with the associated electron flow arrows and the same terminology used in the book for the fundamental organic reaction steps. The reactants and intermediates are shown colliding to create the products of most all the mechanisms in the book. The format of the chemical structures can be toggled between line drawings, stick figures, or ball and stick figures, as desired by the student. Further, the animated mechanisms are accompanied by an audio explanation of the step that is being displayed on the computer screen. These animations emphasize in an easily displayed format the fundamental organic reaction steps and the associated electron flow. This format is particularly useful in the modern age of personal computers and e-readers, because the students can access the mechanisms at anytime.

For Students

❭ **OWL Quick Prep for Organic Chemistry**
Instant Access OWL Quick Prep for Organic Chemistry (90 Days) ISBN 0-495-56027-8

Quick Prep is a self-paced online short course that helps students succeed in organic chemistry. Students who completed Quick Prep through an organized class or self-study averaged almost a full letter grade higher in their subsequent general chemistry course than those who did not. Intended to be taken prior to the start of the semester, Quick Prep is appropriate for both underprepared students and for students who seek a review of basic skills and concepts. Quick Prep is approximately 10 hours of instruction delivered through OWL with no textbook required and can be completed at any time in the student's schedule. Professors can package a printed access card for Quick Prep with the textbook or students can purchase instant access at www.cengagebrain.com. To view an OWL Quick Prep demonstration and for more information, visit **www.cengage.com/chemistry/quickprep**.

❭ **Updated Student Study Guide and Solutions Manual:** By Brent and Sheila Iverson of the University of Texas, Austin, this manual contains detailed solutions to all text problems. An electronic version of this manual is available to students through OWL. ISBN 1-111-42681-3

❭ **Pushing Electrons: A Guide for Students of Organic Chemistry, third edition.** By Daniel P. Weeks, Northwestern University, this paperback workbook is designed to help students learn techniques of electron pushing. ISBN 0-030-20693-6

❭ **New Visit CengageBrain.com** At **www.cengagebrain.com** you can access the course materials described here as well as purchase Cengage products. Search by the ISBNs in this section or by this textbook's ISBN on the back cover. Instructors can log in at **login.cengage.com**.

❭ New **CengageBrain.com App** Now students can prepare for class anytime and anywhere using the CengageBrain.com application developed specifically for the Apple iPhone® and iPod touch®, which allows students to access free study materials—book-specific quizzes, flash cards, related Cengage Learning materials and more—so they can study the way they want, when they want to . . . even on the go. For more information about this complimentary application, please visit **www.cengagebrain.com**.

❭ Updated **Student Companion Site** This site includes a glossary, flashcards, and an interactive periodic table, and a sample of the Study Guide and Student Solutions Manual, which are all accessible from **www.cengagebrain.com**.

For the Instructor

Supporting materials are available to qualified adopters. Please consult your local Cengage Learning sales representative for details. **Go to login.cengage.com**, find this textbook, and choose Instructor's Companion Site to see samples of these materials, request a desk copy, locate your sales representative or download WebCT and Blackboard versions of the Test Bank.

❭ Updated **PowerLecture Instructor's Resource CD/DVD Package** This dual-platform package is a digital library and presentation tool that includes text-specific PowerPoint® lectures, which instructors can customize if desired by importing their own lecture slides or other materials. The package also contains art and tables from the text in a variety of electronic formats, multimedia animations and molecular models to supplement lectures, and ExamView® testing software. With ExamView's friendly interface, instructors can create, deliver, and customize tests based on questions written specifically for this text. ISBN 1-111-42689-9

❭ Updated **Test Bank on PowerLecture** by David M. Collard, Georgia Institute of Technology. A bank of more than 1,000 problems of varying types and difficulties for instructors to use for tests, quizzes, or homework assignments.

Apple, iPhone, iPod touch, and iTunes are trademarks of Apple Inc., registered in the U.S. and other countries

Acknowledgments

This book is the product of collaboration of many individuals, some obvious, others not so obvious. It is with gratitude that we herein acknowledge the contributions of the many.

Lisa Lockwood as executive editor has masterfully guided the revision of the text. Sandi Kiselica has been a rock of support as senior developmental editor. We so appreciate her ability to set challenging but manageable schedules for us and then her constant encouragement as we worked to meet those deadlines. Others at the Cengage Learning organization have helped to shape our words into this text, including, Teresa Trego, production project manager; John Walker, creative director; Stephanie VanCamp, media editor; and Lisa Weber, senior media editor. Patrick Franzen of PreMediaGlobal served as our production editor. Also, many thanks to David Brown of Florida Gulf Coast University who authored the OWL questions for this book.

We are also indebted to the many reviewers of our manuscript who helped shape its contents. With their guidance, we have revised this text to better meet the needs of our and their students.

Sixth Edition

Thomas Albright	University of Houston
Zachary D. Aron	Indiana University
Valerie Ashby	University of North Carolina
B. Mikael Bergdahl	San Diego State University
Robert Boikess	Rutgers University
Jean Chmielewski	Purdue University
Elizabeth Harbron	The College of William and Mary
Arif Karim	University of California, Los Angeles
Susan King	University of California, Irvine
Mark Lipton	Purdue University
Allan Pinhas	University of Cincinnati
Owen Priest	Northwestern University
Jonathan Stoddard	California State University, Fullerton

Fifth Edition

Jon Antilla	University of Southern Florida
Christopher Bielawski	University of Texas
Alan Campion	University of Texas
David Cartrette	South Dakota State University
H. J. Peter de Lijser	California State University, Fullerton
Malcolm Forbes	University of North Carolina
John Grutzner	Purdue University
Robert C. Kerber	SUNY, Stony Brook

Spencer Knapp	Rutgers University
Paul Kropp	University of North Carolina
Deborah Lieberman	University of Cincinnati
James Mack	University of Cincinnati
Felix Ngassa	Grand Valley State University
Milton Orchin	University of Cincinnati
Allan Pinhas	University of Cincinnati
Suzanne Ruder	Virginia Commonwealth University
Laurie Starkey	California State Polytechnic University, Pomona
Qian Wang	University of South Carolina
Alexander Wei	Purdue University
Laurie Witucki	Grand Valley State University
Lei Zhu	Florida State University

Chapter 1

Covalent Bonding and Shapes of Molecules

© Cengage Learning/Charles D. Winters

A model of the structure of diamond, one form of pure carbon. Each carbon is bonded to four other carbons at the corners of a tetrahedron. Inset: a model of fullerene (C_{60}). See the box "Fullerene—A New Form of Carbon."

According to the simplest definition, **organic chemistry** is the study of the compounds of carbon. Perhaps its most remarkable feature is that most organic compounds consist of carbon and only a few other elements—chiefly, hydrogen, oxygen, and nitrogen. Chemists have discovered or made well over ten million compounds composed of carbon and these three other elements. Organic compounds are everywhere around us—in our foods, flavors, and fragrances; in our medicines, toiletries, and cosmetics; in our plastics, films, fibers, and resins; in our paints and varnishes; in our glues and adhesives; in our fuels and lubricants; and, of course, in our bodies and those of all living things.

Let us begin our study of organic chemistry with a review of how the elements of C, H, O, and N combine by sharing electron pairs to form bonds, and ultimately molecules. There is a great deal of material in this chapter, but you have encountered much of it in your previous chemistry courses. However, because all subsequent chapters in this book use this material, it is essential that you understand it and can use it fluently.

1.1 | Electronic Structure of Atoms

An atom contains a small, dense nucleus made of neutrons and positively charged protons. Most of the mass of an atom is contained in its nucleus. The nucleus is surrounded by an extranuclear space containing negatively charged electrons. The nucleus of an atom has a diameter of 10^{-14} to 10^{-15} meters (m). The extranuclear space where its electrons are found is a much larger volume with a diameter of approximately 10^{-10} m (Figure 1.1).

⟁WL Online homework for this chapter may be assigned in OWL for Organic Chemistry, an online learning assessment tool.

1

Figure 1.1
A schematic view of an atom. Most of the mass of an atom is concentrated in its small, dense nucleus.

Shell
A region of space around a nucleus that can be occupied by electrons, corresponding to a principal quantum number.

Quantized
Having discrete values for energy and momentum.

Delocalization
The spreading of electron density over a larger volume of space.

Orbital
A region of space that can hold two electrons.

Orthogonal
Having no net overlap.

Table 1.1	Distribution of Electrons in Shells	
Shell	**Number of Electrons Shell Can Hold**	**Relative Energies of Electrons in These Shells**
4	32	higher
3	18	↑
2	8	
1	2	lower

Shells define the probability of finding an electron in various regions of space relative to the nucleus. The energy of electrons in the shells is quantized. **Quantization** means that only specific values of energy are possible, rather than a continuum of values. These shells only occur at quantized energies in which three important effects balance each other. The first is the electrostatic attraction that the electrons have to the nucleus and that draws them toward the nucleus, the second is the electrostatic repulsion between the electrons, and the third is the wavelike nature of an electron that prefers to be delocalized, thereby spreading the electron density away from the nuclei. **Delocalization** is a term that describes the spreading of electron density over a larger volume of space.

Electron shells are identified by the principal quantum numbers 1, 2, 3, and so forth. Each shell can contain up to $2n^2$ **electrons**, where n is the number of the shell. Thus, the first shell can contain 2 electrons, the second 8 electrons, the third 18 electrons, the fourth 32 electrons, and so on (Table 1.1). Electrons in the first shell are nearest to the positively charged nucleus and are held most strongly by it; these electrons are lowest in energy. Electrons in higher-numbered shells are farther from the positively charged nucleus and are held less strongly.

Shells are divided into subshells designated by the letters s, p, d, and f, and, within these subshells, electrons are grouped in orbitals (Table 1.2). An **orbital** is a region of space that can hold two electrons and has a specific quantized energy. The first shell contains a single orbital called a $1s$ orbital. The second shell contains one s orbital and three p orbitals. The three $2p$ orbitals reflect orthogonal angular momentum states in three-dimensional space. **Orthogonal** in this context results in 90° angles between the orbitals, but in all cases orthogonal also means that the orbitals have no net overlap. As a point of reference, to discuss the $2p$ orthogonal orbitals, we consider them to be directed along the x-, y-, and z-axes and give them designations, $2p_x$, $2p_y$, and $2p_z$. The third shell contains one $3s$ orbital, three $3p$ orbitals, and five $3d$ orbitals. The shapes of s and p orbitals are shown in Figures 1.8 and 1.9, and are described in more detail in Section 1.6B.

Table 1.2	Distribution of Orbitals in Shells
Shell	**Orbitals Contained in That Shell**
3	$3s$, $3p_x$, $3p_y$, $3p_z$, plus five $3d$ orbitals
2	$2s$, $2p_x$, $2p_y$, $2p_z$
1	$1s$

A. Electron Configuration of Atoms

Ground-state electron configuration
The lowest-energy electron configuration for an atom or molecule.

The electron configuration of an atom is a description of the orbitals its electrons occupy. Every atom has an infinite number of possible electron configurations. At this stage, we are concerned primarily with the **ground-state electron configuration**—the electron configuration of lowest energy. We determine the ground-state electron configuration of an atom by using the following three rules.

Rule 1: The Aufbau ("Build-Up") Principle. Orbitals fill in order of increasing energy, from lowest to highest. In this course, we are concerned primarily with the elements of the first, second, and third periods of the Periodic Table. Orbitals fill in the order $1s$, $2s$, $2p$, $3s$, $3p$, and so on.

Rule 2: The Pauli Exclusion Principle. The Pauli exclusion principle requires that only two electrons can occupy an orbital and that their spins must be paired. To understand what it means to have paired spins, recall from general chemistry that just as the earth has a spin, electrons have a quantum mechanical property referred to as spin. And, just as the earth has magnetic north (N) and south (S) poles, so do electrons. As described by quantum mechanics, a given electron can exist in only two different spin states. Two electrons with opposite spins are said to have **paired spins**.

Aufbau principle
Orbitals fill in order of increasing energy, from lowest to highest.

Pauli exclusion principle
No more than two electrons may be present in an orbital. If two electrons are present, their spins must be paired.

The quantum mechanical property of spin gives an electron a tiny magnetic field

When their tiny magnetic fields are aligned N-S, the electron spins are paired

When filling orbitals with electrons, place no more than two in an orbital. For example, with four electrons, the $1s$ and $2s$ orbitals are filled and are written $1s^2 2s^2$. With an additional six electrons, the set of three $2p$ orbitals is filled and is written $2p_x^2 2p_y^2 2p_z^2$. Alternatively, a filled set of three $2p$ orbitals may be written $2p^6$.

Rule 3: Hund's Rule. Hund's rule has two parts. The first part states that when orbitals of equal energy (called degenerate) are available but there are not enough electrons to fill all of them completely, then one electron is added to each orbital before a second electron is added to any one of them. The second part of Hund's rule states that the spins of the single electrons in the degenerate orbitals should be aligned. Recall that electrons have a negative charge; partially filling orbitals as much as possible minimizes electrostatic repulsion between electrons. After the $1s$ and $2s$ orbitals are filled with four electrons, a fifth electron is added to the $2p_x$ orbital, a sixth to the $2p_y$ orbital, and a seventh to the $2p_z$ orbital. Only after each $2p$ orbital contains one electron is a second electron added to the $2p_x$ orbital. Carbon, for example, has six electrons, and its ground-state electron configuration is $1s^2 2s^2 2p_x^1 2p_y^1 2p_z^0$. Alternatively, it may be simplified to $1s^2 2s^2 2p^2$. Table 1.3 shows ground-state electron configurations of the first 18 elements of the Periodic Table.

Chemists routinely write **energy-level diagrams** that pictorially designate where electrons are placed in an electron configuration. For example, the energy-level diagram for the electron configuration of carbon, $1s^2$, $2s^2$, $2p^2$, shows three energy levels, one each for the $1s$, $2s$, and $2p$ orbitals. Moving up in the diagram means higher energy. Electrons in these diagrams are drawn as arrows. The Aufbau principle tells us to place the first four electrons in the $1s$ and $2s$ orbitals, and the Pauli exclusion principle tells us to pair the two electrons in each orbital (shown as arrows with opposing directions). The remaining two electrons are left to go into the $2p$ level, and because there are three such orbitals, the second part of Hund's rule tells us to place these electrons in different orbitals with their spins aligned (shown

Table 1.3	Ground-State Electron Configurations for Elements 1–18		
First Period*	**Second Period**	**Third Period**	
H 1 $1s^1$	Li 3 [He] $2s^1$	Na 11 [Ne] $3s^1$	
He 2 $1s^2$	Be 4 [He] $2s^2$	Mg 12 [Ne] $3s^2$	
	B 5 [He] $2s^2 2p^1$	Al 13 [Ne] $3s^2 3p^1$	
	C 6 [He] $2s^2 2p^2$	Si 14 [Ne] $3s^2 3p^2$	
	N 7 [He] $2s^2 2p^3$	P 15 [Ne] $3s^2 3p^3$	
	O 8 [He] $2s^2 2p^4$	S 16 [Ne] $3s^2 3p^4$	
	F 9 [He] $2s^2 2p^5$	Cl 17 [Ne] $3s^2 3p^5$	
	Ne 10 [He] $2s^2 2p^6$	Ar 18 [Ne] $3s^2 3p^6$	

*Elements are listed by symbol, atomic number, and simplified ground-state electron configuration.

as arrows pointing in the same direction). We will use energy-level diagrams later in this chapter to understand bonding, and throughout the book when discussing relative energies of orbitals.

Energy level
diagram for carbon

Example 1.1

Write the ground-state electron configuration for each element showing the occupancy of each p orbital. For (c) write the energy-level diagram.

(a) Lithium   **(b)** Oxygen   **(c)** Chlorine

Solution

(a) Lithium (atomic number 3): $1s^2\, 2s^1$

(b) Oxygen (atomic number 8): $1s^2\, 2s^2\, 2p_x^2\, 2p_y^1\, 2p_z^1$

(c) Chlorine (atomic number 17): $1s^2\, 2s^2 2p_x^2\, 2p_y^2\, 2p_z^2 3s^2\, 3p_x^2\, 3p_y^2\, 3p_z^1$

Energy level
diagram for chlorine

Problem 1.1

Write and compare the ground-state electron configurations for each pair of elements.

(a) Carbon and silicon   **(b)** Oxygen and sulfur   **(c)** Nitrogen and phosphorus

B. The Concept of Energy

In the discussion of energy-level diagrams, the lines were drawn on the diagram to depict relative energy. All measurements of energy are relative, meaning that we compare an energy state to some reference state. For example, in the energy-level diagram for carbon, the $1s$ level is the reference and the $2s$ and $2p$ levels are placed higher on the diagram relative to it. But, you may be asking, "How is energy defined?"

Energy is the ability to do work. The higher in energy any entity is, the more work it can perform. To understand this, let's imagine an example. You drop a weight to drive a spike into the earth. As you hold the weight above the ground, it is unstable relative to when it is lying on the ground. You expended energy lifting the weight off the ground, and that energy is stored in the weight due to its position. The stored energy is referred to as **potential energy**, because it can potentially be released, if, for example, you let go of the weight. If you hold the weight higher above the ground, it will be increasingly unstable, possess more stored energy, and have a higher potential energy. Now when you drop the weight, it will drive the spike deeper into the earth.

The force that restores the weight to its resting state on the ground is the gravitational attraction of the weight to the earth. Interestingly, the farther the weight is from the earth, the easier it is to take the weight even farther from the earth. As an extreme example, thousands of miles above the earth the weight has incredibly large potential energy and could wreak serious damage to a building if dropped. But, at that distance, it is relatively easy to remove the weight farther from the earth because the gravitational attraction is weak.

We can generalize this example of a weight to chemical structures. Unstable structures have energy in them waiting to be released if given an opportunity. When a species is higher in energy, it has more energy stored. When that energy is released, work can be done. In chemistry, the release of energy is very often harnessed to do work, such as the burning of gasoline to drive the pistons in an internal combustion engine that propels an automobile. However, in chemical reactions carried out in the laboratory the release of energy commonly just heats up the reaction vessel.

With these thoughts in mind, let's return to the energy-level diagram of carbon. In the **ground state** of carbon, the electrons are placed in accordance with the quantum chemistry principles (Aufbau principle, Hund's rule, Pauli exclusion principle, etc.) that dictate the lowest energy form of carbon. If we place the electrons in a different manner (as an example, only one electron in $2s$ and three electrons in $2p$), we would have a higher energy state of carbon, referred to as an **excited state**. All of nature seeks its lowest energy state, and therefore achieving a state in accord with the quantum chemistry principles acts as a restoring force (analogous to gravity on the weight) that will drive the electrons back to their lowest energy state. When the electrons are rearranged back to this ground state, energy is released.

Note that the electrons in the lowest energy orbital, $1s$, are held tightest to the nucleus and are the hardest to remove from the atom. It would take the largest amount of energy to remove these electrons relative to the other electrons. The energy it takes to remove an electron from an atom or molecule is called the **ionization potential**. The $1s$ electrons, therefore, have the highest ionization potential. However, the electrons in the $2p$ levels of carbon are the farthest from the nucleus and are held the weakest. They are the easiest to remove from the atom, and therefore have the lowest ionization potential (called the first ionization potential). This is analogous to the fact that it is easier to remove the weight from the earth the farther it is from the surface.

Energy
The ability to do work.

Potential energy
The energy that can be released if given an opportunity.

Ground state
The lowest energy state of a system.

Excited state
A state of a system at higher energy than the ground state.

First ionization potential
The energy needed to remove the most loosely held electron from an atom or molecule.

C. Lewis Dot Structures

Valence electrons
Electrons in the valence (outermost) shell of an atom.

Valence shell
The outermost occupied electron shell of an atom.

Lewis dot structure
The symbol of an element surrounded by a number of dots equal to the number of electrons in the valence shell of the atom.

Gilbert N. Lewis (1875–1946) introduced the theory of the electron pair that extended our understanding of covalent bonding and of the concept of acids and bases. It is in his honor that we often refer to an "electron dot" structure as a Lewis structure.

When discussing the physical and chemical properties of an element, chemists often focus on the electrons in the outermost shell of the atom because these electrons are involved in the formation of chemical bonds and in chemical reactions. Carbon, for example, with the ground-state electron configuration $1s^2 2s^2 2p^2$, has four outer-shell electrons. Outer-shell electrons are called **valence electrons**, and the energy level in which they are found is called the **valence shell**. To show the outermost electrons of an atom, we commonly use a representation called a **Lewis dot structure**, after the American chemist Gilbert N. Lewis (1875–1946) who devised this notation. A Lewis dot structure shows the symbol of the element surrounded by a number of dots equal to the number of electrons in the outer shell of an atom of that element. In Lewis dot structures, the atomic symbol represents the core; that is, the nucleus and all inner shell electrons. Table 1.4 shows Lewis dot structures for the first 18 elements of the Periodic Table.

The noble gases helium and neon have filled valence shells. The valence shell of helium is filled with two electrons; that of neon is filled with eight electrons. Neon and argon have in common an electron configuration in which the s and p orbitals of their valence shells are filled with eight electrons. The valence shells of all other elements shown in Table 1.4 contain fewer than eight electrons.

For C, N, O, and F in period 2 of the Periodic Table, the valence electrons belong to the second shell. With eight electrons, this shell is completely filled. For Si, P, S, and Cl in period 3 of the Periodic Table, the valence electrons belong to the third shell. This shell is only partially filled with eight electrons; the $3s$ and $3p$ orbitals are fully occupied, but the five $3d$ orbitals can accommodate an additional ten electrons.

Table 1.4	Lewis Dot Structures for Elements 1–18*						
1A	**2A**	**3A**	**4A**	**5A**	**6A**	**7A**	**8A**
H·							He:
Li·	Be:	B:	·C:	·N:	:O:	:F:	:Ne:
Na·	Mg:	Al:	·Si:	·P:	:S:	:Cl:	:Ar:

*These dots represent electrons from the valence shell. They are arranged as pairs or single electrons in accordance with Hund's rule.

1.2 | Lewis Model of Bonding

In 1916, Lewis devised a beautifully simple model that unified many of the observations about chemical bonding and reactions of the elements. He pointed out that the chemical inertness of the noble gases indicates a high degree of stability of the electron configurations of these elements: helium with a valence shell of two electrons ($1s^2$), neon with a valence shell of eight electrons ($2s^2 2p^6$), and argon with a valence shell of eight electrons ($3s^2 3p^6$). The tendency of atoms to react in ways that achieve an outer shell of eight valence electrons is particularly common among second-row elements of Groups 1A–7A (the main-group elements) and is given the special name **octet rule**.

Octet rule
Group 1A–7A elements react to achieve an outer shell of eight valence electrons.

Example 1.2

Show how the loss of an electron from a sodium atom leads to a stable octet.

Solution

The ground-state electron configurations for Na and Na$^+$ are:

$$\text{Na (11 electrons): } 1s^2 2s^2 2p^6 3s^1$$

$$\text{Na}^+ \text{ (10 electrons): } 1s^2 2s^2 2p^6$$

Thus, Na^+ has a complete octet of electrons in its outermost (valence) shell and has the same electron configuration as neon, the noble gas nearest it in atomic number.

Problem 1.2

Show how each chemical change leads to a stable octet.

(a) Sulfur forms S^{2-}. **(b)** Magnesium forms Mg^{2+}.

A. Formation of Chemical Bonds

According to Lewis's model, atoms interact in such a way that each participating atom acquires a completed outer-shell electron configuration resembling that of the noble gas nearest to it in atomic number. Atoms acquire completed valence shells in two ways.

1. An atom may become ionic, that is lose or gain enough electrons to acquire a completely filled valence shell. An atom that gains electrons becomes an **anion** (a negatively charged ion), and an atom that loses electrons becomes a **cation** (a positively charged ion). A positively charged ion and a negatively charged ion attract each other. This attraction can lead to the formation of ionic crystals such as sodium chloride, in which each positive ion is surrounded by negative ions and vice versa, in a definite geometric arrangement that depends on the crystal. When atoms are held together primarily by attraction of oppositely charged ions, we say that an **ionic interaction** exists between them. (This ionic interaction is often referred to as an ionic bond.)

2. An atom may share electrons with one or more other atoms to complete its valence shell. A chemical bond formed by sharing electrons is called a **covalent bond**.

3. Bonds may be partially ionic and partially covalent; these bonds are called **polar covalent bonds**. Polar covalent bonds are defined more precisely in the next section.

Anion
An atom or group of atoms bearing a negative charge.

Cation
An atom or group of atoms bearing a positive charge.

Ionic interaction
Attraction between oppositely charged ions.

Covalent bond
A chemical bond formed between two atoms by sharing one or more pairs of electrons.

B. Electronegativity and Chemical Bonds

How do we estimate the degree of ionic or covalent character in a chemical bond? One way is to compare the electronegativities of the atoms involved. **Electronegativity** is a measure of an atom's attraction for electrons that it shares in a chemical bond with another atom. The most widely used scale of electronegativities (Table 1.5) was devised by Linus Pauling in the 1930s.

On the Pauling scale, fluorine, the most electronegative element, is assigned an electronegativity of 4.0, and all other elements are assigned values in relation to fluorine. As you study the electronegativity values in this table, note that they generally increase from left to right within a period of the Periodic Table and

Electronegativity
A measure of the force of an atom's attraction for electrons.

Linus Pauling (1901–1994) was the first person ever to receive two unshared Nobel prizes. He received the 1954 Nobel Prize in Chemistry for his contributions to our understanding of chemical bonding. He received the 1962 Nobel Peace Prize for his efforts on behalf of international control of nuclear weapons testing.

Table 1.5	Electronegativity Values for Some Atoms (Pauling Scale)

							H 2.1										
1A	**2A**												**3A**	**4A**	**5A**	**6A**	**7A**
Li 1.0	Be 1.5												B 2.0	C 2.5	N 3.0	O 3.5	F 4.0
Na 0.9	Mg 1.2	**3B**	**4B**	**5B**	**6B**	**7B**		**8B**		**1B**	**2B**		Al 1.5	Si 1.8	P 2.1	S 2.5	Cl 3.0
K 0.8	Ca 1.0	Sc 1.3	Ti 1.5	V 1.6	Cr 1.6	Mn 1.5	Fe 1.8	Co 1.8	Ni 1.8	Cu 1.9	Zn 1.6	Ga 1.6	Ge 1.8	As 2.0	Se 2.4	Br 2.8
Rb 0.8	Sr 1.0	Y 1.2	Zr 1.4	Nb 1.6	Mo 1.8	Tc 1.9	Ru 2.2	Rh 2.2	Pd 2.2	Ag 1.9	Cd 1.7	In 1.7	Sn 1.8	Sb 1.9	Te 2.1	I 2.5
Cs 0.7	Ba 0.9	La 1.1	Hf 1.3	Ta 1.5	W 1.7	Re 1.9	Os 2.2	Ir 2.2	Pt 2.2	Au 2.4	Hg 1.9	Tl 1.8	Pb 1.8	Bi 1.9	Po 2.0	At 2.2

■ <1.0 ■ 1.5 – 1.9 ■ 2.5 – 2.9
■ 1.0 – 1.4 ■ 2.0 – 2.4 ■ 3.0 – 4.0

generally decrease from top to bottom within a group. Values increase from left to right because the increasing positive charge on the nucleus results in a greater force of attraction for the atom's valence electrons. Electronegativity decreases from top to bottom because the increasing distance of the valence electrons from the nucleus results in a lower attraction between the nucleus and these electrons.

Let's further analyze the trends in the Periodic Table we just discussed. As you proceed from left to right in a row of the Periodic Table, the atoms get smaller. This contraction occurs because as you go across a row, the electrons are placed in the same shell, but the charge on the nuclei is increasing, thereby pulling the electrons in closer. This means that the orbitals get lower in energy as you move from left to right in the table and the atoms hold their electrons tighter. It therefore takes more energy to remove the electrons from atoms as you move toward the right in the Periodic Table (with some exceptions), meaning that these atoms have a higher first ionization potential.

In contrast, consider adding rather than removing an electron to the atoms. For example, when an electron is added to the halogens (Group 7A), energy is released because these atoms achieve a noble gas configuration. The energy released upon addition of an electron is called the **electron affinity**, which becomes more favorable as you move from left to right in a row of the Periodic Table. In contrast, as you proceed down a column in the Periodic Table, the principal quantum levels increase and the outermost electrons are farther from the nuclei, are held less tightly, and have lower ionization potentials. The atoms also have decreasing electron affinities. Because the electronegativity of an atom reflects its tendency to hold on to and to acquire electrons, the phenomenon arises from a combination of ionization potentials and electron affinities.

When combining the trends of moving from left to right and up to down in the Periodic Table, you can conclude that fluorine must have the most tightly held electrons of any atom that can make bonds. Further, fluorine most tightly holds any electron that it gains during ion formation or covalent bond formation. Hence, fluorine has the highest electronegativity of any atom.

Electron affinity
Energy added or released when an electron is added to an atom or molecule.

Example 1.3

Judging from their relative positions in the Periodic Table, which element in each set is the more electronegative?

(a) Lithium or carbon   (b) Nitrogen or oxygen   (c) Carbon or oxygen

Solution

The elements in these sets are all in the second period of the Periodic Table. Electronegativity in this period increases from left to right.

(a) $C > Li$ (b) $O > N$ (c) $O > C$

Problem 1.3

Judging from their relative positions in the Periodic Table, which element in each set is the more electronegative?

(a) Lithium or potassium   (b) Nitrogen or phosphorus   (c) Carbon or silicon

Formation of Ions

Ions are formed by the transfer of electrons from the valence shell of an atom of lower electronegativity to the valence shell of an atom of higher electronegativity. As a rough guideline, we say that ions will form if the difference in electronegativity between interacting atoms is 1.9 or greater. As an example, ions are formed from sodium (electronegativity 0.9) and fluorine (electronegativity 4.0). In the following equation, we use a single-headed (barbed) curved arrow to show the transfer of one electron from sodium to fluorine.

$$Na\cdot + \cdot\ddot{\underset{\displaystyle\cdot\cdot}{F}}: \longrightarrow Na^+ : \ddot{\underset{\displaystyle\cdot\cdot}{F}}:^-$$

In forming Na^+F^-, the single $3s$ valence electron of sodium is transferred to the partially filled valence shell of fluorine:

$$Na(1s^22s^22p^63s^1) + F(1s^22s^22p^5) \longrightarrow Na^+(1s^22s^22p^6) + F^-(1s^22s^22p^6)$$

As a result of this transfer of one electron, both sodium and fluorine form ions that have the same electron configuration as neon, the noble gas nearest each in atomic number. The attraction between ions leads to a strong crystal lattice with high melting points in ionic salts such as sodium fluoride.

Covalent Bonds

A covalent bond is a chemical bond formed between atoms by the sharing of one or more pairs of electrons to give a noble gas configuration at each atom. The simplest example of a covalent bond occurs in the hydrogen molecule. When two hydrogen atoms bond, the single electrons from each combine to form an electron pair. This shared pair completes the valence shell of each hydrogen. According to the Lewis model, a pair of electrons in a covalent bond functions in two ways simultaneously: it is shared by two atoms and at the same time fills the outer (valence) shell of each. We use a line between the two hydrogens to symbolize the covalent bond formed by the sharing of a pair of electrons.

$$H\cdot + \cdot H \longrightarrow H:H \quad \text{Symbolized } H—H$$

$$\Delta H^0 = -435 \text{ kJ } (-104 \text{ kcal})/\text{mol}$$

In this pairing, a large amount of energy is released, meaning that two hydrogen atoms are unstable relative to H_2. The same amount of energy, called the **bond dissociation enthalpy** (BDE, also known as the bond dissociation energy) would have to be absorbed to break the bond.

Later in the chapter, we see that electrons have both wave and particle character (Section 1.6). When bonds are formed by the sharing of two electrons between adjacent atoms, the system becomes more stable because the wave character of the electrons is stabilized relative to two separate atoms. The wave that represents the electrons in a bond is partially concentrated in the space between the two nuclei, leading to repulsion between these electrons. In contrast, the electrons are attracted to each nucleus, and shield the repulsion between the two positively charged nuclei. The lowering of the energy of the wave character of the electrons along with their added attraction to each nucleus is balanced with the repulsion between the nuclei and between the electrons. This balance results in an optimum internuclear distance called the bond length.

The distance between nuclei participating in a chemical bond is called the **bond length**. Every covalent bond has a characteristic bond length. In H—H, it is 74 pm (picometer; 1 pm = 10^{-12} m). We use SI units of picometers in this book; many chemists still use Å (Ångstroms); 1 Å = 100 pm.

Because each bond requires two electrons, a maximum of four bonds can form with second row atoms. For each unshared pair of electrons on an atom (called a **lone pair**), one fewer bond is possible.

In many situations, filled valence shells can be satisfied only when bonded atoms share more than two electrons. In these cases, multiple covalent bonds form between the same two atoms. For example, four electrons shared between two atoms form a double bond. Six shared electrons form a triple bond.

To summarize, a good way to distinguish covalent bonds from ionic attraction is that covalent bonds have defined geometries and connectivities resulting from the sharing of electrons. In other words, the number and positions of atoms taking part in covalent bonds are defined. In crystals, the positions of the atoms associated through ionic attraction depend on the particular crystal lattice.

Polar Covalent Bonds

Although all covalent bonds involve the sharing of electrons, they differ widely in the degree of sharing. Homonuclear diatomics such as H_2, N_2, O_2, and F_2 share

Bond length
The distance between nuclei in a covalent bond in picometers (pm; 1 pm = 10^{-12} m) or Å (1Å = 10^{-10} m).

Table 1.6	Classification of Chemical Bonds	
Difference in Electronegativity Between Bonded Atoms	**Type of Bond**	
Less than 0.5	Nonpolar covalent	
0.5 to 1.9	Polar covalent	
Greater than 1.9	Ions formed	

the electrons equally between the two atoms, and are said to have nonpolar covalent bonds. Many compounds like HCl and H_2O share the electrons in the bond unequally, and are said to contain polar covalent bonds. The polarity in the bond increases with increasing difference in electronegativity between the bonded atoms (Table 1.6).

A covalent bond between carbon and hydrogen, for example, is classified as **nonpolar covalent** because the difference in electronegativity between these two atoms is $2.5 - 2.1 = 0.4$.

An example of a **polar covalent bond** is that of H—Cl. The difference in electronegativity between chlorine and hydrogen is $3.0 - 2.1 = 0.9$. An important consequence of the unequal sharing of electrons in a polar covalent bond is that the more electronegative atom gains a greater fraction of the shared electrons and acquires a partial negative charge, indicated by the symbol $\delta-$. The less electronegative atom has a smaller fraction of the shared electrons and acquires a partial positive charge, indicated by the symbol $\delta+$. Alternatively, we show the direction of bond polarity by an arrow with the arrowhead pointing toward the negative end and a plus sign on the tail of the arrow at the positive end.

$$\overset{\delta+}{H}\text{—}\overset{\delta-}{Cl} \qquad \overset{\longrightarrow}{H\text{—}Cl}$$

Nonpolar covalent bond
A covalent bond between atoms whose difference in electronegativity is less than approximately 0.5.

Polar covalent bond
A covalent bond between atoms whose difference in electronegativity is between approximately 0.5 and 1.9.

Example 1.4

Classify each bond as nonpolar covalent or polar covalent or state that ions are formed.

(a) O—H **(b)** N—H **(c)** K—Br **(d)** C—Mg

Solution

Based on differences in electronegativity between the bonded atoms, three of these bonds are polar covalent and one involves the formation of ions.

Bond	Difference in Electronegativity	Type of Interaction
(a) O—H	$3.5 - 2.1 = 1.4$	Polar covalent
(b) N—H	$3.0 - 2.1 = 0.9$	Polar covalent
(c) K—Br	$2.8 - 0.8 = 2.0$	Ions formed
(d) C—Mg	$2.5 - 1.2 = 1.3$	Polar covalent

Problem 1.4

Classify each bond as nonpolar covalent or polar covalent or state that ions are formed.

(a) S—H **(b)** P—H **(c)** C—F **(d)** C—Cl

Electronegativity varies somewhat depending on the chemical environment and oxidation state of an atom; therefore, these rules are only guidelines and must be used with caution. Lithium iodide, for example, has a high melting point

(449°C) and boiling point (1180°C) characteristic of ionic compounds. Yet, based on the difference in electronegativity between these two elements of $2.5 - 1.0 = 1.5$ units, we would classify LiI as a polar covalent compound. In particular, carbon in compounds is not atomic carbon, and it often behaves as if its electronegativity is less than shown in Table 1.5. In fact, carbon often behaves as if it were less electronegative than hydrogen.

Example 1.5

Using the symbols $\delta-$ and $\delta+$, indicate the direction of polarity in each polar covalent bond.

(a) C—O   (b) N—H   (c) C—Mg

Solution

Electronegativity is given beneath each atom. The atom with the greater electronegativity has the partial negative charge; the atom with the lesser electronegativity has the partial positive charge.

(a) $\overset{\delta+}{\text{C}}-\overset{\delta-}{\text{O}}$ (b) $\overset{\delta-}{\text{N}}-\overset{\delta+}{\text{H}}$ (c) $\overset{\delta-}{\text{C}}-\overset{\delta+}{\text{Mg}}$

    2.5 3.5   3.0 2.1   2.5 1.2

Problem 1.5

Using the symbols $\delta-$ and $\delta+$, indicate the direction of polarity in each polar covalent bond.

(a) C—N   (b) N—O   (c) C—Cl

The polarity of a covalent bond is measured by a vector quantity called a **bond dipole moment** and is given the symbol μ (Greek mu). Bond dipole moment is defined as the product of the charge, e (either the $\delta+$ or $\delta-$ because each is the same in absolute magnitude), on one of its atoms times the distance, d, separating the two atoms. The SI unit for a dipole moment is the coulomb · meter, but they are commonly reported instead in a derived unit called the debye (D: $1 \text{ D} = 3.34 \times 10^{-30} \text{ C} \cdot \text{m}$). Table 1.7 lists bond dipole moments for the types of covalent bonds we deal with most frequently in this course.

Bond dipole moment (μ)
A measure of the polarity of a covalent bond. It is the product of the charge on either atom of a polar covalent bond times the distance between the nuclei.

Table 1.7	Average Bond Dipole Moments of Selected Covalent Bonds				
Bond	**Bond Dipole (D)**	**Bond**	**Bond Dipole (D)**	**Bond**	**Bond Dipole (D)**
H—C	0.3	C—F	1.4	C—O	0.7
H—N	1.3	C—Cl	1.5	C=O	2.3
H—O	1.5	C—Br	1.4	C—N	0.2
H—S	0.7	C—I	1.2	C≡N	3.5

C. Lewis Structures for Molecules and Polyatomic Ions

The ability to write Lewis structures for molecules and polyatomic ions is a fundamental skill for the study of organic chemistry. The following guidelines will help you do this. As you study these guidelines, look at the examples in Table 1.8.

1. Determine the number of valence electrons in the molecule or ion. To do this, add the number of valence electrons contributed by each atom. For ions, add one electron for each negative charge on the ion, and subtract one electron for

each positive charge on the ion. For example, the Lewis structure for a water molecule, H_2O, must show eight valence electrons: one from each hydrogen and six from oxygen. The Lewis structure for the hydroxide ion, HO^-, must also show eight valence electrons: one from hydrogen, six from oxygen, plus one for the negative charge on the ion.

2. Determine the connectivity (arrangement) of atoms in the molecule or ion. Except for the simplest molecules and ions, this connectivity must be determined experimentally because alternative possibilities can lead to the possibility of **isomers**. Isomers are different compounds with the same molecular formula. For example, there are two different compounds with the molecular formula C_4H_{10}: one with four atoms connected in a row, and one in which there are three in a row and the fourth branches off. We discuss isomers extensively in Section 2.2 and Chapter 3.

Isomers
Different compounds with the same molecular formula.

$$CH_3{-}CH_2{-}CH_2{-}CH_3 \qquad CH_3{-}\overset{\overset{\textstyle CH_3}{|}}{CH}{-}CH_3$$

The two isomers of C_4H_{10}

For some molecules and ions given as examples in the text, you are asked to propose an arrangement of atoms. For most, however, you are given the experimentally determined arrangement.

3. Connect the atoms with single bonds. Then arrange the remaining electrons in pairs so that each atom in the molecule or ion has a complete outer shell. Each hydrogen atom must be surrounded by two electrons. Each atom of carbon, oxygen, nitrogen, and halogen must be surrounded by eight electrons (per the octet rule).

4. Each pair of electrons (**bonding electrons**) shared between two atoms is shown as a single line between the atoms. Each unshared pair of electrons (often called a **lone pair** or **nonbonding electrons**) is shown as a pair of dots.

5. If two atoms only share a single pair of electrons, they form a single bond and a single line is drawn between them. If two pairs of electrons are shared between two atoms, they form a double bond (two lines). If three pairs of electrons are shared between two atoms, they form a triple bond (three lines).

Bonding electrons
Valence electrons involved in forming a covalent bond (i.e., shared electrons).

Nonbonding electrons
Valence electrons not involved in forming covalent bonds. Also called unshared pairs or lone pairs.

Table 1.8 shows Lewis structures, molecular formulas, and names for several compounds. The number of valence electrons each molecule contains is shown in parentheses. Notice that, in these molecules, each hydrogen is surrounded by two valence electrons, and each carbon, nitrogen, oxygen, and chlorine is surrounded by eight valence electrons. Furthermore, each carbon has four bonds, nitrogen has three bonds and one unshared pair of electrons, oxygen has two bonds and two unshared pairs of electrons (lone pairs), and chlorine (and other halogens as well) has one bond and three unshared pairs of electrons.

Table 1.8	Lewis Structures for Several Compounds*		
H$-\overset{..}{\underset{..}{O}}-$H	H$-\overset{..}{\underset{\underset{\textstyle H}{|}}{N}}-$H	H$-\overset{\overset{\textstyle H}{|}}{\underset{\underset{\textstyle H}{|}}{C}}-$H	H$-\overset{..}{\underset{..}{Cl}}:$
H_2O (8)	NH_3 (8)	CH_4 (8)	HCl (8)
Water	**Ammonia**	**Methane**	**Hydrogen chloride**

* *The number of valence electrons is shown in parentheses.*

Example 1.6

Draw Lewis structures, showing all valence electrons, for these molecules.

(a) CO_2 **(b)** CH_3OH **(c)** CH_3Cl

Solution

The number of valence electrons each molecule contains appears under the Lewis structure.

(a)

$$\ddot{O}=C=\ddot{O}$$

Carbon dioxide
(16 valence electrons)

(b)

$$H-\underset{\underset{H}{|}}{\overset{\overset{H}{|}}{C}}-\ddot{O}-H$$

Methanol
(14 valence electrons)

(c)

$$H-\underset{\underset{H}{|}}{\overset{\overset{H}{|}}{C}}-\ddot{\underset{..}{Cl}}:$$

Chloromethane
(14 valence electrons)

Problem 1.6

Draw Lewis structures, showing all valence electrons, for these molecules.

(a) C_2H_6 **(b)** CS_2 **(c)** HCN

D. Formal Charge

Throughout this course, we deal not only with molecules but also with polyatomic cations and anions. Examples of polyatomic cations are the hydronium ion, H_3O^+, and the ammonium ion, NH_4^+. An example of a polyatomic anion is the bicarbonate ion, HCO_3^-. It is important that you are able to determine which atom or atoms in a neutral molecule or polyatomic ion bear a positive or negative charge. The charge on an atom in a molecule or polyatomic ion is called its **formal charge**. To derive a formal charge:

1. Write a correct Lewis structure for the molecule or ion.
2. Assign to each atom all its unshared (nonbonding) electrons and half its shared (bonding) electrons.
3. Compare this number with the number of valence electrons in the neutral, unbonded atom. Remember that for a neutral unbonded atom the positive charge on the nucleus equals the number of its surrounding electrons. So, when there are more positive charges on the nucleus than counterbalancing negative charges from the surrounding electrons, that atom has a positive formal charge. Conversely, if the number of electrons assigned to a bonded atom is greater than that assigned to the unbonded atom, the atom has a negative formal charge.

> **Formal charge**
> The charge on an atom in a polyatomic ion or molecule.

$$\text{Formal charge} = \begin{array}{c}\text{Number of valence}\\ \text{electrons in the neutral,}\\ \text{unbonded atom}\end{array} - \left(\begin{array}{c}\text{All unshared}\\ \text{electrons}\end{array} + \begin{array}{c}\text{One half of all}\\ \text{shared electrons}\end{array}\right)$$

4. The sum of all the formal charges is equal to the total charge on the molecule or ion.

Example 1.7

Draw Lewis structures for these ions, and show which atom in each bears the formal charge.

(a) H_3O^+ **(b)** HCO_3^-

Solution

(a) The Lewis structure for the hydronium ion must show 8 valence electrons: 3 from the three hydrogens, 6 from oxygen, minus 1 for the single positive charge.

An oxygen atom has 6 valence electrons. The oxygen atom in H_3O^+ is assigned 2 unshared electrons and 1 from each shared pair of electrons, giving it a formal charge of $6 - (2 + 3) = +1$.

assigned 5 valence
electrons: formal
charge of +1 on O

$$H-\overset{\cdot\cdot}{\underset{|}{O}}{}^{+}-H$$
$$H$$

(b) The Lewis structure for the bicarbonate ion must show 24 valence electrons: 4 from carbon, 18 from the three oxygens, 1 from hydrogen, plus 1 for the single negative charge. Loss of a hydrogen ion from carbonic acid (Table 1.8) gives the bicarbonate ion. Carbon is assigned 1 electron from each shared pair and has no formal charge $(4 - 4 = 0)$. Two oxygens are assigned 6 valence electrons each and have no formal charges $(6 - 6 = 0)$. The third oxygen is assigned 7 valence electrons and has a formal charge of $6 - (6 + 1) = -1$.

assigned 7 valence
electrons: formal
charge of −1

$$:\overset{\cdot\cdot}{O}:$$
$$\|$$
$$H-\overset{\cdot\cdot}{\underset{\cdot\cdot}{O}}-C-\overset{\cdot\cdot}{\underset{\cdot\cdot}{O}}-H$$

$$:\overset{\cdot\cdot}{O}:$$
$$\|$$
$$H-\overset{\cdot\cdot}{\underset{\cdot\cdot}{O}}-C-\overset{\cdot\cdot}{\underset{\cdot\cdot}{O}}:{}^{-}$$

Carbonic acid, H_2CO_3   **Bicarbonate ion, HCO_3^-**

Problem 1.7

Draw Lewis structures for these ions, and show which atom in each bears the formal charge.

(a) $CH_3NH_3^+$ **(b)** CO_3^{2-} **(c)** OH^-

When writing Lewis structures for molecules and ions, you must remember that elements of the second period, including carbon, nitrogen, oxygen, and fluorine, can accommodate no more than 8 electrons in the four orbitals ($2s$, $2p_x$, $2p_y$, and $2p_z$) of their valence shells. Following are two Lewis structures for nitric acid, HNO_3, each with the correct number of valence electrons (24):

10 electrons in the
valence shell of nitrogen

$$:\overset{\cdot\cdot}{O}:$$
$$\|$$
$$H-\overset{\cdot\cdot}{\underset{\cdot\cdot}{O}}-\overset{+}{N}-\overset{\cdot\cdot}{\underset{\cdot\cdot}{O}}:{}^{-}$$

$$:\overset{\cdot\cdot}{O}:$$
$$\|$$
$$H-\overset{\cdot\cdot}{\underset{\cdot\cdot}{O}}-N=\overset{\cdot\cdot}{\underset{\cdot\cdot}{O}}:$$

An acceptable   Not an acceptable
Lewis structure   Lewis structure

The structure on the left is an acceptable Lewis structure. It shows the required 24 valence electrons, and each oxygen and nitrogen has a completed valence shell of eight electrons. Further, it shows a positive formal charge on nitrogen and a negative formal charge on one of the oxygens. Note that the sum of the formal charges on the acceptable Lewis structure for HNO_3 is zero. The structure on the right is not an acceptable Lewis structure. Although it shows the correct number of valence electrons, it places 10 electrons in the valence shell of nitrogen.

After enough practice counting valence electrons, you will begin to recognize functional groups based on the numbers of bonds and lone pairs on the atoms. For example, hydrogen has one bond and no lone pairs, neutral carbon with an octet has four bonds and no lone pairs, neutral nitrogen with an octet has three bonds and one lone pair, neutral oxygen has two bonds and two lone pairs, and neutral

Drawing Lewis structures from condensed structural formulas is a survival skill for organic chemistry students. There are three steps you should follow to draw a correct structure.

1. From a structural formula, obtain information about which atoms are bonded to each other in a molecule. Connect all of the appropriate atoms with single bonds (single lines) first.

   Example: $CH_3CH_2CH_2COOCH_3$

   Comment: The difficult part of this structure is deciding how to arrange the two oxygen atoms. Using the arrangement shown will produce a stable structure with filled valences for all of the atoms after you have completed Step 3 below. With practice you will begin to recognize common functional groups (Section 1.3) such as the carboxylic ester group (—COOCH$_3$ in this example). If you are unsure, you must draw the different possibilities you are considering, and upon completing the structure, determine which one produces the stable structure with the maximum number of filled valence shells around the atoms.

2. Determine how many electrons have been used for the bonds and how many remain. Add all of the additional valence electrons for each atom that does not already have a filled valence shell due to the single bonds. Remember to assign one electron to each atom taking part in a single bond for the purpose of counting valence electrons around atoms. Make sure to keep track of any formal charges that may be present in the condensed structural formula (the present example has none).

   Comment: Recall that each neutral carbon atom has four valence electrons, and each neutral oxygen atom has six valence electrons. After taking into account all of the single bonds in the molecule, the carbon atom bonded to both oxygen atoms has a single electron left over (4 total electrons − 3 single bonds = 1 electron left over), the oxygen atom bonded only to one carbon atom has five electrons left over (6 total electrons − 1 single bond = 5 electrons left over), and the other oxygen atom has four electrons left over (6 total electrons − 2 single bonds = 4 electrons left over).

3. Add multiple bonds to eliminate unpaired electrons. Draw the remaining nonbonding electrons as lone pairs.

   Comment: The only unpaired electrons were on carbon and oxygen, leading to one new bond being formed.

   The Lewis structure is now complete. The good news is that drawing Lewis structures will get easier with practice.

halogens have one bond and three lone pairs. When counting bonds for this analysis, double bonds count as two bonds, and triple bonds count as three bonds. For atoms with a formal charge, the number of bonds and lone pairs is altered. For example, positively charged nitrogen has four bonds and no lone pairs, positively charged oxygen has three bonds and one lone pair, and positively charged carbon has three bonds and no lone pairs (carbon has an unfilled valence shell).

Negatively charged carbon has three bonds and one lone pair, negatively charged nitrogen has two bonds and two lone pairs, and negatively charged oxygen has one bond and three lone pairs.

The guidelines for formal charge can be generalized. Any atom with an octet plus one bond more than its neutral form has a positive formal charge. Any atom with an octet and one bond less than its neutral form has a negative charge. For example, oxygen is neutral with an octet and two bonds, such as in water. Therefore, anytime that oxygen has an octet and three bonds it will be positive, but with an octet and only one bond it will be negative. We will see in the next section some atoms that have an octet and two bonds more than their neutral form, and hence have a plus two formal charge.

E. Exceptions to the Octet Rule

The Lewis model of covalent bonding focuses on valence electrons and the necessity for each atom other than H participating in a covalent bond to have a completed valence shell of eight electrons. Although most molecules formed by main-group elements (Groups 1A–7A) have structures that satisfy the octet rule, there are important exceptions to this rule.

One group of exceptions consists of molecules containing atoms of Group 3A elements. The following graphic is a Lewis structure for BF_3. In this uncharged covalent compound, boron is surrounded by only six valence electrons. Aluminum trichloride is an example of a compound in which aluminum, the element immediately below boron in Group 3A, has an incomplete valence shell. Because their valence shells are only partially filled, trivalent compounds of boron and aluminum exhibit a high reactivity with compounds that have extra electrons, enabling them to fill their octets (Section 4.7).

6 electrons in the valence shells of boron and aluminum

Boron trifluoride   **Aluminum trichloride**

1.3 | Functional Groups

Functional group
An atom or group of atoms within a molecule that shows a characteristic set of physical and chemical properties.

Carbon combines with other atoms (e.g., H, N, O, S, halogens) to form structural units called **functional groups**. Functional groups are important for three reasons. First, they are the units by which we divide organic compounds into classes. Second, they are sites of characteristic chemical reactions. A particular functional group, in all compounds that contain it, undergoes the same types of chemical reactions. Third, functional groups serve as a basis for naming organic compounds.

We introduce here several of the functional groups we encounter early in this course. At this point, our concern is only with pattern recognition. We shall have more to say about the structure and properties of these functional groups in following chapters. A complete list of the major functional groups we study in this text is presented on the inside front cover.

Alcohol
A compound containing an —OH (hydroxyl) group bonded to a tetrahedral carbon atom.

Hydroxyl group
An —OH group.

A. Alcohols

The functional group of an **alcohol** is an —OH (**hydroxyl**) group bonded to a tetrahedral carbon atom (a carbon having single bonds to four other atoms). Here is the Lewis structure of ethanol.

Ḧ

| | |
—C—Ö—H   H—C—C—Ö—H
| |
Ḧ Ḧ

Functional   **Ethanol**
group   (an alcohol)

We can also represent this alcohol in a more abbreviated form called a condensed structural formula. In a **condensed structural formula**, CH_3 indicates a carbon bonded to three hydrogens, CH_2 indicates a carbon bonded to two hydrogens, and CH indicates a carbon bonded to one hydrogen. Unshared pairs of electrons are generally not shown in a condensed structural formula. Thus the condensed structural formula for the alcohol with molecular formula C_2H_6O is CH_3—CH_2—OH. It is also common to write these formulas in an even more condensed manner, by omitting all single bonds: CH_3CH_2OH.

H H
| |
H—C—C—ÖH CH_3—CH_2—OH CH_3CH_2OH
| |
H H

Lewis structure   Condensed   Fully
structural formula   condensed
structural formula

Alcohols are classified as **primary (1°)**, **secondary (2°)**, or **tertiary (3°)** depending on the number of carbon atoms bonded to the carbon bearing the —OH group.

H   H CH_3
| | |
CH_3—C—OH CH_3—C—OH CH_3—C—OH
| | |
H CH_3 CH_3

A 1° alcohol   A 2° alcohol   A 3° alcohol

Primary (1°)
A compound containing a functional group bonded to a carbon atom bonded to only one other carbon atom and two hydrogens.

Secondary (2°)
A compound containing a functional group bonded to a carbon atom bonded to two other carbon atoms and one hydrogen atom.

Tertiary (3°)
A compound containing a functional group bonded to a carbon atom bonded to three other carbon atoms.

Example 1.8

Draw Lewis structures and condensed structural formulas for the two alcohols with the molecular formula C_3H_8O. Classify each as primary, secondary, or tertiary.

Solution

Begin by drawing the three carbon atoms in a chain. The oxygen atom of the hydroxyl group may be bonded to the carbon chain in two ways: either to an end carbon or to the middle carbon.

C—C—C
|
OH

HO—C—C—C   Same as

Same as   OH
|
C—C—C   C—C—C—OH   C—C—C

The carbon chain   Only two locations for the OH group

Finally, add seven more hydrogens for a total of eight shown in the molecular formula. Show unshared electron pairs on the Lewis structures but not on the condensed structural formulas.

Lewis structures   Condensed structural
                                          formulas

$$H-\overset{\overset{\displaystyle H}{|}}{\underset{\underset{\displaystyle H}{|}}{C}}-\overset{\overset{\displaystyle H}{|}}{\underset{\underset{\displaystyle H}{|}}{C}}-\overset{\overset{\displaystyle H}{|}}{\underset{\underset{\displaystyle H}{|}}{C}}-\overset{..}{\underset{..}{O}}-H$$

$CH_3CH_2CH_2OH$

A primary alcohol

$$H-\overset{\overset{\displaystyle H}{|}}{\underset{\underset{\displaystyle H}{|}}{C}}-\overset{\overset{\displaystyle :\overset{\displaystyle H}{O}:}{|}}{\underset{\underset{\displaystyle H}{|}}{C}}-\overset{\overset{\displaystyle H}{|}}{\underset{\underset{\displaystyle H}{|}}{C}}-H$$

$$\overset{\displaystyle OH}{CH_3CHCH_3}$$

A secondary alcohol

The secondary alcohol, whose common name is isopropyl alcohol, is the major component in rubbing alcohol. We also describe other functional groups such as halides as primary, secondary, and tertiary. For example,

$$H_3C-\overset{\overset{\displaystyle CH_3}{|}}{\underset{\underset{\displaystyle CH_3}{|}}{C}}-\overset{..}{\underset{..}{C}}l:\quad \text{is a tertiary halide.}$$

Problem 1.8

Draw Lewis structures and condensed structural formulas for the four alcohols with the molecular formula $C_4H_{10}O$. Classify each alcohol as primary, secondary, or tertiary.

B. Amines

Amino group
A compound containing a nitrogen atom bonded to one, two, or three carbon atom(s) by single bonds.

Primary (1°) amine
An amine in which nitrogen is bonded to one carbon and two hydrogens.

Secondary (2°) amine
An amine in which nitrogen is bonded to two carbons and one hydrogen.

Tertiary (3°) amine
An amine in which nitrogen is bonded to three carbons.

The functional group of an amine is an **amino group**, a nitrogen atom bonded to one, two, or three carbon atom(s) by single bonds. In a **primary (1°) amine**, nitrogen is bonded to one carbon atom. In a **secondary (2°) amine**, it is bonded to two carbon atoms, and in a **tertiary (3°) amine**, it is bonded to three carbon atoms. Notice that this classification scheme is different from that used with alcohols and halides!

$H-\overset{..}{N}-H$	$CH_3-\overset{..}{N}-H$	$CH_3-\overset{..}{N}-H$	$CH_3-\overset{..}{N}-CH_3$
$\underset{\displaystyle H}{\vert}$	$\underset{\displaystyle H}{\vert}$	$\underset{\displaystyle CH_3}{\vert}$	$\underset{\displaystyle CH_3}{\vert}$
Ammonia	**Methylamine** (a 1° amine)	**Dimethylamine** (a 2° amine)	**Trimethylamine** (a 3° amine)

Example 1.9

Draw condensed structural formulas for the two primary amines with the molecular formula C_3H_9N.

Solution

For a primary amine, draw a nitrogen atom bonded to two hydrogens and one carbon.

C—C—C—N—H   C—C—N—H   CH₃CH₂CH₂—N—H   CH₃CH—N—H

The three carbons may be bonded to nitrogen in two ways.

Add seven hydrogens to give each carbon four bonds and give the correct molecular formula.

Problem 1.9

Draw structural formulas for the three secondary amines with the molecular formula $C_4H_{11}N$.

C. Aldehydes and Ketones

The functional group of both aldehydes and ketones is the **C=O (carbonyl) group**. In formaldehyde, CH_2O, the simplest **aldehyde**, the carbonyl carbon is bonded to two hydrogens. In all other aldehydes, it is bonded to one hydrogen and one carbon. In a condensed structural formula, the aldehyde group may be written showing the carbon-oxygen double bond as —CH=O; alternatively, it may be written —CHO. In a **ketone**, the carbonyl carbon is bonded to two carbon atoms.

Carbonyl group
A C=O group.

Aldehyde
A compound containing a —CHO group.

Ketone
A compound containing a carbonyl group bonded to two carbons.

Functional group

Acetaldehyde
(an aldehyde)

Functional group

Acetone
(a ketone)

Example 1.10

Draw condensed structural formulas for the two aldehydes with the molecular formula C_4H_8O.

Solution

First, draw the functional group of an aldehyde and then add the remaining carbons, which may be bonded to the carbonyl group in two ways. Then, add seven hydrogens to complete the four bonds to each carbon.

$$\underset{\displaystyle \text{O}}{\overset{\displaystyle \text{O}}{CH_3CH_2CH_2\overset{\parallel}{C}H}}$$

or

$$CH_3CH_2CH_2CHO$$

$$\underset{CH_3}{\overset{\displaystyle \text{O}}{CH_3\overset{\parallel}{C}H\overset{\parallel}{C}H}}$$

or

$$\underset{CH_3}{CH_3CHCHO}$$

Problem 1.10

Draw condensed structural formulas for the three ketones with the molecular formula $C_5H_{10}O$.

D. Carboxylic Acids

Carboxylic acid
A compound containing a carboxyl, —COOH, group.

Carboxyl group
A —COOH group.

The functional group of a **carboxylic acid** is a —**COOH** (**carboxyl**: *carb*onyl + hydr*oxyl*) **group**. In a condensed structural formula, a carboxyl group may also be written —CO_2H.

$$-\overset{\displaystyle \overset{\cdot\cdot}{\underset{\cdot\cdot}{O}}}{\underset{}{\overset{\parallel}{C}}}-\overset{\cdot\cdot}{\underset{\cdot\cdot}{O}}-H$$

Functional group

$$CH_3-\overset{\displaystyle O}{\overset{\parallel}{C}}-\overset{\cdot\cdot}{\underset{}{O}}-H$$

Acetic acid
(a carboxylic acid)

Example 1.11

Draw a condensed structural formula for the single carboxylic acid with molecular formula $C_3H_6O_2$.

Solution

The only way the carbon atoms can be bonded is three in a chain, and the —COOH group must be on an end carbon of the chain.

$$CH_3-CH_2-\overset{\displaystyle \overset{\cdot\cdot}{O}}{\overset{\parallel}{C}}-\overset{\cdot\cdot}{\underset{\cdot\cdot}{O}}-H \quad \text{or} \quad CH_3CH_2COOH$$

Problem 1.11

Draw condensed structural formulas for the two carboxylic acids with the molecular formula $C_4H_8O_2$.

E. Carboxylic Esters

A **carboxylic ester**, commonly referred to as an **ester**, is a derivative of a carboxylic acid in which the hydrogen of the carboxyl group is replaced by a carbon-containing group.

Functional group   **Methyl acetate** (an ester)

Carboxylic ester
A derivative of a carboxylic acid in which H of the carboxyl group is replaced by a carbon.

Example 1.12

The molecular formula of methyl acetate is $C_3H_6O_2$. Draw the structural formula of another ester with this same molecular formula.

Solution

There is only one other ester with this molecular formula. Its structural formula is

$$H-\overset{\overset{\displaystyle O}{\|}}{C}-O-CH_2-CH_3$$

(we will usually write this $HCOOC_2H_5$)

Problem 1.12

Draw structural formulas for the four esters with the molecular formula $C_4H_8O_2$.

F. Carboxylic Amides

A **carboxylic amide**, commonly referred to as an **amide**, is a derivative of a carboxylic acid in which the —OH of the carboxyl group is replaced by an amine. As the model shows, the group is planar, something we will explain later.

Functional group   **Dimethylacetamide** (an amide)

Carboxylic amide
A derivative of a carboxylic acid in which the —OH is replaced by an amine.

1.4 | Bond Angles and Shapes of Molecules

In Section 1.2, we used a shared pair of electrons as the fundamental unit of a covalent bond and drew Lewis structures for several molecules and ions containing various combinations of single, double, and triple bonds. We can predict bond angles in these and other molecules and ions in a very straightforward way using a concept referred to as **valence-shell electron-pair repulsion (VSEPR)**. VSEPR is based on the electrons in an atom's valence shell. These valence electrons may be involved in the formation of single, double, or triple bonds, or they may be

VSEPR
A method for predicting bond angles based on the idea that electron pairs repel each other and keep as far apart as possible.

unshared (lone pair). Each combination creates a negatively charged region of space, and, because "like" charges repel each other, the various regions of electron density around an atom will spread out so that each is as far away from the others as possible.

Figure 1.2
A methane molecule, CH_4.
(a) Lewis structure and
(b) shape.

(a)   (b)

We use VSEPR in the following way to predict the shape of a methane molecule, CH_4. The Lewis structure for CH_4 shows a carbon atom surrounded by four regions of electron density, each of which contains a pair of electrons forming a bond to a hydrogen atom. According to VSEPR, the four regions radiate from carbon so that they are as far away from each other as possible. This occurs when the angle between any two pairs of electrons is 109.5°. Therefore, we predict all H—C—H bond angles to be 109.5°, and the shape of the molecule to be **tetrahedral** (Figure 1.2). The H—C—H bond angles in methane have been measured experimentally and found to be 109.5°, identical to those predicted.

We predict the shape of an ammonia molecule, NH_3, in exactly the same manner. The Lewis structure of NH_3 shows nitrogen surrounded by four regions of electron density. Three regions contain single pairs of electrons forming covalent bonds with hydrogen atoms. The fourth region contains an unshared pair of electrons (Figure 1.3). Using VSEPR, we predict that the four regions of electron density around nitrogen are arranged in a tetrahedral manner, that H—N—H bond angles are 109.5°, and that the shape of the molecule is **pyramidal** (like a triangular pyramid). The observed bond angles are 107.3°. This small difference between the predicted and observed angles can be explained by proposing that the unshared pair of electrons on nitrogen repels adjacent electron pairs more strongly than do bonding pairs.

Figure 1.4 shows a Lewis structure and a ball-and-stick model of a water molecule. In H_2O, oxygen is surrounded by four regions of electron density. Two of these regions contain pairs of electrons used to form single covalent bonds to the two hydrogens; the remaining two contain unshared electron pairs. Using VSEPR, we predict that the four regions of electron density around oxygen repel each other and are arranged in a tetrahedral manner. The predicted H—O—H bond angle is 109.5°. Experimental measurements show that the actual bond angle is 104.5°, a value smaller than that predicted. This difference between the predicted and observed bond angles can be explained by proposing, as we did for NH_3, that unshared pairs of electrons repel adjacent pairs more strongly than do bonding pairs. Note that the distortion from 109.5° is greater in H_2O, which has two unshared pairs of electrons, than it is in NH_3, which has only one unshared pair.

A general prediction emerges from this discussion of the shapes of CH_4, NH_3, and H_2O molecules. If a Lewis structure shows four regions of electron density around a central atom, VSEPR predicts a tetrahedral distribution of electron density and bond angles of approximately 109.5°.

In many of the molecules we shall encounter, an atom is surrounded by three regions of electron density. Figure 1.5 shows Lewis structures and ball-and-stick models for formaldehyde, CH_2O, and ethylene, C_2H_4.

According to VSEPR, a double bond is treated as a single region of electron density. In formaldehyde, carbon is surrounded by three regions of electron density: two regions contain single pairs of electrons forming single bonds to hydrogen atoms while the third region contains two pairs of electrons forming a double bond to oxygen. In ethylene, each carbon atom is also surrounded by three regions of

Tetrahedral
A bonding arrangement in which an atom is bonded to four atoms located at the corners of a tetrahedron.

Pyramidal
A bonding arrangement in which an atom is bonded to three atoms in a triangular pyramid.

(a)

H—N̈—H
   |
   H

(b)

Unshared electron pair

107.3°

Figure 1.3

An ammonia molecule, NH_3.
(a) Lewis structure and
(b) shape.

(a)

H—Ö—H

(b)

Unshared electron pairs

104.5°

Figure 1.4

A water molecule, H_2O. (a) Lewis structure and (b) shape.

Formaldehyde

(a) Top view

(b) Side view

Indicates bond is projecting away from viewer

Indicates bond is projecting toward viewer

116.5°
121.8°

Ethylene

(a) Top view

(b) Side view

117.2°
121.4°

Figure 1.5
Shapes of formaldehyde, CH_2O, and ethylene, C_2H_4. Molecules shown from (a) top view and (b) side view. Note that chemists commonly use solid wedges to represent bonds projecting toward the viewer, and broken wedges for bonds projecting away from the viewer.

electron density: two contain single pairs of electrons, and the third contains two pairs of electrons.

Three regions of electron density about an atom are farthest apart when they are coplanar (in the same plane) and make angles of 120° with each other. Thus, the predicted H—C—H and H—C—O bond angles in formaldehyde and the predicted H—C—H and H—C—C bond angles in ethylene are all 120° and the atoms are coplanar. The experimentally measured angles are quite close to this prediction, as shown in Figure 1.5.

In still other types of molecules, a central atom is surrounded by only two regions of electron density. Figure 1.6 shows Lewis structures and ball-and-stick models of carbon dioxide, CO_2, and acetylene, C_2H_2.

In carbon dioxide, carbon is surrounded by two regions of electron density: each contains two pairs of electrons and forms a double bond to an oxygen atom. In acetylene, each carbon is also surrounded by two regions of electron density. One contains a single pair of electrons and forms a single bond to a hydrogen atom, and the other contains three pairs of electrons and forms a triple bond to a carbon atom. In each case, the two regions of electron density are farthest apart if they form a straight line through the central atom and create an angle of 180°. Both carbon dioxide and acetylene are **linear** molecules. Predictions of VSEPR are summarized in Table 1.9.

(a) $\ddot{O}=C=\ddot{O}$

Carbon dioxide

(b) $H—C\equiv C—H$

Acetylene

Figure 1.6
Shapes of (a) carbon dioxide, CO_2, and (b) acetylene, C_2H_2, molecules.

Table 1.9	Predicted Molecular Shapes (VSEPR)			
Regions of Electron Density Around Central Atom	**Predicted Distribution of Electron Density**	**Predicted Bond Angles**	**Examples**	
4	Tetrahedral	109.5°		
3	Trigonal planar	120°		
2	Linear	180°	$\ddot{O}=C=\ddot{O}$ $H—C\equiv C—H$	

Example 1.13

Predict all bond angles in these molecules.

(a) CH_3Cl (b) $CH_2{=}CHCl$

Solution

(a) The Lewis structure for CH_3Cl shows carbon surrounded by four regions of electron density. Therefore, we predict the distribution of electron pairs about carbon to be tetrahedral, all bond angles to be 109.5°, and the shape of CH_3Cl to be tetrahedral. The actual H—C—Cl bond angle is 108°.

(b) The Lewis structure for $CH_2{=}CHCl$ shows each carbon surrounded by three regions of electron density. Therefore, we predict all bond angles to be 120°. The actual C—C—Cl bond angle is 122.4°.

Problem 1.13

Predict all bond angles for these molecules.

(a) CH_3OH (b) PF_3 (c) H_2CO_3

1.5 | Polar and Nonpolar Molecules

We can now combine our understanding of bond polarity (Section 1.2B) and molecular geometry (Section 1.4) to predict the polarity of polyatomic molecules. As we shall see, to be polar, a molecule must have one or more polar bonds. But, as we shall also see, not every molecule with polar bonds is polar.

To predict whether a molecule is polar, we need to determine (1) if the molecule has polar bonds and (2) the arrangement of its atoms in space (using VSEPR, for example). The **molecular dipole moment (μ)** of a molecule is the vector sum of its individual bond dipoles. Sometimes, the bond dipoles exactly cancel each other due to a molecule's geometry. In carbon dioxide, for example, each C—O bond is polar with oxygen, the more electronegative atom, bearing a partial negative charge and with carbon bearing a partial positive charge. Because carbon dioxide is a linear molecule, the vector sum of its two bond dipoles is zero; therefore, the dipole moment of a CO_2 molecule is zero. Boron trifluoride is planar with bond angles of 120°. Although each B—F bond is polar, the vector sum of its bond dipoles is zero, and BF_3 has no dipole moment. Carbon tetrachloride is tetrahedral with bond angles of 109.5°. Although it has four polar C—Cl bonds, the vector sum of its bond dipoles is zero, and CCl_4 also has no dipole moment.

Molecular dipole moment (μ)
The vector sum of individual bond dipoles.

Fullerene—A New Form of Carbon

A favorite chemistry examination question is: What are the elemental forms of carbon? For a long time, the answer was that pure carbon is found in two forms: graphite and diamond. These forms have been known for centuries, and it was generally believed that they are the only forms of carbon having extended networks of C atoms in well-defined structures. But that is not so! The scientific world was startled in 1985 when Richard Smalley of Rice University and Harry W. Kroto of the University of Sussex, UK, and their coworkers announced that they had detected a new form of carbon with the molecular formula C_{60}. They suggested that the molecule has a structure that resembles a soccer ball; it has 12 five-membered rings and 20 six-membered rings arranged such that each five-membered ring is surrounded by 5 six-membered rings. This structure reminded its discoverers of a geodesic dome, a structure invented by the innovative American engineer and philosopher R. Buckminster Fuller. Therefore, the official name of this allotrope of carbon has become *fullerene*. Most chemists refer to it as C_{60} or fullerene; some call it "buckyball." Kroto, Smalley, and Robert F. Curl were awarded the 1996 Nobel Prize in Chemistry for this work. Some other fullerenes, such as C_{70} and C_{84}, have now been isolated and studied as well.

Fullerenes have a rich chemistry. They behave as if they have electron-deficient double bonds. Many different fullerene adducts with a variety of structures have been prepared. Cationic derivatives of these adducts, for example, bind tightly to DNA and can be used to visualize it by electron microscopy.

An important development in this field has been the preparation of single-wall nanotubes, which are based on C_{60} or higher fullerenes and are extended for a very long distance to make long molecules that are hundreds of times stronger than steel and can act as molecular wires. Nanotubes can be used as the probe (a very sharp tip) in atomic force microscopes, which can be used to image single molecules. The nanotubes make the sharpest possible tips because they are of molecular dimensions. Nanotubes are also playing an increasingly important role in the field of nanotechnology, including applications in the area of molecular electronics. Nanotubes can be superconducting or semiconducting, depending on the exact structure of the nanotube.

A fullerene (C_{60}) molecule.

A nanotube.

Yves Rubin, UCLA

Carbon dioxide
$\mu = 0$ D

Boron trifluoride
$\mu = 0$ D

Carbon tetrachloride
$\mu = 0$ D

Other molecules, such as water and ammonia, have polar bonds and dipole moments greater than zero; they are polar molecules. Each O—H bond in a water molecule and each N—H bond in ammonia are polar, with oxygen and nitrogen, the more electronegative atoms, bearing a partial negative charge and each hydrogen bearing a partial positive charge.

The charge densities are easily computed by modern desktop computer programs such as Spartan. Here are electrostatic potential maps that display the computed electronic charge density in water and ammonia. In these models, red represents negative charge and blue represents positive charge. In agreement with the dipole moment diagram and our expectations, the more electronegative atom has substantial negative charge in both molecules. For more information on how to interpret these plots, see Appendix 7.

direction
of dipole
moment
in water

Water
$\mu = 1.85$ D

An electrostatic potential map (elpot) of a water molecule.

direction
of dipole
moment
in ammonia

Ammonia
$\mu = 1.47$ D

An electrostatic potential map (elpot) of an ammonia molecule.

An electrostatic potential map (elpot) of a formaldehyde molecule.

Example 1.14

Which of these molecules are polar? For each that is polar, specify the direction of its dipole moment.

(a) CH_3Cl **(b)** CH_2O **(c)** C_2H_2

Solution

Both chloromethane, CH_3Cl, and formaldehyde, CH_2O, have polar bonds and, because of their geometries, are polar molecules. Because of its linear geometry, acetylene, C_2H_2, has no dipole moment. The experimentally measured dipole moments are shown. The electrostatic potential map (elpot) of formaldehyde clearly shows this charge distribution.

(a)

Chloromethane
$\mu = 1.87$ D

(b)

Formaldehyde
$\mu = 2.33$ D

(c) H—C≡C—H

Acetylene
$\mu = 0$ D

Problem 1.14

Which molecules are polar? For each that is polar, specify the direction of its dipole moment.

(a) CH_2Cl_2 **(b)** HCN   **(c)** H_2O_2

1.6 | Quantum or Wave Mechanics

Thus far in this chapter, we have concentrated on the Lewis model of bonding and on VSEPR. The Lewis model deals primarily with the coordination numbers of atoms (the number of bonds a given atom can form), and VSEPR deals primarily with bond angles and molecular geometries. Although each is useful in its own way, neither gives us any means of accounting in a quantitative or even semiquantitative way for the reasons atoms combine in the first place to form covalent bonds with the liberation of energy. At this point, we need to study an entirely new approach to the theory of covalent bonding, one that provides a means of understanding not only the coordination numbers of atoms and molecular geometries but also the energetics of chemical bonding.

A. Moving Particles Exhibit the Properties of a Wave

The beginning of this new approach to the theory of covalent bonding was provided by Albert Einstein (1879–1955), a German-born American physicist. In 1905, Einstein postulated that light consists of photons of electromagnetic radiation. The energy, E, of a photon is proportional to the frequency, v (Greek nu), of the light. The proportionality constant in this equation is Planck's constant, h.

$$E = hv$$

In 1923, the French physicist Louis de Broglie followed Einstein's lead and advanced the revolutionary idea that if light exhibits properties of particles in motion, then a particle in motion should exhibit the properties of a wave. He proposed that a particle of mass m and speed v has an associated wavelength λ (Greek lambda), given by the equation

$$\lambda = \frac{h}{mv} \qquad \text{(the de Broglie relationship)}$$

Illustrated in Figure 1.7 is a wave such as might result from plucking a guitar string. The mathematical equation that describes this wave is called a **wave equation**. The numerical value(s) of the solution(s) of a wave equation may be positive (corresponding to a wave crest), negative (corresponding to a wave trough), or zero. The sign of the numerical value of the wave equation is called the phase, and changes in sign are referred to as phase changes. Phase changes occur at nodes. A **node** is any point where the value of a solution of a wave equation is zero. A **nodal plane** is any plane perpendicular to the direction of propagation that runs through a node. Shown in Figure 1.7 are three nodal planes.

Node
A point in space where the value of a solution of a wave equation is zero.

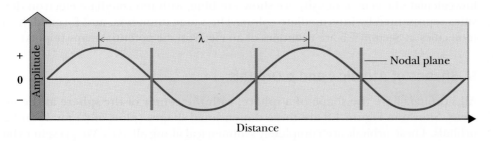

Figure 1.7
Characteristics of a wave associated with a moving particle. Wavelength is designated by the symbol λ.

Erwin Schrödinger built on the idea of de Broglie and in 1926 proposed an equation that could be used to describe the wave properties associated with an electron in an atom or a molecule. **Quantum mechanics (wave mechanics)** is the branch of science that studies particles and their associated waves. Solving the Schrödinger equation gives a set of solutions called **wave functions**. Each wave function ψ (Greek psi) is associated with a unique set of quantum numbers and with a particular atomic or molecular orbital. A wave function occupies three-dimensional space and is called an **orbital**. Each orbital can contain no more than two electrons. The value of ψ^2 is proportional to the probability of finding an electron at a given point in space. Looked at in another way, the value of ψ^2 at any point in space is proportional to the electron density at that point. A plot of electron density (ψ^2) in a given orbital theoretically reaches to infinity but becomes vanishingly small at long distances from the nucleus. Notice that although the value of ψ at any point can be positive or negative, the value of ψ^2 will always be positive in an orbital. In other words, the electron density in two regions of an orbital will be equal if those regions have the same absolute value of ψ, regardless of whether that value is negative or positive. Of course, ψ^2 will be zero at a node. We often represent orbitals as

Quantum mechanics
The branch of science that studies particles and their wavelike properties.

Wave function
A solution to a set of equations that defines the energy of an electron in an atom and the region of space it may occupy.

a solid or mesh with the surface representing the volume within which some amount (such as 95%) of the electron density is contained.

For most aspects of organic chemistry, it is best to consider the wavelike properties of electrons. In this text, we concentrate on wave functions and shapes associated with s and p atomic orbitals because they are the orbitals most often involved in covalent bonding in organic compounds. The wave nature is reflected by the orbital that the electron resides in. We should not think about finding an electron as a particle at a particular location in space, but rather, we should consider the electron density in various regions of space as the square of the amplitude of the orbital at that position in space (see previous paragraph).

When we describe orbital interactions we are referring to interactions of waves. Waves interact constructively or destructively (adding or subtracting). When two waves overlap, positive **phasing** adds constructively with positive phasing, as does negative phasing with negative phasing. However, positive and negative phasing also add destructively, meaning they cancel. For waves on the ocean or on a plucked guitar string, this characteristic is sometimes referred to as waves adding "in phase" and "out of phase."

There is one reason to consider the particle nature of electrons, and that is charge. Negatively charged particles, such as electrons, are attracted to positively charged particles, such as nuclei, while being repulsed by other negative particles, such as other electrons.

It is convenient to add up the electron densities in all of the orbitals in a molecule and then to determine which areas of a molecule have larger and smaller amounts of electron density. In general, the greater electron density is on the more electronegative atoms, especially those with lone pairs. Relative electron density distribution in molecules is important because it allows us to identify sites of chemical reactivity. Many reactions involve an area of relatively high electron density on one molecule reacting with an area of relatively low electron density on another molecule. It is convenient to keep track of overall molecular electron density distributions using computer graphics. This text presents electrostatic potential maps (elpots) in which areas of relatively high calculated electron density are shown in red and areas of relatively low calculated electron density are shown in blue, with intermediate electron densities represented by intermediate colors. The water, ammonia, and formaldehyde molecules in Section 1.5 are examples of an electrostatic potential maps (elpots).

B. Shapes of Atomic *s* and *p* Orbitals

All s orbitals have the shape of a sphere, with the center of the sphere at the nucleus. Shown in Figure 1.8 are three dimensional shapes (plots of ψ) for $1s$ and $2s$ orbitals. These orbitals are completely symmetrical along all axes. We present orbitals that were calculated using the Schrödinger equation, as well as the common cartoons that organic chemists use to represent them. Mesh diagrams are used throughout this book because they allow you to see the inner features of orbitals, such as the change in phase in the $2s$ orbital. You should not consider the lines of the mesh as trajectories of electrons, but rather that the mesh shows a surface within which about 95% of the electron density exists.

Shown in Figure 1.9 are the three-dimensional shapes (plots of ψ) of the three $2p$ orbitals, combined in one diagram to illustrate their relative orientations in space. Each $2p$ orbital consists of two lobes arranged in a straight line with the nucleus in the middle. The three $2p$ orbitals are mutually perpendicular and are designated $2p_x$, $2p_y$, and $2p_z$. The sign of the wave function of a $2p$ orbital is positive in one lobe, zero at the nucleus, and negative in the other lobe. The plus or minus is simply the sign of the mathematical function ψ_{2p} and has no relationship to energy or electron distribution. These signs are shown by blue or red colors; however, these colors should not be confused with the colors used to represent charge density. Recall that the value of ψ^2 is always positive, so the probability of finding electron density in the $(+)$ lobe of a $2p$ orbital (i.e., value of ψ is positive) is the same as that of finding it in the $(-)$ lobe (i.e., value of ψ is negative). Again, electron density is zero at the node.

Figure 1.8

(a) Calculated and
(b) cartoon representations of
the 1s and 2s atomic orbitals
showing an arbitrary boundary
containing about 95% of the
electron density. Note that the 2s
orbital has a phase change within
the spherical electron density.

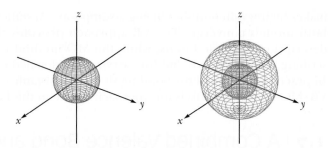

(a) 1s and 2s orbitals computed using the Schrödinger equation

(b) Cartoon representations of 1s and 2s orbitals

Figure 1.9

Three-dimensional
representations of the $2p_x$, $2p_y$,
and $2p_z$ atomic orbitals and their
orientation in space relative to
one another. Nodal planes are
shaded. Note that the lobes of
the computed orbitals (a) are
more spherical in appearance
than the bowling-pin-like
cartoon drawings (b) chemists
routinely use to represent them.

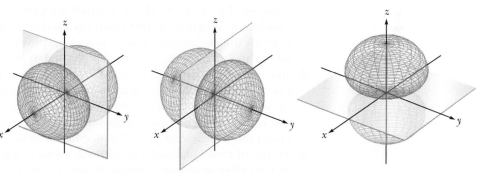

(a) 2p Orbitals computed using the Schrödinger equation

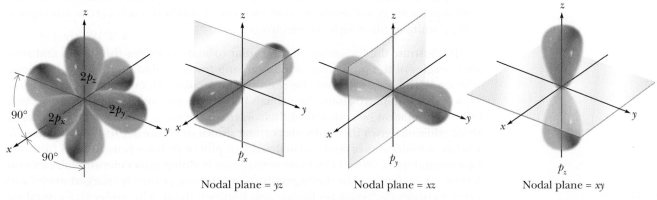

(b) Cartoon representations of 2p orbitals

Besides providing a way to determine the shapes of atomic orbitals, the Schrödinger equation also provides a way, at least approximately, to quantify the energetics of covalent bond formation. These approximations have taken two forms: (1) valence bond (VB) theory and (2) molecular orbital (MO) theory. Both theories of chemical bonding use the methods of quantum mechanics, but each makes slightly different simplifying assumptions. At sufficiently high levels of theory, both models converge. The VB approach provides the most easily visualized description of single bonds, while the MO method is most convenient for describing multiple bonds and for carrying out detailed calculations on computers. In practice, the model for bonding that most organic chemists use is a combined VB/MO theory, and this is the approach we use in this book.

1.7 | A Combined Valence Bond and Molecular Orbital Theory Approach to Covalent Bonding

A. Molecular Orbital Theory; Formation of Molecular Orbitals

Molecular orbital (MO) theory
A theory of chemical bonding in which electrons in molecules occupy molecular orbitals that extend over the entire molecule and are formed by the combination of the atomic orbitals that make up the molecule.

Molecular orbital (MO) theory begins with the hypothesis that electrons in atoms exist in atomic orbitals and assumes that electrons in molecules exist in molecular orbitals. Just as the Schrödinger equation can be used to calculate the energies and shapes of atomic orbitals, molecular orbital theory assumes that the Schrödinger equation can also be used to calculate the energies and shapes of molecular orbitals. Following is a summary of the rules used in applying molecular orbital theory to the formation of covalent bonds.

1. Combination of n atomic orbitals (mathematically adding and subtracting wave functions) forms a set of n molecular orbitals (new wave functions); that is, the number of molecular orbitals formed is equal to the number of atomic orbitals combined. When only MO theory is used to model bonding in organic compounds the molecular orbitals are spread over all atoms in a molecule or ion whose atomic orbitals are properly aligned to overlap with one another.

2. Just like atomic orbitals, molecular orbitals are arranged in order of increasing energy. It is possible to calculate reasonably accurate relative energies of a set of molecular orbitals. Experimental measurements such as those derived from molecular spectroscopy can also be used to provide very detailed information about the relative energies of molecular orbitals.

3. Filling of molecular orbitals with electrons is governed by the same principles as the filling of atomic orbitals. Molecular orbitals are filled beginning with the lowest energy unoccupied molecular orbital (the Aufbau principle). A molecular orbital can accommodate no more than two electrons, and their spins must be paired (the Pauli exclusion principle). When two or more molecular orbitals of equal energy are available, one electron is added to each before any equivalent orbital is filled with two electrons.

To illustrate the formation of molecular orbitals, consider the shapes and relative energies of the molecular orbitals arising from combination of two $1s$ atomic orbitals. With only two atoms the resulting molecular orbitals are simple to visualize, whereas with more atoms the orbitals become increasingly complex (see Section 1.9). Combination by addition of their wave functions (a process referred to as in phase addition) gives the molecular orbital shown in Figure 1.10(a). As with atomic orbitals, a molecular orbital is visualized as a plot of its wave function (ψ) in three-dimensional space. When electrons occupy this bonding molecular orbital, electron density is concentrated in the region between the two positively charged nuclei and serves to offset the repulsive interaction between them. The molecular orbital we have just described is called a sigma bonding molecular orbital and is given the symbol σ_{1s} (pronounced sigma one ess). A **bonding molecular orbital** is one in which electrons have a lower energy than they would in the isolated atomic orbitals. A **sigma (σ) bonding molecular orbital** is one in which electron density lies between the two nuclei, along the axis joining them, and is *cylindrically symmetric* about the axis.

Combination of two $1s$ atomic orbitals by subtraction of their wave functions (a process referred to as out of phase addition) gives the molecular orbital shown in Figure 1.10(b). If electrons occupy this orbital, electron density is concentrated

Bonding molecular orbital
A molecular orbital in which electrons have a lower energy than they would in isolated atomic orbitals.

Sigma (σ) molecular orbital
A molecular orbital in which electron density is concentrated between two nuclei, along the axis joining them, and is cylindrically symmetric.

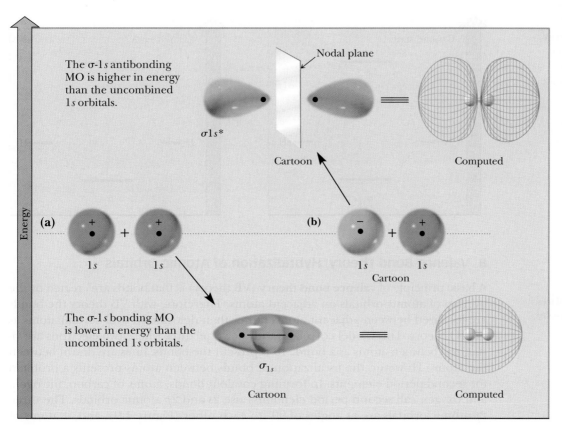

Figure 1.10

Molecular orbitals (plots of ψ) derived from combination of two $1s$ atomic orbitals: (a) combination by addition and (b) combination by subtraction. Electrons in the bonding MO spend most of their time in the region between the two nuclei and bond the atoms together. Electrons in the antibonding MO lead to repulsion between nuclei and decrease bonding.

outside the region between the two nuclei. There is a nodal plane, or plane of zero electron density, between the atoms. This molecular orbital is called a sigma anti-bonding molecular orbital and is given the symbol σ^*_{1s} (pronounced sigma star one ess). An **antibonding molecular orbital** is one in which the electrons in it have a higher energy (are more easily removed) than they would have in the isolated atomic orbitals. Population of this orbital with electrons actually causes repulsion of the nuclei involved. An asterisk (*) is used to indicate that a molecular orbital is antibonding.

The **ground state** of an atom or molecule is its state of lowest energy. In the ground state of a hydrogen molecule, the two electrons occupy the σ_{1s} MO with paired spins. An **excited state** is any electronic state other than the ground state. In the lowest excited state of the hydrogen molecule, one electron occupies the σ_{1s} MO, and the other occupies the σ^*_{1s} MO. There is no net bonding in this excited state, and dissociation will result. Energy-level diagrams of the ground state and the low-est excited state of the hydrogen molecule are shown in Figure 1.11. Under normal circumstances, we do not have to consider antibonding orbitals because they are unoccupied.

A key feature of MO theory is that molecular orbitals extend over entire molecules because all of the orbitals of all of the atoms take part in constructing molecular orbitals. A second key feature is that the molecular orbitals are created by the in phase and out of phase addition (sometimes called subtraction) of all the atomic orbitals that are aligned to overlap on all atoms in a molecule. These two features are ex-tremely powerful when generating quantitative computational models of molecules. However, the full MO description of molecules is not particularly useful for students trying to understand and visualize covalent bonding and structures with σ bonds. To understand and visualize σ bonds in molecules, an approach called valence bond theory (VB theory) is more useful. Next, we describe some of the basic principles of VB theory, and then combine MO and VB theories to arrive at the method most commonly applied to the description of bonding in organic molecules.

Antibonding molecular orbital
A molecular orbital in which electrons have a higher energy than they would in isolated atomic orbitals.

Ground state
The lowest energy state of an atom or molecule.

Excited state
Any electronic state other than the ground state. Will be higher in energy than the ground state.

Figure 1.11

A molecular orbital energy diagram for the hydrogen molecule, H_2. (a) Ground state and (b) lowest excited state.

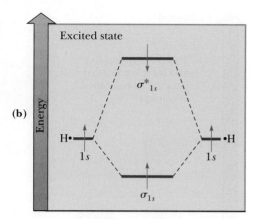

B. Valence Bond Theory; Hybridization of Atomic Orbitals

Valence bond theory
A model of bonding that places electron pairs between adjacent atoms to create bonds

A basic principle of **valence bond theory** (VB theory) is that bonds are created by the overlap of atomic orbitals on adjacent atoms. Therefore, with VB theory the bonds are localized between adjacent atoms rather than delocalized over several atoms as in MO theory. This model correlates with Lewis pictures where two electrons are visualized between atoms as a bond. To represent the bonds, lines are drawn between the atoms. However, the localization of bonds between atoms presents a problem for second-period elements. In forming covalent bonds, atoms of carbon, nitrogen, and oxygen (all second-period elements) use $2s$ and $2p$ atomic orbitals. The three $2p$ atomic orbitals are at angles of 90° to each other (Figure 1.9), and, if atoms of second-period elements used these orbitals to form covalent bonds, we would expect bond angles around each would be approximately 90°. However, we rarely observe bond angles of 90° in organic molecules. What we find instead are bond angles of approximately 109.5° in molecules with only single bonds, 120° in molecules with double bonds, and 180° in molecules with triple bonds, as shown in Table 1.9.

Hybrid orbital
An atomic orbital formed by the combination of two or more atomic orbitals.

To account for the observed bond angles in a way that is intuitive for chemists, Linus Pauling proposed that atomic orbitals for each atom should be thought of as first combining to form new atomic orbitals, called **hybrid orbitals**, which then interact to form bonds by overlapping with orbitals from other atoms. The hybrid orbitals have the bond angles we observe around each atom, so molecular structure and bonding based on the overlap of hybrid orbitals provides an intuitive understanding. Being able to construct organic molecules from the overlap of hybrid orbitals is an essential organic chemistry survival skill.

Hybridization
The combination of atomic orbitals of different types.

Hybrid orbitals are formed by combinations of atomic orbitals, a process called **hybridization**. Mathematically, this is accomplished by combining the wave functions of the $2s$ (ψ_{2s}) and three $2p$ (ψ_{2p_x}, ψ_{2p_y}, ψ_{2p_z}) orbital wave functions. The number of hybrid orbitals formed is equal to the number of atomic orbitals combined. Elements of the second period form three types of hybrid orbitals, designated sp^3, sp^2, and sp, each of which can contain up to two electrons.

sp^3 Hybrid Orbitals—Bond Angles of Approximately 109.5°

sp^3 Hybrid orbital
A hybrid atomic orbital formed by the combination of one s atomic orbital and three p atomic orbitals.

The mathematical combination of the $2s$ atomic orbital and three $2p$ atomic orbitals forms four equivalent sp^3 **hybrid orbitals** described by four new wave functions. Plotting ψ for the four new wave functions gives a three-dimensional visualization of the four sp^3 hybrid orbitals. Each sp^3 hybrid orbital consists of a larger lobe pointing in one direction and a smaller lobe of opposite sign pointing in the opposite direction. The axes of the four sp^3 hybrid orbitals are directed toward the corners of a regular tetrahedron, and sp^3 hybridization results in bond angles of approximately 109.5° (Figure 1.12). Note that each sp^3 orbital has 25% s-character and 75% p-character because those are the percentages of the orbitals combined when constructing them (one $2s$ orbital, three $2p$ orbitals). We will refer back to these percentages several times in the text.

Computed

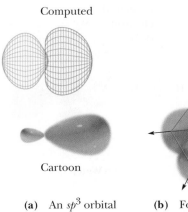

Cartoon

(a) An sp^3 orbital

109.5°

(b) Four sp^3 orbitals in a tetrahedral arrangement

(c)

© Cengage Learning/Charles D. Winters

Figure 1.12

sp^3 Hybrid orbitals.
(a) A single sp^3 hybrid orbital in computed and cartoon form. (b) Three-dimensional cartoon representation of four sp^3 hybrid orbitals centered on the same atom and directed toward the corners of a regular tetrahedron. (c) If four balloons of similar size and shape are tied together, they will naturally assume a tetrahedral geometry.

Atoms with four sp^3 hybrid atomic orbitals are referred to as **sp^3 hybridized**, or as having a hybridization state of sp^3. A diagram depicting this hybridization shows the creation of four orbitals of equal energy.

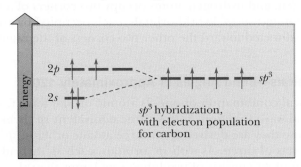

sp^3 hybridization, with electron population for carbon

Population of these orbitals with electrons follows the standard rules for populating atomic orbitals. For example, for C with four valence electrons, each sp^3 orbital contains a single electron. Carbon can therefore make four bonds, one with each sp^3 hybrid. Nitrogen, with five valence electrons, would have a lone pair and three singly occupied sp^3 orbitals. The orbital populations of O and F follow in an analogous fashion.

You must remember that superscripts in the designation of hybrid orbitals tell you how many atomic orbitals have been combined to form the hybrid orbitals. You know that the designation sp^3 represents a hybrid orbital because it shows a combination of s and p orbitals. The superscripts in this case tell you that *one s* atomic orbital and *three p* atomic orbitals are combined in forming the hybrid orbital. Do not confuse this use of superscripts with that used in writing a ground-state electron configuration, as for example $1s^2 2s^2 2p^5$ for fluorine. In the case of a ground-state electron configuration, superscripts tell you the number of electrons in each orbital or set of orbitals.

In Section 1.2, we described the covalent bonding in CH_4, NH_3, and H_2O in terms of the Lewis model, and in Section 1.4 we used VSEPR to predict bond angles of approximately 109.5° in each molecule. Now let us consider the bonding in these molecules in terms of the overlap of hybrid atomic orbitals. To bond with four other atoms with bond angles of 109.5°, carbon uses sp^3 hybrid orbitals. Carbon has four valence electrons, and one electron is placed in each sp^3 hybrid orbital. Each partially filled sp^3 hybrid orbital then overlaps with a partially filled $1s$ atomic orbital of hydrogen to form the four sigma (σ) bonds of methane, and hydrogen atoms occupy the corners of a regular tetrahedron (Figure 1.13). We address how to create and model these σ bonds in Section 1.7C.

In bonding with three other atoms, the five valence electrons of nitrogen are distributed so that one sp^3 hybrid orbital is filled with a pair of electrons (the lone pair) and the other three sp^3 hybrid orbitals have one electron each. Overlapping σ bonds formed by overlap of sp^3 and $1s$ orbitals

$$H-\overset{\overset{\displaystyle H}{|}}{\underset{\underset{\displaystyle H}{|}}{C}}-H$$

Methane

Unshared electron pair

$$H-\overset{\displaystyle \cdot\cdot}{N}-H$$
$$\underset{\displaystyle H}{|}$$

Ammonia

Unshared electron pairs

$$H-\overset{\cdot\cdot}{\underset{\cdot\cdot}{O}}-H$$

Water

Figure 1.13
Orbital overlap pictures of methane, ammonia, and water.

of these partially filled sp^3 hybrid orbitals with $1s$ atomic orbitals of three hydrogen atoms produces an NH_3 molecule (Figure 1.13).

In bonding with two other atoms, the six valence electrons of oxygen are distributed so that two sp^3 hybrid orbitals are filled, and the remaining two have one electron each. Each partially filled sp^3 hybrid orbital overlaps with a $1s$ atomic orbital of hydrogen, and hydrogen atoms occupy two corners of a regular tetrahedron. The remaining two sp^3 hybrid orbitals, each occupied by an unshared pair of electrons, are directed toward the other two corners of the regular tetrahedron (Figure 1.13).

sp^2 Hybrid Orbitals—Bond Angles of Approximately 120°

sp^2 Hybrid orbital
A hybrid atomic orbital formed by the combination of one s atomic orbital and two p atomic orbitals.

The mathematical combination of one $2s$ atomic orbital wave function and two $2p$ atomic orbital wave functions forms three equivalent **sp^2 hybrid orbital** wave functions. Because they are derived from three atomic orbitals, sp^2 hybrid orbitals always occur in sets of three. As with sp^3 orbitals, each sp^2 hybrid orbital (three-dimensional plot of ψ) consists of two lobes, one larger than the other. The axes of the three sp^2 hybrid orbitals lie in a plane and are directed toward the corners of an equilateral triangle; the angle between sp^2 hybrid orbitals is 120°. The third $2p$ atomic orbital (remember $2p_x$, $2p_y$, $2p_z$) is not involved in hybridization (its wave function is not mathematically combined with the other three) and remains as two lobes lying perpendicular to the plane of the sp^2 hybrid orbitals. Figure 1.14 shows three equivalent sp^2 orbitals along with the remaining unhybridized $2p$ atomic orbital. Each sp^2 orbital has 33% s-character and 67% p-character (one $2s$ orbital, two $2p$ orbitals).

An atom possessing three sp^2 hybrid orbitals and a single p atomic orbital is referred to as **sp^2 hybridized**, or as having a hybridization state of sp^2. The

Computed

Cartoon

(a) An sp^2 orbital

(b) Three sp^2 orbitals

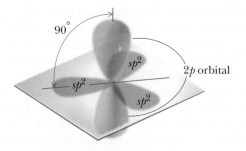

(c) A $2p$ orbital perpendicular to three sp^2 orbitals.

Figure 1.14
sp^2 Hybrid orbitals and a single p orbital on an sp^2 hybridized atom. (a) A single sp^2 hybrid orbital in computed and cartoon form. (b) Three sp^2 hybrized orbitals in a trigonal planar arrangement. (c) The lone p orbital.

energy-level diagram shows the creation of the three hybrid orbitals and a remaining $2p$ orbital. With C, all four orbitals contain a single electron. Therefore, a carbon atom makes three bonds with the sp^2 hybrids and one bond with a p orbital. With N, a lone pair is residing in an sp^2 hybrid orbital, and with O, two lone pairs are residing in sp^2 orbitals. Verify the orbital population of O for yourself by placing six valence electrons into the hybridized orbital diagram shown.

In Section 1.2, we covered a few apparent exceptions to the octet rule, where boron compounds such as BF_3 are common examples. Analogously, VSEPR tells us that BH_3 is trigonal planar, with 120° H—B—H bond angles. Therefore, sp^2 hybridization is the appropriate descriptor for B in such structures. Boron has three valence electrons, and one electron is placed in each sp^2 hybrid orbital. Each partially filled sp^2 hybrid overlaps with a $1s$ hydrogen orbital, containing one electron, to form three B—H σ bonds (Figure 1.15). The unhybridized $2p$ atomic orbital is empty. Note that BH_3 is a highly reactive molecule, and generally dimerizes in solution (see Section 6.4).

sp Hybrid Orbitals—Bond Angles of Approximately 180°

The mathematical combination of one $2s$ atomic orbital and one $2p$ atomic orbital produces two equivalent sp **hybrid orbital** wave functions. Because they are derived from two atomic orbitals, sp hybrid orbitals always occur in sets of two. The three-dimensional plot of ψ shows that the two sp hybrid orbitals lie at an angle of 180°. The axes of the unhybridized $2p$ atomic orbitals are perpendicular to each other and to the axis of the two sp hybrid orbitals. In Figure 1.16, sp hybrid orbitals are shown on the x-axis, and unhybridized $2p$ orbitals are on the y- and z-axes. Each sp orbital has 50% s-character and 50% p-character because those are the percentages of the orbitals combined when constructing them (one $2s$ orbital, one $2p$ orbital).

(a)

(b)

σ bonds formed by the overlap of sp^2 and $1s$ orbitals   Empty $2p$ obital

Borane

Figure 1.15
Covalent bond formation in borane, BH_3. (a) Lewis structure for borane. (b) Orbital overlap picture of borane.

sp Hybrid orbital
A hybrid atomic orbital formed by the combination of one s atomic orbital and one p atomic orbital.

Figure 1.16
sp Hybrid orbitals and two $2p$ orbitals on an sp hybridized atom. (a) A single sp hybrid orbital in computed and cartoon forms. (b) Two sp hybrid orbitals in a linear arrangement. (c) The two $2p$ orbitals in perpendicular orientations to the sp hybrid orbitals.

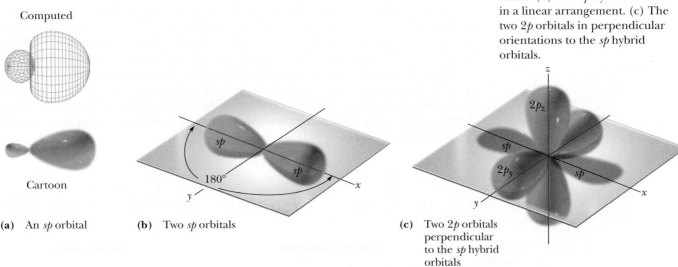

Computed

Cartoon

(a) An sp orbital

(b) Two sp orbitals

(c) Two $2p$ orbitals perpendicular to the sp hybrid orbitals

You may have noticed by now that the cartoon representations of sp^3, sp^2, and sp orbitals all look the same. Although the computed sp orbital is the most spherical and the sp^3 orbital is the most p-like with sp^2 between, chemists often do not attempt to render these differences in drawings.

An atom possessing two sp hybrid orbitals and two $2p$ orbitals is called **sp hybridized**, and the energy-level diagram for a carbon atom is shown below.

C. An Analysis of S and P

Compounds containing S and P are often drawn with more bonds than are allowed by the octet rule and therefore the hybridization states of these atoms are difficult to assign. For example, compare trimethylphosphine to phosphoric acid. The P in trimethylphosphine has three bonds and a lone pair of electrons and is sp^3 hybridized. However, a common depiction of phosphoric acid has five bonds to P and is a well-accepted representation. The creation of five bonds has been explained by invoking the use of $3d$ orbitals on phosphorus to accommodate the additional bond. However, the use of $3d$ orbitals for bonding in such structures is currently in debate. An alternative depiction assumes sp^3 hybridization and gives the P an octet and a positive formal charge (the symbol $\equiv$ means an equivalent structure drawn two different ways). The oxygen involved in the double bond of the common depiction has one bond and a negative formal charge in the alternative depiction.

Trimethylphosphine
(obeys octet rule)

Phosphoric acid

Sulfur, another third-period element, likewise is commonly depicted with varying numbers of bonds. Because S is below O in the Periodic Table we can predict it will be neutral with two bonds and an octet and is sp^3 hybridized (see hydrogen sulfide). Dimethylsulfoxide is commonly drawn with ten valence electrons and a double bond to oxygen, but is now considered as the alternative charge-separated structure with a positive formal charge on an sp^3 hybridized S and negative formal charge on O. Lastly, sulfuric acid has a similar dichotomy, either being drawn as having 12 valence electrons on S or the increasingly accepted picture with an sp^3 S having an octet with a plus two formal charge.

Hydrogen sulfide

Dimethyl sulfoxide

Sulfuric acid

Many biochemical structures, including DNA, are in part made up of derivatives of phosphoric acid. These derivatives are referred to as phosphoesters. The P atoms in these structures are nearly always depicted with five bonds. However, as just described, the modern view of such structures involves charge separation and sp^3 hybridization at P. Because the five-bond representation is historically the most widely spread depiction, this depiction is how we render such structures throughout this book. Yet, you should keep in mind that the alternative is now considered more correct.

3'-Adenine monophosphate

Common depiction   More correct

Glucose 6-phosphate

Common depiction   More correct

D. Combining Valence Bond (VB) and Molecular Orbital (MO) Theories: The Creation of σ and π Bonding and Antibonding Orbitals

At the beginning of this section (1.7A), we noted that the molecular orbitals of H_2 are created by adding and subtracting wave functions for the atomic $1s$ orbitals on the individual hydrogen atoms. We also noted that the exercise of adding and subtracting atomic orbitals such as $1s$, $2s$, and $2p$ to create molecular orbitals is one principle of MO theory, where the resulting molecular orbitals are spread across atoms in the entire molecule. Yet, orbitals spread across many atoms in a molecule are often difficult to visualize. Therefore, it is convenient to consider orbitals to be localized between adjacent atoms as in valence bond theory, or slightly delocalized over three or more atoms (as is done in Section 1.9). The most common model for bonding in organic compounds uses the MO theory notions of addition and subtraction to create the molecular orbitals, but also includes some principles from valence bond theory.

Valence bond theory views bonding as arising from electron pairs localized between adjacent atoms. These pairs of electrons create bonds. Further, organic chemists commonly use the atomic orbitals involved in the three hybridization states of atoms (sp^3, sp^2, and sp) to create the orbitals that hold these electrons because doing so allows the resulting orbitals to match the experimentally determined geometries around the atoms. Therefore, hybridization is also a VB theory concept. But how do we make the orbitals that contain the electrons and that reside between adjacent atoms? This is where we turn back to MO theory.

To create orbitals that are localized between adjacent atoms we add and subtract (also called in and out of phase addition) the atomic orbitals on the adjacent atoms, which are aligned to overlap with each other. For example, let's consider methane, CH_4 (Figure 1.17). The sp^3 hybrid orbitals point at the $1s$ hydrogen orbitals, and therefore we add and subtract these atomic orbitals to create the molecular orbitals. As with H_2, one resulting molecular orbital is lower in energy than the two separate atomic orbitals, and is called the bonding σ orbital. The other resulting

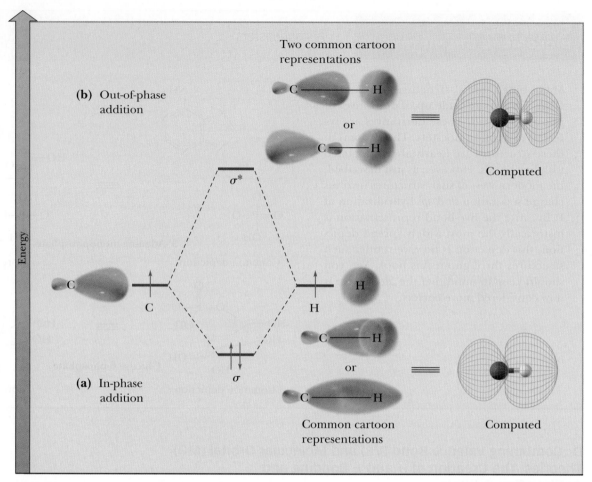

Two common cartoon
representations

(b) Out-of-phase
addition

Computed

σ^*

Energy

C

C

H

H

σ

(a) In-phase
addition

or

Common cartoon
representations

Computed

Figure 1.17

Molecular orbital mixing diagram for the creation of any C—H σ bond. (a) In-phase addition of a C hybrid orbital (albeit sp^3, sp^2, or sp) with a H $1s$ orbital forms a σ orbital that is lower in energy than the two starting orbitals. When the resulting orbital is populated with two electrons, a σ bond results. (b) Addition of the orbitals in an out-of-phase manner (meaning reversing the phasing of one of the starting orbitals) leads to an antibonding σ^* orbital.

molecular orbital is higher in energy than the two atomic orbitals and is antibonding. Only the lower-energy orbital is populated with electrons in methane. Population of the σ bonding orbital results in what we call a σ bond between the C and the H. Each of the four C—H bonds in methane is created in the manner discussed here. Also, although we created this picture for the C—H bonds in methane, we will view all C—H bonds in other organic structures in the same way. In other words, even with sp^2 and sp hybridized carbons, we think of their C—H bonds as looking similar to those in Figure 1.17.

It is important to note in Figures 1.17 and 1.10, as well as in the orbital mixing diagrams (Figures 1.18, 1.21, and 1.25), that the energy of the antibonding orbital goes up further than the drop experienced by the bonding orbital. This is the reason that population of the antibonding orbital with electrons leads to cleavage of the bond. Early, less accurate theoretical methods for modeling bonding found the increase and decrease in energy of these respective orbitals to be identical, but the bonding approach described here correctly predicts the relative energies.

An identical approach used to create C—H σ bonds is used to create C—C σ bonds. For example, whenever a C—C bond exists in an organic structure we consider the overlap of hybrid orbitals on the two carbons. As shown in Figure 1.18, the overlap of two sp^3 hybrid orbitals on the individual carbons creates σ bonding and antibonding molecular orbitals. Only the bonding orbital is populated with electrons, thereby creating a carbon-carbon σ bond. We consider all C—C σ bonds to consist of orbitals similar to those given in Figure 1.18.

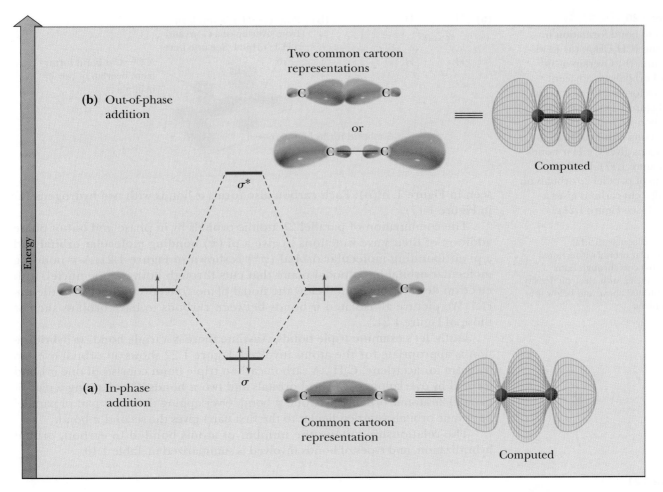

(b) Out-of-phase addition

Two common cartoon representations

Computed

(a) In-phase addition

Common cartoon representation

Computed

Energy

Figure 1.18

Molecular orbital mixing diagram for the creation of any C—C σ bond. (a) In-phase addition of two C hybrid orbitals (albeit sp^3, sp^2, or sp orbital) forms a σ orbital that is lower in energy than the two starting orbitals. When the resulting orbital is populated with two electrons, a σ bond results. (b) Addition of the orbitals in an out-of-phase manner (meaning reversing the phasing of one of the starting orbitals) leads to an antibonding σ^* orbital.

(a)

(b) A C—C σ bond forms from overlap of two sp^3 orbitals

6 C—H σ bonds form from overlap of a C sp^3 with H $1s$ orbitals on hydrogen atoms

Figure 1.19

(a) Lewis structure for ethane (CH_3CH_3). (b) Overlap of sp^3 hybrid orbitals on adjacent carbons forms a C—C σ bond (see Figure 1.18), and overlap of carbon sp^3 hybrid orbitals with hydrogen $1s$ orbitals gives C—H σ bonds (see Figure 1.17).

For example, the simplest two-carbon compound ethane (CH_3CH_3) contains one C—C σ bond and six C—H σ bonds. As shown in Figure 1.19, we consider these bonds to arise from overlap of H $1s$ and C sp^3 orbitals, while the actual bonding orbitals appear as shown in Figures 1.17 and 1.18.

Let's now examine compounds with a double bond. Anytime there is a double bond, sp^2 hybridization should be considered for the atoms involved. For example, second-period elements use a combination of an sp^2 hybrid orbital and the unhybridized $2p$ atomic orbital to form double bonds. Consider ethylene, C_2H_4, whose Lewis structure is shown in Figure 1.20(a). A σ bond between the carbons in ethylene is formed by overlapping sp^2 hybrid orbitals along a common axis as

Figure 1.20

Covalent bond formation in ethylene (CH_2CH_2). (a) Lewis structure. (b) Overlap of sp^2 hybrid orbitals on adjacent carbons forms a C—C σ bond (see Figure 1.18), and overlap of carbon sp^2 hybrid orbitals on carbons with $1s$ orbitals on hydrogens gives C—H σ bonds (see Figure 1.17). Further, overlap of parallel $2p$ orbitals on the adjacent carbons gives a π bond (see Figure 1.21).

Pi (π) molecular orbital
A molecular orbital with a nodal plane that cuts through both atomic nuclei, with electron density concentrated above and below the nodal plane.

(a) H, H C=C H, H

(b) Four C—H σ bonds form from overlap of a C sp^2 and a H $1s$ orbital. See one here.

A C—C σ bond forms from overlap of two sp^2 orbitals

A π bond forms between these two $2p$ orbitals.

seen in Figure 1.18(b). Each carbon also forms σ bonds with two hydrogens (as in Figure 1.17).

The combination of parallel $2p$ atomic orbitals by in phase and out of phase addition of their wave functions to give a **pi (π) bonding molecular orbital** and a pi antibonding molecular orbital (π^*) is shown in Figure 1.21. A π bonding molecular orbital has a nodal plane that cuts through both atomic nuclei, with electron density above and below the nodal plane concentrated between the nuclei. We picture all isolated π bonds between carbons to have orbitals such as those of Figure 1.21.

Lastly, let's examine triple bonds. Anytime there is a triple bond, sp hybridization is appropriate for the atoms involved. Figure 1.22 shows an orbital overlap diagram for acetylene, C_2H_2. A carbon-carbon triple bond consists of one σ bond formed by overlapping sp hybrid orbitals and two π bonds. Overlapping a pair of parallel $2p$ atomic orbitals gives one π bond. Overlapping the other pair of parallel $2p$ atomic orbitals (perpendicular to the first pair) gives the second π bond.

The relationship among the number of atoms bonded to carbon, orbital hybridization, and types of bonds involved is summarized in Table 1.10.

Figure 1.21

Molecular orbital mixing diagram for the creation of any C—C π bond. (a) Addition of two p atomic orbitals in phase leads to a π orbital that is lower in energy than the two separate starting orbitals. When populated with two electrons the π orbital gives a π bond. (b) Addition of the p orbitals in an out-of-phase manner (meaning a reversal of phasing in one of the starting orbitals) leads to a π^* orbital. Population of this orbital with one or two electrons leads to weakening or cleavage of the π bond respectively.

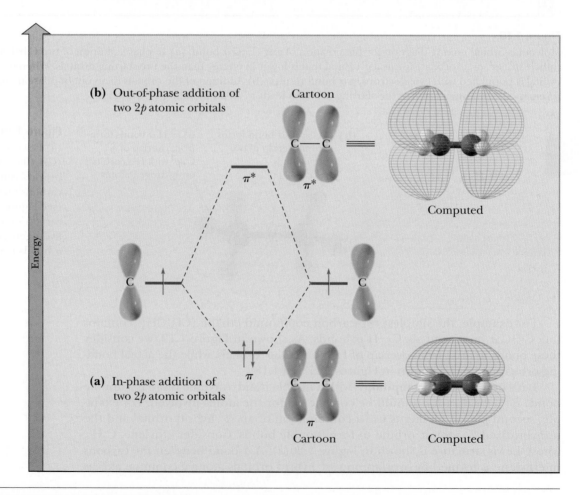

(b) Out-of-phase addition of two $2p$ atomic orbitals

Cartoon

π^*

C—C

π^*

Computed

Energy

C

C

(a) In-phase addition of two $2p$ atomic orbitals

π

C—C

π

Cartoon

Computed

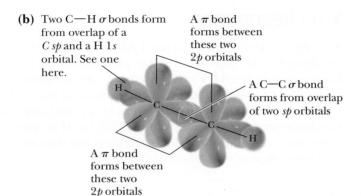

(a) H—C≡C—H

(b) Two C—H σ bonds form from overlap of a C sp and a H 1s orbital. See one here.

A π bond forms between these two 2p orbitals

A C—C σ bond forms from overlap of two sp orbitals

A π bond forms between these two 2p orbitals

Figure 1.22
Covalent bond formation in acetylene. (a) Lewis structure. (b) Overlap of sp hybrid orbitals on adjacent carbons forms a C—C σ bond (see Figure 1.18), and overlap of carbon sp hybrid orbitals with hydrogen 1s orbitals gives C—H σ bonds (see Figure 1.17). Further, overlap of parallel 2p orbitals on the adjacent carbons gives a C—C π bond (see Figure 1.21). Two such π bonds exist in acetylene.

Example 1.15

Describe the bonding in 1-methoxypropanone ($CH_3OCH_2COCH_3$) in terms of (a) hybridization of C and O, (b) type of bonds between C and O, and (c) type of orbitals that hold the lone electron pairs on O.

Solution

(a)

(b)

(c)

Problem 1.15

Describe the bonding in these molecules in terms of hybridization of C and N, and the types of bonds between carbon and nitrogen, and if there are any lone pairs, describe what type of orbital contains these electrons.

(a) $CH_3CH{=}CH_2$

(b) CH_3NH_2

Table 1.10	Covalent Bonding of Carbon				
Groups Bonded to Carbon	Orbital Hybridization	Predicted Bond Angles	Types of Bonds to Carbon	Example	Name
4	sp^3	109.5°	Four σ bonds	H—C—C—H (ethane structure)	**Ethane**
3	sp^2	120°	Three σ bonds and one π bond	C=C (ethylene structure)	**Ethylene**
2	sp	180°	Two σ bonds and two π bonds	H—C≡C—H	**Acetylene**

1.8 | Resonance

As chemists developed a deeper understanding of covalent bonding in organic compounds, it became obvious that, for a great many molecules and ions, no single Lewis structure provides a truly accurate representation. For example, Figure 1.23 shows three Lewis structures for the carbonate ion, CO_3^{2-}, each of which shows carbon bonded to three oxygen atoms by a combination of one double bond and two single bonds. Each Lewis structure implies that one carbon-oxygen bond is different from the other two. However, this is not the case. Experiments showed that all three carbon-oxygen bonds are identical.

Figure 1.23

(a–c) Three Lewis structures for the carbonate ion.

The problem for chemists, then, was how to describe the structure of molecules and ions for which no single Lewis structure is adequate and yet still retain Lewis structures. As an answer to this problem, Linus Pauling proposed the theory of resonance.

A. Theory of Resonance

The theory of **resonance** was developed primarily by Pauling in the 1930s. According to this theory, many molecules and ions are best described by writing two or more Lewis structures and considering the real molecule or ion to be a composite of these structures. Individual Lewis structures are called **contributing structures**. They are also sometimes referred to as **resonance structures** or **resonance contributors**. We show that the real molecule or ion is a **resonance hybrid** of the various contributing structures by interconnecting them with **double-headed arrows**. Do not confuse the double-headed arrow with the double arrow used to show chemical equilibrium. As we explain shortly, resonance structures are not in equilibrium with each other.

Three contributing structures for the carbonate ion are shown in Figure 1.24. These three contributing structures are said to be equivalent; they have identical patterns of covalent bonding.

The use of the term "resonance" for this theory of covalent bonding might suggest to you that bonds and electron pairs are constantly changing back and forth from one position to another over time. This notion is not at all correct. The carbonate ion, for example, has one and only one real structure. The problem is ours—how do we draw that one real structure? The resonance method is a way to describe the real structure and at the same time retain Lewis structures with electron-pair bonds. Thus, although we realize that the carbonate ion is not accurately represented by any one contributing structure shown in Figure 1.24, we

Resonance
A theory that many molecules are best described as a hybrid of several Lewis structures.

Contributing structures
Representations of a molecule or ion that differ only in the distribution of valence electrons.

Resonance hybrid
A molecule, ion, or radical described as a composite of a number of contributing structures.

Double-headed arrow
A symbol used to show that structures on either side are resonance contributing structures.

Figure 1.24

(a–c) The carbonate ion represented as a resonance hybrid of three equivalent contributing structures. Curved arrows show the redistribution of valence electrons between one contributing structure and the next. (d) An electrostatic potential map of a carbonate ion shows that the negative charge is distributed equally among the three oxygens.

Notice in Figure 1.24 that the only difference among contributing structures (a), (b), and (c) is the position of valence electrons. To generate one resonance structure from another, chemists use a symbol called a **curved arrow**. *The arrow indicates where a pair of electrons originates (the tail of the arrow) and where it is positioned in the next structure (the head of the arrow).*

A curved arrow is nothing more than a bookkeeping symbol for keeping track of electron pairs, or, as some call it, **electron pushing**. Do not be misled by its simplicity. Electron pushing will help you see the relationship among contributing structures. Later in the course, it will help you follow bond-breaking and bond-forming steps in organic reactions. Stated directly, electron pushing is a survival skill in organic chemistry.

Following are contributing structures for the nitrite and acetate ions. Curved arrows show how the contributing structures are interconverted. For each ion, the contributing structures are equivalent.

Nitrite ion
(equivalent contributing structures)

Acetate ion
(equivalent contributing structures)

A common mistake is to use curved arrows to indicate the movement of atoms or positive charges. This is never correct. Curved arrows must be used only to show the repositioning of electron pairs.

When you draw curved arrows to indicate the creation of a new contributing structure, the arrows should always start on either a double (or triple) bond or a lone pair of electrons, as seen in the examples given above. Further, the arrows should either end at an atom that can accept a bond or should create a lone pair of electrons on an atom. Often when a new bond to an atom is created, one of the existing bonds to that atom must break so as not to exceed the octet rule. In the two examples given above, the central N of nitrite and C of acetate both acquire one bond and break one bond when you redistribute valence electrons between the contributing structures.

Curved arrow
A symbol used to show the redistribution of valence electrons in resonance contributing structures or reaction mechanisms, symbolizing movement of two electrons.

continue to represent it by one of these for convenience. We understand, of course, that what is intended is the resonance hybrid.

Example 1.16

Draw the contributing structure indicated by the curved arrow(s). Be certain to show all valence electrons and all formal charges.

Solution

Problem 1.16

Draw the contributing structure indicated by the curved arrows. Be certain to show all valence electrons and all formal charges.

(a) $H-\overset{\overset{\displaystyle :\ddot{O}:}{\|}}{C}-\ddot{O}:^-$ ⟷

(b) $H-\overset{\overset{\displaystyle :\ddot{O}:}{\|}}{C}-\ddot{O}:^-$ ⟷

(c) $CH_3-\overset{\overset{\displaystyle :\ddot{O}:}{\|}}{C}-\ddot{O}-CH_3$ ⟷

B. Rules for Writing Acceptable Contributing Structures

Certain rules must be followed in writing acceptable contributing structures:

1. All contributing structures must have the same number of valence electrons.
2. All contributing structures must obey the rules of covalent bonding; no contributing structure may have more than two electrons in the valence shell of hydrogen or more than eight electrons in the valence shell of a second-period element. Third-period elements, such as phosphorus and sulfur, may be drawn indicating up to 12 electrons in their valence shells (however see Section 1.7C).
3. The positions of all nuclei must be the same in all contributing structures; that is, contributing structures differ only in the distribution of valence electrons.
4. All contributing structures must have the same number of paired and unpaired electrons.

Example 1.17

Which sets are valid pairs of contributing structures?

(a) $CH_3-\overset{\overset{\displaystyle :O:}{\|}}{C}-CH_3$ and $CH_3-\overset{\overset{\displaystyle :\ddot{O}:^-}{|}}{\underset{+}{C}}-CH_3$

(b) $CH_3-\overset{\overset{\displaystyle :O:}{\|}}{C}-CH_3$ and $CH_2=\overset{\overset{\displaystyle :\ddot{O}-H}{|}}{C}-CH_3$

Solution

(a) These are valid contributing structures. They differ only in the distribution of valence electrons.

(b) These are not valid contributing structures. They differ in the connectivity of their atoms.

Problem 1.17

Which sets are valid pairs of contributing structures?

(a) $CH_3-\overset{\displaystyle \ddot{O}:}{\underset{\ddot{O}:^-}{C}}$ and $CH_3-\overset{\displaystyle \ddot{O}:^-}{\underset{\ddot{O}:^-}{C^+}}$

(b) $CH_3-\overset{\displaystyle \ddot{O}:}{\underset{\ddot{O}:^-}{C}}$ and $CH_3-\overset{\displaystyle \ddot{O}:}{\underset{\ddot{O}:}{C^-}}$

C. Estimating the Relative Importance of Contributing Structures

Not all structures contribute equally to a resonance hybrid. We describe four ways to predict which structure contributes more to the hybrid. But before examining these preferences, let's consider how to think about the fact that contributing structures may contribute unequally to many resonance hybrids. Suppose you combined yellow and blue paints to make green paint. This is analogous to writing two structures that contribute equally to the final resonance hybrid. The contributing structures do not interconvert back and forth, but rather are used to represent only one structure—just as the green paint you mixed is not yellow one moment and blue at another moment. Continuing with our paint analogy, how might you describe two different shades of green? One color might have a greater contribution of yellow while the other a greater contribution of blue. This is analogous to the fact that many molecules have a greater contribution of one contributing structure. Importantly, there is only one actual way that the electrons are distributed in the molecule, but the limitations of how we draw molecules with dot pairs and lines require us to rely on examining two or more depictions—just as many colors of paint mixed in different proportions give only one final color.

The following preferences will help you to estimate the relative importance of the various contributing structures. In fact, we can rank structures by the number of these preferences they follow. Those that follow the most preferences contribute the most to the hybrid, and any structure that violates all four of these preferences can be ignored.

Preference 1: Filled Valence Shells

Structures in which all atoms have filled valence shells (completed octets) contribute more than those in which one or more valence shells are unfilled. For example, the following are the contributing structures for Example 1.16(c) and its solution.

$$CH_3-\overset{+}{\underset{..}{O}}{=}C-H \quad \longleftrightarrow \quad CH_3-\overset{..}{\underset{..}{O}}-\overset{+}{C}-H$$
$$\quad\quad\quad | \quad\quad\quad\quad\quad\quad\quad\quad\quad\quad | $$
$$\quad\quad\quad H \quad\quad\quad\quad\quad\quad\quad\quad\quad H $$

Greater contribution:	Lesser contribution:
both carbon and oxygen have	carbon has only six electrons
complete valence shells	in its valence shell

Preference 2: Maximum Number of Covalent Bonds

Structures with a greater number of covalent bonds contribute more than those with fewer covalent bonds. In the illustration for preference 1, the structure on the left has eight covalent bonds and makes the greater contribution to the hybrid. The structure on the right has only seven covalent bonds. Yet, this rule should never lead you to write structures that exceed eight valence electrons (some exceptions exist, such as S and P discussed in Section 1.7C).

$$CH_3-\overset{+}{\underset{..}{O}}{=}C-H \quad \longleftrightarrow \quad CH_3-\overset{..}{\underset{..}{O}}-\overset{+}{C}-H$$
$$\quad\quad\quad | \quad\quad\quad\quad\quad\quad\quad\quad\quad\quad | $$
$$\quad\quad\quad H \quad\quad\quad\quad\quad\quad\quad\quad\quad H $$

Greater contribution:	Lesser contribution:
eight covalent bonds	seven covalent bonds

Preference 3: Least Separation of Unlike Charges

Structures that involve separation of unlike charges contribute less than those that do not involve charge separation because separation of charges costs energy.

$$\overset{:O:}{\underset{\|}{}} \quad\quad\quad\quad \overset{:\overset{..}{O}:^-}{\underset{|}{}}$$
$$CH_3-C-CH_3 \quad \longleftrightarrow \quad CH_3-\overset{}{\underset{+}{C}}-CH_3$$

Greater contribution:	Lesser contribution:
no separation of	separation of unlike
unlike charges	charges

Preference 4: Negative Charge on a More Electronegative Atom

Structures that carry a negative charge on a more electronegative atom contribute more than those with the negative charge on a less electronegative atom. Conversely, structures that carry a positive charge on a less electronegative atom contribute

An electrostatic potential map of an acetone molecule.

more than those that carry the positive charge on a more electronegative atom. Following are three contributing structures for acetone:

(a)
Lesser
contribution

(b)
Greater
contribution

(c)
Should not
be drawn

Structure (b) makes the largest contribution to the hybrid. Structure (a) contributes less because it involves separation of unlike charge and because carbon has an incomplete octet. Nevertheless, on structure (a), the more electronegative O atom has the negative charge and the less electronegative C atom has the positive charge. Structure (c) violates all four preference rules and should not be drawn, and arrow 2 on structure (b) can be ignored.

It is important to realize that if resonance structures contribute unequally, the actual structure of the hybrid most resembles the structure that contributes most! The electrostatic potential map of acetone shows the negative charge (red) on oxygen and the positive charge (blue) on carbon in agreement with the results we derive from the resonance treatment.

Example 1.18

Estimate the relative contribution of the members in each set of contributing structures.

(a)   **(b)**

Solution

(a) The structure on the right makes a greater contribution to the hybrid because it places the negative charge on oxygen, the more electronegative atom.

(b) The structures are equivalent and make equal contributions to the hybrid.

Problem 1.18

Estimate the relative contribution of the members in each set of contributing structures.

(a)   **(b)**

To emphasize, do not confuse resonance contributing structures with equilibration among different species. A molecule described as a resonance hybrid is not equilibrating among individual electron configurations of the contributing structures. Rather, the molecule has only one structure, which is best described as a hybrid of its various contributing structures.

The contributing structures for acetone drawn in the preceding section give us insight into the charge distribution of the molecule. In the lesser contributing structure, there is a positive charge on the carbon and a negative charge on the oxygen, as emphasized in the electrostatic potential map shown for acetone. We could come to the same conclusion by recognizing that the C=O bond is polar, with partial positive and negative charges on C and O respectively (as we saw in Section 1.5). Given the insights into bonding orbitals developed in the previous section, we should now be able to use our hybrid VB/MO theory to visualize this polarization. In other words, the σ and π orbitals discussed for ethylene in Section 1.7C should be slightly perturbed for acetone to reflect the polarization in the C=O that is absent in a C=C. Here, only the π bond of C=O will be analyzed, because the lesson for the σ bond is identical.

Figure 1.25 shows the MO theory mixing diagram for the creation of the π and π^* orbitals of C=O. The only difference between this diagram and that of Figure 1.21 is that the energy level of the O $2p$ orbital is placed lower than the C $2p$ orbital. The $2p$ orbitals for O are lower in energy because O is more electronegative than C. In MO theory, when the starting atomic orbitals are not equal in energy, the resulting molecular orbitals most resemble (have a greater contribution from) the atomic orbital that is closest in energy. Hence, the bonding π orbital has a larger contribution from the O, while the antibonding π orbital has a larger contribution from the C. Because only the bonding orbital is occupied, the electron density in the bond is concentrated on O. In contrast, in the antibonding orbital, there is considerable orbital character on C, thereby placing a partial positive charge on this atom (a fact that will become important when we start to analyze reactions on C=O groups). Hence, the simple polarization concepts (Section 1.5), the notion of resonance (Section 1.8), and the VB/MO model of bonding (Section 1.7) all lead to the same conclusions.

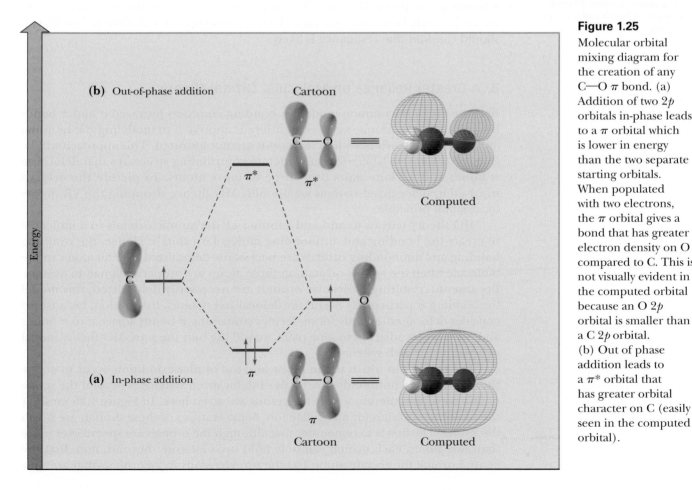

Figure 1.25
Molecular orbital mixing diagram for the creation of any C—O π bond. (a) Addition of two $2p$ orbitals in-phase leads to a π orbital which is lower in energy than the two separate starting orbitals. When populated with two electrons, the π orbital gives a bond that has greater electron density on O compared to C. This is not visually evident in the computed orbital because an O $2p$ orbital is smaller than a C $2p$ orbital. (b) Out of phase addition leads to a π^* orbital that has greater orbital character on C (easily seen in the computed orbital).

1.9 | Molecular Orbitals for Delocalized Systems

A. Resonance Revisited

As discussed in Section 1.8, resonance gives us an approach to depicting bonding when more than one Lewis structure is possible. The real bonding and electron distribution in the molecule is a weighted average of the various contributing structures. For example, let's again examine the acetate anion, which has a 50/50 contribution from two equivalent contributing structures. In the electron pushing that is used to show the interconversion of the two contributing structures, a lone pair on the oxygen is used to create a π bond to the central carbon while the π bond becomes a lone pair. This is an example of charge **delocalization**, which is always a stabilizing effect. Chemists commonly write a structure that is meant to depict this charge delocalization using dashed lines and a $-1/2$ charge on each oxygen. The dashed line in the average structure implies four electrons (a lone pair and a π bond) involved in bonding between the O, C, and O.

Delocalization
The spreading of charge and/or electron density over a larger volume of space.

Acetate ion

Such delocalization occurs for systems that are referred to as conjugated. **Conjugation** means a lack of an intervening atom between π bonds or between π bonds and lone-pair electrons. Whenever you encounter a structure that has conjugated double bonds or lone-pair electrons conjugated with double bonds, you should consider that resonance is likely.

Conjugation
Lack of atoms between π bonds or between π bonds and lone-pair electrons.

B. A Greater Reliance on Molecular Orbital Theory

Because the most common model for bonding considers localized σ and π bonds between adjacent atoms, we need a different approach to modeling the bonding and antibonding orbitals when the bonds are not localized. This approach will be true for any system where we can generate contributing structures that delocalize π bonds and/or lone pairs over three or more atoms. To picture the orbitals involved in delocalized systems we use only MO theory, dropping the VB theory localization principle.

MO theory tells us to add and subtract *all* the atomic orbitals in a molecule to create the bonding and antibonding molecular orbitals. Hence, the resulting bonding and antibonding orbitals are necessarily delocalized over all atoms in the molecule when the atomic orbitals on those atoms are properly aligned to overlap. Because the resulting molecular orbitals are necessarily delocalized, this model for bonding is particularly useful for delocalized systems. In addition, because we consider delocalization only for molecules containing π bonds adjacent to π bonds and/or π bonds adjacent to lone pairs, we will use only the pure MO theory model of bonding for such systems.

The manner in which the in phase and out of phase addition occur to create the bonding and antibonding orbitals of delocalized systems is beyond the scope of this book, and therefore only the results are given here. In Figure 1.26 we show the molecular orbitals for an acetate ion. Some features of these orbitals are worth discussing. The first is to remember that although the orbitals are spread over more than two atoms, each orbital can only hold two electrons. Second, note that the central orbital for acetate anion has the two electrons in $2p$ orbitals that are only

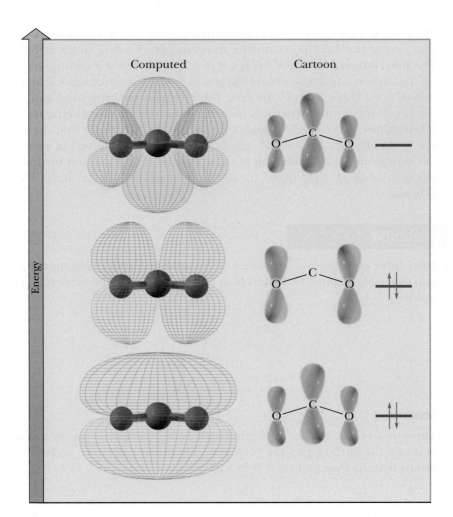

Figure 1.26
An example of molecular orbitals for delocalized systems: the three π orbitals of acetate anion. Only the lowest two are populated with electrons.

Computed   Cartoon

on the oxygens. This electron distribution is in agreement with the contributing structures of acetate, which place the lone pairs on only the oxygens and $-1/2$ of a charge on each oxygen.

C. Hybridization Considerations in Light of Resonance and MO Theory

Given the discussions of delocalization and MO theory in the last two sections, we need to reconsider the assignment of hybridization to various atoms. For example, what is the hybridization of each oxygen in acetate? If you consider only an individual contributing structure as indicated below, the conclusion would be that one oxygen is sp^3 and one is sp^2. However, the other contributing structure would flip-flop the hybridization conclusion.

$$CH_3 - C \quad \longleftrightarrow \quad CH_3 - C$$

Examining the π bonds in the contributing structures allows the correct hybridization assignment to be made. For an atom to be involved in π bonding it must have a $2p$ orbital. Of the two hybridization choices, only sp^2 atoms have

$2p$ orbitals. Therefore, in acetate both oxygens are sp^2 hybridized. This means that two of the lone pairs on the oxygens are residing in sp^2 hybrid orbitals. The third lone pair on each oxygen is in a $2p$ orbital that is mixing with $2p$ orbitals on the carbon and the alternate oxygen to create the three delocalized molecular orbitals of Figure 1.24(a). In fact, the electron pushing used to interconvert the two contributing structures of acetate shows lone pairs being used to create π bonds (see Section 1.9A), and it should not be surprising that these lone pairs are in $2p$ orbitals. In general, whenever atoms are involved in resonance that has one or more contributing structures involving a π bond to that atom, the atoms must be sp^2 hybridized (or on rare occasions, sp hybridized, but never sp^3 hybridized).

Example 1.19

Draw three contributing structures of the following amide and state the hybridization of the highlighted O, C, and N. In which orbitals do the three lone pairs drawn reside?

Solution

There is a reasonable contributing structure (C), which places a double bond between the carbon and the nitrogen. Therefore, all three atoms are sp^2. This means that the lone pair on N is in a $2p$ orbital.

A.   B.   C.

Problem 1.19

Draw three contributing structures of the following compound (called guanidine) and state the hybridization of the four highlighted atoms. In which orbitals do the three lone pairs drawn reside?

guanidine

1.10 | Bond Lengths and Bond Strengths in Alkanes, Alkenes, and Alkynes

Values for bond lengths and bond strengths (bond dissociation enthalpies) for ethane, ethylene, and acetylene are given in Table 1.11.

Table 1.11	Bond Lengths and Bond Strengths for Ethane, Ethylene, and Acetylene				
Name	**Formula**	**Bond**	**Bond Orbital Overlap**	**Bond Length (pm)**	**Bond Strength [kJ (kcal)/mol]**
Ethane	H H | | H—C—C—H | | H H	C—C	sp^3–sp^3	153.2	376 (90)
		C—H	sp^3–$1s$	111.4	422 (101)
Ethylene	H H \ / C=C / \ H H	C—C	sp^2–sp^2, $2p$–$2p$	133.9	727 (174)
		C—H	sp^2–$1s$	110.0	464 (111)
Acetylene	H—C≡C—H	C—C	sp–sp, two $2p$–$2p$	121.2	966 (231)
		C—H	sp–$1s$	109.0	566 (133)

As you study Table 1.11, note the following points:

1. Carbon-carbon triple bonds are shorter than carbon-carbon double bonds, which in turn are shorter than carbon-carbon single bonds. This order of bond lengths exists because there are three versus two versus one bond holding the carbon atoms together.

2. The C—H bond in acetylene is shorter than that in ethylene, which in turn is shorter than that in ethane. The relative lengths of these C—H bonds are determined by the percent *s*-character in the hybrid orbital of carbon forming the σ bond with hydrogen. The greater the percent *s*-character of a hybrid orbital, the closer electrons in it are held to the nucleus and the shorter the bond because *s* electrons are on average closer to the nucleus than *p* electrons. The relative lengths of C—H single bonds correlate with the fact that the percent *s*-character in an *sp* orbital is 50%, in an sp^2 orbital is 33.3%, and in an sp^3 orbital is 25%. Also, because electrons in *s* orbitals are bound more tightly than those in *p* orbitals, the more *s*-character in a bond, the stronger it is. This is the first of several times we will refer to the percent *s*-character in hybrid orbitals. It has a large influence on a variety of molecular properties.

3. There is a correlation between bond length and bond strength; the shorter the bond, the stronger it is. A carbon-carbon triple bond is the shortest C—C bond; it is also the strongest. The carbon-hydrogen bond in acetylene is the shortest; it is also the strongest.

4. Although a C—C double bond is stronger than a C—C single bond, it is not twice as strong. By the same token, a C—C triple bond is stronger than a C—C single bond, but it is not three times as strong. These differences arise because the overlap of orbitals lying on the same axis and forming σ bonds is more efficient (gives a greater bond strength) than overlap of orbitals lying parallel to each other and forming π bonds.

Section 1.1 Electronic Structure of Atoms

Problems: 1.1, 1.21, 1.22, 1.23

- Atoms consist of a small, dense nucleus and electrons distributed about the nucleus in regions of space called **shells**.
  - Each shell can contain as many as $2n^2$ electrons, where n is the number of the shell.
  - Each principal energy level is subdivided into regions of space called **orbitals**. The **valence shell** is the outermost occupied shell, and it contains the **valence electrons**. Valence electrons are important because they take part in chemical bonding.
- The **Lewis dot structure** of an atom shows the symbol of the atom surrounded by a number of dots equal to the number of electrons in the valence shell of the atom.

Section 1.2 Lewis Model of Bonding

Problem: 1.2

- According to the Lewis model of covalent bonding, atoms bond together in such a way that each atom participating in a chemical bond acquires a completed valence-shell electron configuration resembling that of the noble gas nearest it in atomic number.
  - Anions and cations attract each other but do not form bonds with defined directionality.
  - A **covalent bond** is a chemical bond formed by the sharing of electron pairs between adjacent atoms.
  - The tendency of main-group elements (Groups 1A–7A) to achieve an outer shell of eight valence electrons is called the **octet rule**.
- **Electronegativity** is a measure of the force of attraction by an atom for electrons it shares in a chemical bond with another atom.
  - A **nonpolar covalent bond** is a covalent bond in which the difference in electronegativity of the bonded atoms is less than 0.5.
  - A **polar covalent bond** is a covalent bond in which the difference in electronegativity of the bonded atoms is between 0.5 and 1.9.

Problems: 1.3–1.5,
1.24–1.26, 1.33–1.37, 1.66

    - In a polar covalent bond, the more electronegative atom bears a partial negative charge ($\delta-$) and the less electronegative atom bears a partial positive charge ($\delta+$)
    - A polar bond has a bond dipole moment equal to the product of the absolute value of the partial charge times the distance between the dipolar charges (the bond length).
- An acceptable **Lewis structure** for a molecule or an ion must show (1) the correct connectivity of atoms, (2) the correct number of valence electrons, (3) no more than two electrons in the outer shell of hydrogen and no more than eight electrons in the outer shell of any second-period element, and (4) all formal charges.

Problems: 1.6, 1.7,
1.27, 1.32, 1.64, 1.65

  - There are some apparent exceptions to the octet rule: neutral compounds of boron and aluminum can have only six valence electrons.

Section 1.3 Functional Groups

Problems: 1.8–1.12,
1.41–1.47

- **Functional groups** are characteristic structural units by which we divide organic compounds into classes and that serve as a basis for nomenclature.
  - Functional groups are also sites of chemical reactivity. A particular functional group generally undergoes the same types of chemical reactions in whatever compound it occurs.

Section 1.4 Bond Angles and Shapes of Molecules

Problems: 1.13, 1.38–1.40,
1.67, 1.68

- Bond angles of molecules and polyatomic ions can be predicted using Lewis structures and **valence-shell electron-pair repulsion (VSEPR)**.
  - For atoms surrounded by four regions of electron density, VSEPR predicts bond angles of 109.5°; for atoms surrounded by three regions of electron density, it

predicts bond angles of 120°; and for two regions of electron density, it predicts bond angles of 180°.

Section 1.5 Polar and Nonpolar Molecules

The **dipole moment** of a molecule is the vector sum of its bond dipoles.

Problems: 1.14, 1.49, 1.50

Section 1.6 Quantum or Wave Mechanics

- **Quantum mechanics** is the branch of science that studies particles and their associated waves. It provides a way to determine the shapes of atomic orbitals and to quantify the energetics of covalent bond formation.

Section 1.7 A Combined Valence Bond and Molecular Orbital Theory Approach to Covalent Bonding

- According to **molecular orbital (MO)** theory, combination of n atomic orbitals gives n molecular orbitals.
  - Molecular orbitals are divided into sigma (σ) and pi (π) bonding and anti-bonding molecular orbitals. These orbitals are arranged in order of increasing energy, and their order of filling with electrons is governed by the same rules as for filling atomic orbitals.
  - Although useful for quantitative calculations on computers, molecular orbital theory alone does not provide an intuitive understanding of σ bonds in complex molecules.

Problem: 1.61

- For an intuitive understanding of bonding in molecules, we use molecular orbital theory concepts (in phase and out of phase addition of overlapping orbitals to give bonding and antibonding orbitals) in combination with valence bond theory.
  - **Valence bond theory** involves the combination of atomic orbitals on each atom in a process called **hybridization**, and the resulting atomic orbitals are called **hybrid orbitals**.

Problems: 1.15, 1.55, 1.62

  - The combination of one $2s$ atomic orbital and three $2p$ atomic orbitals produces four equivalent sp^3 **hybrid orbitals**, each directed toward a corner of a regular tetrahedron at angles of 109.5°.
  - The combination of one $2s$ atomic orbital and two $2p$ atomic orbitals produces three equivalent sp^2 **hybrid orbitals**, the axes of which lie in a plane at angles of 120°.
  - The combination of one $2s$ atomic orbital and one $2p$ atomic orbital produces two equivalent sp **hybrid orbitals**, the axes of which lie at an angle of 180°.
- S and P atoms are commonly depicted with more than 8 valence electrons, invoking participation of $3d$ electrons.
  - Recent calculations reveal that in many of these cases the S and P atoms are best thought of a sp^3 hybridized with a formal charge.
- In the combined molecular orbital theory/valence bond theory approach, bonding in organic molecules is thought of as the in phase addition (overlapping to create bonding orbitals) and out of phase addition, also referred to as subtraction (to create antibonding orbitals) of the hybridized (and possibly any unhybridized $2p$) atomic orbitals on adjacent atoms.
  - All C—C, C—O, and C—N single bonds are sigma (σ) bonds formed by the overlapping of hybrid orbitals.
  - All C—H, O—H, N—H single bonds are sigma (σ) bonds formed by overlapping hybrid orbitals on C, O, or N with the $1s$ orbital of H.
  - All C=C, C=O, C=N, N=N, and N=O double bonds are a combination of one sigma (σ) bond formed by overlapping hybrid orbitals and one pi (π) bond formed by overlapping parallel, unhybridized $2p$ orbitals.
  - All C≡C and C≡N triple bonds are a combination of one sigma (σ) bond formed by the overlap of sp hybrid orbitals and two pi (π) bonds formed by the overlap of two sets of parallel, unhybridized $2p$ orbitals.

Problems: 1.15, 1.57–1.61, 1.63, 1.71, 1.72, 1.76

Section 1.8 Resonance

- According to the **theory of resonance**, molecules and ions for which no single Lewis structure is adequate are best described by writing two or more **contributing structures**. The real molecule or ion is a **resonance hybrid** of the various contributing structures.

Problems: 1.16–1.18, 1.51–1.55, 1.69, 1.70, 1.74

  - The most important contributing structures have (1) filled valence shells, (2) a maximum number of covalent bonds, (3) the least separation of unlike charges, and (4) any negative charge on the more electronegative atom and/or any positive charge on the less electronegative atom. Double-headed arrows are drawn between contributing structures to describe the hybrid.
  - Curved arrows show the manner in which valence electrons are redistributed from one contributing structure to the next. Use of curved arrows in this way is commonly referred to as **electron pushing**. Curved arrows always show movement of electron pairs, never atoms.

Section 1.9 Molecular Orbitals for Delocalized Systems

- Most examples of molecules described by more than one resonance contributing structure have charge and/or electron **delocalization**, a stabilizing effect in which charge and/or electrons are spread over more than two atoms.
  - Delocalization occurs in molecules that have **conjugation**. A π bond and lone pair of electrons without an intervening atom are conjugated.

Problems: 1.19, 1.73, 1.75, 1.77

  - The delocalized electrons are in molecular orbitals formed from overlapping $2p$ orbitals on three or more adjacent atoms.
    - If atoms are taking part in delocalization as described by resonance contributing structures, they must have $2p$ orbitals, so they must be sp^2 hybridized, or in rare cases sp hybridized.

Section 1.10 Bond Lengths and Bond Strengths in Alkanes, Alkenes, and Alkynes

- The greater the number of bonds between two atoms, the shorter the bond length and the greater the bond strength.
  - Carbon-carbon triple bonds are shorter and stronger than carbon-carbon double bonds, which are shorter and stronger than carbon-carbon single bonds.
- The more *s*-character the hybridized orbital taking part in a bond, the shorter and stronger it is.

◥WL **Questions and problems for this chapter are available in OWL. Use the OWL access card that came with this textbook to access assigned questions and problems for this chapter.**

Chapter 2

Alkanes and Cycloalkanes

© L. Lefkowitz/Taxi/Getty

Bunsen burners burn natural gas, which is primarily methane with small amounts of ethane, propane, and butane (Section 2.9A). *Inset:* a model of methane, the major component of natural gas.

In this chapter, we begin our study of organic compounds with the physical and chemical properties of alkanes, the simplest types of organic compounds. Actually, alkanes are members of a larger group of organic compounds called hydrocarbons. A **hydrocarbon** is a compound composed of only carbon and hydrogen. Figure 2.1 shows the four classes of hydrocarbons, along with the characteristic pattern of bonding between the carbon atoms in each.

Alkanes are **saturated hydrocarbons**; that is, they contain only carbon-carbon single bonds. In this context, *saturated* means that each carbon has the maximum number of hydrogens bonded to it. We often refer to alkanes as **aliphatic hydrocarbons** because the physical properties of the higher members of this class resemble those of the long carbon-chain molecules we find in animal fats and plant oils (Greek: *aleiphar*, fat or oil).

A hydrocarbon that contains one or more carbon-carbon double bond, triple bond, or benzene ring is classified as an **unsaturated hydrocarbon**. We study alkanes (saturated hydrocarbons) in this chapter. We study alkenes and alkynes (both unsaturated hydrocarbons) in Chapters 5, 6, and 7, and we study arenes (also unsaturated hydrocarbons) in Chapters 21 and 22.

2.1 | The Structure of Alkanes

Methane (CH_4) and ethane (C_2H_6) are the first two members of the alkane family. Figure 2.2 shows Lewis structures and molecular models for these molecules. The shape of methane is tetrahedral and all H—C—H bond angles are 109.5°. Each carbon atom in ethane is also tetrahedral, and all bond angles are approximately 109.5°.

☼WL **Online homework for this chapter may be assigned in OWL for Organic Chemistry.**

Figure 2.1
The four classes of
hydrocarbons.

Figure 2.2
Methane and ethane. Lewis
structures and ball-and-stick
models.

Although the three-dimensional shapes of larger alkanes are more complex than those of methane and ethane, the four bonds about each carbon are still arranged in a tetrahedral manner, and all bond angles are approximately 109.5°.

The next three alkanes are propane, butane, and pentane. In the following representations, these hydrocarbons are drawn first as condensed structural formulas that show all carbons and hydrogens. They are also drawn in an even more abbreviated form called a **line-angle formula**. In a line-angle formula, each vertex and line ending represents a carbon atom. Although we do not show hydrogen atoms in line-angle formulas, we assume that they are there in sufficient numbers to give each carbon four bonds.

Line-angle formula
An abbreviated way to draw structural
formulas in which vertices and line
endings represent carbons.

Ball-and-stick model

Line-angle formula

Structural formula

$CH_3CH_2CH_3$
Propane

$CH_3CH_2CH_2CH_3$
Butane

$CH_3CH_2CH_2CH_2CH_3$
Pentane

We can write structural formulas for alkanes in still another abbreviated form. The structural formula of pentane, for example, contains three CH_2 (methylene) groups in the middle of the chain. We can collect them and write the structural formula of pentane as $CH_3(CH_2)_3CH_3$. Table 2.1 gives the names and molecular

Table 2.1		Names, Molecular Formulas, and Condensed Structural Formulas for the First 20 Alkanes with Unbranched Chains			

Name	Molecular Formula	Condensed Structural Formula	Name	Molecular Formula	Condensed Structural Formula
Methane	CH_4	CH_4	Undecane	$C_{11}H_{24}$	$CH_3(CH_2)_9CH_3$
Ethane	C_2H_6	CH_3CH_3	Dodecane	$C_{12}H_{26}$	$CH_3(CH_2)_{10}CH_3$
Propane	C_3H_8	$CH_3CH_2CH_3$	Tridecane	$C_{13}H_{28}$	$CH_3(CH_2)_{11}CH_3$
Butane	C_4H_{10}	$CH_3(CH_2)_2CH_3$	Tetradecane	$C_{14}H_{30}$	$CH_3(CH_2)_{12}CH_3$
Pentane	C_5H_{12}	$CH_3(CH_2)_3CH_3$	Pentadecane	$C_{15}H_{32}$	$CH_3(CH_2)_{13}CH_3$
Hexane	C_6H_{14}	$CH_3(CH_2)_4CH_3$	Hexadecane	$C_{16}H_{34}$	$CH_3(CH_2)_{14}CH_3$
Heptane	C_7H_{16}	$CH_3(CH_2)_5CH_3$	Heptadecane	$C_{17}H_{36}$	$CH_3(CH_2)_{15}CH_3$
Octane	C_8H_{18}	$CH_3(CH_2)_6CH_3$	Octadecane	$C_{18}H_{38}$	$CH_3(CH_2)_{16}CH_3$
Nonane	C_9H_{20}	$CH_3(CH_2)_7CH_3$	Nonadecane	$C_{19}H_{40}$	$CH_3(CH_2)_{17}CH_3$
Decane	$C_{10}H_{22}$	$CH_3(CH_2)_8CH_3$	Eicosane	$C_{20}H_{42}$	$CH_3(CH_2)_{18}CH_3$

formulas of the first 20 alkanes. Note that the names of all these alkanes end in -*ane*. We will have more to say about naming alkanes in Section 2.3.

Alkanes have the general molecular formula C_nH_{2n+2}. Thus, given the number of carbon atoms in an alkane, we can determine the number of hydrogens in the molecule and its molecular formula. For example, decane, with 10 carbon atoms, must have $(2 \times 10) + 2 = 22$ hydrogen atoms and a molecular formula of $C_{10}H_{22}$.

2.2 | Constitutional Isomerism in Alkanes

Constitutional isomers are compounds that have the same molecular formula but different structural formulas. By "different structural formulas," we mean that constitutional isomers differ in the kinds of bonds they have (single, double, or triple) and/or in the connectivity of their atoms.

For the molecular formulas CH_4, C_2H_6, and C_3H_8, only one connectivity is possible. For the molecular formula C_4H_{10}, two connectivities are possible. In one of these, named butane, the four carbons are bonded in a chain; in the other, named 2-methylpropane, three carbons are bonded in a chain with the fourth carbon as a branch on the chain.

Constitutional isomers
Compounds with the same molecular formula but a different connectivity of their atoms.

$CH_3CH_2CH_2CH_3$

Butane
(bp -0.5°C)

CH_3
|
CH_3CHCH_3

2-Methylpropane
(bp -11.6°C)

Butane and 2-methylpropane are constitutional isomers; they are different compounds and have different physical and chemical properties. Their boiling points, for example, differ by approximately 11°C.

In Section 1.3, we encountered several examples of constitutional isomers. We saw, for example, that there are two alcohols with the molecular formula C_3H_8O, two aldehydes with the molecular formula C_4H_8O, and two carboxylic acids with the molecular formula $C_4H_8O_2$.

To determine whether two or more structural formulas represent constitutional isomers (that is, different compounds with the same molecular formula), write the molecular formula of each and then compare them. All compounds that have the same molecular formula, but different structural formulas (different connectivities of their atoms), are constitutional isomers.

Example 2.1

Do the condensed formulas in each pair represent the same compound or constitutional isomers?

(a) $CH_3CH_2CH_2CH_2CH_2CH_3$ and $CH_3CH_2CH_2$ (each is C_6H_{14})
                                              $CH_2CH_2CH_3$

(b) $\overset{\displaystyle CH_3 \quad CH_3}{CH_3CHCH_2CH}$ and $\overset{\displaystyle CH_3}{CH_3CH_2CHCHCH_3}$ (each is C_7H_{16})
                       CH_3 CH_3

Solution

(a) The molecules are drawn here as both condensed structural formulas and line-angle formulas. Each formula has an unbranched chain of six carbons; the two are identical and represent the same compound.

$$\overset{123456}{CH_3CH_2CH_2CH_2CH_2CH_3} \quad \text{and} \quad \overset{123}{CH_3CH_2CH_2}$$

(b) Each formula has a chain of five carbons with two $-CH_3$ branches. Although the branches are identical, they are at different locations on the chains; these formulas represent constitutional isomers.

Problem 2.1

Do the line-angle formulas in each pair represent the same compound or constitutional isomers?

(a) and   **(b)** and

Example 2.2

Write line-angle formulas for the five constitutional isomers with the molecular formula C_6H_{14}.

Solution

In solving problems of this type, you should devise a strategy and then follow it. Here is one such strategy. First, draw a line-angle formula for the constitutional isomer with all six carbons in an unbranched chain. Then, draw line-angle formulas for all constitutional isomers with five carbons in a chain and one carbon as a branch on the chain. Finally, draw line-angle formulas for all constitutional isomers with four carbons in a chain and two carbons as branches.

| Six carbons in an unbranched chain | Five carbons in a chain; with one carbon as a branch | Four carbons in a chain; two carbons as branches |

No constitutional isomers with only three carbons in the longest chain are possible for C_6H_{14}.

Problem 2.2

Draw line-angle formulas for the three constitutional isomers with the molecular formula C_5H_{12}.

The ability of carbon atoms to form strong bonds with other carbon atoms results in a staggering number of constitutional isomers. As the table shows, there are 3 constitutional isomers with the molecular formula C_5H_{12}, 75 constitutional isomers with the molecular formula $C_{10}H_{22}$, and almost 37 million with the molecular formula $C_{25}H_{52}$.

Thus, for even a small number of carbon and hydrogen atoms, a very large number of constitutional isomers is possible. Because constitutional isomers have different chemical properties, a rich diversity of chemistry is possible within these sets.

Carbon Atoms	Constitutional Isomers
1	0
5	3
10	75
15	4,347
25	36,797,588

2.3 | Nomenclature of Alkanes and the IUPAC System

A. The IUPAC System

Ideally, every organic compound should have a name from which its structural formula can be drawn. For this purpose, chemists have adopted a set of rules established by the International Union of Pure and Applied Chemistry (IUPAC).

The IUPAC name of an alkane with an unbranched chain of carbon atoms consists of two parts: (1) a prefix that indicates the number of carbon atoms in the chain and (2) the suffix *-ane* to show that the compound is a saturated hydrocarbon. Table 2.2 gives the prefixes used to show the presence of 1 to 20 carbon atoms.

The first four prefixes listed in Table 2.2 were chosen by the IUPAC because they were well established in the language of organic chemistry. In fact, they were well established even before there were hints of the structural theory underlying

| Table 2.2 | Prefixes Used in the IUPAC System to Show the Presence of 1 to 20 Carbon Atoms in an Unbranched Chain |

Prefix	Number of Carbon Atoms	Prefix	Number of Carbon Atoms
meth-	1	undec-	11
eth-	2	dodec-	12
prop-	3	tridec-	13
but-	4	tetradec-	14
pent-	5	pentadec-	15
hex-	6	hexadec-	16
hept-	7	heptadec-	17
oct-	8	octadec-	18
non-	9	nonadec-	19
dec-	10	eicos-	20

the discipline. For example, the prefix *but-* appears in the name *butyric acid,* a compound of four carbon atoms formed by air oxidation of butter (Latin: *butyrum,* butter). Prefixes to show five or more carbons are derived from Greek or Latin numbers. See Table 2.1 for the names, molecular formulas, and condensed structural formulas for the first 20 unbranched alkanes.

The IUPAC name of an alkane with a branched chain consists of a parent name that indicates the longest chain of carbon atoms in the compound and substituent names that indicate the groups bonded to the parent chain.

a methyl
substituent

$$CH_3$$
$$CH_3CH_2CH_2CHCH_2CH_2CH_2CH_3$$

4-Methyloctane

Alkyl group
A group derived by removing a hydrogen from an alkane; given the symbol R—.

A substituent group derived from an alkane by the removal of a hydrogen atom is called an **alkyl group**, and is commonly represented by the symbol R—. We name alkyl groups by dropping the *-ane* from the name of the parent alkane and adding the suffix *-yl.* The substituent derived from methane, for example, is methyl, CH_3— and that derived from ethane is ethyl, CH_3CH_2—.

The rules of the IUPAC system for naming alkanes follow:

1. The name for an alkane with an unbranched chain of carbon atoms consists of a prefix showing the number of carbon atoms in the chain and the ending *-ane.*
2. For branched-chain alkanes, select the longest chain of carbon atoms as the parent chain; its name becomes the root name.
3. Give each substituent on the parent chain a name and a number. The number shows the carbon atom of the parent chain to which the substituent is bonded. Use a hyphen to connect the number to the name.

$$CH_3$$
$$CH_3CHCH_3$$

2-Methylpropane

4. If there is one substituent, number the parent chain from the end that gives the substituent the lower number.

$$CH_3$$
$$CH_3CH_2CH_2CHCH_3$$

2-Methylpentane   (not 4-Methylpentane)

5. If there are two or more identical substituents, number the parent chain from the end that gives the lower number to the substituent encountered first. The number of times the substituent occurs is indicated by a prefix *di-, tri-, tetra-, penta-, hexa-,* and so on. A comma is used to separate position numbers.

2,4-Dimethylhexane   (not 3,5-Dimethylhexane)

6. If there are two or more different substituents, list them in alphabetical order, and number the chain from the end that gives the lower number to the substituent encountered first. If there are different substituents in equivalent positions on opposite ends of the parent chain, give the substituent of lower alphabetical order the lower number.

3-Ethyl-5-methylheptane   (not 3-Methyl-5-ethylheptane)

7. The prefixes *di-*, *tri-*, *tetra-*, and so on are not included in alphabetizing. Alphabetize the names of the substituents first, and then insert these prefixes. In the following example, the alphabetizing parts are *ethyl* and *methyl*, not ethyl and dimethyl.

4-Ethyl-2,2-dimethylhexane
(not 2,2-Dimethyl-4-ethylhexane)

8. Where there are two or more parent chains of identical length, choose the parent chain with the greater number of substituents.

3-Ethyl-2-methylhexane   (not 3-Isopropylhexane)

Substituents are named following this same set of rules. Those with unbranched chains are named by dropping -*ane* from the name of the parent alkane and replacing it with -*yl*. Thus, unbranched alkyl substituents are named *methyl, ethyl, propyl, butyl, pentyl,* and so forth. Substituents with branched chains are named according to rules 2 and 3. The IUPAC names and structural formulas for unbranched and branched alkyl groups containing one to five carbon atoms are given in Table 2.3. Also given in parentheses are common names for the alkyl substituents. Their common names are so deeply entrenched in organic chemistry that in the official IUPAC nomenclature system, it is acceptable to use either the formal IUPAC name (such as 1-methylethyl) or the common name (in this case, isopropyl) for the alkyl substituents given in Table 2.3.

Table 2.3	Names for Alkyl Groups with One to Five Carbons. Common Names and Their Abbreviations Are Given in Parentheses		
Name	**Condensed Structural Formula**	**Name**	**Condensed Structural Formula**
Methyl (Me)	—CH_3	1,1-Dimethylethyl (*tert*-butyl, *t*-Bu)	CH_3 | —CCH_3 | CH_3
Ethyl (Et)	—CH_2CH_3		
Propyl (Pr)	—$CH_2CH_2CH_3$		
1-Methylethyl (isopropyl, iPr)	—$CHCH_3$ | CH_3	Pentyl	—$CH_2CH_2CH_2CH_2CH_3$
		3-Methylbutyl (isopentyl)	—$CH_2CH_2CHCH_3$ | CH_3
Butyl (Bu)	—$CH_2CH_2CH_2CH_3$		
2-Methylpropyl (isobutyl, iBu)	—CH_2CHCH_3 | CH_3	2-Methylbutyl	—$CH_2CHCH_2CH_3$ | CH_3
1-Methylpropyl (*sec*-butyl, *s*-Bu)	—$CHCH_2CH_3$ | CH_3	2,2-Dimethylpropyl (neopentyl)	CH_3 | —CH_2CCH_3 | CH_3

Example 2.3

Write the IUPAC and common names for these alkanes.

(a) (b)

Solution

Number the longest chain in each compound from the end of the chain toward the substituent that is encountered first. For (a), the longest chain is four carbons (a butane) with a methyl group on carbon 2. For (b), the longest chain is seven carbons (a heptane), with substituents on carbons 2 and 4.

(a)

2-Methylbutane

(b)

2-Methyl-4-(1-methylethyl)heptane or
4-Isopropyl-2-methylheptane

Problem 2.3

Write IUPAC names for these alkanes.

(a) (b)

B. Common Names

In an alternative system known as common nomenclature, the total number of carbon atoms in an alkane, regardless of their arrangement, determines the name. The first three alkanes are methane, ethane, and propane. All alkanes with the molecular formula C_4H_{10} are called butanes, all those with the molecular formula C_5H_{12} are called pentanes, all those with the molecular formula C_6H_{14} are called hexanes, and so forth. The fact that an alkane chain is unbranched is sometimes indicated by the prefix *n-* (normal); an example is *n*-pentane for $CH_3CH_2CH_2CH_2CH_3$. For branched-chain alkanes beyond propane, *iso-* indicates that one end of an otherwise unbranched chain terminates in a $(CH_3)_2CH—$ group, and *neo-* indicates that it terminates in $—C(CH_3)_3$. Following are examples of common names.

$$CH_3CH_2CH_2CH_3 \qquad CH_3\overset{\overset{\displaystyle CH_3}{|}}{C}HCH_3 \qquad CH_3CH_2CH_2CH_2CH_3 \qquad CH_3CH_2\overset{\overset{\displaystyle CH_3}{|}}{C}HCH_3 \qquad CH_3\overset{\overset{\displaystyle CH_3}{|}}{\underset{\underset{\displaystyle CH_3}{|}}{C}}CH_3$$

| Butane | Isobutane | Pentane | Isopentane | Neopentane |

This system of common names has no good way of naming other branching patterns; for more complex alkanes, it is necessary to use the more flexible IUPAC system of nomenclature.

In this text, we concentrate on IUPAC names. However, we also use common names, especially when the common name is used almost exclusively in everyday discussions among chemists. When both IUPAC and common names are given in the text, we give the IUPAC name first, followed by the common name in parentheses. In this way, you should have no doubt about which name is which.

C. The IUPAC System—A General System of Nomenclature

The naming of alkanes and cycloalkanes in Section 2.3A illustrates the application of the IUPAC system of nomenclature to a specific class of organic compounds. Now let us describe the general approach of the IUPAC system. The name we give to any compound with a chain of carbon atoms consists of three parts: a **prefix**, an **infix** (a modifying element inserted into a word), and a **suffix**. Each part provides specific information about the structure of the compound.

1. The prefix indicates the number of carbon atoms in the parent chain. Prefixes that show the presence of 1 to 20 carbon atoms in an unbranched chain are given in Table 2.2.
2. The infix indicates the nature of the carbon-carbon bonds in the parent chain.

Infix	Nature of Carbon-Carbon Bonds in the Parent Chain
-an-	all single bonds
-en-	one or more double bonds
-yn-	one or more triple bonds

3. The suffix indicates the class of compound to which the substance belongs.

Suffix	Class of Compound
-e	hydrocarbon
-ol	alcohol
-al	aldehyde
-amine	amine
-one	ketone
-oic acid	carboxylic acid

Example 2.4

Following are IUPAC names and structural formulas for four compounds:

(a) $CH_2=CHCH_3$ (b) CH_3CH_2OH (c) $CH_3CH_2CH_2CH_2\overset{\overset{O}{\|}}{C}OH$ (d) $HC\equiv CH$

     Propene   Ethanol   Pentanoic acid   Ethyne

Divide each name into a prefix, an infix, and a suffix, and specify the information about the structural formula that is contained in each part of the name.

Solution

(a) prop-en-e ← a hydrocarbon
     ⌐a carbon-carbon double bond
     ⌐three carbon atoms

(b) eth-an-ol ← an alcohol
     ⌐only carbon-carbon single bonds
     ⌐two carbon atoms

(c) pent-an-oic acid ← a carboxylic acid
     ⌐only carbon-carbon single bonds
     ⌐five carbon atoms

(d) eth-yn-e ← a hydrocarbon
     ⌐a carbon-carbon triple bond
     ⌐two carbon atoms

Problem 2.4

Combine the proper prefix, infix, and suffix, and write the IUPAC name for each compound:

(a) $CH_3\overset{\overset{\displaystyle O}{\|}}{C}CH_3$
(b) $CH_3(CH_2)_3\overset{\overset{\displaystyle O}{\|}}{C}H$
(c) [cyclopentane ring]—OH
(d) [cycloheptene ring]

D. Classification of Carbon and Hydrogen Atoms

We classify a carbon atom as primary ($1°$), secondary ($2°$), tertiary ($3°$), or quaternary ($4°$), depending on the number of carbon atoms bonded to it. A carbon bonded to one carbon atom is a primary carbon; a carbon bonded to two carbon atoms is a secondary carbon, and so forth. For example, propane contains two primary carbons and one secondary carbon; 2-methylpropane contains three primary carbons and one tertiary carbon; and 2,2,4-trimethylpentane contains five primary carbons, one secondary carbon, one tertiary carbon, and one quaternary carbon.

two $1°$ carbons   a $3°$ carbon   a $4°$ carbon

$CH_3-CH_2-CH_3$ $CH_3-CH-CH_3$ $CH_3-C-CH_2-CH-CH_3$

a $2°$ carbon

Propane   **2-Methylpropane**   **2,2,4-Trimethylpentane**

Hydrogens are also classified as primary, secondary, or tertiary depending on the type of carbon to which each is bonded. Those bonded to a primary carbon are classified as primary hydrogens, those bonded to a secondary carbon are secondary hydrogens, and those bonded to a tertiary carbon are tertiary hydrogens.

2.4 | Cycloalkanes

A hydrocarbon that contains carbon atoms joined to form a ring is called a **cyclic hydrocarbon**. When all carbons of the ring are saturated, the hydrocarbon is called a **cycloalkane**.

Cycloalkane
A saturated hydrocarbon that contains carbons joined to form a ring.

A. Structure and Nomenclature

Cycloalkanes of ring sizes from 3 to over 30 are found in nature, and, in principle, there is no limit to ring size. Five-membered rings (cyclopentanes) and six-membered rings (cyclohexanes) are especially common and will receive special attention. Figure 2.3 shows structural formulas of cyclopropane, cyclobutane, cyclopentane, and cyclohexane. When writing structural formulas for cycloalkanes, chemists rarely show all carbons and hydrogens. Rather, they use line-angle formulas to represent cycloalkane rings. Each ring is represented by a regular polygon that has the same number of sides as there are carbon atoms in the ring. For example, chemists represent cyclobutane by a square, cyclopentane by a pentagon, and cyclohexane by a hexagon.

Figure 2.3
Examples of cycloalkanes.

Cyclopropane
C_3H_6

Cyclobutane
C_4H_8

Cyclopentane
C_5H_{10}

Cyclohexane
C_6H_{12}

Cycloalkanes contain two fewer hydrogen atoms than an alkane with the same number of carbon atoms and have the general formula C_nH_{2n}. For example, compare the molecular formulas of cyclohexane, C_6H_{12}, and hexane, C_6H_{14}.

To name a cycloalkane, prefix the name of the corresponding open-chain alkane with *cyclo-*, and name each substituent on the ring. If there is only one substituent on the cycloalkane ring, there is no need to give it a number. If there are two substituents, number the ring by beginning with the substituent of lower alphabetical order. If there are three or more substituents, number the ring so as to give the substituents the lowest set of numbers, and list them in alphabetical order.

Example 2.5

Write the molecular formula and the IUPAC name for each cycloalkane.

(a) (b) (c)

Solution

(a) First replace each vertex and line terminus with a carbon and then add hydrogens as necessary to give each carbon four bonds. The molecular formula of this compound is C_8H_{16}. Because there is only one substituent on the ring, there is no need to number the atoms of the ring. This compound's name is (1-methylethyl) cyclopentane. The substituent also could be named isopropyl, giving the alternative IUPAC name isopropylcyclopentane.

$$C_8H_{16}$$

(b) The two substituents are ethyl and 1,1-dimethylethyl, and the IUPAC name of the cycloalkane is 1-ethyl-4-(1,1-dimethylethyl)cyclohexane. The substituents also could be named ethyl and *tert*- butyl, giving the cycloalkane the alternative IUPAC name 1-*tert*-butyl-4-ethylcyclohexane. Its molecular formula is $C_{12}H_{24}$.

(c) Number the ring to give the three substituents the lowest set of numbers and then list them in alphabetical order. The name of this compound is 2-ethyl-1,4-dimethylcyclohexane, and its molecular formula is $C_{10}H_{20}$.

Problem 2.5

Write the molecular formula, IUPAC name, and common name for each cycloalkane.

(a) (b) (c)

B. Bicycloalkanes

An alkane that contains two rings that share two carbon atoms in common is classified as a **bicycloalkane**. The shared carbon atoms are called **bridgehead carbons**, and the carbon chain connecting them is called a **bridge**. The general formula of a bicycloalkane is C_nH_{2n-2}. Figure 2.4 shows three examples of bicycloalkanes along with the IUPAC and common name of each.

Bicycloalkane
An alkane containing two rings that share two carbons.

Figure 2.4

Examples of bicycloalkanes.

Bicyclo[4.3.0]nonane
(Hydrindane)

Bicyclo[4.4.0]decane
(Decalin)

Bicyclo[2.2.1]heptane
(Norbornane)

Example 2.6

Write the general formula for an alkane, a cycloalkane, and a bicycloalkane. How do these general formulas differ?

Solution

General formulas are C_nH_{2n+2} for an alkane, C_nH_{2n} for a cycloalkane, and C_nH_{2n-2} for a bicycloalkane. Each general formula in this series has two fewer hydrogens than the previous member of the series.

Problem 2.6

Write molecular formulas for each bicycloalkane, given its number of carbon atoms.

(a) Hydrindane (9 carbons) **(b)** Decalin (10 carbons) **(c)** Norbornane (7 carbons)

The IUPAC has a set of rules for naming bicyclealkanes, but they are beyond the scope of this text. We will continue to refer to most bicycloalkanes by their common names.

Example 2.7

Following are structural formulas and common names for three bicyclic compounds. Write the molecular formula of each compound.

(a) α-Pinene **(b)** Camphor **(c)** Caryophyllene

Solution

(a) The molecular formula of α-pinene is $C_{10}H_{16}$. α-Pinene is a major component, often as high as 65% by volume, of pine oil and turpentine.
(b) The molecular formula of camphor is $C_{10}H_{16}O$. Camphor, obtained from the camphor tree, *Cinnamonium camphora*, is used in the manufacture of certain plastics, lacquers, and varnishes.
(c) The molecular formula of caryophyllene is $C_{15}H_{24}$. Caryophyllene is one of the fragrant components of oil of cloves.

Problem 2.7

Following are the structural formulas and names of four bicycloalkanes. Write the molecular formula of each compound. Which of these compounds are constitutional isomers?

(a) Thujane   (b) Carane   (c) Pinane   (d) Bornane

2.5 | Conformations of Alkanes and Cycloalkanes

Structural formulas are useful for showing the connectivity of atoms in a molecule. However, they usually do not show three-dimensional shapes. As chemists try to understand more and more about the relationships between structure and the chemical and physical properties of compounds, it becomes increasingly important to know more about the three-dimensional shapes of molecules.

In this section, we ask you to look at molecules as three-dimensional objects and to visualize not only bond angles, but also distances between various atoms and groups of atoms within the molecules. We also describe intramolecular strain, which we divide into three types: torsional strain, steric strain, and angle strain. We urge you to build models (either physically or with desktop modeling programs such as Chem3D or Spartan) of the molecules discussed in this section so that you become comfortable in dealing with them as three-dimensional objects, and understand fully the origins of the various types of intramolecular strain.

Strain, a key concept in organic chemistry, is a measure of the energy stored in a compound due to a structural distortion. Chemicals are physical entities, each possessing an optimal structure, just as does any macroscopic object. If you perturb the optimal structure, you put a strain on the system. A tree branch bending in the wind is a strained form of the branch. A bent branch is at higher energy than an undisturbed branch, and it releases that energy when relaxing between wind gusts. In this section, we describe strained forms of alkanes that interconvert with more optimal structures. The interconversion between the strained and relaxed structures results from collisions and thermal motions. Some structures are permanently strained, and do not readily have a pathway to release that strain. Analogously, a compressed spring is strained, and unless the compression is removed, the spring is stuck in the strained form. Energy is stored in the compressed spring waiting to be released if given an opportunity. The opportunity to release permanently stored strain in a chemical structure comes from chemical reactions, and near the end of this chapter we analyze combustion (burning) as a means of measuring this kind of strain.

A. Alkanes

Alkanes of two or more carbons can be twisted into a number of different three-dimensional arrangements of their atoms by rotating about one or more carbon-carbon bonds. Any three-dimensional arrangement of atoms that results from rotation about single bonds is called a **conformation**. Figure 2.5(a) shows a ball-and-stick model of a **staggered conformation** of ethane. In this conformation, the three C—H bonds on one carbon are as far apart as possible from those bonds on the adjacent carbon. Figure 2.5(b), called a **Newman projection**, (named for Melvin Newman, of Ohio State University, who developed these projections) is a shorthand way to represent this conformation of ethane. In a Newman projection, a molecule is viewed down the axis of a C—C bond. The three atoms or groups of atoms on the carbon nearer your eye are shown on lines extending from the center

Strain
An instability within a structure associated with higher internal energy.

Conformation
Any three-dimensional arrangement of atoms in a molecule that results from rotation about a single bond.

Staggered conformation
A conformation about a carbon-carbon single bond in which the atoms or groups on one carbon are as far apart as possible from atoms or groups on an adjacent carbon.

Newman projection
A way to view a molecule by looking along a carbon-carbon single bond to help evaluate the relative orientations of attached groups.

Figure 2.5

A staggered conformation of ethane. (a) Ball-and-stick models and (b) Newman projection.

(a) Side view — Turned almost end-on — End view

(b) Newman projection

of the circle at angles of 120°. The three atoms or groups of atoms on the carbon farther from your eye are shown on lines extending from the circumference of the circle, also at angles of 120°. Remember that bond angles about each carbon in ethane are approximately 109.5° and not 120°, as this Newman projection might suggest. The three lines in front represent bonds directed toward you, whereas the three lines in back point away from you.

Figure 2.6

An eclipsed conformation of ethane. (a, b) Ball-and-stick models and (c) Newman projection.

(a) Side view (b) Turned almost end-on (c) Newman projection

Eclipsed conformation
A conformation about a carbon-carbon single bond in which the atoms or groups on one carbon are as close as possible to the atoms or groups on an adjacent carbon.

Dihedral angle
The angle created by two intersecting planes.

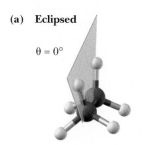

(a) Eclipsed

$\theta = 0°$

(b) Staggered

$\theta = 60°$

Figure 2.7
Dihedral angles in ethane. (a) An eclipsed conformation and (b) a staggered conformation.

Figure 2.6 shows a ball-and-stick model and a Newman projection for an **eclipsed conformation** of ethane. In this conformation, the three C—H bonds on one carbon are as close as possible to the three C—H bonds on the adjacent carbon. In other words, hydrogen atoms on the back carbon are eclipsed by the hydrogen atoms on the front carbon (just as the sun is eclipsed when the moon passes in front of it). Different conformations are often called **conformational isomers** or **conformers**.

If we are to discuss energy relationships among conformations, it is convenient to define the term *dihedral angle*. A **dihedral angle**, θ (Greek theta), is the angle created by two intersecting planes, each plane defined by three atoms. In the Newman projection of the eclipsed conformation of ethane in Figure 2.7(a), two H—C—C planes are shown. The angle at which these planes intersect (the dihedral angle) is 0°. A staggered conformation in which the dihedral angle of the two H—C—C planes is 60° is illustrated in Figure 2.7(b).

In principle, there are an infinite number of conformations of ethane that differ only in the degree of rotation about the carbon-carbon single bond. Because there is a small energy barrier between conformations, rotation is not completely free. As we shall see, the lowest energy (the most stable) conformation of ethane is a staggered conformation. The highest energy (the least stable) conformation is an eclipsed conformation. At room temperature, ethane molecules undergo collisions with sufficient energy so that the energy barrier between extreme conformations can be crossed and rotation about the carbon-carbon single bond from one conformation to another occurs rapidly.

The difference in energy between an eclipsed conformation and a staggered conformation of ethane is approximately 12.6 kJ (3.0 kcal)/mol (approximately 4.2 kJ (1.0 kcal)/mol for each eclipsed H/H), and is referred to as torsional strain. **Torsional strain** (also called eclipsed-interaction strain) arises when nonbonded atoms separated by three bonds are forced from a staggered conformation to an eclipsed conformation. In ethane, for example, torsional strain occurs when pairs of hydrogens H(4)-H(6), H(5)-H(8), and H(3)-H(7) on adjacent carbons are forced into eclipsed positions. The models shown here represent only one of the three different but equivalent eclipsed conformations of ethane.

Ethane
staggered conformation

Ethane
eclipsed conformation

Figure 2.8 shows the relationship between energy and dihedral angle for the conformations of ethane. All energy diagrams in this book use *energy* as a vertical axis. Several types of energy—potential energy, Gibbs free energy, and enthalpy—are important in various contexts. The diagrams for all these types of energy are often nearly indistinguishable, and in most cases "energy" will do for the concepts we are introducing. When it is necessary to be more precise about the type of energy, we will do so and explain why.

Torsional strain
Strain that arises when nonbonded atoms separated by three bonds are forced from a staggered conformation to an eclipsed conformation. Torsional strain is also called eclipsed-interaction strain.

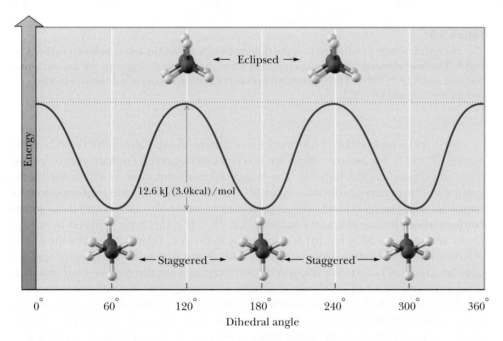

Figure 2.8
The energy of ethane as a function of dihedral angle. The eclipsed conformations are approximately 12.6 kJ (3.0 kcal)/mol higher in energy than the staggered conformations.

There has been disagreement over the years as to the origin of the torsional strain in the eclipsed conformations of ethane. It was originally thought that this strain is the result of the repulsion between eclipsed hydrogen nuclei; they are separated by 255 pm in a staggered conformation, but by only 235 pm in an eclipsed conformation. Alternatively, it has been held that the torsional strain is the result of repulsion between the electron clouds of the adjacent C—H bonds. Theoretical molecular orbital calculations, however, suggest that the energy difference between the conformational extremes arises not from destabilization of the eclipsed conformation, but rather from stabilization of the staggered conformation. This stabilization of the staggered conformation arises because of a small donor-acceptor interaction (donation of electron density from a filled orbital into an empty acceptor orbital) between a filled C—H bonding MO of one carbon and the empty or unfilled C—H antibonding MO on the adjacent carbon with which it is aligned. This donor-acceptor stabilization is lost when a staggered conformation is converted to an eclipsed conformation. Thus, although the fact remains that eclipsed ethane is higher in energy than staggered ethane, the reason for this energy difference is still a subject of study among chemists.

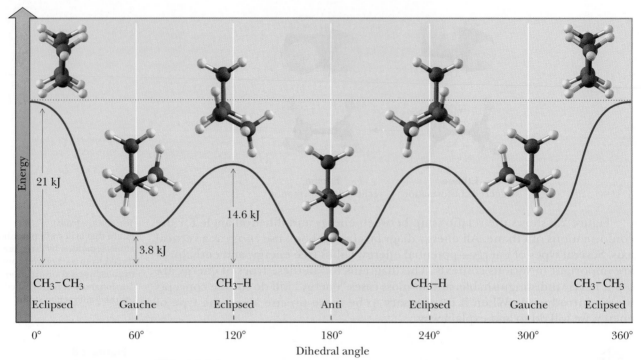

Figure 2.9
The energy of butane as a function of the dihedral angle about the bond between carbons 2 and 3. The lowest energy conformation occurs when the two methyl groups are the maximum distance apart ($\theta = 180°$). The highest energy conformation occurs when the two methyl groups are eclipsed ($\theta = 0°$).

Anti conformation
A conformation about a single bond in which two groups on adjacent carbons lie at a dihedral angle of 180°.

Gauche conformation
A conformation about a single bond of an alkane in which two groups on adjacent carbons lie at a dihedral angle of 60°.

Angle strain
The strain that arises when a bond angle is either compressed or expanded compared to its optimal value.

Steric strain
The strain that arises when nonbonded atoms separated by four or more bonds are forced closer to each other than their atomic (contact) radii would allow. Steric strain is also called nonbonded interaction strain, or van der Waals strain.

Next let us look at the conformations of butane viewed along the bond between carbons 2 and 3. For butane, there are two types of staggered conformations and two types of eclipsed conformations. The staggered conformation in which the methyl groups are the maximum distance apart ($\theta = 180°$) is called the **anti conformation**; the staggered conformation in which they are closer together ($\theta = 60°$) is called the **gauche conformation**. In one eclipsed conformation ($\theta = 0°$), methyl is eclipsed by methyl. In the other ($\theta = 120°$), methyl is eclipsed by hydrogen. Figure 2.9 shows the energy relationships for rotation from $-180°$ to $180°$. Note that both the gauche and anti conformations of butane are staggered conformations, yet the gauche conformations are approximately 3.8 kJ (0.9 kcal)/mol higher in energy than the anti conformation.

In dealing with the relative stabilities of the various conformations of butane, we again encounter torsional strain. In addition, we encounter two other types of strain: angle strain and steric strain. **Angle strain** results when a bond angle in a molecule is either expanded or compressed compared to its optimal values. Strain can also occur when bond lengths are forced to become shorter or longer than normal. In general, bond stretching is not as easy as bond bending. **Steric strain** (also called nonbonded interaction or van der Waals strain) results when nonbonded atoms separated by four or more bonds are forced closer to each other than their atomic (contact) radii allow—that is, when they are forced to smash into each other. The total of all types of strain can be calculated by molecular mechanics programs; angles and bond lengths are chosen in an iterative procedure so that an optimum geometry with the lowest total energy is calculated. Such routines are included with popular desktop programs such as Chem3D and Spartan. Molecular mechanics calculations carried out with these or other programs can determine the lowest energy arrangement of atoms in a given conformation of a molecule, a process referred to as energy minimization.

Let us illustrate the origins of both angle strain and steric strain by comparing the anti ($\theta = 180°$) and eclipsed ($\theta = 0°$) conformations of butane. In an energy-minimized conformation of anti butane, the C—C—C bond angle is 111.9° and all H—C—H bond angles are between 107.4° and 107.9°.

Now consider the eclipsed conformation. Figure 2.10(a) shows models of an eclipsed butane conformation in which C—C—C bond angles are set at 111.9°, the same value as in the energy-minimized anti conformation. Notice that in this eclipsed conformation, hydrogens H(5)-H(13) and H(5)-H(14) are forced to within 178 pm of each other, a distance closer than their contact radii would allow. When the energy of eclipsed butane is minimized [Figure 2.10(b)], the distance between these same pairs of hydrogens increases to 231 pm (which reduces steric strain). At the same time, the C(4)-C(3)-C(2) and C(3)-C(2)-C(1) bond angles increase from 111.9° to 119.0° (which increases angle strain).

Space-filling model, top view

Space-filling model, top view

(a) Ball-and-stick model, side view

(b) Ball-and-stick model, side view

Figure 2.10

Eclipsed conformations of butane. (a) Non-energy-minimized and (b) energy-minimized eclipsed conformations. The calculated difference in energy between the non-energy-minimized and energy-minimized eclipsed conformations is 3.6 kJ (0.86 kcal)/mol.

Thus, the energy-minimized eclipsed conformation represents a balance between a decrease in steric strain and an increase in angle strain. The calculated strain in the energy-minimized eclipsed conformation of butane is 21 kJ (5.0 kcal)/mol, relative to the staggered conformation.

An energy-minimized gauche conformation of butane (Figure 2.11) is approximately 3.8 kJ (0.90 kcal)/mol higher in energy than the anti, staggered conformation. This difference in energy is caused almost entirely by the steric strain (nonbonded interaction strain) between the two methyl groups. At any given instant, a larger number of butane molecules are in the anti conformation than in the gauche conformation, and the number of molecules in the eclipsed conformation is vanishingly small. The percentage of the anti conformation present at 20°C is about 70%.

Butane

Figure 2.11

One of two equivalent energy-minimized gauche conformations of butane.

Although the two gauche conformations ($\theta = 60°$ and $300°$) have equal energies, they are not identical. They are related by reflection; that is, one gauche conformation is the reflection of the other, just as your right hand is the reflection of your left hand. The conformations with eclipsed —CH$_3$ and —H groups ($\theta = 120°$ and $240°$) are also related by reflection. We shall have more to say about objects and their mirror reflections in Chapter 3.

Example 2.8

Following is the structural formula of 1,2-dichloroethane:

$$\text{Cl—C—C—Cl}$$

with H, H on top and H, H on bottom of the two carbons.

1,2-Dichloroethane

(a) Draw Newman projections for all staggered conformations formed by rotation from 0° to 360° about the carbon-carbon single bond.
(b) Which staggered conformation(s) has (have) the lowest energy? Which has (have) the highest energy?
(c) Which, if any, of these staggered conformations are related by reflection?

Solution

(a) If we take the dihedral angle when the chlorines are eclipsed as a reference point, staggered conformations occur at dihedral angles 60°, 180°, and 300°.

Gauche	Anti	Gauche
$\theta = 60°$	$\theta = 180°$	$\theta = 300°$

(b) We predict that the anti conformation (dihedral angle $\theta = 180°$) has the lowest energy. The two gauche conformations (dihedral angle $\theta = 60°$ and $300°$) are of higher but equal energy. We are not given data in the problem to calculate the actual energy differences.
(c) The two gauche conformations are related by reflection.

Problem 2.8

For 1,2-dichloroethane:

(a) Draw Newman projections for all eclipsed conformations formed by rotation from 0° to 360° about the carbon-carbon single bond.
(b) Which eclipsed conformation(s) has (have) the lowest energy? Which has (have) the highest energy?
(c) Which, if any, of these eclipsed conformations are related by reflection?

A word of caution. Although we talk about eclipsed along with staggered conformations, this can be misleading. Of all the conformations of butane, the eclipsed conformations are at the highest points on the energy profile. The gauche and anti conformations are in energy troughs. As a result, butane molecules spend their time in the anti and gauche conformations, and only fleetingly pass through the eclipsed conformations when interconverting between anti and gauche. A further nuance is that C—H and C—C bonds have vibrational motions at room temperature, and these vibrations occur simultaneously with bond rotations. The reality is that hydrocarbons are exceedingly complex and dynamic structures in solution, undergoing constant interconversions between low-energy conformations through combined vibrations and rotations.

B. Cycloalkanes

The structures and energies of cyclic alkanes are highly dependent on the size of their rings. Of major importance is **small ring strain**, which exists in ring sizes below six carbons and is due to the C—C—C bond angles not being able to achieve the optimal tetrahedral angle of 109.5°. Further, just as alkanes have rapidly interconverting conformations with varying degrees of torsional strain along their C—C single bonds, so do cyclic alkanes of four carbons or more. The interconversions involve rotations along C—C single bonds of the rings. However, the bonds can only rotate so far without breaking the ring, and hence the rotations are limited to certain angles.

i. Cyclopropane

The observed C—C—C bond angles in cyclopropane are 60° (Figure 2.12), a value considerably smaller than the bond angle of 109.5° predicted for sp^3 hybridized carbon atoms. This compression from the optimal bond angle introduces a considerable angle strain. Furthermore, because cyclopropane is planar, there are six pairs of fully eclipsed C—H bonds, which introduce considerable torsional strain. The combined angle and torsional strain energy in cyclopropane is approximately 116 kJ (27.7 kcal)/mol. Because of its extreme degree of intramolecular strain, cyclopropane and its derivatives undergo several ring-opening reactions not seen with larger cycloalkanes.

ii. Cyclobutane

Nonplanar or puckered conformations are favored in all cycloalkanes larger than cyclopropane. If cyclobutane were planar [Figure 2.13(a)], all C—C—C bond angles would be 90° (which would maximize angle strain), and there would be eight pairs of eclipsed hydrogen interactions (which would maximize torsional strain). Rotations along the C—C bonds can a slightly relieve this strain. Puckering of the ring [Figure 2.13(b)] alters the strain energy in two ways: (1) it decreases the torsional strain associated with eclipsed interactions, and (2) it increases further the angle strain caused by the compression of C—C—C bond angles. Because the decrease in torsional strain is greater than the increase in angle strain, puckered cyclobutane is more stable than planar cyclobutane. In the conformation of lowest energy, the measured C—C—C bond angles are 88°, and the strain energy in cyclobutane is approximately 110 kJ (26.3 kcal)/mol. Just like butane, cyclobutane is not static but undergoes interconversion between the puckered conformations.

Small ring strain
A strain associated with ring sizes below six arises from nonoptimal bond angles.

(a)

pairs of eclipsed
C-H interactions

(b)

60°

Figure 2.12
Cyclopropane. (a) Structural formula and (b) ball-and-stick model.

(a) Hypothetical planar conformation   (b) Puckered or butterfly conformations

Figure 2.13
Cyclobutane. (a) In the planar conformation, there are eight pairs of eclipsed C—H interactions. (b) The energy is a minimum in the puckered (butterfly) conformation.

iii. Cyclopentane

If cyclopentane were to adopt a planar conformation, all C—C—C bond angles would be 108° [Figure 2.14(a)]. This angle differs only slightly from the tetrahedral angle of 109.5°; consequently, there would be little angle strain in this conformation. In a planar conformation, however, there are 10 pairs of fully eclipsed C—H bonds creating a torsional strain of approximately 42 kJ (10 kcal)/mol. To relieve at least a part of this torsional strain, the ring twists by rotations along the C—C bonds into the **"envelope" conformation** shown in Figure 2.14(b). In this conformation, four carbon atoms are in a plane, and the fifth bends out of the plane, rather like an envelope with its flap bent upward. Cyclopentane exists as a dynamic equilibrium of five equivalent envelope conformations in which each carbon atom alternates as the out-of-plane atom. Note that the average C—C—C bond angle

Figure 2.14

Cyclopentane. (a) In the planar conformation, there are 10 pairs of eclipsed C—H interactions. (b) The most stable conformation is a puckered or envelope conformation.

Hypothetical planar conformation   Puckered envelope conformation

is reduced to 105° (increasing angle strain). In the envelope conformation, the number of eclipsed C—H interactions is also reduced (decreasing torsional strain). Thus, the envelope conformation should be thought of as an energetic compromise that balances a decreased torsional strain with an increased angle strain. The total strain energy in puckered cyclopentane is approximately 27 kJ (6.5 kcal)/mol.

iv. Cyclohexane

Cyclohexane adopts a number of puckered conformations that interconvert via C—C rotations, the most stable of which is the **chair conformation** (Figure 2.15). In this conformation, all C—C—C bond angles are 110.9° (minimizing angle strain), and all hydrogens on adjacent carbons are staggered with respect to one another (minimizing torsional strain). In addition, no two atoms are close enough to each other for nonbonded interaction strain to exist. Thus, there is little strain in a chair conformation of cyclohexane.

The C—H bonds in a chair conformation of cyclohexane are arranged in two different orientations. Six C—H bonds are called **axial bonds**, and the other six are called **equatorial bonds**. One way to visualize the difference between these two types of bonds is to imagine an axis through the center of the chair, perpendicular to the floor (Figure 2.16).

Equatorial bonds are approximately perpendicular to the imaginary axis and form an equator about the ring. Equatorial bonds alternate, first slightly up and then slightly down as you move from one carbon of the ring to the next. Axial bonds are parallel to the imaginary axis. Three axial bonds point straight up; the other three axial bonds point straight down. Axial bonds also alternate, first up and then down as you move from one carbon of the ring to the next. Notice also that if the

Chair conformation
The most stable conformation of a cyclohexane ring; all bond angles are close to 109.5°, and all bonds on adjacent carbons are staggered.

Axial bond
A bond to a chair conformation of cyclohexane that extends from the ring parallel to the imaginary axis through the center of the ring; a bond that lies roughly perpendicular to the equator of the ring.

Equatorial bond
A bond to a chair conformation of cyclohexane that extends from the ring roughly perpendicular to the imaginary axis through the center of the ring; a bond that lies roughly along the equator of the ring.

Figure 2.15

A chair conformation of cyclohexane.

View from the side   View from above

Figure 2.16

A chair conformation of cyclohexane, showing axial and equatorial C—H bonds. The plane of the ring is defined by four carbons; the fifth carbon is above the plane, and the sixth carbon is below it.

(a) Ball-and-stick model showing all 12 hydrogens

(b) The six equatorial C—H bonds shown in red

(c) The six axial C—H bonds shown in blue

axial bond on a carbon points upward, then the equatorial bond on that carbon points slightly downward. Conversely, if the axial bond on a particular carbon points downward, then the equatorial bond on that carbon points slightly upward.

There are many other nonplanar conformations of cyclohexane, two of which, a **boat** and a **twist-boat**, are shown in Figure 2.17.

Boat conformation
A nonplanar conformation of a cyclohexane ring in which carbons 1 and 4 of the ring are bent toward each other.

Figure 2.17
Boat and twist-boat conformations of cyclohexane.

"Flagpole" H atoms

Boat conformation   Twist-boat conformation

You can visualize interconversion of chair and boat conformations by twisting about two carbon-carbon bonds as illustrated in Figure 2.18. A boat conformation is considerably less stable than a chair conformation because of the torsional strain associated with four pairs of eclipsed C—H interactions and the steric strain between the two "flagpole" hydrogens. The difference in energy between chair and boat conformations is approximately 27 kJ (6.5 kcal)/mol.

Twist-boat conformation
A nonplanar conformation of a cyclohexane ring that is twisted from and slightly more stable than a boat conformation.

(a) Chair   **(b) Boat**
These hydrogens are   "Flagpole" H are now
very far apart.   almost in contact.

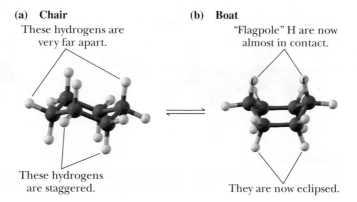

These hydrogens   They are now eclipsed.
are staggered.

Figure 2.18
Interconversion of (a) a chair conformation to (b) a boat conformation produces one set of flagpole steric interactions and four sets of eclipsed hydrogen interactions.

Some of the strain in the boat conformation can be relieved by a slight twisting of the ring to a twist-boat conformation. It is estimated by computer modeling that a twist-boat conformation is favored over a boat conformation by approximately 6.3 kJ (1.5 kcal)/mol. Figure 2.19 shows an energy diagram for the interconversion between chair, twist-boat, and boat conformations. The large difference in energy between chair and boat or twist-boat conformations means that, at room temperature, molecules in the chair conformation make up more than 99.99% of the equilibrium mixture.

Twist this   Twist this
carbon up   carbon down

Chair conformation   Boat conformation   Chair conformation

For cyclohexane, the two equivalent chair conformations can be interconverted by twisting one chair first into a boat and then into the alternative chair.

Figure 2.19
Energy diagram for interconversion of chair, twist-boat, and boat conformations of cyclohexane. The chair conformation is the most stable because angle, torsional, and steric strain are at a minimum.

When one chair is converted to the other, a change occurs in the relative orientations in space of the hydrogen atoms bonded to each carbon. All hydrogen atoms axial in one chair become equatorial in the other and vice versa (Figure 2.20). The conversion of one chair conformation of cyclohexane to the other occurs rapidly at room temperature.

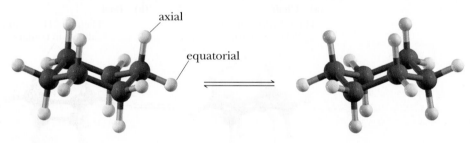

Figure 2.20
Interconversion of alternative chair conformations of cyclohexanes. All C—H bonds equatorial in one chair are axial in the alternative chair and vice versa.

Example 2.9

Following is a chair conformation of cyclohexane showing one methyl group and one hydrogen.

(a) Indicate by a label whether each group is equatorial or axial.
(b) Draw the alternative chair conformation and again label each group as axial or equatorial.

Solution

One chair   The alternative chair

Problem 2.9

Following is a chair conformation of cyclohexane with the carbon atoms numbered 1 through 6.

(a) Draw hydrogen atoms that are above the plane of the ring on carbons 1 and 2 and below the plane of the ring on carbon 4.

(b) Which of these hydrogens are equatorial? Which are axial?

(c) Draw the alternative chair conformation. Which hydrogens are equatorial? Which are axial? Which are above the plane of the ring? Which are below it?

If we replace a hydrogen atom of cyclohexane with an alkyl group, the group occupies an equatorial position in one chair and an axial position in the alternative chair. This means that the two chairs are no longer equivalent and no longer of equal stability.

A convenient way to describe the relative stabilities of chair conformations with equatorial and axial substituents is in terms of a type of steric strain called **diaxial (axial-axial) interaction**. The term diaxial interaction refers to the steric strain between an axial substituent and an axial hydrogen (or another group) on the same side of a cyclohexane ring. The axial positions on the same side of the ring are extremely close to each other, and any groups larger than hydrogen atoms will introduce steric strain between the larger group and the other two axial hydrogen atoms. Because this type of steric strain originates between groups on carbons 1 and 3 of a cyclohexane ring, it is often called a 1,3-diaxial interaction.

Consider methylcyclohexane (Figure 2.21). When —CH$_3$ is axial, it is parallel to the axial C—H bonds on carbons 3 and 5. Thus, for axial methylcyclohexane, there are two unfavorable methyl-hydrogen diaxial interactions. No such unfavorable interactions exist when the methyl group is in an equatorial position. For methylcyclohexane, the equatorial methyl conformation is favored over the axial methyl conformation by approximately 7.28 kJ (1.74 kcal)/mol.

Given the difference in strain energy between the axial and equatorial conformations of methylcyclohexane, we can calculate the ratio of the two conformations at equilibrium using the equation that relates the change in Gibbs free energy ($\Delta G°$) for an equilibrium, the equilibrium constant (K_{eq}), and the temperature (T) in kelvins. R, the universal gas constant, has the value 8.314 J (1.987 cal)$\cdot K^{-1} \cdot$mol^{-1}.

$$\Delta G° = -RT \ln K_{eq}$$

Substituting the value of -7.28 kJ/mol (axial methyl $\rightarrow$ equatorial methyl) for $\Delta G°$ and solving the equation gives a value of 18.9 for the equilibrium constant at room temperature (25°C = 298 K).

$$\ln K_{eq} = \frac{-(-7280 \text{ J} \cdot \text{mol}^{-1})}{8.314 \text{ J} \cdot \text{K}^{-1} \cdot \text{mol}^{-1} \times 298 \text{ K}} = 2.939$$

$$K_{eq} = \frac{18.9}{1} = \frac{\text{equatorial}}{\text{axial}}$$

Diaxial interaction
Refers to the steric strain arising from interaction between an axial substituent and an axial hydrogen (or other group) on the same side of a chair conformation of a cyclohexane ring.

Diaxial interactions

Figure 2.21
Two chair conformations of methylcyclohexane. The steric strain introduced by two diaxial interactions makes the axial methyl conformation less stable by approximately 7.28 kJ (1.74 kcal)/mol.

You will be asked to draw chair conformations of cyclohexane often because these conformations allow you to identify which substituents are axial and which are equatorial. Although drawing chair conformations takes practice, following a few simple guidelines will make you an expert at drawing even complicated substitution patterns.

Step 1: Start by drawing two sets of parallel lines, each set at a slight angle.

Step 2: Complete each chair by drawing the ends connected to the parallel lines, in each case making one end tip up, the other down.

Step 3: Draw the axial bonds as vertical lines that are in the direction of the larger angle at each ring atom.

Axial bonds

Step 4: Draw the equatorial bonds using the bonds of the ring as guides for the angles. This is the tricky part. For the chair conformation on the left, the equatorial bonds on carbons 2 and 5 are parallel to the ring bonds between carbons 3–4 and 1–6 (the two ring bonds drawn in red). The equatorial bonds of carbons 1 and 4 are parallel to the bonds between carbons 2–3 and 5–6 (the two ring bonds drawn in green), and the equatorial bonds of carbons 3 and 6 are parallel to the bonds between carbons 1–2 and 4–5 (the two bonds drawn in purple). Similarly, for the alternative chair on the right, the equatorial bonds on carbons 3 and 6 are parallel to the ring bonds drawn in red, the equatorial bonds of carbons 2 and 5 are parallel to the ring bonds drawn in green, and the equatorial bonds of carbons 1 and 4 are parallel to the ring bonds drawn in purple.

Equatorial bonds

Table 2.4 | ΔG° (Axial-Equatorial) for Monosubstituted Cyclohexanes at 25°C

axial ⟶ equatorial

	$-\Delta G°$			$-\Delta G°$	
Group	**kJ/mol**	**kcal/mol**	**Group**	**kJ/mol**	**kcal/mol**
C≡N	0.8	0.19	NH_2	5.9	1.41
F	1.0	0.24	COOH	5.9	1.41
C≡CH	1.7	0.41	$C=CH_2$	7.1	1.70
I	1.9	0.45	CH_3	7.28	1.74
Cl	2.2	0.53	CH_2CH_3	7.3	1.75
Br	2.4	0.57	$CH(CH_3)_2$	9.0	2.15
OH	3.9	0.93	$C(CH_3)_3$	21.0	5.00

Thus, at any given instant at room temperature, a much larger number of methylcyclohexane molecules have the methyl group in the equatorial conformation rather than in the axial conformation. The percentage of equatorial is (equatorial/(equatorial + axial) × 100%); that is, (18.9/19.9) × 100% = about 95%.

Table 2.4 shows the difference in free energy between axial and equatorial substituents for several monosubstituted cyclohexanes. Notice that as the size of the alkyl substituent increases, the preference for conformations with the group equatorial increases. With a group as large as *tert*-butyl, the energy of the axial conformer becomes so large that the equatorial conformation is approximately 4000 times more abundant at room temperature than the axial conformation. In fact, a chair with an axial *tert*-butyl group is so unstable that, if a *tert*-butyl group is forced into an axial position, the ring adopts a twist-boat conformation.

As shown by the free energy values given in Table 2.4, the preference for the equatorial position among the halogens increases in the order F < I < Cl < Br. Yet the size of the halogen atoms increases in the order F < Cl < Br < I. This anomaly occurs because the C—I bond is so long that the center of the iodine atom is too far from the axial hydrogen to interact with it strongly.

Example 2.10

Label all methyl-hydrogen (CH_3/H) diaxial interactions in the following chair conformation of 1, 2, 4-trimethylcyclohexane.

1,2,4-Trimethylcyclohexane

Solution

There are four methyl-hydrogen 1,3-diaxial interactions in this example; each axial methyl group has two sets of 1,3-diaxial interactions with parallel hydrogen atoms on the same side of the ring. The equatorial methyl group has no diaxial interactions.

Problem 2.10

Draw the alternative chair conformation for the trisubstituted cyclohexane given in Example 2.10. Label all CH_3/H 1,3-diaxial interactions in this chair conformation.

Example 2.11

Calculate the ratio of the diequatorial to diaxial conformation of this disubstituted cyclohexane at 25°C.

Solution

For these two chair conformations, $\Delta G°$ (2 axial CH_3 → 2 equatorial CH_3) = 2 × (−7.28 kJ/mol) = −14.56 kJ (3.5 kcal)/mol. Substituting this value in the equation $\Delta G° = -RT \ln K_{eq}$ gives a ratio of 357:1.

$$\ln K_{eq} = \frac{-(-14{,}560\,\text{J/mol})}{8.314\,\text{J} \cdot \text{K}^{-1} \cdot \text{mol}^{-1} \times 298\,\text{K}} = 5.877$$

$$K_{eq} = \frac{357}{1} = \frac{\text{diequatorial chair conformation}}{\text{diaxial chair conformation}}$$

Thus, at any given instant at room temperature, approximately $357/358 \times 100\%$ = 99.7% of the molecules of this compound are in the diequatorial chair conformation.

Problem 2.11

Draw a chair conformation of 1,4-dimethylcyclohexane in which one methyl group is equatorial and the other is axial. Draw the alternative chair conformation, and calculate the ratio of the two conformations at 25°C.

2.6 | *Cis,Trans* Isomerism in Cycloalkanes and Bicycloalkanes

Stereoisomers
Compounds that have the same molecular formula, the same connectivity of their atoms, but a different orientation of their atoms in space.

In this section we introduce the concept of stereoisomerism. **Stereoisomers** are compounds that have (1) the same molecular formula, (2) the same connectivity of their atoms, (3) but a different orientation of their atoms in space. Recall that constitutional isomers, the only other type of isomers we have studied so far, also have the same molecular formula but a different connectivity. The difference between constitutional isomers and stereoisomers are summarized in Figure 2.22.

We begin our study of stereoisomers with the study of *cis,trans* isomerism in cycloalkanes.

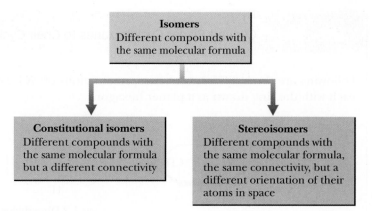

Figure 2.22
Relationship between stereoisomers and constitutional isomers.

A. *Cis,Trans* Isomerism in Cycloalkanes

Cycloalkanes with substituents on two or more carbons of the ring show a type of stereoisomerism called ***cis,trans* isomerism**, which we can illustrate by considering 1,2-dimethylcyclopentane. In the following structural formulas, the cyclopentane ring is drawn as a regular pentagon viewed through the plane of the ring. Carbon-carbon bonds of the ring projecting toward you are shown as heavy lines.

Cis,trans **isomers**
Stereoisomers that have the same connectivity but a different arrangement of their atoms in space as a result of the presence of either a ring or a carbon-carbon double bond.

cis-1,2-Dimethyl-
cyclopentane

trans-1,2-Dimethyl-
cyclopentane

In one isomer of 1,2-dimethylcyclopentane, the methyl groups are on the same side of the ring; in the other, they are on opposite sides of the ring. The prefix *cis* (Latin: on the same side) indicates that the substituents are on the same side of the ring; the prefix *trans* (Latin: across) indicates that they are on opposite sides of the ring. The *cis* isomer cannot be converted to the *trans* isomer, and vice versa, without breaking and reforming one or more bonds, a process that does not occur at or near room temperature. The *cis* isomer is approximately 7.1 kJ (1.7 kcal)/mol higher in energy (less stable) than the *trans* isomer because of the steric strain of the methyl groups on adjacent carbons in the *cis* isomer.

Alternatively, the cyclopentane ring can be viewed as a regular pentagon seen from above, with the ring in the plane of the page. Substituents on the ring then either project toward you (they project up, above the plane of the page) and are shown by solid wedges, or they project away from you (they project down, below the plane of the page) and are shown by broken wedges. In the following structural formulas, only the two methyl groups are shown; hydrogen atoms of the ring are not shown.

Cis
A prefix meaning on the same side.

Trans
A prefix meaning across from.

H_3C CH_3 H_3C CH_3

cis-1,2-Dimethyl-
cyclopentane

trans-1,2-Dimethyl-
cyclopentane

We say that 1,2-dimethylcyclopentane has two stereocenters. A **stereocenter** is an atom, most commonly carbon, about which exchange of two groups produces

Stereocenter
An atom, most commonly carbon, about which exchange of two groups produces a different stereoisomer.

Following are three different stereorepresentations of *cis*-1,2-dimethylcyclohexane, each with the ring drawn as a planar hexagon.

cis-1,2-Dimethylcyclohexane
(planar hexagon representations)

Students often find it difficult to convert substituted cyclohexanes from a planar hexagon representation such as these to a chair conformation. A good rule of thumb is that "up is up and down is down." If a substituent is *up* in a planar hexagon representation, place it *up* on the same carbon of the chair conformation. If a substituent is *down* on a planar hexagon representation, place it *down* on the same carbon of the chair conformation. Note that up or down on a chair conformation may be axial or equatorial, depending on which ring carbon you are considering. For *cis*-1,2-dimethylcyclohexane on which both methyl groups are up in the planar hexagon representation, the two methyl groups are also up in a chair conformation. Each of the alternative chair conformations has one methyl group axial and one equatorial. It is generally helpful to draw the hydrogen atoms bonded to the ring carbons bearing substituents, to make it absolutely clear which positions are equatorial and which are axial.

cis-1,2-Dimethylcyclohexane
(alternative chair conformations)

a different stereoisomer. Both carbons 1 and 2 of 1,2-dimethylcyclopentane, for example, are stereocenters; in this molecule, exchange of H and CH_3 groups at either stereocenter converts a *trans* isomer to a *cis* isomer, or vice versa. Alternatively, we refer to the stereoisomers of 1,2-dimethylcyclobutane as having either a *cis* or a *trans* configuration. **Configuration** refers to the arrangement of atoms about a stereocenter. We say, for example, that exchange of groups at either stereocenter in the *cis* configuration gives a stereoisomer with the *trans* configuration.

Configuration
Refers to the arrangement of atoms about a stereocenter.

Example 2.12

Which cycloalkanes show *cis,trans* isomerism? For each that does, draw the *cis* and *trans* isomers.

(a) Methylcyclopentane   (b) 1,1-Dimethylcyclopentane   (c) 1,3-Dimethylcyclobutane

Solution

(a) Methylcyclopentane does not show *cis,trans* isomerism. It has only one substituent on the ring.

(b) 1,1-Dimethylcyclobutane does not show *cis,trans* isomerism. Because both methyl groups are bonded to the same carbon, only one arrangement is possible for them.

(c) 1,3-Dimethylcyclobutane shows *cis,trans* isomerism. In the following structural formulas, cyclobutane is drawn as a planar ring viewed first from the side, and then from above.

cis-1,3-Dimethylcyclobutane   **trans-1,3-Dimethylcyclobutane**

Problem 2.12

Which cycloalkanes show *cis,trans* isomerism? For each that does, draw both isomers.

(a)   (b)   (c)

Two *cis,trans* isomers are possible for 1,4-dimethylcyclohexane. For the purposes of determining the number of *cis,trans* isomers in substituted cycloalkanes, it is adequate to draw the cycloalkane ring as a planar polygon as is done here.

trans-1,4-Dimethylcyclohexane   **cis-1,4-Dimethylcyclohexane**

We can also draw the *cis* and *trans* isomers of 1,4-dimethylcyclohexane as non-planar chair conformations. In working with alternative chair conformations, it is helpful to remember that all groups axial in one chair become equatorial in the alternative chair, and vice versa. In one chair conformation of *trans*-1,4-dimethylcyclohexane, the two methyl groups are axial; in the alternative chair conformation, they are equatorial. Of these chair conformations, the one with both methyl groups equatorial is more stable by approximately 14.6 kJ (3.5 kcal)/mol and makes up the large majority of a sample of *trans*-1,4-dimethylcyclohexane.

trans-1,4-Dimethylcyclohexane

The alternative chair conformations of *cis*-1,4-dimethylcyclohexane are of equal energy. In one chair, one methyl group is equatorial and the other is axial.

In the alternative chair, the orientations in space of the methyl groups are reversed. The result is that a collection of *cis*-1,4-dimethylcyclohexane molecules is composed of rapidly equilibrating alternative chairs in equal proportions.

***cis*-1,4-Dimethylcyclohexane**
(these conformations are of equal stability)

Example 2.13

Following is a chair conformation of 1,3-dimethylcyclohexane.

(a) Is this a chair conformation of *cis*-1,3-dimethylcyclohexane or of *trans*-1,3-dimethylcyclohexane?
(b) Draw the alternative chair conformation of this compound. Of the two chair conformations, which is the more stable?
(c) Draw a planar hexagon representation of the isomer shown in this example.

Solution

(a) The isomer shown is *cis*-1,3-dimethylcyclohexane; the two methyl groups are on the same side of the ring.

(b)

Diequatorial conformation   Diaxial conformation
(more stable)   (less stable)

(c)

Problem 2.13

Following is a planar hexagon representation for one isomer of 1,2,4-trimethylcyclohexane. Draw the alternative chair conformations of this compound, and state which of the two is the more stable.

Example 2.14

Here is one *cis,trans* isomer of 2,4-dimethylcyclohexanol. Complete the alternative chair conformations on the right.

Solution

For (a), the CH_3 group on carbon 2 must be below the plane of the ring, which on this carbon is axial. The CH_3 group on carbon 4 must be above the plane of the ring, which on this carbon is equatorial. (b) The methyl group on carbon 2 is equatorial; the methyl group on carbon 4 is axial.

Problem 2.14

Here is one *cis,trans* isomer of 3,5-dimethylcyclohexanol. Complete the alternative chair conformations.

B. *Cis,Trans* Isomerism in Bicycloalkanes

By far the most common bicycloalkanes, and the ones we concentrate on in this section, are decalin and hydrindane (Section 2.4B).

Decalin   **Hydrindane**

Two stereoisomers of decalin and hydrindane are possible depending on whether the two hydrogen atoms at the ring junction are *trans* or *cis* to each other. If we draw conformations for the six-membered rings in the two decalins, we see that each ring can exist in its more stable chair conformation. In *trans*-decalin, the hydrogens at the ring junction are axial to both rings; that is, the ring-junction hydrogen above the plane of the rings is axial to ring A and to ring B. Likewise, the ring-junction hydrogen below the plane of the ring is axial to both rings. The situation is different in *cis*-decalin. Each ring-junction hydrogen is axial to one ring but equatorial to the other ring.

Axial to both rings

trans-Decalin

Equatorial to ring A
and axial to ring B

cis-Decalin

Let us look more closely at *trans*-decalin, by far the more common stereoisomer of decalin. An important feature of this bicycloalkane is that each ring is locked into one chair conformation; neither ring can invert to its alternative chair. This means, for example, that if an —OH group is equatorial in a decalinol (a decalin alcohol), it remains equatorial; it cannot become axial because the cyclohexane ring is locked into this one conformation. Likewise, if an —OH group is axial, it remains axial.

Suppose you are given the structural formula on the left for the decalinol. Can you tell from looking at this structure whether the —OH group is axial or equatorial? You can't tell directly, but you can figure it out. Remember that, in *trans*-decalin, the H atoms at the ring junctions are axial to each ring. Remember also that in a chair cyclohexane, axial is up on one carbon, down on the next, up on the next, and so on. Therefore, if you start with the axial group at either ring junction and work your way from one carbon to the next until you come to the carbon bearing the —OH group, you come to the conclusion that the —OH on the structural formula is equatorial to ring A.

A good example of the occurrence of these types of ring systems is in the steroids, all of which contain a carbon skeleton consisting of three six-membered rings and one five-membered ring connected as shown here. This ring system is present in both animal and plant steroids. Steroids are present in human metabolism as cholesterol, steroid hormones, and bile acids (Section 26.4).

The steroid nucleus

Nature is by no means limited to carbon atoms in six-membered rings. Tetrodotoxin, one of the most potent toxins known, is composed of a set of interconnected six-membered rings, each in a chair conformation. All but one of these rings have atoms other than carbon in them. Tetrodotoxin is produced in the liver and ovaries of many species of *Tetraodontidae*, especially the puffer fish, so called because it inflates itself to an almost spherical spiny ball when it is alarmed. The puffer fish is evidently a species that is highly preoccupied with defense, but the Japanese are not put off. They regard the puffer, called "fugu" in Japanese, as a delicacy. To serve it in a public restaurant, a chef must be registered as sufficiently skilled in removing the toxic organs so as to make the flesh safe to eat.

Symptoms of tetrodotoxin poisoning begin with attacks of severe weakness, progressing to complete paralysis and eventual death. Tetrodotoxin blocks sodium ion channels, which are essential for neurotransmission. This prevents communication between neurons and muscle cells and results in the fatal symptoms described.

A puffer fish with its body inflated.

© David Fleetham/Alamy

Tetrodotoxin

Following are two stereorepresentations for cholestanol. In the conformational representation on the right, notice that all ring junctions are *trans,* all groups at each ring-junction atom are axial to the ring, and the —OH group on ring A is equatorial.

Cholestanol

Another type of bicycloalkane is a six-membered ring in which an added CH_2 group forms a bridge between carbons 1 and 4. You can view and draw this molecule from any number of perspectives. What becomes obvious if you view it from the side, as in (c), is that the one-carbon bridge locks the six-membered ring into a boat conformation. Notice that, even though (a) and (b) show the carbon skeleton of the molecule, it is not obvious from them that a locked boat conformation is embedded in the molecule. The lesson here is that it is essential to draw a molecule as a three-dimensional shape to best reveal what you want to show.

one carbon bridge between carbons
1 and 4 of the six-membered ring

(a)   **(b)**   **(c)**   **Camphor**

An example of a natural product containing this bicyclic skeleton is camphor.

Another example of a carbon skeleton that contains several six-membered rings, all of which are locked into chair conformations, is adamantane (c). To understand how the carbon skeleton of adamantane can be constructed, imagine that you (a) start with a chair cyclohexane, (b) add the three axial bonds on the top side of the ring, and (c) then connect each of the axial bonds to a CH group. You now have adamantane, a compound first isolated from petroleum. Amantadine, a 1° amino derivative of adamantane, is an antiviral agent used to treat influenza A.

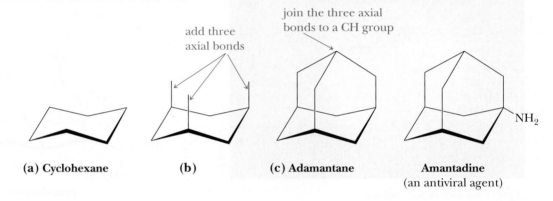

add three
axial bonds

join the three axial
bonds to a CH group

(a) Cyclohexane   **(b)**   **(c) Adamantane**   **Amantadine**
(an antiviral agent)

2.7 | Physical Properties of Alkanes and Cycloalkanes

The most important property of alkanes and cycloalkanes is their almost complete lack of polarity. As we saw in Section 1.2B, the difference in electronegativity between carbon and hydrogen is 2.5 − 2.1 = 0.4 on the Pauling scale, and, given this small difference, we classify a C—H bond as nonpolar covalent. Therefore, alkanes are nonpolar compounds, and only weak interactions exist between their molecules.

A. Boiling Points

The boiling points of alkanes are lower than those of almost any other type of compound of the same molecular weight. In general, both boiling and melting points of alkanes increase with increasing molecular weight (Table 2.5).

Alkanes containing 1 to 4 carbons are gases at room temperature, and those containing 5 to 17 carbons are colorless liquids. High-molecular-weight alkanes (those with 18 or more carbons) are white, waxy solids. Several plant waxes are high-molecular-weight alkanes. The wax found in apple skins, for example, is an unbranched alkane with molecular formula $C_{27}H_{56}$. Paraffin wax, a mixture of high-molecular-weight alkanes, is used for wax candles, in lubricants, and to seal home-canned jams, jellies, and other preserves. Petrolatum, so named because it is derived from petroleum refining, is a liquid mixture of high-molecular-weight alkanes. Sold as mineral oil and Vaseline, petrolatum is used as an ointment base in pharmaceuticals and cosmetics and as a lubricant and rust preventative.

Table 2.5	Physical Properties of Some Unbranched Alkanes			
Name	Condensed Structural Formula	Melting Point (°C)	Boiling Point (°C)	Density of Liquid* (g/mL at 0°C)
Methane	CH_4	−182	−164	(a gas)
Ethane	CH_3CH_3	−183	−88	(a gas)
Propane	$CH_3CH_2CH_3$	−190	−42	(a gas)
Butane	$CH_3(CH_2)_2CH_3$	−138	0	(a gas)
Pentane	$CH_3(CH_2)_3CH_3$	−130	36	0.626
Hexane	$CH_3(CH_2)_4CH_3$	−95	69	0.659
Heptane	$CH_3(CH_2)_5CH_3$	−90	98	0.684
Octane	$CH_3(CH_2)_6CH_3$	−57	126	0.703
Nonane	$CH_3(CH_2)_7CH_3$	−51	151	0.718
Decane	$CH_3(CH_2)_8CH_3$	−30	174	0.730

*For comparison, the density of H_2O is 1 g/mL at 4°C.

B. Dispersion Forces and Interactions Among Alkane Molecules

Methane is a gas at room temperature and atmospheric pressure. It can be converted to a liquid if cooled to −164°C, and to a solid if further cooled to −182°C. The fact that methane (or any other compound, for that matter) can exist as a liquid or a solid depends on the existence of intermolecular forces of attraction between the molecules of each pure compound. Although the forces of attraction between particles are generally electrostatic in nature, they vary widely in their relative strengths. The strongest attractive forces are between ions, for example between Na^+ and Cl^- in NaCl [787 kJ (188 kcal)/mol]. Dipole-dipole interactions and hydrogen bonding [8–42 kJ (2–10 kcal)/mol] are weaker attractive forces. We shall have more to say about these intermolecular attractive forces in Chapter 10 when we discuss the physical properties of alcohols, compounds containing polar O—H groups.

   Dispersion forces [0.08 − 8 kJ (0.02 − 2 kcal)/mol] are the weakest intermolecular attractive forces. The existence of dispersion forces accounts for the fact that low-molecular-weight, nonpolar substances, such as hydrogen (H_2), neon (Ne), and methane (CH_4), can be liquefied. To visualize the origin of dispersion forces, it is necessary to think in terms of instantaneous distributions of electron density rather than average distributions. Consider, for example, neon, which is a gas at room temperature and 1 atm pressure. It can be liquefied when cooled to −246°C. From the heat of vaporization, it can be calculated that the neon-neon attractive interaction in the liquid state is approximately 0.3 kJ (0.07 kcal)/mol. We account for this intermolecular attractive force in the following way. Over time, the distribution of electron density in a neon atom is symmetrical, and there is no dipole moment [Figure 2.23(a)]. However, at any instant, there is a nonzero probability that its electron density will be polarized (shifted) more toward one part of the atom than toward another. This temporary polarization creates a temporary dipole moment, which in turn induces temporary dipole moments in adjacent atoms [Figure 2.23(b)].

Dispersion forces
Very weak intermolecular forces of attraction resulting from the interaction between temporary induced dipoles.

Figure 2.23
Dispersion forces. (a) The distribution of electron density averaged over time in a neon atom is symmetrical, and there is no net polarity. (b) Temporary polarization of one neon atom induces temporary polarization in adjacent atoms. Electrostatic attractions between temporary dipoles are called dispersion forces.

Electrostatic attraction between temporary dipoles

(a)   (b)

There is electrostatic attraction between the temporary dipoles, which is the source of the small attraction referred to as dispersion forces.

The strength of dispersion forces depends on how easily an electron cloud can be polarized. Electrons in smaller atoms and molecules are held closer to their nuclei and, therefore, are not easily polarized. Electrons in larger atoms and molecules are more easily polarized. For this reason, the strength of dispersion forces tends to increase with increasing molecular mass and size. Intermolecular attractive forces between Cl_2 molecules and between Br_2 molecules are estimated to be 2.9 kJ (0.7 kcal)/mol and 4.2 kJ (1.0 kcal)/mol, respectively.

Dispersion forces are inversely proportional to the sixth power of the distance between interacting atoms or molecules. For them to be important, the interacting atoms or molecules must be in virtual contact with one another.

Because interactions between alkane molecules consist only of these very weak dispersion forces, boiling points of alkanes are lower than those of almost any other type of compound of the same molecular weight. As the number of atoms and the molecular weight of alkanes increase, there is more opportunity for dispersion forces between their molecules and boiling points increase.

C. Melting Points and Density

Melting points of alkanes also increase with increasing molecular weight. The increase, however, is not as regular as that observed for boiling points because the packing of molecules into ordered patterns of solids changes as molecular size and shape change.

The average density of the alkanes listed in Table 2.5 is about 0.7 g/mL; that of higher-molecular-weight alkanes is about 0.8 g/mL. All liquid and solid alkanes are less dense than water (1.0 g/mL) and, therefore, float on it.

D. Constitutional Isomers Have Different Physical Properties

Alkanes that are constitutional isomers are different compounds and have different physical and chemical properties. Listed in Table 2.6 are boiling points, melting points, and densities of the five constitutional isomers with molecular formula C_6H_{14}. The boiling point of each of the branched-chain isomers of C_6H_{14} is lower than that of hexane itself, and the more branching there is, the lower the boiling point. These differences in boiling point are related to molecular shape in the following way. The only forces of attraction between alkane molecules are dispersion forces. As branching increases, the shape of an alkane molecule becomes more compact, and its surface area decreases. As surface area decreases, contact among adjacent molecules decreases, the strength of dispersion forces decreases, and boiling points also decrease. Thus, for any group of alkane constitutional isomers, the least branched isomer usually has the highest boiling point, and the most branched isomer usually has the lowest boiling point.

Table 2.6	Physical Properties of the Isomeric Alkanes of Molecular Formula C_6H_{14}			
Name	Boiling Point (°C)	Melting Point (°C)	Density (g/mL)	
Hexane	68.7	−95	0.659	Hexane
2-Methylpentane	60.3	−154	0.653	
3-Methylpentane	63.3	−118	0.664	
2,3-Dimethylbutane	58.0	−129	0.661	
2,2-Dimethylbutane	49.7	−98	0.649	2,2-Dimethylbutane

Example 2.15

Arrange the alkanes in each set in order of increasing boiling point.

(a) Butane, decane, and hexane
(b) 2-Methylheptane, octane, and 2,2,4-trimethylpentane

Solution

(a) All of these compounds are unbranched alkanes. As the number of carbon atoms in the chain increases, dispersion forces between molecules increase, as does the amount of energy required to put the molecules into motion. Therefore, the larger the unbranched alkane, the higher are the boiling points. Decane has the highest boiling point, and butane has the lowest.

Butane	**Hexane**	**Decane**
bp −0.5°C	bp 69°C	bp 174°C

(b) These three alkanes are constitutional isomers with molecular formula C_8H_{18}. Their relative boiling points depend on the degree of branching. 2,2,4-Trimethylpentane, the most highly branched isomer, has the smallest surface area and the lowest boiling point. Octane, the unbranched isomer, has the largest surface area and the highest boiling point.

2,2,4-Trimethylpentane	**2-Methylheptane**	**Octane**
(bp 99°C)	(bp 118°C)	(bp 126°C)

Problem 2.15

Arrange the alkanes in each set in order of increasing boiling point.

(a) 2-Methylbutane, 2,2-dimethylpropane, and pentane
(b) 3,3-Dimethylheptane, 2,2,4-trimethylhexane, and nonane

2.8 | Reactions of Alkanes

Alkanes and cycloalkanes are quite unreactive toward most reagents, a behavior consistent with the fact that they are nonpolar compounds and contain only strong sigma bonds. Under certain conditions, however, alkanes and cycloalkanes do react with O_2 and with the halogens Cl_2 and Br_2. At this point, we present only their combustion with oxygen. We discuss their reaction with halogens in Chapter 8.

A. Oxidation

The oxidation of alkanes by O_2 to give carbon dioxide and water is by far their most economically important reaction. Oxidation of saturated hydrocarbons is the basis for their use as energy sources for heat [natural gas, liquefied petroleum gas (LPG), and fuel oil] and power (gasoline, diesel fuel, and aviation fuel). Following are balanced equations for the complete oxidation of methane (the major component of natural gas) and propane (the major component of LPG).

$$CH_4 + 2O_2 \longrightarrow CO_2 + 2H_2O \qquad \Delta H° = -890.4 \text{ kJ } (-212.8 \text{ kcal})/\text{mol}$$

Methane

$$CH_3CH_2CH_3 + 5O_2 \longrightarrow 3CO_2 + 4H_2O \qquad \Delta H° = -2220 \text{ kJ } (-530.6 \text{ kcal})/\text{mol}$$

Propane

Heat of combustion

Standard heat of combustion is the heat released when one mole of a substance in its standard state (gas, liquid, solid) is oxidized completely to carbon dioxide and water, and is given the symbol $\Delta H°$.

In this and all other hydrocarbon oxidations, the energy of the products is less than that of the reactants, with the difference in energy being given off as the **heat of combustion**. The heat of combustion is the energy of the products minus that of the reactants, and for the combustion of methane, for example, it is $(-890.4 - 0) = -890.4$ kJ (-212.8 kcal)/mol.

Table 2.7	Heats of Combustion of Four Constitutional Isomers of C_8H_{18}	
Hydrocarbon	**Structural Formula**	**$\Delta H°$ [kJ/mol (kcal/mol)]**
Octane		-5470.6 (-1307.5)
2-Methylheptane		-5465.6 (-1306.3)
2,2-Dimethylhexane		-5458.4 (-1304.6)
2,2,3,3-Tetramethylbutane		-5451.8 (-1303.0)

B. Heats of Combustion and Relative Stability of Alkanes and Cycloalkanes

One important use of heats of combustion is to give us information about the relative stabilities of isomeric hydrocarbons. To illustrate, consider the heats of combustion of the four constitutional isomers given in Table 2.7. All four compounds undergo combustion according to this equation.

$$C_8H_{18} + \frac{25}{2} O_2 \longrightarrow 8CO_2 + 9H_2O$$

We see that octane has the largest (most negative) heat of combustion. As branching increases, the $\Delta H°$ decreases (becomes less negative). Of these four isomers, the isomer with four methyl branches has the lowest (least negative) heat of combustion. Therefore, we conclude that branching increases the stability of an alkane.

Figure 2.24 is a graphical analysis of the data in Table 2.7. Because all four compounds give the same products on oxidation, the only differences among them are their relative energies.

Figure 2.24

Heats of combustion in kJ (kcal)/mol of four isomeric alkanes with molecular formula C_8H_{18}.

Heat of combustion, ΔH^0 kJ(kcal)/mol

-5470.6 (-1307.5) -5465.6 (-1306.3) -5458.4 (-1304.6) -5451.8 (-1303.0)

$8 CO_2 + 9 H_2O$

As we saw in Section 2.5B, there is considerable strain in small-ring cycloalkanes. We can measure this strain by measuring the heat of combustion versus the ring size. It has been determined by measurement of the heats of combustion of a series of unbranched alkanes that the average heat of combustion per methylene (CH_2) group is 658.7 kJ (157.4 kcal)/mol. Using this value, we can calculate a predicted heat of combustion for each cycloalkane. Strain energy is the difference between the predicted and actual heats of combustion. These results are displayed graphically in Figure 2.25.

Figure 2.25
Strain energy of cycloalkanes as a function of ring size.

We see that cyclopropane has the largest strain energy of any cycloalkane, which is consistent with the extreme compression of its C—C—C bond angles from 109.5° to 60°. Cyclobutane and cyclopentane each have less strain, and cyclohexane, as expected, has zero strain. What is perhaps surprising is the presence of strain in rings of from 7 to 13 carbon atoms. This strain is primarily the result of torsional and steric strain caused by the fact that these rings are constrained to conformations that cannot achieve ideal bond and torsional angles.

The large amount of strain suffered by the small cycloalkanes means that these structures are unstable relative to the larger cycloalkanes, such as cyclohexane. During the combustion process, all the strain is released because stable entities (CO_2 and H_2O) are formed. The strain in cyclopropane relative to cyclohexane is released as more heat (per carbon) during burning. In other words, cyclopropane burns hotter because considerable energy is stored in its unstable structure.

2.9 | Sources and Importance of Alkanes

The three major sources of alkanes throughout the world are the fossil fuels, namely natural gas, petroleum, and coal. These fossil fuels account for approximately 85% of the total energy consumed in the United States. Nuclear electric power, hydroelectric power, and renewable energy sources such as solar and wind power, make up most of the remaining 15%. In addition, these fossil fuels provide the bulk of the raw materials for the organic chemicals used worldwide.

A. Natural Gas

Natural gas consists of approximately 90–95% methane, 5–10% ethane, and a mixture of other relatively low-boiling alkanes, chiefly propane, butane, and 2-methylpropane. The current widespread use of ethylene as the organic chemical industry's most important building block is largely the result of the ease with which ethane can be separated from natural gas and cracked into ethylene. **Cracking**

A petroleum refinery.

is a process whereby a saturated hydrocarbon is converted into an unsaturated hydrocarbon plus H_2. Ethane is cracked by heating it in a furnace at 800 to 900°C for a fraction of a second. The production of ethylene in the United States in 1997 was 51.1 billion pounds, making it the number one organic compound produced by the U.S. chemical industry, on a weight basis. The bulk of ethylene produced in this manner is used to create organic polymers, as described in Chapter 29.

$$CH_3CH_3 \xrightarrow{800-900°C} CH_2{=}CH_2 + H_2$$

Ethane   **Ethene**
               (Ethylene)

B. Petroleum

Petroleum is a thick, viscous, liquid mixture of thousands of compounds, most of them hydrocarbons, formed from the decomposition of marine plants and animals. Petroleum and petroleum-derived products fuel automobiles, aircraft, and trains. They provide most of the greases and lubricants required for the machinery of our highly industrialized society. Furthermore, petroleum, along with natural gas, provides nearly 90% of the organic raw materials for the synthesis and manufacture of synthetic fibers, plastics, detergents, drugs, dyes, and a multitude of other products.

It is the task of the petroleum refining industry to produce usable products with a minimum of waste, from the thousands of different hydrocarbons in the liquid mixture. The various physical and chemical processes for this purpose fall into two broad categories: separation processes, which separate the complex mixture into various fractions, and reforming processes, which alter the molecular structure of the hydrocarbon components themselves.

The fundamental separation process in refining petroleum is fractional distillation (Figure 2.26). Practically all crude petroleum that enters a refinery goes to distillation units, where it is heated to temperatures as high as 370 to 425°C and separated into fractions. Each fraction contains a mixture of hydrocarbons that boils within a particular range.

Figure 2.26
Fractional distillation of petroleum. The lighter, more volatile fractions are removed from higher up the column; the heavier, less volatile fractions are removed from lower down.

Gasoline is a complex mixture of C_6 to C_{12} hydrocarbons. The quality of gasoline as a fuel for internal combustion engines is expressed in terms of an octane rating. Engine knocking occurs when a portion of the air-fuel mixture explodes prior to the piston reaching the top of its stroke (usually as a result of heat developed during compression) and independently of ignition by the spark plug. The resulting shockwave of the piston against the cylinder wall reverberates, creating a characteristic metallic "pinging" sound. Two compounds were selected as reference fuels. One of these, 2,2,4-trimethylpentane (isooctane), has very good antiknock properties and was assigned an octane rating of 100. The name *isooctane* is a trivial name, its only relation to 2,2,4-trimethylpentane is that both compounds have eight carbons. Heptane, the other reference compound, has poor antiknock properties and was assigned an octane rating of 0.

Heptane
(octane rating 0)

2,2,4-Trimethylpentane
(octane rating 100)

The **octane rating** of a particular gasoline is the percent of isooctane in a mixture of isooctane and heptane that has antiknock properties equivalent to that of the test gasoline. For example, the antiknock properties of 2-methylhexane are the same as those of a mixture of 42% isooctane and 58% heptane; therefore, the octane rating of 2-methylhexane is 42. Octane itself has an octane rating of -20, which means that it produces even more engine knocking than heptane.

The two most common reforming processes are cracking, as illustrated by the thermal conversion of ethane to ethylene (Section 2.9A), and catalytic reforming. Catalytic reforming is illustrated by the conversion of hexane first to cyclohexane and then to benzene.

Hexane   **Cyclohexane**   **Benzene**

C. Coal

To understand how coal can be used as a raw material for the production of organic compounds, it is necessary to discuss synthesis gas. **Synthesis gas** is a mixture of carbon monoxide and hydrogen in varying proportions, depending on the means by which it is manufactured. Synthesis gas is prepared by passing steam over hot coal; it is also prepared by partial oxidation of methane with oxygen.

$$C \ + H_2O \xrightarrow{\text{heat}} CO + H_2$$

Coal

$$CH_4 \ + \frac{1}{2}O_2 \xrightarrow{\text{catalyst}} CO + 2H_2$$

Methane

Two important organic compounds produced today almost exclusively from carbon monoxide and hydrogen are methanol and acetic acid. In the production of methanol, the ratio of carbon monoxide to hydrogen is adjusted to 1:2 and the mixture is passed over a catalyst at elevated temperature and pressure.

$$CO + 2H_2 \xrightarrow{\text{catalyst}} CH_3OH$$

Methanol

Treatment of methanol, in turn, with carbon monoxide over a different catalyst gives acetic acid.

$$CH_3OH + CO \xrightarrow{\text{catalyst}} CH_3\overset{\displaystyle O}{\overset{\|}{C}}OH$$

Methanol **Acetic acid**

Because the processes for making methanol and acetic acid directly from carbon monoxide are commercially proven, it is likely that the decades ahead will see the development of routes to other organic chemicals from coal via methanol.

Summary

Section 2.1 The Structure of Alkanes

Problems: 2.16–2.18

- A **hydrocarbon** is a compound composed only of carbon and hydrogen.
  - Saturated hydrocarbons (alkanes and cycloalkanes) contain only C—C single bonds. Alkanes have the general formula C_nH_{2n+2}.

Section 2.2 Constitutional Isomerism in Alkanes

Problems: 2.1, 2.2, 2.19–2.23, 2.25

- **Constitutional isomers** have the same molecular formula but a different connectivity of their atoms.

Section 2.3 Nomenclature of Alkanes and the IUPAC System

- Alkanes are named according to a systematic set of rules developed by the International Union of Pure and Applied Chemistry (IUPAC). To name an alkane:
  - The main chain is identified, which is the longest carbon chain. The alkyl group substituents are identified and named, each one ending in -yl.

Problems: 2.24, 2.26–2.28

  - The main chain is numbered to give the first substituent encountered the lowest number and each substituent is assigned a number from the main chain.
  - The name is constructed by listing the substituents with their numbers in alphabetical order, followed by the main chain name ending in -ane.
- There is an older common nomenclature for naming molecules that is still used for many common alkanes, but only the IUPAC system can name all molecules, even the very complex ones.
- The IUPAC name of a compound consists of three parts:
  - A **prefix** that tells the number of carbon atoms in the parent chain.
  - An **infix** that tells the nature of the carbon-carbon bonds in the parent chain.
  - A **suffix** that tells the class to which the compound belongs.

Problems: 2.24, 2.29–2.31

- A carbon atom is classified as *primary (1°), secondary (2°), tertiary (3°)*, or *quaternary (4°)* depending on the number of carbon atoms bonded to it.

Section 2.4 Cycloalkanes

- A saturated hydrocarbon that contains carbon atoms bonded to form a ring is called a **cycloalkane**.
- To name a cycloalkane, name and locate each substituent on the ring and prefix the name of the open-chain alkane by *cyclo-*.

Problems: 2.28–2.29

- Five-membered rings and six-membered rings are especially abundant in the biological world.

Section 2.5 Conformations of Alkanes and Cycloalkanes

- A **conformation** is any three-dimensional arrangement of the atoms of a molecule resulting from rotations about the single bonds.
  - One convention for showing conformations is the **Newman projection**.
  - A **dihedral angle** is the angle created by two intersecting planes.
    - For ethane, **staggered conformations** occur at dihedral angles of 60°, 180°, and 300°. **Eclipsed conformations** occur at dihedral angles of 0°, 120°, and 240°.
    - For butane, viewed along the C_2—C_3 bond, the staggered conformation of dihedral angle 180° is called an **anti conformation**; the staggered **conformations** of dihedral angle 60° and 300° are called **gauche conformations**.
    - The anti conformation of butane is lower in energy than the gauche conformations by approximately 2.8 kJ (0.9 kcal)/mol.
- **Intramolecular strain** is of three types:
  - **Torsional strain** (also called eclipsed interaction strain) arises when nonbonded atoms separated by three bonds are forced from a staggered conformation to an eclipsed conformation.
  - **Angle strain** arises from creation of either abnormally large or abnormally small bond angles.
  - **Steric strain** (also called nonbonded interaction or van der Waals strain) arises when nonbonded atoms separated by four or more bonds are forced abnormally close to each other.

  Problems: 2.8, 2.32–2.35

  Problems: 2.37, 2.39, 2.45–2.47

- The relationship between the change in Gibbs free energy, temperature in kelvins, and an equilibrium constant is given by the equation $\Delta G° = -RT \ln K_{eq}$.
- In all cycloalkanes larger than cyclopropane, nonplanar conformations are favored.
  - The lowest-energy conformation of cyclopentane is an **envelope conformation**.
  - The lowest-energy conformations of cyclohexane are two interconvertible **chair conformations**.
    - In a chair conformation, six bonds are **axial** and six bonds are **equatorial**.
    - Bonds axial in one chair are equatorial in the alternative chair.
    - Boat and twist-boat conformations are higher in energy than chair conformations.
    - Groups larger than H are less stable in the axial position because of 1,3-diaxial steric interactions.
    - The more stable chair conformation of a substituted cyclohexane is the one that minimizes diaxial steric interactions and has large groups equatorial.

  Problems: 2.9–2.11, 2.36, 2.38, 2.45, 2.46, 2.48–2.52, 2.61–2.65

Section 2.6 *Cis, Trans* Isomerism in Cycloalkanes and Bicycloalkanes

- **Stereoisomers** are compounds that have the same connectivity but different orientation of their atoms in space.
  - A **stereocenter** is an atom (most commonly a carbon atom) about which exchange of two groups produces a different stereoisomer.
  - **Configuration** refers to the arrangement of atoms or groups of atoms bonded to a stereocenter.
  - *Cis,trans* isomers have the same molecular formula and the same connectivity of their atoms, but the arrangement of their atoms in space cannot be interconverted by rotation about single bonds.
    - *Cis* substituents are on the same side of the ring.
    - *Trans* substituents are on opposite sides of the ring.
    - Most cycloalkanes with substituents on two or more carbons show *cis,trans* isomerism.

  Problems: 2.12–2.14, 2.40–2.44

Section 2.7 Physical Properties of Alkanes and Cycloalkanes

- Alkanes are nonpolar compounds, and the only forces of attraction between their molecules are dispersion forces, which are weak electrostatic interactions between temporary induced dipoles of adjacent atoms or molecules.

Problems: 2.15, 2.53–2.55

– Low-molecular-weight alkanes are gases at room temperature and atmospheric pressure.
– Higher-molecular-weight alkanes are liquids. Very-high-molecular-weight alkanes are solids.
– Among a set of alkane constitutional isomers, the least branched isomer generally has the highest boiling point; the most branched isomer generally has the lowest boiling point.

Section 2.8 Reactions of Alkanes

Problems: 2.57–2.60

• As determined by heats of combustion, strain in cycloalkanes varies with ring size.
• Cyclohexane, which has the most common ring size among organic compounds, is strain free.

Section 2.9 Sources and Importance of Alkanes

• **Natural gas** consists of 90–95% methane with lesser amounts of ethane and other low-molecular-weight hydrocarbons.
• **Petroleum** is a liquid mixture of literally thousands of different hydrocarbons.
  – The most important processes in petroleum refining are fractional distillation, catalytic cracking, and catalytic reforming.
• **Synthesis gas**, a mixture of carbon monoxide and hydrogen, can be derived from natural gas, coal, or petroleum.

⭕WL Questions and problems for this chapter are available in OWL. Use the OWL access card that came with this textbook to access assigned questions and problems for this chapter.

Key Reactions

1. Oxidation of Alkanes (Section 2.8A) Oxidation of alkanes to carbon dioxide and water is the basis for their use as energy sources of heat and power.

$$CH_3CH_2CH_3 + 5O_2 \longrightarrow 3CO_2 + 4H_2O \qquad \Delta H° = -2220 \text{ kJ } (-530.6 \text{ kcal})/\text{mol}$$

Chapter 3

Stereoisomerism and Chirality

Bob Nichols/ARS/USDA

Tartaric acid (Section 3.4B) is found in grapes and other fruits, both free and as its salts. Inset: a model of the *R,R* stereoisomer of tartaric acid.

The study of molecules as three-dimensional objects is called **stereochemistry**. The ability to visualize stereochemical relationships is a survival skill in organic chemistry. The chemistry and properties of the key molecules of biochemistry are critically dependent upon stereochemistry. Our goal in this chapter is to expand your awareness of molecules as three-dimensional objects. We suggest you purchase a set of models (or, if you prefer, a computer modeling program such as Chem3D or Spartan). Alternatively, you may have access to a computer lab with a modeling program. Use your models frequently as an aid in visualizing the spatial concepts in this and later chapters.

3.1 | Chirality—The Handedness of Molecules

A **mirror image** is the reflection of an object in a mirror. When you look in a mirror, you see a reflection, or mirror image, of yourself. Now suppose your mirror image became a three-dimensional object. We could then ask, "What is the relationship between you and your mirror image?" To clarify what we mean by *relationship* we might instead ask, "Can your reflection be superposed on (placed on top of) the original 'you' in such a way that every detail of the reflection corresponds exactly to the original?" The answer is that you and your mirror image are not superposable if details are included. For example, if you have a ring on the little finger of your right hand, your mirror image has the ring on the little finger of its left hand. If you part your hair on your right side, it will be parted on the left side in your reflection. You and your reflection are different objects. You cannot *exactly* superpose one on the other.

Stereochemistry
The study of three-dimensional arrangements of atoms in molecules.

⚲WL **Online homework for this chapter may be assigned in OWL for Organic Chemistry.**

The horns of this African gazelle show chirality and are mirror images of each others.

William H. Brown

Chiral

From the Greek, *cheir*, hand; an object that is not superposable on its mirror image; an object that has handedness.

Achiral

An object that lacks chirality; an object that has no handedness.

Plane of symmetry

An imaginary plane passing through an object dividing it so that one half is the mirror image of the other half.

Center of symmetry

A point so situated that identical components of an object are located on opposite sides and equidistant from that point along any axis passing through it.

Objects that are not superposable on their mirror images are said to be **chiral** (pronounced ki-ral, to rhyme with spiral; from the Greek: *cheir*, hand). That is, they show handedness. Chirality is encountered in three-dimensional objects of all sorts. Your left hand is chiral and so is your right hand (they are approximately mirror images of each other). A spiral binding on a notebook is chiral. A machine screw with a right-handed thread is chiral. A ship's propeller is chiral. As you examine objects around you, you will undoubtedly conclude that the vast majority of them are chiral as well.

The contrasting situation to chirality occurs when an object and its mirror image are superposable. An object and its mirror image are superposable if one of them can be oriented in space so that all its features (corners, edges, points, designs, etc.) correspond exactly to those in the other member of the pair. If this can be done, the object and its mirror image are identical; the original object is **achiral** (that is, lacks chirality). Examples of objects lacking chirality are an undecorated cup, a regular tetrahedron, a cube, and a perfect sphere.

Molecules are objects of the size of several to dozens of Angstroms, yet as discussed in the previous two chapters, they have definite shapes and geometries. These properties make some molecules chiral objects. To figure out if a molecule is chiral, it is actually simpler to first determine if it is achiral. We look for symmetries in a molecule to see if it is achiral. If these symmetries are missing, we can conclude that the molecule is chiral.

An object or molecule will be achiral if it has one or more of certain elements of symmetry. The most common such elements in organic compounds are the plane and center of symmetry. As we shall see, any object or molecule with either of these symmetry elements is achiral and can be superposed on its mirror image. A **plane of symmetry** is an imaginary plane passing through an object or molecule dividing it such that one half is the reflection of the other half. The cube shown in Figure 3.1 has several planes of symmetry. Both the beaker and the compound bromochloromethane have a single plane of symmetry. A **center of symmetry** is a point so situated that identical components of the object or molecule are located equidistant and on opposite sides from the point along any axis passing through that point. The cube shown in Figure 3.1 has a center of symmetry as does the cyclobutane. Because it has a center of symmetry, the cyclobutane is identical to its mirror image, and is achiral.

Objects or molecules that lack both of these symmetry elements are chiral. We can illustrate the chirality of an organic molecule by considering 1-bromo-1-chloroethane. Figure 3.2 shows three-dimensional representations and ball-and-stick models for 1-bromo-1-chloroethane and its mirror image. This molecule has

Figure 3.1

Symmetry in objects. A cube has several planes of symmetry and a center of symmetry. The beaker and CH_2BrCl each have a single plane of symmetry. The cyclobutane has a center of symmetry.

Figure 3.2
Stereorepresentations of
1-bromo-1-chloroethane and
its mirror image.

neither a plane nor a center of symmetry. A model of 1-bromo-1-chloroethane can be turned and rotated in any direction in space, but as long as bonds are not broken and rearranged, only two of the four groups bonded to the central carbon of one molecule can be made to coincide with those of its mirror image. Because 1-bromo-1-chloroethane and its mirror image are nonsuperposable, they are chiral objects.

3.2 | Stereoisomerism

Isomers are different compounds with the same molecular formula. Thus far, we have encountered three types of isomers. Constitutional isomers (Section 2.2) have the same molecular formula but a different connectivity of atoms in their molecules. Examples of pairs of constitutional isomers are pentane and 2-methylbutane, and 1-pentene and cyclopentane.

Pentane
(C_5H_{12}) and   **2-Methylbutane**
(C_5H_{12}) **1-Pentene**
(C_5H_{10}) and   **Cyclopentane**
(C_5H_{10})

A second type of isomerism is stereoisomerism. **Stereoisomers** have the same molecular formula and the same connectivity but different orientations of their atoms in space. One example of stereoisomerism we have seen thus far is that of *cis,trans* isomers in cycloalkanes (Section 2.6), which arise because substituents on a ring are locked into one of two orientations in space with respect to one another by the ring. Isomers of this type are called **configurational isomers** because they differ by the configuration of substituents on an atom.

Configurational isomers (*cis,trans* isomers)

H_3C◢◯◣CH_3 and H_3C◢◯⊲CH_3

cis-**1,4-Dimethylcyclohexane**   *trans*-**1,4-Dimethylcyclohexane**

In configurational isomers, the positions of the atoms cannot interchange so as to be identical simply by rotations along single bonds. In the *cis*- and *trans*-1,4-dimethylcyclohexane isomers just presented, the methyl groups are identically attached to the cyclohexane except that they are on the same or opposite sides of the ring respectively and their interconversion would require breaking of a bond between the cyclohexane ring and a methyl group.

Chirality can arise from the ability of a molecule to exist as configurational isomers. The *cis*- and *trans*-1,4-dimethyl cyclohexane molecules are not chiral because they both possess a plane of symmetry that passes through the methyl groups. However, let's re-examine 1-bromo-1-chloroethane in Figure 3.2. This molecule exists in two configurations that are nonsuperposable mirror images. Hence, these are stereoisomers that are chiral objects, and they are configurational isomers.

Stereoisomers
Isomers that have the same molecular formula and the same connectivity of their atoms but a different orientation of their atoms in space.

Configurational isomers
Isomers that differ by the configuration of substituents on an atom.

Enantiomers
Stereoisomers that are
nonsuperposable mirror images of
each other; refers to a relationship
between pairs of objects.

Chiral center
A tetrahedral atom, most commonly
carbon, that is bonded to four
different groups.

Stereocenter
An atom about which exchange of
two groups produces a stereoisomer.
Chiral centers are one type
of stereocenter.

Such objects, or molecules, are called enantiomers. **Enantiomers** are nonsuperposable mirror images. Note that the terms "chiral" and "achiral" refer to objects, while the term "enantiomer" refers to the relationship between a pair of objects.

The most common cause of chirality in organic molecules is a tetrahedral atom, most commonly carbon, bonded to four different groups. A carbon atom with four different groups bonded to it lacks the two key symmetry elements and is called a **chiral center**. The carbon atom of 1-bromo-1-chloroethane bearing the —Cl, —H, —CH₃, and —Br groups is a chiral center.

The term **stereocenter** is also used to describe a carbon bonded to four different groups but is broader. A stereocenter is an atom about which exchange of two groups produces a stereoisomer. For example, both carbons in *cis*-2-butene and *trans*-2-butene are stereocenters. The C-2 in 2-butanol is a stereocenter as well because exchange of the OH and H creates a stereoisomer, an enantiomer in this case.

So far we have seen that configurational isomers can exist as chiral and achiral objects. When the isomers are chiral, they are enantiomers of one another, just as right and left hands are enantiomers. In contrast, *cis*- and *trans*-1,4-dimethylcyclohexane and *trans*- and *cis*-2-butene are configurational isomers, but they are not chiral objects.

How To	Draw Chiral Molecules

It is worthwhile to notice that there are several different ways to represent the three-dimensional structure of chiral molecules on a two-dimensional page. For example, following are four different representations of one enantiomer of 2-butanol.

Representation (1) shows the tetrahedral geometry of the chiral center. We can turn (1) slightly in space and tip it a bit to place the carbon framework in the plane of the paper and give representation (2). In (2), we still have two groups in the plane of the paper, one coming toward us, and one going away from us. As discussed in Chapter 2, these are represented by wedged and dashed bonds, respectively. For an even more abbreviated representation, we can turn (2) into the line-angle formula (3). Although we don't normally show hydrogens in a line-angle formula, we do in (3) just to remind ourselves that the fourth group on this chiral center is really there and that it is H. Finally, we can carry the abbreviation a step further and write this enantiomer of 2-butanol as (4). Here we omit the H on the chiral center. We know it must be there because carbon needs four bonds, and we know it must be behind the plane of the paper because the OH is in front. Clearly the abbreviated formulas (3) and (4) are the easiest to write, and we will rely on these representations throughout the remainder of the text. When you need to write three-dimensional representations of chiral centers, try to keep the carbon framework in the plane of the paper and the other two atoms or groups of atoms on the chiral center toward and away from you. Often, it is important to draw both enantiomers of a chiral molecule. An easy way to do this is to interchange two of the groups bonded to the chiral center. Using representation (4) as a model, here are two different representations for its enantiomers.

(4)   Enantiomer of (4)   Enantiomer of (4)

Therefore, they cannot exist as enantiomers. Instead, we call these isomers diastereomers.

Stereocenters

trans-2-Butene   swap   **cis-2-Butene**

swap   A stereocenter, also a chiral center

HO H   →   H OH

Diasteromers are stereoisomers that are not mirror images of one another. The term "disastereomers" refers to a relationship between a pair of objects. It is important to note that diastereomers can be chiral or achiral objects but that enantiomers must be chiral objects. We will see in Section 3.4 that diastereomers will arise whenever there are two or more stereocenters in a molecule, as is the case with 1,4-dimethylcyclohexane and 2-butene.

Diastereomers
Stereoisomers that are not mirror images of each other; refers to relationships among two or more objects.

Example 3.1

Each molecule has one chiral center. Draw stereorepresentations for the enantiomers of each.

(a) Cl   (b) Cl

Solution

You will find it helpful to view models of enantiomer pairs from different perspectives as is done in these representations. As you work with these pairs of enantiomers, notice that each has a tetrahedral carbon atom bonded to four different groups, which makes the molecule chiral.

Problem 3.1

Each molecule has one chiral center. Draw stereorepresentations for the enantiomers of each.

(a) (b)

In all the molecules studied so far, chirality arises because of the presence of a tetrahedral carbon chiral center. Chiral centers are not limited to carbon. Following are stereorepresentations of a chiral cation in which the chiral center is nitrogen. We discuss the chirality of nitrogen centers in more detail in Section 23.3. Enantiomers of tetrahedral silicon, phosphorus, and germanium compounds have also been isolated.

Another form of stereoisomerism is **conformational isomerism**. In Section 2.5, we discussed that conformational changes occur within molecules via rotations along single bonds. For example, recall that butane exists in gauche and anti forms, as shown below using Newman projections. These are not configurational isomers because bonds do not need to be broken in order to interchange atoms; instead the bonds only need to be rotated. Clearly they are stereoisomers because the atoms are arranged differently in space.

Gauche   Anti   Gauche

Stereoisomers are either diastereomers or enantiomers and therefore we can also apply these terms to conformational isomers. The gauche and anti forms of butane are diastereomers because they are not mirror images. The two gauche forms of butane are enantiomers because they are mirror images and not superposable (see the reflection in the mirror given). Hence, these forms of butane are chiral. However, butane is *not* a chiral molecule because these three isomers interconvert very rapidly at room temperature and because they interconvert through the intermediacy of the anti isomer, which is achiral (refer back to Figure 2.9 to see the interconversion). The anti isomer is achiral because there is a plane of symmetry when all four carbons are planar.

This discussion highlights an important point: chirality can be present in molecules without chiral centers. This condition occurs via conformational isomerism, but it is less common than chirality arising from configurational isomerism. When, unlike in butane, the barrier to interconversion is large and the enantiomers cannot interconvert at ambient temperature through a planar form, the molecule will be chiral and the enantiomers can be separated. An example is the following substituted biphenyl. Because of the large groups on the rings, there are substantial nonbonded interactions in the planar conformer and the twisted forms shown have far lower energies. The nonbonded interactions result in a very high barrier to rotation around the carbon-carbon single bond connecting the rings, and rotation is very slow. Although this molecule has no chiral center, the mirror images are not superposable, and the molecule is chiral. Isomers of this sort, which lack a chiral center but do not interconvert because of hindered rotation, are called **atropisomers**.

Atropisomers
Enantiomers that lack a chiral center and differ because of hindered rotation.

3.3 | Naming Chiral Centers—The *R,S* System

So far, we have discussed the fact that enantiomers exist. We have not considered the question of which isomer is which, that is, the **absolute configuration**, or which is the right-handed enantiomer and which is the left. For a given sample of a pure enantiomer, the correct arrangement must be determined by experiment. Experimental determination of absolute configuration can be accomplished by using x-ray analysis of a derivative that has a chiral center with a known absolute configuration. In biological molecules, many absolute configurations were determined by comparison to absolute configurations of the chiral center in glyceraldehyde.

A system for designating the absolute configuration of a chiral center was devised in the late 1950s by R. S. Cahn and C. K. Ingold in England and V. Prelog in Switzerland and is named after them. The system, also called the ***R,S* system**, has been incorporated into the IUPAC rules of nomenclature. The orientation of groups about a chiral center is specified using a set of priority rules.

Absolute configuration
Which of the two possible isomers an enantiomer is (i.e., whether it is the right- or left-handed isomer).

***R,S* System**
A set of rules for specifying absolute configuration about a chiral center; also called the Cahn-Ingold-Prelog system.

R
From the Latin, *rectus*, straight, correct; used in the *R,S* convention to show that the order of priority of groups on a chiral center is clockwise.

S
From the Latin, *sinister*, left; used in the *R,S* convention to show that the order of priority of groups on a chiral center is counterclockwise.

Priority Rules

1. Each atom bonded to the chiral center is assigned a priority. Priority is based on atomic number; the higher the atomic number, the higher the priority. Following

are several substituents arranged in order of increasing priority. The atomic number of the atom determining priority is shown in parentheses.

$$\begin{matrix} (1) & (6) & (7) & (8) & (16) & (17) & (35) & (53) \\ -H, & -CH_3, & -NH_2, & -OH, & -SH, & -Cl, & -Br, & -I \end{matrix}$$

Increasing priority ⟶

2. If priority cannot be assigned on the basis of the atoms bonded directly to the chiral center (because of a tie, that is, the same first atom on more than one substituent), look at the next set of atoms and continue until a priority can be assigned. Priority is assigned at the first point of difference. Following is a series of groups arranged in order of increasing priority. The atomic number of the atom on which the assignment of priority is based is shown above it.

$$\begin{matrix} (1) & (6) & (7) & (8) & (17) \\ -CH_2-H & -CH_2-CH_3 & -CH_2-NH_2 & -CH_2-OH & -CH_2-Cl \end{matrix}$$

Increasing priority ⟶

If two carbons have substituents of the same priority, priority is assigned to the carbon that has more of these substituents. Thus, $-CHCl_2 > -CH_2Cl$.

3. Atoms participating in a double or triple bond are considered to be bonded to an equivalent number of similar "phantom" atoms (shown here highlighted) by single bonds; that is, atoms of the double bond are duplicated, and atoms of a triple bond are triplicated. The phantom atoms are bonded to no other atoms.

4. *Note:* priority assignment is made at the *first point of difference* between groups. A common mistake is to assume that larger groups must always have higher priority, but this might not necessarily be the case. For example, a $-CH_2Cl$ group has priority over a $-CH_2CH_2CH_2CH_3$ group because the Cl atom is the *first point of difference*.

Example 3.2

Assign priorities to the groups in each set.

(a) $\overset{\displaystyle O}{\overset{\|}{-C}}OH$ and $\overset{\displaystyle O}{\overset{\|}{-C}}H$ 　(b) $-\overset{\displaystyle H}{\underset{\displaystyle H}{C}}=\overset{}{\underset{\displaystyle H}{C}}$ and $-\overset{\displaystyle CH_3}{\underset{\displaystyle H}{C}}-CH_3$

Solution

(a) The first point of difference is O of the —OH in the carboxyl group compared to —H in the aldehyde group. The carboxyl group has a higher priority.

first point of difference
along path of higher priority

Carboxyl group
(higher priority)

Aldehyde group
(lower priority)

(b) Carbon 1 in each group has the same pattern of atoms; namely C(C,C,H) (i.e., carbon bonded to two carbons and a hydrogen). For the vinyl group, bonding at carbon 2 is C(C,H,H). For the isopropyl group, at carbon 2 it is C(H,H,H). The vinyl group is higher in priority than is the isopropyl group.

first point of difference
along path of higher priority

Vinyl group
(higher priority)

Isopropyl group
(lower priority)

Problem 3.2

Assign priorities to the groups in each set.

(a) —CH_2OH and —CH_2CH_2OH **(b)** —CH_2OH and —$CH=CH_2$
(c) —CH_2OH and —$C(CH_3)_3$

| **How To** | Assign *R* or *S* Configuration to a Chiral Center |

1. Locate the chiral center, identify its four substituents, and assign a priority from 1 (highest) to 4 (lowest) to each substituent.
2. Orient the molecule in space so that the group of lowest priority (4) is directed away from you, analogously to the steering column of a car. The three groups of higher priority (1–3) then project toward you, as would the spokes of the steering wheel.
3. Read the three groups projecting toward you in order from highest priority (1) to lowest priority (3).
4. If the groups are read in a clockwise direction, the configuration is designated as **R** (Latin: *rectus*, straight, correct); if they are read in a counterclockwise direction, the configuration is **S** (Latin: *sinister*, left). You can also visualize this as follows: turning the steering wheel to the right (down the order of priority) equals *R*; turning it to the left equals *S*.

Group of lowest priority
points away from you

Example 3.3

Assign an *R* or *S* configuration to the chiral center in each molecule.

(a) (b) (c)

Solution

View each molecule through the chiral center along the bond from the chiral center toward the group of lowest priority. In (a), the order of priority is Cl > CH$_2$CH$_3$ > CH$_3$ > H; the configuration is *S*. In (b), the order of priority is Cl > CH=CH > CH$_2$ > H; the configuration is *R*. In (c), the order of priority is OH > CH$_2$COOH > CH$_2$CH$_2$OH > CH$_3$; the configuration is *R*.

(a)

(b)

(c)

Problem 3.3

Assign an *R* or *S* configuration to the chiral center in each molecule.

(a) (b) (c) H—C—OH

3.4 | Acyclic Molecules with Two or More Stereocenters

When two or more stereocenters exist in a molecule, multiple stereoisomers are possible. When stereoisomerism was first introduction in Section 2.2, we noted that both double-bond carbons in *trans*-2-butene are stereocenters because the swapping of a methyl and a hydrogen on either carbon creates the stereoisomer *cis*-2-butene. More double bonds lead to more possible stereoisomers. For example, the molecule 2,4-heptadiene can exist as four stereoisomers. This situation gets even more complicated when the stereocenters are also chiral centers.

2,4-heptadiene

A. Enantiomers and Diastereomers

We have now seen several examples of molecules with one chiral center and verified that, for each, two stereoisomers (one pair of enantiomers) are possible. Now let us consider molecules with two or more chiral centers. To generalize, for a molecule with n chiral centers, the maximum number of stereoisomers possible is 2^n. We have already seen that, for a molecule with one chiral center, $2^1 = 2$ stereoisomers are possible. For a molecule with two chiral centers, $2^2 = 4$ stereoisomers are possible; for a molecule with three chiral centers, $2^3 = 8$ stereoisomers are possible, and so forth.

Let us begin our study of molecules with multiple chiral centers by considering 2,3,4-trihydroxybutanal, a molecule with two chiral centers, shown here highlighted.

$$\text{HOCH}_2\text{—CH—CH—CHO}$$
$$\text{OH} \quad \text{OH}$$

2,3,4-Trihydroxybutanal

The maximum number of stereoisomers possible for this molecule is $2^2 = 4$, each of which is drawn in Figure 3.3. One of these pairs is called erythrose, and the other, threose.

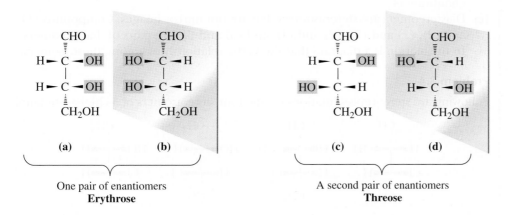

Figure 3.3

The four stereoisomers of 2,3,4-trihydroxybutanal, a compound with two chiral centers.

Stereoisomers (a) and (b) are nonsuperposable mirror images and are, therefore, a pair of enantiomers. Stereoisomers (c) and (d) are also nonsuperposable mirror images and are a second pair of enantiomers. One way to describe the four stereoisomers of 2,3,4-trihydroxybutanal is to say that they consist of two pairs of enantiomers. Enantiomers (a) and (b) of 2,3,4-trihydroxybutanal are given the names (2R,3R)-erythrose and (2S,3S)-erythrose; enantiomers (c) and (d) are given the names (2R,3S)-threose and (2S,3R)-threose. Note that all of the chiral centers in a molecule are reversed in its enantiomer. The molecule with the 2R,3S configuration is the enantiomer of the molecule with 2S,3R, and the molecule with 2S,3S is the enantiomer of the molecule with 2R,3R. Erythrose and threose belong to the class of compounds called carbohydrates, which we discuss in Chapter 25. Erythrose is found in erythrocytes (red blood cells), hence the derivation of its name.

We have specified the relationship between (a) and (b) and between (c) and (d); each represents a pair of enantiomers. What is the relationship between (a) and (c), between (a) and (d), between (b) and (c), and between (b) and (d)? The answer is that they are diastereomers. Recall that diastereomers are stereoisomers that are not mirror images (enantiomers). As we see in this example, molecules with at least two chiral centers can have diastereomers.

Example 3.4

Following are stereorepresentations for the four stereoisomers of 1,2,3-butanetriol. *R* and *S* configurations are given for the chiral centers in (1) and (4).

(1)   (2)   (3)   (4)

(a) Write the IUPAC names for each compound showing the *R* or *S* configuration of each chiral center.
(b) Which molecules are enantiomers?
(c) Which molecules are diastereomers?

Solution

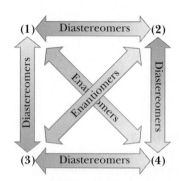

(a) (1) (2*S*,3*S*)-1,2,3-Butanetriol   (2) (2*S*,3*R*)-1,2,3-Butanetriol
    (3) (2*R*,3*S*)-1,2,3-Butanetriol   (4) (2*R*,3*R*)-1,2,3-Butanetriol
(b) Enantiomers are stereoisomers that are nonsuperposable mirror images. As you see from their configurations, compounds (1) and (4) are one pair of enantiomers, and compounds (2) and (3) are a second pair of enantiomers.
(c) Diastereomers are stereoisomers that are not mirror images. Compounds (1) and (2), (1) and (3), (2) and (4), and (3) and (4) are pairs of diastereomers. In the margin is a diagram that shows the relationships among these isomers.

Problem 3.4

Following are stereorepresentations for the four stereoisomers of 3-chloro-2-butanol.

(1)   (2)   (3)   (4)

(a) Assign an *R* or *S* configuration of each chiral center.
(b) Which compounds are enantiomers?
(c) Which compounds are diastereomers?

B. Meso Compounds

Certain molecules containing two or more chiral centers have special symmetry properties that reduce the number of stereoisomers to fewer than the maximum number predicted by the 2^n rule. One such molecule is 2,3-dihydroxybutanedioic acid, more commonly named tartaric acid.

$$HO\overset{\text{O}}{\underset{}{C}}-CH-CH-C\overset{\text{O}}{\underset{}{}}OH$$
$$\quad\quad\quad OH \quad OH$$

2,3-Dihydroxybutanedioic acid
(Tartaric acid)

Tartaric acid is a colorless, crystalline compound. During the fermentation of grape juice, potassium bitartrate (one carboxyl group is present as a potassium salt,

Figure 3.4
Stereoisomers of tartaric acid. One pair of enantiomers and one meso compound.

(a)   (b)   (c)   (d)

A pair of enantiomers   A meso compound

—COO⁻K⁺) deposits as a crust on the sides of wine casks. When collected and purified, it is sold commercially as cream of tartar.

In tartaric acid, carbons 2 and 3 are chiral centers, and the maximum number of stereoisomers possible is $2^2 = 4$; these stereorepresentations are drawn in Figure 3.4. Structures (a) and (b) are nonsuperposable mirror images and are, therefore, a pair of enantiomers. Structures (c) and (d) are also mirror images, but they are superposable. To see this, imagine that you first rotate (d) by 180° in the plane of the paper, lift it out of the plane of the paper, and place it on top of (c). If you do this mental manipulation correctly, you find that (d) is superposable on (c). Therefore, (c) and (d) are not different molecules; they are the same molecule, just drawn here in a different orientation. Because (c) and its mirror image are superposable, (c) is achiral.

Another way to determine that (c) is achiral is to see that it has a plane of symmetry that bisects the molecule in such a way that the top half is the reflection of the bottom half. Thus, even though (c) has two chiral centers, it is an achiral object (Section 3.2).

The stereoisomer of tartaric acid represented by (c) or (d) is called a **meso compound**. A meso compound contains two or more chiral centers but is achiral. To be a meso compound, a molecule must also have chiral isomers. We can now answer the question: How many stereoisomers are there of tartaric acid? The answer is three: one meso compound and one pair of enantiomers. Note that the meso compound is a diastereomer of each member of the pair of enantiomers.

From this example, we can make this generalization about meso compounds: They have an internal mirror plane (or center of symmetry). Commonly, there are two chiral centers, each with the same four groups: one is *R* the other *S*.

Notice that the stereoisomers in Figure 3.4 are shown in only one conformation. Because conformational isomers in acyclic systems interconvert extremely rapidly, we need consider only the most symmetric conformers. Other conformers may be rotated to give the most symmetric one for determination of symmetry properties (see Problem 3.5).

It is sometimes helpful to identify stereochemical relationships among molecules based on assignment of configuration to all the chiral centers. In some cases, this can be easier than searching for and comparing the most symmetric conformations. Two stereoisomers will be enantiomers if all of the chiral centers present are reversed, and the stereoisomers will be diastereomers if only some of the chiral centers are reversed.

Enantiomers have identical physical and chemical properties in an achiral environment. Examples of achiral environments are solvents that have no chiral centers such as H_2O, CH_3CH_2OH, or CH_2Cl_2. The enantiomers of tartaric acid (Table 3.1), for example, have the same melting point, the same boiling point, the same solubility in water and other common solvents, the same value of pK_a, and they undergo the same acid-base reactions. The enantiomers of tartaric acid do, however, differ in optical activity (the ability to rotate the plane of plane-polarized light), which we discuss in Section 3.7B. Diastereomers have different physical and chemical properties, even in achiral environments. Meso tartaric acid has different physical properties from those of the enantiomers and can be separated from them by methods such as crystallization.

Meso compound
An achiral compound possessing two or more chiral centers that also has chiral isomers.

Table 3.1 — Some Physical Properties of the Stereoisomers of Tartaric Acid

	(R,R)-Tartaric Acid	(S,S)-Tartaric Acid	Meso Tartaric Acid
Specific rotation*	+12.7	−12.7	0
Melting point (°C)	171–174	171–174	146–148
Density at 20°C (g/cm³)	1.7598	1.7598	1.660
Solubility in water at 20°C (g/100 mL)	139	139	125
pK_1 (25°C)	2.98	2.98	3.23
pK_2 (25°C)	4.34	4.34	4.82

*Specific rotation is discussed in Section 3.7B.

Example 3.5

Following are stereorepresentations for the three stereoisomers of 2,3-butanediol. The carbons are numbered beginning from the left, as shown in (1).

(1)   (2)   (3)

(a) Assign an R or S configuration to each chiral center.
(b) Which are enantiomers?
(c) Which is the meso compound?
(d) Which are diastereomers?

Solution

(a) (1) (2R,3R)-2,3-Butanediol   (2) (2R,3S)-2,3-Butanediol
    (3) (2S,3S)-2,3-Butanediol
(b) Compounds (1) and (3) are enantiomers.
(c) Compound (2) is a meso compound. Note that compound (2) can be drawn in two symmetric conformations, one of which has a plane of symmetry and the other of which has a center of symmetry.

Plane of symmetry   Center of symmetry

(d) (1) and (2) are diastereomers; (2) and (3) are also diastereomers.

Problem 3.5

Following are four Newman projection formulas for tartaric acid.

| (1) | (2) | (3) | (4) |

(a) Which represent the same compound?
(b) Which represent enantiomers?
(c) Which represent a meso compound?
(d) Which are diastereomers?

C. Fischer Projection Formulas

Glyceraldehyde contains a chiral center and therefore exists as a pair of enantiomers.

(R)-Glyceraldehyde   **(S)-Glyceraldehyde**

Chemists use two-dimensional representations called **Fischer projections** to show the configuration of molecules with multiple chiral centers, especially carbohydrates. To write a Fischer projection, draw a three-dimensional representation of the molecule oriented so that the vertical bonds from the chiral center are directed away from you and the horizontal bonds from it are directed toward you. Then write the molecule as a two-dimensional figure with the chiral center indicated by the point at which the bonds cross.

Fischer projection
A two-dimensional projection of a molecule; in these projections, groups on the right and left are by convention in front, while those at the top and bottom are to the rear.

(R)-Glyceraldehyde
(three-dimensional
representation)

(R)-Glyceraldehyde
(Fischer projection)

The horizontal segments of this Fischer projection represent bonds directed toward you, and the vertical segments represent bonds directed away from you. The only atom in the plane of the paper is the chiral center. Because a Fischer projection implies that the groups to each side are in front and those at top and bottom are behind the plane of the paper, rotations of these drawings by 90° are not permissible.

Example 3.6

Draw a Fischer Projection of (2R,3R) erythrose (Figure 3.3).

Solution

$$
\begin{array}{c}
\text{CHO} \\
\text{H} \long!\!\longrightarrow \text{OH} \\
\text{H} \longrightarrow \text{OH} \\
\text{CH}_2\text{OH}
\end{array}
$$

(2R, 3R)-Erythrose
(2,3,4-Trihydroxybutanal)

Problem 3.6

Give a complete stereochemical name for the following compound, which is a 1,2,3-butanetriol.

$$
\begin{array}{c}
\text{CH}_2\text{OH} \\
\text{HO} \longrightarrow \text{H} \\
\text{HO} \longrightarrow \text{H} \\
\text{CH}_3
\end{array}
$$

3.5 | Cyclic Molecules with Two or More Stereocenters

In this section, we concentrate on derivatives of cyclopentane and cyclohexane containing two stereocenters. We can analyze stereoisomerism in cyclic compounds in the same way as in acyclic compounds.

A. Disubstituted Derivatives of Cyclopentane

Let us start with 2-methylcyclopentanol, a compound with two chiral centers. We predict a maximum of $2^2 = 4$ stereoisomers. Both the *cis* isomer and the *trans* isomer are chiral: The *cis* isomer exists as one pair of enantiomers, and the *trans* isomer exists as a second pair of enantiomers. The *cis* and *trans* isomers are stereoisomers that are not mirror images of each other, that is, they are diastereomers.

cis-2-Methylcyclopentanol
(a pair of enantiomers)

trans-2-Methylcyclopentanol
(a pair of enantiomers)

1,2-Cyclopentanediol also has two chiral centers; therefore, the 2^n rule predicts a maximum of $2^2 = 4$ stereoisomers. As seen in the following stereodrawings, only three stereoisomers exist for this compound. The *cis* isomer is achiral (meso) because it and its mirror image are superposable. An alternative way to identify the *cis* isomer as achiral is to notice that it possesses a plane of symmetry that bisects the molecule into two mirror halves. The *trans* isomer is chiral and exists as a pair of enantiomers.

cis-1,2-Cyclopentanediol
(a meso compound)

trans-1,2-Cyclopentanediol
(a pair of enantiomers)

Example 3.7

How many stereoisomers exist for 3-methylcyclopentanol?

Solution

There are four stereoisomers of 3-methylcyclopentanol. The *cis* isomer exists as one pair of enantiomers; the *trans* isomer, as a second pair of enantiomers.

cis-3-Methylcyclopentanol
(a pair of enantiomers)

trans-3-Methylcyclopentanol
(a pair of enantiomers)

Problem 3.7

How many stereoisomers exist for 1,3-cyclopentanediol?

B. Disubstituted Derivatives of Cyclohexane

Cyclohexane derivatives exist in chair conformations. However, when evaluating disubstituted cyclohexane derivatives for symmetry, it is helpful to analyze structures drawn flat. As an example of a disubstituted cyclohexane, let us consider the methylcyclohexanols. 4-Methylcyclohexanol has two stereocenters, but they are not chiral centers because the carbons of the stereocenters do not have four different groups attached. 4-Methylcyclohexanol therefore exists as two diastereomers, a pair of *cis,trans* isomers. Both of these isomers are achiral. In each, a plane of symmetry runs through the —CH₃ and —OH groups and the carbons bonded to them.

trans-4-Methylcyclohexanol
(achiral, plane of symmetry)

cis-4-Methylcyclohexanol
(achiral, plane of symmetry)

3-Methylcyclohexanol has two chiral centers and exists as $2^2 = 4$ stereoisomers. The *cis* isomer exists as one pair of enantiomers; the *trans* isomer, as a second pair of enantiomers.

cis-3-Methylcyclohexanol
(a pair of enantiomers)

trans-3-Methylcyclohexanol
(a pair of enantiomers)

Similarly, 2-methylcyclohexanol has two chiral centers and exists as $2^2 = 4$ stereoisomers. The *cis* isomer exists as one pair of enantiomers; the *trans* isomer, as a second pair of enantiomers.

Example 3.8

How many stereoisomers exist for 1,3-Cyclohexanediol?

Solution

1,3-Cyclohexanediol has two chiral centers and, according to the 2^n rule, has a maximum of $2^2 = 4$ stereoisomers. The *trans* isomer of this compound exists as a pair of enantiomers. The *cis* isomer has a plane of symmetry and is a meso compound. Therefore, although the 2^n rule predicts a maximum of four stereoisomers for 1,3-cyclohexanediol, only three exist: one meso compound and one pair of enantiomers.

cis-1,3-Cyclohexanediol
(meso)

trans-1,3-Cyclohexanediol
(a pair of enantiomers)

Problem 3.8

How many stereoisomers exist for 1,4-cyclohexanediol?

Examining the planar structures indicates that 1,2-cyclohexanediol has three stereoisomers. The *trans* isomer exists as a pair of enantiomers, while the *cis* isomer is a meso compound because it has a plane of symmetry. Interestingly, the two alternative chair conformations of *cis*-1,2-cyclohexanediol are each chiral and each is the mirror image of the other. However, because they are the same energy and they interconvert rapidly at room temperature, *cis*-1,2-cyclohexanediol is effectively meso as predicted by analysis of the planar structures.

trans-1,2-Cyclohexanediol
(a pair of enantiomers)

cis-1,2-Cyclohexanediol
(achiral, plane of symmetry)

chair-chair
interconversion

cis-1,2-Cyclohexanediol
chair conformations

3.6 | Tying All the Terminology Together

One of the major challenges in learning stereochemistry is mastering all of the terminology. In this and the previous chapters, we have presented several terms that describe the various kinds of isomers. Figure 3.5 summarizes the different isomers we have discussed so far, as well as the relationships among them. It delineates a step-by-step series of questions to guide you through the decision-making process for stereochemical terminology along with some of the same chemical examples that have been used thus far.

As a review, let's summarize Figure 3.5. Isomers occur when chemicals have the same molecular formula, but stereoisomers occur when the compounds have the same

Figure 3.5
Relationships among isomers along with some chemical examples.

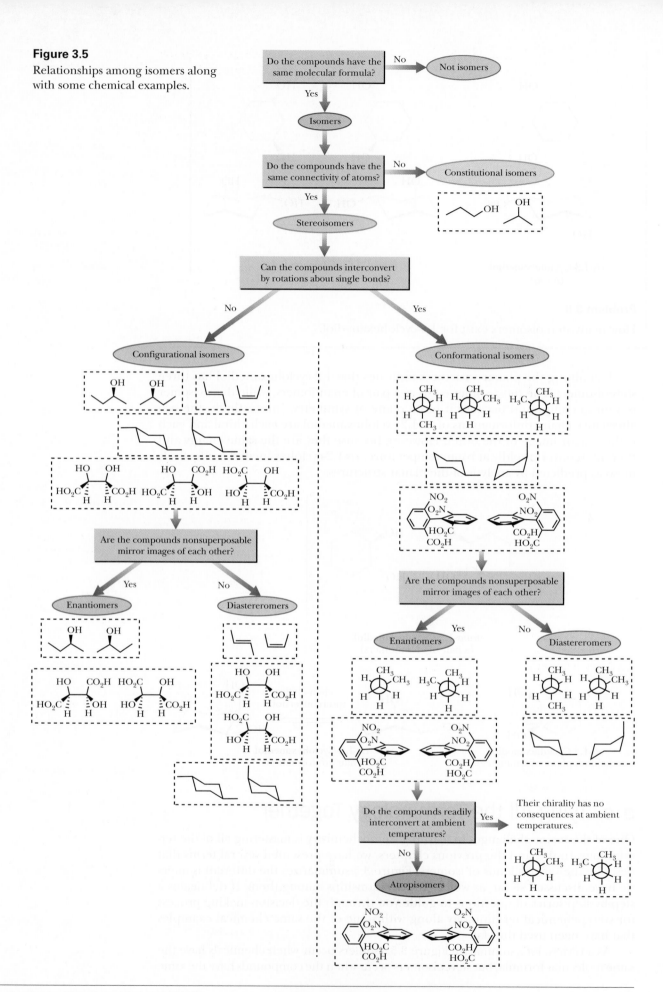

connectivity of their atoms but those atoms are arranged differently in space. Conformational isomers are a form of stereoisomers that interconvert by rotations along single bonds. Sometimes these isomers are nonsuperposable mirror images and are thus enantiomers, or if not, they are diastereomers. If there is restricted rotation such that the enantiomers cannot interconvert at ambient temperatures, then they are called atropisomers. When the enantiomers interconvert rapidly, there are no chemical consequences of their isomeric relationship. So, it is generally ignored.

When the stereoisomers are not interconverted by rotations along single bonds, they are referred to as configurational isomers. These isomers are either nonsuperposable mirror images and are thus enantiomers, or they are diastereomers.

3.7 | Optical Activity—How Chirality Is Detected in the Laboratory

As we have already established, enantiomers are different compounds, and thus we must expect that they differ in some properties. One property that differs between enantiomers is their effect on the rotation of the plane of polarized light. Each member of a pair of enantiomers rotates the plane of polarized light in opposite directions, and for this reason, enantiomers are said to be **optically active**.

The phenomenon of optical activity was discovered in 1815 by the French physicist Jean Baptiste Biot. To understand how it is detected in the laboratory, we must first understand something about plane-polarized light and a polarimeter, the device used to detect optical activity.

Optically active
Refers to a compound that rotates the plane of polarized light.

A. Plane-Polarized Light

Ordinary light consists of waves vibrating in all planes perpendicular to its direction of propagation (Figure 3.6). Certain materials such as calcite or a Polaroid sheet (a plastic film containing properly oriented crystals of an organic substance embedded in it) selectively transmit light waves vibrating only in parallel planes. Such radiation is said to be **plane polarized**.

Plane-polarized light is the vector sum of left and right circularly polarized light that propagates through space as left- and right-handed helices. These two forms of light are enantiomers, and because of their opposite handedness, each component interacts in an opposite way with chiral molecules. The result of this interaction is that each member of a pair of enantiomers rotates the plane of polarized light in an opposite direction.

Plane-polarized light
Light oscillating in only a single plane.

B. Polarimeters

A **polarimeter** consists of a monochromatic light source, a polarizing filter and an analyzing filter (each made of calcite or Polaroid film), and a sample tube (Figure 3.7). If the sample tube is empty, the intensity of the light reaching the eye is at its minimum when the polarizing axes of the two filters are at right angles. If the analyzing filter is turned either clockwise or counterclockwise, more light is transmitted.

The ability of molecules to rotate the plane of polarized light can be observed using a polarimeter in the following way. First, a sample tube filled with solvent is

Polarimeter
An instrument for measuring the ability of a compound to rotate the plane of polarized light.

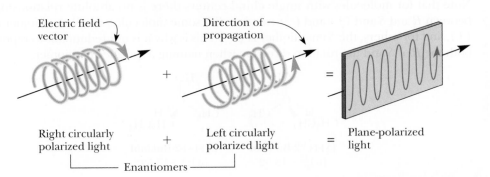

Electric field vector

Direction of propagation

Right circularly polarized light + Left circularly polarized light = Plane-polarized light

Enantiomers

Figure 3.6
Plane-polarized light is a mixture of left and right circularly polarized light.

Figure 3.7

Schematic diagram of a polarimeter with its sample tube containing a solution of an optically active compound. The analyzing filter has been turned clockwise by α degrees to restore the dark field to the observer.

Light source

Polarizing filter

Plane of polarized light

Filled sample tube

Analyzing filter is turned until field is dark

$0°$

α

Observed rotation

The number of degrees through which a compound rotates the plane of polarized light.

Dextrorotatory

Refers to a substance that rotates the plane of polarized light to the right.

Levorotatory

Refers to a substance that rotates the plane of polarized light to the left.

Specific rotation

Observed rotation of the plane of polarized light when a sample is placed in a tube 1.0 dm in length and at a concentration of 1 g/mL. For a pure liquid, concentration is expressed in g/mL (density).

placed in the polarimeter. The analyzing filter is adjusted so that the field is dark (the easiest position for the human eye to observe). This position of the analyzing filter is taken as 0°. When a solution of an optically active compound is placed in the sample tube, some light now passes through the analyzing filter; the optically active compound has rotated the plane of polarized light from the polarizing filter so that it is now no longer at an angle of 90° to the analyzing filter. The analyzing filter is then rotated to restore darkness in the field of view. The number of degrees, α, through which the analyzing filter must be rotated to restore darkness to the field of view is called the **observed rotation**. If the analyzing filter must be turned to the right (clockwise) to restore darkness, we say that the compound is **dextrorotatory** (Latin: *dexter*, on the right side). If the analyzing filter must be turned to the left (counterclockwise), we say that the compound is **levorotatory** (Latin: *laevus*, on the left side).

The magnitude of the observed rotation for a particular compound depends on its concentration, the length of the sample tube, the temperature, the solvent, and the wavelength of the light used. **Specific rotation**, $[\alpha]$, is defined as the observed rotation at a specific cell length and sample concentration.

$$\text{Specific rotation} = [\alpha]_\lambda^T = \frac{\text{Observed rotation (degrees)}}{\text{Length (dm)} \times \text{Concentration}}$$

The standard cell length is 1 decimeter (1 dm or 10 cm). Concentration is expressed in grams per milliliter (g/mL). Because specific rotation depends on temperature (T, in degrees Celsius) and wavelength λ of light, these variables are designated, respectively, as superscript and subscript. The light source most commonly used in polarimetry is the sodium D line ($\lambda = 589$ nm), the line responsible for the yellow color of sodium-vapor lamps.

In reporting either observed or specific rotation, a dextrorotatory compound is indicated with a plus sign in parentheses, ($+$), and a levorotatory compound with a minus sign in parentheses, ($-$). For any pair of enantiomers, one enantiomer is dextrorotatory, and the other is levorotatory. For each member of the pair, the absolute value of the specific rotation is exactly the same, but the sign is opposite. Following are the specific rotations of the enantiomers of 2-butanol at 25°C using the D line of sodium. Note that for molecules with single chiral centers there is no absolute relationship between R and S and ($+$) and ($-$) rotation. For some molecules, the R enantiomer is ($+$), and for others, the S enantiomer is ($+$). This is why it is not redundant to report both the R or S and the rotation direction when naming a specific enantiomer.

A polarimeter is used to measure the rotation of plane-polarized light as it passes through a sample.

© Richard Megna/Fundamental Photos

(S)-(+)-2-Butanol
$[\alpha]_D^{25} +13.52$

(R)-(−)-2-Butanol
$[\alpha]_D^{25} -13.52$

Example 3.9

A solution is prepared by dissolving 400 mg of testosterone, a male sex hormone, in 10.0 mL of ethanol and placing it in a sample tube 10.0 cm in length. The observed rotation of this sample at 25°C using the D line of sodium is +4.36°. Calculate the specific rotation of testosterone.

Solution

The concentration of testosterone is 400 mg/10.0 mL = 0.0400 g/mL. The length of the sample tube is 1.00 dm. Inserting these values in the equation for calculating specific rotation gives

$$\text{Specific rotation} = \frac{\text{Observed rotation (degrees)}}{\text{Length (dm)} \times \text{Concentration (g/mL)}} = \frac{+4.36°}{1.00 \times 0.0400} = +109$$

Problem 3.9

The specific rotation of progesterone, a female sex hormone, is +172. Calculate the observed rotation for a solution prepared by dissolving 300 mg of progesterone in 15.0 mL of dioxane and placing it in a sample tube 10.0 cm long.

C. Racemic Mixtures

An equimolar mixture of two enantiomers is called a **racemic mixture**, a term derived from the name "racemic acid" (Latin: *racemus,* a cluster of grapes). Racemic acid is the name originally given to an equimolar mixture of the enantiomers of tartaric acid (Table 3.1). Because a racemic mixture contains equal numbers of dextrorotatory and levorotatory molecules, its specific rotation is zero. Alternatively, we say that a racemic mixture is optically inactive. A racemic mixture is indicated by adding the prefix (±) to the name of the compound [or sometimes the prefix (*d,l*)].

Racemic mixture
A mixture of equal amounts of two enantiomers.

D. Optical Purity (Enantiomeric Excess)

When dealing with a pair of enantiomers, it is essential to have a means of describing the composition of that mixture and the degree to which one enantiomer is in excess relative to its mirror image. One way of describing the composition of a mixture of enantiomers is by its percent **optical purity**, a property that can be observed directly. Optical purity is the specific rotation of a mixture of enantiomers divided by the specific rotation of the enantiomerically pure substance when they are at the same concentration.

Optical purity
The specific rotation of a mixture of enantiomers divided by the specific rotation of the enantiomerically pure substance (expressed as a percent). Optical purity is numerically equal to enantiomeric excess, but experimentally determined.

$$\text{Percent optical purity} = \frac{[\alpha]_{\text{sample}}}{[\alpha]_{\text{pure enantiomers}}} \times 100$$

An alternative way to describe the composition of a mixture of enantiomers is by its **enantiomeric excess (ee)**, which is the difference in the number of moles of each enantiomer in a mixture compared to the total number of moles of both. Enantiomeric excess in percent is calculated by taking the difference in the percentage of each enantiomer.

Enantiomeric excess (ee)
The difference between the percentage of two enantiomers in a mixture.

$$\text{Enantiomeric excess (ee)} = \%R - \%S$$

For example, if a mixture consists of 75% of the *R* enantiomer and 25% of the *S* enantiomer, then the enantiomeric excess of the *R* enantiomer is 50%. Enantiomeric excess and optical purity are numerically identical.

Example 3.10

Figure 3.9 presents a scheme for separation of the enantiomers of mandelic acid. The specific rotation of optically pure (S)-(−)-mandelic acid is −158. Suppose that, instead of isolating pure (S)-(−)-mandelic acid from this scheme, the sample is a mixture of enantiomers with a specific rotation of −134. For this sample, calculate the following.

(a) The enantiomeric excess of this sample of (S)-(−)-mandelic acid.
(b) The percentage of (S)-(−)-mandelic acid and of (R)-(+)-mandelic acid in the sample.

Solution

(a) The enantiomeric excess of (S)-(−)-mandelic acid is 84.8%.

$$\text{Enantiomeric excess} = \frac{-134}{-158} \times 100 = 84.8\%$$

(b) This sample is 84.8% (S)-(−)-mandelic acid and 15.2% (R,S)-mandelic acid. The (R,S)-mandelic acid is 7.6% S enantiomer and 7.6% R enantiomer. The sample, therefore, contains 92.4% of the S enantiomer and 7.6% of the R enantiomer. We can check these values by calculating the observed rotation of a mixture containing 92.4% (S)-(−)-mandelic acid and 7.6% (R)-(+)-mandelic acid as follows:

Specific rotation = 0.924 × (−158) + 0.076 × (+158) = −146 + 12 = −134

which agrees with the experimental specific rotation.

Problem 3.10

One commercial synthesis of naproxen (the active ingredient in Aleve and a score of other over-the-counter and prescription nonsteroidal anti-inflammatory drug preparations) gives the enantiomer shown in 97% enantiomeric excess.

Naproxen
(a nonsteroidal anti-inflammatory drug)

(a) Assign an R or S configuration to this enantiomer of naproxen.
(b) What are the percentages of R and S enantiomers in the mixture?

3.8 | The Significance of Chirality in the Biological World

Except for inorganic salts and a relatively few low-molecular-weight organic substances, the molecules in living systems, both plant and animal, are chiral. Although these molecules can exist as a number of stereoisomers, almost invariably only one stereoisomer is found in nature. This occurrence is a consequence of the fact that their natural syntheses are catalyzed by enzymes, which are also chiral. Of course, instances do occur in which more than one stereoisomer is found, but these rarely exist together in the same biological system.

A. Chirality in Enzymes

Let us look more closely at the chirality of enzymes. An illustration is chymotrypsin, an enzyme in the intestines of animals, which catalyzes the hydrolysis of proteins during digestion. Chymotrypsin, like all proteins, is composed of a long molecular chain of amino acids that folds up into the active enzyme. Human chymotrypsin has 268 chiral centers that result from the amino acids, so the maximum number of stereoisomers possible is 2^{268}, a staggeringly large number, almost beyond comprehension.

Fortunately, because each chiral amino acid is only present as a single stereoisomer, only one of the possible stereoisomers of chymotrypsin is produced. Because enzymes are chiral substances and are present as single stereoisomers, most either produce or react only with substances that are single stereoisomers (if chiral).

As an interesting illustration of this enzyme chirality principle, Stephen B. Kent, then at The Scripps Research Institute in La Jolla, California, synthesized the enantiomer of a natural enzyme by using the enantiomers of the amino acids normally found in proteins as building blocks. As expected, this synthetic enzyme enantiomer only catalyzed a reaction with a substrate that was the enantiomer of the chiral substrate utilized by the natural enzyme.

B. How an Enzyme Distinguishes Between a Molecule and Its Enantiomer

Enzymes are chiral catalysts. Some are completely specific for the catalysis of the reaction of only one particular compound, whereas others are less specific and catalyze similar reactions of a family of compounds. An enzyme catalyzes a biological reaction of molecules by first positioning them at a binding site on its surface. These molecules may be held at the binding site by a combination of hydrogen bonds, electrostatic attractions, dispersion forces, or even temporary covalent bonds.

An enzyme with a specific binding site for a molecule with a chiral center can distinguish between a molecule and its enantiomer or one of its diastereomers. Assume, for example, that an enzyme involved in catalyzing a reaction of glyceraldehyde has a binding site with groups that interact with —H, —OH, and —CHO. Assume further that the binding sites are arranged in the enzyme binding site as shown in Figure 3.8. The enzyme can distinguish (R)-$(+)$-glyceraldehyde (the natural or biologically active form) from its enantiomer (S)-$(-)$-glyceraldehyde because the natural enantiomer can be bound to the binding site with three groups interacting with their appropriate binding sites; the other enantiomer can, at best, bind to only two of these sites.

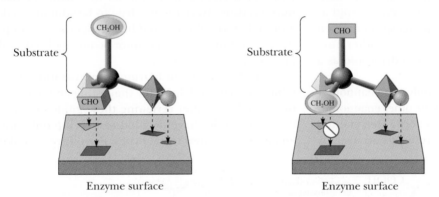

This enantiomer of glyceraldehyde fits the three specific binding sites on the enzyme surface.

This enantiomer of glyceraldehyde does not fit the same binding sites.

Figure 3.8
A schematic diagram of an enzyme surface capable of interacting with (R)-$(+)$-glyceraldehyde at three binding sites, but with (S)-$(-)$-glyceraldehyde at only two of these sites.

Because interactions between molecules in living systems take place in a chiral environment, it should be no surprise that a molecule and its enantiomer or diastereomers have different physiological properties. The tricarboxylic acid (TCA) cycle, for example, produces and then metabolizes only (S)-$(+)$-malic acid. Because only one enantiomer is produced, both the production and metabolism of (S)-$(+)$-malic acid are said to be enantioselective.

$$\text{HOOC} \overset{\text{COOH}}{\underset{\text{H} \quad \text{OH}}{\diagup}}$$

(S)-$(+)$-Malic acid

That interactions between molecules in the biological world are highly enantio-selective is not surprising, but just how these interactions are accomplished at the molecular level with such precision and efficiency is a great puzzle that is an active area of scientific research. Scientists in the field of bioorganic chemistry seek to better understand and harness the incredible power of biological molecules, especially enzymes, for use in organic chemistry applications.

CONNECTIONS TO BIOLOGICAL CHEMISTRY

Chiral Drugs

Some of the common drugs used in human medicine, for example aspirin (Section 18.5B), are achiral. Others are chiral and sold as single enantiomers. The penicillin and erythromycin classes of antibiotics and the drug captopril are all chiral drugs. Captopril, which is very effective for the treatment of high blood pressure and congestive heart failure, was developed in a research program designed to discover effective inhibitors for angiotensin-converting enzyme (ACE). It is manufactured and sold as the S,S stereoisomer. A large number of chiral drugs, however, are sold as racemic mixtures. The popular analgesic ibuprofen (the active ingredient in Motrin and many other non-aspirin analgesics) is an example of a drug sold as a racemic mixture.

For racemic drugs, most often only one enantiomer exerts the beneficial effect, whereas the other enantiomer either has no effect or may even exert a detrimental effect. Thus, enantiomerically pure drugs are usually

dopamine. This compound does not cross the blood-brain barrier to the required site of action in the brain. What is administered, instead, is the prodrug, a compound that is not active by itself but is converted in the body to an active drug. 3,4-Dihydroxyphenylalanine is such a prodrug. It crosses the blood-brain barrier and then undergoes decarboxylation catalyzed by the enzyme dopamine decarboxylase. This enzyme is specific for the S enantiomer, which is commonly known as L-DOPA. It is essential, therefore, to administer the enantiomerically pure prodrug. Were the prodrug to be administered in a racemic form, there could be a dangerous buildup of the R enantiomer, which cannot be metabolized by the enzymes present in the brain.

The U.S. Food and Drug Administration has established new guidelines for the testing and marketing of chiral drugs. After reviewing these guidelines, many drug companies have decided to develop only single enantiomers of new chiral drugs. For this reason, there

Captopril

(S)-Ibuprofen

(S)-(–)-3,4-Dihydroxyphenylalanine
(L-DOPA)
$[\alpha]_D^{13} -13.1$

enzyme-catalyzed decarboxylation →

Dopamine + CO_2

more effective than their racemic counterparts. For example, only the S enantiomer of ibuprofen is biologically active. Interestingly, the body converts the inactive R enantiomer to the active S enantiomer. Another good example is the drug dihydroxyphenylalanine used in the treatment of Parkinson's disease. The active drug is

has been a tremendous recent interest in developing stereoselective synthetic methods that yield a single enantiomer or diastereomer. In addition to regulatory pressure, there are patent considerations. If a company has patents on a racemic drug, a new patent can often be taken out on one of its enantiomers.

In organisms ranging from bacteria to humans, the 20 most common amino acids share the familiar structural motif of a central carbon atom bonded to a hydrogen atom, an amino group, and a carboxyl group. In addition, 19 of the 20 amino acids have an additional group of atoms, referred to as the "side chain," bonded to the central carbon. The central carbon atom of these 19 is a chiral center, having the configuration shown here. At neutral pH, amino acids exist as an internal salt (commonly called a zwitterion) as a result of the relative acid strength of the carboxyl group and base strength of the amino group, respectively (see Section 4.4).

Side chain

$$H \overset{\text{C*}}{\underset{H_3N^+}{\cdots}} COO^-$$

Ionized (zwitterion) form

Proteins are long chains of amino acids covalently bonded together by amide bonds (Section 1.3F) formed between the carboxyl group of one amino acid and the amino group of another amino acid. Because they are made of pure amino acid stereoisomers, proteins themselves are single stereoisomers despite having several hundred or more chiral centers. These chains fold into complex three-dimensional structures capable of a remarkable array of chemical functions.

Proteins are one of the key functional components of all living cells. This means that the chemistry of living systems is chiral, and that the presence of single stereoisomers is the rule rather than the exception for biological molecules. Another of the key biochemical building blocks, the carbohydrates, have multiple chiral centers, adding to the chiral nature of living systems.

For most proteins, $n = 150-750$

The amino acid side chains encompass a variety of functional groups including benzene rings, alkyl groups, amino groups, thiol (—SH) groups, and carboxyl groups. One amino acid, namely glycine, has no side chain; only another hydrogen is bonded to the central carbon.

According to the Cahn-Ingold-Prelog priority rules, 18 of the 19 side chain groups are lower priority than the amino and carboxyl groups, so these 18 have the *S* configuration. The exception is cysteine in which the side chain is a —CH₂—SH group. Because of the sulfur atom, the —CH₂—SH group is of higher priority than the carboxyl group, and the configuration of cysteine is *R*. Two of the amino acid side chains themselves have a chiral center (isoleucine and threonine). All 19 chiral amino acids are found in nature as single stereoisomers the vast majority of the time.

$$CH_2SH$$
$$H \overset{\text{C*}}{\underset{H_3N^+}{\cdots}} COO^-$$

Cysteine

You will encounter a different type of stereochemical designation when it comes to amino acids. In this designation, stereochemistry is related to the structure of glyceraldehyde, a molecule with one chiral center (Section 3.4C). The chiral center in the 19 common chiral amino acids is structurally related to the glyceraldehyde stereoisomer that rotates plane-polarized light in the levorotatory or (−) direction, so this glyceraldehyde *and all the amino acids structurally related to it* are designated "L." The common 19 chiral amino acids are often referred to as L-amino acids in biochemistry texts and literature. Note that this does *not* mean that each of the 19 common amino acids rotates plane polarized light in the levorotatory or (−) direction; they do not. Many students can be confused by this. The "L" designation refers *only* to the *structural* relationship of the common 19 chiral amino acids to the levorotatory (−) isomer of glyceraldehyde. For example, the specific rotation of L-serine is +14.45 in 1 *M* HCl, but −6.83 in water. The change in the ionization of the carboxyl group causes a change in specific rotation, but not a change in configuration of the chiral center.

Although relatively rare, amino acids with the opposite configuration are found in some organisms. These amino acids are structurally related to the dextrorotatory or (+) isomer of glyceraldehyde, so they are referred to as D amino acids. The D-amino acids are the enantiomers of corresponding L-amino acids.

L-Amino acids
(common)

D-Amino acids
(rare)

3.9 | Separation of Enantiomers—Resolution

The separation of a racemic mixture into its enantiomers is called **resolution**.

Resolution

Separation of a racemic mixture into its enantiomers.

A. Resolution by Means of Diastereomeric Salts

One general scheme for separating enantiomers requires chemical conversion of a pair of enantiomers into two diastereomers with the aid of an enantiomerically pure chiral resolving agent. This chemical resolution is successful because the diastereomers thus formed are different compounds, have different physical properties, and often can be separated by physical means (most commonly fractional crystallization or column chromatography) and purified. The final step in this scheme for resolution is chemical conversion of the separated diastereomers back to the individual enantiomers and recovery of the chiral resolving agent.

A reaction that lends itself to chemical resolution is salt formation because it is readily reversible.

$$\text{RCOOH} \quad + \quad \text{:B} \rightleftharpoons \text{RCOO}^-\text{HB}^+$$

Carboxylic acid   Base   Salt

Chiral bases available from plants are often used as chiral resolving agents for racemic acids. More commonly, chemists now use commercially available chiral amines such as (R,S)-1-phenylethylamine.

(R)-1-Phenylethanamine   **(S)-1-Phenylethanamine**

The resolution of mandelic acid by way of its diastereomeric salts with the natural chiral base cinchonine is illustrated in Figure 3.9. Racemic mandelic acid and optically pure (+)-cinchonine (Cin) are dissolved in boiling water, giving a solution of a pair of diastereomeric salts. Diastereomers have different solubilities, and when the solution cools, the less-soluble diastereomeric salt crystallizes. This salt is collected and purified by further recrystallization. The filtrates, richer in the more soluble diastereomeric salt, are concentrated to give this salt, which is also purified by further recrystallization. The purified diastereomeric salts are treated with aqueous HCl to precipitate the nearly pure enantiomers of mandelic acid. Cinchonine remains in the aqueous solution as its water-soluble hydrochloride salt.

Optical rotations and melting points of racemic mandelic acid, cinchonine, the purified diastereomeric salts, and the pure enantiomers of mandelic acid are

Figure 3.9

Resolution of mandelic acid.

given in Figure 3.9. Note the following two points: (1) The diastereomeric salts have different specific rotations and different melting points. (2) The enantiomers of mandelic acid have identical melting points and have specific rotations that are identical in magnitude but opposite in sign.

Resolution of a racemic base with a chiral acid is carried out in a similar way. Acids that are commonly used as chiral resolving agents are (+)-tartaric acid, (−)-malic acid, and (+)-camphoric acid (Figure 3.10). These and other naturally occurring chiral resolving agents are produced in plant and animal systems as single enantiomers.

B. Enzymes as Resolving Agents

In their quest for enantiomerically pure compounds, organic chemists have developed several new techniques for preparation of enantiomerically pure materials. One approach is to use enzymes. Enzymes are themselves chiral so they can produce single enantiomer products. A class of enzymes used for this purpose is the esterases, which catalyze the hydrolysis of esters to give an alcohol and a carboxylic acid.

The ethyl esters of naproxen crystallize in two enantiomeric crystal forms; one containing the *R* ester and the other containing the *S* ester. Each is insoluble in water. Chemists then use an esterase in alkaline solution to hydrolyze selectively the *S* ester to the (*S*)-carboxylic acid, which goes into solution as the sodium salt. The *R* ester is unaffected by these conditions. Filtering the alkaline solution recovers

Figure 3.10
Some carboxylic acids used as chiral resolving agents.

(2R,3R)-(+)-Tartaric acid

$[\alpha]_D^{20}$ +12.5

(S)-(−)-Malic acid

$[\alpha]_D^{20}$ −27

(1S,3R)-(+)-Camphoric acid

$[\alpha]_D^{20}$ +46.5

the crystals of the *R* ester. After crystals of the *R* ester are removed, the alkaline solution is acidified to give enantiomerically pure (*S*)-naproxen.

The recovered *R* ester is racemized (converted to an *R,S* mixture) and treated again with the esterase. Thus, by recycling the *R* ester, all the racemic ester is converted to (*S*)-naproxen.

Ethyl ester of (S)-naproxen
1. Esterase
2. HCl, H₂O

NaOH, H₂O

Ethyl ester of (R)-naproxen
(Not affected by the esterase)

(S)-Naproxen

Chemists are no longer limited to using only natural enzymes for resolution of chiral molecules. Using genetic engineering techniques, it is now possible to develop new esterases or other enzymes in the laboratory that react specifically with single enantiomers of important substrates. Scientists have also engineered new properties into enzymes such as better stability at higher temperature or enhanced solubility in organic solvent/water mixtures, increasing the practical value of enzymes for organic chemistry.

C. Resolution by Means of Chromatography on a Chiral Substrate

Chromatography
A separation method involving passing a vapor or solution mixture through a column packed with a material with different affinities for different components of the mixture.

Chromatography is a term used to describe the purification of molecules in which a sample to be purified interacts with a solid material, and different components of the sample separate based on their different relative interactions with the solid material. Separation can be accomplished using the sample in either the gas phase (usually for analytical purposes) or the liquid phase (analytical or preparative separations are possible). The solid material is packed into a column and a solvent (or, in gas chromatography, a gas) passes down the column, carrying the more weakly bound components of the mixture with it more rapidly than the more tightly bound ones.

A common method of resolving enantiomers today is chromatography using a chiral column packing material. Each enantiomer interacts differently with the chiral molecules of the packing material, and the elution time will (in principle at least) be different for the two enantiomers. A wide variety of chiral column packings have been developed for this purpose.

Summary

Section 3.1 Chirality—The Handedness of Molecules

- Molecules that are not superposable on their mirror images are said to be **chiral**.
  - A **mirror image** is the reflection of an object in a mirror.
  - **Chirality** is a property of an object as a whole, not of a particular atom.
  - An **achiral** object lacks chirality; that is, it is an object with a superposable mirror image. Almost all achiral objects possess at least one plane or center of symmetry.
    - A **plane of symmetry** is an imaginary plane passing through an object dividing it such that one half is the reflection of the other half.
    - A **center of symmetry** is a point so situated that identical components of the object are located on opposite sides and equidistant from the point along any axis passing through that point.

Problems: 3.11, 3.12, 3.33, 3.34

Section 3.2 Stereoisomerism

- **Isomers** are molecules that have the same molecular formula but possess different structures because of different connectivity of atoms and/or spatial arrangement of atoms.
- **Constitutional isomers** are molecules that have the same molecular formula but have a different connectivity of atoms.
- **Stereoisomers** are isomers that have the same connectivity of atoms but have a different spatial arrangement of their atoms.
  - **Conformational isomers** are stereoisomers that interconvert by rotation along single bonds.
  - **Configurational isomers** are stereoisomers that do not interconvert at room temperature either by virtue of having to break single or double bonds to interconvert.
  - **Enantiomers** are stereoisomers that are nonsuperposable mirror images.
  - **Diastereomers** are stereoisomers that are not mirror images.
  - **Atropisomers** are enantiomers that lack a chiral center and differ because of hindered rotation along a single bond.
- A **stereocenter** is an atom about which exchange of two groups produces a stereoisomer.
- A **chiral center** is a tetrahedral atom with four different groups bonded to it.

Problems: 3.1, 3.13–3.15, 3.17, 3.19, 3.36–3.38

Section 3.3 Naming Chiral Centers—The *R,S* System

- The **absolute configuration** at any chiral center can be specified by the *R,S* **system**.
  - To apply this convention:
    (1) Each atom or group of atoms bonded to the chiral center is assigned a priority and numbered from highest priority to lowest priority.
    (2) The molecule is oriented in space so that the group of lowest priority is directed away from the observer.
    (3) The remaining three groups are read in order from highest priority to lowest priority. If the order of groups is clockwise, the configuration is *R* (Latin: *rectus*, right, correct). If the order is counterclockwise, the configuration is *S* (Latin: *sinister*, left).
  - To invert the configuration at a chiral center (*R* to *S* or *vice versa*), switch the location of any two groups bonded to the chiral center.

Problems: 3.2, 3.3, 3.16, 3.20–3.24, 3.35

Section 3.4 Acyclic Molecules with Two or More Stereocenters

- For a molecule with n chiral centers, the maximum number of stereoisomers possible is 2^n.
  - Certain molecules have special symmetry properties that reduce the number of stereoisomers to fewer than that predicted by the 2^n rule. Two stereoisomers are enantiomers if *all* of the chiral centers have the opposite configuration (*R* vs. *S* or vice versa).
  - If even one chiral center is the same between two stereoisomers, they are diastereomers.
  - A **meso compound** contains two or more chiral centers bonded in such a way that it is achiral. Meso compounds usually have an internal plane of symmetry.
- **Fischer projections** are two-dimensional projections of chiral molecules useful when evaluating stereochemistry.
  - In Fischer projections, groups on the right and left are by convention in front.
  - Groups on the top and bottom are by convention behind.
- The type of stereoisomer relationship dictates the relationships of properties between stereoisomers.
  - Enantiomers have identical physical and chemical properties in achiral environments but different properties in chiral environments such as in the presence of plane-polarized light. They also have different properties in the presence of chiral reagents and enzymes as chiral catalysts.
  - Diastereomers have different physical and chemical properties even in achiral environments.

Problems: 3.4, 3.5, 3.25–3.28, 3.30–3.32

Problem: 3.6

Section 3.5 Cyclic Molecules with Two or More Stereocenters

Problems: 3.7, 3.8

- When evaluating symmetry in **ring structures** with chiral centers such as cyclohexane derivatives, it is helpful to evaluate the planar representations.

Section 3.6 Tying All the Terminology Together

- There are two overarching classes of isomers: constitutional isomers and stereoisomers.
  - Stereoisomerism can arise either from configurational or conformational isomerism.
  - Configurational isomers are either diastereomers or enantiomers (if the compound is chiral).
  - Conformational *isomers* are either diastereomers or enantiomers (if the compound is chiral). Enantiomers that do not readily interconvert by bond rotations are **atropisomers**.

Section 3.7 Optical Activity—How Chirality Is Detected in the Laboratory

- Plane-polarized light is used to evaluate stereoisomers.
  - Light that vibrates in parallel planes is said to be plane-polarized.
  - Plane-polarized light contains equal components of left and right circularly polarized light.
  - A **polarimeter** is an instrument used to detect and measure the magnitude of optical activity.
- A compound is said to be **optically active** if it rotates the plane of plane-polarized light.
  - **Observed rotation** is the number of degrees the plane of plane-polarized light is rotated.
  - **Specific rotation** is the observed rotation measured in a cell 1 dm long and at a sample concentration of 1 g/mL.
    - A compound is **dextrorotatory** if the analyzing prism must be turned clockwise to restore the zero point.
    - A compound is **levorotatory** if the analyzing prism must be turned counterclockwise to restore the zero point.

Problems: 3.9, 3.29

- Each member of a pair of enantiomers rotates the plane of polarized light an equal number of degrees but opposite in direction.
- A **racemic mixture** is a mixture of equal amounts of two enantiomers and has a specific rotation of zero.
- Percent **optical purity** (identical to **enantiomeric excess**) is defined as the specific rotation of a mixture of enantiomers divided by the specific rotation of the pure enantiomer times 100.
- **Meso compounds** are not chiral so they are optically inactive.

Problems: 3.10, 3.39, 3.40

Section 3.8 The Significance of Chirality in the Biological World

- Biological molecules are chiral, and are usually found as single enantiomers in living systems.
  - Enzymes are made of chiral amino acids and they catalyze biological reactions in a chiral way by first positioning the molecule or molecules at binding sites and holding them there by a combination of hydrogen bonds, electrostatic attractions, dispersion forces, and, sometimes, covalent bonds.
  - An enzyme with specific binding sites for three of the four groups on a chiral center can distinguish between a molecule and its enantiomer.
  - Almost all enzyme-catalyzed reactions are **stereoselective**.

Section 3.9 Separation of Enantiomers—Resolution

- **Resolution** is the experimental process of separating a mixture of enantiomers into the two pure enantiomers.
  - A common chemical means of resolving organic compounds is to treat the racemic mixture with a chiral resolving agent that converts the mixture of enantiomers into a pair of diastereomers. The diastereomers are separated based on differences in their physical properties; each diastereomer is then converted to a pure stereoisomer, uncontaminated by its enantiomer.
  - Enzymes are also used as resolving agents because of their ability to catalyze a reaction of one enantiomer but not that of its mirror image.
  - **Chromatography** on a chiral substrate is also an effective separation method.

⭕WL Questions and problems for this chapter are available in OWL. Use the OWL access card that came with this textbook to access assigned questions and problems for this chapter.

Chapter 4
Acids and Bases

© Emilio Ereza/Alamy

Citrus fruits are sources of citric acid. Lemon juice, for example, contains 5–8% citric acid. Inset: a model of citric acid.

A great many organic reactions either are acid-base reactions or involve catalysis by an acid or base at some stage. Of the reactions involving acid catalysis, some use proton-donating acids, such as H_2SO_4, H_3O^+, and $CH_3CH_2OH_2^+$. Others use Lewis acids, such as BF_3 and $AlCl_3$. It is essential, therefore, that you have a good grasp of the fundamentals of acid-base chemistry. In this and following chapters, we study the acid-base properties of the major classes of organic compounds.

4.1 | Arrhenius Acids and Bases

The first useful definition of acids and bases was put forward by Svante Arrhenius in 1884. According to the original Arrhenius definition, an acid is a substance that dissolves in water to produce H^+ ions. A base is a substance that dissolves in water to produce OH^- ions. Today we know that it is not accurate to refer to an acid as giving rise to simple H^+ ions. An H^+ ion immediately combines with a water molecule to give a hydronium ion, H_3O^+. Hydration of the hydronium ion itself gives the ion $H_5O_2^+$. Modern research suggests that the monohydrated and dihydrated forms (H_3O^+ and $H_5O_2^+$) are the major hydrated forms present in aqueous solution, and they are present in approximately equal concentrations. Throughout this text, we will represent a proton dissolved in aqueous solution as a hydronium ion, H_3O^+, although it is an oversimplification.

$$H^+(aq) + H_2O(l) \longrightarrow H_3O^+(aq)$$

Hydronium ion

⬥WL Online homework for this chapter may be assigned in OWL for Organic Chemistry.

$$H_3O^+(aq) + H_2O(l) \rightleftharpoons H_5O_2^+(aq)$$

Hydronium ion

We therefore must modify the Arrhenius acid definition to take into account these interactions of H^+ with water molecules. Apart from this modification, the Arrhenius definitions of an acid and a base are still valid and useful today, as long as we are talking about aqueous solutions.

4.2 | Brønsted-Lowry Acids and Bases

In 1923 the Danish chemist Johannes Brønsted and the English chemist Thomas Lowry independently proposed the following definitions: an **acid** is a proton donor, a **base** is a proton acceptor, and an acid-base reaction is a proton-transfer reaction.

Brønsted-Lowry acid
A proton donor.

Brønsted-Lowry base
A proton acceptor.

A. Conjugate Acid-Base Pairs Differ by a Proton

According to the Brønsted-Lowry definitions, any pair of molecules or ions that can be interconverted by transfer of a proton is called a **conjugate acid-base pair**. When an acid transfers a proton to a base, the acid is converted to its **conjugate base**. When a base accepts a proton, it is converted to its **conjugate acid**. Fundamental to these definitions is the fact that the members of a conjugate acid-base pair differ by a proton.

We can illustrate the relationships among conjugate acid-base pairs by examining the reaction of hydrogen chloride with water to form chloride ion and hydronium ion.

Conjugate base
The species formed when an acid transfers a proton to a base.

Conjugate acid
The species formed when a base accepts a proton from an acid.

```
          ┌──── conjugate acid-base pair ────┐
    ┌─ conjugate acid-base pair ─┐
    ↓                            ↓           ↓              ↓
  HCl(aq)  +  H₂O(l)    ⇌    Cl⁻(aq)   +   H₃O⁺(aq)
```

Hydrogen chloride	**Water**	**Chloride ion**	**Hydronium ion**
(acid)	(base)	(conjugate base of HCl)	(conjugate acid of H₂O)

The acid HCl donates a proton and is converted to its conjugate base Cl^-. The base H_2O accepts a proton and is converted to its conjugate acid H_3O^+. The members of each conjugate acid-base pair differ only by a proton.

We can show the transfer of a proton from an acid to a base by using a **curved arrow**, the same curved arrow symbol that we used in Section 1.8 to show the relocation of electron pairs among resonance contributing structures. For acid-base reactions, we write the Lewis structure of each reactant and product, showing all valence electrons on reacting atoms. We then use curved arrows to show the change in position of electron pairs during the reaction. The tail of the curved arrow is located at an electron pair, either a lone pair or a bonding pair, as the case may be. The head of the curved arrow shows the new location of the electron pair. For example, a curved arrow originating at a lone pair and pointing to an adjacent atom indicates formation of a new bond, while an arrow originating at a bonding electron pair and pointing toward a previously bonded atom indicates a breaking of that bond (see Section 6.2D for a more thorough discussion of the use of curved arrows).

relocating this electron
pair forms a new O—H bond

$$H-\ddot{O}: + H-\ddot{C}l: \rightleftharpoons H-\overset{+}{\underset{H}{O}}-H + :\ddot{C}l:^-$$

relocating this electron
pair breaks the H—Cl bond

In this equation, the curved arrow on the left shows that an unshared pair of electrons on oxygen changes position to form a new covalent bond with hydrogen. The curved arrow on the right shows that the H—Cl bond breaks and that its electron pair is given entirely to chlorine to form a chloride ion. Thus, in the reaction of HCl with H_2O, a proton is transferred from HCl to H_2O. In the process, an O—H bond forms and an H—Cl bond breaks. It is important to keep in mind that arrows indicate movement of electrons, not atoms.

The most common mistake when writing curved arrows is to draw arrows that point backward. In the following example the arrow originates at the H—Cl bond and terminates at an oxygen atom. As drawn, this notation means that the two electrons of the H—Cl bond are moving with the hydrogen, and hence H:$^-$ (hydride anion) is being transferred rather than a H$^+$ (proton). Appendix 10 contains more examples of common mistakes.

$$\text{wrong}$$

$$H-\overset{\cdot\cdot}{\underset{|}{O}}: \quad + \quad H-Cl$$
$$H$$

We have illustrated the application of the Brønsted-Lowry definitions using water as a reactant. The Brønsted-Lowry definitions, however, do not require water as a reactant. Consider the following reaction between acetic acid and ammonia.

$$\text{conjugate acid-base pair}$$
$$\text{conjugate acid-base pair}$$

$$CH_3COOH + NH_3 \rightleftharpoons CH_3COO^- + NH_4^+$$

Acetic acid	Ammonia	Acetate ion	Ammonium ion
(acid)	(base)	(conjugate base of acetic acid)	(conjugate acid of ammonia)

We can use curved arrows to show how this reaction takes place. The curved arrow on the right shows that the unshared pair of electrons on nitrogen becomes shared between N and H to form a new H—N bond. At the same time that the H—N bond forms, the O—H bond breaks, and the electron pair of the O—H bond moves entirely to oxygen to form —O$^-$ of the acetate ion, as depicted by the curved arrow on the left. The result of these two electron-pair shifts is the transfer of a proton from an acetic acid molecule to an ammonia molecule.

$$CH_3-\overset{\overset{\displaystyle :O:}{\|}}{C}-\overset{\cdot\cdot}{O}-H \quad + \quad \overset{H}{\underset{H}{\overset{|}{N}}}-H \rightleftharpoons CH_3-\overset{\overset{\displaystyle :O:}{\|}}{C}-\overset{\cdot\cdot}{\underset{\cdot\cdot}{O}}:^- + H-\overset{H}{\underset{H}{\overset{+|}{N}}}-H$$

Acetic acid	Ammonia	Acetate ion	Ammonium ion
(proton donor)	(proton acceptor)		

Example 4.1

For each conjugate acid-base pair, identify the first species as an acid or base and the second species as its conjugate base or conjugate acid. In addition, draw Lewis structures for each species, showing all valence electrons and any formal charges.

(a) CH_3OH, $CH_3OH_2^+$ (b) H_2CO_3, HCO_3^- (c) CH_3NH_2, $CH_3NH_3^+$

Solution

(a) CH₃—Ö—H   CH₃—Ö⁺—H
    Base   H
                  Conjugate
                  acid

(b) H—Ö—C(=O)—Ö—H   H—Ö—C(=O)—Ö⁻
        Acid   Conjugate
                  base

(c) CH₃—N̈—H   CH₃—N⁺—H (with H top and bottom)
       H
    Base   Conjugate
             acid

Problem 4.1

For each conjugate acid-base pair, identify the first species as an acid or base and the second species as its conjugate acid or conjugate base. In addition, draw Lewis structures for each species, showing all valence electrons and any formal charges.

(a) H_2SO_4, HSO_4^- (b) NH_3, NH_2^- (c) CH_3OH, CH_3O^-

Example 4.2

Write these reactions as proton-transfer reactions. Label which reactant is the acid and which is the base, which product is the conjugate base of the original acid, and which is the conjugate acid of the original base. In addition, write Lewis structures for each reactant and product, and use curved arrows to show the flow of electrons in each reaction.

(a) $H_2O + NH_4^+ \longrightarrow H_3O^+ + NH_3$
(b) $CH_3CH_2OH + NH_2^- \longrightarrow CH_3CH_2O^- + NH_3$

Solution

(a) Water is the base (proton acceptor), and ammonium ion is the acid (proton donor).

    Base   Acid   Conjugate acid   Conjugate base
                                 of H_2O of NH_4^+

(b) Ethanol is the acid (proton donor), and amide ion (NH_2^-) is the base (proton acceptor).

   Acid   H
            Base   Conjugate base   Conjugate acid
                       of CH_3CH_2OH of NH_2^-

Problem 4.2

Write these reactions as proton-transfer reactions. Label which reactant is the acid and which is the base, which product is the conjugate base of the original acid, and which is the conjugate acid of the original base. In addition, write Lewis structures for each reactant and product, and use curved arrows to show the flow of electrons in each reaction.

(a) $CH_3SH + OH^- \rightleftharpoons CH_3S^- + H_2O$

(b) $CH_2{=}O + HCl \rightleftharpoons CH_2{=}OH^+ + Cl^-$

B. Brønsted-Lowry Bases with Two or More Receptor Sites

Thus far, we have dealt with Brønsted-Lowry bases that have only one site that can act as a proton acceptor in an acid-base reaction. Many organic compounds have two or more such sites. In the following discussion, we restrict our attention to compounds containing a carbonyl group in which the carbonyl carbon is bonded to either an oxygen or a nitrogen. The principle we develop here is an extremely important one and applies to other types of molecules as well. The more stable charged species is the one in which the charge is more delocalized. Relative charge delocalization can often be understood by considering resonance (Section 1.8). Let us consider first the potential sites for proton transfer to an oxygen atom of a carboxylic acid such as acetic acid. Proton transfer to the carbonyl oxygen gives cation A, and proton transfer to the hydroxyl oxygen gives cation B.

A
(protonation on the carbonyl oxygen)

B
(protonation on the hydroxyl oxygen)

We now examine each cation and determine which is the more stable (lower in energy). For cation A, we can write three contributing structures. Two of these place the positive charge on oxygen, and one places it on carbon.

A-1
(C and O have complete octets)

A-2
(C has incomplete octet)

A-3
(C and O have complete octets)

Of these three structures, A-1 and A-3 make the greater contributions to the hybrid because all atoms in each have complete octets; A-2 makes a lesser contribution because its carbonyl carbon has an incomplete octet. Thus, on protonation of the carbonyl oxygen, the positive charge is delocalized over three atoms with the greater share of it being on the two equivalent oxygen atoms. (The two oxygens were not equivalent before protonation, but they are now.)

Protonation on the hydroxyl oxygen gives cation B for which we can write two resonance contributing structures.

B-1

B-2
(charge separation and adjacent positive charges)

Of these, B-2 makes at best only a minor contribution to the hybrid because of the adjacent positive charges; therefore, the charge on this cation is, in effect, localized on the hydroxyl oxygen.

From this analysis of cations A and B, we see that protonation of a carboxyl group occurs preferentially on the carbonyl oxygen because this cation has greater delocalization of the positive charge.

Example 4.3

The functional group created when the —OH of a carboxyl group is replaced by an NH_2 group is called an amide (Section 1.3F). Draw the structural formula of acetamide, which is derived from acetic acid, and determine if proton transfer to the amide group from HCl occurs preferentially on the amide oxygen or the amide nitrogen.

Solution

Following is a Lewis structure for acetamide and its two possible protonated forms.

$$CH_3-C(:O:)-N(H)-H + HCl \rightleftharpoons CH_3-C(\overset{+}{O}-H)=N(H)-H \quad \text{or} \quad CH_3-C(:O:)-\overset{+}{N}(H)-H + Cl^-$$

Acetamide (an amide)	A (protonation on the amide oxygen)	B (protonation on the amide nitrogen)

Of the three contributing structures that can be drawn for cation A, structures A-1 and A-3 make the greater contributions to the hybrid because all atoms in each have complete octets; of these two contributors, A-3 has the positive charge on the less electronegative atom and, therefore, makes a greater contribution than A-1. The result is that the positive charge in cation A is delocalized over three atoms, the greater share of it being on nitrogen and oxygen.

A-1 (complete octets; positive charge on more electronegative atom)	A-2 (incomplete octet for C)	A-3 (complete octets; positive charge on less electronegative atom)

Only two contributing structures can be drawn for cation B. Of these, B-2 requires creation and separation of unlike charges and places positive charges on adjacent atoms. It therefore makes little contribution to the hybrid. Thus, the positive charge in cation B is essentially localized on the amide nitrogen.

B-1 (protonation on the amide nitrogen)	B-2 (adjacent plus charges; charge separation)

From this analysis, we conclude that as in the acetic acid example, proton transfer to the carbonyl oxygen of the amide group gives the more delocalized, and hence, the more stable cation, A.

Problem 4.3

Following is a structural formula for guanidine, the compound by which migratory birds excrete excess metabolic nitrogen. The hydrochloride salt of this compound is a white crystalline powder, freely soluble in water and ethanol.

(a) Write a Lewis structure for guanidine showing all valence electrons.

(b) Does proton transfer to guanidine occur preferentially to one of its —NH$_2$ groups (cation A) or to its =NH group (cation B)? Explain.

$$\underset{\textbf{Guanidine}}{\overset{\overset{\displaystyle NH}{\|}}{H_2N-C-NH_2}} + HCl \;\rightleftharpoons\; \underset{A}{\overset{\overset{\displaystyle NH}{\|}}{H_2N-C-NH_3^+}} \quad\text{or}\quad \underset{B}{\overset{\overset{\displaystyle NH_2^+}{\|}}{H_2N-C-NH_2}} + Cl^-$$

C. π Electrons as Brønsted-Lowry Bases

Thus far we have considered proton transfer to atoms having a nonbonding pair of electrons. Proton-transfer reactions also occur with compounds having π electrons (e.g., the π electrons of carbon-carbon double and triple bonds). The π electrons of the carbon-carbon double bond of 2-butene, for example, react with strong acids such as H$_2$SO$_4$, H$_3$PO$_4$, HCl, HBr, and HI by proton transfer to form a new carbon-hydrogen bond.

$$CH_3-CH=CH-CH_3 + H-\ddot{B}r: \;\rightleftharpoons\; CH_3-\overset{+}{C}-\underset{\underset{H}{|}}{\overset{\overset{H}{|}}{C}}-CH_3 + :\ddot{\underset{..}{B}r}:^-$$

2-Butene　　　　　　　　　　**sec-Butyl cation**
　　　　　　　　　　　　　　　　(a 2° carbocation)

The result of this proton-transfer reaction is formation of a **carbocation**, a species in which one of its carbons has only six electrons in its valence shell and carries a charge of +1. Because the carbon bearing the positive charge in the *sec*-butyl cation has only two other carbons bonded to it, it is classified as a secondary (2°) carbocation. We will study the formation, structure, and reactions of carbocations in detail in Chapter 6.

Example 4.4

The acid-base reaction between 2-methyl-2-butene and HI can in principle form two carbocations. Write chemical equations for the formation of each carbocation and use curved arrows to show the proton transfer in each reaction.

$$\underset{\textbf{2-Methyl-2-butene}}{CH_3-\underset{\underset{CH_3}{|}}{C}=CH-CH_3}$$

Solution

Proton transfer to carbon 3 of this alkene gives a tertiary (3°) carbocation. Proton transfer to carbon 2 gives a secondary (2°) carbocation.

$$CH_3-\underset{\underset{CH_3}{|}}{C}=CH-CH_3 + H-\ddot{I}:$$

$$\longrightarrow H_3C-\overset{+}{C}-\underset{\underset{H_3C}{|}\,\underset{H}{|}}{\overset{\overset{H}{|}}{C}}-CH_3 + :\ddot{\underset{..}{I}}:^-$$

A 3° carbocation

$$\longrightarrow H_3C-\underset{\underset{H_3C}{|}}{\overset{\overset{H}{|}}{C}}-\overset{+}{\underset{\underset{H}{|}}{C}}-CH_3 + :\ddot{\underset{..}{I}}:^-$$

A 2° carbocation

Problem 4.4

Write an equation to show the proton transfer between each alkene or cycloalkene and HCl. Where two carbocations are possible, show each.

(a) $CH_3CH_2CH\!=\!CHCH_3$

2-Pentene

(b)

Cyclohexene

4.3 | Acid Dissociation Constants, pK_a, and the Relative Strengths of Acids and Bases

Any quantitative measure of the acidity of organic acids or bases involves measuring the equilibrium concentrations of the various components in an acid-base equilibrium. The strength of an acid is then expressed by an equilibrium constant. The dissociation (ionization) of acetic acid in water is given by the following equation.

$$CH_3\overset{\displaystyle O}{\overset{\displaystyle \|}{C}}OH \ + \ H_2O \ \rightleftharpoons \ CH_3\overset{\displaystyle O}{\overset{\displaystyle \|}{C}}O^- \ + \ H_3O^+$$

Acetic acid   **Water**   **Acetate ion**   **Hydronium ion**

We can write an equilibrium expression for the dissociation of this or any other uncharged acid in a more general form; dissociation of the acid, HA, in water gives an anion, A^-, and the hydronium ion, H_3O^+. The equilibrium constant for this ionization is

$$HA + H_2O \rightleftharpoons A^- + H_3O^+$$

$$K_{eq} = \frac{[H_3O^+][A^-]}{[HA][H_2O]}$$

Because water is the solvent for this reaction and its concentration changes very little when HA is added to it, we can treat the concentration of water as a constant equal to 1000 g/L or approximately 55.6 mol/L. We can then combine these two constants (K_{eq} and the concentration of water) to define a new constant called an **acid dissociation constant**, given the symbol K_a.

Acid dissociation constant
Equal to the equilibrium constant (K_{eq}) for an acid dissociation reaction multiplied by the concentration of water $[H_2O]$.

$$K_a = K_{eq}[H_2O] = \frac{[H_3O^+][A^-]}{[HA]}$$

Because dissociation constants for most acids, including organic acids, are numbers with negative exponents, acid strengths are often expressed as **pK_a** ($-\log_{10}K_a$). The pK_a for acetic acid is 4.76, which means that acetic acid is a weak acid because the major species present in aqueous solution is undissociated CH_3COOH. Table 4.1 gives names, molecular formulas, and values of pK_a for some organic and inorganic acids. As you study the information in this table, note the following relationships:

$$pK_a = -\log_{10}K_a$$

- The larger the value of pK_a, the weaker the acid.
- The smaller the value of pK_a, the stronger the acid.
- The weaker the acid, the stronger its conjugate base.
- The stronger the acid, the weaker its conjugate base.

Table 4.1 pKₐ Values for Some Organic and Inorganic Acids

	Acid	Formula	pK_a	Conjugate Base	
Weaker acid	Ethane	CH_3CH_3	51	$CH_3CH_2^-$	Stronger conjugate base
	Ethylene	$CH_2=CH_2$	44	$CH_2"CH^-$	
	Ammonia	NH_3	38	NH_2^-	
	Hydrogen	H_2	35	H^-	
	Acetylene	$HC\equiv CH$	25	$HC\equiv C^-$	
	Ethanol	CH_3CH_2OH	15.9	$CH_3CH_2O^-$	
	Water	H_2O	15.7	HO^-	
	Methylammonium ion	$CH_3NH_3^+$	10.64	CH_3NH_2	
	Bicarbonate ion	HCO_3^-	10.33	CO_3^{2-}	
	Phenol	C_6H_5OH	9.95	$C_6H_5O^-$	
	Ammonium ion	NH_4^+	9.24	NH_3	
	Hydrogen sulfide	H_2S	7.04	HS^-	
	Carbonic acid	H_2CO_3	6.36	HCO_3^-	
	Benzoic acid	C_6H_5COOH	4.19	$C_6H_5COO^-$	
	Hydrogen fluoride	HF	3.2	F^-	
	Phosphoric acid	H_3PO_4	2.1	$H_2PO_4^-$	
	p-Toluenesulfonic acid	$CH_3C_6H_4SO_3H$	0.7	$CH_3C_6H_4SO_3^-$	
	Nitric acid	HNO_3	−1.5	NO_3^-	Weaker conjugate base
Stronger acid	Hydronium ion	H_3O^+	−1.74	H_2O	
	Sulfuric acid	H_2SO_4	−5.2	HSO_4^-	
	Hydrogen chloride	HCl	−7	Cl^-	
	Hydrogen bromide	HBr	−8	Br^-	
	Hydrogen iodide	HI	−9.9	I^-	

Example 4.5

For each value of pK_a, calculate the corresponding value of K_a. Which compound is the stronger acid?

(a) Ethanol, $pK_a = 15.9$ **(b)** Carbonic acid, $pK_a = 6.36$

Solution

Recall that $pK_a = -\log_{10}K_a$ so calculating $K_a = 10^{-pK_a}$ values gives:

(a) For ethanol, $K_a = 1.3 \times 10^{-16}$ **(b)** For carbonic acid, $K_a = 4.4 \times 10^{-7}$

Because the value of pK_a for carbonic acid is smaller than that for ethanol, carbonic acid is the stronger acid, and ethanol is the weaker acid.

Problem 4.5

For each value of K_a, calculate the corresponding value of pK_a. Which compound is the stronger acid?

(a) Acetic acid, $K_a = 1.74 \times 10^{-5}$ **(b)** Chloroacetic acid, $K_a = 1.38 \times 10^{-3}$

    The pK_a values can also be used to readily estimate the equilibrium constants (K_a). For example, if the pK_a of an acid is near zero, then the equilibrium constant for the reaction of that acid protonating water is near 1. Negative pK_a values correlate to acids with equilibrium constants greater than 1, while positive pK_a values are for acids with equilibrium constants less than 1. Each single unit difference between pK_a values represents a tenfold increase or decrease in the strength of the acids being compared.

Values of pK_a in aqueous solution in the range 2 to 12 can be measured quite accurately. Values of pK_a smaller than 2 are less accurate because very strong acids, such as HCl, HBr, and HI, are completely ionized in water, and the only acid present in solutions of these acids is H_3O^+. For acids too strong to be measured accurately in water, less basic solvents such as acetic acid or mixtures of water and sulfuric acid are used. Although none of the halogen acids, for example, is completely ionized in acetic acid, HI shows a greater degree of ionization in this solvent than either HBr or HCl; therefore, HI is the strongest acid of the three. Values of pK_a greater than 12 are also less precise. For bases too strong to be measured in aqueous solution, more basic solvents such as liquid ammonia and dimethyl sulfoxide are used. Because different solvent systems are used to measure relative strengths at either end of the acidity scale, pK_a values smaller than 2 and greater than 12 should be used only in a qualitative way when comparing them with values in the middle of the scale.

4.4 | The Position of Equilibrium in Acid-Base Reactions

We know from the value of pK_a for an acid whether an aqueous solution of the acid contains more molecules of the undissociated acid or its ions. A negative pK_a value indicates that the majority of molecules of the acid are dissociated in water, while a positive value indicates that most acid molecules remain undissociated in water. HCl, for example, a strong acid with a pK_a of -7, is almost completely dissociated at equilibrium in aqueous solution, and the major species present are H_3O^+ and Cl^-.

$$HCl + H_2O \rightleftharpoons H_3O^+ + Cl^- \qquad pK_a = -7$$

For acetic acid on the other hand, which is a weak acid with a pK_a of 4.76, the major species present at equilibrium in aqueous solution are CH_3COOH molecules.

$$CH_3\overset{\overset{\text{O}}{\|}}{C}OH + H_2O \rightleftharpoons CH_3\overset{\overset{\text{O}}{\|}}{C}O^- + H_3O^+ \qquad pK_a = 4.$$

In these acid-base reactions, water is the base. But what if we have a base other than water as the proton acceptor, or what if we have an acid other than hydrogen chloride or acetic acid as the proton donor? How do we determine quantitatively or even qualitatively which species are present at equilibrium; that is, how do we determine where the position of equilibrium lies?

Let us take as an example the acid-base reaction of acetic acid and ammonia to form acetate ion and ammonium ion.

$$CH_3\overset{\overset{\text{O}}{\|}}{C}OH \;+\; NH_3 \;\rightleftharpoons\; CH_3\overset{\overset{\text{O}}{\|}}{C}O^- \;+\; NH_4^+$$

Acetic acid　　**Ammonia**　　　**Acetate ion**　　**Ammonium ion**

There are two acids present in this equilibrium, acetic acid and ammonium ion. There are also two bases present, ammonia and acetate ion. A way to analyze this equilibrium is to view it as a competition between the two bases, ammonia and acetate ion, for a proton. The question we then ask is "Which of these is the stronger base?" The information we need to answer this question is in Table 4.1. We first determine which conjugate acid is the stronger acid and couple this with the fact that the stronger the acid, the weaker its conjugate base. From Table 4.1, we see that acetic acid, pK_a 4.76, is the stronger acid, which means that CH_3COO^- is the weaker base. Conversely, ammonium ion, pK_a 9.24, is the weaker acid, which means than NH_3 is the stronger base. We can now label the relative strengths of each acid and base.

$$CH_3\overset{\overset{\text{O}}{\|}}{C}OH \;+\; NH_3 \;\rightleftharpoons\; CH_3\overset{\overset{\text{O}}{\|}}{C}O^- \;+\; NH_4^+$$

Acetic acid　　　**Ammonia**　　　**Acetate ion**　　　**Ammonium ion**

pK_a 4.76 (stronger acid)　　(stronger base)　　(weaker base)　　pK_a 9.24 (weaker acid)

Not only can we use values of pK_a for each acid to estimate the position of equilibrium, but we can also use them to calculate an equilibrium constant for the acid-base reaction. A series of mathematical manipulations allows us to do so. For example, consider the following acid-base reaction:

$$HA \quad + \quad B \quad \rightleftharpoons \quad A^- \quad + \quad BH^+$$

| Acid | Base | Conjugate base of HA | Conjugate acid of B |

The equilibrium constant for this reaction is

$$K_{eq} = \frac{[A^-][BH^+]}{[HA][B]}$$

Multiplying the right-hand side of the equation by $[H_3O^+]/[H_3O^+]$ gives a new expression, which upon rearrangement becomes the K_a of acid HA divided by the K_a of acid BH^+:

$$K_{eq} = \frac{[A^-][BH^+]}{[HA][B]} \times \frac{[H_3O^+]}{[H_3O^+]} = \frac{[A^-][H_3O^+]}{[HA]} \times \frac{[BH^+]}{[B][H_3O^+]} = \frac{K_{HA}}{K_{HB^+}}$$

Upon taking the logarithm of each side of this equation, we arrive at this expression:

$$pK_{eq} = pK_{HA} - pK_{BH^+}$$

Thus, if we know the pK_a values of each acid in the equilibria, we can calculate the equilibrium constant for the acid-base reaction using these steps:

1. Look up the pK_a values of the acid and conjugate acid involved in the reactants and the products, respectively, of an acid-base reaction.
2. Subtract the pK_a of the conjugate acid (product) from the pK_a of the acid (reactant).
3. Because $pK_{eq} = -\log_{10}K_{eq}$, we take the antilog of $-pK_{eq}$ and to arrive at K_{eq}.

Example: Let's calculate the K_{eq} for the reaction of acetic acid with ammonia.

$$CH_3CO_2H \quad + \quad NH_3 \quad \rightleftharpoons \quad CH_3CO_2^- \quad + \quad NH_4^+$$

$pK_a = 4.76$ $pK_a = 9.24$

Following Steps 1 through 3, we have $pK_{eq} = 4.76 - 9.24 = -4.48$, so $K_{eq} = 3.0 \times 10^4$.

Because we know that acid-base reactions are favored when the stronger acid reacts with the stronger base to give the weaker acid and the weaker base, we can conclude that the equilibrium for the reaction between acetic acid and ammonia lies to the right. Using the mathematical approach just developed we can calculate that the preference to the right is 3.0×10^4. This means that if we started with equal amounts of acetic acid and ammonia, the reaction would prefer the products 30,000 to 1.

The most important concept is the following:

In an acid-base reaction, the position of equilibrium always favors reaction of the stronger acid and stronger base to form the weaker acid and weaker base.

Thus, at equilibrium, the major species present are the weaker acid and weaker base. Therefore, in the reaction between acetic acid and ammonia, the equilibrium lies to the right, and the major species present are acetate ion and ammonium ion.

The pH of living cells is generally between 7.0 and 8.5, a range often referred to as physiological pH. At physiological pH, several of the common functional groups found in biological molecules are ionized because they are either acids that are deprotonated or bases that are protonated. A good rule of thumb is that an acid will be substantially deprotonated if its pK_a is two or more units lower than the pH of the solution. A carboxyl group (the functional group of carboxylic acids) is present in all amino acids, as well as in the side chains of glutamic acid and aspartic acid. The pK_a values for carboxylic acids are typically between 4 and 5. At pH values of 7 or above, carboxylic acids are essentially fully deprotonated and therefore anionic.

$$R-\overset{\overset{\displaystyle O}{\|}}{C}-OH \qquad R-\overset{\overset{\displaystyle O}{\|}}{C}-O^-$$

Carboxylic acid   Form present at
pK_a 4–5   physiological pH

Another functional group found in biomolecules is the phosphodiester group. This group is found as part of the backbone of nucleic acids such as DNA and RNA. The pK_a values for phosphodiesters are between 1 and 3.

$$R-O-\overset{\overset{\displaystyle OH}{|}}{\underset{\underset{\displaystyle O}{\|}}{P}}-O-R' \qquad R-O-\overset{\overset{\displaystyle O^-}{|}}{\underset{\underset{\displaystyle O}{\|}}{P}}-O-R'$$

Phosphodiester   Form present
pK_a 1–3   at physiological pH

At physiological pH values, the phosphodiester group will be present in its anion form. Therefore, the backbones of nucleic acids (in which there is a repeating pattern of phosphodiester groups) will be polyanionic, a factor that has a major influence on their overall properties.

There are also basic functional groups in biological molecules. A good rule of thumb here is that a base will be substantially protonated if the pK_a of its conjugate acid is two or more units higher than the pH of the solution. Two important examples include amino groups and guanidino groups. The pK_a values for the conjugate acids of amines and guanidines are about 9 to 11 and 13 to 14, respectively. As a result,

these groups are protonated and therefore positively charged at physiological pH.

$$R-NH_2 \qquad\qquad R-NH_3^+$$

An amine group pK_a 9–11
                            Form present at
                            physiological pH

$$R-NH-\overset{\overset{\displaystyle NH}{\|}}{C}-NH_2 \qquad R-NH-\overset{\overset{\displaystyle NH_2^+}{\|}}{C}-NH_2$$

A guanidine group pK_a 13–14
                            Form present at
                            physiological pH

An interesting case is the imidazole group, which comprises the side chain of the amino acid histidine. An imidazole group, whose conjugate acid has a pK_a between 6 and 7, will be present at physiological pH as a mixture of protonated and deprotonated forms. This ability to exist in both forms can be significant in situations in which proton transfer reactions are important for the function of the protein containing an imidazole group.

$$\text{(imidazole)}-R \;+\; H_3O^+ \;\rightleftharpoons\; \text{(imidazole-H}^+)-R \;+\; H_2O$$

Imidazole group pK_a 6–7
                            Conjugate acid of
                            imidazole

Finally, no highlight section on the acid-base properties of biological molecules would be complete without a discussion of amino acids. At physiological pH, both the amino and carboxyl groups are ionized. Free amino acids are found in a number of situations in organisms, for example as neurotransmitters in mammals.

$$\underset{H_2N}{\overset{\displaystyle R}{\underset{\displaystyle}{\overset{|}{\underset{H}{C}}}}}\diagdown COOH \qquad \underset{^+H_3N}{\overset{\displaystyle R}{\overset{|}{\underset{H}{C}}}}\diagdown COO^-$$

An amino acid contains   Ionized or zwitterion
an amino group and a   form present at
carboxyl group   physiological pH

Example 4.6

Predict the position of equilibrium, and calculate the equilibrium constant, K_{eq}, for each acid-base reaction.

(a) C_6H_5OH $\quad + \quad$ HCO_3^- $\quad \rightleftharpoons \quad$ $C_6H_5O^-$ $\quad + \quad$ H_2CO_3

$\quad$ **Phenol** $\qquad\qquad$ **Bicarbonate ion** $\qquad\quad$ **Phenoxide ion** $\quad$ **Carbonic acid**

(b) $HC{\equiv}CH$ $\quad + \quad$ NH_2^- $\quad \rightleftharpoons \quad$ $HC{\equiv}C^-$ $\quad + \quad$ NH_3

$\quad$ **Acetylene** $\qquad\qquad$ **Amide ion** $\qquad\qquad$ **Acetylide ion** $\qquad$ **Ammonia**

Solution

(a) Carbonic acid is the stronger acid; the position of this equilibrium lies to the left. Phenol does not transfer a proton to bicarbonate ion to form carbonic acid.

$$C_6H_5OH + HCO_3^- \rightleftharpoons C_6H_5O^- + H_2CO_3 \qquad pK_{eq} = 9.95 - 6.36 = 3.59$$

$\quad$ pK_a 9.95 $\qquad\qquad\qquad\qquad\quad$ pK_a 6.36 $\qquad$ $K_{eq} = 10^{-3.59} = 2.57 \times 10^{-4}$
$\quad$ Weaker acid $\qquad\qquad\qquad\qquad$ Stronger acid

(b) Acetylene is the stronger acid; the position of this equilibrium lies to the right.

$$HC{\equiv}CH + NH_2^- \rightleftharpoons HC{\equiv}C^- + NH_3 \qquad pK_{eq} = 25 - 38 = -13$$

$\qquad$ pK_a 25 $\qquad\qquad\qquad\qquad$ pK_a 38 $\qquad$ $K_{eq} = 10^{13}$
$\qquad$ Stronger acid $\qquad\qquad\qquad$ Weaker acid

Problem 4.6

Predict the position of equilibrium, and calculate the equilibrium constant, K_{eq}, for each acid-base reaction.

(a) CH_3NH_2 $\quad + \quad$ CH_3COOH $\quad \rightleftharpoons \quad$ $CH_3NH_3^+$ $\quad + \quad$ CH_3COO^-

$\quad$ **Methylamine** $\qquad$ **Acetic acid** $\qquad\quad$ **Methylammonium** $\quad$ **Acetate**
$\qquad\qquad\qquad\qquad\qquad\qquad\qquad\qquad\qquad$ **ion** $\qquad\qquad\quad$ **ion**

(b) $CH_3CH_2O^-$ $\quad + \quad$ NH_3 $\quad \rightleftharpoons \quad$ CH_3CH_2OH $\quad + \quad$ NH_2^-

$\quad$ **Ethoxide ion** $\qquad$ **Ammonia** $\qquad\qquad$ **Ethanol** $\qquad\qquad$ **Amide ion**

4.5 | Thermochemistry and Mechanisms of Acid-Base Reactions

The transfer of a proton from an acid to a base is among the simplest of all chemical reactions that are discussed in this book. It is therefore an excellent reaction to examine as a means of introducing the concepts of thermochemistry and reaction mechanisms.

Reaction mechanism
A step-by-step description of how a chemical reaction occurs.

A **reaction mechanism** describes in detail how a reaction occurs. It describes which bonds are broken and which ones are formed, as well as the order and relative rates of the various bond-breaking and bond-forming steps. A complete mechanism ideally describes the positions of all atoms and the energy of the entire system during each and every instant of the reaction. This ideal, however, is rarely ever achieved in practice.

Thermochemistry
The study of the energy of chemical structures.

Thermochemistry is the study of the energy of the entire system at each and every instant of a reaction. It is therefore part of the information gained when deciphering a complete mechanism. Before examining the mechanisms and thermochemistry of acid-base reactions, we must examine what causes chemical reactions to occur, and introduce some basic energy principles that are key to chemical reactivity.

A. Thermal Reactions and Transition States

Most chemical reactions occur via collisions. Imagine placing two chemicals, A—H and B⁻, in a reaction vessel along with a solvent. Compounds A—H and B⁻ will move through the solvent by jostling around, hitting and bouncing off individual solvent molecules. On occasion, A—H and B⁻ will collide with one another. The velocity at which the molecules move through the vessel is proportional to their kinetic energy ("kinetic" meaning motion). Higher-energy collisions occur between molecules possessing larger kinetic energy. During collision, the structures of the molecule contort and flex, and collisions of higher energy lead to larger distortions in structure. Analogously, a head-on collision is more damaging between automobiles moving at 65 mph than at 35 mph. At higher temperatures, the energy of the collisions is greater because the molecules are moving more rapidly. Reactions that result by virtue of the kinetic energy put into a reaction vessel due to temperature are called **thermal reactions**.

During the collisions, the kinetic energy of the reactants is converted into potential energy. The potential energy is stored in the chemical structures in the form of the structural strains, and the energy is released as the molecules again adopt their optimal geometries (see Section 2.5 for a discussion of strain). This situation is analogous to jumping up and down on the branch of a tree. Let's say the most stable geometry of the branch is horizontal. Each jump (collision) strains the branch by bending away from the optimal horizontal geometry. If you jump hard enough, the branch will break. Similarly, if molecules collide with enough force, bonds will break. Further, some collisions cause distortions that lead to rearrangements of bonds.

During a collision process that yields a reaction, a structure forms that is called the **transition state** and is also commonly called the activated complex. It has a particular geometry possessing partially broken and partially formed bonds, and it is so strained that it transitions to new structures that are less strained. The structure of a bent tree branch that finally cannot take any more strain and therefore snaps is a good analogy to the transition state in chemical reactions. We will talk more about the transition state in the next section.

Transition state
The highest energy point on a reaction coordinate diagram. The chemical structure at this point is commonly called the activated complex.

B. Reaction Coordinate Diagrams and Thermochemistry

Chemists use **reaction coordinate diagrams** to show the changes in energy for the molecules involved in a chemical reaction. Energy is measured on the vertical axis and the change in position of the atoms during the reaction is represented on the horizontal axis, called the **reaction coordinate**. The reaction coordinate is not a time axis; rather it shows how far the reaction has progressed in terms of structural changes. Figure 4.1 shows a reaction coordinate diagram for the reaction of acid H—A and base B⁻ (reactants) to form H—B and A⁻ (products). Wells on the plots represent stable structures that have lifetimes, while other points along the curve represent unstable structures that cannot be isolated. The transition state exists at the highest energy point on a reaction coordinate diagram. Transition states last less than one picosecond. The example given in Figure 4.1 occurs in one step, meaning that bond breaking in the reactants and bond formation in the products occur simultaneously upon collision. Every point along the reaction coordinate that is not reactants or products represents a chemical structure that has characteristics of both the reactants and products. Positions along the x-axis close to the reactants represent structures more resembling the reactants, while positions close to the products resemble the products. Any collisions that raise the energy of the reactants (moving toward the products from the reactants in the diagram) but do not raise the energy enough to achieve the transition state simply result in glancing blows. In contrast, most collisions that possess enough energy to achieve the transition state result in passing of the chemical structures over this point. The resulting entities relax to the structure of the products with a release of energy.

Reaction coordinate diagram
A graph showing the energy changes that occur during a chemical reaction; energy is plotted on the vertical axis while the reaction progress is plotted along the horizontal axis.

Figure 4.1

A reaction coordinate diagram for a one-step reaction between H—A and B⁻. The dashed lines in the transition state indicate that the H—A bond is partially broken and the new H—B bond is partially formed. On completion of the reaction, the H—A bond is fully broken and the H—B bond is fully formed. The energy of the reactants is higher than that of the products. The energy axis in this diagram is Gibbs free energy ($G°$).

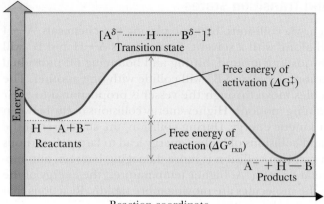

Gibbs free energy change, $\Delta G°$
The energy that dictates the position of chemical equilibria and rates of chemical reactions. It is a thermodynamic function of enthalpy, entropy, and temperature, given by the equation $\Delta G° = \Delta H° - T\Delta S°$. The position of the equilibria for the reaction favors the product(s) if $\Delta G° < 0$ and favors the reactant(s) if $\Delta G° > 0$.

Thermodynamics is the study of the relative energies between any two entities on a reaction coordinate diagram that are shown in wells, such as the energy between the reactants and products in Figure 4.1. The energy that chemists plot on the vertical axis of a reaction coordinate diagram can vary. For reactions at constant pressure, the Gibbs free energy ($G°$) is often used. The naught superscript (°) indicates standard states (298 K and 1 atm). This kind of energy controls the rates and equilibrium of reactions. For example, a **change in Gibbs free energy** ($\Delta G°$ or $\Delta G°_{rxn}$) between the reactants and products is related to the equilibrium constant through this equation:

$$\Delta G° = -RT \ln K_{eq},$$

where $R = 8.31$ J/K mol and T is the temperature in kelvins.

Example 4.7

Calculate the $\Delta G°$ for a reaction with a K_{eq} of 1.0×10^3 at 298 K.

Solution

$\Delta G° = -(8.31 \text{ J/K mol}) \times 298 \text{ K} \times \ln(1.0 \times 10^3) = -17.1$ kJ/mol (-4.09 kcal/mol)

Problem 4.7

Calculate K_{eq} for a reaction with $\Delta G° = -17.1$ kJ/mol (-4.09 kcal/mol) at 328 K. Compare this value to the 1×10^3 seen at 298 K.

The Gibbs free energy difference between the reactants and the transition state, $\Delta G^{\ddagger}$, is called the Gibbs **free energy of activation** (Figure 4.1). It controls the rate of the reaction and the ability of the reactants to achieve the transition state. Because transition states are higher in energy than the reactants, $\Delta G^{\ddagger}$ is always a positive value. If it is easy to achieve the transition state, $\Delta G^{\ddagger}$ will be a small value and the reaction will be fast, but if it is difficult to achieve the transition state, $\Delta G^{\ddagger}$ will be relatively large and the reaction will be slow. **Kinetics** is the study of the rates of chemical reactions. Being able to surmount the barrier means that there are obtainable collisions that transition the reactants to the products (that is, a pathway for the reaction exists).

A fundamental principle of nature is that all systems will spontaneously attain a lower energy state if a pathway is available. To analyze reactions, chemists often look at the difference between the energy of the products ($G°_P$) and the energy of the reactants ($G°_R$).

$$\Delta G° = G°_P - G°_R$$

If the products are more stable (a lower numerical energy value), the change in energy is negative. A negative $\Delta G°$ (called an **exergonic** reaction) is favorable, while a positive $\Delta G°$ (called an **endergonic** reaction) is unfavorable. The more negative the value of $\Delta G°$, the greater the driving force for the reaction to occur. Lastly, there are some simple qualitative relationships between Gibbs free energy changes and equilibrium constants for favorable and unfavorable reactions; when $\Delta G°$ is greater than 0, K_{eq} is less than 1 (unfavorable); when $\Delta G° = 0$, $K_{eq} = 1$; when $\Delta G°$ is less that 0, K_{eq} is greater than 1 (favorable).

The change in Gibbs free energy is a function of the change in two terms: enthalpy and entropy, as expressed by the Gibbs-Helmholtz equation:

$$\Delta G° = \Delta H° - T\Delta S°.$$

Enthalpy ($H°$) is the energy contained in chemical bonds and solvation. The difference in enthalpy between reactants and products is called the **heat of reaction** ($\Delta H°$). It reflects differences in bond strengths and solvation. If the bonds formed in the product(s) are stronger than the bonds in the reactant(s), heat will be released, the solution will warm, and the reaction is called **exothermic** ($\Delta H°$ is negative). Conversely, if the bonds formed in the product(s) are weaker, heat will be absorbed from the solution so it cools, and the reaction is **endothermic** ($\Delta H°$ is positive). Although $\Delta G°$ dictates rates and equilibria, enthalpy is commonly plotted along the Y-axes of reaction coordinate diagrams. This is because enthalpy reflects the intrinsic stability of the chemical structures involved in the reaction. Structures that have stronger bonds, and/or that are better solvated, are more stable and thereby have lower enthalpy.

Entropy (S) measures disorder versus order, and disorder is favorable. It is a fundamental principle of nature that, given time, everything degrades (the second law of Thermodynamics). This is because all systems become more stable as the number of freely moving particles increases and the chaotic movement of the particles increases. Hence, when molecules fragment, and/or when molecular vibrations, such as bond rotations, increase their freedom of movement, entropy becomes more favorable. A favorable change in entropy ($\Delta S°$) is defined as a positive value and represents an increase in disorder, whereas an increase in order is reflected by a negative value. Because entropy is multiplied by temperature in the Gibbs-Helmholtz equation, and bond cleavage is associated with a positive $\Delta S°$, high temperature often leads to a negative $\Delta G°$ and spontaneous bond breaking. This occurs even if the reaction is endothermic by virtue of the breaking of bonds. Cracking of petroleum (Section 2.9B) is an example of a reaction that occurs with an increase in entropy. The relationship among $\Delta G°$, $\Delta H°$, and $\Delta S°$ and the position of equilibrium in chemical reactions are summarized in Table 4.2.

It is important to note that there is commonly a correlation between the Gibbs free energy of reaction ($\Delta G°$) and the Gibbs free energy of activation ($\Delta G^{\ddagger}$). Often,

Exergonic reaction
A reaction in which the Gibbs free energy of the products is lower than that of the reactants. The position of equilibrium for an exergonic reaction favors products.

Endergonic reaction
A reaction in which the Gibbs free energy of the products is higher than that of the reactants. The position of equilibrium for an endergonic reaction favors starting materials.

Enthalpy change, $\Delta H°$
The difference in total bond strengths and solvation between various points under comparison on a reaction coordinate diagram.

Heat of reaction, $\Delta H°$
The difference in the enthalpy between the reactants and products.

Exothermic reaction
A reaction in which the enthalpy of the products is lower than the enthalpy of the reactants; heat will be released.

Endothermic reaction
A reaction in which the enthalpy of the products is higher than the enthalpy of the reactants; heat will be absorbed.

Table 4.2	Relationship Among $\Delta G°$, $\Delta H°$, $\Delta S°$	
	$\Delta S° < 0$	$\Delta S° > 0$
$\Delta H° > 0$	$\Delta G° > 0$; the position of equilibrium favors reactants	At higher temperatures when $T\Delta S° < \Delta H°$ and $\Delta G° < 0$, the position of equilibrium favors products
$\Delta H° < 0$	At lower temperatures when $T\Delta S° < \Delta H°$ and $\Delta G° < 0$; the position of equilibrium favors products	$\Delta G° < 0$; the position of equilibrium favors products

(a)

Computed

(b)

Cartoon

Figure 4.2
Calculated mesh diagram of an antibonding O—H orbital and a cartoon representation. Both pictures display that the orbital is primarily on the hydrogen.

more favorable reactions occur more quickly. Yet, there is no strict rule that correlates kinetics with thermodynamics, and many exceptions are seen. For example, diamond is a much more stable form of carbon than graphite, but we do not observe graphite spontaneously transforming into diamond. There is a strong energy preference to do so ($\Delta G° \ll 0$), but there is no accessible mechanism or pathway for this pathway to occur ($\Delta G^\ddagger \gg 0$), hence the reaction is impossibly slow.

C. Mechanism and Thermochemistry of Acid-Base Reactions

We will use the concepts just set forth in many subsequent chapters of this book. Yet, now we can use them to better understand acid-base chemistry. Let's first focus on issues that affect the rate of an acid-base reaction.

For an acid to transfer its proton to a base, the two entities must collide. Moreover, they must collide with a specific geometry. The trajectory of the base (B^-) must be toward the H of H—A, not toward the A group. Further, it makes sense that there would be an optimal trajectory for the approach of H—A, and B^- that will lead to the lowest-energy pathway for passing the proton from the acid to the base. In this case a linear approach is optimal. The transition state has a linear A—H—B bond angle and the proton is partially shared between both A and B (look back at Figure 4.1).

This optimal trajectory for the collision of H—A and B^- can be rationalized by examination of an A—H antibonding orbital, and Figure 4.2 shows a specific example of this kind of orbital for an O—H bond of an alcohol (ROH). The orbital has greatest magnitude on H and it lies along the O—H axis. This orbital will be most efficiently filled with a lone pair of electrons from the base colliding with the H along the H—O axis, which will lead to cleavage of the O—H bond. As depicted by the arrow pushing that describes the mechanism, the lone pair on the base can be considered as the electron pair that forms the new H—B bond, while the electron pair in the O—H bond can be considered as the electrons that create the lone pair on O^-.

$$R—\overset{..}{\underset{..}{O}}—H \; + \; :B^- \longrightarrow [R—\overset{..}{\underset{..}{O}}{}^{\delta-}\text{·······}H\text{·······}B^{\delta-}] \longrightarrow R—\overset{..}{\underset{..}{O}}{:}^- + H—B$$

The following sections of this chapter focus on predicting relative acidities, which is an analysis of thermodynamics. The focus will be on enthalpy because it measures the intrinsic stabilities of the acids and bases on both sides of the equilibria. We do not consider entropy because an acid and a base exist in both the reactants and the products, and therefore the number of molecules does not change during the reaction. Hence, enthalpy is a good predictor for acid-base reaction equilibria.

4.6 | Molecular Structure and Acidity

Now let us examine in some detail the relationships between molecular structure and acidity. The overriding principle in determining the relative acidities of uncharged organic acids is the stability of the conjugate base anion, A^-, resulting from loss of a proton; the more stable the anion, the greater the acidity of the acid, HA. As we discuss in this section, ways to stabilize A^- include

- Having the negative charge on a more electronegative atom
- Having the negative charge on a larger atom
- Delocalizing the negative charge as described by resonance contributing structures
- Spreading the negative charge onto electron-withdrawing groups by the inductive effect (polarization of sigma bonds)
- Having the negative charge in an orbital with more s character

A. Electronegativity of the Atom Bearing the Negative Charge

Let us consider the relative acidities of the following series of hydrogen acids, all of which are in the same period of the Periodic Table.

Acid					Conjugate base
Methanol pK_a ~16	$CH_3-\overset{..}{\underset{..}{O}}-H$			$CH_3-\overset{..}{\underset{..}{O}}:^-$	**Methoxide ion**
Methylamine pK_a ~38	$CH_3-\overset{\mid}{\underset{\mid}{N}}-H$ H			$CH_3-\overset{\mid}{N}:^-$ H	**Methylamide ion**
Ethane pK_a ~51	$CH_3-\overset{\overset{H}{\mid}}{\underset{\underset{H}{\mid}}{C}}-H$	Increasing acidity ↑	Increasing basicity ↓	$CH_3-\overset{\overset{H}{\mid}}{\underset{\underset{H}{\mid}}{C}}:^-$	**Ethyl anion**

The pK_a value for ethane is given in Table 4.1, but values for methylamine and methanol are not. We can, however, make good guesses about the pK_a values of these acids by reasoning that the nature of the alkyl group bonded to nitrogen or oxygen has only a relatively small effect on the acidity of the hydrogen bonded to the heteroatom (in organic chemistry, an atom other than carbon). Therefore, we estimate that the pK_a of methylamine is approximately the same as that of ammonia (pK_a 38), and that the pK_a of methanol is approximately the same as that of ethanol (pK_a 15.9). As we see, ethane is the weakest acid in this series, and ethyl anion is the strongest conjugate base. Conversely, methanol is the strongest acid, and methoxide ion is the weakest conjugate base.

The relative acidity within a period of the Periodic Table is related to the electronegativity of the atom in the anion that bears the negative charge. The greater the electronegativity of this atom, the more strongly its electrons are held, and the more stable the anion is. Conversely, the smaller the electronegativity of this atom, the less tightly its electrons are held, and the less stable the anion is. Oxygen, the most electronegative of the three atoms compared, has the largest electronegativity (3.5 on the Pauling scale), and methanol forms the most stable anion. Carbon, the least electronegative of the three (2.5 on the Pauling scale), forms the least stable anion. Because methanol forms the most stable anion, it is the strongest acid in this series. Ethane is the weakest acid in the series.

It is essential to understand that this argument based on electronegativity applies only to acids within the same period (row) of the Periodic Table. Anions of atoms within the same period have approximately the same size, and their energies of solvation are approximately the same.

B. Size of the Atom Bearing the Negative Charge

To illustrate how the acidity of hydrogen acids varies within a group (column) of the Periodic Table, let us compare the acidities of methanol and methanethiol, CH_3SH. We estimated in the previous section that the pK_a of methanol is around 16, which is pretty close to its measure pK_a value of 15.5. We can estimate the pK_a of methanethiol in the following way. The pK_a of hydrogen sulfide, H_2S, is given in Table 4.1 as 7.04. If we assume that substitution of a methyl group for a hydrogen makes only a slight change in acidity, we might estimate that the pK_a of methanethiol is approximately 7.0. Actually, the pK_a of methanethiol is around 10.4, the difference being a result of the different way in which the methanethiolate anion is solvated versus HS—. Nevertheless, methanethiol is the stronger acid, and methanethiolate ion is the weaker conjugate base.

$$\text{CH}_3\text{—}\overset{..}{\underset{..}{\text{S}}}\text{—H} \quad + \quad \text{CH}_3\text{—}\overset{..}{\underset{..}{\text{O}}}\text{:}^{-} \quad \rightleftharpoons \quad \text{CH}_3\text{—}\overset{..}{\underset{..}{\text{S}}}\text{:}^{-} \quad + \quad \text{CH}_3\text{—}\overset{..}{\underset{..}{\text{O}}}\text{—H}$$

Methanethiol	**Methoxide**	**Methanethiolate**	**Methanol**
pK_a 10.4	**ion**	**ion**	pK_a 15.5
(stronger acid)	(stronger base)	(weaker base)	(weaker acid)

The relative acidity of these two hydrogen acids, and, in fact, any set of hydrogen acids within a group of the Periodic Table, is related to the size of the atom bearing the negative charge. Size increases from top to bottom within a group because the valence electrons are in increasingly higher principal energy levels. This means that (1) they are farther from the nucleus, and (2) they occupy a larger volume of space. Because sulfur is below oxygen in the Periodic Table, it is larger than oxygen. Accordingly, the negative charge on sulfur in methanethiolate ion is spread over a larger volume of space (more delocalized, Section 4.2B); therefore, the CH_3S^- anion is more stable. The negative charge on oxygen in methoxide ion is confined to a smaller volume of space; therefore, the CH_3O^- anion is less stable.

CH_3S^- $\qquad\qquad$ CH_3O^-

We see this same trend in the strength of the halogen acids, HF, HCl, HBr, and HI, which increase in strength from HF (the weakest) to HI (the strongest). Of their anions, iodide ion is the largest; its charge is delocalized over the largest volume of space and, therefore, is the most stable. HI is the strongest of the halogen acids. Conversely, fluoride ion is the smallest anion; its charge is the most concentrated, and fluoride ion is the least stable. HF is, therefore, the weakest acid of the halogen acids.

This size trend that we use to rationalize that HI is the strongest and HF the weakest acid of the halogen acids runs contrary to what would be predicted from electronegativity arguments. Based on electronegativity, fluoride has the most stable negative charge because fluorine is the most electro-negative element. Therefore, you would predict that HF should be the strongest acid. This is one example, of many in chemistry, where there are opposing trends that predict reactivity, and one of the trends dominates what we observe experimentally.

We will return to the relative sizes of these ions in Chapter 9 at which point we will discuss the nature of their solvation in polar solvents. We will see that, for the ions of the same charge, the smaller the ion, the greater its solvation, and that the degree of solvation has a profound effect on their relative reactivities.

C. Delocalization of Charge in the Anion

Carboxylic acids are weak acids. Values of pK_a for most unsubstituted carboxylic acids fall within the range 4 to 5. The pK_a for acetic acid, for example, is 4.76. Values of pK_a for most alcohols, compounds that also contain an —OH group, fall within the range 15 to 18; the pK_a for ethanol, for example, is 15.9. Thus, alcohols are slightly weaker acids than water (pK_a = 15.7).

We account for the greater acidity of carboxylic acids compared with alcohols using the resonance model and looking at the relative stabilities of the alkoxide and carboxylate ions. Our guideline is this: the more stable the conjugate base anion, the farther the position of equilibrium is shifted toward the right, and the more acidic the compound is.

Here we take the acid ionization of an alcohol as a reference equilibrium.

$$CH_3CH_2\overset{..}{\underset{..}{O}}-H + H_2O \rightleftharpoons CH_3CH_2\overset{..}{\underset{..}{O}}{:}^- + H_3O^+$$

An alcohol   An alkoxide ion

Elpot of ethoxide ion

In the alkoxide anion, the negative charge is localized on oxygen. In contrast, ionization of a carboxylic acid gives an anion for which we can write two equivalent contributing structures that result in delocalization of the negative charge of the conjugate base anion. Because of this delocalization of negative charge, a carboxylate anion is more stable than an alkoxide anion. Therefore a carboxylic acid is a stronger acid than an alcohol.

Equivalent contributing structures; the carboxylate anion is stabilized by delocalization of the negative charge

Elpot of acetate ion

D. Inductive Effect and Electrostatic Stabilization of the Anion

We see an example of the **inductive effect** in alcohols in the fact that an electronegative substituent adjacent to the carbon bearing the —OH group increases the acidity of the alcohol. Compare, for example, the acidities of ethanol and 2,2,2-trifluoroethanol. The acid dissociation constant for 2,2,2-trifluoroethanol is larger than that of ethanol by more than three orders of magnitude, which means that the 2,2,2-trifluoroethoxide ion is considerably more stable than the ethoxide ion.

Inductive effect
The polarization of the electron density of a covalent bond caused by the electronegativity of a nearby atom.

Ethanol
pK_a 15.9

2,2,2-Trifluoroethanol
pK_a 12.4

We account for the increased stability of the 2,2,2-trifluoroethoxide ion in the following way. Fluorine is more electronegative than carbon (4.0 versus 2.5); therefore, the C—F bond has a significant dipole, indicated in the following figure by symbols to show the partial charges. There is an attractive stabilization by the interaction of the negatively charged oxygen and the partial positive charge on the carbon bearing the fluorines, which results in stabilization of the trifluoroethoxide ion.

Another way to think about the relative acidity of these two compounds is to realize that some of the negative charge from the oxygen atom of the anion has been delocalized onto the electronegative fluorine atoms, thereby stabilizing the charged species through delocalization of the charge.

The partial positive charge helps neutralize the negative charge on oxygen

Elpot of trifluoroethoxide ion

Stabilization by the inductive effect falls off rapidly with increasing distance of the electronegative atom(s) from the site of the negative charge. Compare, for example, the pK_a values of alcohols substituted with fluorine on carbons 2 versus 3 versus 4. When fluorine atoms are more than two carbons away from the carbon bearing the —OH group, they have almost no effect on acidity.

CF_3-CH_2-OH	$CF_3-CH_2-CH_2-OH$	$CF_3-CH_2-CH_2-CH_2-OH$
2,2,2-Trifluoroethanol	**3,3,3-Trifluoro-1-propanol**	**4,4,4-Trifluoro-1-butanol**
(pK_a 12.4)	(pK_a 14.6)	(pK_a 15.4)

We also see the operation of the inductive effect in the acidity of halogen-substituted carboxylic acids.

$$Cl-CH_2-\overset{\overset{\displaystyle O}{\|}}{C}-OH \qquad CH_3-\overset{\overset{\displaystyle O}{\|}}{C}-OH$$

Chloroacetic acid
pK_a 2.86

Acetic acid
pK_a 4.76

The pK_a values of these two acids indicate that chloroacetic acid is approximately two orders of magnitude more acidic than acetic acid. In the case of chloroacetate anion, the negative charge is stabilized by electrostatic interaction between the partial negative charges on the oxygens and the partial positive charge on the carbon bearing the chlorine atom.

The partial positive charge helps neutralize the negative charge on the oxygens

In the hybrid, charge is distributed equally between the two oxygens

Elpot of chloroacetate ion

As was the case with halogen substitution and the acidity of alcohols, the acid-enhancing effect of halogen substitution in carboxylic acids falls off rapidly with increasing distance between the point of substitution and the carboxyl group.

Butanoic acid	**4-Chlorobutanoic acid**	**3-Chlorobutanoic acid**	**2-Chlorobutanoic acid**
pK_a 4.82	pK_a 4.52	pK_a 3.98	pK_a 2.83

E. Hybridization and the Percent *s* Character of the Atom Bearing the Negative Charge

To see the effect of hybridization, we consider the case of two or more conjugate base anions, each with the same charge and same element bearing the charge. The only difference is the hybridization of the atom bearing the negative charge. The acidity of a hydrogen bound to a carbon atom of an alkane, an alkene, and an alkyne is of special importance to us.

One of the major differences between the chemistry of alkynes and that of alkenes and alkanes is that a hydrogen bonded to a triply bonded carbon atom is sufficiently acidic that it can be removed by a strong base, such as sodium amide, $NaNH_2$ or sodium hydride, NaH. Table 4.3 gives pK_a values for an alkyne, alkene, and alkane. Also given for comparison are values for ammonia and water.

We account for the greater acidity of alkynes in the following way. The lone pair of electrons on a carbon anion lies in a hybrid orbital: an sp orbital for an alkyne anion, an sp^2 orbital for an alkene anion, and an sp^3 orbital for an alkane anion. An sp orbital has 50% *s* character; an sp^2 orbital, 33%, and an sp^3 orbital, 25% (Section 1.7). Electrons in an *s* orbital are lower in energy than those in a *p* orbital; that is, they are held more tightly to the nucleus. Therefore, the more *s* character in a hybrid orbital, the more electronegative the atom will be, and the more acidic a hydrogen bonded to it will be (and the more stable the anion will be). Of the three types of compounds, the carbon in an alkyne (sp hybridized with 50% *s* character) is the most electronegative. Therefore, an alkyne anion is the most stable of the series, and an alkyne is the strongest acid of the series. By similar reasoning, the alkane carbon (sp^3 hybridized with 25% *s* character) is the least electronegative, and an alkane is the weakest acid of the series. An alkene anion, with 33% *s* character, is intermediate.

Table 4.3	Acidity of Alkanes, Alkenes, and Alkynes		
Weak Acid		**Conjugate Base**	**pK_a**
Water	HO—H	HO⁻	15.7
Alkyne	HC≡C—H	HC≡C⁻	25
Ammonia	H_2N—H	H_2N⁻	38
Alkene	CH_2=CH—H	CH_2=CH⁻	44
Alkane	CH_3CH_2—H	CH_3CH_2⁻	51

Increasing acidity →

4.7 | Lewis Acids and Bases

Gilbert N. Lewis, who proposed that covalent bonds are formed by sharing one or more pairs of electrons (Section 1.2A), further generalized the theory of acids and bases to include a group of substances not included in the Brønsted-Lowry concept. According to the Lewis definition, an acid is a species that can form a new covalent bond by accepting a pair of electrons; a base is a species that can form a new covalent bond by donating a pair of electrons. In the following general equation, the **Lewis acid**, A, accepts a pair of electrons in forming the new covalent bond and acquires a negative formal charge. The **Lewis base**, :B, donates the pair of electrons and acquires a positive formal charge.

Lewis acid
Any molecule or ion that can form a new covalent bond by accepting a pair of electrons.

Lewis base
Any molecule or ion that can form a new covalent bond by donating a pair of electrons.

Note that, although we speak of a Lewis base as "donating" a pair of electrons, the term is not fully accurate. Donating in this case does not mean that the electron pair under consideration is removed completely from the valence shell of the base. Rather, donating means that the electron pair becomes shared with another atom to form a covalent bond. Charged species can also be Lewis acids and Lewis bases. In these cases, Lewis acids have a positive charge and Lewis bases have a negative charge. If both reactants are equally but oppositely charged, a Lewis acid-Lewis base reaction will result in the formation of a new covalent bond in which the product has no charge.

$$A^+ \quad + \quad :B^- \quad \rightleftharpoons \quad A{-}B$$

<table>
<tr><td>Lewis acid</td><td>Lewis base</td><td>New covalent bond</td></tr>
<tr><td>(electron pair</td><td>(electron pair</td><td>formed in this Lewis</td></tr>
<tr><td>acceptor)</td><td>donor)</td><td>acid-base reaction</td></tr>
</table>

An example of this type of a Lewis acid-base reaction is that of a carbocation (a Lewis acid) with bromide ion, a Lewis base. The *sec*-butyl cation, for example, reacts with bromide ion to form 2-bromobutane.

sec-Butyl cation   **Bromide**   **2-Bromobutane**
(a carbocation)   **ion**   (racemic)

The Lewis concept of acids and bases includes proton-transfer reactions; all Brønsted-Lowry bases (proton acceptors) are also Lewis bases, and all Brønsted-Lowry acids (proton donors) are also Lewis acids. The Lewis model, however, is more general in that it is not restricted to proton-transfer reactions.

Consider the reaction that occurs when boron trifluoride gas is dissolved in diethyl ether.

Diethyl ether   **Boron trifluoride**   A BF$_3$-ether complex
(a Lewis base)   (a Lewis acid)

Elpot of   Elpot of BF$_3$
diethyl ether

Boron, a Group 3A element, has three electrons in its valence shell, and, after forming single bonds with three fluorine atoms to give BF$_3$, boron still has only six electrons in its valence shell. Because it has an empty orbital in its valence shell and can accept two electrons into it, boron trifluoride is electron deficient and, therefore, a Lewis acid. In forming the O—B bond, the oxygen atom of diethyl ether (a Lewis base) donates an electron pair, and boron accepts the electron pair. The reaction between diethyl ether and boron trifluoride is classified as an acid-base reaction according to the Lewis model, but because there is no proton transfer involved, it is not classified as an acid-base reaction by the Brønsted-Lowry model. Said another way, all Brønsted-Lowry acids are **protic acids**; Lewis acids may be protic acids or they may be **aprotic acids**.

As a final note we make a tie to terminology that will be used extensively in future chapters. In our analyses of many organic chemical reactions we will show that high-electron-density regions on molecules or ions react with low-electron-density regions of other molecules or ions, quite often resulting in the formation of a new covalent bond. We call the electron-rich species a **nucleophile** (nucleus-loving), meaning that it is seeking a region of low electron density (such as an atomic nucleus). We call the low-electron-density species an **electrophile** (electron-loving), meaning that it is seeking a region of high electron density. Nucleophiles are therefore analogous to Lewis bases and electrophiles are analogous to Lewis acids. Although chemists use the terms interchangeably, nucleophile and electrophile are most commonly used in kinetics discussions, while Lewis acid-base terminology is mostly used in thermodynamics discussions. We will talk about these connections in more detail throughout the book.

Protic acid
An acid that is a proton donor in an acid-base reaction.

Aprotic acid
An acid that is not a proton donor; an acid that is an electron pair acceptor in a Lewis acid-base reaction.

Example 4.8

Write an equation for the reaction between each Lewis acid-base pair, showing electron flow by means of curved arrows.

(a) $BF_3 + NH_3 \longrightarrow$ **(b)** $(CH_3)_2 CH^+ + Cl^- \longrightarrow$

Solution

(a) BF_3 has an empty orbital in the valence shell of boron and is the Lewis acid. NH_3 has an unshared pair of electrons in the valence shell of nitrogen and is the Lewis base. In this example, each of these atoms takes on a formal charge; the resulting structure, however, has no net charge.

Lewis acid   Lewis base

(b) The trivalent carbon atom in the isopropyl cation has an empty orbital in its valence shell and is, therefore, the Lewis acid. Chloride ion is the Lewis base.

Lewis acid   Lewis base

Problem 4.8

Write an equation for the reaction between each Lewis acid-base pair, showing electron flow by means of curved arrows.

(a) $(CH_3CH_2)_3B + OH^- \longrightarrow$ **(b)** $CH_3Cl + AlCl_3 \longrightarrow$

Section 4.1 Arrhenius Acids and Bases

By the Arrhenius definitions, an acid is a substance that dissolves in water to produce H^+ (H_3O^+) ions and a base is a substance that dissolves in water to produce OH^- ions.

Section 4.2 Brønsted-Lowry Acids and Bases

- A **Brønsted-Lowry acid** is a proton donor and a **Brønsted-Lowry base** is a proton acceptor.
- Any pair of molecules or ions that can be interconverted by transfer of a proton is called a **conjugate acid-base pair**.
  - Neutralization of an acid by a base is a proton-transfer reaction in which the acid is transformed into its **conjugate base** and the base is transformed into its **conjugate acid**.
- The more stable charged species is the one in which the charge is more delocalized.
  - Brønsted-Lowry bases with two or more receptor sites will be protonated on the site that gives the more delocalized charge.
- π electrons can act as Brønsted-Lowry bases to give cations, a reaction that is important in the chemistry of alkenes.

Section 4.3 Acid Dissociation Constants, pK_a, and the Relative Strengths of Acids and Bases

- A strong acid or strong base is one that is completely ionized in water and a weak acid or weak base is one that is only partially ionized in water.
  - The **acid dissociation constant K_a** is a quantitative measure of acid strength.
  - Acid strengths are generally reported as **pK_a** values, which are equal to $-\log_{10}K_a$.
  - Stronger acids have larger K_a values and therefore smaller pK_a values.
  - Among the most common weak organic acids are carboxylic acids, compounds that contain the —COOH (carboxyl) group.
    - The value of K_a for acetic acid, a representative carboxylic acid, is 1.74×10^{-5} M; the value of pK_a for acetic acid is 4.76.

Section 4.4 The Position of Equilibrium in Acid-Base Reactions

- The **position of equilibrium** in an acid-base reaction favors reaction of the stronger acid (lower pK_a value) with the stronger base to form the weaker acid (higher pK_a value) and the weaker base.
  - For quantitative calculations, the pK_{eq} for an acid-base reaction equals the difference of the pK_a for the stronger acid (pK_{aHA}) and the protonated form of the stronger base (pK_{aBH^+})
    - $pK_{eq} = pK_{a(HA)} - pK_{a(BH^+)}$
    - K_{eq} for an acid-base reaction is equal to $10^{(-pK_{eq})}$.
  - A good rule of thumb is that an acid will be substantially *deprotonated* if its pK_a is two or more units lower than the pH of an aqueous solution. A base will be *protonated* if the pK_a of its conjugate acid is two or more units higher than the pH of an aqueous solution.

    - Carboxylic acids and phosphodiesters are generally deprotonated and anionic at pH 7–8, while amines and guanidinium groups are generally found in their protonated and positively charged forms.

Section 4.5 Thermochemistry and Mechanisms of Acid-Base Reactions

- A **reaction mechanism** describes in detail how a reaction occurs.
  - Most chemical reactions occur via collisions.
  - The reactants must collide with the proper orientation and enough energy to reach the **transition state** and proceed to products.
  - A reaction can be described by a **reaction coordinate diagram**, which is a plot of energy versus the progress of the reaction.

- **Thermochemistry** is the study of energy of the entire system at each moment of a reaction.
  - **Thermodynamics** is the study of the relative energies between any two states in wells on a reaction coordinate diagram.
    - If products are more stable than reactants, the overall thermodynamics are favorable for reaction and the step is **exergonic**.
    - If products are less stable than reactants, the overall thermodynamics are not favorable for reaction and the step is **endergonic**.
  - **Kinetics** is the study of rates of chemical reactions.
    - The lower the **free energy of activation** (energy difference between the transition state and starting state), the faster the rate of the corresponding reaction step, and *vice versa.*
  - **Enthalpy** is the energy contained within chemical bonds and solvation.
    - If bonds formed in a product are stronger than those broken in the starting materials, heat is given off and the reaction is **exothermic**. If bonds formed in a product are weaker than those broken in the starting materials, heat is absorbed and the reaction is **endothermic**.
  - **Entropy** measures chaos versus order, and chaos is favorable.
- For acid-base reactions to occur, the acid and base must collide with a geometry in which the proton to be transferred is between the proton-donor and proton-acceptor atoms, more or less in a linear geometry.

Section 4.6 Molecular Structure and Acidity

- The acidity of an acid is determined by the stability of the anion formed on deprotonation, according to the rule that more acidic molecules form more stable anions upon deprotonation. Factors that influence the stability of an anion are:
  - Electronegativity of the atom bearing the negative charge because more electronegative atoms are more stable as anions.
  - Size of the atom bearing the negative charge because larger atoms can more easily accommodate a negative charge (it is spread over a larger area).
  - Delocalization of charge in the anion, usually described by resonance contributing structures because greater delocalization of charge is stabilizing.
  - The inductive effect because adjacent electronegative atoms such as the halogens will stabilize a nearby negative charge.
  - The hybridization of the atom bearing the negative charge because the greater the percentage of *s* character of an atom, the more stable it will be as an anion.

Problems: 4.34, 4.35, 4.38, 4.41, 4.42, 4.46

Section 4.7 Lewis Acids and Bases

- A **Lewis acid** is a species that can form a new covalent bond by accepting a pair of electrons; a **Lewis base** is a species that can form a new covalent bond by donating a pair of electrons.
  - All Brønsted-Lowry acids (proton donors) are also Lewis acids and all Brønsted-Lowry bases (proton acceptors) are also Lewis bases, but the Lewis acid-base model is far more general in that it applies to reactions beyond just proton transfers.

Problems: 4.7, 4.25–4.28, 4.30, 4.36, 4.39

☉WL Questions and problems for this chapter are available in OWL. Use the OWL access card that came with this textbook to access assigned questions and problems for this chapter.

1. **Proton-Transfer Reaction (Section 4.2)**   A proton-transfer reaction involves transfer of a proton from a proton donor (a Brønsted-Lowry acid) to a proton acceptor (a Brønsted-Lowry base).

$$\underset{\text{Proton donor}}{H-\overset{+}{\underset{|}{\overset{|}{O}}}-H} + \underset{\text{Proton acceptor}}{:\overset{|}{\underset{|}{N}}-H} \longrightarrow H-\overset{|}{\underset{|}{O}}: + H-\overset{+}{\underset{|}{N}}-H$$

2. **Position of Equilibrium in an Acid-Base Reaction (Section 4.4)**   The stronger acid reacts with the stronger base to give a weaker acid and a weaker base. K_{eq} for this equilibrium can be calculated from pK_a values for the two acids.

$$\underset{\substack{\text{p}K_a\ 4.76 \\ \text{(stronger acid)}}}{CH_3COH} + CN^- \rightleftharpoons \underset{}{CH_3CO^-} + \underset{\substack{\text{p}K_a\ 9.31 \\ \text{(weaker acid)}}}{HCN}$$

$$\text{p}K_{eq} = 4.76 - 9.31 = -4.55$$
$$K_{eq} = 3.55 \times 10^4$$

3. **Lewis Acid-Base Reaction (Section 4.7)**   A Lewis acid-base reaction involves sharing an electron pair between an electron pair donor (a Lewis base) and an electron pair acceptor (a Lewis acid).

$$:\overset{}{\underset{}{O}}: + \ \underset{F}{\overset{F}{B}}-F \longrightarrow \overset{+}{\underset{}{O}}-\underset{F}{\overset{F}{B}}-F$$

Chapter 5

Alkenes: Bonding, Nomenclature, and Properties

© Mark Muench/Stone/Getty Images

Haze in the Blue Ridge Mountains. This haze is caused by light-scattering from the aerosol caused by the photooxidation of isoprene and other hydrocarbons emitted by trees and other plants. Many naturally occurring hydrocarbons are formed from isoprene units (see Section 5.4).

Unsaturated hydrocarbon
A hydrocarbon containing one or more carbon-carbon double or triple bonds. The three most important classes of unsaturated hydrocarbons are alkenes, alkynes, and arenes.

An **unsaturated hydrocarbon** contains one or more carbon-carbon double or triple bonds. The term "unsaturation" indicates that there are fewer hydrogens bonded to carbon than in an alkane, C_nH_{2n+2}. The three most important classes of unsaturated hydrocarbons are alkenes, alkynes, and arenes. **Alkenes** contain a carbon-carbon double bond and, with one double bond and no rings, have the general formula C_nH_{2n}. Alkynes contain a carbon-carbon triple bond and, with one triple bond and no rings, have the general formula C_nH_{2n-2}. The simplest alkene is ethylene, and the simplest alkyne is acetylene.

Ethylene
(an alkene)

Side view

End view

Acetylene
(an alkyne)

Side view

End view

In this chapter, we study the structure, nomenclature, and physical properties of alkenes. Alkynes are discussed separately in Chapter 7.

ⓌWL Online homework for this chapter may be assigned in OWL for Organic Chemistry.

Arene
A term used to classify benzene and its derivatives.

Arenes are a third class of unsaturated hydrocarbons. The Lewis structure of benzene, the simplest arene, is shown below.

Benzene
(an arene)

Top view

Side view

Just as a group derived by removal of an H from an alkane is called an alkyl group and given the symbol R—(Section 2.3A), a group derived by removal of an H from an arene is called an **aryl group** and given the symbol **Ar—**.

Aryl group (Ar—)
A group derived from an arene by removal of an H.

Phenyl group
A group derived by removing an H from benzene; abbreviated C_6H_5— or Ph—.

When the aryl group substituent on a parent chain is a benzene ring, it is given the special name **phenyl group**. You might think that, when present as a substituent, benzene would become benzyl, just as ethane becomes ethyl. This is not so! "Phene" is a now-obsolete name for benzene, and, although this name is no longer used, a derivative has persisted in the name "phenyl." Following is a structural formula for the phenyl group and two alternative representations for it. Throughout this text, we will represent benzene by a hexagon with three inscribed double bonds. It is also common to represent it by a hexagon with an inscribed circle. We explain the reasons for the alternative representations in Chapter 21.

Benzene   An alternative representation for benzene C_6H_5—   Ph—

Alternative representations for the phenyl group

The chemistry of benzene and its derivatives is quite different from that of alkenes and alkynes, but, even though we do not study the chemistry of arenes until Chapters 21 and 22, we will show structural formulas of compounds containing aryl groups before that time. The three double bonds in a six-membered ring create a special stabilization called aromaticity, which lowers the reactivity of benzene relative to other alkenes. What you need to remember at this point is that an aryl group is not chemically reactive under any of the conditions we describe in Chapters 6 through 20.

5.1 | Structure of Alkenes

A. Shapes of Alkenes

Using valence-shell electron-pair repulsion (Section 1.4) for a carbon-carbon double bond, we predict a value of 120° for the bond angles about each carbon. The observed H—C—C bond angle in ethylene is 121.1°, close to that predicted. In other alkenes, deviations from the predicted angle of 120° may be somewhat larger because of the strain introduced by nonbonded interactions created by groups bonded to the carbons of the double bond. The C—C—C bond angle in propene, for example, is 123.9°.

121.1°

Ethylene

123.9°

Propene

How To	Calculate the Index of Hydrogen Deficiency

Valuable information about the structural formula of an unknown compound can be obtained by inspecting its molecular formula. In addition to learning the number of atoms of carbon, hydrogen, oxygen, nitrogen, and so forth in a molecule of the compound, we can also determine what is called its index of hydrogen deficiency. For each ring and π bond, the molecular formula has two fewer hydrogens. The **index of hydrogen deficiency** is the sum of the number of rings and π bonds in a molecule. It is determined by comparing the number of hydrogens in the molecular formula of a compound whose structure is to be determined ($H_{molecule}$) with the number of hydrogens in a reference alkane of the same number of carbon atoms ($H_{reference}$).

Index of hydrogen deficiency
The sum of the number of rings and π bonds in a molecule.

$$\text{Index of hydrogen deficiency} = \frac{(H_{reference} - H_{molecule})}{2}$$

1. The molecular formula of a reference acyclic alkane is C_nH_{2n+2} (Section 2.1).
2. To compare the molecular formula for a compound containing elements besides carbon and hydrogen, write the formula of the reference hydrocarbon with the same number of carbon atoms, and make the following adjustments to the number of hydrogen atoms in the unknown.

   a. Replace each monovalent atom of a Group 7 element (F, Cl, Br, I) with one hydrogen; halogen substitutes for hydrogen and reduces the number of hydrogens by one per halogen. The general formula of an acyclic monochloroalkane, for example, is $C_nH_{2n+1}Cl$; the general formula of the corresponding acyclic alkane is C_nH_{2n+2}.

   b. No correction is necessary for the addition of divalent atoms of Group 6 elements (O, S, Se). Insertion of a divalent Group 6 element into a hydrocarbon does not change the number of hydrogens.

   c. For each atom of a trivalent Group 5 element (N, P, As) present, add one hydrogen, because insertion of a trivalent Group 5 element adds one hydrogen to the molecular formula. The general molecular formula for an acyclic alkylamine, for example, is $C_nH_{2n+3}N$.

B. Carbon-Carbon Double Bond Orbitals

In Section 1.7C, we described the formation of a carbon-carbon double bond in terms of the overlap of atomic orbitals. A carbon-carbon double bond consists of one σ bond and one π bond (Figure 5.1). Each carbon of the double bond uses its three sp^2 hybrid orbitals to form σ bonds to three atoms. The unhybridized $2p$ atomic orbitals, which lie perpendicular to the plane created by the axes of the three sp^2 hybrid orbitals, combine to form two π molecular orbitals: one bonding

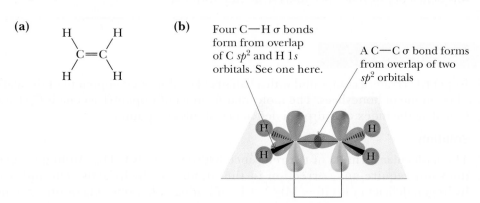

(a)

(b) Four C—H σ bonds form from overlap of C sp^2 and H $1s$ orbitals. See one here.

A C—C σ bond forms from overlap of two sp^2 orbitals

A π bond forms between these two $2p$ orbitals.

Figure 5.1
Covalent bonding in ethylene.
(a) Lewis structure and
(b) orbital overlap model showing the σ and π bonds.

and the other antibonding. For the unhybridized $2p$ orbitals to be parallel, thus giving maximum overlap, the two carbon atoms of the double bond and the four bonded atoms must lie in a plane.

It takes approximately 264 kJ (63 kcal)/mol to break the π bond in ethylene, that is, to rotate one carbon by 90° with respect to the other where zero overlap occurs between $2p$ orbitals on adjacent carbons (Figure 5.2). This energy is considerably greater than the thermal energy available at room temperature; consequently, rotation about a carbon-carbon double bond does not occur under normal conditions. You might compare rotation about a carbon-carbon double bond, such as in ethylene, with that about a carbon-carbon single bond, such as in ethane (Section 2.5A). Whereas rotation about the carbon-carbon single bond in ethane is relatively free [rotation barrier approximately 12.5 kJ (3.0 kcal)/mol], rotation about the carbon-carbon double bond in ethylene is severely restricted.

(a)   **(b)**

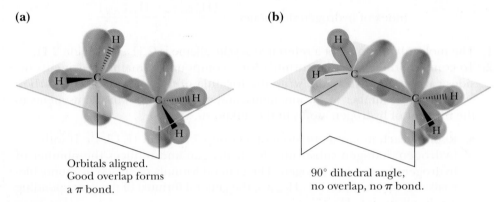

Orbitals aligned.
Good overlap forms
a π bond.

90° dihedral angle,
no overlap, no π bond.

Figure 5.2
Restricted rotation about a carbon-carbon double bond. (a) Orbital overlap model showing the π bond. (b) The π bond is broken by rotating the plane of one H—C—H group by 90° with respect to the plane of the other H—C—H group.

Example 5.1

Calculate the index of hydrogen deficiency for 1-hexene, C_6H_{12}, and account for this deficiency by reference to its structural formula.

Solution

The molecular formula of the reference acyclic alkane of six carbon atoms is C_6H_{14}. The index of hydrogen deficiency of 1-hexene $(14 - 12)/2 = 1$ and is accounted for by the one π bond in 1-hexene.

Problem 5.1

Calculate the index of hydrogen deficiency of cyclohexene, C_6H_{10}, and account for this deficiency by reference to its structural formula.

Example 5.2

Isopentyl acetate, a compound with a banana-like odor, is a component of the alarm pheromone of honeybees. The molecular formula of isopentyl acetate is $C_7H_{14}O_2$. Calculate the index of hydrogen deficiency of this compound.

Solution

The molecular formula of the reference hydrocarbon is C_7H_{16}. Adding oxygens does not require any correction in the number of hydrogens. The index of hydrogen deficiency is $(16 - 14)/2 = 1$, indicating either one ring or one π bond. Following is the structural formula of isopentyl acetate. It contains one π bond, in this case, a carbon-oxygen π bond.

Isopentyl acetate

Problem 5.2

The index of hydrogen deficiency of niacin is 5. Account for this index of hydrogen deficiency by reference to the structural formula of niacin.

Nicotinamide (Niacin)

C. *Cis,Trans* Isomerism in Alkenes

Because of restricted rotation about a carbon-carbon double bond, any alkene in which each carbon of the double bond has two different groups bonded to it shows ***cis,trans* isomerism**. For example, 2-butene has two stereoisomers. In *cis*-2-butene, the two methyl groups are on one side of the double bond and the two hydrogens are on the other side. In *trans*-2-butene, the two methyl groups are on opposite sides of the double bond. These two compounds cannot be converted into one another at room temperature because of the restricted rotation about the double bond; they are different compounds (diastereomers), with different physical and chemical properties.

Cis,trans **isomers**
Isomers that have the same order of attachment of their atoms but a different arrangement of their atoms in space owing to the presence of either a ring (Section 2.6) or a carbon-carbon double bond (Section 5.1C).

steric strain

***cis*-2-Butene**
mp −139°C, bp 4°C

***trans*-2-Butene**
mp −106°C, bp 1°C

Cis alkenes with double bonds in open chains are less stable than their *trans* isomers because of steric strain between alkyl substituents on the same side of the double bond, as can be seen in space-filling models of the *cis* and *trans* isomers of 2-butene. This is the same type of strain that results in the preference for equatorial methylcyclohexane over axial methylcyclohexane (Section 2.5B). *Trans*-2-butene is more stable than the *cis* isomer by about 4.2 kJ (1.0 kcal)/mol because of the sum of steric strain and angle strain that results from the two methyls moving apart. At the energy minimum, the C═C—CH₃ angle for *cis*-2-butene is about 127°.

5.2 | Nomenclature of Alkenes

Alkenes are named using the IUPAC system, but, as we shall see, some are usually referred to by their common names.

A. IUPAC Names

To form IUPAC names for alkenes, change the *-an-* infix of the parent alkane to *-en-* (Section 2.3C). Hence, CH_2═CH_2 is named ethene, and CH_3CH═CH_2 is named propene. In higher alkenes, where isomers exist that differ in location of the double bond, a numbering system must be used. According to the IUPAC system,

1. Number the longest carbon chain that contains the double bond in the direction that gives the carbon atoms of the double bond the lowest possible numbers.
2. Indicate the location of the double bond by the number of its first carbon.

3. Name branched or substituted alkenes in a manner similar to alkanes.

4. Number the carbon atoms, locate and name substituent groups, locate the double bond, and name the main chain.

$$CH_3CH_2CH_2CH_2CH{=}CH_2 \qquad CH_3CH_2\overset{\underset{\displaystyle CH_3}{|}}{C}HCH_2CH{=}CH_2 \qquad CH_3\overset{\underset{\displaystyle CH_3}{|}}{C}HCH_2\overset{\underset{\displaystyle CH_2CH_3}{|}}{C}{=}CH_2$$

1-Hexene   **4-Methyl-1-hexene**   **2-Ethyl-4-methyl-1-pentene**

Note that there is a chain of six carbon atoms in 2-ethyl-4-methyl-1-pentene. However, because the longest chain that contains the double bond has only five carbons, the parent hydrocarbon is pentane, and the molecule is named as a disubstituted 1-pentene.

Example 5.3

Write the IUPAC name of each alkene.

(a)   **(b)**

Solution

(a) 4-Ethyl-3,3-dimethyl-1-octene   **(b)** 2-Methyl-2-butene

Problem 5.3

Write the IUPAC name of each alkene.

(a)   **(b)**

B. Common Names

Some alkenes, particularly those of low molecular weight, are known almost exclusively by their common names, as illustrated by the common names of these alkenes.

$$CH_2{=}CH_2 \qquad CH_3CH{=}CH_2 \qquad CH_3\overset{\underset{\displaystyle CH_2}{\|}}{\underset{\displaystyle }{C}}{=}CH_2$$

Methylene
A $CH_2{=}$ group.

Vinyl
A $CH_2{=}CH{-}$ group.

Allyl
A $CH_2{=}CHCH_2{-}$ group.

IUPAC name:	**Ethene**	**Propene**	**2-Methylpropene**
Common name:	**Ethylene**	**Propylene**	**Isobutylene**

Furthermore, the common names **methylene** (a CH_2 group), **vinyl,** and **allyl** are often used to show the presence of the following alkenyl groups:

Alkenyl Group	IUPAC Name	Common Name	Example	IUPAC Name (Common Name)
$CH_2{=}$	**Methylidene**	**Methylene**	$H_2C{=}$	**Methylidenecyclopentane** (Methylenecyclopentane)
$CH_2{=}CH{-}$	**Ethenyl**	**Vinyl**	$CH_2{=}CH{-}$	**Ethenylcyclopentane** (Vinylcyclopentane)
$CH_2{=}CHCH_2{-}$	**2-Propenyl**	**Allyl**	$CH_2{=}CHCH_2{-}$	**2-Propenylcyclopentane** (Allylcyclopentane)

C. Systems for Designating Configuration in Alkenes

The *Cis,Trans* System

The most common method for specifying the configuration in alkenes uses the prefixes *cis* and *trans*. There is no doubt which isomers are intended by the names *trans*-3-hexene and *cis*-3-hexene. For more complex alkenes, the orientation of the atoms of the parent chain determines whether the alkene is *cis* or *trans*. On the right is a structural formula for the *cis* isomer of 3,4-dimethyl-2-pentene. In this example, carbon atoms of the main chain (carbons 1 and 4) are on the same side of the double bond and, therefore, this alkene is *cis*.

cis
In *cis,trans*-alkene nomenclature, it refers to molecules in which the carbon atoms of the main chain are on the same side of the double bond.

trans
In *cis,trans*-alkene nomenclature, it refers to molecules in which the carbon atoms of the main chain are on opposite sides of the double bond.

trans-**3-Hexene**   *cis*-**3-Hexene**   *cis*-**3,4-Dimethyl-2-pentene**

Example 5.4

Name each alkene and show the configuration about each double bond using the *cis,trans* system.

(a)

(b)

Solution

(a) The chain contains seven carbon atoms and is numbered from the end that gives the lower number to the first carbon of the double bond. Its name is *trans*-3-heptene.

(b) The longest chain contains eight carbon atoms and is numbered from either end of the chain so that the first carbon of the double bond is carbon 4 of the chain. Its name is *cis*-4-octene.

Problem 5.4

Which alkenes show *cis,trans* isomerism? For each alkene that does, draw the *trans* isomer.

(a) 2-Pentene   (b) 2-Methyl-2-pentene   (c) 3-Methyl-2-pentene

---

The *E,Z* System

Because the *cis,trans* system becomes confusing with tri- and tetrasubstituted alkenes and is not detailed enough to name all alkenes, chemists developed the **E,Z system**. This system uses the priority rules of the *R,S* system (Section 3.3) to assign priority to the substituents on each carbon of a double bond. Using these rules, we decide which group on each carbon has the higher priority. If the groups of higher priority are on the same side of the double bond, the configuration of the alkene is **Z** (German: *zusammen*, together). If they are on opposite sides of the double bond, the alkene is **E** (German: *entgegen*, opposite).

E,Z system
A system to specify the configuration of groups about a carbon-carbon double bond.

Z
From the German, *zusammen*, together. Specifies that groups of higher priority on the carbons of a double bond are on the same side.

E
From the German, *entgegen*, opposite. Specifies that groups of higher priority on the carbons of a double bond are on opposite sides.

Z (*zusammen*)   *E* (*entgegen*)

Throughout this text, we use the *cis,trans* system for alkenes in which each carbon of the C=C bond bears a hydrogen. We use the *E,Z* system in all other cases. The *E,Z* system should always be used if confusion is possible.

Example 5.5

Name each alkene and specify its configuration by the *E,Z* system.

(a) (b) (c)

Solution

(a) The group of higher priority on carbon 2 is methyl; that of higher priority on carbon 3 is isopropyl. Because the groups of higher priority are on the same side of the double bond, the alkene has the *Z* configuration. Its name is (*Z*)-3,4-dimethyl-2-pentene.
(b) Groups of higher priority on carbons 2 and 3 are —Cl and —CH_2CH_3. Because these groups are on opposite sides of the double bond, the configuration of this alkene is *E*, and its name is (*E*)-2-chloro-2-pentene.
(c) The groups of higher priority are on opposite sides of the double bond; the configuration is *E*. The name of this bromoalkene is (*E*)-1-bromo-4-isopropyl-5-methyl-4-octene.

higher priority
on carbon 5

higher priority
on carbon 4

Problem 5.5

Name each alkene and specify its configuration by the *E,Z* system.

(a) (b) (c)

D. Cycloalkenes

In naming **cycloalkenes**, the carbon atoms of the ring double bond are numbered 1 and 2 in the direction that gives the substituent encountered first the smaller number.

3-Methylcyclopentene   **4-Ethyl-1-methylcyclohexene**   **1,6-Dimethylcyclohexene**

Example 5.6

Write the IUPAC name of each cycloalkene.

(a) (b) (c)

Solution

(a) 3,3-Dimethylcyclohexene **(b)** 1,2-Dimethylcyclopentene
(c) 4-(1-Methylethyl)-1-methylcyclohexene

Problem 5.6

Write the IUPAC name of each cycloalkene.

(a) **(b)** **(c)**

E. *Cis,Trans* Isomerism in Cycloalkenes

Following are structural formulas for four cycloalkenes:

Cyclopentene   **Cyclohexene**   **Cycloheptene**   **Cyclooctene**

In these representations, the configuration about each double bond is *cis*. Is it possible to have a *trans* configuration in these and larger cycloalkenes? To date, *trans*-cyclooctene is the smallest *trans* cycloalkene that has been prepared in pure form and is stable at room temperature. Yet, even in this *trans* cycloalkene, there is considerable angle strain; the double bond's $2p$ orbitals make an angle of 44° to each other. *Cis*-cyclooctene is more stable than its *trans* isomer by 38 kJ (9.1 kcal)/mol. Note that the *trans* isomer is chiral even though it has no chiral center.

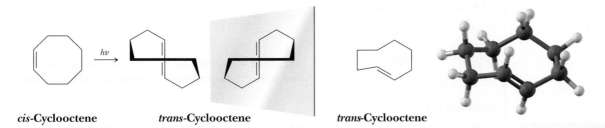

cis-**Cyclooctene**   *trans*-**Cyclooctene**   *trans*-**Cyclooctene**

F. Bridgehead Alkenes

Another manner in which alkenes become strained is when one of the carbons in the double bond is placed at a bridgehead carbon of a bicyclic ring system. A bridgehead carbon is one that is contained in both rings of the hydrocarbon. For example, the carbons shown below with arrows are at bridgeheads. In norborene, the double bond does not contain the bridgehead carbon, but in the other illustration, the double bond is at a bridgehead. This arrangement of a double bond imparts considerable strain because the alkene cannot be planar while the rest of the carbons in the bicycle system span the rings.

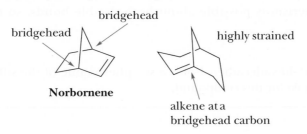

bridgehead
bridgehead   highly strained

Norbornene

alkene at a
bridgehead carbon

G. Dienes, Trienes, and Polyenes

For alkenes containing two or more double bonds, the infix -en- is changed to -adien-, -atrien-, and so on. Those that contain several double bonds are also referred to more generally as polyenes (Greek: *poly*, many). Following are examples of three dienes.

1,4-Pentadiene   **2-Methyl-1,3-butadiene (Isoprene)**   **1,3-Cyclopentadiene**

H. *Cis,Trans* Isomerism in Dienes, Trienes, and Polyenes

Thus far we have considered *cis,trans* isomerism in alkenes containing only one carbon-carbon double bond. For an alkene with one carbon-carbon double bond that can show *cis,trans* isomerism, two stereoisomers are possible. For an alkene with **n** carbon-carbon double bonds, each of which can show *cis,trans* isomerism, 2^n stereoisomers are possible.

Example 5.7

How many stereoisomers are possible for 2,4-heptadiene?

Solution

This molecule has two carbon-carbon double bonds, each of which shows *cis,trans* isomerism. As shown in this table, there are $2^2 = 4$ stereoisomers. Two of these are drawn on the right.

Double Bond	
C2—C3	**C4—C5**
trans	*trans*
trans	*cis*
cis	*trans*
cis	*cis*

(2E,4E)-2,4-Heptadiene
trans,trans-2,4-Heptadiene

(2E,4Z)-2,4-Heptadiene
trans,cis-2,4-Heptadiene

Problem 5.7

Draw structural formulas for the other two stereoisomers of 2,4-heptadiene.

Example 5.8

How many stereoisomers are possible for 10,12-hexadecadien-1-ol?

$$CH_3(CH_2)_2CH{=}CHCH{=}CH(CH_2)_8CH_2OH$$

10,12-Hexadecadien-1-ol

Solution

Cis,trans isomerism is possible about both double bonds, so there are four stereoisomers.

Problem 5.8

(10E,12Z)-10,12-hexadecadien-1-ol is a sex pheromone of the silkworm. Draw a structural formula for this compound.

Silkworms spinning cocoons at a silk farm. Japan.

© Paul Chesley/Stone/Getty Images

An example of a biologically important compound for which a number of *cis,trans* isomers are possible is vitamin A. There are four carbon-carbon double bonds in the chain of carbon atoms bonded to the substituted cyclohexene ring, and each has the potential for *cis,trans* isomerism. There are $2^4 = 16$ stereoisomers possible for this structural formula. Vitamin A is the all-*E* isomer.

Vitamin A (retinol)

5.3 | Physical Properties of Alkenes

Alkenes are nonpolar compounds, and the only attractive forces between their molecules are dispersion forces (Section 2.7B). Their physical properties, therefore, are similar to those of alkanes. Alkenes of two, three, and four carbon atoms are gases at room temperature. Those of five or more carbons are colorless liquids that are less dense than water. Alkenes are insoluble in water but soluble in one another, in other nonpolar organic liquids, and in ethanol. Table 5.1 lists physical properties of some alkenes.

Chemical Connections

The Case of the Iowa and New York Strains of the European Corn Borer

Although humans communicate largely by sight and sound, chemicals are the primary means of communication for the vast majority of other species in the animal world. Often, communication within a species is specific for one of two or more configurational isomers. For example, a member of a given species may respond to a *cis* isomer of a chemical but not to the *trans* isomer. Or, alternatively, it might respond to a quite precise blend of *cis* and *trans* isomers but not to other blends of these same isomers.

Several groups of scientists have studied the components of the sex pheromones of both the Iowa and the New York strains of the European corn borer. Females of these closely related species secrete the sex attractant 11-tetradecenyl acetate. Males of the Iowa strain show maximum response to a mixture containing 96% of the *cis* isomer and 4% of the *trans* isomer. When the pure *cis* isomer is used alone, males are only weakly attracted. Males of the New York strain show an entirely different response. They respond maximally to a mixture containing 3% of the *cis* isomer and 97% of the *trans* isomer.

The European corn borer, Pyrausta nubilalis

There is evidence that optimum response to a narrow range of stereoisomers as we see here is widespread in nature and that a great many insects maintain species isolation for mating and reproduction by the stereochemistry of their pheromones.

cis-**11-Tetradecenyl acetate**

trans-**11-Tetradecenyl acetate**

Table 5.1	Physical Properties of Some Alkenes		
Name	**Structural Formula**	**mp (°C)**	**bp (°C)**
Ethylene	$CH_2=CH_2$	−169	−104
Propene	$CH_3CH=CH_2$	−185	−47
1-Butene	$CH_3CH_2CH=CH_2$	−185	−6
1-Pentene	$CH_3CH_2CH_2CH=CH_2$	−138	30
***cis*-2-Pentene**	$\begin{array}{c} H_3C \qquad CH_2CH_3 \\ C=C \\ H \qquad\quad H \end{array}$	−151	37
***trans*-2-Pentene**	$\begin{array}{c} H_3C \qquad H \\ C=C \\ H \qquad\; CH_2CH_3 \end{array}$	−156	36
2-Methyl-2-butene	$\begin{array}{c} CH_3 \\ | \\ CH_3C=CHCH_3 \end{array}$	−134	39

5.4 | Naturally Occurring Alkenes— Terpene Hydrocarbons

Terpene
A compound whose carbon skeleton can be divided into two or more units identical with the carbon skeleton of isoprene.

A **terpene** is a compound whose carbon skeleton can be divided into two or more units that are identical with the carbon skeleton of isoprene. Carbon 1 of an isoprene unit is called the head; carbon 4 is called the tail. Terpenes are formed by bonding the tail of one isoprene unit to the head of another. This is called the **isoprene rule**.

$$\begin{array}{c} CH_3 \\ | \\ CH_2=C-CH=CH_2 \end{array}$$

2-Methyl-1,3-butadiene
(Isoprene)

A study of terpenes provides a glimpse of the wondrous diversity that nature can generate from a simple carbon skeleton. Terpenes also illustrate an important principle of the molecular logic of living systems, namely, that in building large molecules, small subunits are bonded together enzymatically by an iterative process and then modified by subsequent precise enzyme-catalyzed reactions. Chemists use the same principles in the laboratory, but our methods do not have the precision and selectivity of the enzyme-catalyzed reactions of living systems.

Probably the terpenes most familiar to you, at least by odor, are components of the so-called essential oils obtained by steam distillation or ether extraction of various parts of plants. Essential oils contain the relatively low-molecular-weight substances that are in large part responsible for characteristic plant fragrances. Many essential oils, particularly those from flowers, are used in perfumes.

One example of a terpene obtained from an essential oil is myrcene, $C_{10}H_{16}$, a component of bayberry wax and oils of bay and verbena. Myrcene is a triene with a parent chain of eight carbon atoms and two one-carbon branches [Figure 5.3(a)].

Head-to-tail bonds between isoprene units are vastly more common in nature than are the alternative head-to-head or tail-to-tail patterns. Figure 5.4 shows structural formulas of five more terpenes, all derived from two isoprene units. Geraniol has the same carbon skeleton as myrcene. In the last four terpenes of Figure 5.4, the carbon atoms present in myrcene and geraniol are cross-linked to give cyclic structures. To help you identify the points of cross linkage and ring formation, the carbon atoms of the geraniol skeleton are numbered 1 through 8. This numbering pattern is used in the remaining terpenes to show points of cross-linking. In both limonene

Figure 5.3
Myrcene. (a) Structural formula and (b) ball-and-stick model.

(a)   (b)

Bond joining the head of one isoprene unit to the tail of the second

Bond joining the head of one isoprene unit to the tail of the second.

Geraniol
(rose and other flowers)

Limonene
(oil of lemon and orange)

Menthol
(peppermint)

α-Pinene
(turpentine)

Camphor
(camphor tree)

Figure 5.4
Five terpenes, each divisible into two isoprene units.

and menthol, a carbon-carbon bond is present between carbons 1 and 6. In α-pinene, carbon-carbon bonds are present between carbons 1 and 6 and between carbons 4 and 7. In camphor, they are between carbons 1 and 6 and between 3 and 7.

Vitamin A (Section 5.2H), a terpene with the molecular formula $C_{20}H_{30}O$, consists of four isoprene units linked head-to-tail and cross-linked at one point to form a six-membered ring.

Vitamin A (retinol)

The synthesis of substances in living systems is a fascinating area of research and one of the links between organic chemistry and biochemistry. However tempting it might be to propose that nature synthesizes terpenes by joining together molecules of isoprene, this is not quite the way it is done. We will discuss the creation of new carbon-carbon bonds during terpene biosynthesis in Section 26.4B.

CONNECTIONS TO BIOLOGICAL CHEMISTRY

The Importance of *Cis* Double Bonds in Fats Versus Oils

Fats and oils are very similar in that both are triesters of glycerol, hence the name triglyceride. Hydrolysis of a triglyceride in aqueous base followed by acidification gives glycerol and three carboxylic acids. Because these carboxylic acids can be derived from fats, they are called fatty acids:

$$
\begin{array}{l}
\quad\quad O \\
\quad\quad \| \\
O \quad CH_2OCR \\
\| \quad | \\
R'COCH \quad\quad O \\
\quad | \quad\quad \| \\
\quad CH_2OCR''
\end{array}
\xrightarrow[\text{2. HCl, H}_2\text{O}]{\text{1. NaOH, H}_2\text{O}}
\begin{array}{l}
CH_2OH \\
| \\
HOCH \\
| \\
CH_2OH
\end{array}
+
\begin{array}{l}
RCOOH \\
R'COOH \\
R''COOH
\end{array}
$$

A triglyceride   1,2,3-Propanetriol   Fatty acids
(a triester of glycerol)   (Glycerol)

The most common fatty acids have between 12 and 24 carbon atoms in an unbranched chain.

The main difference between fats and oils is the temperature at which they melt. Fats are solids or semisolids at or near room temperature, while oils are liquids. The

occurring fatty acids have *cis* configurations. The triglycerides of animal fats are richer in saturated fatty acids, whereas the triglycerides of plant oils (for example, corn, soybean, canola, olive, and palm oils) are richer in unsaturated fatty acids.

The carbon-carbon single bonds of saturated fatty acid alkyl chains exist largely in the staggered, anti-conformation, so they can pack together relatively well, and are held together by dispersion forces (Section 2.8B). As a result, both saturated fatty acids and the triglycerides

Stearic acid
mp 70°C
(a saturated C_{18} fatty acid)

Oleic acid
mp 13°C
(a monounsaturated *cis* C_{18} fatty acid)

Linolenic acid
mp −17°C
(a polyunsaturated *cis* C_{18} fatty acid)

different physical properties of fats and oils result from the presence of different fatty acids.

Fatty acids with no double bonds are referred to as saturated fatty acids, those with a single double bond are called monounsaturated fatty acids, and those with more than one double bond are called polyunsaturated fatty acids. The double bonds in almost all naturally

derived from them are solids at room temperature. However, the *cis* double bonds place a considerable "kink" in the chains of monounsaturated and polyunsaturated fatty acid chains limiting their packing, decreasing surface contact and dispersion forces, thereby increasing fluidity.

As a result, monounsaturated and polyunsaturated fatty acids and the triglycerides composed of them are

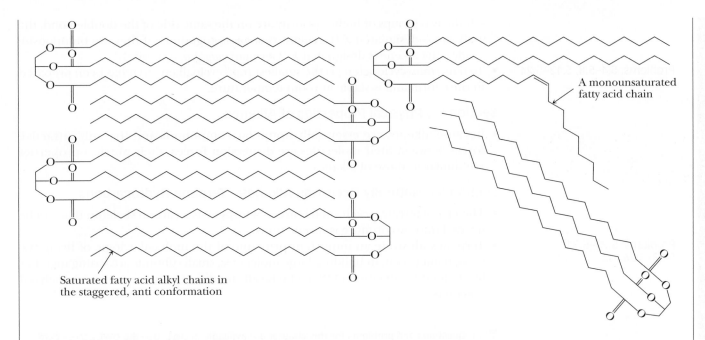

A monounsaturated
fatty acid chain

Saturated fatty acid alkyl chains in
the staggered, anti conformation

liquid at room temperature. Overall, oils have fatty acids with more *cis* double bonds. Fats have fewer *cis* double bonds and more saturated fatty acids. For example, butter has a high content of saturated fats and is a solid at room temperature, whereas salad oil (from plant oils) has a high content of polyunsaturated fatty acids and is liquid, even at freezing temperatures. Olive oil, which has a high content of the monounsaturated fatty acid oleic acid (hence the name of oleic acid), will solidify in the refrigerator.

Summary

Section 5.1 Structure of Alkenes

- An **alkene** is an **unsaturated hydrocarbon** that contains a carbon-carbon double bond. The general formula of an alkene is C_nH_{2n}.
  - A carbon-carbon double bond consists of one σ bond formed by the overlap of sp^2 hybrid orbitals and one π bond formed by the overlap of parallel $2p$ orbitals.
  - The strength of the π bond in ethylene is approximately 264 kJ (63 kcal)/mol, which is considerably weaker than the carbon-carbon σ bond.
  - The structural feature that makes *cis,trans* isomerism possible in alkenes is lack of rotation about the two carbons of the double bond.

  Problems: 5.9–5.12, 5.25–5.28

- **Index of hydrogen deficiency** is the sum of the number of π bonds and rings in a molecule.

  Problems: 5.1, 5.2, 5.32

Section 5.2 Nomenclature of Alkenes

- According to the IUPAC system, the presence of a carbon-carbon double bond is shown by changing the infix of the parent alkane from *-an-* to *-en-*.
  - For compounds containing two or more double bonds, the infix *-en-* is changed to *-adien-*, *-atrien-*, and so on.
  - The names **methylene**, **vinyl**, and **allyl** are commonly used to show the presence of $=CH_2$, $—CH=CH_2$, and $—CH_2CH=CH_2$ groups, respectively.
- Whether an alkene is *cis* or *trans* is determined by the orientation of the main carbon chain about the double bond. The configuration of a carbon-carbon double bond is specified more precisely by the *E,Z system* using the same set of priority rules used for the *R,S* system.

  Problem: 5.3

Problems: 5.4–5.8, 5.13–5.24

– If the two groups of higher priority are on the same side of the double bond, the alkene is designated **Z** (German: *zusammen*, together); if they are on opposite sides, the alkene is designated **E** (German: *entgegen*, opposite).

• To date, *trans*-cyclooctene is the smallest *trans* cycloalkene that has been prepared in pure form and is stable at room temperature.

Section 5.3 Physical Properties of Alkenes

• Because alkenes are essentially nonpolar compounds and the only attractive forces between their molecules are **dispersion forces**, their physical properties are similar to those of alkanes.

Section 5.4 Naturally Occurring Alkenes—Terpene Hydrocarbons

Problems: 5.29–5.31, 5.33–5.36

• The characteristic structural feature of a **terpene** is a carbon skeleton that can be divided into two or more isoprene units.

• Terpenes illustrate an important principle of the molecular logic of living systems, namely that, in building large molecules, small subunits are strung together by an iterative process and then chemically modified by precise enzyme-catalyzed reactions.

⬛WL Questions and problems for this chapter are available in OWL. Use the OWL access card that came with this textbook to access assigned questions and problems for this chapter.

Chapter 6

Reactions of Alkenes

Image copyright Pichugin Dmitry 2010. Used under license from Shutterstock.com

These Kayaks are fabricated from polyethylene. Inset: a model of ethylene, the alkene monomer from which polyethylene is derived.

Over 10 million organic compounds have been discovered or made by organic chemists! It would seem to be an almost impossible task to learn the chemical properties of this many compounds. Fortunately, the study of organic compounds is not as formidable a task as you might think. Although organic compounds can undergo a wide variety of chemical reactions, only certain portions of their structures are changed in any particular reaction. As we will see in this chapter, the same functional group, in whatever organic molecule we find it, undergoes the same types of chemical reactions. Therefore, you do not have to study the chemical reactions of even a fraction of the 10 million known organic compounds. Instead, you need only to identify a few characteristic types of functional groups and then study the chemical reactions that each undergoes.

6.1 | Reactions of Alkenes—An Overview

The most characteristic reaction of alkenes is **addition** to the carbon-carbon double bond in such a way that the π bond is broken and, in its place, σ bonds form to two new atoms or groups of atoms. Table 6.1 gives several examples of reactions at a carbon-carbon double bond along with the descriptive name(s) associated with each. Some of these reactions are treated separately under oxidations (Section 6.5) and reductions (Section 6.6), but are included in this table because they are formally additions.

In the following sections, we study these alkene reactions in considerable detail, with particular attention to the mechanism by which each occurs.

◆WL **Online homework for this chapter may be assigned in OWL for Organic Chemistry.**

Table 6.1 | Characteristic Alkene Addition Reactions

Reaction	Descriptive Name(s)
$\text{C=C} + \text{HCl (HX)} \longrightarrow$ product with H and Cl (X) added	Hydrochlorination (hydrohalogenation)
$\text{C=C} + \text{H}_2\text{O} \longrightarrow$ product with H and OH added	Hydration
$\text{C=C} + \text{Br}_2\ (\text{X}_2) \longrightarrow$ product with (X) Br and Br (X) added	Bromination (halogenation)
$\text{C=C} + \text{Br}_2\ (\text{X}_2) \xrightarrow{\text{H}_2\text{O}}$ product with HO and Br (X) added	Bromo(halo)hydrin formation
$\text{C=C} + \text{Hg(OAc)}_2 \xrightarrow{\text{H}_2\text{O}}$ product with HgOAc and HO added	Oxymercuration
$\text{C=C} + \text{BH}_3 \longrightarrow$ product with H and BH$_2$ added	Hydroboration
$\text{C=C} + \text{OsO}_4 \longrightarrow$ product with HO and OH added	Diol formation (oxidation)
$\text{C=C} + \text{H}_2 \longrightarrow$ product with H and H added	Hydrogenation (reduction)

Addition reaction

A reaction in which two atoms or ions react with a double bond, forming a compound with the two new groups bonded to the carbons of the original double bond.

Another addition reaction of alkenes is the formation of chain-growth polymers (Greek: *poly,* many, and *meros,* part). In the presence of certain catalysts called initiators, many alkenes form polymers made by the addition of monomers (Greek: *mono,* one, and *meros,* part) to a growing polymer chain as illustrated by the formation of polyethylene from ethylene.

$$n\,\text{CH}_2\text{=CH}_2 \xrightarrow{\text{initiator}} \left(\text{CH}_2\text{CH}_2\right)_n$$

In alkene polymers of industrial and commercial importance, n is a large number, typically several thousand. We discuss this alkene reaction in Chapter 29.

6.2 | Organic Reactions Involving Reactive Intermediates

A. Reaction Coordinate Diagrams

In Chapter 4, we introduced reaction mechanisms and reaction coordinate diagrams. These concepts were explained in the context of acid-base chemistry, which involves the transfer of a proton from one chemical entity to another occurring in one single step. Although many chemical transformations take place in one step, many others involve two or more steps.

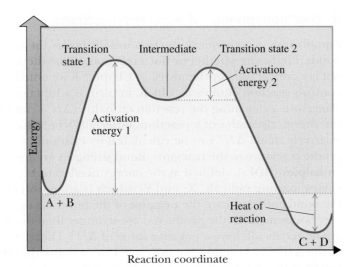

Figure 6.1
An energy diagram for a two-step reaction involving formation of a reaction intermediate. The energy of the reactants is higher than that of the products, and energy is released in the conversion of A + B to C + D.

In a reaction that occurs in two or more steps, each step has its own transition state and activation energy. Figure 6.1 shows an energy diagram for the conversion of reactants to products in two steps. A **reactive intermediate** corresponds to an energy minimum between two transition states, in this case between transition states 1 and 2. Because the energies of the intermediates we describe in this chapter are higher than that of either reactants or products, they are highly reactive and usually cannot be isolated. However, some have significant lifetimes and can be observed directly using special experimental strategies.

Reactive intermediate
A high-energy species, formed between two successive reaction steps, that lies in an energy minimum between the two transition states.

The most common reactive intermediates in organic chemistry involve abnormal bonding at a carbon atom. In this chapter, the mechanisms involve *carbocation* intermediates where carbon has three bonds and only six valence electrons, and is positively charged (see Section 6.2A). In future chapters, we will encounter radical intermediates where carbon has three bonds and only seven valence electrons. When carbon has three bonds but retains a full octet, it is negatively charged and is referred to as a carbanion. All three intermediates—carbocations, radicals, and carbanions—are significantly higher in energy than either the reactants or the products, and therefore a reaction coordinate diagram analogous to Figure 6.1 applies.

$$
\begin{array}{ccc}
\underset{\text{Carbocation}}{\overset{R}{\underset{R}{\overset{|}{\underset{}{R-\overset{+}{C}}}}}} &
\underset{\text{Radical}}{\overset{R}{\underset{R}{\overset{|}{\underset{}{R-\overset{\cdot}{C}}}}}} &
\underset{\text{Carbanion}}{\overset{R}{\underset{R}{\overset{|}{\underset{}{R-\overset{-}{\underset{\cdot\cdot}{C}}}}}}}
\end{array}
$$

Figure 6.2
The reactive intermediates common in organic chemistry transformations.

The slowest step in a multistep reaction, called the **rate-determining step**, is the step that crosses the highest energy barrier. In the two-step sequence shown in Figure 6.1, Step 1 crosses the higher energy barrier and is, therefore, the rate-determining step.

Rate-determining step
The step in a multistep reaction sequence that crosses the highest energy barrier.

Each step of a chemical reaction involves the crossing of a peak on an energy diagram. The peaks represent chemical structures on the energy surfaces that are transitions between stable structures. Reaction intermediates are represented in wells (troughs) on the energy surfaces. The peaks on the surface are referred to as **transition states** (recall Section 4.5A).

It is important to make a distinction between reaction intermediates, such as carbocations, and transition states. *Reaction intermediates* have lifetimes longer than the time that it takes for a bond to vibrate, whereas *transition states* have lifetimes only on the order of the lifetime of a bond vibration (in the femtosecond range, or billionths of a second, 10^{-9} s). For most practical purposes, the chemical structures of transition states can be considered as having no measurable lifetime, whereas reaction intermediates have distinct measurable lifetimes, even though they may be extremely short.

B. Thermodynamics of Addition Reactions

Equilibrium in a chemical reaction usually favors the side with stronger chemical bonds. Predicting whether or not a reaction will be thermodynamically favorable or not is a critical issue in chemistry. In Chapter 4, we noted that the Gibbs free energy controls reaction equilibria, which, in turn, is a function of enthalpy and entropy. Therefore, calculating the reaction enthalpy ($\Delta H°$) can be used in part to predict the thermodynamics of a reaction, especially when the reaction entropy ($\Delta S°$) is relatively small. $\Delta H°$ can be calculated by examining the bond strengths in the products relative to the reactants. Bond strengths are reported as **bond dissociation enthalpies** (BDEs), defined as the energy needed to break any bond, X—Y, into its corresponding radicals, X· and Y·, which is an endothermic process. Therefore, the bond enthalpies are the negative of the bond strengths. Reactions in which the new bonds made in the product(s) are stronger than the bonds broken in the starting materials will have a negative value of $\Delta H°$. Therefore, to get a rough estimate of the $\Delta H°$ of a reaction, we only need to examine the BDEs in the reactants and products that have undergone a change; and in fact, average bond enthalpies work remarkably well for rough estimates. A table of average BDEs is given in Appendix 3. To calculate $\Delta H°$ for the reaction, we subtract the BDEs of the products from the BDEs of the reactants:

$$\Delta H°_{rxn} = BDEs_R - BDEs_P$$

In this chapter, we examine the mechanisms, regiochemistry, and stereochemistry of the addition of water, HCl, HBr, Cl_2, Br_2, as well as other reagents to alkenes. In Problem 6.14 a table of BDEs is given. Using this table, let's calculate the reaction enthalpy of just one of the addition reactions covered in this chapter—for example, the addition of bromine to an alkene:

			543 kJ(130 kcal)/mol
			−602 kJ(-144 kcal)/mol
			−59 kJ(-14 kcal)/mol
351 kJ (84 kcal)/mol	192 kJ (46 kcal)/mol	2x 301 kJ (72 kcal)/mol	

This reaction involves cleavage of a C=C π bond and a Br—Br σ bond and the formation of two C—Br σ bonds. We can estimate the BDE of the C=C π bond by taking the difference between the BDE for $CH_2=CH_2$ (727 kJ/mol) and $CH_3—CH_3$ (376 kJ/mol) which equals 351 kJ/mol. Other important BDEs for this calculation are Br—Br (192 kJ/mol) and C—Br (310 kJ/mol). Therefore, $\Delta H°_{rxn}$ is

$$\Delta H°_{rxn} = (351 \text{ kJ/mol} + 192 \text{ kJ/mol}) - (301 \text{ kJ/mol} + 301 \text{ kJ/mol})$$

$$= -59 \text{ kJ/mol}$$

C=C π bond   Br—Br bond   C—Br bond   C—Br bond

The reaction is therefore exothermic, because stronger bonds are made in the product compared to those broken in the starting materials. If the entropy is favorable or near zero, the reaction will be exergonic. In other words, the reaction is thermodynamically favorable and will spontaneously occur (assuming a relatively low activation energy).

C. Developing a Reaction Mechanism

To develop an understanding of a reaction mechanism, chemists begin by first using a combination of experience and intuition to propose several sets of steps or mechanisms, each of which might account for the overall chemical transformation. Next, chemists test each proposed mechanism by designing and carrying out experiments that provide experimental observations that will allow them to exclude those mechanisms that are not consistent with the facts. Modern computational methods now allow detailed predictions of mechanistic pathways more accurately than ever before, but the results must still be compared with experiment for confirmation.

A mechanism becomes generally established by excluding reasonable alternatives and by showing that it is consistent with every test that can be devised. This, of course, does not mean that a generally accepted mechanism is a completely accurate description of the chemical events, only that it is the best mechanism chemists have been able to devise. It is important to keep in mind that, as new experimental evidence is obtained, it may be necessary to modify a generally accepted mechanism or possibly even discard it in favor of an alternative.

Before we go on to consider reactions and reaction mechanisms, we might ask why chemists go to the trouble of establishing them and why you must spend time learning about them. One reason is very practical: mechanisms provide a framework within which to organize a great deal of descriptive chemistry. For example, with insight into how reagents add to particular alkenes, it is possible to generalize and then to predict how the same reagents might add to other alkenes. A second reason lies in the intellectual satisfaction derived from constructing models that accurately reflect the behavior of chemical systems. Finally, to a creative scientist, a mechanism is a tool to be used in the search for new information and new understanding. A mechanism consistent with all that is known about a reaction can be used to make predictions about chemical interactions as yet unexplored, and experiments can be designed to test these predictions. Thus, reaction mechanisms provide a way not only to organize knowledge but also to extend it.

D. The Correct Use of Arrows to Indicate Electron Movement

The ability to write an organic reaction mechanism properly is key to student success in organic chemistry classes. Organic chemists use a technique called **electron pushing** (sometimes called arrow pushing) to depict the flow or movement of electrons during chemical reactions. Arrow pushing helps chemists keep track of the way in which electrons and their associated atoms redistribute as bonds are made and broken. The first essential rule to keep in mind is the following:

Electron Pushing
The use of arrows in organic chemistry mechanisms to indicate the flow or movement of electrons.

First Rule: Arrows are used to indicate movement of *electrons.*

A regular arrow (double-sided arrowhead) is used to indicate the movement of two electrons, while a line with a single-sided arrowhead (sometimes called a "fish hook arrow") is used for single electron movement involved with radical reactions that are first described in Chapter 8.

The majority of reactions that will be discussed in this book involve movement of pairs of electrons, so they are represented by double-sided arrowheads.

Arrow pushing was first introduced in Section 1.8A in the discussion of resonance contributing structures. Recall that when comparing two or more contributing structures, an arrow was used to show how two electrons (lines representing bonds or pairs of dots representing lone pairs) could be redistributed within a single chemical structure to create an alternative Lewis line structure representation of the bonding. By convention, arrows are used to keep track of *all* pairs of electrons that are in different locations in the two different contributing Lewis line structures, shown here for the acetate anion and benzene molecule.

Acetate anion contributing structures

Benzene contributing structures

Keep in mind that in the case of resonance, 1) the atoms do not move between contributing structures, and 2) the electrons are not actually moving. The true chemical structure should be thought of as a hybrid of the contributing Lewis line structures. It is worth pointing out that when used with contributing structures, arrows generally indicate only the interconversion of π bonds and lone pairs (acetate anion) or just π bonds (benzene), not the formation or breaking of σ bonds.

In chemical reactions, both electrons and atoms change positions as both π and σ bonds are formed and broken. Arrow pushing is used to keep track of the movement of all electrons involved with each step of the overall transformation. Because electrons are located in orbitals surrounding atoms, when bonds are formed or broken, the movement of electrons between orbitals is necessarily accompanied by the movement of the associated atoms, which leads to the second rule of arrow pushing when depicting chemical reaction mechanisms:

Second Rule: Arrows are never used to indicate the movement of atoms directly. The arrows only show atom movement indirectly as a consequence of electron movement when covalent bonds are made and broken.

We have already used arrow pushing to show proton transfer several times in Chapter 4. The example below shows the transfer of a proton from the relatively acidic acetic acid molecule to the relatively basic hydroxide anion. We show this process with one arrow (labeled "a" in the diagram) that starts at a lone pair of electrons on the basic oxygen atom of the hydroxide anion, then points to the acidic H atom of acetic acid to indicate formation of the new bond being made. A second arrow originates at the line representing the breaking O—H bond and points to the O atom to denote creation of a lone pair (arrow "b"). In this reaction, the proton is being transferred between molecules, and the arrows indicate movement of the electrons involved.

Correct use of arrows to
indicate electron movement
during a reaction

A common mistake beginning students make is that they will erroneously write an arrow pointing *from* the H of the acetic acid *to* the O atom of the hydroxide anion. This is wrong, because such an arrow would be indicating the H *atom* movement directly, not *electron* movement! Other common mistakes in arrow pushing are given in Appendix 10.

Incorrect arrow because it is pointing in the wrong direction! Never use arrows to indicate atom movement directly

E. Electron Sources and Sinks: How to *Predict* What Will Occur in an Organic Reaction Mechanism

Combined with the arrows shown for the contributing structures described previously, we have now seen all three of the situations illustrated by arrows with double-sided arrowheads, namely the redistribution of π bonds and/or lone pairs, formation of a new σ bond (generally from a lone pair or π bond), and breaking of a σ bond (generally to form a new lone pair or sometimes a new π bond). Often, as in the case of the acetic acid–hydroxide ion reaction, more than one arrow is used in a given mechanism step. Now that you have seen all of the important types of arrows, we can point out the most important common feature between them:

Third Rule: Arrows always start at an **electron source** and end at an **electron sink**.

electron source → electron sink

An **electron source** is a bond or a lone pair of electrons. It is most commonly a π bond or a lone pair on an atom of relatively high electron density, or a σ-bond that must break during a reaction. An **electron sink** is an atom on a molecule or ion that can accept a new bond or lone pair of electrons.

Learning to identify the characteristic sources and sinks in different functional groups is the key to learning organic chemistry reaction mechanisms. For example, for arrows that depict the formation of a new σ bond between two molecules, the electron source is often readily identified as being a lone pair on the most electron rich atom of one molecule or ion, and the electron sink is readily identified as the most electron poor atom of the other molecule or ion. Thus, the prediction of many of the most important electron sources and sinks comes down to lessons concerning the differences in electronegativity between atoms that were presented in Section 1.2, which allow you to identify the most electron rich and electron poor atoms in molecules. As an aid to your analysis, the red and blue colors of the various electrostatic surface maps given throughout this book indicate the negative (electron rich) and positive (electron poor) regions of molecules. We will have more to say about this reactivity pattern a little bit later.

This leads us to another commonly encountered type of process that deserves mention. As you will see in this and many later chapters, making a new bond to an electron sink often requires the simultaneous breaking of one of the bonds present at the sink atom to avoid overfilling its valence orbitals, a situation referred to as hypervalence.

Fourth Rule: Bond breaking will occur to avoid overfilling valence (hypervalence) on an atom serving as an electron sink.

In these cases, the electron source for the arrow is the bond being broken, and the sink is an atom able to accommodate the electrons as a lone pair, generally an electronegative atom such as an O atom or a halogen. If an ion is created, that ion is often stabilized by resonance delocalization or other stabilizing interactions.

Electron source
A bond or lone pair of electrons, usually an area of relatively high electron density, that serves as the origin of a mechanism arrow. An electron source characteristically interacts with an electron sink in an organic mechanism.

Electron sink
An atom that accepts a new bond or lone pair of electrons. An electron sink is what an arrow points toward in an organic chemistry mechanism.

Returning to the proton transfer reaction between acetic acid and hydroxide, we can now summarize our analysis of this simple one-step mechanism.

Viewed in the context of the third rule, when considering the arrow used to make a new σ bond (arrow "a"), the hydroxide O atom is the electron source (most negatively charged atom) and the acetic acid H atom is the electron sink (atom with highest partial positive charge). This is illustrated using the electrostatic molecular surfaces shown below the reaction equation. The O atom of hydroxide ion has the greatest localized negative charge as indicated by the most intense red color and the acetic acid proton being transferred has the most intense positive charge character indicated by the most intense blue color. In order to avoid overfilling the valence of the H atom during the reaction (fourth rule), the O—H bond of acetic acid must be broken (arrow "b"). In so doing, the acetate ion is formed. Note that the acetate ion is stabilized by resonance delocalization.

Based on our analysis of the reaction between acetic acid and the hydroxide anion, you should now appreciate that the transfer of a proton (a so-called Brønsted acid-base reaction) is really just a special case of the common pattern of reactivity between an electron source (the base) and the proton as an electron sink, combined with breaking a bond to satisfy valence and create a relatively stable ion.

The addition or removal of protons during chemical reactions is so common that proton transfer steps are referred to by name directly, and we will use phrases such as "add a proton" or "take a proton away" when referring to them. However, proton transfer reactions are not the only case in which we use special names to describe a particular type of common reaction that involves arrows between electron sources and electron sinks.

As briefly mentioned in Section 4.7, a broader terminology is applied to the very common case of reactions in which new σ bonds form between electron rich and electron poor regions of molecules. **Nucleophiles** (meaning nucleus seeking) are molecules that have lone pairs or relatively electron rich π bonds that act as electron sources for arrows making new bonds. **Electrophiles** (meaning electron seeking) are molecules with relatively electron poor atoms that serve as sinks for these arrows. Analogously, a molecule, or region of a molecule, that is a source for such an arrow is called nucleophilic, while a molecule or region of a molecule that is a sink for these arrows is referred to as being electrophilic. Based on this description, it should be clear that nucleophiles are analogous to Lewis bases and

Nucleophile
From the Greek meaning nucleus loving, seeking a region of low electron density. An electron source that can donate a pair of electrons to form a new covalent bond; alternatively, a Lewis base.

Electrophile
From the Greek meaning electron loving, seeking a region of high electron density. An electron sink that can accept a pair of electrons to form a new covalent bond; alternatively, a Lewis acid.

electrophiles are analogous to Lewis acids. Chemists use these terms interchange-ably, although nucleophile and electrophile are more commonly used in kinetics discussions while Lewis acid and Lewis base are more commonly used in discus-sions about reaction thermodynamics. We will use all of these terms throughout the rest of the book.

It is helpful to summarize the appropriate use of key terms associated with arrow pushing and reaction mechanisms. The terms "source" and "sink" are used to identify the start and end of each reaction mechanism arrow, which is indicating the change in location of electron pairs. The terms "nucleophile" and "electrophile" (as well as "Lewis base" and "Lewis acid") are used to describe molecules based on their chemical reactivity and propensity to either donate or receive electrons when they interact. Protons can be thought of as a specific type of electrophile, and for reactions in which a proton is transferred, the nucleo-phile is called a base.

Example 6.1

The following two sets of reactions, ((a) and (b)), show possibilities for arrow pushing in individual reaction steps. Identify which is wrong and explain why. Next, using arrow pushing correctly, label which molecule is the nucleophile and which is the electrophile.

(a)

(b)

Solution

In each case the first arrow pushing scenario is wrong. The arrows shown below with stars over them do not start at a source of electrons, but rather they start at positions of relative positive charge, which is incorrect. In the correct arrow pushing, the arrow labeled "a" depicts the interaction of a region of relative high negative charge (a π-bond or lone pair) with an atom of relatively high partial positive charge on the other reactant. Therefore, the molecule acting as the source for the σ bond-forming arrow "a" is the nucleophile while the molecule containing the sink atom is the electrophile.

Incorrect   **Correct**

The arrow labeled "b" is needed to satisfy valence, and is not considered when defining the nucleophile and electrophile.

Problem 6.1

The following two sets of reactions, ((a) and (b)), show possibilities for arrow pushing in individual reaction steps. Identify which is wrong and explain why. Next, using the correct arrow pushing, label which molecule is the nucleophile and which is the electrophile.

(a)

$$:\ddot{I}:^- \quad H_3C-\ddot{C}l: \longrightarrow :\ddot{I}-CH_3 + :\ddot{C}l:^-$$

$$:\ddot{I}:^- \quad H_3C-\ddot{C}l: \longrightarrow :\ddot{I}-CH_3 + :\ddot{C}l:^-$$

(b)

$$\begin{array}{c} H_3C \\ \diagdown \\ C-CH_3 \\ \diagup \\ H_3C \end{array} \overset{+}{} :\ddot{C}l:^- \longrightarrow \begin{array}{c} H_3C \\ \diagdown \\ C-\ddot{C}l: \\ \diagup \\ H_3C \end{array}$$

$$\begin{array}{c} H_3C \\ \diagdown \\ C-CH_3 \\ \diagup \\ H_3C \end{array} \overset{+}{} :\ddot{C}l:^- \longrightarrow \begin{array}{c} H_3C \\ \diagdown \\ C-\ddot{C}l: \\ \diagup \\ H_3C \end{array}$$

F. Putting it All Together: It Comes Down to a Multiple Choice Situation

In the sections and chapters that follow, many different reaction mechanisms will be described in a stepwise fashion. Each arrow can be classified according to one of the three overall situations we have already encountered: (i) redistribution of π bonds and/or lone pairs, (ii) formation of a new σ bond from a lone pair or π bond, (iii) breaking a σ bond to give a new lone pair or π bond. When learning new mechanisms, first focus on the overall transformation that takes place. It might be a reaction in which atoms or groups are added (an addition reaction), a reaction in which atoms or groups are removed (an elimination reaction), a reaction in which atoms or groups replace an atom or group (a substitution reaction), or other processes we will encounter. Often, the overall process is composed of multiple steps. Once you have the overall process in mind, it is time to think about the individual steps that convert starting material(s) into product(s). Predicting complete multi-step mechanisms, then, comes down to learning how to predict the individual steps.

Understanding, as opposed to memorizing, mechanisms is critical to mastering organic chemistry. Although the mechanisms you encounter throughout the course may seem entirely different, they are actually related in fundamental ways. In fact, almost all of the organic reaction mechanisms you will learn are composed of only a few different individual elements (steps) that are put together in various combinations. Your job is to learn these individual mechanism elements, and then understand how to assemble them into the steps of the correct mechanism for an overall reaction.

Fortunately, there are a surprisingly small number of different types of characteristic mechanism elements (patterns of arrows) to be considered when trying to predict individual steps of even complex chemical reactions. For this reason, you should view the prediction of each step in an organic mechanism as essentially a *multiple choice* situation in which your most common choices are the following four:

1. **Make a new bond between a nucleophile (source for an arrow) and an electrophile (sink for an arrow)** Use this element when there is a nucleophile present in the solution as well as an electrophile suitable for reaction to occur.

$$H-\ddot{O}-H \quad + \quad CH_3\overset{+}{C}HCH_3 \quad \rightleftharpoons \quad CH_3CHCH_3$$

Water	**Isopropyl cation**	Oxonium ion
(a nucleophile)	intermediate	intermediate
	(an electrophile)	

2. **Break a bond so that relatively stable molecules or ions are created** Use this element when there is no suitable nucleophile-electrophile or proton transfer reaction, but breaking a bond can create neutral molecules or relatively stable ions, or both.

$$
\underset{\substack{\text{2-Bromo-2-methylpropane}}}{\overset{\overset{\displaystyle :\ddot{Br}:}{|}}{CH_3CCH_3}} \quad \rightleftharpoons \quad \underset{\substack{\textit{tert}\text{-Butyl cation}}}{\overset{+}{CH_3CCH_3}} \quad + \quad \underset{\substack{\text{Bromide ion}}}{:\ddot{Br}:^-}
$$

2-Bromo-2-methylpropane

tert-Butyl cation
intermediate
(stabilized by inductive
and hyperconjugation
effects. See Section 6.3A)

Bromide ion
(relatively stable anion)

3. **Add a proton** Use this element when there is no suitable nucleophile-electrophile reaction, but there is a strong acid present or the molecule has a strongly basic functional group.

Ethyl acetate
(a carboxylic ester)

Hydronium ion
(a strong acid)

A protonated
intermediate

Water

4. **Take a proton away** Use this element when there is no suitable nucleophile-electrophile reaction, but the molecule has a strongly acidic proton or there is a strong base present.

Oxonium ion
intermediate
(strong acid)

Water
(can act as a base)

2-Propanol

Hydronium ion

The situation is even simpler than you might expect because 1 and 2 are the functional reverse of each other, as are 3 and 4 in many cases. Many times, more than one of the four choices occurs simultaneously in the same mechanistic step.

These four mechanistic elements are the foundation upon which you will build an understanding of organic reaction mechanisms. They are far and away the most common; however, they are not the only mechanistic elements you will see. Once in a while, you will encounter a reaction with a unique mechanistic element specific to the particular molecules involved. For example, Section 6.4 describes a mechanistic element unique to hydroboration reactions. In addition, we will describe other important mechanistic elements such as *1,2 shifts* (Section 6.3C) that occur in several reactions.

In the following sections and chapters of the book, you will learn important properties of the different functional groups that allow you to deduce the appropriate choices for the individual steps in reaction mechanisms. To help you accomplish this, as new mechanisms are introduced throughout the rest of the book, we will label each mechanistic step as one of the common ones when appropriate, emphasizing the analogous features between even complex mechanisms. When you are able to predict which of the common mechanistic elements is(are) the most appropriate for a given step in a mechanism, you will then be able to push electrons correctly without relying on memorization. At that point, you will have taken a major step toward mastering organic chemistry!

6.3 | Electrophilic Additions

We begin our introduction to the chemistry of alkenes by examining five common addition reactions. We first study some of the experimental observations about each addition reaction and then look at its mechanism. Through the study of these particular reactions and their mechanisms, you will develop a general understanding of how alkenes undergo addition reactions.

The overall reaction of electrophilic species adding to π bonds is called **electrophilic addition**. As the different electrophilic addition reactions are described in the next sections of this chapter, you will notice that the key step in these mechanisms involves the π electrons of the alkene reacting as a nucleophile with an electrophilic species in a bond-making step. In order to emphasize that the π bond is the source for a bond-making arrow, we will classify this mechanistic element as **make a bond between a nucleophile (π bond) and an electrophile**.

At this point, it is appropriate to ask what makes a molecule or ion an electrophile? As defined previously, electrophiles are "electron-loving" chemical species. Hence, electrophiles react with nucleophiles, which are molecular species that are electron rich. The formation of a bond via nucleophilic attack on an electrophile is one of organic chemistry's fundamental reactions.

There are at least three characteristic features that can make a molecule or ion electrophilic. The most common of these features is a region of low electron density, reflected by a partial or full positive charge. The full or partial positive charge aids reaction with an electron rich region of a nucleophile through electrostatic attraction. A second common feature of some electrophiles is the lack of an octet on an atom. The electrophiles listed below fit one or both of these criteria. We will examine some of their reactions with alkene nucleophiles in this chapter and with other nucleophiles in future chapters.

Electrophilic addition
A common type of reaction with alkenes in which an electrophilic species adds to a π bond.

A positive charge with an acidic hydrogen	A partial positive charge with an acidic hydrogen	A positive charge with a Lewis acidic mercury atom	A boron atom lacking an octet	A positive charge on a carbon atom that is lacking an octet

A third characteristic that can make a molecule or ion electrophilic is a relatively weak bond to an atom that can depart as a stable ion or molecule. These electrophilic species can be considered as electron seeking because their reactions with nucleophiles create stronger bonds and therefore more stable molecules. In such cases, there is often no partial or full positive charge on the electrophilic atom. Molecular halogens (X_2) are good examples. Their bonds are weak relative to the ones they form upon reaction with a nucleophile; this process is further aided by release of a stable halogen anion (X^-). Another example of a molecule that is electrophilic for these reasons is methyl iodide (we examine its reactions with nucleophiles in Chapter 9). There is little to no charge polarity in methyl iodide. However, the carbon is electrophilic because its bond to iodine is weak, and reaction with a nucleophile produces the stable iodide (I^-) anion.

A weak bond between chlorine atoms	A weak bond between bromine atoms	A weak bond between a carbon and iodine atom

A. Addition of Hydrogen Halides

The hydrogen halides HCl, HBr, and HI add to alkenes to give haloalkanes (alkyl halides). These additions may be carried out either with the pure reagents (neat) or in the presence of a polar solvent such as acetic acid. HCl reacts sluggishly compared to the other two acids. Addition of HBr to ethylene gives bromoethane (ethyl bromide):

$$CH_2{=}CH_2 + \boxed{HBr} \longrightarrow \overset{\boxed{H}}{\underset{}{CH_2}}{-}\overset{\boxed{Br}}{\underset{}{CH_2}}$$

Ethylene   **Bromoethane**
                                            (Ethyl bromide)

Addition of HBr to propene gives 2-bromopropane (isopropyl bromide); hydrogen adds to carbon 1 of propene, and bromine adds to carbon 2. If the orientation of addition were reversed, 1-bromopropane (propyl bromide) would be formed. The observed result is that 2-bromopropane is formed to the virtual exclusion of 1-bromopropane. We say that addition of HBr to propene is highly regioselective. A **regioselective reaction** is a reaction in which one direction of bond forming or breaking occurs in preference to all other directions of bond forming or breaking.

Regioselective reaction
An addition or substitution reaction in which one of two or more possible products is formed in preference to all others that might be formed.

$$CH_3CH{=}CH_2 + \boxed{HBr} \longrightarrow \overset{\boxed{Br}\ \boxed{H}}{CH_3CH{-}CH_2} + \overset{\boxed{H}\ \boxed{Br}}{CH_3CH{-}CH_2}$$

Propene   **2-Bromopropane**   **1-Bromopropane**
                                                  (not observed)

This regioselectivity was noted by Vladimir Markovnikov who made the generalization known as **Markovnikov's rule**: in the addition of H—X to an alkene, hydrogen adds to the double-bonded carbon that has the greater number of hydrogens already bonded to it. Although Markovnikov's rule provides a way to predict the products of many alkene addition reactions, it does not explain why one product predominates over other possible products.

Markovnikov's rule
In the addition of HX, H_2O, or ROH to an alkene, hydrogen adds to the carbon of the double bond having the greater number of hydrogens.

Example 6.2

Name and draw a structural formula for the product of each alkene addition reaction:

(a) $\overset{CH_3}{\underset{}{CH_3C}}{=}CH_2 + HI \longrightarrow$ (b) [cyclopentene with CH₃] $+ HCl \longrightarrow$

Solution

Using Markovnikov's rule, we predict that 2-iodo-2-methylpropane is the product in (a) and that 1-chloro-1-methylcyclopentane is the product in (b).

(a) $\overset{CH_3}{\underset{I}{CH_3CCH_3}}$ (b) [cyclopentane ring with Cl and CH₃]

  2-Iodo-2-methylpropane   **1-Chloro-1-methylcyclopentane**

Problem 6.2

Name and draw a structural formula for the product of each alkene addition reaction:

(a) [cyclohexene]—$CH_3 + HBr \longrightarrow$ (b) [cyclohexylidene]$=CH_2 + HI \longrightarrow$

Chemists account for the addition of HX to an alkene by arrow pushing and a two-step mechanism, which we illustrate by the reaction of 2-butene with hydrogen bromide to give 2-bromobutane. We will first look at this two-step mechanism in overview and then go back and study each step in detail.

Mechanism Electrophilic Addition of HBr to 2-Butene

Step 1: Make a new bond between a nucleophile (π bond) and an electrophile—add a proton. Addition begins with the transfer of a proton from H—Br to 2-butene, as shown by the two curved arrows on the left side of the equation. The first curved arrow shows that the π bond of the alkene breaks and its electron pair forms a new covalent bond with the hydrogen atom of H—Br. The second curved arrow shows that the polar covalent bond in H—Br breaks and its electron pair moves to bromine to form a bromide ion. The result of this step is the formation of an organic cation and bromide ion.

$$CH_3CH\!=\!\!CHCH_3 \;+\; \overset{\delta+}{H}\!-\!\overset{\delta-}{\ddot{B}r}: \;\underset{\text{determining}}{\overset{\text{slow, rate}}{\rightleftharpoons}}\; CH_3\overset{+}{C}H\!-\!\overset{\overset{\displaystyle H}{|}}{C}HCH_3 \;+\; :\!\ddot{B}r\!:^{\!-}$$

(a nucleophile) (an electrophile)   *sec*-**Butyl cation**
                                                     (a 2° carbocation
                                                     intermediate)

Step 2: Make a new bond between a nucleophile and an electrophile. Reaction of the *sec*-butyl cation (an electrophile) with bromide ion (a nucleophile) completes the valence shell of carbon and gives 2-bromobutane.

$$:\!\ddot{B}r\!:^{\!-} \;+\; CH_3\overset{+}{C}HCH_2CH_3 \;\overset{\text{fast}}{\longrightarrow}\; CH_3\overset{\overset{\displaystyle :\ddot{B}r:}{|}}{C}HCH_2CH_3$$

Bromide ion   *sec*-**Butyl cation**   **2-Bromobutane**
(a nucleophile)   (an electrophile)   (a racemic mixture)

Now that we have looked at this two-step mechanism in overview, let us go back and look at the individual steps in more detail. There is a great deal of important organic chemistry embedded in them.

The reaction commences when the electron-rich π bond of the alkene (the electron source) interacts with the relatively electropositive H atom of H—Br (an electron sink) to form a new bond. The π bond is relatively electron-rich because the π-bonding electron density is above and below the bond axis, not between the positively charged atomic nuclei as is the case with σ-bonding electron density. Most of the reactions of alkenes described in this chapter belong to this category, with the alkene π bond acting as a nucleophile and interacting with a variety of electrophiles. We refer to this mechanistic step as **make a bond between a nucleophile (π bond) and an electrophile** to emphasize that the source of the bond-making arrow is the alkene π bond. Because a proton is the electrophile in this case, we can also think of this step as **addition of a proton**. HBr can be identified as an electrophile because it has an area of low electron density and therefore partial positive charge on the H atom; breaking the H—Br bond leads to a stable Br anion.

Step 1, the addition of a proton to the alkene, results in formation of a cationic intermediate. One carbon atom in this intermediate has only six electrons in its valence shell and carries a charge of +1. A species containing a positively charged carbon atom is called a **carbocation** (*carbon* + *cation*). Such carbon-containing cations are also called *carbonium ions* and *carbenium ions*. Carbocations are classified as primary (1°), secondary (2°), or tertiary (3°), depending on the number of carbon atoms bonded to the carbon bearing the positive charge. All carbocations are electrophiles as well as Lewis acids (Section 4.7).

Carbocation
A species in which a carbon atom has only six electrons in its valence shell and bears a positive charge.

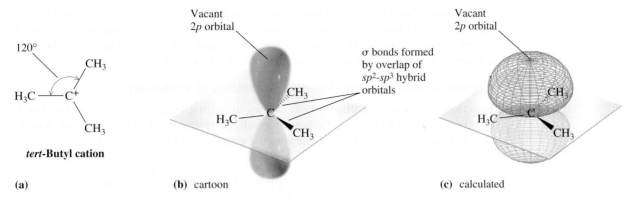

(a) **(b)** cartoon **(c)** calculated

Figure 6.3
The structure of the *tert*-butyl cation. **(a)** Lewis structure, **(b)** cartoon, and **(c)** a calculated structure indicating a vacant 2*p* orbital.

In a carbocation, the carbon bearing the positive charge is bonded to three other atoms and, as predicted by VSEPR, the three bonds about the cationic carbon are coplanar and form bond angles of approximately 120°. According to the orbital hybridization model, the electron-deficient carbon of a carbocation uses sp^2 hybrid orbitals to form σ bonds to the three attached groups. The unhybridized 2*p* orbital lies perpendicular to the σ bond framework and contains no electrons. Figure 6.3 shows a Lewis structure, a cartoon, and a calculated structure indicating the vacant 2*p* orbital for the *tert*-butyl cation.

Figure 6.4 shows an energy diagram for the two-step reaction of 2-butene with HBr. The slower, rate-determining step (the one that crosses the higher energy barrier) is Step 1, which leads to the formation of the 2° carbocation intermediate. This carbocation intermediate lies in an energy minimum between the transition states for Steps 1 and 2. The carbocation intermediate (an electrophile) reacts within microseconds, or faster, with bromide ion (a nucleophile) in a nucleophile-electrophile combination to give 2-bromobutane. Note that the energy level for 2-bromobutane (the product) is lower than the energy level for 2-butene and HBr (the reactants). Thus, in this electrophilic addition reaction, energy is released; the reaction is exergonic.

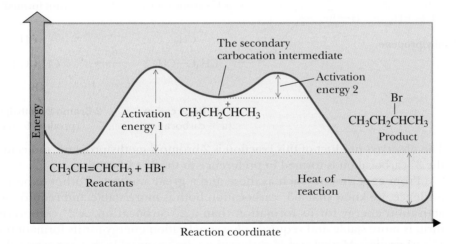

Reaction coordinate

Figure 6.4
An energy diagram for the two-step addition of HBr to 2-butene. The reaction is exergonic.

The reaction involves a collision with the appropriate trajectory for transfer of the electrophile to the alkene. As with any collision in everyday life, strain results from physical deformations introduced into the colliding partners. In this case, the collision must occur with enough energy to both contort the structure of the alkene toward that of a carbocation and to break the HBr bond. The physical structure of the transition state is the lowest energy contorted geometry that gives way to the carbocation intermediate.

The second step of the reaction is much easier than the first step because the reactant is now a carbocation. Carbocations are highly reactive electrophiles because of their positive charge and the lack of an octet at carbon, and they commonly have lifetimes of microseconds or less. Because the highly electrophilic cation intermediate is so unstable, the energy necessary in the collision with the nucleophilic bromide to surmount the second barrier is very low. As with a highly unstable object in the macroscopic world (such as a book teetering on the edge of a desk), only a small disturbance is necessary to give the intermediate a pathway to a lower energy state (such as tapping the book to cause it to fall).

Regioselectivity and the Relative Stabilities of Carbocations

Reaction of HX and an alkene can, at least in principle, give two different carbocation intermediates because the proton could be transferred to either of the doubly bonded carbon atoms, as illustrated by the reaction of HBr with propene.

$CH_3CH_2\text{—}CH_2$ $\longrightarrow$ $CH_3CH_2CH_2\ddot{B}r:$
Propyl cation   **1-Bromopropane**
(a 1° carbocation)   (not formed)

CH_3CH=CH_2 + H—$\ddot{B}r:$

Propene

$CH_3\overset{+}{C}H\text{—}CH_3$ $\longrightarrow$ CH_3CHCH_3
Isopropyl cation   **2-Bromopropane**
(a 2° carbocation)   (product formed)

The observed product is 2-bromopropane, indicating that the 2° carbocation intermediate is formed in preference to the 1° carbocation intermediate.

Similarly, in the reaction of HBr with 2-methylpropene, proton transfer to the carbon-carbon double bond might form either the isobutyl cation (a 1° carbocation) or the *tert*-butyl cation (a 3° carbocation).

$CH_3CH\text{—}CH_2$ $\longrightarrow$ $CH_3CHCH_2\ddot{B}r:$
Isobutyl cation   **1-Bromo-2-methylpropane**
(a 1° carbocation)   (not formed)

$CH_3\overset{CH_3}{\underset{}{C}}$=$CH_2$ + H—$\ddot{B}r:$

2-Methylpropene

$CH_3\overset{+}{C}\text{—}CH_3$ $\longrightarrow$ CH_3CCH_3
tert-Butyl cation   **2-Bromo-2-methylpropane**
(a 3° carbocation)   (product formed)

The observed product of this reaction is 2-bromo-2-methylpropane, indicating that the 3° carbocation is formed in preference to the 1° carbocation.

From experiments such as these and a great amount of other experimental evidence, we know that a 3° carbocation both is more stable and requires a lower activation energy for its formation than a 2° carbocation. A 2° carbocation, in turn, is more stable and requires a lower activation energy for its formation than a 1° carbocation. Methyl and 1° carbocations are so unstable they are rarely observed in solution. It follows, then, that a more stable carbocation intermediate forms faster than a less stable carbocation intermediate. Following is the order of stability of four types of alkyl carbocations.

Methyl cation   **Ethyl cation**   **Isopropyl cation**   **tert-Butyl cation**
(Methyl)   (1°)   (2°)   (3°)

Increasing carbocation stability

Now that we know the order of stability of carbocations, how do we account for this order? As we saw during the discussion of anion stability in Section 4.5, a system bearing a charge (either positive or negative) is more stable if the charge is delocalized. Using this principle, we can explain the order of stability of carbocations if we assume that alkyl groups bonded to a positively charged carbon release electron density toward that carbon and thereby help delocalize the positive charge on the cation. We account for the electron-releasing ability of alkyl groups bonded to a cationic carbon by two effects: the inductive effect and hyperconjugation.

The Inductive Effect

The inductive effect operates in the following way. The electron deficiency of the cationic carbon exerts an electron-withdrawing **inductive effect** that polarizes electrons from adjacent sigma bonds toward it. In this way, the positive charge of the cation is not localized on the trivalent carbon but rather is delocalized over nearby atoms. The larger the volume over which the positive charge is delocalized, the greater the stability of the cation. Thus, as the number of alkyl groups bonded to the cationic carbon increases, the stability of the cation increases. Figure 6.5 illustrates the electron-withdrawing inductive effect of the positively charged carbon and the resulting delocalization of charge. According to quantum mechanical calculations, the charge on carbon in the methyl cation is approximately $+0.645$, and the charge on each of the hydrogen atoms is $+0.118$. Thus, even in the methyl cation, the positive charge is not localized entirely on carbon. Rather, it is delocalized over the volume of space occupied by the entire ion. Delocalization of charge is even much more extensive in the *tert*-butyl cation.

Inductive effect
The polarization of the electron density of a covalent bond resulting from the electronegativity of a nearby atom.

(a) H_3C^+ (b) $(CH_3)_3C^+$

Figure 6.5
Electrostatic potential plots showing the distribution of the positive charge in (a) the methyl cation and (b) the *tert*-butyl cation. Electron donation by the methyl groups decreases the positive charge (blue area) on the central carbon of the *tert*-butyl cation.

Hyperconjugation

The second related effect by which alkyl groups stabilize carbocations is hyperconjugation. **Hyperconjugation** involves partial overlapping of the σ-bonding orbital of an adjacent C—H or C—C bond of the alkyl group with the vacant $2p$ orbital of the cationic carbon (Figure 6.6). In other words, some electron density of the alkyl group C—H or C—C bond is mixed into the $2p$ orbital. As shown in Figure 1.25 for analogous mixing, the result is lower and higher energy orbitals, the first filled and the second empty, respectively. In the case of hyperconjugation, the populated orbital remains primarily C—H bonding (Figure 6.6b) and the empty orbital remains primarily $2p$ (Figure 6.6c), but Figure 6.6b and c both show the delocalization that is the hallmark of hyperconjugation. The net result of hyperconjugation is an increase of electron density on the cationic carbon, thereby delocalizing the positive charge onto the adjacent alkyl groups. As more alkyl groups are bonded to a cationic carbon, the hyperconjugation effect becomes stronger and the carbocation becomes more stable.

Hyperconjugation
Interaction of electrons in a sigma-bonding orbital with the vacant $2p$ orbital of an adjacent positively charged carbon.

Figure 6.6
Hyperconjugation.

(a)
Cartoon diagram of the ethyl cation. There is delocalization of C—H bonding electrons into the empty $2p$ orbital of the positively charged carbon. The orbital that results from this delocalization contains two electrons (see **(b)**).

(b)
A calculated orbital populated with two electrons. This orbital is delocalized over the C—H bond and the $2p$ orbital of the positively charged carbon of the ethyl cation.

(c)
Calculated empty orbital residing mostly on the positively charged carbon of ethyl cation. Some empty orbital character exists on the methyl group also.

Example 6.3

Arrange these carbocations in order of increasing stability.

 (a)   **(b)**   **(c)**

Solution

Carbocation (a) is 2°, (b) is 3°, and (c) is 1°. In order of increasing stability they are c < a < b.

Problem 6.3

Arrange these carbocations in order of increasing stability.

(a) **(b)** **(c)**

Example 6.4

Propose a mechanism for the addition of HI to methylidenecyclohexane to give 1-iodo-1-methylcyclohexane. Which step in your mechanism is rate determining?

Solution

Propose a two-step mechanism similar to that proposed for the addition of HBr to propene.

Step 1: **Make a new bond between a nucleophile (π bond) and an electrophile—add a proton.** A rate-determining addition of a proton from HI to the carbon-carbon double bond gives a 3° carbocation intermediate.

Methylidenecyclohexane
(Methylenecyclohexane)

slow, rate determining

A 3° carbocation
intermediate

Step 2: **Make a new bond between a nucleophile and an electrophile.** Reaction of the 3° carbocation intermediate (an electrophile) with iodide ion (a nucleophile) completes the valence shell of carbon and gives the product.

fast

1-Iodo-1-methylcyclohexane

Problem 6.4

Propose a mechanism for addition of HI to 1-methylcyclohexene to give 1-iodo-1-methylcyclohexane. Which step in your mechanism is rate determining?

B. Addition of Water: Acid-Catalyzed Hydration

In the presence of an acid catalyst, most commonly concentrated sulfuric acid, water adds to an alkene to give an alcohol. The addition of water is called **hydration.**

Hydration
The addition of water.

$$CH_3CH=CH_2 + H_2O \xrightarrow{H_2SO_4} CH_3CH-CH_2$$
$$\overset{\displaystyle OH\ \ H}{|\ \ \ |}$$

Propene　　　　　**2-Propanol**

$$CH_3\overset{\displaystyle CH_3}{\underset{}{C}}=CH_2 + H_2O \xrightarrow{H_2SO_4} CH_3\overset{\displaystyle CH_3}{\underset{HO\ \ H}{C}}-CH_2$$

2-Methylpropene　　　　**2-Methyl-2-propanol**

In the case of simple alkenes, H adds to the carbon of the double bond with the greater number of hydrogens, and OH adds to the carbon with the fewer hydrogens. Thus, H—OH adds to alkenes in accordance with Markovnikov's rule.

Example 6.5

Draw a structural formula for the product of the acid-catalyzed hydration of 1-methylcyclohexene.

Solution

$$+ H_2O \xrightarrow{H_2SO_4}$$

1-Methylcyclohexene　　　　**1-Methylcyclohexanol**

Problem 6.5

Draw a structural formula for the product of each alkene hydration reaction:

(a) (structure) + H₂O $\xrightarrow{\text{H}_2\text{SO}_4}$ **(b)** (structure) + H₂O $\xrightarrow{\text{H}_2\text{SO}_4}$

The mechanism for acid-catalyzed hydration of alkenes is quite similar to what we already proposed for addition of HCl, HBr, and HI to alkenes and is illustrated by conversion of propene to 2-propanol. As you study this mechanism, note that it is consistent with the fact that acid is a catalyst; an H_3O^+ is consumed in Step 1 but another is generated in Step 3.

Mechanism   Acid-Catalyzed Hydration of Propene

Step 1: **Make a new bond between a nucleophile (π bond) and an electrophile— add a proton.**   Addition of a proton to propene from H_3O^+ gives a 2° carbocation intermediate.

$$CH_3CH=CH_2 + H-\overset{\cdot\cdot}{\underset{\underset{H}{|}}{O}}-H \underset{\text{determining}}{\overset{\text{slow, rate}}{\rightleftharpoons}} CH_3\overset{+}{C}HCH_3 + :\overset{\cdot\cdot}{\underset{\underset{H}{|}}{O}}-H$$

A 2° carbocation intermediate

Step 2: **Make a new bond between a nucleophile and an electrophile.**   The 2° carbocation intermediate (an electrophile) completes its valence shell by forming a new covalent bond with an unshared pair of electrons of the oxygen atom of water (a nucleophile) and gives an **oxonium ion.**

$$CH_3\overset{+}{C}HCH_3 + :\overset{\cdot\cdot}{\underset{\underset{H}{|}}{O}}-H \overset{\text{fast}}{\rightleftharpoons} CH_3\overset{\overset{\overset{\displaystyle H\underset{\displaystyle }{\overset{+}{\underset{\displaystyle O}{\diagup}}}H}{|}}{}}{C}HCH_3$$

An oxonium ion

Step 3: **Take a proton away.**   Loss of a proton from the relatively acidic oxonium ion to water gives the alcohol and generates a new acid catalyst.

$$CH_3\overset{\overset{\displaystyle H\underset{\displaystyle }{\overset{+}{\underset{\displaystyle O}{\diagup}}}H}{|}}{C}HCH_3 + :\overset{\cdot\cdot}{\underset{\underset{H}{|}}{O}}-H \overset{\text{fast}}{\rightleftharpoons} CH_3\overset{\overset{\displaystyle H\underset{\displaystyle }{\overset{\cdot\cdot}{\underset{\displaystyle O}{\diagup}}}:}{|}}{C}HCH_3 + H-\overset{\cdot\cdot}{\underset{\underset{H}{|}}{\overset{+}{O}}}-H$$

Example 6.6

Propose a mechanism for the acid-catalyzed hydration of methylidenecyclohexane to give 1-methylcyclohexanol. Which step in your mechanism is rate determining?

Solution

Propose a three-step mechanism similar to that for the acid-catalyzed hydration of propene.

Step 1: **Make a new bond between a nucleophile (π bond) and an electrophile—add a proton.**   Proton transfer from the acid catalyst to the alkene gives a 3° carbocation intermediate. Formation of the 3° carbocation intermediate is rate determining.

Oxonium ion
An ion in which oxygen bears a positive charge.

A 3° carbocation
intermediate

Step 2: Make a new bond between a nucleophile and an electrophile. Reaction of the 3° carbocation intermediate (an electrophile) with water (a nucleophile) completes the valence shell of carbon and gives an oxonium ion.

An oxonium ion

Step 3: Take a proton away. Proton transfer from the oxonium ion to water gives the alcohol and generates a new acid catalyst.

Problem 6.6

Propose a mechanism for the acid-catalyzed hydration of 1-methylcyclohexene to give 1-methylcyclohexanol. Which step in your mechanism is rate determining?

C. Carbocation Rearrangements

As we have seen in the preceding discussions, the expected product of electrophilic addition to a carbon-carbon double bond involves rupture of the π bond and formation of two new σ bonds in its place. In addition of HCl to 3,3-dimethyl-1-butene, however, only 17% of 2-chloro-3,3-dimethylbutane, the expected product, is formed. The major product is 2-chloro-2,3-dimethylbutane, a compound with a different connectivity of its atoms compared with that in the starting alkene. We say that formation of 2-chloro-2,3-dimethylbutane involves a **rearrangement**. Typically, either an alkyl group or a hydrogen migrates, each with its bonding electrons, from an adjacent atom to the electron-deficient atom. In the rearrangements we examine in this chapter, migration is to an adjacent electron-deficient carbon atom bearing a positive charge.

Rearrangement
A change in connectivity of the atoms in a product compared with the connectivity of the same atoms in the starting material.

3,3-Dimethyl-1-butene   **2-Chloro-3,3-dimethylbutane**   **2-Chloro-2,3-dimethylbutane**
(the expected product; 17%)   (the major product; 83%)
(a racemic mixture)

Formation of the rearranged product in this reaction can be accounted for by the following mechanism, the key step of which is a type of rearrangement called a **1,2-shift**. In the rearrangement shown in Step 2, the migrating group is a methyl group with its bonding electrons. A 1,2 shift represents a new mechanistic choice in addition to the four choices presented in Section 6.2F. A 1,2 shift is far less common than the other choices.

1,2-Shift
A type of rearrangement in which an atom or group of atoms moves with its bonding electrons from one atom to an adjacent electron-deficient atom.

Mechanism Carbocation Rearrangement in the Addition of HCl to an Alkene

Step 1: **Make a new bond between a nucleophile (π bond) and an electrophile—add a proton.** Proton transfer to the alkene gives a 2° carbocation intermediate.

A 2° carbocation
intermediate

Step 2: **1,2 Shift.** Migration of a methyl group with its bonding electrons from an adjacent carbon to the positively charged carbon of the 2° carbocation gives a more stable 3° carbocation. In this rearrangement, the major movement is that of the bonding electron pair with the methyl group following.

A 2° carbocation
intermediate

A 3° carbocation
intermediate

Step 3: **Make a new bond between a nucleophile and an electrophile.** Reaction of the 3° carbocation intermediate (an electrophile) with chloride ion (a nucleophile) gives the rearranged product.

The driving force for this rearrangement is the fact that the less stable 2° carbocation is converted to a more stable 3° carbocation. From the study of this and other carbocation rearrangements, we find that 2° carbocations rearrange to more stable 2° or 3° carbocations. They rearrange in the opposite direction only under special circumstances such as where the relief of ring strain provides added driving force.

Rearrangements also occur in the acid-catalyzed hydration of alkenes, especially when the carbocation formed in the first step can rearrange to a more stable carbocation. For example, acid-catalyzed hydration of 3-methyl-1-butene gives 2-methyl-2-butanol. In this example, the group that migrates is a hydrogen with its bonding pair of electrons, in effect a hydride ion H:⁻.

$$
\underset{\textbf{3-Methyl-1-butene}}{\overset{\overset{\displaystyle CH_3}{|}}{CH_3CHCH=CH_2}} + H_2O \xrightarrow{H_2SO_4} \underset{\textbf{2-Methyl-2-butanol}}{\overset{\overset{\displaystyle CH_3}{|}}{\underset{\underset{\displaystyle OH}{|}}{CH_3CCH_2CH_3}}}
$$

Example 6.7

Propose a mechanism for the acid-catalyzed hydration of 3-methyl-1-butene to give 2-methyl-2-butanol.

Solution

Following is a four-step mechanism for the formation of 2-methyl-2-butanol.

Step 1: **Make a new bond between a nucleophile (π bond) and an electrophile—add a proton.** Proton transfer from H_3O^+, the acid catalyst, to the double bond of the alkene gives a 2° carbocation intermediate.

A 2° carbocation
intermediate

Step 2: **1,2 Shift.** The less stable 2° carbocation rearranges to a more stable 3° carbocation by migration of a hydrogen with its pair of bonding electrons (in effect, a hydride ion).

A 2° carbocation   A 3° carbocation
intermediate   intermediate

Step 3: **Make a new bond between a nucleophile and an electrophile.** Reaction of the 3° carbocation intermediate (an electrophile) with water (a nucleophile) completes the valence shell of carbon and gives an oxonium ion.

An oxonium ion

Step 4: **Take a proton away.** Proton transfer from the oxonium ion to water gives the product and generates a new H_3O^+ to continue the hydration reaction.

2-Methyl-2-butanol

Problem 6.7

The acid-catalyzed hydration of 3,3-dimethyl-1-butene gives 2,3-dimethyl-2-butanol as the major product. Propose a mechanism for the formation of this alcohol.

3,3-Dimethyl-1-butene   **2,3-Dimethyl-2-butanol**

D. Addition of Bromine and Chlorine

Chlorine, Cl_2, and bromine, Br_2, react with alkenes at room temperature by adding halogen atoms to the two carbon atoms of the double bond with formation of two new carbon-halogen bonds. Fluorine, F_2, adds to alkenes, but, because its reactions are very fast and difficult to control, this reaction is not a useful laboratory procedure. Iodine, I_2, also adds, but the reaction is not preparatively useful. Halogenation with bromine or chlorine is generally carried out either with the pure reagents or by mixing them in an inert solvent such as CH_2Cl_2.

$$CH_3CH = CHCH_3 + \boxed{Br_2} \xrightarrow[CH_2Cl_2]{} CH_3CH - CHCH_3$$

2-Butene   **2,3-Dibromobutane**

The addition of bromine or chlorine to a cycloalkene gives a *trans*-dihalocycloalkane formed as a racemic mixture. The addition of bromine to cyclohexene, for example, gives *trans*-1,2-dibromocyclohexane:

Cyclohexene   ***trans*-1,2-Dibromocyclohexane**
(a racemic mixture)

At first glance, the two *trans* enantiomers may appear to be the same structure. However, there is no plane of symmetry or center of symmetry so they are both chiral. You should make molecular models of both enantiomers and convince yourself that they are indeed nonsuperposable mirror images, and not just the same molecule viewed from a different perspective.

We discuss the addition stereochemistry of Cl_2 and Br_2 to alkenes in more detail in Section 6.7. For now, it is sufficient to point out that these reactions proceed with anti (from the opposite side or face) addition of halogen atoms; that is, they occur with **anti stereoselectivity**.

Reaction of bromine with an alkene is a particularly useful qualitative test for the presence of a carbon-carbon double bond. If we dissolve bromine in dichloromethane, the solution is red. Both alkenes and dibromoalkanes are colorless. If we now mix a few drops of the bromine solution with an alkene, a dibromoalkane is formed, and the red solution becomes colorless.

Stereoselective reaction
A reaction in which one stereoisomer is formed in preference to all others. A stereoselective reaction may be enantioselective or diastereoselective, as the case may be.

Anti stereoselectivity
The addition of atoms or groups of atoms to opposite faces of a carbon-carbon double bond.

Example 6.8

Complete these reactions, showing the stereochemistry of the product.

(a) $+ Br_2 \xrightarrow[CH_2Cl_2]{}$ **(b)** $+ Cl_2 \xrightarrow[CH_2Cl_2]{}$

Solution

Addition of both Br_2 and Cl_2 occurs with anti stereoselectivity, which means that the halogen atoms are *trans* to each other in each product.

(a) $+ Br_2 \xrightarrow[CH_2Cl_2]{}$

***trans*-1,2-Dibromocyclopentane**
(a racemic mixture)

(b)

(a racemic mixture)

Problem 6.8

Complete these reactions.

(a) $CH_3\overset{\displaystyle CH_3}{\underset{\displaystyle CH_3}{C}}CH{=}CH_2 + Br_2 \xrightarrow{CH_2Cl_2}$ **(b)**

$+ Cl_2 \xrightarrow{CH_2Cl_2}$

Anti Stereoselectivity and Bridged Halonium Ion Intermediates

We explain the addition of bromine and chlorine to alkenes and its anti stereoselectivity by the following two-step mechanism.

Mechanism Addition of Bromine with Anti Stereoselectivity

Step 1: **Make a new bond between a nucleophile (π bond) and an electrophile.** Reaction is initiated by interaction of the π electrons of the alkene with bromine (or chlorine as the case may be) that acts likes an electrophile to form an intermediate in which bromine bears a positive charge. A bromine atom bearing a positive charge is called a **bromonium ion**, and the cyclic structure of which it is a part is called a **bridged bromonium ion**.

Although a bridged bromonium ion may look odd because it has two bonds to bromine, it is nevertheless an acceptable Lewis structure. Calculation of formal charge places a positive charge on bromine. This intermediate is a hybrid of three resonance-contributing structures. Because of the planarity of the atoms forming the π bond, the bridged bromonium ion can form with equal probability on the top or bottom face of the alkene. In this step, the Br_2 or Cl_2 acts as an electrophile because the halogen-halogen bond is weak, and reaction leads to the departure of the stable X^- anion.

Step 2: **Make a new bond between a nucleophile and an electrophile.** Attack of bromide ion (a nucleophile) on carbon from the side opposite the bromonium ion (an electrophile) opens the three-membered ring to give the anti product. The bromines are not only anti but also in the same plane (coplanar). Thus, we call this an anti-coplanar attack. Attack by Br^- can occur at either carbon of the bromonium ion.

Anti (coplanar) orientation of added bromine atoms

Newman projection of the product

Addition of chlorine or bromine to cyclohexene and its derivatives gives a *trans* diaxial product because only axial positions on adjacent atoms of a cyclohexane ring are anti and coplanar. The initial *trans* diaxial conformation of the product is in equilibrium with the *trans* diequatorial conformation, and, in simple derivatives of cyclohexane, the *trans* diequatorial conformation is more stable and predominates. Because the original bromonium ion can form on either face of the double bond with equal probability, both *trans* enantiomers are formed as a racemic mixture.

(1S,2S)-1,2-Dibromo-cyclohexane

(1R,2R)-1,2-Dibromo-cyclohexane

Trans diaxial conformations (less stable)

Trans diequatorial conformations (more stable)

E. Addition of HOCl and HOBr

Treating an alkene with Br_2 or Cl_2 in the presence of water results in addition of OH and Br, or OH and Cl, to the carbon-carbon double bond to give a **halohydrin**.

Halohydrin
A compound containing a halogen atom and a hydroxyl group on adjacent carbons; those containing Br and OH are bromohydrins, and those containing Cl and OH are chlorohydrins.

Propene

1-Chloro-2-propanol
(a chlorohydrin)
(a racemic mixture)

Addition of HOCl and HOBr is regioselective (halogen adds to the less substituted carbon atom) and anti stereoselective. Both the regioselectivity and anti stereoselectivity are illustrated by the addition of HOBr to 1-methylcyclopentene. Bromine and the hydroxyl group add anti to each other with Br bonding to the less substituted carbon and OH bonding to the more substituted carbon. Dimethyl sulfoxide (DMSO) is used as a cosolvent with water to assist the solubility of the alkene.

1-Methylcyclopentene   **2-Bromo-1-methylcyclopentanol**
(a racemic mixture)

To account for the regioselectivity and anti stereoselectivity of halohydrin reactions, chemists propose a three-step mechanism.

Mechanism Halohydrin Formation and Its Anti Stereoselectivity

Step 1: Make a new bond between a nucleophile (π bond) and an electrophile.
Reaction of the π electrons of the alkene with bromine (an electrophile) gives a bridged bromonium ion intermediate. This intermediate has some of the character of a carbocation (to account for the regioselectivity) and some of the character of a halonium ion (to account for the anti stereoselectivity). The secondary carbocation makes a substantial contribution to the structure of the resonance hybrid; the primary carbocation is higher in energy and makes little contribution.

This carbocation is a major contributing structure

The bridged bromonium ion plays an important role

This structure plays little role

Step 2: Make a new bond between a nucleophile and an electrophile. Attack of H_2O (a nucleophile) on the more substituted carbon of the bridged bromonium ion (an electrophile) opens the three-membered ring.

In the case of a bromonium ion derived from a symmetrical alkene, both carbons are attacked by H_2O with equal probability. In the case of unsymmetrical alkenes, as for example that derived from 2-methylpropene, there is preferential opening of the cyclic bromonium ion intermediate by attack of H_2O on the more substituted carbon of the alkene. At first glance, this may seem counterintuitive because the more substituted carbon might be considered less accessible to a nucleophile. However, the experimentally observed preferential attack at the more substituted carbon atom can be explained by a combination of two factors working together.

Carbocation Character

As the accompanying electrostatic potential maps show, there is more carbocation character on the more substituted carbon, which directs attack of the nucleophile preferentially to this carbon. Recall that alkyl groups stabilize carbocations, explaining the greater carbocation character at the more substituted carbon.

Activation Energy To Reach the Ring-Opening Transition State

As the accompanying electrostatic potential maps also show, the carbon-halogen bond to the more substituted carbon of the halonium ion is longer than the bond to the less substituted carbon. This difference in bond lengths in the transition state means that the ring-opening transition state can be reached more easily by attack at the more substituted carbon.

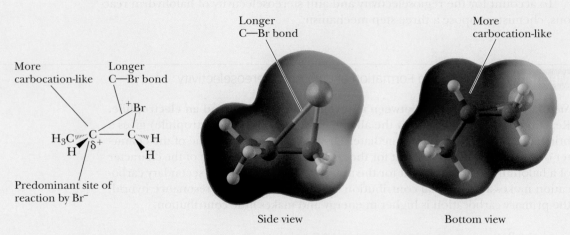

Bridged bromonium ion from propene.

Step 3: Take a proton away. Proton transfer to water completes the reaction.

Example 6.9

Draw the structure of the bromohydrin formed by treating 2-methylpropene with Br_2/H_2O.

Solution

Addition is regioselective, with —OH adding to the more substituted carbon and —Br adding to the less substituted carbon.

$$CH_3-\underset{\underset{\displaystyle CH_3}{|}}{C}=CH_2 \; + \; Br_2 \; + \; H_2O \; \longrightarrow \; CH_3-\underset{\underset{\displaystyle HO}{|}}{\overset{\overset{\displaystyle CH_3}{|}}{C}}-\underset{\underset{\displaystyle Br}{|}}{CH_2} \; + \; HBr$$

2-Methylpropene

1-Bromo-2-methyl-2-propanol
(achiral)

Problem 6.9

Draw the structure of the chlorohydrin formed by treating 1-methylcyclohexene with Cl_2/H_2O.

F. Oxymercuration-Reduction

The hydration of an alkene can be accomplished by treating it with mercury(II) acetate (mercuric acetate) in water followed by reduction of the resulting organomercury compound with sodium borohydride, $NaBH_4$. In the following structural formulas for mercury(II) acetate, the acetate group is written in full in the first formula and abbreviated as AcO in the other formulas.

$$CH_3CO-Hg-OCCH_3 \quad AcO-Hg-OAc \quad \text{or} \quad Hg(OAc)_2$$

Mercury(II) acetate
(Mercuric acetate)

The result of oxymercuration followed by sodium borohydride reduction is Markovnikov addition of H—OH to an alkene.

1-Hexene → **(S)-2-Hexanol** + **(R)-2-Hexanol**
(a racemic mixture)

Reagents: 1. Hg(OAc)₂, H₂O 2. NaBH₄

We discuss this reaction in two stages, first oxymercuration and then reduction.

Oxymercuration, the addition of mercury(II) to one carbon of the double bond and oxygen to the other, is illustrated by the first step in the two-step conversion of 1-hexene to 2-hexanol. Oxymercuration is regioselective: HgOAc becomes bonded to the less substituted carbon of the alkene, and OH of water becomes bonded to the more substituted carbon.

Sidebar:

Oxymercuration-reduction
A method for converting an alkene to an alcohol. The alkene is treated with mercury(II) acetate followed by reduction with sodium borohydride.

1-Hexene + **Mercury(II) acetate** + H₂O ⟶ **An organomercury compound** (a racemic mixture) + **Acetic acid** (CH₃COH)

Reduction of the organomercury compound by sodium borohydride, NaBH₄, replaces HgOAc by H.

(a racemic mixture) → **2-Hexanol** (a racemic mixture) + **Acetic acid** (CH₃COH) + Hg

Oxymercuration of 3,3-dimethyl-1-butene followed by NaBH₄ reduction gives racemic 3,3-dimethyl-2-butanol exclusively and illustrates a very important feature of this reaction sequence: It occurs without rearrangement.

3,3-Dimethyl-1-butene → **(R)-3,3-Dimethyl-2-butanol** + **(S)-3,3-Dimethyl-2-butanol**
(a racemic mixture)

Reagents: 1. Hg(OAc)₂, H₂O 2. NaBH₄

You might compare the product of oxymercuration-reduction of 3,3-methyl-1-butene with the product formed by acid-catalyzed hydration of the same alkene (Section 6.3C). In the former, no rearrangement occurs. In the latter, the major product is 2,3-dimethyl-2-butanol, a compound formed by rearrangement. The fact that no rearrangement occurs during oxymercuration-reduction indicates that at no time is a free carbocation intermediate formed.

The stereoselectivity of the oxymercuration step is illustrated by the reaction of mercury(II) acetate with cyclopentene.

Cyclopentene

(Anti addition of OH and HgOAc)
(racemic)

The fact that the oxymercuration step is both regioselective and anti stereo-selective has led chemists to propose the following mechanism, which is closely analogous to that for the addition of Br_2 and Cl_2 to an alkene (Section 6.3D).

Mechanism Oxymercuration-Reduction of an Alkene

Step 1: Break a new bond to give stable molecules or ions. Dissociation of mercury(II) acetate gives $AcOHg^+$ (an electrophile) and acetate anion.

$$AcO-Hg-OAc \longrightarrow AcO-Hg^+ + AcO^-$$

(an electrophile)

Step 2: Make a new bond between a nucleophile (π bond) and an electrophile. Attack of the alkene π bond onto $AcOHg^+$ (an electrophile) forms a bridged mercurinium ion intermediate.

An open carbocation
intermediate
(a minor contributor)

A bridged mercurinium
ion intermediate
(the major contributor)

An open carbocation
intermediate
(a negligible contributor)

This intermediate closely resembles a bridged bromonium ion intermediate (Section 6.3D). However, in the bridged mercurinium ion intermediate, the two π electrons of the carbon-carbon double bond form a ring containing three atoms bonded by two electrons. The open cation structure with the positive charge on the 2° carbon is a minor contributing structure to the resonance hybrid. The open cation contributor with the positive charge on the 1° carbon is a negligible contributor.

The fact that oxymercuration occurs without rearrangement indicates that the intermediate formed in Step 2 is not a true carbocation but rather a resonance hybrid largely with the character of a bridged mercurinium ion intermediate.

Step 3: Make a new bond between a nucleophile and an electrophile. Anti attack of water (a nucleophile) on the bridged mercurinium ion intermediate (an electrophile) occurs at the more substituted carbon to open the three-membered ring.

Anti addition of
HgOAc and HOH

Proton transfer from this product to water completes oxymercuration of the alkene. We account for the regioselectivity just as we did for the regioselectivity of halohydrin formation in Section 6.3E. Of the two carbons of the mercurinium ion intermediate, the more substituted carbon has a greater degree of partial positive charge and is attacked by the nucleophile, H_2O. In addition, computer modeling indicates that the carbon-mercury bond to the more substituted carbon of the bridged mercurinium ion intermediate is longer than the one to the less substituted carbon, which means that the ring-opening transition state is reached more easily by attack at the more substituted carbon.

Step 4: Reduction of the C—HgOAc bond to a C—H bond gives the final product and metallic mercury. The mechanism of this step is beyond the scope of this chapter and will not be discussed in detail, which is why we have not drawn arrows to indicate movement of electrons. The mechanism is thought to involve radicals (unpaired electrons). The key from our point of view is that the stereochemistry of any chiral carbon bearing the Hg atom is scrambled in this reduction step. The net result is that even though the OH group and the Hg atom add with anti stereochemistry in the oxymercuration step, following reduction of Hg to H, the products reflect a mixture of syn and anti addition of the H atom and OH group.

6.4 | Hydroboration-Oxidation

The net reaction from hydroboration and subsequent oxidation of an alkene is hydration of a carbon-carbon double bond. Because hydrogen is added to the more substituted carbon of the double bond and —OH to the less substituted carbon, we refer to the regiochemistry of hydroboration and subsequent oxidation as non-Markovnikov hydration:

Hydroboration-oxidation of alkenes is a valuable laboratory method for the regioselective and stereoselective hydration of alkenes. Furthermore, this sequence of reactions occurs without rearrangement.

Hydroboration is the addition of borane, BH_3, to an alkene to form a trialkylborane. The overall reaction occurs in three successive steps. BH_3 reacts with one molecule of alkene to form an alkylborane, then with a second molecule of alkene to form a dialkylborane, and finally with a third molecule of alkene to form a trialkylborane.

Hydroboration-oxidation
A method for converting an alkene to an alcohol. The alkene is treated with borane (BH_3) to give a trialkylborane, which is then oxidized with alkaline hydrogen peroxide to give the alcohol.

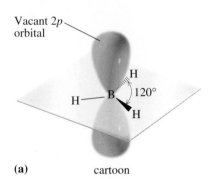

Vacant 2*p* orbital

(a)   cartoon

(b)   calculated

(a) Cartoon and **(b)** Calculated orbitals and bond angles in BF_3.

Syn stereoselective
The addition of atoms or groups of atoms to the same face of a carbon–carbon double bond.

Borane cannot be prepared as a pure compound because it dimerizes to diborane, B_2H_6, a toxic gas that ignites spontaneously in air.

$$2\ BH_3 \rightleftharpoons B_2H_6$$

Borane   **Diborane**

However, BH_3 forms stable Lewis acid-base complexes with ethers. Borane is most often used as a commercially available solution of BH_3 in THF.

Tetrahydrofuran   **BH₃ · THF**
(THF)

    Boron, atomic number 5, has three electrons in its valence shell. To bond with three other atoms, boron uses sp^2 hybrid orbitals. The unoccupied $2p$ orbital of boron is perpendicular to the plane created by boron and the three other atoms to which it is bonded. An example of a stable, trivalent boron compound is boron trifluoride, BF_3, a planar molecule with F—B—F bond angles of 120° (Section 1.2E). Because of the vacant $2p$ orbital in the valence shell of boron, BH_3, BF_3, and all other trivalent compounds of boron are electrophiles. These compounds of boron resemble carbocations, except that, unlike carbocations, they are electrically neutral. BH_3 is a planar molecule with H—B—H bond angles of 120° (see margin).
    Addition of borane to alkenes is regioselective and stereoselective.

- Regioselective: upon addition of borane to an unsymmetrical alkene, boron becomes bonded predominantly to the less substituted carbon of the double bond. **(non-Markovnikov)**.
- Stereoselective: hydrogen and boron add from the same face of the double bond; that is, the reaction is **syn** (from the same side) **stereoselective**.

Both the regioselectivity and syn stereoselectivity are illustrated by hydroboration of 1-methylcyclopentene.

1-Methylcyclopentene   (Syn addition of BH₃)
(*R* = 2-methylcyclopentyl)
(a racemic mixture)

    Note that other textbooks will refer to the regioselectivity of the hydroboration–oxidation reaction as anti-Markovnikov, while we prefer the term non-Markovnikov. We do not favor the use of anti-Markovnikov because it can be confusing to students due to the syn, not anti, stereochemistry of addition observed with this reaction.

| Mechanism | Hydroboration |

Step 1: The addition of borane to an alkene is initiated by coordination of the vacant $2p$ orbital of boron with the electron pair of the π bond. This coordination is a Lewis acid-base interaction, analogous to the coordination of THF to borane shown above. We designate this coordination with an arrow from the alkene to the boron.

$$CH_3CH_2CH_2CH{=}CH_2 \xrightleftharpoons{\ HBR_2\ } CH_3CH_2CH_2CH{=}CH_2 \quad H{-}B\overset{R}{\underset{R}{\diagdown}}$$

Step 2: **Make a new bond between a nucleophile (π bond) and an electrophile with simultaneous bond formation to H.** After coordination, the boron adds to the less substituted carbon of the alkene, thereby placing the hydrogen on the more substituted carbon, via a cyclic, four-centered transition state. Boron and hydrogen add simultaneously and from the same face of the double bond (syn addition).

$$\underset{\delta-}{H}\!-\!\underset{\delta+}{B} \qquad\qquad \overset{H}{\underset{|}{}}\ \overset{B}{\underset{|}{}}$$

$$CH_3CH_2CH_2CH{=}CH_2 \longrightarrow CH_3CH_2CH_2CH{-}CH_2$$

We account for the regioselectivity by a combination of steric and electronic factors. In terms of steric effects, boron, the larger part of the reagent, adds selectively to the less hindered carbon of the double bond, and hydrogen, the smaller part of the reagent, adds to the more hindered carbon. It is believed that the observed regioselectivity is due largely to these steric effects.

Electronic effects probably also influence the regioselectivity. The electronegativity of hydrogen (2.1) is slightly greater than that of boron (2.0); hence, there is a small degree of polarity (approximately 5%) to each B—H bond, with boron bearing a partial positive charge and hydrogen a partial negative charge. It is proposed that there is some degree of carbocation character in the transition state and that the partial positive charge is on the more substituted carbon.

$$\underset{\delta-}{H}\!-\!\underset{\delta+}{B} \qquad\qquad \underset{\delta-}{H}\text{-}\text{-}\underset{}{B}$$

$$CH_3CH_2CH_2CH{=}CH_2 \longrightarrow CH_3CH_2CH_2\underset{\delta+}{CH}{=}\!=\!CH_2$$

Starting reagents	Transition state
	(some carbocation character exists in the transition state)

Trialkylboranes are rarely isolated. Rather, they are converted directly to other products formed by substitution of another atom (H, O, N, C, or halogen) for boron. One of the most important reactions of trialkylboranes is with hydrogen peroxide in aqueous sodium hydroxide. Hydrogen peroxide is an oxidizing agent and, under these conditions, oxidizes a trialkylborane to an alcohol and sodium borate, Na_3BO_3.

Mechanism Oxidation of a Trialkylborane by Alkaline Hydrogen Peroxide

Step 1: **Make a new bond between a nucleophile and an electrophile.** Donation of a pair of electrons from a hydroperoxide ion (a nucleophile) to the boron atom of the trialkylborane (an electrophile) gives an intermediate in which boron has a filled valence shell and bears a negative formal charge.

$$HOOH + OH^- \rightleftharpoons HOO^- + HOH$$

Hydrogen peroxide	Hydroperoxide ion

negative formal charge on boron

$$\underset{|}{\overset{R}{\underset{R}{R{-}B}}} + \ddot{\underset{..}{O}}{-}\ddot{\underset{..}{O}}{-}H \rightleftharpoons \underset{|}{\overset{R}{\underset{R}{R{-}B}}}{-}\ddot{\underset{..}{O}}{-}\ddot{\underset{..}{O}}{-}H$$

A trialkylborane (an electrophile)	**Hydroperoxide ion** (a nucleophile)

Step 2: 1,2 Shift. Rearrangement of an R group with its pair of bonding electrons to an adjacent oxygen (a 1,2-shift) results in ejection of hydroxide ion.

$$R-\overset{\underset{|}{R}}{\overset{-}{B}}-\overset{..}{\overset{..}{O}}-\overset{..}{\overset{..}{O}}-H \longrightarrow R-\overset{\underset{|}{R}}{B}-\overset{..}{\overset{..}{O}}: + \ ^-\overset{..}{O}-H$$

Two more reactions with hydroperoxide ion followed by rearrangements give a trialkylborate.

$$R-\overset{\underset{|}{R}}{B}-\overset{..}{\overset{..}{O}}: \ \xrightarrow[\text{NaOH}]{H_2O_2} \ \xrightarrow[\text{NaOH}]{H_2O_2} \ R-\overset{..}{\overset{..}{O}}-\overset{\overset{..}{\overset{|}{O}-R}}{B}-\overset{..}{\overset{..}{O}}-R$$

A trialkylborate
(a triester of boric acid)

Step 3: Reaction of the trialkylborate with aqueous NaOH gives the alcohol and sodium borate.

$$(RO)_3B \quad + \ 3\,NaOH \longrightarrow 3\,ROH + \quad Na_3BO_3$$

A trialkylborate   **Sodium borate**

Hydrogen peroxide oxidation of a trialkylborane is stereoselective in that the configuration of the alkyl group is retained; whatever the position of boron in relation to other groups in the trialkylborane, the OH group by which it is replaced occupies the same position. Thus, the net result of hydroboration-oxidation of an alkene is syn stereoselective addition of H and OH to a carbon-carbon double bond, combined with non-Markovnikov regioselectivity.

Example 6.10

Draw structural formulas for the alcohol formed by hydroboration-oxidation of each alkene.

(a) $CH_3\overset{\underset{|}{CH_3}}{C}{=}CHCH_3$ (b)

Solution

(a)   +   (b)   +

(a racemic mixture)   (a racemic mixture)

Problem 6.10

Draw structural formulas for the alkene that gives each alcohol upon hydroboration-oxidation.

(a)   (b)

6.5 | Oxidation

We begin this section with a review of the definitions of oxidation and reduction. We then consider two common oxidation reactions of alkenes.

Oxidation is the loss of electrons, and **reduction** is the gain of electrons. In the following reactions, propene is transformed into two different compounds by reactions we study in this and the following section. The first reaction involves reduction and the second involves oxidation. These equations, however, do not specify what reagents are necessary to bring about the particular transformation. Each does specify, however, that the carbon atoms of the products are derived from those of propene.

Oxidation
The loss of electrons. Alternatively, either the loss of hydrogens, the gain of oxygens, or both.

Reduction
The gain of electrons. Alternatively, either the gain of hydrogen, loss of oxygens, or both.

$$CH_3CH=CH_2 \quad \text{Propene}$$

reduction → CH_3CH-CH_2 (with H, H) **Propane**

oxidation → CH_3CH-CH_2 (with OH, OH) **1,2-Propanediol**

Example 6.11

Use a balanced half-reaction to show that each transformation involves an oxidation.

(a) $CH_3CH_2CH_2OH \longrightarrow CH_3CH_2CH$ (with =O)

(b) $CH_3CH=CH_2 \longrightarrow CH_3CH + HCH$ (each with =O)

Solution

First complete a material balance and then a charge balance.

(a) $CH_3CH_2CH_2OH \longrightarrow CH_3CH_2CH + 2\,H^+ + 2e^-$

(b) $CH_3CH=CH_2 + 2\,H_2O \longrightarrow CH_3CH + HCH + 4\,H^+ + 4e^-$

The first transformation is a two-electron oxidation; the second is a four-electron oxidation. To bring each about requires an oxidizing agent.

Problem 6.11

Use a balanced half-reaction to show that each transformation involves a reduction.

(a) cyclohexanone → cyclohexanol

(b) $CH_3CH_2COH \longrightarrow CH_3CH_2CH_2OH$ (with =O)

As an alternative way to recognize oxidation/reduction, recall from your course in general chemistry that oxidation and reduction can be defined in terms of the loss or gain or oxygens or hydrogens. For organic compounds:

oxidation: The addition of O to and/or removal of H from a carbon atom
reduction: The removal of O from and/or addition of H to a carbon atom

Example 6.12

Tell which of these transformations are oxidations and which are reductions based on whether there is addition or removal of O or H.

(a) $CH_3CH_2OH \longrightarrow CH_3CH$ (with =O)

(b) $CH_3CCH_2COH \longrightarrow CH_3CHCH_2COH$ (with =O, =O → OH, =O)

(c) $CH_3CH=CH_2 \longrightarrow CH_3CH_2CH_3$

Solution

(a) Oxidation; there is a loss of two hydrogens.
(b) Reduction; there is a gain of two hydrogens.
(c) Reduction; there is a gain of two hydrogens.

Problem 6.12

Tell which of these transformations are oxidations and which are reductions based on whether there is addition or removal of O or H.

A. OsO₄—Oxidation of an Alkene to a Glycol

Glycol, Vicinal diol
A compound with hydroxyl (—OH) groups on adjacent carbons.

Osmium tetroxide, OsO_4, and certain other transition metal oxides are effective oxidizing agents for the conversion of an alkene to a 1,2-diol (a **glycol**, also known as a **vicinal diol**). OsO_4 is highly electrophilic and adds in a single step to the alkene. Oxidation of an alkene by OsO_4 is syn stereoselective; it involves syn addition of an OH group to each carbon of the double bond. For example, oxidation of cyclopentene gives *cis*-1,2-cyclopentanediol. Note that both *cis* and *trans* isomers are possible for this glycol but that only the *cis* glycol forms in this oxidation.

A cyclic osmate   ***cis*-1,2-Cyclopentanediol**
                         (a *cis* glycol)

The syn stereoselectivity of the osmium tetroxide oxidation of an alkene is accounted for by the formation of a cyclic osmate in such a way that the five-membered osmium-containing ring is bonded in a *cis* configuration to the original alkene. Osmates can be isolated and characterized. Usually, the osmate is treated directly with a reducing agent, such as $NaHSO_3$, which cleaves the osmium-oxygen bonds to give a *cis* glycol and reduced forms of osmium.

The drawbacks of OsO_4 are that it is both expensive and highly toxic. One strategy to circumvent the high cost is to use it in catalytic amounts along with a stoichiometric amount of another oxidizing agent whose purpose is to reoxidize the reduced forms of osmium and, thus, recycle the osmium reagent. Secondary oxidizing agents commonly used for this purpose are hydrogen peroxide and *tert*-butyl hydroperoxide. When this procedure is used, there is no need for a reducing step using $NaHSO_3$.

$$CH_3 \mid$$
$$CH_3COOH \mid$$
HOOH CH_3

Hydrogen   *tert*-Butyl hydroperoxide
peroxide   **(*t*-BuOOH)**

One way to tell if these or other transformations involve oxidation, reduction, or neither is to use the method of balanced half-reactions.

To write a balanced half-reaction:

1. Write a half-reaction showing the organic reactant(s) and product(s).
2. Complete a material balance; that is, balance the number of atoms on each side of the half-reaction. To balance the number of oxygens and hydrogens for a reaction that takes place in acid solution, use H_2O for oxygens and then H^+ for hydrogens. For a reaction that takes place in basic solution, use H_2O and OH^-.
3. Complete a charge balance; that is, balance the charge on both sides of the half-reaction. To balance the charge, add electrons, e^-, to one side or the other. The equation completed in this step is a balanced half-reaction.

If electrons appear on the right side of the balanced half reaction, the reactant gives up electrons and is oxidized. If electrons appear on the left side of a balanced half reaction, the reactant has gained electrons and is reduced. If no electrons appear in the balanced half reaction, then the transformation involves neither oxidation nor reduction. Let us apply these steps to the transformation of propene to propane.

Step 1:   Half-reaction $CH_3CH{=}CH_2 \longrightarrow CH_3CH_2CH_3$

Step 2:   Material balance $CH_3CH{=}CH_2 + 2\,H^+ \longrightarrow CH_3CH_2CH_3$

Step 3:   Balanced half-reaction $CH_3CH{=}CH_2 + 2\,H^+ + 2e^- \longrightarrow CH_3CH_2CH_3$

Because two electrons appear on the left side of the balanced half-reaction (Step 3), conversion of propene to propane is a two-electron reduction. To bring it about requires use of a reducing agent.

A balanced half-reaction for the transformation of propene to 1,2-propanediol requires two electrons on the right side of the equation for a charge balance; this transformation is a two-electron oxidation.

$$
\underset{\textbf{Propene}}{CH_3CH{=}CH_2} + 2\,H_2O \longrightarrow \underset{\textbf{1,2-Propanediol}}{CH_3\overset{\overset{\displaystyle OH}{|}}{CH}{-}\overset{\overset{\displaystyle OH}{|}}{CH_2}} + 2\,H^+ + 2e^-
$$

Balanced half-reaction:

Following is a balanced half-reaction for the transformation of propene to 2-propanol.

$$
\text{Balanced half-reaction:} \quad \underset{\textbf{Propene}}{CH_3CH{=}CH_2} + H_2O \longrightarrow \underset{\textbf{2-Propanol}}{CH_3\overset{\overset{\displaystyle OH}{|}}{CH}CH_3}
$$

Because no electrons are required to achieve an electrical balance in the half-reaction, conversion of propene to 2-propanol is neither oxidation nor reduction; this reaction can be brought about by acid-catalyzed hydration of propene.

It is important to realize that this strategy for recognizing oxidation and reduction is only that, a strategy. In no way does it give any indication of how a particular oxidation or reduction might be carried out in the laboratory. For example, the balanced half-reaction for the transformation of propene to propane requires $2H^+$ and $2e^-$. Yet by far the most common laboratory procedure for reducing propene to propane does not involve H^+ at all; rather it involves molecular hydrogen, H_2, and a transition metal catalyst (Section 6.6).

B. Ozone—Cleavage of a Carbon-Carbon Double Bond (Ozonolysis)

Treating an alkene with ozone, O_3, followed by a suitable work-up cleaves the carbon-carbon double bond and forms two carbonyl ($C=O$) groups in its place. Once again, as with all reagents discussed in this chapter, O_3 is strongly electrophilic. This reaction is noteworthy because it is one of the very few organic reactions that breaks C—C bonds. The alkene is dissolved in an inert solvent, such as CH_2Cl_2, and a stream of ozone is bubbled through the solution. The products isolated from ozonolysis depend on the reaction conditions. Hydrolysis of the reaction mixture with water yields hydrogen peroxide, an oxidizing agent that can bring about further oxidations. To prevent side reactions caused by reactive peroxide intermediates, a weak reducing agent, most commonly dimethyl sulfide, $(CH_3)_2S$, is added during the work-up to reduce peroxides to water.

$$CH_3C{=}CHCH_2CH_3 \xrightarrow[\text{2. } (CH_3)_2S]{\text{1. } O_3} CH_3\overset{O}{\overset{\|}{C}}CH_3 \; + \; H\overset{O}{\overset{\|}{C}}CH_2CH_3$$

2-Methyl-2-pentene   **Propanone**   **Propanal**
(a ketone)   (an aldehyde)

The initial product of reaction of an alkene with ozone is an adduct called a molozonide, which rearranges under the conditions of the reaction to an isomeric compound called an ozonide. Low-molecular-weight ozonides are explosive and are rarely isolated. They are treated directly with a weak reducing agent to give the carbonyl-containing products.

$$CH_3CH{=}CHCH_3 \xrightarrow{O_3} \left[CH_3CH{-}CHCH_3 \right] \longrightarrow \; \text{An ozonide} \xrightarrow{(CH_3)_2S} 2\ CH_3\overset{O}{\overset{\|}{C}}H$$

2-Butene   A molozonide   An ozonide   **Acetaldehyde**

To understand how an ozonide is formed, we must first examine the structure of ozone. We can write this molecule as a hybrid of four contributing structures, all of which show separation of unlike charge.

It is not possible to write a Lewis structure for ozone without separation of charges.

Mechanism   Formation of an Ozonide

Step 1: Ozone reacts with the alkene with both electrophilic and nucleophilic character in a single, simultaneous step.

A molozonide

Step 2: Break a bond to give stable molecules or ions. Relocating valence electrons in the molozonide results in cleavage of one carbon-carbon and one oxygen-oxygen bond. The resulting fragments then recombine to form an ozonide.

Step 3: Reduction of the ozonide and cleavage results in the final carbonyl fragments.

Example 6.13

Draw structural formulas for the products of the following ozonolysis reactions, and name the new functional groups formed in each oxidation.

(a) $\dfrac{1.\ O_3}{2.\ (CH_3)_2S}$ **(b)** $\dfrac{1.\ O_3}{2.\ (CH_3)_2S}$

Solution

(a)

Propanal	**2-Methylpropanal**
(an aldehyde)	(an aldehyde)

(b)

an aldehyde   a ketone

5-Oxohexanal
(a ketoaldehyde)

Problem 6.13

What alkene with the molecular formula C_6H_{12}, when treated with ozone and then dimethyl sulfide, gives the following product(s)?

(a)

(only product)

(b)

(equal moles of each)

(c)

(only product)

6.6 | Reduction

Most alkenes are reduced quantitatively by molecular hydrogen, H_2, in the presence of a transition metal catalyst to give an alkane. Yields are usually quantitative or nearly so.

$+\ H_2\ \xrightarrow[\text{25°C, 3 atm}]{\text{Pd}}$ $\Delta H° = -119.5\ \text{kJ}\ (-28.6\ \text{kcal})/\text{mol}$

Cyclohexene   **Cyclohexane**

Although the addition of hydrogen to an alkene is exothermic, reduction is immeasurably slow in the absence of a catalyst. Commonly used transition metal catalysts include platinum, palladium, ruthenium, and nickel. Because the conversion of an alkene to an alkane involves reduction by hydrogen in the presence of a catalyst, the process is called **catalytic reduction** or, alternatively, **catalytic hydrogenation**.

Monosubstituted and disubstituted carbon-carbon double bonds react readily at room temperature under a few atmospheres (atm) pressure of hydrogen. Trisubstituted carbon-carbon double bonds require slightly elevated temperatures and pressures of up to 100 psi (pounds per square inch). Tetrasubstituted carbon-carbon double bonds are difficult to reduce and may require temperatures up to 275°C and hydrogen pressures of 1000 psi. Recall that 1 atm is equal to about 15 psi.

The metal catalyst is used as a finely powdered solid or may be supported on some inert material, such as finely powdered charcoal or alumina. The reaction is usually carried out by dissolving the alkene in ethanol or another nonreacting organic solvent, adding the solid catalyst, and then shaking the mixture under hydrogen gas at pressures of from 1 to 50 atm. Alternatively, the metal may be complexed with certain organic molecules and used in the form of a soluble complex (Section 6.7C).

Catalytic reduction is stereoselective, with the vast majority proceeding by syn addition of hydrogens to the carbon-carbon double bond.

Parr Instrument Co., Moline, IL

Parr shaker-type hydrogenation apparatus.

A. Mechanism of Catalytic Reduction

The transition metals used in catalytic hydrogenation are able to adsorb large quantities of hydrogen onto their surfaces, probably by forming metal-hydrogen σ bonds. Similarly, alkenes are also adsorbed on metal surfaces with formation of carbon-metal bonds. Addition of hydrogen atoms to the alkene occurs in two steps (Figure 6.7).

Figure 6.7
The addition of hydrogen to an alkene involving a transition metal catalyst. **(a)** Hydrogen and the alkene are adsorbed on the metal surface and **(b)** one hydrogen atom is transferred to the alkene forming a new C—H bond. The other carbon remains adsorbed on the metal surface. **(c)** A second C—H bond is formed, and the alkane is desorbed.

Under some experimental conditions, particularly with tetrasubstituted double bonds, some percentage of the product may appear to be formed by anti addition of hydrogen. Catalytic reduction of 1,2-dimethylcyclohexene, for example, yields predominantly *cis*-1,2-dimethylcyclohexane. Along with the *cis* isomer are formed lesser amounts of *trans*-1,2-dimethylcyclohexane as a racemic mixture.

1,2-Dimethyl-cyclohexene + H₂ →(Pt) **70% to 85% *cis*-1,2-Dimethyl-cyclohexane** + **30% to 15% *trans*-1,2-Dimethyl-cyclohexane** (a racemic mixture)

If addition of hydrogens is syn stereoselective, how then do we account for the formation of a *trans* product? It has been found that before a second hydrogen can be delivered from the metal surface to complete the reduction, there is transfer of a hydrogen from a carbon atom adjacent to the original double bond to the metal surface along with a newly added hydrogen. This hydrogen transfer, in effect, reverses the first step and forms a new alkene that is isomeric with the original alkene. As shown in the following equation, 1,2-dimethylcyclohexene undergoes isomerization on the metal surface to 1,6-dimethylcyclohexene. This alkene then leaves the metal surface. When it is later readsorbed and reduced, hydrogens are still added to it with syn stereoselectivity, but not necessarily from the same side as the original hydrogen.

1,2-Dimethylcyclohexene   **1,6-Dimethylcyclohexene**

B. Heats of Hydrogenation and the Relative Stabilities of Alkenes

The heat of hydrogenation of an alkene is its heat of reaction, $\Delta H°$, with hydrogen to form an alkane. Table 6.2 lists heats of reaction for the catalytic hydrogenation of several alkenes.

Three important points are derived from the information given in Table 6.2.

1. The reduction of an alkene to an alkane is an exothermic process. This observation is consistent with the fact that, during hydrogenation, there is net conversion of a H—H bond and a weaker π bond to stronger σ bonds; that is, one σ bond (H—H) and one π bond (C=C) are broken, and two new σ bonds (C—H) form. For a comparison of the relative strengths of σ and π bonds, refer to Section 1.10.

2. Heats of hydrogenation depend on the degree of substitution of the carbon-carbon double bond; the greater the substitution, the lower the heat of hydrogenation. Compare, for example, heats of hydrogenation of ethylene (no substituents), propene (one substituent), 1-butene (one substituent), and the *cis* and *trans* isomers of 2-butene (two substituents).

Table 6.2	Heats of Hydrogenation of Several Alkenes	
Name	**Structural Formula**	**$\Delta H°$ [kJ(kcal)/mol]**
Ethylene	CH_2CH_2	$-137\ (-32.8)$
Propene	$CH_3CH{=}CH_2$	$-126\ (-30.1)$
1-Butene	$CH_3CH_2CH{=}CH_2$	$-127\ (-30.3)$
cis-**2-Butene**	$H_3C\diagdown\quad CH_3$ / $C{=}C$ / $H\diagup\quad H$	$-119.7\ (-28.6)$
trans-**2-Butene**	$H_3C\diagdown\quad H$ / $C{=}C$ / $H\diagup\quad CH_3$	$-115.5\ (-27.6)$
2-Methyl-2-butene	$H_3C\diagdown\quad CH_3$ / $C{=}C$ / $H_3C\diagup\quad H$	$-113\ (-26.9)$
2,3-Dimethyl-2-butene	$H_3C\diagdown\quad CH_3$ / $C{=}C$ / $H_3C\diagup\quad CH_3$	$-111\ (-26.6)$

Ethylene

trans-**2-Butene**

2,3-Dimethyl-2-butene

3. The heat of hydrogenation of a *trans* alkene is lower than that of the isomeric *cis* alkene. Compare, for example, the heats of hydrogenation of *cis*-2-butene and *trans*-2-butene. Because reduction of each alkene gives butane, any difference in heats of hydrogenation must be caused by a difference in relative energy between the two alkenes (Figure 6.8). The alkene with the lower (less negative) value of $\Delta H°$ is the more stable.

Figure 6.8

Heats of hydrogenation of *cis*-2-butene and *trans*-2-butene. *Trans*-2-Butene is more stable than *cis*-2-butene by 4.2 kJ (1.0 kcal)/mol.

We explain the greater stability of *trans* alkenes relative to *cis* alkenes in terms of steric strain (Section 2.5A), which can be visualized using spacefilling models (Figure 6.9). In *cis*-2-butene, the two —CH₃ groups are sufficiently close to each other that there is steric strain caused by repulsion between the two methyl groups. This repulsion is reflected in the larger heat of hydrogenation (decreased stability) of *cis*-2-butene compared with *trans*-2-butene [approximately 4.2 kJ (1.0 kcal)/mol]. Thus hydrogenation allows measurement of the strain energy of *cis*-2-butene directly.

CONNECTIONS TO BIOLOGICAL CHEMISTRY

Trans Fatty Acids: What They Are and How to Avoid Them

Fats and oils were introduced in Chapter 5 (Connections to Biological Chemistry). Fats are added to processed foods to provide a desirable firmness along with a moist texture and pleasant taste. To supply the demand for dietary fats of the appropriate consistency, the *cis* double bonds of vegetable oils are partially hydrogenated by using hydrogen in the presence of a Ni or other transition metal catalyst. The greater the extent of hydrogenation, the higher the melting point of the triglyceride. By controlling the degree of hydrogenation, a triglyceride with a melting point below room temperature can be converted to a semisolid or even a solid product.

Unfortunately, because of the reversible interaction of a carbon-carbon double bond with the Ni catalyst, some of the double bonds remaining in the triglyceride may be isomerized from the less stable *cis* configuration to the more stable *trans* configuration. Recall that a key step in catalytic hydrogenation involves the cleavage of the alkene π bond and the bonding of its two carbons to the surface of the transition metal catalyst (Figure 6.7). This process is reversible, thus allowing equilibration between the *cis* and *trans* configurations.

The oils used for frying in fast-food restaurants are usually partially hydrogenated plant oils and, thus, contain substantial amounts of *trans* fatty acids that are transferred to the foods fried in them. Other major sources of *trans* fatty acids in the diet include stick margarine, certain commercial bakery products, creme-filled cookies, chips, frozen breakfast foods, and cake mixes.

Recent studies have shown that consuming a significant amount of *trans* fatty acids can lead to serious health problems related to serum cholesterol levels. Low overall serum cholesterol and a decreased ratio of low-density lipoprotein (LDL) cholesterol to high-density lipoprotein (HDL) cholesterol are associated with good overall cardiovascular health. High serum cholesterol and an elevated ratio of LDL cholesterol to HDL cholesterol are linked to a high incidence of cardiovascular disease, especially atherosclerosis. Research has indicated that diets high in either saturated fatty acids or *trans* fatty acids raise the ratio of serum LDL cholesterol to HDL cholesterol and substantially increase the risk of cardiovascular disease.

The Food and Drug Administration requires that processed foods list the content of *trans* fatty acids. A diet low in saturated and *trans* fatty acids is recommended, along with more fish, whole grains, fruits, and vegetables. The recommendation is also for daily exercise, which is tremendously beneficial, regardless of diet.

(a)

(b)

Figure 6.9

Space-filling models of **(a)** *cis*-2-butene and **(b)** *trans*-2-butene. Steric clashing shown as a yellow shading.

cis-**2-Butene**
Nonbonded interaction strain

trans-**2-Butene**
No nonbonded interaction strain

In addition to the steric reason that makes a *trans*-alkene more stable than a *cis*-alkene, there is an electronic reason for the fact that increased alkyl-group substitution on an alkene imparts stability. Due to the lower energy of $2s$ atomic orbitals relative to $2p$ atomic orbitals, the increased s character of sp^2 hybridized orbitals makes these orbitals more electron withdrawing than sp^3 orbitals. Therefore, the carbons involved in double bonds are slightly electron withdrawing. Hence, in a fashion analogous to the electron donation from alkyl groups that stabilizes carbocations, alkyl groups also donate electrons to alkenes, thereby imparting increased stability to the alkenes. The more alkyl groups, the greater the stabilization.

6.7 | Molecules Containing Chiral Centers as Reactants or Products

As the structure of an organic compound is altered in the course of a reaction, one or more chiral centers, usually at carbon, may be created, inverted, or destroyed. In Section 6.7A, we consider two alkene addition reactions in which a chiral molecule is created in an achiral environment. In doing so, we will illustrate the point that an optically active compound (that is, an enantiomerically pure compound or even an enantiomerically enriched compound) can never be produced from achiral starting materials reacting in an achiral environment. Then, in Section 6.7B, we consider the reaction of achiral starting materials reacting in a chiral environment, in this case in the presence of a chiral catalyst. We shall see that an enantiomerically pure product may be produced from achiral reagents if the reaction takes place in a chiral environment.

A. Reaction of Achiral Starting Materials in an Achiral Environment

The addition of bromine to 2-butene (Section 6.3D) gives 2,3-dibromobutane, a molecule with two chiral centers. Three stereoisomers are possible for this compound: a meso compound and a pair of enantiomers (Section 3.4). We now ask the following questions.

- What is the stereochemistry of the product; is it one enantiomer, a pair of enantiomers, the meso compound, or a mixture of all three stereoisomers?
- Is it optically active or optically inactive?

A partial answer is that the stereochemistry of the product formed depends on the configuration of the alkene. Let us first examine the addition of bromine to *cis*-2-butene.

The reaction of bromine with *cis*-2-butene on either face of the planar part of the molecule gives the same bridged bromonium ion intermediate (Figure 6.10). Although this intermediate has two chiral centers, it has a plane of symmetry and is, therefore, meso. Attack of Br$^-$ on this meso intermediate from the side opposite that of the bromonium ion bridge gives a pair of enantiomers. Attack of bromide ion on carbon 2 gives the (2*S*,3*S*) enantiomer; attack on carbon 3 gives the (2*R*,3*R*) enantiomer. Attack of bromide ion occurs at equal rates at each carbon; therefore, the enantiomers are formed in equal amounts, and 2,3-dibromobutane is obtained as a racemic mixture (Figure 6.10). Thus, the product is chiral, but because it is a racemic mixture, it is optically inactive.

Figure 6.10

Anti stereoselective addition of bromine to *cis*-2-butene gives 2,3-dibromobutane as a racemic mixture. The product is chiral, but because it is formed as a racemic mixture, it is optically inactive.

Step 1: **Make a new bond between a nucleophile (π bond) and and an electrophile** Reaction of *cis*-2-butene with bromine forms bridged bromonium ions which are meso and identical.

cis-**2-Butene**
(achiral)

(meso bridged bromonium ion intermediate)

Step 2: **Make a new bond between a nucleophile and an electrophile** Attack of bromide at carbons 2 and 3 occurs with equal probability to give enantiomeric products in a racemic mixture.

Attack of bromide ion at carbon 3

(*2R,3R*)-**2,3-Dibromobutane**

Attack of bromide ion at carbon 2

(*2S,3S*)-**2,3-Dibromobutane**

Product is a racemic mixture

We have described reaction of Br—Br on one side of the carbon-carbon double bond. Reaction on the opposite side followed by opening of the resulting bromonium ion intermediate produces the same two stereoisomers as a racemic mixture.

The addition of Br_2 to *trans*-2-butene leads to two enantiomeric bridged bromonium ion intermediates. Attack by Br⁻ at either carbon atom of either bromonium ion intermediate gives the meso product, which is achiral and therefore optically inactive (Figure 6.11).

As we have seen in this discussion, the stereochemistry of the product formed by the addition of bromine to 2-butene depends on the stereochemistry of the starting alkene; the addition of halogen to *cis*-2-butene gives a racemic mixture, whereas the addition of halogen to *trans*-2-butene gives a meso product. Accordingly, we say that the addition of Br_2 or Cl_2 to an alkene is stereospecific. A **stereospecific reaction** is a special type of stereoselective reaction in which the stereochemistry of the product depends on the stereochemistry of the starting material.

In Section 6.5A, we studied oxidation of alkenes by osmium tetroxide in the presence of peroxides. This oxidation results in syn stereoselective hydroxylation of the alkene to form a glycol. In the case of cycloalkenes, the product is a *cis* glycol. The first step in each oxidation involves formation of a cyclic osmate and is followed immediately by reaction with water to give a glycol.

Stereospecific reaction

A special type of stereoselective reaction in which the stereochemistry of the product is dependent on the stereochemistry of the starting material.

cis-**2-Butene**
(achiral)

$\xrightarrow[\text{ROOH}]{OsO_4}$

(*2S,3R*)-**2,3-Butanediol**

(*2R,3S*)-**2,3-Butanediol**

identical; a meso compound

trans-2-Butene
(achiral)

OsO₄ / ROOH

(2S,3S)-2,3-Butanediol

(2R,3R)-2,3-Butanediol

a pair of
enantiomers;
a racemic
mixture

Syn hydroxylation of *cis*-2-butene gives meso-2,3-butanediol; because the meso compound is achiral, the product is optically inactive. Syn hydroxylation of *trans*-2-butene gives racemic 2,3-butanediol. Because the diol is formed as a racemic mixture, the product of the oxidation of the *trans* alkene is also optically inactive. Thus, the osmium tetroxide oxidation of an alkene is stereospecific; the stereochemistry of the product depends on the stereochemistry of the starting alkene.

Notice that the stereochemistry of the product of the osmium tetroxide oxidation of *trans*-2-butene is opposite to that formed on the addition of bromine to *trans*-2-butene. Osmium tetroxide oxidation gives the glycol as a pair of enantiomers forming a racemic mixture. Addition of bromine to *trans*-2-butene gives the dibromoalkane as a meso compound. A similar difference is observed between the stereochemical outcomes of these reactions with *cis*-2-butene. The difference in outcomes occurs because bromination of an alkene involves anti addition, whereas oxidation by osmium tetroxide involves syn addition.

In this section, we have seen two examples of reactions in which achiral starting materials give chiral products. In each case, the chiral product is formed as a racemic mixture (which is optically inactive) or as a meso compound (which is also

Step 1: **Make a new bond between a nucleophile (π bond) and an electrophile** Reaction of *trans*-2-butene with bromine forms bridged bromonium ions which are enantiomers

trans-2-Butene
(achiral)

Br₂

(bridged bromonium ion intermediates;
a pair of enantiomers)

Figure 6.11

Anti stereoselective addition of bromine to *trans*-2-butene gives meso-2,3-dibromobutane.

Step 2: **Make a new bond between a nucleophile and an electrophile** Attack of bromide on either carbon of either enantiomer gives meso-2,3-dibromobutane

Attack of bromide ion at carbon 3

(2R,3S)-2,3-Dibromobutane
(meso)

Attack of bromide ion at carbon 2

(2S,3R)-2,3-Dibromobutane

Identical

optically inactive). These results illustrate a very important point about the creation of chiral molecules: an optically active product (that is, an enantiomerically pure compound or even an enantiomerically enriched compound) can never be produced from achiral starting materials and achiral reagents reacting under achiral conditions. Although the molecules of the product may be chiral, the product is always optically inactive (either meso or a pair of enantiomers).

We will encounter many reactions throughout the remainder of this course where achiral starting materials are converted into chiral products under achiral reaction conditions. For convenience, we often draw just one of the enantiomeric products, but you must always keep in mind that, under these experimental conditions, the product will always be racemic and therefore optically inactive.

B. Reaction of a Chiral Starting Material in an Achiral Environment

Let us consider the bromination of (*R*)-4-*tert*-butylcyclohexene. Recall that, in derivatives of cyclohexane in which interconversion between one chair conformation and the other is not possible or is severely restricted, the *trans* diaxial product is isolated. If a cyclohexane ring contains a bulky alkyl group, such as *tert*-butyl (Section 2.6B), then the molecule exists overwhelmingly in a conformation in which the *tert*-butyl group is equatorial. Reaction of bromine with enantiomerically pure (*R*)-4-*tert*-butylcyclohexene occurs at both faces of the six-membered ring. Because bromine atoms must add in an axial manner, each bromonium ion intermediate reacts with bromide ion to give the same product. In the favored chair conformation of this product, *tert*-butyl is equatorial, the bromine atoms remain axial, and only a single diastereomer is formed.

(*R*)-4-*tert*-**Butylcyclohexene**

(1*S*,2*S*,4*R*)-1,2-Dibromo-4-*tert*-**butylcyclohexane**

In effect, the presence of the bulky *tert*-butyl group controls the orientation of the two bromine atoms added to the ring.

C. Reaction of Achiral Starting Materials in a Chiral Environment

The reduction of a carbon-carbon double bond can be carried out using hydrogen in the presence of a transition metal catalyst. Because hydrogen atoms are delivered to either face of the double bond with equal probability, if a new chiral center is created, equal amounts of both the *R* and *S* configurations will be produced.

Within the last three decades, chemists have discovered ways to embed transition metal hydrogenation catalysts in chiral molecules with the result that hydrogen can be delivered to only one face of the alkene. In catalytic reductions where a new chiral center is formed, a large enantiomeric excess of one enantiomer may be formed, and the reaction is said to be **enantioselective**. The most widely used of these chiral hydrogenation catalysts involve the chiral ligand

Enantioselective reaction
A reaction that produces one enantiomer in preference to the other.

2,2-bis-(diphenylphosphanyl)-1,1′-binaphthyl, more commonly known as BINAP. BINAP has been resolved into its *R* and *S* enantiomers. The fact that BINAP can be resolved depends on restricted rotation about the single bond joining the two naphthalene rings. The two enantiomers are atropisomers (Section 3.2).

BINAP

(*S*)-(–)-BINAP
$[\alpha]_D^{25}$ –223

(*R*)-(+)-BINAP
$[\alpha]_D^{25}$ +223

Treating either enantiomer of BINAP with ruthenium chloride forms a complex in which ruthenium is bound as a complex ion in the chiral environment of the larger BINAP molecule. This complex is soluble in dichloromethane, CH_2Cl_2, and can be used as a homogeneous hydrogenation catalyst.

$$(R)\text{-BINAP} + RuCl_3 \longrightarrow (R)\text{-BINAP-RuCl}_3$$

Using (*R*)-BINAP-RuCl₃ as a hydrogenation catalyst, (*S*)-naproxen is formed in greater than 98% enantiomeric excess. (*S*)-Naproxen is the antiinflammatory and pain reliever in Aleve and several other over-the-counter medications. Note that, in this reduction, neither the benzene rings of the naphthyl group nor the carboxyl group are reduced. We will have more to say about the reduction of these groups in later chapters.

(*S*)-Naproxen
(ee > 98%)

BINAP-RuCl₃ complexes are somewhat specific for the types of carbon-carbon double bond reductions they catalyze. To be reduced, the double bond must have some kind of neighboring functional group that serves as a directing group during the reduction. The most common of these directing groups are the carboxyl group of carboxylic acids and esters, and the hydroxyl group of alcohols. As seen in the following example, only the carbon-carbon double bond nearer the —OH group is reduced. Geraniol, as the name might suggest, is a natural product isolated from rose and geranium oils. It is also present in citronella and lemon grass oils. With this chiral catalyst, only the *R* enantiomer is formed from the (*S*)-BINAP-RuCl₃ complex, and only the *S* enantiomer is formed from the (*R*)-BINAP-RuCl₃ complex.

(*E*)-3,7-Dimethyl-2,6-octadien-1-ol
(Geraniol)

(*S*)-BINAP-RuCl₃

(*R*)-3,7-Dimethyl-6-octen-1-ol

(*R*)-BINAP-RuCl₃

(*S*)-3,7-Dimethyl-6-octen-1-ol

Section 6.1 Reactions of Alkenes—An Overview

- The most characteristic reaction of alkenes is an **addition reaction** to the carbon-carbon double bond, breaking the π bond and producing two new σ bonds to two new atoms or groups of atoms.

Section 6.2 Organic Reactions Involving Reactive Intermediates

- A **reaction mechanism** is a detailed description of how and why a chemical reaction occurs as it does, which bonds are broken and which new ones are formed, the order in which the various bond-breaking and bond-forming steps take place and their relative rates, the role of the solvent if the reaction takes place in solution, and the role of the catalyst if one is present.
  - A **reactive intermediate** corresponds to an energy minimum between two transition states.
  - In a multistep reaction, the **rate-determining step** is the step that crosses the highest energy barrier.
- The most common reactive intermediates in organic chemistry are those that lack an octet at carbon or that are carbon-based anions.
  - In particular, common reactive intermediates are **carbocations** (positive charge on a carbon with only six valence electrons), **carbon radicals** (one unpaired electron, so only seven valence electrons on carbon), and **carbanions** (three bonds and one lone pair on carbon).
- **Bond dissociation enthalpy** (BDE) is the amount of energy required to break a bond into two radicals. BDEs can be used to estimate whether a reaction will be favorable (exothermic) or unfavorable (endothermic).
  - This analysis assumes that entropy remains relatively constant throughout the reaction.
  - The change in enthalpy for a reaction ($\Delta H°$) can be estimated as the difference between the BDEs of bonds broken minus the BDEs of bonds made during the reaction.
  - A reaction is exothermic if the bonds made in the product(s) are stronger than the bonds broken in the reactant(s).
- **Arrow pushing** is used by chemists to indicate the mechanisms of reactions. There are four rules to be followed with arrow pushing:
  - Arrows indicate movement of electrons
  - Arrows never indicate movement of atoms directly, only indirectly when associated with electrons
  - Arrows point from **electron sources** (areas of relatively high electron density) to **electron sinks** (an atom that can accept a new bond or lone pair)
  - Breaking a bond usually occurs so that a relatively stable ion or neutral molecule is created
- When the electron source is a base and the electron sink is a proton, a proton transfer reaction takes place, a very common situation in mechanisms of reactions.
- For bond forming mechanism steps that do not involve proton transfers, the electron source is called a **nucleophile** and the electron sink is called an **electrophile**.
- A **nucleophile** (literally nucleus loving) is any atom, molecule, or ion that can donate a pair of electrons to form a new covalent bond. Alternatively, a nucleophile is a Lewis base.
  - Nucleophiles contain an electron rich atom or bond.
  - A lone pair, or π bond, is usually the source for a bond-forming arrow on a nucleophile.
- An **electrophile** (literally electron loving) is any atom, molecule, or ion that can accept a pair of electrons to form a new covalent bond. Alternatively, an electrophile is a Lewis acid. An electrophile will have one or more of the following attributes:
  - A region of full or partial positive charge
  - The lack of a full octet
  - A weak bond that breaks to make a stable ion such as X^-

Problem: 6.13

- The individual steps of mechanisms can be thought of as a multiple choice situation, in which most of the steps encountered in even complex mechanisms are one of four choices:
  - Make a new bond between a nucleophile (electron rich source) and an electrophile (electron poor sink).
  - Break a bond so that a stable molecule or ion is created
  - Add a proton
  - Take a proton away

Section 6.3 Electrophilic Additions

- Addition reactions to alkenes occur with a variety of **electrophiles.**
- The key step in electrophilic addition reactions is the making of a new bond when the alkene π bond reacts as a nucleophile with an electrophile.
  - This is the rate-determining step in the mechanism.
  - The electrophile can be a variety of species including a proton.
- This bond-making step creates a **carbocation** intermediate when the electrophile is a proton.
  - A carbocation is a positively charged ion that contains a carbon atom with only six electrons in its valence shell.
  - Carbocations are planar with bond angles of approximately 120° about the positive carbon.
  - The order of stability of carbocations is 3°> 2° > 1° > methyl.
  - Carbocations are stabilized by the electron-releasing **inductive effect** of alkyl groups bonded to the cationic carbon and by **hyperconjugation.** Hyperconjugation is a stabilizing interaction that involves overlap of σ bonding electron density from adjacent alkyl groups with the empty $2p$ orbital of the cationic carbon atom.
  - Carbocations can **rearrange.**
    - The driving force for **carbocation rearrangement** is conversion to a more stable 2° or 3° carbocation.
    - Rearrangement is by a 1,2-shift in which an atom or group of atoms with its bonding electrons moves from an adjacent atom to an electron-deficient atom.
- A **regioselective reaction** is a reaction in which one direction of bond forming or bond breaking occurs in preference to all other directions.
  - According to **Markovnikov's rule,** in the addition of HX or H_2O to an unsymmetrical alkene, hydrogen adds to the carbon of the double bond having the greater number of hydrogens, and X or OH adds to the more substituted carbon (fewer H atoms).
  - Markovnikov's rule can be explained by realizing that two different carbocations can be formed from the reaction between an unsymmetrical alkene and an electrophile. Formation of the more stable carbocation will predominate, and will lead to the product predicted by Markovikov's rule.

Problems: 6.1–6.8, 6.14–6.35, 6.43, 6.44, 6.48–6.52

- A **stereoselective reaction** is a reaction in which one stereoisomer is formed in preference to all others that might be formed.
  - A stereoselective reaction may be **enantioselective** or **diastereoselective,** depending on whether the product stereoisomers are enantiomers or diastereomers.
  - Addition of new atoms or groups of atoms from opposite faces (or sides) of a double bond is called **anti addition.** These reactions often occur with **anti stereoselectivity.** In cyclic systems, anti addition is equivalent to *trans* coplanar addition.
  - **Syn addition** is the addition of atoms or groups of atoms to the same face (or side) of a double bond.

Section 6.4 Hydroboration-Oxidation

Problems: 6.9, 6.36, 6.37, 6.43, 6.44

- Non-Markovnikov hydration of an alkene can be achieved by **hydroboration-oxidation,** because the key step in this reaction sequence involves a four-membered transition state featuring syn addition, not a carbocation intermediate.

Section 6.5 Oxidation

Problems: 6.10–6.12, 6.38–6.40, 6.43, 6.44

- **Oxidation** is the loss of electrons; **reduction** is the gain of electrons.
  - Alternatively, oxidation is the gain of oxygen atoms and/or loss of hydrogen atoms.

Section 6.6 Reduction

- **Reduction** is the loss of oxygen atoms and/or the gain of hydrogen atoms.
- From **heats of hydrogenation** of a series of alkenes, we conclude that in general,
  - The greater the degree of substitution of a carbon-carbon double bond, the more stable the alkene.

Problems: 6.41–6.44

  - A *trans* alkene is more stable than an isomeric *cis* alkene.

Section 6.7 Molecules Containing Chiral Centers as Reactants or Products

- For reactions in which chiral centers are created, only racemic mixtures are formed from achiral reagents in achiral media.
- A **stereospecific reaction** is a reaction in which the stereochemistry of the product depends on the stereochemistry of the starting material.

Problems: 6.45–6.47

  - All stereospecific reactions are stereoselective, but not all stereoselective reactions are stereospecific, as you will see later.
  - Optically active products can never be formed by the reaction of achiral starting materials in an achiral environment.
  - Optically active products may be formed, however, by the reaction of achiral starting materials in a chiral environment.

OWL Questions and problems for this chapter are available in OWL. Use the OWL access card that came with this textbook to access assigned questions and problems for this chapter.

Key Reactions

1. **Addition of HX (Section 6.3A)**   HX is used to convert alkenes to haloalkanes in an electrophilic addition reaction. The two-step mechanism involves initial protonation of the alkene π bond to form a carbocation, which reacts with X^- to give the product haloalkane. The X atom becomes bonded to the more highly substituted atom of the alkene, so it follows Markovnikov regioselectivity (derived from the preference for forming the more stable carbocation intermediate). Carbocation rearrangements are possible.

See Appendix 8 for a summary of stereochemical terms.

2. **Acid-Catalyzed Hydration (Section 6.3B)**   Water, in the presence of an acid catalyst, converts an alkene into an alcohol. The mechanism involves initial protonation of the alkene π bond to give a carbocation, which reacts with the nucleophile water to create a second intermediate, which loses a proton to give the alcohol product. The OH group becomes bonded to the more highly substituted carbon of the alkene, so the reaction displays Markovnikov regioselectivity (derived from the preference for forming the more stable carbocation intermediate). Carbocation rearrangements are possible.

3. **Addition of Bromine and Chlorine (Section 6.3D)** Cl_2 or Br_2 is used to convert an alkene into a vicinal dihalide. The mechanism involves attack by the alkene π bond on one atom of X_2 to give a bridged halonium ion intermediate (a cation) that is, in turn, attacked by X^- from the backside to give the vicinal dihalide. Rearrangements do not occur. The reaction displays anti addition stereoselectivity because of the halonium ion intermediate. The reaction is stereospecific because Z alkenes give different products than E alkenes.

(a racemic mixture)

4. **Addition of HOCl and HOBr (Section 6.3E)** Cl_2 or Br_2, in the presence of H_2O, is used to convert alkenes to halohydrins. The mechanism involves attack by the alkene π bond on one atom of X_2 to give a bridged halonium ion intermediate (a cation) that is, in turn, attacked by H_2O from the backside to give a new intermediate, which loses a proton to give the halohydrin. Rearrangements are not observed and the reaction displays anti addition stereoselectivity because of the halonium ion intermediate. The reaction gives Markovnikov regioselectivity in that the —OH group becomes bonded to the more highly substituted alkene carbon. This regioselectivity is derived from attack by H_2O on the carbon of the halonium ion intermediate capable of accommodating more carbocation character. The reaction is stereospecific because Z alkenes give different products than isomeric E alkenes.

(a racemic mixture)

5. **Oxymercuration-Reduction (Section 6.3F)** Oxymercuration-reduction is used to convert alkenes to alcohols. The mechanism involves reaction of the alkene π bond with an $HgOAc^+$ to give a bridged mercurinium ion intermediate (a cation) that is, in turn, attacked by H_2O from the backside to give a new intermediate, which loses a proton. In a second step, $NaBH_4$ is added to replace the Hg atom with H. The first step is anti stereoselective because HgOAc and OH add from opposite faces of the alkene. However, the $NaBH_4$ reduction step scrambles the stereochemistry as H replaces Hg, so the overall process is scrambled anti and syn addition. The reaction displays Markovnikov regioselectivity in that the —OH group becomes bonded to the more highly substituted alkene carbon. This regioselectivity is derived from attack by H_2O on the carbon of the mercurinium ion intermediate capable of accommodating more carbocation character. Rearrangements are not observed.

6. **Hydroboration-Oxidation (Section 6.4)** BH_3 followed by basic H_2O_2 converts alkenes into alcohols with non-Markovnikov regioselectivity (the OH adds to the less substituted alkene carbon) and syn stereoselectivity without rearrangement. The reaction is stereospecific because *cis* alkenes give different products than *trans* alkenes. The mechanism involves coordination of the alkene π bond to the vacant $2p$ orbital of borane, followed by a four-membered ring transition state, which simultaneously adds H to the more substituted carbon and boron to the less substituted alkene carbon. The basic peroxide replaces the boron with OH.

(a racemic mixture)

7. **Oxidation to a Vicinal Diol by OsO₄ (Section 6.5A)** OsO_4 followed by $NaHSO_3$ converts alkenes into vicinal diols with syn stereoselectivity and without rearrangement. The reaction is stereospecific because *cis* alkenes give different products than *trans* alkenes. The mechanism involves formation of a cyclic osmate ester.

A cyclic osmate   **cis-1,2-Cyclopentanediol**
(a *cis* glycol)

8. **Oxidation by Ozone (Section 6.5B)**   Ozone, O_3, followed by dimethylsulfide cleaves the carbon-carbon double bond of alkenes and gives two carbonyl groups in its place. The mechanism involves formation of a molozonide intermediate that rearranges to an ozonide intermediate, which is reduced using the dimethyl sulfide to give the carbonyl products.

9. **Addition of H_2; Catalytic Reduction (Section 6.6)**   In the presence of an appropriate transition metal, H_2 reduces alkenes to alkanes with syn stereoselectivity and without rearrangement. The reaction can be stereospecific because *Z* alkenes can give different products than *E* alkenes. The mechanism involves cleavage of the alkene π bond through adsorption onto the surface of the transition metal. This step is followed by formation of two carbon-hydrogen bonds from the same face of the alkane molecule as it desorbs.

10. **Enantioselective Reduction (Section 6.7)**   The most useful chiral hydrogenation catalysts involve a chiral phosphorus-containing ligand complexed with either ruthenium or rhodium.

Chapter 7

Alkynes

Cutting with an oxyacetylene torch.
Inset: a model of acetylene.

Adam Crowley/Getty Images

In this chapter, we continue our discussion of the chemistry of carbon-carbon π bonds as we now consider the chemistry of alkynes. Because alkenes and alkynes are similar in that the multiple bond in each is a combination of σ and π bonds, both types of functional groups undergo similar chemical reactions. Alkynes undergo electrophilic additions of X_2, HX, and H_2O. They undergo hydroboration-oxidation, and the carbon-carbon triple bond can be reduced first to a double bond and then to a single bond. An important reaction of alkynes that is not typical of alkenes is the conversion of terminal alkynes to their alkali metal salts, which are good nucleophiles and, therefore, valuable building blocks for the construction of larger molecules through the formation of carbon-carbon bonds.

7.1 | Structure of Alkynes

The functional group of an **alkyne** is a carbon-carbon triple bond. The simplest alkyne is ethyne, C_2H_2, more commonly named acetylene (Figure 7.1). Acetylene is a linear molecule; all bond angles are 180°. The carbon-carbon bond length in acetylene is 121 pm (1.21 Å) (Figure 7.1).

By comparison, the length of the carbon-carbon double bond in ethylene is 134 pm (1.34 Å), and that of the carbon-carbon single bond in ethane is 153 pm (1.53 Å). Thus, triple bonds are shorter than double bonds, which, in turn, are shorter than single bonds. The bond dissociation enthalpy of the carbon-carbon triple bond in acetylene [966 kJ (231 kcal)/mol] is considerably larger than that for the carbon-carbon double bond in ethylene [727 kJ (174 kcal)/mol] and the carbon-carbon

Alkyne
An unsaturated hydrocarbon that contains one or more carbon-carbon triple bonds.

◖WL **Online homework for this chapter may be assigned in OWL for Organic Chemistry.**

227

Figure 7.1
Acetylene. **(a)** Lewis structure,
(b) ball-and-stick model, and
(c) space-filling model.

(a) H—C≡C—H

(b)

180°

121 pm

(c)

Acetylene

single bond in ethane [376 kJ (90 kcal)/mol]. The difference in bond dissociation enthalpies between the carbon-carbon triple bond in acetylene and the carbon-carbon double bond in ethylene is only 239 kJ (57 kcal)/mol. This difference indicates that a π bond in an alkyne is weaker than a π bond in an alkene.

A triple bond is described in terms of the overlap of sp hybrid orbitals of adjacent carbon atoms to form a σ bond, the overlap of parallel $2p_y$ orbitals to form one π bond, and the overlap of parallel $2p_z$ orbitals to form a second π bond (Figure 1.22). In acetylene, each carbon-hydrogen bond is formed by the overlap of a $1s$ orbital of hydrogen with an sp orbital of carbon. Because of the 50% s-character of the acetylenic C—H bond, it is unusually strong (see Table 1.11 and related text).

7.2 | Nomenclature of Alkynes

A. IUPAC Names

According to the rules of the IUPAC system, the infix -*yn*- is used to show the presence of a carbon-carbon triple bond (Section 2.3). Thus, HC≡CH is named ethyne, and CH₃C≡CH is named propyne. The IUPAC system retains the name acetylene; therefore, there are two acceptable names for HC≡CH, ethyne and acetylene. Of these names, acetylene is used much more frequently.

There is no need to use a number to locate the position of the triple bond in ethyne and propyne; there is only one possible location for it in each compound. For larger molecules, number the longest carbon chain that contains the triple bond from the end that gives the triply bonded carbons the lower numbers. Show the location of the triple bond by the number of its first carbon. If a hydrocarbon chain contains more than one triple bond, we use the infixes -*adiyn*-, -*atriyn*-, and so forth.

3-Methyl-1-butyne   **6,6-Dimethyl-3-heptyne**   **1,6-Heptadiyne**

Example 7.1

Write the IUPAC name of each compound.

(a)   **(b)** Cl   **(c)**

Solution

(a) 2-Pentyne   **(b)** 3-Chloropropyne   **(c)** 2,6-Octadiyne

Problem 7.1

Write the IUPAC name of each compound.

(a)   **(b)**   **(c)** Br

B. Common Names

Common names for alkynes are derived by prefixing the names of substituents on the carbon-carbon triple bond to the word *acetylene*. Note in the third example that, when a carbon-carbon double bond (indicated by *-en-*) and a carbon-carbon triple bond (indicated by *-yn-*) are both present in the same molecule, the IUPAC rules specify that the location of the double bond takes precedence in numbering the compound.

	$CH_3C{\equiv}CH$	$CH_3C{\equiv}CCH_3$	$CH_2{=}CHC{\equiv}CH$
IUPAC name:	**Propyne**	**2-Butyne**	**1-Buten-3-yne**
Common name:	**Methylacetylene**	**Dimethylacetylene**	**Vinylacetylene**

Alkynes in which the triple bond is between carbons 1 and 2 are commonly referred to as **terminal alkynes**. Examples of terminal alkynes are propyne and 1-butyne.

Terminal alkyne
An alkyne in which the triple bond is between carbons 1 and 2.

Example 7.2

Write the common name of each alkyne.

(a)   (b)   (c)

Solution

(a) Isobutylmethylacetylene (b) (S)-*sec*-Butylacetylene (c) *tert*-Butylacetylene

Problem 7.2

Write the common name of each alkyne.

(a)   (b)   (c)

The smallest cycloalkyne that has been isolated is cyclooctyne. This molecule is quite unstable and polymerizes rapidly at room temperature. The C—C≡C bond angle in cyclooctyne is calculated to be approximately 155°, indicating a high degree of angle strain. Cyclononyne (Figure 7.2) has also been prepared and is stable at room temperature. The calculated C—C≡C bond angles in this cycloalkyne are approximately 160°, which still represents a considerable distortion from the optimal 180°. You can see the distortion of the C—C≡C bond angles in the accompanying optimized molecular model. You can also see the degree to which C—C and C—H bonds on adjacent carbons are staggered, thus minimizing torsional strain.

Figure 7.2
Structure of cyclononyne.

7.3 | Physical Properties of Alkynes

The physical properties of alkynes are similar to those of alkanes and alkenes with analogous carbon skeletons. The lower-molecular-weight alkynes are gases at room temperature. Those that are liquids at room temperature have densities less than 1.0 g/mL (less dense than water). Listed in Table 7.1 are melting points, boiling points, and densities of several low-molecular-weight alkynes. Because alkynes, like alkanes and alkenes, are nonpolar compounds, they are insoluble in water and other polar solvents. They are soluble in each other and in other nonpolar organic solvents.

Table 7.1	Physical Properties of Some Low-Molecular-Weight Alkynes			
Name	Formula	Melting Point (°C)	Boiling Point (°C)	Density at 20°C (g/mL)
Ethyne	$HC{\equiv}CH$	−81	−84	(a gas)
Propyne	$CH_3C{\equiv}CH$	−102	−23	(a gas)
1-Butyne	$CH_3CH_2C{\equiv}CH$	−126	8	(a gas)
2-Butyne	$CH_3C{\equiv}CCH_3$	−32	27	0.691
1-Pentyne	$CH_3(CH_2)_2C{\equiv}CH$	−90	40	0.690
1-Hexyne	$CH_3(CH_2)_3C{\equiv}CH$	−132	71	0.716
1-Octyne	$CH_3(CH_2)_5C{\equiv}CH$	−79	125	0.746
1-Decyne	$CH_3(CH_2)_7C{\equiv}CH$	−36	174	0.766

7.4 | Acidity of 1-Alkynes

One of the major differences between the chemistry of alkynes and that of alkenes or alkanes is that a hydrogen bonded to a triply bonded carbon atom of a terminal alkyne is sufficiently acidic that it can be removed by a strong base, such as sodium amide $NaNH_2$ (Table 4.1), to give an acetylide anion.

$$H{-}C{\equiv}C{-}H + {:}NH_2^- \rightleftharpoons H{-}C{\equiv}C{:}^- + {:}NH_3 \qquad K_{eq} = 10^{13}$$

pK_a 25   (Stronger   (Weaker   pK_a 38
(Stronger   base)   base)   (Weaker
acid)   acid)

Other strong bases commonly used to form acetylide anions are sodium hydride and lithium diisopropylamide (LDA).

$$Na^+H{:}^- \qquad\qquad [(CH_3)_2CH]_2\ddot{N}{:}^-Li^+$$

Sodium hydride   Lithium diisopropylamide
(LDA)

Because water is a stronger acid than acetylene, the hydroxide ion is not a strong enough base to convert a terminal alkyne to an alkyne anion.

$$H{-}C{\equiv}C{-}H + {:}\ddot{O}H^- \rightleftharpoons H{-}C{\equiv}C{:}^- + H{-}\ddot{O}H \qquad K_{eq} = 10^{-9.3}$$

pK_a 25   (Weaker   (Stronger   pK_a 15.7
(Weaker   base)   base)   (Stronger
acid)   acid)

The pK_a values for alkene and alkane hydrogens are so large (they are so weakly acidic) that neither the commonly used alkali metal hydroxides nor sodium hydride, sodium amide, or lithium diisopropylamide are strong enough bases to remove a proton from alkanes or alkenes.

7.5 | Preparation of Alkynes

A. Alkylation of Acetylide Anions with Methyl and 1° Haloalkanes

As we have already seen, an acetylide anion is a strong base. An acetylide anion is also a nucleophile; it has an unshared pair of electrons that it can donate to another atom to form a new covalent bond. In this instance, an acetylide anion donates its unshared pair of electrons to the carbon of a methyl or primary haloalkane, and, in so doing, the acetylide nucleophile replaces the halogen atom. This type of reaction is called a nucleophilic substitution. For example, treating sodium acetylide with 1-bromobutane gives 1-hexyne.

$$HC\equiv C:^- Na^+ + \quad \xrightarrow[\text{substitution}]{\text{nucleophilic}} \quad + \; Na^+Br^-$$

| **Sodium acetylide** | **1-Bromobutane** | **1-Hexyne** |

Because an alkyl group (in this case a butyl group) is added to a molecule, this type of reaction is also called an **alkylation reaction**. We limit our discussion of nucleophilic substitution in this chapter to the reactions of acetylide anions with methyl and 1° haloalkanes. We discuss the scope and limitations of nucleophilic substitution in more general terms in Chapter 9. For reasons we discuss in Chapter 9, alkylation of acetylide anions is practical only with methyl and primary halides.

Because of the ready availability of acetylene and the ease with which it is converted to a good nucleophile, alkylation of an acetylide anion is the most convenient laboratory method for the synthesis of terminal alkynes, meaning one with the carbon-carbon triple bond at the end of a carbon chain. The process of alkylation can be repeated, and a terminal alkyne can, in turn, be converted to an alkyne with the carbon-carbon triple bond in the middle of a carbon chain. Such an alkyne is usually referred to as an **internal alkyne**. An important feature of this reaction is that new carbon-carbon bonds are made, allowing the construction of larger carbon backbones from smaller ones.

Alkylation reaction
Any reaction in which a new carbon-carbon bond to an alkyl group is formed.

Internal alkyne
A carbon-carbon triple bond in the middle of a carbon chain.

$$CH_3CH_2C\equiv C^- Na^+ + CH_3CH_2{-}Br \xrightarrow[\text{substitution}]{\text{nucleophilic}} CH_3CH_2C\equiv CCH_2CH_3 + Na^+Br^-$$

| **Sodium butynide** | **Bromoethane** | **3-Hexyne** |

B. Alkynes from Alkenes

To prepare an alkyne from an alkene, the alkene is first treated with 1 mole of either bromine (Br_2) or chlorine (Cl_2) to give a dihaloalkane (Section 6.3D). Treating the dihaloalkane with 2 moles of a strong base such as sodium amide ($NaNH_2$) in liquid ammonia [$NH_3(l)$] brings about two successive **dehydrohalogenations**. Recall that addition of HX to an alkene is called hydrohalogenation; removal of HX from a haloalkane is called dehydrohalogenation. The removal of atoms from adjacent carbons to form an alkene is also called an elimination reaction and is discussed fully in Chapter 9. The following example shows the conversion of 2-butene to 2-butyne.

Dehydrohalogenation
The removal of HX from a molecule.

| *trans*-2-Butene | 2,3-Dibromobutane (meso) |

$$\underset{\substack{\text{2,3-Dibromobutane} \\ \text{(meso)}}}{\underset{\underset{\underset{\text{H}}{|}}{\underset{\text{H}_3\text{C}}{\overset{\text{Br}}{\underset{|}{\text{C}}}}} - \underset{\underset{\text{Br}}{|}}{\overset{\overset{\text{H}}{|}}{\text{C}}}\text{CH}_3}{}} + 2\text{ NaNH}_2 \xrightarrow[-33°C]{\text{NH}_3(l)} \underset{\text{2-Butyne}}{\text{CH}_3\text{C}\equiv\text{CCH}_3} + 2\text{ NaBr} + 2\text{ NH}_3$$

Given the ease of converting alkenes to dihaloalkanes and then to alkynes, alkenes are versatile starting materials for the preparation of alkynes.

With a strong base such as sodium amide, both dehydrohalogenations occur readily. However, with weaker bases, such as sodium hydroxide or potassium hydroxide in ethanol, it is often possible to stop the reaction after the first dehydrohalogenation and isolate the haloalkene.

$$\underset{\substack{\text{2,3-Dibromobutane} \\ \text{(meso)}}}{} + \text{KOH} \xrightarrow{\text{ethanol}} \underset{\substack{\text{(E)-2-Bromo-2-butene}}}{} + \text{KBr} + \text{H}_2\text{O}$$

In practice, it is much more common to use a stronger base and go directly to the alkyne.

The following equations show the conversion of 1-hexene to 1-hexyne. Note that 3 moles of sodium amide are used in this sequence. Two moles are required for the double dehydrohalogenation reaction, which gives 1-hexyne. As soon as any 1-hexyne (a weak acid, pK_a 25) forms, it reacts with sodium amide (a strong base) to give an alkyne salt. Thus, a third mole of sodium amide is required to complete the dehydrohalogenation of the remaining bromoalkene. Addition of water (a weak acid) or aqueous acid completes the sequence and gives 1-hexyne.

$$\underset{\text{1-Hexene}}{\text{CH}_3(\text{CH}_2)_3\text{CH}=\text{CH}_2} \xrightarrow{\text{Br}_2} \underset{\substack{\text{1,2-Dibromohexane} \\ \text{(a racemic mixture)}}}{\text{CH}_3(\text{CH}_2)_3\overset{\overset{\text{Br}}{|}}{\text{CH}}\overset{\overset{\text{Br}}{|}}{\text{CH}}_2} \xrightarrow[-2\text{ HBr}]{2\text{ NaNH}_2} \underset{\text{1-Hexyne}}{[\text{CH}_3(\text{CH}_2)_3\text{C}\equiv\text{CH}]} \xrightarrow{\text{NaNH}_2}$$

$$\underset{\text{Sodium salt of 1-hexyne}}{\text{CH}_3(\text{CH}_2)_3\text{C}\equiv\text{C}^-\text{Na}^+} \xrightarrow{\text{H}_2\text{O}} \underset{\text{1-Hexyne}}{\text{CH}_3(\text{CH}_2)_3\text{C}\equiv\text{CH}}$$

In dehydrohalogenation of a haloalkene with at least one hydrogen on each adjacent carbon, a side reaction occurs, namely the formation of an allene.

$$\underset{\substack{\text{A haloalkene} \\ \text{(a vinylic halide)}}}{\text{R}\overset{\overset{\text{H}}{|}}{\underset{\underset{\text{R}}{|}}{\text{C}}}-\overset{\overset{\text{X}}{|}}{\text{C}}=\overset{\overset{\text{H}}{|}}{\text{C}}\text{R}} \xrightarrow[-\text{HX}]{\text{NaNH}_2} \begin{cases} \underset{\text{An allene}}{\overset{\overset{\text{R}}{}\quad\overset{\text{H}}{}}{\underset{\text{R}\quad\text{R}}{\text{C}=\text{C}=\text{C}}}} \\ \\ \underset{\text{An alkyne}}{\text{R}\overset{\overset{\text{H}}{|}}{\underset{\underset{\text{R}}{|}}{\text{C}}}\text{C}\equiv\text{CR}} \end{cases}$$

An **allene** has two adjacent carbon-carbon double bonds; that is, it contains a C=C=C functional group. The simplest allene is 1,2-propadiene, commonly named allene. In it, each end carbon is sp^2 hybridized, and the middle carbon

Allene
The compound $CH_2=C=CH_2$. Any compound that contains adjacent carbon-carbon double bonds; that is, any molecule that contains a C=C=C functional group.

is *sp* hybridized. Each carbon-carbon σ bond is formed by the overlap of *sp* and *sp²* hybrid orbitals. One π bond is formed by the overlap of parallel $2p_y$ orbitals; the other, by the overlap of parallel $2p_z$ orbitals. The two π-bonding molecular orbitals are in planes perpendicular to each other, as are the two H—C—H groups.

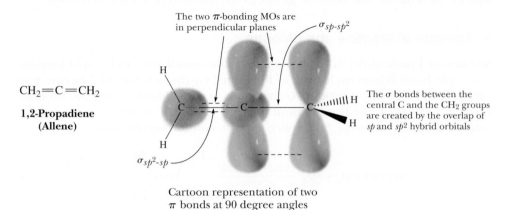

CH₂=C=CH₂

1,2-Propadiene (Allene)

The two π-bonding MOs are in perpendicular planes

$\sigma\,sp\text{-}sp^2$

The σ bonds between the central C and the CH₂ groups are created by the overlap of *sp* and *sp²* hybrid orbitals

$\sigma sp^2\text{-}sp$

Cartoon representation of two π bonds at 90 degree angles

Most allenes are less stable than their isomeric alkynes. For example, allene itself is less stable by 6.7 kJ (1.6 kcal)/mol than its constitutional isomer propyne, and 1,2-butadiene is less stable than 2-butyne by 16.7 kJ (4.0 kcal)/mol.

$$CH_2{=}C{=}CH_2 \longrightarrow CH_3C{\equiv}CH \qquad \Delta H^\circ = -6.7 \text{ kJ } (-1.6 \text{ kcal})/\text{mol}$$

$$CH_2{=}C{=}CHCH_3 \longrightarrow CH_3C{\equiv}CCH_3 \qquad \Delta H^\circ = -16.7 \text{ kJ } (-4.0 \text{ kcal})/\text{mol}$$

Because of their lower stability relative to isomeric alkynes, allenes are generally only minor products of alkyne-forming dehydrohalogenation reactions.

Example 7.3

Show how you might convert 1-pentene to 1-pentyne.

Solution

This synthesis can be done in two steps: Treating 1-pentene with 1 mole of bromine gives 1,2-dibromopentane. Treating this dibromoalkane with 3 moles of sodium amide followed by H₂O gives 1-pentyne.

Br₂ →

Br

1. 3 NaNH₂
2. H₂O

1-Pentene   **1,2-Dibromopentane**
(a racemic mixture)

1-Pentyne

Problem 7.3

Draw a structural formula for an alkene and dichloroalkane with the given molecular formula that yields the indicated alkyne by each reaction sequence.

(a) $C_6H_{12} \xrightarrow{Cl_2} C_6H_{12}Cl_2 \xrightarrow{2\ NaNH_2}$

(b) $C_7H_{14} \xrightarrow{Cl_2} C_7H_{14}Cl_2 \xrightarrow{2\ NaNH_2}$

7.6 | Electrophilic Addition to Alkynes

Analogous to alkenes, a characteristic reaction of alkynes is a π bond acting as a nucleophile to make a new bond with an electrophile. As a result, alkynes undergo many of the same electrophilic additions as alkenes. In this section, we study the addition of bromine and chlorine as well as the addition of hydrogen halides.

A. Addition of Bromine and Chlorine

Addition of 1 mole of Br_2 to an alkyne gives a dibromoalkene. Addition of bromine to a triple bond is stereoselective, as illustrated by the reaction of 2-butyne with 1 mole of Br_2. The major product corresponds to anti addition of the two bromine atoms. Carrying out the bromination in acetic acid with added bromide ion, for example LiBr, increases the preference for anti addition significantly.

$$CH_3C{\equiv}CCH_3 + Br_2 \xrightarrow[\text{anti addition}]{CH_3COOH, \; LiBr}$$

2-Butyne　　　　　　　　　　　　**(E)-2,3-Dibromo-2-butene**

Alkynes similarly undergo addition of Cl_2, although less stereoselectively than with Br_2.

Addition of bromine to alkynes follows much the same type of mechanism as it does for addition to alkenes (Section 6.3D), namely, formation of a bridged bromonium ion intermediate, which is then attacked by bromide ion from the face opposite that occupied by the positively charged bromine atom.

Addition of a second mole of Br_2 gives a tetrabromoalkane.

$$ + Br_2 \longrightarrow CH_3\overset{\overset{\displaystyle Br}{|}}{\underset{\underset{\displaystyle Br}{|}}{C}}-\overset{\overset{\displaystyle Br}{|}}{\underset{\underset{\displaystyle Br}{|}}{C}}CH_3$$

(E)-2,3-Dibromo-2-butene　　　　**2,2,3,3-Tetrabromobutane**

B. Addition of Hydrogen Halides

Alkynes add either 1 or 2 moles of HBr and HCl, depending on the ratios in which the alkyne and halogen acid are mixed.

$$CH_3C{\equiv}CH \xrightarrow{HBr} CH_3\overset{\overset{\displaystyle Br}{|}}{C}{=}CH_2 \xrightarrow{HBr} CH_3\overset{\overset{\displaystyle Br}{|}}{\underset{\underset{\displaystyle Br}{|}}{C}}CH_3$$

Propyne　　　　**2-Bromopropene**　　　　**2,2-Dibromopropane**

As shown in this equation, addition of both the first and second moles of HBr is regioselective. Addition of hydrogen halides follows Markovnikov's rule (Section 6.3A); hydrogen adds to the carbon that has the greater number of hydrogens. We can account for this regioselectivity of addition of HX by a two-step mechanism for each addition.

Step 1: Make a new bond between a nucleophile (π bond) and an electrophile—add a proton Proton transfer from HBr to the alkyne gives a **vinylic carbocation**; the more stable 2° vinylic carbocation is formed in preference to the less stable 1° vinylic carbocation.

Vinylic carbocation
A carbocation in which the positive charge is on one of the carbons of a carbon-carbon double bond.

$$CH_3C\equiv CH + H\text{—}\overset{..}{\underset{..}{Br}}: \longrightarrow CH_3\overset{+}{C}=CH_2 \ + :\overset{..}{\underset{..}{Br}}:^-$$

A 2° vinylic
carbocation

Step 2: Make a new bond between a nucleophile and an electrophile Reaction of the vinylic carbocation (an electrophile) with bromide ion (a nucleophile) gives the vinylic bromoalkene.

$$CH_3\overset{+}{C}=CH_2 + :\overset{..}{\underset{..}{Br}}:^- \longrightarrow CH_3\overset{\overset{:\overset{..}{\underset{..}{Br}}:}{|}}{C}=CH_2$$

2-Bromopropene

Alkynes are considerably less reactive toward most electrophilic additions than are alkenes. The major reason for this difference is the instability of the sp-hybridized vinylic carbocation intermediate formed from an alkyne compared with the sp^2-hybridized alkyl carbocation formed from an alkene.

In the addition of the second mole of HX, Step 1 is the reaction of the electron pair of the remaining π bond with HBr to form a carbocation. Of the two possible carbocations, the one with the positive charge on the carbon bearing the halogen is favored because of delocalization of the positive charge through resonance.

1° Carbocation

Resonance-stabilized 2° carbocation

Addition of 1 mole of HCl to acetylene gives chloroethene (vinyl chloride), a compound of considerable industrial importance.

$$HC\equiv CH + HCl \longrightarrow CH_2=CHCl$$

Ethyne   **Chloroethene**
(Acetylene)   **(Vinyl chloride)**

Vinyl chloride is the monomer in the production of the polymer poly(vinyl chloride), abbreviated PVC.

$$n\ CH_2=CHCl \xrightarrow{\text{catalyst}} \overset{\overset{\text{Cl}}{|}}{-(CH_2CH-)_n}$$

Vinyl chloride   **Poly(vinyl chloride)**
(PVC)

PVC dominates much of the plumbing and construction market for plastics. Approximately 67% of all pipe, fittings, and conduit, along with 42% of all plastics used in construction at the present time, are fabricated from PVC. We will describe the synthesis of this polymer and its properties in Chapter 29. Our purpose here is to describe the synthesis of vinyl chloride.

At one time, hydrochlorination of acetylene was the major source of vinyl chloride. As the cost of production of acetylene increased, however, manufacturers of vinyl chloride sought other routes to this material. The starting material chosen was ethylene, which can be converted to vinyl chloride in two steps: Treating ethylene with chlorine gives 1,2-dichloroethane, which, when heated in the presence of charcoal or other catalyst, loses a molecule of HCl to form vinyl chloride.

$$CH_2{=}CH_2 \xrightarrow{Cl_2} \underset{\substack{|\quad| \\ CH_2CH_2}}{\overset{Cl \quad Cl}{}} \xrightarrow[catalyst]{heat} CH_2{=}CHCl \ + \ HCl$$

Ethylene **1,2-Dichloroethane** **Vinyl chloride**

Chlorine atoms of the byproduct are recycled by passing the HCl mixed with air over a copper(II) catalyst, which results in the oxidation of HCl to $2Cl_2$.

$$4\,HCl \ + \ O_2 \xrightarrow{CuCl_2} 2\,H_2O \ + \ 2\,Cl_2$$

We described the production of vinyl chloride first from acetylene and then from ethylene to illustrate an important point about industrial organic chemistry. The aim is to produce a desired chemical from readily available and inexpensive starting materials by reactions in which byproducts can be recycled. All chemical companies now support this objective to minimize both costs and production of materials that require disposal or can harm the environment.

7.7 | Hydration of Alkynes to Aldehydes and Ketones

The elements of H_2O can be added to the carbon-carbon triple bond of an alkyne by the same two reactions used for the hydration of alkenes, namely hydroboration-oxidation and acid-catalyzed hydration. Even though the reagents are similar, the products from hydration of alkenes and alkynes are quite different.

A. Hydroboration-Oxidation

Borane adds readily to an internal alkyne as illustrated by its reaction with 3-hexyne.

3-Hexyne
A trialkenylborane
(R = *cis*-3-hexenyl group)

Notice that the hydroboration of an internal alkyne stops after the addition of 1 mole of borane. The product is a trialkenylborane (the infix -*enyl*- shows the presence of a carbon-carbon double bond on the carbon bonded to boron). As with hydroboration of alkenes (Section 6.4), hydroboration of alkynes is stereoselective; it involves syn addition of hydrogen and boron.

Terminal alkynes also react regioselectively with borane to form trialkenylboranes. In practice, however, the reaction is difficult to stop at this stage because the alkenyl group reacts further with borane to undergo a second hydroboration.

It is possible to prevent the second hydroboration step and, in effect, stop the reaction at the alkenylborane stage by using a sterically hindered disubstituted borane. One of the most widely used of these is di-*sec*-isoamyl borane, (sia)$_2$BH, prepared by treating borane with two equivalents of 2-methyl-2-butene (amyl is an older common name for pentyl).

2-Methyl-2-butene   **Di-*sec*-isoamylborane [(sia)$_2$BH]**

Reaction of this sterically hindered dialkylborane with a terminal alkyne results in a single hydroboration and formation of an alkenylborane.

1-Octyne   An alkenylborane

Just as with hydroboration of unsymmetrical alkenes, the addition of (sia)$_2$BH to a carbon-carbon triple bond of a terminal alkene is regioselective; boron adds to the less substituted carbon.

Treatment of an alkenylborane with hydrogen peroxide in aqueous sodium hydroxide gives a product that corresponds to hydration of an alkyne; that is, it corresponds to addition of H to one carbon of the triple bond and OH to the other as illustrated by the hydroboration-oxidation of 2-butyne.

$K_{eq} = 6.7 \times 10^6$

2-Butyne   **2-Buten-2-ol** (an enol)   **2-Butanone** (a ketone)   (for keto-enol tautomerism)

The initial product of hydroboration-oxidation of an alkyne is an **enol**, a compound containing a hydroxyl group bonded to a carbon of a carbon-carbon double bond. The name enol is derived from the fact that it is both an alkene (-*en*-) and an alcohol (-*ol*). To this point, hydroboration-oxidation of alkynes is identical to that of alkenes (Section 6.4).

Enols are in equilibrium with a constitutional isomer formed by migration of a hydrogen atom from oxygen to carbon and rearrangement of the carbon-carbon double bond to form a carbon-oxygen double bond. As can be seen from the value of K_{eq}, 2-butanone (the keto form) is much more stable than its enol. Keto forms in general are more stable than enol forms because (1) a C=O π bond is generally stronger than a C=C π bond, whereas (2) C—H and O—H σ bonds generally have similar bond strengths.

The keto and enol forms of 2-butanone are said to be tautomers. **Tautomers** are constitutional isomers that are in equilibrium with each other and differ only in the location of a hydrogen atom or other atom and a double bond relative to a

Enol
A compound containing a hydroxyl group bonded to a doubly bonded carbon atom.

Tautomers
Constitutional isomers in equilibrium with each other that differ in the location of a hydrogen atom and a double bond relative to a heteroatom, most commonly O, N, or S.

NDRB

heteroatom, most commonly O, N, and S. This type of isomerism is called tautomerism. Because the type of tautomerism we are dealing with in this section involves keto (from ketone) and enol forms, it is commonly called **keto-enol tautomerism**. We discuss keto-enol tautomerism in more detail in Section 16.9.

Hydroboration of a terminal alkyne using disiamylborane followed by oxidation in alkaline hydrogen peroxide also gives an enol that, in this case, is in equilibrium with the more stable aldehyde. Thus, hydroboration-oxidation of a terminal alkyne gives an aldehyde.

Keto-enol tautomerism
A type of isomerism involving keto (from ketone) and enol tautomers.

1-Octyne   **An enol**   **Octanal**

Example 7.4

Hydroboration-oxidation of 2-pentyne gives a mixture of two ketones, each with the molecular formula $C_5H_{10}O$. Propose structural formulas for these two ketones and for the enol from which each is derived.

Solution

Because each carbon of the triple bond in 2-pentyne has the same degree of substitution, very little regioselectivity occurs during hydroboration. Two enols are formed, and the isomeric ketones are formed from them.

2-Pentyne   **2-Pentene-3-ol**   **2-Pentene-2-ol**
                 (an enol)   **(an enol)**

3-Pentanone   **2-Pentanone**

Problem 7.4

Draw a structural formula for a hydrocarbon with the given molecular formula that undergoes hydroboration-oxidation to give the indicated product.

B. Acid-Catalyzed Hydration

In the presence of concentrated sulfuric acid and Hg(II) salts as catalysts, alkynes undergo the addition of water in a reaction analogous to the oxymercuration of alkenes (Section 6.3F). The Hg(II) salts most often used for this purpose are HgO, $HgSO_4$, or $Hg(OAc)_2$. For terminal alkynes, addition of water follows Markovnikov's rule; hydrogen adds to the carbon atom of the triple bond bearing the hydrogen.

NDRB

The resulting enol is in equilibrium with the more stable keto form, so the product isolated is a ketone (an aldehyde in the case of acetylene itself).

$$CH_3C{\equiv}CH + H_2O \xrightarrow[\text{HgSO}_4]{\text{H}_2\text{SO}_4} CH_3\overset{\overset{\displaystyle OH}{|}}{C}{=}CH_2 \rightleftharpoons CH_3\overset{\overset{\displaystyle O}{\parallel}}{C}CH_3$$

Propyne   **Propen-2-ol**   **Propanone**
                (an enol)   **(Acetone)**

The mechanism of this reaction is illustrated by the hydration of propyne to give propanone (acetone).

Mechanism   HgSO$_4$/H$_2$SO$_4$ Catalyzed Hydration of an Alkyne

Step 1: **Make a new bond between a nucleophile (π bond) and an electrophile**
Attack of the C—C triple bond on the Hg^{2+}(an electrophile) gives a bridged mercurinium ion intermediate.

A bridged mercurinium
ion intermediate

Step 2: **Make a new bond between a nucleophile and an electrophile**   Attack of water (a nucleophile) on the bridged mercurinium ion intermediate (an electrophile) from the side opposite the bridge opens the three-membered ring.

Because the 2° vinylic cation structure makes a greater contribution to the hybrid than the 1° vinylic cation structure, attack of water occurs preferentially at the more substituted carbon, which accounts for the observed regioselectivity of the reaction.

Step 3: **Take a proton away**   Proton transfer to solvent gives an organomercury enol.

Step 4: **Keto-enol tautomerism**   Tautomerism of the enol gives the keto form.

Enol form   Keto form

Step 5: Add a proton Proton transfer to the carbonyl group of the ketone gives an oxonium ion.

$$H\overset{+}{\underset{\underset{H}{|}}{\ddot{O}}}-H + CH_3-\overset{\overset{\textstyle :\ddot{O}:}{\|}}{C}-CH_2-Hg^+ \longrightarrow CH_3-\overset{\overset{\textstyle \overset{H}{\underset{}{\overset{+}{\ddot{O}}}}}{\|}}{C}-CH_2-Hg^+ + H_2\ddot{O}:$$

Steps 6 and 7: Break a bond to give stable molecules or ions followed by keto-enol tautomerism Loss of Hg^{2+} from the oxonium ion gives the enol form of the final product. Tautomerism of the enol gives the ketone.

$$CH_3-\overset{\overset{\textstyle \overset{H}{\underset{}{\overset{+}{\ddot{O}}}}}{\|}}{C}-CH_2-Hg^+ \longrightarrow Hg^{2+} + CH_3-\overset{\overset{\textstyle \overset{H}{\underset{}{\ddot{O}}}}{|}}{C}=CH_2 \rightleftharpoons CH_3-\overset{\overset{\textstyle :\ddot{O}:}{\|}}{C}-CH_3$$

$$\qquad\qquad\qquad\qquad\qquad\qquad\qquad\text{(Enol form)}\qquad\qquad\text{(Keto form)}$$

Example 7.5

Show reagents to bring about the following conversions.

(a) (b)

Solution

(a) Hydration of this monosubstituted alkyne using a mercuric ion catalyst gives an enol that is in equilibrium with the more stable keto form.

(b) Hydroboration using disiamylborane followed by treatment with alkaline hydrogen peroxide gives an enol that is in equilibrium with the more stable aldehyde.

Problem 7.5

Hydration of 2-pentyne gives a mixture of two ketones, each with the molecular formula $C_5H_{10}O$. Propose structural formulas for these two ketones and for the enol from which each is derived.

7.8 | Reduction of Alkynes

Three types of reactions are used to convert alkynes to alkenes and alkanes: catalytic reduction, hydroboration-protonolysis, and dissolving-metal reduction.

A. Catalytic Reduction

Treatment of an alkyne with H_2 in the presence of a transition metal catalyst, most commonly palladium, platinum, or nickel, results in the addition of 2 moles of H_2 to the alkyne and its conversion to an alkane. Catalytic reduction of an alkyne can be brought about at or slightly above room temperature and with moderate pressures of hydrogen gas.

$$CH_3C\equiv CCH_3 + 2\ H_2 \xrightarrow[\text{3 atm}]{\text{Pd, Pt, or Ni}} CH_3CH_2CH_2CH_3$$

2-Butyne   **Butane**

Reduction of an alkyne occurs in two stages: first addition of 1 mole of H_2 to form an alkene and then addition of the second mole to the alkene to form the alkane. In most cases, it is not possible to stop the reaction at the alkene stage.

However, by careful choice of catalyst, it is possible to stop the reduction after the addition of 1 mole of hydrogen. The catalyst most commonly used for this purpose consists of finely powdered palladium metal deposited on solid calcium carbonate that has been specially modified with lead salts. This combination is known as the **Lindlar catalyst**. Reduction (hydrogenation) of alkynes over a Lindlar catalyst is stereoselective; **syn addition** of two hydrogen atoms to the carbon-carbon triple bond gives a *cis* alkene.

Lindlar catalyst
Finely powdered palladium metal deposited on solid calcium carbonate that has been specially modified with lead salts. Its particular use is as a catalyst for the reduction of an alkyne to a *cis* alkene.

$$CH_3C\equiv CCH_3 + H_2 \xrightarrow{\substack{\text{Lindlar}\\\text{catalyst}}} \underset{\substack{| \\ H}}{\overset{\substack{H_3C \\ |}}{C}}=\underset{\substack{| \\ H}}{\overset{\substack{CH_3 \\ |}}{C}}$$

2-Butyne   *cis*-**2-Butene**

Because addition of hydrogen in the presence of the Lindlar catalyst is stereoselective for syn addition, it has been proposed that reduction proceeds by simultaneous or nearly simultaneous transfer of two hydrogen atoms from the surface of the metal catalyst to the alkyne. We presented a similar mechanism in Section 6.6A for the catalytic reduction of an alkene to an alkane.

B. Hydroboration-Protonolysis

As we have just seen in Section 7.7A, internal alkynes react with borane to give a trialkenylborane. Treating a trialkenylborane with a carboxylic acid, such as acetic acid, results in stereoselective replacement of boron by hydrogen: a *cis* alkenyl group bonded to a boron is converted to a *cis* alkene.

A trialkenylborane   *cis*-**3-Hexene**

The net effect of hydroboration of an internal alkyne followed by treatment with acetic acid is reduction of the alkyne to a *cis* alkene. Thus, hydroboration-protonolysis and catalytic reduction over a Lindlar catalyst provide alternative schemes for conversion of an alkyne to a *cis* alkene.

C. Dissolving-Metal Reduction

Alkynes can also be reduced to alkenes by using either sodium or lithium metal in liquid ammonia or in low-molecular-weight primary or secondary amines. The alkali metal is the reducing agent and, in the process, is oxidized to M^+, which dissolves as a metal salt in the solvent for the reaction. Reduction of an alkyne to an alkene by lithium or sodium in liquid ammonia, $NH_3(l)$, is stereoselective; it involves mainly **anti addition** of two hydrogen atoms to the triple bond.

$$R{-}\!\!\equiv\!\!{-}R' + 2\,Na + 2\,NH_3 \xrightarrow[NH_3(l)]{} \underset{\underset{R}{}}{\overset{\overset{H\quad R'}{}}{C{=}C}}\underset{H}{} + 2\,NaNH_2$$

4-Octyne   ***trans*-4-Octene**

Thus, by the proper choice of reagents and reaction conditions, it is possible to reduce an alkyne to either a *cis* alkene (by catalytic reduction or hydroboration-protonolysis) or to a *trans* alkene (by dissolving-metal reduction).

The stereoselectivity of alkali metal reduction of alkynes to alkenes can be accounted for by the following mechanism. As you study this mechanism, note that it involves two one-electron reductions and two proton-transfer reactions. The stereochemistry of the alkene is determined in Step 3. Adding the four steps and canceling species that appear on both sides of the equation gives the overall equation for the reaction.

Mechanism   Reduction of an Alkyne by Sodium in Liquid Ammonia

Step 1: A one-electron reduction of the alkyne gives an alkenyl radical anion; that is, an ion containing an unpaired electron on one carbon and a negative charge on an adjacent carbon (note that we use a single-headed arrow to show the repositioning of single electrons).

$$R{-}C{\equiv}C{-}R + \cdot Na \longrightarrow R{-}\overset{\cdot}{C}{=}\overset{..^-}{C}{-}R \longleftrightarrow R{-}\overset{..^-}{C}{=}\overset{\cdot}{C}{-}R + Na^+$$

A resonance-stabilized alkenyl radical anion

Step 2: **Add a proton**   The alkenyl radical anion (a very strong base) abstracts a proton from a molecule of ammonia (under these conditions, a weak acid) to give an alkenyl radical.

$$R{-}\overset{\cdot}{C}{=}\overset{..^-}{C}{-}R + H{-}\overset{..}{\underset{H}{N}}{-}H \longrightarrow \underset{R}{\overset{}{C}}{=}\underset{H}{\overset{R}{C}} + \overset{..}{\underset{H}{:N}}{-}H$$

An alkenyl   **Amide**
radical   **anion**

Step 3: A one-electron reduction of the alkenyl radical gives an alkenyl anion. The *trans* alkenyl anion is more stable than its *cis* isomer, and the stereochemistry of the final product is determined in this step.

$$Na\cdot + \overset{R}{\underset{R}{C}}=\overset{\cdot}{\underset{H}{C}} \longrightarrow Na^+ + \overset{R}{\underset{R}{\overset{\cdot\cdot}{C}}}=\overset{-}{\underset{H}{C}}$$

An alkenyl anion

Step 4: Add a proton A second proton-transfer reaction completes the reduction and gives the *trans* alkene.

$$H-\overset{H}{\underset{H}{\overset{\cdot\cdot}{N}}}-H + \overset{R}{\underset{R}{\overset{\cdot\cdot}{C}}}=\overset{\cdot\cdot}{\underset{H}{C}} \longrightarrow H-\overset{H}{\underset{H}{\overset{\cdot\cdot}{N}}} + \overset{R}{\underset{R}{C}}=\overset{R}{\underset{H}{C}}$$

Amide anion A *trans* alkene

7.9 | Organic Synthesis

We have now seen how to prepare both terminal and internal alkynes from acetylene and substituted acetylenes, and we have seen several common reactions of alkynes, including addition (HX, X_2, and H_2O), hydroboration-oxidation, and reduction. Now let us move a step farther to consider what might be called the art of **organic synthesis**.

A. Retrosynthetic Analysis

Synthesis is generally the most important objective of organic chemists, applicable to the preparation of compounds for use as pharmaceuticals, agrochemicals, plastics, elastomers, or textile fibers. A successful synthesis must provide the desired product in a maximum yield and with a maximum control of stereochemistry at all stages of the synthesis. Furthermore, there is an increasing desire to develop "green" syntheses, that is, syntheses that do not produce or release byproducts harmful to the environment.

Our goal in this section is to develop an ability to plan a successful synthesis. The best strategy is to work backwards from the desired product. First, we analyze the target molecule in the following way.

1. Count the carbon atoms of the carbon skeleton of the target molecule. Determining how to build the carbon skeleton from available starting materials is often the most challenging part of a synthesis. If you must add carbons, you need to consider what carbon-carbon bond-forming reactions are available to you. At this stage in the course, you have only one such reaction, namely alkylation of acetylide anions with methyl or primary halides (Section 7.5).
2. Analyze the functional groups. What are they and how can they be changed to facilitate formation of the carbon skeleton? How can they then be changed to give the final set of functional groups in the desired product? Regiospecificity and stereospecificity can be important considerations.
3. Now work backwards; surprisingly, it is often easier to identify the last step in a synthesis rather than trying to start from the beginning. This process is referred to as **retrosynthetic analysis**. The idea is to start with the product structure and work backwards, one step at a time, until you reach a point at which you can create your synthetic intermediate from the starting materials. If you cannot do this using your first proposed route, try working backwards using a different sequence of reactions. Synthesis problems are creative in the sense that they force

Organic synthesis
A series of reactions by which a set of organic starting materials is converted to a more complicated structure.

Retrosynthesis
A process of reasoning backwards from a target molecule to a suitable set of starting materials.

you to create entirely new ways to connect the reactions you have been taught. To be good at synthesis, you must first be very familiar with the reactions, including how to carry out those reactions on new molecules. The best way to keep track of the reactions you are learning is an Organic Chemistry Roadmap. You should work all of the end of chapter problems under the "Organic Chemistry Roadmap" heading and become familiar with Appendix 11.

Target Molecule: *cis*-3-Hexene

As readily available starting materials, we use acetylene and haloalkanes.

Target molecule:

cis-3-Hexene

Analysis

We note there are six carbons in the product and only two in acetylene. We will need to construct the carbon skeleton through carbon-carbon bond formation with haloalkanes totaling four additional carbon atoms. The functional group in the product is a *cis* carbon-carbon double bond, which can be prepared by catalytic reduction of a carbon-carbon triple bond using the Lindlar catalyst (Section 7.8A). We then disconnect the carbon skeleton into possible starting materials, which we can later reconnect by known reactions. In the example here, we disconnect at the two carbon-carbon single bonds adjacent to the triple bond. These bonds can be formed during the synthesis by alkylation of the acetylide dianion using two haloalkanes (Section 7.5A), each with two carbon atoms (e.g., bromoethane). We use an open arrow to symbolize a step in a retrosynthesis.

disconnect
here

$:C\equiv C:^-$ + 2 CH_3CH_2Br

cis-3-Hexene   **3-Hexyne**   **Acetylide dianion**   **Bromoethane**

Synthesis

Our starting materials for this synthesis of *cis*-3-hexene are acetylene and bromoethane, both readily available compounds. This synthesis is carried out in five steps as follows.

$HC\equiv CH$ 1. $NaNH_2$ 2. CH_3CH_2Br →   3. $NaNH_2$ 4. CH_3CH_2Br →   5. H_2 Lindlar catalyst   →

Acetylene   **1-Butyne**   **3-Hexyne**   **cis-3-Hexene**

Target Molecule: 2-Heptanone

2-Heptanone is responsible for the "peppery" odor of cheeses of the Roquefort type. As readily available starting materials, we again use acetylene and haloalkanes.

Target molecule:

$$\overset{O}{\overset{\|}{\diagup}}\diagdown\diagup\diagdown$$

2-Heptanone

Analysis

We note that there are seven carbons in the product and only two in acetylene. We will need to construct the carbon skeleton through carbon-carbon bond formation with haloalkanes totaling five carbon atoms. The functional group in the target molecule is a ketone, which we can prepare by hydration of a carbon-carbon triple bond. Hydration of 1-heptyne gives only 2-heptanone, whereas hydration of 2-heptyne gives a mixture of 2-heptanone and 3-heptanone. Therefore, we choose a functional group interconversion via 1-heptyne.

An acid-catalyzed hydration gives a mixture of 2-heptanone and 3-heptanone.

2-Heptyne

O

2-Heptanone

Acid-catalyzed hydration gives 2-heptanone

1-Heptyne

$HC\equiv\overset{-}{C}:$ + Br

Acetylide anion　　**1-Bromopentane**

Synthesis

This synthesis can be carried out in three steps as follows.

$HC\equiv CH$　$\xrightarrow[\text{2. Br}]{\text{1. NaNH}_2}$　**1-Heptyne**　$\xrightarrow[\text{H}_2\text{SO}_4, \text{ HgSO}_4]{\text{3. H}_2\text{O}}$　

O

2-Heptanone

Example 7.6

How might the scheme for the synthesis of 2-heptanone be modified so that the product is heptanal?

Solution

Steps 1 and 2 are the same and give 1-heptyne. Instead of acid-catalyzed hydration of 1-heptyne, treat the alkyne with $(sia)_2BH$ followed by alkaline hydrogen peroxide (Section 7.7).

$HC\equiv CH$　$\xrightarrow[\text{2. Br}]{\text{1. NaNH}_2}$　**1-Heptyne**　$\xrightarrow[\text{4. H}_2\text{O}_2/\text{NaOH}]{\text{3. (sia)}_2\text{BH}}$　

H

O　**Heptanal**

Problem 7.6

Show how the synthetic scheme in Example 7.6 might be modified to give the following.

(a) 1-Heptanol　　　**(b)** 2-Heptanol

B. Reactions of Functional Groups in Complex Molecules

A defining feature of organic chemistry is that a functional group often reacts in the same fashion whether it is in a relatively simple molecule or in a more complex molecule that contains numerous other functional groups. As a result, the synthesis of increasingly complex molecules can be carried out in a rational manner, provided that the reactivities of all the functional groups present are taken into account.

Fexofenadine, sold under the trade name Allegra, is a powerful nonsedating antihistamine. An efficient synthesis of fexofenadine published by a team from McGill University in 1994 involved the conversion of an alkyne to a ketone in acidic solution, using HgO (Section 7.7B). The ketone function ended up on the carbon adjacent to the benzene ring because positive charge in the intermediate is stabilized at this position by resonance delocalization. This ketone product is converted to fexofenadine through two additional reactions we will not discuss here. Although the McGill synthesis is efficient, the use of toxic HgO makes this overall synthesis inappropriate for commercial use, and an alternative approach is used when the product is prepared for sale. Fexofenadine has one chiral center and is sold as a racemic mixture.

Fexofenadine
(racemic)

This example illustrates two important points about how functional group analysis is used to develop systematic syntheses of complex molecules. First, the alkyne was the only functional group to react under the conditions used, and it reacted as expected based on comparison to simpler alkyne molecules such as those discussed in this chapter. You will learn about the reactivities of many functional groups in this course, and in each case they can be used to predict what happens in complex molecules that contain them. Secondly, although they do not react directly, other functional groups often have an influence over the regiochemistry or stereochemistry of a reaction. The benzene ring in the present example did not react directly with the HgO reagent but did influence the regiochemistry of the product.

Summary

Problem: 7.7

Section 7.1 Structure of Alkynes

- **Alkynes** contain one or more carbon-carbon triple bonds.
  - The triple bond is a combination of one σ bond formed by the overlap of *sp* hybrid orbitals and two π bonds formed by the overlap of two sets of parallel $2p$ orbitals.

Section 7.2 Nomenclature of Alkynes

- According to IUPAC nomenclature, the infix –*yn*- is used to show the presence of a carbon-carbon triple bond.
  - We name the main chain as the longest one that contains the triple bond, and assign numbers from the end that gives the carbon atoms of the triple bond the lower set of numbers.

Problems: 7.1, 7.2

- The IUPAC system retains the name acetylene.
- Common names are derived by adding a prefix for the substituent to the word *acetylene*.

Section 7.3 Physical Properties of Alkynes

- The physical properties of alkynes are similar to those of alkanes and alkenes of comparable carbon skeletons.

Section 7.4 Acidity of 1-Alkynes

- The pK_a values of terminal alkynes are approximately 25; they are less acidic than water and alcohols but more acidic than alkanes, alkenes, and ammonia.
  - The hydrogen atom bonded to a carbon-carbon triple bond is sufficiently acidic so that it can be removed by a strong base, most commonly sodium amide ($NaNH_2$), sodium hydride (NaH), or lithium diisopropylamide (LDA).

Problem: 7.10

Section 7.5 Preparation of Alkynes

- Alkynes can be prepared through alkylation of acetylide anions with methyl or primary halides.
  - This reaction is important because it creates a carbon-carbon bond. Starting with acetylene, even unsymmetrical alkynes can be made using two sequential deprotonation–**alkylation reactions**.
- Alkynes can be made from alkenes through a two-step sequence of halogenation with X_2, followed by double dehydrohalogenation.
  - **Allenes** are sometimes seen as a byproduct of the double dehydrohalogenation.

Problems: 7.3, 7.8, 7.9, 7.15, 7.19, 7.22, 7.26

Section 7.6 Electrophilic Addition to Alkynes

- Alkynes add Br_2 to give dibromoalkenes via anti addition stereochemistry.
  - A second mole of Br_2 can be added to give a tetrabromoalkane.
  - Addition of Cl_2 occurs with less stereoselectivity than the addition of Br_2.
- Alkynes add two moles of HCl or HBr, with both the halogens adding to the same carbon.
  - For terminal alkynes, Markovnikov's rule is followed and the hydrogen ends up on the terminal carbon atom.

Problems: 7.11, 7.16

Section 7.7 Hydration of Alkynes to Aldehydes and Ketones

- Alkynes react with borane [BH_3 or $(sia)_2BH$] followed by basic peroxide to give aldehydes or ketones.
  - The sterically hindered $(sia)_2BH$ gives regioselective reaction with 1-alkynes leading to aldehydes. The reaction initially produces an **enol**.
- The functional group of an enol is an —OH group on a carbon atom of an alkene.
  - The enol form is in equilibrium with the keto form, namely an aldehyde or ketone, and the equilibrium almost always lies far on the side of the keto form.
  - **Tautomers** are constitutional isomers that are in equilibrium with each other but differ in the location of a hydrogen and a double bond relative to a heteroatom, most commonly O, N, and S.
  - **Keto-enol tautomerism** is the most common type of tautomerism we encounter in this course.
- Alkynes react with water in the presence of H_2SO_4 and Hg(II) salts to give ketones, or acetaldehyde in the case of acetylene, via an enol intermediate.

Section 7.8 Reduction of Alkynes

- Alkynes react with two moles of H_2 in the presence of a transition metal catalyst to give alkanes in a reaction that does not ordinarily stop at the alkene stage.
- The **Lindlar catalyst** is a deactivated hydrogenation catalyst that stops alkyne hydrogenation at the alkene stage, allowing conversion of alkynes to *cis* alkenes. Hydroboration followed by an acid workup (instead of basic peroxide) also gives *cis* alkenes from alkynes. Hydroboration–protonolysis is an alternative way to prepare a *cis* alkene from an alkyne.
- Dissolving metal reduction gives *trans* alkenes from alkynes.

Problem: 7.17

Section 7.9 Organic Synthesis

- When planning an **organic synthesis** involving several reactions, it is best to use the following systematic approach:
  1. Count the carbons in the products versus the starting material so you know what fragments must be added or subtracted.
  2. Analyze the functional groups.
  3. Work backwards (**retrosynthesis**).
- A **functional group** often reacts in the same fashion whether it is in a relatively simple molecule or a more complex molecule that contains numerous other functional groups.
  - Although they do not react directly, other functional groups often have an influence over the regiochemistry or stereochemistry of a reaction.

Problems: 7.6, 7.14, 7.20, 7.21–7.32

Problem: 7.33

ⓌWL Questions and problems for this chapter are available in OWL. Use the OWL access card that came with this textbook to access assigned questions and problems for this chapter.

Key Reactions

1. **Acidity of Terminal Alkynes (Section 7.4)**   Treatment of terminal alkynes (pK_a 25) with a strong base, most commonly $NaNH_2$, NaH, or lithium diisopropylamine (LDA), gives an acetylide salt.

$$HC \equiv CH + Na^+NH_2^- \longrightarrow HC \equiv C^-Na^+ + NH_3$$

2. **Alkylation of Acetylide Anions (Section 7.5A)**   Acetylide anions are nucleophiles and will displace halide ion from methyl and 1° haloalkanes.

$$HC \equiv C{:}^-Na^+ + \underset{}{\diagdown\!\diagup\!\diagdown}Br \longrightarrow \underset{}{\diagdown\!\diagup\!\diagdown\!\diagup\!\equiv} + Na^+Br^-$$

3. **Synthesis of an Alkyne from an Alkene (Section 7.5B)**   Treating an alkene with Br_2 or Cl_2 gives a dihaloalkane. Treating the dihaloalkane with $NaNH_2$ or other strong base results in two successive dehydrohalogenations to give an alkyne.

$$RCH{=}CHR \xrightarrow{Br_2} \underset{\substack{|\;\;\;\;\;| \\ Br\;\;Br}}{RCH{-}CHR} \xrightarrow[NH_3(l)]{2\,NaNH_2} RC \equiv CR$$

4. **Addition of Br_2 and Cl_2 (Section 7.6A)**   Addition of 1 mole of Br_2 or Cl_2 is anti stereoselective; anti addition of halogen to an alkyne gives an (E)-dihaloalkene. Addition of a second mole of halogen gives a tetrahaloalkane.

$$CH_3C \equiv CCH_3 \xrightarrow{Cl_2} \underset{\substack{Cl \;\;\;\;\;CH_3}}{\overset{\substack{H_3C \;\;\;\;\;Cl}}{C{=}C}} \xrightarrow{Cl_2} \underset{\substack{|\;\;\;| \\ Cl\;\;Cl}}{CH_3{-}\overset{\substack{Cl\;\;Cl}}{C{-}C}{-}CH_3}$$

5. **Addition of HX (Section 7.6B)**   Addition of HX is regioselective. Reaction by way of a vinylic carbocation intermediate follows Markovnikov's rule. Addition of 2 HX gives a geminal dihaloalkane.

$$CH_3C \equiv CH \xrightarrow{HBr} \underset{\substack{|\\Br}}{CH_3\overset{\substack{Br}}{C}{=}CH_2} \xrightarrow{HBr} \underset{\substack{|\\Br}}{CH_3\overset{\substack{Br}}{C}CH_3}$$

6. **Keto-Enol Tautomerism (Section 7.7A)**   In an equilibrium between a keto form and an enol form, the keto form generally predominates.

$$\underset{\text{Enol form}}{CH_3CH{=}\overset{\substack{OH\\|}}{C}CH_3} \rightleftharpoons \underset{\text{Keto form}}{CH_3CH_2\overset{\substack{O\\||}}{C}CH_3}$$

7. Hydroboration-Oxidation (Section 7.7A) Hydroboration of an internal alkyne is syn stereoselective. Oxidation of the resulting trialkenylborane by H_2O_2/NaOH gives an enol that is in equilibrium, through keto-enol tautomerism, with a ketone.

Hydroboration of a terminal alkyne using a hindered dialkylborane followed by oxidation of the resulting trialkenylborane with H_2O_2/NaOH and then keto-enol tautomerism gives an aldehyde.

8. Acid-Catalyzed Hydration (Section 7.7B) Acid-catalyzed addition of water in the presence of Hg(II) salts is regioselective. Keto-enol tautomerism of the resulting enol gives a ketone.

$$CH_3C{\equiv}CH + H_2O \xrightarrow[HgSO_4]{H_2SO_4} CH_3\overset{OH}{\underset{|}{C}}{=}CH_2 \rightleftharpoons CH_3\overset{O}{\overset{\|}{C}}CH_3$$

9. Catalytic Reduction (Section 7.8A) Reaction of an alkyne with 2 moles of H_2 under moderate pressure in the presence of a transition metal catalyst at room temperature gives an alkane.

Catalytic reduction of an alkyne in the presence of the Lindlar catalyst is syn stereoselective; Lindlar reduction of an internal alkyne gives a *cis* alkene.

10. Hydroboration-Protonolysis (Section 7.8B) Hydroboration of an alkyne followed by protonolysis also converts an alkyne to a *cis* alkene.

11. Reduction Using Na or Li Metal in NH₃(*l*) (Section 7.8C) Alkali metal reduction is stereoselective: anti addition of hydrogens to an internal alkyne gives a *trans* alkene. The mechanism involves a radical mechanism with two sequential single-electron transfers from the Na or Li metal, each followed by deprotonation of the amine solvent.

Chapter 8

Haloalkanes, Halogenation, and Radical Reactions

Many common objects such as these pipes are made of poly(vinyl chloride). *Inset:* a model of chloroethene (vinyl chloride).

Image copyright Christina Richards 2010. Used under license from Shutterstock.com

Compounds containing a halogen atom covalently bonded to an sp^3 hybridized carbon atom are named haloalkanes, or, in the common system of nomenclature, alkyl halides. Several haloalkanes are important laboratory and industrial solvents. In addition, haloalkanes are invaluable building blocks for organic synthesis because of the variety of ways in which the halogen may be substituted with other groups or eliminated to produce π bonds.

In this chapter, we begin with the structure and physical properties of haloalkanes. We then study radical halogenation of alkanes as a vehicle to introduce an important type of reaction mechanism, namely the mechanism of radical chain reactions. Reactions of oxygen with alkenes and a radical mechanism for HBr addition to alkenes complete the chapter.

8.1 | Structure

Haloalkane (alkyl halide)
A compound containing a halogen atom covalently bonded to an sp^3 hybridized carbon atom. Given the symbol R—X.

Haloalkene (vinylic halide)
A compound containing a halogen bonded to one of the carbons of a carbon-carbon double bond.

Haloarene (aryl halide)
A compound containing a halogen atom bonded to a benzene ring. Given the symbol Ar—X.

The general symbol for a **haloalkane** is R—X, where —X may be —F, —Cl, —Br, or —I. If a halogen is bonded to a doubly bonded carbon of an alkene, the compound belongs to a class called **haloalkenes**. If it is bonded to a benzene ring, the compound belongs to a class called **haloarenes**, which have the general symbol Ar—X.

A haloalkane
(an alkyl halide)

A haloalkene
(an alkenyl
or vinylic halide)

A haloarene
(an aryl halide)

⬤WL Online homework for this chapter may be assigned in OWL for Organic Chemistry.

8.2 | Nomenclature

A. IUPAC System

IUPAC names for haloalkanes are derived by naming the parent alkane according to the rules given in Section 2.3A.

- The parent chain is numbered from the direction that gives the first substituent encountered the lowest number, whether it is a halogen or an alkyl group. If two groups could have the same lowest number from the end of the chain, give the group of lower alphabetical order the lower number. An example is 2-bromo-4-methylpentane.
- Halogen substituents are indicated by the prefixes *fluoro-*, *chloro-*, *bromo-*, and *iodo-* and are listed in alphabetical order with other substituents.
- The location of each halogen atom on the parent chain is given by a number preceding the name of the halogen.
- In haloalkenes, numbering the parent hydrocarbon is determined by the location of the carbon-carbon double bond. Numbering is done in the direction that gives the carbon atoms of the double bond and substituents the lowest set of numbers.

(S)-2-Bromo-4-methylpentane **(1S,2S)-2-Chlorocyclohexanol** **(R)-4-Bromocyclohexene**

B. Common Names

Common names of haloalkanes and haloalkenes consist of the common name of the alkyl group followed by the name of the halide as a separate word. Hence, the name **alkyl halide** is a common name for this class of compounds. In the following examples, the IUPAC name of the compound is given first, and then its common name is given in parentheses:

(R)-2-Bromobutane   **Chloroethene**   **3-Chloropropene**
((R)-sec-Butyl bromide) **(Vinyl chloride)** **(Allyl chloride)**

Several polyhaloalkanes are important solvents and are generally referred to by their common names. Dichloromethane (methylene chloride) is the most widely used haloalkane solvent. Compounds of the type CHX_3 are called **haloforms**. The common name for $CHCl_3$, for example, is chloroform. The common name methyl chloroform for the compound CH_3CCl_3 derives from this name. Methyl chloroform and trichloroethylene (trichlor) were once common solvents for industrial cleaning. Because they are somewhat toxic and cause environmental problems, they have been phased out.

Haloform
A compound of the type CHX_3 where X is a halogen.

CH_2Cl_2	$CHCl_3$	CH_3CCl_3	$CCl_2{=}CHCl$
Dichloromethane	**Trichloromethane**	**1,1,1-Trichloroethane**	**Trichloroethylene**
(Methylene chloride)	**(Chloroform)**	**(Methyl chloroform)**	**(Trichlor)**

Hydrocarbons in which all hydrogens are replaced by halogens are commonly called perhaloalkanes or perhaloalkenes. Perchloroethylene, commonly known as perc, is a dry cleaning solvent that is also being phased out.

Perchloroethane   Perfluoropropane   Perchloroethylene

Example 8.1

Write the IUPAC name and, where possible, the common name of each compound. Show stereochemistry where relevant.

(a)   (b)   (c)

Solution

(a) 1-Bromo-2,2-dimethylpropane. Its common name is neopentyl bromide.
(b) (*E*)-4-Bromo-3,4-dimethyl-2-pentene.
(c) (*S*)-2-Bromooctane.

Problem 8.1

Write the IUPAC name, and where possible, the common name of each compound. Show stereochemistry where relevant.

(a)   (b)   (c)   (d)

8.3 | Physical Properties of Haloalkanes

A. Polarity

Fluorine, chlorine, and bromine are all more electronegative than carbon (Table 1.5), and, as a result, C—X bonds with these atoms are polarized with a partial negative charge on halogen and a partial positive charge on carbon. Table 8.1 shows that each of the halomethanes has a substantial dipole moment. The electrostatic potential map of fluoromethane shows the large charge separation in this compound caused by the dipole.

The magnitude of a dipole moment depends on the size of the partial charges, the distance between them, and the polarizability of the three pairs of unshared electrons on each halogen. For the halomethanes, the dipole moment increases as the electronegativity of the halogen and the bond length increase. These two

Table 8.1	Dipole Moments (Gas Phase) of Halomethanes		
Halomethane	Electronegativity of Halogen	Carbon-Halogen Bond Length (pm)	Dipole Moment (debyes, D)
CH_3F	4.0	139	1.85
CH_3Cl	3.0	178	1.87
CH_3Br	2.8	193	1.81
CH_3I	2.5	214	1.62

trends run counter to each other, with the net effect that chloromethane has the largest dipole moment of the series. The experimental dipole moments also show that the electronegativity of carbon is less than the standard Pauling value of 2.5 in many compounds (as we mentioned in Section 1.2B), or CH_3I would have no dipole moment.

B. Boiling Point

Electrostatic potential map of fluoromethane.

Haloalkanes are associated in the liquid state by a combination of attractive dipole-dipole, dipole-induced dipole, and induced dipole-induced dipole (dispersion) forces. These forces are grouped together under the term **van der Waals forces**, in honor of J. D. van der Waals, the 19th century Dutch physicist. As atoms or molecules are brought closer and closer, van der Waals attractive forces are overcome by repulsive forces between the electron clouds of adjacent atoms. The energy minimum is where the net attraction is the strongest. Nonbonded interatomic and intermolecular distances at these minima can be measured by X-ray crystallography of solid compounds, and each atom and group of atoms can be assigned an atomic or molecular radius called a **van der Waals radius**. Van der Waals radii for selected atoms and groups of atoms are given in Table 8.2.

Notice from Table 8.2 that the van der Waals radius of fluorine is only slightly greater than that of hydrogen and that, among the halogens, only iodine has a larger van der Waals radius than methyl.

van der Waals forces
A group of intermolecular attractive forces including dipole-dipole, dipole-induced dipole, and induced dipole-induced dipole (dispersion) forces.

van der Waals radius
The minimum distance of approach to an atom that does not cause nonbonded interaction strain.

Table 8.2	van der Waals Radii (pm) for Selected Atoms and Groups of Atoms					
H	F	Cl	Br	CH_2	CH_3	I
120	135	180	195	200	200	215

Increasing van der Waals radius

Boiling points of several low-molecular-weight haloalkanes and the alkanes from which they are derived are given in Table 8.3. Several trends are to be noticed from these data.

1. As with hydrocarbons, constitutional isomers with branched chains have lower boiling points than their unbranched-chain isomers (Section 2.7D). Compare, for example, the boiling points of unbranched-chain 1-bromobutane (butyl bromide, bp 100°C) with the more branched and compact 2-bromo-2-methylpropane (*tert*-butyl bromide, bp 72°C). Branched-chain constitutional isomers have lower boiling points because they have a more spherical shape and, therefore, decreased surface area, leading to smaller van der Waals forces between their molecules.
2. For an alkane and haloalkane of comparable size and shape, the haloalkane has a higher boiling point. Compare, for example, the boiling points of ethane (bp −89°C) and bromomethane (bp 4°C). Although both molecules are roughly the same size and have roughly the same effective contact area, the boiling point of bromomethane is considerably higher. This difference in boiling points is due to the dipole moment in bromomethane, as well as the greater **polarizability** of bromine compared to methyl. Recall from Section 2.7B that the strength of dispersion forces, the weakest of all intermolecular forces, depends on the polarizability of electrons, which, in turn, depends on how tightly they are held by the nucleus. Unshared electron pairs have a higher polarizability than electrons shared in a covalent bond. In addition, the farther electrons are from the nucleus, the less tightly they are held and the greater their polarizability. Therefore, the larger the halogen, the greater its polarizability.

Polarizability
A measure of the ease of distortion of the distribution of electron density about an atom or group in response to interaction with other molecules or ions. Fluorine, which has a high electronegativity, holds its electrons tightly and has a very low polarizability. Iodine, which has a lower electronegativity and holds its electrons less tightly, has a very high polarizability.

Table 8.3

Table 8.3	Boiling Points of Some Low-Molecular-Weight Alkanes and Haloalkanes					
		Boiling Point (°C)				
Alkyl Group	**Name**	**H**	**F**	**Cl**	**Br**	**I**
CH_3—	**Methyl**	−161	−78	−24	4	43
CH_3CH_2—	**Ethyl**	−89	−37	13	38	72
$CH_3(CH_2)_2$—	**Propyl**	−45	3	46	71	102
$(CH_3)_2CH$—	**Isopropyl**	−45	−11	35	60	89
$CH_3(CH_2)_3$—	**Butyl**	0	32	77	100	130
$CH_3CH_2(CH_3)CH$—	*sec*-**Butyl**	0	25	67	90	119
$(CH_3)_2CHCH_2$—	**Isobutyl**	−1	16	68	91	120
$(CH_3)_3C$—	*tert*-**Butyl**	−1	12	51	72	98
$CH_3(CH_2)_4$—	**Pentyl**	36	63	108	129	157
$CH_3(CH_2)_5$—	**Hexyl**	69	92	134	155	181

3. The boiling points of fluoroalkanes are comparable to those of hydrocarbons of similar molecular weight. Compare, for example, the boiling points of hexane (MW 86.2, bp 69°C) and 1-fluoropentane (MW 90.1, bp 63°C) and the boiling points of 2-methylpropane (MW 58.1, bp −1°C) and 2-fluoropropane (MW 62.1, bp −11°C). This low boiling point is attributable to the small size of fluorine, the tightness with which its electrons are held, and their particularly low polarizability.

C. Density

The densities of liquid haloalkanes are greater than those of hydrocarbons of comparable molecular weight because of the halogens' large mass-to-volume ratio. A bromine atom and a methyl group have almost identical van der Waals radii, but bromine has a mass of 79.9 atomic mass units (amu) compared with 15 amu for methyl. Table 8.4 gives densities for some low-molecular-weight haloalkanes that are liquid at 25°C. The densities of all liquid bromoalkanes and iodoalkanes are greater than that of water.

Although the densities of liquid chloroalkanes (Table 8.4) are less than that of water, further substitution of chlorine for hydrogen increases the density to the point where di- and polychloroalkanes have a greater density than water (Table 8.5). These compounds sink in water and form the lower layer when mixed because they are immiscible with water.

Table 8.4	Densities of Some Low-Molecular-Weight Haloalkane			
		Density of Liquid (g/mL) at 25°C		
Alkyl Group	**Name**	**Cl**	**Br**	**I**
CH_3—	**Methyl**	—	—	2.279
CH_3CH_2—	**Ethyl**	—	1.460	1.936
$CH_3(CH_2)_2$—	**Propyl**	0.891	1.354	1.749
$(CH_3)_2CH$—	**Isopropyl**	0.862	1.314	1.703
$CH_3(CH_2)_3$—	**Butyl**	0.886	1.276	1.615
$(CH_3)_3C$—	*tert*-**Butyl**	0.842	1.221	1.545

Table 8.5	Density of Polyhalomethanes			
		Density of Liquid (g/mL) at 25°C		
Haloalkane	X =	Cl	Br	I
CH_2X_2		1.327	2.497	3.325
CHX_3		1.483	2.890	4.008
CX_4		1.594	3.273	4.23

D. Bond Lengths and Bond Strengths

With the exception of C—F bonds, C—X bonds are weaker than C—H bonds as measured by bond dissociation enthalpies (BDE), which is a measure of bond strength (Section 6.2B). A table of BDE values for many bonds is given in Appendix 3. Recall that bond dissociation enthalpy is defined as the amount of energy required to break a bond homolytically into two radicals in the gas phase at 25°C.

$$A \overset{\frown}{}\!\!\overset{\frown}{} B \longrightarrow A\cdot + \cdot B$$

A **radical**, sometimes called a free radical, is any chemical species that contains one or more unpaired electrons. Radicals are produced from a molecule by cleavage of a bond in such a way that each atom or fragment participating in the bond retains one electron, a process called **homolytic bond cleavage**. In the more common **heterolytic bond cleavage**, a bond breaks in such a way that one of the species retains both electrons. We use **fishhook arrows** to show the change in position of single electrons and to indicate a homolytic mechanism.

This reaction used to determine BDE is a "virtual" one because it can't actually be carried out in most cases. Instead, the extensive tables of BDEs are collected from thermochemical data on heats of combustion, hydrogenation, and other reactions. The useful thing about these data is that by adding and subtracting them, heats of reaction can be calculated with confidence for reactions that have never been measured.

C—X BDEs are tabulated in Table 8.6. As the size of the halogen atom increases, the C—X bond length increases, and its strength decreases. These relationships between bond strength and bond length help us to understand the difference in the ease with which haloalkanes undergo reactions that involve carbon-halogen bond breaking. Fluoroalkanes, for example, with the strongest and shortest C—X bonds, are highly resistant to bond breaking under most conditions. This characteristic inertness is one of the factors that make perfluoroalkanes such as Teflon such useful materials.

Radical
Any chemical species that contains one or more unpaired electrons.

Homolytic bond cleavage
Cleavage of a bond so that each fragment retains one electron, producing radicals.

Heterolytic bond cleavage
Cleavage of a bond so that one fragment retains both electrons and the other has none.

Fishhook arrow
A barbed, curved arrow used to show the change in position of a single electron.

Table 8.6	Average Bond Dissociation Enthalpies for C—H and C—X Bonds	
Bond	Bond Length (pm)	Bond Dissociation Enthalpy [kJ (kcal)/mol]
C—H	109	414 (99)
C—F	142	464 (111)
C—Cl	178	355 (85)
C—Br	193	309 (78)
C—I	214	228 (57)

8.4 | Preparation of Haloalkanes by Halogenation of Alkanes

As we saw in Sections 6.3A and 6.3D, haloalkanes can be prepared by the addition of HX and X_2 to alkenes. They are also prepared by replacement of the —OH group of alcohols by halogen (Section 10.5). Many of the simpler, low-molecular-weight haloalkanes are prepared by the halogenation of alkanes, illustrated here by treating 2-methylpropane with bromine at an elevated temperature.

$$
\begin{array}{ccccccc}
& CH_3 & & & & CH_3 & \\
& | & & & & | & \\
CH_3CH & + & Br_2 & \xrightarrow{\text{heat}} & CH_3CBr & + & HBr \\
& | & & & & | & \\
& CH_3 & & & & CH_3 &
\end{array}
$$

2-Methylpropane   **2-Bromo-2-methylpropane**
(Isobutane)   **(*tert*-Butyl bromide)**

Halogenation of alkanes is common with Br_2 and Cl_2. Fluorine, F_2, is seldom used because its reactions with alkanes are so exothermic that they are difficult to control and can actually cause C—C bond cleavage and even explosions. Iodine, I_2, is seldom used because the reaction is endothermic and the position of equilibrium favors alkane and I_2 rather than iodoalkane and HI.

If a mixture of methane and chlorine gas is kept in the dark at room temperature, no detectable change occurs. If, however, the mixture is heated or exposed to light, a reaction begins almost at once with the evolution of heat. The products are chloromethane and hydrogen chloride. What occurs is a **substitution** reaction, in this case, substitution of a methane hydrogen atom by a chlorine atom and the production of an equivalent amount of hydrogen chloride.

Substitution
A reaction in which an atom or group of atoms in a compound is replaced by another atom or group of atoms.

$$ CH_4 + Cl_2 \xrightarrow{\text{heat}} CH_3Cl + HCl $$

Methane   **Chloromethane**
                           (Methyl chloride)

If chloromethane is allowed to react with more chlorine, further chlorination produces a mixture of dichloromethane (methylene chloride), trichloromethane (chloroform), and tetrachloromethane (carbon tetrachloride). Notice that in the last equation, the reagent Cl_2 is placed over the reaction arrow and the equivalent amount of HCl formed is not shown. Placing reagents over reaction arrows and omitting byproducts is commonly done to save space.

$$ CH_3Cl + Cl_2 \xrightarrow{\text{heat}} CH_2Cl_2 + HCl $$

Chloromethane   **Dichloromethane**
(Methyl chloride)   **(Methylene chloride)**

$$ CH_2Cl_2 \xrightarrow[\text{heat}]{Cl_2} CHCl_3 \xrightarrow[\text{heat}]{Cl_2} CCl_4 $$

Dichloromethane   **Trichloromethane**   **Tetrachloromethane**
(Methylene chloride)   **(Chloroform)**   **(Carbon tetrachloride)**

It is possible to prepare chloromethane or tetrachloromethane in relatively pure form by this reaction. In the case of chloromethane, a large excess of methane is used; for tetrachloromethane, a large excess of chlorine drives the reaction to complete halogenation. The other chlorinated methanes can be separated by distillation of partially chlorinated mixtures.

Treating ethane with bromine gives bromoethane (ethyl bromide).

$$ CH_3CH_3 + Br_2 \xrightarrow{\text{heat}} CH_3CH_2Br + HBr $$

Ethane   **Bromoethane**
                          (Ethyl bromide)

In all cases, monosubstituted products are only obtained using an excess of ethane.

A. Regioselectivity

Treating propane with bromine gives a mixture consisting of approximately 92% of 2-bromopropane and 8% of 1-bromopropane:

$$CH_3CH_2CH_3 + Br_2 \xrightarrow[\text{or light}]{\text{heat}} CH_3\overset{\overset{\displaystyle Br}{|}}{C}HCH_3 + CH_3CH_2CH_2Br + HBr$$

Propane   **2-Bromopropane**   **1-Bromopropane**
                                       (92%)   (8%)

Propane contains eight hydrogens; one set of six equivalent primary hydrogens and one set of two secondary hydrogens (Section 2.3D). The hydrogens in each set are equivalent because of rapid bond rotation about C—C single bonds. Substitution of bromine for a primary hydrogen gives 1-bromopropane; substitution of bromine for a secondary hydrogen gives 2-bromopropane. If there were random substitution of any one of the eight hydrogens in propane, we would predict that the isomeric bromopropanes would be formed in the ratio of 6:2, or 75% 1-bromopropane and 25% 2-bromopropane. In fact, in the bromination of propane, substitution of a secondary hydrogen rather than a primary hydrogen is strongly favored. 2-Bromopropane is the major product and the reaction is highly regioselective.

Product Distribution	$CH_3CH_2CH_2Br$	$CH_3\overset{\overset{Br}{|}}{C}HCH_3$
Prediction based on ratio of six 1° H to two 2° H	75%	25%
Experimental observation	8%	92%

Other experiments have shown that substitution at a tertiary hydrogen is favored over both secondary and primary hydrogens. For example, monobromination of 2-methylpentane is very regioselective and gives almost exclusively 2-bromo-2-methylpentane.

2-Methylpentane   **2-Bromo-2-methylpentane**

The reaction of bromine with an alkane occurs in the order 3° > 2° > 1° hydrogen. Chlorination of alkanes is also regioselective, but much less so than bromination. For example, treatment of propane with chlorine gives a mixture of approximately 57% 2-chloropropane and 43% 1-chloropropane.

Propane   **2-Chloropropane**   **1-Chloropropane**
                                   (57%)   (43%)

Thus, we can conclude that, although both bromine and chlorine are regioselective in hydrogen replacement in the order 3° > 2° > 1°, regioselectivity is far greater for bromination than for chlorination. From data on product distribution, it has been determined that regioselectivity per hydrogen for bromination is approximately 1600:80:1, whereas it is only about 5:4:1 for chlorination. We will discuss reasons for this difference in Section 8.5.

Example 8.2

Name and draw structural formulas for all monobromination products formed by treating 2-methylpropane with Br_2. Predict the major product based on the regioselectivity of the reaction of Br_2 with alkanes.

$$\text{2-Methylpropane} + Br_2 \xrightarrow[\text{heat}]{\text{light or}} \text{monobromoalkanes} + HBr$$

2-Methylpropane

Solution

2-Methylpropane has nine equivalent primary hydrogens and one tertiary hydrogen. Substitution of bromine for a primary hydrogen gives 1-bromo-2-methylpropane; substitution for the tertiary hydrogen gives 2-bromo-2-methylpropane. Given that the regioselectivity per hydrogen of bromination for $3° > 2° > 1°$ hydrogens is approximately 1600:80:1, it is necessary to correct for the number of hydrogens: nine primary and one tertiary. The result is that 99.4% of the product is 2-bromo-2-methylpropane and 0.6% is 1-bromo-2-methylpropane.

$$\text{2-Methylpropane} + Br_2 \xrightarrow[\text{or light}]{\text{heat}} \text{2-Bromo-2-methylpropane} + \text{1-Bromo-2-methylpropane} + HBr$$

2-Methylpropane **2-Bromo-2-methylpropane** (major product) **1-Bromo-2-methylpropane**

$$\text{Predicted \% 2-bromo-2-methylpropane} = \frac{1 \times 1600}{(1 \times 1600) + (9 \times 1)} \times 100 = 99.4\%$$

Problem 8.2

Name and draw structural formulas for all monochlorination products formed by treatment of 2-methylpropane with Cl_2. Predict the major product based on the regioselectivity of the reaction of Cl_2 with alkanes.

B. Energetics

We can learn a lot about these reactions by careful consideration of the energetics of each step. A selection of C—H BDE values is given in Table 8.7.

Note that the BDE values for saturated hydrocarbons depend on the type of hydrogen being abstracted and are in the order methane $> 1° > 2° > 3°$. This order can best be understood by comparison to carbocations. In general, interactions

Table 8.7	Bond Dissociation Enthalpies for Selected C—H Bonds			
Hydrocarbon	**Radical**	**Name of Radical**	**Type of Radical**	**$\Delta H°$ [kJ(kcal)/mol]**
$CH_2{=}CHCH_2{-}H$	$CH_2{=}CHCH_2\cdot$	Allyl	Allylic	372 (89)
$C_6H_5CH_2{-}H$	$C_6H_5CH_2\cdot$	Benzyl	Benzylic	376 (90)
$(CH_3)_3C{-}H$	$(CH_3)_3C\cdot$	*tert*-Butyl	3°	405 (97)
$(CH_3)_2CH{-}H$	$(CH_3)_2CH\cdot$	Isopropyl	2°	414 (99)
$CH_3CH_2{-}H$	$CH_3CH_2\cdot$	Ethyl	1°	422 (101)
$CH_3{-}H$	$CH_3\cdot$	Methyl	Methyl	439 (105)
$CH_2{=}CH{-}H$	$CH_2{=}CH\cdot$	Vinyl	Vinylic	464 (111)

Radical stability →

that are stabilizing to carbocations are also stabilizing to radicals. This makes sense because radicals have only seven electrons in the valence shell of the carbon bearing the radical. Therefore, like carbocations, the more alkyl groups the carbon radicals have, the more stable they are. Note also that an sp^2 C—H bond is particularly strong, as mentioned in Section 1.10.

Chemists calculate the heat of any reaction (enthalpy change) by comparing the bond strengths in the reactants and products. The heat of a reaction equals the sum of the BDEs of the bonds broken minus the sum of the BDEs of the bonds formed:

$$\Delta H° = \Sigma BDEs(\text{broken}) - \Sigma BDEs(\text{formed})$$

This equation makes sense because energy must be added to break bonds while it is released when bonds are formed.

Using data from Table 8.7 and Appendix 3, we can calculate the enthalpy of reaction, $\Delta H°$, for the halogenation of methane. Energy is required to break CH_3—H and Cl—Cl bonds [439 and 247 kJ (105 and 59 kcal)/mol, respectively]. Energy is released in making the CH_3—Cl and H—Cl bonds [−351 and −431 kJ (−84 and −103 kcal)/mol, respectively]. Based upon these enthalpies, we calculate that chlorination of methane to form chloromethane and hydrogen chloride liberates 96 kJ (23 kcal)/mol.

$$CH_4 + Cl_2 \longrightarrow CH_3Cl + HCl \qquad \Delta H^0 = -96 \text{ kJ/mol } (-23 \text{ kcal/mol})$$

BDE, kJ/mol   +439   +247   −351   −431

Example 8.3

Using the table of bond dissociation enthalpies in Appendix 3, calculate $\Delta H°$ for bromination of propane to give 2-bromopropane and hydrogen bromide.

Solution

Under each molecule is given the enthalpy for breaking or forming each corresponding bond. The calculated heat of reaction is −71 kJ (−17 kcal)/mol.

$$\underset{+414}{CH_3\overset{H}{\underset{|}{C}}HCH_3} + \underset{+192}{Br—Br} \longrightarrow \underset{-309}{CH_3\overset{Br}{\underset{|}{C}}HCH_3} + \underset{-368}{H—Br} \qquad \Delta H^0 = -71 \text{ kJ/mol } (-17 \text{ kcal/mol})$$

BDE, kJ/mol

Problem 8.3

Using the table of bond dissociation enthalpies in Appendix 3, calculate $\Delta H°$ for bromination of propane to give 1-bromopropane and hydrogen bromide.

8.5 | Mechanism of Halogenation of Alkanes

From detailed studies of the conditions and products for halogenation of alkanes, chemists have concluded that these reactions occur by a type of mechanism called a radical chain mechanism.

A. Formation of Radicals

Following are four reactions that result in homolytic cleavage to give radicals. Note that in the first three cases the bond being broken is between atoms with at least two lone pairs. Lone pair repulsion is one factor that weakens sigma bonds. BDEs of these reactions from Appendix 3 are shown on the right of each equation.

Dibenzoyl peroxide   **Benzoyl radicals**

$\Delta H^0 = +130$ kJ/mol (+31 kcal/mol)

$$CH_3CH_2\ddot{O}-\ddot{O}CH_2CH_3 \xrightarrow{80° C} CH_3CH_2\ddot{O}\cdot + \cdot\ddot{O}CH_2CH_3$$

Diethyl peroxide   **Ethoxy radicals**

$\Delta H^0 = +159$ kJ/mol (+38 kcal/mol)

$$:\ddot{Cl}-\ddot{Cl}: \xrightarrow[\text{light}]{\text{heat or}} :\ddot{Cl}\cdot + \cdot\ddot{Cl}:$$

Chlorine   **Chlorine atoms**

$\Delta H^0 = +247$ kJ/mol (+59 kcal/mol)

$$CH_3-CH_3 \xrightarrow{\text{heat}} CH_3\cdot + \cdot CH_3$$

Ethane   **Methyl radicals**

$\Delta H^0 = +376$ kJ/mol (+90 kcal/mol)

Energy to cause bond cleavage and generation of radicals can be supplied either by light or heat. The energy of visible and ultraviolet radiation (wavelength from 200 to 700 nm) falls in the range of 585 to 167 kJ (140 to 40 kcal)/mol and is of the same order of magnitude as the bond dissociation enthalpies of halogen-halogen covalent bonds. The bond dissociation enthalpy of Br_2 is 192 kJ (46 kcal)/mol; that for Cl_2 is 247 kJ (59 kcal)/mol. Dissociation of these halogens can also be brought about by heating at temperatures above 350°C.

Oxygen-oxygen single bonds in peroxides (ROOR) and hydroperoxides (ROOH) have dissociation enthalpies in the range of 146 to 209 kJ (35 to 50 kcal)/mol, and compounds containing these bonds are cleaved to radicals at considerably lower temperatures than those required for rupture of carbon-carbon bonds. Diethyl peroxide, for example, begins to dissociate to ethoxy radicals at 80°C. Dissociation of ethane into two methyl radicals occurs only at very high temperature.

B. A Radical Chain Mechanism

To account for the products formed from halogenation of alkanes, chemists propose a radical chain mechanism involving three types of steps: (1) **chain initiation**, (2) **chain propagation**, and (3) **chain termination**. We illustrate radical halogenation of alkanes by the reaction of chlorine with ethane.

Mechanism	Radical Chlorination of Ethane

Chain initiation involves formation of radicals from nonradical species. Chlorine is homolytically dissociated by heat or light.

Step 1: $$:\ddot{Cl}-\ddot{Cl}: \xrightarrow[\text{or light}]{\text{heat}} :\ddot{Cl}\cdot + \cdot\ddot{Cl}:$$

Chain propagation involves reaction of a radical and a molecule to form a new radical. The chlorine atom formed in Step 1 attacks the alkane, removing a hydrogen atom in another homolytic reaction.

Step 2: $CH_3\ddot{C}H_2—H + \cdot\ddot{C}l: \longrightarrow CH_3CH_2\cdot + H—\ddot{C}l:$

Step 3: $CH_3CH_2\cdot + :\ddot{C}l—\ddot{C}l: \longrightarrow CH_3CH_2—\ddot{C}l: + \cdot\ddot{C}l:$

Chain termination involves destruction of radicals. The first three possible chain termination steps involve coupling of radicals to form a new covalent bond. The fourth chain termination step, called disproportionation, involves transfer of a hydrogen atom from the beta position of one radical to another radical and formation of an alkane and an alkene.

Step 4: $CH_3CH_2\cdot + \cdot CH_2CH_3 \longrightarrow CH_3CH_2—CH_2CH_3$

Step 5: $CH_3CH_2\cdot + \cdot\ddot{C}l: \longrightarrow CH_3CH_2—\ddot{C}l:$

Step 6: $:\ddot{C}l\cdot + \cdot\ddot{C}l: \longrightarrow :\ddot{C}l—\ddot{C}l:$

Step 7: $CH_3CH_2\cdot + CH_2\overset{\displaystyle H}{—}CH_2\cdot \longrightarrow CH_3CH_3 + CH_2{=}CH_2$

Initiation

The characteristic feature of a **chain initiation** step is formation of radicals from nonradical compounds. In the case of chlorination of ethane, chain initiation is by thermal or light-induced homolysis of the Cl—Cl bond to give two chlorine radicals.

Chain initiation
A step in a chain reaction characterized by the formation of reactive intermediates (radicals, anions, or cations) from nonradical or noncharged molecules.

Chain Propagation

The characteristic feature of a **chain propagation** step is reaction of a radical and a molecule to give a new radical. A chlorine atom, also called a chlorine radical, is consumed in Step 2, but an ethyl radical is produced. Similarly, an ethyl radical is consumed in Step 3, but a chlorine radical is produced. Steps 2 and 3 can repeat thousands of times as long as neither radical is removed by a different reaction.

A second characteristic feature of chain propagation steps is that, when added together, they give the observed stoichiometry of the reaction. Adding Steps 2 and 3 and canceling structures that appear on both sides of the equation gives the balanced equation for the radical chlorination of ethane (Figure 8.1).

The number of times a cycle of chain propagation steps repeats is called **chain length**. Chain lengths can range from a few to many thousand, depending on the relative rates of reactions and the concentrations of various species.

Chain propagation
A step in a chain reaction characterized by the reaction of a reactive intermediate and a molecule to give a new reactive intermediate and a new molecule.

Chain length
The number of times the cycle of chain propagation steps repeats in a chain reaction.

Chain Termination

A characteristic feature of a **chain termination** step is destruction of radicals. Among the most important chain termination reactions during halogenation of alkanes are radical couplings, illustrated by Steps 4, 5, 6, and disproportionation illustrated by Step 7 in the mechanism for the halogenation of ethane. One of these termination steps gives the product ethyl chloride. However, the loss of ethyl and chlorine radicals terminates two chains and therefore stops many dozens to thousands of subsequent product-forming propagation steps.

Note that chain termination steps are usually relatively rare compared to chain propagation steps in radical chain reactions. This is because at any one time, the concentration of radical species is very low, making a collision between two radicals a relatively rare event. The relatively rare occurrence of chain termination steps explains why there can be chain lengths of many thousands of steps in a radical chain reaction.

Chain termination
A step in a chain reaction that involves destruction of reactive intermediates.

Figure 8.1

The cycle of chain propagation steps from the halogenation of ethane radical chain mechanism.

CH_3CH_2 + Cl_2 $\longrightarrow$ CH_3CH_2Cl + HCl

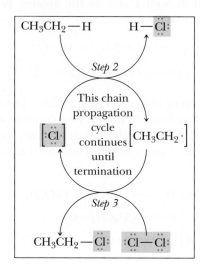

A major concern with chlorofluorocarbons is destruction of the ozone layer (mentioned in "Chemical Connections: Freons" in this chapter). The ozone consuming reactions are thought to occur through a radical chain mechanism. The production of one radical species derived from a CFC can destroy a large number of ozone molecules before any termination steps occur.

The structures, geometries, and relative stabilities of simple alkyl radicals are similar to those of alkyl carbocations. Methyl radical is planar, and all other radicals are nearly so, with bond angles near 120° about the carbon with the unpaired electron. This geometry indicates that carbon is sp^2 hybridized and that the unpaired electron occupies the unhybridized $2p$ orbital. As mentioned, the order of stability of alkyl radicals, like alkyl carbocations, is 3° > 2° > 1° methyl.

C. Energetics of Chain Propagation Steps

After the radical chain is initiated, the heat of reaction is derived entirely from the heat of reaction of the individual chain propagation steps. In Step 2 of radical chlorination of ethane, for example, energy is required to break the CH_3CH_2—H bond [422 kJ (101 kcal)/mol], but energy is released on formation of the H—Cl bond [−431 kJ (−103 kcal)/mol]. Similarly, energy is required in Step 3 to break the Cl—Cl bond [247 kJ (59 kcal)/mol], but energy is released on formation of the CH_3CH_2—Cl bond [−355 kJ (−85 kcal)/mol]. We see that, just as the sum of the chain propagation steps for radical halogenation gives the observed stoichiometry, the sum of the heats of reaction for each propagation step is equal to the observed heat of reaction:

Bond Dissociation Enthalpies for Chain Propagation Steps

	Reaction Step				$\Delta H°$, kJ/mol (kcal/mol)
Step 2:	CH_3CH_2—H + ·Cl $\longrightarrow$ CH_3CH_2· + H—Cl				
	+422			−431	−9 (−2)
Step 3:	CH_3CH_2· + Cl—Cl $\longrightarrow$ CH_3CH_2—Cl + ·Cl				
		+247	−355		−108 (−26)
Sum:	CH_3CH_2—H + Cl—Cl $\longrightarrow$ CH_3CH_2—Cl + H—Cl				−117 (−28)

Of all the fluoroalkanes, **chlorofluorocarbons (CFCs)** manufactured under the trade name **Freons** are the most widely known. CFCs are nontoxic, nonflammable, odorless, and noncorrosive and seemed to be ideal replacements for the hazardous compounds such as ammonia and sulfur dioxide formerly used as heat-transfer agents in refrigeration systems. Among the CFCs most widely used for this purpose were trichlorofluoromethane (CCl_3F, Freon-11) and dichlorodifluoromethane (CCl_2F_2, Freon-12). They are particularly desirable refrigerants because of their low boiling points.

The CFCs found wide use as industrial cleaning solvents to prepare surfaces for coatings, to remove cutting oils and waxes from millings, and to remove protective coatings. CFCs were also used as propellants for aerosol sprays.

Concern about the environmental impact of CFCs arose in the 1970s when it was shown that more than 4.5×10^5 kg/yr of these compounds were being emitted into the atmosphere. Then, in 1974 Sherwood Rowland of the University of California, Irvine, and Mario Molina, now at the University of California, San Diego, announced their theory, which has since been amply confirmed, of ozone (O_3) destruction by these compounds. When released into the air, CFCs escape to the lower atmosphere, but because of their inertness, they do not decompose there. Slowly they find their way to the stratosphere. What makes CFCs so dangerous is that a single CFC molecule in the stratosphere can lead to the destruction of many thousands of molecules of ozone through a radical chain mechanism (Section 8.5B). The process is initiated when the sun's unfiltered UV light in the stratosphere causes homolytic cleavage of the CFC to give a Cl radical. The Cl radical reacts with ozone in the first chain propagation step to give O_2 and ClO. The ClO reacts with an O atom (created when UV light hits O_2 or O_3 in the stratosphere) to give O_2 and another Cl radical, the latter of which reacts with a new ozone molecule, thus continuing the radical chain reaction. Not only is an ozone molecule destroyed for each cycle of the chain propagation steps, but the O atom consumed might have created a new O_3 molecule had it not been intercepted by the ClO.

The problem for humans and all other living things on the planet is that ozone acts as a shield for the Earth against short-wavelength ultraviolet radiation from the sun. Scientists believe that an increase in short-wavelength ultraviolet radiation reaching the Earth will lead to the destruction of certain crops and agricultural species, and even to an increased incidence of skin cancer in light-skinned individuals.

The results of this concern were that in 1987 most countries subscribed to the so-called Montreal Protocol, which set limits on the production and use of ozone-depleting CFCs and urged a complete phaseout of their production by the year 1996. This phaseout has resulted in enormous costs and is not yet complete in developing countries. The fact that an international agreement on the environment that set limits on the production of any substance could be reached is indeed amazing and bodes well for the health of the planet. Rowland, Molina, and Paul Crutzen, a Dutch chemist at the Max Planck Institute for Chemistry in Germany, were awarded the 1995 Nobel prize in chemistry for their work on this topic.

The chemical industry has responded by developing less-ozone-depleting alternatives to CFCs, among which are the hydrofluorocarbons (HFCs) and hydrochlorofluorocarbons (HCFCs). These compounds are much more chemically reactive in the atmosphere than the Freons and are destroyed before reaching the stratosphere. However, they tend to act as "greenhouse gases" and may contribute to global warming. For this reason, they are likely to be replaced in turn.

HFC-134a   **HCFC-123**

We must not assume, however, that haloalkanes are introduced into the environment only by human action. It is estimated, for example, that annual production of bromomethane from natural sources is 2.7×10^8 kg, largely from marine algae, giant kelp, and volcanoes. Furthermore, global emission of chloromethane is estimated to be 4.5×10^9 kg/yr, most of it from terrestrial and marine biomass. These haloalkanes, however, have only short atmospheric lifetimes, and only a tiny fraction of them reach the stratosphere. The CFCs are the problem; they have longer atmospheric lifetimes, reach the stratosphere, and do their damage there.

Example 8.4

Using the table of bond dissociation enthalpies in Appendix 3, calculate $\Delta H°$ for each propagation step in the radical bromination of propane to give 2-bromopropane and HBr.

Solution

Here are the two chain propagation steps along with bond dissociation enthalpies for the bonds broken and the bonds formed. The first chain propagation step is endothermic, the second is exothermic, and the overall reaction is exothermic by 71 kJ (17 kcal)/mol.

		ΔH^0
Step 2:	$\overset{\text{H}}{\underset{|}{\text{CH}_3\text{CHCH}_3}} + \text{Br} \cdot \longrightarrow \text{CH}_3\overset{\cdot}{\text{CHCH}_3} + \text{H—Br}$	+46 kJ/mol (+11 kcal/mol)
Step 3:	$\text{CH}_3\overset{\cdot}{\text{CHCH}_3} + \text{Br—Br} \longrightarrow \overset{\text{Br}}{\underset{|}{\text{CH}_3\text{CHCH}_3}} + \text{Br} \cdot$	−117 kJ/mol (−28 kcal/mol)
Sum:	$\overset{\text{H}}{\underset{|}{\text{CH}_3\text{CHCH}_3}} + \text{Br—Br} \longrightarrow \overset{\text{Br}}{\underset{|}{\text{CH}_3\text{CHCH}_3}} + \text{H—Br}$	−71 kJ/mol (−17 kcal/mol)

Problem 8.4

Write a pair of chain propagation steps for the radical bromination of propane to give 1-bromopropane, and calculate $\Delta H°$ for each propagation step and for the overall reaction.

D. Regioselectivity of Bromination Versus Chlorination: Hammond's Postulate

The regioselectivity in halogenation of alkanes can be accounted for in terms of the relative stabilities of radicals (3° > 2° > 1° > methyl). As we will see, the energy of the transition state reflects the energy of the radicals; more stable radical products are formed with a lower activation energy, thus faster. But how do we account for the greater regioselectivity in bromination of alkanes compared with chlorination of alkanes? To do so, we need to consider **Hammond's postulate**, a refinement of transition state theory proposed in 1955 by George Hammond, then at Iowa State University. According to this postulate:

> The structure of the transition state for an exothermic reaction step is reached relatively early in the reaction, so it resembles the reactants of that step more than the products. Conversely, the structure of the transition state for an endothermic reaction step is reached relatively late, so it resembles the products of that step more than the reactants.

It is important to realize that we cannot observe a transition state directly. Until the advent of modern computational theory, we could only infer its existence, structure, and stability from experiment. Hammond's postulate gives us a reasonable way of deducing something about the structure of a transition state by examining things we can observe: the structure of reactants and products and heats of reaction. Hammond's postulate applies equally well to multistep reactions. The transition state of any exothermic step in a multistep sequence looks more like the starting material(s) of that step; the transition state of any endothermic step in the sequence looks more like the product(s) of that step. Thus changes in starting material energy affect the transition state of an exothermic reaction more than

Hammond's postulate
The structure of the transition state for an exothermic step looks more like the reactants of that step than the products. Conversely, the structure of the transition state for an endothermic step looks more like the products of that step than the reactants.

(a) Highly exothermic reaction

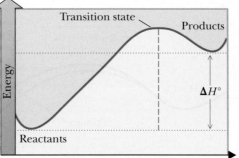

(b) Highly endothermic reaction

Figure 8.2

Hammond's postulate. Energy diagrams for two one-step reactions. In the exothermic reaction, the transition state occurs early, and its structure resembles that of the reactants. In the endothermic reaction, the transition state occurs late, and its structure resembles that of the products.

changes in product energy. The converse is true for an endothermic reaction. Today, we can carry out high-level computations that reveal details of complex transition states and strongly support the validity of Hammond's postulate.

Shown in Figure 8.2 are energy diagrams for a highly exothermic reaction and a highly endothermic reaction, each occurring in one step.

Now let us apply Hammond's postulate to explain the relative regioselectivities of chlorination versus bromination of alkanes. In applying this postulate, we deal with the rate-determining step of the reaction, which, in radical halogenation of alkanes, is the abstraction of a hydrogen atom by a halogen radical. Given in Table 8.8 are heats of reaction, $\Delta H°$, for the hydrogen abstraction step in chlorination and bromination of the different hydrogens of 2-methylpropane (isobutane). Also given under the formulas of isobutane, HCl, and HBr are bond dissociation enthalpies for the bonds broken (1° and 3° C—H) and formed (H—Cl and H—Br) in each step. Because the 3° radical is more stable than the 1° radical, the BDE for the 3° H is lower than that of a 1° H by about 17 kJ (4 kcal)/mol and the difference in $\Delta H°$ between 1° and 3° for the two reactions is just this amount.

Table 8.8	Enthalpies of Reaction for Hydrogen Abstraction in 2-Methylpropane
Reaction Step	ΔH^0 [kJ (kcal)/mol]

Chlorination

$$\text{(isobutane with H)} + \cdot Cl \longrightarrow \text{(primary radical)} \cdot + \; H—Cl \qquad -9(-2)$$

+422(101)   Primary radical   −431(−103)

$$\text{(isobutane, 3° H)}—H + \cdot Cl \longrightarrow \text{(tertiary radical)} \cdot + \; H—Cl \qquad -26(-6)$$

+405(97)   Tertiary radical   −431(−103)

⎫ 17(4)

Bromination

$$\text{(isobutane with H)} + \cdot Br \longrightarrow \text{(primary radical)} \cdot + \; H—Br \qquad +54(+13)$$

+422(101)   −368(−88)

$$\text{(isobutane, 3° H)}—H + \cdot Br \longrightarrow \text{(tertiary radical)} \cdot + \; H—Br \qquad +37(+9)$$

+405(97)   −368(−88)

⎫ 17(4)

Figure 8.3

Transition states and energetics for hydrogen abstraction in the radical chlorination and bromination of 2-methylpropane (isobutane). The product is the intermediate radical, R·.

(a) Chlorination

(b) Bromination

Abstraction of hydrogen by chlorine is exothermic, which, according to Hammond's postulate, means that the transition state for H abstraction by Cl· is reached early in the course of the reaction [Figure 8.3(a)]. Therefore, the structure of the transition state for this step resembles the reactants, namely the alkane and a chlorine atom, not the product radicals. As a result, there is relatively little radical character on carbon in this transition state, and regioselectively in radical chlorination is only slightly influenced by the relative stabilities of radical intermediates. Products are determined more by whether a chlorine atom happens to collide with a 1°, 2°, or 3° H.

In Section 8.4A, we were given the fact that the selectivity for abstraction of a 3° H compared to a 1° H in chlorination is 5:1; this ratio directly reflects the relative reaction rates of these hydrogens with chlorine atoms. Using this ratio of reaction rates and the relationship between $\Delta G^{\ddagger}$ and rate constants, we can calculate that the difference in activation energies, $\Delta\Delta G^{\ddagger}$, for the abstraction of a 3° H versus a 1° H is about 4 kJ (1 kcal)/mol. However, we can calculate from the primary and tertiary C—H bond dissociation enthalpies [Figure 8.3(a)] that $\Delta\Delta H°$ for the two reactions is about 17 kJ (4 kcal)/mol. Thus the difference in product stabilities is only slightly reflected in the transition states and the resulting reaction rates.

Contrast this reaction with bromination [Figure 8.3(b)]. For bromination, the selectivity of 3° H to 1° H is 1600:1, which corresponds to $\Delta\Delta G^{\ddagger}$ of approximately 18 kJ (4.2 kcal)/mol. The $\Delta\Delta H°$ for the formation of the primary and tertiary radicals is the same in bromination and in chlorination (it is just the difference in BDE of the C—H bonds). But the rate-determining step for bromination, because it is endothermic, has a transition state much more like the product radical, and the transition state reflects nearly all the energy difference of the primary and tertiary radicals. The later transition state, and the correspondingly larger $\Delta\Delta G^{\ddagger}$ (which causes a large difference in reaction rates), is the reason for the much larger regioselectivity in radical bromination than in radical chlorination.

BDE values are $\Delta H°$ and not $\Delta G°$. Recall that $\Delta G = \Delta H - T\Delta S$. Because we are dealing with similar reactions, we can assume that entropy differences between them are nearly zero and, therefore, $\Delta\Delta G° \approx \Delta\Delta H°$ and $\Delta\Delta G^{\ddagger} \approx \Delta\Delta H^{\ddagger}$, which allows us to make these comparisons.

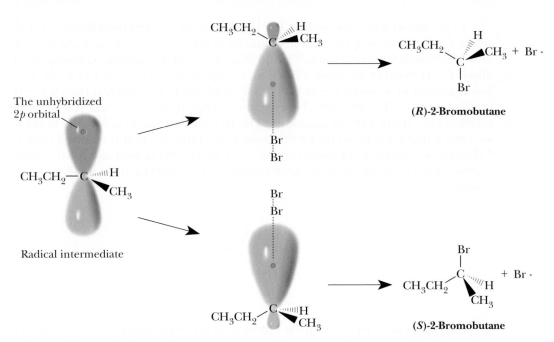

Figure 8.4
Radical bromination of butane.

(R)-2-Bromobutane

(S)-2-Bromobutane

E. Stereochemistry of Radical Halogenation

When radical halogenation produces a chiral center or takes place at a hydrogen on an existing chiral center, the product is an equal mixture of R and S enantiomers. Consider, for example, radical bromination of butane, which produces 2-bromobutane.

$$CH_3CH_2CH_2CH_3 + Br_2 \xrightarrow[\text{or light}]{\text{heat}} CH_3CH_2\overset{\overset{\displaystyle Br}{|}}{C}HCH_3 + HBr$$

Butane   **2-Bromobutane**
(racemic)

In this example, both of the starting materials are achiral, and, as is true for any reaction of achiral starting materials taking place in an achiral environment that gives a chiral product (Section 6.7A), the product is a racemic mixture (Figure 8.4).

In the case of the *sec*-butyl radical, the carbon bearing the unpaired electron is sp^2 hybridized, and the unpaired electron lies in the unhybridized $2p$ orbital. Reaction of the alkyl radical intermediate with halogen in the second chain propagation step occurs with equal probability from either face to give an equal mixture of the R and S configurations at the newly created chiral center.

8.6 | Allylic Halogenation

We saw in Section 6.3D that propene and other alkenes react with Br_2 and Cl_2 at room temperature by addition to the carbon-carbon double bond. If, however, propene and one of these halogens are allowed to react at a high temperature, an entirely different reaction takes place, namely substitution of a halogen occurs at the **allylic carbon** (the carbon next to a carbon-carbon double bond). We illustrate **allylic substitution** by the reaction of propene with chlorine at high temperature:

$$CH_2{=}CHCH_3 + Cl_2 \xrightarrow{350°C} CH_2{=}CHCH_2Cl + HCl$$

Propene   **3-Chloropropene**
(Allyl chloride)

Allylic carbon
A carbon adjacent to a carbon-carbon double bond.

Allylic substitution
Any reaction in which an atom or group of atoms is substituted for another atom or group of atoms at an allylic carbon.

A comparable reaction takes place when propene is treated with bromine at an elevated temperature.

To predict which of the various C—H bonds in propene is most likely to break when a mixture of propene and bromine or chlorine is heated, we need to look at bond dissociation enthalpies. We find that the bond dissociation enthalpy of an allylic C—H bond in propene (Table 8.7) is approximately 92 kJ (22 kcal)/mol less than that of a vinylic C—H bond and 50 kJ (12 kcal)/mol less than a C—H bond of ethane. The allyl radical is even more stable than a 3° radical; this unusual stability also applies to carbocations. The reason that the allylic C—H bond is so weak is discussed in Section 8.6B. Note from Table 8.7 that the benzyl radical $C_6H_5CH_2\cdot$ is stabilized in exactly the same way as the allyl radical and for the same reason; benzylic compounds undergo many of the same reactions as allylic compounds (Section 21.5).

Treating propene with bromine or chlorine at elevated temperatures illustrates a very important point about organic reactions: it is often possible to change the product(s) by changing the mechanism through a change in reaction conditions. Under the high temperatures used in this reaction, the concentration of bromine radicals becomes much higher than at room temperature; this greatly accelerates the substitution reaction, which occurs by the radical halogenation mechanism. At room temperature, there are far fewer radicals, and electrophilic addition is observed.

$$CH_2=CHCH_3 + Br_2 \xrightarrow{\text{high temp.}} CH_2=CHCH_2Br + HBr \qquad \text{(allylic substitution)}$$

Propene   **3-Bromopropene**

$$CH_2=CHCH_3 + Br_2 \xrightarrow[CH_2Cl_2]{\text{room temp.}} CH_2CHCH_3 \qquad \text{(electrophilic addition)}$$

Propene   Br   Br

1,2-Dibromopropane
(racemic)

A very useful way to carry out allylic bromination in the laboratory at or slightly above room temperature is to use the reagent *N*-bromosuccinimide (NBS) in dichloromethane (CH_2Cl_2). Reaction between an alkene and NBS is most commonly initiated by light. This reaction involves a net double substitution: a bromine in NBS and a hydrogen in the alkene exchange places.

Cyclohexene   **N-Bromosuccinimide**   **3-Bromocyclohexene**   **Succinimide**
                (NBS)   (racemic)

A. Mechanism of Allylic Halogenation

Allylic bromination and chlorination proceed by a radical chain mechanism involving the same type of chain initiation, chain propagation, and chain termination steps involved in the radical halogenation of alkanes.

Chain Initiation Chain initiation involves formation of radicals from NBS by light-induced homolytic cleavage of the N—Br bond in NBS. This step is analogous to the homolytic dissociation of chlorine to chlorine radicals.

Step 1:

$$\text{(N—Br:)} \xrightarrow{h\nu} \text{(N·)} + \cdot\ddot{\text{B}}\text{r:}$$

Chain Propagation Chain propagation involves the formation of products. Reaction of a radical and a nonradical gives a new radical. (Both radicals formed in the initiation can abstract hydrogen atoms. We show only the Br· reaction.) In the first propagation step, a bromine atom abstracts an allylic hydrogen (the weakest C—H bond in propene) to produce an allyl radical. The allyl radical, in turn, reacts with a bromine molecule to form allyl bromide and a new bromine atom.

Step 2: $CH_2{=}CHCH_2{-}H + \cdot\ddot{\text{B}}\text{r:} \longrightarrow CH_2{=}CHCH_2\cdot + H{-}\ddot{\text{B}}\text{r:}$

Allylic hydrogen   *Allyl Radical*

Step 3: $CH_2{=}CHCH_2\cdot + :\ddot{\text{B}}\text{r}{-}\ddot{\text{B}}\text{r:} \longrightarrow CH_2{=}CHCH_2{-}\ddot{\text{B}}\text{r:} + \cdot\ddot{\text{B}}\text{r:}$

Note that, as always, this combination of chain propagation steps adds up to the observed stoichiometry. This reaction is exactly like halogenation of alkanes, but is strongly regioselective for the allylic hydrogen because of its weak bond.

Chain Termination—The Destruction of Radicals Propagation of the chain reaction continues until termination steps produce nonradical products and thus stop further reaction.

Step 4: $:\ddot{\text{B}}\text{r}\cdot + \cdot\ddot{\text{B}}\text{r:} \longrightarrow :\ddot{\text{B}}\text{r}{-}\ddot{\text{B}}\text{r:}$

Step 5: $CH_2{=}CHCH_2\cdot + \cdot\ddot{\text{B}}\text{r:} \longrightarrow CH_2{=}CHCH_2{-}\ddot{\text{B}}\text{r:}$

Step 6: $CH_2{=}CHCH_2\cdot + \cdot CH_2CH{=}CH_2 \longrightarrow CH_2{=}CHCH_2{-}CH_2CH{=}CH_2$

The Br_2 necessary for Step 2 is formed by reaction of product HBr with NBS.

Step 7:

$$\text{(N—Br} + \text{HBr)} \longrightarrow \text{(N—H} + Br_2\text{)}$$

Bromine formed in this step then reacts with an allyl radical to continue the chain propagation reactions (Step 3).

The mechanism we described for allylic bromination by NBS poses the following problem. NBS is the indirect source of Br_2, which then takes part in chain propagation. But if Br_2 is present in the reaction mixture, why does it not react instead with the carbon-carbon double bond by electrophilic addition? In other words, why is the observed reaction allylic substitution rather than addition to the double bond? The answer is that the rates of the chain propagation steps are much faster than the rate of electrophilic addition of bromine to the alkene when radicals are present. Furthermore, the concentration of Br_2 is very low throughout the course of the reaction, which slows the rate of electrophilic addition.

B. Structure of the Allyl Radical

The allyl radical can be represented as a hybrid of two contributing structures. Here fishhook arrows show the redistribution of single electrons between contributing structures. Note three π electrons take part in this resonance.

$$\overset{\frown}{CH_2}=CH\overset{\curvearrowleft}{\cdot}CH_2 \longleftrightarrow \dot{C}H_2-CH=CH_2$$

(Equivalent contributing structures)

The position of the radical electron in the two contributing structures predicts that radical reactivity will occur at carbons 1 and 3 but not at carbon 2. This result is indeed experimentally observed.

The eight atoms of the allyl radical lie in a plane, and all bond angles are approximately 120°. Each carbon atom is sp^2 hybridized, and the three $2p$ orbitals participating in resonance delocalization of the radical are parallel to one another as shown in Figure 8.5. Like charged systems, in which a delocalized charge is more stable than a localized one, delocalized unpaired electron density leads to more stable structures than localized unpaired electron density.

Because of the radical-stabilizing influence of this electron delocalization, it is reasonable to expect that the BDE of an allylic C—H is significantly weaker than that of a primary C—H. In fact, based on bond dissociation enthalpies, we conclude that an allyl radical is even more stable than a 3° alkyl radical. Note that because of the larger amount of s character in its carbon sp^2 hybrid orbital, a vinylic C—H bond is stronger (has a larger bond dissociation enthalpy) than any sp^3 C—H bond and is never abstracted in homolytic reactions.

According to the molecular orbital description, the conjugated system of the allyl radical involves the formation of three molecular orbitals by overlap of three $2p$ atomic orbitals (Figure 8.5). The lowest energy MO has zero nodes, the next MO has one node, and the highest energy MO has two nodes. The molecular orbital of intermediate energy in this case leads to neither net stabilization nor destabilization and is, therefore, called a nonbonding MO. The lowest π MO is at a lower energy than the isolated $2p$ orbitals.

In the lowest energy (ground) state of the allyl radical, two electrons of the π system lie in the π-bonding MO, and the third lies in the π-nonbonding MO; the π-antibonding MO is unoccupied. Because the lowest π MO is at a lower energy

Figure 8.5
Molecular orbital model of covalent bonding in the allyl radical. Combination of three $2p$ atomic orbitals gives three π molecular MOs. The lowest, a π-bonding MO, has zero nodes. The next in energy, a π-nonbonding MO, has one node, and the highest in energy, a π-antibonding MO, has two nodes.

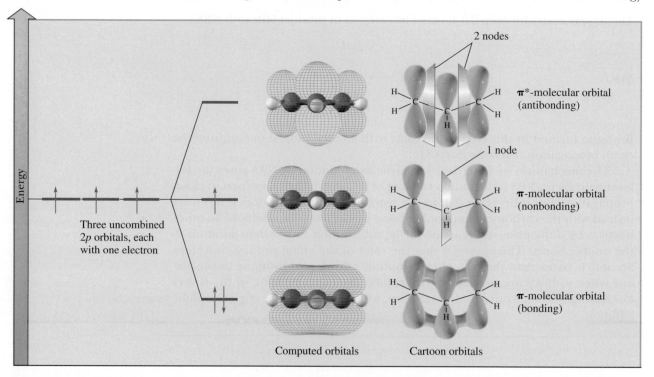

than the isolated $2p$ atomic orbitals, putting two electrons in this MO releases considerable energy, which accounts for the stability of the allyl radical.

The lone electron of the allyl radical is associated with the π-nonbonding MO, which places electron density on carbons 1 and 3 only. This localization is shown clearly in the unpaired electron density map in Figure 8.6. Thus both the resonance model and molecular orbital theory are consistent in predicting radical character on carbons 1 and 3 of the allyl radical but no radical character on carbon 2, consistent with the experimental observation that reaction occurs at carbons 1 and 3 of an allylic radical. Importantly, when there is a difference, the reaction will occur to generate the alkene product that is the most stable, in other words with the more highly substituted double bond.

Figure 8.6
Unpaired electron spin density map for the allyl radical. Unpaired electron density (green cones) appears only on carbons 1 and 3.

Example 8.5

Account for the fact that allylic bromination of 1-octene by NBS gives these products.

1-Octene → (NBS / CH_2Cl_2) → **3-Bromo-1-octene** (racemic, 17%) + **(E)-1-Bromo-2-octene** (83%)

Solution

The rate-determining step of this radical chain mechanism is hydrogen abstraction from the allylic position on 1-octene to give a 2° allylic radical. This radical is stabilized by delocalization of the two π electrons and the unpaired electron. Reaction of the radical at carbon 1 gives the major product. Reaction at carbon 3 gives the minor product. The more substituted (and more stable) alkene isomer predominates.

1-Octene + ·Br → [resonance structures] + HBr

→ (Br₂) → **3-Bromo-1-octene** (racemic)

+ → (Br₂) → **1-Bromo-2-octene** (major product) + ·Br + ·Br

The reason for this regioselectivity seems to be that the resonance contributor of the allylic radical with the more substituted double bond dominates. This hypothesis is borne out by the unpaired electron spin density, here calculated for the radical formed from 1-butene (Figure 8.7). The terminal carbon has the highest spin density (blue).

Problem 8.5

Given the solution to Example 8.5, predict the structure of the product(s) formed when 3-hexene is treated with NBS.

Figure 8.7
Unpaired electron spin density map of the radical formed from 1-butene. Spin (blue) is on carbons 1 and 3, but more is on carbon 1.

Autoxidation
Air oxidation of materials such as unsaturated fatty acids

8.7 | Radical Autoxidation

One of the most important destructive reactions for materials, foods, and also for living systems is called **autoxidation**; that is, oxidation requiring oxygen and no other reactant. This reaction takes place by a radical chain mechanism very similar to that for allylic bromination. If you open a bottle of cooking oil that has stood for a long time, you will notice the hiss of air entering the bottle because there is a negative pressure caused by the consumption of oxygen by autoxidation of the oil.

Many plants contain polyunsaturated fatty acid esters in their leaves or seeds. However, in the natural state, they are protected against autoxidation by a variety of agents, one of the most important of which is α-tocopherol (vitamin E, Section 26.6C). This compound is a phenol (Section 21.4). The characteristic of phenols that makes them effective as protective agents against autoxidation is their O—H bond, which is even weaker than the doubly allylic C—H bond (see Appendix 3). Vitamin E reacts preferentially with the initial peroxy radical to give a highly delocalized and therefore stabilized phenoxy radical, which is very unreactive and survives to scavenge another peroxy radical, ROO·. The resulting peroxide is relatively stable. The antioxidant thus removes two molecules of peroxy radical and stops the radical chain oxidation dead in its tracks.

Vitamin E

Peroxy radical

A hydroperoxide

A phenoxy radical

A peroxide group

A peroxide derived from vitamin E

BHT → **BHT radical** + ROOH

Because vitamin E is removed in the processing of many food products, similar phenols such as BHT (Problem 22.23) are often added to foods to prevent spoilage by autoxidation. The BHT radical is unreactive because of the steric hindrance provided by the adjacent *tert*-butyl groups. Formation of the BHT radical on the phenolic O atom therefore stops the autoxidation chain propagation analogous to vitamin E. These compounds are all called **radical inhibitors** because of their ability to terminate radical chains before significant damage to the food occurs. Radical inhibitors are often referred to as preservatives. Similar compounds are also added to many materials, such as plastics and rubber, to protect them against autoxidation.

Similar oxidative degradation of materials in low-density lipoproteins (LDL, Section 26.4A) deposited in arteriosclerotic plaques on the walls of arteries leads to cardiovascular disease in humans. Many effects of aging and possibly cancer in humans and damage to materials such as rubber and plastic occur by similar mechanisms.

Cooking oil contains **polyunsaturated fatty acid esters**. (See Section 5.4 "Connections to Biological Chemistry: The Importance of *Cis* Double Bonds in Fats Versus Oils," as well as Section 26.1.) The most common of these compounds have chains of 16 or 18 carbons containing 1,4-diene functional groups. (Both double bonds are *cis*; the nature of R_1 and R_2 need not concern us at this stage.) The hydrogens on the CH_2 group between the double bonds are doubly allylic; that is, they are allylic with respect to both double bonds. As you might expect, the radical formed by abstraction of one of these hydrogens is unusually stable because it is even more delocalized (described by resonance contributing structures) than an allylic radical. An allylic C—H bond is much weaker than a corresponding alkane C—H bond, and the doubly allylic C—H is even weaker.

Autoxidation begins when a radical initiator, X·, formed either by light activation of an impurity in the oil or by thermal decomposition of peroxide impurities, abstracts a doubly allylic hydrogen to form a radical. This radical is delocalized through resonance with both double bonds (1 and 2 in the following structure).

> **Radical inhibitor**
> A compound such as a phenol that selectively reacts with radicals to remove them from a chain reaction and terminate the chain.

A polyunsaturated fatty acid ester (R—H)

This radical reacts with oxygen, itself a (very unreactive) diradical, to form a peroxy radical, which then reacts with the CH_2 of another 1,4-diene fatty acid ester (R—H) to give a new radical (R·) and a hydroperoxide. Hydroperoxides are formed on both sides by reactions with the resonance hybrid; only one is shown. The new radical reacts again with oxygen, causing a radical chain reaction in which hundreds of molecules of fatty acid ester are oxidized for each initiator radical.

Peroxy radical   A hydroperoxide

The ultimate fate of the peroxide, and some of the peroxy radical as well, is complex. Some autoxidation products degrade to short-chain aldehydes and carboxylic acids with unpleasant "rancid" smells familiar to anyone who has smelled old cooking oil or aged foods that contain polyunsaturated oils. It has been suggested that some products of autoxidation of oils are toxic and/or carcinogenic. Oils lacking the 1,4-diene structure are much less easily oxidized.

Example 8.6

What products would you expect from the following reaction? Indicate the major one and specify stereochemistry if relevant.

Solution

The major product has the more substituted double bond. (Both are racemic.)

Major   Minor
(racemic)   (racemic)

Problem 8.6

Show the products of the following reaction and indicate the major one.

$$CH_3(CH_2)_4CH_2CH{=}CH_2 \xrightarrow{\text{O}_2, \text{ initiator}}$$

8.8 | Radical Addition of HBr to Alkenes

When the addition of hydrogen halides to alkenes was first studied systematically in the 1930s, chemists observed that the addition of HBr sometimes gives Markovnikov addition, and sometimes gives non-Markovnikov addition. These two modes of addition of HBr are illustrated for 2-methylpropene (isobutylene).

Markovnikov addition

2-Methylpropene + HBr → [no peroxides] → 2-Bromo-2-methylpropane

Non-Markovnikov addition

2-Methylpropene + HBr → [peroxides] → 1-Bromo-2-methylpropane

The puzzle was solved in 1933 when it was discovered that non-Markovnikov products are observed only in the presence of peroxides or other sources of radicals. In the absence of radicals, addition of HBr gives only the expected Markovnikov product.

To account for the products of HBr addition to alkenes in the presence of peroxides, chemists proposed a radical chain mechanism like the one for halogenation (Section 8.6A). In the following mechanism, the source of initiating radicals is a dialkyl peroxide, which is frequently present as an impurity in the solvent or alkene.

Mechanism Radical-Initiated Non-Markovnikov Addition of HBr to Alkenes

Chain Initiation

Homolytic cleavage of a dialkyl peroxide is induced by light or heat to give two alkoxy radicals. An alkoxy radical then reacts with HBr by hydrogen abstraction to give an alcohol and a bromine radical.

Step 1: R—Ö—Ö—R ⟶ R—Ö· + ·Ö—R

A dialkyl peroxide   Two alkoxy radicals

Step 2: R—Ö· + H—Br· ⟶ R—Ö—H + ·Br·

Bromine radical

Chain Propagation

A bromine radical adds to the carbon-carbon double bond regioselectively to give the more substituted (and more stable) carbon radical. The carbon radical, in turn, reacts with a molecule of HBr to give the bromoalkane and generate a new bromine radical. Note that in each propagation step, one radical is consumed, but another is formed.

Step 3:

A 3° radical

Step 4:

1-Bromo-2-methylpropane

Chain Termination

The most important chain termination steps are the combination of a carbon radical with a bromine radical, and the combination of two bromine radicals. Each of these steps destroys one or both of the radical intermediates in the chain:

Step 5: $\ddot{Br}\cdot + \cdot\ddot{Br}: \longrightarrow :\ddot{Br}—\ddot{Br}:$

Step 6:

The observed non-Markovnikov regioselectivity of radical addition of HBr to an alkene is a combination of a steric factor and an electronic factor. First, a bromine radical attacks the less hindered carbon of the double bond (the steric factor). Second, as mentioned in Section 8.4C, the relative stabilities of radicals parallel those of carbocations (Section 6.3A).

Because the intermediate in Step 3 is a radical, it does not preserve stereo-chemistry in Step 4 when relevant.

The addition of the bromine atom to the alkene could occur at either C of the double bond, but it is dominated by addition that gives the more stable radical. In the mechanism just described, the two choices are the formation of a primary or tertiary radical. Because a tertiary carbon radical is more stable than a primary radical, the regiochemistry ends up as non-Markovnikov. Recall that the polar addition of HBr to an alkene is regioselective (Section 6.3A), with bromine adding to the more substituted carbon (Markovnikov addition). There is an important similarity between the polar mechanism (Section 6.3A) and the radical mechanism. The regiochemistry of each reaction is dominated by the reactions that proceed through the most stable reactive intermediates, which are in both cases tertiary, as shown below. This pair of alkene additions illustrates how the products of a reaction often can be altered by a change in experimental conditions and a change in mechanism, in this case, a change from a polar mechanism to a radical mechanism. The non-Markovnikov creation of a haloalkane can be very useful in organic synthesis.

Radical mechanism

Polar mechanism

Interestingly, the non-Markovnikov addition occurs only with HBr. Both HCl and HI always add to alkenes according to Markovnikov's rule. The reason for the differences in reactivity between HCl, HBr, and HI is a subtle balance of bond strengths. As just discussed, the mechanism of non-Markovnikov addition involves hydrogen radical abstraction from the H—X bond and addition of a halogen radical to the alkene. The bond strength in HCl is too strong for the abstraction of an H atom, while the bond is sufficiently weak enough in HBr and HI. However, with HI, the addition of iodine atom to the double bond does not occur because the π bond is stronger than a C—I bond. With HCl and HBr, the C—X bond strengths exceed the π bond, and therefore these reactions are favorable. In summary, only HBr can add to alkenes via a radical mechanism, whereas all HX acids can add via mechanisms involving carbocations (Chapter 6).

Example 8.7

Predict the product of the following reaction:

Solution

Mixture of *cis* and *trans*; the radical intermediate does not preserve stereochemistry

Problem 8.7

Predict the major product of the following reaction:

Summary

Section 8.1 Structure

- **Haloalkanes** contain a halogen covalently bonded to an sp^3 hybridized carbon.

Section 8.2 Nomenclature

- In the IUPAC system, halogen atoms are named fluoro-, chloro-, bromo-, and iodo-, and are listed in alphabetical order with other substituents. In the common system, they are named as **alkyl halides**.
- **Haloalkenes** contain a halogen covalently bonded to an sp^2 hybridized carbon of an alkene. In the common system, they are named alkenyl or vinylic halides.
- **Haloarenes** contain a halogen atom covalently bonded to a benzene ring.

Problems: 8.1, 8.8, 8.9

Section 8.3 Physical Properties of Haloalkanes

- The **van der Waals radius** of fluorine is only slightly greater than that of hydrogen, and, among the other halogens, only iodine has a larger van der Waals radius than methyl.

- Among alkanes and chloro-, bromo-, and iodoalkanes of comparable size and shape, the haloalkanes have the higher boiling points predominantly because of the greater **polarizability** of the unshared electrons of the halogen atom.
  - Polarizability refers to the distortion of the distribution of electron density around an atom that is interacting with another atom or ion.
  - The electron density on larger, less electronegative atoms is more polarizable than that of electrons on more electronegative atoms with smaller atomic radii.
- Boiling points of fluoroalkanes are generally comparable to those of alkanes of similar size and shape because of the uniquely low polarizability of the valence electrons of fluorine.

Problems: 8.10–8.12

- The density of liquid haloalkanes is greater than that of hydrocarbons of comparable molecular weight because of the halogen's larger mass-to-volume ratio.

Section 8.4 Preparation of Haloalkanes by Halogenation of Alkanes

- **Free radical halogenation** of alkanes uses Cl_2 or Br_2 and light or heat to produce haloalkanes.
  - Free radical halogenation is regioselective, and replaces H atoms with the halogen in the order allylic $>3°>2°>1°>$ methyl.
    - This order of reactivity can be predicted on the basis of radical intermediate stabilities, which follow the same order. Radical intermediates are analogous to carbocations in the sense that they are stabilized by the same interactions, namely resonance delocalization and hyperconjugation with attached alkyl groups.

Problems: 8.2, 8.13–8.15, 8.20, 8.33–8.35

- Regioselectivity of halogenation is greater for bromination than for chlorination.
- Bond dissociation enthalpies (BDEs) are the enthalpies of homolytic bond cleavage for a given type of bond.
  - The overall energetics of a reaction can be calculated by adding all the BDEs of the bonds broken, minus the BDEs of the bonds made in the reaction.

Problems: 8.3, 8.17, 8.27

  - Exothermic reactions (that is, ones favored to give products) have overall negative values calculated in this way.

Section 8.5 Mechanism of Halogenation of Alkanes

- A **radical chain mechanism** consists of three types of steps: chain initiation, chain propagation, and chain termination.
  - In **chain initiation**, radicals are formed from nonradical compounds.
  - In a **chain propagation** step, a radical and a molecule react to give a new radical and a new molecule.
    - When summed, chain propagation steps give the observed stoichiometry of the reaction.
    - **Chain length** is the number of times a cycle of chain propagation steps repeats.
  - In a **chain termination** step, radicals are destroyed.
- Simple alkyl radicals are planar or almost so with bond angles of 120° about the carbon with the unpaired electron.

Problems: 8.4, 8.16, 8.21, 8.32

- Heats of reaction for a radical reaction and for individual chain initiation, propagation, and termination steps can be calculated from BDEs.
- According to **Hammond's postulate**, the structure of the transition state of an exothermic reaction step occurs early and looks more like the reactants of that step than like the products; thus changes in the reactants have a large effect on the rate.
  - Conversely, the structure of the transition state of an endothermic reaction step occurs later and looks more like the products of that step than like the reactants, and changes in products have a large effect on the rate. Hammond's postulate accounts for the fact that bromination of an alkane is more regioselective than chlorination.

– For both bromination and chlorination of alkanes, the rate-determining step is hydrogen abstraction to form an alkyl radical. Hydrogen abstraction is endothermic for bromination and exothermic for chlorination.

Problems: 8.18, 8.19

Section 8.6 Allylic Halogenation

• **Allylic substitution** is any reaction in which an atom or group of atoms is substituted for another atom or group of atoms at a carbon adjacent to a carbon-carbon double bond.
  – **Allylic halogenation** proceeds by a radical chain mechanism.
• Because of delocalization of unpaired electron density through resonance, the allyl radical is more stable than the *tert*-butyl radical.

Problems: 8.5, 8.22–8.26, 8.36

Section 8.7 Radical Autoxidation

• **Autoxidation** of unsaturated compounds is an important process in aging and degradation of materials.

Problems: 8.5, 8.29, 8.31

Section 8.8 Radical Addition of HBr to Alkenes

• Addition of HBr to alkenes in the presence of peroxides can lead to non-Markovnikov addition because under these conditions a free-radical chain mechanism operates.

Problems: 8.7, 8.30

⚙WL Questions and problems for this chapter are available in OWL. Use the OWL access card that came with this textbook to access assigned questions and problems for this chapter.

Key Reactions

1. **Chlorination and Bromination of Alkanes (Section 8.4)** Chlorination and bromination of alkanes are regioselective in the order 3° H > 2° H > 1° H. Bromination has a higher regioselectivity than chlorination. The mechanism involves a radical chain process.

$$CH_3CH_2CH_3 + Br_2 \xrightarrow[\text{or light}]{\text{heat}} \underset{92\%}{CH_3\overset{\overset{\displaystyle Br}{|}}{C}HCH_3} + \underset{8\%}{CH_3CH_2CH_2-Br}$$

2. **Allylic Bromination (Section 8.6)** These reactions occur at high temperatures (heat is the radical initiator) using the halogens themselves. Bromination using *N*-bromosuccinimide (NBS) is initiated by light. The mechanism involves a radical chain process, with a resonance delocalized allyl radical intermediate.

(racemic)

3. **Autoxidation (Section 8.7)** Autoxidation involves reaction of a CH bond, especially an allylic one, with oxygen under radical initiation conditions. The primary product is a hydroperoxide. The mechanism involves a radical chain process in which resonance-delocalized allylic radical intermediates react with molecular oxygen to give a peroxy radical that continues the radical chain.

(racemic)

4. HBr Addition to Alkenes Under Radical Conditions   Non-Markovnikov addition of HBr to alkenes occurs by a radical mechanism in the presence of peroxides, in which a Br· reacts with the π bond of the alkene to create a radical intermediate that abstracts H· from H—Br to continue the chain process. The regioselectivity of the products is the opposite of the ordinary Markovnikov addition products that form under polar conditions, and hence it is a useful alternative to polar addition of HBr in organic synthesis.

Chapter 9

Nucleophilic Substitution and β-Elimination

Hydroxide ion reacts with bromomethane (upper models) to give methanol and bromide ion (lower models) by an S_N2 mechanism (Section 9.3).

Nucleophilic substitution refers to any reaction in which an electron-rich **nucleophile** (meaning "nucleus loving") (Nu:) replaces a **leaving group** (Lv). Viewed in the context of the mechanism elements first described in Section 6.2F, nucleophilic substitution is a combination of making a new bond between a nucleophile and an electrophile, and breaking a bond so that relatively stable molecules or ions are created. All nucleophiles are electron sources and can also be considered as Lewis bases (Section 4.7). As first discussed in Chapter 6, with the exception of radical reactions (Chapter 8), essentially every reaction you will study involves a reaction of a Lewis acid (that can also be considered as a good electron sink) with a Lewis base. In these reactions, the Lewis base, which is electron rich, reacts with the Lewis acid, which is electron poor. The Lewis acid is called an **electrophile** (meaning "electron loving"). The leaving group (Lv) can be a halide (X) or other electronegative group that can form a stable anion or other stable species. It should be noted that not all nucleophiles and leaving groups covered in this chapter are negatively charged. Further, not all nucleophiles react with all electrophiles; recognizing which do and which do not is part of what you should take from this chapter. Here is a general equation for a nucleophilic substitution reaction.

Nucleophilic substitution
Any reaction in which one nucleophile is substituted for another at a tetravalent carbon atom.

Nucleophile
From the Greek, meaning nucleus loving. A molecule or ion that donates a pair of electrons to another atom or ion to form a new covalent bond; a Lewis base.

Leaving group (Lv)
The group that is displaced in a substitution or that is lost in an elimination.

Electrophile
From the Greek, meaning electron loving. A molecule or ion that accepts a pair of electrons from another atom or molecule in a reaction; a Lewis acid.

Nucleophile   Electrophile
(Lewis base)   (Lewis acid)

◉WL Online homework for this chapter may be assigned in OWL for Organic Chemistry.

An example of this reaction that you have already studied is the alkylation of terminal alkynes (Section 7.5A). Another is the reaction of hydroxide ion with chloromethane. In this reaction, chloromethane is the electrophile. Because of the electronegativity of chlorine, there is a partial positive charge on the carbon. An electrostatic potential map shows the negative electron density on HO⁻ (the nucleophile) interacting with the partial positive charge on the methyl group.

$$HO:^- \quad + \quad \delta^+ CH_3 \overset{\frown}{-} \overset{..}{\underset{..}{Cl}} : \delta^- \longrightarrow CH_3 \overset{..}{-} \overset{..}{O}H \quad + \quad :\overset{..}{\underset{..}{Cl}} :^-$$

Nucleophile   Electrophile
(Lewis base)   (Lewis acid)

Electrostatic potential map showing the nucleophile (OH⁻) reacting at its negative (red) end with the electrophilic carbon (blue) in the reaction of hydroxide with chloromethane.

Nucleophiles are also Brønsted bases (Section 4.2), although some are very weak ones. The stronger ones can remove protons as well as attack at carbon centers. A reaction in which a halide and a hydrogen on the neighboring (β) carbon are removed is called a **β-elimination**. Nucleophilic substitution and base-promoted β-elimination are therefore competing reactions. For example, ethoxide ion reacts with bromocyclohexane as a nucleophile to give ethoxycyclohexane (cyclohexyl ethyl ether) and as a Brønsted base to give cyclohexene and ethanol.

β-Elimination
A reaction in which a molecule, such as HCl, HBr, HI, or HOH, is split out or eliminated from adjacent carbons.

as a nucleophile, ethoxide ion attacks this carbon

nucleophilic substitution

OCH₂CH₃

+ Na⁺ :Br:⁻

+ CH₃CH₂O:⁻Na⁺

A nucleophile and a base

β-elimination
− (HBr)

+ CH₃CH₂OH + Na⁺ :Br:⁻

as a Brønsted base, ethoxide ion attacks this hydrogen

In this chapter, we study substitution and β-elimination. By using these reactions, we can convert haloalkanes to compounds with other functional groups including alcohols, ethers, thiols, sulfides, amines, nitriles, alkenes, and alkynes. Nucleophilic substitution and β-elimination open entirely new areas of organic chemistry and are a major method of interconverting functional groups. One of the most challenging aspects of the study of these reactions is deciding whether substitution or elimination is likely to prevail, and this will be the major focus of the last part of the chapter.

9.1 | Nucleophilic Substitution in Haloalkanes

Nucleophilic substitution is one of the most important reactions of haloalkanes and can lead to a wide variety of new functional groups, many of which are illustrated in Table 9.1. Some of these reactions proceed smoothly at room temperature; others occur only at elevated temperatures, as we will see in later sections. As you study the entries in this table, note these points:

1. If the nucleophile is negatively charged, as, for example, OH⁻ and HC≡C⁻, then in a substitution reaction the atom donating the pair of electrons becomes neutral in the product.
2. If the nucleophile is uncharged, as, for example, NH₃ and CH₃OH, then in the substitution reaction, the atom donating the pair of electrons becomes positively charged in the initial product.
3. In the middle of the table are two reactions involving N≡C⁻ and HC≡C⁻ nucleophiles. In these nucleophilic substitution reactions, the products have

new carbon-carbon bonds, as we have seen for alkynes in Section 7.5A. The formation of new carbon-carbon bonds is important in organic chemistry because it provides a means of extending a molecular carbon skeleton.

Table 9.1 Some Nucleophilic Substitution Reactions

Reaction: $Nu:^- + CH_3\ddot{B}r: \longrightarrow CH_3Nu + :\ddot{B}r:^-$

Nucleophile	Product	Class of Compound Formed
$:\ddot{O}H \longrightarrow CH_3\ddot{O}H$		An alcohol
$:\ddot{O}R \longrightarrow CH_3\ddot{O}R$		An ether
$:\ddot{S}H \longrightarrow CH_3\ddot{S}H$		A thiol (a mercaptan)
$:\ddot{S}R \longrightarrow CH_3\ddot{S}R$		A sulfide (a thioether)
$:C\equiv CH \longrightarrow CH_3C\equiv CH$		An alkyne
$:C\equiv N: \longrightarrow CH_3C\equiv N:$		A nitrile
$:\ddot{I}:^- \longrightarrow CH_3\ddot{I}:$		An alkyl iodide
$:\ddot{N}=\overset{+}{N}=\ddot{N}:^- \longrightarrow CH_3-\ddot{N}=\overset{+}{N}=\ddot{N}:^-$		An alkyl azide
$:NH_3 \longrightarrow CH_3NH_3^+$		An alkylammonium ion
$:\underset{\underset{H}{\mid}}{\ddot{O}}-H \longrightarrow CH_3\overset{+}{\underset{\underset{H}{\mid}}{\ddot{O}}}-H$		An alcohol (after proton is taken away)
$:\underset{\underset{H}{\mid}}{\ddot{O}}-CH_3 \longrightarrow CH_3\overset{+}{\underset{\underset{H}{\mid}}{\ddot{O}}}-CH_3$		An ether (after proton is taken away)

Example 9.1

Complete these nucleophilic substitution reactions. In each reaction, show all electron pairs on both the nucleophile and the leaving group.

(a) $\text{\textbackslash}\diagup\diagdown\diagup\text{Br} + CH_3O^-Na^+ \xrightarrow[\text{methanol}]{}$

(b) $\text{\textbackslash}\diagup\diagdown\diagup\diagdown\text{Cl} + NH_3 \xrightarrow[\text{ethanol}]{}$

Solution

(a) Methoxide ion is the nucleophile and bromide is the leaving group.

$\diagdown\diagup\diagdown\diagup\ddot{B}r: + CH_3\ddot{O}:^-Na^+ \xrightarrow[\text{methanol}]{} \diagdown\diagup\diagdown\diagup\ddot{O}CH_3 + Na^+:\ddot{B}r:^-$

| **1-Bromobutane** | **Sodium methoxide** | **1-Methoxybutane (Butyl methyl ether)** | **Sodium bromide** |

(b) Ammonia is the nucleophile and chloride is the leaving group.

$\diagdown\diagup\diagdown\diagup\ddot{C}l: + :NH_3 \xrightarrow[\text{ethanol}]{} \diagdown\diagup\diagdown\diagup NH_3^+ :\ddot{C}l:^-$

| **1-Chlorobutane** | **Ammonia** | **Butylammonium chloride** |

Problem 9.1

Complete the following nucleophilic substitution reactions. In each reaction, show all electron pairs on both the nucleophile and the leaving group.

(a) [cyclohexane with Cl] $+ CH_3CO^-Na^+$ (with C=O) $\xrightarrow{\text{ethanol}}$

(b) [structure with I] $+$ [structure] S^-Na^+ $\xrightarrow{\text{acetone}}$

(c) [structure with Br] $+ H_3C{-}C{\equiv}C^-Li^+$ $\xrightarrow{\text{dimethyl sulfoxide}}$

9.2 Mechanisms of Nucleophilic Aliphatic Substitution

On the basis of experimental observations developed over a 70-year period, two limiting mechanisms for nucleophilic substitutions have been proposed, called S_N2 and S_N1. A fundamental difference between them is the timing of bond breaking between carbon and the leaving group and of bond forming between carbon and the nucleophile.

A. S_N2 Mechanism

At one extreme, bond breaking and bond forming occur simultaneously. Thus departure of the leaving group is assisted by the incoming nucleophile. This mechanism is designated **S_N2**, where S stands for *S*ubstitution, N for *N*ucleophilic, and 2 for a **bimolecular reaction**. This type of substitution reaction is classified as bimolecular because both the haloalkane and the nucleophile are involved in the rate-determining step.

Bimolecular reaction
A reaction in which two species are involved in the rate-determining step.

S_N2 reaction
A bimolecular nucleophilic substitution reaction.

Mechanism An S_N2 Reaction

Make a new bond between a nucleophile and an electrophile and simultaneously break a bond to give stable molecules or ions. The nucleophile attacks the reactive center from the side opposite the leaving group; that is, an **S_N2 reaction** involves backside attack of the nucleophile. In this diagram, the dashed lines in the transition state represent partially formed or broken bonds.

Transition state with simultaneous bond breaking and bond forming

---

As noted in the mechanism just given, the nucleophile attacks from the backside. Backside attack by the nucleophile is facilitated in two ways. First, because of the polarization of the C—Br bond, the carbon atom has a partial positive charge and therefore attracts the electron-rich nucleophile (as shown for methyl chloride at the beginning of the chapter). Second, the electron density of the nucleophile entering from the backside assists in breaking the C—Br bond, thereby helping the bromide leave. The electron density of the nucleophile attacking from the backside can be thought of as populating the antibonding molecular orbital of the C—Br bond, weakening the C—Br bond as the new C—O σ bond becomes stronger. This anti-bonding C—Br orbital has most of its character on the backside of the C. Therefore, upon collision with a nucleophile, the most effective way to fill this orbital is by collision from the backside of the carbon, which breaks the C—Br bond on the other side of the carbon. Other reaction geometries are higher in energy because they do not produce an efficient orbital overlap that leads to weakening of the C—Lv bond. Backside attack has important stereochemical consequences, as we will see in Section 9.3A.

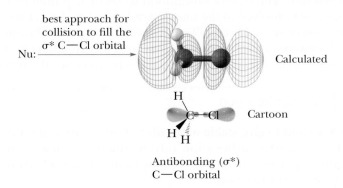

Figure 9.1 shows an energy diagram for an S_N2 reaction. Because there is a single step in the S_N2 mechanism, the energy diagram has one energy barrier that corresponds to a single transition state. Recall that transition states are fleeting structures with essentially no lifetime. They are not intermediates but instead represent transitions between two structures that lie in wells (troughs) on energy surfaces such as Figure 9.1.

The collision between the nucleophile and the electrophile must occur with enough energy to surmount the barrier to the reaction. This energy barrier is present because of the distortion from optimal bonding arrangements that occurs at the transition state. At the transition state of an S_N2 reaction the C is distorted into a trigonal bipyramidal geometry with one bond partially forming and one bond partially breaking. Hence, the transition state structure is considerably strained relative to the reactant and product and is a hybrid structure that is transitioning between the reactant and product. A good way to think about this is that the transition state has higher internal energy due to the structural distortions caused by the collision—like the strain that takes place when two rubber balls collide, deform,

Figure 9.1

An energy diagram for an S_N2 reaction. There is one transition state (no reactive intermediate).

9.2 Mechanisms of Nucleophilic Aliphatic Substitution   **285** and then bounce off one another. The difference is that in chemistry, the collisions can occur with sufficient energy to cause bond breaking and forming, leading to a new structure.

B. S$_N$1 Mechanism

The other limiting mechanism is called the **S$_N$1 reaction**. In this mechanism, bond breaking between carbon and the leaving group is entirely completed before bond forming with the nucleophile begins. In the designation S$_N$1, 1 stands for **unimolecular reaction**. This type of substitution is classified as unimolecular because only the haloalkane is involved in the rate-determining step. An S$_N$1 mechanism is illustrated by the solvolysis of 2-bromo-2-methylpropane (*tert*-butyl bromide) in methanol to form 2-methoxy-2-methylpropane (*tert*-butyl methyl ether) and HBr. In this reaction, the nucleophile (methanol) is also the solvent, hence the name **solvolysis**.

S$_N$1 reaction
A unimolecular nucleophilic substitution reaction.

Unimolecular reaction
A reaction in which only one species is involved in the rate-determining step.

Solvolysis
A nucleophilic substitution in which the solvent is also the nucleophile.

Mechanism   An S$_N$1 Reaction

Step 1: **Break a bond to give stable molecules or ions.**   Ionization of the C—Lv (Lv=Br) bond forms a carbocation intermediate. Because no nucleophile is assisting the departure of the halide anion, this is the relatively slow, rate-determining step of the reaction.

A carbocation intermediate;
its shape is trigonal planar

Step 2: **Make a new bond between a nucleophile and an electrophile.**   Reaction of the carbocation intermediate (an electrophile) with methanol (a nucleophile) gives an oxonium ion. Attack of the nucleophile occurs with approximately equal probability from either face of the planar carbocation intermediate.

Nucleophile   Electrophile   An oxonium ion
(Lewis base) (Lewis acid)

Step 3: **Take a proton away.**   Proton transfer from the oxonium ion to methanol completes the reaction and gives *tert*-butyl methyl ether.

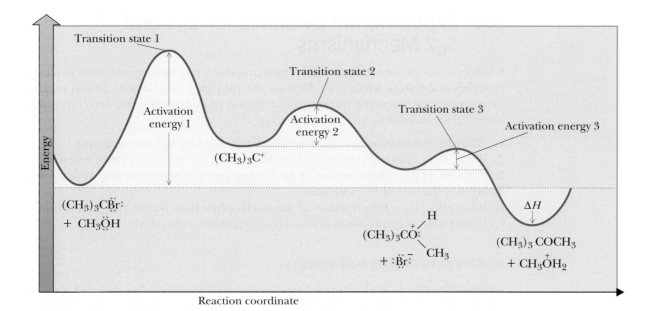

In an S_N1 reaction, the rate-determining step is the cleavage of the C—Lv bond to form a carbocation intermediate, shown as the structure in the second well of Figure 9.2. As we presented in Chapter 6, carbocations are two electrons shy of an octet and quickly react with nucleophiles, such as the solvent methanol. In the example of Figure 9.2, after reaction of the carbocation with methanol, the structure created has a proton on oxygen. A transition state exists on the energy diagram for each individual step. The last step in this three-step mechanism is a proton-transfer reaction following the S_N1 reaction.

As discussed, the rate-determining step of an S_N1 reaction involves unimolecular cleavage of the haloalkane to a carbocation and a halide anion. One can envision this occurring due to collisions with the solvent. Recall that the reactant is dissolved in a solvent, and that there is continual thermal motion consisting of translation and tumbling of both the reactant and solvent molecules. The haloalkane is continually being jostled around within the solvent, being hit by the solvent from all directions. When one of these collisions is of high enough energy to distort the haloalkane into a geometry in which the bond to the leaving group is almost completely broken, the transition state for departure of the leaving group can be achieved and the S_N1 mechanism enabled. This is in contrast to the S_N2 mechanism, where a collision with the nucleophile from the backside initiates the reaction.

Figure 9.2

An energy diagram for an S_N1 reaction. There are three transition states before the final product is created. The first is for formation of the carbocation intermediate, the second is for the reaction of the carbocation with methanol to give an oxonium ion, and the third is for taking off the proton. Step 1 crosses the highest energy barrier and therefore is rate determining.

C. Key Mechanistic Differences Between S_N2 and S_N1 Reactions

Now that we have introduced the two most dominant mechanisms for substitution reactions on alkyl halides (R—Lv), it is worthwhile to point out some of the key differences. First, an S_N2 reaction involves a single step and therefore has no intermediates. As with all chemical reactions, however, it has a transition state. In contrast, an S_N1 reaction has two steps (or three steps when a proton is removed as the last step), each with a transition state. Importantly, an intermediate carbocation is formed. The single-step versus the two-step mechanisms for S_N2 and S_N1, respectively, along with the positive charge on the carbocation intermediate involved in an S_N1 reaction, are key factors that influence the preference of one mechanism over the other. Let us now examine the experimental evidence on which these two contrasting mechanisms are based and learn what structural features cause one mechanism to dominate over the other.

9.3 | Experimental Evidence for S_N1 and S_N2 Mechanisms

It is important to note that chemists often propose a mechanism and then perform experimental tests to validate or disprove the proposed mechanism. In this regard, we consider the following questions as a means of contrasting the two commonly observed mechanisms.

1. What are the kinetics and stereochemistry of S_N2 and S_N1 mechanisms?
2. What effect does the structure of the haloalkane have on the rate of reaction?
3. What effect does the structure of the leaving group have on the rate of reaction?
4. What is the role of the solvent?
5. What effect does the structure of the nucleophile have on the rate of reaction?
6. Under what conditions are skeletal rearrangements observed?

A. Kinetics and Stereochemistry

Chemists routinely perform two experiments to distinguish S_N2 and S_N1 mechanisms. The first involves performing a kinetic analysis, which means that the rates of reactions are followed as the concentration of reactants is changed. The second experiment is to run the substitution reaction with an alkyl-Lv structure, where the C bonded to the Lv is a chiral center, and then examine the stereochemistry of the products. Because these two approaches are commonly applied for studying substitution reactions, we group them together here.

The kinetic order of nucleophilic substitutions can be studied by measuring the effect on rate of varying the concentrations of haloalkane and nucleophile. Those reactions whose rate is dependent only on the concentration of haloalkane are classified as S_N1; those reactions whose rate is dependent on the concentration of both haloalkane and nucleophile are classified as S_N2.

S_N1 Kinetics

Because the transition state for formation of the carbocation intermediate in an S_N1 mechanism involves only the haloalkane and not the nucleophile and this step is rate determining, it is a unimolecular process. The result is a first-order reaction. In this instance, the rate of reaction is expressed as the rate of disappearance of the starting material, 2-bromo-2-methylpropane. The rate has no dependence on the concentration of the nucleophile. We can conclude that any substitution reaction whose rate depends only upon [R-Lv] proceeds via an S_N1 mechanism.

2-Bromo-2-methylpropane
(*tert*-Butyl bromide)

$$\text{Rate} = -\frac{d[(CH_3)_3CBr]}{dt} = k[(CH_3)_3CBr]$$

2-Methoxy-2-methylpropane

S_N2 Kinetics

By contrast, there is only one step in the S_N2 mechanism. For the reaction of OH^- and CH_3Br, for example, both species must collide and are present in the transition state; that is, the reaction is bimolecular. The reaction between CH_3Br and NaOH to give CH_3OH and NaBr is second order: it is first order in CH_3Br and first order in OH^-, so doubling the concentration of either increases the rate by a factor of two. Doubling both of the reactants increases the rate by a factor of four. When the rate of a substitution reaction depends upon both [R-Lv] and [Nu⁻] we conclude that an S_N2 mechanism is occurring.

$$CH_3\ddot{B}r\colon + \ Na^+\colon\!\ddot{O}H^- \longrightarrow CH_3\ddot{O}H + Na^+\colon\!\ddot{B}r\colon^-$$

Bromomethane   **Methanol**

$$Rate = -\frac{d[CH_3Br]}{dt} = k[CH_3Br][OH^-]$$

Example 9.2

The reaction of *tert*-butyl bromide with azide ion (N_3^-) in methanol is a typical S_N1 reaction. What happens to the rate of the reaction if $[N_3^-]$ is doubled?

Solution

The rate remains the same because the nucleophile concentration does not appear in the rate equation for an S_N1 reaction.

Problem 9.2

The reaction of bromomethane with azide ion (N_3^-) in methanol is a typical S_N2 reaction. What happens to the rate of the reaction if $[N_3^-]$ is doubled?

S_N1 Stereochemistry

Experiments in which nucleophilic substitution takes place at a chiral center provide us with information about the stereochemical course of the reaction. One of the compounds studied to determine the stereochemistry of an S_N1 reaction utilized the following chloroalkane. When either enantiomer of this molecule undergoes nucleophilic substitution by an S_N1 pathway, the product is racemic. The reason is that ionization of this secondary chloride forms an achiral carbocation. Attack of the nucleophile can occur from either side of the planar carbocation carbon, resulting in enantiomeric products. The R and S enantiomers are formed in equal amounts, and in this case the product is a racemic mixture.

R enantiomer   Planar carbocation   *S* enantiomer   *R* enantiomer
(achiral)

A racemic mixture

The S_N1 mechanism just described results in complete racemization. Although examples of complete racemization have been observed, it is common to find only partial racemization, with the predominant product being the one with inversion of configuration at the chiral center. Although bond breaking between carbon and the leaving group is complete, the leaving group (chloride ion in this example) remains associated for a short time with the carbocation in an ion pair.

Approach of the nucleophile from this side is less hindered. $\longrightarrow$

$$R_1$$
$$\underset{H\ R_2}{\overset{|}{C^+}}\ :\overset{..}{\underset{..}{Cl}}:^-$$

Approach of the nucleophile from this side is hindered $\longleftarrow$ by chloride ion, which remains associated with the carbocation as an ion pair.

To the extent that the leaving group remains associated with the carbocation as an ion pair, it hinders approach of the nucleophile from that side of the carbocation. The result is that somewhat more than 50% of the product is formed by attack of the nucleophile from the side of the carbocation opposite that of the leaving group. Whenever we observe partial to complete racemization of stereochemistry, we conclude that an S_N1 mechanism is operative.

S_N2 Stereochemistry

Every S_N2 reaction proceeds with backside attack by the nucleophile and therefore inversion of configuration. This was shown in an ingenious experiment designed by the English chemists E. D. Hughes and C. K. Ingold. They studied the exchange reaction between enantiomerically pure 2-iodooctane and iodine-131, a radioactive isotope of iodine. Iodine-127, the naturally occurring isotope of iodine, is stable and does not undergo radioactive decay. Here acetone is the solvent.

2-Iodooctane

Hughes and Ingold first demonstrated that the reaction is second order: first order in 2-iodooctane and first order in iodide ion. Therefore, the reaction proceeds by an S_N2 mechanism. They observed further that the rate of racemization of enantiomerically pure 2-iodooctane is exactly twice the rate of incorporation of iodine-131. This observation must mean, they reasoned, that each displacement of iodine-127 by iodine-131 proceeds with inversion of configuration, as illustrated in the following equation. Note that the reaction was run to only a low percent conversion in order to minimize further reactions from the product; such a result would invert stereochemistry again.

(S)-2-Iodooctane   **(R)-2-Iodooctane**

Substitution with inversion of configuration in one molecule cancels the rotation of one molecule that has not reacted so that, for each molecule undergoing inversion, one racemic pair is formed. Inversion of configuration in 50% of the molecules results in 100% racemization. Whenever complete inversion of configuration is found in a substitution reaction we conclude that an S_N2 mechanism is occurring.

Example 9.3

Complete these S_N2 reactions, showing the configuration of each product.

(a) [structure] $+ CH_3COO^-Na^+ \longrightarrow$ (b) [structure] $+ CH_3NH_2 \longrightarrow$

Solution

S$_N$2 reactions occur with inversion of configuration at the chiral center. In (a), the starting material is the *(R,R)* isomer; the product is the *(R,S)* isomer. In (b), the starting material is the *(R)* enantiomer; the product is the *(S)* enantiomer.

(a) [structure: cyclopentane ring with CH$_3$ and OCCH$_3$ (ester, O double bond), labeled *R* and *S*]

(b) CH$_3$—N$^+$—H with H on top, Br$^-$, pentyl chain below, labeled *S*

Problem 9.3

Complete these S$_N$2 reactions, showing the configuration of each product.

(a) [structure: cyclohexane with Br and H$_3$C substituents] + Na$^+$N$_3^-$ $\longrightarrow$

(b) [structure: benzene ring with CHBr CH$_3$ group] + CH$_3$S$^-$Na$^+$ $\longrightarrow$

B. Structure of the Alkyl Portion of the Haloalkane

The rates of S$_N$1 reactions are governed mainly by electronic factors, namely the relative stabilities of carbocation intermediates. The rates of S$_N$2 reactions, on the other hand, are governed mainly by steric factors, and their transition states are particularly sensitive to *bulky groups* at the site of reaction. The ability of groups, because of their size, to hinder access to a reaction site within a molecule is called **steric hindrance**.

S$_N$1 Considerations

Relative Stabilities of Carbocations

Let us first consider the effect of the alkyl group of the haloalkane on S$_N$1 reactions. As discussed, the rate-determining step of an S$_N$1 mechanism is formation of a carbocation, and therefore the stability of the resulting carbocation is a dominant consideration.

As we learned in Section 6.3A, 3° carbocations are the most stable (lowest activation energy for their formation) due to hyperconjugation, whereas 1° carbocations are the least stable (highest activation energy for their formation). In fact, 1° carbocations are so unstable that they rarely are ever formed in solution. Because carbocations are high-energy intermediates, the transition states for their formation are very similar to the carbocation in energy (Hammond postulate, Section 8.5D). Therefore, 3° haloalkanes are most likely to react by carbocation formation; 2° haloalkanes are the next most likely to react by carbocation formation, while methyl and 1° haloalkanes react in this manner only when they are specially stabilized.

Allylic carbocations, like allylic radicals (Section 8.6), have a double bond next to the electron deficient carbon. The allyl cation is the simplest allylic carbocation. Because the allyl cation has only one substituent on the carbon bearing the positive charge, it is a primary allylic carbocation. Allylic carbocations are considerably more stable than comparably substituted alkyl carbocations because delocalization is associated with the resonance interaction between the positively charged carbon and the adjacent π bond. The allyl cation, for example, can be represented as a hybrid of two equivalent contributing structures. The result is that the positive charge appears only on carbons 1 and 3, as shown in the accompanying electrostatic potential map.

$$CH_2{=}CH{-}\overset{+}{C}H_2 \longleftrightarrow \overset{+}{C}H_2{-}CH{=}CH_2$$

Allyl cation
(a hybrid of two equivalent contributing structures)

Steric hindrance
The ability of groups, because of their size, to hinder access to a reaction site within a molecule.

Allylic carbocation
A carbocation in which an allylic carbon bears the positive charge.

Allylic
Next to a carbon-carbon double bond.

Recall that, in general, a distributed charge in a molecule is more stabilizing than a more localized charge. It has been determined experimentally that the double bond of one adjacent vinyl group provides approximately as much stabilization as two alkyl groups. Thus, the allyl cation and 2° isopropyl cation are of comparable stability.

Electrostatic potential map for the allyl cation. The positive charge (blue) is on carbons 1 and 3.

These cations are of comparable stability

$$CH_2=CH-\overset{+}{C}H_2 \qquad CH_3-\overset{+}{C}H-CH_3$$

1° Allylic cation   2° Alkyl cation

The classification of allylic cations as 1°, 2°, and 3° is determined by the location of the positive charge in the more important contributing structure. Following are examples of 2° and 3° allylic carbocations.

$$CH_2=CH-\overset{+}{C}H-CH_3 \qquad CH_2=CH-\overset{+}{\underset{\underset{\textstyle CH_3}{|}}{C}}-CH_3$$

2° Allylic cation   3° Allylic cation

Benzylic carbocation
A carbocation in which a carbon attached to a benzene ring bears the positive charge.

Benzylic carbocations show approximately the same stability as allylic carbocations. Both are stabilized by resonance delocalization of the positive charge due to adjacent π bonds. It can be written as $C_6H_5-CH_2^+$.

Benzyl cation
(a benzylic carbocation)

In Section 6.3A, we presented the order of stability of methyl, 1°, 2°, and 3° carbocations. We can now expand this order to include 1°, 2°, and 3° allylic, as well as 1°, 2°, and 3° benzylic carbocations.

$$\text{methyl} < 1° \text{ alkyl} < \left\{\begin{array}{l} 2° \text{ alkyl} \\ 1° \text{ allylic} \\ 1° \text{ benzylic} \end{array}\right\} < \left\{\begin{array}{l} 3° \text{ alkyl} \\ 2° \text{ allylic} \\ 2° \text{ benzylic} \end{array}\right\} < \left\{\begin{array}{l} 3° \text{ allylic} \\ 3° \text{ benzylic} \end{array}\right\}$$

Increasing stability of carbocations ⟶

Thus, in summary, S_N1 mechanisms should be considered for allyl and benzylic haloalkanes, even if they are primary haloalkanes.

Lastly, we note that S_N1 reactions rarely occur with sp^2 carbons and never occur on sp carbons. The carbocations derived from sp^2 C—X or sp C—X bonds are too unstable to form.

Example 9.4

Write an additional resonance contributing structure for each carbocation and state which of the two makes the greater contribution to the resonance hybrid. Classify each contributing structure as a 1°, 2°, or 3° allylic cation.

(a)   (b)

Solution

The additional resonance contributing structure in each case is a 3° allylic cation. The contributing structure having the greater degree of substitution on the positively charged carbon makes the greater contribution to the hybrid.

(a)   Greater contribution   (b)   Greater contribution

Problem 9.4

Write an additional resonance contributing structure for each carbocation and state which of the two makes the greater contribution to the resonance hybrid.

(a) =CH₂   (b)

S_N2 Considerations

Steric Hindrance

In an S_N2 mechanism the nucleophile must approach the substitution center and begin to form a new covalent bond to it while the leaving group is departing. If we compare the ease of approach to the substitution center of a 1° haloalkane with that of a 3° haloalkane, we see that the approach is considerably easier for bromoethane than for *tert*-butyl bromide. Two hydrogen atoms and one alkyl group screen the backside of the substitution center of a 1° haloalkane. In contrast, three alkyl groups screen the backside of the 3° haloalkane.

Bromoethane
(Ethyl bromide)

2-Bromo-2-methylpropane
(*tert*-Butyl bromide)

Given the competition between electronic and steric factors, we find that tertiary haloalkanes react by an S_N1 mechanism because 3° carbocation intermediates are relatively stable and tertiary haloalkanes are very hindered toward backside attack. Hence, 3° haloalkanes are never observed to react by an S_N2 mechanism. Halomethanes and primary haloalkanes have little crowding around the reaction site and react by an S_N2 mechanism; they are never observed to react by an S_N1 mechanism because methyl and primary carbocations are so unstable. Secondary haloalkanes may react by either S_N1 or S_N2 mechanisms, depending on the nucleophile and solvent. The competition between electronic and steric factors and their effects on relative rates of nucleophilic substitution reactions of haloalkanes are summarized in Figure 9.3.

Figure 9.3

Effect of steric and electronic factors in competition between S_N1 and S_N2 reactions of haloalkanes. Methyl and primary haloalkanes react only by the S_N2 mechanism; they do not react by S_N1. Tertiary haloalkanes do not react by S_N2; they react only by S_N1. Secondary haloalkanes may be made to react by either S_N1 or S_N2 mechanisms depending on the solvent and the choice of nucleophile, which are topics addressed later in this chapter.

We see a similar effect of steric hindrance on S_N2 reactions in molecules with branching at the β-carbon. The carbon bearing the halogen in a haloalkane is called the α-carbon and the next carbon is called the β-carbon. Table 9.2 shows relative rates of S_N2 reactions on a series of primary bromoalkanes. In these data, the rate of nucleophilic substitution of bromoethane is taken as a reference and is given the value 1.0. As CH_3 branches are added to the β-carbon, the relative rate of reaction decreases. Compare the relative rates of bromoethane (no β-branch) with that of 1-bromo-2,2-dimethylpropane (neopentyl bromide), a compound with three β-branches. The rate of S_N2 substitution of this compound is only 10^{-5} that of bromoethane. For all practical purposes, primary halides with three β-branches do not undergo S_N2 reactions.

As shown in Figure 9.4, the carbon of the C—Br bond in bromoethane is unhindered and open to attack by a nucleophile in an S_N2 reaction. On the other hand, three β-methyl groups screen the corresponding carbon in neopentyl bromide. Thus, although the carbon bearing the leaving group is primary, approach of the nucleophile is so hindered that the rate of S_N2 reaction of neopentyl bromide is greatly reduced compared to bromoethane.

Table 9.2	Effect of β-Branches on the Rate of S_N2 Reactions			
Alkyl bromide	Br / β	Br / β	Br / β	Br / β
β-Branches	0	1	2	3
Relative rate	1.0	4.1×10^{-1}	1.2×10^{-3}	1.2×10^{-5}

CH₃CH₂Br

$$CH_3CCH_2Br$$ with CH₃ above and CH₃ below the central C

Free access →

Blocked access ┄→

Bromoethane
(Ethyl bromide)

1-Bromo-2,2-dimethylpropane
(Neopentyl bromide)

Figure 9.4
The effect of β-branching in S_N2 reactions on a primary haloalkane. With bromoethane, attack of the nucleophile is unhindered. With 1-bromo-2,2-dimethylpropane, the three β-branches block approach of the nucleophile to the backside of the C—Br bond, thus drastically reducing the rate of S_N2 reaction of this compound.

Lastly, we note that S_N2 reactions never occur on sp^2 or sp hybridized carbons. Hence, you should never consider performing a substitution on a vinyl halide (C=C—X), aryl-halide, or alkynyl-halide (C≡C—X).

C. The Leaving Group

In the transition state for nucleophilic substitution on a haloalkane, the **leaving group** develops a partial negative charge in both S_N1 and S_N2 reactions; therefore, the ability of a group to function as a leaving group is related to how stable it is as an anion. The most stable anions, and therefore the best leaving groups, are the (weak) conjugate bases of strong acids. Thus we can use the information on the relative strengths of organic and inorganic acids in Table 4.1 to determine which anions are the best leaving groups. This order is shown here.

Reactivity as a leaving group ⟵

rarely function as leaving groups

$$I^- > Br^- > Cl^- \sim H_2O \gg F^- > CH_3CO^- > HO^- > CH_3O^- > NH_2^-$$
(with O double-bonded above the CH₃C of CH₃CO⁻)

Stability of anion; strength of conjugate acid ⟵

The best leaving groups in this series are the halides, I^-, Br^-, and Cl^-. Hydroxide ion, OH^-, methoxide ion, CH_3O^-, and amide ion, NH_2^-, are such poor leaving groups that they rarely, if ever, are displaced in nucleophilic aliphatic substitution.

D. The Solvent

Solvents dissolve reactants and provide the medium in which reactions take place. The solvent plays a role of paramount importance in substitution reactions. It has an influence on the rates of both S_N1 and S_N2 reactions, on relative nucleophilicity (see the next section), and on the balance between S_N1 and S_N2 mechanisms for secondary haloalkanes. Before analyzing the effect of the solvent on these properties we must first examine solvent characteristics.

Common solvents can be divided into two groups: protic and aprotic. Furthermore, solvents are classified as polar and nonpolar based on their **dielectric constant**. The greater the value of the dielectric constant of a solvent, the better it solvates, and so the smaller the interaction between ions of opposite charge dissolved in it. We say that a solvent is a **polar solvent** if it has a dielectric constant of 15 or greater. A solvent

Dielectric constant
A measure of a solvent's ability to insulate opposite charges from one another.

Table 9.3	Common Protic Solvents	
Solvent	**Structure**	**Dielectric Constant (25°C)**
Water	H_2O	79
Formic acid	HCOOH	59
Methanol	CH_3OH	33
Ethanol	CH_3CH_2OH	24
Acetic acid	CH_3COOH	6

is a **nonpolar solvent** if it has a dielectric constant of less than 5. Solvents with a dielectric constant between 5 and 15 are borderline.

The common **protic solvents** for nucleophilic substitution reactions are water, low-molecular-weight alcohols, and low-molecular-weight carboxylic acids (Table 9.3). Each of these has a partially negatively charged oxygen bonded to a partially positively charged hydrogen atom. Protic solvents solvate ionic substances by electrostatic interactions between anions and the partially positively charged hydrogens of the solvent and also between cations and partially negatively charged atoms of the solvent. By our guideline, water, formic acid, methanol, and ethanol are classified as **polar protic solvents**. Because of its smaller dielectric constant, acetic acid is classified as a moderately polar protic solvent.

The **aprotic solvents** most commonly used for nucleophilic substitution reactions are given in Table 9.4. Of these, dimethyl sulfoxide (DMSO), acetonitrile, *N,N*-dimethylformamide (DMF), and acetone are classified as **polar aprotic solvents**. Dichloromethane and tetrahydrofuran (THF) are moderately polar aprotic solvents. Diethyl ether, toluene, and hexane are classified as nonpolar aprotic solvents.

Effect of Solvent on S_N1 Reactions

Nucleophilic substitution by an S_N1 pathway involves creation and separation of opposite charges in the transition state of the rate-determining step. For this

Protic solvent
A solvent that is a hydrogen-bond donor; the most common protic solvents contain —OH groups. Common protic solvents are water, low-molecular-weight alcohols such as ethanol, and low-molecular-weight carboxylic acids.

Aprotic solvent
A solvent that cannot serve as a hydrogen-bond donor; nowhere in the molecule is there a hydrogen bonded to an atom of high electronegativity. Common aprotic solvents are dichloromethane, diethyl ether, and dimethyl sulfoxide.

Table 9.4	Common Aprotic Solvents	
Solvent	**Structure**	**Dielectric Constant**
Polar		
Dimethyl sulfoxide (DMSO)	$\overset{\displaystyle O}{\overset{\|}{CH_3SCH_3}}$	48.9
Acetonitrile	$CH_3C{\equiv}N$	37.5
N,N-Dimethylformamide (DMF)	$\overset{\displaystyle O}{\overset{\|}{HCN(CH_3)_2}}$	36.7
Acetone	$\overset{\displaystyle O}{\overset{\|}{CH_3CCH_3}}$	20.7
Moderately Polar		
Dichloromethane	CH_2Cl_2	9.1
Tetrahydrofuran (THF)	(cyclic ether structure) O	7.6
Nonpolar		
Diethyl ether	$CH_3CH_2OCH_2CH_3$	4.3
Toluene	(benzene ring)—CH_3	2.3
Hexane	$CH_3(CH_2)_4CH_3$	1.9

Increasing solvent polarity

reason, the rate of S_N1 reactions depends on both the ability of the solvent to keep opposite charges separated and its ability to stabilize both positive and negative sites by solvation. The solvents that best solvate charges are polar protic solvents such as H_2O, low-molecular-weight alcohols such as methanol and ethanol, and, to a lesser degree, low-molecular-weight carboxylic acids such as formic acid and acetic acid. As seen in Table 9.5, the rate of solvolysis of 2-chloro-2-methylpropane (*tert*-butyl chloride) increases by a factor of 10^5 when the solvent is changed from ethanol to water because water better solvates the carbocation and chloride anion.

The increased rate of an S_N1 reaction in higher polarity solvents can also be explained by an analysis of reaction coordinate diagrams. Figure 9.5a shows curves for two different solvents. The higher polarity solvent will generally better solvate the reactants, intermediates, and products, and therefore the entire curve is lower in energy with the higher polarity solvent. However, the increased solvation is greatest for the intermediate carbocation, which lowers the barrier to its formation and thereby accelerates the reaction.

The Effect of Solvent on S_N2 Reactions

The most common type of S_N2 reaction involves a negatively charged nucleophile and a negatively charged leaving group. The central carbon atom has a partial positive charge in the starting material, while in the transition state it may have either a smaller or larger positive charge depending on the conditions, but the negative charge from the nucleophile is dispersed across the adding nucleophile and the departing leaving group.

Transition state

The stronger the solvation of the nucleophile, the greater the energy required to remove the nucleophile from its solvation shell to reach the transition state, and hence the lower the rate of the S_N2 reaction.

Table 9.5	Rates of an S_N1 Reaction as a Function of Solvent	

$$CH_3CCl + ROH \xrightarrow{\text{solvolysis}} CH_3COR + HCl$$

Solvent	$\dfrac{k_{\text{(solvent)}}}{k_{\text{(ethanol)}}}$
Water	100,000
80% water: 20% ethanol	14,000
40% water: 60% ethanol	100
Ethanol	1

Increasing polarity of solvent

Increasing rate of solvolysis

Figure 9.5

Energy diagrams for substitution reactions in different polarity solvents. **(a)** An S_N1 reaction run in two different solvents. The higher polarity solvent better solvates all species but has the greatest difference in solvation for the intermediate carbocation. The reaction is therefore faster in the higher polarity solvent. **(b)** An S_N2 reaction run in two different solvents. The higher polarity solvent better solvates all the species but has the greatest difference in solvation for the anionic nucleophile reactant. The reaction is therefore faster in the lower polarity solvent.

(a)

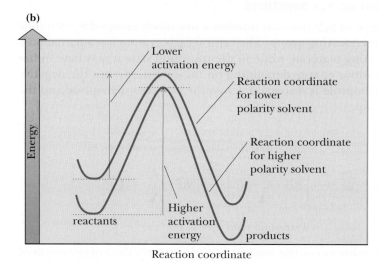

(b)

Polar aprotic solvents can solvate cations very well, but they solvate anions (nucleophiles) relatively poorly, because they cannot donate a hydrogen bond to an anion, as can a protic solvent. For this reason, nucleophiles are freer and more reactive in polar aprotic solvents than in protic solvents, so the rates of S_N2 reactions are dramatically accelerated, often by several orders of magnitude compared to the same reaction in protic solvents.

These effects can be seen pictorially by comparing reaction coordinate diagrams for an S_N2 reaction in solvents of different polarity. Figure 9.5b shows that all species in the reaction are better solvated in the more polar solvent. But, because the anionic nucleophile reactant is particularly well solvated (lower starting energy) in the more polar solvent, it is more reactive in the less polar solvent. Consequently, the activation energy is lower in the less polar solvent, and therefore the reaction is faster.

Table 9.6 shows ratios of rate constants for the S_N2 reaction of 1-bromobutane with sodium azide as a function of solvent. The rate of reaction in methanol is taken as a reference and assigned a relative rate of 1. Although chemists may prefer to use a polar aprotic solvent for an S_N2 reaction because it will be completed faster than when using a polar protic solvent, you should realize that the S_N2 mechanism is viable in all solvents that dissolve the nucleophile and electrophile. Only polar protic solvents are viable for the S_N1 mechanism.

Nucleophilicity
A kinetic property measured by the rate at which a nucleophile causes nucleophilic substitution on a reference compound under a standardized set of experimental conditions.

E. Structure of the Nucleophile

Nucleophilicity is a kinetic property and is measured by relative rates of reaction. Relative nucleophilicities for a series of nucleophiles are established by measuring

Table 9.6	Rates of an S_N2 Reaction as a Function of Solvent

Solvent Type	Solvent		$\dfrac{k_{(solvent)}}{k_{(methanol)}}$
Polar aprotic	Acetonitrile	$CH_3C{\equiv}N$	5000
	DMF	$(CH_3)_2NCHO$	2800
	DMSO	$(CH_3)_2S{=}O$	1300
Polar protic	Water	H_2O	7
	Methanol	CH_3OH	1

the rate at which each displaces a leaving group from a haloalkane; for instance, the rate at which each nucleophile displaces bromide ion from bromoethane in ethanol at 25°C. Here is a reaction using ammonia as the nucleophile.

$$CH_3CH_2\overset{..}{\underset{..}{Br}}: + :NH_3 \xrightarrow[\text{Ethanol}]{25°C} CH_3CH_2\overset{+}{N}H_3 + :\overset{..}{\underset{..}{Br}}:^-$$

From these studies, we can then make correlations between the structure of a nucleophile and its relative nucleophilicity. Listed in Table 9.7 are the types of nucleophiles we deal with most commonly in this text and their nucleophilicity in alcohol or water. The more rapidly a nucleophile reacts with a substrate in an S_N2 reaction, the more nucleophilic it is, by definition.

Because all nucleophiles are Brønsted bases as well (see Chapter 4), we also study correlations between nucleophilicity and basicity. **Basicity** and nucleophilicity are often related because they both involve a lone pair of electrons making a bond to another atom. In the case of a base, the lone pair makes a bond to a proton, while with a nucleophile the lone pair most commonly creates a bond to an electrophilic carbon. In general, sterically unhindered strong bases are good nucleophiles.

Basicity
An equilibrium property measured by the position of equilibrium in an acid-base reaction, such as for example the acid-base reaction between ammonia and water.

Table 9.7	Common Nucleophiles and Their Relative Nucleophilicities in Alcohol or Water

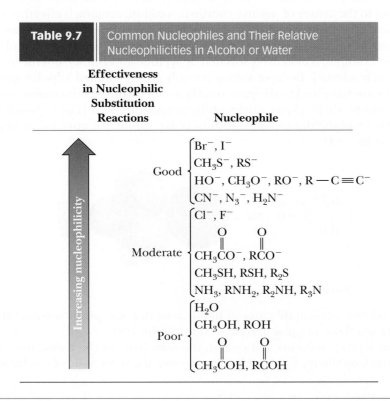

For example, oxygen anions such as hydroxide and methoxide (CH_3O^-) are good nucleophiles because they are also strong bases. Weaker oxygen bases are similarly weaker nucleophiles. As an example, carboxylate anions (RCO_2^-) are classified as moderate nucleophiles. Because of this trend with basicity, we can confidently conclude that anionic atoms are more nucleophilic than their neutral counterparts. Hence, neutral oxygen species such as water, alcohols, and carboxylic acids are weak nucleophiles. As a rough guideline for oxygen and amine nucleophilicities, we consider those nucleophiles that have conjugate acids with pK_a's above 11 to be **strong nucleophiles**, around 11 to be **moderate nucleophiles**, and below 11 to be **weak nucleophiles**. This guideline classifies amines as moderate nucleophiles, as shown in Table 9.7.

But upon inspection of Table 9.7, you might be asking, "Why are the extremely weak bases iodide and bromide anions good nucleophiles?" Also, the table shows that anionic sulfur species, as well as cyanide and azide, are good nucleophiles even though none of these species are particularly strong bases. As we now describe, nucleophilicity is complex, and depends upon solvent and shape, not just base strength.

Solvation Effects on Nucleophilicities

The solvent in which nucleophilic substitutions are carried out has a marked effect on relative nucleophilicities. For a fuller understanding of the role of the solvent, let us consider nucleophilic substitution reactions carried out in polar aprotic solvents and in polar protic solvents. An organizing principle for substitution reactions is the following:

All other factors being equal, the stronger the interaction of the nucleophile with the solvent, the lower its nucleophilicity.

Conversely, the less the nucleophile interacts with solvent, the greater its nucleophilicity.

The most commonly used polar aprotic solvents (DMSO, acetone, acetonitrile, and DMF) are very effective in solvating cations (in addition to the attraction to the negative end of the dipole, the lone pairs on oxygen and nitrogen act as Lewis bases) but are not nearly as effective in solvating anions. Consider, for example, acetone. Because the negative end of its dipole and the lone pairs on oxygen can come close to the center of positive charge in a cation, acetone is effective in solvating cations. The positive end of its dipole, however, is shielded by the two methyl groups and is therefore less effective in solvating anions. The sodium ion of sodium iodide, for example, is effectively solvated by acetone and DMSO, but the iodide ion is only poorly solvated. Because anions are only poorly solvated in polar aprotic solvents, they are freer and participate readily in nucleophilic substitution reactions. In these solvents their relative nucleophilicities parallel their relative basicities. The relative nucleophilicities of halide ions in polar aprotic solvents, for example, are $F^- > Cl^- > Br^- > I^-$.

Solvation of NaI in acetone.

The relative nucleophilicities of halide ions in polar protic solvents are quite different from those in polar aprotic solvents (Table 9.8).

In polar protic solvents, iodide ion, the least basic of the halide ions, has the greatest nucleophilicity. Conversely, fluoride ion, the most basic of the halide ions,

Table 9.8	Relative Nucleophilicities of Halide Ions in Polar Aprotic and Protic Solvents

Solvent	Increasing nucleophilicity
Polar aprotic	$I^- < Br^- < Cl^- < F^-$
Polar protic	$F^- < Cl^- < Br^- < I^-$

has the smallest nucleophilicity. The reason for this reversal of correlation between nucleophilicity and basicity lies in the degree of solvation of anions in protic solvents compared with aprotic solvents, and in polarizability trends.

- In polar aprotic solvents, anions are only weakly solvated and, therefore, relatively free to participate in nucleophilic substitution reactions, and basicity dictates nucleophilicity.
- In polar protic solvents, anions are highly solvated by hydrogen bonding with solvent molecules and therefore are less free to participate in nucleophilic substitution reactions, and polarizability dictates nucleophilicity.

The negative charge on the fluoride ion, the smallest of the halide ions, is concentrated in a small volume, and the very tightly held solvent shell formed by a polar protic solvent constitutes a barrier between fluoride ion and substrate. The fluoride ion must be at least partially removed from its tightly held solvation shell before it can participate in nucleophilic substitution. The negative charge on the iodide ion, the largest and most polarizable of the halide ions, is far less concentrated, the solvent shell is less tightly held, and iodide is considerably freer to participate in nucleophilic substitution reactions.

Recall that the order of polarizability is $F^- < Cl^- < Br^- < I^-$, which reflects the greater ability of electron clouds to undergo perturbations during chemical reactions such as S_N2 displacements. Iodide, being the most polarizable, makes it the best nucleophile in a polar protic solvent.

Effect of Shape on Nucleophilicity

As described above, the nucleophile in an S_N2 reaction attacks the backside of a C—Lv bond. That backside attack can have varying degrees of steric hindrance depending upon the R-group of the haloalkane. Not surprisingly, nucleophiles that are shaped like bullets or spears can better penetrate past the steric hindrance, and are in general better nucleophiles. Two prime examples are azide and cyanide, both of which are cylindrically shaped anions. Although neither are particularly basic nor polarizable, both are excellent nucleophiles. In contrast, when an otherwise good nucleophile, such as an alkoxide, is large and bulky, its ability to be a nucleophile diminishes. For example, whereas ethoxide (EtO^-) is an excellent nucleophile, *tert*-butoxide (t-BuO^-) is not a nucleophile.

F. Skeletal Rearrangement

As we saw in Section 6.3C, skeletal rearrangement is typical of reactions involving a carbocation intermediate that can rearrange to a more stable one. Because there is little or no carbocation character at the substitution center, S_N2 reactions are free of rearrangement. In contrast, S_N1 reactions often proceed with rearrangement. An example of an S_N1 reaction involving rearrangement is solvolysis of 2-chloro-3-phenylbutane in methanol, a polar protic solvent and a weak nucleophile. The major substitution product is the ether with a rearranged structure. The chlorine atom in the starting material is on a 2° carbon, but the methoxyl group in the product is on the adjacent 3° carbon.

2-Chloro-3-phenylbutane
(racemic)

2-Methoxy-2-phenylbutane
(racemic)

As seen in the following mechanism, reaction is initiated by heterolytic cleavage of the carbon-chlorine bond to form a 2° carbocation, which rearranges to a considerably more stable 3° carbocation by shift of a hydrogen with its pair of electrons (a hydride ion) from the adjacent benzylic carbon. Note that not only is the rearranged carbocation tertiary (hyperconjugation stabilization), but it is also benzylic (stabilization by resonance delocalization).

Mechanism Rearrangement During Solvolysis of 2-Chloro-3-phenylbutane

Step 1: **Break a bond to give stable molecules or ions.** Ionization of the C—Cl bond gives a 2° carbocation intermediate.

A 2° carbocation

Step 2: **1,2 Shift.** Migration of a hydrogen atom with its bonding electrons (a hydride ion) gives a more stable 3° benzylic carbocation intermediate.

A 3° benzylic
carbocation

Step 3: **Make a new bond between a nucleophile and an electrophile.** Reaction of the 3° benzylic carbocation intermediate (an electrophile) with methanol (a nucleophile) forms an oxonium ion.

An oxonium ion

Step 4: **Take a proton away.** Proton transfer to solvent (in this case, methanol) gives the final product.

In general, migration of a hydrogen atom or alkyl group with its bonding electrons occurs when a more stable carbocation can be formed.

Table 9.9 | Summary of S_N1 Versus S_N2 Reactions of Haloalkanes

Type of Alkyl Halide	S_N2	S_N1
Methyl CH_3X	**S_N2 is favored.**	**S_N1 does not occur.** The methyl cation is so unstable that it is never observed in solution.
Primary RCH_2X	**S_N2 is favored.**	**S_N1 rarely occurs.** Primary cations are so unstable that they are rarely formed in solution (allylic and benzylic cations can be exceptions).
Secondary R_2CHX	**S_N2 is favored** in aprotic solvents with good nucleophiles.	**S_N1 is favored** in protic solvents with poor nucleophiles. Carbocation rearrangements may occur.
Tertiary R_3CX	**S_N2 does not occur** because of steric hindrance around the reaction center.	**S_N1 is favored** because of the ease of formation of tertiary carbocations.
Substitution at a chiral center	**Inversion of configuration.** The nucleophile attacks the chiral center from the side opposite the leaving group.	**Racemization is favored.** The carbocation intermediate is planar, and attack of the nucleophile occurs with equal probability from either side. There is often some net inversion of configuration.

Now that we have considered the many factors involved in substitution reactions, we present an overview useful to predict the type of mechanism that dominates under certain reaction conditions (Table 9.9). Examples of how to use this summary are given in Section 9.4.

9.4 | Analysis of Several Nucleophilic Substitution Reactions

Predictions about the mechanism for a particular nucleophilic substitution reaction must be based on considerations of the structure of the haloalkane, the nucleophile, the leaving group, and the solvent. Following are five nucleophilic substitution reactions and an analysis of the factors that favor an S_N1 or S_N2 mechanism for each and the products that result from the mechanism used. Note that in the following examples we ignore competing elimination because it has not been discussed yet.

Nucleophilic Substitution 1

(*R*)-2-Chlorobutane + CH_3OH/H_2O ⟶ ?

The mixture of methanol and water is a polar protic solvent and a good ionizing solvent in which to form carbocations. 2-Chlorobutane ionizes in this solvent to form a fairly stable 2° carbocation intermediate. Both water and methanol are poor nucleophiles. From this analysis, we predict that reaction occurs primarily by an S_N1 mechanism. Ionization of the 2° chloroalkane gives a carbocation intermediate, which then reacts with either water or methanol as the nucleophile to give the observed products. Each product is formed as an approximately 50:50 mixture of *R* and *S* enantiomers.

(R)-2-Chlorobutane   **2-Butanol**   **2-Methoxybutane**
(racemic)   (racemic)

Nucleophilic Substitution 2

This is a primary bromoalkane with two beta branches in the presence of a cyanide ion, a good nucleophile. Dimethyl sulfoxide (DMSO), a polar aprotic solvent, is a particularly good solvent in which to carry out nucleophile-assisted substitution reactions because of its good ability to solvate cations (in this case, Na^+), and its poor ability to solvate anions (in this case, CN^-). From this analysis, we predict that this reaction occurs by an S_N2 mechanism.

Nucleophilic Substitution 3

(R)-2-Bromobutane

Bromine is a good leaving group, and it is on a 2° carbon. The methylsulfide ion is a good nucleophile. Acetone, a polar aprotic solvent, is a good medium in which to carry out S_N2 reactions but a poor medium in which to carry out S_N1 reactions. From this analysis, we predict that this reaction occurs by an S_N2 mechanism and that the product is the *S* enantiomer.

(R)-2-Bromobutane   **(S)-2-Methylsulfanylbutane**

Nucleophilic Substitution 4

Ionization of the carbon-bromine bond forms a resonance-stabilized 2° allylic carbocation. Acetic acid is a poor nucleophile, which reduces the likelihood of an S_N2 reaction. Further, acetic acid is a moderately polar protic (hydroxylic) solvent that favors S_N1 reaction. From this analysis, we predict that this reaction occurs by an S_N1 mechanism and both enantiomers of the product are observed.

(R)-3-Bromocyclo-   **(R)-3-Acetoxy-**   **(S)-3-Acetoxy-**
hexene   **cyclohexene**   **cyclohexene**

Br + (C₆H₅)₃P $\xrightarrow{\text{toluene}}$?

The bromoalkane is primary, and bromine is a good leaving group. Trivalent compounds of phosphorus, a third-row element, are moderate nucleophiles. Toluene is a nonpolar aprotic solvent. Given the combination of a primary halide, a good leaving group, a moderate nucleophile, and a nonpolar aprotic solvent, we predict the reaction occurs by an S_N2 pathway.

Br + (C₆H₅)₃P $\xrightarrow{\text{toluene}}$ $\overset{+}{P}$(C₆H₅)₃ Br⁻

Example 9.5

Write the expected substitution product(s) for each reaction and predict the mechanism by which each product is formed.

(a) [cyclopentenyl chloride] + CH₃OH $\xrightarrow{\text{methanol}}$ **(b)** [structure with I] + CH₃CO⁻Na⁺ $\xrightarrow{\text{DMSO}}$

S enantiomer   *R* enantiomer

Solution

(a) This 2° allylic chloride is treated with methanol, a poor nucleophile and a polar protic solvent. Ionization of the carbon-chlorine bond forms a secondary allylic cation that is stabilized by resonance delocalization. Therefore, we predict reaction by an S_N1 mechanism and formation of the product as a roughly racemic mixture.

[cyclopentenyl chloride] + CH₃OH $\xrightarrow[\text{methanol}]{S_N1}$ [OCH₃ product] + [OCH₃ product] + HCl

S enantiomer   *S* enantiomer   *R* enantiomer
                       (~50%)   (~50%)

(b) Iodide is a good leaving group on a moderately accessible secondary carbon. Acetate ion dissolved in a polar aprotic solvent is a moderate nucleophile. We predict substitution by an S_N2 pathway with inversion of configuration at the chiral center.

[structure with I] + CH₃CO⁻Na⁺ $\xrightarrow[\text{DMSO}]{S_N2}$ [OCCH₃ product] + Na⁺I⁻

R enantiomer   *S* enantiomer

Problem 9.5

Write the expected substitution product(s) for each reaction and predict the mechanism by which each product is formed.

(a) [cyclohexane structure with Br] + Na⁺SH⁻ $\xrightarrow{\text{acetone}}$ **(b)** [structure with Cl] + HCOH $\longrightarrow$

(R)-2-Chlorobutane

9.5 | β-Elimination

All nucleophiles have an electron pair that can take part in a reaction, either as a lone pair or sometimes as a π-bond. This means that all nucleophiles are also bases, because any pair of electrons can accept a proton. Hence, chemists are routinely confronted with competing reactions that depend upon a balance between the basicity and nucleophilicity of the reactants we use. As mentioned in the introduction to this chapter, the **β-elimination reaction** is the competing process to substitution that we observe. Viewed in the context of the mechanistic elements described in Section 6.2F, β-elimination is the combination of take a proton away and break a bond to give stable molecules or ions.

Dehydrohalogenation
Removal of —H and —X from adjacent carbons; a type of β-elimination.

Here, we study a type of β-elimination called **dehydrohalogenation**. In the presence of base, halogen is removed from one carbon of a haloalkane and hydrogen is removed from an adjacent carbon to form an alkene.

A haloalkane   Strong base   An alkene

Strong bases promote β-elimination reactions. Strong bases that serve effectively in β-eliminations of haloalkanes are OH⁻, OR⁻, NH₂⁻, and acetylide anions. Following are three examples of base-promoted β-elimination reactions. In the first example, the base is shown as a reactant. In the second and third examples, the base is a reactant but is shown over the reaction arrow. Note that the solvent used is commonly the conjugate acid of the base used in the elimination.

1-Bromodecane   **Potassium**   **1-Decene**
                       ***tert*-butoxide**

2-Bromo-2-   **2-Methyl-2-butene**   **2-Methyl-1-butene**
methylbutane   (major product)

1-Bromo-1-methylcyclopentane   **1-Methylcyclopentene**   **Methylenecyclopentane**
                                    (major product)

In the second and third illustrations, there are two nonequivalent β-carbons, each bearing a hydrogen; therefore, two alkenes are possible. In each case, the major product of these and most other β-elimination reactions is the more substituted (and therefore the more stable) alkene (Section 6.6B). Formation of the more substituted alkene in an elimination is common, but it is not always the outcome. When the more substituted alkene is the dominant product, the reaction is said to follow **Zaitsev's rule**, or to undergo **Zaitsev elimination**.

Zaitsev's rule
A rule stating that the major product of a β-elimination reaction is the most stable alkene; that is, it is the alkene with the greatest number of substituents on the carbon-carbon double bond.

Example 9.6

Predict the β-elimination product(s) formed when each bromoalkane is treated with sodium ethoxide in ethanol. If two or more products might be formed, predict which is the major product.

(a)

Br

(racemic)

(b) Br

Solution

(a) There are two nonequivalent β-carbons in this bromoalkane, and two alkenes are possible. 2-Methyl-2-butene, the more substituted alkene, is the major product.

2-Methyl-2-butene   3-Methyl-1-butene
(major product)

(b) There is only one β-carbon in this bromoalkane, and only one alkene is possible.

3-Methyl-1-butene

Problem 9.6

Predict the β-elimination product(s) formed when each chloroalkane is treated with sodium ethoxide in ethanol. If two or more products might be formed, predict which is the major product.

(a)   (b)   (c)

9.6 | Mechanisms of β-Elimination

There are two limiting mechanisms for β-eliminations. A fundamental difference between them is the timing of the bond-breaking and bond-forming steps. Recall that we made the same statement about the two limiting mechanisms for nucleophilic substitution reactions (Section 9.3).

A. E1 Mechanism

E1

A unimolecular β-elimination reaction.

At one extreme, breaking of the C—Lv bond to give a carbocation is complete before any reaction occurs with the base to lose a hydrogen and form the carbon-carbon double bond. This mechanism is designated an **E1 reaction**, where E stands for elimination and 1 stands for *uni*molecular. One species, in this case the haloalkane, is involved in the rate-determining step. The mechanism of an E1 reaction is illustrated here by the reaction of 2-bromo-2-methylpropane to form 2-methylpropene.

Mechanism   E1 Reaction of 2-Bromo-2-methylpropane

Step 1: **Break a bond to give stable molecules or ions.**   Rate-determining ionization of the C—Br bond gives a carbocation intermediate.

$$CH_3-\underset{\underset{:\ddot{Br}:}{|}}{\overset{\overset{CH_3}{|}}{C}}-CH_3 \xrightarrow[\text{determining}]{\text{slow, rate}} CH_3-\overset{\overset{CH_3}{|}}{\underset{+}{C}}-CH_3 + :\ddot{Br}:^-$$

(A carbocation intermediate)

Step 2: **Take a proton away.**   Proton transfer from the carbocation intermediate to solvent (in this case methanol) gives the alkene.

$$\underset{H_3C}{\overset{H}{\diagdown}}\ddot{O}: + H-CH_2-\overset{\overset{CH_3}{|}}{\underset{+}{C}}-CH_3 \xrightarrow{\text{fast}} \underset{H_3C}{\overset{H}{\diagdown}}\ddot{O}{}^{\pm}H + CH_2=\overset{\overset{CH_3}{|}}{C}-CH_3$$

In an E1 mechanism, one transition state exists for the formation of the carbocation in Step 1 and a second exists for the loss of a hydrogen in Step 2 (Figure 9.6). Formation of the carbocation intermediate in Step 1 crosses the higher energy barrier and is the rate-determining step. This reaction competes with S$_N$1 substitution. E1 and S$_N$1 almost always occur together.

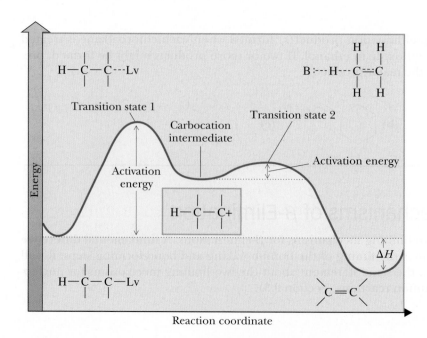

Figure 9.6

An energy diagram for an E1 reaction showing two transition states and one carbocation intermediate.

B. E2 Mechanism

At the other extreme of elimination mechanisms is a concerted process. In an **E2 reaction**, here illustrated by the reaction of 2-bromobutane with sodium ethoxide, proton transfer to the base, formation of the carbon-carbon double bond, and ejection of the bromide ion occur simultaneously; all bond-breaking and bond-forming steps are concerted. Because the base removes a β-hydrogen at the same time that the C—Br bond is broken to form a halide ion, the transition state has considerable double-bond character (Figure 9.7).

E2
A bimolecular β-elimination reaction.

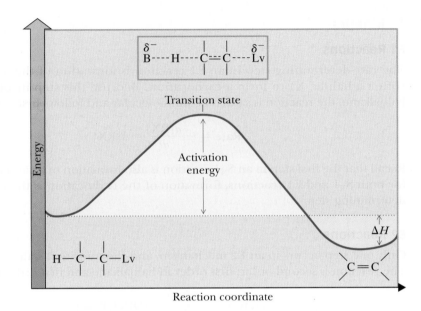

Figure 9.7
An energy diagram for an E2 reaction. There is considerable double-bond character in the transition state.

Mechanism   E2 Reaction of 2-Bromopropane

Take a proton away and simultaneously break a bond to give stable molecules or ions.   Bond breaking and bond forming are concerted; that is, they occur simultaneously.

$$CH_3CH_2\overset{..}{\underset{..}{O}}{:}^- \ + \ H-\underset{\underset{CH_3}{|}}{\overset{\overset{CH_3}{|}}{CH}}-CH-\overset{..}{\underset{..}{Br}}{:} \ \longrightarrow \ CH_3CH_2\overset{..}{\underset{..}{O}}-H \ + \ CH_3CH=CHCH_3 \ + \ {:}\overset{..}{\underset{..}{Br}}{:}^-$$

   This mechanism is designated E2, where E stands for Elimination and 2 stands for Bimolecular; both the haloalkane and the base are involved in the transition state for the rate-determining step.

   Although in principle any base can be made to induce an E2 reaction under appropriate experimental conditions, chemists commonly employ particularly strong bases such as hydroxide, alkoxides, and amide anions (NR_2^-). These bases have conjugate acids with pK_a's above 11. When we use other bases whose conjugate acid pK_a's are near or below 11, such as carboxylates, thiolates, and cyanide (just a few examples), the intention is to effect a substitution reaction via using these reactants as nucleophiles. Therefore, one simplifying aspect of the competition between substitution and elimination is to consider an E2 pathway only when hydroxide, alkoxides, acetylides, and amide anions are used.

9.7 | Experimental Evidence for E1 and E2 Mechanisms

As we examine some of the experimental evidence on which these two contrasting mechanisms are based, we consider the following questions:

1. What are the kinetics of base-promoted β-eliminations?
2. Where two or more alkenes are possible, what factors determine the ratio of the possible products?
3. What is the stereoselectivity?

A. Kinetics

E1 Reactions

The rate-determining step in an E1 reaction is ionization of the leaving group (often a halide, X) to form a carbocation. Because this step involves only the haloalkane, the reaction is said to be unimolecular and follows first-order kinetics.

$$\text{Rate} = -\frac{d[RX]}{dt} = k[RX]$$

Recall that the first step in an S_N1 reaction is also formation of a carbocation. Thus, for both S_N1 and E1 reactions, formation of the carbocation is the first and rate-determining step.

E2 Reactions

Only one step occurs in an E2 mechanism, and the transition state is bimolecular. The reaction is second order: first order in haloalkane and first order in base.

$$\text{rate} = -\frac{d[RX]}{dt} = k[RX][\text{Base}]$$

B. Regioselectivity

E1 Reactions

The major product in E1 reactions is the more stable alkene; that is, the alkene with the more highly substituted carbon-carbon double bond (Zaitsev's rule). After the carbocation is formed in the rate-determining step of an E1 reaction, it may lose a hydrogen to complete β-elimination, or it may rearrange to a more stable carbocation and then lose a hydrogen.

E2 Reactions

For E2 reactions using strong bases and in which the leaving group is a halide ion, the major regioisomeric product is also that formed following Zaitsev's rule, unless special steric relations apply (Section 9.7C). Double-bond character is so highly developed in the transition state that the relative stability of possible alkenes commonly determines which regioisomer is the major product. Thus, the transition state of lowest energy is commonly that leading to the most highly substituted alkene. For similar reasons, *trans* double bonds predominate over *cis* double bonds in the products when either is possible. Note that E2 elimination at a 2° carbon predominates over S_N2 reaction with the strongly basic alkoxide ions.

| 2-Bromohexane (racemic) | $\xrightarrow[\text{CH}_3\text{OH}]{\text{CH}_3\text{O}^-\text{Na}^+}$ | 2-Hexene (74%) (*cis* and *trans*) | + | 1-Hexene (26%) |

With larger, sterically hindered bases such as *tert*-butoxide, however, where isomeric alkenes are possible, the major product is often the less substituted alkene because reaction occurs primarily at the most accessible H atom. Sterically hindered bases like *tert*-butoxide are also noteworthy because the steric hindrance prevents them from reacting as nucleophiles, even with primary alkyl halides.

C. Stereoselectivity

The stereochemistry of E2 reactions is controlled by a conformational effect. The lowest-energy transition state of an E2 reaction is commonly the one in which the —Lv and —H are oriented anti and coplanar (at a dihedral angle of 180°) to one another. The reason for this preferred geometry is that it allows for proper orbital overlap between the base, the proton being removed, and the departing leaving group. Remembering the anti and coplanar geometry requirement is important because it allows prediction of alkene stereochemistry in E2 reactions, namely whether *E* or *Z* products are produced.

—H and —Lv are anti and coplanar
(dihedral angle 180°)

This is shown more clearly in a Newman projection with a bromide as the leaving group:

As with the required backside attack associated with an S_N2 reaction, there is an orbital-based reason for the anti and coplanar arrangement of the —H and —Lv involved in an E2 reaction. The following diagram shows a filled C—H σ bonding molecular orbital aligned with the empty C—Lv σ antibonding molecular orbital.

Filled C—H
σ orbitals

Empty C—Lv
σ^* orbital

As the strong base removes the proton, we consider the two electrons in the C—H orbital filling the antibonding C—Lv orbital and thereby breaking the C—Lv bond. An anti and coplanar arrangement of the C—H and C—Lv leads to proper phasing in the resulting π bond. When the H and Lv are aligned as shown, a collision of the base along the C—H bond leads to the lowest-energy E2 pathway.

For example, treatment of 1,2-dibromo-1,2-diphenylethane with sodium methoxide in methanol gives 1-bromo-1,2-diphenylethylene. The meso isomer of 1,2-dibromo-1,2-diphenylethane gives *(E)*-1-bromo-1,2-diphenylethylene, whereas the racemic mixture of 1,2-dibromo-1,2-diphenylethane gives *(Z)*-1-bromo-1,2-diphenylethylene. We use the anti coplanar requirement of the transition state to account for the stereospecificity of these E2 β-eliminations:

$$C_6H_5\overset{Br}{\underset{|}{C}}H-\overset{Br}{\underset{|}{C}}HC_6H_5 + CH_3O^-Na^+ \xrightarrow[CH_3OH]{E2} \underset{Br}{\overset{C_6H_5}{}}C=C\overset{C_6H_5}{\underset{H}{}} + CH_3OH + Na^+Br^-$$

meso-1,2-Dibromo-1,2-diphenylethane

***(E)*-1-Bromo-1,2-diphenylethylene**

$$C_6H_5\overset{Br}{\underset{|}{C}}H-\overset{Br}{\underset{|}{C}}HC_6H_5 + CH_3O^-Na^+ \xrightarrow[CH_3OH]{E2} \underset{Br}{\overset{C_6H_5}{}}C=C\overset{H}{\underset{C_6H_5}{}} + CH_3OH + Na^+Br^-$$

racemic-1,2-Dibromo-1,2-diphenylethane

***(Z)*-1-Bromo-1,2-diphenylethylene**

Because it is preferred for an E2 reaction that —H and —Lv be anti and coplanar, it is important to identify the reactive conformation of a haloalkane starting material. Following is a stereorepresentation of the meso isomer of 1,2-dibromo-1,2-diphenylethane, drawn to show the plane of symmetry.

Mechanism   E2 Reaction of meso-1,2-Dibromo-1,2-diphenylethane

Take a proton away and simultaneously break a bond to give stable molecules or ions. Clockwise rotation of the left carbon by 60° brings —H and —Br into the required anti and coplanar relationship.

Plane of symmetry

Meso isomer

rotate left carbon by 60°

—H and —Br are now anti and coplanar as required for E2

***(E)*-1-Bromo-1,2-diphenylethylene**

E2 reaction on this conformation gives only the *(E)*-alkene, as shown below in a Newman projection. The —H and —Br are anti and coplanar. All bond-breaking and bond-forming steps are concerted.

***(E)*-1-Bromo-1,2-diphenylethylene**

E2 reaction of either enantiomer of the racemic mixture of 1,2-dibromo-1,2-diphenylethane gives only the *(Z)*-alkene as predicted by analysis of the proper anti and coplanar conformations.

**Take a proton away and simultaneously break a bond to give stable molecules
or ions.** The —H and —Br are anti and coplanar. All bond-breaking and bond-
forming steps are concerted.

The required *trans* and coplanar transition state geometry can also be used to
predict the regiochemistry of E2 elimination in halocyclohexanes such as chloro-
cyclohexanes. In these molecules, anti and coplanar correspond to *trans*, diaxial.
Consider the E2 reaction of the *cis* isomer of 1-chloro-2-isopropylcyclohexane. The
major product is 1-isopropylcyclohexene, the more substituted cycloalkene.

cis-1-**Chloro-2-isopropylcyclohexane**   1-**Isopropylcyclohexene** (*R*)-3-**Isopropylcyclohexene**
                                                   (major product)

Mechanism E2 Reaction of *cis*-1-Chloro-2-isopropylcyclohexane

**Take a proton away and simultaneously break a bond to give stable molecules
or ions.** In the more stable chair conformation of the *cis* isomer, the considerably larger isopropyl group is equatorial and the smaller chlorine is axial. In
this chair conformation, —H on carbon 2 and —Cl on carbon 1 are anti and
coplanar. Concerted E2 elimination gives 1-isopropylcyclohexene, a trisubstituted alkene, as the major product. Note that —H on carbon 6 and —Cl are also
anti and coplanar. Dehydrohalogenation of this combination of —H and —Cl
gives 3-isopropylcyclohexene, a disubstituted, and therefore less stable, alkene.

The formation of the 1-isomer as the major product is in agreement with Zaitsev's rule. However, Zaitsev's rule can be counteracted by the anti coplanar arrangement of —H and —Lv (Example 9.7).

1-Isopropylcyclohexene

The factors favoring E1 or E2 elimination are summarized in Table 9.10.

Table 9.10	Summary of E1 Versus E2 Reactions for Haloalkanes	
Alkyl Halide	**E1**	**E2**
Primary RCH_2X	E1 not observed. Primary carbocations are so unstable that they are never observed in solution.	E2 is favored if elimination is observed. Usually requires sterically hindered strong base.
Secondary R_2CHX	Main reaction with weak bases such as H_2O, ROH.	Main reaction with strong bases such as OH^- and OR^-.
Tertiary R_3CX	Main reaction with weak bases such as H_2O, ROH.	Main reaction with strong bases such as OH^- and OR^-.

Example 9.7

From *trans*-1-chloro-2-isopropylcyclohexane, only 3-isopropylcyclohexene, the less substituted alkene, is formed. Using conformational analysis, explain why this product is observed. Also, will the E2 reaction with *trans*-1-chloro-2-isopropylcyclohexane or *cis*-1-chloro-2-isopropylcyclohexane occur faster under the same basic conditions?

trans-1-Chloro-2-isopropylcyclohexane

(*R*)-3-Isopropylcyclohexene

Solution

Step 1: In the more stable chair conformation of the *trans* isomer, both isopropyl and chlorine are equatorial. In this conformation, the hydrogen atom on carbon 2 is *cis* to the chlorine atom. One of the hydrogen atoms on carbon 6 is *trans* to —Cl, but it is not anti and coplanar. Therefore, the reaction is not favored from this conformation. In the alternative, less stable chair conformation of the *trans* isomer, both isopropyl and chlorine are axial. In this conformation, the axial hydrogen in carbon 6 is anti and coplanar to chlorine, and E2 β-elimination can occur to give 3-isopropylcyclohexene. Thus, even though the diaxial conformation is less stable, the reaction goes through this conformation because it is the only one with an anti-coplanar arrangement of the Cl and a β-H; consequently the non-Zaitsev product is formed.

More stable chair
(no H is anti and
coplanar to Cl)

Less stable chair
(H on carbon 6 is
anti and coplanar to Cl)

Step 2: E2 reaction can take place now that an —H and —Cl are anti and coplanar. This reaction doesn't follow the Zaitsev rule because the mechanism of the reaction requires the anti arrangement.

(R)-3-Isopropylcyclohexene

The rate at which the *cis* isomer undergoes E2 reaction is considerably greater than the rate for the *trans* isomer. We can account for this observation in the following manner. The more stable chair conformation of the *cis* isomer has —H and —Cl anti and coplanar, and the activation energy for the reaction is that required to reach the E2 transition state. The more stable chair conformation of the *trans* isomer, however, cannot undergo anti elimination. To react, it must first be converted to the less stable chair, and the transition state for elimination is correspondingly higher in energy because of the axial isopropyl group.

Problem 9.7

1-Chloro-4-isopropylcyclohexane exists as two stereoisomers: one *cis* and one *trans*. Treatment of either isomer with sodium ethoxide in ethanol gives 4-isopropylcyclohexene by an E2 reaction.

1-Chloro-4-
isopropylcyclohexane

4-Isopropylcyclohexene

The *cis* isomer undergoes E2 reaction several orders of magnitude faster than the *trans* isomer. How do you account for this experimental observation?

Although *much* rarer than an anti and coplanar arrangement of the C—H and C—Lv bonds in an E2 reaction, a syn and coplanar arrangement of these bonds can also lead to E2. Such an arrangement means that the C—H and C—Lv bonds are eclipsed, and therefore only certain constrained ring systems have this geometry. As an example, the following reaction occurs via elimination of H_a rather than H_b because the C—H_a bond is aligned with the C—Cl bond while the C—H_b bond is gauche to the C—Cl bond.

Syn and coplanar

9.8 | Substitution Versus Elimination

Nucleophilic substitution and β-elimination often compete with each other, and the ratio of products formed by these reactions depends on the relative rates of the two reactions. In this section, we consider factors that influence this competition.

A. S$_N$1 Versus E1 Reactions

Reactions of secondary and tertiary haloalkanes in polar protic solvents give mixtures of substitution and elimination products. In both reactions, Step 1 is the formation of a carbocation intermediate. This step is then followed by one or more characteristic carbocation reactions: (1) loss of a hydrogen (E1) to give an alkene, (2) reaction with solvent (S$_N$1) to give a substitution product, or (3) rearrangement followed by reaction (1) or (2). In polar protic solvents, the products formed depend only on the structure of the particular carbocation. For example, *tert*-butyl chloride and *tert*-butyl iodide in 80% aqueous ethanol both react with solvent giving the same mixture of substitution and elimination products. Because iodide ion is a better leaving group than chloride ion, *tert*-butyl iodide reacts over 100 times faster than *tert*-butyl chloride. Yet the ratio of products is the same because the intermediate *tert*-butyl cation is the same.

It is difficult to predict the ratio of substitution to elimination products for first-order reactions of haloalkanes. For the majority of cases, however, S$_N$1 predominates over E1 when weak bases are used.

B. S$_N$2 Versus E2 Reactions

It is considerably easier to predict the ratio of substitution to elimination products for second-order reactions of haloalkanes with reagents that act both as nucleophiles and bases. The guiding principles are:

1. Branching at the α-carbon or β-carbon(s) increases steric hindrance about the α-carbon and significantly retards S$_N$2 reactions. Conversely, branching at the α-carbon or β-carbon(s) increases the rate of E2 reactions because of the increased stability of the alkene product.
2. The greater the nucleophilicity of the attacking reagent, the greater the S$_N$2-to-E2 ratio. Conversely, the greater the basicity of the attacking reagent, the greater the E2-to-S$_N$2 ratio.

Attack of base on a β-hydrogen by E2 is only slightly affected by branching at the α-carbon; alkene formation is accelerated.

S_N2 attack of a nucleophile is impeded by branching at the α- and β-carbons.

A second point involves a relative comparison of nucleophilicity to basicity. It is often difficult to definitively predict in advance if nucleophilicity will or will not outcompete basicity, thereby favoring or not favoring S_N2 versus E2. This competition is particularly important with secondary haloalkane (see below). However, there is a general guideline that is reasonably predictive for secondary haloalkanes. If a nucleophile/base has a conjugate acid with a pK_a below 11 and is a good nucleophile, then an S_N2 reaction will dominate. If the pK_a of the conjugate acid of the nucleophile/base is above 11, the basic character will usually outcompete the nucleophilic character, and an E2 reaction will dominate. A particularly good example of this phenomenon is the comparison between a thiolate anion (RS^-) and an alkoxide anion (RO^-). Thiolates are excellent nucleophiles in polar protic media, and the pK_as of their conjugate acids (thiols, RSH) are in the range of 10 to 12. However, alkoxides are also excellent nucleophiles, but the pK_as of their conjugate acids (alcohols, ROH) are much higher, in the range of 16 to 18. Hence, of a reaction with the same haloalkane in the same polar protic solvent, the percent of S_N2 will be greater for a thiolate nucleophile/base, while the percent of E2 will be greater for the reaction performed with an alkoxide nucleophile/base.

Temperature is another factor that influences the balance between S_N2 and E2 reactions. In general, higher temperatures result in increasing extents of elimination at the expense of substitution. The reason derives from differences in the number of products compared to the number of reactants. Elimination reactions involve the creation of increasing numbers of molecules because a base and an R—Lv react to give the conjugate acid of the base, an alkene, and the free leaving group. In contrast, substitution reactions do not change the number of molecules because a nucleophile and an R—Lv react to give the substituted product and the free leaving group. The more particles formed in a reaction the more entropically favored is that reaction. Hence, because elimination reactions create more particles than substitution reactions, they are more entropically favored, which will be reflected in differences in the energies of the transition states for these two reactions. Recall that $\Delta G = \Delta H - T\Delta S$, and therefore entropy effects become more accentuated at higher temperatures because the $T\Delta S$ term becomes increasingly important. Thus, when two or more reactions are in competition, at higher temperatures the reactions with the more favorable entropies will increase at the expense of those with less favorable entropies. With the competition between S_N2 and E2, the more favorable entropy for elimination results in an increase at higher temperatures.

S_N2

$$Nu{:}^- \ + \ R{-}Lv \longrightarrow Nu{-}R \ + \ Lv{:}^-$$

E2

$$Base{:}^- \ + \ R{-}Lv \longrightarrow Base{-}H \ + \ alkene \ + \ Lv{:}^-$$

C. Putting It All Together

In this chapter, we examined S_N2, S_N1, E2, and E1 mechanisms and how they compete with each other depending upon the alkyl group, the leaving group, the solvent, and the nucleophile. We also examined solvent effects upon nucleophilicity. Nature does not always have clear-cut rules, but here we summarize guidelines that chemists use to predict the outcome of reactions between haloalkanes and various nucleophiles and bases.

Figure 9.8 shows a flow chart that allows you to predict the major product of substitution or elimination reactions. Use the chart as a guide to the following discussion. Alternatively, you can follow the discussion by referring to Table 9.11.

(a) R-group is primary and sterically unhindered: no S_N1 or E1

(b) R-group is secondary

(c) R-group is tertiary: No S_N2

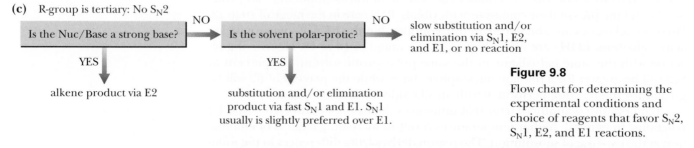

Figure 9.8
Flow chart for determining the experimental conditions and choice of reagents that favor S_N2, S_N1, E2, and E1 reactions.

Table 9.11	Summary of Substitution Versus Elimination Reactions of Haloalkanes	
Halide	**Reaction**	**Comments**
Methyl CH$_3$X	S_N2	S_N1 reactions of methyl halides are never observed. The methyl cation is so unstable that it is not observed in common solvents.
Primary RCH$_2$X	S_N2	The main reaction with good nucleophiles/ weak bases such as I$^-$ and CH$_3$COO$^-$.
	E2	The main reaction with strong, bulky bases such as (CH$_3$)$_3$CO$^-$.
	S_N1/E1	Primary cations are rarely formed in solution, and, therefore, S_N1 and E1 reactions of primary halides are unlikely.
Secondary R$_2$CHX	S_N2	The main reaction with bases/nucleophiles where pK_a of the conjugate acid is 11 or less, as for example I$^-$ and CH$_3$COO$^-$.
	E2	The main reaction with bases/nucleophiles where the pK_a of the conjugate acid is 11 or greater, as for example OH$^-$ and CH$_3$CH$_2$O$^-$.
	S_N1/E1	Common in reactions with weak nucleophiles in polar protic solvents, such as water, methanol, and ethanol.
Tertiary R$_3$CX	E2	Main reaction with strong bases such as HO$^-$ and RO$^-$.
	S_N1/E1	Main reactions with poor nucleophiles/weak bases if the solvent is polar protic.
	S_N2	S_N2 reactions of tertiary halides are never observed because of the extreme crowding around the 3° carbon.

We first classify the haloalkane as (a) primary (RCH_2X), (b) secondary (R_2CHX), or (c) tertiary (R_3CX). We consider 1° carbons that are not sterically hindered in Part (a) because this covers most 1° cases. If the 1° carbon is sterically hindered, such as that in the neopentyl group (R = $(CH_3)_3CCH_2X$, recall Section 9.3B), we treat it as if it were a secondary carbon (b) that cannot undergo an elimination reaction. The flow chart does not show what happens if the alkyl group is methyl (CH_3X), because the only possible outcome is an S_N2 reaction irrespective of the structure of the nucleophile, the leaving group, and the solvent (Table 9.11 puts CH_3X at the top). Recall that methyl cations are too unstable to form (S_N1 is ruled out) and there is only one carbon, meaning that elimination to create a double bond is impossible (E1 and E2 are ruled out).

We next examine the structure of the nucleophile. Because all nucleophiles are bases, we refer to them as Nuc/Base.

(a) Primary alkyl groups:

- Because primary carbocations are too unstable to form, S_N1 or E1 mechanisms are not possible.
- If the Nuc/Base is a strong base and sterically hindered, it will *not* be a good nucleophile and E2 is the major pathway. A common example is the use of *tert*-butoxide ion as the Nuc/Base. Amide anions are exceptions. Although they are not hindered, they are so basic that E2 dominates.
- If the Nuc/Base is a strong base and not sterically hindered, we next consider if it is a good nucleophile. Examples of strong bases that are also good nucleophiles are hydroxide, acetylide, and methoxide. Weak bases are defined as those that have conjugate acids with pK_a's below 11. Examples of weak bases that are good nucleophiles are thiolate (RS^-), cyanide (NC^-), iodide (I^-), and azide (N_3^-) anions. Even moderate nucleophiles that are weak bases, such as unhindered amines (NR_3) and phosphines (PR_3), participate in efficient S_N2 reactions. Hence, with any of the basic or weakly basic good-to-moderate nucleophiles, we find products predominantly from S_N2 pathways. However, some accompanying E2 mechanism is likely with strong bases.
- Finally, if the Nuc/Base is neither a good nor moderate nucleophile, we are likely to get S_N2 and E2 in a ratio that is difficult to predict, or to get no reaction. Examples are water, alcohols, and carboxylic acids.

(b) Secondary alkyl groups:

- If the Nuc/Base is a strong base, whether or not it is hindered, E2 will dominate. Strong bases are defined as those that have conjugate acids with pK_a's above 11, such as hydroxide, alkoxides, acetylides, and H_2N^-.
- With a weak base that is a good to moderate nucleophile, S_N2 will dominate. Examples are those nucleophiles that have conjugate acids with pK_a's below 11. However, because the alkyl group is secondary, the S_N2 reaction may be sluggish, and potentially S_N1 and E2/E1 elimination pathways may compete to a small extent.
- When the nucleophile is not good, we need to examine the solvent. In a polar protic solvent, often with gentle warming, we can induce S_N1 and E1 pathways. The extent that substitution or elimination occur is hard to predict.
- Finally, if the solvent is neither polar nor protic, and the Nuc/Base is neither a strong base nor a good nucleophile, all four reaction pathways are possible, and it is difficult to predict which will dominate, or whether a reaction will occur at all.

(c) Tertiary alkyl groups:

- We start by noting that an S_N2 mechanism cannot occur on a tertiary alkyl group.
- If the Nuc/Base is a strong base, E2 will dominate.
- Because S_N2 is not possible, we do not have to consider if the Nuc/Base is a strong or weak nucleophile. Hence, this question is not relevant to predicting the dominant reaction pathway. Therefore, instead we check the solvent.

- If the solvent is polar and protic, we can induce S_N1 and E1 pathways, often by applying heat. Whether substitution or elimination dominates is hard to predict.
- Finally, if the Nuc/Base is not a strong base, and the solvent is not polar and protic, S_N1, E2, and E1 reaction pathways are possible, and it is difficult to predict which will dominate, or whether a reaction will occur at all.

9.9 | Analysis of Several Competitions Between Substitutions and Eliminations

Now that we have outlined a step-by-step process with which to make predictions about relative extents of S_N2, S_N1, E2, and E1 competing mechanism, we'll examine some specific examples in the same manner as we did in Section 9.4. Following are five examples of reactions between a haloalkane and a Nuc/Base in specific solvents, along with an analysis of the factors that favor the various mechanistic pathways. We start with examples that have clear-cut predictions and move to those that are more challenging. In some cases. we also cover predictions for the reactions under slightly different experimental conditions.

Competition 1

The haloalkane is primary, and therefore S_N1 and E1 cannot occur. The Nuc/Base is a strong base because the pK_a of its conjugate acid is several units above 11 (pK_a HOEt = 15.9). But, due to its basicity and because it is not sterically hindered, ethoxide is also a good nucleophile. Hence, S_N2 will dominate over E2.

Competition 2

The haloalkane is tertiary, and therefore S_N2 cannot occur. The Nuc/Base is a weak base. The solvent is polar protic, and therefore S_N1 and E1 mechanisms will occur. S_N1 or E2 will generally dominate.

If this same reaction were performed with NaOEt in the ethanol, E2 would have been the dominant pathway because ethoxide is a strong base.

Competition 3

The haloalkane is secondary. Because the Nuc/Base has a pK_a of its conjugate acid far below 11 (pK_a of HOAc = 4.7), it is a weak base, and hence there will be little to no E2. However, acetate is a moderate nucleophile. Hence, S_N2 is the best prediction.

Because the solvent is polar protic, there could be a minor extent of $S_N1/E1$. If the sodium acetate were left out of the reaction and it were heated, the prediction would be $S_N1/E1$.

Competition 4

The haloalkane is secondary. The Nuc/Base has a pK_a of its conjugate acid near or slightly below 11 (HCN, p$K_a = 9.3$), and hence is a moderate to weak base. However, cyanide anion is an excellent nucleophile. Consequently, S_N2 will dominate over E2. Furthermore, the solvent DMF (dimethylforamide) is polar and aprotic and also supports S_N2 or E2, but it does not assist S_N1 or E1. Because the reactant is chiral, the S_N2 inverts the configuration.

Competition 5

The haloalkane is tertiary, and therefore S_N2 cannot occur. The Nuc/Base is a weak base (pK_a HN$_3$ = 4.9), and therefore E2 is not obvious. However, the solvent is not protic but is simply polar. Therefore, S_N1 and E1 are not going to be favored. This is a case that is difficult to predict using Figure 9.8 or Table 9.11.

However, the lack of a polar protic solvent means that E2 is most likely, even with the weak base. The E2 occurs with an anti and coplanar arrangement of the Br and H that are eliminated, giving an *E* alkene. Any substitution from an S_N1 pathway would lead to racemization of the chiral center that possessed the leaving group.

Example 9.8

Predict whether each reaction proceeds predominantly by substitution (S_N1 or S_N2) or elimination (E1 or E2) or whether the two compete. Write structural formulas for the major organic product(s).

Solution

(a) A 3° haloalkane is heated with a strong base/good nucleophile. Elimination by an E2 reaction predominates to give 2-methyl-2-butene as the major product.

(b) Reaction of a 1° haloalkane with this moderate nucleophile/weak base gives substitution by an S_N2 reaction.

$$\text{(isopentyl)–Br} + (C_2H_5)_3N \xrightarrow[\text{CH}_2\text{Cl}_2]{30°C} \text{(isopentyl)–}\overset{+}{N}(C_2H_5)_3 \; Br^-$$

Problem 9.8

Predict whether each reaction proceeds predominantly by substitution (S_N1 or S_N2) or elimination (E1 or E2) or whether the two compete. Write structural formulas for the major organic product(s).

(a)

$$\underset{\text{Br}}{\text{(sec-alkyl bromide)}} + CH_3O^-Na^+ \xrightarrow[\text{methanol}]{}$$

(b)

$$\underset{}{\text{(cyclohexyl with Cl and CH}_3\text{)}} + Na^+I^- \xrightarrow[\text{acetone}]{}$$

(c)

$$C_6H_5\underset{}{\diagup}\text{Br} + Na^+CN^- \xrightarrow[\text{methanol}]{}$$

An important take-home lesson from this chapter is that understanding key transition state or reactive intermediate geometries as well as relative transition state energies allows the prediction of product stereochemistry and regiochemistry. Backside attack in S_N2 reactions, the anti and coplanar geometry of the H atom and leaving group in E2 reactions, and the presence of carbocation intermediates in S_N1 reactions are important examples of reaction geometries that dictate stereochemistry. Understanding the relative energies of alternative possible transition states is also important. In the case of β-elimination reactions, relative transition state energies provide a rationale for Zaitsev's rule of regiochemistry. As you go through the rest of this book, try to learn key features of reaction mechanisms that dictate the stereochemistry and regiochemistry of reaction products. You should think of mechanisms as more than just electron pushing: they involve three-dimensional molecular interactions with associated relative energies that control the formation of products.

9.10 | Neighboring Group Participation

So far, we have considered two limiting mechanisms for nucleophilic substitutions that focus on the degree of covalent bonding between the nucleophile and the substitution center during departure of the leaving group. In an S_N2 mechanism, the leaving group is assisted in its departure by the nucleophile. In an S_N1 mechanism, the leaving group is not assisted in this way. An essential criterion for distinguishing between these two pathways is the order of reaction. Nucleophile-assisted substitutions are second order: first order in RX and first order in nucleophile. Nucleophile-unassisted substitutions are first order: first order in RX and zero order in nucleophile.

Chemists recognize that certain nucleophilic substitutions have the kinetic characteristics of first-order (S_N1) substitution but, in fact, involve two successive displacement reactions. A characteristic feature of a great many of these reactions is the presence of an internal nucleophile, most commonly sulfur, nitrogen, or oxygen, on the carbon atom beta to the leaving group. This neighboring nucleophile participates in the departure of the leaving group to give an intermediate, which then reacts with an external nucleophile to complete the reaction.

The mustard gases are one group of compounds that react by participation of a neighboring group. The characteristic structural feature of a mustard gas is a two-carbon chain, with a halogen on one carbon and a divalent sulfur or trivalent nitrogen on the other carbon (S-C-C-Lv or N-C-C-Lv). An example of a mustard gas is *bis*(2-chloroethyl)sulfide, a poison gas used extensively in World War I and at one time, at least, manufactured by Iraq. This compound is a deadly vesicant (blistering agent) and quickly causes conjunctivitis and blindness.

ClCH$_2$CH$_2$SCH$_2$CH$_2$Cl

bis(2-Chloroethyl)sulfide
(a sulfur mustard gas)

$$\overset{\displaystyle CH_3}{\underset{\displaystyle |}{ClCH_2CH_2-N-CH_2CH_2Cl}}$$

bis(2-Chloroethyl)methylamine
(a nitrogen mustard gas)

Bis(2-chloroethyl)sulfide and *bis*(2-chloroethyl)methylamine are not gases at all. They are oily liquids with a high vapor pressure, hence the designation "gas." Nitrogen and sulfur mustards react very rapidly with moisture in the air and in the mucous membranes of the eye, nose, and throat to produce HCl, which then burns and blisters these sensitive tissues. What is unusual about the reactivity of the mustard gases is that they react very rapidly with water, a very poor nucleophile.

Mustard gases also react rapidly with other nucleophiles, such as those in biological molecules, which makes them particularly dangerous chemicals. Of the two steps in the mechanism of the hydrolysis of a sulfur mustard, the first is the slower and is rate-determining. As a result, the rate of reaction is proportional to the concentration of the sulfur mustard but independent of the concentration of the external nucleophile. Thus, although this reaction has the kinetic characteristics of an S$_N$1 reaction, it actually involves two successive S$_N$2 displacement reactions.

Mechanism Hydrolysis of a Sulfur Mustard—Participation by a Neighboring Group

***Step 1:* Make a new bond between a nucleophile and an electrophile and simultaneously break a bond to give stable molecules or ions.** The reason for the extremely rapid hydrolysis of the sulfur mustards is neighboring group participation by sulfur in the ionization of the carbon-chlorine bond to form a cyclic sulfonium ion. This is the rate-determining step of the reaction; although it is the slowest step, it is much faster than reaction of a typical primary chloroalkane with water. At this point, you should review halogenation of alkenes (Sections 6.3D and 6.3F) and compare the cyclic halonium ions formed there with the cyclic sulfonium ion formed here.

A cyclic
sulfonium ion

***Step 2:* Make a new bond between a nucleophile and an electrophile.** The cyclic sulfonium ion contains a highly strained three-membered ring, and reacts rapidly with an external nucleophile to open the ring followed by proton transfer to H$_2$O to give H$_3$O$^+$. In this S$_N$2 reaction, H$_2$O is the nucleophile and sulfur is the leaving group.

Step 3: Take a proton away. Proton transfer to water completes the reaction.

The net effect of these reactions is nucleophilic substitution of Cl by OH.

We continue to use the terms S_N2 and S_N1 to describe nucleophilic substitution reactions. You should realize, however, that these designations do not adequately describe all nucleophilic substitution reactions.

Example 9.9

Write a mechanism for the hydrolysis of the nitrogen mustard *bis*(2-chloroethyl) methylamine.

Solution

Following is a three-step mechanism.

Step 1: Make a new bond between a nucleophile and an electrophile and simultaneously break a bond to give stable molecules or ions An internal S_N2 reaction in which ionization of the C—Cl bond is assisted by the neighboring nitrogen atom to form a highly strained three-membered ring.

Step 2: Make a new bond between a nucleophile and an electrophile Reaction of the cyclic ammonium ion with water opens the three-membered ring. In this S_N2 reaction, H_2O is the nucleophile and nitrogen is the leaving group.

Step 3: Take a proton away Proton transfer to the basic nitrogen completes the reaction.

Problem 9.9

Knowing what you do about the regioselectivity of S_N2 reactions, predict the product of hydrolysis of this compound.

Autopsies of soldiers killed by sulfur mustards in World War I revealed, among other things, very low white blood cell counts and defects in bone marrow development. From these observations, it was realized that sulfur mustards have profound effects on rapidly dividing cells. This became a lead observation in the search

nucleophilicity, but the resulting compound was not sufficiently soluble in water for intravenous injection. The solubility problem was solved by adding a carboxyl group. When the carboxyl group was added directly to the aromatic ring, however, the resulting compound was too stable and, therefore, not biologically active.

Mechlorethamine   (An aziridinium ion intermediate)

for less toxic alkylating agents for use in treatment of cancers, which have rapidly dividing cells. Attention turned to the less reactive nitrogen mustards. One of the first compounds tested was mechlorethamine. As with other mustards, the reaction of mechlorethamine with nucleophiles is rapid because of the formation of an aziridinium ion.

Adding a propyl bridge (chlorambucil) or an aminoethyl bridge (melphalan) between the aromatic ring and the carboxyl group solved both the solubility problem and the reactivity problem. Note that melphalan is chiral. It has been demonstrated that the R and S enantiomers have approximately equal therapeutic potency.

Nucleophilicity of nitrogen acceptable, but the compound is too insoluble in water for intravenous injection

Solubility in water is acceptable, but nucleophilicity of nitrogen reduced and compound is unreactive

Mechlorethamine undergoes very rapid reaction with water (hydrolysis) and with other nucleophiles, so much so that within minutes after injection into the body, it has completely reacted. The problem for the chemist, then, was to find a way to decrease the nucleophilicity of nitrogen while maintaining reasonable water solubility. Substitution of phenyl for methyl reduced the

The clinical value of the nitrogen mustards lies in the fact that they undergo reaction with certain nucleophilic sites on the heterocyclic aromatic amine bases in DNA (see Chapter 28). For DNA, the most reactive nucleophilic site is N-7 of guanine. Next in reactivity is N-3 of adenine, followed by N-3 of cytosine.

Chlorambucil   **Melphalan**
(racemic)

The nitrogen mustards are bifunctional alkylating agents; one molecule of nitrogen mustard undergoes reaction with two molecules of nucleophile. Guanine alkylation leaves one free reactive alkylating group, which can react with another base, giving cross links that lead to miscoding during DNA replication. The therapeutic value of the nitrogen mustards lies in their ability to disrupt normal base pairing. This prevents replication of the cells, and the rapidly dividing cancer cells are more sensitive than normal cells.

Summary

Problem: 9.1

Section 9.1 Nucleophilic Substitution in Haloalkanes

- **Nucleophilic substitution** is any reaction in which a nucleophile replaces another electron-rich group called a **leaving group (Lv)**.
  – A nucleophile (Nu:$^-$) is an electron-rich molecule or ion that donates a pair of electrons to another atom or ion to form a new covalent bond.

Section 9.2 Mechanisms of Nucleophilic Aliphatic Substitution

- There are two limiting mechanisms for nucleophilic substitution, namely **S$_N$2** and **S$_N$1**.
  – In the S$_N$2 reaction mechanism, bond forming and bond breaking occur simultaneously.
    – S$_N$2 reactions are bimolecular because both nucleophile and haloalkane concentrations influence reaction rate.
    – The nucleophile must approach the carbon-leaving group (C—Lv) bond from the backside in order to populate the C—Lv antibonding orbital and allow reaction.
  – In the S$_N$1 mechanism, the leaving group departs first in the rate-determining step, leaving a carbocation intermediate that reacts with the nucleophile in a second step.
    – S$_N$1 reactions are unimolecular because only the haloalkane concentration influences reaction rate.

Section 9.3 Experimental Evidence for S$_N$1 and S$_N$2 Mechanisms

- The S$_N$2 reaction can be identified based on kinetics of the reaction and stereochemistry of the products.
  – Because an S$_N$2 reaction is bimolecular, doubling the concentration of either haloalkane or nucleophile will double the rate of the reaction.
  – Because backside attack geometry is required, reaction at a chiral center results in inversion of configuration.
- The S$_N$1 reaction can also be identified based on kinetics of the reaction and stereochemistry of the products.
  – Because an S$_N$1 reaction is unimolecular, doubling the concentration of only the haloalkane can double the rate of the reaction.

- Because a planar and achiral carbocation intermediate is formed that can be attacked with roughly equal probability from either face, reaction at a chiral center results in racemization of stereochemistry.
  - The chiral center is often not completely racemized because the leaving group forms an ion pair with the carbocation intermediate, partially blocking one face.
- The structure of the haloalkane influences the reaction rate and mechanism.
  - Haloalkanes that can form more stable carbocations react faster if an S_N1 mechanism occurs.
    - Because S_N1 reactions involve carbocations, rearrangements (1,2 shifts) can occur if they lead to more stable carbocation intermediates.
  - Steric hindrance on the backside of the C—Lv bond of a haloalkane slows down or possibly prevents an S_N2 mechanism.
- The more stable the anion produced upon reaction, the better the **leaving group ability**.
- **Solvent** properties can have an important influence on reaction mechanisms.
  - **Protic solvents** are hydrogen-bond donors. The most common protic solvents are those containing —OH groups.
  - **Aprotic solvents** cannot serve as hydrogen-bond donors. Common aprotic solvents are acetone, diethyl ether, dimethyl sulfoxide, and N,N-dimethylformamide.
  - **Polar solvents** interact strongly with ions and polar molecules.
  - **Nonpolar solvents** do not interact strongly with ions and polar molecules.
  - The **dielectric constant** is the most commonly used measure of solvent polarity.
  - **Solvolysis** is a nucleophilic substitution reaction in which the solvent is the nucleophile.
  - **Polar protic solvents** accelerate S_N1 reactions by stabilizing the charged carbocation intermediate.
  - **Polar aprotic solvents** accelerate S_N2 reactions because they do not interact strongly with the nucleophile.
- **Nucleophiles** are categorized as good, moderate, or poor.
  - **Good nucleophiles** are generally anions. **Moderate nucleophiles** are generally neutral, with one or more available lone pairs. **Poor nucleophiles** are generally polar protic solvents.
  - All things being equal, the stronger the interaction of a nucleophile with solvent, the lower the nucleophilicity.
  - Small nucleophiles with very little steric hindrance are better nucleophiles for S_N2 reactions.

Problems: 9.2–9.4, 9.10–9.13, 9.15–9.22, 9.24–9.36

Section 9.4 Analysis of Several Nucleophilic Substitution Reactions

- Methyl or primary haloalkanes react through S_N2 mechanisms because of an absence of steric hindrance and lack of carbocation stability.
- Secondary haloalkanes react through an S_N2 mechanism in aprotic solvents with good nucleophiles, but through an S_N1 mechanism in protic solvents with poor nucleophiles. E1 is usually less when S_N1 occurs.
- Tertiary haloalkanes react through an S_N1 mechanism because the steric hindrance disfavors S_N2 backside attack, and the attached alkyl groups stabilize a carbocation.

Problems: 9.5, 9.14, 9.23

Section 9.5 β-Elimination

- A **β-elimination reaction** involves removal of atoms or groups of atoms from adjacent carbon atoms.
  - **Dehydrohalogenation** is a β-elimination reaction that involves loss of an H and a halogen atom from adjacent carbons to create an alkene from a haloalkane.
  - β-Elimination to give the more highly substituted alkene is called **Zaitsev elimination**.

Problem: 9.6

Section 9.6 Mechanisms of β-Elimination

- The two limiting mechanisms for β-elimination reactions are the **E1** and **E2** mechanisms.

- In the E1 mechanism, the leaving group departs to give a carbocation, then a proton is taken off an adjacent carbon atom by base to create the product alkene.
    - E1 reactions are unimolecular because only the haloalkane concentration influences the rate of the reaction.
- In the E2 mechanism, the halogen departs at the same time that an H atom is removed by base from an adjacent carbon atom to create the product alkene.
    - E2 reactions are bimolecular because both the haloalkane and base concentrations influence the rate of the reaction.

Section 9.7 Experimental Evidence for E1 and E2 Mechanisms

- E2 reactions are stereoselective in that the lowest energy transition state is the one in which the leaving group and H atoms that depart are oriented anti and coplanar.
    - This anti and coplanar requirement determines whether E or Z alkenes are produced. For cyclohexane derivatives, both the leaving group and departing H atom must be axial.
- Both E1 and E2 reactions are regioselective, favoring formation of the more stable (Zaitsev) product alkene (as long as Lv and H can be oriented anti and coplanar).
    - The more stable alkene is generally the more highly substituted alkene.

Problems: 9.7, 9.37–9.42

Section 9.8 Substitution Versus Elimination

- When deciding which substitution or elimination mechanism dominates a reaction, analyze the structure of the haloalkane, the choice of the solvent, and the relative base strength of the nucleophile.
- Methyl or primary haloalkanes do not react through E1 or S_N1 mechanisms.
    - S_N2 is favored for all nucleophiles except for exceptionally strong bases (H_2N^-) or sterically hindered ones (*tert*-butoxide), which cause E2 to predominate.
- Secondary haloalkanes can react through any of the mechanisms.
    - If the nucleophile is a strong base (conjugate acids with pK_a's above 11 such as hydroxide, alkoxides, acetylides, and H_2N^-), E2 predominates.
    - Weak bases (conjugate acids with pK_a's below 11) that are good or moderate nucleophiles (see Table 9.7) react predominantly by an S_N2 mechanism.
    - Poor nucleophiles (that are polar protic solvents) react through a combination $S_N1/E1$ pathways, the exact ratio of which is hard to predict.
- Tertiary haloalkanes cannot react by an S_N2 mechanism.
    - If the nucleophile is a strong base (conjugate acids with pK_a's above 11 such as hydroxide, alkoxides, acetylides, and H_2N^-), E2 predominates.
    - For other nucleophiles in a polar protic solvent, reaction is through a combination $S_N1/E1$ pathways, the exact ratio of which is hard to predict.

Problems: 9.8, 9.43–9.62

Section 9.9 Analysis of Several Competitions Between Substitutions and Eliminations

- Predicting whether substitution or elimination reactions will dominate is a matter of following the logic given in Section 9.8.
    - Either Table 9.11 or the flowchart given in Figure 9.8 (summarized just above) will lead to a successful analysis of the majority of reactions that organic chemists perform.

Section 9.10 Neighboring Group Participation

- Certain nucleophilic displacements that have the kinetic characteristic of S_N1 reactions (first order in haloalkane and zero order in nucleophile) involve two successive S_N2 reactions.
    - Many such reactions involve participation of a neighboring nucleophile.
    - The mustard gases are one group of compounds whose nucleophilic substitution reactions involve neighboring group participation.

Problem: 9.9

☞WL **Questions and problems for this chapter are available in OWL. Use the OWL access card that came with this textbook to access assigned questions and problems for this chapter.**

1. **Nucleophilic Aliphatic Substitution: S$_N$2 (Section 9.3)**   S$_N$2 reactions occur in one step; departure of the leaving group is assisted by the incoming nucleophile, and both nucleophile and leaving group are involved in the transition state. The nucleophile may be negatively charged as in the first example or neutral as in the second example.

S$_N$2 reactions result in inversion of configuration at the reaction center. They are accelerated more in polar aprotic solvents than polar protic solvents. The relative rates of S$_N$2 reactions are governed by steric factors, namely the degree of crowding around the site of reaction.

2. **Nucleophilic Aliphatic Substitution: S$_N$1 (Section 9.3)**   An S$_N$1 reaction occurs in two steps. Step 1 is a slow, rate-determining ionization of the C—Lv bond to form a carbocation intermediate followed in Step 2 by rapid reaction of the carbocation intermediate with a nucleophile to complete the substitution. Reaction at a chiral center gives largely racemization, often accompanied with a slight excess of inversion of configuration. Reactions often involve carbocation rearrangements and are accelerated by polar protic solvents. S$_N$1 reactions are governed by electronic factors, namely the relative stabilities of carbocation intermediates. The following reaction involves an S$_N$1 reaction, with a hydride shift.

3. **β-Elimination: E1 (Sections 9.6, 9.7)**   An E1 reaction occurs in two steps: slow, rate-determining breaking of the C—Lv bond to form a carbocation intermediate followed by rapid proton transfer to solvent to form an alkene. An E1 reaction is first order in haloalkane and zero order in base. Skeletal rearrangements are common.

4. **β-Elimination: E2 (Sections 9.6, 9.7)**   An E2 reaction occurs in one step: simultaneous reaction with base to remove a hydrogen, formation of the alkene, and departure of the leaving group. Elimination is stereoselective, requiring an anti and coplanar arrangement of the groups being eliminated.

5. **Neighboring Group Participation (Section 9.10)**   Neighboring group participation is characterized by first-order kinetics and participation of an internal nucleophile in departure of the leaving group, as in hydrolysis of a sulfur or nitrogen mustard gas. The mechanism for their solvolysis involves two successive nucleophilic displacements.

Chapter 10
Alcohols

Fermentation vats of wine grapes at the Beaulieu Vineyards, California. *Inset:* a model of ethanol.

© Cephas Picture Library/Alamy

In this chapter, we study the physical and chemical properties of alcohols, a class of compounds containing the —OH (hydroxyl) group. We also study thiols, a class of compounds containing the —SH (sulfhydryl) group.

$$CH_3CH_2OH \qquad CH_3CH_2SH$$

Ethanol   **Ethanethiol**
(an alcohol)   (a thiol)

Ethanol is the additive in E85, the alcohol in alcoholic beverages, and an important industrial solvent. Ethanethiol, like all other low-molecular-weight thiols, has a stench; such smells from skunks, rotten eggs, and sewage are caused by thiols or H_2S.

Alcohols are important because they can be converted into many other types of compounds, including alkenes, haloalkanes, aldehydes, ketones, carboxylic acids, and esters. Not only can alcohols be converted to these compounds, but these compounds can also be converted to alcohols. Thus, alcohols play a central role in the interconversion of organic functional groups.

Hydroxyl groups are found in carbohydrates and certain amino acids. Following are two representations for glucose, the most abundant organic compound in nature. On the left is a Fischer projection showing the configuration of all chiral centers. On the right is a cyclic structure, the predominant form in which this molecule exists in both the solid form and in solution. The amino acid L-serine is one of the 20 amino acid building blocks of proteins.

🦉WL **Online homework for this chapter may be assigned in OWL for Organic Chemistry.**

D-Glucose
(Fischer projection)

β-D-Glucopyranose

L-Serine

Because sulfur and oxygen are both Group 6 elements, thiols and alcohols undergo many of the same types of reactions. Sulfur, a third-row element, however, can undergo some reactions that are not possible for alcohols. In addition, sulfur's electronegativity and basicity are less than those of oxygen.

10.1 | Structure and Nomenclature of Alcohols

A. Structure

The functional group of an alcohol is an **—OH (hydroxyl) group** (Section 1.3A) bonded to an sp^3-hybridized carbon. The oxygen atom of an alcohol is also sp^3 hybridized. Two sp^3 hybrid orbitals of oxygen form σ bonds to atoms of carbon and hydrogen, and the remaining two sp^3 hybrid orbitals each contain an unshared pair of electrons. Figure 10.1 shows a Lewis structure and a ball-and-stick model of methanol, CH_3OH, the simplest alcohol. The measured C—O—H bond angle in methanol is 108.9°, very close to the perfectly tetrahedral angle of 109.5°.

(a)

(b)

Figure 10.1
Methanol, CH_3OH. **(a)** Lewis structure and **(b)** ball-and-stick model.

B. Nomenclature

In the IUPAC system, the longest chain of carbon atoms containing the —OH group is selected as the parent alkane and numbered from the end closer to —OH. To show that the compound is an alcohol, change the suffix *-e* of the parent alkane to *-ol* (Section 2.3), and use a number to show the location of the —OH group. The location of the —OH group takes precedence over alkyl groups and halogen atoms in numbering the parent chain. For cyclic alcohols, numbering begins with the carbon bearing the —OH group. Because the —OH group is understood to be on carbon 1 of the ring, there is no need to give its location a number. In complex alcohols, the number for the hydroxyl group is often placed between the infix and the suffix. Thus, for example, 2-methyl-1-propanol and 2-methylpropan-1-ol are both acceptable names.

Common names for alcohols are derived by naming the alkyl group bonded to —OH and then adding the word *alcohol*. Here are IUPAC names and, in parentheses, common names for several low-molecular-weight alcohols.

1-Propanol
(Propyl alcohol)

2-Propanol
(Isopropyl alcohol)

1-Butanol
(Butyl alcohol)

(S)-2-Butanol
((S)-*sec*-Butyl alcohol)

2-Methyl-1-propanol
(Isobutyl alcohol)

2-Methyl-2-propanol
(*tert*-Butyl alcohol)

Example 10.1

Write IUPAC names for these alcohols.

(a) (b)

Solution

(a) (R)-4-Methyl-2-pentanol.

(b) (1R,2R)-2-Methylcyclohexanol. Note that the designation of the configuration as R,R specifies not only the absolute configuration of each chiral center but also that the —CH₃ and —OH groups are *trans* to each other on the ring. The alcohol can also be named *trans*-2-methylcyclohexanol and, while this name specifies that the hydroxyl and methyl groups are *trans* to each other, it does not specify the absolute configuration of either group.

Problem 10.1

Write IUPAC names for these alcohols, and include the configuration for (a).

(a) (b)

We classify alcohols as **primary (1°), secondary (2°), or tertiary (3°)**, depending on whether the —OH group is on a primary, secondary, or tertiary carbon.

$$R{-}\underset{\underset{H}{|}}{\overset{\overset{H}{|}}{C}}{-}OH \qquad R{-}\underset{\underset{H}{|}}{\overset{\overset{R'}{|}}{C}}{-}OH \qquad R{-}\underset{\underset{R''}{|}}{\overset{\overset{R'}{|}}{C}}{-}OH$$

Primary (1°)   Secondary (2°)   Tertiary (3°)

Example 10.2

Classify each alcohol as primary, secondary, or tertiary.

(a) (b) $CH_3\underset{\underset{CH_3}{|}}{\overset{\overset{CH_3}{|}}{C}}OH$ (c)

Solution

(a) Secondary (2°)   (b) Tertiary (3°)   (c) Primary (1°)

Problem 10.2

Classify each alcohol as primary, secondary, or tertiary.

(a) (b) (c) (d)

Diol
A compound containing two hydroxyl groups.

Triol
A compound containing three hydroxyl groups.

In the IUPAC system, a compound containing two hydroxyl groups is named as a **diol**, one containing three hydroxyl groups as a **triol**, and so on. In IUPAC names for diols, triols, and so on, the final -*e* (the suffix) of the parent alkane name is retained, as for example in the name 1,2-ethanediol. As with many organic compounds, common names for certain diols and triols have persisted. Compounds containing hydroxyl groups on adjacent carbons are often referred to as **glycols** (Section 6.5). Ethylene glycol and propylene glycol are synthesized from ethylene and propylene, respectively, hence their common names.

1,2-Ethanediol
(Ethylene glycol)

1,2-Propanediol
(Propylene glycol)

1,3-Propanediol

1,2,3-Propanetriol
(Glycerol, glycerine)

Compounds containing —OH and C=C groups are often referred to as **unsaturated alcohols** because of the presence of the carbon-carbon double bond. In the IUPAC system, the double bond is shown by changing the infix of the parent alkane from -*an*- to -*en*- (Section 2.3), and the hydroxyl group is shown by changing the suffix of the parent alkane from -*e* to -*ol*. Numbers must be used to show the location of both the carbon-carbon double bond and the hydroxyl group. The parent alkane is numbered to give the —OH group the lowest possible number; that is, the group shown by a suffix (in this case, -*ol*) takes precedence over the group shown by an infix (in this case, -*en*-).

Example 10.3

Write IUPAC names for these unsaturated alcohols.

(a) CH$_2$=CHCH$_2$OH   (b) HO

Solution

(a) 2-Propen-1-ol. Its common name is allyl alcohol.
(b) (*E*)-2-Hexen-1-ol (*trans*-2-Hexen-1-ol)

Problem 10.3

Write IUPAC names for these unsaturated alcohols.

(a)   OH   (b)

Dipole-dipole interaction
The attraction between the positive end of one dipole and the negative end of another.

Hydrogen bonding
The attractive interaction between a hydrogen atom bonded to an atom of high electronegativity (most commonly O or N) and a lone pair of electrons on another atom of high electronegativity (again, most commonly O or N).

10.2 | Physical Properties of Alcohols

Because of the presence of the polar —OH group, alcohols are polar compounds, with partial positive charges on carbon and hydrogen and a partial negative charge on oxygen (Figure 10.2).

The attraction between the positive end of one dipole and the negative end of another is called **dipole-dipole interaction**. When the positive end of one of the dipoles is a hydrogen atom bonded to O or N (atoms of high electronegativity) and the negative end of the other dipole is an O or N atom, the attractive interaction between dipoles is particularly strong and is given the special name of **hydrogen bonding**. The length of a hydrogen bond in water is 177 pm, about 80% longer than an O—H covalent bond. The strength of a hydrogen bond in water is approximately 21 kJ (5 kcal)/mol. For comparison, the strength of the O—H covalent bond in water is approximately 498 kJ (118 kcal)/mol. As can be seen by comparing these numbers, an O---H hydrogen bond is considerably weaker than an O—H covalent bond. The presence of a large number of hydrogen bonds in liquid water, however, has an important cumulative effect on the physical properties of water. Because of hydrogen bonding, extra energy is required to separate each water molecule from its neighbors, hence the relatively high boiling point of water.

Figure 10.2
Polarity of the C—O—H bond in an alcohol.

Figure 10.3

The association of ethanol molecules in the liquid state by hydrogen bonding. Each O—H can participate in up to three hydrogen bonds (one through hydrogen and two through oxygen). Only two of the three possible hydrogen bonds per molecule are shown.

Hydrogen bonding

Similarly, there is extensive hydrogen bonding between alcohol molecules in the pure liquid. Figure 10.3 shows the association of ethanol molecules by hydrogen bonding between the partially negative oxygen atom of one ethanol molecule and the partially positive hydrogen atom of another ethanol molecule.

Table 10.1 lists the boiling points and solubilities in water for several groups of alcohols and hydrocarbons of similar molecular weight. Of the compounds compared in each group, the alcohols have the higher boiling points because more energy is needed to overcome the attractive forces of hydrogen bonding between their polar —OH groups. The presence of additional hydroxyl groups in a molecule further increases the extent of hydrogen bonding, as can be seen by comparing the boiling points of hexane (bp 69°C), 1-pentanol (bp 138°C), and 1,4-butanediol (bp 230°C), all of which have approximately the same molecular weight. Because of increased dispersion forces between larger molecules, boiling points of all types of compounds, including alcohols, increase with increasing molecular weight. Compare, for example, the boiling points of ethanol (bp 78°C), 1-propanol (bp 97°C), 1-butanol (bp 117°C), and 1-pentanol (bp 138°C).

| Table 10.1 | Boiling Points and Solubilities in Water of Five Groups of Alcohols and Hydrocarbons of Similar Molecular Weight | | | |

Structural Formula	Name	Molecular Weight (g/mol)	Boiling Point (°C)	Solubility in Water
CH_3OH	Methanol	32	65	Infinite
CH_3CH_3	Ethane	30	−89	Insoluble
CH_3CH_2OH	Ethanol	46	78	Infinite
$CH_3CH_2CH_3$	Propane	44	−42	Insoluble
$CH_3CH_2CH_2OH$	1-Propanol	60	97	Infinite
$CH_3CH_2CH_2CH_3$	Butane	58	0	Insoluble
$CH_3CH_2CH_2CH_2OH$	1-Butanol	74	117	8 g/100 g
$CH_3CH_2CH_2CH_2CH_3$	Pentane	72	36	Insoluble
$HOCH_2CH_2CH_2CH_2OH$	1,4-Butanediol	90	230	Infinite
$CH_3CH_2CH_2CH_2CH_2OH$	1-Pentanol	88	138	2.3 g/100 g
$CH_3CH_2CH_2CH_2CH_2CH_3$	Hexane	86	69	Insoluble

Hydrogen bonds have directionality in that the donor and acceptor groups must be oriented appropriately with respect to each other for hydrogen bonding to occur. Important hydrogen bond donors in biological molecules include —OH groups (proteins, carbohydrates) and —NH groups (proteins, nucleic acids). Important hydrogen bond acceptors are any N or O with a lone pair of electrons, such as C=O groups (proteins, carbohydrates, nucleic acids), —OH groups (proteins, carbohydrates), and COO⁻ groups (proteins).

With directionality comes the potential for hydrogen bonds to organize molecules at many levels ranging from the folding of biological molecules to the specific binding and recognition between a pharmaceutical and its receptor. The drug atorvastatin (Lipitor) is used to treat high cholesterol. Cholesterol is synthesized in the liver from the two-carbon acetyl group of acetyl coenzyme A (acetyl-CoA). A key intermediate in the sequence of reactions leading to the synthesis of cholesterol is a six-carbon molecule named mevalonate (Section 26.4B). Atorvastatin specifically binds to, and blocks the action of, HMG-CoA reductase, a key enzyme in the biosynthesis of mevalonate. Atorvastatin binds to this enzyme in preference to the large number of other potential enzyme targets because (1) the drug has a shape complementary to the catalytic cavity (the active site) of HMG-CoA reductase (Figure 1), and (2) it can form at least nine specific hydrogen bonds with functional groups at the active site on the enzyme (Figure 2).

The complementary shape and pattern of hydrogen bonding ensure that atorvastatin binds to HMG-CoA reductase and inhibits its ability to catalyze the formation of mevalonate. The hallmark of this and other effective drugs is their ability to bind strongly with their intended target molecules, while at the same time not interacting with other molecules that could lead to unwanted side effects.

Figure 1
A space-filling model of the cholesterol-lowering drug atorvastatin (Lipitor) bound to the active site of its enzyme target HMG-CoA reductase (shown as a yellow surface). The shape of the drug is complementary to the active site of the enzyme.

Figure 2
Hydrogen bonding (shown in red) between atorvastatin and the functional groups at the active site of the enzyme HMG-CoA reductase. The nine hydrogen bonds (shown in red), many of which involve hydroxyl groups on atorvastatin or the enzyme surface, help to provide the specificity that directs the binding of the drug to its target enzyme.

The effect of hydrogen bonding in alcohols is illustrated dramatically by comparing the boiling points of ethanol (bp 78°C) and its constitutional isomer dimethyl ether (bp −24°C). The difference in boiling point between these two compounds is caused by the presence of a polar O—H group in the alcohol, which is capable of forming intermolecular hydrogen bonds. This hydrogen bonding increases the attractive forces between molecules of ethanol, and, thus, ethanol has a higher boiling point than dimethyl ether.

$$CH_3CH_2OH \qquad CH_3OCH_3$$

Ethanol   **Dimethyl ether**
bp 78°C   bp −24°C

Because alcohols can interact by hydrogen bonding with water, they are more soluble in water than alkanes and alkenes of comparable molecular weight. Methanol, ethanol, and 1-propanol are soluble in water in all proportions. As molecular weight increases, the physical properties of alcohols become more like those of hydrocarbons of comparable molecular weight. Higher-molecular-weight alcohols are much less soluble in water because of the increase in size of the hydrocarbon portion of their molecules.

Example 10.4

Following are three alcohols with the molecular formula $C_4H_{10}O$. Their boiling points, from lowest to highest, are 82.3°C, 99.5°C, and 117°C. Which alcohol has which boiling point?

1-Butanol   **2-Butanol**   **2-Methyl-2-propanol**

Solution

Boiling points of these constitutional isomers depend on the strength of intermolecular hydrogen bonding. The primary —OH group of 1-butanol is the most accessible for intermolecular hydrogen bonding; this alcohol has the highest boiling point, 117°C. The tertiary —OH group of 2-methyl-2-propanol is the least accessible for intermolecular hydrogen bonding; this alcohol has the lowest boiling point, 82.3°C.

Problem 10.4

Arrange these compounds in order of increasing boiling point.

Heptane   **1,5-Pentanediol**   **1-Hexanol**

Example 10.5

Arrange these compounds in order of increasing solubility in water.

Hexane   **1,4-Butanediol**   **1-Pentanol**

Solution

Hexane, C_6H_{14}, a nonpolar hydrocarbon, has the lowest solubility in water. Both 1-pentanol and 1,4-butanediol are polar compounds due to the presence of —OH groups, and each interacts with water molecules by hydrogen bonding. Because 1,4-butanediol has more sites within its molecules for hydrogen bonding than 1-pentanol, the diol is more soluble in water than 1-pentanol. The water solubilities of these compounds are given in Table 10.1.

Insoluble   2.3 g/100 mL   Infinitely soluble

Problem 10.5

Arrange these compounds in order of increasing solubility in water.

HO〜〜〜   HO〜〜   Cl〜〜Cl

1-Butanol   **1-Propanol**   **1,2-Dichloroethane**

10.3 | Acidity and Basicity of Alcohols

Alcohols can function as both weak acids (proton donors) and weak bases (proton acceptors). Table 10.2 lists the acid ionization constants for several low-molecular-weight alcohols.

In dilute aqueous solution, only methanol (pK_a 15.5) is more acidic than water.

$$CH_3\ddot{O}-H + \ddot{O}-H \rightleftharpoons CH_3\ddot{O}:^- + H-\overset{+}{\underset{H}{\ddot{O}}}-H \qquad K_a = \frac{[CH_3O^-][H_3O^+]}{[CH_3OH]} = 10^{-15.5}$$

$$pK_a = 15.5$$

Ethanol has about the same acidity as water. Higher-molecular-weight, water-soluble alcohols are slightly weaker acids than water. Thus, although alcohols have some acidity, they are not strong enough acids to react with weak bases such as sodium bicarbonate or sodium carbonate. (At this point, it would be wise to review Section 4.4, which discusses the position of equilibrium in acid-base reactions.)

For simple alcohols, such as methanol and ethanol, acidity depends primarily on the degree of solvation and stabilization of the alkoxide ion by water molecules. The negatively charged oxygen atoms of the methoxide and ethoxide ions are almost as accessible for solvation as the hydroxide ion is; therefore, these alcohols are about as acidic as water. As the bulk of the alkyl group bonded to oxygen increases, the ability of water molecules to solvate the alkoxide ion decreases. 2-Methyl-2-propanol (*tert*-butyl alcohol) is a weaker acid than either methanol or ethanol, primarily because of the bulk of the *tert*-butyl group, which reduces solvation of the *tert*-butoxide anion by surrounding water molecules.

In the presence of strong acids, the oxygen atom of an alcohol is a base and reacts with an acid by proton transfer to form an oxonium ion.

$$CH_3CH_2-\ddot{O}-H + H-\overset{+}{\underset{H}{\ddot{O}}}-H \xrightarrow{H_2SO_4} CH_3CH_2-\overset{+}{\underset{H}{\ddot{O}}}-H + \ddot{O}-H$$

Ethanol   **Hydronium ion**   **Ethyloxonium ion**
                     (pK_a −1.7) pK_a −2.4
                     (weaker acid)   (stronger acid)

Table 10.2	pK_a Values for Selected Alcohols in Dilute Aqueous Solution*		
Compound	**Structural Formula**	**pK_a**	
Hydrogen chloride	HCl	−7	Stronger acid
Acetic acid	CH_3COOH	4.8	
Methanol	CH_3OH	15.5	
Water	H_2O	15.7	↑
Ethanol	CH_3CH_2OH	15.9	
2-Propanol	$(CH_3)_2CHOH$	17	Weaker acid
2-Methyl-2-propanol	$(CH_3)_3COH$	18	

*Also given for comparison are pK_a values for water, acetic acid, and hydrogen chloride.

To summarize, when trying to predict the mechanisms of reactions involving an hydroxy group, you need to keep in mind that it is both a weak acid and a weak base, so consider adding a proton or taking a proton away in the initial steps of mechanisms when there is a strong acid or base present, respectively. In addition, an important mechanistic theme in many of the reactions of alcohols is that the —OH group, a poor leaving group, reacts with protons or a variety of strong electrophiles to create —$\overset{+}{O}H_2$ or analogous group, a much better leaving group, enabling subsequent substitution or elimination reactions to take place.

10.4 | Reaction of Alcohols with Active Metals

Alcohols react with Li, Na, K, and other active metals to liberate hydrogen and form metal alkoxides. In this oxidation/reduction reaction, Na is oxidized to Na^+, and H^+ is reduced to H_2.

$$2\ CH_3OH + 2\ Na \longrightarrow 2\ CH_3O^-Na^+ + H_2$$

Sodium methoxide
(MeO⁻Na⁺)

To name a metal alkoxide, name the cation first, followed by the name of the anion. The name of the anion is derived from the prefix showing the number of carbon atoms and their arrangement (*meth-*, *eth-*, *isoprop-*, tert-*but-*, and so on) followed by the suffix *-oxide*.

Alkoxide ions are nearly the same or somewhat stronger bases than the hydroxide ion. In addition to sodium methoxide, the following metal salts of alcohols are commonly used in organic reactions requiring a strong base in a nonaqueous solvent, as for example sodium ethoxide in ethanol and potassium *tert*-butoxide in 2-methyl-2-propanol (*tert*-butyl alcohol).

$$CH_3CH_2O^-Na^+ \qquad CH_3\overset{\overset{\displaystyle CH_3}{|}}{\underset{\underset{\displaystyle CH_3}{|}}{C}}O^-K^+$$

Sodium ethoxide   **Potassium *tert*-butoxide**
(EtO⁻Na⁺)   **(*t*-BuO⁻K⁺)**

Alcohols can also be converted to salts by reaction with bases stronger than alkoxide ions. One such base is sodium hydride, NaH. Hydride ion, $H:^-$, the conjugate base of H_2, is an extremely strong base.

$$CH_3CH_2OH + Na^+H^- \longrightarrow CH_3CH_2O^-Na^+ + H_2$$

Ethanol   **Sodium hydride**   **Sodium ethoxide**

Reactions of sodium hydride with compounds containing acidic hydrogens are irreversible and driven to completion by the formation of H_2, which is given off as a gas.

Example 10.6

Write a balanced equation for the reaction of cyclohexanol with sodium metal.

Solution

Problem 10.6

Predict the position of equilibrium for this acid-base reaction.

$$CH_3CH_2O^-Na^+ + CH_3\overset{\overset{\displaystyle O}{\|}}{C}OH \rightleftharpoons CH_3CH_2OH + CH_3\overset{\overset{\displaystyle O}{\|}}{C}O^-Na^+$$

10.5 | Conversion of Alcohols to Haloalkanes and Sulfonates

Conversion of an alcohol to a haloalkane involves substitution of halogen for —OH at a saturated carbon. The most common reagents for this conversion are the halogen acids (HCl, HBr, and HI) and certain inorganic halides (PBr_3, $SOCl_2$, and $SOBr_2$).

A. Reaction with HCl, HBr, and HI

Tertiary alcohols react rapidly with HCl, HBr, and HI. Mixing a low-molecular-weight, water-soluble tertiary alcohol with concentrated hydrochloric acid for a few minutes at room temperature results in conversion of the alcohol to a chloroalkane.

$$\text{>}\!-\!\text{OH} \;+\; \text{HCl} \;\xrightarrow{25°C}\; \text{>}\!-\!\text{Cl} \;+\; H_2O$$

2-Methyl-2-propanol   **2-Chloro-2-methylpropane**

Reaction is evident by formation of a water-insoluble chloroalkane that separates from the aqueous layer. Low-molecular-weight, water-soluble primary and secondary alcohols are unreactive under these conditions.

Water-insoluble tertiary alcohols are converted to tertiary halides by bubbling gaseous HX through a solution of the alcohol dissolved in diethyl ether or tetrahydrofuran (THF).

$$\text{1-Methylcyclohexanol} \;+\; \text{HCl} \;\xrightarrow[\text{ether}]{0°C}\; \text{1-Chloro-1-methylcyclohexane} \;+\; H_2O$$

1-Methylcyclohexanol   **1-Chloro-1-methyl-cyclohexane**

Water-insoluble primary and secondary alcohols react only slowly under these conditions.

Primary and secondary alcohols are converted to bromoalkanes and iodoalkanes by treatment with hydrobromic and hydroiodic acids. For example, when heated to reflux with concentrated HBr, 1-butanol is converted smoothly to 1-bromobutane.

$$\text{1-Butanol} + \text{HBr} \xrightarrow[\text{reflux}]{H_2O} \text{1-Bromobutane} + H_2O$$

1-Butanol   **1-Bromobutane**

Many secondary alcohols give at least some rearranged product, evidence for the formation of carbocation intermediates during their reaction. For example, treating 3-pentanol with HBr gives 3-bromopentane as the major product, along with some 2-bromopentane.

$$\text{3-Pentanol} + \text{HBr} \xrightarrow{\text{heat}} \text{3-Bromopentane} + \text{2-Bromopentane} + H_2O$$

3-Pentanol   **3-Bromopentane**   **2-Bromopentane**
                        (major product)   (a product of rearrangement, racemic)

Primary alcohols with extensive β–branching give large amounts of a product derived from rearrangement. For example, treatment of 2,2-dimethyl-1-propanol (neopentyl alcohol) with HBr gives a rearranged product almost exclusively.

2,2-Dimethyl-1-propanol　　　　　　　　**2-Bromo-2-methylbutane**
(a product of rearrangement)

Based on observations of the relative ease of reaction of alcohols with HX ($3° > 2° > 1°$) and the occurrence of rearrangements, chemists propose an S_N1 mechanism for the conversion of tertiary and secondary alcohols to haloalkanes by concentrated HX, with the formation of a carbocation intermediate.

Mechanism　Reaction of a 3° Alcohol with HBr—An S_N1 Reaction

Step 1: **Add a proton.**　While we often show HBr as the acid present in solution, the actual acid involved in this reaction is H_3O^+ formed by dissociation of HBr in aqueous solution.

Rapid and reversible proton transfer from H_3O^+ to the —OH group of the alcohol gives an oxonium ion, which converts —OH, a poor leaving group, into —$\overset{+}{O}H_2$, a better leaving group.

2-Methyl-2-propanol　　　　　　　　　　**An oxonium ion**
(*tert*-Butyl alcohol)

Step 2: **Break a bond to give stable molecules or ions.**　Loss of water gives a 3° carbocation intermediate.

A 3° carbocation
intermediate

Step 3: **Make a new bond between a nucleophile and an electrophile.**　Reaction of the 3° carbocation (an electrophile) with bromide ion (a nucleophile) gives the alkyl halide.

2-Bromo-2-methylpropane
(*tert*-Butyl bromide)

Primary alcohols with no β-branching react with HX by an S_N2 mechanism. In the rate-determining step, halide ion reacts at the carbon bearing the oxonium ion to displace H_2O and form the C—X bond.

Step 1: Add a proton. Rapid and reversible proton transfer gives an oxonium ion, which transforms —OH, a poor leaving group, into —OH$_2$, a better leaving group.

An oxonium ion

Step 2: Make a new bond between a nucleophile and an electrophile and simultaneously break a bond to give stable molecules or ions. Nucleophilic displacement of H$_2$O by Br$^-$ gives the bromoalkane.

For primary alcohols with extensive β-branching, such as 2,2-dimethyl-1-propanol (neopentyl alcohol), it is difficult if not impossible for reaction to occur by direct displacement of H$_2$O from the primary carbon. Furthermore, formation of a 1° carbocation is also difficult, if not impossible. Instead, primary alcohols with extensive β-branching react by a mechanism involving formation of a 3° carbocation intermediate by simultaneous loss of H$_2$O and migration of an alkyl group, as illustrated by the conversion of 2,2-dimethyl-1-propanol to 2-chloro-2-methylbutane. Because the rate-determining step of this transformation involves only one reactant, namely the protonated alcohol, it is classified as an S$_N$1 reaction.

| Mechanism | Rearrangement upon Treatment of Neopentyl Alcohol with HCl |

Step 1: Add a proton. Rapid and reversible proton transfer gives an oxonium ion. This step converts —OH, a poor leaving group, into —OH$_2$, a better leaving group.

2,2-Dimethyl-1-propanol   An oxonium ion

Step 2: 1,2 Shift and simultaneously break a bond to give stable molecules or ions. Two changes take place simultaneously in this step; the C—O bond breaks and a methyl group with its pair of bonding electrons migrates to the site occupied by the departing H$_2$O group. The result of these changes is loss of H$_2$O and the formation of a 3° carbocation.

A 3° carbocation
intermediate

Step 3: Make a new bond between a nucleophile and an electrophile. Reaction of the 3° carbocation (an electrophile) with chloride ion (a nucleophile) gives the 3° haloalkane.

2-Chloro-2-methylbutane

In summary, preparation of haloalkanes by treatment of ROH with HX is most useful for primary and tertiary alcohols. The central theme in all these reactions is that protonation of —OH, a very poor leaving group, transforms it into —$\overset{+}{O}H_2$, a better leaving group so that an S_N1 or S_N2 reaction can take place with a halide nucleophile. Because of the possibility of rearrangement, this process is less useful for secondary alcohols (except for simple cycloalkanols) and for primary alcohols with extensive branching on the β-carbon.

B. Reaction with Phosphorous Tribromide

An alternative method for the synthesis of bromoalkanes from primary and secondary alcohols is through the use of phosphorous tribromide, PBr_3.

| 2-Methyl-1-propanol (Isobutyl alcohol) | Phosphorus tribromide | 1-Bromo-2-methylpropane (Isobutyl bromide) | Phosphorous acid |

This method of preparation of bromoalkanes takes place under milder conditions than treatment with HBr. Although rearrangement sometimes occurs with PBr_3, the extent is considerably less than that with HBr, especially when the reaction mixture is kept at or below 0°C.

Mechanism Reaction of a Primary Alcohol with PBr_3

Conversion of an alcohol to a bromoalkane takes place in two steps.

Step 1: Make a new bond between a nucleophile and an electrophile and simultaneously break a bond to give stable molecules or ions. Nucleophilic displacement on phosphorus by the oxygen atom of the alcohol gives a protonated dibromophosphite group, which converts —OH, a poor leaving group, into a good leaving group.

Step 2: Make a new bond between a nucleophile and an electrophile and simultaneously break a bond to give stable molecules or ions. Nucleophilic displacement of the protonated dibromophosphite group by bromide ion gives the bromoalkane.

The other two bromine atoms on phosphorus are replaced in similar reactions, giving three moles of RBr and one mole of phosphorous acid, H_3PO_3.

C. Reaction with Thionyl Chloride and Thionyl Bromide

The most widely used reagent for the conversion of primary and secondary alcohols to chloroalkanes is thionyl chloride, $SOCl_2$. Yields are high, and rearrangements are seldom observed. The byproducts of this conversion are HCl and SO_2.

$$\text{1-Heptanol} + SOCl_2 \xrightarrow{\text{pyridine}} \text{1-Chloroheptane} + SO_2 + HCl$$

1-Heptanol **Thionyl chloride** **1-Chloroheptane**

Similarly, thionyl bromide, $SOBr_2$ can be used to convert an alcohol to a bromoalkane.

Reactions with these reagents are most commonly carried out in the presence of pyridine (Section 23.1) or a tertiary amine such as triethylamine, Et_3N. The function of the amine (a weak base) is twofold. First, it catalyzes the reaction by forming a small amount of the alkoxide in equilibrium. The alkoxide is more reactive than the alcohol as a nucleophile. In addition, the amine neutralizes the HCl or HBr generated during the reaction and, in this way, prevents unwanted side reactions.

The 3° amine promotes formation of an alkoxide:

$$R-\ddot{O}-H + :NEt_3 \rightleftharpoons R-\ddot{O}:^- + H-\overset{+}{N}Et_3$$

It also neutralizes the HCl formed during the reaction:

$$Et_3N: + H-\ddot{C}l: \longrightarrow Et_3\overset{+}{N}-H \quad :\ddot{C}l:^-$$

Triethylammonium chloride

A particular value of thionyl halides is that their reaction with alcohols is stereoselective; it occurs with inversion of configuration. Reaction of thionyl chloride with (S)-2-octanol, for example, in the presence of a tertiary amine occurs with inversion of configuration and gives (R)-2-chlorooctane.

$$\text{(S)-2-Octanol} + SOCl_2 \xrightarrow{\text{3° amine}} \text{(R)-2-Chlorooctane} + SO_2 + HCl$$

(S)-2-Octanol **Thionyl chloride** **(R)-2-Chlorooctane**

A key feature of the reaction of an alcohol with thionyl chloride is the formation of an alkyl chlorosulfite, which converts OH^-, a poor leaving group, into a chlorosulfite that now contains a good leaving group. If the reaction between the alcohol and thionyl chloride is carried out at 0°C or below, the alkyl chlorosulfite can be isolated.

An alkyl chlorosulfite

Nucleophilic displacement of this leaving group by chloride ion gives the product.

D. Formation of Aryl and Alkyl Sulfonates

As we have just seen, alcohols react with thionyl chloride to form alkyl chlorosulfites. Alcohols also react with compounds called sulfonyl chlorides to form alkylsulfonates. Sulfonyl chlorides are derived from sulfonic acids, compounds that are very strong acids, comparable in strength to sulfuric acid.

A sulfonyl chloride   A sulfonic acid   A sulfonate anion
(a very strong acid)   (a very weak base and stable anion; a very good leaving group)

What is important for us at this point is that a sulfonate anion is a very weak base and stable anion, and therefore a very good leaving group in nucleophilic substitution reactions.

Two of the most commonly used sulfonyl chlorides are *p*-toluenesulfonyl chloride (abbreviated tosyl chloride, TsCl) and methanesulfonyl chloride (abbreviated mesyl chloride, MsCl). Treating ethanol with *p*-toluenesulfonyl chloride in the presence of pyridine gives ethyl *p*-toluenesulfonate (ethyl tosylate). Pyridine is added to catalyze the reaction and to neutralize the HCl formed as a byproduct. Cyclohexanol is converted to cyclohexyl methanesulfonate (cyclohexyl mesylate) by a similar reaction of cyclohexanol with methanesulfonyl chloride.

Ethanol   **p-Toluenesulfonyl chloride**   **Ethyl p-toluenesulfonate (Ethyl tosylate)**

Cyclohexanol   **Methanesulfonyl chloride**   **Cyclohexyl methanesulfonate (Cyclohexyl mesylate)**

In formation of either a tosylate or a mesylate, the reaction involves breaking the O—H bond of the alcohol; it does not affect the C—O bond in any way. If the carbon bearing the —OH group is a chiral center, sulfonate ester formation takes place with retention of configuration.

A particular advantage of sulfonate esters is that, through their use, a hydroxyl group, a very poor leaving group, can be converted to a tosylate or mesylate group, often shown as OTs and OMs, respectively. Both are very good leaving groups readily displaced by nucleophilic substitution.

Ethyl p-toluenesulfonate   **p-Toluenesulfonate anion (a good leaving group)**

Following is a two-step sequence for conversion of (*S*)-2-octanol to (*R*)-2-octyl acetate via a tosylate. The first step involves cleavage of the O—H bond and proceeds with retention of configuration at the chiral center. The second step involves S$_N$2 nucleophilic displacement of tosylate by acetate ion and proceeds with inversion of configuration at the chiral center.

(S)-2-Octanol   **Tosyl chloride**   **(S)-2-Octyl tosylate**

Sodium acetate   **(S)-2-Octyl tosylate** $\xrightarrow[\text{ethanol}]{\text{S}_N2}$ **(R)-2-Octyl acetate**   + Na$^+$OTs$^-$

Example 10.7

Show how to convert *trans*-4-methylcyclohexanol to *cis*-1-iodo-4-methylcyclohexane via a tosylate.

Solution

Treat the alcohol with *p*-toluenesulfonyl chloride in pyridine to form a tosylate with retention of configuration. Then treat the tosylate with sodium iodide in acetone. The S$_N$2 reaction with inversion of configuration gives the product. Because of the requirement for backside attack by the I$^-$ nucleophile, the tosylate group must be in the axial position to react. Backside attack is not possible for an equatorial leaving group on a cyclohexane ring. The molecule must undergo a ring flip before the I$^-$ displacement reaction can occur.

Problem 10.7

Show how to convert (*R*)-2-pentanol to (*S*)-2-pentanethiol via a tosylate:

(R)-2-Pentanol   **(S)-2-Pentanethiol**

10.6 | Acid-Catalyzed Dehydration of Alcohols

An alcohol can be converted to an alkene by **dehydration**; that is, by the elimination of a molecule of water from adjacent carbon atoms. Dehydration is most often brought about by heating the alcohol with either 85% phosphoric acid or concentrated sulfuric acid. Primary alcohols are the most difficult to dehydrate and generally require heating in concentrated sulfuric acid at temperatures as high as 180°C. Secondary alcohols undergo acid-catalyzed dehydration at somewhat lower temperatures. Acid-catalyzed dehydration of tertiary alcohols often requires temperatures only slightly above room temperature.

Dehydration
Elimination of water.

$$CH_3CH_2OH \xrightarrow[180°C]{H_2SO_4} CH_2{=}CH_2 + H_2O$$

Cyclohexanol   **Cyclohexene**

$$\underset{\substack{\text{2-Methyl-2-propanol}\\(\textit{tert}\text{-Butyl alcohol})}}{\underset{\overset{|}{\text{CH}_3}}{\overset{\overset{\text{CH}_3}{|}}{\text{CH}_3\text{COH}}}} \xrightarrow[\text{50°C}]{\text{H}_2\text{SO}_4} \underset{\substack{\text{2-Methylpropene}\\(\text{Isobutylene})}}{\overset{\overset{\text{CH}_3}{|}}{\text{CH}_3\text{C}{=}\text{CH}_2}} + \text{H}_2\text{O}$$

Thus, the ease of acid-catalyzed dehydration of alcohols is in this order:

$$1° \text{ alcohol } < \ 2° \text{ alcohol } < \ 3° \text{ alcohol}$$

Ease of dehydration of alcohols ➤

When isomeric alkenes are obtained in acid-catalyzed dehydration of an alcohol, the alkene having the greater number of substituents on the double bond (the more stable alkene) generally predominates (Zaitsev's rule, Section 9.5).

$$\underset{\substack{\text{2-Butanol}\\(\text{racemic})}}{\overset{\overset{\text{OH}}{|}}{\text{CH}_3\text{CH}_2\text{CHCH}_3}} \xrightarrow[\text{heat}]{85\% \text{ H}_3\text{PO}_4} \underset{\substack{(E \text{ and } Z)\text{-2-Butene}\\(80\%)}}{\text{CH}_3\text{CH}{=}\text{CHCH}_3} + \underset{\substack{\text{1-Butene}\\(20\%)}}{\text{CH}_3\text{CH}_2\text{CH}{=}\text{CH}_2} + \text{H}_2\text{O}$$

Example 10.8

Draw structural formulas for the alkenes formed on acid-catalyzed dehydration of each alcohol. Where isomeric alkenes are possible, predict which alkene is the major product.

(a) 3-Methyl-2-butanol (racemic)   (b) 2-Methylcyclopentanol (racemic)

Solution

(a) Elimination of H_2O from carbons 2-3 gives 2-methyl-2-butene; elimination from carbons 1-2 gives 3-methyl-1-butene. 2-Methyl-2-butene, with three alkyl groups (three methyl groups) on the double bond, is the major product (Zaitsev rule). 3-Methyl-1-butene, with only one alkyl group (an isopropyl group) on the double bond, is the minor product. A small amount of 2-methyl-1-butene is formed by rearrangement.

3-Methyl-2-butanol
(racemic)   **2-Methyl-2-butene**
(major product)   **3-Methyl-1-butene**

(b) The major product, 1-methylcyclopentene, has three alkyl substituents on the double bond. 3-Methylcyclopentene has only two substituents on the double bond.

2-Methylcyclopentanol
(racemic)   **1-Methylcyclopentene**
(major product)   **3-Methylcyclopentene**
(racemic)

Problem 10.8

Draw structural formulas for the alkenes formed by acid-catalyzed dehydration of each alcohol. Where isomeric alkenes are possible, predict which is the major product.

(a) 2-Methyl-2-butanol   (b) 1-Methylcyclopentanol

Dehydration of primary and secondary alcohols is often accompanied by rearrangement. Acid-catalyzed dehydration of 3,3-dimethyl-2-butanol, for example, gives a mixture of two alkenes, each of which is the result of a rearrangement.

3,3-Dimethyl-2-butanol
(racemic)

$\xrightarrow[140°-170°C]{H_2SO_4}$

2,3-Dimethyl-2-butene
(80%)

+

2,3-Dimethyl-1-butene
(20%)

Based on the relative rates of dehydration of alcohols (3° > 2° > 1°) and the prevalence of rearrangement, particularly among primary and secondary alcohols, chemists propose a three-step mechanism for acid-catalyzed dehydration of secondary and tertiary alcohols. This mechanism involves formation of a carbocation in the rate-determining step and, therefore, is classified as an E1 mechanism.

Mechanism Acid-Catalyzed Dehydration of 2-Butanol—An E1 Reaction

Step 1: Add a proton. Proton transfer from H_3O^+ to the OH group of the alcohol gives an oxonium ion; —OH, a poor leaving group, is converted to $-OH_2^+$, a better leaving group.

rapid and reversible

Step 2: Break a bond to give stable molecules or ions. Breaking of the C—O bond and loss of H_2O gives a 2° carbocation intermediate.

slow, rate determining

$+ H_2\ddot{O}:$

A 2° carbocation intermediate

Step 3: Take a proton away. Proton transfer from a carbon adjacent to the positively charged carbon to H_2O gives the alkene. In this step, the electrons of the C—H σ bond become the electrons of the π bond.

rapid and reversible

We account for acid-catalyzed dehydration accompanied by rearrangement in the same manner we did for the rearrangements that accompany S_N1 and E1 reactions (Section 9.3F). Rearrangement occurs through formation of a carbocation intermediate followed by migration of an atom or group, with its bonding pair of electrons, from the β-carbon to the carbon bearing the positive charge.

3,3-Dimethyl-2-butanol
(racemic)

$\xrightarrow[-H_2O]{H^+}$

A 2° carbocation intermediate

A 3° carbocation intermediate

$\xrightarrow{H_2O}$

2,3-Dimethyl-2-butene

$+ H_3O^+$

$\xrightarrow{H_2O}$

2,3-Dimethyl-1-butene

$+ H_3O^+$

The driving force for rearrangements of this type is conversion of a less stable carbocation to a more stable one. Proton transfer to H_2O then gives the alkenes. As in other cases of acid-catalyzed dehydration of alkenes, the Zaitsev rule applies, and the more substituted alkene predominates.

Primary alcohols with little or no β-branching undergo acid-catalyzed dehydration to give a terminal alkene and rearranged alkenes. Acid-catalyzed dehydration of 1-butanol, for example, gives only 12% of 1-butene. The major product is a mixture of the *trans* and *cis* isomers of 2-butene.

1-Butanol   *trans*-2-Butene *cis*-2-Butene 1-Butene
          (56%)   (32%) (12%)

We account for the formation of these products by a combination of E1 and E2 mechanisms.

Mechanism Acid-Catalyzed Dehydration of an Unbranched Primary Alcohol

Step 1: **Add a proton.** Proton transfer from H_3O^+ to the OH group gives an oxonium ion.

1-Butanol

Step 2: **Take a proton away and simultaneously break a bond to give stable molecules or ions.** Simultaneous proton transfer to solvent and loss of H_2O gives the carbon-carbon double bond of the terminal alkene.

1-Butene

Step 3: **1,2 Shift and simultaneously break a bond to give stable molecules or ions.** Simultaneous shift of a hydride ion from the β-carbon to the α-carbon and loss of H_2O gives a carbocation intermediate.

A 2° carbocation

Step 4: **Take a proton away.** Transfer of a proton from a carbon adjacent to the carbocation to solvent gives the rearranged alkenes.

trans-2-Butene *cis*-2-Butene

Example 10.9

Propose a mechanism to account for this acid-catalyzed dehydration.

(racemic)

Solution

Proton transfer to the OH group to form an oxonium ion (**add a proton**) followed by loss of H_2O (**break a bond to give stable molecules or ions**) gives a 2° carbocation intermediate. Migration of a methyl group with its bonding pair of electrons from the adjacent carbon to the positively charged carbon gives a more stable 3° carbocation (**1,2-shift**). Proton transfer from this intermediate to a base, here shown as H_2O, gives the observed product (**take a proton away**).

A 2° carbocation intermediate   A 3° carbocation intermediate

Problem 10.9

Propose a mechanism to account for this acid-catalyzed dehydration.

(racemic)

In Section 6.3B we discussed the acid-catalyzed hydration of alkenes to give alcohols. In the present section, we have discussed the acid-catalyzed dehydration of alcohols to give alkenes. In fact, both the alkene hydration and the alcohol dehydration reactions are reversible and represent different directions of the same process. The following equilibrium exists.

An alkene   An alcohol

Large amounts of water (use of dilute aqueous acid) favor alcohol formation, whereas scarcity of water (use of concentrated acid) or experimental conditions where water is removed (heating the reaction mixture above 100°C) favor alkene formation. Thus, depending on experimental conditions, it is possible to use the hydration-dehydration equilibrium to prepare either alcohols or alkenes in high yields.

    This hydration-dehydration equilibrium illustrates a very important principle in the study of reaction mechanisms—the **principle of microscopic reversibility**. According to this principle, the sequence of transition states and reactive intermediates (that is, the mechanism) for any reversible reaction must be the same, but in reverse order, for the backward reaction as for the forward reaction.

Principle of microscopic reversibility
This principle states that the sequence of transition states and reactive intermediates in the mechanism of any reversible reaction must be the same, but in reverse order, for the reverse reaction as for the forward reaction.

As an illustration of the principle of microscopic reversibility, notice that the mechanism we presented in this section for the acid-catalyzed dehydration of 2-butanol to give 2-butene is exactly the reverse of that presented in Section 6.3B for the acid-catalyzed hydration of 2-butene to give 2-butanol.

10.7 | The Pinacol Rearrangement

Compounds containing hydroxyl groups on two adjacent carbon atoms are called vicinal *diols*, or alternatively, *glycols*. Such compounds can be synthesized by a variety of methods, including oxidation of alkenes by OsO_4 (Section 6.5A). The products of acid-catalyzed dehydration of glycols are quite different from those of acid-catalyzed dehydration of alcohols. For example, treating 2,3-dimethyl-2,3-butanediol (commonly called pinacol) with concentrated sulfuric acid gives 3,3-dimethyl-2-butanone (commonly called pinacolone):

2,3-Dimethyl-2,3-butanediol
(Pinacol)

3,3-Dimethyl-2-butanone
(Pinacolone)

Note two features of this reaction: (1) it involves dehydration of a glycol to form a ketone, and (2) it involves migration of a methyl group from one carbon to an adjacent carbon. Acid-catalyzed conversion of pinacol to pinacolone is an example of a type of reaction called the **pinacol rearrangement**. We account for the conversion of pinacol to pinacolone by a four-step mechanism.

Mechanism The Pinacol Rearrangement of 2,3-Dimethyl-2,3-butanediol (Pinacol)

Step 1: **Add a proton.** Proton transfer from the acid catalyst to one of the —OH groups gives an oxonium ion, which converts —OH, a poor leaving group, into —OH₂, a better leaving group.

An oxonium ion

Step 2: **Break a bond to give stable molecules or ions.** Loss of H_2O from the oxonium ion gives a 3° carbocation intermediate.

A 3° carbocation
intermediate

Step 3: **1,2 Shift.** Migration of a methyl group from the adjacent carbon with its bonding electrons gives a new, more stable resonance-stabilized cation intermediate. Of the two contributing structures we can draw for it, the one on the right makes the greater contribution because, in it, both carbon and oxygen have complete octets of valence electrons (Section 1.8).

Cation intermediate stabilized
by charge delocalization

Step 4: Take a proton away. Proton transfer to solvent gives pinacolone.

The pinacol rearrangement is general for all glycols. In the rearrangement of pinacol, a symmetrical diol, equivalent carbocations are formed no matter which —OH becomes protonated and leaves. Studies of unsymmetrical vicinal diols reveal that the —OH group that becomes protonated and leaves is the one that gives rise to the more stable carbocation. For example, treatment of 2-methyl-1,2-propanediol with cold concentrated sulfuric acid gives a 3° carbocation. Subsequent migration of hydride ion (H: ⁻) followed by transfer of a proton from the new cation to solvent gives 2-methylpropanal.

| 2-Methyl-1,2-propanediol | A 3° carbocation intermediate | | 2-Methylpropanal |

Example 10.10

Predict the product formed by treating each vicinal diol with H_2SO_4.

(a)

(b)

(racemic)

Solution

(a) *Step 1:* **Add a proton followed by break a bond to give stable molecules or ions.**
Protonation of the 3° hydroxyl group followed by loss of H_2O gives a 3° carbocation intermediate.

(racemic)   A 3° carbocation
                                    intermediate

Step 2: 1,2 Shift. Migration of a hydride ion from the adjacent carbon gives a resonance-stabilized cation intermediate.

A resonance-stabilized cation intermediate

The contributing structure on the right has filled valence shells on both carbon and oxygen and, therefore, makes the greater contribution to the hybrid.

Step 3: Take a proton away. Proton transfer from the resonance-stabilized cation intermediate to water completes the reaction to give 2-methylbutanal.

2-Methylbutanal
(racemic)

(b) Protonation of either hydroxyl group followed by loss of water gives a 3° carbocation. The group that then migrates is a CH_2 group of the five-membered ring, and the product is a bicyclic ketone.

(racemic)   A 3° carbocation intermediate   **Spiro[4.5]decan-6-one**

The product belongs to the class of compounds called spiro compounds, in which two rings share only one carbon atom.

Problem 10.10

Propose a mechanism to account for the following transformation.

10.8 | Oxidation of Alcohols

Oxidation of a primary alcohol gives an aldehyde or a carboxylic acid, depending on experimental conditions. Secondary alcohols are oxidized to ketones. Tertiary alcohols are not oxidized. Following is a series of transformations in which a primary alcohol is oxidized first to an aldehyde and then to a carboxylic acid. The fact that each transformation involves oxidation is indicated by the symbol O in brackets over the reaction arrow:

A primary   An aldehyde   A carboxylic
alcohol   acid

A. Chromic Acid

A reagent commonly used in the laboratory for the oxidation of a primary alcohol to a carboxylic acid is chromic acid, H_2CrO_4. Chromic acid is prepared by dissolving either chromium(VI) oxide or potassium dichromate in aqueous sulfuric acid:

$$CrO_3 \ + \ H_2O \ \xrightarrow{H_2SO_4} \ H_2CrO_4$$

Chromium(VI) **Chromic acid**
oxide

$$K_2Cr_2O_7 \ \xrightarrow{H_2SO_4} \ H_2Cr_2O_7 \ \xrightarrow{H_2O} \ 2\ H_2CrO_4$$

Potassium **Chromic acid**
dichromate

A solution of chromic acid in aqueous sulfuric acid is known as the **Jones reagent**.

Because of the low solubility of most organic compounds in water, their oxidation by chromic acid is commonly carried out by dissolving them in acetone and then adding a stoichiometric amount of Jones reagent to complete the oxidation.

Oxidation of 1-hexanol, for example, using chromic acid in the mixed solvent of aqueous sulfuric acid and acetone gives hexanoic acid in high yield.

1-Hexanol   **Hexanal**   **Hexanoic acid**
(not isolated)

These experimental conditions are more than sufficient to oxidize the intermediate aldehyde to a carboxylic acid.

Secondary alcohols are oxidized to ketones by chromic acid:

2-Isopropyl-5-methyl-   **2-Isopropyl-5-methyl-**
cyclohexanol   **cyclohexanone**
(Menthol)   **(Menthone)**

Tertiary alcohols are resistant to oxidation because the carbon bearing the —OH is already bonded to three carbon atoms and, therefore, cannot form an additional carbon-oxygen bond.

1-Methylcyclopentanol

Thus, the prerequisite for the oxidation of an alcohol to an aldehyde or ketone is at least one H on the carbon bearing the OH group.

Step 1: Reaction of the alcohol and chromic acid gives an alkyl chromate by a mechanism similar to that for the formation of a carboxylic ester (Section 17.7). There is no change in oxidation state of either carbon or chromium as a result of this step.

Cyclohexanol   **An alkyl chromate**

Step 2: **Take a proton away and simultaneously break bonds to give stable molecules or ions.**   Reaction of the alkyl chromate with a base, here shown as a water molecule, results in cleavage of a C—H bond, formation of the carbonyl group, and reduction of chromium(VI) to chromium(IV).

Cyclohexanone

This step is the oxidation-reduction step; carbon undergoes a two-electron oxidation and chromium(VI) undergoes a two-electron reduction to chromium(IV). Chromium(IV) then participates in further oxidations by a similar mechanism and eventually is transformed to Cr(III).

We have shown that in aqueous chromic acid, a primary alcohol is oxidized first to an aldehyde and then to a carboxylic acid. In the second step, it is not the aldehyde that is oxidized, but rather the aldehyde hydrate formed by addition of a molecule of water to the aldehyde carbonyl group (hydration). An —OH of the aldehyde hydrate reacts with chromic acid to complete the oxidation of the aldehyde to a carboxylic acid.

The hydration step is critical because it converts the aldehyde carbonyl into two hydroxyl groups. Chromic acid can only react with an —OH to form an alkyl chromate ester, not a carbonyl. The hydration process is catalyzed by acid, and although equilibrium favors the aldehyde, enough hydrate is made to complete the reaction. We will study the hydration of aldehydes and ketones in more detail in Section 16.7.

An aldehyde   An aldehyde   An alkyl   A carboxylic
                               hydrate   chromate ester   acid

B. Pyridinium Chlorochromate

The form of Cr(VI) most commonly used for oxidation of a primary alcohol to an aldehyde is prepared by dissolving CrO_3 in aqueous HCl and adding pyridine to precipitate **pyridinium chlorochromate (PCC)** as a solid.

$$CrO_3 + HCl + \text{(pyridine)} \longrightarrow \text{(pyridinium chlorochromate)} \quad ClCrO_3{}^-$$

pyridinium ion
chlorochromate ion

Pyridine   **Pyridinium chlorochromate**
(PCC)

PCC oxidations are carried out in aprotic solvents, most commonly dichloromethane, CH_2Cl_2.

PCC not only is selective for the oxidation of primary alcohols to aldehydes but also has little effect on carbon-carbon double bonds or other easily oxidized functional groups. In the following example, geraniol, a primary terpene alcohol, is oxidized to geranial without affecting either carbon-carbon double bond.

Geraniol   **Geranial**

PCC does not oxidize aldehydes further because the PCC reagent is not used in water but rather in an organic solvent, usually CH_2Cl_2. Without water, the product aldehyde is not in equilibrium with the aldehyde hydrate. Recall that only an —OH of an aldehyde hydrate is susceptible to further oxidation by Cr(VI), not an aldehyde carbonyl. Both PCC and H_2CrO_4 can be used for the oxidation of a 2° alcohol to a ketone.

Example 10.11

Draw the product of treating each alcohol with PCC.

(a) 1-Hexanol   **(b)** 2-Hexanol   **(c)** Cyclohexanol

Solution

1-Hexanol, a primary alcohol, is oxidized to hexanal. 2-Hexanol, a secondary alcohol, is oxidized to 2-hexanone. Cyclohexanol, a secondary alcohol, is oxidized to cyclohexanone.

(a) **Hexanal**   (b) **2-Hexanone**   (c) **Cyclohexanone**

Problem 10.11

Draw the product of treating each alcohol in Example 10.11 with chromic acid.

Potassium dichromate oxidation of ethanol to acetic acid is the basis for the original breath alcohol screening test used by law enforcement agencies to determine a person's blood alcohol content. The test is based on the difference in color between the dichromate ion (reddish orange) in the reagent and the chromium

ken off, a mouthpiece is fitted to one end, and the other end is inserted into the neck of a plastic bag. The person being tested then blows into the mouthpiece until the plastic bag is inflated.

As breath containing ethanol vapor passes through the tube, reddish orange dichromate is reduced to

$$CH_3CH_2OH \;+\; Cr_2O_7^{2-} \;\xrightarrow[\text{H}_2\text{O}]{\text{H}_2\text{SO}_4}\; CH_3\overset{\displaystyle O}{\overset{\|}{C}}OH \;+\; Cr^{3+}$$

| **Ethanol** | **Dichromate ion** (reddish orange) | | **Acetic acid** | **Chromium(III) ion** (green) |

(III) ion (green) in the product. Thus, color change from reddish orange to green can be used as a measure of the quantity of ethanol present in a sample of a person's breath.

In its simplest form, a breath alcohol screening test consists of a sealed glass tube containing a potassium dichromate-sulfuric acid reagent impregnated on silica gel. To administer the test, the ends of the tube are bro-

green chromium (III). The concentration of ethanol in the breath is then estimated by measuring how far the green color extends along the length of the tube. When the green color extends beyond the halfway point, the person is judged to have a sufficiently high blood alcohol content to warrant further, more precise testing.

The Breathalyzer, a more accurate testing device, operates on the same principle as the simplified screening test. In a Breathalyzer test, a measured volume of breath is bubbled through a solution of potassium dichromate in aqueous sulfuric acid, and the color change is measured spectrophotometrically.

These tests measure alcohol in the breath. The legal definition of being under the influence of alcohol, however, is based on blood alcohol content, not breath alcohol content. The chemical correlation between these two measurements is that air deep within the lungs is in equilibrium with blood passing through the pulmonary arteries, and an equilibrium is established between blood alcohol and breath alcohol. It has been determined by tests in a person drinking alcohol that 2100 mL of breath contains the same amount of ethanol as 1.00 mL of blood.

Glass tube containing potassium dichromate–sulfuric acid coated on silica gel particles

Person forces breath through mouthpiece into the tube

As person blows into the tube, the plastic bag becomes inflated

C. Periodic Acid Oxidation of Glycols

Periodic acid, H_5IO_6 (or alternatively $HIO_4 \cdot 2H_2O$), is a white crystalline solid, mp 122°C. Its major use in organic chemistry is for the cleavage of a glycol to two carbonyl groups. In the process, periodic acid is reduced to iodic acid.

$$\begin{array}{c} -\!\overset{|}{C}\!-\!OH \\ -\!\overset{|}{C}\!-\!OH \end{array} + HIO_4 \longrightarrow \begin{array}{c} -\overset{|}{C}\!\!=\!\!O \\ -\overset{|}{C}\!\!=\!\!O \end{array} + HIO_3 + H_2O$$

| **Periodic acid** | | **Iodic acid** |

For example, periodic acid oxidizes *cis*-1,2-cyclohexanediol to hexanedial.

| *cis*-1,2-Cyclohexanediol | Periodic acid | Hexanedial | Iodic acid |

Mechanism Oxidation of a Glycol by Periodic Acid

Step 1: Reaction of the glycol with periodic acid gives a five-membered cyclic periodate.

A cyclic periodate

There is no change in the oxidation state of either iodine or the glycol as a result of formation of the cyclic periodate.

Step 2: Break bonds to give stable molecules or ions.   Redistribution of valence electrons within the cyclic periodate gives HIO_3 and two carbonyl groups. A result of this electron redistribution is an oxidation of the organic component and a reduction of the iodine-containing component.

HIO_3
(Iodic acid)

This mechanism is consistent with the fact that HIO_4 oxidations are restricted to glycols that can form a five-membered cyclic periodate. Any glycol that cannot form such a cyclic periodate is not oxidized by periodic acid. Following are structural formulas for two isomeric decalindiols. Only the *cis* glycol can form a cyclic periodate with periodic acid, and only the *cis* glycol is oxidized by this reagent.

The *trans* isomer is unreactive toward periodic acid

The *cis* isomer forms a cyclic periodate and is cleaved

It should come as no surprise that biological systems do not use agents like potassium dichromate or the oxides of other transition metals for the oxidation of alcohols to aldehydes and ketones, or for the oxidation of aldehydes to carboxylic acids. What biological systems use instead is NAD$^+$, a compound derived from niacin, one of the B complex vitamins.

nitrogen. The key concept here is that NAD$^+$ is a two-electron oxidizing agent, and in the process undergoes a two-electron reduction to NADH.

NAD$^+$ serves as an oxidizing agent in a wide variety of enzyme-catalyzed reactions, two of which are shown here. The oxidation of ethanol to acetaldehyde is the first step in the reaction by which the liver detoxifies

A pyridine ring

**Nicotinic acid
(Niacin; Vitamin B6)**

The business end of NAD$^+$

An amide group

The plus sign in NAD$^+$ represents this charge on nitrogen

**Nicotinamide adenine
dinucleotide (NAD$^+$)**

In converting niacin to NAD$^+$, the body converts the carboxyl group to an amide group (the —OH of the carboxyl group is replaced by —NH$_2$). In addition, the nitrogen of the six-membered ring is bonded to the sugar ribose, which is in turn bonded to a molecule called adenosine diphosphate (ADP). We will study ribose and ADP in Chapters 25 and 28. For now, we abbreviate the combination of ribose and ADP as Ad. The function of the Ad portion of the molecule is to position NAD$^+$ on the surface of the enzyme in the proper orientation relative to the molecule it is to oxidize.

When NAD$^+$ functions as a biological oxidizing agent, it is reduced to NADH. In this transformation, NAD$^+$ gains one H and loses the positive charge on its

$+ \, H^+ + 2e^-$ $\quad \underset{\text{oxidation}}{\overset{\text{reduction}}{\rightleftharpoons}} \quad$

NAD$^+$
(oxidized form)

NADH
(reduced form)

ethanol. The oxidation of lactate to pyruvate is one step in the process by which the body derives energy from the oxidation of carbohydrates. Lactate is the end-product of anaerobic (without oxygen) glycolysis.

$$\text{CH}_3\text{CH}_2\text{OH} + \text{NAD}^+ \xrightarrow{\underset{\text{dehydrogenase}}{\text{alcohol}}} \text{CH}_3\overset{\displaystyle O}{\overset{\|}{\text{CH}}} + \text{NADH} + \text{H}^+$$

Ethanol

**Ethanal
(Acetaldehyde)**

$$\underset{\textbf{Lactate}}{\overset{\displaystyle \text{OH}}{\underset{|}{\text{CH}_3\text{CHCOO}^-}}} + \text{NAD}^+ \xrightarrow{\underset{\text{dehydrogenase}}{\text{lactate}}} \underset{\textbf{Pyruvate}}{\text{CH}_3\overset{\displaystyle O}{\overset{\|}{\text{CCOO}^-}}} + \text{NADH} + \text{H}^+$$

Following is a mechanism for the oxidation of an alcohol by alcohol dehydrogenase and NAD$^+$.

Mechanism Oxidation of an Alcohol by NAD$^+$

Arrow①: A basic group, B$^-$, on the enzyme removes H$^+$ from the —OH group.

Arrow②: The H—O σ bond breaks as a C=O bond forms.

Arrow③: Transfer of a hydride ion from the carbon bearing the —OH group to NAD$^+$ creates the new C—H bond in NADH. This is the oxidation-reduction step: The alcohol is oxidized, and NAD$^+$ is reduced.

Arrows④–⑤: Electrons within the ring flow to the positively charged nitrogen.

An electron pair is added to nitrogen.

NAD$^+$ **NADH**

Because enzymes are chiral catalysts, NAD$^+$ oxidation takes place in a chiral environment, with the result that hydride transfer to NAD$^+$ is stereoselective; some enzymes catalyze the addition of hydride ion to the top face of NAD$^+$; others, to the bottom face. In the case of alcohol dehydrogenase, the hydride ion is transferred to the top face.

As we will see in Section 16.11, NADH can in turn reverse the process and transfer a hydride ion stereoselectively to the carbonyl group of an aldehyde or ketone, thus reducing these types of molecules to either primary or secondary alcohols respectively.

Example 10.12

What products are formed when each glycol is treated with HIO$_4$?

(a)

(b)

(racemic)

Solution

The bond between the carbons bearing the —OH groups is cleaved and each —OH group is converted to a carbonyl group.

(a) **(b)**

Problem 10.12

α-Hydroxyketones and α-hydroxyaldehydes are also oxidized by treatment with periodic acid.

$$\xrightarrow[\text{2. HIO}_4]{\text{1. H}_2\text{O}}$$

(racemic)

It is not the α-hydroxyketone or aldehyde, however, that undergoes reaction with periodic acid but the hydrate formed by addition of water to the carbonyl group of the α-hydroxyketone or aldehyde. Write a mechanism for the oxidation of this α-hydroxyaldehyde by HIO_4.

10.9 | Thiols

A. Structure

Thiol
A compound containing an —SH (sulfhydryl) group bonded to an sp^3 hybridized carbon.

Mercaptan
A common name for a thiol; that is, any compound that contains an —SH (sulfhydryl) group.

The functional group of a **thiol** is an —**SH** (**sulfhydryl**) group bonded to an sp^3 hybridized carbon. Figure 10.4 shows a Lewis structure and a ball-and-stick model of methanethiol, CH_3SH, the simplest thiol. The C—S—H bond angle in methanethiol is 100.3°. By way of comparison, the H—S—H bond angle in H_2S is 93.3°. If a sulfur atom were bonded to two other atoms by fully hybridized sp^3 hybrid orbitals, bond angles about sulfur would be approximately 109.5°. If, instead, a sulfur atom were bonded to two other atoms by unhybridized $3p$ orbitals, bond angles would be approximately 90°. The fact that the C—S—H bond angle in methanethiol is 100.3° and the H—S—H bond angle in H_2S is 93.3° indicates that there is considerably more p-character (and hence less s-character) in the bonding orbitals of divalent sulfur than there is in those of divalent oxygen.

B. Nomenclature

In the older literature, thiols are often referred to as **mercaptans**, which literally means mercury capturing. They react with Hg^{2+} in aqueous solution to give sulfide salts as insoluble precipitates. Thiophenol, C_6H_5SH, for example, gives $(C_6H_5S)_2Hg$.

In the IUPAC system, thiols are named by selecting as the parent alkane the longest chain of carbon atoms that contains the —SH group. To show that the compound is a thiol, retain the final -*e* in the name of the parent alkane and add the suffix -*thiol*. The location of the —SH group takes precedence over alkyl groups and halogens in numbering the parent chain.

In the IUPAC system, —OH takes precedence over —SH in both numbering and naming. In compounds containing these two functional groups, an —SH group is indicated by the IUPAC prefix *sulfanyl-*. Alternatively, it may be indicated by the common-name prefix *mercapto-*.

(a)

(b)

100.3°

Figure 10.4
Methanethiol, CH_3SH. (a) Lewis structure and (b) ball-and-stick model.

Common names for simple thiols are derived by naming the alkyl group bonded to —SH and adding the word *mercaptan.*

1-Butanethiol
(Butyl mercaptan)

2-Methyl-1-propanethiol
(Isobutyl mercaptan)

2-Sulfanylethanol
(β-Mercaptoethanol)

Example 10.13

Write names for these thiols.

(a) **(b)**

Solution

(a) 1-Pentanethiol (pentyl mercaptan)
(b) (*E*)-2-Butene-1-thiol (*trans*-2-butene-1-thiol)

Problem 10.13

Write IUPAC names for these thiols.

(a) **(b)**

C. Physical Properties

The most outstanding physical characteristic of low-molecular-weight thiols is their stench. The scent of skunks is primarily the result of the following two thiols.

2-Butene-1-thiol

3-Methyl-1-butanethiol
(Isopentyl mercaptan)

A blend of low-molecular-weight thiols is added to natural gas as odorants. The most common of these is 2-methyl-2-propanethiol (*tert*-butyl mercaptan) because it is the most resistant to oxidation and has the greatest soil penetration. 2-Propanethiol is also used for this purpose, usually as a blend with 2-methyl-2-propanethiol.

2-Methyl-2-propanethiol
(*tert*-Butyl mercaptan)

2-Propanethiol
(Isopropyl mercaptan)

Because of the very low polarity of the S—H bond, thiols show little association by hydrogen bonding. Consequently, they have lower boiling points and are less soluble in water and other polar solvents than alcohols of comparable molecular weights. Table 10.3 gives names and boiling points for three low-molecular-weight thiols. Shown for comparison are boiling points for alcohols that contain the same number of carbon atoms.

In Section 10.2, we illustrated the importance of hydrogen bonding in alcohols by comparing the boiling points of ethanol (bp 78°C) and its constitutional isomer dimethyl ether (bp −24°C). By comparison, the boiling point of ethanethiol is 35°C, and that of its constitutional isomer dimethyl sulfide is 37°C.

$$CH_3CH_2SH \qquad\qquad CH_3SCH_3$$

Ethanethiol   **Dimethyl sulfide**
bp 35°C   bp 37°C

The fact that the boiling points of these constitutional isomers are almost identical indicates that little or no association by hydrogen bonding occurs between thiol molecules.

Table 10.3	Boiling Points of Thiols and Alcohols of the Same Number of Carbon Atoms		
Thio	**bp (°C)**	**Alcohol**	**bp (°C)**
Methanethiol	6	**Methanol**	65
Ethanethiol	35	**Ethanol**	78
1-Butanethiol	98	**1-Butanol**	117

D. Thiols in Biological Molecules

The thiol group is found in the amino acid L-cysteine. L-Cysteine is important because the thiol groups of pairs of cysteines are oxidized to disulfide bonds (Section 10.9G), which are a major factor in stabilizing the three-dimensional structure of protein molecules. The thiol group of cysteine functions as a nucleophile in certain enzyme mechanisms. In addition, it binds the metal in certain metal-containing enzymes.

(R)-2-Amino-3-mercaptopropanoic acid
(L-Cysteine)

E. Preparation

The most common preparation of thiols, RSH, depends on the high nucleophilicity of the hydrosulfide ion, HS⁻ (Section 9.3E). Sodium hydrosulfide is prepared by bubbling H_2S through a solution of NaOH in water or aqueous ethanol. Reaction of HS⁻ with a haloalkane gives a thiol.

$$CH_3(CH_2)_8CH_2I + Na^+SH^- \xrightarrow[\text{ethanol}]{S_N2} CH_3(CH_2)_8CH_2SH + Na^+I^-$$

1-Iododecane   **Sodium**   **1-Decanethiol**
hydrosulfide

In practice, the scope and limitations of this reaction are governed by the limitations of the S_N2 reaction and by competition between substitution and β-elimination (Section 9.8). The reaction is most useful for preparation of thiols from primary halides. Yields are lower from secondary halides because of the competing β-elimination (E2) reaction. With tertiary halides, β-elimination (E2) predominates, and the alkene formed by dehydrohalogenation is the major product.

In a commercial application of thiol formation by this nucleophilic substitution, the sodium salt of thioglycolic acid is prepared by the reaction of sodium hydrosulfide and sodium iodoacetate.

$$Na^+ SH^- \;+\; ICH_2\overset{\displaystyle O}{\overset{\displaystyle \|}{C}}O^- Na^+ \;\xrightarrow{S_N2}\; HSCH_2\overset{\displaystyle O}{\overset{\displaystyle \|}{C}}O^- Na^+ \;+\; Na^+ I^-$$

Sodium   **Sodium iodoacetate**   **Sodium mercaptoacetate**
hydrosulfide   **(Sodium thioglycolate)**

The sodium and ammonium salts of thioglycolic acid are used in cold waving of hair. These compounds work by breaking the disulfide bonds of hair proteins, which maintain the overall structure of hair (Section 27.6C). The disulfide bonds are subsequently reformed by oxidation (Section 10.9G) in a second step. The calcium salt of thioglycolic acid is used as a depilatory; that is, it is used to remove body hair.

$$(HSCH_2COO^-)_2Ca^{2+}$$

Calcium mercaptoacetate
(Calcium thioglycolate)

F. Acidity

Hydrogen sulfide is much more acidic than water.

$$H_2O + H_2O \rightleftharpoons HO^- + H_3O^+ \qquad pK_a = 15.7$$

$$H_2S + H_2O \rightleftharpoons HS^- + H_3O^+ \qquad pK_a = 7.0$$

Similarly, thiols are stronger acids than alcohols. Compare, for example, the pK_a values of ethanol and ethanethiol in dilute aqueous solution.

$$CH_3CH_2OH + H_2O \rightleftharpoons CH_3CH_2O^- + H_3O^+ \qquad pK_a = 15.9$$

$$CH_3CH_2SH + H_2O \rightleftharpoons CH_3CH_2S^- + H_3O^+ \qquad pK_a = 10.5$$

The greater acidity of thiols compared to alcohols can be explained by the fact that sulfur (a third-period element) is larger than oxygen (a second-period element). The negative charge on an alkylsulfide ion (RS^-) is delocalized over a larger area and is therefore more stable than the negative charge on an alkoxide ion (RO^-).

Thiols are sufficiently strong acids so that when dissolved in aqueous sodium hydroxide, they are converted completely to alkylsulfide salts.

$$CH_3CH_2SH \quad + \quad Na^+OH^- \longrightarrow \quad CH_3CH_2S^-Na^+ \quad + \quad H_2O$$

$$pK_a = 10.5 \qquad\qquad\qquad\qquad\qquad\qquad pK_a = 15.7$$
(Stronger acid) (Stronger base)   (Weaker base)   (Weaker acid)

To name salts of thiols, give the name of the cation first, followed by the name of the alkyl group to which is attached the suffix -*sulfide*. For example, the sodium salt derived from ethanethiol is named sodium ethylsulfide.

G. Oxidation

Many of the chemical properties of thiols stem from the fact that the sulfur atom of a thiol is oxidized easily to several higher oxidation states, the most common of which are shown in the following flow diagram.

Each oxidation requires a specific oxidizing agent to avoid over-oxidation. There are other oxidation states, too, but they are not very stable. Note that, as drawn, the valence shell of sulfur appears to contain 8 electrons in a thiol and a disulfide, 10 electrons in a sulfinic acid, and 12 electrons in a sulfonic acid. Recall that, although the matter is still being debated, recent calculations indicate that in sulfinic acids and sulfonic acids, it is best to consider each S—O bond to be a single bond, with a negative charge on O and one or two positive charges on S, respectively (Section 1.2E).

The most common oxidation-reduction reaction of sulfur compounds in biological systems is interconversion between a thiol and a disulfide. The functional group of a disulfide is an —S—S— group.

$$2 \text{ RSH} + \text{I}_2 \longrightarrow \text{RSSR} + 2 \text{ HI}$$

A thiol   A disulfide

Thiols are also oxidized to disulfides by molecular oxygen

$$2 \text{ RSH} + \tfrac{1}{2}\text{O}_2 \longrightarrow \text{RSSR} + \text{H}_2\text{O}$$

A thiol   A disulfide

In fact, thiols are so susceptible to oxidation that they must be protected from contact with air during storage.

The disulfide bond is an important structural feature stabilizing the tertiary structure of many proteins (Section 27.6C).

Recall from Table 9.7 that thiols are moderate nucleophiles and thiolates are good nucleophiles. As such, they rank as better nucleophiles in both their neutral and anionic states as compared to alcohols and alkoxides, respectively. The reason that thiols and thiolates are better nucleophiles is that the sulfur atom is more polarizable than oxygen, and increased polarizability enhances nucleophilicity. Also, because the pK_a values for thiols are generally less than 11, thiolate anions react with secondary alkyl halides primarily through an S_N2 substitution mechanism rather than an E2 elimination. The reverse is true for the significantly more basic alkoxides.

Summary

Section 10.1 Structure and Nomenclature of Alcohols

- The functional group of an alcohol is an —**OH (hydroxyl) group** bonded to an sp^3 hybridized carbon.
- Alcohols are classified as **1°**, **2°**, or **3°** depending on whether the —OH group is bonded to a **primary**, **secondary**, or **tertiary** carbon.
- IUPAC names of alcohols are derived by changing the suffix of the parent alkane from *-e* to *-ol.*
  - The chain is numbered from the direction that gives the carbon bearing the —OH the lower number.
  - In compounds containing other functional groups of higher precedence, the presence of —OH is indicated by the prefix *hydroxy-.*

Problems: 10.1–10.3, 10.14–10.17

- Common names for alcohols are derived by naming the alkyl group bonded to —OH and adding the word *alcohol.*

Section 10.2 Physical Properties of Alcohols

- Alcohols are polar compounds with oxygen bearing a partial negative charge and both the carbon and the hydroxyl hydrogen bearing partial positive charges.
  - Because of intermolecular association by **hydrogen bonding**, the boiling points of alcohols are higher than those of hydrocarbons of comparable molecular weight.
  - Because of increased dispersion forces, the boiling points of alcohols increase with increasing molecular weight.

Problems: 10.4–10.5, 10.18–10.24, 10.47, 10.48

  - Alcohols interact with water by hydrogen bonding and therefore are more soluble in water than hydrocarbons of comparable molecular weight.

Section 10.3 Acidity and Basicity of Alcohols

- Alcohols can function as both weak acids (proton donors) and weak bases (proton acceptors).
  - The pK_a of most alcohols is in the 16–18 range, similar to water. Loss of a proton from an alcohol produces an alkoxide anion, a relatively strong base.
  - In the presence of a strong acid, the —OH group can be protonated. Protonation of the —OH group produces the good leaving group —OH$_2^+$, and this is a common mechanistic theme for alcohol reactions.

Problems: 10.25–10.28, 10.49

Section 10.4 Reaction of Alcohols with Active Metals

- Alcohols react with active metals such as Li, Na, and K to liberate H$_2$ and form metal alkoxides.

Problems: 10.6, 10.29, 10.30

Section 10.5 Conversion of Alcohols to Haloalkanes and Sulfonates

- Alcohols can be converted to haloalkanes by reaction with HCl, HBr, and HI, or PBr$_3$, SOCl$_2$ or SOBr$_2$.
  - The HX reactions work best for tertiary alcohols, where they involve a carbocation intermediate.
  - PBr$_3$, SOCl$_2$, or SOBr$_2$ react well with secondary and primary alcohols. These reactions involve an S$_N$2 displacement, so inversion of stereochemistry is seen when the OH group is bonded to a chiral center.
- Alcohols react with sulfonyl chlorides to give alkyl sulfonates. The sulfonate group is a good leaving group analogous to a halogen atom.

Problems: 10.7, 10.29–10.31, 10.33, 10.35, 10.40, 10.43–10.45, 10.50–10.52, 10.55, 10.57, 10.59

Section 10.6 Acid-Catalyzed Dehydration of Alcohols

- Treatment of alcohols with strong acid leads to **dehydration**, which is elimination of water from adjacent carbon atoms to give an alkene.
  - Zaitsev's rule is followed, that is, the predominant product is the most stable (usually the most highly substituted) alkene. The mechanism is the reverse of acid catalyzed hydration of an alkene.
- According to the **principle of microscopic reversibility**, the sequence of transition states and reactive intermediates (that is, the mechanism) for any reversible reaction must be the same, but in reverse order, for the reverse reaction as for the forward reaction.

Problems: 10.8, 10.9, 10.29–10.32, 10.34, 10.35, 10.37, 10.40, 10.42, 10.45, 10.46

Section 10.7 The Pinacol Rearrangement

- Some compounds containing OH groups on two adjacent carbon atoms (sometimes called glycols) undergo a characteristic reaction in acid to generate a rearranged aldehyde or ketone product.

Problems: 10.10, 10.36, 10.50

Section 10.8 Oxidation of Alcohols

- Primary alcohols are oxidized by aqueous H$_2$CrO$_4$ to give carboxylic acids in a process that involves initial aldehyde formation, followed by conversion to an aldehyde hydrate that is further oxidized to the carboxylic acid.
- Primary alcohols are oxidized to aldehydes using PCC, a reagent that is not used in water, thereby precluding hydrate formation and any further oxidation.
- Secondary alcohols can be oxidized to ketones using either PCC or H$_2$CrO$_4$.
- Tertiary alcohols are not oxidized.
- Glycols can be oxidized by periodic acid to give two carbonyl species along with cleavage of the C—C bond between the —OH groups.

Problems: 10.11, 10.12, 10.29–10.31, 10.35, 10.40, 10.41, 10.45, 10.50, 10.51, 10.53, 10.54, 10.56

Section 10.9 Thiols

- A **thiol** is a sulfur analog of an alcohol; it contains an **—SH (sulfhydryl) group** in place of an —OH group.
- Thiols are named in the same manner as alcohols, but the suffix *-e* of the parent alkane is retained and *-thiol* added.
  - In compounds containing other functional groups of higher precedence, the presence of —SH is indicated by the prefix *mercapto-*.

- Common names for thiols are derived by naming the alkyl group bonded to —SH and adding the word *mercaptan*.
- Because the S—H bond is almost nonpolar, the physical properties of thiols are more like those of hydrocarbons of comparable molecular weight.
- Thiols are much more acidic than alcohols.
- Thiols are oxidized to give disulfides, sulfinic acids, and sulfonic acids.

Problem: 10.13

OWL Questions and problems for this chapter are available in OWL. Use the OWL access card that came with this textbook to access assigned questions and problems for this chapter.

Key Reactions

1. Acidity of Alcohols (Section 10.3) In dilute aqueous solution, methanol and ethanol are comparable in acidity to water; secondary and tertiary alcohols are somewhat weaker acids.

$$CH_3OH + H_2O \rightleftharpoons CH_3O^- + H_3O^+ \qquad pK_a = 15.5$$

2. Reaction with Active Metals (Section 10.4) Alcohols react with Li, Na, K, and other active metals to form metal alkoxides, which are nearly the same or somewhat stronger bases than the alkali metal hydroxides such as NaOH and KOH.

$$2\,CH_3OH + 2\,Na \longrightarrow 2\,CH_3O^-Na^+ + H_2$$

3. Reaction with HCl, HBr, and HI (Section 10.5A) Primary alcohols react by an S_N2 mechanism. Strong acid protonates the —OH group, converting it to the good leaving group, —OH_2^+, setting up either S_N2 or S_N1 reactions.

$$\text{\raisebox{0pt}{$\diagdown\!\diagup\!\diagdown$}OH} + HBr \xrightarrow{\text{reflux}} \text{\raisebox{0pt}{$\diagdown\!\diagup\!\diagdown$}Br} + H_2O$$

Tertiary alcohols react by an S_N1 mechanism with formation of a carbocation intermediate.

$$\text{$>$—OH} + HCl \xrightarrow{25°C} \text{$>$—Cl} + H_2O$$

Secondary alcohols may react by an S_N2 or an S_N1 mechanism, depending on the alcohol and experimental conditions. Primary alcohols with extensive β-branching react by an S_N1 mechanism involving formation of a rearranged carbocation.

$$\text{\raisebox{0pt}{$\diagup\!\!\!\times\!\diagdown$}OH} + HBr \longrightarrow \text{\raisebox{0pt}{$\diagup\!\!\underset{Br}{\diagdown\!\!\diagup}\!\diagdown$}} + H_2O$$

4. Reaction with PBr₃ (Section 10.5B) Alcohols react with PBr_3 to initially displace Br^- to give a protonated dibromophosphite intermediate, which is displaced from the backside by Br^- to give the bromoalkane. Although some rearrangement may occur with this reagent, it is less likely than in the reaction of an alcohol with HBr.

$$\overset{\text{OH}}{\underset{|}{3\,CH_3CHCH_3}} + PBr_3 \longrightarrow \overset{\text{Br}}{\underset{|}{3\,CH_3CHCH_3}} + H_3PO_3$$

5. Reaction with SOCl₂ and SOBr₂ (Section 10.5C) Alcohols react with $SOCl_2$ to initially give an alkyl chlorosulfite intermediate, which is displaced from the backside by Cl^- to give the chloroalkane. This is often the method of choice for converting a primary or secondary alcohol to an alkyl chloride or alkyl bromide.

$$\text{\raisebox{0pt}{$\diagdown\!\diagup\!\diagdown$}OH} + SOCl_2 \xrightarrow{\text{pyridine}} \text{\raisebox{0pt}{$\diagdown\!\diagup\!\diagdown$}Cl} + SO_2 + HCl$$

6. Acid-Catalyzed Dehydration (Section 10.6) When isomeric alkenes are possible, the major product is generally the more substituted alkene (Zaitsev's rule). Rearrangements are common with secondary alcohols and also with primary alcohols with extensive β-branching. The mechanism is the reverse of acid-catalyzed hydration of an alkene, involving initial protonation of the —OH group followed by loss of water to give a carbocation, which loses a proton to give an alkene.

$$\underset{\substack{\text{2-Butanol}}}{CH_3CH_2\overset{\displaystyle OH}{\underset{\displaystyle |}{CH}}CH_3} \xrightarrow[\text{heat}]{85\% \ H_3PO_4} \underset{\substack{\text{2-Butene}\\(80\%)}}{CH_3CH=CHCH_3} + \underset{\substack{\text{1-Butene}\\(20\%)}}{CH_3CH_2CH=CH_2} + H_2O$$

7. Pinacol Rearrangement (Section 10.7) Dehydration of a glycol involves formation of a carbocation intermediate, rearrangement, and loss of H$^+$ to give an aldehyde or ketone.

8. Oxidation of a Primary Alcohol to a Carboxylic Acid (Section 10.8A) A primary alcohol is oxidized to a carboxylic acid by chromic acid. The mechanism involves initial formation of an alkyl chromate intermediate, followed by reaction with base to remove a proton, generating the carbonyl group of an aldehyde and simultaneously reducing the chromium (VI) to chromium (IV). An initially formed aldehyde adds water, generating an aldehyde hydrate, which is oxidized according to the same mechanism to give the carboxylic acid.

9. Oxidation of a Secondary Alcohol to a Ketone (Section 10.8A and 10.8B) A secondary alcohol is oxidized to a ketone by chromic acid or by PCC.

10. Oxidation of a Primary Alcohol to an Aldehyde (Section 10.8B) The oxidation of a primary alcohol to an aldehyde is most conveniently carried out using pyridinium chlorochromate (PCC). Because there is no water, the aldehyde does not form the hydrate, and the oxidation reaction stops at the aldehyde stage.

11. Oxidative Cleavage of a Glycol (Section 10.8C) HIO$_4$ reacts with a glycol to form a five-membered cyclic periodate intermediate that undergoes carbon-carbon bond cleavage to form two carbonyl groups.

12. Acidity of Thiols (Section 10.9F) Thiols are weak acids, pK_a 10–11, but considerably stronger than alcohols, pK_a 15.5–18.

$$CH_3CH_2SH + H_2O \rightleftharpoons CH_3CH_2S^- + H_3O^+ \qquad pK_a = 10.5$$

13. Oxidation of Thiols to Disulfides (Section 10.9G) Oxidation by weak oxidizing agents such as O$_2$ and I$_2$ gives disulfides.

$$2 \ RSH + \tfrac{1}{2}O_2 \longrightarrow RSSR + H_2O$$

Chapter 11

Ethers, Epoxides, and Sulfides

The discovery that inhaling ethers could make a patient insensitive to pain revolutionized the practice of medicine. Inset: A model of isoflurane, $CF_3CHClOCHF_2$, a halogenated ether widely used as an inhalation anesthetic in both human and veterinary medicine.

© Alan Levenson/Getty Images

In this chapter, we first discuss the structure, nomenclature, physical properties, and chemical properties of ethers and then compare their physical properties with those of isomeric alcohols. Next, we study the preparation and chemical properties of a group of cyclic ethers called epoxides. As we shall see, their most important reactions involve nucleophilic substitution. This chapter continues the discussion of S_N1 and S_N2 reaction mechanisms begun in Chapter 9 and continued into Chapter 10.

11.1 | Structure of Ethers

Ether
A compound containing an oxygen atom bonded to two carbon atoms.

The functional group of an **ether** is an atom of oxygen bonded to two carbon atoms. Figure 11.1 shows a Lewis structure and a ball-and-stick model of dimethyl ether, CH_3OCH_3, the simplest ether. In dimethyl ether, two sp^3 hybrid orbitals of oxygen form σ bonds to the two carbon atoms. The other two sp^3 hybrid orbitals of oxygen each contain an unshared pair of electrons. The C—O—C bond angle in dimethyl ether is 110.3°, a value close to the tetrahedral angle of 109.5°.

In still other ethers, the ether oxygen is bonded to sp^2 hybridized carbons. In ethoxyethene (ethyl vinyl ether), for example, the ether oxygen is bonded to one sp^3 and one sp^2 hybridized carbon.

$$CH_3CH_2—O—CH=CH_2$$
Ethoxyethene
(Ethyl vinyl ether)

OWL Online homework for this chapter may be assigned in OWL for Organic Chemistry.

11.2 | Nomenclature of Ethers

In the IUPAC system, ethers are named by selecting the longest carbon chain as the parent alkane and naming the —OR group bonded to it as an **alkoxy group**. Common names are derived by listing the alkyl groups bonded to oxygen in alphabetical order and adding the word "ether." Following are the IUPAC names and, in parentheses, the common names for three low-molecular-weight ethers.

Alkoxy group
An —OR group, where R is an alkyl group.

$CH_3CH_2OCH_2CH_3$ (structure with OH and OCH$_2$CH$_3$) CH_3OCCH_3 (with CH$_3$ groups)

Ethoxyethane   **(1R,2R)-2-Ethoxycyclohexanol**   **2-Methoxy-2-methylpropane**
(Diethyl ether)   **(trans-2-Ethoxycyclohexanol)**   **(tert-Butyl methyl ether)**

(a)

$H—C—\overset{..}{\underset{..}{O}}—C—H$

(b)

110.3°

Figure 11.1
Dimethyl ether, CH_3OCH_3.
(a) Lewis structure and
(b) ball-and-stick model.

Chemists almost invariably use common names for low-molecular-weight ethers. For example, although ethoxyethane is the IUPAC name for $CH_3CH_2OCH_2CH_3$, it is rarely called that; rather, it is called diethyl ether, ethyl ether, or even more commonly, simply ether. The abbreviation for *tert*-butyl methyl ether is MTBE, after the common name methyl *tert*-butyl ether (incorrectly alphabetized, as you will recognize).

Three other ethers deserve special mention. 2-Methoxyethanol and 2-ethoxyethanol, more commonly known as Methyl Cellosolve and Cellosolve, are good polar protic solvents in which to carry out organic reactions and are also used commercially in some paint strippers. *Di*ethylene *gly*col di*me*thyl *e*ther, more commonly known by its acronym, diglyme, is a common solvent for hydroboration and $NaBH_4$ reductions.

2-Methoxyethanol   **2-Ethoxyethanol**   **Diethylene glycol dimethyl ether**
(Methyl Cellosolve)   **(Cellosolve)**   **(Diglyme)**

Cyclic ethers are given special names. The presence of an oxygen atom in a saturated ring is indicated by the prefix *ox-*, and ring sizes from three to six are indicated by the endings *-irane, -etane, -olane,* and *-ane,* respectively. Several of these smaller ring cyclic ethers are more often referred to by their common names, here shown in parentheses. Numbering of the atoms of the ring begins with the oxygen atom. These compounds and others in which there is a heteroatom (noncarbon atom) in the ring are called **heterocycles**. Heterocycles containing O or N atoms are particularly common in organic chemistry, and several clinically important pharmaceuticals are heterocycles.

Heterocycle
A cyclic compound whose ring contains more than one kind of atom. Oxirane (ethylene oxide), for example, is a heterocycle whose ring contains two carbon atoms and one oxygen atom.

Oxirane   **Oxetane**   **Oxolane**   **Oxane**   **1,4-Dioxane**
(Ethylene oxide)   **(Tetrahydrofuran)**   **(Tetrahydropyran)**

Example 11.1

Write IUPAC and common names for these ethers.

(a) $CH_3\overset{CH_3}{\underset{CH_3}{C}}OCH_2CH_3$ **(b)** (cyclohexyl—O—cyclohexyl)   **(c)** $CH_2{=}CHOCH_3$

Solution

(a) 2-Ethoxy-2-methylpropane. Its common name is *tert*-butyl ethyl ether.
(b) Cyclohexoxycyclohexane. Its common name is dicyclohexyl ether.
(c) Methoxyethene. Its common name is methyl vinyl ether.

Problem 11.1

Write IUPAC and common names for these ethers.

(a)   (b)   (c)

11.3 | Physical Properties of Ethers

Ethers are polar molecules in which oxygen bears a partial negative charge and each attached carbon bears a partial positive charge (Figure 11.2). However, only weak dipole-dipole interactions exist between ether molecules in the liquid state. Consequently, boiling points of ethers are much lower than those of alcohols of comparable molecular weight (Table 11.1) and are close to those of hydrocarbons of comparable molecular weight (Table 2.5).

Figure 11.2

Although ethers are polar compounds, there are only weak dipole-dipole interactions between their molecules in the liquid state.

Only very weak dipole-dipole interaction

Because ethers cannot act as hydrogen bond donors, they are much less soluble in water than alcohols. However, they can act as hydrogen bond acceptors (Figure 11.3), which makes them more water-soluble than hydrocarbons of comparable molecular weight and shape (compare data in Tables 2.5 and 11.1).

Figure 11.3

Ethers are hydrogen bond acceptors only. They are not hydrogen bond donors.

Hydrogen bond

Table 11.1	Boiling Points and Solubilities in Water of Some Ethers and Alcohols of Comparable Molecular Weight			

Structural Formula	Name	Molecular Weight	bp (°C)	Solubility in Water
CH_3CH_2OH	Ethanol	46	78	Infinite
CH_3OCH_3	Dimethyl ether	46	−24	7.8 g/100 g
$CH_3CH_2CH_2CH_2OH$	1-Butanol	74	117	7.4 g/100 g
$CH_3CH_2OCH_2CH_3$	Diethyl ether	74	35	8.0 g/100 g
$HOCH_2CH_2CH_2CH_2OH$	1,4-Butanediol	90	230	Infinite
$CH_3CH_2CH_2CH_2CH_2OH$	1-Pentanol	88	138	2.3 g/100 g
$CH_3OCH_2CH_2OCH_3$	Ethylene glycol dimethyl ether	90	84	Infinite
$CH_3CH_2CH_2CH_2OCH_3$	Butyl methyl ether	88	71	Slight

Example 11.2

Arrange these compounds in order of increasing solubility in water.

Ethylene glycol   **Diethyl ether**   **Hexane**
dimethyl ether

Solution

Water is a polar solvent. Hexane, a nonpolar hydrocarbon, has the lowest solubility in water. Both diethyl ether and ethylene glycol dimethyl ether are polar compounds because of the presence of the polar C—O—C bond, and each interacts with water as a hydrogen bond acceptor.

Of these three compounds, ethylene glycol dimethyl ether is the most soluble in water because it has more sites for hydrogen bonding (a total of four lone pairs on two O atoms) than diethyl ether.

Insoluble   8g/100g water   Soluble in all proportions

Problem 11.2

Arrange these compounds in order of increasing boiling point.

11.4 | Preparation of Ethers

A. Williamson Ether Synthesis

The most common general method for the synthesis of ethers, the **Williamson ether synthesis**, involves nucleophilic displacement of a halide ion or other good leaving group by an alkoxide ion.

Sodium	Iodomethane	2-Methoxypropane
isopropoxide	(Methyl iodide)	(Isopropyl methyl ether)

In planning a Williamson ether synthesis, it is essential to use a combination of reactants that maximizes nucleophilic substitution and minimizes any competing β-elimination (E2, Section 9.6B). Yields of ether are highest when the halide to be displaced is on a methyl or a primary carbon. Yields are low in the displacement from secondary halides (because of competing β-elimination), and the Williamson ether synthesis fails altogether with tertiary halides (because β-elimination by an E2 mechanism is the exclusive reaction). For example, *tert*-butyl methyl ether can be prepared by the reaction of potassium *tert*-butoxide and bromomethane. Note that bromomethane is the only haloalkane with little enough steric hindrance to react with the highly hindered potassium *tert*-butoxide. Even primary haloalkanes would not react to give a high yield of the corresponding *tert*-butyl ether.

Potassium	Bromomethane	2-Methoxy-2-methylpropane
tert-butoxide	(Methyl bromide)	(*tert*-Butyl methyl ether)

Williamson ether synthesis
A general method for the synthesis of dialkyl ethers by an S$_N$2 reaction between a haloalkane and an alkoxide ion.

With the alternative combination of sodium methoxide and 2-bromo-2-methyl-propane, no ether is formed; 2-methylpropene, formed by dehydrohalogenation, is the only product.

$$CH_3\underset{\underset{CH_3}{|}}{\overset{\overset{CH_3}{|}}{C}}Br \quad + \quad CH_3O^-Na^+ \quad \xrightarrow{E2} \quad CH_3\underset{\underset{CH_3}{|}}{C}=CH_2 \quad + \quad CH_3OH + Na^+Br^-$$

2-Bromo-2-   **Sodium**   **2-Methylpropene**
methylpropane   **methoxide**

Example 11.3

Show the combination of alcohol and haloalkane that can best be used to prepare each ether by the Williamson ether synthesis.

(a) [structure]   (b) [structure with OEt]

Solution

(a) Treat 2-propanol with sodium metal to form sodium isopropoxide. Then treat this metal alkoxide with 1-bromobutane.

[reaction structure] Br + Na⁺ ⁻O[structure] $\xrightarrow{S_N2}$ [structure] O[structure] + Na⁺Br⁻

1-Bromobutane   **Sodium**
             isopropoxide

The alternative combination of sodium butoxide and 2-bromopropane gives considerably more elimination product.

(b) Treat (*S*)-2-butanol with sodium metal to form the sodium alkoxide. This reaction involves only the O—H bond and does not affect the chiral center. Then, treat this sodium alkoxide with a haloethane, for example ethyl iodide (EtI), to give the desired product.

[structure OH] $\xrightarrow[\text{2. EtI}]{\text{1. Na}}$ [structure OEt]

     (*S*)-2-**Butanol**   (*S*)-2-**Ethoxyethane**

An alternative synthesis is to convert the (*R*)-2-butanol to its tosylate (Section 10.5D) followed by treatment with sodium ethoxide.

[structure OH] $\xrightarrow[\text{2. EtO}^-\text{Na}^+]{\text{1. TsCl, pyridine}}$ $CH_3CH=CHCH_3$ + [structure OEt]

   (*R*)-2-**Butanol**   **Product of E2**   (*S*)-2-**Ethoxybutane**
                                        (major product)   **(product of S$_N$2 reaction;**
                                                       a minor product)

This synthesis, however, gives only a low yield of the desired product. Recall from Section 9.8C that when a 2° halide or by analogy a 2° tosylate is treated with a strong base/good nucleophile, E2 is the major reaction.

Problem 11.3

Show how you might use the Williamson ether synthesis to prepare each ether.

(a) [structure]   (b) [structure]

B. Acid-Catalyzed Dehydration of Alcohols

Diethyl ether and several other commercially available ethers are synthesized on an industrial scale by the acid-catalyzed dehydration of primary alcohols. Intermolecular dehydration of ethanol for example, gives diethyl ether.

$$2\ CH_3CH_2OH \xrightarrow[140°C]{H_2SO_4} CH_3CH_2OCH_2CH_3 + H_2O$$

Ethanol   **Diethyl ether**

Mechanism   Acid-Catalyzed Intermolecular Dehydration of a Primary Alcohol

Step 1: **Add a proton.**   Proton transfer from the acid catalyst to the hydroxyl group gives an oxonium ion, which converts —OH, a poor leaving group, into —OH$_2^+$, a better leaving group.

An oxonium ion

Step 2: **Make a new bond between a nucleophile and an electrophile and simultaneously break a bond to give stable molecules or ions.**   Nucleophilic displacement of H$_2$O by the OH group of a second alcohol molecule gives a new oxonium ion.

A new oxonium ion

Step 3: **Take a proton away.**   Proton transfer from the new oxonium ion to H$_2$O completes the reaction.

Note that the acid is a catalyst in this reaction. One proton is used in Step 1 but another is generated in Step 3.

Yields of ethers from the acid-catalyzed intermolecular dehydration of alcohols are highest for symmetrical ethers formed from unbranched primary alcohols. Examples of symmetrical ethers formed in good yield by this method are dimethyl ether, diethyl ether, and dibutyl ether. From secondary alcohols, yields of ether are lower because of competition from acid-catalyzed dehydration (Section 10.6). In the case of tertiary alcohols, dehydration to an alkene is the only reaction.

Example 11.4

Explain why this reaction does not give a good yield of ethyl hexyl ether.

Solution

From this reaction, we expect a mixture of three ethers: diethyl ether, ethyl hexyl ether, and dihexyl ether.

Problem 11.4

Show how ethyl hexyl ether might be prepared by a Williamson ether synthesis.

C. Acid-Catalyzed Addition of Alcohols to Alkenes

Under suitable conditions, alcohols can be added to the carbon-carbon double bond of an alkene to give an ether. The usefulness of this method of ether synthesis is limited to the interaction of alkenes that form stable carbocations and methanol or primary alcohols. An example is the commercial synthesis of *tert*-butyl methyl ether (MTBE). 2-Methylpropene and methanol are passed over an acid catalyst to give the ether.

$$CH_3C{=}CH_2 + CH_3OH \xrightarrow[\text{catalyst}]{\text{acid}} CH_3\overset{\displaystyle CH_3}{\underset{\displaystyle CH_3}{C}}OCH_3$$

2-Methoxy-2-methylpropane
(***tert*-Butyl methyl ether**)

Mechanism   Acid-Catalyzed Addition of an Alcohol to an Alkene

Step 1: Make a new bond between a π bond and an electrophile—add a proton.
Proton transfer from the acid catalyst to the alkene gives a carbocation intermediate.

$$CH_3C{=}CH_2 + H{-}\overset{+}{\underset{H}{\ddot{O}}}{-}CH_3 \rightleftharpoons CH_3\overset{+}{C}CH_3 + :\overset{\ }{\underset{H}{\ddot{O}}}{-}CH_3$$

Step 2: Make a new bond between a nucleophile and an electrophile.   Reaction of the carbocation intermediate (an electrophile and a Lewis acid) with the alcohol (a nucleophile and a Lewis base) gives an oxonium ion.

$$CH_3\overset{+}{C}CH_3 + H\ddot{O}CH_3 \rightleftharpoons CH_3\underset{\overset{+}{\underset{H}{\overset{|}{O}}}\diagdown CH_3}{C}CH_3$$

Step 3: Take a proton away.   Proton transfer to solvent (in this case methanol) completes the reaction.

$$CH_3{-}\ddot{O}{-}H + CH_3\underset{\overset{+}{\underset{H}{\overset{|}{O}}}\diagdown CH_3}{C}CH_3 \rightleftharpoons CH_3{-}\overset{+}{\ddot{O}}{-}H + CH_3\underset{\overset{\ }{\underset{CH_3}{\overset{|}{:O}}}}{C}CH_3$$

At one time MTBE was added to gasoline under a mandate from the Environmental Protection Agency to add "oxygenates," which make gasoline burn more smoothly (it raises the octane number) and lower exhaust emissions. As an octane-improving additive, MTBE is superior to ethanol (the additive in ethanol blend fuels such as E10 and E85). A blend of 15% MTBE with gasoline improves octane rating by approximately 5 units. Unfortunately, because MTBE is much more soluble in water than gasoline, it has gotten into the water table in many places, in some cases because of leaky gas station storage tanks. It has been detected in lakes, reservoirs, and water supplies, in some cases at concentrations that exceed limits for both "taste and odor" and human health. Consequently, its use as a gasoline additive is being phased out.

11.5 | Reactions of Ethers

Ethers resemble hydrocarbons in their resistance to chemical reaction. They do not react with oxidizing agents such as potassium dichromate or potassium permanganate. They are stable toward even very strong bases, and, except for tertiary alkyl ethers, they are not affected by most weak acids at moderate temperatures. Because of their good solubilizing properties and general inertness to chemical reaction, ethers are excellent solvents in which to carry out many organic reactions.

A. Acid-Catalyzed Cleavage by Concentrated HX

Cleavage of dialkyl ethers requires both a strong acid and a good nucleophile, hence the use of concentrated aqueous HI (57%) or HBr (48%). Dibutyl ether, for example, reacts with hot concentrated HBr to give two molecules of 1-bromobutane.

$$\text{(structure)} \quad + 2\,HBr \xrightarrow{\text{heat}} 2 \quad \text{(structure)} Br + H_2O$$

Dibutyl ether   **1-Bromobutane**

Concentrated HCl (38%) is far less effective in cleaving dialkyl ethers, primarily because Cl^- is a weaker nucleophile in water than either I^- or Br^-.

The mechanism of acid-catalyzed cleavage of dialkyl ethers depends on the nature of the carbons bonded to oxygen. If both carbons are primary, cleavage involves an S_N2 reaction in which a halide ion is the nucleophile. Otherwise cleavage is by an S_N1 reaction.

Mechanism   Acid-Catalyzed Cleavage of a Dialkyl Ether

Step 1: Add a proton.   Proton transfer from the acid catalyst to the oxygen atom of the ether gives an oxonium ion.

$$CH_3CH_2-\ddot{O}-CH_2CH_3 + H-\overset{+}{\underset{H}{\ddot{O}}}-H \underset{\text{reversible}}{\overset{\text{fast and}}{\rightleftharpoons}} CH_3CH-\overset{+}{\underset{H}{\ddot{O}}}-CH_2CH_3 + :\underset{H}{\ddot{O}}-H$$

An oxonium ion

Step 2: Make a new bond between a nucleophile and an electrophile and simultaneously break a bond to give stable molecules or ions.   Nucleophilic displacement by halide ion on the primary carbon cleaves the C—O bond; the leaving group is CH_3CH_2OH, a weak base and a poor nucleophile.

$$:\ddot{Br}:^- + CH_3CH_2-\overset{+}{\underset{H}{\ddot{O}}}-CH_2CH_3 \xrightarrow{S_N2} CH_3CH_2-\ddot{Br}: + :\underset{H}{\ddot{O}}-CH_2CH_3$$

This cleavage produces one molecule of bromoalkane and one molecule of alcohol. In the presence of excess concentrated HBr, the alcohol is converted to a second molecule of bromoalkane by another S_N2 process (Section 9.2).

Tertiary, allylic, and benzylic ethers are particularly susceptible to cleavage by acid, often under quite mild conditions. Tertiary butyl ethers, for example, are cleaved by aqueous HCl at room temperature. Proton transfer from the acid to the oxygen atom of the ether produces an oxonium ion, which then cleaves to produce a particularly stable 3°, allylic, or benzylic carbocation. Reaction of the carbocation with Cl⁻ completes the reaction.

A *tert*-butyl ether   A 3° carbocation
                                         intermediate

Example 11.5

Account for the fact that treating most methyl ethers with concentrated HI gives CH_3I and ROH as the initial major products rather than CH_3OH and RI, as illustrated by the following reaction.

Solution

The first step is protonation of the ether oxygen to give an oxonium ion. Cleavage is by an S_N2 pathway on the less hindered methyl carbon.

site of attack by nucleophile

Problem 11.5

Account for the fact that treatment of *tert*-butyl methyl ether with a limited amount of concentrated HI gives methanol and *tert*-butyl iodide rather than methyl iodide and *tert*-butyl alcohol.

Example 11.6

Draw structural formulas for the major products of each reaction.

(a) + HBr ⟶   (b) + HBr
        (Excess)   (1.0 mol)

Solution

(a) Cleavage on either side of the ether oxygen by an S_N2 mechanism gives an alcohol and a bromoalkane. Reaction of the alcohol then gives a second molecule of bromoalkane.

(b) Proton transfer to the ether oxygen followed by cleavage gives a 3° carbocation intermediate, which may then (1) react with bromide ion to give a bromoalcohol or (2) lose a proton to give an unsaturated alcohol.

+ HBr ⟶

(1.0 mol)   A 3° carbocation
                 intermediate

+ Br⁻
S_N1 HO   Br
                4-Bromo-4-methylpentan-l-ol

− H⁺
E1   HO
                4-Methylpent-3-en-1-ol

Problem 11.6

Draw structural formulas for the major products of each reaction.

(a) $\overset{\displaystyle CH_3}{\underset{\displaystyle CH_3}{CH_3C\!-\!OCH_3}}$ + HBr $\longrightarrow$ (b) [cyclic ether structure] + HBr $\longrightarrow$

              (Excess)   (Excess)

B. Ether Safety Alert: Flammability and Formation of Hydroperoxides

Two hazards must be avoided when working with diethyl ether and other low-molecular-weight ethers. First, the commonly used ethers have low boiling points and are highly flammable, a dangerous combination. Consequently, open flames and electric appliances with sparking contacts must be avoided where ethers are being used (laboratory refrigerators and ovens are frequent causes of ignition). Because diethyl ether is so volatile (its boiling point is 35°C), it should be used in a fume hood to prevent the build-up of vapors and possible explosion. Second, anhydrous ethers react with molecular oxygen at a C—H bond adjacent to the ether oxygen to form **hydroperoxides**, which are dangerous because they are explosive.

Hydroperoxide
A compound containing an —OOH group.

Diethyl ether   A hydroperoxide

Hydroperoxidation proceeds by a radical chain mechanism (see "Chemical Connections: Antioxidants" in Chapter 8). Rates of hydroperoxide formation increase dramatically if the C—H bond adjacent to oxygen is secondary, as for example in diisopropyl ether, because of favored generation of a relatively stable 3° radical intermediate next to oxygen.

Diisopropyl ether   A hydroperoxide

This hydroperoxide precipitates from solution as a waxy solid and is particularly dangerous.

Hydroperoxides in ethers can be detected by shaking a small amount of the ether with an acidified 10% aqueous solution of potassium iodide, KI, or by using starch iodine paper with a drop of acetic acid. Peroxides oxidize iodide ion to iodine, I_2, which gives a yellow color to the solution. Hydroperoxides can be removed by treating them with a reducing agent. One effective procedure is to shake the hydroperoxide-contaminated ether with a solution of iron (II) sulfate in dilute aqueous sulfuric acid. You should never use ethers past their expiration date, and you should properly dispose of them before then.

11.6 | Silyl Ethers as Protecting Groups

When dealing with organic compounds containing two or more functional groups, it is often necessary to protect one functional group (to prevent its reaction) while carrying out a reaction at another functional group. Suppose, for example, that you wish to convert 4-pentyn-1-ol to 4-heptyn-1-ol.

4-Pentyn-1-ol   **4-Heptyn-1-ol**

The new carbon-carbon bond can be formed by treating the acetylide anion (Section 7.5) of 4-pentyn-1-ol with bromoethane. 4-Pentyn-1-ol, however, contains two acidic hydrogens, one on the hydroxyl group (pK_a 16–18) and the other on the carbon-carbon triple bond (pK_a 25). Treatment of this compound with one equivalent of $NaNH_2$ forms the alkoxide anion (the —OH group is the stronger acid) rather than the acetylide anion.

Alkoxide anion

To carry out the synthesis of 4-heptyne-1-ol, we must first protect the —OH group to prevent its reaction with sodium amide. A good **protecting group** is

- Easily added to the sensitive functional group.
- Resistant to the reagents used to transform the unprotected functional group or groups.
- Easily removed to regenerate the original functional group.

Chemists have devised protecting groups for most functional groups, and we will encounter several of them in this text. In this section, we concentrate on the most common type of hydroxyl-protecting group, namely silyl ethers.

Silicon is in Group 4A of the Periodic Table, immediately below carbon. Like carbon, silicon also forms tetravalent compounds such as the following.

Silicon dioxide   **Silane**   **Tetramethylsilane**   **Chlorotrimethylsilane**

An —OH group can be converted to a silyl ether by treating it with a trialkylsilyl chloride in the presence of an amine base. For example, treating an alcohol with chlorotrimethylsilane in the presence of a tertiary amine, such as triethylamine or pyridine, gives a trimethylsilyl ether.

Chlorotri-   A trimethylsilyl   **Triethylammonium**
methylsilane   ether   **chloride**

The function of the tertiary amine is to catalyze the reaction by forming some of the more nucleophilic alkoxide ion and to neutralize the HCl formed during the reaction. Tertiary amines are not nucleophilic because the three attached alkyl groups provide steric hindrance. Tertiary amines can react with protons and act as a base, but will not react with electrophiles such as chlorotrimethylsilane.

Replacement of one of the methyl groups of the trimethylsilyl group by *tert*-butyl gives the *tert*-butyldimethylsilyl (TBDMS) group, which is considerably more stable than the trimethylsilyl group. Other common silyl protecting groups are the triethylsilyl (TES) and triisopropylsilyl (TIPS) groups

| **Trimethylsilyl chloride** (**TMSCl**) | **Triethylsilyl chloride** (**TESCl**) | **t-Butyldimethylsilyl chloride** (**TBDMSCl**) | **Triisopropylsilyl chloride** (**TIPSCl**) |

Protecting group
An unreactive group reversibly created for the purpose of preventing a functional group from potentially reacting to give an unwanted product or products.

Silyl ethers are unaffected by most oxidizing and reducing agents, and are stable to most nonaqueous acids and bases. The TBDMS group is stable in aqueous solution within the pH range 2 to 12, which makes it one of the most widely used —OH protecting groups. Silyl ether blocking groups are most easily removed by treatment with fluoride ion, generally in the form of tetrabutylammonium fluoride, $Bu_4N^+F^-$.

$$RCH_2O-Si\!\!\!< \ + \ F^- \ \xrightarrow[\substack{\text{THF} \\ \text{(The solvent)}}]{Bu_4N^+F^-} \ RCH_2OH \ + \ F-Si\!\!\!<$$

A TBDMS-protected
alcohol

This cleavage of the protecting group depends on the fact that a silicon-fluorine σ bond is considerably stronger (582 kJ/mol) than a silicon-oxygen σ bond (368 kJ/mol). In fact, the Si—F σ bond is one of the strongest σ bonds known. The large difference in bond strengths between Si—O and Si—F bonds drives the silyl ether cleavage reaction to completion.

We can use a silyl ether in the following way to convert 4-pentyn-1-ol to 4-heptyn-1-ol. Treating 4-pentyn-1-ol with chlorotrimethylsilane in the presence of triethylamine gives the trimethylsilyl ether. Treatment of the terminal alkyne with sodium amide followed by bromoethane forms the new carbon-carbon bond. Subsequent removal of the TMS protecting group with tetrabutylammonium fluoride gives the desired 4-heptyn-1-ol.

4-Pentyn-1-ol

4-Heptyn-1-ol

Example 11.7

Compare the polarity of the C—Cl bond in $(CH_3)_3C$—Cl with the polarity of the Si—Cl bond in $(CH_3)_3Si$—Cl.

Solution

The difference in electronegativity between carbon and chlorine is $3.0 - 2.5 = 0.5$; that between silicon and chlorine is $3.0 - 1.8 = 1.2$; a Si—Cl bond is more polar than a C—Cl bond.

Problem 11.7

The trimethylsilyl protecting group is easily removed in aqueous solution containing a trace of acid. Propose a mechanism for this reaction. (Note that a TBDMS protecting group is stable under these conditions because of the greater steric crowding around silicon created by the *t*-butyl group.)

11.7 | Epoxides: Structure and Nomenclature

Although **epoxides** are technically classed as ethers, we discuss them separately because of their exceptional chemical reactivity compared with other ethers. Simple epoxides are named as derivatives of oxirane, the parent epoxide. Where the epoxide is a part of another ring system, it is named using the prefix *epoxy*.

Epoxide
A cyclic ether in which oxygen is one atom of a three-membered ring.

Oxirane
(Ethylene oxide)

cis-2,3-Dimethyloxirane
(*cis*-2-Butene oxide)

1,2-Epoxycyclohexane
(Cyclohexene oxide)

Common names of epoxides are derived by giving the name of the alkene from which the epoxide is formally derived followed by the word "oxide" an example is *cis*-2-butene oxide.

11.8 | Synthesis of Epoxides

A. Ethylene Oxide

Ethylene oxide, one of the few epoxides synthesized on an industrial scale, is prepared by passing a mixture of ethylene and air (or oxygen) over a silver catalyst.

$$2\ CH_2{=}CH_2 + O_2 \xrightarrow{\ Ag\ } 2\ H_2C\underset{O}{\diagdown\!\!\diagup}CH_2$$

Oxirane
(Ethylene oxide)

This method only works well for the production of oxirane from ethylene. As a result, other methods have been developed for the production of epoxides.

B. Internal Nucleophilic Substitution in Halohydrins

One method for the preparation of epoxides from alkenes involves (1) treating the alkene with chlorine or bromine in water to form a chlorohydrin or bromohydrin followed by (2) treating the halohydrin with a base to bring about intramolecular displacement of Cl⁻. These steps convert propene first to 1-chloro-2-propanol, and then to methyloxirane (propylene oxide).

$$CH_3CH{=}CH_2 \xrightarrow{Cl_2,\ H_2O} \overset{\displaystyle OH}{\underset{\displaystyle Cl}{CH_3CH{-}CH_2}} \xrightarrow{NaOH,\ H_2O} CH_3CH\underset{O}{\diagup\!\!\diagdown}CH_2$$

Propene

1-Chloro-2-propanol
(a chlorohydrin)
(racemic)

Methyloxirane
(Propylene oxide)
(racemic)

We studied the reaction of alkenes with chlorine or bromine in water to form halohydrins in Section 6.3E and saw that it is both regioselective and stereoselective (for an alkene that shows *cis-trans* isomerism, it is also stereospecific). Conversion of a halohydrin to an epoxide with base is stereoselective as well and can be viewed as an internal S$_N$2 reaction. Hydroxide ion or other base abstracts a proton from the halohydrin hydroxyl group to form an alkoxide ion, a good nucleophile, which then displaces halogen on the adjacent carbon. As with all S$_N$2 reactions, attack of the nucleophile is from the backside of the C—X bond and causes inversion of configuration at the site of substitution.

$$:\!\overset{\displaystyle \cdot\cdot}{O}\!:^{-} \underset{:\overset{\cdot\cdot}{Cl}:}{\overset{\diagup}{C}{-}C} \xrightarrow{\text{internal }S_N2} \overset{:\overset{\cdot\cdot}{O}:}{C{-}C} + :\overset{\cdot\cdot}{\underset{\cdot\cdot}{Cl}}:^{-}$$

An epoxide

Note that this displacement of halide by the alkoxide ion can also be viewed as an intramolecular variation of the Williamson ether synthesis (Section 11.4A). In this case, the displacing alkoxide and leaving halide ions are on adjacent carbon atoms.

Example 11.8

Conversion of an alkene to a halohydrin and internal displacement of a halide ion by an alkoxide ion are both stereoselective. Use this information to demonstrate that the configuration of the alkene is preserved in the epoxide. As an illustration, show that reaction of *cis*-2-butene by this two-step sequence gives *cis*-2,3-dimethyloxirane (*cis*-2-butene oxide).

Solution

Addition of HOCl to an alkene occurs by stereoselective anti addition of —OH and —Cl to the double bond (Section 6.3E). The conformation of this product is also the conformation necessary for stereoselective backside displacement of the halide ion by alkoxide ion. Thus, a *cis* alkene gives a *cis* disubstituted oxirane, and the transformation from alkene to epoxide is stereospecific.

cis-2-Butene

Cl$_2$, H$_2$O

Two chlorohydrins
(as a pair of
enantiomers)

NaOH, H$_2$O | internal S$_N$2

This epoxide has two chiral centers but is achiral; it is a meso compound

cis-2,3-Dimethyloxirane
(achiral; a meso compound)

Problem 11.8

Consider the possibilities for stereoisomerism in the bromohydrin and epoxide formed from *trans*-2-butene.

(a) How many stereoisomers are possible for the bromohydrin? Which of the possible bromohydrin stereoisomers are formed by treating *trans*-2-butene with bromine in water?

(b) How many stereoisomers are possible for the epoxide? Which of the possible stereoisomers is/are formed in this two-step sequence?

C. Oxidation of Alkenes with Peroxycarboxylic Acids

The most common laboratory method for the synthesis of epoxides from alkenes is oxidation with a peroxycarboxylic acid (a peracid). Three of the most widely used peroxyacids are *meta*-chloroperoxybenzoic acid (MCPBA), the magnesium salt of monoperoxyphthalic acid (MMPP), and peroxyacetic acid.

**meta-Chloroperoxybenzoic acid
(MCPBA)**

**Magnesium monoperoxyphthalate
(MMPP)**

**Peroxyacetic acid
(Peracetic acid)**

Following is a balanced equation for the epoxidation of cyclohexene by a peroxycarboxylic acid, RCO_3H. In the process, the peroxycarboxylic acid is reduced to a carboxylic acid.

| Cyclohexene | A peroxy-carboxylic acid | | 1,2–Epoxycyclohexane (Cyclohexene oxide) | A carboxylic acid |

For an alkene that shows *cis-trans* isomerism, epoxidation is also stereospecific: the stereochemistry of the product depends on the stereochemistry of the starting alkene. Epoxidation of *cis*-2-butene, for example, yields only the meso compound *cis*-2,3-dimethyloxirane, and epoxidation of *trans*-2-butene yields only the enantiomers of *trans*-2,3-dimethyloxirane.

trans-2-Butene

trans-2,3-Dimethyloxirane
(A pair of enantiomers)

A mechanism for epoxidation by a peroxyacid must take into account the following facts. (1) The reaction takes place in nonpolar solvents, which means that the reaction cannot involve the formation of ions or any species with significant separation of unlike charges. (2) The reaction is stereospecific, with complete retention of the alkene configuration, which means that even though the π bond of the carbon-carbon double bond is broken, at no time is there free rotation about the remaining σ bond. Following is a mechanism consistent with these observations.

Mechanism   Epoxidation of an Alkene by RCO_3H

The bond-making and bond-breaking steps are thought to be concerted, or nearly so.

The concerted formation of the two C—O bonds of the epoxide ensures that the reaction is stereospecific; that is, that *cis* alkenes give *cis* epoxides and *trans* alkenes give *trans* epoxides.

D. Sharpless Asymmetric Epoxidation

One of the most useful organic reactions discovered in the last several decades is the titanium-catalyzed asymmetric epoxidation of primary allylic alcohols developed by Professor Barry Sharpless, then at Stanford University. The reagent consists of *tert*-butyl hydroperoxide, titanium tetraisopropoxide [Ti(O-iPr)$_4$], and diethyl tartrate. Recall from Section 3.4A that tartaric acid has two chiral centers and exists as three stereoisomers: a pair of enantiomers and a meso compound. The form of tartaric acid used in the Sharpless epoxidation is either pure (+)-diethyl tartrate or its enantiomer, (−)-diethyl tartrate. The *tert*-butyl hydroperoxide is the oxidizing agent and must be present in molar amounts. Titanium tetraisopropoxide and diethyl tartrate combine to make the active catalyst and are present in lesser amounts, generally 5 to 10 mole percent.

What is remarkable about the Sharpless epoxidation is that it is stereospecific based on the diethyl tartrate added; either enantiomer of an epoxide can be produced depending on which enantiomer of diethyl tartrate is used. If the (−)-enantiomer is used, the product is enantiomer A. If the (+)-enantiomer is used, the product is enantiomer B.

(2S,3S)-(−)-Diethyl tartrate

(2R,3R)-(+)-Diethyl tartrate

When predicting the stereochemistry of a Sharpless epoxidation product, you will find it helpful to draw the allylic alcohol in the same orientation each time, for example, as shown in the margin.

When drawn in this manner, the (−)-tartrate catalyzes delivery of the epoxide oxygen from the top face of the alkene; the (+)-tartrate catalyzes its delivery from the bottom face.

The mechanism of this catalyzed epoxidation has been studied in detail and involves formation of a chiral complex in which the carbonyl oxygen of diethyl tartrate displaces one of the isopropoxide groups on titanium. When the R—OOH oxidizing agent is added, it displaces a second isopropoxide group. Finally, the oxygen of the allylic alcohol displaces a third isopropoxide group. Thus, although neither the alkene nor the ROOH oxidizing agent is chiral, both are now held in a fixed stereochemical relationship to the other in the chiral environment created by the diethyl tartrate-titanium complex. In this chiral environment, oxygen is now delivered to either the top face or the bottom face of the alkene, depending on which enantiomer of diethyl tartrate is present.

For their pioneering work in the field of enantioselective synthesis, Sharpless (along with William Knowles and Ryoji Noyori) received the 2001 Nobel Prize in chemistry.

With (–)-diethyl tartrate, oxygen is delivered to the top face

With (+)-diethyl tartrate, oxygen is delivered to the bottom face

Example 11.9

Draw the expected products of Sharpless epoxidation of each allylic alcohol using (+)-diethyl tartrate as the chiral catalyst.

(a) [structure with OH]   (b) [structure with OH]

Solution

(a)

(b) In this solution, the carbon skeleton of the allylic alcohol is first reoriented to match the orientation in the template, the epoxidation is completed, and then the carbon skeleton is reoriented to match the original drawing.

Problem 11.9

Draw the expected products of Sharpless epoxidation of each allylic alcohol using (+)-diethyl tartrate as the chiral catalyst.

(a) **(b)**

11.9 | Reactions of Epoxides

Because of the strain associated with the three-membered ring, epoxides undergo a variety of ring-opening reactions, the characteristic feature of which is nucleophilic substitution at one of the carbons of the epoxide ring with the oxygen atom as the leaving group.

Characteristic reaction of epoxides:

A. Acid-Catalyzed Ring Opening

In the presence of an acid catalyst, such as sulfuric acid, epoxides are hydrolyzed to glycols. As an example, acid-catalyzed hydrolysis of oxirane gives 1,2-ethanediol (ethylene glycol).

$$H_2C\overset{}{\underset{O}{\diagdown\diagup}}CH_2 \ + \ H_2O \ \xrightarrow{H^+} \ HOCH_2CH_2OH$$

Oxirane	1,2-Ethanediol
(Ethylene oxide)	(Ethylene glycol)

Mechanism Acid-Catalyzed Hydrolysis of an Epoxide

Step 1: Add a proton. Proton transfer from the acid catalyst to oxygen of the epoxide gives a bridged oxonium ion intermediate.

Step 2: Make a new bond between a nucleophile and an electrophile and simultaneously break a bond to give stable molecules or ions. Backside attack of H_2O on the protonated epoxide (a bridged oxonium ion) opens the three-membered ring.

Step 3: Take a proton away. Proton transfer to solvent completes the formation of the glycol.

Attack of a nucleophile on a protonated epoxide shows an anti stereoselectivity typical of S_N2 reactions; the nucleophile attacks anti to the leaving hydroxyl group, and the —OH groups in the glycol thus formed are anti. As a result, hydrolysis of an epoxycycloalkane yields a *trans*-1,2-cycloalkanediol.

1,2-Epoxycyclopentane
(Cyclopentene oxide)
(achiral)

***trans*-1,2-Cyclopentanediol**
(a racemic mixture)

Note the similarity in ring opening of this bridged oxonium ion intermediate and the bridged halonium ion intermediate in electrophilic addition of halogens or X_2/H_2O to an alkene (Sections 6.3D and 6.3E). In each case, the intermediate is a three-membered ring with a heteroatom bearing a positive charge, and attack of the nucleophile is anti to the leaving group.

Because there is some carbocation character developed in the transition state for an acid-catalyzed epoxide ring opening, attack of the nucleophile on unsymmetrical epoxides occurs preferentially at the carbon better able to bear a partial positive charge.

1-Methyl-1,2-epoxycyclohexane

2-Methoxy-2-methylcyclohexanol

The stereochemistry of acid-catalyzed ring openings is S_N2-like in that attack of the nucleophile is from the side opposite the bridged oxonium ion intermediate. The regiochemistry, however, is S_N1-like. Because of the partial carbocation character of the transition state, attack of the nucleophile on the oxonium ion intermediate occurs preferentially at the more substituted carbon. That is, attack occurs at the site better able to bear the partial positive charge that develops on carbon in the transition state in analogy to attack on a bridged bromonium ion.

At this point, let us compare the stereochemistry of the glycol formed by acid-catalyzed hydrolysis of an epoxide with that formed by oxidation of an alkene with osmium tetroxide (Section 6.5A). Each reaction sequence is stereoselective but gives a different stereoisomer. Acid-catalyzed hydrolysis of cyclopentene oxide gives *trans*-1,2-cyclopentanediol; osmium tetroxide oxidation of cyclopentene gives *cis*-1, 2-cyclopentanediol. Thus, a cycloalkene can be converted to either a *cis* glycol or a *trans* glycol by the proper choice of reagents.

trans-1,2-Cyclopentanediol
(formed as a racemic mixture)

cis-1,2-Cyclopentanediol
(achiral)

B. Nucleophilic Ring Opening

Ethers are not normally susceptible to reaction with nucleophiles. Epoxides, however, are different. Because of the strain associated with a three-membered ring, epoxides undergo ring-opening reactions with a variety of nucleophiles. Good nucleophiles attack an epoxide ring by an S_N2 mechanism and show an S_N2-like regioselectivity; that is, the nucleophile attacks at the less hindered carbon. Following is an equation for the reaction of methyloxirane (propylene oxide) with sodium methoxide in methanol.

$$CH_3CH \underset{O}{\overset{}{\diagdown\diagup}} CH_2 + CH_3OH \xrightarrow[S_N2]{CH_3O^-Na^+} CH_3CH-CH_2-OCH_3$$

$$\underset{OH}{\overset{|}{}}$$

Mechanism   Nucleophilic Opening of an Epoxide Ring

Step 1: **Make a new bond between a nucleophile and an electrophile and simultaneously break a bond to give stable molecules or ions.**   Backside attack of the nucleophile on the less hindered carbon of the highly strained epoxide opens the ring and displaces O⁻.

$$CH_3CH \underset{:O:}{\overset{}{\diagdown\diagup}} CH_2 + CH_3\ddot{O}:^- Na^+ \xrightarrow{S_N2} CH_3CHCH_2\ddot{O}CH_3$$

$$\underset{:\ddot{O}:^- Na^+}{\overset{|}{}}$$

Step 2: **Add a proton.**   Proton transfer completes the reaction.

$$CH_3\ddot{O}-H + CH_3CHCH_2\ddot{O}CH_3 \longrightarrow CH_3\ddot{O}:^- Na^+ + CH_3CHCH_2\ddot{O}CH_3$$

$$\underset{:\ddot{O}:^- Na^+}{\overset{|}{}} \qquad\qquad\qquad \underset{:\ddot{O}H}{\overset{|}{}}$$

The nucleophilic ring opening of epoxides is also stereoselective; as expected of an S_N2 reaction, attack of the nucleophile is anti to the leaving group. An illustration is the reaction of cyclohexene oxide with sodium methoxide in methanol to give *trans*-2-methoxycyclohexanol.

Cyclohexene oxide → **trans-2-Methoxycyclohexanol** (a racemic mixture)

The value of epoxides lies in the number of nucleophiles that bring about ring opening and the combinations of functional groups that can be prepared from them. The most important of these ring-opening reactions are summarized in the following chart.

$HSCH_2\overset{\displaystyle CH_3}{CHOH}$

A β-mercaptoalcohol

$HOCH_2\overset{\displaystyle CH_3}{CHOH}$

A glycol

$HC\equiv CCH_2\overset{\displaystyle CH_3}{CHOH}$

A β-alkynylalcohol

Na^+SH^-/H_2O

H_2O/H_3O^+

1. $HC\equiv C^-Na^+$
2. H_2O

$H_2C\!\!-\!\!\overset{\displaystyle CH_3}{CH}$ with O

Methyloxirane

$N\equiv C^-Na^+/H_2O$

NH_3

$N\equiv CCH_2\overset{\displaystyle CH_3}{CHOH}$

A β-hydroxynitrile

$H_2NCH_2\overset{\displaystyle CH_3}{CHOH}$

A β-aminoalcohol

The reactions with a terminal alkyne anion or the cyanide anion are particularly noteworthy because a new carbon–carbon bond is formed in each reaction.

Finally, treatment with $LiAlH_4$ reduces an epoxide to an alcohol. Lithium aluminum hydride is similar to sodium borohydride, $NaBH_4$, in that it is a donor of hydride ion, $H:^-$, which is both a strong base and a good nucleophile. In the reduction of a substituted epoxide by $LiAlH_4$, attack of the hydride ion occurs preferentially at the less hindered carbon of the epoxide, an observation consistent with S_N2 reactivity.

1. $LiAlH_4$
2. H_2O

(R)-Phenyloxirane
((R)-Styrene oxide)

(S)-1-Phenylethanol

11.10 | Ethylene Oxide and Epichlorohydrin: Building Blocks in Organic Synthesis

Ethylene oxide is a valuable building block for organic synthesis because each carbon of its two-carbon skeleton has a functional group. Following is a flow chart illustrating some of the functional groups and types of molecules that can be generated from this building block. The key to recognizing a structural unit derived from ethylene oxide is the presence of an $Nu\!-\!CH_2\!-\!CH_2\!-\!OH$ group. In cases where the $-OH$ group is subsequently modified by replacement with another nucleophile, you will find the group $Nu\!-\!CH_2\!-\!CH_2\!-\!Nu$. The nucleophiles most widely used in modification of the $-OH$ group are ammonia, 1° amines, and 2° amines (Section 1.3B). We have seen all of these reactions before, but not in this form.

As you study this flow chart, notice the following points.

- Reactions (1) and (8) use carbon nucleophiles to open the three-membered ring and thus form new carbon-carbon bonds.
- Reaction (2) is a catalytic reduction of the carbon-nitrogen triple bond to a 1° amine. Just as a carbon-carbon triple bond can be reduced to a carbon-carbon single bond by hydrogen in the presence of a transition metal catalyst, a carbon-nitrogen triple bond can be similarly reduced.
- Reactions (3) and (4) are openings of the epoxide ring by nitrogen nucleophiles.
- Reaction (5) is an intramolecular acid-catalyzed dehydration of a 1,5-diol to give a cyclic ether.
- Reaction (7) involves two successive S_N2 reactions to form a nitrogen-containing ring.

An example of a compound, part of which is derived from the two-carbon skeleton of ethylene oxide, is the local anesthetic procaine. Here is a retrosynthetic analysis of procaine.

Procaine

Ethylene oxide   **Diethylamine**

The hydrochloride salt of procaine is marketed under the trade name Novocaine. We will show how to complete the synthesis of procaine when we study the derivatives of carboxylic acids in Chapter 18.

The epoxide epichlorohydrin is also a valuable synthetic building block because each of its three carbons contains a reactive functional group.

Epichlorohydrin

Epichlorohydrin is an oily liquid, bp 118°C. It is insoluble in water and nonpolar hydrocarbon solvents, but soluble in polar aprotic solvents such as diethyl ether and dichloromethane. Epichlorohydrin is synthesized industrially by the following series of three reactions.

Step 1: Allylic halogenation by a radical chain mechanism (Section 8.6A).

$$\text{Propene} + Cl_2 \xrightarrow{500°C} \text{Cl} \diagup\!\!\diagdown\!\!\diagup + HCl$$

Propene   **3-Chloropropene**
(Allyl chloride)

Step 2: Treating the haloalkene with chlorine in water gives a chlorohydrin (Section 6.3E).

$$Cl\diagup\!\!\diagdown\!\!\diagup + Cl_2/H_2O \longrightarrow Cl\diagdown\!\!\underset{OH}{\diagup}\!\!\diagdown Cl + HCl$$

Step 3: Treating the chlorohydrin with calcium hydroxide brings about an internal S_N2 reaction and gives epichlorohydrin.

$$Cl\diagdown\!\!\underset{OH}{\diagup}\!\!\diagdown Cl + Ca(OH)_2 \longrightarrow Cl\diagup\!\!\diagdown\!\!\overset{O}{\triangle} + CaCl_2$$

3-Chloro-1,2-epoxypropane
(Epichlorohydrin)
(racemic)

The characteristic structural feature of a product derived from epichlorohydrin is a three-carbon unit with —OH on the middle carbon, and a carbon, nitrogen, oxygen, or sulfur nucleophile bonded to the two end carbons.

$$Cl\diagup\!\!\diagdown\!\!\overset{O}{\triangle} \xrightarrow{Nu:^-} Nu\diagup\!\!\diagdown\!\!\overset{O}{\triangle} \xrightarrow{Nu:^-} Nu\diagdown\!\!\underset{OH}{\diagup}\!\!\diagdown Nu$$

Epichlorohydrin

An example of a compound that contains the three-carbon skeleton of epichlorohydrin is nadolol, a β-adrenergic blocker with vasodilating activity.

Nadolol
(racemic)

A nucleophile derived
by removal of the acidic
H from an —OH group

The nitrogen
nucleophile
of a 1° amine

Members of this class of compounds have received enormous clinical attention because of their effectiveness in treating hypertension (high blood pressure), migraine headaches, glaucoma, ischemic heart disease, and certain cardiac arrhythmias. Shown in this retrosynthetic analysis are the two nucleophiles used in the synthesis of nadolol. We will show how to complete the synthesis of nadolol when we study the chemistry of benzene and its derivatives in Chapters 21 and 22.

We are not concerned at this stage with how these nucleophiles are generated or, if there are two nucleophiles used, which nucleophile is added first or reacts at which site. Our concern with ethylene oxide and epichlorohydrin at this stage of the course is only that you recognize the structural features in a target molecule that might be derived from these building blocks. Call it pattern recognition if you will. Later, after we study the chemistry of other functional groups, we will discuss in detail the chemistry of how the target molecules are synthesized in the laboratory.

11.11 | Crown Ethers

Crown ether
A cyclic polyether derived from ethylene glycol and substituted ethylene glycols.

In the early 1960s, Charles Pedersen of DuPont discovered a family of cyclic polyethers derived from ethylene glycol and substituted ethylene glycols. Compounds of this structure are named **crown ethers** because one of their most stable conformations resembles the shape of a crown. These ethers are named by the system devised by Pedersen. The parent name *crown* is preceded by a number describing the size of the ring and followed by a number describing the number of oxygen atoms in the ring, as for example, 18-crown-6.

| **18-Crown-6** (a cyclic hexamer) | Space-filling model, viewed from above | Ball-and-stick model, viewed through an edge | Electrostatic potential map showing the electron-rich interior and the nonpolar exterior |

For his work, Pedersen shared the 1987 Nobel Prize for chemistry with Donald J. Cram of the United States and Jean-Marie Lehn of France.

The most significant structural feature of crown ethers is that the diameter of the cavity created by the repeating oxygen atoms of the ring is comparable to the diameter of alkali metal ions. The diameter of the cavity in 18-crown-6, for example, is approximately the diameter of a potassium ion. When a potassium ion is inserted into the cavity of 18-crown-6, the unshared electron pairs on the six oxygens of the crown ether are close enough to the potassium ion to provide very effective solvation for K^+:

Diameter of cavity created by six oxygens is approximately 260 to 320 pm

Diameter of K^+ is 266 pm

Ion	Diameter (pm)
Li^+	136
Na^+	194
K^+	266
Rb^+	294
Mg^{2+}	164
Ca^{2+}	286

A complex of K^+ and 18-crown-6

18-Crown-6 forms somewhat weaker complexes with rubidium ion (a somewhat larger ion) and with sodium ion (a somewhat smaller ion). It does not coordinate to any appreciable degree with lithium ion (a considerably smaller ion). 12-Crown-4, however, with its smaller cavity, does form a strong complex with lithium ion.

The cavity of a crown ether is a polar region, and the unshared pairs of electrons on the oxygen atoms lining the cavity provide effective solvation for alkali metal ions. The outer surface of the crown is nonpolar and hydrocarbon-like, and, thus, crown ethers and their alkali metal ion complexes dissolve readily in nonpolar organic solvents.

Crown ethers have proven to be particularly valuable because of their ability to cause inorganic salts to dissolve in nonpolar aprotic organic solvents such as methylene chloride, hexane, and benzene. Potassium permanganate, for example, does not dissolve in benzene. If 18-crown-6 is added to benzene, the solution takes on the purple color characteristic of permanganate ion. The crown-potassium ion complex is soluble in benzene and brings permanganate ion into solution with it. The resulting "purple benzene" is a valuable reagent for the oxidation of water-insoluble organic compounds.

Crown ethers have also proven valuable in nucleophilic displacement reactions. The cations of potassium salts, such as KF, KCN, or KN_3, are very tightly bound within the solvation cavity of 18-crown-6 molecules. The anions, however, are only weakly solvated, and because of the geometry of cation binding within the cavity of the crown, only loose ion pairing occurs between the anion and cation. Thus, in nonpolar aprotic solvents, these anions are without any appreciable solvent shell and are, therefore, highly reactive as nucleophiles. The nucleophilicity of F^-, CN^-, N_3^-, and other anions in nonpolar aprotic solvents containing an 18-crown-6 equals and often exceeds their nucleophilicity in polar aprotic solvents such as DMSO and acetonitrile.

11.12 | Sulfides

A. Nomenclature

To derive the IUPAC name of a **sulfide** (also called a **thioether**), select the longest carbon chain as the parent alkane and name the sulfur-containing substituent as an *alkylsulfanyl* group. To derive a common name, list the groups bonded to sulfur and add the word sulfide to show the presence of the —S— group:

Sulfide
The sulfur analog of an ether; a molecule containing a sulfur atom bonded to two carbon atoms. Sulfides are also called thioethers.

Ethylsulfanylethane
(Diethyl sulfide)

2-Ethylsulfanylpropane
(Ethyl isopropyl sulfide)

The functional group of a **disulfide** is an —S—S— group. IUPAC names of disulfides are derived by selecting the longest carbon chain as the parent alkane and indicating the disulfide-containing substituent as an *alkyldisulfanyl* group. Common names of disulfides are derived by listing the names of the groups bonded to sulfur and adding the word *disulfide:*

Disulfide
A molecule containing an —S—S— group.

Ethyldisulfanylethane
(Diethyl disulfide)

B. Preparation of Sulfides

Symmetrical sulfides, RSR (also called symmetrical thioethers), are prepared by treating one mole of Na_2S (where S^{2-} is the nucleophile) with two moles of haloalkane.

$$2 \text{ RX} + Na_2S \rightarrow RSR + 2 \text{ NaX}$$
A Sulfide

This same reaction can also be used to prepare five- and six-membered cyclic sulfides. Treating a 1,4-dihaloalkane with Na_2S gives a five-membered cyclic sulfide; treating a 1,5-dihaloalkane with Na_2S gives a six-membered ring.

1,4-Dichlorobutane

Thiolane
(Tetrahydrothiophene)

1,5-Dichloropentane   **Thiane**
(Tetrahydrothiopyran)

Unsymmetrical sulfides, RSR′, are prepared by converting a thiol to a sodium salt with either sodium hydroxide or sodium ethoxide and then allowing the salt to react with a haloalkane.

$$CH_3(CH_2)_8CH_2S^-Na^+ + CH_3I \xrightarrow{S_N2} CH_3(CH_2)_8CH_2SCH_3 + Na^+I^-$$

Sodium 1-decanethiolate   **1-Methylsulfanyldecane**
(Decyl methyl sulfide)

This method of thioether formation is the sulfur analog of the Williamson ether synthesis (Section 11.4A).

Note that all these reactions leading to sulfides (thioethers) are direct applications of nucleophilic substitution reactions (Chapter 9).

C. Oxidation of Sulfides

Many of the properties of sulfides stem from the fact that divalent sulfur is a reducing agent; it is easily oxidized to two higher oxidation states. Treatment of a sulfide with one mole of 30% aqueous hydrogen peroxide at room temperature gives a sulfoxide, as illustrated by oxidation of methyl phenyl sulfide to methyl phenyl sulfoxide. Several other oxidizing agents, including sodium periodate, $NaIO_4$, also bring about the same conversion. Treatment of a sulfoxide with $NaIO_4$ brings about its oxidation to a sulfone.

Methyl phenyl sulfide   **Methyl phenyl sulfoxide**   **Methyl phenyl sulfone**

Dimethyl sulfoxide (DMSO) is manufactured on an industrial scale by air oxidation of dimethyl sulfide in the presence of oxides of nitrogen.

Dimethyl sulfide   **Dimethyl sulfoxide**

Summary

Problems: 11.1, 11.10, 11.11

Section 11.1 Structure of Ethers

- An **ether** contains an atom of oxygen bonded to two carbon atoms.

Section 11.2 Nomenclature of Ethers

- In the IUPAC name, the parent chain is named and the —OR group is named as an alkoxy substituent.
- Common names for ethers are derived by naming the two groups bonded to oxygen followed by the word ether. Heterocyclic ethers have an oxygen atom as one of the members of a ring.

Section 11.3 Physical Properties of Ethers

- Ethers are weakly polar compounds and associate by weak dipole-dipole interactions and dispersion forces.
  - The boiling points of ethers are close to those of hydrocarbons of comparable molecular weight but much lower than those of the corresponding alcohols.
  - Because ethers are hydrogen bond acceptors, they are more soluble in water than are hydrocarbons of comparable molecular weight.

Problems: 11.2, 11.12–11.14

Section 11.4 Preparation of Ethers

- Ethers are prepared using the **Williamson ether synthesis**, which is an S_N2 reaction between an alkoxide and a methyl or primary alkyl halide (secondary or tertiary alkyl halides give too much unwanted E2 elimination).
- Symmetrical ethers can be prepared through dehydration of an alcohol in strong acid.
- Ethers can be prepared through the acid-catalyzed addition of methyl or primary alcohols to alkenes that can form a stable carbocation upon protonation, via a mechanism analogous to acid-catalyzed hydration of an alkene.

Problems: 11.3, 11.4, 11.15, 11.16, 11.41, 11.42, 11.43

Section 11.5 Reactions of Ethers

- Ethers are cleaved by concentrated HX via a mechanism that involves protonation of the ether oxygen followed by S_N1 or S_N2 substitution with X^-, the exact mechanism being determined by the structure of the ether.
- When handling ethers, special precautions are necessary.
  - Common ethers are highly volatile and flammable. They are oxidized in the presence of O_2 to explosive hydroperoxides, so they must be disposed of prior to their expiration date.

Problems: 11.5, 11.6

Section 11.6 Silyl Ethers as Protecting Groups

- **Protecting groups** are used to temporarily prevent a functional group from reacting.
- Silyl ethers are used to protect alcohol groups.
  - The silyl ether is prepared from the alcohol and a silyl chloride.
  - Silyl ethers are stable to oxidations, reductions, nonaqueous acid, and nonaqueous base.
  - Silyl ethers are removed by treatment with tetrabutylammonium fluoride, a reaction that exploits the extraordinarily strong Si-F bond.

Problems: 11.7, 11.17, 11.18, 11.38, 11.44

Section 11.7 Epoxides: Structure and Nomenclature

- **Epoxides** are three-membered ring, cyclic ethers.
  - In IUPAC nomenclature, epoxides are named as oxirane derivatives, unless they are part of another ring system and given the prefix *epoxy*-.
  - In common nomenclature, epoxides are named from the alkene from which they are derived, followed by the word "oxide."

Problems: 11.10, 11.11

Section 11.8 Synthesis of Epoxides

- Epoxides are synthesized from alkenes using peroxycarboxylic acids (RCO_3H), in a reaction involving a concerted five-membered ring transition state.
- Epoxides can be synthesized from internal nucleophilic substitution of halohydrins, which can be derived from hydrohalogenation of alkenes.
  - The reaction involves deprotonation of the alcohol, followed by backside attack by the alkoxide anion on the adjacent C—X bond.
- One of the most useful reactions discovered in the last several decades is the **Sharpless epoxidation** of primary allylic alcohols to give single enantiomers of chiral epoxides in a predictable fashion.
  - This reaction uses a titanium catalyst, a peroxide, and the enantiomers of diethyl tartrate.
  - Analyzing the steric environment created by the groups on the alkene allows the predominant product enantiomer to be predicted when either enantiomer of diethyl tartrate is used.

Problems: 11.8, 11.9, 11.27–11.29, 11.45

Section 11.9 Reactions of Epoxides

- Epoxides undergo acid-catalyzed ring opening to add weak nucleophiles such as water and alcohols to give diols or ether-alcohols, respectively, with anti stereochemistry.
    - The mechanism involves protonation of the epoxide oxygen to give a cation intermediate (analogous to a bridged halonium ion), followed by backside nucleophilic attack on the carbon more able to accept a positive charge (in unsymmetrical epoxides).
- Epoxides react with strong nucleophiles such as hydroxide, alkoxides, and ammonia or amines via an S_N2 mechanism at the less hindered carbon of unsymmetrical epoxides.
    - The attack is from the back of the C—O bond.
    - The observed regiochemistry is often the opposite of that seen with acid-catalyzed ring opening, providing access to both regioisomer products of unsymmetrical epoxides.

Problems: 11.19–11.26, 11.30–11.38, 11.43

Section 11.10 Ethylene Oxide and Epichlorohydrin: Building Blocks in Organic Synthesis

- **Ethylene oxide** is a valuable two-carbon building block for synthesis because after nucleophilic attack, both of its carbons contain a functional group.
    - Ammonia or primary or secondary amines are commonly used nucleophiles, but other sulfur or oxygen nucleophiles can also be used in the reaction.
    - In more complex molecules, the ethylene oxide building blocks can be recognized by the Nu—CH_2—CH_2—OH pattern, or Nu—CH_2—CH_2—Nu if the OH group is exchanged for another nucleophile.
- **Epichlorohydrin** is a valuable three-carbon building block for synthesis because after nucleophilic attack, all three of its carbons contain a functional group.
    - In more complex molecules, epichlorohydrin building blocks can be recognized by the Nu—CH_2—CHOH—CH_2—Nu pattern.
    - Ammonia or primary or secondary amines are commonly used nucleophiles, but other sulfur or oxygen nucleophiles can also be used.

Problems: 11.19–11.23, 11.39–11.40

Section 11.11 Crown Ethers

- **Crown ethers** are cyclic polyethers having 12 or more atoms in a ring.
    - The cavity of a crown ether is a polar region, and the unshared pairs of electrons on the ether oxygens can solvate alkali metal ions.
    - The cavity of **18-crown-6**, for example, has approximately the same diameter as the potassium ion, explaining the strong interaction between the two.
    - The outer surface of a crown ether is nonpolar and hydrocarbon-like.
    - Crown ethers are valuable for their ability to cause ionic compounds to dissolve in nonpolar organic solvents.

Problem: 11.22

Section 11.12 Sulfides

- **Sulfides (thioethers)** are named as alkylsulfanylalkanes. Common names for sulfides are derived by naming the two groups bonded to sulfur followed by the word *sulfide*.
- A **disulfide** (R-S-S-R) is named by choosing the longer alkyl group as the parent chain, and indicating the disulfide-containing portion as an alkyldisulfanyl group. Common names are derived by naming the two groups bonded to each sulfur, followed by the word *disulfide*.
- Symmetrical sulfides are created by reacting Na_2S with two moles of haloalkane. Unsymmetrical sulfides are synthesized by reacting the sodium salt of a thiol with a haloalkane.
- Sulfides can be oxidized with H_2O_2 by addition of one oxygen atom to sulfur to make a **sulfoxide**, and further oxidized with $NaIO_4$ to add a second oxygen atom to sulfur to make a **sulfone**.

◖WL Questions and problems for this chapter are available in OWL. Use the OWL access card that came with this textbook to access assigned questions and problems for this chapter.

1. **Williamson Ether Synthesis (Section 11.4A)** The Williamson ether synthesis is a general method for the synthesis of dialkyl ethers by an S_N2 reaction between a haloalkane and an alkoxide ion.

$$CH_3\underset{\underset{CH_3}{|}}{\overset{\overset{CH_3}{|}}{C}}O^-K^+ + CH_3Br \xrightarrow{S_N2} CH_3\underset{\underset{CH_3}{|}}{\overset{\overset{CH_3}{|}}{C}}OCH_3 + K^+Br^-$$

Yields are highest with methyl, 1° alkyl halides, and 1° allylic halides. They are considerably lower with 2° halides because of competition from E2 elimination. The Williamson ether synthesis reaction fails altogether with 3° halides.

2. **Acid-Catalyzed Dehydration of Alcohols (Section 11.4B)** Yields are highest for symmetrical ethers formed from unbranched primary alcohols. The mechanism involves protonation of an —OH group, followed by displacement of the H_2O leaving group by a second alcohol molecule acting as a nucleophile, followed by loss of a proton to give the ether.

$$2\ CH_3CH_2OH \xrightarrow[140°C]{H_2SO_4} CH_3CH_2OCH_2CH_3 + H_2O$$

3. **Acid-Catalyzed Addition of Alcohols to Alkenes (Section 11.4C)** Proton transfer to the alkene generates a carbocation. Nucleophilic addition of an alcohol to the carbocation followed by proton transfer to the solvent gives the ether.

$$CH_3\underset{}{\overset{\overset{CH_3}{|}}{C}}{=}CH_2 + CH_3OH \xrightarrow[\text{catalyst}]{\text{acid}} CH_3\underset{\underset{CH_3}{|}}{\overset{\overset{CH_3}{|}}{C}}OCH_3$$

4. **Acid-Catalyzed Cleavage of Dialkyl Ethers (Section 11.5A)** Cleavage of ethers requires both a strong acid and a good nucleophile, hence the use of concentrated HBr and HI.

$$\langle\ \rangle{-}OCH_3 + HI\ (\text{excess}) \longrightarrow \langle\ \rangle{-}I + CH_3I + H_2O$$

The mechanism involves initial protonation of the ether oxygen.

Cleavage of primary and secondary alkyl ethers is by an S_N2 pathway. Cleavage of tertiary alkyl ethers is by an S_N1 pathway.

5. **Reaction of Alcohols with Chloro-tert-butyldimethylsilane (Section 11.6B)** The *tert*-butyldimethylsilyl (t-BuMe$_2$Si—) group is used to protect primary and secondary alcohols.

$$RCH_2OH + Cl{-}\underset{|}{\overset{|}{Si}}{\Big\langle} \xrightarrow{\text{pyridine}} RCH_2O\underset{|}{\overset{|}{Si}}{\Big\langle}$$

Chloro-*tert*-butyl dimethylsilane A *tert*-butyldimethylsilyl ether

The protecting group is removed by treating the silyl ether with fluoride ion to regenerate the original alcohol.

6. **Oxidation of Alkenes by Peroxycarboxylic acids (Section 11.8B)** Three commonly used peroxycarboxylic acid oxidizing agents are *meta*-chloroperoxybenzoic acid, the magnesium salt of monoperoxyphthalic acid, and peroxyacetic acid. Each reagent oxidizes an alkene to an epoxide. The mechanism is a concerted rearrangement of electrons involving a five-membered ring, the π bond of an alkene, and the terminal O atom of the peroxyacid to give an epoxide and carboxylic acid.

$$\langle\ \rangle + R\overset{\overset{O}{\|}}{C}OOH \longrightarrow \langle\ \rangle{\cdots}O + R\overset{\overset{O}{\|}}{C}OH$$

7. Synthesis of Epoxides from Halohydrins (Section 11.8C)
Formation of the halohydrin and the following intramolecular S_N2 reaction are both stereoselective (the configuration of the alkene is retained in the epoxide) and stereospecific (for alkenes that show *cis,trans* isomerism, the configuration of the epoxide depends on the configuration of the alkene).

8. Sharpless Asymmetric Epoxidation (Section 11.8D)
Oxidation of the carbon-carbon double bond of a 1° allylic alcohol by *tert*-butyl hydroperoxide in the presence of a chiral catalyst consisting of either (+)- or (−)-diethyl tartrate and titanium tetraisopropoxide gives an enantiomerically pure epoxide. The enantiomer formed depends on which enantiomer of diethyl tartrate is used in the catalyst.

A 1° allylic   *tert*-Butyl
alcohol   **hydroperoxide**

9. Acid-Catalyzed Hydrolysis of Epoxides (Section 11.9A)
Hydrolysis of an epoxide derived from a cycloalkene gives a *trans* glycol. The reaction involves initial protonation of the epoxide O atom, followed by nucleophilic attack of water, then loss of a proton to give the *trans* diol. The reaction also occurs with alcohol nucleophiles, and when there is a difference, reaction of the nucleophile occurs predominantly at the more substituted carbon of the protonated epoxide.

(racemic)

10. Nucleophilic Ring Opening of Epoxides (Section 11.9B)
Attack on the epoxide is regioselective with the nucleophile attacking the less substituted carbon of the epoxide.

11. Reduction of an Epoxide to an Alcohol (Section 11.9B)
Regioselective hydride ion transfer from lithium aluminum hydride to the less hindered carbon of the epoxide gives an alcohol.

(racemic)   (racemic)

12. Oxidation of Sulfides (Section 11.12C)
Oxidation of a sulfide gives either a sulfoxide or a sulfone, depending on the oxidizing agent and experimental conditions. Air oxidation of dimethyl sulfide is a commercial route to dimethyl sulfoxide, a polar aprotic solvent.

Chapter 12

Infrared Spectroscopy

© Chris Taylor/CSIRO/Photo Researchers, Inc.

A scientist working with a Fourier transform infrared spectrometer. Inset: a model of 3-methyl-2-butanone. For an IR spectrum of this compound, see Figure 12.2.

Determination of molecular structure is one of the central themes of organic chemistry. For this purpose, chemists today rely almost exclusively on instrumental methods, four of which we discuss in this text. We begin in this chapter with infrared (IR) spectroscopy. Then, in Chapters 13 and 14, we introduce nuclear magnetic resonance (NMR) spectroscopy and mass spectrometry (MS). A brief introduction to ultraviolet-visible spectroscopy is contained in Chapter 20, as part of our discussion of conjugated systems.

12.1 | Electromagnetic Radiation

Gamma rays, x-rays, ultraviolet light, visible light, infrared radiation, microwaves, and radio waves are all types of **electromagnetic radiation**. Because electromagnetic radiation behaves as a wave traveling at the speed of light, it can be described in terms of its wavelength and its frequency.

Table 12.1 summarizes **wavelengths (λ)**, **frequencies**, and energies of various regions of the electromagnetic spectrum. The wavelengths of visible light fall in the range 400–700 nm. The infrared region of the electromagnetic spectrum (which can be felt as heat but not seen) falls in the range 0.7–300 μm.

Frequency, the number of full cycles of a wave that pass a given point in a second, is given the symbol ν (Greek nu) and is reported in **hertz (Hz)**, which has the units s^{-1}. Wavelength and frequency are inversely proportional, and one can be calculated from the other using the following relationship:

$$\lambda\nu = c$$

Electromagnetic radiation
Light and other forms of radiant energy.

Wavelength (λ)
The distance between consecutive peaks on a wave.

Frequency
The number of full cycles of a wave that pass a given point in a second; it is given the symbol ν (Greek nu) and reported in hertz (Hz), which has the units s^{-1}.

Hertz (Hz)
The unit in which frequency is measured: s^{-1} (read "per second").

⦿WL **Online homework for this chapter may be assigned in OWL for Organic Chemistry.**

Table 12.1 | Wavelengths, Frequencies, and Energies of Some Regions of the Electromagnetic Spectrum

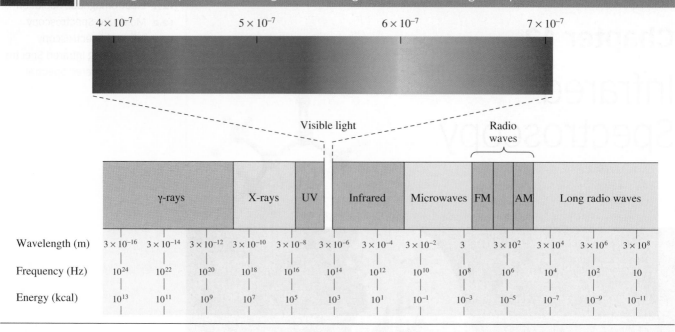

Wavelength (m)	3×10^{-16}	3×10^{-14}	3×10^{-12}	3×10^{-10}	3×10^{-8}	3×10^{-6}	3×10^{-4}	3×10^{-2}	3	3×10^{2}	3×10^{4}	3×10^{6} 3×10^{8}
Frequency (Hz)	10^{24}	10^{22}	10^{20}	10^{18}	10^{16}	10^{14}	10^{12}	10^{10}	10^{8}	10^{6}	10^{4}	10^{2} 10
Energy (kcal)	10^{13}	10^{11}	10^{9}	10^{7}	10^{5}	10^{3}	10^{1}	10^{-1}	10^{-3}	10^{-5}	10^{-7}	10^{-9} 10^{-11}

where c is the velocity of light, 3.00×10^{8} m/s. For example, consider the infrared radiation of wavelength 1.5×10^{-5} m (15 μm). The frequency of this radiation is 2.0×10^{13} Hz.

$$\nu = \frac{3 \times 10^{8}\,\text{m}}{\text{s} \cdot 1.5 \times 10^{-5}\,\text{m}} = 2.0 \times 10^{13}\,\text{Hz}$$

An alternative way to describe electromagnetic radiation is in terms of its properties as a stream of particles called **photons**. The energy in a mole of photons is related to the frequency of the radiation by the equations

$$E = h\nu = h\frac{c}{\lambda}$$

where E is the energy in kJ (kcal)/mol and h is Planck's constant, 3.99×10^{-13} kJ (9.537×10^{-14} kcal) $\cdot$ s $\cdot$ mol^{-1}.

Wavelength is usually expressed in the SI base unit of meters. Other derived units commonly used to express wavelength are given in Table 12.2.

Molecular spectroscopy
The study of which frequencies of radiation are absorbed or emitted by a particular substance and the correlation of these frequencies with details of molecular structure.

Table 12.2 | Common Units Used to Express Wavelength (λ)

Unit	Relation to Meter
Meter (m)	—
Millimeter (mm)	1 mm = 10^{-3} m
Micrometer (μm)	1 μm = 10^{-6} m
Nanometer (nm)	1 nm = 10^{-9} m
Angstrom (Å)	1 Å = 10^{-10} m

12.2 | Molecular Spectroscopy

An atom or molecule can be made to undergo a transition from energy state E_1 to a higher energy state E_2 by irradiating the atom or molecule with electromagnetic radiation corresponding to the energy difference between states E_1 and E_2 as illustrated schematically in Figure 12.1. When the atom or molecule returns from state E_2 to state E_1, an equivalent amount of energy is emitted.

Molecular spectroscopy is the experimental process of measuring which frequencies of radiation are absorbed or emitted by a particular substance and then attempting to correlate patterns of energy absorption or emission with details of molecular structure. Table 12.3 summarizes the regions of the electromagnetic spectrum of most interest to us and the relationships of each to changes in atomic and molecular energy levels.

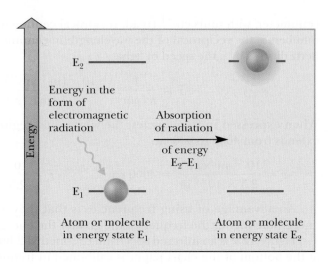

Figure 12.1
Absorption of energy in the form of electromagnetic radiation causes an atom or molecule in energy state E_1 to change to a higher energy state E_2.

Table 12.3	Types of Energy Transitions Resulting from Absorption of Energy from Three Regions of the Electromagnetic Spectrum		
Region of Electromagnetic Spectrum	**Frequency (hertz)**	**Type of Spectroscopy**	**Absorption of Electromagnetic Radiation Results in Transitions Between**
Radio frequency	$3 \times 10^7 - 9 \times 10^8$	Nuclear magnetic resonance	Nuclear spin levels
Infrared	$3 \times 10^{11} - 2.5 \times 10^{14}$	Infrared	Vibrational energy levels
Ultraviolet-visible	$2.5 \times 10^{14} - 1.5 \times 10^{15}$	Ultraviolet-visible	Electronic energy levels

12.3 | Infrared Spectroscopy

Infrared spectroscopy provides a direct method of detecting certain functional groups in a molecule. In this chapter, we first develop a basic understanding of the theory behind infrared spectroscopy, and then we concentrate on the interpretation of spectra and the information they can provide about details of molecular structure. Infrared spectroscopy depends on the absorption of infrared light in the wavelength range 2.5×10^{-6} to 2.5×10^{-5} m by vibrating bonds within molecules. Infrared spectroscopy is useful to organic chemists not only for the determination of molecular structure but also for many other applications. For example, infrared spectroscopy is used to identify certain illegal substances in forensic science, and a type of infrared frequency measurement is used as an important probe of atmospheric and mineral composition in space exploration programs such as the recent Mars rover missions.

A. The Vibrational Infrared Spectrum

Organic molecules are flexible. As we discussed in Chapter 2, atoms and groups of atoms can rotate about single covalent bonds. In addition, covalent bonds can stretch and bend just as if their atoms were joined by flexible springs. **Infrared spectroscopy**, also called **IR spectroscopy**, probes stretching and bending vibrations of organic molecules.

The **vibrational infrared region**, which extends from 2.5×10^{-6} to 2.5×10^{-5} m in wavelength, is used for infrared spectroscopy. Radiation in this region is most commonly referred to by its frequency in **wavenumbers**, $\bar{\nu}$, the number of waves per

Infrared (IR) spectroscopy
A spectroscopic technique in which a compound is irradiated with infrared radiation, absorption of which causes covalent bonds to change from a lower vibration state to a higher one. Infrared spectroscopy is particularly valuable for determining the kinds of functional groups are present in a molecule.

Vibrational infrared region
The portion of the infrared region that extends from 4000 to 400 cm^{-1}.

Wavenumbers $\bar{\nu}$
The frequency of electromagnetic radiation expressed as the number of waves per centimeter, with units cm^{-1} (read: reciprocal centimeters).

centimeter, with units cm^{-1} (read: reciprocal centimeters). The frequency in wavenumbers is the reciprocal of the wavelength in centimeters, or the frequency (ν) in hertz divided by c, the speed of light.

$$\bar{\nu} = \frac{1}{\lambda\ (\text{cm})} = \frac{10^{-2}\ (\text{m} \cdot \text{cm}^{-1})}{\lambda\ (\text{m})} = \frac{\nu}{c}$$

When expressed in frequencies, the vibrational region of the infrared spectrum extends from 4000 to 400 cm^{-1}.

$$\bar{\nu} = \frac{10^{-2}\ \text{m} \cdot \text{cm}^{-1}}{2.5 \times 10^{-6}\ \text{m}} = 4000\ \text{cm}^{-1} \qquad \bar{\nu} = \frac{10^{-1}\ \text{m} \cdot \text{cm}^{-1}}{2.5 \times 10^{-5}\ \text{m}} = 400\ \text{cm}^{-1}$$

An advantage of using frequencies is that they are directly proportional to energy; the higher the frequency, the greater the energy of the radiation.

Figure 12.2 is an infrared spectrum of 3-methyl-2-butanone. The horizontal axis at the bottom of the chart paper is calibrated in frequency (wavenumbers, cm^{-1}); that at the top is calibrated in wavelength (micrometers, μm). The frequency scale is often divided into two or more regions. For all spectra reproduced in this text, it is divided into three linear regions: 4000–2200 cm^{-1} 2200–1000 cm^{-1}, and 1000–400 cm^{-1}. The vertical axis measures transmittance (the fraction of light transmitted), with 100% at the top and 0% at the bottom. Thus, the baseline for an infrared spectrum (100% transmittance of radiation through the sample, 0% absorption) is at the top of the chart paper, and absorption of radiation corresponds to a trough or valley. Strange as it may seem, we commonly refer to infrared absorptions as peaks, even though they are conventionally displayed pointing downward.

The spectrum in Figure 12.2 was recorded using a neat sample, which means the pure liquid. A few drops of the liquid are placed between two sodium chloride discs and spread to give a thin film through which infrared radiation is then passed. Liquid and solid samples may also be dissolved in carbon tetrachloride or another solvent with minimal infrared absorption and analyzed in a liquid sampling cell. Still another way to obtain the infrared spectrum of a solid is to mix it with potassium bromide and then compress the mixture into a thin transparent wafer, which is placed in the beam of the spectrophotometer. Both NaCl and KBr are transparent to infrared radiation and can be used as windows and optics for infrared spectroscopy. Note that KBr and NaCl are ionic solids, and thus they do not have covalent bonds to absorb infrared radiation. For this reason, they do not interfere with the IR spectrum of the sample. These materials are very easily damaged by moisture. Infrared spectra of gas samples are determined using specially constructed gas-handling cells.

Figure 12.2

Infrared spectrum of 3-methyl-2-butanone.

Example 12.1

Some infrared spectrophotometers are calibrated to record spectra on an ordinate that is linear in wavelength (μm), whereas others record them on an ordinate that is linear in frequency (cm^{-1}). Carry out the following conversions (note the convenient formula for converting between μm and cm^{-1}: $\bar{\nu}\lambda = 10^4$).

(a) 7.05 μm to cm^{-1} **(b)** 3.35 μm to cm^{-1} **(c)** 3280 cm^{-1} to μm

Solution

(a) 1418 cm^{-1} **(b)** 2985 cm^{-1} **(c)** 3.05 μm

Problem 12.1

Which is higher in energy?

(a) Infrared radiation of 1715 cm^{-1} or of 2800 cm^{-1}?
(b) Radio-frequency radiation of 300 MHz or of 60 MHz?

B. Molecular Vibrations

Atoms joined by covalent bonds are not permanently fixed in one position but rather undergo continual vibrations relative to each other. The energies associated with these vibrations are quantized, which means that, within a molecule, only specific vibrational energy levels are allowed. The energies associated with transitions between vibrational energy levels in most covalent molecules correspond to frequencies in the infrared region, 4000–400 cm^{-1}.

For a molecule to absorb this radiation, its vibration must result in a *substantial change in the bond dipole moment*. If two opposite charges are connected by a spring, a change in distance between the charges corresponds to a change in dipole moment. In general, the greater the bond dipole, the greater the change in dipole moment caused by a vibration. Any vibration that leads to a substantial change in dipole moment is said to be **infrared active**. The greater the change is, the more intense the absorption will be. Covalent bonds whose vibration does not result in a change in bond dipole moment, for example, as a result of symmetry in the molecule, are said to be infrared inactive. The carbon-carbon double and triple bonds in symmetrically substituted alkenes and alkynes, for example 2,3-dimethyl-2-butene and 2-butyne, do not absorb infrared radiation because vibration does not result in a substantial bond dipole change.

Infrared active
Any molecular vibration that leads to a substantial change in dipole moment is observed in an IR spectrum.

2,3-Dimethyl-2-butene   **2-Butyne**

For a nonlinear molecule containing n atoms, $3n - 6$ allowed fundamental vibrations exist. For a molecule as simple as ethanol, CH_3CH_2OH, there are 21 fundamental vibrations, and for hexanoic acid, $CH_3(CH_2)_4COOH$, there are 54. Thus, even for relatively simple molecules, a large number of vibrational energy levels exist, and the patterns of energy absorption for these and larger molecules are very complex. For linear molecules with n atoms, there are $3n - 5$ allowed fundamental vibrations.

The simplest vibrational motions in molecules giving rise to absorption of infrared radiation are stretching and bending motions. Figure 12.3 illustrates the fundamental stretching and bending vibrations for a methylene group.

A different technique called **Raman spectroscopy** is complementary to infrared spectroscopy in that infrared-inactive vibrations are seen in Raman spectra, while Raman-inactive vibrations are the ones that are infrared active. A more complete description of Raman spectroscopy is beyond the scope of this text, but it is a very useful technique for studying certain molecules.

Raman spectroscopy
A vibrational molecular spectroscopy that is complementary to infrared (IR) spectroscopy in that infrared inactive vibrations are seen in Raman spectra.

Figure 12.3

Fundamental stretching and
bending vibrations for a
methylene group.

| Symmetric stretching | Scissoring | Rocking |

| Asymmetric stretching | Wagging | Twisting |

Stretching vibrations   **Bending vibrations**

C. Characteristic Absorption Patterns

Analysis of the modes of vibration for a molecule is very complex because all the atoms contribute to the vibrational modes. However, we can make useful generalizations about where absorptions due to particular vibrational modes will appear in an infrared spectrum by considering each individual bond and ignoring other bonds in the molecule. As a simplifying assumption, let us consider two covalently bonded atoms as two vibrating masses connected by a spring. The total energy is proportional to the frequency of vibration. The frequency of a stretching vibration is given by the following equation, which is derived from Hooke's law for a vibrating spring.

$$\bar{\nu} = 4.12 \sqrt{\frac{K}{\mu}}$$

Here $\bar{\nu}$ is the frequency of the vibration in wavenumbers (cm^{-1}); K is the force constant of the bond, a measure of the bond's strength, in dynes per centimeter; and μ is the "reduced mass" of the two atoms, $(m_1 m_2)/(m_1 + m_2)$, where m is the mass of the atoms.

Force constants for single, double, and triple bonds are approximately 5, 10, and 15×10^5 dynes/cm, respectively, thus approximately in the ratio $1 : 2 : 3$. Using the value for the force constant for a single bond, we calculate the frequency for the stretching vibration of a single bond between ^{12}C and ^{1}H as follows:
For ^{12}C—^{1}H stretching:

$$\text{Reduced mass} = 12 \times 1/(12 + 1) = 0.923 \text{ g/atom}$$

and

$$\bar{\nu} = 4.12 \sqrt{\frac{K}{\mu}} = 4.12 \sqrt{\frac{5 \times 10^5}{0.923}} = 3032 \text{ cm}^{-1}$$

The experimentally determined value for the frequency of an alkyl C—H stretching vibration is approximately 3000 cm^{-1}. Given the simplifying assumptions made in this calculation and the fact that the value of the force constant for a single bond is an average value, the agreement between the calculated value and the experimental value is remarkably good. Although frequencies calculated in this manner can be close to the experimental values, they are generally not accurate enough for precise determination of molecular structure.

Hooke's law predicts that the *position* of the absorption of a stretching vibration in an IR spectrum depends both on the strength of the vibrating bond and on the masses of the atoms connected by the bond. The stronger the bond is and the lighter the atoms are, the higher the frequency of the stretching vibration will be. As we saw earlier, the *intensity* of an absorption depends primarily on the change in dipole of the vibrating bond.

Example 12.2

Calculate the stretching frequency in wavenumbers for a carbon-carbon double bond. Assume that each carbon is the most abundant isotope, namely ^{12}C.

Solution

Assume a force constant of 10×10^5 dynes per centimeter for C=C. The calculated frequency is 1682 cm^{-1}, a value close to the experimental value of 1650 cm^{-1}.

$$\bar{\nu} = 4.12 \sqrt{\frac{10 \times 10^5}{12 \times 12/(12 + 12)}} = 1682 \text{ cm}^{-1}$$

Problem 12.2

Without doing the calculation, which member of each pair do you expect to occur at the higher frequency?

(a) C=O or C=C stretching   (b) C=O or C—O stretching
(c) C≡C or C=O stretching   (d) C—H or C—Cl stretching

Detailed interpretation of most infrared spectra is difficult because of the complexity of vibrational modes. In addition to the fundamental vibrational modes we have described, other types of absorptions occur, resulting in so-called overtone and coupling peaks that are usually quite weak.

To one skilled in the interpretation of infrared spectra, the absorption patterns can yield an enormous amount of information about chemical structure. However, we have neither the time nor the need to develop this level of competence. The value of infrared spectra for us is that they can be used to determine the presence or absence of certain functional groups. A carbonyl group, for example, typically shows strong absorption at approximately 1630–1820 cm^{-1}. The position of absorption for a particular carbonyl group depends on whether it is an aldehyde, ketone, carboxylic acid, or ester, or, if it is in a ring, it depends on the size of the ring. In this chapter, we discuss how structural variations, such as ring size or other factors, affect this value.

D. Correlation Tables

Data on absorption patterns of functional groups are collected in tables called **correlation tables**. Table 12.4 lists infrared absorptions for the types of bonds and functional groups we examine most often. A cumulative correlation table can be found in Appendix 6. In these tables, the intensity of a particular absorption is often referred to as strong (s), medium (m), or weak (w). In general, bonds between C and O where the electronegativity difference is largest have the largest dipole moments and tend to give the strongest infrared absorptions.

Organic chemists pay most attention to the region from 3500 to 1500 cm^{-1} because the stretching and bending vibrations for most functional groups are found

Table 12.4	Infrared Stretching Frequencies of Selected Functional Groups	
Bond	**Stretching Frequency (cm^{-1})**	**Intensity**
O—H	3200–3650	Weak to strong (strongest when H-bonded)
N—H	3100–3550	Medium
C—H	2700–3300	Weak to medium
C=C	1600–1680	Weak to medium
C=O	1630–1820	Strong
C—O	1000–1250	Strong

Fingerprint region
Vibrations in the region 1500 to 400 cm^{-1} of IR spectra are complex and difficult to analyze but are characteristic for different molecules.

in this region. Vibrations in the region 1500 to 400 cm^{-1} are more complex and far more difficult to analyze. Because even slight variations in molecular structure and absorption patterns are most obvious in this region, it is often called the **fingerprint region**. If two compounds have even slightly different structures, the differences in their infrared spectra are most clearly discernible in this region.

Example 12.3

What functional group is most likely present if a compound shows IR absorption at these frequencies?

(a) 1705 cm^{-1} **(b)** 2950 cm^{-1}

Solution

(a) A C=O group   **(b)** An aliphatic C—H group

Problem 12.3

A compound shows strong, very broad IR absorption in the region 3300–3600 cm^{-1} and strong, sharp absorption at 1715 cm^{-1}. What functional group accounts for both of these absorptions?

Example 12.4

Propanone and 2-propen-1-ol are constitutional isomers. Show how to distinguish between them by IR spectroscopy.

$$\underset{\substack{\textbf{Propanone}\\ \textbf{(Acetone)}}}{CH_3-\overset{\displaystyle O}{\overset{\|}{C}}-CH_3} \qquad \underset{\substack{\textbf{2-Propen-1-ol}\\ \textbf{(Allyl alcohol)}}}{CH_2=CH-CH_2-OH}$$

Solution

Only propanone shows strong absorption in the C=O stretching region, 1630–1820 cm^{-1}. Alternatively, only 2-propen-1-ol shows strong absorption in the O—H stretching region, 3200–3650 cm^{-1}.

Problem 12.4

Propanoic acid and methyl ethanoate are constitutional isomers. Show how to distinguish between them by IR spectroscopy.

$$\underset{\textbf{Propanoic acid}}{CH_3CH_2\overset{\displaystyle O}{\overset{\|}{C}}OH} \qquad \underset{\substack{\textbf{Methyl ethanoate}\\ \textbf{(Methyl acetate)}}}{CH_3\overset{\displaystyle O}{\overset{\|}{C}}OCH_3}$$

12.4 | Interpreting Infrared Spectra

A. Alkanes

Infrared spectra of alkanes are usually simple with few peaks, the most common of which are given in Table 12.5.

Figure 12.4 is an infrared spectrum of decane. The strong peak with multiple splitting between 2850 and 3000 cm^{-1} is characteristic of alkane C—H stretching; it is strong in this spectrum because there are so many C—H bonds and no other functional groups. The other prominent peaks correspond to methylene bending at 1465 cm^{-1} and methyl bending at 1380 cm^{-1}.

Table 12.5 Infrared Absorptions of Alkanes, Alkenes, Alkynes, and Arenes

Hydrocarbon	Vibration	Frequency (cm^{-1})	Intensity
Alkane			
C—H	Stretching	2850–3000	Medium
CH$_2$	Bending	1450–1475	Medium
CH$_3$	Bending	1375 and 1450	Weak to medium
C—C	(Not useful for interpretation—too many bands)		
Alkene			
C—H	Stretching	3000–3100	Weak to medium
C=C	Stretching	1600–1680	Weak to medium
Alkyne			
C—H	Stretching	3300	Medium to strong
C≡C	Stretching	2100–2250	Weak
Arene			
C—H	Stretching	3030	Weak to medium
C=C	Stretching	1450–1600	Medium
C—H	Bending	690–900	Strong

Figure 12.4
Infrared spectrum of decane.

B. Alkenes

An easily recognized alkene absorption is the vinylic C—H stretching slightly to the left of 3000 cm^{-1} (i.e., higher frequency). Also characteristic of alkenes is C=C stretching at 1600–1680 cm^{-1}. This vibration, however, is often weak and difficult to observe: the more symmetrical the alkene is, the weaker the absorption will be. Both vinylic C—H stretching and C=C stretching can be seen in the infrared spectrum of cyclohexene (Figure 12.5). Also visible are the aliphatic C—H stretching near 2900 cm^{-1} and methylene bending near 1475 cm^{-1}.

C. Alkynes

Alkyne ≡C—H stretching occurs near 3300 cm^{-1}, at higher frequency than for either alkyl —C—H or vinylic =C—H stretching. This peak is usually sharp and strong. The (sp-$1s$) C—H bond is unusually strong and therefore has a higher

Figure 12.5

Infrared spectrum of cyclohexene.

force constant than alkene (sp^2-1s) C—H bonds, which in turn absorb at higher frequency than the even weaker alkane (sp^3-1s) C—H bonds. Recall that the greater the percent s character of an atom's hybridization, the more tightly held the electrons, and accordingly the stronger (and shorter) the bonds to atoms such as H. Therefore, the sp-1s C—H bonds (50% s character on carbon) are the strongest, and the sp^3-1s C—H bonds (25% s character on carbon) are the weakest, with the sp^2-1s C—H bonds in between. Also observed in terminal alkynes is absorption near 2150 cm^{-1} owing to C≡C stretching. Both of these peaks can be seen in the infrared spectrum of 1-octyne (Figure 12.6). For internal alkynes, the C≡C stretching absorption is often very weak or completely absent (in symmetric alkynes) because stretching of this bond results in little or no change in the bond dipole moment (Section 12.3B).

Figure 12.6

Infrared spectrum of 1-octyne.

D. Arenes (Benzene and Its Derivatives)

Aromatic rings show a medium to weak peak in the C—H stretching region at approximately 3030 cm^{-1} characteristic of (sp^2-1s) =C—H bonds. In addition, aromatic rings show strong absorption in the region 690–900 cm^{-1} as a result of out-of-plane C—H bending. Finally, these compounds show several absorptions owing to C=C stretching between 1450 and 1600 cm^{-1}. Actually, these are complex vibrational modes of the entire ring. Some modes involve all atoms moving in and out (breathing), whereas others involve some atoms moving in and others moving out. The intensities of these peaks can vary depending on the symmetry of ring substitution patterns. Each of these characteristic absorption patterns can be seen in the infrared spectrum of toluene (Figure 12.7).

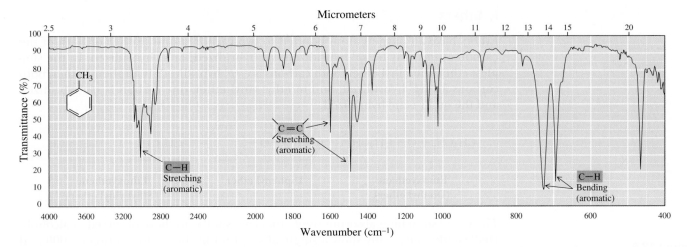

Figure 12.7
Infrared spectrum of toluene.

E. Alcohols

Both the position of the O—H stretching absorption and its intensity depend on the extent of hydrogen bonding. Under conditions where there is extensive hydrogen bonding between alcohol molecules (in pure alcohol or in concentrated solutions of the alcohol), the O—H stretching absorption occurs as a broad peak at 3200–3500 cm^{-1}. The variety of hydrogen-bonded states in different molecules leads to this broadening. The "free" O—H stretch near 3650 cm^{-1} is seen only in very dilute solution in non-hydrogen-bonding solvents. The C—O stretching absorption appears in the range 1000–1250 cm^{-1} (Table 12.6).

Table 12.6	Infrared Absorptions of Alcohols	
Bond	**Frequency (cm^{-1})**	**Intensity**
O—H (free)	3600–3650	Weak
O—H (hydrogen bonded)	3200–3500	Medium, broad
C—O	1000–1250	Medium

Shown in Figure 12.8 is an infrared spectrum of neat 1-hexanol. The hydrogen-bonded O—H stretching appears as a broad band of strong intensity centered at 3340 cm^{-1}. The C—O stretching appears at 1058 cm^{-1}, a value characteristic of primary alcohols.

Figure 12.8
Infrared spectrum of 1-hexanol (neat, salt plates).

F. Ethers

Ethers have an oxygen atom bonded to two carbon atoms. Either or both of the carbon atoms may be sp^3 hybridized, sp^2 hybridized, or sp hybridized. In the simplest ether, dimethyl ether, both carbons are sp^3 hybridized. In diphenyl ether, both carbons are sp^2 hybridized, and in ethyl vinyl ether, one carbon is sp^3 hybridized and the other is sp^2 hybridized.

$$CH_3CH_2-O-CH_2CH_3$$

Ethoxyethane
(Diethyl ether)

Diphenyl ether

$$CH_3CH_2-O-CH=CH_2$$

Ethoxyethene
(Ethyl vinyl ether)

The C—O stretching absorptions of ethers are similar to those observed in alcohols. Dialkyl ethers typically show a single absorption in the region between 1000 and 1250 cm^{-1} as can be seen in the infrared spectrum of dibutyl ether (Figure 12.9).

Figure 12.9
Infrared spectrum of dibutyl ether.

Aromatic ethers (compounds in which the ether oxygen is bonded to one or more benzene rings) and vinyl ethers (compounds in which the ether oxygen is bonded to one or more sp^2 hybridized carbon of a C=C bond) typically show two C—O stretching vibrations, one at either end of the range for C—O stretching. Anisole (Figure 12.10), for example, shows C—O stretching vibrations at 1050 cm^{-1} (sp^3 C—O) and 1250 cm^{-1} (sp^2 C—O). Ethers in which one of the bonds is attached to an sp^2 hybridized carbon typically also have a band in the region between 1200 and 1250 cm^{-1}.

Figure 12.10
Infrared spectrum of anisole.

Figure 12.11

Infrared spectrum of 1-butanamine, a primary amine.

The presence or absence of O—H stretching at 3200 to 3500 cm^{-1} for hydrogen-bonded O—H can be used to distinguish between an ether and an isomeric alcohol. A C—O stretching absorption is also present in esters (see Section 12.4J). In this case, the presence or absence of C=O stretching can be used to distinguish between an ether and an ester.

G. Amines

The most important and readily observed infrared absorptions of primary and secondary amines are the result of N—H stretching vibrations and appear in the region 3300–3500 cm^{-1}. Like O—H bonds, N—H bonds become broader and shift to longer wavelength when they take part in hydrogen bonding. Primary amines have two bands in this region: one caused by symmetric stretching and the other by asymmetric stretching. The two N—H stretching absorptions characteristic of a primary amine can be seen in the IR spectrum of 1-butanamine (Figure 12.11). Secondary amines give only one absorption in this region. Tertiary amines have no N—H and, therefore, are transparent in this region of the infrared spectrum.

H. Aldehydes and Ketones

Aldehydes and ketones show characteristic strong infrared absorption between 1630 and 1820 cm^{-1} associated with the stretching vibration of the carbon-oxygen double bond. The stretching vibration for the carbonyl group of menthone occurs at 1705 cm^{-1} (Figure 12.12). The large difference in electronegativity between carbon and oxygen, leading to the correspondingly large dipole moment of the carbonyl group, is the reason for the strong intensity of this IR absorption.

Figure 12.12

Infrared spectrum of menthone.

Because few other bond vibrations absorb energy between 1630 and 1820 cm^{-1}, absorption in this region of the spectrum is a reliable means for confirming the presence of a carbonyl group. Because several different functional groups contain a carbonyl group, it is often not possible to tell from absorption in this region alone whether the carbonyl-containing compound is an aldehyde, a ketone, a carboxylic acid (Section 12.4I), an ester (Section 12.4J), an amide (Section 12.4J), or an anhydride (Section 12.4J). However, other absorptions such as the C—O stretch in esters can help distinguish these groups. Aldehydes frequently have a weak but very distinctive absorption at 2720 cm^{-1} caused by the stretching of the C—H of the CHO group.

The position of the C=O stretching vibration is quite sensitive to the molecular environment of the carbonyl group, as illustrated by comparing these cycloalkanones. Cyclohexanone, which has very little angle strain shows absorption at 1715 cm^{-1}. As ring size decreases and angle strain increases, the C=O absorption shifts to a higher frequency as shown in the following series.

| 1715 cm^{-1} | 1745 cm^{-1} | 1780 cm^{-1} | 1850 cm^{-1} |

The presence of an adjacent carbon-carbon double bond or benzene ring in **conjugation** with the carbonyl group results in a shift of the C=O absorption to a lower frequency, as seen by comparing the carbonyl stretching frequencies of the following molecules. Conjugation occurs when the π electrons of adjacent π bonds interact with each other due to overlap of the unhybridized 2p orbitals and is a consequence of the special properties of extended π systems (Chapter 20). Conjugation also moves the C=C stretch to the right (lower frequency) and increases its intensity.

Conjugation
A situation in which two multiple bonds are separated by a single bond. Alternatively, a series of overlapping 2p orbitals. 1,3-Butadiene, for example, is a conjugated diene, and 3-butene-2-one is a conjugated ketone.

| 1717 cm^{-1} | 1690 cm^{-1} | 1700 cm^{-1} |

I. Carboxylic Acids

A carboxyl group gives rise to two characteristic absorptions in an infrared spectrum. One of these occurs in the region 1700–1725 cm^{-1} and is associated with the stretching vibration of the carbonyl group. This range of absorption is essentially the same as that for the carbonyl group of aldehydes and ketones, but it is usually broader in the case of the carboxyl carbonyl because of intermolecular hydrogen bonding with the carboxyl O—H of another molecule. The other infrared absorption characteristic of a carboxyl group is a peak between 2500 and 3300 cm^{-1} owing to the stretching vibration of the O—H group, which often overlaps the C—H stretching absorptions. The O—H absorption is generally very broad as a result of hydrogen bonding between molecules of the carboxylic acid. Both C=O and O—H stretches can be seen in the infrared spectrum of pentanoic acid in Figure 12.13.

J. Derivatives of Carboxylic Acids

In Chapter 18, you will learn about several important derivatives of carboxylic acids. The most important are amides, esters, anhydrides, and nitriles.

| An amide | An ester | An acid anhydride | A nitrile |

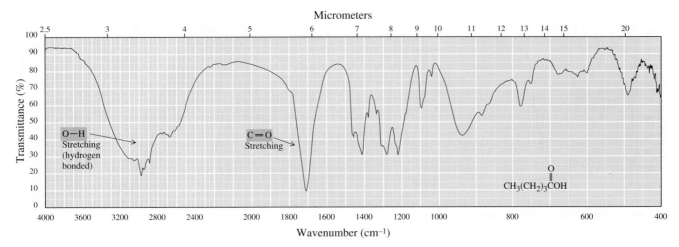

Figure 12.13

Infrared spectrum
of pentanoic acid.

Figure 12.14

Infrared spectrum of
N-methylpropanamide
(a secondary amide).

The functional group of a carboxylic ester, most commonly referred to as simply an ester, is a carbonyl group bonded to an —OR group.

$$\underset{\textbf{Ethyl ethanoate} \atop \textbf{(Ethyl acetate)}}{CH_3\overset{\overset{\displaystyle O}{\|}}{C}-OCH_2CH_3} \qquad \underset{\textbf{Isopropyl pentanoate}}{CH_3(CH_2)_3\overset{\overset{\displaystyle O}{\|}}{C}-OCH(CH_3)_2}$$

Note that one of the carbons of the C—O—C group is sp^2 hybridized and the other is sp^3 hybridized.

Esters display strong C=O stretching absorption in the region between 1735 and 1800 cm^{-1}. As in ketones, this band is shifted to higher frequency in smaller rings and to lower frequency by adjacent double bonds. In addition, esters also display strong C—O stretching absorptions in the region 1000–1100 cm^{-1} for the sp^3 C—O stretch and 1200–1250 cm^{-1} for the sp^2 C—O stretch (Figure 12.15). Ethers in which one of the carbons bonded to oxygen is sp^2 hybridized also show this band.

The most important infrared absorption of carboxylic acids and their derivatives is the result of the C=O stretching vibration. Infrared spectroscopic data for these derivatives and the other carbonyl-containing compounds are summarized in Table 12.7.

The carbonyl stretching of amides occurs at 1630–1680 cm^{-1}, at a lower frequency than for other carbonyl compounds. This observation can be explained by a weakened carbonyl C=O bond as described through the three most important resonance-contributing structures of amides, two of which indicate single bond

Figure 12.15

Infrared spectrum of ethyl butanoate.

character (see Example 1.19 and the Connections to Biological Chemistry box entitled "The Unique Structure of Amides" in Section 18.1). Primary and secondary amides show N—H stretching in the region 3200–3400 cm^{-1}; primary amides (RCONH$_2$) usually show two N—H absorptions, whereas secondary amides (RCONHR) show only a single N—H absorption (Figure 12.14).

Anhydrides have two carbonyl stretching absorptions, one near 1760 cm^{-1} and the other near 1810 cm^{-1}, the result of symmetric and unsymmetric stretching

Table 12.7 | Infrared Absorptions of Molecules Containing Carbonyl Groups

Carbonyl Group		Vibration	Frequency (cm^{-1})	Intensity
$\overset{O}{\underset{}{\overset{\|}{RCR'}}}$	Ketones (Section 16.1) C=O	Stretching	1630–1820	Strong
$\overset{O}{\underset{}{\overset{\|}{RCH}}}$	Aldehydes (Section 16.1) C=O C—H	Stretching Stretching	1630–1820 2720	Strong Weak
$\overset{O}{\underset{}{\overset{\|}{RCOH}}}$	Carboxylic Acids (Section 17.1) C=O O—H	Stretching Stretching	1700–1725 2500–3300	Strong Strong (broad)
$\overset{O}{\underset{}{\overset{\|}{RCNH_2}}}$	Amides (Section 18.1D) C=O N—H (1° amides have two N—H stretches) (2° amides have one N—H stretch)	Stretching Stretching	1630–1680 3200, 3400	Strong Medium
$\overset{O}{\underset{}{\overset{\|}{RCOR'}}}$	Carboxylic Esters (Section 18.1C) C=O sp^2 C—O sp^3 C—O	Stretching Stretching Stretching	1735–1800 1200–1250 1000–1100	Strong Strong Strong
$\overset{O\ \ O}{\underset{}{\overset{\|\ \ \|}{RCOCR}}}$	Acid Anhydrides (Section 18.1B) C=O C—O	Stretching Stretching	1740–1760 and 1800–1850 900–1300	Strong Strong
RC≡N	Nitriles (Section 18.1E) C≡N	Stretching	2200–2250	Medium

vibrations of the entire anhydride functional group. In addition, anhydrides display strong C—O stretching absorption in the region 900–1300 cm^{-1}. Nitriles can be distinguished by a medium C≡N stretching absorption at 2200–2250 cm^{-1}.

12.5 | Solving Infrared Spectral Problems

The following steps provide a systematic approach to solving IR problems.

Step 1: Check the region around 3000 cm^{-1}. Absorption in this region is caused by C—H stretching. Absorption is generally to the right of 3000 cm^{-1} for sp^3 C—H stretching of alkanes and to the left for the sp^2 C—H stretching of alkenes and aromatic rings.

Step 2: Is there a strong, broad band in the region of 3500 cm^{-1}? If yes, then the molecule contains an —OH group either of an alcohol or a carboxylic acid. If there is no absorption around 1700 cm^{-1} (caused by a carbonyl group), then the functional group is an —OH group of an alcohol. If there is a peak around 1700 cm^{-1}, then the functional group may be a carboxyl group. One or two peaks in the 3500 cm^{-1} region at somewhat lower frequency than for —OH may indicate a 2° or 1° amine, respectively.

Step 3: Is there a sharp peak in the region 1630–1820 cm^{-1}? If yes, there is a C=O group present. This peak will probably be the strongest peak in the spectrum. If no peak is present in this region, there is no C=O present. The type of carbonyl-containing functional group can often be determined by looking for the presence or absence of an aldehyde C—H stretch at 2720 cm^{-1}, a carboxyl O—H stretch around 3500 cm^{-1}, and a C—O stretch around 1000–1250 cm^{-1}.

Finally, here is a note of caution on interpreting infrared spectra. Even though it is possible to obtain a great deal of valuable information about a compound from its infrared spectrum, it is often very difficult to determine its structure based solely on this information, so other types of information should be sought.

Summary

Section 12.1 Electromagnetic Radiation

- **Electromagnetic radiation** behaves as a wave traveling at the speed of light and is described in terms of its **wavelength** and **frequency**.
  - The wavelengths of visible light fall in the range 400–700 nm.
  - The wavelengths of infrared radiation fall in the range 2–15 μm.
  - Wavelength and frequency are inversely proportional.

Section 12.2 Molecular Spectroscopy

- An atom or molecule can be made to undergo a transition from energy state E_1 to a higher energy state E_2 by irradiating the atom or molecule with electromagnetic radiation corresponding to the energy difference between states E_1 and E_2.
- **Molecular spectroscopy** is the experimental process of measuring which frequencies of radiation are absorbed or emitted by a particular substance and then attempting to correlate patterns of energy absorption or emission with details of molecular structure.

Section 12.3 Infrared Spectroscopy

- **Infrared spectroscopy** provides a direct method for detecting certain functional groups in a molecule.
  - In infrared spectroscopy, absorption of energy correlates to characteristic stretching and bending vibrations of functional groups in molecules.
- The **vibrational infrared region** extends from 7.5×10^{-7} m to 1×10^{-3} m in wavelength.

– Radiation in the vibrational infrared region is commonly referred to by its frequency in **wavenumbers**, or the number of waves per centimeter (cm^{-1}, read: reciprocal centimeters).

– The vibrational infrared spectrum extends from 4000 to 400 cm^{-1}.

• To be **infrared active**, a bond must be polar, and its vibration must cause a substantial change in the dipole moment.

– In general, the larger the dipole moment of the bond, the stronger the absorption seen in an infrared spectrum.

– There are $3n-6$ allowed fundamental vibrations for a nonlinear molecule containing n atoms.

– The simplest vibrations that give rise to absorption of infrared radiation are stretching and bending vibrations. Stretching may be symmetrical or asymmetrical.

– The frequency of vibration for an infrared-active bond can be estimated using **Hooke's law** for the vibration of a simple harmonic oscillator such as a vibrating spring. Hooke's law predicts that the frequency of vibration increases when (1) the bond strength increases and (2) the reduced mass of the vibrating system decreases.

Problems: 12.1–12.4, 12.8, 12.9

• A **correlation table** is a list of the absorption patterns of functional groups.

– The intensity of a peak is referred to as strong (s), medium (m), or weak (w).

– Bending and stretching vibrations characteristic of most functional groups appear in the region from 3500 to 1500 cm^{-1}.

– The region 1500 to 400 cm^{-1} is called the **fingerprint region**, and little useful information concerning functional groups is found in this region.

Section 12.4 Interpreting Infrared Spectra

• **Alkanes** give infrared spectra that are simple, with few useful peaks, although CH stretching (2850–3000 cm^{-1}) and bending (1375–1450, 1450–1475 cm^{-1}) peaks are seen.

• The C—H stretch of **alkenes** (3000–3100 cm^{-1}) and terminal **alkynes** (3300 cm^{-1}) are useful for identification of these functional groups.

– The C=C stretches (1600–1680 cm^{-1}) for alkenes and the C≡C stretches (2100–2250 cm^{-1}) for alkynes are useful if they are strong enough to see.

• For **arenes**, C—H bending (3030 cm^{-1}), —C=C— stretching (1450–1600 cm^{-1}) and C—H bending (690–900 cm^{-1}) are useful peaks.

• For **alcohols**, the O—H (3200–3650 cm^{-1}, depending on hydrogen bonding) and C—O (1000–1250 cm^{-1}) stretches are most useful, while for **ethers**, only the C—O stretches (1000–1250 cm^{-1}) are seen.

• For **amines**, the N—H stretch (3300–3500 cm^{-1}) is useful, with primary amines having two peaks and secondary amines only one.

• For carbonyl species, the strong C=O stretch (1600–1850 cm^{-1}) is characteristic for **aldehydes** (1630–1820 cm^{-1}), **ketones** (1630–1820 cm^{-1}), **carboxylic acids** (1700–1725 cm^{-1}), **amides** (1630–1680 cm^{-1}), **carboxylic esters** (1735–1800 cm^{-1}), and **acid anhydrides** (1740–1760 cm^{-1}).

– The presence of other stretches such as C—O stretches for carboxylic esters (1000–1100 cm^{-1} and 1200–1250 cm^{-1}) and anhydrides (900–1300 cm^{-1}), a C—H stretch (2720 cm^{-1}) for aldehydes, an O—H stretch (2500–3300 cm^{-1}) for carboxylic acids, and N—H stretch (3200 and 3400 cm^{-1}) for amides is also used to distinguish these functional groups.

Problems: 12.10, 12.11

• For **nitriles**, the C≡N stretch (2200–2250 cm^{-1}) is useful.

Section 12.5 Solving Infrared Spectral Problems

• A systematic approach to solving IR problems involves:

– Analyzing the 3000 region to see if there are alkane C—H stretches (to the right of 3000 cm^{-1} or alkene/arene to the left of 3000 cm^{-1}).

– Checking for characteristic broad O—H or N—H peaks (3500 cm^{-1})

Problems: 12.5–12.7

– Checking for strong carbonyl peaks (1630–1820 cm^{-1}), then the C—O (1000–1250 cm^{-1}) or other peaks that might accompany the carbonyl in certain functional groups.

ⓌWL Questions and problems for this chapter are available in OWL. Use the OWL access card that came with this textbook to access assigned questions and problems for this chapter.

Chapter 13

Nuclear Magnetic Resonance Spectroscopy

Paul Shambroom/Science Source/Photo Researchers, Inc.

Magnetic resonance imaging is a useful medical diagnostic tool. Inset: a model of methyl acetate. For a ^{1}H-NMR spectrum of methyl acetate, see Figure 13.5.

In this chapter, we concentrate on absorption of radio-frequency radiation by nuclei and the resulting transitions between energy levels, better known as **nuclear magnetic resonance (NMR) spectroscopy**. Felix Bloch and Edward Purcell, both of the United States, first detected the phenomenon of nuclear magnetic resonance in 1946. They shared the 1952 Nobel Prize for physics. Nuclear magnetic resonance (NMR) spectroscopy was developed in the late 1950s and, within a decade, became the single most important technique available to chemists for the determination of molecular structure. Nuclear magnetic resonance spectroscopy gives us information about the number and types of atoms in a molecule: for example, about hydrogens using ^{1}H-NMR spectroscopy, and about carbons using ^{13}C-NMR spectroscopy. It can also give us substantial information about the connectivity of the atoms and, in many cases, can allow determination of the structure of a molecule with no additional information.

13.1 | Nuclear Spin States

You are already familiar from general chemistry with the concepts that (1) an electron has a spin quantum number of $\frac{1}{2}$, with allowed values of $+\frac{1}{2}$ and $-\frac{1}{2}$, and that (2) a moving charge has an associated magnetic field. In effect, an electron behaves as if it is a tiny bar magnet and has a magnetic moment. The same effect holds for certain atomic nuclei.

Any atomic nucleus that has an odd mass number, an odd atomic number, or both, also has a spin and a resulting nuclear magnetic moment. The allowed

⊙WL Online homework for this chapter may be assigned in OWL for Organic Chemistry.

Table 13.1	Spin Quantum Numbers and Allowed Nuclear Spin States for Selected Isotopes of Elements Common to Organic Compounds									
Element	^{1}H	^{2}H	^{12}C	^{13}C	^{14}N	^{15}N	^{16}O	^{19}N	^{31}P	^{32}S
Nuclear spin quantum number (I)	$\frac{1}{2}$	1	0	$\frac{1}{2}$	1	$\frac{1}{2}$	0	$\frac{1}{2}$	$\frac{1}{2}$	0
Number of spin states	2	3	1	2	3	2	1	2	2	1

nuclear spin states are determined by the spin quantum number, I, of the nucleus. A nucleus with spin quantum number I has $2I + 1$ spin states. Our focus in this chapter is on nuclei of ^{1}H and ^{13}C, isotopes of the two elements most common to organic compounds. Each has a **nuclear spin quantum number** of $\frac{1}{2}$ and therefore has $2(\frac{1}{2}) + 1 = 2$ allowed spin states. Quantum numbers and allowed nuclear spin states for these nuclei and those of other elements common to organic compounds are shown in Table 13.1. Note that ^{12}C, ^{16}O, and ^{32}S each have a spin quantum number of zero and only one allowed nuclear spin state; these nuclei are inactive in NMR spectroscopy.

13.2 | Orientation of Nuclear Spins in an Applied Magnetic Field

Within a sample containing ^{1}H and ^{13}C atoms, the orientations of the nuclear magnetic moments associated with their nuclear spins are completely random. When placed between the poles of a powerful magnet of field strength B_0, however, interactions between the nuclear spins and the applied magnetic field are quantized, with the result that only certain orientations of nuclear magnetic moments are allowed. For ^{1}H and ^{13}C nuclei, only two orientations are allowed as illustrated in Figure 13.1. By convention, nuclei with spin designated as $+\frac{1}{2}$ are aligned with the applied magnetic field and are in the lower energy state; nuclei with spin designated as $-\frac{1}{2}$ are aligned against the applied magnetic field and are in the higher energy state.

The most important NMR physical concept from the point of view of molecular structure determination is that **the difference in energy between nuclear spin states for a given nucleus is proportional to the strength of the magnetic field experienced by that nucleus (Figure 13.2)**. We will come back to this concept several times in this chapter. At an applied field strength of 7.05 T, which is readily available with present-day superconducting electromagnets, the difference in energy between nuclear spin states for ^{1}H is approximately 0.120 J (0.0286 cal)/mol (corresponding to electromagnetic radiation of 300 MHz). At 7.05 T, the energy difference in nuclear spin states for ^{13}C nuclei is approximately 0.030 J (0.00715 cal)/mol (corresponding to

Note: the SI unit for magnetic field strength is the **tesla (T)**. A unit still in common use, however, is the gauss (G). Values of T and G are related by the equation $1\,T = 10^4\,G$.

Figure 13.1
^{1}H and ^{13}C nuclei with spin $+\frac{1}{2}$ are aligned with the applied magnetic field, B_0, and are in the lower spin energy state; those with spin $-\frac{1}{2}$ are aligned against the applied magnetic field and are in the higher spin energy state.

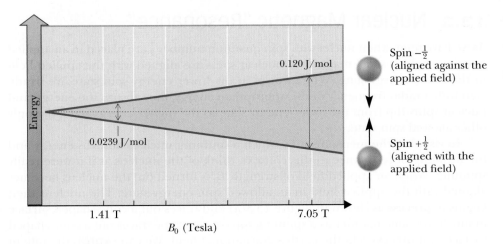

Figure 13.2
The energy difference between the allowed nuclear spin states increases linearly with applied field strength. Values shown here are for ^1H nuclei.

Spin $-\frac{1}{2}$ (aligned against the applied field)

Spin $+\frac{1}{2}$ (aligned with the applied field)

0.120 J/mol

0.0239 J/mol

1.41 T

7.05 T

B_0 (Tesla)

radiation of 75 MHz). Advanced commercial instruments now operate at fields more than three times this value; the operating frequencies are proportional to the field. Sensitivities are more than proportionally higher!

To put these values for nuclear spin energy levels in perspective, energies for transitions between vibrational energy levels observed in infrared (IR) spectroscopy are 8 to 63 kJ (2 to 15 kcal)/mol. Those between electronic energy levels in ultraviolet-visible spectroscopy (see Section 20.3) are 167 to 585 kJ (40 to 140 kcal)/mol. Nuclear transitions involve only small energies, on the order of a few hundredths of a calorie!

Example 13.1

Calculate the ratio of nuclei in the higher spin state to those in the lower spin state, N_h/N_l, for ^1H at 25°C in an applied field strength of 7.05 T.

Solution

Use the equation given in Section 2.6B for the relationship between the difference in energy states and equilibrium constant. In this problem, this relationship has the form

$$\Delta G^0 = -RT \ln \frac{N_h}{N_l}$$

The difference in energy between the higher and lower nuclear spin states in an applied field of 7.05 T is approximately 0.120 J/mol, and the temperature is 25 + 273 = 298 K. Substituting these values in this equation gives

$$\ln \frac{N_h}{N_l} = \frac{-\Delta G^0}{RT} = \frac{-0.120 \text{ J} \cdot \text{mol}^{-1}}{8.314 \text{ J} \cdot \text{K}^{-1} \cdot \text{mol}^{-1} \times 298 \text{ K}} = -4.843 \times 10^{-5}$$

$$\frac{N_h}{N_l} = 0.9999516 = \frac{1.000000}{1.000048}$$

From this calculation, we determine that, for every 1,000,000 hydrogen atoms in the higher energy state in this applied field, there are 1,000,048 in the lower energy state. The excess population of the lower energy state under these conditions is only 48 per million! What is important about this number is that the strength of an NMR signal is proportional to the population difference. As you will see, the greater this difference in populations, the stronger the signal will be, because more spins are flipping.

Problem 13.1

Calculate the ratio of nuclei in the higher spin state to those in the lower spin state, N_h/N_l, for ^{13}C at 25°C in an applied field strength of 7.05 T. The difference in energy between the higher and lower nuclear spin states in this applied field is approximately 0.030 J (0.00715 cal)/mol.

13.3 | Nuclear Magnetic "Resonance"

As we have seen, when nuclei with spin quantum number $\frac{1}{2}$ are placed in an applied magnetic field, a small majority of nuclear spins are aligned with the applied field in the lower energy state. When nuclei in the lower energy spin state are irradiated with a radio frequency of the appropriate energy, they absorb the energy, and nuclear spins flip from the lower energy state to the higher energy state, the only other allowed spin state.

To visualize the mechanism by which a spinning nucleus absorbs energy and the meaning of resonance in this context, think of the nucleus as if it were really spinning. When an applied field of strength B_0 is turned on, the nucleus becomes aligned with the applied field in an allowed spin energy state. The nucleus then begins to **precess** as shown in Figure 13.3(a) and traces out a cone-shaped surface in much the same manner as a spinning top or gyroscope traces out a cone-shaped surface as it precesses in the earth's gravitational field. We can express the rate of precession as a frequency in hertz.

If the precessing nucleus is irradiated with electromagnetic radiation at exactly the precession frequency, then the two frequencies couple, energy is absorbed, and the nuclear spin "flips" from spin state $+\frac{1}{2}$ (with the applied field) to spin state $-\frac{1}{2}$ (against the applied field) as illustrated in Figure 13.3(b). For ^1H in an applied magnetic field of 7.05 T, the frequency of precession is approximately 300 MHz. For ^{13}C in the same field, it is approximately 75 MHz. **Resonance** in this context is the absorption of electromagnetic radiation by a precessing nucleus and the resulting flip of its nuclear spin from the lower energy state to the higher energy state. The spectrometer detects this absorption of electromagnetic radiation and records it as a **signal**. The process is quantized, so that only electromagnetic radiation of precisely the correct frequency causes a nuclear spin to flip. Electromagnetic radiation of a frequency that is too low or too high is not absorbed.

If we were dealing with ^1H nuclei isolated from all other atoms and electrons, any combination of applied field and electromagnetic radiation that produces a signal for one hydrogen nucleus would produce a signal for all hydrogen nuclei. In other words, hydrogens would be indistinguishable. Hydrogens in an organic molecule, however, are not isolated; they are surrounded by electron density.

A key physical principle for NMR is that circulating electrons induce a magnetic field. This principle, first described by Michael Faraday and his law of electromagnetic induction, also explains electromagnets and electric motors. The direction of electron movement dictates the orientation of the induced magnetic field. Of equal importance to NMR, the converse is also true. An applied magnetic field induces electrons to circulate, and the orientation of the field dictates the direction of circulation. You will learn more details of these relationships in

Resonance in NMR spectroscopy
The absorption of electromagnetic radiation by a precessing nucleus and the resulting "flip" of its nuclear spin from the lower energy state to the higher energy state.

Signal
A recording in an NMR spectrum of a nuclear magnetic resonance.

Figure 13.3

The origin of nuclear magnetic "resonance." **(a)** Precession of a spinning nucleus in an applied magnetic field. **(b)** Absorption of electromagnetic radiation occurs when the frequency of radiation is equal to the frequency of precession.

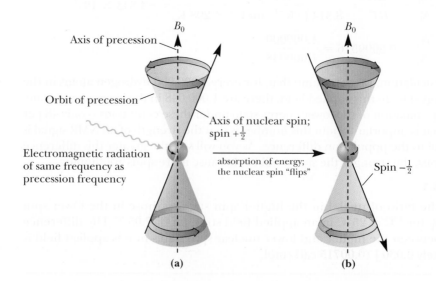

your physics classes. The important point for our purposes is that an applied magnetic field induces the electron density in a molecule to circulate. The spin states of underlying nuclei are, in turn, influenced to a small but measurable degree by the magnetic field created by the induced electron density circulation. The circulation of electron density in a molecule in an applied magnetic field is called a **diamagnetic current**.

It turns out that a molecule's σ-bonding electron density is induced to circulate in a direction that creates a small magnetic field that directly *opposes* the applied magnetic field. As a result of the opposing magnetic fields, the nuclei within the circulating electron density experience a magnetic field that is slightly *smaller* than the actual applied field. In other words, nuclei underneath circulating σ-bonding electron density are *shielded* to a small degree from the applied magnetic field. This nuclear **shielding** is called **diamagnetic shielding**. Although the diamagnetic shielding created by circulating electron density is generally orders of magnitude weaker than the applied fields used in NMR spectroscopy, it is nonetheless significant enough to be measured.

The degree of shielding depends on several factors, which we will take up later. Nevertheless, what we need to establish now is that as the shielding becomes greater, the net magnetic field present at a nucleus becomes smaller, so the energy of electromagnetic radiation required to bring that nucleus into resonance (i.e., "flip its spin") also decreases. Energy is proportional to electromagnetic radiation frequency, and resonance frequencies are plotted on an NMR spectrum. Putting all of these ideas together, we see that a nucleus that is more shielded will come into resonance at lower frequency than a nucleus that is less shielded. **Deshielding** is the term commonly used to express the concept of less shielding. The relationship of *greater* shielding leading to resonance at *lower* frequency is fundamental to understanding NMR spectra, but most students find it challenging to remember correctly. Make sure you have a clear understanding of this concept before going on.

The difference in resonance frequencies caused by differing amounts of shielding is called **chemical shift**. The differences in resonance frequencies among the various hydrogen nuclei within a molecule attributable to shielding/deshielding are generally very small. The difference between the resonance frequencies of hydrogens in chloromethane compared with those in fluoromethane, for example, under an applied field of 7.05 T is only 360 Hz. Considering that the radio-frequency radiation used at this applied field is approximately 300 MHz, the difference in resonance frequencies between these two sets of hydrogens is only slightly greater than 1 **part per million** (1 **ppm**) compared with the irradiating frequency.

$$\frac{360 \text{ Hz}}{300 \times 10^6 \text{ Hz}} = \frac{1.2}{10^6} = 1.2 \text{ ppm}$$

It is customary to measure the resonance frequencies of individual nuclei relative to the resonance frequency of nuclei in a reference compound. The reference compound now universally accepted for ¹H-NMR and ¹³C-NMR spectroscopy is tetramethylsilane (TMS), which is assigned a chemical shift of 0.00 ppm by convention.

$$H_3C-\underset{\underset{CH_3}{|}}{\overset{\overset{CH_3}{|}}{Si}}-CH_3$$

Tetramethylsilane (TMS)

To standardize reporting of NMR data for both ¹H and ¹³C spectra, the chemical shift (δ), in parts per million, is defined as the frequency shift from either the hydrogens or the carbons in TMS divided by the operating frequency of the spectrometer. Thus, by definition, chemical shift is independent of the operating frequency of the spectrometer. On the chart paper used to record NMR spectra,

Diamagnetic current in NMR
The circulation of electron density in a molecule in an applied magnetic field.

Shielding in NMR
Also called diamagnetic shielding; the term refers to the reduction in magnetic field strength experienced by a nucleus underneath electron density induced to circulate when the molecule is placed in a strong magnetic field.

Chemical shift (δ)
The shift in parts per million of an NMR signal relative to the signal of TMS.

chemical shift values are shown in increasing order to the left of the TMS signal. Having values increase to the left is backward compared to the way you normally see scientific values plotted (with larger values to the right), so make sure you recognize this very unusual feature of NMR spectra.

13.4 | An NMR Spectrometer

The essential elements of an NMR spectrometer are a powerful magnet, a radio-frequency generator, a radio-frequency detector, and a sample tube (Figure 13.4).

The sample is dissolved in a solvent, most commonly carbon tetrachloride (CCl_4), deuterochloroform ($CDCl_3$), or deuterium oxide (D_2O), which have no ^1H atoms and do not interfere in ^1H spectra. The sample cell is a small glass tube suspended in the magnetic field and set spinning on its long axis to ensure that all parts of the sample experience a homogeneous applied magnetic field. In the simplest form of the ^1H-NMR experiment, the absorption of electromagnetic radiation is measured as different ^1H nuclei are excited from their $+\frac{1}{2}$ spin states to their $-\frac{1}{2}$ spin states. The frequencies at which the absorptions occur are in the radio-frequency region of the electromagnetic spectrum. The observed absorption frequencies are plotted as peaks relative to the TMS standard on a ppm scale.

Modern Fourier transform NMR (FT-NMR) spectrometers can increase the power of the NMR technique significantly. An FT-NMR spectrometer operates in the following way. The magnetic field is held constant, and the sample is irradiated with a short pulse (approximately 10^{-5} s) of radio-frequency energy that flips the spins of all susceptible nuclei simultaneously. As each nucleus returns to its equilibrium state, it emits a sine wave at the frequency of its resonance. The intensity of the sine wave decays with time and falls to zero as nuclei return to their equilibrium state. A computer records this intensity-versus-time information and then uses a mathematical algorithm called a Fourier transform (FT) to convert it to intensity-versus-frequency information. An FT-NMR spectrum can be recorded in less than two seconds. A particular advantage of FT-NMR spectroscopy is that a large number of spectra (as many as several thousand per sample) can be recorded and digitally summed to give a time-averaged spectrum. Instrumental electronic noise is random and partially cancels out when spectra are time-averaged, but sample signals accumulate and become much stronger relative to the electronic noise. The net result is that good NMR spectra can be obtained with very little sample. All NMR spectra shown in this text were recorded and displayed using FT techniques.

Figure 13.4

Schematic diagram of a nuclear magnetic resonance spectrometer.

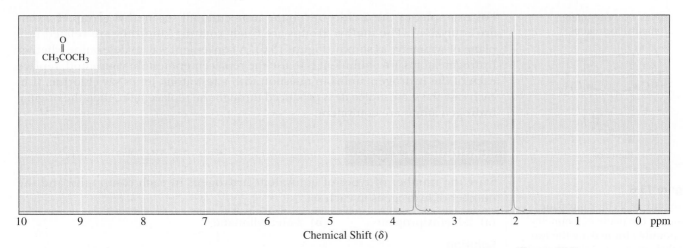

Figure 13.5
^{1}H-NMR spectrum of methyl acetate.

Figure 13.5 shows a 300 MHz ^{1}H-NMR spectrum of methyl acetate. The lower axis is δ, in parts per million. The small signal at δ 0 is caused by the hydrogens of the TMS reference, a small amount of which was added to the sample. The remainder of the spectrum consists of two signals: one for the three hydrogens on the methyl adjacent to oxygen and one for the three hydrogens on the methyl adjacent to the carbonyl group. It is not our purpose at the moment to determine which hydrogens give rise to which signal but only to recognize the form in which an NMR spectrum is recorded and the origin of the calibration marks.

Here is a note on terminology. If a signal is shifted toward the left on the chart paper (larger chemical shift), we say that it is shifted **downfield**. A downfield shift corresponds to decreased shielding around a nucleus, that is, deshielding. If a signal is shifted toward the right (smaller chemical shift), we say that it is shifted **upfield** and therefore corresponds to increased shielding around a nucleus. Students often remember these relationships through the simple expression "more downfield = more deshielded."

Downfield
A signal of an NMR spectrum that is shifted toward the left (larger chemical shift) on the chart paper.

Upfield
A signal of an NMR spectrum that is shifted toward the right (smaller chemical shift) on the chart paper.

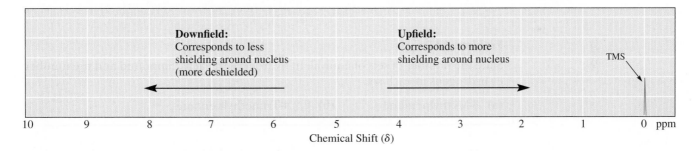

13.5 | Equivalent Hydrogens

Given the structural formula of a compound, how do we know how many signals to expect in its ^{1}H-NMR spectrum? The answer is that equivalent hydrogens give the same ^{1}H-NMR signal; nonequivalent hydrogens give different ^{1}H-NMR signals. **Equivalent hydrogens** have the same chemical environment. H atoms are equivalent (in the same chemical environment) if either of the two following conditions exist:

Equivalent hydrogens
Hydrogens that have the same chemical environment.

1. They are bonded to the same sp^3 hybridized carbon atom, and that carbon atom can rotate freely at room temperature. The rapid bond rotation means that, on average, the H atoms bonded to the same carbon atom see the same chemical environment and are therefore equivalent. For example, all three H atoms on a freely rotating —CH$_3$ group are equivalent, and both H atoms on a freely rotating —CH$_2$— group are usually equivalent, although see Section 13.10 for the exception that occurs when —CH$_2$— groups are near a chiral center.
2. They are related by symmetry, namely a plane or point of symmetry in a molecule.

a set of six equivalent
methyl hydrogens

Plane of
symmetry

Figure 13.6
Structure of 2-chloropropane
showing the plane of symmetry
responsible for making the two
methyl groups, and therefore
the six methyl group H atoms,
equivalent.

For example, in the 2-chloropropane molecule, all six methyl group H atoms are equivalent. The methyl groups are freely rotating, and they are related by a plane of symmetry as shown (Figure 13.6). A convenient way to determine whether hydrogen atoms are equivalent is to use the replacement test. In your mind, replace each of the hydrogen atoms in question with a test atom, let us say chlorine. If in each case the same molecule is produced through the replacement, then the original hydrogen atoms are equivalent.

Example 13.2

State the number of types of equivalent hydrogens in each compound and the number of hydrogens in each set.

(a) 2-Methylpropane   (b) 2-Methylbutane

Solution

(a) 2-Methylpropane contains two sets of equivalent hydrogens: a set of nine equivalent primary hydrogens and one tertiary hydrogen.

(b) 2-Methylbutane contains four sets of equivalent hydrogens. Nine primary hydrogens are in this molecule: one set of three and one set of six. To see that there are two sets, note that replacement by chlorine of any hydrogen in the set of three gives 1-chloro-3-methylbutane. Replacement by chlorine of any hydrogen in the set of six gives 1-chloro-2-methylbutane. In addition, the molecule contains a set of two equivalent secondary hydrogens and one tertiary hydrogen.

Problem 13.2

State the number of sets of equivalent hydrogens in each compound and the number of hydrogens in each set.

(a) 3-Methylpentane   (b) 2,2,4-Trimethylpentane

Here are four organic compounds, each of which has one set of equivalent hydrogens and gives one signal in its ^{1}H-NMR spectrum.

$$CH_3\overset{\overset{\displaystyle O}{\|}}{C}CH_3 \qquad ClCH_2CH_2Cl \qquad \qquad \qquad \overset{H_3C}{\underset{H_3C}{>}}C=C\overset{CH_3}{\underset{CH_3}{<}}$$

Propanone   **1,2-Dichloroethane**   **Cyclopentane**   **2,3-Dimethyl-2-butene**
(Acetone)

Molecules with two or more sets of equivalent hydrogens give rise to a different resonance signal for each set, as illustrated by these four compounds.

$$\overset{\overset{\displaystyle Cl}{|}}{CH_3}CHCl \qquad \qquad =O \qquad \overset{Cl}{\underset{H}{>}}C=C\overset{CH_3}{\underset{H}{<}}$$

1,1-Dichloroethane   **Cyclopentanone**   **(Z)-1-Chloropropene**   **Cyclohexene**
(2 signals)   (2 signals)   (3 signals)   (3 signals)

You should see immediately that valuable information about molecular structure can be obtained simply by counting the number of signals in the ^{1}H-NMR spectrum of a compound. Consider, for example, the two constitutional isomers of molecular formula $C_2H_4Cl_2$. The compound 1,2-dichloroethane has one set of equivalent hydrogens and one signal in its ^{1}H-NMR spectrum. Its constitutional isomer 1,1-dichloroethane has two sets of equivalent hydrogens and two signals in its ^{1}H-NMR spectrum. Thus, simply counting signals allows you to distinguish between these two compounds.

Example 13.3

Each compound gives only one signal in its ^{1}H-NMR spectrum. Propose a structural formula for each compound.

(a) C_2H_6O (b) $C_3H_6Cl_2$ (c) C_6H_{12} (d) C_4H_6

Solution

(a) CH_3OCH_3 (b) $CH_3\overset{\displaystyle Cl}{\underset{\displaystyle Cl}{C}}CH_3$ (c) ⬡ or $\underset{H_3C}{\overset{H_3C}{}}C=C\underset{CH_3}{\overset{CH_3}{}}$ (d) $CH_3C{\equiv}CCH_3$

Problem 13.3

Each compound gives only one signal in its ^{1}H-NMR spectrum. Propose a structural formula for each compound.

(a) C_3H_6O (b) C_5H_{10} (c) C_5H_{12} (d) $C_4H_6Cl_4$

13.6 | Signal Areas

We have just seen that the number of signals in a ^{1}H-NMR spectrum gives us information about the number of sets of equivalent hydrogens. The relative areas of these signals provide additional information. As a spectrum is being run, the instrument's computer numerically measures the area under each signal. In the spectra shown in this text, this information is displayed in the form of a line of integration superposed on the original spectrum. The vertical rise of the line of integration over each signal is proportional to the area under that signal, which, in turn, is proportional to the number of equivalent hydrogens giving rise to that signal. Figure 13.7 shows an integrated ^{1}H-NMR spectrum of *tert*-butyl acetate, $C_6H_{12}O_2$. The spectrum shows signals at δ 1.44 and 1.95. The integrated signal heights are $23 + 67$, or 90 chart divisions, which correspond to 12 hydrogens. From these numbers, we calculate that $(23/90) \times 12$ or

Figure 13.7

^{1}H-NMR spectrum of *tert*-butyl acetate showing the integration. The total vertical rise of 90 chart divisions corresponds to 12 hydrogens, 9 in one set and 3 in the other.

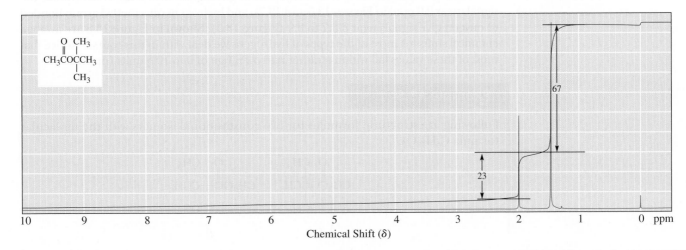

3 hydrogens are in one set, and (67/90) × 12 or 9 hydrogens are in the second set. An alternative way of indicating integration is a numerical readout given over each signal, and you may see this on many ¹H-NMR spectra you encounter.

Example 13.4

Following is a ¹H-NMR spectrum for a compound of molecular formula $C_9H_{10}O_2$. From the integration, calculate the number of hydrogens giving rise to each signal.

Solution

The total vertical rise in the line of integration is 88 chart divisions and corresponds to 10 hydrogens. From these numbers, we calculate that 44/88 × 10, or 5, of the hydrogens give rise to the signal at δ 7.34. By similar calculations, the signals at δ 5.08 and 2.06 correspond to two hydrogens and three hydrogens, respectively.

Problem 13.4

The line of integration of the two signals in the ¹H-NMR spectrum of a ketone with the molecular formula $C_7H_{14}O$ rises 62 and 10 chart divisions, respectively. Calculate the number of hydrogens giving rise to each signal, and propose a structural formula for this ketone.

13.7 | Chemical Shift

The **chemical shift** for a signal in a ¹H-NMR spectrum can give valuable information about the type of hydrogens giving rise to that signal. Hydrogens on methyl groups bonded to sp^3-hybridized carbons, for example, give signals near δ 0.8 to 1.0. Hydrogens on methyl groups bonded to a carbonyl carbon give signals near δ 2.1 to 2.3 (notice the signals near 2.0 ppm in Figures 13.5 and 13.7), and hydrogens on a methyl group bonded to oxygen give signals near δ 3.7 to 3.9 (Figure 13.5). Shown in Figure 13.8 are average chemical shifts for most of the types of hydrogens we deal with in this course. Notice that most of these values fall within a rather narrow range of 0 to 10 δ units (ppm).

Example 13.5

Following are structural formulas for two constitutional isomers with the molecular formula $C_6H_{12}O_2$.

$$
\begin{array}{cc}
\underset{\text{(1)}}{\overset{\displaystyle \text{O} \quad \text{CH}_3}{\underset{\displaystyle \text{CH}_3}{\text{CH}_3\text{COCCH}_3}}}
&
\underset{\text{(2)}}{\overset{\displaystyle \text{O} \quad \text{CH}_3}{\underset{\displaystyle \text{CH}_3}{\text{CH}_3\text{OC}-\text{CCH}_3}}}
\end{array}
$$

(a) Predict the number of signals in the ¹H-NMR spectrum of each isomer.
(b) Predict the ratio of areas of the signals in each spectrum.
(c) Show how you can distinguish between these isomers on the basis of chemical shift.

Solution

(a) The ¹H-NMR spectrum of each consists of two signals (b) in the ratio 9:3, or 3:1. (c) Distinguish between these constitutional isomers by the chemical shift of the single —CH₃ group. The hydrogens of CH₃O are more deshielded (appear farther downfield) than the hydrogens of CH₃C=O. See Figure 13.8 for approximate values for each chemical shift. Experimentally determined values are shown in red.

$$\delta\ 1.95 \qquad \underset{CH_3}{\underset{|}{\overset{O\quad CH_3}{\overset{||\quad |}{CH_3COCCH_3}}}} \qquad \delta\ 1.44 \qquad\qquad \delta\ 3.67 \qquad \underset{CH_3}{\underset{|}{\overset{O\quad CH_3}{\overset{||\quad |}{CH_3OC-CCH_3}}}} \qquad \delta\ 1.20$$

(1)   (2)

Please see Figure 13.7 for the spectrum of *tert*-butyl acetate (1).

Problem 13.5

Following are two constitutional isomers with the molecular formula $C_4H_8O_2$.

$$\overset{O}{\overset{||}{CH_3CH_2OCCH_3}} \qquad\qquad \overset{O}{\overset{||}{CH_3CH_2COCH_3}}$$

(1)   (2)

(a) Predict the number of signals in the ¹H-NMR spectrum of each isomer.
(b) Predict the ratio of areas of the signals in each spectrum.
(c) Show how you can distinguish between these isomers on the basis of chemical shift.

The chemical shift of a particular type of hydrogen depends primarily on the extent of shielding it experiences. Shielding, in turn, depends on three factors: (1) the electronegativity of nearby atoms, (2) the hybridization of the adjacent atoms, and (3) the magnetic induction within an adjacent π bond. Let us consider these factors one at a time.

Figure 13.8
Average values of chemical shifts of representative types of hydrogens. These values are approximate. Other atoms or groups in the molecules may cause signals to appear outside of these ranges.

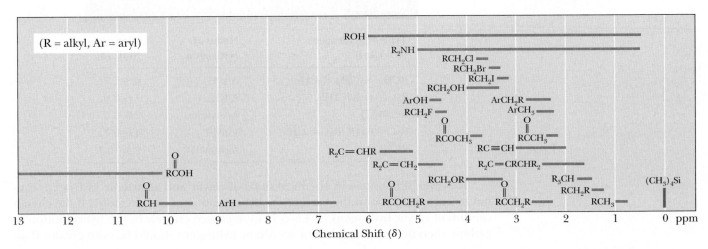

A. Electronegativity of Nearby Atoms

As illustrated in Table 13.2 for the chemical shift of methyl hydrogens in the series $CH_3—X$, the greater the electronegativity of X, the greater the chemical shift will be. The effect of an electronegative substituent falls off quickly with distance. The effect of an electronegative substituent two atoms away is only about 10% of that when it is on the adjacent atom. The effect of an electronegative substituent three atoms away is almost negligible.

Electronegativity and chemical shift are related in the following way: the presence of an electronegative atom or group reduces electron density on atoms bonded to it and therefore the shielding. This effect deshields nearby nuclei and causes them to resonate farther downfield, that is, with a larger chemical shift.

Table 13.2	Dependence of Chemical Shift of CH_3X on the Electronegativity of X	
$CH_3—X$	Electronegativity of X	Chemical Shift (δ) of Methyl Hydrogens
CH_3F	4.0	4.26
CH_3OH	3.5	3.47
CH_3Cl	3.1	3.05
CH_3Br	2.8	2.68
CH_3I	2.5	2.16
$(CH_3)_4C$	2.1	0.86
$(CH_3)_4Si$	1.8	0.00 (by definition)

B. Hybridization of Adjacent Atoms

Hydrogens bonded to an sp^3-hybridized carbon typically have signals at δ 0.8 to 1.7. Vinylic hydrogens (those on a carbon of a carbon-carbon double bond) are considerably deshielded and their signals appear at δ 4.6 to 5.7 (Table 13.3). Part of the explanation for the greater deshielding of vinylic hydrogens compared with alkyl hydrogens lies in the hybridization of carbon. Because a σ-bonding orbital of an sp^2-hybridized carbon has more s-character than a σ-bonding orbital of an sp^3-hybridized carbon (33% compared with 25%), an sp^2-hybridized carbon atom is more electronegative. Vinylic hydrogens are deshielded by this electronegativity effect and their nuclei resonate farther downfield relative to alkyl hydrogens. Similarly, signals for acetylene and aldehyde hydrogens also appear farther downfield compared with alkyl hydrogens.

Table 13.3	The Effect of Hybridization on Chemical Shift	
Type of Hydrogen (R = alkyl)	Name of Hydrogen	Chemical Shift (δ)
RCH_3, R_2CH_2, R_3CH	Alkyl	0.8–1.7
$R_2C=C(R)CHR_2$	Allylic	1.6–2.6
$RC\equiv CH$	Acetylenic	2.0–3.0
$R_2C=CHR, R_2C=CH_2$	Vinylic	4.6–5.7
$RCHO$	Aldehydic	9.5–10.1

However, differences in chemical shifts of vinylic and acetylenic hydrogens cannot be accounted for on the basis of the hybridization of carbon alone. If the chemical shift of vinylic hydrogens (δ 4.6–5.7) were caused entirely by the hybridization of carbon, then the chemical shift of acetylenic hydrogens should be even greater than that of vinylic hydrogens. Yet the chemical shift of acetylenic hydrogens is only δ 2.0 to 3.0. It seems that either the chemical shift of acetylenic hydrogens is abnormally small or the chemical shift of vinylic hydrogens is abnormally large. In either case, another factor must be contributing to the magnitude of the chemical shift. Theoretical and experimental evidence suggest that the chemical shifts of hydrogens bonded to π-bonded carbons are influenced not only by the relative electronegativities of the sp^2- and sp-hybridized carbon atoms but also by magnetic induction from π bonds.

C. Diamagnetic Effects from π Bonds

To understand the influence of π bonds on the chemical shift of an acetylenic hydrogen, imagine that the carbon-carbon triple bond is oriented as shown in Figure 13.9 with respect to the applied field. Because of magnetic induction, the applied field induces a circulation of the π electrons, which in turn produces an induced magnetic field. Given the geometry of an alkyne and the cylindrical nature of its π electron cloud, the induced magnetic field is shielding in the vicinity of the acetylenic hydrogen. Therefore, lower frequency electromagnetic radiation is required to make an acetylenic hydrogen nucleus resonate; the local magnetic field induced in the π bonds shifts the signal of an acetylenic hydrogen upfield to a smaller δ value.

The effect of the induced circulation of π electrons on a vinylic hydrogen (Figure 13.10) is opposite to that on an acetylenic hydrogen. The direction of the induced magnetic field in the π bond of a carbon-carbon double bond is parallel to the applied field in the region of the vinylic hydrogens. The induced magnetic field deshields vinylic hydrogens and, thus, shifts their signal downfield to a larger δ value. The presence of the π electrons in the carbonyl group has a similar effect on the chemical shift of the hydrogen of an aldehyde group.

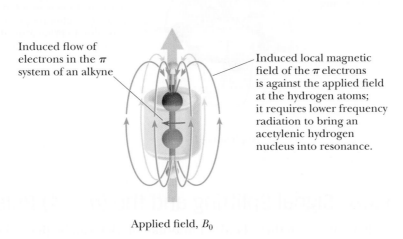

Induced flow of electrons in the π system of an alkyne

Induced local magnetic field of the π electrons is against the applied field at the hydrogen atoms; it requires lower frequency radiation to bring an acetylenic hydrogen nucleus into resonance.

Applied field, B_0

Figure 13.9

A magnetic field induced in the π bonds of a carbon-carbon triple bond shields an acetylenic hydrogen and shifts its signal upfield.

Induced circulation of π electrons in an alkene

Induced local magnetic field of the π electrons reinforces the applied field at the hydrogen atoms; it requires higher frequency radiation to bring a vinylic hydrogen nucleus into resonance.

Applied field, B_0

Figure 13.10

A magnetic field induced in the π bond of a carbon-carbon double bond deshields vinylic hydrogens and shifts their signals downfield.

The bottom line when comparing acetylinic versus vinylic hydrogens is that the hybridization effect, which would predict a larger chemical shift (greater deshielding) for the acetylinic hydrogen signals, is more than overcome by the strongly shielding diamagnetic effect of the cylindrically oriented alkyne π bonds and the deshielding diamagnetic effect of the alkene π bond. The net result is that acetylinic hydrogens have signals that appear less downfield from TMS as compared to the signals for vinylic hydrogens.

The effects of the π electrons in benzene are even more dramatic than in alkenes. All six hydrogens of benzene are equivalent, and its ^{1}H-NMR spectrum is a sharp singlet at δ 7.27. Hydrogens bonded to a substituted benzene ring appear in the region δ 6.5 to 8.5. Few other hydrogens absorb in this region; thus, aryl hydrogens are quite easily identifiable by their distinctive chemical shifts, as much as 2 ppm higher than comparably substituted alkenes.

That aryl hydrogens absorb even farther downfield than vinylic hydrogens is accounted for by the existence of a **ring current**, a special property of aromatic rings (Figure 13.11). When the plane of an aromatic ring tumbles in an applied magnetic field, the applied field causes the π electrons to circulate around the ring, giving rise to the so-called ring current. This induced ring current has associated with it a magnetic field that opposes the applied field in the middle of the ring but reinforces the applied field on the outside of the ring. Thus, given the position of aromatic hydrogens relative to the induced ring current, they are deshielded and come into resonance at a larger chemical shift.

Ring current
An applied magnetic field causes the π electrons of an aromatic ring to circulate, giving rise to the so-called ring current and an associated magnetic field that opposes the applied field in the middle of the ring but reinforces the applied field on the outside of the ring.

Figure 13.11
The magnetic field induced by circulation of π electrons in an aromatic ring deshields the hydrogens of the aromatic ring and shifts their signal downfield.

Induced circulation of π electrons in a benzene ring

Induced local magnetic field of the circulating π electrons reinforces the applied field at the hydrogen atoms; it requires higher frequency radiation to bring aromatic hydrogen nuclei into resonance

Applied field

13.8 | Signal Splitting and the (n + 1) Rule

We have now seen three kinds of information that can be derived from examination of a ^{1}H-NMR spectrum.

1. From the number of signals, we can determine the number of sets of equivalent hydrogens.
2. From integration of signal areas, we can determine the relative numbers of hydrogens giving rise to each signal.
3. From the chemical shift of each signal, we derive information about the types of hydrogens in each set.

A fourth kind of information can be derived from the splitting pattern of each signal. Consider, for example, the ^{1}H-NMR spectrum of 1,1-dichloroethane shown in Figure 13.12. This molecule contains two sets of equivalent hydrogens. According to what we have learned so far, we predict two signals with relative areas 3:1 corresponding to the three hydrogens of the —CH$_3$ group and the one hydrogen of the —CHCl$_2$ group, respectively. You see from the spectrum, however, that there are, in fact, six peaks. These peaks are named by how the signal is split: two peaks are a

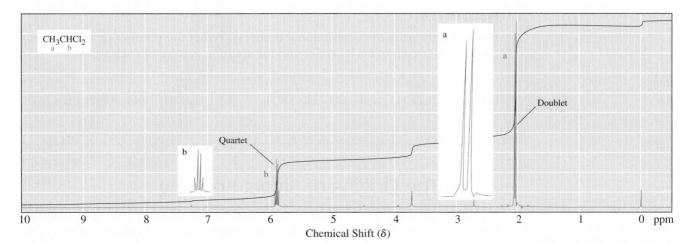

Chemical Shift (δ)

Figure 13.12
¹H-NMR spectrum of
1,1-dichloroethane.

doublet, three peaks are a triplet, and so on. The grouping of two peaks at δ 2.0 is the signal for the three hydrogens of the —CH₃ group, and the grouping of four peaks at δ 5.9 is the signal for the single hydrogen of the —CHCl₂ group. We say that the CH₃ resonance at δ 2.1 is split into a doublet and that the CH resonance at δ 5.9 is split into a quartet.

In many situations, the degree of **signal splitting** can be predicted on the basis of the **(n + 1) rule**. According to this rule, if a hydrogen has n hydrogens nonequivalent to it but equivalent among themselves on the same or adjacent atom(s), its ¹H-NMR signal is split into (n + 1) peaks.

Let us apply the (n + 1) rule to the analysis of the spectrum of 1,1-dichloroethane. The three hydrogens of the —CH₃ group have one nonequivalent neighbor hydrogen (n = 1), and, therefore, their signal is split into a doublet. The single hydrogen of the —CHCl₂ group has a set of three nonequivalent neighbor hydrogens (n = 3), and its signal is split into a quartet.

(n + 1) rule
If a hydrogen has n hydrogens nonequivalent to it but equivalent among themselves on the same or adjacent atom(s), its ¹H-NMR signal is split into (n + 1) peaks.

For these hydrogens, n = 1; their signal is split into (1 + 1) or two peaks—a **doublet**.

For this hydrogen, n = 3; its signal is split into (3 + 1) or four peaks—a **quartet**.

$$CH_3—CH—Cl$$
$$\quad\quad\quad |$$
$$\quad\quad\quad Cl$$

Example 13.6

Predict the number of signals and the splitting pattern of each signal in the ¹H-NMR spectrum of each molecule.

(a) $CH_3\overset{O}{\overset{||}{C}}CH_2CH_3$ (b) $CH_3CH_2\overset{O}{\overset{||}{C}}CH_2CH_3$ (c) $CH_3\overset{O}{\overset{||}{C}}CH(CH_3)_2$

Solution

The sets of equivalent hydrogens in each molecule are labeled *a*, *b*, and *c*. Molecule (a) has three sets of equivalent hydrogens; its ¹H-NMR spectrum shows a singlet, a quartet, and a triplet in the ratio 3 : 2 : 3. Molecule (b) has two sets of equivalent hydrogens; its ¹H-NMR spectrum shows a triplet and a quartet in the ratio 3 : 2. Molecule (c) has three sets of equivalent hydrogens; its ¹H-NMR spectrum shows a singlet, a septet, and a doublet in the ratio 3 : 1 : 6.

Following are pairs of constitutional isomers. Predict the number of signals in the ^{1}H-NMR spectrum of each isomer and the splitting pattern of each signal.

$$\text{(a)} \quad CH_3OCH_2\overset{\overset{\displaystyle O}{\|}}{C}CH_3 \quad \text{and} \quad CH_3CH_2\overset{\overset{\displaystyle O}{\|}}{C}OCH_3 \qquad \text{(b)} \quad CH_3\overset{\overset{\displaystyle Cl}{|}}{\underset{\underset{\displaystyle Cl}{|}}{C}}CH_3 \quad \text{and} \quad ClCH_2CH_2CH_2Cl$$

13.9 | The Origins of Signal Splitting

^{1}H-NMR signal splitting can be understood by considering **spin-spin coupling** between ^{1}H nuclei, an interaction in which nuclear spins of adjacent atoms influence each other. A common type of spin-spin coupling involves the H atoms on two C atoms that are bonded to each other. These H atoms are three bonds apart and are referred to as **vicinal hydrogens**. Coupling between vicinal hydrogens is called **vicinal coupling**, illustrated in Figure 13.13. A **coupling constant (J)** is the separation on an NMR spectrum between adjacent peaks in a split signal and is a quantitative measure of the influence of adjacent nuclei. The magnitude of a coupling constant is expressed in hertz (Hz); for protons in ^{1}H-NMR spectroscopy, the coupling constant is generally in the range 0 to 18 Hz. The values of J depends only on interactions with other nuclei within a molecule, and so it is independent of the applied field strength.

Nuclear spin and hence the chemical shift of the atom labeled H_a in Figure 13.14 is influenced by the vicinal atom H_b, whose nuclear spin might be aligned with or against an applied magnetic field in a ^{1}H-NMR spectrometer. Because of spin-spin coupling, alignment of the H_b nuclear spin *with* the applied magnetic field results in a slightly different chemical shift of the signal for H_a compared to the situation in which the H_b nuclear spin is aligned *against* the applied magnetic field. Across the population of molecules in a sample, there will be similar numbers of molecules

Spin-spin coupling

An interaction in which nuclear spins of adjacent atoms influence each other and lead to the splitting of NMR signals.

Vicinal Hydrogens

H atoms on two C atoms that are bonded to each other.

Vicinal Coupling

Coupling between nuclei of vicinal H atoms.

Coupling constant (J)

The separation on an NMR spectrum (in hertz) between adjacent peaks in a multiplet and a quantitative measure of the influence of the spin-spin coupling with adjacent nuclei.

Figure 13.13

Vicinal H atoms and the vicinal spin-spin coupling that occurs between them.

Figure 13.14

Illustration of spin-spin coupling that gives rise to signal splitting in ^{1}H-NMR spectra.

Figure 13.15
The quartet-triplet ^{1}H-NMR
signals of 3-pentanone showing
the original trace and a scale
expansion to show the signal
splitting pattern more clearly.

having each spin alignment for H$_b$. Any single molecule gives rise to a single signal for H$_a$, but the spectrum of the entire sample shows both. The result is that the signal for the H$_a$ atom appears in the spectrum as a **doublet**. In this hypothetical example, the signal for H$_b$ will also be split into a similar doublet owing to H$_a$ because the effect operates in both directions.

The coupling constant for two vicinal hydrogens on adjacent sp^3 hybridized carbon atoms is approximately 7 Hz. For a spectrometer operating at 300 MHz, a coupling constant of 7 Hz corresponds to only 0.023 ppm. Because peaks with this and comparable values of J are so narrowly spaced, splitting patterns from spectra taken at 300 MHz and higher are often very difficult to see by inspection of the spectra. It is, therefore, common practice to retrace certain signals in expanded form so that splitting patterns are easier to observe (Figure 13.15).

Given in Table 13.4 are approximate values for coupling constants for different types of hydrogens.

Table 13.4	Approximate Values of J for Compounds Containing Alkyl and Alkenyl Groups

$\begin{array}{cc} H_a & H_b \\ \mid & \mid \\ -C-C- \\ \mid & \mid \end{array}$			
6–8 Hz	8–14 Hz	0–5 Hz	0–5 Hz
11–18 Hz	5–10 Hz	0–5 Hz	8–11 Hz

A. Predicting Peak Intensities

As stated previously, in the general case, n equivalent H atoms will cause signal splitting into $n + 1$ peaks. We can now understand this rule and the relative intensities of these peaks by analyzing all possible spin state combinations. There are $n + 1$ different spin state combinations of n spins aligning with or against an applied magnetic field. The probability of a molecule having a given set of spins is proportional to the number of possible spin alignments giving rise to that spin state. The arrows in Figure 13.16 are particularly helpful in understanding this very important concept, with each arrow representing the spin alignment of a ^{1}H nucleus. If there is just one H$_b$ nucleus to consider, there are only two possibilities (↑ or ↓), both

Figure 13.16

The origins of signal splitting patterns. Each arrow represents an H_b nuclear spin orientation.

Observed splitting in signal of H_a

of roughly equal probability, leading to a doublet with a 1:1 ratio of peaks for the signal of H_a (left). Two equivalent H_b nuclei can have three different possible combinations that occur in a 1:2:1 ratio (middle), while three equivalent H_b nuclei can have four possible combinations that occur in a 1:3:3:1 ratio (right).

Alternatively, the ratio of peak areas in any multiplet can be derived from a mathematical mnemonic device called **Pascal's triangle** (Figure 13.17). Here is a note of caution in counting the number of peaks in a multiplet: if the signal of a particular hydrogen is of low intensity compared with others in the spectrum, it may not be possible to distinguish some of the smaller side peaks because of electronic noise in the baseline.

B. Physical Basis for the (n + 1) Rule

Coupling of nuclear spins is mediated through intervening bonds. The extent of coupling is related to a number of factors, including the number of bonds between the H atoms in question. H atoms with more than three bonds between them generally do not exhibit noticeable coupling, although longer range coupling can be seen in some cases. As described in detail above, a common type of coupling is between vicinal H atoms.

An important factor in vicinal coupling is the dihedral angle α between the C—H σ bonds and whether or not it is fixed. As described quantitatively by the Karplus equation, named after Martin Karplus, coupling is maximized when the angle α is $0°$ and $180°$ and is minimized when α is $90°$.

Figure 13.17

Pascal's triangle. As illustrated by the highlighted entries, each entry is the sum of the values immediately above it to the left and the right.

Bonds that rotate rapidly at room temperature do not have a fixed angle between adjacent C—H bonds, so an average angle and an average coupling are observed. This latter concept is important for the interpretation of ^{1}H-NMR spectra for alkanes and other flexible molecules.

It should be noted here that all the nuclei of adjacent hydrogens couple. It is only when coupling is between nonequivalent hydrogens that signal splitting results; coupling between equivalent hydrogens, whether they are on the same or adjacent carbons, does not produce signal splitting.

C. More Complex Splitting Patterns

So far, we have concentrated on spin-spin coupling with only one other nonequivalent set of H atoms. However, more complex situations often arise in which the nuclei of a set of H atoms are coupled to the nuclei of more than one set of nonequivalent H atoms in molecules that do not have rapid bond rotation. In these situations, the coupling from adjacent nonequivalent sets of H atom nuclei *combines* to give more complex signal splitting patterns. Use of a tree diagram is helpful in understanding splitting in these cases. In a tree diagram, the different couplings are applied sequentially.

For example, the atom labeled H_b in Figure 13.18 is adjacent to nonequivalent atoms H_a and H_c on either side, so the resulting coupling will give rise to a so-called doublet of doublets, in other words, a signal with four peaks. Here, the signal for H_b is split into a doublet with coupling constant J_{ab} by H_a, and this doublet is split into a doublet of doublets with coupling constant J_{bc} by H_c. If there were no other H atoms in the molecule to be considered, then the signal for H_a would be a doublet with coupling constant J_{ab} and the signal for H_c would be a doublet with coupling constant J_{bc}. This analysis assumes that the H_a-H_b and H_b-H_c coupling constants, J_{ab} and J_{bc}, are different. If J_{ab} and J_{bc} are equal, the peaks overlap, a situation discussed in Section 13.9F.

Figure 13.18
Coupling that arises when H_b is split by two different nonequivalent H atoms H_a and H_c. This analysis assumes no other coupling in the molecule and that $J_{ab} \neq J_{bc}$.

If H_c is a set of two equivalent H atoms and H_a is still a single H atom, then the observed coupling would be a doublet of triplets, in other words, a signal with six peaks. Again we are assuming that $J_{ab} \neq J_{bc}$. The tree diagram in Figure 13.19 shows the complex pattern that results from this type of splitting. If there were no other H atoms in the molecule to be considered, then the signal for H_a would be a doublet with coupling constant J_{ab} and the signal for H_c would be a doublet with coupling constant J_{bc}. The peaks for H_a and H_b would each integrate to a relative value of one H atom, while the peaks for H_c would integrate to a relative value of 2 H atoms.

In the general case, a signal will be split into $(n + 1) \times (m + 1)$ peaks for an H atom that is coupled to a set of n H atoms with one coupling constant, and to a set of m H atoms with another coupling constant. Note that, in a tree diagram, you get the same splitting patterns no matter which order the two coupling constants are analyzed.

Figure 13.19
Complex coupling that arises when H_b is split by H_a and two equivalent atoms H_c. Again, this analysis assumes no other coupling in the molecule and that $J_{ab} \neq J_{bc}$.

D. Bond Rotation

Because the angle between C—H bonds determines the extent of coupling in a molecule, bond rotation is a key parameter. In alkanes and other molecules with relatively free rotation about C—C σ bonds, H atoms bonded to the same C atom

Figure 13.20

Geminal coupling that occurs when two H atoms on the same carbon atom are not equivalent. This is most common in unsymmetrical alkenes and cyclic molecules.

in —CH_2— and —CH_3 groups are generally equivalent because of the rapid bond rotation. An exception is when a —CH_2— is adjacent to a chiral center, a situation that is discussed in Section 13.10. However, when there is restricted bond rotation, as in alkenes or cyclic structures, H atoms bonded to the same C atom may not be equivalent, especially if the molecule is not symmetrical. Nonequivalent ^{1}H nuclei on the *same* C atom will couple to each other and cause splitting. This is referred to as **geminal coupling** (Figure 13.20). Geminal coupling constants are generally small, on the order of 0 to 5 Hz.

Because of the restricted rotation about C=C bonds, the alkenyl (vinylic) H atoms of unsymmetrical alkenes are not equivalent; in other words, they are in unique chemical environments. For example, ethyl propenoate (ethyl acrylate) is an unsymmetrical terminal alkene; therefore, the three alkenyl H atoms are nonequivalent (Figure 13.21). As a result, their nuclei couple with each other. In alkenes, *trans* coupling generally results in larger coupling constants (J_{trans} = 11–18 Hz) compared to *cis* coupling (J_{cis} = 5–10 Hz), with geminal coupling being by far the smallest (J_{gem} = 0–5 Hz). Unless a high-resolution spectrum is taken, the geminal coupling constant is so small that it is often difficult to see in terminal alkenes. In the spectrum of ethyl propenoate, the geminal coupling is only visible upon close inspection of the signals labeled a and c. You should be able to recognize the characteristic ethyl group pattern of a quartet integrating to two H atoms (—CH_2—, H_d) and a triplet integrating to three H atoms (—CH_3, H_e). Tree diagrams are provided in Figure 13.22 to help decipher patterns of the alkenyl signals.

Figure 13.21

300 MHz ^{1}H-NMR spectrum of ethyl propenoate.

Figure 13.22

Tree diagrams for the complex coupling seen for the alkenyl H atoms in the ^{1}H-NMR spectrum of ethyl propenoate.

Figure 13.23

300 MHz ^{1}H-NMR spectrum of 2-methyl-2-vinyl oxirane. The two H atoms on the oxirane ring are nonequivalent, so they exhibit geminal coupling.

Cyclic structures often exhibit restricted rotation about their C—C σ bonds and can have constrained conformations. The result is that the two H atoms on —CH_2— groups in cyclic molecules can be nonequivalent, leading to complex spin-spin coupling. Substituted epoxides such as 2-methyl-2-vinyloxirane provide a good example (Figure 13.23). The two H atoms on the three-membered epoxide ring are nonequivalent. H_d is *cis* to the vinyl group and *trans* to the methyl group, while H_e is the reverse. Because they are in different chemical environments, they are nonequivalent and exhibit geminal coupling (Figure 13.24). The geminal coupling constant is small but discernable in the spectrum because the signals for both H_d and H_e are doublets. Vinyl H atom H_a is split by both H_b (*trans* coupling) and H_c (*cis* coupling), giving rise to a doublet of doublets, or four peaks. H_b is split by H_a (*trans* coupling) along with H_c, the latter geminal coupling constant is so small that it is barely discernable at this resolution. H_c is split by H_a (*cis* coupling) as well as H_b (geminal coupling). The singlet near 1.5 ppm that integrates to three H atoms is the methyl group labeled f.

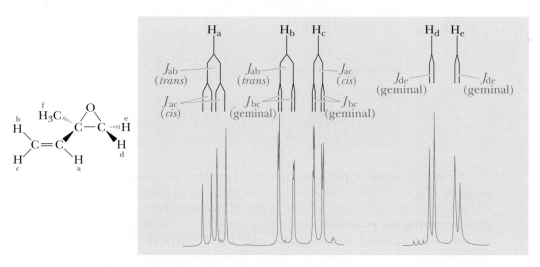

Figure 13.24

Tree diagrams that indicate the complex coupling seen in ^{1}H-NMR signals for the vinyl group and the oxirane ring H atoms of 2-methyl-2-vinyloxirane.

E. Coincidental Overlap

Here is a word of caution: quite often, because peaks can overlap by coincidence, there are fewer *distinguishable* peaks in a signal than predicted. Coincidental peak overlap can occur in any molecule, but it is especially common with flexible alkyl chains. In addition, some coupling constants are so small that peak splitting is hard to see in a spectrum. Thus, the predicted number of peaks using the $(n + 1) \times (m + 1)$ rule should be considered the maximum that *might* be observed. Detailed analysis using extremely high resolution spectrometers is often required to distinguish all the peaks in a highly split signal. You should note also that the types of splitting patterns we have described are applicable only when the separation between coupled signals is much greater than the coupling constant. When this is not the case, spectra can become much more complex.

F. Complex Coupling in Flexible Molecules

Coupling in molecules having unrestricted bond rotation is often simplified to give only $m + n + 1$ peaks, not the expected $(n + 1) \times (m + 1)$. In other words, the number of peaks actually observed for a signal is the number of adjacent hydrogens + 1, no matter how many different sets of equivalent H atoms this represents. The explanation is that bond rotation averages the coupling constants throughout molecules with freely rotating bonds and tends to make them very similar, in the 6 to 8 Hz range, for H atoms on freely rotating sp^3 hybridized C atoms.

Very similar or identical coupling constants simplify splitting patterns. For example, in the hypothetical unsymmetrical molecule depicted in Figure 13.25, the central H_b atoms are coupled to both H_a atoms as well as both H_c atoms. If $J_{ab} \neq J_{bc}$, this would lead to a triplet of triplets or nine peaks in the signal for H_b. However, if the coupling constants are identical so that $J_{ab} = J_{bc}$, the splitting pattern overlaps considerably to generate only five peaks in the signal for H_b. In the general case, simplification because of very similar or identical J values gives a number of peaks equal to the number of adjacent H atoms + 1, regardless of patterns of equivalence.

Figure 13.25

Simplification of signal splitting that occurs when coupling constants are the same.

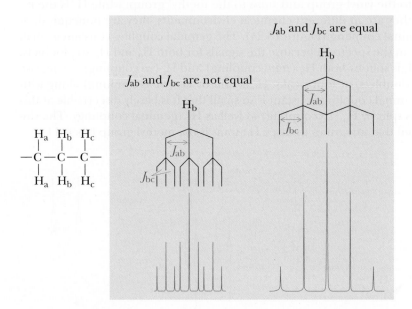

A good example of peak overlap occurs in the spectrum of 1-chloro-3-iodo-propane (Figure 13.26). The signal for the H atoms of the central —CH_2— group (labeled c on the spectrum) is split by the H atoms on both of the other —CH_2— groups, raising the possibility of splitting into $3 \times 3 = 9$ peaks. However, because the values for J_{ab} and J_{bc} are so similar, only $4 + 1 = 5$ peaks are distinguishable in the spectrum for the H_c signal as a result of peak overlap.

Another common example is the kind of splitting of the signal for the central —CH_2— in a —CH_2—CH_2—CH_3 group, such as occurs in the molecule 1-chloropropane, Cl—CH_2—CH_2—CH_3 (Figure 13.27). A maximum of $3 \times 4 = 12$ peaks would be possible for the central —CH_2— signal (labeled b in Figure 13.27), but because the coupling constants are very similar, only $5 + 1 = 6$ peaks are distinguishable.

G. Fast Exchange

Hydrogen atoms bonded to oxygen or nitrogen atoms can exchange with each other faster than the time it takes to acquire a ^{1}H-NMR spectrum. This process is greatly facilitated by even traces of acid or base in a sample. Important affected

Figure 13.26
300 MHz ^{1}H-NMR spectrum of 1-chloro-3-iodopropane

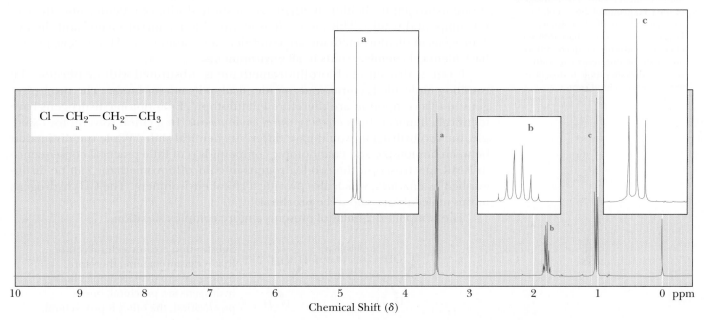

functional groups include carboxylic acids, alcohols, amines, and amides. Fast exchange has two important consequences. First, signals for exchanging H atoms are generally broad singlets that do not take part in splitting with other signals. Second, the signal will disappear altogether if D_2O or a deuterated alcohol is added to the sample because the H atoms will be replaced with D atoms, which are ^{1}H-NMR silent. This latter phenomenon can be used to identify signals from exchangeable H atoms by taking spectra with and without added D_2O. Note that these same exchangeable H atoms also can take part in hydrogen bonds, the presence of which can alter chemical shift in a concentration-dependent fashion.

Figure 13.27
300 MHz ^{1}H-NMR spectrum of 1-chloropropane.

13.10 | Stereochemistry and Topicity

The discussion of the number of equivalent hydrogens given in the beginning of this chapter is slightly oversimplified because stereochemistry can affect chemical shift. Depending on the symmetry of the molecule, otherwise equivalent atoms may be **homotopic**, **enantiotopic**, or **diastereotopic**. The simplest way to visualize the topicity of a molecule (that is, which of the these classes it falls into) is by mentally substituting one of the atoms or groups of atoms by an isotope, and then deciding

Homotopic groups
Atoms or groups on an atom that give an achiral molecule when one of the groups is replaced by an isotope. The hydrogens of the CH_2 group of propane, for example, are homotopic. Replacing either one of them with deuterium gives 2-deuteropropane, which is achiral. Homotopic groups have identical chemical shifts under all conditions.

Enantiotopic groups

Atoms or groups on an atom that give a chiral center when one of the groups is replaced by an isotope. A pair of enantiomers results. The hydrogens of the CH_2 group of ethanol, for example, are enantiotopic. Replacing one of them by deuterium gives (R)-1-deuteroethanol; replacing the other gives (S)-1-deuteroethanol. Enantiotopic groups have identical chemical shifts in achiral environments but different chemical shifts in chiral environments.

Diastereotopic groups

Atoms or groups on an atom that are bonded to an atom that is bonded to two nonidentical groups, one of which contains a chiral center. When one of the atoms or groups is replaced by an isotope, a new chiral center is created and a set of diastereomers results. The hydrogens of the CH_2 group of 2-butanol, for example, are diastereotopic. Diastereotopic groups have different chemical shifts under all conditions, although the differences are only seen for diastereotopic hydrogens very close to a chiral center.

whether the resulting compound would be (a) the same or (b) different from its mirror image or whether (c) diastereomers are possible. Depending on the outcome of the test, the atoms or groups are homotopic, enantiotopic, or diastereotopic, respectively.

Consider the following molecules.

Substitution does not create a chiral center; therefore hydrogens are homotopic.

Dichloromethane (achiral)   achiral

Substitution produces a chiral center; therefore, hydrogens are enantiotopic. Both hydrogens are prochiral; one is pro-R-chiral, the other is pro-S-chiral.

Chlorofluoromethane (achiral)   chiral

If one hydrogen in dichloromethane is substituted with one deuterium, an achiral compound results. This molecule is identical to its mirror image, and the two hydrogens in dichloromethane are equivalent and homotopic. Homotopic groups have identical chemical shifts in all environments.

If one hydrogen of chlorofluoromethane is substituted with deuterium, the resulting molecule is chiral, and not identical to its mirror image. The two hydrogens in this compound are therefore enantiotopic. Enantiotopic hydrogens have identical chemical shifts *except in chiral environments*. In a chiral solvent, for example, the two hydrogens would have different chemical shifts. While the distinction between homotopic and enantiotopic compounds is of little practical consequence in NMR spectroscopy, the two hydrogens in chlorofluoromethane can be distinguished by enzymes, which also provide a chiral environment. The CH_2 hydrogens in this molecule are said to be **prochiral**.

The compound 2-butanol presents a more complex situation.

Substitution creates a chiral center that is diastereomeric; therefore, hydrogens are diastereotopic. Both hydrogens are prochiral; one is pro-R-chiral, the other is pro-S-chiral.

2-Butanol (chiral)   chiral

Molecules such as 3-methyl-2-butanol are even more complicated. If a hydrogen on one of the methyl groups on carbon-3 of 3-methyl-2-butanol is substituted with a deuterium, a new chiral center is created. Because there is already one chiral center, diastereomers are now possible. Thus, the methyl groups on carbon-3 of 3-methyl-2-butanol are diastereotopic. Diastereotopic hydrogens have different chemical shifts under all conditions, which can lead to unexpected complexity in spectra of simple compounds. The ^{1}H-NMR spectrum of 3-methyl-2-butanol is shown in Figure 13.28. The methyl groups on carbon-3 are nonequivalent and give two doublets rather than one doublet of twice the intensity, which would be expected if they were equivalent.

Any molecule with a chiral center near two otherwise identical groups on a carbon with a third substituent has the potential for diastereotopicity. Of course, like any other nonequivalent groups, diastereotopic groups may have very similar or accidentally identical chemical shifts. Generally, the shift differences fall off rapidly with increasing distance from the chiral center.

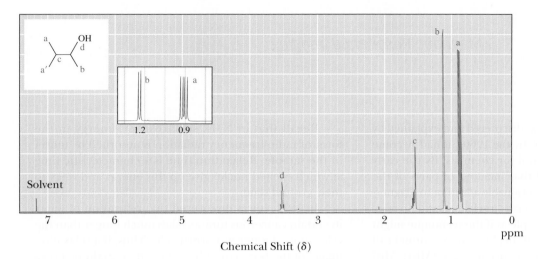

Figure 13.28
^{1}H-NMR spectrum of 3-methyl-2-butanol (500 MHz). The methyl groups on carbon-3 (c) are diastereotopic and therefore nonequivalent. They appear as two doublets.

Example 13.7

Indicate whether the highlighted hydrogens in the following compounds are homotopic, enantiotopic, or diastereotopic.

(a) (b) CH_3CH_2Br (c)

Solution

(a) Diastereotopic (near a chiral center). These hydrogens will have different chemical shifts.
(b) Enantiotopic. These hydrogens will have the same chemical shift except in chiral environments.
(c) Enantiotopic [see part (b)].

Problem 13.7

Following is a ^{1}H-NMR spectrum of 2-butanol. Explain why the CH_2 protons appear as a complex multiplet rather than as a simple quintet.

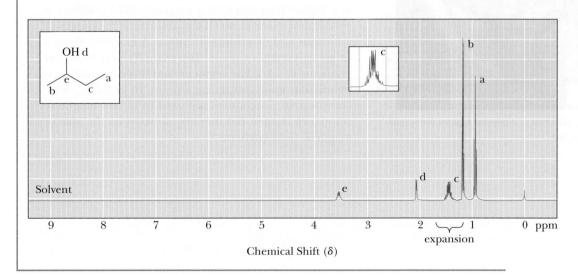

Chemical Shift (δ)

The NMR phenomenon was discovered and explained by physicists in the 1950s and by the 1960s, it had become an invaluable analytical tool for chemists. By the early 1970s, scientists realized that imaging of parts of the body using NMR could be a valuable addition to diagnostic medicine. Because the term "nuclear magnetic resonance" sounds to many people as if the technique might involve radioactive material, health care personnel call the technique magnetic resonance imaging (MRI). MRI has become so important, that in 2003, the Nobel Prize for medicine or physiology was awarded to Paul Lauterbur and Peter Mansfield for their discoveries that led to practical MRI.

The body contains several nuclei that, in principle, could be used for MRI. Of these, hydrogens, most of which come from the hydrogens of water, triglycerides (Section 26.1), and membrane phospholipids (Section 26.5) give the most useful signals. Phosphorus MRI is also used in diagnostic medicine.

Computer-enhanced MRI scan of a normal human brain with pituitary gland highlighted.

Scott ???/Photo Researchers, Inc.

Recall that in NMR spectroscopy, energy in the form of radio-frequency radiation is absorbed by nuclei in the sample. Relaxation time is a characteristic time at which excited nuclei give up this energy and relax to their ground state.

In 1971, it was discovered that relaxation of water in certain cancerous tumors takes much longer than the relaxation of water in normal cells. Thus, if a relaxation image of the body could be obtained, it might be possible to identify tumors at an early stage. Subsequent work demonstrated that many tumors can be identified in this way. Another important application of MRI is in the examination of the brain and spinal cord. White and gray matter are easily distinguished by MRI, which is useful in the study of such diseases as multiple sclerosis. Magnetic resonance imaging and x-ray imaging are, in many cases, complementary. The hard, outer layer of bone is essentially invisible to MRI but shows up extremely well in x-ray images, whereas soft tissue is nearly transparent to x-rays but shows up in MRI.

The key to any medical imaging technique is to know which part of the body gives rise to which signal. In MRI, spatial information is encoded using magnetic field gradients. Recall that a linear relationship exists between the frequency at which a nucleus resonates and the intensity of the magnetic field. In ^{1}H-NMR spectroscopy, we use a homogeneous magnetic field, in which all equivalent hydrogens absorb at the same radio frequency and have the same chemical shift. In MRI, the patient is placed in a magnetic field gradient that can be varied from place to place. Nuclei in the weaker magnetic field gradient absorb at a lower frequency. Nuclei elsewhere in the stronger magnetic field absorb at a higher frequency. In a magnetic field gradient, a correlation exists between the absorption frequency of a nucleus and its position in space. A gradient along a single axis images a plane. Two mutually perpendicular gradients image a line segment, and three mutually perpendicular gradients image a point. In practice, more complicated procedures are used to obtain magnetic resonance images, but they are all based on the idea of magnetic field gradients.

13.11 | ^{13}C-NMR

The development of ^{13}**C-NMR** spectroscopy lagged behind ^{1}H-NMR spectroscopy primarily because of two problems. One is the particularly low natural abundance of ^{13}C (only 1.1%) and the resulting weak signal. The second problem is that the magnetic moment of ^{13}C is considerably smaller than that of ^{1}H, which causes the population of the higher and lower nuclear spin states to differ by much less than for ^{1}H. Taken in combination, these two factors mean that ^{13}C-NMR signals in natural samples (those not artificially enriched with carbon-13) are only about 10^{-4} times the strength of ^{1}H-NMR signals. Even though ^{1}H-NMR spectroscopy became a routine analytical tool in the mid-1960s, it was not until 20 years later with the development of FT-NMR techniques that ^{13}C-NMR spectroscopy became widely available as a routine analytical tool.

As with ^{1}H-NMR spectra, splitting patterns in ^{13}C-NMR spectra are also explained according to the $(n + 1)$ rule. Because, in natural abundance, only 1.1% of carbon atoms are ^{13}C, almost all ^{13}C atoms in a molecule have only magnetically inactive ^{12}C next to them; therefore, ^{13}C—^{13}C signal splitting is not normally observed. However, the signal from a ^{13}C nucleus is split by the hydrogens bonded to it. The signal for a ^{13}C atom with three attached hydrogens is split to a quartet, that for an atom of ^{13}C with two attached hydrogens is split to a triplet, and so on. The ^{13}C—H signal splitting provides important information about the number of hydrogen atoms bonded to carbon. The disadvantage of ^{13}C—H signal splitting is that coupling constants of between 100 and 250 Hz are common. Coupling constants of this magnitude correspond to 1.33 to 3.33 ppm at 75 MHz, which means that the overlap among signals can be significant and that splitting patterns are very often difficult to determine. In addition, there are smaller but significant couplings from hydrogens that are not directly bonded to the carbon, but are separated by two or three bonds. This extensive splitting causes the already weak signals of the ^{13}C to split into many smaller peaks that are easily lost in the noise. For this reason, the most common mode of operation of a ^{13}C-NMR spectrometer is a hydrogen-decoupled mode. (See Problem 13.25 for an interesting problem on the use of coupling constants to determine orbital hybridization.)

In the hydrogen-decoupled mode, the sample is irradiated with two different radio frequencies. The first radio frequency is used to excite the ^{13}C nuclei. The second is a broad spectrum of frequencies that causes all hydrogens in the molecule to undergo rapid transitions among their nuclear spin states. On the time scale of a ^{13}C-NMR spectrum, each hydrogen is in a time average of the two states, with the result that ^{1}H-^{13}C spin-spin interactions are not observed. The term for this process is spin-spin decoupling. In a hydrogen-decoupled spectrum, all ^{13}C signals appear as singlets. The hydrogen-decoupled ^{13}C-NMR spectrum of 1-bromobutane (Figure 13.29) consists of four singlets.

Figure 13.29
Hydrogen-decoupled ^{13}C-NMR spectrum of 1-bromobutane.

CH$_3$CH$_2$CH$_2$CH$_2$Br

Chemical Shift (δ)

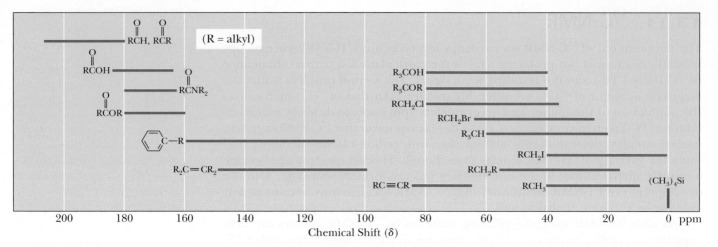

Figure 13.30
^{13}C-NMR chemical shifts of representative groups. These values are approximate. Other atoms or groups in the molecules may cause signals to appear outside of these ranges.

Figure 13.30 shows approximate chemical shifts for ^{13}C-NMR. Notice how much wider the range of chemical shifts is for ^{13}C-NMR than for ^{1}H-NMR. Most chemical shifts for ^{1}H-NMR fall within a rather narrow range of 0 to 10 ppm; however, those for ^{13}C-NMR cover 0 to 220 ppm. Because of this expanded scale, it is very unusual to find any two nonequivalent carbons in the same molecule with identical chemical shifts. Most commonly, each different type of carbon within a molecule has a distinct signal clearly resolved from all other signals.

Notice further that the chemical shift of carbonyl carbons is quite distinct from those of sp^3-hybridized carbons and of other types of sp^2-hybridized carbons. The presence or absence of a carbonyl carbon is quite easy to recognize in a ^{13}C-NMR spectrum. Note that signals from sp^2-hybridized carbons fall in a distinctive range of 100 to 160 ppm.

A great advantage of ^{13}C-NMR spectroscopy is that it is possible to count the number of types of carbon atoms in a molecule. There is one caution here, however. Because of certain complications, including the long relaxation times of ^{13}C nuclei, it is generally not possible to determine the number of carbons of each type by integration of signal areas.

Example 13.8

Predict the number of signals in a proton-decoupled ^{13}C-NMR spectrum of each compound.

(a) $CH_3\overset{\displaystyle O}{\overset{\|}{C}}OCH_3$ (b) $CH_3CH_2CH_2\overset{\displaystyle O}{\overset{\|}{C}}CH_3$ (c) $CH_3CH_2\overset{\displaystyle O}{\overset{\|}{C}}CH_2CH_3$

Solution

Following are the number of signals in the proton-decoupled spectrum of each compound, along with the chemical shifts of each signal. The chemical shifts of the carbonyl carbons are quite distinctive (Figure 13.30) and in these examples occur at δ 171.37, 208.85, and 211.97.

(a) Methyl acetate: three signals (δ 171.37, 51.53, and 20.63)
(b) 2-Pentanone: five signals (δ 208.85, 45.68, 29.79, 17.35, and 13.68)
(c) 3-Pentanone: three signals (δ 211.97, 35.45, and 7.92)

Problem 13.8

Explain how to distinguish between the members of each pair of constitutional isomers based on the number of signals in the proton-decoupled ^{13}C-NMR spectrum of each member.

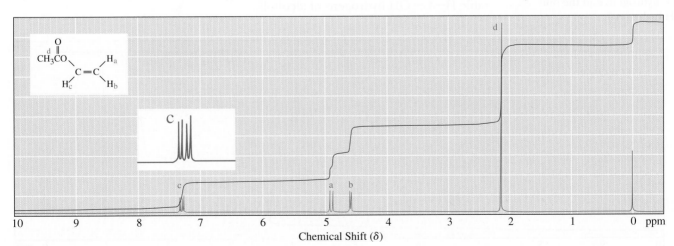

13.12 | Interpretation of NMR Spectra

A. Alkanes

All hydrogens in alkanes are in very similar chemical environments; therefore, ^{1}H-NMR chemical shifts of alkane hydrogens fall within a narrow range of δ 0.8 to 1.7. Chemical shifts for alkane carbons in ^{13}C-NMR spectroscopy fall within the considerably wider range of δ 10 to 60.

B. Alkenes

The chemical shifts of vinylic hydrogens are larger than those of alkane hydrogens and typically fall in the range δ 4.6 to 5.7. Vinylic hydrogens are deshielded by the sp^2-hybridized carbons of the double bond (Section 13.9B) and the local magnetic field induced in the π bond of alkenes (Section 13.9C). The splitting pattern observed in the ^{1}H-NMR spectrum of vinyl acetate (Figure 13.31) is typical of monosubstituted alkenes. The singlet at δ 2.12 represents the three hydrogens of the methyl group. The terminal vinylic hydrogens appear at δ 4.58 and δ 4.90. The internal vinylic hydrogen, which normally appears in the range δ 5.0 to 5.7, is shifted farther downfield to δ 7.30 as a result of deshielding by the adjacent electronegative oxygen atom of the ester.

As shown in Table 13.4, coupling constants are generally larger for *trans* vinylic hydrogens (11–18 Hz) than for *cis* vinylic hydrogens (5–10 Hz), and it is often possible to distinguish between *cis* and *trans* alkenes by an analysis of their coupling constants. It is also possible to distinguish between vicinal hydrogens and geminal hydrogens (=CH$_2$), the latter having small coupling constants, generally in the 0 to 5 Hz range.

Figure 13.31
^{1}H-NMR spectrum of vinyl acetate.

The signal of each vinylic hydrogen in vinyl acetate is predicted to be a doublet of doublets. The signal for H_c, for example, is split to a doublet by coupling with H_a and further split to a doublet of doublets by coupling with H_b. For H_a and H_b, the geminal coupling is so small that it is not visible at this resolution, so their signals appear as doublets. Higher resolution would reveal the geminal coupling.

The sp^2-hybridized carbons of alkenes give ^{13}C-NMR signals in the range δ 100 to 150 ppm (Figure 13.30), which is considerably downfield from sp^3-hybridized carbons.

C. Alcohols

The chemical shift of a hydroxyl hydrogen in a ^{1}H-NMR spectrum is variable and depends on the purity of the sample, the solvent, the concentration, and the temperature. It often appears in the range δ 3.0 to 4.0, but depending on experimental conditions, it may appear as far upfield as δ 0.5. Hydrogens on the carbon bearing the —OH group are deshielded by the electron-withdrawing inductive effect of the oxygen atom, and their signals also typically appear in the range δ 3.4 to 4.0. Shown in Figure 13.32 is the ^{1}H-NMR spectrum of 1-propanol.

Signal splitting between the hydrogen on O—H and its neighbors on the adjacent —CH$_2$— group is seen in the ^{1}H-NMR spectrum of 1-propanol. However, this splitting is rarely seen. The reason is that most samples of alcohol contain traces of acid, base, or other impurities that catalyze the transfer of the hydroxyl proton from the oxygen of one alcohol molecule to the oxygen of another alcohol molecule. This fast exchange decouples the hydroxyl proton from all other nuclei in the molecule (Section 13.9G). For this same reason, the hydroxyl proton does not usually split the signal of any α-hydrogens.

hydrogen bonding

These protons have been exchanged.

Figure 13.32

^{1}H-NMR spectrum of 1-propanol. The hydroxyl hydrogen appears at δ 3.18 as a narrowly spaced triplet. The signal of hydrogens on carbon 1 of 1-propanol appears as a quartet (labeled in the spectra as a multiplet) at δ 3.56 (split by the two CH$_2$ hydrogens and the one OH hydrogen).

D. Ethers

The most distinctive feature of the ^{1}H-NMR spectra of ethers is the chemical shift of hydrogens on the carbons bonded to the ether oxygen. Signals for this type of hydrogen fall in the range δ 3.3 to 4.0, which corresponds to a downfield shift of approximately 2.4 units compared with their normal position in alkanes. The chemical shifts of H—C—O— hydrogens in ethers are similar to those seen for comparable H—C—OH hydrogens of alcohols.

E. Aldehydes and Ketones

The signal for an aldehyde hydrogen typically appears between δ 9.5 and δ 10.1 in a ¹H-NMR spectrum. Because almost nothing else absorbs in this region, it is very useful for identification. Hydrogens on an α-carbon (carbon directly adjacent to carbonyl) of an aldehyde or ketone appear around δ 2.2 to 2.6. The carbonyl carbons of aldehydes and ketones have characteristic positions in the ¹³C-NMR between δ 180 and δ 215 (and can be distinguished from carboxylic acid derivatives, which absorb at higher field).

F. Carboxylic Acids and Esters

Signals for hydrogens on the α-carbon to a carboxyl group in acids and esters appear in a ¹H-NMR spectrum in the range δ 2.0 to 2.6. The hydrogen of a carboxyl group gives a very distinctive signal in the range δ 10 to 13, downfield of most other types of hydrogens, even farther downfield than that of an aldehyde hydrogen (δ 9.5–10.1), and serves to distinguish carboxyl hydrogens from most other types of hydrogens. The ¹H-NMR signal for the carboxyl hydrogen of 2-methylpropanoic acid, for example, appears at δ 13.2 and is shown at the left in Figure 13.33.

The ¹³C resonance of the carboxyl carbon in acids and esters appears in the range δ 165 to 185 and at a distinctly higher field than that in ketones. Hydrogens α to an ester oxygen are strongly deshielded and resonate between δ 3.7 and δ 4.7, more downfield than in alcohols and ethers.

G. Amines

The chemical shifts of amine hydrogens, like those of hydroxyl hydrogens (Section 13.12C), vary between δ 0.5 and δ 5.0, depending on experimental conditions due to hydrogen bonding. As in alcohols, exchange is fast enough that spin-spin splitting between amine hydrogens and hydrogens on adjacent α-carbons is averaged. Thus, amine hydrogens generally appear as broad singlets. Coupling to ¹⁴N (beyond the scope of this text) causes these signals to broaden. Hydrogens α to the amine nitrogen appear around δ 2.5 ppm, about 1 ppm higher than for hydrogens α to oxygen in ethers and alcohols.

Carbons bonded to nitrogen appear in the ¹³C-NMR spectrum approximately 20 ppm lower than in alkanes of comparable structure, but about 20 ppm above carbons attached to oxygen in ethers or alcohols.

Here is a final word. We have barely scratched the surface of what NMR can do. Something called the nuclear Overhauser enhancement (NOE) can determine distances between atoms in molecules that are near each other in three-dimensional space even if there are more than three bonds separating them. In addition, using modern instruments, spectra that examine multiple parameters simultaneously can

Figure 13.33
¹H-NMR spectrum of 2-methylpropanoic acid (isobutyric acid).

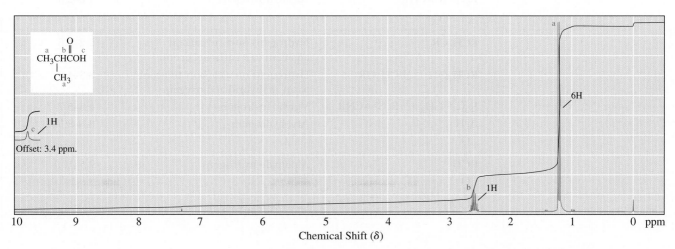

be produced to yield immense amounts of information about even very complicated molecules. Such spectra are plotted on more than one axis so they are referred to as multidimensional spectra. Multidimensional spectra are used to deduce structure and conformation of molecules ranging from small organic molecules to large biological macromolecules such as proteins, DNA, and RNA.

How To	Solve NMR Spectral Problems

One of the first steps in determining molecular structure is establishing the molecular formula. In the past, this task was most commonly done by elemental analysis, combustion analysis to determine percent composition, molecular weight determination, and so forth. More commonly today, molecular weight and molecular formula are determined by mass spectrometry (Chapter 14). In the examples that follow, we assume that the molecular formula of any unknown compound has already been determined, and we proceed from that point using spectral analysis to determine a structural formula.

A. Molecular Formula

Valuable information about the structural formula of an unknown compound can be obtained by inspecting its molecular formula, which gives its index of hydrogen deficiency. Refer to Chapter 5 to review this technique. A molecular formula can often be obtained from the mass spectrum (see Chapter 14).

B. From an ^{1}H-NMR Spectrum to a Structural Formula

The following steps may prove helpful as a systematic approach to solving spectral problems.

Step 1: Molecular formula and index of hydrogen deficiency. Examine the molecular formula, calculate the index of hydrogen deficiency, and deduce what information you can about the presence or absence of rings or π bonds.

Step 2: Number of signals. Count the number of signals to determine the number of sets of equivalent hydrogens present in the compound.

Step 3: Integration. Use the integration and the molecular formula to determine the numbers of hydrogens present in each set.

Step 4: Pattern of chemical shifts. Compare signal chemical shifts with reference tables to determine which functional groups may be present. Keep in mind that these are broad ranges and that signals of each type may be shifted either farther upfield or farther downfield, depending on the details of the molecular structure.

Types of Hydrogens	Descriptive Name	Chemical Shift (δ)
RCH$_3$ RCH$_2$R R$_3$CH	Alkyl hydrogens	0.8–1.7
R$_2$C=CRCHR$_2$	Allylic hydrogens	1.6–2.6
$\overset{\displaystyle O}{\overset{\displaystyle \|}{\text{RCH}_2\text{CR}}}$	Hydrogens on a sp^3 carbon adjacent to a carbonyl group	2.2–2.6
RCH$_2$OH RCH$_2$OR	Hydrogens on a carbon adjacent to an sp^3-hybridized oxygen	3.3–4.0
R$_2$C=CH$_2$ R$_2$C=CHR	Vinylic hydrogens	4.6–5.7
ArH	Aryl hydrogens	6.5–8.5
$\overset{\displaystyle O}{\overset{\displaystyle \|}{\text{RCH}}}$	Aldehyde hydrogens	9.5–10.1
$\overset{\displaystyle O}{\overset{\displaystyle \|}{\text{RCOH}}}$	Carboxyl hydrogens	10–13

Step 5: Signal splitting patterns. Examine splitting patterns for information about the number of nearest nonequivalent hydrogen neighbors.

Step 6: Structural formula. Construct possible molecules from the functional groups present, their relative signal integrations, and any other information you are given, especially the molecular formula and other spectra. Confirm the correct structure by making sure all the available information matches.

Spectral Problem 1: Molecular formula C₅H₁₀O

Analysis of Spectral Problem 1

Step 1: Molecular formula and index of hydrogen deficiency. The reference compound is C_5H_{12}; therefore, the index of hydrogen deficiency is 1 and the molecule contains either one ring or one π bond.

Step 2: Number of signals. There are two signals (a triplet and a quartet) and, therefore, two sets of equivalent hydrogens.

Step 3: Integration. From the integration, the hydrogens in each set are in the ratio 3:2. Because there are 10 hydrogens, 6H must give rise to the signal at δ 1.07, and 4H must give rise to the signal at δ 2.42.

Step 4: Pattern of chemical shifts. The signal at δ 1.07 is in the alkyl region and, based on its chemical shift, most probably indicates a methyl group. No signal occurs between δ 4.6 and δ 5.7; there are no vinylic hydrogens. If a carbon-carbon double bond is in the molecule, there are no hydrogens on it (that is, it is tetrasubstituted). The chemical shift of the four protons at δ 2.42 is consistent with two CH_2 groups next to a carbonyl group.

Step 5: Signal splitting patterns. The methyl signal at δ 1.07 is split into a triplet (t); it must have two neighbors, indicating $—CH_2CH_3$. The signal at δ 2.42 is split into a quartet (q); it must have three neighbors. An ethyl group accounts for these two signals. No other signals occur in the spectrum; therefore, there are no other types of hydrogens in the molecule.

Step 6: Structural formula. Put this information together to arrive at the following structural formula. The chemical shift of the methylene group ($–CH_2–$) at δ 2.42 is consistent with an alkyl group adjacent to a carbonyl group.

$$CH_3—CH_2—\overset{\displaystyle O}{\overset{\displaystyle \|}{C}}—CH_2—CH_3$$

δ 2.42 (q)
δ 1.07 (t)

3-Pentanone

Spectral Problem 2: Molecular formula $C_7H_{14}O$

Analysis of Spectral Problem 2

Step 1: Molecular formula and index of hydrogen deficiency. The index of hydrogen deficiency is 1; the compound contains one ring or one π bond.

Step 2: Number of signals. There are three signals and, therefore, three sets of equivalent hydrogens.

Step 3: Integration. Reading from right to left, there are 9, 3, and 2 hydrogens in these signals.

Step 4: Pattern of chemical shifts. The signal at δ 1.01 is characteristic of a methyl group adjacent to an sp^3 hybridized carbon. The signals at δ 2.11 and δ 2.32 are characteristic of alkyl groups adjacent to a carbonyl group.

Step 5: Signal splitting pattern. All signals are singlets (s). Therefore, none of the groups has hydrogens on neighboring carbons.

Step 6: Structural formula. The compound is 4,4-dimethyl-2-pentanone.

4,4-Dimethyl-2-pentanone

Summary

Section 13.1 Nuclear Spin States

- Absorption of electromagnetic energy leads to spectroscopic tools that are important for the determination of structures of organic molecules.
  - These absorptions can be from electronic transitions: UV-visible spectra (Section 20.2), vibrations (IR spectra, Chapter 12), and nuclear magnetic spin transitions (NMR spectra at radio frequencies).
- Nuclei of ^1H and ^{13}C have a **nuclear spin quantum number** of $\frac{1}{2}$ and allowed nuclear spin states of $+\frac{1}{2}$ and $-\frac{1}{2}$.

Section 13.2 Orientation of Nuclear Spins in an Applied Magnetic Field

- In the presence of an applied magnetic field, B_0, nuclei with spin $+\frac{1}{2}$ are aligned with the applied field and are in the lower energy state; nuclei with spin $-\frac{1}{2}$ are aligned against the applied field and are in the higher energy state.

- The key relationship for NMR is that the difference in energy between the $+\frac{1}{2}$ and $-\frac{1}{2}$ nuclear spin states is proportional to the strength of the magnetic field experienced by a given nucleus.

Problem: 13.1

Section 13.3 Nuclear Magnetic "Resonance"

- When placed in a powerful magnetic field, there is a small population bias for the ^1H and ^{13}C nuclei to be aligned with the magnetic field, and they **precess**.
- **Resonance** is the absorption of electromagnetic radiation by a precessing nucleus and the resulting flip of its nuclear spin from the lower energy spin state to the higher energy spin state.
  - The electromagnetic radiation must be of precisely the same frequency as the nuclear precession frequency for absorbance of energy to occur and thus achieve resonance.
  - For typical experiments, this precession frequency corresponds to electromagnetic radiation in the **radio-frequency** range.
- The experimental conditions required to cause nuclei to resonate are affected by the local chemical and magnetic environments.
  - Electrons around a hydrogen or carbon create local magnetic fields that shield the nuclei of these atoms from the applied field.
    - Any factor that increases the exposure of nuclei to an applied field is said to **deshield** them and shifts their signal downfield to a larger δ value.
    - Conversely, any factor that decreases the exposure of nuclei to an applied field is said to **shield** them and shifts their signal upfield to a smaller δ value.

Problems: 13.10, 13.11

Section 13.4 An NMR Spectrometer

- An NMR spectrometer records resonance as a **signal** and plots the irradiation frequencies where absorption takes place scaled by the strength of the applied magnetic field using the units of **parts per million (ppm)**.
  - For ^1H-NMR, ppm is a convenient scale because differences in resonant frequencies for different ^1H nuclei in molecules are about 1 millionth of the overall frequency of electromagnetic frequency used to cause the resonance.

Section 13.5 Equivalent Hydrogens

- **Equivalent hydrogens** within a molecule are in identical chemical environments.
  - In a molecule, hydrogens are equivalent if they are bonded to the same sp^3 carbon atom and there is free bond rotation involving that carbon atom in the molecule.
    - A lack of free bond rotation, such as that which occurs in small rings or with alkenes, can make hydrogens bonded to the same carbon atom nonequivalent.
  - There might be overall symmetry in the molecule making sets of hydrogens equivalent.

Problems: 13.2, 13.3

Section 13.6 Signal Areas

- The area of a ^1H-NMR signal is proportional to the number of equivalent hydrogens giving rise to that signal.

Problem: 13.4

Section 13.7 Chemical Shift

- **Chemical shift**, δ, is defined as the frequency shift from tetramethyl silane (TMS) divided by the operating frequency of the spectrometer.
  - The resonance signals in ^1H-NMR spectra are reported according to how far they are shifted from the resonance signal of the 12 equivalent hydrogens in TMS.
  - The resonance signals in ^{13}C-NMR spectra are reported according to how far they are shifted from the resonance signal of the four equivalent carbons in TMS.

- The chemical shift of a particular set of equivalent hydrogens depends primarily on three factors:
  - Nearby electronegative atoms have a deshielding effect.
  - In general, the higher the percent of s-character in a hybrid orbital, the greater the deshielding effect on itself and adjacent atoms.
  - The induced local magnetic fields in π bonds either add to, or subtract from, the applied field, depending on the geometry of the induced electron density circulation relative to the atom in question.

Problem: 13.6

Section 13.8 Signal Splitting and the (*n* + 1) Rule

- According to the **(*n* + 1) rule**, if a hydrogen has n hydrogens nonequivalent to it but equivalent among themselves on the same or adjacent atom(s), its ^{1}H-NMR signal will be split into $(n + 1)$ peaks.
  - Splitting patterns are commonly referred to as singlets (s), doublets (d), triplets (t), quartets (q), quintets, and multiplets (m).
  - The relative intensities of peaks in a multiplet can be predicted from an analysis of spin combinations for adjacent hydrogens or from the mnemonic device called **Pascal's triangle**.

Problem: 13.26

  - A **coupling constant (*J*)** is the distance between adjacent peaks in a multiplet and is reported in hertz (Hz). The value of J depends only on internal fields within a molecule and is independent of the spectrometer field.

Section 13.9 The Origins of Signal Splitting

- ^{1}H-NMR signals are split because the spin state ($+\frac{1}{2}$ versus $-\frac{1}{2}$) of nuclei of nonequivalent hydrogens no more than three bonds away influence the net magnetic field experienced by a given nucleus, an interaction known as **spin-spin coupling**.
- When a hydrogen nucleus is coupled to more than one set of adjacent hydrogen nuclei, the couplings combine.
  - In the general case, if a hydrogen nucleus is coupled to a set of n hydrogen nuclei on one side and a set of m hydrogen nuclei on the other, the signal will be split into a maximum of $(n + 1)(m + 1)$ peaks.
    - In molecules that are rigid, for example alkenes or cyclic molecules, all the $(n + 1)(m + 1)$ peaks can often be seen.

Problems: 13.7, 13.27

    - However, because coupling constants can be similar, especially in flexible molecules, this splitting can simplify to a number of observed peaks that is equal to the number of adjacent H atoms + 1, regardless of patterns of equivalence.

Section 13.10 Stereochemistry and Topicity

- Groups of atoms in which substitution of one atom by an isotope creates an achiral molecule are called **homotopic**.
  - Homotopic groups always have identical chemical shifts. Those in which such substitution produces a chiral molecule are **enantiotopic**. Enantiotopic groups have identical chemical shifts, except in a chiral environment.
- Molecules in which substitution produces diastereomers are called **diastereotopic**.
  - Diastereotopic atoms are nonequivalent in all environments so they have different chemical shifts. These differences can lead to complex splitting of the signals of diastereotopic H atoms, especially those adjacent to a chiral center.

Section 13.11 ^{13}C-NMR

Problems: 13.8, 13.27

- ^{13}**C-NMR** is like ^{1}H-NMR, except the nuclear spins of ^{13}C nuclei are being analyzed.
  - ^{13}C-NMR spectra are commonly recorded in a hydrogen-decoupled instrumental mode. In this mode, all ^{13}C signals appear as singlets.
  - The number of different signals in a ^{13}C-NMR spectrum tell you how many nonequivalent carbon atoms are in a molecule.
  - ^{13}C-NMR chemical shifts tell you what kind of carbon atoms are present.

Section 13.12 Interpretation of NMR Spectra

- Four important types of structural information can be obtained from a ^{1}H-NMR spectrum.
  - From the number of signals, we can determine the number of sets of equivalent hydrogens.
  - From the integration of signal areas, we can determine the relative numbers of hydrogens in each set.
  - From the chemical shift of each signal, we can derive information about the chemical environment of the hydrogens in each set.
  - From the splitting pattern of each signal, we can derive information about the number and chemical equivalency of hydrogens on the same and adjacent carbon atoms, in other words the connectivities between different groups on the molecule.

Problems:
13.09, 13.12–13.13,
13.15–13.24, 13.28

OWL Questions and problems for this chapter are available in OWL. Use the OWL access card that came with this textbook to access assigned questions and problems for this chapter.

Chapter 14

Mass Spectrometry

© Mark Muench/Stone/Getty Images

Some mass spectrometers can be quite sophisticated like the one shown to the left. For a partial mass spectrum of the neurotransmitter dopamine, see Figure 14.2. *Inset:* a model of dopamine

Mass spectrometry
An analytical technique for measuring the mass-to-charge ratio (m/z) of ions.

Mass spectrometry is an analytical technique for measuring the mass-to-charge ratio (m/z) of ions, originally positive ions, now both positive and negative. The principles of mass spectrometry were first recognized in 1898. In 1911, J. J. Thomson recorded the first mass spectrum, that of neon, and discovered that this element can be separated into a more abundant isotope, ^{20}Ne, and a less abundant isotope, ^{22}Ne. Using improved instrumentation, F. W. Aston showed that most of the naturally occurring elements are mixtures of isotopes. It was found, for example, that approximately 75% of chlorine atoms in nature are ^{35}Cl, and 25% are ^{37}Cl. Mass spectrometry did not come into common use until the 1950s, when commercial instruments that offered high resolution, reliability, and relatively inexpensive maintenance became available. Today, mass spectrometry is our most valuable analytical tool for the determination of accurate molecular masses. Furthermore, extensive information about the molecular formula and structure of a compound can be obtained from analysis of its mass spectrum. Mass spectrometry is becoming increasingly important in biochemistry as well; sequencing of proteins using this technique alone allows protein structures to be determined on a virtually single-cell scale.

14.1 | A Mass Spectrometer

A mass spectrometer (Figure 14.1) is designed to do three things:

1. Convert neutral atoms or molecules into a beam of positive or negative ions.
2. Separate the ions on the basis of their mass-to-charge (m/z) ratio.
3. Measure the relative abundance of each type of ion.

OWL Online homework for this chapter may be assigned in OWL for Organic Chemistry.

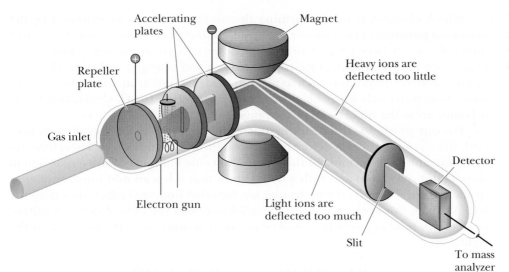

From this information, we can determine both the molecular mass and the molecular formula of an unknown compound. In addition, we can obtain valuable clues about the molecular structure of the compound.

There are many types of mass spectrometers; we have space in this text to describe in detail only the simplest. In the first-generation spectrometers, a vaporized sample in an evacuated ionization chamber is bombarded with high-energy electrons that cause electrons to be stripped from molecules of the sample, giving positively charged ions. Increasingly, radical anions (in which an extra electron has been added to a molecule) are studied; these are beyond the scope of this text. Positive ions are accelerated by a series of negatively charged accelerator plates into an analyzing chamber inside a magnetic (electric in some spectrometers) field perpendicular to the direction of the ion beam. The magnetic field causes the ion beam to curve. The radius of curvature of each ion depends on the charge on the ion (z), its mass (m), the accelerating voltage, and the strength of the magnetic field. A **mass spectrum** is a plot of relative ion abundance versus m/z ratio.

Samples of gases and volatile liquids can be introduced directly into the ionization chamber. Because the interior of a mass spectrometer is kept at a high vacuum, volatile liquids and even some solids are vaporized. For less volatile liquids and solids, the sample may be placed on the tip of a heated probe that is then inserted directly into the ionization chamber. Another extremely useful method for introducing a sample into the ionization chamber is to link a gas chromatograph (GC) or liquid chromatograph (LC) directly to the mass spectrometer. These machines can separate complex mixtures of molecules into pure fractions. Each fraction eluted from the chromatograph enters directly into the ionization chamber of the mass spectrometer, enabling mass determination.

Once in the ionization chamber, molecules of the sample are bombarded with a stream of high-energy electrons emitted from a hot filament and are then accelerated by an electric field to energies of approximately 70 eV [1 eV = 96.49 kJ (23.05 kcal)/mol]. Collisions between molecules of the sample and these high-energy electrons result in loss of electrons from sample molecules to form positive ions. A **molecular ion, $M^{+\cdot}$**, is the species formed by removal of a single electron from a molecule. A molecular ion belongs to a class of ions called **radical cations**. When methane, for example, is bombarded with high-energy electrons, an electron is dislodged from a molecule to give a molecular ion at m/z 16.

$$
\begin{array}{c}
\text{H} \\
| \\
\text{H}\!-\!\text{C}\!-\!\text{H} + e^- \\
| \\
\text{H}
\end{array}
\longrightarrow
\left[
\begin{array}{c}
\text{H} \\
| \\
\text{H}\!-\!\text{C}\!-\!\text{H} \\
| \\
\text{H}
\end{array}
\right]^{+\cdot}
+ \ 2e^-
$$

Molecular ion
(a radical cation)

Molecular ion ($M^{+\cdot}$)
The radical cation formed by removal of a single electron from a parent molecule in a mass spectrometer.

Radical cation
A species formed when a neutral molecule loses one electron; it contains both an odd number of electrons and a positive charge.

Ionization potential (IP)
The minimum energy required to remove an electron from an atom or molecule to a distance where there is no electrostatic interaction between the resulting ion and electron.

Which electron is lost in forming the molecular ion is determined by the **ionization potential** of the atom or molecule. Ionization potentials for most organic molecules are between 8 and 15 eV. The potentials are at the lower end of this range both for nonbonding electrons of oxygen and nitrogen and for π electrons in unsaturated compounds such as alkenes, alkynes, and aromatic hydrocarbons. Ionization potentials for σ electrons, such as those of C—C, C—H, and C—O σ bonds, are at the higher end of the range.

For our purposes, it doesn't matter which electron is lost because, in general, the radical cation is delocalized throughout the molecule. Therefore, we write the molecular formula of the parent molecule in brackets with a plus sign to show that it is a cation and with a dot to show that it has an odd number of electrons. See, for example, the molecular ion for ethyl isopropyl ether, shown on the left. At times, however, we will find it useful to depict the radical cation localized in a certain position to better understand its reactions, as in the formula on the right.

$$[CH_3CH_2OCH(CH_3)_2]^{+\cdot} \qquad CH_3CH_2\overset{\cdot+}{\underset{\cdot\cdot}{O}}CH(CH_3)_2$$

As we shall see in Section 14.2D, a molecular ion can undergo fragmentation to form a variety of smaller cations (which themselves may undergo further fragmentation), as well as radicals and smaller molecules. Only charged fragments are detected.

After molecular ions and their fragments form, a positively charged repeller plate directs the ions toward a series of negatively charged accelerator plates, producing a rapidly traveling ion beam. The ion beam is then focused by one or more slits and passes into a mass analyzer where it enters a magnetic field perpendicular to the direction of the ion beam. The magnetic field causes the ion beam to curve. Cations with larger values of m/z are deflected less than those with smaller m/z values. By varying either the accelerating voltage or the strength of the magnetic field, cations of the same m/z ratio can be focused on a detector, where the ion current is recorded. Modern detectors are capable of detecting single ions and of scanning a desired mass-to-charge region in a few tenths of a second or less.

A **mass spectrum** is a plot of the relative abundance of each cation versus mass-to-charge ratio. The peak resulting from the most abundant cation is called the **base peak** and is assigned an arbitrary intensity of 100. The relative abundances of all other cations in a mass spectrum are reported as percentages of the base peak. Figure 14.2 shows a partial mass spectrum of dopamine, a neurotransmitter in the brain's caudate nucleus, a center involved with coordination and integration of fine muscle movement. A deficiency of dopamine is an underlying biochemical defect in Parkinson's disease.

As can be seen in the following table, the number of peaks recorded depends on the sensitivity of the detector. If we record all peaks with intensity equal to or greater than 0.5% of the base peak, as in Figure 14.2, we find 45 peaks for dopamine. If we record all peaks with intensity equal to or greater than 0.05% of the base peak, we find 120 peaks. (Table 14.1)

Mass spectrum
A plot of the relative abundance of ions versus their mass-to-charge (m/z) ratio.

Base peak
The peak caused by the most abundant ion in a mass spectrum; the most intense peak. It is assigned an arbitrary intensity of 100.

Table 14.1	Number of Peaks Recorded in a Mass Spectrum of Dopamine
Peak Intensity Relative to Base Peak (%)	**Number of Peaks Recorded**
>5	8
>1	31
>0.5	45
>0.05	120

Figure 14.2
A partial mass spectrum of dopamine showing all peaks with intensity equal to or greater than 0.5% of the base peak.

The technique we have described is called **electron ionization mass spectrometry (EI-MS)**. This technique was the first developed and for a time was the one most widely used. It is limited, however, to relatively low-molecular-weight compounds that are vaporized easily in the evacuated ionization chamber. In recent years, a revolution in ionization techniques has extended the use of mass spectrometry to very high molecular-weight compounds and others that cannot be vaporized directly. Among the new techniques is fast-atom bombardment (FAB), which uses high-energy particles, such as xenon atoms accelerated to keV energies, to bombard a dispersion of a compound in a nonvolatile matrix, producing ions of the compound and expelling them into the gas phase. A second technique is matrix-assisted laser desorption ionization mass spectrometry (MALDI), which uses photons from an energetic laser for the same purpose. A third technique is chemical ionization (CI), which uses gas-phase acid-base reactions to produce ions. CI is particularly useful for identifying the molecular mass of a base (Brønsted-Lowry or Lewis) as its conjugate acid, MH^+. In addition, electrospray ionization mass spectrometry (ESI-MS) has become increasingly popular.

In ESI-MS, a solution of the analyte is introduced directly (for example, from a liquid chromatograph) through a charged capillary into a high-vacuum chamber. After entry into the vacuum chamber, the analyte exists in small charged droplets that rapidly evaporate due to the vacuum, leaving the charge concentrated on the analyte molecules. The m/z ratios of the charged molecules are then determined. ESI is a relatively gentle method of generating charged molecular ions in a vacuum. ESI-MS is therefore particularly effective for the ionization of biological macromolecules, large molecules such as polysaccharides (Chapter 25), proteins (Chapter 27), and nucleotides (Chapter 28), allowing determination of their molecular mass and major fragments without the complications caused by the unavoidable overfragmentation seen with harsher ionization methods. MALDI is also a relatively gentle ionization technique that is now commonly used for the ionization of biological macromolecules.

Many modern mass spectrometers use a quadrupole mass analyzer, which consists of four parallel rods. A fixed DC current and alternating RF potentials are applied to the rods. A sample is ionized and then accelerated into the space between the four rods in a direction that is parallel to the rods' long axes. The ions will take an oscillating path due to the alternating RF potentials, which are modulated so that only ions with the correct m/z ratio will have a stable trajectory and make it to the detector at the end of the chamber. The detector is tuned to scan different m/z ratios and to develop the entire spectrum. Multiple quadrupole detectors can be linked in series to gain even greater levels of sensitivity and precision. In these linked detectors, individual peaks isolated from the first detector are injected into the second in order to observe fragments of the fragments. Identification of fragments is greatly improved by this arrangement, which is often called tandem mass spectrometry.

14.2 | Features of a Mass Spectrum

To understand a mass spectrum, we need to understand the relationships between mass spectra and resolution, the presence of isotopes, and the fragmentation of molecules and molecular ions in both the ionization chamber and the analyzing chamber.

A. Resolution

An important operating characteristic of a mass spectrometer is its **resolution**, that is, how well it separates ions of different mass. **Low-resolution mass spectrometry** refers to instruments capable of distinguishing among ions of different nominal mass; that is, ions that differ by one or more atomic mass units (amu). **High-resolution mass spectrometry** refers to instruments capable of distinguishing among ions that differ in precise mass by as little as 0.0001 amu.

Resolution
In mass spectrometry, a measure of how well a mass spectrometer separates ions of different mass.

Low-resolution mass spectrometry
Instrumentation that is capable of separating only ions that differ in mass by 1 or more amu.

High-resolution mass spectrometry
Instrumentation that is capable of separating ions that differ in mass by as little as 0.0001 amu.

To illustrate, compounds with the molecular formulas C_3H_6O and C_3H_8O have nominal masses of 58 and 60 and can be resolved by low-resolution mass spectrometry. The compounds C_3H_8O and $C_2H_4O_2$, however, have the same nominal mass of 60 and cannot be distinguished by low-resolution mass spectrometry. If we calculate the precise mass of each compound using the data in Table 14.1, we see that they differ by 0.03642 amu and can be distinguished by high-resolution mass spectrometry. Observation of a molecular ion with a mass of 60.058 or 60.021 would establish the identities of C_3H_8O and $C_2H_4O_2$, respectively.

Molecular Formula	Nominal Mass	Precise Mass
C_3H_8O	60	60.05754
$C_2H_4O_2$	60	60.02112

B. The Presence of Isotopes

In the mass spectrum of dopamine (Figure 14.2), the molecular ion appears at m/z 153. If you look more closely at this mass spectrum, you see a small peak at m/z 154, from an ion one amu heavier than the molecular ion of dopamine. This peak is actually the sum of four separate peaks, each of amu 154 and each corresponding to the presence in the ion of a single heavier isotope of H, C, N, or O in dopamine. Because this peak corresponds to an ion one amu heavier than the molecular ion, it is called an M + 1 peak. We are concerned in this section primarily with M + 1 and M + 2 peaks.

Virtually all the elements common to organic compounds, including H, C, N, O, S, Cl, and Br, are mixtures of isotopes. Exceptions are fluorine, phosphorus, and iodine, which occur in Nature exclusively as ^{19}F, ^{31}P, and ^{127}I. Table 14.2 shows average atomic weights for the elements most common to organic compounds along with the masses and relative abundances in Nature of the stable isotopes of each. In this table, the relative abundances are tabulated according to the number of atoms of the heavier isotope per 100 atoms of the most abundant isotope. Naturally

Table 14.2	Precise Masses and Natural Abundances of Isotopes Relative to 100 Atoms of the Most Abundant Isotope			
Element	Atomic Weight	Isotope	Precise Mass (amu)	Relative Abundance
Hydrogen	1.0079	^1H	1.00783	100
		^2H	2.01410	0.016
Carbon	12.011	^{12}C	12.0000	100
		^{13}C	13.0034	1.11
Nitrogen	14.007	^{14}N	14.0031	100
		^{15}N	15.0001	0.38
Oxygen	15.999	^{16}O	15.9949	100
		^{17}O	16.9991	0.04
		^{18}O	17.9992	0.20
Sulfur	32.066	^{32}S	31.9721	100
		^{33}S	32.9715	0.78
		^{34}S	33.9679	4.40
Chlorine	35.453	^{35}Cl	34.9689	100
		^{37}Cl	36.9659	31.98
Bromine	79.904	^{79}Br	78.9183	100
		^{81}Br	80.9163	98.0

occurring carbon, for example, is 98.90% ^{12}C and 1.10% ^{13}C. Thus, there are 1.11 atoms of carbon-13 in nature for every 100 atoms of carbon-12.

$$1.10 \times \frac{100}{98.90} = 1.11 \text{ atoms } ^{13}\text{C per 100 atoms } ^{12}\text{C}$$

Example 14.1

Calculate the precise mass of each ion to five significant figures. Unless otherwise indicated, use the mass of the most abundant isotope of each element.

(a) $[CH_2Cl_2]^{\ddagger}$ (b) $[^{13}CH_2Cl_2]^{\ddagger}$ (c) $[CH_2Cl^{37}Cl]^{\ddagger}$

Solution

(a) $12.0000 + 2(1.00783) + 2(34.9689) = 83.953$ (b) 84.957   (c) 85.951

Problem 14.1

Calculate the nominal mass of each ion. Unless otherwise indicated, use the mass of the most abundant isotope of each element.

(a) $[CH_3Br]^{\ddagger}$ (b) $[CH_3^{81}Br]^{\ddagger}$ (c) $[^{13}CH_3Br]^{\ddagger}$

C. Relative Abundance of M, M + 2, and M + 1 Peaks

The most common elements giving rise to significant M + 2 peaks are chlorine, bromine, and oxygen. Chlorine in Nature is 75.77% ^{35}Cl and 24.23% ^{37}Cl. Thus, a ratio of M to M + 2 peaks of approximately 3:1 indicates the presence of a single chlorine atom in the compound. Similarly, bromine in Nature is 50.5% ^{79}Br and 49.5% ^{81}Br; a ratio of M to M + 2 of approximately 1:1 indicates the presence of a single bromine atom in the compound. The contribution of ^{18}O is only 0.2%, but it makes the major contribution to the M + 2 peak in compounds containing only C, H, N, and O. Sulfur is the only other element common to organic compounds that gives a significant M + 2 peak.

Let us use pentane, C_5H_{12}, to illustrate the relationship between M and M + 1 peaks. Pentane has a nominal mass of 72, and its molecular ion appears at m/z 72. In any sample of pentane, there is a probability that there will be a molecule in which one of the atoms of carbon is ^{13}C, the heavier isotope of carbon. This molecule has a nominal mass of 73, and its molecular ion will appear at m/z 73. Similarly, there is a probability that there will be a molecule in which one of the atoms of hydrogen is the heavier isotope of hydrogen, namely deuterium, ^{2}H. The probability of each of these isotope substitutions occurring is related to the natural abundance of each isotope in the following way.

% (M + 1) = Σ (% abundance of heavier isotope × number of atoms in the formula)

Using this formula, we calculate that the relative intensity of the M + 1 peak for pentane is

(M + 1) = Σ (1.11 × 5C + 0.016 × 12H) = 5.55 + 0.19 = 5.74% of molecular ion peak

Notice that the M + 1 peak for pentane is almost entirely from ^{13}C. The same is true for other compounds containing only C and H. Because M + 1 peaks are relatively low in intensity compared to the molecular ion peak and often difficult to measure with any precision, they are not useful for accurate determinations of molecular formulas. M + 1 and M + 2 peaks, however, can be useful for getting a rough idea of the number of carbons, oxygens, sulfurs, and halogens. For example, the spectrum of chloroethane has peaks at m/z 64 and 66 (corresponding to $C_2H_5^{35}Cl$ and $C_2H_5^{37}Cl$, respectively), in a characteristic 3:1 ratio. Figures 14.3 and 14.4 illustrate this for a chloro and a bromo compound.

In contrast, 1-bromopropane has peaks at 122 and 124 for $C_3H_7^{79}Br$ and $C_3H_7^{81}Br$ in about a 1:1 ratio. This ratio of M and M + 2 peaks is characteristic of

Figure 14.3

Mass spectrum of chloroethane.

Figure 14.4

Mass spectrum of 1-bromopropane.

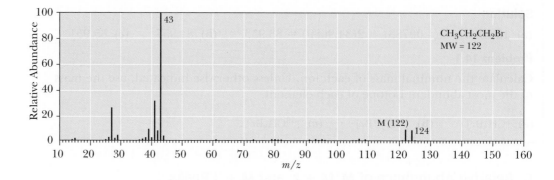

monobrominated compounds. The M and M + 2 peaks in chlorides and bromides in their ratios of 3:1 and 1:1, respectively, are very distinctive and allow almost immediate identification of monochloro and monobromo compounds.

D. Fragmentation of Molecular Ions

To attain high efficiency of molecular ion formation and to give reproducible mass spectra, it is common to use electrons with energies of 70 eV [approximately 6750 kJ (1600 kcal)/mol] in EI. This energy is sufficient not only to dislodge one or more electrons from a molecule but also to cause extensive fragmentation because it is well in excess of bond dissociation enthalpies in organic molecules. These fragments may be unstable as well and, in turn, break apart into even smaller fragments.

$$
[A\text{—}B]^{+} \quad
\begin{array}{l}
\longrightarrow \quad A\cdot \;+\; B^{+} \\[4pt]
 \text{Radical} \quad \text{Cation} \\[10pt]
\longrightarrow \quad A^{+} \;+\; \cdot B \\[4pt]
 \text{Cation} \quad \text{Radical}
\end{array}
$$

not detected by mass spectrometry

Molecular ion (a radical cation)

The molecular ions for some compounds have a sufficiently long lifetime in the analyzing chamber that they are observed in the mass spectrum, sometimes as the base (most intense) peak. Molecular ions of other compounds have a shorter lifetime and are present in low abundance or not at all. As a result, the mass spectrum of a compound ionized with one of the harsher ionization methods such as EI or CI typically (but not always) consists of a peak for the molecular ion and series of peaks for fragment ions. The fragmentation pattern and relative abundances of ions are unique for each compound under a given set of ionizing conditions and are characteristic of that compound. Fragmentation patterns give us valuable information about molecular structure.

A great deal of the chemistry of ion fragmentation can be understood in terms of the formation and relative stabilities of carbocations in solution. Where fragmentation occurs forming new carbocations, the mode of fragmentation that gives the most stable carbocation is favored. Thus, the probability of fragmentation to form a new carbocation increases in the following familiar order.

$$CH_3^+ < 1° < 2° \cong 1° \text{ allylic/benzylic} < 3° \cong 2° \text{ allylic/benzylic} < 3° \text{ allylic/benzylic}$$

Increasing hydrocarbon carbocation stability →

Molecular rearrangements are also characteristic of certain types of functional groups.

14.3 | Interpreting Mass Spectra

Chemists often use mass spectra primarily for the determination of molecular mass and molecular formula. Very rarely do they attempt a full interpretation of a mass spectrum, which can be very time consuming, difficult, and dependent on the experimental details of the ionization. The mass spectrum of dopamine (Figure 14.2), for example, contains at least 45 peaks with intensity equal to or greater than 0.5% of the intensity of the base peak. We have neither the need nor the time to attempt to interpret this level of complexity. Rather, we concentrate in this section on the fragmentation mechanisms giving rise to major peaks.

As we now look at typical mass spectra of the classes of organic compounds we have seen so far, keep the following two points in mind. They provide valuable information about the molecular composition of an unknown compound.

1. The only elements giving rise to significant M + 2 peaks are ^{18}O (0.2%), ^{34}S (4.40%), ^{37}Cl (32%), and ^{81}Br (98%). If no large M + 2 peak is present, then these elements are absent.
2. Is the mass of the molecular ion odd or even? According to the **nitrogen rule**, if a compound has an odd number of nitrogen atoms, its molecular ion will appear at an odd m/z value. Conversely, if a compound has an even number of nitrogen atoms (including zero), its molecular ion will appear at an even m/z value. This rule is most helpful when there is an odd number of nitrogens. You may need additional experimental information to establish the presence of an even number of nitrogens.

Nitrogen rule
A rule stating that the molecular ion of a compound with an odd number of nitrogen atoms has at an odd m/z ratio; if there are zero or an even number of nitrogen atoms, the molecular ion has an even m/z ratio.

A. Alkanes

Two rules will help you interpret the mass spectra of alkanes. (1) Fragmentation tends to occur toward the middle of unbranched chains rather than at the ends. (2) The differences in energy among allylic, benzylic, tertiary, secondary, primary, and methyl carbocations in the gas phase are much greater than the differences among comparable radicals. Therefore, where alternative modes of fragmentation are possible, the more stable carbocation tends to form in preference to the more stable radical.

Unbranched alkanes fragment to form a series of cations differing by 14 amu (a CH_2 group), with each fragment formed by a one-bond cleavage having an odd mass number. The mass spectrum of octane (Figure 14.5), for example, shows a peak for the molecular ion (m/z 114), as well as peaks for $C_6H_{13}^+$ (m/z 85), $C_5H_{11}^+$ (m/z 71), $C_4H_9^+$ (m/z 57), $C_3H_7^+$ (m/z 43), and $C_2H_5^+$ (m/z 29). These correspond to loss of ethyl, propyl, and butyl. Fragmentation of the CH_2—CH_3 bond is not observed; there is no peak corresponding to a methyl cation (m/z 15), nor is there one corresponding to a heptyl cation (loss of methyl, m/z 99). In mass spectrometry, fragmentations are shown by lines through the bond that is cleaved with an angled part toward the fragment that bears the charge.

Fragmentation of branched-chain alkanes leads preferentially to the formation of secondary and tertiary carbocations, and because these cations are more easily formed than methyl and primary carbocations, extensive fragmentation is likely. For this reason, the molecular ion of branched-chain hydrocarbons is often very weak or absent entirely from the spectrum. The molecular ion corresponding to m/z 114 is not observed, for example, in the mass spectrum of the highly

Figure 14.5

Mass spectrum of octane.

Figure 14.6

Mass spectrum of 2,2,4-trimethylpentane. The peak for the molecular ion is of such low intensity that it does not appear in this spectrum.

branched 2,2,4-trimethylpentane (Figure 14.6). The base peak for this hydrocarbon is at $m/z\,57$, which corresponds to the *tert*-butyl cation ($C_4H_9^+$). Other prominent peaks are at $m/z\,43$ (isopropyl cation) and $m/z\,41$ (allyl cation [$CH_2=CHCH_2^+$]).

Sometimes peaks that seem to defy the rules of chemical logic we have encountered so far occur in a mass spectrum. For example, the prominent peak at $m/z\,29$ in the mass spectrum of 2,2,4-trimethylpentane (Figure 14.6) is consistent with the ethyl cation, $CH_3CH_2^+$. There is, however, no ethyl group in the parent molecule! This cation must be formed by some combination of fragmentation and rearrangement quite beyond anything that we have seen up to this point; such rearrangements are common at the high energies of electron-impact mass spectra. Exploration of this point is beyond the scope of this text.

The most common fragmentation patterns of cycloalkanes are loss of side chains and loss of ethylene, $CH_2=CH_2$. The peak at $m/z\,69$ in the mass spectrum of methylcyclopentane (Figure 14.7) is the result of the loss of the one-carbon side chain to give the cyclopentyl cation, $C_5H_9^+$. The base peak at $m/z\,56$ is caused by the loss of ethylene and corresponds to a cation of molecular formula $C_4H_8^+$. Note here that one-carbon cleavages of alkanes and cycloalkanes give fragments with odd mass numbers; two-bond cleavages give fragments with even mass numbers.

Example 14.2

The base peak at $m/z\,56$ in the mass spectrum of methylcyclopentane corresponds to loss of ethylene to give a radical cation with the molecular formula $C_4H_8^{+\cdot}$. Propose a structural formula for this radical cation and show how it might be formed.

Solution

Following is a structural formula for a molecular ion that might be formed in the ionizing chamber. In it, a single electron has been dislodged from a carbon-carbon single bond to give a 1° radical and a 2° carbocation. Rearrangement of bonding electrons in this radical cation gives ethylene and a new radical cation.

Molecular ion | Ethylene | A new radical cation
(a radical cation, *m/z* 84) | | (a 2° carbocation, *m/z* 56)

Problem 14.2

Propose a structural formula for the cation at *m/z* 41 observed in the mass spectrum of methylcyclopentane.

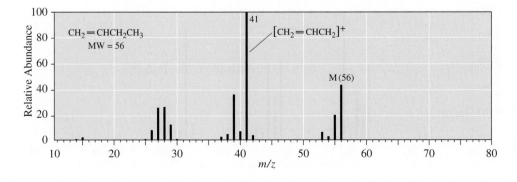

Figure 14.7

Mass spectrum of methylcyclopentane.

B. Alkenes

Alkenes characteristically show a strong molecular ion peak, most probably formed by removal of one π electron from the double bond. Furthermore, they cleave readily to form resonance-stabilized allylic cations, such as the allyl cation seen at *m/z* 41 in the mass spectrum of 1-butene (Figure 14.8).

Cyclohexenes undergo fragmentation to give a 1,3-diene and an alkene in a process that is the reverse of a Diels-Alder reaction (Section 24.6). The terpene limonene, a disubstituted cyclohexene, for example, fragments by a reverse Diels-Alder reaction to give two molecules of 2-methyl-1,3-butadiene (isoprene): one formed as a neutral diene and the other formed as a diene radical cation. Note here that the two-bond cleavage of this hydrocarbon gives fragments with even mass numbers.

Limonene | **A neutral diene** | **A radical cation**
m/z 136 | *m/z* 68 | *m/z* 68

Figure 14.8

Mass spectrum of 1-butene.

C. Alkynes

As with alkenes, alkynes show a strong peak for the molecular ion. Their fragmentation patterns are also similar to those of alkenes. One of the most prominent peaks in the mass spectrum of most alkynes is from the delocalization stabilized 3-propynyl (propargyl) cation (m/z 39) or a substituted propargyl cation.

$$HC\equiv C\overset{\frown}{-}CH_2^+ \longleftrightarrow HC\overset{+}{=}C=CH_2$$

3-Propynyl cation
(Propargyl cation)

Both the molecular ion, m/z 68, and the propargyl cation, m/z 39, are seen in the mass spectrum of 1-pentyne (Figure 14.9). Also seen is the ethyl cation, m/z 29.

Figure 14.9

Mass spectrum of 1-pentyne.

D. Alcohols

The intensity of the molecular ion from primary and secondary alcohols is normally quite low, and there usually is no molecular ion detectable for tertiary alcohols. One of the most common fragmentation patterns for alcohols is loss of a molecule of water to give a peak corresponding to the molecular ion minus 18 (M − 18). Another common pattern is loss of an alkyl group from the carbon bearing the —OH group to form a delocalization stabilized oxonium ion and an alkyl radical. The oxonium ion is particularly stable because of delocalization of charge.

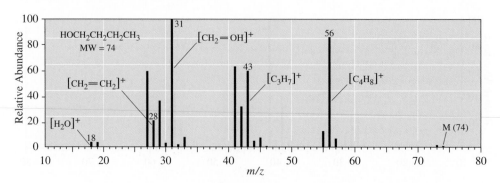

Each of these patterns is found in the mass spectrum of 1-butanol (Figure 14.10). The molecular ion appears at m/z 74. The prominent peak at m/z 56 corresponds to loss of a molecule of water from the molecular ion (M − 18). The base peak at m/z 31 corresponds to cleavage of a propyl group (M − 43) from the carbon bearing the —OH group. The propyl cation is visible at m/z 43.

Figure 14.10

Mass spectrum of 1-butanol.

Example 14.3

A low-resolution mass spectrum of 2-methyl-2-butanol (MW 88) shows 16 peaks. The molecular ion is absent. Account for the formation of peaks at m/z 73, 70, 59, and 55, and propose a structural formula for each cation.

Solution

The peak at m/z 73 (M − 15) corresponds to loss of a methyl radical from the molecular ion. The peak at m/z 59 (M − 29) corresponds to loss of an ethyl radical. Loss of water as a neutral molecule from the molecular ion gives an alkene at m/z 70 (M − 18) as a radical cation. Loss of methyl from this radical cation gives an allylic carbocation at m/z 55.

Problem 14.3

The low-resolution mass spectrum of 2-pentanol shows 15 peaks. Account for the formation of the peaks at m/z 73, 70, 55, 45, 43, and, 41.

E. Aldehydes and Ketones

A characteristic fragmentation pattern of aliphatic aldehydes and ketones is cleavage of one of the bonds to the carbonyl group (α-cleavage). α-Cleavage of 2-octanone, for example, gives carbonyl-containing ions at m/z 43 and 113. α-Cleavage of the aldehyde proton gives an M − 1 peak, which is often quite distinct and provides a useful way to distinguish between an aldehyde and a ketone.

Aldehydes and ketones with a sufficiently long carbon chain show a fragmentation called a McLafferty rearrangement. In a **McLafferty rearrangement** of an aldehyde or ketone, the carbonyl oxygen abstracts a hydrogen five atoms away to give an alkene and a new radical cation. Because McLafferty rearrangements involve cleavage of two bonds, molecules at even m/z give fragments at even m/z.

Mechanism | McLafferty Rearrangement of a Ketone

Reaction occurs through a six-membered ring transition state.

Molecular ion
m/z 86

→ McLafferty rearrangement →

m/z 58

McLafferty rearrangement of 2-octanone, for example, gives 1-pentene and a radical cation at m/z 58, which is the enol of acetone (Section 16.9).

Molecular ion
m/z 128

→ McLafferty rearrangement →

m/z 58

The results of both α-cleavage and McLafferty rearrangement can be seen in the mass spectrum of 2-octanone (Figure 14.11).

F. Carboxylic Acids, Esters, and Amides

The molecular ion peak from a carboxylic acid is generally observed, although it is often very weak. The most common fragmentation patterns are α-cleavage of the carboxyl group to give the ion [COOH]$^+$ at m/z 45 and McLafferty rearrangement. The base peak is very often the result of the McLafferty rearrangement product.

Molecular ion
m/z 88

→ α-cleavage → ·+ [O=C—O—H]$^+$

m/z 45

Figure 14.11

Mass spectrum of 2-octanone. Ions at m/z 43 and 113 result from α-cleavage. The ion at m/z 58 results from McLafferty rearrangement.

$$CH_3CCH_2CH_2CH_2CH_2CH_2CH_3$$
MW = 128

Mechanism | McLafferty Rearrangement of a Carboxylic Acid

Molecular ion
m/z 88

→ McLafferty rearrangement →

m/z 60

Each of these patterns is seen in the mass spectrum of butanoic acid (Figure 14.12).

Figure 14.12

Mass spectrum of butanoic acid. Common fragmentation patterns of carboxylic acids are α-cleavage to give the ion $[COOH]^+$ at m/z 45 and McLafferty rearrangement.

Esters and amides also generally show discernible molecular ion peaks. Like carboxylic acids, their most characteristic fragmentation patterns are α-cleavage and McLafferty rearrangement, both of which can be seen in the mass spectrum of methyl butanoate (Figure 14.13). Peaks at m/z 71 and 59 are the result of α-cleavage.

The peak at m/z 74 is the result of McLafferty rearrangement.

G. Aromatic Hydrocarbons

The mass spectra of most aromatic hydrocarbons show an intense molecular ion peak. The mass spectrum of toluene (Figure 14.14), for example, shows a large molecular ion peak at m/z 92.

The mass spectra of toluene and most other alkylbenzenes show a fragment ion at m/z 91. Although it might seem that the most likely structure for this ion is that of the benzyl cation, experimental evidence suggests a molecular rearrangement to form the more stable tropylium ion (Section 21.2E). In the tropylium ion,

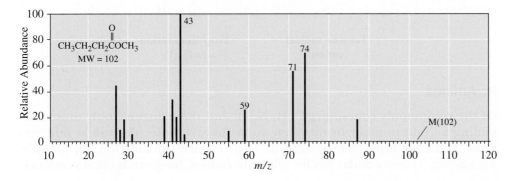

Figure 14.13

Mass spectrum of methyl butanoate. Characteristic fragmentation patterns of esters are α-cleavage and McLafferty rearrangement.

As a result of enormous instrumental advances made during the last few years, mass spectrometry is becoming the method of choice for determining the structures of biological macromolecules, in particular, proteins and DNA. The Nobel Prize in chemistry in 2002 was awarded to John B. Fenn and Koichi Tanaka for the development of the electrospray ionization (ESI) and MALDI methods that have revolutionized structure determination of these molecules. The key advantage of using mass spectrometry for analysis of proteins and nucleic acids is that the entire process can be automated so that information about a large number of molecules can be obtained rapidly. A growing list of DNA sequences at the genome level is now available, and attention is increasingly being turned toward acquiring sequence information at the

analyzer determines the mass of each fragment. Because each amino acid has a slightly different mass (except for two isomeric amino acids leucine and isoleucine), the exact amino acid composition of the fragment can be determined. Powerful computers align overlapping fragments and determine the exact sequence.

For larger proteins, the mass spectrometric method is usually preceded by enzymatic cleavage into fragments. However, in the most advanced systems, the individual fragments do not need to be separated. They are injected into a tandem mass spectrometer (called an MS-MS). The first segment of the instrument separates the individual fragments by mass, and the second segment fragments each peptide and sequences it separately. The MS-MS can also be used to separate, and

protein level with high throughput. Mass spectrometry is playing a major role in this effort.

For example, it is now possible to obtain complete amino acid sequences on polypeptides (proteins, biological polyamides, see Chapter 27) of substantial length. The mass spectrometer cleaves polypeptides into fragments of varying length. Although there are many cleavage modes possible, the main cleavage is at peptide (amide) bonds. Both fragments (from the *N*-terminal and *C*-terminal part) can usually be identified. A mass

then further fragment, the primary ions obtained in the first stage to obtain sequence information directly. Enormous proteins can be sequenced on a picomole or lower concentration level. This method is particularly helpful because it can be used on mixtures (and ultimately, it is believed, on whole cell contents). Similar techniques are also now used for determining the sequence of DNA, the genetic material. Further discussion about applications to macromolecules may be found in Chapters 27 and 28.

---

an aromatic cation, the positive charge is delocalized equally over all seven carbon atoms of the cycloheptatrienyl ring.

Toluene radical cation

Tropylium cation (*m/z* 91)

H. Amines

Of the compounds containing C, H, N, O, and the halogens, only those containing an odd number of nitrogen atoms have a molecular ion at odd *m/z* ratio. Thus, mass spectrometry can be a particularly valuable tool for identifying amines. The molecular ion for aliphatic amines, however, is often very weak.

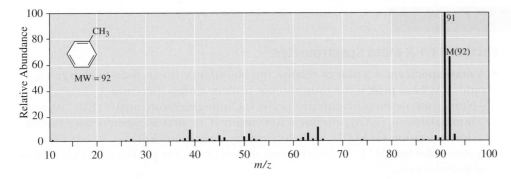

Figure 14.14

Mass spectrum of toluene. Prominent are the intense molecular ion peak at m/z 92 and the tropylium cation at m/z 91.

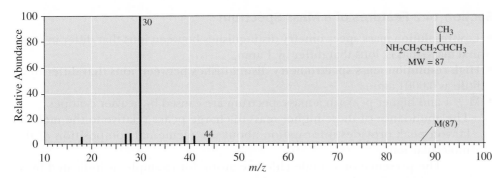

Figure 14.15

Mass spectrum of 3-methyl-1-butanamine (isopentylamine). The most characteristic fragmentation pattern of aliphatic amines is β-cleavage.

The most characteristic fragmentation of amines, and the one that often gives the base peak, is β-cleavage. Where alternative possibilities for β-cleavage exist, it is generally the largest R group that is lost. In contrast to nitrogen-free molecules, single bond fragments from compounds that have one or an odd number of nitrogens give compounds with even mass. The most prominent peak in the mass spectrum of 3-methyl-1-butanamine (Figure 14.15) is from $[CH_2{=}NH_2]^+$, m/z 30, from β-cleavage. β-Cleavage is also characteristic of secondary and tertiary amines. Complex rearrangement and fragmentation processes give the m/z 30 peak as a major fragment even from secondary and tertiary amines.

$$CH_3-\underset{\underset{CH_3}{|}}{CH}-CH_2-CH_2-NH_2 \xrightarrow{\beta\text{-cleavage}} CH_3-\underset{\underset{CH_3}{|}}{CH}-CH_2\cdot + CH_2{=}\overset{+}{N}H_2$$

$$(m/z\ 30)$$

14.4 | Mass Spectrometry in the Organic Synthesis Laboratory and Other Applications

Mass spectrometry, especially when interfaced with separation methods such as gas chromatography (GC-MS) and liquid chromatography (LC-MS), represents a powerful and rapid method of identifying compounds in reaction product mixtures. As a result, GC-MS and LC-MS are taking on increased importance as the primary method of routine molecule identification in the organic synthesis laboratory, both in industrial and academic settings.

Mass spectrometry is also becoming increasingly important for other practical applications. Some devices used to screen luggage in airports use mass spectrometry to identify traces of known explosives. Drug testing of athletes and advanced forensic science uses GC-MS to identify traces of pharmaceuticals or illicit drugs in blood samples. Looking toward the future, with the advent of powerful new-generation quadrupole mass spectrometers that are increasingly small and inexpensive, there will be a dramatic increase in the use of mass spectrometry in the identification of molecules in many aspects of modern life, not just in the research laboratory.

Section 14.1 A Mass Spectrometer

- A **mass spectrum** is a plot of relative ion abundance versus mass-to-charge (m/z) ratio. The **base peak** is the most intense peak in a mass spectrum.
  - Numerous methods of ionization exist, including electron impact (EI), fast-atom bombardment (FAB), chemical ionization (CI), matrix-assisted laser desorption ionization (MALDI), and electrospray ionization (ESI).
- A **molecular ion**, $M^{+\cdot}$ is a radical cation derived from the parent molecule by loss of one electron.

Section 14.2 Features of a Mass Spectrum

- **Low-resolution mass spectrometry** distinguishes between ions that differ in nominal mass, that is, ions that differ by 1 amu.
- **High-resolution mass spectrometry** distinguishes between ions that differ by as little as 0.0001 amu.
- M + 1 and higher peaks in a mass spectrum are caused by heavier isotopes.
  - The abundance of these higher mass-to-charge peaks relative to the molecular ion peak provides information about the elemental composition of the molecular ion.
    - The presence of a single chlorine atom, for example, is indicated by M and M + 2 peaks in a ratio of 3:1.
- The mass spectrum of a compound typically consists of a peak for the molecular ion and a series of peaks for fragment ions.
  - The fragmentation pattern and relative abundances of ions are unique for each compound and are characteristic of that compound.
    - Fragments formed by cleavage of one bond have odd mass if they contain no nitrogen; those formed from the cleavage of two bonds have even mass.
  - Many of the observed fragmentation patterns can be understood in terms of the relative stability of carbocations.
    - Where alternative modes of fragmentation are possible, the more stable carbocation tends to be formed in preference to the more stable radical.

Section 14.3 Interpreting Mass Spectra

- Different functional groups have characteristic fragmentation patterns in mass spectra.
  - **Alkane** mass spectra tend to show fragmentation that occurs toward the middle of unbranched chains, forming a series of fragments differing by 14 amu.
    - Branched chains give fragments at the branches to generate secondary and tertiary cations.
  - **Alkenes** give a strong molecular ion peak and cleave to form allyl cations.
  - **Alkynes** give a strong molecular ion peak and cleave to form the propargyl cation.
  - **Alcohols** give generally weak molecular ions (secondary and tertiary alcohols especially), and often cleave to lose water or they lose an alkyl group bonded to the carbon bonded to the OH group.
  - **Aldehydes and ketones** undergo two kinds of characteristic cleavage.
    - **α-Cleavage** involves cleavage of one of the bonds to the carbonyl group.
    - **McLafferty rearrangement** involves a six-membered ring transition state to give an alkene and new radical cation.

Problems: 14.1, 14.5, 14.8, 14.9, 14.11, 14.12, 14.19–14.23

- **Carboxylic acids, esters, and amides** undergo the same two kinds of characteristic cleavage, α-cleavage and McLafferty rearrangement.
- **Aromatic hydrocarbons** give strong molecular ions and alkyl benzenes often fragment to give the tropylium cation.
- According to the **nitrogen rule**, if a compound has an odd number of nitrogen atoms, its molecular ion will have an odd *m/z* value.

Problems: 14.2–14.4, 14.6, 14.7, 14.10, 14.13–14.18, 14.24–14.37

Problem: 14.38

Section 14.4 Mass Spectrometry in the Organic Synthesis Laboratory and Other Applications

- Mass spectrometry is increasingly used in the forensic laboratory.
- Mass spectral techniques can be used to determine the amino acid sequences of proteins and DNA.

ᗢWL Questions and problems for this chapter are available in OWL. Use the OWL access card that came with this textbook to access assigned questions and problems for this chapter.

Chapter 15

An Introduction to Organometallic Compounds

© Clive Druett: Papilio/CORBIS

A common housefly (*Musca domestica*). Commercially available traps with the attractant muscalure (synthesized via a route using an organometallic compound) combined with a poison are efficient at luring and killing houseflies (Section 15.2B).

Organometallic compound
A compound that contains a carbon-metal bond.

In this chapter, we undertake our first discussion of a broad class of organic compounds called **organometallic compounds**, compounds that contain a carbon-metal bond. In recent years, there has been an enormous explosion in our understanding of their chemistry, particularly as stereospecific (and often enantioselective) reagents for synthetic chemistry. We have already seen one example in the Sharpless enantioselective epoxidation of alkenes (Section 11.8D).

This chapter cannot possibly cover the wealth of organometallic reagents and catalysts that have been developed for synthetic organic chemistry, particularly during the last decade. We focus, therefore, on transformations that are fundamental to synthetic chemistry. Organomagnesium, lithium, and copper reagents have been selected because of their historical importance and their continued use in modern organic synthesis. These reagents are particularly important in the reactions of carbonyl compounds—the focus of the next several chapters. Several more recent reactions of organometallic compounds are discussed in the chapter on C—C bond formation and organic synthesis (Chapter 24).

15.1 | Organomagnesium and Organolithium Compounds

We begin with organomagnesium and organolithium compounds and concentrate on their formation and basicity. We discuss their use in organic synthesis in more detail in later chapters, particularly in Chapters 16 and 18.

OWL Online homework for this chapter may be assigned in OWL for Organic Chemistry.

RMgX   RLi

An organomagnesium   An organolithium
compound   compound
(a Grignard reagent)

A. Formation and Structure

Organomagnesium compounds are among the most readily available, easily prepared, and easily handled organometallics. They are commonly named Grignard reagents after the French chemist Victor Grignard (1871–1935), who was awarded the 1912 Nobel Prize for chemistry for their discovery and application to organic synthesis.

Grignard reagents are typically prepared by the slow addition of an alkyl, aryl, or alkenyl (vinylic) halide to a stirred suspension of a slight excess of magnesium metal in an ether solvent, most commonly diethyl ether or tetrahydrofuran (THF). Organic iodides and bromides generally react very rapidly under these conditions, whereas most organic chlorides react more slowly. Bromides are the most common starting materials for preparation of Grignard reagents. It is common to use the higher-boiling THF (bp 67°C) to prepare Grignard reagents from the less reactive organic halides. Generally there is an induction period at the beginning of the reaction caused by the presence of traces of moisture and a thin oxide coating on the surface of the magnesium. When reaction starts, it is exothermic, and the remaining organic halide is added at a rate sufficient to maintain a gentle reflux of the ether.

Butylmagnesium bromide, for example, is prepared by treating 1-bromobutane in diethyl ether with magnesium metal. Aryl Grignard reagents, such as phenylmagnesium bromide, are prepared in the same manner. These reactions are referred to as oxidative additions, because they result in an increase in the formal oxidation state of magnesium by two, that is, from Mg(0) to Mg(II).

1-Bromobutane   **Butylmagnesium bromide**
(an alkyl Grignard reagent)

Bromobenzene   **Phenylmagnesium
                                        bromide**
(an aryl Grignard reagent)

Although the equation for formation of Grignard reagents looks simple, the mechanism is considerably more complicated and involves radicals. We have no need in this course to discuss the mechanism for their formation. However, we note that for many Grignard reagents there is an equilibrium between monoalkyl and dialkyl magnesium complexes as shown:

$$2 \text{ RMgX} \rightleftharpoons R_2Mg + MgX_2$$

Grignard reagents form on the surface of the metal and dissolve as coordination complexes solvated by ether. In this ether-soluble complex, magnesium acts as a Lewis acid, and the ether acts as a Lewis base (Figure 15.1).

Figure 15.1

Ethylmagnesium bromide
dietherate, a Grignard reagent
solvated by ether.

$$\begin{array}{c}
\text{CH}_3\text{CH}_2 \quad \quad \text{CH}_2\text{CH}_3 \\
\text{O} \\
| \\
\text{CH}_3\text{CH}_2 \longrightarrow \text{Mg} \longleftarrow \text{Br} \\
| \\
\text{O} \\
\text{CH}_3\text{CH}_2 \quad \quad \text{CH}_2\text{CH}_3
\end{array}$$

Organolithium reagents are prepared by treating an alkyl, aryl, or alkenyl halide with two equivalents of lithium metal, as illustrated by the preparation of butyllithium. In this reaction, a solution of 1-chlorobutane in pentane is added to lithium wire at –10°C.

$$\text{Cl} + 2\ \text{Li} \xrightarrow{\text{pentane}} \text{Li} + \text{LiCl}$$

1-Chlorobutane   **Butyllithium**

Organolithium compounds are very reactive as nucleophiles in carbonyl addition reactions even at very low temperatures. They are also powerful and effective bases. For these reasons, they are now widely used in modern synthetic chemistry. However, they react rapidly with atmospheric oxygen and moisture and, therefore, must be used under an inert atmosphere of N_2 or Ar, which decreases their convenience.

The carbon-metal bonds in Grignard and organolithium reagents are best described as polar covalent, with carbon bearing a partial negative charge and the metal bearing a partial positive charge. In their reactions, Grignard and organolithium reagents behave as **carbanions**, which are ions in which carbon has an unshared pair of electrons and bears a negative charge. Shown in Table 15.1 are electronegativity differences (Pauling scale, Table 1.5) between carbon and various metals. From this difference, we can estimate the percent ionic character of each carbon-metal bond.

Carbanion
An ion in which carbon has an
unshared pair of electrons and bears
a negative charge.

Table 15.1	Percent Ionic Character of Some C—M Bonds	
$\overset{\delta-}{\text{C}}\overset{\delta+}{\text{—}}\text{M}$ Bond	Difference in Electronegativity	Percent Ionic Character*
C—Li	2.5 − 1.0 = 1.5	60
C—Mg	2.5 − 1.2 = 1.3	52
C—Al	2.5 − 1.5 = 1.0	40
C—Zn	2.5 − 1.6 = 0.9	36
C—Sn	2.5 − 1.8 = 0.7	28
C—Cu	2.5 − 1.9 = 0.6	24
C—Hg	2.5 − 1.9 = 0.6	24

*Percent ionic character $= \dfrac{E_C - E_M}{E_C} \times 100$

Organolithium and organomagnesium bonds have the highest partial ionic character, whereas those of organocopper and organomercury compounds are lower. These compounds do not behave as salts. Organolithium reagents, for example, which have the highest percent partial ionic character, dissolve in nonpolar hydrocarbon solvents such as pentane because they self-assemble into well-ordered aggregates, $(RLi)_x$, that present a nonpolar surface to the surrounding solvent.

An important aspect of the metals listed in Table 15.1 is that they all have electronegativities that are considerably lower than carbon. This means that the polarity of the carbon-metal bond places a partial negative charge on carbon and a partial positive charge on the metal. The partial negative charge makes the carbon atom both basic and nucleophilic. When Grignard and organolithium reagents are prepared, the carbon atom bearing the halogen is transformed from an electrophilic center (partial positive charge) in the haloalkane, alkene, or arene to a nucleophilic center (partial negative charge) in the organometallic compound. In the structural formula of butylmagnesium bromide on the right, the carbon-magnesium bond is shown as ionic to emphasize its nucleophilic character.

Carbon is an electrophile

Carbon is a nucleophile

$$
\begin{array}{cc}
\overset{H}{\underset{CH_3CH_2CH_2}{H \cdots C - Br}}^{\delta+ \ \delta-} & \overset{H}{\underset{CH_3CH_2CH_2}{H \cdots C : Mg^{2+} \ Br^-}}^{-}
\end{array}
$$

As nucleophiles, these compounds react with the electrophilic carbon atom of the carbonyl groups of aldehydes and ketones (Chapter 16) and of carboxylic esters and acid chlorides (Chapter 18). Herein lies the value of organomagnesium and organolithium reagents in synthetic organic chemistry—as carbon-centered nucleophiles, they enable the formation of new carbon-carbon bonds.

B. Reaction with Proton Acids

Both Grignard and organolithium compounds are very strong bases and react readily with any acid (proton donor) stronger than the alkane from which they are derived. Ethylmagnesium bromide, for example, reacts instantly with water, which donates a proton to give ethane and magnesium salts. This reaction is an example of a much stronger acid and a much stronger base reacting to give a weaker acid and a weaker base (Section 4.4). Ethane is evolved from the reaction mixture as a gas.

$$
\overset{\delta- \quad \delta+}{CH_3CH_2} - MgBr + \overset{\delta+ \ \delta-}{H - OH} \longrightarrow CH_3CH_2 - H + Mg^{2+} + OH^- + Br^- \qquad
\begin{array}{l}
pK_{eq} = -35 \\
K_{eq} = 10^{35}
\end{array}
$$

	pK_a 15.7	pK_a 51	
Stronger base	Stronger acid	Weaker acid	Weaker base

Following are several classes of proton donors that react readily with Grignard and organolithium reagents. Because they react so readily with these types of compounds, Grignard and organolithium compounds cannot be prepared from any organohalogen that also contains one of these functional groups. Nor can they be prepared from any organohalogen compound that also contains a nitro or carbonyl group because they also react with these groups.

R_2NH	$RC \equiv CH$	ROH	HOH	ArOH	RSH	RCOOH
pK_a 38–40	pK_a 25	pK_a 16–18	pK_a 15.7	pK_a 9–10	pK_a 8–9	pK_a 4–5
1° and 2° amines	Terminal alkynes	Alcohols	Water	Phenols	Thiols	Carboxylic acids

Example 15.1

Write an equation for the acid-base reaction between ethylmagnesium iodide and an alcohol. Use curved arrows to show the flow of electrons in this reaction. In addition, show by using appropriate pK_a values that this reaction is an example of a stronger acid and stronger base reacting to give a weaker acid and weaker base.

Solution

The alcohol is the stronger acid, and the partially negatively charged ethyl group is the stronger base.

$$CH_3CH_2-MgI \quad + \quad H-\ddot{O}R \quad \longrightarrow \quad CH_3CH_2-H \quad + \quad MgI^+ \: \ddot{\overset{-}{O}}R$$

| A Grignard reagent (stronger base) | An alcohol pK_a 16–18 (stronger acid) | An alkane pK_a 51 (weaker acid) | A magnesium alkoxide (weaker base) |

Problem 15.1

Explain how these Grignard reagents would react with molecules of their own kind to "self-destruct."

(a)   **(b)**

C. Reaction with Oxiranes

As we saw in Section 11.9, the oxirane ring is so strained that it undergoes ring-opening reactions with a variety of nucleophiles. We can now add Grignard and organolithium reagents to the list of reactive nucleophiles. Butylmagnesium bromide, for example, reacts with oxirane (ethylene oxide) to give a magnesium alkoxide, which, on treatment with aqueous acid, gives 1-hexanol.

Butylmagnesium bromide   **Ethylene oxide**   **A magnesium alkoxide**   **1-Hexanol**

As illustrated in this example, the product of treatment of a Grignard reagent with oxirane followed by protonation of the alkoxide is a primary alcohol with a carbon chain two carbons longer than the original chain. In reaction of a substituted oxirane, the major product corresponds to attack of the Grignard reagent on the less hindered carbon of the three-membered ring in an S_N2-like reaction. Treatment of racemic methyloxirane (propylene oxide) with phenylmagnesium bromide, for example, followed by workup in aqueous acid gives racemic 1-phenyl-2-propanol. The reaction does not work well if one or more of the oxirane carbons is quaternary.

Phenylmagnesium bromide   **Methyloxirane (Propylene oxide)**   **A magnesium alkoxide (racemic)**   **1-Phenyl-2-propanol (racemic)**

(racemic)

Example 15.2

Show how to prepare each alcohol from an organohalogen compound and an oxirane.

(a)

(b)

(racemic mixture
of two pairs of
enantiomers)

Solution

Shown is a retrosynthetic analysis for each compound followed by a synthesis.

(a)

(b)

(a racemic mixture of two
pairs of enantiomers)

Problem 15.2

Recalling the reactions of alcohols from Chapter 10, show how to synthesize each compound from an organohalogen compound and an oxirane, followed by a transformation of the resulting hydroxyl group to the desired oxygen-containing functional group.

(a) (b)

15.2 | Lithium Diorganocopper (Gilman) Reagents

A. Formation and Structure

An important use of organolithium reagents (Section 15.1) is in the preparation of diorganocopper reagents, often called Gilman reagents after Henry Gilman (1893–1986) of Iowa State University who was the first to develop their chemistry.

They are easily prepared by treatment of an alkyl, aryl, or alkenyllithium compound with copper(I) iodide, as illustrated by the preparation of lithium dibutylcopper from butyllithium.

$$2 \; CH_3CH_2CH_2CH_2Li + CuI \xrightarrow[\text{or THF}]{\text{diethyl ether}} (CH_3CH_2CH_2CH_2)_2CuLi + LiI$$

Butyllithium   **Copper(I)**   **Lithium dibutylcopper**
iodide   (a Gilman reagent)

Gilman reagents consist of two organic groups associated with a copper(I) ion giving a negatively charged species, which is the source of the carbon nucleophile. Lithium ion is associated with this negatively charged species as the counter ion.

B. Coupling with Organohalogen Compounds

Gilman reagents are especially valuable for the formation of new carbon-carbon bonds by a coupling reaction with an alkyl chloride, bromide, or iodide (alkyl fluorides are unreactive under these conditions) as illustrated by the following preparation of 2-methyl-1-dodecene. Notice that only one of the Gilman-reagent alkyl groups is transferred in the reaction. Because Gilman reagents are ultimately prepared from halides, this leads to effective coupling of two halides.

1-Iododecane

1. Li, pentane
2. CuI

**Lithium
didecylcopper**   **2-Bromo-
propene**   **2-Methyl-1-dodecene**

This example illustrates the coupling of Gilman reagent, a nucleophile, with a vinylic halide, an electrophile. Vinylic halides are normally quite unreactive toward nucleophilic displacement. Thus, the lithium diorganocopper reaction shown here is unique.

Gilman reagents giving the best yields of coupling products are those prepared from methyl, primary alkyl, allylic, vinylic, and aryl halides via the corresponding organolithium compounds. Yields are lower with secondary and tertiary haloalkanes.

Coupling with a vinylic halide is stereospecific; the configuration of the carbon-carbon double bond is preserved, as illustrated by the synthesis of *trans*-5-tridecene.

(E)-1-Iodo-1-nonene   **Lithium
dibutylcopper**   *(E)*-5-Tridecene

A variation on the preparation of Gilman reagents is to use a Grignard reagent in the presence of a catalytic amount of Cu(I). Zoecon Corporation has developed a synthesis of 150-kg batches of the housefly sex attractant muscalure by treating (*Z*)-1-bromo-9-octadecene with pentylmagnesium bromide in the presence of catalytic amounts of Cu(I). The starting bromoalkene is easily prepared from the readily available (*Z*)-9-octadecenoic acid (oleic acid, Section 26.1). Yields of muscalure are nearly quantitative.

(Z)-1-Bromo-9-octadecene $+ CH_3(CH_2)_4MgBr \xrightarrow[\text{THF}]{\text{CuI}}$ (Z)-9-Tricosene (Muscalure)

The mechanism of these coupling reactions is not fully understood and is the subject of active investigation.

Example 15.3

Show how to bring about each conversion using a lithium diorganocopper reagent.

(a) 1-Bromocyclohexene to 1-methylcyclohexene
(b) 1-Bromo-2-methylpropane to 2,5-dimethylhexane using the bromoalkane as the only source of carbon

Solution

(a) Treat 1-bromocyclohexene with lithium dimethylcopper.

1-Bromocyclohexene $+$ $(CH_3)_2CuLi$ $\longrightarrow$ 1-Methylcyclohexene $+ CH_3Cu + LiBr$

Lithium dimethylcopper

(b) Treat 1-bromo-2-methylpropane with lithium diisobutylcopper, itself prepared from 1-bromo-2-methylpropane.

1. Li, hexane
2. CuI

Lithium diisobutylcopper $+$ 1-Bromo-2-methyl-propane (Isobutyl bromide) $\longrightarrow$ 2,5-Dimethylhexane

Problem 15.3

Show how to bring about each conversion using a lithium diorganocopper reagent.

(a) $\longrightarrow$ (b) $\longrightarrow$

C. Reaction with Oxiranes

The reaction of epoxides with Gilman reagents is an important method for the formation of new carbon-carbon bonds. Like organolithium compounds and Grignard reagents, these compounds bring about regioselective ring opening of substituted epoxides at the less substituted carbon to give alcohols. Treatment of racemic styrene oxide with lithium divinylcopper, for example, followed by work-up in aqueous acid gives racemic 1-phenyl-3-buten-1-ol.

Styrene oxide
(racemic)

1. $(CH_2\!\!=\!\!CH)_2CuLi$
2. H_2O, HCl

1-Phenyl-3-buten-1-ol
(racemic)

Example 15.4

Show two combinations of epoxide and Gilman reagent that can be used to prepare racemic 1-phenyl-3-hexanol.

Solution

The carbon bearing the hydroxyl group must have been one of the carbon atoms of the epoxide ring. The second carbon of the epoxide was either the one to the right of the carbon now bearing the —OH or the one to the left of it. In these solutions, the phenyl group is written C_6H_5—. Either route would be satisfactory.

if the epoxide was
between these carbons

(racemic)

if the epoxide was
between these carbons

OH

C_6H_5

$(C_6H_5CH_2)_2CuLi$ +

C_6H_5

+ $(CH_3CH_2)_2CuLi$

(racemic)   (racemic)

Problem 15.4

Show how to prepare each Gilman reagent in Example 15.4 from an appropriate alkyl halide and each epoxide from an appropriate alkene.

15.3 | Carbenes and Carbenoids

Carbene
A neutral molecule that contains a carbon atom surrounded by only six valence electrons (R_2C:).

A **carbene**, R_2C:, is a neutral molecule in which a carbon atom is surrounded by only six valence electrons. Because they are electron-deficient, carbenes are highly reactive and behave as electrophiles. As we will see, one of their most important types of reactions is with alkenes (nucleophiles) to give cyclopropanes.

A. Methylene

Photolysis
Cleavage by light.

Thermolysis
Cleavage by heating.

The simplest carbene is methylene, CH_2, prepared by **photolysis** (cleavage by light) or **thermolysis** (cleavage on heating) of diazomethane, CH_2N_2, an explosive, toxic gas.

$$H_2C\!\!=\!\!N\!\!=\!\!\ddot{N}: \longleftrightarrow H_2\ddot{C}\!\!-\!\!N\!\!\equiv\!\!N: \xrightarrow{h\nu} H_2C: \;+\; :N\!\!\equiv\!\!N:$$

Methylene
(the simplest
carbene)

In the lowest electronic state of most carbenes, carbon is sp^2 hybridized with the unshared pair of electrons occupying the third sp^2 orbital. The unhybridized $2p$

---

orbital lies perpendicular to the plane created by the three sp^2 orbitals. Note that this orbital description of methylene is very much like that of a carbocation (Section 6.3A). In both species, carbon is sp^2 hybridized with a vacant $2p$ orbital. Methylene in this electronic state combines features of a carbocation and a carbanion in that it has both a vacant p orbital and a lone pair.

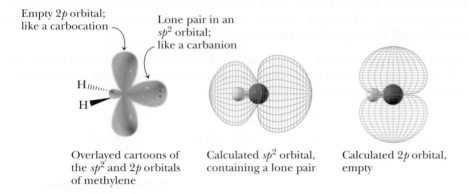

Empty $2p$ orbital; like a carbocation

Lone pair in an sp^2 orbital; like a carbanion

Overlayed cartoons of the sp^2 and $2p$ orbitals of methylene

Calculated sp^2 orbital, containing a lone pair

Calculated $2p$ orbital, empty

Methylene generated in this manner reacts with all C—H and C=C bonds and is so nonselective that it is of little synthetic use.

B. Dichlorocarbene

Although we often think of chlorine atoms as electron-withdrawing substituents, dichlorocarbene is much more stable and chemoselective than free methylene because resonance with the lone pairs on chlorine partially satisfies the electron deficiency on carbon.

$$:\ddot{Cl}—\ddot{C} \longleftrightarrow :\ddot{Cl}—\ddot{C}:^- \longleftrightarrow :\ddot{Cl}=\ddot{C}:^-$$

Dichlorocarbene can be prepared by treating chloroform with potassium *tert*-butoxide, removing the elements of HCl. The resulting carbene reacts cleanly with alkenes to give dichlorocyclopropanes. Addition of a dihalocarbene to an alkene shows syn stereoselectivity.

$$\text{CHCl}_3 \quad + \quad (\text{CH}_3)_3\text{CO}^-\text{K}^+ \quad \longrightarrow \quad \text{Cl}_2\text{C}: \quad + \quad (\text{CH}_3)_3\text{COH} + \text{K}^+\text{Cl}^-$$

Trichloromethane (Chloroform)	**Potassium *tert*-butoxide**	**Dichlorocarbene**	***tert*-Butyl alcohol**

$$\text{Cl}_2\text{C}: \quad +$$

Dichlorocarbene

A dichlorocyclopropane

Reaction of a *cis* alkene with a dihalocarbene gives only a *cis* dihalocyclopropane as illustrated by the reaction of *cis*-3-hexene with dichlorocarbene. Similarly, reaction of a *trans* alkene gives only a *trans* dihalocyclopropane.

$$+ \text{CHCl}_3 \xrightarrow{(\text{CH}_3)_3\text{CO}^-\text{K}^+}$$

(Z)-3-Hexene

(cis)-1,1-Dichloro-2,3-diethylcyclopropane

| Mechanism | Formation of Dichlorocarbene and Its Reaction with Cyclohexene |

Taken together, Steps 1 and 2 result in α-elimination of H and Cl; that is, both atoms are eliminated from the same carbon. We have seen many examples of β-elimination, where hydrogen and a leaving group are eliminated from neighboring carbons. There are very few examples of α-elimination, and they are possible only where no β-hydrogen exists.

Step 1: Take a proton away. Treatment of chloroform, which is somewhat acidic because of its three electron-withdrawing chlorine atoms, with potassium *tert*-butoxide gives the trichloromethide anion.

Trichloromethide anion

Step 2: Break a bond to give stable molecules or ions. Loss of Cl⁻ from CCl₃⁻ gives dichlorocarbene.

Dichlorocarbene

Step 3: Make a bond between a nucleophile (π bond) and an electrophile. Syn addition of dichlorocarbene to cyclohexene gives a dichlorocyclopropane. Compare this step to the electrophilic addition of bromine and chlorine to an alkene first discussed in Section 6.3D.

Dichlorocarbene	Cyclohexene	A dichlorocyclopropane
(an electrophile)	(a nucleophile)	

| Example 15.5 |

Predict the product from the following reaction.

$$CHBr_3 + (CH_3)_3CO^-K^+ \quad + \quad \text{(structure)} \quad \longrightarrow$$

(Z)-3-Methyl-2-pentene

Solution

Bromoform gives dibromocarbene, which reacts stereospecifically with the alkene to give a dibromocyclopropane.

(racemic)

Problem 15.3

Predict the product of the following reaction.

$$\text{(cyclohexylidene)}{=}CH_2 + CHBr_3 + (CH_3)_3CO^-K^+ \longrightarrow$$

C. The Simmons-Smith Reaction

Although methylene prepared from diazomethane itself is not synthetically useful, addition of methylene to an alkene can be accomplished using a reaction first reported by the American chemists Howard Simmons and Ronald Smith. The Simmons-Smith reaction uses diiodomethane and zinc dust activated by a small amount of copper (a so-called "zinc-copper couple") to produce iodomethylzinc iodide, in a reaction reminiscent of a Grignard reaction. Even though we show the **Simmons-Smith reagent** here as ICH_2ZnI, its structure is considerably more complex and not fully understood.

$$CH_2I_2 \quad + \quad Zn(Cu) \xrightarrow[\text{diethyl ether}]{} \quad ICH_2ZnI$$

Diiodomethane   **Zinc-copper couple**   **Iodomethylzinc iodide** (Simmons-Smith reagent)

This organozinc compound reacts with a wide variety of alkenes to give cyclopropanes.

Methylenecyclopentane   **Spiro[4.2]heptane**

(racemic)

Mechanism   The Simmons-Smith Reaction with an Alkene

Although an α-elimination from the Simmons-Smith reagent to give methylene would in principle be possible, the reagent is much more selective than free methylene. Instead, the organozinc compound reacts directly with the alkene by a concerted mechanism to give the cyclopropane-containing product. The Simmons-Smith reagent is an example of a **carbenoid**, a compound that delivers the elements of a carbene without actually producing a free carbene.

Carbenoid
A compound that delivers the elements of a carbene without actually producing a free carbene.

$$\longrightarrow ZnI_2 + H_2C\text{(cyclopropane)}$$

Example 15.6

Draw a structural formula for the product of treating each alkene with the Simmons-Smith reagent.

(a)   (b)

Solution

Reaction at each carbon-carbon double bond forms a cyclopropane ring.

(a)

(b)

(racemic)   (racemic)

Problem 15.6

Show how the following compound could be prepared from any compound containing ten carbons or fewer.

Summary

Section 15.1 Organomagnesium and Organolithium Compounds

- An **organometallic compound** is one that contains a carbon-metal bond.
  - The key feature of many of these reagents is that the carbon of the carbon-metal bond carries a partial negative charge.
  - The partial negative charge on carbon makes it basic and nucleophilic; the latter property can be exploited in organic synthesis in the construction of carbon-carbon bonds.
- Organomagnesium compounds are named **Grignard reagents** after their discoverer, Victor Grignard.
  - Grignard reagents are prepared by reacting alkyl, aryl, or alkenyl halides (chlorides, bromides, and iodides, not fluorides) with a slight excess of magnesium metal in an ether solvent.
  - The carbon-magnesium bond of Grignard reagents is polar covalent, with a partial negative charge on carbon, making it nucleophilc and basic.
- **Organolithium reagents** are prepared by reaction of an alkyl, aryl, or alkenyl halide with two equivalents of lithium metal.
  - The carbon-lithium bond in organolithium compounds is polar covalent, with a partial negative charge on carbon, making it nucleophilic and basic.
- Grignard reagents and organolithium compounds react as **carbon nucleophiles** with a wide range of electrophilic functional groups, including epoxides (and many carbonyl containing species discussed later in the book).
  - For unsymmetrical epoxides, reaction occurs at the less hindered carbon.
  - A new carbon-carbon bond is formed making these reactions very useful for synthesis.
  - The initial product formed is an alkoxide salt that is converted to an alcohol product following an acidic aqueous workup.
- Grignard and organolithium reagents are **very basic**, so they will deprotonate functional groups such as amines, terminal alkynes, alcohols, thiols, and carboxylic acids.

Problems: 15.1, 15.2, 15.11, 15.12, 15.17–15.19, 15.20, 15.23–15.25

Section 15.2 Lithium Diorganocopper (Gilman) Reagents

- Two equivalents of organolithium compounds also react with one equivalent of a copper Cu(I) salt, usually CuI, to give useful reagents called **Gilman reagents**.
  - Gilman reagents react with haloalkanes and haloalkenes to form new carbon-carbon bonds.
  - Gilman reagents also react with oxiranes to give alcohol products with carbon-carbon bond formation following an acidic aqueous workup.

Problems: 15.3, 15.4, 15.7–15.11, 15.19, 15.20, 15.21–15.24

Section 15.3 Carbenes and Carbenoids

- **Carbenes** are neutral molecules that contain a carbon with only six valence electrons, their organometallic-complexed equivalents are called **carbenoids**.
  - Carbenes are sp^2 hybridized, with an empty $2p$ orbital and a lone pair in one sp^2 hybrid orbital.
  - Carbenes are prepared from **photolysis** or **thermolysis** of diazo species such as diazomethane.
  - Dichlorocarbene is prepared from the reactions of chloroform in a strong base.
  - A useful carbenoid is the **Simmons-Smith reagent** prepared from diiodomethane and Zn(Cu).
  - Carbenes such as methylene are too reactive and not used for synthesis.
  - Dichlorocarbene and the Simmons-Smith reagent react stereospecifically with alkenes to give cyclopropanes.

Problems: 15.5, 15.6, 15.12–15.16, 15.23

🦉**WL** Questions and problems for this chapter are available in OWL. Use the OWL access card that came with this textbook to access assigned questions and problems for this chapter.

Key Reactions

1. **Formation of Organomagnesium (Grignard) and Organolithium Compounds (Section 15.1A)** Organomagnesium compounds are prepared by treating an alkyl, aryl, or alkenyl (vinylic) halide with magnesium in diethyl ether or THF. Organolithium compounds are prepared by treating an alkyl, aryl, or alkenyl halide with lithium in pentane or other hydrocarbon solvent.

2. **Reaction of RMgX and RLi with Proton Donors (Section 15.1B)** Both organomagnesium and organolithium compounds are strong bases and react with any proton donor stronger than the alkane from which the organolithium or magnesium compound is derived. Water or other proton donors must be completely excluded during their preparation and use.

$$CH_3CH_2MgBr + H_2O \longrightarrow CH_3CH_2{-}H + Mg(OH)Br$$

3. **Reaction of a Grignard Reagent with an Epoxide (Section 15.1C)** Treatment of a Grignard reagent with an epoxide followed by protonation of the magnesium alkoxide salt in aqueous acid gives an alcohol with its carbon chain extended by two carbon atoms.

4. Formation of Gilman Reagents (Section 15.2A) Lithium diorganocopper (Gilman) reagents are prepared by treating an organolithium compound with copper(I) iodide.

$$2 \text{ CH}_3\text{CH}_2\text{CH}_2\text{CH}_2\text{Li} + \text{CuI} \xrightarrow[\text{or THF}]{\text{ether}} (\text{CH}_3\text{CH}_2\text{CH}_2\text{CH}_2)_2\text{CuLi} + \text{LiI}$$

5. Treatment of a Gilman Reagent with an Alkyl, Aryl, or Alkenyl Halide (Section 15.2B) Coupling of a Gilman reagent with an alkyl, alkenyl, or aryl halide results in formation of a new carbon-carbon bond.

6. Reaction of Dichloro- or Dibromocarbene with an Alkene (Section 15.3B) The dihalocarbene is generated by treatment of CHCl_3 or CHBr_3 with a strong base such as potassium *tert*-butoxide. Addition of the dihalocarbene to an alkene shows syn stereospecificity.

(racemic)

7. The Simmons-Smith Reaction (Section 15.3C) Treatment of CH_2I_2 with a zinc-copper couple generates an organozinc compound, known as the Simmons-Smith reagent, which reacts with alkenes to give cyclopropanes.

(racemic)

Chapter 16

Aldehydes and Ketones

© Omikron/Photo Researchers, Inc.

Rod cells in the human eye. Inset: a model of 11-*cis*-retinal, an oxidized form of vitamin A. For the reaction of 11-*cis*-retinal with opsin to form visual purple, see Section 16.8A.

In this and several of the following chapters, we study the physical and chemical properties of compounds containing the carbonyl group, C=O. Because the carbonyl group is the functional group of aldehydes, ketones, carboxylic acids, and carboxyl derivatives, it is one of the most important functional groups in organic chemistry. The chemical properties of this functional group are straightforward, and an understanding of its few characteristic reaction themes leads very quickly to an understanding of a wide variety of organic reactions.

16.1 | Structure and Bonding

The functional group of an **aldehyde** is a carbonyl group bonded to a hydrogen atom and a carbon atom (Section 1.3C). In methanal (always called formaldehyde), the simplest aldehyde, the carbonyl group is bonded to two hydrogen atoms. In other aldehydes, it is bonded to one hydrogen atom and one carbon atom. Following are Lewis structures for formaldehyde and ethanal (which is always called acetaldehyde).

Methanal	Ethanal	Propanone
(Formaldehyde)	(Acetaldehyde)	(Acetone)

The functional group of a **ketone** is a carbonyl group bonded to two carbon atoms (Section 1.3C). The simplest ketone is propanone, which is always called acetone.

⊙WL **Online homework for this chapter may be assigned in OWL for Organic Chemistry.**

According to valence bond theory, the carbon-oxygen double bond consists of one σ bond formed by the overlap of sp^2 hybrid orbitals of carbon and oxygen and one π bond formed by the overlap of parallel $2p$ orbitals. The two nonbonding pairs of electrons on oxygen lie in the remaining sp^2 hybrid orbitals (Figure 1.25).

16.2 | Nomenclature

A. IUPAC Nomenclature

IUPAC names for aldehydes and ketones follow the familiar pattern of selecting as the parent alkane the longest chain of carbon atoms that contains the functional group. We show the aldehyde group by changing the suffix -e of the parent alkane to -al (Section 2.3). Because the carbonyl group of an aldehyde can only appear at the end of a parent chain and because numbering must start with it as carbon-1, its position is unambiguous; there is no need to use a number to locate it.

For unsaturated aldehydes, the presence of a carbon-carbon double or triple bond is indicated by the infix -en- or -yn-. As with other molecules with both an infix and a suffix, the location of the group corresponding to the suffix determines the numbering pattern.

3-Methylbutanal   **2-Propenal**   **(2E)-3,7-Dimethyl-2,6-octadienal**
                      (Acrolein)   **(Geranial)**

For cyclic molecules in which —CHO is bonded directly to the ring, the molecule is named by adding the suffix -*carbaldehyde* to the name of the ring. The atom of the ring to which the aldehyde group is bonded is numbered 1.

Cyclopentane-   ***trans*-4-Hydroxycyclo-**
carbaldehyde   **hexanecarbaldehyde**

Among the aldehydes for which the IUPAC system retains common names are benzaldehyde and cinnamaldehyde, as well as formaldehyde and acetaldehyde. Note here the alternative ways the phenyl group can be written. In benzaldehyde, it is written as a line-angle formula; in cinnamaldehyde, it is written C_6H_5—.

Benzaldehyde   ***trans*-3-Phenyl-2-propenal**
                    (Cinnamaldehyde)

In the IUPAC system, ketones are named by selecting as the parent alkane the longest chain that contains the carbonyl group and then indicating its presence by changing the suffix from -e to -one (Section 2.3). The parent chain is numbered from the direction that gives the carbonyl carbon the smaller number. The IUPAC system retains the common names acetone, acetophenone, and benzophenone.

Propanone   **Acetophenone**   **Benzophenone**   **1-Phenyl-1-pentanone**
(Acetone)

Benzaldehyde is found in the kernels of bitter almonds. Cinnamaldehyde is found in Ceylon and Chinese cinnamon oils.

© Cengage Learning/Charles D. Winters

Example 16.1

Write IUPAC names for each compound. Specify the configuration of all chiral centers in (a) and (c).

(a)   (b)   (c)

Solution

(a) The parent chain is pentane. The name is (*S*)-3-ethyl-2-hydroxypentanal.
(b) Number the six-membered ring beginning with the carbonyl carbon. The IUPAC name is 3-methyl-2-cyclohexenone.
(c) The name (2*S*,5*R*)-2-isopropyl-5-methylcyclohexanone provides a complete description of the configuration of each chiral center as well as the *trans* relationship between the isopropyl and methyl groups. The common name of this compound is menthone.

Problem 16.1

Write the IUPAC name for each compound. Specify the configuration in (c).

(a)   (b)   (c)

Example 16.2

Write structural formulas for all ketones with the molecular formula $C_6H_{12}O$, and give each its IUPAC name. Which of these ketones are chiral?

Solution

Following are line-angle formulas and IUPAC names for the six ketones with this molecular formula. Only 3-methyl-2-pentanone is chiral; the *R* enantiomer is drawn here.

2-Hexanone   3-Hexanone   (*R*)-3-Methyl-2-pentanone

4-Methyl-2-pentanone   2-Methyl-3-pentanone   3,3-Dimethyl-2-butanone

Problem 16.2

Write structural formulas for all aldehydes with the molecular formula $C_6H_{12}O$, and give each its IUPAC name. Which of these aldehydes are chiral?

B. IUPAC Names for More Complex Aldehydes and Ketones

Order of precedence of functions A ranking of functional groups in order of priority for the purposes of IUPAC nomenclature.

In naming compounds that contain more than one functional group that might be indicated by a suffix, the IUPAC system has established an **order of precedence of functions**. Table 16.1 gives the order of precedence for the functional groups we have studied so far.

Table 16.1	Increasing Order of Precedence of Six Functional Groups			
Functional Group	**Suffix if Higher Priority**	**Prefix if Lower Priority**	**Example When the Functional Group Has Lower Priority**	
Carboxyl	-oic acid	—		
Aldehyde	-al	oxo-	3-Oxopropanoic acid	
Ketone	-one	oxo-	3-Oxobutanoic acid	
Alcohol	-ol	hydroxy-	4-Hydroxybutanoic acid	
Amino	-amine	amino-	3-Aminobutanoic acid	
Sulfhydryl	-thiol	mercapto	2-Mercaptoethanol	

Increasing precedence ↑

Example 16.3

Write the IUPAC name for each compound, being certain to specify configuration where appropriate.

(a) (b) (c)

Solution

(a) 3-Oxobutanal. An aldehyde has higher precedence than a ketone. The presence of the carbonyl group of the ketone is indicated by the prefix *oxo-* (Table 16.1).
(b) (*R*)-2,3-Dihydroxypropanal. Its common name is glyceraldehyde. Glyceraldehyde is the simplest carbohydrate (Section 25.1).
(c) (*R*)-6-Hydroxy-2-heptanone.

Problem 16.3

Write the IUPAC name for each compound.

(a) (b) (c)

C. Common Names

The common name for an aldehyde is derived from the common name of the corresponding carboxylic acid by dropping the word *acid* and changing the suffix *-ic* or *-oic* to *-aldehyde*. Because we have not yet studied common names for carboxylic acids, we are not in a position to discuss common names for aldehydes. We can illustrate how they are derived, however, by reference to a few common names with which you are familiar. The name formaldehyde is derived from formic acid; the name acetaldehyde is derived from acetic acid.

$$\underset{\text{Formaldehyde}}{\overset{\displaystyle O}{\underset{\displaystyle \|}{HCH}}} \qquad \underset{\text{Formic acid}}{\overset{\displaystyle O}{\underset{\displaystyle \|}{HCOH}}} \qquad \underset{\text{Acetaldehyde}}{\overset{\displaystyle O}{\underset{\displaystyle \|}{CH_3CH}}} \qquad \underset{\text{Acetic acid}}{\overset{\displaystyle O}{\underset{\displaystyle \|}{CH_3COH}}}$$

Common names for ketones are derived by naming the two alkyl or aryl groups bonded to the carbonyl group as separate words, followed by the word *ketone*.

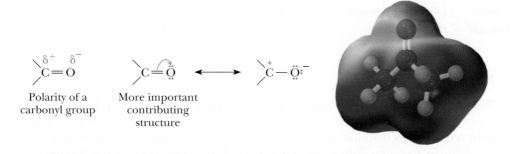

Ethyl isopropyl ketone   **Diethyl ketone**   **Dicyclohexyl ketone**

16.3 | Physical Properties

Oxygen is more electronegative than carbon (3.5 compared with 2.5); therefore, a carbon-oxygen double bond is polar, with oxygen bearing a partial negative charge and carbon bearing a partial positive charge. In addition, the resonance structure shown on the right emphasizes the reactivity of the carbonyl oxygen as a Lewis base and the carbonyl carbon as a Lewis acid. The bond dipole moment of a carbonyl group is 2.3 D (Table 1.7).

An electrostatic potential map for acetone. Note the large negative charge on oxygen (red) and the positive charges (blue) on the three carbons.

Because of the polarity of the carbonyl group, aldehydes and ketones are polar compounds and interact in the pure liquid by dipole-dipole interactions; they have higher boiling points than nonpolar compounds of comparable molecular weight (Table 16.2).

Table 16.2	Boiling Points of Six Compounds of Comparable Molecular Weight		
Name	**Structural Formula**	**Molecular Weight (g/mol)**	**Boiling Point (°C)**
Diethyl ether	$CH_3CH_2OCH_2CH_3$	74	34
Pentane	$CH_3CH_2CH_2CH_2CH_3$	72	36
Butanal	$CH_3CH_2CH_2CHO$	72	76
2-Butanone	$CH_3CH_2COCH_3$	72	80
1-Butanol	$CH_3CH_2CH_2CH_2OH$	74	117
Propanoic acid	CH_3CH_2COOH	74	141

Of the compounds listed in Table 16.2, pentane and diethyl ether have the lowest boiling points. Diethyl ether is a polar molecule, but because of steric hindrance, only weak dipole-dipole interactions exist between its molecules (Section 11.3). Both butanal and 2-butanone are polar molecules, and, because of the intermolecular attraction between their carbonyl groups, their boiling points are higher than those of pentane and diethyl ether. Alcohols (Section 10.2) and carboxylic acids (Section 17.3) are polar molecules that can associate by hydrogen bonding, the strongest intermolecular interaction. The boiling points of 1-butanol and propanoic acid are significantly higher than those of butanal and 2-butanone, compounds whose molecules cannot associate by hydrogen bonding.

The oxygen atoms of the carbonyl groups of aldehydes and ketones act as hydrogen bond acceptors with water; therefore, low-molecular-weight aldehydes and ketones are more soluble in water than are nonpolar compounds of comparable molecular weight. Listed in Table 16.3 are boiling points and solubilities in water for several low-molecular-weight aldehydes and ketones.

Table 16.3	Physical Properties of Selected Aldehydes and Ketones			
IUPAC Name	Common Name	Structural Formula	Boiling Point (°C)	Solubility (g/100 g water)
Methanal	Formaldehyde	$HCHO$	−21	Infinite
Ethanal	Acetaldehyde	CH_3CHO	20	Infinite
Propanal	Propionaldehyde	CH_3CH_2CHO	49	16
Butanal	Butyraldehyde	$CH_3CH_2CH_2CHO$	76	7
Hexanal	Caproaldehyde	$CH_3(CH_2)_4CHO$	129	Slight
Propanone	Acetone	CH_3COCH_3	56	Infinite
2-Butanone	Ethyl methyl ketone	$CH_3COCH_2CH_3$	80	26
3-Pentanone	Diethyl ketone	$CH_3CH_2COCH_2CH_3$	101	5

16.4 | Reactions

One of the most common reaction themes of a carbonyl group is addition of a nucleophile to form a tetrahedral carbonyl addition compound. The carbonyl carbon atom is highly electrophilic due to a significant partial positive charge as well as to the ability to accommodate a new bond through the conversion of the π bond to a lone pair on oxygen of the tetrahedral carbonyl addition compound. These reactions are often referred to as **nucleophilic acyl additions**. In the following general reaction, the nucleophilic reagent is written as Nu:$^-$ to emphasize the presence of its unshared pair of electrons. Notice how the π bond of the carbonyl group breaks as the nucleophile attacks, changing the carbon atom hybridization state, all the while maintaining four bonds to carbon in the tetrahedral carbonyl addition compound.

Nucleophilic acyl addition
A characteristic reaction mechanism of carbonyl-containing compounds such as aldehydes and ketones in which a nucleophile makes a new bond to the electrophilic carbonyl carbon atom.

Tetrahedral carbonyl addition compound

A second common reaction theme of a carbonyl group is reaction with a proton or other Lewis acid to form a resonance-stabilized cation. Protonation increases the electron deficiency of the carbonyl carbon and makes it more reactive toward nucleophiles. This cation then reacts with nucleophiles, followed by removal of a proton, to give a tetrahedral carbonyl addition compound.

Tetrahedral carbonyl
addition compound

Both of these reactivities can be predicted based on the large bond dipole of the carbonyl group. Using acetaldehyde as an example, Lewis acids such as protons are attracted to the partial negative charge (red color of the electrostatic potential map) of the carbonyl oxygen, while electron-rich nucleophiles are attracted to the partial positive charge (blue color of the electrostatic potential map) of the carbonyl carbon.

Site of interaction with
Lewis acids and electrophiles

$:O:$

H_3C H

Site of interaction
with nucleophiles

Carbonyl compounds such as aldehydes and ketones undergo a wide variety of important reactions with most involving nucleophilic acyl addition. The mechanisms for these reactions are similar because almost all of the steps can be described as one of the four mechanistic elements first mentioned in Section 6.2F. When choosing among the four to begin each mechanism, look to see whether or not there is acid or strong base present. If there is a nucleophile, but no acid or base present, the reaction mechanism will begin with the nucleophlie attacking the electrophilic carbonyl carbon atom (make a new bond between a nucleophile and an electrophile). The π bond of the carbonyl breaks as the nucleophile adds. If there is acid present, the proton will add to the carbonyl oxygen atom (add a proton), a process that makes the carbonyl group even more electrophilic. In these cases, the next step will be attack of a nucleophile to the carbonyl carbon atom. In the presence of strong base, carbonyl compounds undergo a third characteristic reaction (Take a proton away) to form a very important species called an enolate anion that is the focus of Chapter 19.

Often, the tetrahedral product of nucleophile addition to a carbonyl is a new chiral center. If none of the starting materials are chiral, then the two enantiomers will be created in equal amounts because the nucleophile will approach the carbonyl from either side with equal probability. As a result, the carbonyl addition product will consist of a racemic mixture.

Approach from the top face

Approach from the bottom face

A new chiral center is created

A racemic mixture

16.5 | Addition of Carbon Nucleophiles

In this section, we examine nucleophilic acyl addition reactions of aldehydes and ketones with the following types of carbon nucleophiles.

RMgX	RLi	$RC\equiv C^-$	$^-C\equiv N$
A Grignard reagent	An organolithium reagent	An anion of a terminal alkyne	Cyanide ion

From the perspective of the organic chemist, addition of a carbon nucleophile is the most important type of carbonyl addition reaction because a new carbon-carbon bond is formed in the process. Each of these reactions follows the same mechanistic two-step pattern of **making a bond between the carbon nucleophile and the electrophilic carbonyl carbon atom** to give the tetrahedral carbonyl addition compound, followed by **adding a proton** to give an —OH group in the product.

A. Addition of Grignard Reagents

The special value of Grignard reagents (Section 15.1) is that they provide excellent ways to form new carbon-carbon bonds. Given the difference in electronegativity between carbon and magnesium ($2.5 - 1.2 = 1.3$), the carbon-magnesium bond of a Grignard reagent is polar covalent with carbon bearing a partial negative charge and magnesium bearing a partial positive charge. A Grignard reagent is therefore a good nucleophile and adds to the carbonyl group of an aldehyde or ketone to form a tetrahedral carbonyl addition compound. The reaction is assisted by the attraction between the partial negative charge on the carbon of the organometallic compound and the partial positive charge on the carbonyl carbon. The alkoxide ions formed in these reactions are strong bases (Section 10.4) and, when treated with an aqueous acid such as HCl or aqueous NH_4Cl during work-up, form alcohols. In the following examples, the magnesium oxygen bond is written $—O^-[MgBr]^+$ to emphasize its ionic character.

Caution: new chiral centers are often created in Grignard reactions with aldehydes or ketones. When neither the aldehyde/ketone nor the Grignard reagent is chiral but the product has a new chiral center, a racemic mixture is formed.

Addition to Formaldehyde Gives a Primary Alcohol

Recall that a Grignard reagent can be considered a carbon nucleophile because there is significant partial negative charge on the carbon attached to the Mg atom. Treatment of a Grignard reagent with formaldehyde followed by protonation in aqueous acid gives a 1° alcohol.

Formaldehyde   A magnesium alkoxide   **1-Propanol** (a 1° alcohol)

Mechanism — Grignard Reagent Reacting with Formaldehyde

Step 1: Make a new bond between a nucleophile and an electrophile. Nucleophilic addition of a Grignard reagent to the electrophilic carbonyl carbon atom gives a tetrahedral carbonyl addition compound.

$$CH_3CH_2\overset{\delta-}{-}\overset{\delta+}{MgBr} \;+\; H\overset{\overset{\overset{\delta-}{\ddot{O}}}{\|}}{\underset{}{C}}{}^{\delta+}H \;\xrightarrow{\text{ether}}\; CH_3CH_2-CH_2 \;\; :\!\ddot{O}\!:^-[MgBr]^+$$

Step 2: Add a proton. In a second step, the chemist adds a dilute acid solution to protonate the alkoxide function of the tetrahedral carbonyl addition compound to give the primary alcohol product.

$$CH_3CH_2-CH_2\;\;:\!\ddot{O}\!:^-[MgBr]^+ \;+\; H-\overset{+}{\underset{H}{\ddot{O}}}-H \;\longrightarrow\; CH_3CH_2-CH_2\;\;\ddot{O}H \;+\; H-\ddot{O}-H$$

It is important to remember that this second step requires the chemist to add the acid after the tetrahedral carbonyl addition compound forms. If the acid were added with the Grignard reagent in a single step, the acid would immediately protonate the Grignard reagent before any further reaction could take place.

Addition to an Aldehyde (Except Formaldehyde) Gives a Secondary Alcohol

Treatment of a Grignard reagent with any other aldehyde followed by protonation in aqueous acid gives a 2° alcohol.

Acetaldehyde
(an aldehyde)

A magnesium
alkoxide

1-Cyclohexylethanol
(a 2° alcohol; racemic)

In this example, the product is chiral and is formed as a racemic mixture.

Addition to a Ketone Gives a Tertiary Alcohol

Treatment of a Grignard reagent with a ketone followed by protonation in aqueous acid gives a 3° alcohol.

**Phenyl-
magnesium
bromide**

Acetone
(a ketone)

A magnesium
alkoxide

2-Phenyl-2-propanol
(a 3° alcohol)

Example 16.4

Racemic 2-phenyl-2-butanol can be synthesized by three different combinations of a Grignard reagent and a ketone. Show each combination.

Solution

In each solution, curved arrows show formation of the new carbon-carbon bond and the alkoxide ion. The new carbon-carbon bond formed by each set of reagents is labeled in the final product.

(a) Ph—MgBr + $\overset{\overset{\displaystyle O:}{\parallel}}{\underset{\underset{\displaystyle CH_3}{|}}{C}}CH_2CH_3$

(b) Ph—$\overset{\overset{\displaystyle O:}{\parallel}}{\underset{\underset{\displaystyle CH_3}{|}}{C}}$ + $\overset{\displaystyle MgBr}{CH_2CH_3}$ ⟶ Ph—$\overset{\overset{\displaystyle :O:^-[MgBr]^+}{}}{\underset{\underset{\displaystyle CH_3}{|}}{C}}$—CH_2CH_3 $\xrightarrow[H_2O]{HCl}$ Ph—$\overset{\overset{\displaystyle :OH}{}}{\underset{\underset{\displaystyle CH_3 (c)}{|}}{C}}$—CH_2CH_3

(a) :ÖH (b)

(c) Ph—$\overset{\overset{\displaystyle O:}{\parallel}}{C}CH_2CH_3$ + $\overset{\displaystyle MgBr}{CH_3}$

2-Phenyl-2-butanol
(produced as a
racemic mixture)

Problem 16.4

Show how these four products can be synthesized from the same Grignard reagent.

(a)
(racemic)

(b)
(racemic)

(c)
(racemic)

(d)
(racemic)

B. Addition of Organolithium Compounds

Organolithium compounds have greater negative charge character on carbon, so they are generally more reactive in nucleophilic acyl addition reactions than organomagnesium compounds and typically give higher yields of products. They are more troublesome to use, however, because they must be prepared and used under an atmosphere of nitrogen or other inert gas. The following synthesis illustrates the use of an organolithium compound to form a sterically hindered tertiary alcohol.

Phenyllithium　　**3,3-Dimethyl-2-butanone**　　A lithium alkoxide
(racemic)　　**3,3-Dimethyl-2-phenyl-2-butanol**
(racemic)

Mechanism Organolithium Reagent Reacting with a Ketone

***Step 1:* Make a new bond between a nucleophile and an electrophile.** Nucleophilic addition of an organolithium reagent to the electrophilic carbonyl carbon atom gives a tetrahedral carbonyl addition compound.

Step 2: Add a proton. In a second step, the chemist adds a dilute acid solution to protonate the alkoxide function of the tetrahedral carbonyl addition compound to give the tertiary alcohol product.

Once again, it is important to remember that this second step requires the chemist to add the acid after the tetrahedral carbonyl addition compound forms. If the acid were added with the organolithium reagent in a single step, the acid would immediately protonate the organolithium reagent before any further reaction could take place.

C. Addition of Anions of Terminal Alkynes

The anion of a terminal alkyne is a nucleophile (Section 7.5) and adds to the carbonyl group of an aldehyde or ketone to form a tetrahedral carbonyl addition compound. In the following example, addition of sodium acetylide to cyclohexanone followed by hydrolysis in aqueous acid gives 1-ethynylcyclohexanol.

| **Sodium acetylide** | **Cyclohexanone** | A sodium alkoxide | **1-Ethynylcyclohexanol** |

Mechanism Alkyne Anion Reacting with a Ketone

Step 1: Make a new bond between a nucleophile and an electrophile. Nucleophilic addition of an alkyne anion to the electrophilic carbonyl carbon atom gives a tetrahedral carbonyl addition compound.

Step 2: Add a proton. In a second step, the chemist adds a dilute acid solution to protonate the alkoxide function of the tetrahedral carbonyl addition compound to give the tertiary alcohol product.

Once again, it is important to remember that this second step requires the chemist to add the acid after the tetrahedral carbonyl addition compound forms. If the acid were added with the alkyne anion in a single step, the acid would immediately protonate the alkyne anion before any further reaction could take place.

These addition compounds (alkynyl alcohols) contain both a hydroxyl group and a carbon-carbon triple bond, each of which can be further modified.

Acid-catalyzed hydration (Section 7.7B) of 1-ethynylcyclohexanol gives an α-hydroxyketone. Alternatively, hydroboration followed by oxidation with alkaline hydrogen peroxide (Section 7.7A) gives a β-hydroxyaldehyde.

This example illustrates two of the most valuable reactions of alkynes: (1) addition of the anion of a terminal alkyne to the carbonyl group of an aldehyde or ketone gives an alkynyl alcohol and (2) hydration of a terminal alkyne gives either an aldehyde or ketone depending on the alkyne and the method of hydration.

D. Addition of Hydrogen Cyanide

Hydrogen cyanide, HCN, adds to the carbonyl group of an aldehyde or ketone to form a tetrahedral carbonyl addition compound called a **cyanohydrin**. For example, HCN adds to acetaldehyde to form acetaldehyde cyanohydrin in 75% yield. We study the naming of compounds containing the nitrile group in Section 18.1E.

Cyanohydrin
A molecule containing an —OH group and a —CN group bonded to the same carbon.

2-Hydroxypropanenitrile
(Acetaldehyde cyanohydrin;
produced as a racemic mixture)

Addition of hydrogen cyanide proceeds by way of cyanide ion. Because HCN is a weak acid, pK_a 9.31, the concentration of cyanide ion in aqueous HCN is too low for cyanohydrin formation to proceed at a reasonable rate. For this reason, cyanohydrin formation is generally carried out by dissolving NaCN or KCN in water and adjusting the pH of the solution to approximately 10.0, giving a solution in which HCN and CN^- are present in comparable concentrations.

Mechanism Formation of a Cyanohydrin

Step 1: **Make a new bond between a nucleophile and an electrophile.** Nucleophilic addition of cyanide ion to the carbonyl carbon gives a tetrahedral carbonyl addition compound.

Step 2: **Add a proton.** Proton transfer from HCN gives the cyanohydrin and generates a new cyanide ion.

For aldehydes and most aliphatic ketones, the position of equilibrium favors cyanohydrin formation. For many aryl ketones (ketones in which the carbonyl carbon is bonded to a benzene ring) and sterically hindered aliphatic ketones, however, the position of equilibrium favors starting materials; cyanohydrin formation is not a useful reaction for these types of compounds. The following synthesis of ibuprofen, for example, failed because the cyanohydrin was formed only in low yield.

4-Isobutylacetophenone   A cyanohydrin   **Ibuprofen**
                                                                    (racemic)

Benzaldehyde cyanohydrin (mandelonitrile) provides an interesting example of a chemical defense mechanism in the biological world. This substance is synthesized by millipedes (*Apheloria corrigata*) and stored in special glands. When a millipede is threatened, the cyanohydrin is released from the storage gland and undergoes enzyme-catalyzed reversal of cyanohydrin formation to produce HCN, which is then released to ward off predators. The quantity of HCN emitted by a single millipede is sufficient to kill a small mouse. Mandelonitrile is also found in bitter almond and peach pits. Its function there is unknown, as is how millipedes survive exposure to hydrogen cyanide.

Benzaldehyde cyanohydrin   **Benzaldehyde**
     (Mandelonitrile)

The value of cyanohydrins as synthetic intermediates lies in the new functional groups into which they can be converted. First, the secondary or tertiary hydroxyl group of the cyanohydrin may undergo acid-catalyzed dehydration to form an unsaturated nitrile. For example, acid-catalyzed dehydration of acetaldehyde cyanohydrin gives acrylonitrile, the monomer from which polyacrylonitrile (Orlon, Table 29.1) is made.

$$CH_3\overset{\overset{\displaystyle OH}{|}}{C}HC\equiv N \xrightarrow[\text{catalyst}]{\text{acid}} CH_2=CHC\equiv N + H_2O$$

2-Hydroxypropanenitrile　　　　　　**Propenenitrile**
(Acetaldehyde cyanohydrin)　　　　　**(Acrylonitrile)**

Second, a nitrile is reduced to a primary amine by hydrogen in the presence of nickel or other transition metal catalyst (Section 18.10C). Catalytic reduction of benzaldehyde cyanohydrin, for example, gives 2-amino-1-phenylethanol.

$$\text{C}_6\text{H}_5\overset{\overset{\displaystyle OH}{|}}{C}HC\equiv N + 2 H_2 \xrightarrow{\text{Ni}} \text{C}_6\text{H}_5\overset{\overset{\displaystyle OH}{|}}{C}HCH_2NH_2$$

Benzaldehyde cyanohydrin　　　　　　**2-Amino-1-phenylethanol**

As we shall see in Section 18.4E, hydrolysis of a nitrile in the presence of an acid catalyst gives a carboxylic acid. Thus, even though nitriles are little used themselves, they are valuable intermediates for the synthesis of other functional groups.

16.6 | The Wittig Reaction

Ylide
A neutral molecule with positive and negative charges on adjacent atoms.

In 1954, Georg Wittig reported a method for the synthesis of alkenes from aldehydes and ketones using compounds called phosphonium **ylides**. For his pioneering study and development of this reaction into a major synthetic tool, Professor Wittig shared the 1979 Nobel Prize for chemistry. (The other recipient was Herbert C. Brown for his studies of hydroboration and the chemistry of organoboron compounds.) A Wittig synthesis is illustrated by the conversion of cyclohexanone to methylenecyclohexane. In this reaction, a C=O double bond is converted to a C=C double bond. A noteworthy aspect of this reaction is that a strong thermodynamic driving force is provided by formation of the very strong P—O bonding interaction in the phosphine oxide product.

$$\text{(cyclohexanone)}=O + Ph_3\overset{+}{P}-\overset{-}{C}H_2 \longrightarrow \text{(cyclohexane)}=CH_2 + Ph_3P=O$$

Cyclohexanone　　**A phosphonium**　　**Methylenecyclohexane**　　**Triphenylphosphine**
　　　　　　　　　　　ylide　　　　　　　　　　　　　　　　　　　　**oxide**

We study the **Wittig reaction** in two stages: first, the formation and structure of phosphonium ylides and, second, the reaction of a phosphonium ylide with the carbonyl group of an aldehyde or ketone to give an alkene. The Wittig reaction is especially valuable as a synthetic tool because it takes place under mild conditions and because the location of the carbon-carbon bond is unambiguously determined. The only disadvantage of the Wittig reaction is that it is subject to steric hindrance. Yields are generally highest with aldehydes that have the least hindered carbonyl group, and are lower with ketones in which the carbonyl group is more hindered.

　　　Phosphorus is the second element in Group 5A of the Periodic Table and, like nitrogen, has five electrons in its valence shell. Examples of trivalent phosphorus compounds are phosphine, PH_3, and triphenylphosphine, Ph_3P. Phosphine is a highly toxic, flammable gas. Triphenylphosphine is a colorless, odorless solid. Because phosphorus is below nitrogen in the Periodic Table, phosphines are weaker bases than amines and also good nucleophiles (Section 9.4E). Treatment of a phosphine with a primary or secondary alkyl halide gives a phosphonium salt by an S_N2 pathway.

$$Ph_3P: \quad + \quad CH_3 \overset{\frown}{-}\overset{\cdots}{I}: \quad \overset{S_N2}{\longrightarrow} \quad Ph_3\overset{+}{P}-CH_3 \quad \overset{:\overset{\cdots}{I}:^-}{}$$

Triphenylphosphine   **Methyltriphenylphosphonium iodide**
(an alkyltriphenylphosphonium salt)

Because phosphines are also weak bases, treatment of a tertiary halide with a phosphine gives largely an alkene by an E2 pathway.

α-Hydrogen atoms on the alkyl group of an alkyltriphenylphosphonium ion are weakly acidic and can be removed by reaction with a very strong base, typically butyllithium (BuLi), sodium hydride (NaH), or sodium amide ($NaNH_2$).

$$CH_3CH_2CH_2CH_2\overset{-}{:}Li^+ \; + \; \overset{\frown}{H}-CH_2-\overset{+}{P}Ph_3\; I^- \; \longrightarrow \; CH_3CH_2CH_2CH_3 \; + \; \overset{-}{:}CH_2-\overset{+}{P}Ph_3 \; + \; LiI$$

Butyllithium   An alkyltriphenyl-   **Butane**   A phosphonium
phosphonium iodide   ylide

The product of removal of a proton from an alkyltriphenylphosphonium ion is a phosphonium ylide. The important feature of a phosphonium ylide is that the deprotonated carbon atom bears considerable partial negative charge, making it a strong carbon-based nucleophile, analogous to species such as Grignard and organolithium reagents. Just like Grignard and organolithium reagents, the deprotonated carbon of phosphonium ylides readily reacts with the electrophilic carbon atom of aldehyde and ketone carbonyl groups.

Mechanism   The Wittig Reaction

Step 1: **Make a new bond between a nucleophile and an electrophile.**   Reaction of a nucleophilic phosphonium ylide with the electrophilic carbonyl carbon of an aldehyde or ketone gives a dipolar intermediate called a **betaine**.

> **Betaine**
> A neutral molecule with nonadjacent positive and negative charges. An example of a betaine is the intermediate formed by addition of a Wittig reagent to an aldehyde or ketone.

Step 2: The betaine collapses to a four-membered oxaphosphetane ring. The name for this four-membered ring system is derived by combining the following: *oxa-* shows that it contains oxygen, *-phosph-* shows that it contains phosphorus, *-et-* shows that it is a four-membered ring, and *-ane* shows only carbon-carbon single bonds in the ring. Oxaphosphetanes can be isolated at low temperature.

An oxaphosphetane

Step 3: **Break bonds to give stable molecules or ions.**   Decomposition of the oxaphosphetane gives triphenylphosphine oxide and an alkene.

$$\overset{:\overset{\cdots}{O}-CR_2}{\underset{Ph_3P-CH_2}{|\quad\quad|}} \longrightarrow Ph_3P=\overset{\cdots}{O} \; + \; R_2C=CH_2$$
Triphenylphosphine   An alkene
oxide

The driving force for a Wittig reaction is the formation of the very strong phosphorus-oxygen bond in triphenylphosphine oxide.

The Wittig reaction is effective with a wide variety of aldehydes and ketones and with ylides derived from a wide variety of primary, secondary, and allylic halides as shown by the following examples.

$$
\diagdown\!\!=\!\!O + Ph_3\overset{+}{P}\diagup\!\!\diagdown\!\!\diagup\!\!\diagdown \longrightarrow \diagdown\!\!=\!\!\diagup\!\!\diagdown\!\!\diagup\!\!\diagdown + Ph_3P\!=\!O
$$

Acetone   **2-Methyl-2-heptene**

$$
Ph\diagup\!\!\overset{O}{\diagdown}\!\!H + Ph_3\overset{+}{P}\diagup\!\!\diagdown \longrightarrow Ph\diagup\!\!\diagdown\!\!=\!\!\diagup + Ph\diagup\!\!\diagup\!\!\diagdown + Ph_3P\!=\!O
$$

Phenylacetaldehyde   **(Z)-1-Phenyl-**   **(E)-1-Phenyl-**
                              2-butene   **2-butene**
                               (87%)   (13%)

$$
Ph\diagup\!\!\overset{H}{\underset{O}{\diagdown}} + Ph_3\overset{+}{P}\diagdown\!\!\overset{O}{\diagup}\!\!OEt \longrightarrow Ph\diagup\!\!\diagdown\!\!\diagup\!\!\overset{O}{\diagdown}\!\!OEt + Ph_3P\!=\!O
$$

Phenylacetaldehyde   **Ethyl (E)-4-phenyl-2-butenoate**
                               (only the *E* isomer is formed)

As illustrated by the second and third examples, some Wittig reactions are *Z* selective, while others are *E* selective. As a general rule, those Wittig reagents with anion-stabilizing substituents, such as a carbonyl group, adjacent to the negative charge are *E* selective. We refer to these ylides as being stabilized. We refer to ylides without an adjacent anion-stabilizing group as being unstabilized; unstabilized ylides are *Z* selective. We can write the following resonance contributing structures for a carbonyl-stabilized ylide.

$$
Ph_3\overset{+}{P}\diagdown\!\!\overset{\overset{..}{:}O}{\diagup}\!\!OEt \longleftrightarrow Ph_3\overset{+}{P}\diagdown\!\!\overset{\overset{..}{:}\overset{-}{O}:}{\diagup}\!\!OEt
$$

Resonance contributing structures for an
ylide stabilized by an adjacent carbonyl group

Stabilization of the ylide through resonance decreases its reactivity, allowing an equilibrium to be established during the product-determining step that favors the more stable *E* isomer.

Because the Wittig reaction is so useful for the preparation of alkenes, chemists have explored several variations of it. One of the most useful of these, known as the Horner-Emmons-Wadsworth modification, uses a phosphonate ester derived from an α-haloester or ketone to generate the Wittig carbanion.

$$
Br-CH_2-\overset{O}{\overset{\|}{C}}-OEt
$$
an α-bromoester
$$
(MeO)_2\overset{O}{\overset{\|}{P}}-CH_2-\overset{O}{\overset{\|}{C}}-OEt + MeBr
$$

An α-phosphonoester

$$
(MeO)_3P:
$$
Trimethylphosphite

$$
Br-CH_2-\overset{O}{\overset{\|}{C}}-R
$$
an α-bromoketone
$$
(MeO)_2\overset{O}{\overset{\|}{P}}-CH_2-\overset{O}{\overset{\|}{C}}-R + MeBr
$$

An α-phosphonoketone

The α-phosphonoesters or ketones used in this variation of the Wittig reaction are formed by two successive S_N2 reactions. Trimethylphosphite is an excellent nucleophile and readily displaces bromine from an α-bromoester or α-bromoketone by an S_N2 reaction. Bromide ion then is the nucleophile in the second S_N2 reaction that generates the α-phosphonoester.

Treatment of a phosphonoester with a strong base followed by an aldehyde or ketone gives an alkene, in this case either an α,β-unsaturated ester, ketone, or aldehyde. A particular advantage of using a phosphonate-stabilized carbanion as the Wittig reagent is that the resulting alkene is either entirely or almost entirely the E isomer; that is, phosphonate-stabilized Wittig reagents are almost exclusively E selective. Another advantage of phosphonate-stabilized ylides is that the by-product, dimethylphosphate anion, is water-soluble and, therefore, easily separated from the desired organic product.

Only the E isomer   **Dimethylphosphate**
is formed   **anion**

Example 16.5

Show how this alkene can be synthesized by a Wittig reaction.

Solution

Starting materials are either cyclopentanone and the triphenylphosphonium ylide derived from bromoethane, or acetaldehyde and the triphenylphosphonium ylide derived from chlorocyclopentane. Either route is satisfactory.

Bromoethane

Chlorocyclopentane

Problem 16.5

Show how each alkene can be synthesized by a Wittig reaction (there are two routes to each).

(a)   **(b)**

16.7 | Addition of Oxygen Nucleophiles

A. Addition of Water: Formation of Carbonyl Hydrates

Nucleophilic acyl addition of water (hydration) to a carbonyl group of an aldehyde or ketone forms a geminal diol, commonly abbreviated gem-diol.

$$\text{C}=\text{O} + \text{H}_2\text{O} \xrightleftharpoons[\text{base}]{\text{acid or}} \text{C}\begin{smallmatrix}\text{OH}\\\text{OH}\end{smallmatrix}$$

Carbonyl group of an aldehyde or ketone

A hydrate (a gem-diol)

A gem-diol is commonly referred to as the hydrate of the corresponding aldehyde or ketone. These compounds are unstable and are rarely isolated. This reaction is catalyzed by acids and bases. The mechanism is identical to that for the addition of alcohols, which is discussed next.

Hydration of an aldehyde or ketone is readily reversible, and the diol can eliminate water to regenerate the aldehyde or ketone. In most cases, equilibrium strongly favors the carbonyl group. For a few simple aldehydes, however, the hydrate is favored. For example, when formaldehyde is dissolved in water at 20°C, the position of equilibrium is such that it is more than 99% hydrated.

$$\begin{smallmatrix}\text{H}\\\text{H}\end{smallmatrix}\text{C}=\text{O} + \text{H}_2\text{O} \rightleftharpoons \begin{smallmatrix}\text{H}\\\text{H}\end{smallmatrix}\text{C}\begin{smallmatrix}\text{OH}\\\text{OH}\end{smallmatrix}$$

Formaldehyde

Formaldehyde hydrate (>99%)

A 37% solution of formaldehyde in water, called formalin, is commonly used to preserve biological specimens.

In contrast, an aqueous solution of acetone consists of less than 0.1% of the hydrate at equilibrium.

$$\text{C}=\text{O} + \text{H}_2\text{O} \rightleftharpoons \text{C}\begin{smallmatrix}\text{OH}\\\text{OH}\end{smallmatrix}$$

Acetone (99.9%)

2,2-Propanediol (0.1%)

B. Addition of Alcohols: Formation of Acetals

Alcohols add to aldehydes and ketones in the same manner as described for water. Nucleophilic acyl addition of one molecule of alcohol to the carbonyl group of an aldehyde or ketone forms a **hemiacetal** (a half-acetal).

Hemiacetal
A molecule containing an —OH and an —OR or —OAr group bonded to the same carbon.

$$\text{C}=\text{O} + \text{H}-\text{OEt} \xrightleftharpoons[\text{base}]{\text{acid or}} \text{C}\begin{smallmatrix}\text{OH}\\\text{OEt}\end{smallmatrix}$$

A hemiacetal

The functional group of a hemiacetal is a carbon bonded to an —OH group and an —OR group.

from an aldehyde

from a ketone

Hemiacetals

Hemiacetals are generally unstable and are only minor components of an equilibrium mixture, except in one very important type of compound. When a hydroxyl group is part of the same molecule that contains the carbonyl group, and a five- or six-membered ring can form, the compound exists almost entirely in the cyclic hemiacetal form. Recall that five- and six-membered rings have relatively little ring strain (Section 6.6B). In the following example, (*S*)-4-hydroxypentanal already has a chiral center, and a new chiral center is created upon formation of the hemiacetal.

(*S*)-4-Hydroxypentanal

Cyclic hemiacetals
(major forms present at equilibrium)

Because of the original chiral center, the product hemiacetals in this example are diastereomers, not enantiomers, and are not necessarily produced in equal amounts.

Simple carbohydrates, all of which are polyhydroxyaldehydes or polyhydroxyketones, exist in solution predominantly as cyclic hemiacetals. Because carbohydrates have several hydroxyl groups, they could potentially form rings of different sizes. Generally, only five- and six-membered cyclic hemiacetals (the most strain-free types of rings) are produced to an appreciable extent. The new chiral center created in carbohydrate cyclic hemiacetals can have either an *R* or an *S* configuration. The carbon atom at the new chiral center of a carbohydrate cyclic hemiacetal is given the special name of **anomeric carbon** and corresponds to the carbonyl carbon atom in the open chain form. The two different cyclic hemiacetals are called **anomers**, and the configuration of each is designated as α or β depending on whether the hemiacetal —OH group is on the same side of the ring as the terminal —CH$_2$OH substituent (β anomer) or on the opposite side (α anomer). Note that owing to the presence of multiple chiral centers, the anomers are diastereomers, not enantiomers.

D-Glucose, the most important carbohydrate in mammalian metabolism, exists as a six-membered cyclic hemiacetal form, as both α and β anomers.

D-Glucose
(open chain form)

β Anomer of D-glucose cyclic hemiacetal
(predominates at equilibrium)

α Anomer of D-glucose cyclic hemiacetal

At equilibrium, the β anomer of D-glucose predominates, because the —OH group of the anomeric carbon is in the more stable equatorial position of the more stable chair conformation. In α-D-glucose, the —OH group on the anomeric carbon is axial. When remembering the names of D-glucose anomers, some students find it helpful to remember the phrase "alpha is axial."

β Anomer of D-glucose
cyclic hemiacetal

α Anomer of D-glucose
cyclic hemiacetal

We discuss the chemistry of carbohydrate cyclic hemiacetals in more detail in Chapter 25.

Formation of hemiacetals is catalyzed by bases such as hydroxide or alkoxide. The function of the catalyst is to remove a proton from the alcohol, making it a better nucleophile.

Mechanism   Base-Catalyzed Formation of a Hemiacetal

Step 1: **Take a proton away.**   Proton transfer from HOR to the base gives the alkoxide.

$$B: + H{-}OR \underset{\text{reversible}}{\overset{\text{fast and}}{\rightleftharpoons}} B{-}H + :\overset{..}{\underset{..}{O}}R$$

Step 2: **Make a new bond between a nucleophile and an electrophile.**   Attack of RO⁻ on the carbonyl carbon gives a tetrahedral carbonyl addition compound.

$$CH_3{-}\overset{\overset{\displaystyle ..}{O}:}{\underset{}{C}}{-}CH_3 + :\overset{..}{\underset{..}{O}}{-}R \rightleftharpoons CH_3{-}\overset{\overset{\displaystyle :O:^-}{|}}{\underset{\underset{\displaystyle :OR}{|}}{C}}{-}CH_3$$

A tetrahedral carbonyl
addition compound

Step 3: **Add a proton.**   Proton transfer from the alcohol to the O⁻ gives the hemiacetal and regenerates the base catalyst.

$$CH_3{-}\overset{\overset{\displaystyle :O:^-}{|}}{\underset{\underset{\displaystyle :OR}{|}}{C}}{-}CH_3 + H{-}\overset{..}{\underset{..}{O}}R \rightleftharpoons CH_3{-}\overset{\overset{\displaystyle :OH}{|}}{\underset{\underset{\displaystyle :OR}{|}}{C}}{-}CH_3 + :\overset{..}{\underset{..}{O}}R$$

Formation of hemiacetals can also be catalyzed by acid, most commonly sulfuric acid, *p*-toluenesulfonic acid, or hydrogen chloride.

Step 1: **Add a proton.**   Using a strong acid such as HCl or sulfuric acid sets up an equilibrium in which the protonated alcohol is preferred. The protonated alcohol is therefore the proton source for subsequent steps in the hemiacetal formation reaction.

$$R-\overset{\cdot\cdot}{\underset{\cdot\cdot}{O}}-H \;+\; H-A \;\rightleftharpoons\; R-\overset{+}{\underset{\underset{H}{|}}{\overset{\cdot\cdot}{O}}}-H \;+\; :A$$

pK$_a$ = −7 to −10   pK$_a$ = −2 to −3

Step 2: **Add a proton.**   Proton transfer from the protonated alcohol to the carbonyl oxygen gives the conjugate acid of the aldehyde or ketone.

$$CH_3-\overset{\overset{\displaystyle :O:}{\|}}{C}-CH_3 \;+\; H-\overset{+}{\underset{\underset{H}{|}}{\overset{\cdot\cdot}{O}}}-R \;\underset{reversible}{\overset{fast\ and}{\rightleftharpoons}}\; CH_3-\overset{\overset{\displaystyle \overset{H}{\underset{|}{\overset{+}{O}}}}{\|}}{C}-CH_3 \;+\; H-\overset{\cdot\cdot}{\underset{\cdot\cdot}{O}}-R$$

In this way, the acid catalyst functions to protonate the carbonyl oxygen and thus renders the carbonyl carbon more electrophilic and more susceptible to attack by the weakly nucleophilic oxygen atom of the alcohol.

Step 3: **Make a new bond between a nucleophile and an electrophile.**   Attack of ROH on the carbonyl carbon gives a tetrahedral carbonyl addition compound.

$$CH_3-\overset{\overset{\displaystyle \overset{H}{\underset{|}{\overset{+}{O}}}}{\|}}{C}-CH_3 \;+\; H-\overset{\cdot\cdot}{\underset{\cdot\cdot}{O}}-R \;\rightleftharpoons\; CH_3-\overset{\overset{\displaystyle \overset{\cdot\cdot}{O}-H}{}}{\underset{\underset{\displaystyle \underset{H\quad R}{\overset{+}{O}}}{|}}{C}}-CH_3$$

A tetrahedral carbonyl
addition compound

Step 4: **Take a proton away.**   Proton transfer from the oxonium ion to an alcohol molecule gives the hemiacetal and regenerates the acid catalyst.

$$CH_3-\overset{\overset{\displaystyle \overset{\cdot\cdot}{O}-H}{}}{\underset{\underset{\displaystyle \underset{H\quad R}{\overset{+}{O}}}{|}}{C}}-CH_3 \;\rightleftharpoons\; CH_3-\overset{\overset{\displaystyle \overset{\cdot\cdot}{O}-H}{}}{\underset{\underset{\displaystyle \overset{\cdot\cdot}{O}-R}{|}}{C}}-CH_3 \;+\; R-\overset{+}{\underset{\underset{H}{|}}{\overset{\cdot\cdot}{O}}}-H$$

R−Ö−H   A hemiacetal

Hemiacetals are often not stable relative to starting materials, but they react further with alcohols to form **acetals** plus a molecule of water. Acetals are considerably more stable than hemiacetals and can be isolated in good yield under the proper conditions.

$$\overset{OH}{\underset{OEt}{\diagup\!\!\!\!\diagdown}} \;+\; H-OEt \;\overset{H^+}{\rightleftharpoons}\; \overset{OEt}{\underset{OEt}{\diagup\!\!\!\!\diagdown}} \;+\; H_2O$$

A hemiacetal   A diethyl acetal

Acetal
A molecule containing two —OR or —OAr groups bonded to the same carbon.

The formation of acetals and its reverse is catalyzed by acids, not by bases, because the OH group cannot be displaced directly by nucleophiles.

The functional group of an acetal is a carbon bonded to two —OR groups.

from an aldehyde

from a ketone

$$R-\overset{\overset{\displaystyle OR'}{|}}{\underset{\underset{\displaystyle H}{|}}{C}}-OR' \qquad R-\overset{\overset{\displaystyle OR'}{|}}{\underset{\underset{\displaystyle R''}{|}}{C}}-OR'$$

Acetals

The mechanism for the acid-catalyzed conversion of a hemiacetal to an acetal is divided into four steps. As you study this mechanism, note that acid is a true catalyst in this reaction. The protonated alcohol is used to add a proton in Step 1, but another protonated alcohol is generated in Step 4. The latter steps of this mechanism are very similar to those for hemiacetal formation.

Mechanism Acid-Catalyzed Formation of an Acetal

Step 1: **Add a proton.** Proton transfer from the protonated alcohol to the hemiacetal OH group gives an oxonium ion.

A hemiacetal   An oxonium ion

Step 2: **Break a bond to give stable molecules or ions.** Loss of water gives a new, resonance-stabilized cation.

A resonance-stabilized cation

Step 3: **Make a new bond between a nucleophile and an electrophile.** Reaction of the resonance-stabilized cation (an electrophile) with methanol (a nucleophile) gives the conjugate acid of the acetal.

A protonated acetal

Step 4: **Take a proton away.** Proton transfer from the protonated acetal to alcohol gives the acetal and generates a new molecule of the acid catalyst.

An acetal

Formation of acetals is often carried out using the alcohol as the solvent and dissolving either dry HCl (hydrogen chloride gas) or *p*-toluenesulfonic acid in the alcohol. Because the alcohol is both a reactant and solvent, it is present in large molar excess, which forces the equilibrium to the right and favors acetal formation. Note that this reaction is completely reversible. Addition of excess water to an acetal causes hydrolysis to the ketone.

An excess of alcohol pushes the equilibrium toward formation of the acetal

Removal of water favors formation of the acetal

$$R-\overset{\overset{\textstyle O}{\|}}{C}-R + 2\ EtOH \ \underset{\text{H}^+}{\rightleftharpoons}\ R-\overset{\overset{\textstyle OEt}{|}}{\underset{\underset{\textstyle R}{|}}{C}}-OEt + H_2O$$

A diethyl acetal

In another experimental technique to force the equilibrium to the right, water is removed from the reaction vessel as an **azeotrope** by distillation using a **Dean-Stark trap** (Figure 16.1). In this method for preparing an acetal, the aldehyde or ketone, alcohol, acid catalyst, and benzene are brought to reflux. The component in this mixture with the lowest boiling point is an azeotrope, bp 69°C, consisting of 91% benzene and 9% water. This vapor is condensed and collected in a side trap where it separates into two layers. At room temperature, the composition of the upper, less dense layer is 99.94% benzene and 0.06% water. The composition of the lower, more dense layer is almost the reverse, 0.07% benzene and 99.93% water. As reflux continues, benzene from the top layer is returned to the refluxing mixture, and water is drawn off at the bottom through a stopcock. A Dean-Stark trap "pumps" water out of the reaction mixture, thus forcing the equilibrium to the right. This same apparatus is used in many other reactions where water needs to be removed, such as formation of enamines (Section 19.4A).

Azeotrope
A liquid mixture of constant composition with a boiling point that is different from that of any of its components.

Vapor rising toward condenser contains 91% benzene and 9% water

Reaction takes place in the benzene layer

Heat source

Water-jacket condenser

H₂O

Condensing vapor

Cold H₂O

Dean-Stark trap

Upper layer contains 99.94% benzene and 0.06% water

Lower layer contains 0.07% benzene and 99.93% water

Stopcock to withdraw lower layer

ACME MANTLE CO.

Figure 16.1
A Dean-Stark trap for removing water by azeotropic distillation with benzene. Toluene or xylene can be used if a higher reaction temperature is desired.

Example 16.6

Show the reaction of the carbonyl group of each aldehyde or ketone with one molecule of alcohol to give a hemiacetal and then with a second molecule of alcohol to give an acetal.

(a) [structure of 2-butanone] + 2 EtOH $\underset{}{\overset{H^+}{\rightleftharpoons}}$ (b) [cyclopentanone] =O + HO〜OH $\overset{H^+}{\rightleftharpoons}$

Note that in part (b), ethylene glycol is a diol, and one molecule of it provides both —OH groups.

Solution

Given are structural formulas of the hemiacetal and then the acetal.

(a) [structure with HO OEt] ⟶ [structure with EtO OEt]

  Hemiacetal   Acetal

(b) [cyclopentane with OH and O—CH2—OH] ⟶ [cyclic acetal spiro structure]

  Hemiacetal   Cyclic acetal

Problem 16.6

Hydrolysis of an acetal in aqueous acid gives an aldehyde or ketone and two molecules of alcohol or one molecule of a diol. Draw the structural formulas for the products of hydrolysis of the following acetals in aqueous acid.

(a) MeO—[benzene ring]—[CH with OMe and OMe]   (b) [dioxolane with methyl]   (c) [tetrahydrofuran ring with OMe]

Like ethers (Section 11.5), acetals are unreactive to bases, hydride-reducing agents such as LiAlH$_4$ and NaBH$_4$, Grignard and other organometallic reagents, oxidizing agents (except, of course, for those involving the use of aqueous acid), and catalytic reduction. This lack of reactivity is because acetals have no sp^2 hybridized electrophilic carbon atom to react with nucleophiles. Because of their lack of reactivity toward these reagents and ready hydrolysis in aqueous acid, acetals are often used to reversibly "protect" the carbonyl groups of aldehydes and ketones while reactions are carried out on other functional groups in the molecule.

C. Acetals as Carbonyl-Protecting Groups

The use of acetals as carbonyl-protecting groups is illustrated by the synthesis of 5-hydroxy-5-phenylpentanal from benzaldehyde and 4-bromobutanal.

[reaction scheme: Benzaldehyde + Br-(CH2)3-CHO (4-Bromobutanal) → 5-Hydroxy-5-phenylpentanal]

Benzaldehyde   **4-Bromobutanal**   **5-Hydroxy-5-phenylpentanal**
(racemic)

One obvious way to form a new carbon-carbon bond between these two molecules is to treat benzaldehyde with the Grignard reagent from 4-bromobutanal. A Grignard reagent formed from 4-bromobutanal, however, reacts immediately with the carbonyl group of another molecule of 4-bromobutanal, during preparation. As a result no significant amount of the Grignard reagent is made. A way to avoid this problem is to protect the carbonyl group of the bromoaldehyde by conversion to an acetal. Cyclic acetals are often used because they are particularly easy to prepare.

A cyclic acetal

Treatment of the protected bromoaldehyde with magnesium in diethyl ether followed by addition of benzaldehyde gives a chiral magnesium alkoxide as a racemic mixture.

A cyclic acetal

A chiral magnesium alkoxide
(produced as a racemic mixture)

Treatment of the magnesium alkoxide with aqueous acid accomplishes two things. First, protonation of the alkoxide anion gives the desired 2° hydroxyl group; second, hydrolysis of the cyclic acetal regenerates the carbonyl group of the aldehyde.

5-Hydroxy-5-phenylpentanal
(formed as a racemic mixture)

D. Tetrahydropyranyl Ethers: Protecting an Alcohol as an Acetal

We have just seen in Section 16.7C that an aldehyde or ketone can be protected by conversion to an acetal. A similar strategy can be used to protect a primary or secondary alcohol. Treatment of the alcohol with dihydropyran in the presence of an acid catalyst, commonly anhydrous HCl or a sulfonic acid, RSO_3H, converts the alcohol into a **tetrahydropyranyl** (THP) **ether**.

Dihydropyran

A tetrahydropyranyl
ether

tetrahydropyranyl
group

Because the **THP** group is an acetal, it is stable in neutral and basic solutions and to most oxidizing and reducing agents. It is removed easily by treatment with dilute aqueous acid to regenerate the original primary or secondary alcohol.

Example 16.7

Write a mechanism for the formation of a THP ether from a primary alcohol RCH_2OH catalyzed by a sulfonic acid $ArSO_3H$.

Solution

Step 1: **Make a new bond between a nucleophile (π bond) and an electrophile—add a proton.** Dihydropyran (a vinyl or enol ether) is weakly basic and is protonated to give a resonance-stabilized cation.

A resonance-stabilized cation

Step 2: **Make a new bond between a nucleophile and an electrophile.** Reaction of the resonance-stabilized cation (an electrophile) with the alcohol (a nucleophile) gives an oxonium ion.

Step 3: **Take a proton away.** Proton transfer from the oxonium ion to $ArSO_3^-$ completes the reaction.

(racemic)

Problem 16.7

Write a mechanism for the acid-catalyzed hydrolysis of a THP ether to regenerate the original alcohol. Into what compound is the THP group converted?

16.8 | Addition of Nitrogen Nucleophiles

A. Ammonia and Its Derivatives

Ammonia, primary aliphatic amines (RNH_2), and primary aromatic amines ($ArNH_2$) react with the carbonyl group of aldehydes and ketones to give an **imine**, often referred to as a **Schiff base**. Imines are usually unstable unless the $C=N$ group is part of an extended system of conjugation (e.g., rhodopsin) and are generally not isolated.

Imine
A compound containing a carbon-nitrogen double bond, $R_2C=NR$; also called a Schiff base.

Schiff base
An alternative name for an imine.

Acetaldehyde   **Aniline**   An imine
(a Schiff base)

Cyclohexanone   **Ammonia**   An imine

Step 1: Make a new bond between a nucleophile and an electrophile. Nucleophilic addition of a primary amine or ammonia gives a tetrahedral carbonyl addition compound with a negative charge on oxygen and a positive charge on nitrogen.

$$CH_3-\overset{\overset{\displaystyle :O:}{\|}}{\underset{\underset{\displaystyle H}{|}}{\underset{\displaystyle H-\ddot{N}-R}{C}}}-CH_3 \rightleftharpoons CH_3-\overset{\overset{\displaystyle :\overset{-}{O}:}{|}}{\underset{\underset{\displaystyle H}{|}}{\underset{\displaystyle H-\overset{+}{N}-R}{C}}}-CH_3$$

Step 2: Take a proton away. Another molecule of amine or ammonia removes a proton to give a neutral nitrogen atom.

$$R-\overset{\underset{\displaystyle H}{|}}{\ddot{N}}-H \quad CH_3-\overset{\overset{\displaystyle :\overset{..}{O}:^{-}}{|}}{\underset{\underset{\displaystyle H}{|}}{\underset{\displaystyle H-\overset{+}{N}-R}{C}}}-CH_3 \rightleftharpoons CH_3-\overset{\overset{\displaystyle :\overset{..}{O}:^{-}}{|}}{\underset{\underset{\displaystyle H}{|}}{\underset{\displaystyle :N-R}{C}}}-CH_3 \quad + \quad R-\overset{\overset{\displaystyle H}{|}}{\underset{\underset{\displaystyle H}{|}}{\overset{+}{N}}}-H$$

Step 3: Add a proton. A proton is added to the oxygen atom to give the neutral aminal intermediate.

$$CH_3-\overset{\overset{\displaystyle :\overset{..}{O}:^{-}}{|}}{\underset{\underset{\displaystyle H}{|}}{\underset{\displaystyle :N-R}{C}}}-CH_3 \quad + \quad H-\overset{\overset{\displaystyle H}{|}}{\underset{\underset{\displaystyle H}{|}}{\overset{+}{N}}}-R \rightleftharpoons CH_3-\overset{\overset{\displaystyle :\overset{..}{O}H}{|}}{\underset{\underset{\displaystyle H}{|}}{\underset{\displaystyle :N-R}{C}}}-CH_3 \quad + \quad R-\overset{\overset{\displaystyle H}{|}}{\underset{\underset{\displaystyle H}{|}}{\ddot{N}}}-H$$

Aminal intermediate

Steps 2 and 3 above are expected to occur more or less at the same time and could take place in either order. The net result of these two steps is the transfer of a proton from one atom on an intermediate to a different site on the same intermediate, although it is unlikely the same proton is transferred because of the other species involved in these steps.

Step 4: Add a proton. The aminal intermediate is protonated on the hydroxyl group. This protonation is not favored but is a necessary step for the mechanism to continue to products.

$$R-\overset{\overset{\displaystyle H}{|}}{\underset{\underset{\displaystyle H}{|}}{\overset{+}{N}}}-H \quad + \quad CH_3-\overset{\overset{\displaystyle :\overset{..}{O}H}{|}}{\underset{\underset{\displaystyle H}{|}}{\underset{\displaystyle :N-R}{C}}}-CH_3 \rightleftharpoons CH_3-\overset{\overset{\displaystyle \overset{H\ \ H}{\underset{\displaystyle }{\overset{+}{O}}}}{|}}{\underset{\underset{\displaystyle H}{|}}{\underset{\displaystyle :N-R}{C}}}-CH_3 \quad + \quad R-\overset{\overset{\displaystyle H}{|}}{\underset{\underset{\displaystyle H}{|}}{\ddot{N}}}-H$$

Step 5: Break a bond to give stable molecules or ions. The $-\overset{+}{O}H_2$ group leaves as water to generate an iminium ion, which is stabilized by resonance delocalization.

Iminium ion that is stabilized by resonance delocalization

Step 6: Take a proton away. The iminium ion is deprotonated to give the imine product.

An imine

Imine formation is sensitive to the amount of acid. Acid is required to protonate the aminal intermediate in Step 4, but too much acid would protonate all of the amine, and thus prevent it from acting as a nucleophile in Step 1. It turns out that a pH of about 4 is optimum for this process to occur.

To give but one example of the importance of imines in biological systems, vitamin A aldehyde (retinal) is bound to the protein opsin in the human retina in the form of an imine. The primary amino group of opsin for this reaction is provided by the side chain of the amino acid lysine (Section 27.1). The imine is called rhodopsin or visual purple.

11-*cis*-Retinal

Rhodopsin
(Visual purple)

Absorption of photons by rhodopsin causes a *cis* to *trans* isomerization of the double bond at carbon 11, and the resulting change in molecular shape leads to creation of a nerve impulse that forms the basis of mammalian vision.

Example 16.8

Write a structural formula for the imine formed in each reaction.

Solution

Given is a structural formula for each imine.

(a) (b)

Problem 16.8

Acid-catalyzed hydrolysis of an imine gives an amine and an aldehyde or ketone. When one equivalent of acid is used, the amine is converted to an ammonium salt. Write structural formulas for the products of hydrolysis of the following imines using one equivalent of HCl.

(a) MeO—⟨benzene ring⟩—CH=NCH_2CH_3 (b) —CH_2N=⟨cyclopentane⟩

Secondary amines react with aldehydes and ketones to form enamines. The name **enamine** is derived from -*en*- to indicate the presence of a carbon-carbon double bond and -*amine* to indicate the presence of an amino group. An example is enamine formation between cyclohexanone and piperidine, a cyclic secondary amine. Water is removed by a Dean-Stark trap (Figure 16.1), which forces the equilibrium to the right.

Enamine
An unsaturated compound derived by the reaction of an aldehyde or ketone and a secondary amine followed by loss of H_2O; R_2C=CR—NR_2.

| Cyclohexanone | Piperidine | An enamine |
| | (a secondary amine) | |

Briefly, the mechanism for formation of an enamine is very similar to that for the formation of an imine. In the first step nucleophilic addition of the secondary amine to the carbonyl carbon of the aldehyde or ketone followed by proton transfer from nitrogen to oxygen gives a tetrahedral carbonyl addition compound. Acid-catalyzed dehydration gives the enamine. It is at this stage that enamine formation differs from imine formation. The nitrogen has no proton to lose. Instead, a proton is lost from the α-carbon of the ketone or aldehyde portion of the molecule in an elimination reaction.

| Phenylacetaldehyde | Piperidine | A tetrahedral carbonyl addition compound | An enamine (a mixture of *E,Z* isomers) |

We will return to the chemistry of enamines and their use in synthesis in Section 19.5.

B. Hydrazine and Related Compounds

Aldehydes and ketones react with **hydrazine** to form compounds called hydrazones as illustrated by treating cyclopentanone with hydrazine.

| | **Hydrazine** | A hydrazone |

The reversible formation of imines from amino groups and carbonyls can play the role of interchanging these groups in biological transformations. For example, **transamination** involves such a shuffling, and is the first reaction in the degradation of amino acids and the last reaction in their synthesis. In transamination, an amino group is transferred from a donor α-amino acid to an acceptor α-ketoacid via imines (referred to in the mechanism given below as aldimines and ketamines to

tissue. Damage to these organs leads to release of transaminases into the blood and determination of serum levels of these enzymes can provide the clinician with valuable information about the extent of heart or liver damage. The two transaminases most commonly assayed for this purpose are serum glutamate oxaloacetate transaminases (SGOT) and serum glutamate pyruvate transaminases (SGPT).

Transaminations serve two vital functions. First, they provide a means to readjust the relative proportions of a

Alanine transaminase:

$$R-\underset{\underset{NH_3^+}{|}}{CH}-CO_2^- + CH_3-\underset{\underset{O}{\parallel}}{C}-CO_2^- \rightleftharpoons R-\underset{\underset{O}{\parallel}}{C}CO_2^- + CH_3-\underset{\underset{NH_3^+}{|}}{CH}-CO_2^-$$

A donor amino acid · · · · · · **pyruvate** · · · · · · · · · · · A new ketoacid · · · · · · **L-Alanine**

Glutamate transaminase:

$$R-\underset{\underset{NH_3^+}{|}}{CH}-CO_2^- + {}^-O_2CCH_2CH_2\underset{\underset{O}{\parallel}}{C}CO_2^- \rightleftharpoons R-\underset{\underset{O}{\parallel}}{C}CO_2^- + {}^-O_2CCH_2CH_2\underset{\underset{NH_3^+}{|}}{CH}CO_2^-$$

A donor amino acid · · · · · · **α-ketoglutarate** · · · · · · A new ketoacid · · · · · · **L-Glutamate**

distinguish their formation from aldehydes and ketones respectively). In the process the acceptor α-ketoacid is transformed into a new amino acid. While several different α-ketoacids participate in transamination reactions, the most important are pyruvate and α-ketoglutarate. During the transamination phase of amino acid synthesis, all amino groups (with the possible exception of those of lysine and threonine) are channeled into either alanine or glutamate.

Transaminations are catalyzed by a specific group of enzymes called amino-transaminases, or more commonly, **transaminases**. While transaminases are found in all cells, their concentrations are particularly high in heart and liver

number of amino acids to meet the particular needs of an organism: in most diets, the amino acid blend does not correspond precisely to what is needed for proper growth and development. Second, transamination serves to collect the nitrogen atoms of all amino acids as glutamate. Glutamate is the central source of nitrogen atoms for biosynthesis.

For catalytic activity, all transaminases require pyridoxal phosphate, a coenzyme derived from pyridoxine, vitamin B$_6$. In its role as a catalyst, this coenzyme undergoes reversible transformations between an aldehyde (pyridoxal phosphate, PLP) and a primary amine (pyridoxamine phosphate, PMP).

Pyrixodine (vitamin B$_6$)

Pyridoxal phosphate (PLP)

Pyridoxine phosphate (PMP)

The transamination reaction begins with reaction of the amino group of an amino acid with the carbonyl group of pyridoxal phosphate to form an aldimine (an imine so named because it is formed between an amino group and an aldehyde). To understand the following step, imagine that it is catalyzed by the presence of a wrap-around enzyme that has on its surface both a basic group (a proton acceptor here represented by Enz:⁻) and an acidic group (a proton donor, here represented by Enz—H). Abstraction of the α proton from the carbon skeleton of the amino acid and redistribution of two pairs of electrons culminating with the abstraction of the proton from an acidic group on the surface of the enzyme gives an isomeric imine here called a ketamine because it is formed by reaction of a primary amine and the carbonyl group of a ketone. Hydrolysis of this ketamine gives a new α-ketoacid and pyridoxine phosphate.

The net effect of this sequence of reactions is transfer of the amino group ($-NH_3^+$) from the α-amino acid to pyridoxal.

Left at the active site of the enzyme is pyridoxamine phosphate, which then combines with a new α-ketoacid to form a second ketamine that then undergoes the reverse of the above sequence to form a new aldimine and eventual hydrolysis to give a new α–amino acid and regenerate pyridoxal phosphate.

This reversible cycle of reactions in effect exchanges amino groups between amino acids and α-ketoacids.

An amino acid

An aldimine

A ketamine

A new α-ketoacid

Pyridoxine phosphate (PMP)

Table 16.4 lists several other derivatives of ammonia and hydrazine that react with aldehydes and ketones to give imines.

The chief value of the nitrogen nucleophiles listed in Table 16.4 is that most aldehydes and ketones react with them to give crystalline solids with sharp melting points. Historically, these derivatives often provided a convenient way to identify liquid aldehydes or ketones. Now, aldehydes and ketones are more readily identified by IR and NMR spectroscopy.

Table 16.4 | Derivatives of Ammonia and Hydrazine Used for Forming Imines

Reagent, H₂N—R	Name of Reagent	Name of Derivative Formed
H₂N—OH	Hydroxylamine	Oxime
H₂N—NH—⟨phenyl⟩	Phenylhydrazine	Phenylhydrazone
H₂N—NH—⟨2,4-dinitrophenyl⟩—NO₂ O₂N	2,4-Dinitrophenylhydrazine	2,4-Dinitrophenylhydrazone
H₂N—NHCNH₂ (with C=O above)	Semicarbazide	Semicarbazone

α-Carbon
A carbon atom adjacent to a carbonyl group.

α-Hydrogen
A hydrogen on a carbon adjacent to a carbonyl group.

Enolate anion
An anion derived by loss of a hydrogen from a carbon alpha to a carbonyl group; the anion of an enol.

Table 16.5

Type of Bond	pK_a
CH₃CH₂O—H	16
CH₃CCH₂—H (with C=O)	20
CH₃C≡C—H	25
CH₂=CH—H	44
CH₃CH₂—H	51

16.9 | Keto-Enol Tautomerism

A. Acidity of α-Hydrogens

A carbon atom adjacent to a carbonyl group is called an **α-carbon**, and hydrogen atoms bonded to it are called **α-hydrogens**.

$$\text{α-hydrogens} \qquad \text{CH}_3-\overset{\text{O}}{\overset{\|}{\text{C}}}-\text{CH}_2-\text{CH}_3 \qquad \text{α-carbons}$$

Because carbon and hydrogen have comparable electronegativities, a C—H bond normally has little polarity. In addition, carbon does not have a high electronegativity (compare it, for example, with oxygen, which has an electronegativity of 3.5), so that an anion based on carbon is relatively unstable. As a result, a hydrogen bonded to carbon usually shows very low acidity. The situation is different, however, for hydrogens alpha to a carbonyl group. α-Hydrogens are more acidic than acetylenic, vinylic, and alkane hydrogens but less acidic than —OH hydrogens of alcohols (Table 16.5).

The greater acidity of α-hydrogens arises because the negative charge on the resulting **enolate anion** is delocalized by resonance, thus stabilizing it relative to an alkane, alkene, or alkyne anion.

$$\text{CH}_3-\overset{\ddot{\text{O}}:}{\overset{\|}{\text{C}}}-\text{CH}_2-\text{H} + :\text{B}^- \rightleftharpoons \text{CH}_3-\overset{\ddot{\text{O}}:}{\overset{\|}{\text{C}}}-\ddot{\text{C}}\text{H}_2^- \longleftrightarrow \text{CH}_3-\overset{:\ddot{\text{O}}^-}{\overset{|}{\text{C}}}=\text{CH}_2 + \text{H}-\text{B}$$

Resonance-stabilized enolate anion

An electrostatic potential map of an enolate anion.

An enolate anion is also stabilized by the electron-withdrawing inductive effect of the electronegative oxygen. Recall that we used these same factors in Section 4.6C to explain the greater acidity of carboxylic acids compared with alcohols.

Example 16.9

Predict the position of the following equilibrium.

Acetophenone

Solution

The pK_a of ethanol is approximately 16 (Table 4.1). Assume that the pK_a of acetophenone is approximately equal to that of acetone, that is, about 20. Ethanol is the stronger acid; therefore, the equilibrium lies to the left.

Weaker acid pK_a 20 — Stronger acid pK_a 16

Problem 16.9

Predict the position of the following equilibrium.

Acetophenone

When an enolate anion reacts with a proton donor, it may do so either on oxygen or on the α-carbon. Protonation of the enolate anion on the α-carbon gives the original molecule in what is called the keto form. Protonation on oxygen gives an enol form. In this way, the keto form of an aldehyde or ketone can be converted into the enol catalyzed by base.

Mechanism Base-Catalyzed Equilibration of Keto and Enol Tautomers

Step 1: **Take a proton away.** A base, B^-, removes an α-hydrogen to give an enolate anion intermediate that is stabilized by resonance delocalization of the negative charge.

Keto form

An enolate anion intermediate that is stabilized by resonance delocalization

Step 2: **Add a proton.** A proton is added to the enolate oxygen anion to give the enol.

Enol form

Enol formation can also be catalyzed by acid. The only difference between the base-catalyzed and acid-catalyzed reactions is the order of proton addition and elimination. In acid-catalyzed reactions, a proton is added first; in base-catalyzed reactions, a proton is removed first.

Mechanism Acid-Catalyzed Equilibration of Keto and Enol Tautomers

***Step 1:* Add a proton.** Rapid and reversible proton transfer from the acid catalyst, H—A, to the carbonyl oxygen gives the conjugate acid of the ketone as a resonance-stabilized oxonium ion.

Keto form · The conjugate acid of the ketone

***Step 2:* Take a proton away.** Proton transfer from the α-carbon to the base, A⁻, gives the enol and generates a new molecule of the acid catalyst.

Enol form

B. The Position of Equilibrium in Keto-Enol Tautomerism

Aldehydes and ketones with at least one α-hydrogen are in equilibrium with their enol forms. We first encountered this type of equilibrium in our study of the hydroboration-oxidation and acid-catalyzed hydration of alkynes in Section 7.7. As we see in Table 16.6, the position of keto-enol equilibrium for simple aldehydes and ketones lies far on the side of the keto form, primarily because carbon-hydrogen single bonds are about as strong as oxygen-hydrogen single bonds but a carbon-oxygen double bond is stronger than a carbon-carbon double bond.

For certain types of molecules, the enol form may be the major form and, in some cases, the only form present at equilibrium. For β-diketones, such as 1,3-cyclohexanedione and 2,4-pentanedione, where an α-carbon lies between two carbonyl groups, the position of equilibrium shifts in favor of the enol form.

1,3-Cyclohexanedione

20%   80%

2,4-Pentanedione (Acetylacetone)

Table 16.6 — The Position of Keto-Enol Equilibrium for Some Simple Aldehydes and Ketones*

Keto Form	Enol Form	% Enol at Equilibrium
$CH_3\overset{\overset{\displaystyle O}{\|\|}}{C}H \rightleftharpoons$	$CH_2=\overset{\overset{\displaystyle OH}{\|}}{C}H$	6×10^{-5}
$CH_3\overset{\overset{\displaystyle O}{\|\|}}{C}CH_3 \rightleftharpoons$	$CH_3\overset{\overset{\displaystyle OH}{\|}}{C}=CH_2$	6×10^{-7}
cyclopentanone $\rightleftharpoons$	cyclopentenol (OH)	1×10^{-6}
cyclohexanone $\rightleftharpoons$	cyclohexenol (OH)	4×10^{-5}

*Data from J. March, *Advanced Organic Chemistry*, 4th ed., Wiley Interscience, New York, 1992, p. 70.

These enols are stabilized by **conjugation** of the π systems of the carbon-carbon double bond and the carbonyl group. The enol of 2,4-pentanedione, an open-chain β-diketone, is further stabilized by intramolecular hydrogen bonding.

Conjugation
A situation that occurs when the electrons of adjacent π bonds interact with each other; that is, when two double bonds are separated by one single bond.

Example 16.10

Write two enol forms for each compound. Which enol of each has the larger concentration at equilibrium?

(a) (b)

Solution

In each case, the major enol form has the more substituted (and more stable) double bond.

(a)

Major enol

(b)

Major enol

Problem 16.10

Draw a structural formula for the keto form of each enol.

(a) (b) (c)

16.10 | Oxidation

A. Oxidation of Aldehydes

Aldehydes are oxidized to carboxylic acids by a variety of common oxidizing agents, including chromic acid and molecular oxygen. In fact, aldehydes are one of the most easily oxidized of all functional groups. Oxidation by chromic acid is illustrated by the conversion of hexanal to hexanoic acid (for the mechanism of this oxidation, review Section 10.8A).

$$\text{CHO} \xrightarrow{\text{H}_2\text{CrO}_4} \text{COOH}$$

Hexanal   **Hexanoic acid**

Aldehydes are also oxidized to carboxylic acids by Ag(I) ion. One laboratory procedure is to shake a solution of the aldehyde in aqueous ethanol or tetrahydrofuran with a slurry of Ag_2O.

$$\text{MeO, HO—CHO} + Ag_2O \xrightarrow[\text{NaOH}]{\text{THF, H}_2\text{O}} \xrightarrow[\text{H}_2\text{O}]{\text{HCl}} \text{MeO, HO—COOH} + 2\,Ag°$$

Vanillin   **Vanillic acid**

Tollens' reagent, another form of Ag(I), is prepared by dissolving silver nitrate in water, adding sodium hydroxide to precipitate Ag(I) as Ag_2O, and then adding aqueous ammonia to redissolve silver(I) as the silver-ammonia complex ion.

$$Ag^+NO_3^- + 2\,NH_3 \xrightleftharpoons{\text{NH}_3,\ \text{H}_2\text{O}} Ag(NH_3)_2{}^+NO_3^-$$

When Tollens' reagent is added to an aldehyde, the aldehyde is oxidized to a carboxylic anion, and Ag(I) is reduced to metallic silver. If this reaction is carried out properly, silver precipitates as a smooth, mirror-like deposit, hence the name silver-mirror test. Ag(I) is rarely used at the present time for the oxidation of aldehydes because of the cost of silver and because other, more convenient methods exist for this oxidation. This reaction, however, is still used for silvering glassware, including mirrors. In this process, formaldehyde or glucose (Section 25.1) is generally used as the aldehyde to reduce Ag(I).

Aldehydes are also oxidized to carboxylic acids by molecular oxygen and by hydrogen peroxide.

$$2 \text{ (benzaldehyde)} + O_2 \longrightarrow 2 \text{ (benzoic acid)}$$

Benzaldehyde   **Benzoic acid**

Reaction with oxygen is a radical chain reaction (Section 8.7). Molecular oxygen is the least expensive and most readily available of all oxidizing agents. On an industrial scale, air oxidation of organic compounds, including aldehydes, is very common. Air oxidation of aldehydes can also be a problem. Aldehydes that are liquid at room temperature are so sensitive to oxidation by molecular oxygen that they must be protected from contact with air during storage. Often this is done by sealing the aldehyde in a container under an atmosphere of nitrogen.

Tollens' reagent
A solution prepared by dissolving Ag_2O in aqueous ammonia; used for selective oxidation of an aldehyde to a carboxylic acid.

© Cengage Learning/Charles D. Winters

A silvered mirror has been deposited in the inside of this flask by the reaction of an aldehyde with Tollens' reagent.

Example 16.11

Draw a structural formula for the product formed by treating each compound with Tollens' reagent followed by acidification with aqueous HCl.

(a) Pentanal   **(b)** Cyclopentanecarbaldehyde

Solution

The aldehyde group in each compound is oxidized to a carboxyl group.

(a)

![structure of pentanoic acid]

Pentanoic acid

(b)

![structure of cyclopentane carboxylic acid]

Cyclopentane carboxylic acid

Problem 16.11

Complete the equations for these oxidations.

(a) Hexanal + $H_2O_2 \longrightarrow$ (b) 3-Phenylpropanal + Tollens' reagent $\longrightarrow$

B. Oxidation of Ketones

In contrast to aldehydes, ketones are oxidized only under rather special conditions. For example, they are not normally oxidized by chromic acid or potassium permanganate. In fact, chromic acid is used routinely to oxidize secondary alcohols to ketones in good yield (Section 10.8A).

Ketones undergo oxidative cleavage, via their enol form, when treated with potassium dichromate, potassium permanganate, and other strong oxidants at higher temperatures and higher concentrations of acid or base. The carbon-carbon double bond of the enol is cleaved to form two carboxyl or ketone groups, depending on the substitution pattern of the original ketone. An important industrial application of this reaction is oxidation by nitric acid of cyclohexanone to hexanedioic (adipic) acid, one of the two monomers required for the synthesis of the polymer nylon 66 (Section 29.5A).

![reaction of cyclohexanone keto form to enol form, then with HNO₃ to hexanedioic acid + oxides of nitrogen]

Cyclohexanone (keto form)   **Cyclohexanone** (enol form)   **Hexanedioic acid (Adipic acid)** + oxides of nitrogen

As the oxidation of cyclohexanone shows, this reaction is most useful for oxidation of symmetrical cycloalkanones, which yield a single product. Most other ketones give mixtures of products because the enol can form on either side of the carbonyl group.

16.11 | Reduction

Aldehydes are reduced to primary alcohols and ketones are reduced to secondary alcohols. In addition, both aldehyde and ketone carbonyl groups can be reduced to methylene groups ($-CH_2-$).

Aldehydes	Can Be Reduced to	Ketones	Can Be Reduced to
RCH ($=O$)	RCH_2OH / RCH_3	RCR' ($=O$)	$RCHR'$ (OH) / RCH_2R'

A. Metal Hydride Reductions

Hydride ion
A hydrogen atom with two electrons in its valence shell; H:—.

By far the most common laboratory reagents for reduction of the carbonyl group of an aldehyde or ketone to a hydroxyl group are sodium borohydride, lithium aluminum hydride (LAH), and their derivatives. These compounds behave as sources of **hydride ion**, a powerful nucleophile, that takes part in a nucleophilic acyl addition reaction.

$$\text{Na}^+ \ \underset{\underset{\text{H}}{|}}{\overset{\overset{\text{H}}{|}}{\text{H}-\text{B}^--\text{H}}} \qquad \text{Li}^+ \ \underset{\underset{\text{H}}{|}}{\overset{\overset{\text{H}}{|}}{\text{H}-\text{Al}^--\text{H}}} \qquad \text{H:}^-$$

| **Sodium borohydride** | **Lithium aluminum hydride (LAH)** | **Hydride ion** |

Lithium aluminum hydride is a very powerful reducing agent; it reduces not only the carbonyl groups of aldehydes and ketones rapidly but also those of carboxylic acids (Section 17.6A) and their functional derivatives (Section 18.10). Sodium borohydride is a less reactive and, therefore, a much more selective reagent, reducing only aldehydes and ketones rapidly. Neither reagent reduces alkenes or alkynes to alkanes.

Reductions using sodium borohydride are most commonly carried out in aqueous methanol, in pure methanol, or in ethanol. The initial product of reduction is a tetraalkyl borate, which, on warming with water, is converted to an alcohol and sodium borate salts. One mole of sodium borohydride reduces four moles of aldehyde or ketone.

$$4 \ \text{R}\overset{\overset{\text{O}}{\parallel}}{\text{C}}\text{H} + \text{NaBH}_4 \xrightarrow{\text{methanol}} (\text{RCH}_2\text{O})_4\text{B}^-\text{Na}^+ \xrightarrow{\text{H}_2\text{O}} 4 \ \text{RCH}_2\text{OH} + \text{borate salts}$$

$$\text{A tetraalkyl borate}$$

Unlike sodium borohydride, lithium aluminum hydride reacts violently with water, alcohols, and other protic solvents to liberate hydrogen gas and form metal hydroxides and alkoxides. Therefore, reductions of aldehydes and ketones using this reagent must be carried out in aprotic solvents, most commonly diethyl ether or tetrahydrofuran. The stoichiometry for lithium aluminum hydride reductions is the same as that for sodium borohydride reductions: one mole of lithium aluminum hydride per four moles of aldehyde or ketone. Because of the formation of gelatinous aluminum salts, aqueous acid or base work-up is usually used to dissolve these salts.

$$4 \ \text{R}\overset{\overset{\text{O}}{\parallel}}{\text{C}}\text{R} + \text{LiAlH}_4 \xrightarrow{\text{ether}} (\text{R}_2\text{CHO})_4\text{Al}^-\text{Li}^+ \xrightarrow[\text{H}^+ \text{ or OH}^-]{\text{H}_2\text{O}} 4 \ \text{R}\overset{\overset{\text{OH}}{|}}{\text{C}}\text{HR} + \text{aluminum salts}$$

$$\text{A tetraalkyl aluminate}$$

Mechanism Sodium Borohydride Reduction of an Aldehyde or Ketone

Step 1: **Make a new bond between a nucleophile and an electrophile.** Nucleophilic addition of a hydride to the electrophilic carbonyl carbon atom gives a tetrahedral carbonyl addition compound.

$$\text{Na}^+ \ \underset{\underset{\text{H}}{|}}{\overset{\overset{\text{H}}{|}}{\text{H}-\text{B}^--\text{H}}} + \text{R}-\overset{\overset{:\ddot{\text{O}}:}{\parallel}}{\text{C}}-\text{R}' \longrightarrow \text{R}-\overset{\overset{:\ddot{\text{O}}-\overline{\text{B}}\text{H}_3 \ \text{Na}^+}{|}}{\underset{\underset{\text{H}}{|}}{\text{C}}}-\text{R}'$$

Step 2: Add a proton. In a second step, the chemist opens the flask and adds water to give the alcohol product.

On the alcohol product, the hydrogen atom bonded to carbon comes from the hydride reducing agent, and the hydrogen atom bonded to oxygen comes from water during hydrolysis of the metal alkoxide salt.

B. Catalytic Reduction

The carbonyl group of an aldehyde or ketone is reduced to a hydroxyl group by hydrogen in the presence of a transition metal catalyst, most commonly finely divided platinum or nickel. Reductions are generally carried out at temperatures from 25 to 100°C and at pressures of hydrogen from 1 to 5 atm. Under such conditions, cyclohexanone is reduced to cyclohexanol.

Cyclohexanone   **Cyclohexanol**

Catalytic reduction of aldehydes and ketones is simple to carry out, yields are generally very high, and isolation of the final product is very easy. A disadvantage is that some other functional groups are also reduced under these conditions, for example, carbon-carbon double and triple bonds.

trans-2-Butenal   1-Butanol
(Crotonaldehyde)

C. Selective Reduction

Using metal hydride reductions, it is possible to reduce an aldehyde or ketone carbonyl group without reducing any C=C bonds that might be present in the same molecule. The metal hydride reducing agents like $LiAlH_4$ and $NaBH_4$ are nucleophilic reagents. Functional groups that are not electrophiles, such as C=C bonds in alkenes, do not react.

The following equations illustrate selective reduction of a carbonyl group in the presence of a carbon-carbon double bond and, alternatively, selective reduction of a carbon-carbon double bond in the presence of a carbonyl group using rhodium on powdered charcoal as a catalyst.

Selective reduction of a carbonyl group:

Selective reduction of a carbon-carbon double bond:

$$RCH{=}CHCR' + H_2 \xrightarrow{Rh} RCH_2CH_2CR'$$

(with carbonyl O on both structures)

Example 16.12

Complete these reductions.

(a) $\xrightarrow[\text{Pt}]{H_2}$ (b) $\xrightarrow{\text{1. NaBH}_4}{\text{2. H}_2\text{O}}$

Solution

The carbonyl group of the aldehyde in (a) is reduced to a primary alcohol, and the carbonyl group of the ketone in (b) is reduced to a secondary alcohol.

(a)   (b)

(racemic)

Problem 16.12

What aldehyde or ketone gives these alcohols on reduction with $NaBH_4$?

(a)   (b)   (c)

(mixture of
three stereoisomers)

D. Reductive Amination

Reductive Amination
A method of preparing substituted amines by treating an aldehyde or ketone with an amine in the presence of, or followed by, a reducing agent.

One of the chief values of imines is that the carbon-nitrogen double bond can be reduced by hydrogen in the presence of a nickel or other transition metal catalyst to a carbon-nitrogen single bond. By this two-step reaction, called **reductive amination**, a primary amine is converted to a secondary amine by way of an imine as illustrated by the conversion of cyclohexylamine to dicyclohexylamine.

Cyclohexanone   **Cyclohexylamine**   (an imine intermediate)   **Dicyclohexylamine**

It is possible to carry out reductive amination in a single step by using a reducing agent that is not powerful enough to reduce the starting aldehyde or ketone, but is strong enough to reduce the more easily reduced imine that is formed. The reducing agent usually used for this purpose is sodium cyanoborohydride, $NaBH_3CN$.

For the reduction of aldehydes and ketones to alcohols, biological systems use NADH, a reagent whose results are equivalent to laboratory hydride-reducing agents. As an example, the final step in alcoholic fermentation—the process by which yeast converts carbohydrates such as glucose to ethanol and carbon dioxide—is the enzyme-catalyzed reduction of acetaldehyde to ethanol.

$$CH_3-\overset{\overset{\displaystyle O}{\|}}{C}-H + NADH + H_3O^+ \xrightarrow[\text{dehydrogenase}]{\text{alcohol}} CH_3CH_2OH + NAD^+ + H_2O$$

Acetaldehyde   **Ethanol**

Alcoholic fermentation is the basis for the brewing of beers and the fermentation of grape sugar in wine making.

As another example, the end product of glycolysis is pyruvate and the reduced coenzyme NADH. In the absence of an adequate supply of oxygen (anaerobic metabolism) to reoxidize NADH to NAD^+ and thereby allow glycolysis to continue, cells use the reduction of pyruvate to lactate as a way to regenerate NAD^+:

$$\overset{\overset{\displaystyle O}{\|}}{\underset{\displaystyle COO^-}{}} + NADH \xrightarrow[\text{dehydrogenase}]{\text{lactate}} \overset{\overset{\displaystyle OH}{|}}{\underset{\displaystyle COO^-}{}} + NAD^+$$

Pyruvate   **(S)-Lactate**

Anyone who exercises to the point of consuming all available oxygen knows the pain and fatigue associated with the buildup of lactate in muscles. With rest and a renewed supply of oxygen, the concentration of lactate decreases rapidly and muscle pain are relieved. Lactate production by anaerobic organisms during fermentation is responsible for the taste of sour milk and the characteristic taste and fragrance of sauerkraut (fermented cabbage).

In still other aldehyde and ketone reductions, biological systems use NADPH as a reducing agent. This molecule, which is a phosphate ester of NADH, functions in the same manner as NADH as a biological reducing agent.

The mechanism for NADH reduction of the carbonyl group of an aldehyde or ketone follows. First the carbonyl-containing compound and NADH are positioned on the surface of the enzyme catalyst in a

highly specific relationship to each other. Then follows a redistribution of valence electrons, one part of which is the transfer of a hydrogen atom with its pair of electrons (in effect a hydride ion) from NADH to the carbonyl compound.

Arrows 1 and 2: Electrons within the ring flow from nitrogen.

Arrow 3: Transfer of a hydride ion from the —CH_2— of the six-membered ring to the carbonyl carbon creates the new C—H bond to the carbonyl carbon.

Arrow 4: The C=O π bond breaks as the new C—H bond forms.

Arrow 5: An acidic group, —BH, on the surface of the enzyme transfers a proton to the newly formed alkoxide ion to complete formation of the hydroxyl group of the product.

The enzyme-catalyzed reduction of pyruvate is completely stereoselective; in muscle tissue, only the *S* enantiomer of lactate is produced. This stereoselectivity arises because the reduction takes place in a chiral

environment created by the enzyme. At the actual reduction step, both pyruvate and NADH are positioned precisely on the chiral surface of the enzyme with the result that the hydride ion from NADH can be delivered only to one face of pyruvate, in this case producing only the *S* enantiomer.

Cyclohexanone   **N-Methylcyclohexylamine**

E. Reduction of a Carbonyl Group to a Methylene Group

Several methods are available for reducing the carbonyl group of an aldehyde or ketone to a methylene group ($-CH_2-$). One of the first discovered was refluxing the aldehyde or ketone with amalgamated zinc (zinc with a surface layer of mercury) in concentrated HCl.

Clemmensen reduction
Reduction of the C=O group of an aldehyde or ketone to a CH$_2$ group using Zn(Hg) and HCl.

This reaction is known as the **Clemmensen reduction** after the German chemist, E. Clemmensen, who developed it in 1912. The mechanism of Clemmensen reduction, although not well understood, involves transfer of electrons from the Zn to reduce the carbonyl group.

Because the Clemmensen reduction requires the use of concentrated HCl, it cannot be used to reduce a carbonyl group in a molecule that also contains acid-sensitive groups, such as a tertiary alcohol that might undergo dehydration, or an acetal that is hydrolyzed so that its resulting carbonyl group is also reduced.

Wolff-Kishner reduction
Reduction of the C=O group of an aldehyde or ketone to a CH$_2$ group using hydrazine and base.

The **Wolff-Kishner reduction**, discovered independently by N. Kishner in 1911 and L. Wolff in 1912 and reported within months of Clemmensen's discovery, is an alternative method for reduction of a carbonyl group to a methylene group. In this reduction, a mixture of the aldehyde or ketone, hydrazine, and concentrated potassium hydroxide is heated at reflux in a high-boiling solvent such as diethylene glycol (bp 245°C).

Mechanism   Wolff-Kishner Reduction

Step 1: Reaction of the carbonyl group of the aldehyde or ketone with hydrazine gives a hydrazone (Section 16.8B).

Step 2: Take a proton away. Hydroxide removes a proton on the terminal nitrogen on the hydrazone.

$$R \atop R' \diagdown C = \ddot{N} - \ddot{N} - H \quad + \quad \ddot{O} - H \quad \rightleftharpoons \quad H - \ddot{O} - H \quad + \quad {R \atop R'} \diagdown C = \ddot{N} - \ddot{N} - H \quad \longleftrightarrow \quad {R \atop R'} \diagdown \overset{-}{C} - \ddot{N} = \ddot{N} - H$$

Step 3: Add a proton. Protonation takes place on the carbon atom to complete the tautomerism process (compare keto-enol tautomerism, Section 16.9).

$$\overset{R}{\underset{R'}{\diagdown}} \overset{-}{C} - \ddot{N} = \ddot{N} - H \quad + \quad H - \ddot{O} - H \quad \rightleftharpoons \quad R - \overset{H}{\underset{R'}{\overset{|}{C}}} - \ddot{N} = \ddot{N} - H \quad + \quad {}^{-}\ddot{O} - H$$

Step 4: Take a proton away. Hydroxide again removes a proton on the terminal nitrogen.

$$R - \overset{H}{\underset{R'}{\overset{|}{C}}} - \ddot{N} = \ddot{N} - H \quad + \quad {}^{-}\ddot{O} - H \quad \rightleftharpoons \quad R - \overset{H}{\underset{R'}{\overset{|}{C}}} - \ddot{N} = \overset{-}{\ddot{N}} \quad + \quad H - \ddot{O} - H$$

Step 5: Break a bond to give stable molecules or ions. Loss of the stable molecule nitrogen drives formation of a highly basic carbanion.

$$R - \overset{H}{\underset{R'}{\overset{|}{C}}} - \ddot{N} = \overset{-}{\ddot{N}} \quad \longrightarrow \quad R - \overset{H}{\underset{R'}{\overset{|}{\overset{-}{C}}}} \quad + \quad :N \equiv N: \quad \textbf{Nitrogen}$$

A carbanion

Step 6: Add a proton. Protonation of the highly basic carbanion gives the final reduced product in which the carbonyl O atom has been replaced with two H atoms.

$$R - \overset{H}{\underset{R'}{\overset{|}{\overset{-}{C}}}} \quad + \quad H - \ddot{O} - H \quad \longrightarrow \quad R - \overset{H}{\underset{R'}{\overset{|}{C}}} - H \quad + \quad {}^{-}\ddot{O} - H$$

Each of the reductions has its special conditions, advantages, and disadvantages. The Clemmensen reduction cannot be used in the presence of groups sensitive to concentrated acid, the Wolff-Kishner reduction cannot be used in the presence of groups sensitive to concentrated base. However, the carbonyl group of almost any aldehyde or ketone can be reduced to a methylene group by one of these methods.

Example 16.13

Complete the following reactions.

(a)

[structure of Vanillin with CHO, OCH₃, OH groups]

Zn/Hg, HCl
heat
→

Vanillin
(from vanilla beans)

(b)

[structure of Camphor]

N₂H₄, KOH
diethylene glycol,
heat
→

Camphor

Solution

Reaction (a) is a Clemmensen reduction, and reaction (b) is a Wolff-Kishner reduction.

(a)

[structure with CH₃, OCH₃, OH groups]

(b)

[bicyclic structure]

Problem 16.13

Complete the following reactions.

(a)

[structure of Civetone, large ring with =O]

Zn/Hg, HCl
heat
→

Civetone
(from the civet cat; used
in perfumery)

(b)

[structure of Citronellal with CHO group]

N₂H₄, KOH
diethylene glycol,
heat
→

Citronellal
(from citronella and
lemon grass oils)

16.12 | Reactions at an α-Carbon

A. Racemization

When enantiomerically pure (either R or S) 3-phenyl-2-butanone is dissolved in ethanol, no change occurs in the optical activity of the solution over time. If, however, a trace of either acid (for example, aqueous or gaseous HCl) or base (for example, sodium ethoxide) is added, the optical activity of the solution begins to decrease gradually and eventually drops to zero. When 3-phenyl-2-butanone is isolated from this solution, it is found to be a racemic mixture. Furthermore, the rate of racemization is proportional to the concentration of acid or base. These observations can be explained by a rate-determining acid- or base-catalyzed formation of an achiral enol intermediate. Tautomerism of the achiral enol to the chiral keto form generates the R and S enantiomers with equal probability.

[structures: (R)-3-Phenyl-2-butanone ⇌ An achiral enol ⇌ (S)-3-Phenyl-2-butanone]

(R)-3-Phenyl-2-butanone   An achiral enol   **(S)-3-Phenyl-2-butanone**

Racemization by this mechanism occurs only at α-carbon chiral centers with at least one α-hydrogen.

B. Deuterium Exchange

When an aldehyde or ketone with one or more α-hydrogens is dissolved in an aqueous solution enriched with D_2O and also containing catalytic amounts of either D^+ or OD^-, exchange of α-hydrogens occurs at a rate that is proportional to the concentration of the acid or base catalyst. We account for incorporation of deuterium by proposing a rate-determining acid- or base-catalyzed enolization followed by incorporation of deuterium as the enol form converts to the keto form.

$$\underset{\textbf{Acetone}}{\overset{O}{\underset{\|}{CH_3CCH_3}}} + 6\ D_2O \underset{\xrightleftharpoons}{\overset{D^+ \text{ or } OD^-}{}} \underset{\textbf{Acetone-}d_6}{\overset{O}{\underset{\|}{CD_3CCD_3}}} + 6\ HOD$$

Deuterium exchange has two values. First, by observing changes in hydrogen ratios before and after deuterium exchange, it is possible to determine the number of exchangeable α-hydrogens in a molecule. Second, exchange of α-hydrogens is a convenient way to introduce an isotopic label into molecules.

In naming compounds, the presence of deuterium is shown by the symbol "d," and the number of deuterium atoms is shown by a subscript following it. In addition to acetone-d_6, more than 225 deuterium-labeled compounds are available commercially, in isotopic enrichments of up to 99.8 atom % D. Among these are

CDCl₃	CD₃COD	CH₃COD	NaBD₄	CH₃CH₂OD
Chloroform-d	**Acetic-d₃ acid-d**	**Acetic acid-d**	**Sodium borodeuteride**	**Ethanol-d**

Deuterated solvents, such as $CDCl_3$, acetone-d_6, and benzene-d_6, are used as solvents in ^{1}H-NMR spectroscopy because they lack protons that might otherwise obscure the spectrum of the compound of interest.

C. α-Halogenation

Aldehydes and ketones with at least one α-hydrogen react at the α-carbon with bromine and chlorine to form α-haloaldehydes and α-haloketones as illustrated by bromination of acetophenone.

Acetophenone + Br₂ → (CH₃COOH) → α-Bromoacetophenone + HBr

Bromination or chlorination at an α-carbon is catalyzed by both acid and base. For acid-catalyzed halogenation, acid generated by the reaction catalyzes further reaction. The slow step of acid-catalyzed halogenation is formation of an enol. This is followed by rapid reaction of the double bond with halogen to give the α-haloketone.

Mechanism Acid-Catalyzed α-Halogenation of a Ketone

Step 1: **Keto-enol tautomerism.**   Acid-catalyzed keto-enol tautomerism gives the enol.

$$R'-\overset{\overset{\displaystyle O}{\|}}{C}-\overset{\overset{\displaystyle H}{|}}{\underset{\underset{\displaystyle R}{|}}{C}}-R \underset{\text{slow}}{\rightleftharpoons} \overset{H-O}{\underset{R'}{\diagdown}}C=C\overset{R}{\diagup}$$

Step 2: **Make a new bond between a π bond and an electrophile.**   Nucleophilic attack of the enol on the electrophilic halogen molecule, X₂, gives the conjugate acid of an α-haloketone.

$$\overset{H-\ddot{O}:}{\underset{R'}{\diagdown}}C=C\overset{R}{\underset{R}{\diagup}} + :\ddot{Br}-\ddot{Br}: \overset{\text{fast}}{\longrightarrow} \overset{H-\overset{+}{\ddot{O}}}{\underset{R'}{\diagdown}}C-\overset{:\ddot{Br}:}{\underset{\underset{R}{|}}{C}}-R + :\ddot{Br}:^-$$

Step 3: **Take a proton away.**   Proton transfer to a base, in this case a molecule of acetic acid, gives the α-haloketone.

$$CH_3-\overset{\overset{\displaystyle :O:}{\|}}{C}-OH + \overset{H-\overset{+}{\ddot{O}}}{\underset{R'}{\diagdown}}C-\overset{:\ddot{Br}:}{\underset{\underset{R}{|}}{C}}-R \overset{\text{fast}}{\longrightarrow} \overset{:\ddot{O}:}{\underset{R'}{\diagdown}}C-\overset{:\ddot{Br}:}{\underset{\underset{R}{|}}{C}}-R + CH_3-\overset{\overset{\displaystyle :\overset{+}{O}-H}{\|}}{C}-OH$$

An α-bromoketone

The slow step in base-promoted α-halogenation is removal of an α-hydrogen by base to form an enolate anion, which then reacts with halogen by nucleophilic displacement to form the final product. This procedure for α-halogenation produces HX as a by-product, and, in order to keep the solution basic, it is necessary to add slightly more than one mole of base per mole of aldehyde or ketone. Because base is a reactant required in equimolar amounts, we say that this reaction is base-promoted rather than base-catalyzed.

Mechanism Base-Promoted α-Halogenation of a Ketone

Step 1: **Take a proton away.**   Proton transfer from the α-carbon to the base gives a resonance-stabilized enolate anion.

$$R'-\overset{\overset{\displaystyle :\ddot{O}}{\|}}{C}-\overset{\overset{\displaystyle H}{|}}{\underset{\underset{\displaystyle R}{|}}{C}}-R + :\ddot{O}H^- \underset{\text{slow}}{\rightleftharpoons} \overset{:\ddot{O}:}{\underset{R'}{\diagdown}}C-C\overset{R}{\underset{R}{\diagup}} \longleftrightarrow \overset{:\ddot{O}:^-}{\underset{R'}{\diagdown}}C=C\overset{R}{\underset{R}{\diagup}} + H_2\ddot{O}:$$

Resonance-stabilized enolate anion

Step 2: **Make a new bond between a nucleophile (π bond) and an electrophile.** Nucleophilic attack of the enolate anion on halogen gives an α-bromoketone.

$$\overset{:\ddot{O}:^-}{\underset{R'}{\diagdown}}C=C\overset{R}{\underset{R}{\diagup}} + :\ddot{Br}-\ddot{Br}: \overset{\text{fast}}{\longrightarrow} \overset{:\ddot{O}:}{\underset{R'}{\diagdown}}C-\overset{:\ddot{Br}:}{\underset{\underset{R}{|}}{C}}-R + :\ddot{Br}:^-$$

A major difference exists between acid-catalyzed and base-promoted α-halogenation. In principle, both can lead to polyhalogenation. In practice, the rate of acid-catalyzed introduction of a second halogen is considerably lower than the rate of the first halogenation because the electronegative α-halogen decreases the basicity of the carbonyl oxygen toward protonation. Thus, it is generally possible to stop acid-catalyzed halogenation at a single substitution. For base-promoted halogenation, each successive halogenation is more rapid than the previous one because introduction of an electronegative halogen atom on an α-carbon further increases the acidity of remaining α-hydrogens; thus, each successive α-hydrogen is removed more rapidly than the previous one. For this reason, base-promoted halogenation is generally not a useful synthetic reaction.

Summary

Section 16.1 Structure and Bonding

- An **aldehyde** contains a carbonyl group bonded to a hydrogen atom and a carbon atom.
- A **ketone** contains a carbonyl group bonded to two carbon atoms.

Section 16.2 Nomenclature

- An aldehyde is named by changing -*e* of the parent alkane to -*al*.
- A ketone is named by changing -*e* of the parent alkane to -*one* and using a number to locate the carbonyl group.
- In naming compounds that contain more than one functional group, the IUPAC system has established an **order of precedence of functions**.
  - A selection of functional groups ranked from highest to lowest in order of precedence: carboxylic acid (highest), aldehyde, ketone, alcohol, amine, thiol (lowest).
  - If the carbonyl group of an aldehyde or ketone is lower in precedence than other functional groups in the molecule, it is indicated by the infix -*oxo*-.

Problems: 16.1–16.3, 16.14, 16.15

Section 16.3 Physical Properties

- Aldehydes and ketones are polar compounds and interact in the pure state by dipole-dipole interactions.
  - They have higher boiling points and are more soluble in water than nonpolar compounds of comparable molecular weight.

Problems: 16.16, 16.17

Section 16.4 Reactions

- An important structural feature of a carbonyl group is the strong dipole moment in which there is a partial negative charge on oxygen and a partial positive charge on carbon.
  - Lewis acids such as protons react with carbonyl groups at the oxygen atom.
  - Nucleophiles react with carbonyl groups at the carbon atom.
  - One of the most common reaction themes of aldehydes and ketones is addition of a nucleophile to the carbonyl carbon to form a tetrahedral carbonyl addition compound.
  - Often, a new chiral center is created by this reaction.
  - When none of the starting materials is chiral, a racemic mixture is formed.
  - Many of these reactions form new carbon-carbon bonds, making this a very important class of reactions in organic synthesis.

Section 16.5 Addition of Carbon Nucleophiles

- **Grignard reagents** add to formaldehyde, aldehydes, and ketones to give primary, secondary, and tertiary alcohols, respectively.
- **Organolithium** and **terminal alkyne anions** add to the carbonyl group of aldehydes and ketones to give alcohols according to a mechanism similar to addition of Grignard reagents.
- **Hydrogen cyanide (HCN)** adds to aldehydes and ketones to give cyanohydrins.

Problems: 16.4, 16.18–16.22, 16.51, 16.52, 16.54–16.56, 16.59, 16.61, 16.62, 16.68, 16.69

Section 16.6 The Wittig Reaction

Problems: 16.5, 16.23–16.28, 16.51, 16.52, 16.53, 16.60, 16.61

- **Wittig reactions** involve addition of phosphonium **ylides** (deprotonated phosphonium salts) with aldehyde and ketone carbonyl groups to give alkenes.
  - *E* and *Z* products are generally both formed.
  - Wittig reagents with anion stabilizing groups (such as carbonyls) adjacent to the negatively charged carbon are more *E* selective.

Section 16.7 Addition of Oxygen Nucleophiles

Problems: 16.6, 16.29–16.37

- Water adds to aldehydes and ketones to give hydrates, which are geminal diols.
  - The reaction is only favorable for simple aldehydes, especially formaldehyde.
- Alcohols add to aldehydes and ketones to give **hemiacetals** (one alcohol added), then **acetals** (two alcohols added).
  - Hemiacetals are only stable when five- and six-membered rings are formed from a carbonyl and OH group on the same molecule.
  - Hemiacetal formation can be catalyzed by either acid or base.
  - Acetal formation can be catalyzed only by acid.
  - The overall process of acetal formation from alcohols and an aldehyde or ketone is acid catalyzed and reversible.
  - The relative ratio of alcohol to water in the reaction determines the ratio of carbonyl to acetal species present at equilibrium.
  - Water is removed from reactions to favor acetal formation using a **Dean-Stark trap**.

Problems: 16.67, 16.70, 16.71

- Carbohydrates are predominantly found in the cyclic hemiacetal form.
  - The **anomeric carbon** is the carbon bonded to two oxygen atoms.
  - The anomeric carbon can be formed as either the *α* or *β* **anomer**.

Problems: 16.7, 16.63, 16.64

- Acetals, usually as five- or six-membered rings, are often used as carbonyl-**protecting groups**.
  - A protecting group reversibly masks the reactivity of a molecule so that unwanted side reactions are prevented.
  - Grignard reagents can be prepared from molecules containing carbonyl groups as long as the carbonyl group is protected as a cyclic acetal.
  - Cyclic acetal protecting groups are removed by adding excess aqueous acid.
  - **Tetrahydropyranyl ethers**, which are cyclic hemiacetals, can be used as a protecting group for alcohols.

Section 16.8 Addition of Nitrogen Nucleophiles

- Ammonia and primary amines add to aldehydes and ketones to give **imines**, sometimes called **Schiff bases**, that have carbon-nitrogen double bonds.

Problems: 16.8, 16.38–16.40, 16.43, 16.44

- Secondary amines, especially cyclic secondary amines such as piperidine, form **enamines** with aldehydes and ketones.
  - Enamines have a carbon-nitrogen single bond with an adjacent carbon-carbon double bond.
- **Hydrazine** and related compounds react with aldehydes and ketones to give analogous products with carbon-nitrogen double bonds.

Section 16.9 Keto-Enol Tautomerism

Problems: 16.9, 16.10, 16.41, 16.42, 16.57, 16.58, 16.65

- The carbon atom adjacent to a carbonyl group is called an **α-carbon**, and a hydrogen bonded to it is called an **α-hydrogen**.
  - The pK_a of an α-hydrogen of an aldehyde or ketone is approximately 20, which makes it less acidic than alcohols but more acidic than terminal alkynes.
  - The α-hydrogen is acidic because the deprotonated **enolate anion** is stabilized by delocalization of charge through resonance.
- Aldehydes and ketones equilibrate between **keto** and **enol forms**.
  - Keto-enol equilibration is catalyzed by both acid and base.
  - Base catalysis of keto-enol equilibration involves an enolate anion intermediate.
  - The keto form is favored for most aldehydes and ketones.
  - Molecules that favor the enol form at equilibrium generally have conjugated enols.

Section 16.10 Oxidation

- Aldehydes are easily oxidized to carboxylic acids using a variety of reagents including chromic acid, silver salts, peroxides, and molecular oxygen, O_2.
- Ketones are not easily oxidized, requiring extremely strong oxidizing agents as well as heat.

Problems: 16.11, 16.43, 16.44

Section 16.11 Reduction

- Aldehydes and ketones are reduced to primary and secondary alcohols, respectively.
  - **Metal hydride** reducing agents such as $LiAlH_4$ and $NaBH_4$ are effective for reducing aldehydes and ketones.
  - Metal hydride reducing agents do not reduce carbon-carbon double bonds.
  - Hydrogenation using H_2 and a transition metal catalyst can be used to reduce aldehyde or ketone carbonyls, although carbon-carbon double bonds in a molecule may also be reduced.
  - Carbon-carbon double bonds are easier to reduce than carbonyls using hydrogenation, so conditions can often be found in which only a carbon-carbon double bond is reduced in the presence of an aldehyde or ketone carbonyl group.
  - Conversely, by using metal hydride reagents, carbonyls can be reduced without reducing carbon-carbon double bonds.

Problems: 16.12, 16.13, 16.43, 16.44, 16.51, 16.52, 16.59

- In **reductive amination**, ketones or aldehydes react with amines in the presence of an appropriate reducing agent such as $NaBH_3CN$ to give substituted amines.
- Aldehyde and ketone carbonyl groups can be reduced to methyl or methylene groups, respectively, using two different complementary reactions.
  - The **Clemmensen reduction** uses amalgamated zinc $Zn(Hg)$ in strong acid.
  - The **Wolff-Kishner reduction** uses hydrazine and base.
  - The Clemmensen reduction is used when the molecule is stable to strong acid, and the Wolff-Kishner reduction is used when the molecule is stable to strong base.

Section 16.12 Reactions at an α-Carbon

- The α-carbon of aldehydes and ketones has special reactivity derived from the keto-enol equilibrium (referred to as **tautomerization**) that occurs in acid or base.
  - α-Carbon chiral centers of aldehydes and ketones racemize in the presence of acid or base through formation of an achiral enol during keto-enol tautomerization.
  - Deuterium can be exchanged for α-hydrogens catalyzed by acid or base through formation of the enol during keto-enol tautomerization.
- Aldehydes or ketones with at least one α-hydrogen react with halogens in acid or base to form α-haloaldehydes and α-haloketones.

Problems: 16.45–16.50, 16.66

⬤WL Questions and problems for this chapter are available in OWL. Use the OWL access card that came with this textbook to access assigned questions and problems for this chapter.

Key Reactions

1. **Reaction with Grignard Reagents (Section 16.5A)** Treating formaldehyde with a Grignard reagent followed by hydrolysis gives a primary alcohol. Similar treatment of any other aldehyde gives a secondary alcohol. Treatment of a ketone gives a tertiary alcohol. The mechanism involves an initial attack of the nucleophilic carbon of the Grignard reagent at the electrophilic carbon of the carbonyl to give the alkoxide Mg salt. The flask is opened and the acid is added to protonate the alkoxide, and produce the alcohol.

2. Reaction with Organolithium Reagents (Section 16.5B)
Reactions of aldehydes and ketones with organolithium reagents are similar to those with Grignard reagents.

$$\text{(ketone)} \xrightarrow[\text{2.HCl, H}_2\text{O}]{\text{1.C}_6\text{H}_5\text{Li}} \text{(alcohol, } C_6H_5 \text{)}$$

(produced as a
racemic mixture)

The mechanism involves an initial attack of the nucleophilic carbon of the organolithium reagent at the electrophilic carbon of the carbonyl to give the alkoxide Li salt. The flask is opened and the acid is added to protonate the alkoxide, thereby producing the alcohol product.

3. Reaction with Anions of Terminal Alkynes (Section 16.5C)
Treating an aldehyde or ketone with the alkali metal salt of a terminal alkyne followed by hydrolysis gives an α-alkynylalcohol.

$$\text{(cyclohexanone)} \xrightarrow[\text{2.HCl, H}_2\text{O}]{\text{1.HC}\equiv\text{C}^-\text{Na}^+} \text{(OH, C}\equiv\text{CH)}$$

4. Reaction with HCN to Form Cyanohydrins (Section 16.5D)
For aldehydes and most sterically unhindered aliphatic ketones, equilibrium favors formation of the cyanohydrin. For aryl ketones, equilibrium favors starting materials, and little cyanohydrin is obtained. The mechanism involves addition of the CN anion to the carbonyl carbon, followed by protonation of the resulting alkoxide.

$$C_6H_5\overset{\text{O}}{\overset{\|}{\text{C}}}\text{H} + \text{HC}\equiv\text{N} \xrightarrow{\text{NaCN}} C_6H_5\overset{\text{OH}}{\overset{|}{\text{C}}}\text{HC}\equiv\text{N}$$

(Formed as a
racemic mixture)

5. The Wittig Reaction (Section 16.6)
Treating an aldehyde or ketone with a triphenylphosphonium ylide gives a betaine intermediate, which rearranges to an oxaphosphetane intermediate, which in turn fragments to give triphenylphosphine oxide and an alkene.

$$\text{(cyclohexanone)}=\text{O} + \text{Ph}_3\overset{+}{\text{P}}-\overset{-}{\text{CH}_2} \longrightarrow \text{(cyclohexane)}=\text{CH}_2 + \text{Ph}_3\text{P}=\text{O}$$

6. The Horner-Emmons-Wadsworth Modification of the Wittig Reaction (Section 16.6)
This modification of the original Wittig reaction uses a phosphonate ester derived from an α-haloester, aldehyde, or ketone to generate the Wittig carbanion and shows very high E selectivity.

$$(\text{MeO})_2\overset{\text{O}}{\overset{\|}{\text{P}}}-\text{CH}_2-\overset{\text{O}}{\overset{\|}{\text{C}}}-\text{OEt} \xrightarrow[\text{2.}]{\text{1. strong base}} \text{(alkene)} \overset{\text{O}}{\overset{\|}{\text{C}}}\text{OEt} + \text{MeO}-\overset{\text{O}}{\overset{\|}{\underset{\text{OMe}}{\text{P}}}}-\text{O}^-$$

7. Hydration (Section 16.7A)
The degree of hydration is greater for aldehydes than for ketones.

$$\text{H}\overset{\text{O}}{\overset{\|}{\text{C}}}\text{H} + \text{H}_2\text{O} \rightleftharpoons \text{H}\overset{\text{OH}}{\underset{\text{H}}{\overset{|}{\text{C}}}}\text{OH}$$

(>99%)

8. **Addition of Alcohols to Form Hemiacetals (Section 16.7B)**   Hemiacetals are only minor components of an equilibrium mixture of aldehyde or ketone and alcohol, except where the —OH and the C=O are parts of the same molecule and a five- or six-membered ring can form. The reaction is catalyzed by acid or base. The mechanism involves protonation of the carbonyl oxygen atom that facilitates attack by an alcohol at the carbonyl carbon, followed by loss of a proton to give the hemiacetal.

9. **Addition of Alcohols to Form Acetals (Section 16.7B)**   Formation of acetals is catalyzed by acid. Acetals are stable to water and aqueous base but are hydrolyzed in aqueous acid. Acetals are valuable as carbonyl-protecting groups. The mechanism for conversion of a hemiacetal to an acetal involves protonation of the OH group of the hemiacetal followed by loss of water to give a resonance-stabilized cation, which is attacked by the second molecule of alcohol.

10. **Addition of Ammonia and Its Derivatives: Formation of Imines (Section 16.8A)** Addition of ammonia or a primary amine to the carbonyl group of an aldehyde or ketone forms a tetrahedral carbonyl addition compound. Loss of water from this intermediate gives an imine. The mechanism for imine formation involves an initial attack of the nucleophilic nitrogen atom on the carbonyl carbon atom, followed by proton transfer to the OH, creating an H_2O group that then departs.

11. **Addition of Secondary Amines: Formation of Enamines (Section 16.8A)**   Addition of a secondary amine to the carbonyl group of an aldehyde or ketone forms a tetrahedral carbonyl addition intermediate. Acid-catalyzed dehydration of this intermediate gives an enamine.

12. **Addition of Hydrazine and Its Derivatives (Section 16.8B)**   Treating an aldehyde or ketone with hydrazine gives a hydrazone.

Derivatives of hydrazine react similarly.

13. **Keto-Enol Tautomerism (Section 16.9B)**   The keto form predominates at equilibrium, except for those aldehydes and ketones in which the enol is stabilized by resonance or hydrogen bonding.

14. Oxidation of an Aldehyde to a Carboxylic Acid (Section 16.10A) The aldehyde group is among the most easily oxidized functional groups. Oxidizing agents include H_2CrO_4, $KMnO_4$, Ag_2O, Tollens' reagent, H_2O_2, and O_2.

15. Metal Hydride Reduction (Section 16.11A) Both $LiAlH_4$ and $NaBH_4$ are selective in that neither reduces isolated carbon-carbon double or triple bonds. The mechanism for metal hydride reducing agents involves initial nucleophilic attack by an H^- equivalent, followed by protonation of the alkoxide ion to give the alcohol.

(racemic)

16. Catalytic Reduction (Section 16.11B) Catalytic reduction of the carbonyl group of an aldehyde or ketone to a hydroxyl group is simple to carry out and yields of the alcohols are high. A disadvantage of this method is that some other functional groups, including carbon-carbon double and triple bonds, may also be reduced.

17. Reductive Amination (Section 16.11D) Ketones or aldehydes react with amines in the presence of an appropriate reducing agent such as $NaBH_3CN$ to give substituted amines. The mechanism involves initial reaction of the carbonyl and amine to form an imine intermediate that is reduced by the $NaBH_3CN$ to give the substituted amine.

18. Clemmensen Reduction (Section 16.11E) Reduction of the carbonyl group of an aldehyde or ketone using amalgamated zinc in the presence of concentrated hydrochloric acid gives a methylene group.

19. Wolff-Kishner Reduction (Section 16.11E) Formation of a hydrazone followed by treatment with base, commonly KOH in diethylene glycol or potassium *tert*-butoxide in dimethyl sulfoxide, reduces the carbonyl group of an aldehyde or ketone to a methylene group. The mechanism involves initial hydrazone formation, followed by base-catalyzed tautomerization, deprotonation, loss of N_2, and reprotonation of the carbanion intermediate to give the fully reduced product.

20. Deuterium Exchange at an α-Carbon (Section 16.12B) Acid- or base-catalyzed deuterium exchange at an α-carbon involves formation of an enol or enolate anion intermediate:

$$CH_3\overset{\displaystyle O}{\overset{\|}{C}}CH_3 + 6D_2O \underset{\displaystyle}{\overset{DCl}{\rightleftharpoons}} CD_3\overset{\displaystyle O}{\overset{\|}{C}}CD_3 + 6HOD$$

21. Halogenation at an α-Carbon (Section 16.12C) The rate-determining step in acid-catalyzed α-halogenation is the formation of an enol. In base-promoted α-halogenation, it is formation of an enolate anion. Acid catalyzed α-halogenation involves reaction of the halogen with the enol form at the carbonyl compound. Base promoted α-halogenation involves reaction of the enolate anion intermediate with the halogen.

Chapter 17

Carboxylic Acids

The active ingredients in these two nonprescription pain relievers are derivatives of arylpropanoic acids. Inset: a model of (S)-ibuprofen. See Chemical Connections: "From Willow Bark to Aspirin and Beyond."

© Charles D. Winters/Cengage Learning

The most important chemical property of carboxylic acids, another class of organic compounds containing the carbonyl group, is their acidity. Furthermore, carboxylic acids form numerous important derivatives, including esters, amides, anhydrides, and acid halides. In this chapter, we study carboxylic acids and in Chapter 18 we study their derivatives.

17.1 | Structure

The functional group of a carboxylic acid is the **carboxyl group** (Section 1.3D), so named because it is made up of a *carb*onyl group and a hydr*oxyl* group. Following is a Lewis structure of the carboxyl group as well as three alternative representations for it.

$$-\overset{\overset{\displaystyle \ddot{O}:}{\|}}{\underset{\displaystyle \ddot{O}-H}{C}} \qquad -\overset{\overset{\displaystyle O}{\|}}{\underset{\displaystyle O-H}{C}} \qquad -COOH \qquad -CO_2H$$

Alternative representations of a carboxyl group

The general formula for an aliphatic carboxylic acid is RCOOH; the general formula for an aromatic carboxylic acid is ArCOOH.

OWL Online homework for this chapter may be assigned in OWL for Organic Chemistry.

17.2 | Nomenclature

A. IUPAC System

The IUPAC name of a carboxylic acid is derived from that of the longest carbon chain that contains the carboxyl group by dropping the final *-e* from the name of the parent alkane and adding the suffix *-oic* followed by the word "acid" (Section 2.3C). The chain is numbered beginning with the carbon of the carboxyl group. Because the carboxyl carbon is understood to be carbon 1, there is no need to give it a number. The IUPAC system retains the common names formic acid and acetic acid, which are always used to refer to these acids.

HCOOH	CH₃COOH	
Methanoic acid	**Ethanoic acid**	**3-Methylbutanoic acid**
(Formic acid)	**(Acetic acid)**	**(Isovaleric acid)**

If the carboxylic acid contains a carbon-carbon double or triple bond, change the infix from *-an-* to *-en-* or *-yn-*, as the case may be, to indicate the presence of the multiple bond and show the location of the multiple bond by a number.

Propenoic acid — **(Acrylic acid)**   ***trans*-2-Butenoic acid** — **(Crotonic acid)**   ***trans*-3-Phenylpropenoic acid** — **(Cinnamic acid)**

In the IUPAC system, a carboxyl group takes precedence over most other functional groups (Table 16.1), including hydroxyl groups, amino groups, and the carbonyl groups of aldehydes and ketones. As illustrated in the following examples, an —OH group is indicated by the prefix *hydroxy-*; an —NH₂ group, by *amino-*; and the C=O group of an aldehyde or ketone, by *oxo-*.

(*R*)-5-Hydroxyhexanoic acid   **5-Oxohexanoic acid**   **4-Aminobutanoic acid**

Dicarboxylic acids are named by adding the suffix *-dioic acid* to the name of the carbon chain that contains both carboxyl groups. The numbers of the carboxyl carbons are not indicated because they can be only at the ends of the parent chain. Following are IUPAC and common names for several important aliphatic dicarboxylic acids.

Ethanedioic acid — **(Oxalic acid)**   **Propanedioic acid** — **(Malonic acid)**

Butanedioic acid — **(Succinic acid)**   **Pentanedioic acid** — **(Glutaric acid)**   **Hexanedioic acid** — **(Adipic acid)**

Image copyright Wong Hock Weng 2010. Used under license from Shutterstock.com

Formic acid was first obtained in 1670 from the destructive distillation of ants, whose Latin genus is Formica. *It is one of the components of the venom injected by stinging ants.*

The name *oxalic acid* is derived from one of its sources in the biological world, namely, plants of the genus *Oxalis*, one of which is rhubarb. Adipic acid is one of the two monomers required for the synthesis of the polymer nylon 66 (Section 29.5A). A mnemonic phrase for remembering the common names for the dicarboxylic acids oxalic through adipic is *Oh my, such good apples.*

A carboxylic acid containing a carboxyl group bonded to a cycloalkane ring is named by giving the name of the ring and adding the suffix -*carboxylic acid*. The atoms of the ring are numbered beginning with the carbon bearing the —COOH group.

2-Cyclohexenecarboxylic acid

trans-**1,3-Cyclopentanedicarboxylic acid**

The simplest aromatic carboxylic acid is benzoic acid. Derivatives are named by using numbers to show the location of substituents relative to the carboxyl group. Certain aromatic carboxylic acids have common names by which they are more usually known. For example, 2-hydroxybenzoic acid is more often called salicylic acid, a name derived from the fact that this aromatic carboxylic acid was first isolated from the bark of the willow, a tree of the genus *Salix*.

Benzoic acid

2-Hydroxybenzoic acid (Salicylic acid)

1,2-Benzenedicarboxylic acid (Phthalic acid)

1,4-Benzenedicarboxylic acid (Terephthalic acid)

Aromatic dicarboxylic acids are named by adding the words *dicarboxylic acid* to "benzene," for example, 1,2-benzenedicarboxylic acid and 1,4-benzenedicarboxylic acid. Each is more usually known by its common name: phthalic acid and terephthalic acid, respectively. Terephthalic acid is one of the two organic components required for the synthesis of the textile fiber known as Dacron polyester, or Dacron (Section 29.5B).

B. Common Names

Aliphatic carboxylic acids, many of which were known long before the development of structural theory and IUPAC nomenclature, are named according to their source or for some characteristic property. Table 17.1 lists several of the unbranched aliphatic carboxylic acids found in the biological world along with their common name and Latin or Greek derivation. Those of 16, 18, and 20 carbon atoms are particularly abundant in fats and oils (Section 26.1) and in the phospholipid components of biological membranes (Section 26.5).

When common names are used, the Greek letters α, β, γ, δ, and so forth are often added as a prefix to locate substituents. The α-position in a carboxylic acid is the one next to the carboxyl group; an α-substituent in a common name is equivalent to a 2-substituent in an IUPAC name. GABA is an inhibitory neurotransmitter in the central nervous system of humans. Alanine is one of the 20 protein-derived amino acids.

4-Aminobutanoic acid (γ-Aminobutyric acid, GABA)

(S)-2-Aminopropanoic acid [(S)-α-Aminopropionic acid; L-Alanine]

Table 17.1

Table 17.1 | Several Aliphatic Carboxylic Acids—Their Common Names and Derivations

Structure	IUPAC Name	Common Name	Derivation
HCOOH	Methanoic acid	Formic acid	Latin: *formica*, ant
CH$_3$COOH	Ethanoic acid	Acetic acid	Latin: *acetum*, vinegar
CH$_3$CH$_2$COOH	Propanoic acid	Propionic acid	Greek: *propion*, first fat
CH$_3$(CH$_2$)$_2$COOH	Butanoic acid	Butyric acid	Latin: *butyrum*, butter
CH$_3$(CH$_2$)$_3$COOH	Pentanoic acid	Valeric acid	Latin: *valeriana*, a flowering plant
CH$_3$(CH$_2$)$_4$COOH	Hexanoic acid	Caproic acid	Latin: *caper*, goat
CH$_3$(CH$_2$)$_6$COOH	Octanoic acid	Caprylic acid	Latin: *caper*, goat
CH$_3$(CH$_2$)$_8$COOH	Decanoic acid	Capric acid	Latin: *caper*, goat
CH$_3$(CH$_2$)$_{10}$COOH	Dodecanoic acid	Lauric acid	Latin: *laurus*, laurel
CH$_3$(CH$_2$)$_{12}$COOH	Tetradecanoic acid	Myristic acid	Greek: *myristikos*, fragrant
CH$_3$(CH$_2$)$_{14}$COOH	Hexadecanoic acid	Palmitic acid	Latin: *palma*, palm tree
CH$_3$(CH$_2$)$_{16}$COOH	Octadecanoic acid	Stearic acid	Greek: *stear*, solid fat
CH$_3$(CH$_2$)$_{18}$COOH	Eicosanoic acid	Arachidic acid	Greek: *arachis*, peanut

In common names, the presence of a ketone carbonyl in a substituted carboxylic acid is indicated by the prefix *keto-*, illustrated by the common name β-ketobutyric acid. This substituted carboxylic acid is also named acetoacetic acid. In deriving this name, 3-oxobutanoic acid is regarded as a substituted acetic acid. In the common nomenclature, the substituent is named an **aceto group**, CH$_3$CO—.

Aceto group
A CH$_3$CO— group; also called an acetyl group.

3-Oxobutanoic acid
(β-Ketobutyric acid;
Acetoacetic acid)

Acetyl group
(an aceto group)

Example 17.1

Write the IUPAC name for each carboxylic acid.

(a)

(b) (racemic)

(c)

(d)

Solution

(a) (Z)-9-Octadecenoic acid (oleic acid)
(b) *trans*-2-Hydroxycyclohexanecarboxylic acid
(c) (R)-2-Hydroxypropanoic acid [(R)-lactic acid]
(d) Chloroacetic acid

Problem 17.1

Each of these carboxylic acids has a well-recognized common name. A derivative of glyceric acid is an intermediate in glycolysis. Maleic acid is an intermediate in the tricarboxylic acid (TCA) cycle. Mevalonic acid is an intermediate in the biosynthesis of steroids. Write the IUPAC name for each compound. Be certain to specify configuration.

(a) HO⟍⟍COOH / OH — **Glyceric acid**

(b) COOH⟍⟍COOH — **Maleic acid**

(c) HO CH₃ / HO⟍⟍COOH — **Mevalonic acid**

17.3 | Physical Properties

In the liquid and solid states, carboxylic acids are associated by hydrogen bonding into dimers, as shown here for acetic acid in the liquid state.

Hydrogen bond

$$H_3C-C \quad \underset{\delta^-}{O} \cdots \underset{\delta^+}{H}-O \quad \underset{\delta^+}{O}-H \cdots \underset{\delta^-}{O} \quad C-CH_3$$

Electrostatic potential map
of the acetic acid dimer

Hydrogen bond

Carboxylic acids have significantly higher boiling points than other types of organic compounds of comparable molecular weight, such as alcohols, aldehydes, and ketones. For example, butanoic acid (Table 17.2) has a higher boiling point than either 1-pentanol or pentanal. The higher boiling points of carboxylic acids result from their polarity and from the fact that they form very strong intermolecular hydrogen bonds.

Carboxylic acids also interact with water molecules by hydrogen bonding through both the carbonyl and hydroxyl groups. Because of greater hydrogen-bonding interactions, carboxylic acids are more soluble in water than alcohols, ethers, aldehydes, and ketones of comparable molecular weight. The solubility of a carboxylic acid in water decreases as its molecular weight increases. We account

Table 17.2	Boiling Points and Solubilities in Water of Selected Carboxylic Acids, Alcohols, and Aldehydes of Comparable Molecular Weight			
Structure	**Name**	**Molecular Weight (g/mol)**	**Boiling Point (°C)**	**Solubility (g/100 g H₂O)**
CH_3COOH	**Acetic acid**	60.1	118	Infinite
$CH_3CH_2CH_2OH$	**1-Propanol**	60.1	97	Infinite
CH_3CH_2CHO	**Propanal**	58.1	48	16.0
$CH_3(CH_2)_2COOH$	**Butanoic acid**	88.1	163	Infinite
$CH_3(CH_2)_3CH_2OH$	**1-Pentanol**	88.1	137	2.3
$CH_3(CH_2)_3CHO$	**Pentanal**	86.1	103	Slight
$CH_3(CH_2)_4COOH$	**Hexanoic acid**	116.2	205	1.0
$CH_3(CH_2)_5CH_2OH$	**1-Heptanol**	116.2	176	0.2
$CH_3(CH_2)_5CHO$	**Heptanal**	114.1	153	0.1

for this trend in the following way. A carboxylic acid consists of two regions of distinctly different polarity: a polar **hydrophilic** carboxyl group and, except for formic acid, a nonpolar **hydrophobic** hydrocarbon chain. The hydrophilic carboxyl group increases water solubility; the hydrophobic hydrocarbon chain decreases water solubility.

Hydrophilic
From the Greek, meaning water loving.

Hydrophobic
From the Greek, meaning water fearing.

Chemical Connections

From Willow Bark to Aspirin and Beyond

The first drug developed for widespread use was aspirin, one of today's most common pain relievers. Americans alone consume approximately 80 billion tablets of aspirin a year! The story of the development of this modern pain reliever goes back more than 2000 years. In 400 BCE, the Greek physician Hippocrates recommended chewing bark of the willow tree to alleviate the pain of childbirth and to treat eye infections.

The active component of willow bark was found to be salicin, a compound composed of salicyl alcohol bonded to a unit of β-D-glucose (Section 25.1). Hydrolysis of salicin in aqueous acid gives salicyl alcohol, which can be oxidized to salicylic acid. Salicylic acid proved to be an even more effective reliever of pain, fever, and inflammation than salicin, without its extremely bitter taste. Unfortunately, salicylic acid causes severe irritation of the mucous membrane lining of the stomach.

Salicin   **Salicyl alcohol**   **Salicylic acid**

In the search for less irritating but still effective derivatives of salicylic acid, chemists at the Bayer division of I. G. Farben in Germany in 1883 prepared acetylsalicylic acid and gave it the name aspirin.

Salicylic acid   **Acetylsalicylic acid (Aspirin)**

Aspirin proved to be less irritating to the stomach than salicylic acid and also more effective in relieving the pain and inflammation of rheumatoid arthritis. Aspirin, however, is still irritating to the stomach and frequent use can cause duodenal ulcers in susceptible persons.

In the 1960s, in a search for even more effective and less irritating analgesics and antiinflammatory drugs, chemists at the Boots Pure Drug Company in England synthesized a series of compounds related in structure to salicylic acid. Among them, they discovered an even more potent compound, which they named ibuprofen. Soon thereafter, Syntex Corporation in the United States developed naproxen, the active ingredient in Aleve. Each compound has one chiral center and can exist as a pair of enantiomers. For each drug, the physiologically active form is the *S* enantiomer.

(S)-Naproxen

(S)-Ibuprofen

In the 1960s, it was discovered that aspirin acts by inhibiting cyclooxygenase (COX), a key enzyme in the conversion of arachidonic acid to prostaglandins (Section 26.3). With this discovery, it became clear why only one enantiomer of ibuprofen and naproxen is active: only the *S* enantiomer of each has the correct handedness to bind to COX and inhibit its activity.

Recently it was recognized that there are actually two cyclooxygenases; one is more important for the inflammation pathway, and the other affects the stability of blood vessels. Aspirin and other nonsteroidal antiinflammatory drugs (NSAIDs) inhibit both, which is why they can cause gastrointestinal bleeding. New drugs (Celebrex, for example) have been developed that inhibit only the inflammatory enzyme pathway and are remarkably effective for suppression of inflammation (for example, in arthritis) without the gastrointestinal side effects. However, drugs like Celebrex have been found to have an increased risk of heart attack or stroke.

hydrophobic (nonpolar) tail

hydrophilic (polar) head

Decanoic acid
(0.2 g/100 mL H₂O)

Electrostatic potential map of decanoic acid

The first four aliphatic carboxylic acids (formic, acetic, propanoic, and butanoic acids) are infinitely soluble in water because the hydrophobic character of the hydrocarbon chain is more than counterbalanced by the hydrophilic character of the carboxyl group. As the size of the hydrocarbon chain increases relative to the size of the hydrophilic group, water solubility decreases. The solubility of hexanoic acid is 1.0 g/100 g H₂O, while that of decanoic acid is only 0.2 g/100 g H₂O.

One other physical property of carboxylic acids must be mentioned. The liquid carboxylic acids from propanoic acid to decanoic acid have extremely foul odors, about as bad as those of thiols, though different. Butanoic acid is found in stale perspiration, is a major component of "locker room odor," and provides the characteristic odor of regurgitated milk. Pentanoic acid smells even worse, and goats, which secrete C_6, C_8, and C_{10} acids, are famous for their unpleasant odors.

17.4 | Acidity

A. Acid Ionization Constants

Carboxylic acids are weak acids. Values of K_a for most unsubstituted aliphatic and aromatic carboxylic acids fall within the range 10^{-4} to 10^{-5}. The value of K_a for acetic acid, for example, is 1.74×10^{-5}. Its pK_a is 4.76.

$$CH_3COOH + H_2O \rightleftharpoons CH_3COO^- + H_3O^+ \quad K_a = \frac{[CH_3COO^-][H_3O^+]}{[CH_3COOH]} = 1.74 \times 10^{-5}$$

$$pK_a = 4.76$$

As we discussed in Sections 4.6C and 4.6D, the greater acidity of carboxylic acids (pK_a 4–5) compared with alcohols (pK_a 16–18) is because of delocalization of the negative charge of the carboxylate anion through resonance, and because of the electron-withdrawing inductive effect of the carbonyl group. There is no comparable resonance or inductive stabilization of alkoxide ions.

We saw in Section 4.6D that substitution at the α-carbon of an atom or group of atoms of higher electronegativity than carbon further increases the acidity of carboxylic acids by the inductive effect. Compare, for example, the acidity of acetic acid (pK_a 4.76) and chloroacetic acid (pK_a 2.86). To see the effects of multiple halogen substitution, compare the values of pK_a for acetic acid with its mono-, di-, and trichloro derivatives. A single chlorine substituent increases acid strength by nearly 100. Trichloroacetic acid, the strongest of the three acids, is a stronger acid than H_3PO_4.

Formula:	CH₃COOH	ClCH₂COOH	Cl₂CHCOOH	Cl₃CCOOH
Name:	Acetic acid	Chloroacetic acid	Dichloroacetic acid	Trichloroacetic acid
pK_a:	4.76	2.86	1.48	0.70

Increasing acid strength →

The trend found in the above chart for the acidity of the substituted carboxylic acids is best understood by an analysis of the stability of the anionic conjugate bases that are created. The electron withdrawing ability of the electronegative chlorine atoms stabilizes the nearby anionic charge by electron withdrawal through σ-bonds. The strength of this inductive effect is proportional to the number of chlorine atoms present, explaining why trichloroacetic acid is the strongest acid in the series.

Inductive effect spreads
the negative charge
onto the chlorine atoms.

We also see an example of the inductive effect in a comparison of the relative acidities of benzoic acid and acetic acid. Because of the stronger electron-withdrawing inductive effect of the sp^2 hybridized carbon of the benzene ring compared with the sp^3 hybridized carbon of the methyl group, benzoic acid is a stronger acid than acetic acid; its K_a is approximately four times that of acetic acid.

Benzoic acid
pK_a 4.19

Acetic acid
pK_a 4.76

Example 17.2

Which is the stronger acid in each pair?

(a) Benzoic acid   or   4-Nitrobenzoic acid

(b) 2-Hydroxypropanoic acid (Lactic acid)   or   Propanoic acid

Solution

(a) 4-Nitrobenzoic acid (pK_a 3.42) is a considerably stronger acid than benzoic acid (pK_a 4.19) because of the electron-withdrawing inductive effect of the nitro group. This inductive effect makes the anionic conjugate base of 4-nitrobenzoic acid more stable than that of benzoic acid and hence leads to the greater acidity.

(b) 2-Hydroxypropanoic acid (pK_a 3.08) is a stronger acid than propanoic acid (pK_a 4.87) because of the electron-withdrawing inductive effect of the adjacent hydroxyl oxygen. This effect stabilizes the anionic conjugate base of 2-hydroxypropionic acid.

Problem 17.2

Which is the stronger acid in each pair?

(a) CH_3COOH or CH_3SO_3H

Acetic acid   Methanesulfonic acid

(b)

2-Oxopropanoic acid (Pyruvic acid)   or   Propanoic acid

One final point about carboxylic acids: when a carboxylic acid dissolves in an aqueous solution, the form of the carboxylic acid present depends on the pH of the solution in which it is dissolved. Consider typical carboxylic acids, which have pK_a values in the range of 4.0 to 5.0. When the pH of the solution is equal to the pK_a of the carboxylic acid (that is, the pH of the solution is in the range 4.0 to 5.0), the acid and its anion (its conjugate base) are present in equal concentrations. If the pH of the solution is adjusted to 2.0 or lower by the addition of a strong acid, the carboxylic acid then is present in solution almost entirely as RCOOH. If, on the other hand, the pH of the solution is adjusted to 7.0 or higher, the carboxylic acid is present almost entirely as its anion. Thus, even in a neutral solution (pH 7.0), a carboxylic acid is present predominantly as its anion.

$$R-\overset{\overset{\textstyle O}{\|}}{C}-OH \underset{H^+}{\overset{OH^-}{\rightleftharpoons}} R-\overset{\overset{\textstyle O}{\|}}{C}-OH + R-\overset{\overset{\textstyle O}{\|}}{C}-O^- \underset{H^+}{\overset{OH^-}{\rightleftharpoons}} R-\overset{\overset{\textstyle O}{\|}}{C}-O^-$$

| predominant species when the pH of the solution is 2.0 or less | present in equal concentrations when the pH of the solution is equal to the pK_a of the acid | predominant species when the pH of the solution is 7.0 or greater |

A carboxylate anion has a negative charge, so in biological systems, molecules with substantial numbers of carboxylate anions have considerable negative charge. Because of the hydrophilic character of carboxylate anions, molecules with a large number of them are highly water-soluble.

B. Reaction with Bases

All carboxylic acids, whether soluble or insoluble in water, react with NaOH, KOH, and other strong bases to form water-soluble salts.

Benzoic acid   +   NaOH $\xrightarrow{H_2O}$ **Sodium benzoate**
(slightly soluble in water)   (60 g/100 mL water)

Carboxylic acids also form water-soluble salts with ammonia and amines:

Benzoic acid   +   NH$_3$ $\xrightarrow{H_2O}$ **Ammonium benzoate**
(slightly soluble in water)   (20 g/100 mL water)

Carboxylic acids react with sodium bicarbonate and sodium carbonate to form water-soluble sodium salts and carbonic acid (a weaker acid). Carbonic acid, in turn, decomposes to give water and carbon dioxide, which evolves as a gas.

$$CH_3COOH + NaHCO_3 \longrightarrow CH_3COO^-Na^+ + \boxed{H_2CO_3}$$
$$\boxed{H_2CO_3} \longrightarrow CO_2 + H_2O$$
$$\overline{CH_3COOH + NaHCO_3 \longrightarrow CH_3COO^-Na^+ + CO_2 + H_2O}$$

Salts of carboxylic acids are named in the same manner as the salts of inorganic acids; the cation is named before the anion. The name of the anion is derived from the name of the carboxylic acid by dropping the suffix -*ic acid* and adding the suffix -*ate*.

Example 17.3

Complete each acid-base reaction and name the carboxylic salt formed.

(a) $\diagdown\diagup\diagdown$ COOH + NaOH $\longrightarrow$ **(b)** $\diagup\diagdown$ (OH) COOH + NaHCO$_3$ $\longrightarrow$

(*S*)-Lactic acid

Solution

Each carboxylic acid is converted to its sodium salt. In (b), carbonic acid is formed; it decomposes to carbon dioxide and water.

(a) $\diagdown\diagup\diagdown$ COOH + NaOH $\longrightarrow$ $\diagdown\diagup\diagdown$ COO$^-$Na$^+$ + H$_2$O

   Butanoic acid   **Sodium butanoate**

(b) $\diagup\diagdown$ (OH) COOH + NaHCO$_3$ $\longrightarrow$ $\diagup\diagdown$ (OH) COO$^-$Na$^+$ + H$_2$O + CO$_2$

   (*S*)-**Lactic acid**   **Sodium (*S*)-lactate**

Problem 17.3

Write equations for the reaction of each acid in Example 17.3 with ammonia, and name the carboxylic salt formed.

A consequence of the water solubility of carboxylic acid salts is that water-insoluble carboxylic acids can be converted to water-soluble ammonium or alkali metal salts and extracted into aqueous solution. The salt, in turn, can be transformed back to the free carboxylic acid by addition of HCl, H$_2$SO$_4$, or other strong acid. These reactions allow an easy separation of carboxylic acids from water-insoluble non-acidic compounds.

Figure 17.1
Flow chart for the separation of benzoic acid from benzyl alcohol.

Shown in Figure 17.1 is a flow chart for the separation of benzoic acid, a water-insoluble carboxylic acid, from benzyl alcohol, a nonacidic compound. First, the mixture of benzoic acid and benzyl alcohol is dissolved in diethyl ether. When the ether solution is shaken with aqueous NaOH or other strong base, benzoic acid is converted to its water-soluble sodium salt. Then the ether and aqueous phases are separated. The ether solution is distilled, yielding first diethyl ether (bp 35°C) and then benzyl alcohol (bp 205°C). The aqueous solution is acidified with HCl, and benzoic acid precipitates as a crystalline solid (mp 122°C) and is recovered by filtration.

17.5 | Preparation of Carboxylic Acids

We have seen how carboxylic acids are prepared by oxidation of primary alcohols (Section 10.8) and aldehydes (Section 16.10A). We mention an additional method involving a Grignard reagent.

Treating a Grignard reagent with carbon dioxide gives the magnesium salt of a carboxylic acid, which, on protonation with aqueous acid, gives a carboxylic acid.

Carbon dioxide A magnesium carboxylate **Cyclopentane-carboxylic acid**

Thus, carbonation of a Grignard reagent is a convenient way to convert an alkyl or aryl halide to a carboxylic acid.

17.6 | Reduction

The carboxyl group is one of the organic functional groups most resistant to reduction. It is not affected by catalytic hydrogenation under conditions that easily reduce aldehydes and ketones to alcohols and that reduce alkenes and alkynes to alkanes. The most common reagent for the reduction of carboxylic acids to primary alcohols is the very powerful reducing agent lithium aluminum hydride (Section 16.11A).

A. Lithium Aluminum Hydride

Lithium aluminum hydride, $LiAlH_4$ (LAH), reduces a carboxylic acid to a primary alcohol in excellent yield, although heating is required. LAH is usually dissolved in diethyl ether or tetrahydrofuran (THF). When carboxylic acids react with $LiAlH_4$, the initial product is a tetraalkoxy aluminate ion, which is then treated with water to give the primary alcohol and lithium and aluminum hydroxides. These hydroxides are insoluble in diethyl ether and THF and are removed by filtration. Evaporation of the solvent then yields the primary alcohol.

3-Cyclopentene-carboxylic acid **4-Hydroxymethylcyclopentene**

Alkenes are generally not affected by metal hydride reducing agents. These reagents function as hydride ion donors, that is, as nucleophiles, and alkenes are not attacked by nucleophiles.

Yearly production of acetic acid in the United States is approximately 10^7 kg, a volume that ranks it at the top of the list of organic chemicals manufactured by the U.S. chemical industry. The first industrial synthesis of acetic acid was commercialized in 1916 in Canada and Germany, using acetylene as a feedstock. The process involved two stages: (1) hydration of acetylene to acetaldehyde, followed by (2) oxidation of acetaldehyde to acetic acid by molecular oxygen, catalyzed by cobalt(III) acetate.

$$HC\equiv CH + H_2O \xrightarrow[HgSO_4]{H_2SO_4} \left[\begin{matrix} OH \\ | \\ CH_2=CH \end{matrix} \right] \longrightarrow CH_3\overset{O}{\overset{||}{C}}H \xrightarrow[Co^{3+}]{O_2} CH_3\overset{O}{\overset{||}{C}}OH$$

Acetylene   Enol of   **Acetaldehyde**   **Acetic acid**
                    acetaldehyde

The technology of producing acetic acid from acetylene is simple, yields are high, and these factors made this procedure the major route to acetic acid for over 50 years. Acetylene was prepared by the reaction of calcium carbide with water. Calcium carbide, in turn, was prepared by heating calcium oxide (from limestone, $CaCO_3$) with coke (from coal) to between 2000 and 2500°C in an electric furnace.

$$CaO + 2C \xrightarrow{2500°C} CaC_2 \xrightarrow{2 H_2O} HC\equiv CH + Ca(OH)_2$$

Calcium   **Calcium**   **Acetylene**
oxide   **carbide**

This preparation of calcium carbide requires enormous amounts of energy so, as the cost of energy rose, acetylene ceased to be an economical feedstock from which to manufacture acetic acid.

As an alternative feedstock, chemists turned to ethylene, already available in huge quantities from the refining of natural gas and petroleum. The process of producing acetic acid from ethylene depends on the fact, known since 1894, that, in the presence of catalytic amounts of Pd^{2+} and Cu^{2+} salts, ethylene is oxidized by molecular oxygen to acetaldehyde.

$$2 CH_2=CH_2 + O_2 \xrightarrow[\text{(Wacker process)}]{Pd^{2+}, Cu^{2+}} 2 CH_3\overset{O}{\overset{||}{C}}H$$

The first chemical plant to use ethylene oxidation for the manufacture of acetaldehyde was built in Germany by Wacker-Chemie in 1959, and the process itself became known as the **Wacker process**.

In another approach to the synthesis of acetic acid, chemists turned to a route based on carbon monoxide, a readily available raw material. The carbonylation of methanol is exothermic. The challenge was to find a catalytic system that would bring about this reaction.

In 1973, the Monsanto Company in the United States developed a process for the carbonylation of methanol in the presence of small amounts of soluble rhodium(III) salts, HI, and H_2O.

$$CH_3OH + CO \xrightarrow[HI, H_2O]{Rh(CO)_2I_2^-} CH_3\overset{O}{\overset{||}{C}}OH \qquad \Delta H^0 = -138 \text{ kJ} (33 \text{ kcal})/\text{mol}$$

In the reduction of a carboxyl group by lithium aluminum hydride, the first hydride ion reacts with the carboxyl hydrogen to give H_2. The resulting carboxylate anion reacts with a hydride ion at the carbonyl carbon atom, with the assistance of Li^+ and Al complexes acting as Lewis acids. Following hydride reaction, departure of an oxygen atom as an aluminum oxide species produces an intermediate aldehyde. The aldehyde immediately reacts under the reaction conditions (Section 16.11A) to give a tetraalkoxy aluminate ion that hydrolyzes in an aqueous workup to yield the final product alcohol. Following are balanced equations for treatment of a carboxylic acid with $LiAlH_4$ to form a tetraalkoxy aluminate ion, followed by its hydrolysis in water.

$$4 \text{ RСOH} + 3 \text{ LiAlH}_4 \longrightarrow \left[4 \text{ RСH} \right] \longrightarrow [(\text{RCH}_2\text{O})_4\text{Al}]\text{Li} + 2 \text{ LiAlO}_2 + 4 \text{ H}_2$$

$$\downarrow 4\text{H}_2\text{O}$$

$$4 \text{ RCH}_2\text{OH} + \text{LiOH} + \text{Al(OH)}_3$$

In the reduction of a carboxyl group, two hydrogens from LiAlH_4 are delivered to the carbonyl group. The hydrogen on the hydroxyl group of the product is provided by water or by aqueous acid during work-up. The mechanism of lithium aluminum hydride reduction of carboxyl derivatives is presented in Section 18.10.

B. Selective Reduction of Other Functional Groups

Because carboxyl groups are not affected by the conditions of catalytic hydrogenation, which normally reduce aldehydes, ketones, alkenes, and alkynes, it is possible to selectively reduce these functional groups to alcohols or alkanes in the presence of carboxyl groups.

5-Oxohexanoic acid → **5-Hydroxyhexanoic acid** (racemic)

We saw in Section 16.11A that both LiAlH_4 and NaBH_4 reduce aldehydes and ketones to alcohols. Only LiAlH_4, however, reduces carboxyl groups. Thus, it is possible to reduce an aldehyde or ketone carbonyl group selectively in the presence of a carboxyl group by using the less reactive NaBH_4 as the reducing agent. An example is the selective reduction of the following ketoacid to a hydroxyacid.

5-Oxo-5-phenylpentanoic acid → **5-Hydroxy-5-phenylpentanoic acid** (racemic)

17.7 | Esterification

A. Fischer Esterification

Fischer esterification

The process of forming an ester by heating at reflux a carboxylic acid and an alcohol in the presence of an acid catalyst, commonly H_2SO_4, ArSO_3H, or HCl.

Esters can be prepared by treating a carboxylic acid with an alcohol in the presence of an acid catalyst, most commonly H_2SO_4, ArSO_3H, or gaseous HCl. Conversion of a carboxylic acid and alcohol to an ester is given the special name **Fischer esterification** after the German chemist, Emil Fischer (1852–1919). An example of Fischer esterification is the treatment of acetic acid with ethanol in the presence of concentrated sulfuric acid, which gives ethyl acetate and water.

Ethanoic acid (Acetic acid) + **Ethanol** (Ethyl alcohol) ⇌ **Ethyl ethanoate** (Ethyl acetate) + H_2O

Acid-catalyzed esterification is reversible, and generally the quantities of both carboxylic acid and alcohol remaining at equilibrium are appreciable. If, for example, 60.1 g (1.00 mol) of acetic acid and 60.1 g (1.00 mol) of 1-propanol are heated under reflux in the presence of a few drops of concentrated sulfuric acid until equilibrium is reached, the reaction mixture contains approximately 0.67 mol each of propyl acetate and water and 0.33 mol each of acetic acid and 1-propanol. Thus, at equilibrium, about 67% of the carboxylic acid and alcohol are converted to the desired ester. The mechanism of Fischer esterification involves nucleophilic acyl substitution, which is a major focus of the next chapter. Consequently, the mechanism of this reaction will be covered in Chapter 18.

	Acetic acid	1-Propanol	Propyl acetate	
Initial:	1.00 mol	1.00 mol	0.00 mol	0.00 mol
Equilibrium:	0.33 mol	0.33 mol	0.67 mol	0.67 mol

By control of reaction conditions, it is possible to use Fischer esterification to prepare esters in high yields. If the alcohol is inexpensive compared with the carboxylic acid, a large excess of it can be used to drive the equilibrium to the right and achieve a high conversion of carboxylic acid to its ester. Alternatively, water can be removed by azeotropic distillation and a Dean-Stark trap (Figure 16.1).

Example 17.4

Complete the equation for each Fischer esterification reaction.

Solution

Following is the structural formula for the ester produced in each reaction.

Problem 17.4

Complete the equation for each Fischer esterification.

B. Formation of Methyl Esters Using Diazomethane

Treating a carboxylic acid with diazomethane, usually in ether solution, converts the carboxylic acid under mild conditions and in very high yield to its methyl ester.

$$\underset{\text{RCOH}}{\overset{\text{O}}{\|}} + CH_2N_2 \xrightarrow{\text{ether}} \underset{\text{RCOCH}_3}{\overset{\text{O}}{\|}} + N_2$$

Diazomethane   A methyl ester

Pyrethrin is a natural insecticide obtained from the powdered flower heads of several species of *Chrysanthemum*,

Pyrethrin I

particularly *C. cinerariaefolium*. The active substances in pyrethrum, principally pyrethrins I and II, are contact poisons for insects and cold-blooded vertebrates.

Because their concentrations in the pyrethrum powder used in chrysanthemum-based insecticides are nontoxic to plants and higher animals, pyrethrum powder has found wide use in household and livestock sprays as well as in dusts for edible plants. Pyrethrins I and II are esters of chrysanthemic acid.

Although pyrethrum powders are effective insecticides, the active substances in them are destroyed rapidly in the environment. In an effort to develop synthetic compounds as effective as these natural insecticides but with greater biostability, chemists prepared a series of esters related in structure to chrysanthemic acid. Among the synthetic pyrethrenoids now in common use in household and agricultural products are the following.

Permethrin

Bifenthrin

Diazomethane, a potentially explosive, toxic **yellow gas**, is best represented as a hybrid of two resonance contributing structures.

Diazomethane
(a resonance hybrid of two
important contributing structures)

Because of the hazards associated with the use of diazomethane, it is used only where other means of preparation of methyl esters are too harsh, and, even then, it is used only in small quantities.

Mechanism Formation of a Methyl Ester Using Diazomethane

The reaction of a carboxylic acid with diazomethane occurs in two steps.

***Step 1:* Take a proton away.** Proton transfer from the carboxyl group to diazomethane gives a carboxylate anion and methyldiazonium cation.

A carboxylate   **Methyldiazonium**
anion   **cation**

Step 2: Make a new bond between a nucleophile and an electrophile. Nucleophilic displacement of N_2, an extraordinarily good leaving group, gives the methyl ester.

$$R-\overset{\overset{\displaystyle :O:}{\|}}{C}-\overset{..}{\underset{..}{O}}:^- + \,^+CH_3-N\equiv N: \xrightarrow{S_N2} R-\overset{\overset{\displaystyle :O:}{\|}}{C}-\overset{..}{\underset{..}{O}}-CH_3 + :N\equiv N:$$

17.8 | Conversion to Acid Chlorides

The functional group of an acid halide is a carbonyl group bonded to a halogen atom. Among the acid halides, acid chlorides are the most frequently used in the laboratory and in industrial organic chemistry.

Functional group of an acid halide   **Acetyl chloride**   **Benzoyl chloride**

We study the nomenclature, structure, and characteristic reactions of acid halides in Chapter 18. Here, we are concerned only with their synthesis from carboxylic acids.

Chemical Connections

Esters as Flavoring Agents

Flavoring agents are the largest class of food additives. At the present time, over a thousand synthetic and natural flavors are available. The majority of these are concentrates or extracts from the material whose flavor is desired. These flavoring agents are often complex mixtures of tens to hundreds of compounds. A number of flavoring agents, many of them esters, however, are synthesized industrially. Many of these synthetic flavoring agents are major components of the natural flavors, and adding only one or a few of them is sufficient to make ice cream, soft drinks, or candy taste naturally flavored.

Structure	Name	Flavor
	Ethyl formate	Rum
	(3-Methyl)butyl acetate (Isopentyl acetate)	Banana
	Octyl acetate	Orange
	Methyl butanoate	Apple
	Ethyl butanoate	Pineapple
	Methyl 2-aminobenzoate (Methyl anthranilate)	Grape

Acid chlorides are most often prepared by treating a carboxylic acid with thionyl chloride, the same reagent used to convert an alcohol to a chloroalkane (Section 10.5C).

$$\text{Butanoic acid} + \text{SOCl}_2 \longrightarrow \text{Butanoyl chloride} + \text{SO}_2 + \text{HCl}$$

Butanoic acid   Thionyl chloride   Butanoyl chloride

The mechanism for the reaction of thionyl chloride with a carboxylic acid to form an acid chloride is similar to that presented in Section 10.5C for the conversion of an alcohol to a chloroalkane, and involves initial chlorosulfite formation, followed by nucleophilic attack of chloride ion on the carbonyl carbon to give a tetrahedral carbonyl addition intermediate, which decomposes to give the acid chloride, SO_2, and chloride ion.

Example 17.5

Complete the equation for each reaction.

(a) [structure] OH + SOCl₂ ⟶   **(b)** [structure] OH + SOCl₂ ⟶

Solution

Following are the products of each reaction.

(a) [structure] Cl + SO₂ + HCl   **(b)** [structure] Cl + SO₂ + HCl

Problem 17.5

Complete the equation for each reaction.

(a) [structure with COOH and OCH₃] + SOCl₂ ⟶   **(b)** [cyclohexanol structure] + SOCl₂ ⟶

17.9 | Decarboxylation

A. β-Ketoacids

Decarboxylation
Loss of CO_2 from a carboxyl group.

Decarboxylation is the loss of CO_2 from the carboxyl group of a molecule. Almost any carboxylic acid, heated to a very high temperature, undergoes thermal decarboxylation.

$$\underset{\text{heat}}{\overset{\text{decarboxylation}}{\text{RCOH} \xrightarrow{\hspace{2cm}}}} \text{RH} + \text{CO}_2$$

Most carboxylic acids, however, are quite resistant to moderate heat and melt or even boil without decarboxylation. Exceptions are carboxylic acids that have a carbonyl group β to the carboxyl group. This type of carboxylic acid undergoes decarboxylation quite readily on mild heating. For example, warming 3-oxobutanoic acid brings about its decarboxylation to give acetone and carbon dioxide.

$$\text{3-Oxobutanoic acid} \xrightarrow{\text{warm}} \text{Acetone} + \text{CO}_2$$

3-Oxobutanoic acid
(Acetoacetic acid)   **Acetone**

Decarboxylation on moderate heating is a unique property of 3-oxocarboxylic acids (β-ketoacids) and is not observed with other classes of ketoacids.

Mechanism Decarboxylation of a β-Ketocarboxylic Acid

Step 1: Redistribution of six electrons in a cyclic six-membered transition state gives carbon dioxide and an enol.

Step 2: Keto-enol tautomerism (Section 16.9B) of the enol gives the more stable keto form of the product.

(A cyclic six-membered   Enol of   A ketone
transition state)   a ketone

A hydrogen bond between the carboxyl hydrogen atom and the β-carbonyl oxygen promotes the reaction by favoring a conformation on the path to the six-membered ring transition state. Through this conformational stabilization, the molecules have a much higher probability of undergoing reaction, and the reaction occurs rapidly at moderate temperatures.

Hydrogen bond

Reactive
conformation

An important example of decarboxylation of a β-ketoacid in the biological world occurs during the oxidation of foodstuffs in the tricarboxylic acid (TCA) cycle. One of the intermediates in this cycle is oxalosuccinic acid, which undergoes spontaneous decarboxylation to produce α-ketoglutaric acid. Only one of the three carboxyl groups of oxalosuccinic acid has a carbonyl group in the β-position to it, and this carboxyl group is lost as CO_2.

Only this carboxyl
has a C=O
β to it.

Oxalosuccinic acid   **α-Ketoglutaric acid**

B. Malonic Acid and Substituted Malonic Acids

The presence of a ketone or aldehyde carbonyl group β to the carboxyl group is sufficient to facilitate decarboxylation. In the more general reaction, decarboxylation is facilitated by the presence of any carbonyl group at the β-position, including that of a carboxyl group or ester. Malonic acid and substituted

3-Oxobutanoic acid (acetoacetic acid) and its reduction product, 3-hydroxybutanoic acid, are synthesized in the liver from acetyl-CoA, a product of the metabolism of fatty acids and certain amino acids. 3-Hydroxybutanoic acid and 3-oxobutanoic acid are known collectively as ketone bodies.

3-Oxobutanoic acid
(Acetoacetic acid)

3-Hydroxybutanoic acid
(β-Hydroxybutyric acid)

The concentration of ketone bodies in the blood of healthy, well-fed humans is approximately 0.01 mM/L. However, in persons suffering from starvation or diabetes mellitus, the concentration of ketone bodies may increase to as much as 500 times normal. Under these conditions, the concentration of acetoacetic acid increases to the point that it undergoes spontaneous decarboxylation to form acetone and carbon dioxide. Acetone is not metabolized by humans and is excreted through the kidneys and the lungs. The odor of acetone is responsible for the characteristic "sweet smell" of the breath of severely diabetic patients.

malonic acids, for example, undergo thermal decarboxylation, as illustrated by the decarboxylation of malonic acid when it is heated slightly above its melting point of 135–137°C.

$$HOCCH_2COH \xrightarrow{140°-150°C} CH_3COH + CO_2$$

Propanedioic acid
(Malonic acid)

The mechanism of decarboxylation of malonic acids is very similar to what we have just seen for the decarboxylation of β-ketoacids. In each case, formation of a cyclic, six-membered transition state involving rearrangement of three electron pairs gives the enol form of a carboxylic acid, which is tautomerized to the carboxylic acid.

Mechanism   Decarboxylation of a β-Dicarboxylic Acid

Step 1: Redistribution of six electrons in a cyclic six-membered transition state gives carbon dioxide and the enol form of a carboxyl group.

Step 2: Keto-enol tautomerism (Section 16.9B) of the enol gives the more stable keto form of the carboxyl group.

A cyclic six-membered
transition state

Enol of a
carboxylic acid

A carboxylic
acid

Example 17.6

Each of these carboxylic acids undergoes thermal decarboxylation.

(a)

(b)

(c)

Draw a structural formula for the enol intermediate and final product formed in each reaction.

Solution

Following is a structural formula for the enol intermediate and the final product of each decarboxylation.

(a)

(b)

(c)

Problem 17.6

Account for the observation that the following β-ketoacid can be heated for extended periods at temperatures above its melting point without noticeable decarboxylation.

Summary

Section 17.1 Structure

- A **carboxylic acid** (—COOH) contains a carbonyl group bonded to an —OH group.

Section 17.2 Nomenclature

- IUPAC names of carboxylic acids are derived from the parent alkane by dropping the suffix -e and adding -oic acid.
  - The carbon with the carboxyl group is understood to be carbon 1, so there is no need to give it a number.
  - The carboxyl group takes precedence over most functional groups.
  - **Dicarboxylic acids** are named as -dioic acids and the parent chain is the one that contains both carboxyl groups.

Problems: 17.1, 17.7–17.14

Section 17.3 Physical Properties

- A carboxyl group is polar and, in the liquid and solid states, carboxylic acids are associated by hydrogen bonding into dimers.
  - Carboxylic acids have higher boiling points and are more soluble in water than alcohols, aldehydes, ketones, and ethers of comparable molecular weight.
- A carboxylic acid consists of two regions of distinctly different polarity; a polar **hydrophilic** carboxyl group, which increases solubility in water, and a nonpolar **hydrophobic** hydrocarbon chain, which decreases solubility in water.
  - The low-molecular-weight carboxylic acids are infinitely soluble in water because the hydrophilic carboxyl group more than counterbalances the hydrophobic hydrocarbon chain.
  - As the size of the carbon chain increases, however, the hydrophobic group becomes dominant, and solubility in water decreases.

Problems: 17.15–17.17

Section 17.4 Acidity

Problems: 17.2, 17.3, 17.25–17.32

- Values of pK_a for aliphatic carboxylic acids are in the range 4.0–5.0.
  - The greater acidity of carboxylic acids compared with alcohols is explained by charge delocalization through resonance in a carboxylate anion relative to an alkoxide ion and the electron-withdrawing inductive effect of the carbonyl group.
  - Electron-withdrawing substituents near the carboxyl group increase its acidity.

Section 17.5 Preparation of Carboxylic Acids

Problems: 17.18–17.24, 17.42, 17.48

- Carboxylic acids can be prepared by oxidation of primary alcohols and aldehydes.
- Treating a Grignard reagent with carbon dioxide (CO_2) gives the magnesium salt of a carboxylic acid, which, on protonation with aqueous acid, gives a carboxylic acid.

Section 17.6 Reduction

- Lithium aluminum hydride ($LiAlH_4$) reduces a carboxylic acid to a primary alcohol, although heating is required.

Problems: 17.32, 17.33, 17.34, 17.51

  - Other reducing agents such as catalytic hydrogenation and $NaBH_4$ cannot reduce a carboxylic acid, so these can be used to reduce other functional groups without affecting a carboxyl group in the same molecule.

Section 17.7 Esterification

Problems: 17.4, 17.32, 17.35–17.39, 17.43–17.45, 17.48–17.50

- **Fischer esterification** is the preparation of an ester by treating a carboxylic acid with an alcohol in the presence of an acid catalyst such as sulfuric acid. Fischer esterification is reversible.
- Treating a carboxylic acid with diazomethane (CH_2N_2) gives a methyl ester in high yield.

Section 17.8 Conversion to Acid Chlorides

Problems: 17.5, 17.32, 17.47

- Acid chlorides are prepared from a carboxyl group by treatment with thionyl chloride ($SOCl_2$).

Section 17.9 Decarboxylation

Problems: 17.6, 17.40, 17.41

- Carboxylic acids with a carbonyl β to the carboxyl group undergo **decarboxylation** (loss of CO_2) upon heating.
  - The reaction is important for β-keto acids as well as malonic acid derivatives.

⨀WL **Questions and problems for this chapter are available in OWL. Use the OWL access card that came with this textbook to access assigned questions and problems for this chapter.**

Key Reactions

1. Acidity of Carboxylic Acids (Section 17.4A)   Values of pK_a for most unsubstituted aliphatic and aromatic carboxylic acids are within the range pK_a 4–5.

$$CH_3\overset{O}{\overset{\|}{C}}OH + H_2O \rightleftharpoons CH_3\overset{O}{\overset{\|}{C}}O^- + H_3O^+ \qquad K_a = 1.74 \times 10^{-5}$$

The presence of electron-withdrawing groups near the carboxyl group decreases its pK_a (increases its acidity).

2. Reaction of Carboxylic Acids with Bases (Section 17.4B)   Carboxylic acids form water-soluble salts with alkali metal hydroxides, carbonates, and bicarbonates, as well as with ammonia and aliphatic and aromatic amines.

$$\langle\!\!\bigcirc\!\!\rangle\!\!-COOH + NaOH \xrightarrow{H_2O} \langle\!\!\bigcirc\!\!\rangle\!\!-COO^-\,Na^+ + H_2O$$

$$\langle\!\!\bigcirc\!\!\rangle\!\!-COOH + NH_3 \xrightarrow{H_2O} \langle\!\!\bigcirc\!\!\rangle\!\!-COO^-\,NH_4^+$$

3. Carbonation of a Grignard reagent (Section 17.5)   Adding CO_2 to a Grignard reagent followed by acidification provides a useful route to carboxylic acids.

Cyclopentane-carboxylic acid

4. Industrial Preparation of Acetic Acid by the Carbonylation of Methanol (Section 17.5)

$$CO + 2\ H_2 \xrightarrow{\text{catalyst}} CH_3OH \xrightarrow[\substack{\text{rhodium(III)} \\ \text{HI, H}_2\text{O}}]{\text{CO}} CH_3\overset{\displaystyle O}{\overset{\|}{C}}OH$$

5. Reduction by Lithium Aluminum Hydride (Section 17.6A)   Lithium aluminum hydride reduces a carboxyl group to a primary alcohol.

6. Fischer Esterification (Section 17.7A)   An ester can be prepared by treating a carboxylic acid with an alcohol in the presence of an acid catalyst.

Fischer esterification is reversible. To achieve high yields of ester, it is necessary to force the equilibrium to the right. One way to accomplish this is to use an excess of alcohol; another is to remove water by azeotropic distillation using a Dean-Stark trap.

7. Reaction with Diazomethane (Section 17.7B)   Diazomethane is used to form methyl esters from carboxylic acids. The mechanism involves protonation of the diazomethane carbon atom by the carboxylic acid to make a methyldiazonium cation, followed by attack of the resulting carboxylate on the methyldiazonium cation to give the methyl ester and N_2.

Because diazomethane is explosive and poisonous, it is used only when other means of preparing methyl esters are not suitable.

8. Conversion to Acid Halides (Section 17.8)   Acid chlorides, the most common and widely used of the acid halides, are prepared by treating a carboxylic acid with thionyl chloride. The mechanism, similar to that of the conversion of alcohols to chloroalkanes, involves initial chlorosulfite formation, followed by nucleophilic attack of chloride ion on the carbonyl carbon to give a tetrahedral carbonyl addition intermediate, which decomposes to give the acid chloride, SO_2, and chloride ion.

9. **Decarboxylation of β-Ketoacids (Section 17.9A)** β-Ketoacids decarboxylate upon heating. The mechanism involves redistribution of electrons in a six-membered transition state to give CO_2 and the enol of a ketone, which tautomerizes to give a ketone. The reaction is facilitated by a hydrogen bond between the carboxyl hydrogen atom and β-carbonyl oxygen.

10. **Decarboxylation of β-Dicarboxylic Acids (Section 17.9B)** The mechanism of decarboxylation of a β-dicarboxylic acid is similar to that for decarboxylation of a β-ketoacid.

Chapter 18

Functional Derivatives of Carboxylic Acids

© SCIMAT/Science Source/Photo Researchers, Inc.

Colored scanning electron micrograph of *Penicillium s.* fungus. The stalklike objects are condiophores to which are attached numerous round condia. The condia are the fruiting bodies of the fungus. Inset: a model of amoxicillin. See Chemical Connections: "The Penicillins and Cephalosporins: β-Lactam Antibiotics."

I n this chapter, we study five classes of organic compounds, each related to the carboxyl group: acid halides, acid anhydrides, esters, amides, and nitriles.

Under the general formula of each functional group is an illustration to show you how the group is formally related to a carboxylic acid. Formal loss of —OH from a carboxyl group and H— from H—Cl, for example, gives an acid chloride. Similarly, loss of —OH from a carboxyl group and H— from ammonia gives an amide. For illustrative purposes, we show each of these reactions as a formal loss of water. However, as we will see in this chapter, some actual mechanisms do not involve a step in which an H_2O molecule is lost.

ᗜWL **Online homework for this chapter may be assigned in OWL for Organic Chemistry.**

561

18.1 | Structure and Nomenclature

A. Acid Halides

Acyl group
An RCO— or ArCO— group.

The functional group of an **acid halide** (acyl halide) is an **acyl group (RCO—)** bonded to a halogen atom. Acid chlorides are the most common acid halides.

| An acyl group | Ethanoyl chloride (Acetyl chloride) | Benzoyl chloride | Hexanedioyl chloride (Adipoyl dichloride) |

Acid halides are named by changing the suffix -*ic acid* in the name of the parent carboxylic acid to -*yl halide*.

Similarly, replacement of —OH in a sulfonic acid by chlorine gives a derivative called a **sulfonyl chloride**. Following are structural formulas for two sulfonic acids and the acid chloride derived from each.

Methanesulfonic acid **Methanesulfonyl chloride (MsCl)** ***p*-Toluenesulfonic acid** ***p*-Toluenesulfonyl chloride (Tosyl chloride, TsCl)**

B. Acid Anhydrides

Carboxylic Anhydrides

The functional group of a **carboxylic anhydride** is two acyl groups bonded to an oxygen atom. These compounds are called **acid anhydrides** because they are formally derived from two carboxylic acids by the loss of water. An anhydride may be symmetrical (two identical acyl groups), or it may be mixed (two different acyl groups). Anhydrides are named by replacing the word *acid* in the name of the parent carboxylic acid with the word *anhydride*.

Acetic anhydride   **Benzoic anhydride**

Cyclic anhydrides are named from the dicarboxylic acids from which they are derived. Here are the cyclic anhydrides derived from succinic acid, maleic acid, and phthalic acid.

Succinic anhydride **Maleic anhydride** **Phthalic anhydride**

Phosphoric Anhydrides

Because of the special importance of anhydrides of phosphoric acid in biological chemistry, we include them here to show their similarity with the anhydrides of carboxylic acids. The functional group of a **phosphoric anhydride** is two phosphoryl groups bonded to an oxygen atom. Here are structural formulas for two anhydrides of phosphoric acid and the ions derived by ionization of each acidic hydrogen.

Diphosphoric acid
(Pyrophosphoric acid)

Diphosphate ion
(Pyrophosphate ion)

Triphosphoric acid

Triphosphate ion

C. Esters

Esters of Carboxylic Acids

The functional group of a **carboxylic ester** is an acyl group bonded to —OR or —OAr. Both IUPAC and common names of esters are derived from the names of the parent carboxylic acids. The alkyl or aryl group bonded to oxygen is named first, followed by the name of the acid in which the suffix *-ic acid* is replaced by the suffix *-ate*.

$$CH_3COCH_2CH_3$$

Ethyl ethanoate
(Ethyl acetate)

Diethyl propanedioate
(Diethyl malonate)

Lactones: Cyclic Esters

Cyclic esters are called **lactones**. The IUPAC system has developed a set of rules for naming these compounds. Nonetheless, the simplest lactones are still named by dropping the suffix *-ic acid* or *-oic acid* from the name of the parent carboxylic acid and adding the suffix *-olactone*. The location of the oxygen atom in the ring is indicated by a number if the IUPAC name of the acid is used, or by a Greek letter α, β, γ, δ, ε, and so forth, if the common name of the acid is used.

Lactone
A cyclic ester.

(S)-3-Butanolactone
((S)-β-Butyrolactone)

4-Butanolactone
(γ-Butyrolactone)

6-Hexanolactone
(ε-Caprolactone)

Esters of Phosphoric Acid

Phosphoric acid has three —OH groups and forms mono-, di-, and triesters, which are named by giving the name(s) of the alkyl or aryl group(s) bonded to oxygen followed by the word *phosphate,* as for example dimethyl phosphate. In more complex phosphoric esters, it is common to name the organic molecule and then indicate the presence of the phosphoric ester using either the word *phosphate* or

Cocaine is an alkaloid present in the leaves of the South American coca plant *Erythroxylon coca*. It was first isolated in 1880, and soon thereafter its property as a local anesthetic was discovered. Cocaine was introduced into medicine and dentistry in 1884 by two young Viennese physicians, Sigmund Freud and Karl Koller. Unfortunately, the use of cocaine can create a dependence, as Freud himself observed when he used it to wean a colleague from morphine and thereby produced one of the first documented cases of cocaine addiction.

Cocaine

After determining cocaine's structure, chemists could ask, "How is the structure of cocaine related to its anesthetic effects? Can the anesthetic effects be separated from the habituation effect?" If these questions could be answered, it might be possible to prepare synthetic drugs with the structural features essential for the anesthetic activity but without those giving rise to the undesirable effects. Chemists focused on three structural features of cocaine: its benzoic ester, its basic nitrogen atom, and something of its carbon skeleton. This search resulted in 1905 in the synthesis of procaine, which almost immediately replaced cocaine in dentistry and surgery. Lidocaine was introduced in 1948 and today is one of the most widely used local anesthetics. More recently, other members of the "caine" family of local anesthetics have been introduced, for example etidocaine. All of these local anesthetics are administered as their water-soluble hydrochloride salts.

Procaine
(Novocain)

Lidocaine
(Xylocaine)

Etidocaine
(Duranest; racemic)

Cocaine reduces fatigue, permits greater physical endurance, and gives a feeling of tremendous confidence and power. In some of the Sherlock Holmes stories, the great detective injects himself with a 7% solution of cocaine to overcome boredom.

Thus, seizing on clues provided by nature, chemists have been able to synthesize drugs far more suitable for a specific function than anything known to be produced by nature itself.

the prefix *phospho-*. On the right are two phosphoric esters, each of special importance in the biological world.

Dimethyl phosphate

Glyceraldehyde 3-phosphate

Pyridoxal 5-phosphate

Phosphoenolpyruvate

Glyceraldehyde 3-phosphate is an intermediate in glycolysis, the metabolic pathway by which glucose is converted to pyruvate. Pyridoxal phosphate is one of the metabolically active forms of vitamin B_6. Each of these esters is shown as it is ionized

In 1933, a disgruntled farmer delivered a pail of unclotted blood to the laboratory of Dr. Karl Link at the University of Wisconsin and tales of cows bleeding to death from minor cuts. Over the next couple of years, Link and his collaborators discovered that when cows are fed moldy clover, their blood clotting is inhibited, and they bleed to death from minor cuts and scratches. From the moldy clover they isolated the anticoagulant dicoumarol, a substance that delays or prevents blood clotting. Dicoumarol exerts its anticoagulation effect by interfering with vitamin K activity. Within a few years after its discovery, dicoumarol became widely used to treat victims of heart attack and others at risk for developing blood clots.

Dicoumarol is a derivative of coumarin, a lactone that gives sweet clover its pleasant smell. Coumarin, which does not interfere with blood clotting, is converted to dicoumarol as sweet clover becomes moldy.

In a search for even more potent anticoagulants, Link developed warfarin (named for the Wisconsin Alumni Research Foundation), now used primarily as a rat poison. When rats consume it, their blood fails to clot, and they bleed to death. Warfarin is also used as a blood anticoagulant in humans. The *S* enantiomer shown here is more active than the *R* enantiomer. The commercial product is sold as a racemic mixture. The synthesis of racemic warfarin is described in Problem 19.59.

(S)-Warfarin
(a synthetic anticoagulant)

Coumarin
(from sweet clover)

as sweet clover becomes moldy

Dicumarol
(an anticoagulant)

at pH 7.4, the pH of blood plasma; the two hydroxyl groups of these phosphoryl groups are ionized giving each a charge of -2. The molecular backbones of both DNA and RNA contain phosphoric diesters in each repeating unit.

D. Amides and Imides

The functional group of an **amide** is an acyl group bonded to a nitrogen atom. Amides are named by dropping the suffix *-oic acid* from the IUPAC name of the parent acid, or *-ic acid* from its common name, and adding *-amide*. If the nitrogen atom of an amide is bonded to an alkyl or aryl group, the group is named, and its location on nitrogen is indicated by *N-*. Two alkyl or aryl groups on nitrogen are indicated by *N,N-di-*. *N,N-*Dimethylformamide (DMF) is a widely used polar aprotic solvent (Section 9.3D).

CH_3CNH_2

$CH_3C-N\langle{}^H_{CH_3}$

$H-C-N\langle{}^{CH_3}_{CH_3}$

Acetamide
(a 1° amide)

***N*-Methylacetamide**
(a 2° amide)

***N,N*-Dimethyl-formamide (DMF)**
(a 3° amide)

Lactam
A cyclic amide.

Cyclic amides are given the special name **lactam**. Their names are derived in a manner similar to those of lactones, with the difference that the suffix *-lactone* is replaced by *-lactam*.

(S)-3-Butanolactam   **6-Hexanolactam**
((S)-β-Butyrolactam)   **(ε-Caprolactam)**

Imide
A functional group in which two acyl groups, RCO— or ArCO—, are bonded to a nitrogen atom.

The functional group of an **imide** is two acyl groups bonded to nitrogen. Both succinimide and phthalimide are cyclic imides.

Succinimide   Phthalimide

Example 18.1

Write the IUPAC name for each compound.

Solution

Given first is the IUPAC name and then, in parentheses, the common name.
(a) Methyl 3-methylbutanoate (methyl isovalerate, from isovaleric acid)
(b) Ethyl 3-oxobutanoate (ethyl β-ketobutyrate, from β-ketobutyric acid)
(c) Hexanediamide (adipamide, from adipic acid)
(d) Phenylethanoic anhydride (phenylacetic anhydride, from phenylacetic acid)

Problem 18.1

Draw a structural formula for each compound.
(a) *N*-Cyclohexylacetamide
(b) 1-Methylpropyl methanoate
(c) Cyclobutyl butanoate
(d) *N*-(1-Methylheptyl)succinimide
(e) Diethyl adipate
(f) 2-Aminopropanamide

The penicillins were discovered in 1928 by the Scottish bacteriologist Sir Alexander Fleming. As a result of the brilliant experimental work of Sir Howard Florey, an Australian pathologist, and Ernst Chain, a German chemist who fled Nazi Germany, penicillin G was introduced into the practice of medicine in 1943. For their pioneering work in developing one of the most effective antibiotics of all time, Fleming, Florey, and Chain were awarded the 1945 Nobel Prize in medicine or physiology.

The mold from which Fleming discovered penicillin was *Penicillium notatum,* a strain that gives a relatively low yield of penicillin. It was replaced in commercial production of the antibiotic by *P. chrysogenum,* a strain cultured from a mold found growing on a grapefruit in a market in Peoria, Illinois.

The structural feature common to all penicillins is a β-lactam ring fused to a five-membered thiazolidine ring.

The penicillins differ in the group bonded to the acyl carbon

Amoxicillin
(a β-lactam antibiotic)

The penicillins owe their antibacterial activity to a common mechanism that inhibits the biosynthesis of a vital part of bacterial cell walls.

Soon after the penicillins were introduced into medical practice, penicillin-resistant strains of bacteria began to appear and have since proliferated. One approach to combating resistant strains is to synthesize newer, more effective penicillins. Among those developed are ampicillin, methicillin, and amoxicillin. Another approach is to search for newer, more effective β-lactam antibiotics. At the present time, the most effective of these are the cephalosporins, the first of which was isolated from the fungus *Cephalosporium acremonium.*

The cephalosporins differ in the group bonded to the acyl carbon and the side chain of the thiazine ring

**Cephalexin
(Keflex)**

The cephalosporin antibiotics have an even broader spectrum of antibacterial activity than the penicillins and are effective against many penicillin-resistant bacterial strains. However, resistance to the cephalosporins is now also widespread.

A common mechanism of resistance in bacteria involves their production of a specific enzyme, called a β-lactamase, that catalyzes the hydrolysis of the β-lactam ring, which is common to all penicillins and cephalosporins. Several compounds have been found that inhibit this enzyme, and now drugs based on these compounds can be taken in combination with penicillins and cephalosporins to restore their effectiveness when resistance is known to be a problem. The commonly prescribed formulation called Augmentin is a combination of a β-lactamase inhibitor and a penicillin. It is used as a second line of defense against childhood ear infections when resistance is suspected. Most children know it as the white liquid with a banana taste.

E. Nitriles

The functional group of a **nitrile** is a cyano (C≡N) group bonded to a carbon atom. IUPAC names follow the pattern alkanenitrile: for example, ethanenitrile. Common names are derived by dropping the suffix *-ic* or *-oic acid* from the name of the parent carboxylic acid and adding the suffix *-onitrile.*

Nitrile
A compound containing a —C≡N (cyano) group bonded to a carbon atom.

$$CH_3C\equiv N \qquad \langle\bigcirc\rangle-C\equiv N \qquad \langle\bigcirc\rangle-CH_2C\equiv N$$

Ethanenitrile   **Benzonitrile**   **Phenylethanenitrile**
(Acetonitrile)   **(Phenylacetonitrile)**

18.2 | Acidity of Amides, Imides, and Sulfonamides

Following are structural formulas of a primary amide, a sulfonamide, and two cyclic imides, along with pK_a values for each.

Acetamide   **Benzenesulfonamide**   **Succinimide**   **Phthalimide**
pK_a 15–17 pK_a 10 pK_a 9.7 pK_a 8.3

Values of pK_a for amides of carboxylic acids are in the range of 15–17, which means that they are comparable in acidity to alcohols. Amides show no evidence of acidity in aqueous solution; that is, water-insoluble amides do not react with aqueous solutions of NaOH or other alkali metal hydroxides to form water-soluble salts.

Imides (pK_a 8–10) are considerably more acidic than amides and readily dissolve in 5% aqueous NaOH by forming water-soluble salts. We account for the acidity of imides in the same manner as for the acidity of carboxylic acids (Section 17.4), namely the imide anion is stabilized by delocalization of its negative charge. The more important contributing structures for the anion formed by ionization of an imide delocalize the negative charge on nitrogen and the two carbonyl oxygens.

A resonance-stabilized anion

Sulfonamides derived from ammonia and primary amines are also sufficiently acidic to dissolve in aqueous solutions of NaOH or other alkali metal hydroxides by forming water-soluble salts. The pK_a of benzenesulfonamide is approximately 10. We account for the acidity of sulfonamides in the same manner as for imides, namely the resonance stabilization of the resulting anion.

Benzenesulfonamide   A resonance-stabilized anion

Example 18.2

Phthalimide is insoluble in water. Will phthalimide dissolve in aqueous NaOH?

Solution

Phthalimide is the stronger acid, and NaOH is the stronger base. The position of equilibrium, therefore, lies to the right. Phthalimide dissolves in aqueous NaOH by forming a water-soluble sodium salt.

$$pK_{eq} = -7.4$$
$$K_{eq} = 2.5 \times 10^7$$

pK_a 8.3 (stronger acid)   (stronger base)   (weaker base) pK_a 15.7 (weaker acid)

Problem 18.2

Will phthalimide dissolve in aqueous sodium bicarbonate?

CONNECTIONS TO BIOLOGICAL CHEMISTRY

The Unique Structure of Amide Bonds

Amides have structural characteristics that are unique among carboxylic acid derivatives. In the late 1930s, Linus Pauling discovered that the bond angles about the nitrogen atom of an amide bond in proteins are close to 120°; the amide nitrogen is trigonal planar and sp^2 hybridized. We know that amides are best represented as a hybrid of three resonance contributing structures (see Section 1.9C).

resonance hybrid indicates the presence of a restricted bond rotation about the C—N bond. The measured C—N bond rotation barrier in amides is approximately 63–84 kJ (15–20 kcal)/mol, large enough so that, at room temperature, rotation about the C—N bond is restricted. In addition, because the lone pair on nitrogen is delocalized into the π bond, it is not as available for interacting with protons and other Lewis acids. Thus,

This contributing structure
places a double bond
between C and N

The fact that the six atoms of an amide bond are planar with bond angles of 120° means that the resonance structure on the right makes a significant contribution to the hybrid, and that the hybrid looks very much like this third structure. Inclusion of the third contributing structure explains why the amide nitrogen is sp^2 hybridized and therefore trigonal planar. Also, the presence of a partial double bond (π bond) in the

amide nitrogens are not basic. In fact, in acid solution, amides are protonated on the carbonyl oxygen atom, rather than on the nitrogen (review Example 4.2). Finally, delocalization of the nitrogen lone pair reduces the electrophilic character (partial positive charge) on the carbonyl carbon, thus reducing the susceptibility of amides to nucleophilic attack.

Amides protonate here

Less electrophilic than other carbonyls

Large rotation barrier due to partial double bond

N atom is sp^2 hybridized and non-basic

All of the atoms in the box are in the same plane

The amide —NH group is a good hydrogen bond donor, while the amide carbonyl is a good hydrogen bond acceptor, allowing both primary and secondary amides to form strong hydrogen bonds.

As we will see in Chapter 27, the ability of amides to participate in both intermolecular and intramolecular hydrogen bonding is an important factor in determining the three-dimensional structure of polypeptides and proteins.

18.3 | Characteristic Reactions

In this and subsequent sections, we examine the interconversions of various carboxylic acid derivatives. All these reactions begin with formation of a tetrahedral carbonyl addition intermediate (make a new bond between a nucleophile and an electrophile).

A. Nucleophilic Acyl Addition

The first step of this reaction is exactly analogous to the addition of alcohols to aldehydes and ketones (Section 16.7B). This reaction can be carried out under basic conditions, in which a negatively charged nucleophile adds directly to the carbonyl carbon. The tetrahedral carbonyl addition intermediate formed then adds a proton from a proton donor, HA. The result of this reaction is nucleophilic acyl addition.

Nucleophilic acyl addition (basic conditions):

A carboxylic acid derivative

Tetrahedral carbonyl addition intermediate

Addition product

As with aldehydes and ketones, this reaction can also be catalyzed by acid, in which case protonation (add a proton) of the carbonyl oxygen precedes the attack of the nucleophile.

Nucleophilic acyl addition (acidic conditions):

A carboxylic acid derivative

Tetrahedral carbonyl addition intermediate

B. Nucleophilic Acyl Substitution

For functional derivatives of carboxylic acids, the fate of the tetrahedral carbonyl addition intermediate is quite different from that of aldehydes and ketones; the intermediate collapses to expel the leaving group (Lv) and regenerate the carbonyl group (break a bond to give stable molecules or ions). The result of this addition-elimination sequence is **nucleophilic acyl substitution**.

Nucleophilic acyl substitution A reaction in which a nucleophile bonded to the carbon of an acyl group is replaced by another nucleophile.

Nucleophilic acyl substitution (basic conditions):

Tetrahedral carbonyl addition intermediate

Substitution product

The major difference between nucleophilic acyl addition and nucleophilic acyl substitution is that aldehydes and ketones do not have a group that can leave as a relatively stable anion. They undergo only nucleophilic addition. The four carboxylic acid derivatives we study in this chapter have a leaving group, Lv, that can leave as a relatively stable anion or as a neutral species. Neutral molecules commonly serve as nucleophiles in this reaction, mainly when it is carried out under acid-catalyzed conditions. When these reactions are catalyzed by acid, protonation precedes nucleophilic attack, and similarly protonation precedes leaving group departure. We will see this sequence numerous times in this chapter.

Nucleophilic acyl substitution (acidic conditions):

C. Relative Reactivity

The four carboxylic acid derivatives that are the focus of this chapter have the relative reactivity toward nucleophilic acyl substitution as shown below. The differences in this trend are dramatic. For example, at common ambient temperatures and neutral pH, acid halides will react with water within seconds to minutes, while anhydrides will do so over minutes to hours. Esters, however, do not react with water at appreciable rates under these conditions, taking many years to hydrolyze; amides take centuries to react. Hence, acid halides and acid anhydrides are so reactive that they are not found in nature, whereas esters and amides are universally present.

There are two effects that lead to this trend. One is relative leaving group ability. We show below the leaving groups as anions in order to illustrate an important point: the weaker the base (that is, the more stable the anion), the better the leaving group (Figure 18.1). The weakest base in the series and the best leaving group is the halide ion; acid halides are the most reactive toward nucleophilic acyl substitution. The strongest base and the poorest leaving group is the amide ion; amides are the least reactive toward nucleophilic acyl substitution.

Figure 18.1
Anion leaving group ability and basicity.

$$R_2N^- \quad RO^- \quad \overset{\overset{\textstyle O}{\|}}{RCO^-} \quad X^-$$

Increasing leaving ability

Increasing basicity

The second effect derives from the relative resonance stabilization of the carboxylic acid derivatives. As shown below, each derivative can be written with contributing structures that will be stabilizing to some extent. The second contributing structure that we show for each carboxylic acid derivative has a positive charge on the carbonyl carbon. This structure reflects the electrophilicity of these carbons. However, for each derivative, it is the other contributing structures that reflect the relative resonance stabilization of the derivatives.

Let's start with an analysis of the acid chloride. The third contributing structure for an acid chloride has a carbon to chlorine double bond whose π-bond is weak due to poor orbital overlap between the differentially sized p-orbitals on these two atoms. Further, there is a positive charge on the electronegative chlorine atom. Both of these factors make this a poor contributing structure for the acid chloride. An acid anhydride has five contributing structures; the last two shown place positive charges on the central oxygen. However, these positive charges are adjacent to an electron-withdrawing carbonyl group. Hence, these two contributing structures are not very reasonable depictions of an acid anhydride. But, the analogous contributing structure for an ester places the positively charged oxygen near an electron-donating alkyl group, which stabilizes this charge. Accordingly, this contributing structure is a reasonable depiction of an ester; it is stabilizing, and it lowers the susceptibility of the carbonyl carbon to nucleophilic attack. Lastly, the third contributing structure for an amide has a positive charge on the less electronegative nitrogen (relative to oxygen as with an ester), making this an even more reasonable structure and thereby increasingly stabilizing. In fact, the C=N double bond character of an amide is significant. This increased stability makes the amide the least susceptible to nucleophilic attack.

Acid chloride contributing structures

Acid anhydride contributing structures

Ester contributing structures

Amide contributing structures

Taken together, the combined effects of leaving group ability and susceptibility to nucleophilic attack reinforce each other, thereby resulting in the order of reactivity given below.

| Amide | < | Ester | < | Acid anhydride | < | Acid halide |

Increasing reactivity toward nucleophilic acyl substitution

D. Catalysis

The reactivity of acid halides and acid anhydrides is high enough that the common nucleophiles used to interconvert the carboxylic acid derivatives will react directly with these species without any catalysis. However, esters and amides are so stable that some form of acid or base catalysis is required. Acid catalysis is used to increase the electrophilicity of the carboxylic acid derivatives and to facilitate leaving group departure. Placing a proton on the carbonyl oxygen creates significantly more positive charge on the carbonyl carbon making it more susceptible to nucleophilic attack. In addition, placing a proton on the leaving group makes it more readily depart as a stable molecule.

Base is used to increase nucleophilicity by converting a neutral nucleophile to an anionic nucleophile, for example ethanol to sodium ethoxide. In addition, under basic conditions, the tetrahedral addition intermediates are negatively charged and therefore more apt to expel a negatively charged leaving group. We will see detailed mechanisms involving both acid and base in this chapter.

E. Fischer Esterification Revisited

Now that we have introduced the general steps involved in nucleophilic acyl substitution, let's turn to Fischer esterification, a reaction from the previous chapter. This reaction occurs via nucleophilic acyl substitution and therefore is an excellent introduction to the mechanisms used to interconvert the carboxylic acid derivatives described in the remainder of this chapter. Recall that Fischer esterification is the acid-catalyzed reaction of a carboxylic acid with an alcohol to create an ester.

Ethanoic acid　　　**Ethanol**　　　　**Ethyl ethanoate**
(Acetic acid)　　**(Ethyl alcohol)**　　　**(Ethyl acetate)**

The acid catalysis is used to enhance the electrophilicity of the carboxylic acid toward nucleophilic attack by the alcohol (Step 1 of the following Mechanism box). Although the acid added may be H_2SO_4 or HCl or another acid, the actual catalyst that initiates the reaction is the conjugate acid of the alcohol, ROH_2^+, used in the esterification. The next steps are nucleophilic attack followed by deprotonation, and along with Step 1, are analogous to the acid-catalyzed reaction of aldehydes and ketones with alcohols to form hemiacetals (Section 16.7B). After protonation of the leaving group (Step 4), the leaving group departure takes place (Step 5), followed by a final deprotonation.

All of the intermediates are either neutral or positively charged because the reaction is carried out in acidic solution. A common mistake made by students is to "mix media" in mechanisms, that is, combine acidic and basic intermediates in the same reaction. Proton transfer reactions are exceedingly fast, so a strong acid and a strong base could never be found in the same reaction. A good rule of thumb is that reactions carried out in acidic media will have neutral or positively charged intermediates, while reactions carried out in basic media will be neutral or negatively charged.

Mechanism　Fischer Esterification

Step 1: **Add a proton.**　The reaction begins with protonation, which increases the electrophilicity of the carboxylic acid carbonyl carbon.

Step 2: **Make a new bond between a nucleophile and an electrophile.** The alcohol adds to the carbonyl carbon atom.

Step 3: Take a proton away. Deprotonation gives a tetrahedral addition intermediate.

Step 4: Add a proton. Placing a proton on an —OH converts it to —OH_2^+; this process allows the much better leaving group water to depart.

Step 5: Break a bond to give stable molecules or ions. Water departs as a leaving group.

Step 6: Take a proton away. A final deprotonation gives the ester product and regenerates the acid catalyst.

18.4 | Reaction with Water: Hydrolysis

A. Acid Chlorides

Low-molecular-weight acid chlorides react very rapidly with water to form carboxylic acids and HCl.

$$CH_3\overset{O}{\overset{\|}{C}}Cl + H_2O \longrightarrow CH_3\overset{O}{\overset{\|}{C}}OH + HCl$$

Acetyl chloride

Higher molecular-weight acid halides are less soluble and, consequently, react less rapidly with water.

Mechanism Hydrolysis of an Acid Chloride

Acid chlorides are so reactive that hydrolysis does not require acid or base catalysis, and therefore the steps in the mechanism do not involve putting on or taking off protons prior to the nucleophilic attack and/or the leaving group departure.

Step 1: **Make a new bond between a nucleophile and an electrophile.** Water attacks the carbonyl carbon directly to give a tetrahedral addition intermediate.

Step 2: **Take away a proton.** Removal of a proton is rapid.

Step 3: **Break a bond to give stable molecules or ions.** Expulsion of the chloride anion leaving group yields the carboxylic acid product.

This reaction creates the very strong acid HCl (H_3O^+ and Cl^-). Chemists commonly add a weak base, such as pyridine, to neutralize the acid that is created.

B. Acid Anhydrides

Anhydrides are generally less reactive than acid chlorides. However, the lower molecular-weight anhydrides also react readily with water to form two molecules of carboxylic acid.

$$CH_3COCCH_3 + H_2O \longrightarrow CH_3COH + HOCCH_3$$

Acetic anhydride

As with the hydrolysis of acid chlorides, the hydrolysis of acid anhydrides will occur without an added acid or base catalyst (although sometimes acid is used), and therefore the mechanism is similar to that given above. The acid-catalyzed mechanism is analogous to that with esters, discussed in the very next section.

C. Esters

Esters are hydrolyzed very slowly at neutral pH, even when heated to reflux. Hydrolysis becomes considerably more rapid, however, when they are heated to reflux in aqueous acid or base. The mechanism of acid-catalyzed hydrolysis highlights the logic and key steps involved in many of the mechanisms discussed in this chapter. Therefore, let's analyze this reaction in great detail with a "How To" box.

HOW TO	Write Mechanisms for Interconversions of Carboxylic Acid Derivatives

In Figure 18.2, we will see that acid chlorides react with water, carboxylic acids, alcohols, and amines. Anhydrides undergo reactions with water, alcohols, and amines. Esters undergo reactions with water and amines, and lastly, amides undergo reactions with water. Considering that this is a list of ten reactions, each of which can be performed with the addition of acid or base (the acid chloride and anhydride reactions don't require acid or base), there are nearly twenty different reactions for interconversions of carboxylic acids and their functional groups. Combining the four most common mechanistic elements you have seen throughout this book will allow you to write each mechanism without resorting to memorization.

1. Make a new bond between a nucleophile and an electrophile
2. Break a bond to give stable molecules or ions
3. Add a proton
4. Take a proton away

Because the mechanisms for many of the reactions discussed in this chapter are relatively long, these steps may be used repetitively. To put each step together in the proper sequence, we recommend examining each reaction with regard to the following three principles.

  I. First figure out which bonds must break and form throughout the mechanism.

 II. Avoid mixed media errors. In other words, when writing a mechanism for a reaction occurring in strongly basic media (contains hydroxide or alkoxides) do not create any intermediates that are strong acids (R_2OH^+ structures). Similarly, when writing a mechanism for a reaction occurring in strongly acidic media (contains hydronium or protonated alcohols ROH_2^+) do not create any intermediates that are highly basic (hydroxide, alkoxides, amide anions). (See Appendix 10 for a greater discussion of mixed media errors).

III. Analyze each intermediate you write in your mechanism to conclude when nucleophilic additions, leaving group departure, and proton transfers are feasible.

Acid-catalyzed ester hydrolysis

Let's put the logic together to construct the mechanism for the hydrolysis of an ester in acidic water. Examination of the hydrolysis of an ester shows that the OR′ group has been replaced with OH; thus an OR′ group has to depart as a stable molecule or ion, and an OH group must be a nucleophile at some point during the mechanism (Principle I). Given this, we start considering possible steps to write, thinking of each step almost as a multiple-choice situation among the four mechanistic elements.

Step 1: If water added directly to the carbonyl (make a bond), we would create an anionic oxygen on the ester carbonyl. Because the reaction is carried out in acid and the anionic oxygen is basic, this would constitute a mixed media error, and is therefore a mistake (Principle II). The OR' group cannot depart from an sp^2 carbon (break a bond) because it would leave as an alkoxide and we are in acidic media (Principle II). There are no protons that can be removed (take a proton away; Principle III). Hence, by process of elimination the first step must be protonation of the ester to make structure B. Therefore, **Add a proton**.

Step 2: Structure B still has no leaving group that can depart (break a bond) given the acid media. We cannot protonate a second time (add a proton) because that would create a dication, and if we take off the proton (take a proton away), that simply leads back to A. So, by process of elimination we predict that there must be nucleophilic addition to give C. Therefore, **Make a bond between a nucleophile and an electrophile.**

Step 3: From structure C, water could depart as a leaving group (break a bond), but that simply regenerates B. No nucleophilic attack is possible on C (make a bond) because the carbon is tertiary (cannot undergo S_N2 attack), and we should not put on another proton (add a proton) because that would again create a dication. Thus, again by process of elimination, we conclude we must take a proton off to give D. Therefore, **Take a proton away.**

Step 4: A leaving group cannot depart directly from D (break a bond) because it would be either a hydroxide or an alkoxide and we are in acidic media (Principle II). Hence, the leaving group must be protonated first, giving E. Therefore, **Add a proton.**

Step 5: Protonation in Step 4 allows for the leaving group to depart and the creation of F. Therefore, **Break a bond to give stable molecules or ions.**

Step 6: F finally just needs to lose a proton to give the product G. Therefore, **Take a proton away.**

Using the three principles of logic and four possible steps presented above should allow you to write a reasonable mechanism for all the carboxylic acid and carboxylic acid derivative interconversions discussed in this chapter, as well as many other mechanisms in past and future chapters.

Microscopic Reversibility

We have now discussed Fischer esterification (formation of an ester in an acidic solution of an alcohol) and the hydrolysis of an ester in acidic water. When we discussed Fischer esterification, we pointed out that it is an equilibrium reaction. Ester hydrolysis in aqueous acid is also an equilibrium reaction. The two reactions proceed via the same nucleophilic addition/elimination mechanism, except that they are the reverse of each other. As first introduced in Section 10.6, the principle of microscopic reversibility states that for any reversible reaction, the sequence of intermediates and transition states must be the same but in reverse order for the backward versus forward reaction. In general, the reverse of protonation (Add a proton) is deprotonation (Take away a proton). The reverse of nucleophilic attack (Make a bond between a nucleophile and an electrophile) is leaving group departure (Break a bond to give stable molecules or ions).

With Fischer esterification and ester hydrolysis, we can see the principle of microscopic reversibility by comparing the Mechanism box for Fischer esterification (Section 18.3D) and the How To box immediately above. First, note that they both have six overall steps. Now let's examine the corresponding steps. In the following analysis, we will compare within parentheses the structures lettered with Roman numerals in the Fischer Esterification Mechanism box to the capital letters in the How To box, respectively.

The esterification starts with a protonation of the carbonyl oxygen, while the hydrolysis ends with a deprotonation of a carbonyl oxygen (I = G). The second step of esterification is nucleophilic attack on the carbonyl carbon, while the second to last step of hydrolysis is leaving group departure to create a carbonyl (II = F). The third step of esterification is to remove a proton of the nucleophile that added, while the third to the last step of hydrolysis is to protonate what will be the leaving group (III = E). The fourth step of esterification is to protonate what will be the leaving group, while third step of hydrolysis is to deprotonate what was the nucleophile (IV = D). It is important to note at this point that the third step of esterification creates the same neutrally charged tetrahedral intermediate via deprotonation that the third step of hydrolysis supplies a proton to. The fifth step of esterification is leaving group departure, while the second step of hydrolysis is nucleophilic attack (V = C). The last step of esterification is the deprotonation of the carbonyl oxygen, while the first step of hydrolysis is to add a proton to the carbonyl oxygen (VI = B). By using the principle of microscopic reversibility, you should be able to write the mechanism of any reverse reaction once you know and understand the forward reaction.

Saponification

Hydrolysis of esters may also be carried out using hot aqueous base, such as aqueous NaOH.

Saponification

Hydrolysis of an ester in aqueous NaOH or KOH to an alcohol and the sodium or potassium salt of a carboxylic acid.

Hydrolysis of esters in aqueous base is often called **saponification**, a reference to the use of this reaction in the manufacture of soaps (Section 26.2A) through hydrolysis of triglyceride ester groups. Although the carbonyl carbon of an ester is not strongly electrophilic, hydroxide ion is a good nucleophile and adds to the carbonyl carbon to form a tetrahedral carbonyl addition intermediate, which in turn collapses to give a carboxylic acid and an alkoxide ion. The carboxylic acid reacts with the alkoxide ion or other base present to form a carboxylate anion. Thus, each mole of ester hydrolyzed requires one mole of base.

Mechanism Hydrolysis of an Ester in Aqueous Base (Saponification)

Step 1: Make a new bond between a nucleophile and electrophile. Addition of hydroxide ion to the carbonyl carbon of the ester gives a tetrahedral carbonyl addition intermediate.

Tetrahedral carbonyl addition intermediate

Step 2: Break a bond to give stable molecules or ions. Collapse of this intermediate gives a carboxylic acid and an alkoxide ion.

Step 3: Take a proton away. Proton transfer between the carboxyl group and the alkoxide ion gives the carboxylate anion. This strongly exothermic acid-base reaction drives the whole reaction to completion.

There are two major differences between hydrolysis of esters in aqueous acid and aqueous base.

1. For hydrolysis of an ester in aqueous acid, acid is required in only catalytic amounts. For hydrolysis in aqueous base, base is required in stoichiometric amounts because it is a reactant, not a catalyst.
2. Hydrolysis of an ester in aqueous acid is reversible, but hydrolysis in aqueous base is irreversible because a carboxylate anion (weakly electrophilic, if at all) is not attacked by ROH (a weak nucleophile).

Other acid derivatives react with base in an identical manner to esters.

Example 18.3

Complete and balance equations for the hydrolysis of each ester in aqueous sodium hydroxide. Show all products as they are ionized under these conditions.

Solution

The products of hydrolysis of (a) are benzoic acid and 2-propanol. In aqueous NaOH, benzoic acid is converted to its sodium salt. Therefore, one mole of NaOH is required for hydrolysis of one mole of this ester. Compound (b) is a diester of ethylene glycol. Two moles of NaOH are required for its hydrolysis.

(a) Ph—C(=O)—O⁻Na⁺   +   (2-propanol)   **(b)** 2 (acetate) O⁻Na⁺ + HO—CH₂CH₂—OH

| **Sodium benzoate** | **2-Propanol** (**Isopropyl alcohol**) | **Sodium acetate** | **1,2-Ethanediol** (**Ethylene glycol**) |

Problem 18.3

Complete and balance equations for the hydrolysis of each ester in aqueous solution; show each product as it is ionized under the indicated experimental conditions.

(a) (benzene ring with COOCH₃, COOCH₃) + NaOH $\xrightarrow{H_2O}$ **(b)** (diketo-ester structure) OEt + H₂O $\xrightarrow{HCl}$

D. Amides

Compared to esters, amides require considerably more vigorous conditions for hydrolysis in both acid and base. Amides undergo hydrolysis in hot aqueous acid to give a carboxylic acid and an ammonium ion. Hydrolysis is driven to completion by the acid-base reaction between ammonia or the amine and acid to form an ammonium salt. One mole of acid is required per mole of amide.

(structure)—C(=O)—NH₂ + H₂O + HCl $\xrightarrow[\text{heat}]{H_2O}$ (structure)—C(=O)—OH + NH₄⁺Cl⁻
       Ph   Ph

| (**R**)-2-Phenylbutanamide | (**R**)-2-Phenylbutanoic acid |

In aqueous base, the products of amide hydrolysis are a carboxylate salt and ammonia or an amine. Hydrolysis in aqueous base is driven to completion by the acid-base reaction between the resulting carboxylic acid and base to form a salt. One mole of base is required per mole of amide.

$$CH_3\overset{O}{\overset{\|}{C}}NH\text{—}\bigcirc + NaOH \xrightarrow[\text{heat}]{H_2O} CH_3\overset{O}{\overset{\|}{C}}O^-Na^+ + H_2N\text{—}\bigcirc$$

| **N-Phenylethanamide** (**N-Phenylacetamide, Acetanilide**) | **Sodium acetate**   **Aniline** |

The steps in the mechanism for the hydrolysis of amides in aqueous acid are similar to those for the hydrolysis of esters in aqueous acid.

S_N2

Although an addition/elimination sequence involving the formation of a tetrahedral carbonyl addition intermediate is the most common mechanism for the hydrolysis of esters, alternative pathways are followed in special cases. One such case occurs with methyl esters in basic conditions. Recall that S_N2 reactions are most favorable with CH_3Lv (where Lv = leaving group) relative to 1°, 2°, and 3° alkyl groups. With methyl esters an S_N2 mechanism has a lower energy transition state than those involved in the addition/elimination sequence, and therefore the S_N2 pathway dominates.

Make a new bond between a nucleophile and an electrophilie and simultaneously break a bond to give stable molecules and ions

S_N1

Another special case occurs in acidic media when the alkyl group bonded to the oxygen can form an especially stable carbocation. In these cases. protonation of the carbonyl oxygen is followed by cleavage of the O—C bond to give a carboxylic acid and a carbocation. Benzyl and *tert*-butyl esters readily undergo this type of ester hydrolysis in acid. The carbocation is then trapped by water to create an alcohol. This is an S_N1 reaction in which the leaving group is a carboxylic acid.

Step 1: Add a proton

Step 2: Break a bond to give stable molecules or ions

Step 3: Make a bond between a nucleophile and an electrophile

Step 4: Take a proton away

Step 1: **Add a proton.** Protonation of the carbonyl oxygen gives a resonance-stabilized cation intermediate.

Cation stabilized by resonance delocalization

The role of the proton in this step is to protonate the carbonyl oxygen to increase the electrophilic character of the carbonyl carbon.

Step 2: **Make a new bond between a nucleophile and an electrophile.** Addition of water to the carbonyl carbon.

Step 3: **Take a proton away/add a proton.** Proton transfer between the O and N atoms gives a carbonyl addition intermediate. It is assumed that a solvent molecule accepts the acidic proton on the O atom, and a hydronium ion donates the proton to the N atom, although the exact timing of these events may be different for different molecules in the flask.

Tetrahedral carbonyl addition intermediate

Step 4: **Break a bond to make stable molecules or ions.** Note that the leaving group in this step is a neutral amine (a weaker base), a far better leaving group than an amide ion (a much stronger base).

Step 5: **Take a proton away.** Proton transfer between the very acidic protonated carbonyl and relatively basic amine gives the carboxylic acid and ammonium ion products.

The mechanism for the hydrolysis of amides in aqueous base is more complex than that for the hydrolysis of esters in aqueous base because the amide anion is such a poor leaving group.

Mechanism Hydrolysis of an Amide in Aqueous Base

Step 1: **Make a new bond between a nucleophile and an electrophile.** Addition of hydroxide ion to the carbonyl carbon gives a tetrahedral carbonyl addition intermediate.

Tetrahedral carbonyl
addition intermediate

Step 2: **Take a proton away.** The accepted mechanism involves the creation of a dianionic tetrahedral intermediate, which has enough negative charge to expel the amide anion.

Step 3: **Break a bond to give stable molecules or ions/add a proton.** The amide anion has little to no lifetime in water because it is so basic, and therefore will be instantly protonated by water upon its formation, or potentially during its expulsion (as shown here).

Example 18.4

Write equations for the hydrolysis of these amides in concentrated aqueous HCl. Show all products as they exist in aqueous HCl, and the number of moles of HCl required for hydrolysis of each amide.

(a) $CH_3CN(CH_3)_2$ (b)

Solution

(a) Hydrolysis of *N,N*-dimethylacetamide gives acetic acid and dimethylamine. Dimethylamine, a base, is protonated by HCl to form dimethylammonium ion and is shown in the balanced equation as dimethylammonium chloride. One mole of HCl is required per mole of amide.

$$CH_3CN(CH_3)_2 + H_2O + HCl \xrightarrow{heat} CH_3COH + (CH_3)_2NH_2^+Cl^-$$

(b) Hydrolysis of this δ-lactam gives the protonated form of 5-aminopentanoic acid. One mole of HCl is required per mole of amide.

$$\text{(structure)} + H_2O + HCl \xrightarrow[heat]{} HO \qquad NH_3^+Cl^-$$

Problem 18.4

Complete equations for the hydrolysis of the amides in Example 18.4 in concentrated aqueous NaOH. Show all products as they exist in aqueous NaOH and the number of moles of NaOH required for hydrolysis of each amide.

E. Nitriles

The cyano group of a nitrile is hydrolyzed in aqueous acid to a carboxyl group and ammonium ion as shown in the following equation.

$$\text{Ar}-CH_2C{\equiv}N + 2 H_2O + H_2SO_4 \xrightarrow[heat]{H_2O} \text{Ar}-CH_2COH + NH_4^+HSO_4^-$$

Phenylacetonitrile   **Phenylacetic acid**   **Ammonium hydrogen sulfate**

In hydrolysis of a cyano group in aqueous acid, protonation of the nitrogen atom gives a cation that reacts with water to give an imidic acid (the enol of an amide). Keto-enol tautomerism of the imidic acid gives an amide. The amide is then hydrolyzed, as already described, to a carboxylic acid and an ammonium ion.

$$R{-}C{\equiv}N + H_2O \xrightarrow{H^+} \underset{\substack{\text{An imidic acid} \\ \text{(enol of an amide)}}}{R{-}\overset{OH}{\underset{}{C}}{=}NH} \rightleftharpoons \underset{\text{An amide}}{R{-}\overset{O}{\underset{}{C}}{-}NH_2}$$

The reaction conditions required for acid-catalyzed hydrolysis of a cyano group are typically more vigorous than those required for hydrolysis of an amide, and in the presence of excess water, a cyano group is hydrolyzed first to an amide and then to a carboxylic acid. It is possible to stop at the amide by using sulfuric acid as a catalyst and one mole of water per mole of nitrile. Selective hydrolysis of a nitrile to an amide, however, is not a good method for the preparation of amides. They are better prepared from acid chlorides, acid anhydrides, or esters.

Hydrolysis of a cyano group in aqueous base gives a carboxylate anion and ammonia. The reaction is driven to completion by the acid-base reaction between the carboxylic acid and base to form a carboxylate anion. Acidification of the reaction mixture during workup converts the carboxylate anion to the carboxylic acid.

$$CH_3(CH_2)_9C{\equiv}N + H_2O + NaOH \xrightarrow[\text{heat}]{H_2O} CH_3(CH_2)_9\overset{\displaystyle O}{\overset{\|}{C}}O^-Na^+ + NH_3 \xrightarrow[H_2O]{HCl} CH_3(CH_2)_9\overset{\displaystyle O}{\overset{\|}{C}}OH + NaCl + NH_4Cl$$

Undecanenitrile   **Sodium undecanoate**   **Undecanoic acid**

Mechanism   Hydrolysis of a Cyano Group to an Amide in Aqueous Base

Hydrolysis of a cyano group in aqueous base involves initial formation of the anion of an imidic acid, which, after proton transfer from water, undergoes keto-enol tautomerism to give an amide. The amide is then hydrolyzed by aqueous base, as we have seen earlier, to the carboxylate anion and ammonia.

Step 1: **Make a new bond between a nucleophile and an electrophile.** Hydroxide adds to the electrophilic C atom of the cyano group.

A nitrile

Step 2: **Add a proton.** Proton transfer from water gives an imidic acid.

An imidic acid

Step 3: **Keto-enol tautomerism.** Tautomerism of the imidic acid gives the amide.

An imidic acid   An amide

The acid-catalyzed reaction proceeds similarly; the only difference is in the order of proton transfers.

Hydrolysis of nitriles is a valuable route to the synthesis of carboxylic acids from primary or secondary haloalkanes. In this route, one carbon in the form of a cyano group (Table 8.1) is added to a carbon chain and then converted to a carboxyl group.

$$CH_3(CH_2)_8CH_2Cl \xrightarrow[\substack{\text{ethanol,} \\ \text{water}}]{KCN} CH_3(CH_2)_9C\equiv N \xrightarrow[\text{heat}]{H_2SO_4, H_2O} CH_3(CH_2)_9\overset{\displaystyle O}{\overset{\|}{C}}OH$$

1-Chlorodecane   **Undecanenitrile**   **Undecanoic acid**

Hydrolysis of cyanohydrins, which are obtained by the addition of HCN to an aldehyde or ketone (Section 16.5D), provides a valuable route to α-hydroxycarboxylic acids, as illustrated by the synthesis of mandelic acid.

Benzaldehyde   **Benzaldehyde cyanohydrin**   **2-Hydroxyphenylacetic acid**
                             (Mandelonitrile)   **(Mandelic acid)**
                                (racemic)   **(racemic)**

Example 18.5

Show how to bring about the following conversions using as one step the hydrolysis of a cyano group.

(a)   **4-Chloroheptane**   **2-Propylpentanoic acid**
                                (Valproic acid)

Solution

(a) Treatment of 4-chloroheptane with KCN in aqueous ethanol gives a nitrile. Hydrolysis of the cyano group in aqueous sulfuric acid gives the product.

This synthesis can also be accomplished by conversion of the chloroalkane to a Grignard reagent followed by carbonation and hydrolysis in aqueous acid.

(b) Treatment of cyclohexanone with HCN/KCN in aqueous ethanol gives a cyanohydrin. Hydrolysis of the cyano group in concentrated sulfuric acid gives the carboxyl group of the product.

Problem 18.5

Synthesis of nitriles by nucleophilic displacement of halide from an alkyl halide is practical only with primary and secondary alkyl halides. It fails with tertiary alkyl halides. Why? What is the major product of the following reaction?

18.5 | Reaction with Alcohols

A. Acid Halides

An acid halide reacts with an alcohol to give an ester.

Butanoyl chloride   Cyclohexanol   Cyclohexyl butanoate

Because acid halides are so reactive toward even weak nucleophiles such as alcohols, no catalyst is necessary for these reactions.

In cases in which the alcohol or resulting ester is sensitive to acid, the reaction can be carried out in the presence of a tertiary amine to neutralize the HCl as it is formed. The amines most commonly used for this purpose are pyridine and triethylamine.

Et_3N

Pyridine   Triethylamine

When used for this purpose, each amine is converted to its hydrochloride salt. Pyridine, for example, is converted to pyridinium chloride, as illustrated by its use in the synthesis of isoamyl benzoate.

| **Benzoyl chloride** | **3-Methyl-1-butanol (Isoamyl alcohol)** | **Pyridine** | **3-Methylbutyl benzoate (Isoamyl benzoate)** | **Pyridinium chloride** |

Sulfonic acid esters are prepared by the reaction of an alkane- or arenesulfonyl chloride with an alcohol or phenol. Two of the most common sulfonyl chlorides are *p*-toluenesulfonyl chloride, abbreviated TsCl, and methanesulfonyl chloride, abbreviated MsCl (Section 18.1A).

p-Toluenesulfonyl chloride (Tosyl chloride; TsCl)   (R)-2-Octanol   (R)-2-Octyl p-toluenesulfonate ((R)-2-Octyl tosylate)

As discussed in Section 10.5D, a special value of *p*-toluenesulfonic (tosylate) and methanesulfonic (mesylate) esters is that, in forming them, an —OH is converted from a poor leaving group (hydroxide ion) in nucleophilic displacement to an excellent leaving group, the *p*-toluenesulfonate (tosylate) or methanesulfonate (mesylate) anions.

B. Acid Anhydrides

Acid anhydrides react with alcohols to give one mole of ester and one mole of a carboxylic acid.

$$\underset{\substack{\text{Acetic anhydride}}}{\text{CH}_3\overset{\text{O}}{\overset{\|}{\text{C}}}\text{O}\overset{\text{O}}{\overset{\|}{\text{C}}}\text{CH}_3} + \underset{\substack{\text{Ethanol}}}{\text{HOCH}_2\text{CH}_3} \longrightarrow \underset{\substack{\text{Ethyl acetate}}}{\text{CH}_3\overset{\text{O}}{\overset{\|}{\text{C}}}\text{OCH}_2\text{CH}_3} + \underset{\substack{\text{Acetic acid}}}{\text{CH}_3\overset{\text{O}}{\overset{\|}{\text{C}}}\text{OH}}$$

Phthalic anhydride + **2-Butanol (sec-Butyl alcohol) (racemic)** → **1-Methylpropyl hydrogen phthalate (sec-Butyl hydrogen phthalate) (racemic)**

Thus, the reaction of an alcohol with an anhydride is a useful method for the synthesis of esters. This reaction is catalyzed by acids and by tertiary amines.

Aspirin is synthesized on an industrial scale by the reaction of acetic anhydride and salicylic acid.

2-Hydroxybenzoic acid (Salicylic acid) + **Acetic anhydride** → **Acetylsalicylic acid (Aspirin)** + **Acetic acid**

C. Esters

Transesterification
Exchange of the —OR or —OAr group of an ester for another —OR or —OAr group.

Esters react with alcohols in an acid-catalyzed reaction called **transesterification**. For example, it is possible to convert methyl acrylate to butyl acrylate by heating the methyl ester with 1-butanol in the presence of an acid catalyst.

Methyl propenoate (Methyl acrylate) (bp 81°C) + **1-Butanol (bp 117°C)** $\overset{\text{HCl}}{\rightleftharpoons}$ **Butyl propenoate (Butyl acrylate) (bp 147°C)** + **Methanol (bp 65°C)**

The acids most commonly used for transesterification are HCl as a gas bubbled into the reaction medium and *p*-toluenesulfonic acid.

Transesterification is an equilibrium reaction and can be driven in either direction by control of experimental conditions. For example, in the reaction of methyl acrylate with 1-butanol, transesterification is carried out at a temperature slightly above the boiling point of methanol (the lowest boiling component in the mixture). Methanol distills from the reaction mixture, thus shifting the position of equilibrium in favor of butyl acrylate. Conversely, reaction of butyl acrylate with a large excess of methanol shifts the equilibrium to favor formation of methyl acrylate.

Example 18.6

Complete the following transesterification reactions (the stoichiometry of each is given in the problem).

(a) [structure: ester] + MeOH $\xrightarrow{H^+}$ (b) [structure: diester] + 2MeOH $\xrightarrow{H^+}$

Solution

(a) [structure: methyl ester] OMe + HO [chain] $_8$ (b) 2 [structure] OMe + HO [chain] OH

Problem 18.6

Complete the following transesterification reaction (the stoichiometry is given in the equation).

2 [structure: benzoate ester] OCH$_3$ + HO [chain] OH $\xrightarrow{H^+}$

D. Amides

Amides, the least reactive of the functional derivatives of carboxylic acids, do not react with alcohols. Thus, the reaction of an amide with an alcohol cannot be used to prepare an ester.

18.6 | Reactions with Ammonia and Amines

A. Acid Halides

Acid halides react readily with ammonia and 1° and 2° amines to form amides. For complete conversion of an acid halide to an amide, two equivalents of ammonia or amine are used, one to form the amide and one to neutralize the hydrogen halide formed.

[structure] Cl + 2 NH$_3$ $\longrightarrow$ [structure] NH$_2$ + NH$_4{}^+$Cl$^-$

| **Hexanoyl chloride** | **Ammonia** | **Hexanamide** | **Ammonium chloride** |

Mechanism Reaction of an Acid Chloride and Ammonia

Step 1: **Make a new bond between a nucleophile and an electrophile.** Ammonia adds to the carbonyl carbon.

[mechanism structure:]

R—C—Cl: $\rightleftharpoons$ R—C—Cl:

with :N—H / H adding to give H—N$^+$—H

Step 2: **Take a proton away.** Proton transfer gives a tetrahedral carbonyl addition intermediate.

$$R-\overset{\overset{\displaystyle :\overset{..}{\overset{..}{O}}{}^{\!-}}{|}}{\underset{\underset{\displaystyle H}{\overset{\displaystyle |}{\overset{+}{N}}-H}}{C}}-\overset{..}{\underset{..}{Cl}}: \quad \rightleftharpoons \quad R-\overset{\overset{\displaystyle :\overset{..}{\overset{..}{O}}:{}^{\!-}}{|}}{\underset{\underset{\displaystyle H}{\overset{\displaystyle |}{:N}-H}}{C}}-\overset{..}{\underset{..}{Cl}}: \quad + \quad H-\overset{\overset{\displaystyle H}{\overset{|}{+}}}{\underset{\underset{\displaystyle H}{|}}{N}}-H$$

Tetrahedral carbonyl
addition intermediate

Step 3: **Break a bond to give stable molecules or ions.** The tetrahedral carbonyl addition intermediate then expels the chloride as a leaving group.

$$R-\overset{\overset{\displaystyle :\overset{..}{\overset{..}{O}}:{}^{\!-}}{|}}{\underset{\underset{\displaystyle H}{\overset{\displaystyle |}{:N}-H}}{C}}-\overset{..}{\underset{..}{Cl}}: \quad \rightleftharpoons \quad R-\overset{\overset{\displaystyle :O:}{\|}}{C}-\overset{..}{\underset{\underset{\displaystyle H}{|}}{N}}-H \quad + \quad :\overset{..}{\underset{..}{Cl}}:{}^{\!-}$$

B. Acid Anhydrides

Acid anhydrides react with ammonia and 1° and 2° amines to form amides. As with acid halides, two moles of amine are required; one mole to form the amide and one mole to neutralize the carboxylic acid byproduct.

$$\overset{\overset{\displaystyle O}{\|}}{CH_3C}\overset{\overset{\displaystyle O}{\|}}{O}CCH_3 \quad + \quad 2\,NH_3 \quad \longrightarrow \quad \overset{\overset{\displaystyle O}{\|}}{CH_3C}NH_2 \quad + \quad \overset{\overset{\displaystyle O}{\|}}{CH_3C}O^-NH_4{}^+$$

Acetic anhydride **Ammonia**   **Ethanamide**   **Ammonium acetate**
                                            (Acetamide)

Alternatively, if the amine used to make the amide is expensive, a non-nucleophilic tertiary amine such as triethylamine may be used to neutralize the carboxylic acid.

C. Esters

Esters react with ammonia and with 1° and 2° amines to form amides.

$$Ph\overset{\overset{\displaystyle O}{\|}}{\diagup}OEt \quad + \quad NH_3 \quad \longrightarrow \quad Ph\overset{\overset{\displaystyle O}{\|}}{\diagup}NH_2 \quad + \quad EtOH$$

Ethyl phenylacetate   **Phenylacetamide**   **Ethanol**

Because an alkoxide anion is a poor leaving group compared with either a halide or a carboxylate ion, esters are less reactive toward ammonia, 1° amines, and 2° amines than are acid halides or acid anhydrides. The reaction often requires heating or high concentrations of amine, or both.

D. Amides

Amides do not react with ammonia or primary or secondary amines.

Example 18.7

Complete the following reactions (the stoichiometry of each reaction is given in the equation).

(a) [structure: ethyl butanoate] OEt + NH$_3$ ⟶ **(b)** EtO–CO–OEt + 2 NH$_3$ ⟶

Ethyl butanoate   **Diethyl carbonate**

Solution

(a) [structure] NH$_2$ + EtOH   **(b)** H$_2$N–CO–NH$_2$ + 2 EtOH

Butanamide   **Urea**

Problem 18.7

Complete and balance equations for the following reactions (the stoichiometry of each reaction is given in the equation).

(a) CH$_3$CO—⟨benzene ring⟩—OCCH$_3$ + 2 NH$_3$ ⟶   **(b)** [cyclic lactone structure] + NH$_3$ ⟶

18.7 | Reaction of Acid Chlorides with Salts of Carboxylic Acids

Acid chlorides react with salts of carboxylic acids to give anhydrides. Most commonly used are the sodium or potassium salts.

CH$_3$CCl   +   Na^{+-}OC—⟨benzene ring⟩   ⟶   CH$_3$COC—⟨benzene ring⟩   +   Na$^+$Cl$^-$

Acetyl chloride   **Sodium benzoate**   **Acetic benzoic anhydride**

Reaction of an acid halide with a carboxylate anion of a carboxylic acid is a particularly useful method for synthesis of mixed anhydrides.

18.8 | Interconversion of Functional Derivatives

We have seen throughout the past several sections that acid chlorides are the most reactive toward nucleophilic acyl substitution, followed by acid anhydrides and esters, while the least reactive are amides. Carboxylate anions are negatively charged and therefore repel nucleophiles; the resonance in these species is quite stabilizing. Both of these factors make carboxylate anions essentially inert to nucleophilic acyl substitution (hence, we have not even examined them to this point in the chapter). Another useful way to think about the reactions of the functional derivatives of carboxylic acids is summarized in Figure 18.2.

Any functional group lower in this figure can be prepared from any functional group above it by treatment with an appropriate oxygen or nitrogen nucleophile. An acid chloride, for example, can be converted to an acid anhydride, an ester, an amide, or a carboxylic acid. Acid anhydrides, esters, and amides, however, do not react with chloride ion to give acid chlorides.

Figure 18.2

Reactivities of carboxyl derivatives toward nucleophilic acyl substitution. A more reactive derivative may be converted to a less reactive derivative by treatment with an appropriate reagent. Treatment of a carboxylic acid with thionyl chloride (the acid chloride of sulfurous acid) converts it to the more reactive acid chloride. Carboxylic acids are about as reactive as esters under acidic conditions, but they are converted to unreactive carboxylates under basic conditions.

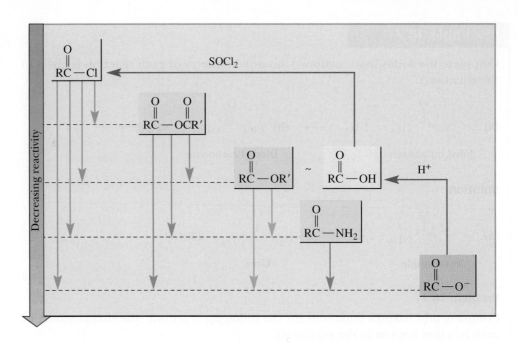

18.9 | Reactions with Organometallic Compounds

A. Grignard Reagents

Treating a formic ester with two moles of a Grignard reagent followed by hydrolysis of the magnesium alkoxide salt in aqueous acid gives a secondary alcohol.

$$\underset{\substack{\text{An ester of}\\\text{formic acid}}}{\text{HCOCH}_3} + 2\ \text{RMgX} \longrightarrow \underset{\substack{\text{magnesium}\\\text{alkoxide}\\\text{salt}}}{} \xrightarrow{\text{H}_2\text{O, HCl}} \underset{\substack{\text{A 2°}\\\text{alcohol}}}{\text{HC}-\text{R}} + \text{CH}_3\text{OH}$$

Treating an ester other than a formate with a Grignard reagent gives a tertiary alcohol in which two of the groups bonded to the carbon bearing the —OH group are the same.

$$\underset{\substack{\text{An ester of any acid}\\\text{other than formic acid}}}{\text{CH}_3\text{COCH}_3} + 2\ \text{RMgX} \longrightarrow \underset{\substack{\text{magnesium}\\\text{alkoxide}\\\text{salt}}}{} \xrightarrow{\text{H}_2\text{O, HCl}} \underset{\substack{\text{A 3°}\\\text{alcohol}}}{\text{CH}_3\text{C}-\text{R}} + \text{CH}_3\text{OH}$$

Reaction of an ester with a Grignard reagent involves formation of two successive tetrahedral carbonyl addition intermediates.

Mechanism Reaction of an Ester with a Grignard Reagent

Step 1: **Make a new bond between a nucleophile and an electrophile.** The reaction begins with addition of one mole of Grignard reagent to the carbonyl carbon to form a tetrahedral carbonyl addition intermediate.

The reaction mechanism at the top shows:

$$H_3C-\overset{\overset{\displaystyle :O:}{\|}}{C}-\overset{..}{\underset{..}{O}}-CH_3 \ + \ R-MgX \ \longrightarrow \ H_3C-\overset{\overset{\displaystyle :\overset{..}{O}:^-[MgX]^+}{|}}{\underset{\underset{\displaystyle R}{|}}{C}}-\overset{..}{\underset{..}{O}}-CH_3$$

Tetrahedral carbonyl
addition intermediate

Step 2: Break a bond to give stable molecules or ions. Because an alkoxide ion is a moderately good leaving group from a tetrahedral carbonyl addition intermediate, this intermediate collapses to give a ketone and a magnesium alkoxide salt. To this point in the mechanism, we have nucleophilic acyl substitution.

$$H_3C-\overset{\overset{\displaystyle :\overset{..}{O}:^-[MgX]^+}{|}}{\underset{\underset{\displaystyle R}{|}}{C}}-\overset{..}{\underset{..}{O}}-CH_3 \ \longrightarrow \ H_3C-\overset{\overset{\displaystyle :O:}{\|}}{C}-R \ + \ H_3C-\overset{..}{\underset{..}{O}}:^-[MgX]^+$$

A ketone

Step 3: Make a new bond between a nucleophile and an electrophile. The ketone reacts with a second mole of Grignard reagent to form a second tetrahedral carbonyl addition compound.

$$H_3C-\overset{\overset{\displaystyle :\overset{..}{O}:}{\|}}{C}-R \ + \ R-MgX \ \longrightarrow \ H_3C-\overset{\overset{\displaystyle :\overset{..}{O}:^-[MgX]^+}{|}}{\underset{\underset{\displaystyle R}{|}}{C}}-R$$

A ketone   Tetrahedral carbonyl
addition intermediate

Step 4: Add a proton. The chemist adds aqueous acid to the reaction and the resulting hydrolysis gives a tertiary alcohol. These last two steps constitute nucleophilic acyl addition.

$$H_3C-\overset{\overset{\displaystyle :\overset{..}{O}:^-[MgX]^+}{|}}{\underset{\underset{\displaystyle R}{|}}{C}}-R \quad H-\overset{..+}{\underset{\underset{\displaystyle H}{|}}{O}}-H \quad \underset{\text{(Added to flask)}}{\longrightarrow} \quad H_3C-\overset{\overset{\displaystyle :\overset{..}{O}\diagup^H}{|}}{\underset{\underset{\displaystyle R}{|}}{C}}-R$$

A 3° alcohol

It is important to realize that it is not possible to use RMgX and an ester to prepare a ketone; the intermediate ketone is more reactive than the ester and reacts immediately with the Grignard reagent to give a tertiary alcohol.

Example 18.8

Complete each Grignard reaction.

(a) $H-\overset{\overset{\displaystyle O}{\|}}{C}-OCH_3 \ \xrightarrow[\text{2. } H_2O, \text{ HCl}]{\text{1. } 2 \ \diagdown\diagup MgBr}$

(b) $\diagup\diagdown\overset{\overset{\displaystyle O}{\|}}{C}-OCH_3 \ \xrightarrow[\text{2. } H_2O, \text{ HCl}]{\text{1. } 2 \ \text{(phenyl)}MgBr}$

Solution

Sequence (a) gives a secondary alcohol, and sequence (b) gives a tertiary alcohol.

(a) [structure: carbon chain with OH on a central carbon]

(b) [structure: carbon chain with OH on carbon bearing two Ph groups]

Problem 18.8

Show how to prepare each alcohol by treating an ester with a Grignard reagent.

(a) [structure: bis(cyclopentyl)methanol with OH, —CH—]

(b) [structure: alcohol with OH, Ph, and two allyl groups]

B. Organolithium Compounds

Organolithium compounds are even more powerful nucleophiles than Grignard reagents and react with esters to give the same types of secondary and tertiary alcohols as shown for Grignard reagents, often in higher yields.

$$
\underset{\text{RCOCH}_3}{\overset{\text{O}}{\|}} \quad \xrightarrow[\text{2. H}_2\text{O, HCl}]{\text{1. 2 R'Li}} \quad \text{R}-\underset{\underset{\text{R'}}{|}}{\overset{\overset{\text{OH}}{|}}{\text{C}}}-\text{R'}
$$

C. Lithium Diorganocuprates

Acid chlorides react readily with lithium diorganocopper (Gilman) reagents (Section 15.2) to give ketones, as illustrated by the conversion of pentanoyl chloride to 2-hexanone. The reaction is carried out at $-78°C$ in either diethyl ether or tetrahydrofuran. Following hydrolysis in aqueous acid, the ketone is isolated in good yield.

[structure: pentanoyl chloride] $\xrightarrow[\text{2. H}_2\text{O}]{\text{1. (CH}_3)_2\text{CuLi, ether, } -78°C}$ [structure: 2-hexanone]

Pentanoyl chloride   **2-Hexanone**

Notice that, under these conditions, the ketone does not react further. This contrasts with the reaction of an ester with a Grignard reagent or organolithium compound, where the intermediate ketone reacts with a second mole of the organometallic compound to give an alcohol. The reason for this difference in reactivity is that the tetrahedral carbonyl addition intermediate in a diorganocuprate reaction is stable at $-78°C$; it survives until the workup causes it to decompose to the ketone, at which point the Gilman reagent has been destroyed.

R_2CuLi reagents react readily only with the very reactive acid chlorides; they do not react with aldehydes, ketones, esters, amides, acid anhydrides, or nitriles. The following compound contains both an acid chloride and an ester group. When treated with lithium dimethylcopper, only the acid chloride reacts.

[structure: methyl ester + acid chloride compound] $\xrightarrow[\text{2. H}_2\text{O}]{\text{1. (CH}_3)_2\text{CuLi, ether, } -78°C}$ [structure: methyl ester + ketone product]

Example 18.9

Show how to bring about each conversion in good yield.

(a)   (b)

Solution

(a) Treat the acid chloride with lithium dimethylcopper followed by H_2O.

(b) Treat the carboxylic acid with thionyl chloride to form the acid chloride, followed by treatment with lithium diallylcopper and then aqueous acid.

Problem 18.9

Show how to bring about each conversion in good yield.

(a) PhCOOH $\longrightarrow$ (b) $CH_2{=}CHCl \longrightarrow$

18.10 | Reduction

Most reductions of carbonyl compounds, including aldehydes and ketones, are now accomplished by transfer of hydride ions from boron or aluminum hydrides. We have already seen the use of sodium borohydride to reduce the carbonyl group of aldehydes and ketones to hydroxyl groups (Section 16.11A) and the use of lithium aluminum hydride to reduce not only aldehyde and ketone carbonyl groups but also carboxyl groups to hydroxyl groups (Section 17.6A).

A. Esters

Lithium aluminum hydride reduces an ester to two alcohols; the alcohol derived from the acyl group is primary and is usually the objective of the reduction.

Methyl(*S*)-2-phenylpropanoate   (*S*)-2-Phenyl-1-propanol Methanol

Mechanism | Reduction of an Ester by Lithium Aluminum Hydride

As you study this mechanism, note that Steps 1 and 3 are closely analogous to the reaction of Grignard reagents with an ester, with the exception that a hydride ion rather than a carbanion is being donated to the carbonyl carbon.

Step 1: Make a new bond between a nucleophile and an electrophile. Nucleophilic addition of hydride ion to the carbonyl carbon gives a tetrahedral carbonyl addition intermediate. The hydride ion is not free but is donated by the AlH_4^- ion.

$$R-\overset{\overset{\displaystyle :\ddot{O}:}{\|}}{C}-\ddot{\underset{..}{O}}-CH_3 \ + \ H{-}AlH_3^- \ \longrightarrow \ R-\overset{\overset{\displaystyle :\ddot{O}:^-}{|}}{\underset{\underset{\displaystyle H}{|}}{C}}-\ddot{\underset{..}{O}}-CH_3 \ + \ AlH_3$$

Tetrahedral carbonyl
addition intermediate

Step 2: Break a bond to give stable molecules or ions. Collapse of this intermediate by loss of alkoxide ion gives a new carbonyl-containing compound.

$$R-\overset{\overset{\displaystyle :\ddot{O}:^-}{|}}{\underset{\underset{\displaystyle H}{|}}{C}}-\ddot{\underset{..}{O}}-CH_3 \ \longrightarrow \ R-\overset{\overset{\displaystyle :O:}{\|}}{C}-H \ + \ ^-\ddot{\underset{..}{O}}-CH_3$$

An aldehyde

Step 3: Make a new bond between a nucleophile and an electrophile. Nucleophilic addition of a second hydride ion to the newly formed carbonyl group gives an alkoxide ion.

$$R-\overset{\overset{\displaystyle :O:}{\|}}{C}-H \ + \ H{-}AlH_3^- \ \longrightarrow \ R-\overset{\overset{\displaystyle :\ddot{O}:^-}{|}}{\underset{\underset{\displaystyle H}{|}}{C}}-H \ + \ AlH_3$$

Tetrahedral carbonyl
addition intermediate

Step 4: Add a proton. The chemist adds water to the reaction and the resulting hydrolysis gives a primary alcohol.

$$R-\overset{\overset{\displaystyle :\ddot{O}:^-}{|}}{\underset{\underset{\displaystyle H}{|}}{C}}-H \quad \overset{H{-}\ddot{\underset{..}{O}}{-}H}{\underset{\substack{(Added \\ to\ flask)}}{\longrightarrow}} \quad R-\overset{\overset{\displaystyle :\ddot{O}{-}H}{|}}{\underset{\underset{\displaystyle H}{|}}{C}}-H \ + \ ^-\ddot{\underset{..}{O}}-H$$

A 1° alcohol

Sodium borohydride is not normally used to reduce esters because the reaction is very slow. Because of this lower reactivity of sodium borohydride toward esters, it is possible to reduce the carbonyl group of an aldehyde or ketone to a hydroxyl group with this reagent without reducing an ester or carboxyl group in the same molecule.

(racemic)

Reduction of an ester to a primary alcohol can be viewed as two successive hydride ion transfers, as shown in the mechanism we have just presented. Chemists wondered if it might be possible to modify the structure of the reducing agent so as to reduce an ester to an aldehyde and no further. A useful modified hydride-reducing agent developed for this purpose is diisobutylaluminum hydride (DIBALH).

$[(CH_3)_2CHCH_2]_2AlH$

Diisobutylaluminum hydride (DIBALH)

DIBALH reductions are typically carried out in toluene or hexane at $-78°C$ (dry ice/acetone temperature) followed by warming to room temperature and addition of aqueous acid to hydrolyze the aluminum salts and liberate the aldehyde. Reduction of esters using DIBALH has become a valuable method for the synthesis of aldehydes, as illustrated by the conversion of methyl hexanoate to hexanal.

Methyl hexanoate   **Hexanal**

If a DIBALH reduction of an ester is carried out at room temperature, the ester is reduced to a primary alcohol. At low temperature, the tetrahedral carbonyl addition intermediate does not eliminate alkoxide ion, and the more reactive aldehyde is not formed until after workup, when the hydride ion has been destroyed. Thus, temperature control is critical for the selective reduction of an ester to an aldehyde.

B. Amides

Lithium aluminum hydride reduction of amides can be used to prepare 1°, 2°, or 3° amines, depending on the degree of substitution of the amide.

Octanamide   **1-Octanamine**

N,N-Dimethylbenzamide   **N,N-Dimethylbenzylamine**

The mechanism for the reduction of an amide to an amine is shown here divided into four steps.

Step 1: **Make a new bond between a nucleophile and an electrophile.** Hydride ion adds to the carbonyl carbon.

$$R-\overset{\overset{\displaystyle :O:}{\|}}{\underset{\underset{\displaystyle H}{|}}{C}}-\overset{\cdot\cdot}{N}-H \;+\; H-AlH_3^- \longrightarrow R-\overset{\overset{\displaystyle :\overset{-}{\underset{\cdot\cdot}{O}:}}{|}}{\underset{\underset{\displaystyle H}{|}}{C}}-\overset{\underset{\displaystyle H}{|}}{\overset{\cdot\cdot}{N}}-H \;+\; AlH_3$$

Tetrahedral carbonyl
addition intermediate

Step 2: **Make a new bond between a nucleophile and an electrophile.** A Lewis acid-base reaction between —O⁻ (a Lewis base) and AlH_3 (a Lewis acid) forms an oxygen-aluminum bond.

$$R-\overset{\overset{\displaystyle :\overset{-}{\underset{\cdot\cdot}{O}:}}{|}}{\underset{\underset{\displaystyle H}{|}}{C}}-\overset{\underset{\displaystyle H}{|}}{\overset{\cdot\cdot}{N}}-H \;+\; AlH_3 \longrightarrow R-\overset{\overset{\displaystyle :\overset{\cdot\cdot}{O}}{|}\diagdown^{AlH_3^-}}{\underset{\underset{\displaystyle H}{|}}{C}}-\overset{\underset{\displaystyle H}{|}}{\overset{\cdot\cdot}{N}}-H$$

Step 3: **Break a bond to give stable molecules or ions.** Rearrangement of electron pairs ejects H_3AlO^{2-} and generates an iminium ion. Because aluminum hydroxides are somewhat acidic, H_3AlO^{2-} is a reasonably good leaving group.

$$R-\overset{\overset{\displaystyle :\overset{\cdot\cdot}{O}}{|}\diagdown^{AlH_3^-}}{\underset{\underset{\displaystyle H}{|}}{C}}-\overset{\underset{\displaystyle H}{|}}{\overset{\cdot\cdot}{N}}-H \longrightarrow R-\overset{\underset{\displaystyle H}{|}}{C}=\overset{+}{\underset{\underset{\displaystyle H}{|}}{N}}-H \;+\; [:\overset{\cdot\cdot}{\underset{\cdot\cdot}{O}}-AlH_3]^{2-}$$

An iminium ion

Step 4: **Make a new bond between a nucleophile and an electrophile.** In the final step the iminium ion adds a second hydride ion to complete the reduction.

$$R-\overset{\underset{\displaystyle H}{|}}{C}=\overset{+}{\underset{\underset{\displaystyle H}{|}}{N}}-H \;+\; H-AlH_3^- \longrightarrow R-\overset{\overset{\displaystyle H}{|}}{\underset{\underset{\displaystyle H}{|}}{C}}-\overset{\underset{\displaystyle H}{|}}{\overset{\cdot\cdot}{N}}-H \;+\; AlH_3$$

A 1° amine

Example 18.10

Show how to bring about each conversion.

(a) $C_6H_5\overset{\overset{\displaystyle O}{\|}}{C}OH \longrightarrow C_6H_5CH_2-N\,\big\rangle$ **(b)** ⬡—$\overset{\overset{\displaystyle O}{\|}}{C}OH \longrightarrow$ ⬡—CH_2NHCH_3

Solution

The key in each part is to convert the carboxylic acid to an amide and then to reduce the amide with $LiAlH_4$. The amide can be prepared by treating the carboxylic acid with $SOCl_2$ to give the acid chloride (Section 17.8) and then treating the acid chloride with an amine (Section 18.6A). Alternatively, the carboxylic acid can be converted to an ethyl ester by Fischer esterification (Section 18.3E), and the ester can then be treated with an amine to give the amide. Solution (a) uses the acid chloride route, and solution (b) uses the ester route.

(a) $C_6H_5\overset{O}{\overset{\|}{C}}OH$ $\xrightarrow{SOCl_2}$ $C_6H_5\overset{O}{\overset{\|}{C}}Cl$ $\xrightarrow{\overset{HN\underset{\smile}{\frown}}{}}$ $C_6H_5\overset{O}{\overset{\|}{C}}-N\underset{\smile}{\frown}$ $\xrightarrow[\text{2. H}_2\text{O}]{\text{1. LiAlH}_4}$ $C_6H_5CH_2-N\underset{\smile}{\frown}$

(b) ⬡—$\overset{O}{\overset{\|}{C}}OH$ $\xrightarrow[\text{2. CH}_3\text{NH}_2]{\text{1. CH}_3\text{CH}_2\text{OH, H}^+}$ ⬡—$\overset{O}{\overset{\|}{C}}NHCH_3$ $\xrightarrow[\text{4. H}_2\text{O}]{\text{3. LiAlH}_4}$ ⬡—CH_2NHCH_3

Problem 18.10

Show how to convert hexanoic acid to each amine.

(a) ~~~~~NMe₂ **(b)** ~~~~~$\overset{}{\underset{H}{N}}$⟨

Example 18.11

Show how to convert phenylacetic acid to each compound.

(a) Ph$\overset{O}{\overset{\|}{\diagup}}$OMe **(b)** Ph$\overset{O}{\overset{\|}{\diagup}}$NH₂ **(c)** Ph⤳NH₂ **(d)** Ph⤳OH

Solution

Prepare methyl ester (a) by Fischer esterification (Section 18.3E) of phenylacetic acid with methanol. Then treat this ester with ammonia to prepare amide (b). Alternatively, treat phenylacetic acid with thionyl chloride (Section 17.8) to give an acid chloride. Then treat this acid chloride with ammonia to give amide (b). Reduction of the amide (b) by LiAlH₄ gives the primary amine (c). Similar reduction of either phenylacetic acid or ester (a) gives alcohol (d).

Problem 18.11

Show how to convert (*R*)-2-phenylpropanoic acid to each compound.

(a) Ph⤳OH

(*R*)-2-Phenyl-1-propanol

(b) Ph⤳NH₂

(*R*)-2-Phenyl-1-propanamine

C. Nitriles

Lithium aluminum hydride reduces the cyano group of a nitrile to a primary amino group.

$$CH_3CH{=}CH(CH_2)_4C{\equiv}N \xrightarrow[\text{2. } H_2O]{\text{1. LiAlH}_4} CH_3CH{=}CH(CH_2)_4CH_2NH_2$$

6-Octenenitrile   **6-Octen-1-amine**

Reduction of cyano groups is useful for the preparation of primary amines only.

Summary

Section 18.1 Structure and Nomenclature

- An **acyl group** is R—C—
- The most important carboxylic acid derivatives are acid halides, acid anhydrides, esters, amides, and nitriles.
- **Acid halides** have an acyl group bonded to a halogen, usually Cl.
  - Acid halides are named by replacing the *-ic acid* of the parent acid with *-yl halide.*
- An **acid anhydride** has two acyl groups bonded to an O atom.
  - Acid anhydrides are named by replacing the word *acid* of the parent acid with *anhydride.*
- **Esters** have an acyl group bonded to an —OR group.
  - The alkyl or aryl group bonded to the O atom is named first, followed by the name of the parent acid in which *-ic acid* is replaced by *-ate.*
  - **Lactones** are cyclic esters in which the acyl carbon and O atom are part of the ring.

Problems: 18.1, 18.12–18.18

- **Amides** have an acyl group bonded to an N atom.
  - If the N atom has one or two groups attached, the groups are named after the designator *N-*, then *-ic acid* of the parent acid is replaced by *-amide.*
  - **Lactams** are cyclic amides in which the acyl carbon and N atom are part of the ring.
  - In amides, the N atom is sp^2 hybridized owing to a third significant resonance contributing structure that places a double bond between the acyl C and N atoms.
  - The acyl C—N bond has partial double bond character and a correspondingly large rotation barrier.
  - The amide acyl C atom is less electrophilic than that of the other carboxylic acid derivatives.
  - The amide N atom is not basic.
- **Nitriles** have a cyano group bonded to carbon.
  - Nitriles are named by replacing *-oic acid* of the parent acid with *-onitrile.*

Section 18.2 Acidity of Amides, Imides, and Sulfonamides

Problem: 18.12

- **Imides** have two acyl groups attached to the same N atom.
- Imides are considerably more acidic than amides due to delocalization of the negative charge of the *N*-deprotonated anion over both carbonyls.

Section 18.3 Characteristic Reactions

- The characteristic reaction of carboxylic acid derivatives is **nucleophilic acyl substitution**.
  - A strong nucleophile adds directly to the electrophilic acyl carbon, breaking the C=O π bond, thereby creating a tetrahedral addition intermediate.
  - The reaction with weaker nucleophiles can be catalyzed by acid, in which case the acyl O atom is protonated first.

– The tetrahedral intermediate collapses by losing a leaving group and reforming the C—O π bond.
  – More stable anions are better anionic leaving groups.

  – Leaving group ability increases in the order $H_2N^- < RO^- < R\overset{\overset{\textstyle O}{\|}}{C}-O^- < Cl^-$.
  – Stability imparted by resonance increases in the order of acid halide < acid anhydride < ester < amide.
  – The carboxylic acid derivatives increase in reactivity in the order amides < esters < acid anhydrides < acid chlorides.
  – The mechanism of Fischer esterification involves initial protonation of the carbonyl group oxygen, followed by nucleophilic attack of an alcohol on the carbonyl carbon, and then a proton transfer. Loss of water gives the ester.

Section 18.4 Reaction with Water: Hydrolysis

- Acid chlorides and acid anhydrides react spontaneously with water to give a carboxylic acid and HCl, or two molecules of carboxylic acid, respectively.
  – The reaction is catalyzed by acid, but the reaction will occur without added acid because the acid produced in the reaction catalyzes the process.
- Acid or base catalysis is needed for the hydrolysis of esters and amides but is not required for acid halides and acid anhydrides.
    – The role of an acid is to increase the electrophilicity of the carbonyl and to protonate the leaving group in order to facilitate its departure.
    – The role of a base is to improve nucleophilicity and to facilitate leaving group departure by creating anionic tetrahedral intermediates.
- Esters react with water in the presence of an acid catalyst to produce a carboxylic acid and alcohol. The mechanism is exactly the reverse of Fischer esterification, explaining why Fischer esterification is an equilibrium process.
- The principle of microscopic reversibility states that the sequence of intermediates and transition states is the same, except in reverse, for the forward and reverse pathways of equilibrium reactions.
- Esters hydrolyze in base in a process called **saponification** because it has been used to hydrolyze triglyceride ester groups in soap manufacturing.
- Amides hydrolyze in either acid or base, and in both cases the reaction is stoichiometric, not catalytic, and requires more vigorous conditions than even esters.
- Nitriles hydrolyze in strong acid to give a carboxylic acid and ammonium ion. They hydrolyze in strong base to give a carboxylate ion and an amine.

Section 18.5 Reaction with Alcohols

- Acid chlorides react with alcohols to give esters and HCl. This reaction is analogous to the formation of sulfonic esters discussed in Chapter 9. When the product ester is acid sensitive, a base such as a tertiary amine is used to neutralize the HCl as it is formed.
- Acid anhydrides react with alcohols to give one molecule of ester and one molecule of carboxylic acid.
- Esters react with alcohols in an acid-catalyzed reaction called **transesterification**, an equilibrium process in which one ester —OR group is exchanged for another.
- Amides are not reactive enough to react with alcohols.

Problems: 18.6, 18.19, 18.20, 18.27–18.30, 18.36–18.38, 18.40, 18.42, 18.48, 18.53, 18.56, 18.60–18.62, 18.68, 18.69

Section 18.6 Reactions with Ammonia and Amines

- Acid chlorides react with two equivalents of ammonia, 1°, and 2° amines, to form an amide and one equivalent of an ammonium chloride.
- Acid anhydrides react with two equivalents of ammonia, 1°, and 2° amines, to form an amide and one equivalent of an ammonium carboxylate salt.
- Esters react slowly with ammonia, 1°, and 2° amines to form an amide and an alcohol.

Section 18.7 Reaction of Acid Chlorides with Salts of Carboxylic Acids

- Acid chlorides react with carboxylate anions to give acid anhydrides and a chloride salt.

Section 18.8 Interconversion of Functional Derivatives

- The general rule is that you can make any less reactive carboxylic acid derivative from any more reactive carboxylic acid derivative and the appropriate oxygen or nitrogen nucleophile.
  - Acid chlorides can be used to make any of the other carboxylic acid derivatives.
  - Because acid chlorides can be made from carboxylic acids using $SOCl_2$, and all the carboxylic acid derivatives can be hydrolyzed, it is possible to interconvert any of the carboxylic acid derivatives.
- The mechanisms shown in the chapter are combinations of the following four steps.
  - Putting on a proton (Add a proton)
  - Taking off a proton (Take a proton away)
  - Attack of a nucleophile on an sp^2 carbon to give a tetrahedral addition intermediate. (Make a new bond between a nucleophile and electrophile)
  - Departure of a leaving group from an sp^3 carbon atom. (Break a bond to give stable molecules or ions)
- Do not to mix media in reaction mechanisms.
  - Do not create a strong base in a reaction run in acid.
  - Do not create a strong acid in a reaction run in base.

Section 18.9 Reactions with Organometallic Compounds

- Esters add two molecules of a Grignard reagent, and following aqueous acid workup, give an alcohol product, in which two groups bonded to the carbon bearing the —OH group are the same.
  - Formate esters give a secondary alcohol; all other esters give a tertiary alcohol.
  - Organolithium reagents can be used in place of the Grignard reagents to carry out the same reaction.
- Acid chlorides react with Gilman reagents to give ketones.

Problems: 18.8, 18.9, 18.19, 18.20, 18.32, 18.41

Section 18.10 Reduction

- Esters react with lithium aluminum hydride followed by aqueous acid to form two alcohols.
  - Sodium borohydride reacts very slowly with esters, and is not used for this purpose.
- Diisobutylaluminum hydride (DIBALH) reacts with esters at low temperatures to give aldehydes. At higher temperatures it reacts all the way to the alcohol.
- Amides react with lithium aluminum hydride to give amines.
- Lithium aluminum hydride reduces the cyano group of nitriles to a primary (1°) amine.

Problems: 18.10, 18.11, 18.19, 18.20, 18.23–18.27, 18.32, 18.33, 18.43, 18.48, 18.52–18.54

⊌WL **Questions and problems for this chapter are available in OWL. Use the OWL access card that came with this textbook to access assigned questions and problems for this chapter.**

Key Reactions

1. **Acidity of Imides (Section 18.2)**   Imides (pK_a 8–10) dissolve in aqueous NaOH by forming water-soluble sodium salts.

Insoluble in water   A water-soluble
                              sodium salt

Imides are more acidic than amides because the imide anion is stabilized by delocalization of the negative charge onto the two carbonyl oxygens.

2. Acidity of Sulfonamides (Section 18.2) Sulfonamides (pK_a 9–10) dissolve in aqueous NaOH by forming water-soluble salts. The sulfonamide anion is stabilized by delocalization of the negative charge onto the two O atoms.

$$C_6H_5SO_2N(H)(CH_2)_4CH_3 + NaOH \longrightarrow C_6H_5SO_2N^-Na^+(CH_2)_4CH_3 + H_2O$$

Insoluble in water   A water-soluble salt

3. Hydrolysis of an Acid Chloride (Section 18.4A) Low-molecular-weight acid chlorides react vigorously with water.

$$CH_3CCl + H_2O \longrightarrow CH_3COH + HCl$$

Higher molecular-weight acid chlorides react less rapidly.

4. Hydrolysis of an Acid Anhydride (Section 18.4B) Acid anhydrides react readily with water.

$$CH_3COCCH_3 + H_2O \longrightarrow CH_3COH + HOCCH_3$$

The mechanism involves protonation of the acyl oxygen, attack by water to create the tetrahedral addition intermediate, transfer of a proton to the acyl oxygen of the leaving group, and departure of the leaving carboxylic acid.

5. Hydrolysis of an Ester (Section 18.4C) Esters are hydrolyzed only in the presence of acid or base. Acid is a catalyst. Base is required in an equimolar amount. In acid, the mechanism involves protonation of the acyl oxygen, attack by water to create the tetrahedral addition intermediate, transfer of a proton to the oxygen of the —OR group, and departure of the leaving alcohol.

$$CH_3CO\text{-}C_6H_{11} + NaOH \xrightarrow{H_2O} CH_3CO^-Na^+ + HO\text{-}C_6H_{11}$$

In base, the mechanism involves direct addition of the strong nucleophile HO$^-$ to give the tetrahedral addition intermediate, followed by collapse to give a carboxylic acid and an alkoxide, which transfers a proton from the relatively acidic carboxylic acid to the relatively basic alkoxide to give the carboxylate ion and alcohol as final products.

6. Hydrolysis of an Amide (Section 18.4D) Either acid or base is required in an amount equivalent to that of the amide. In acid, the mechanism is similar to that for esters, except the departing amine is basic and reacts with a proton to give an ammonium ion product. This last step consumes a proton, explaining why the process is not catalytic in acid.

$$\text{(PhCH(CH_2CH_3)C(O)NH_2)} + H_2O + HCl \xrightarrow[\text{heat}]{H_2O} \text{(PhCH(CH_2CH_3)C(O)OH)} + NH_4^+Cl^-$$

$$CH_3CNH\text{-}C_6H_5 + NaOH \xrightarrow[\text{heat}]{H_2O} CH_3CO^-Na^+ + H_2N\text{-}C_6H_5$$

In base, the mechanism for amide hydrolysis is more complex than with esters, involving initial attack by HO$^-$, deprotonation to give a dianionic tetrahedral intermediate, followed by loss of amide ion, which is immediately protonated to give the amine product.

7. Hydrolysis of a Nitrile (Section 18.4E) Either acid or base is required in an amount equivalent to that of the nitrile. In acid, the mechanism involves an initial protonation of the nitrile N atom, followed by attack by water to give an imidic acid that tautomerizes to give an amide, and the rest proceeds the same as for amide hydrolysis in acid.

$$\text{C}_6\text{H}_5-\text{CH}_2\text{C}\equiv\text{N} + 2\,\text{H}_2\text{O} + \text{H}_2\text{SO}_4 \xrightarrow[\text{heat}]{\text{H}_2\text{O}} \text{C}_6\text{H}_5-\text{CH}_2\text{COH} + \text{NH}_4{}^+\text{HSO}_4{}^-$$

$$\text{CH}_3(\text{CH}_2)_9\text{C}\equiv\text{N} + \text{H}_2\text{O} + \text{NaOH} \xrightarrow[\text{heat}]{\text{H}_2\text{O}} \text{CH}_3(\text{CH}_2)_9\text{CO}^-\text{Na}^+ + \text{NH}_3$$

In base, the mechanism involves an initial attack of HO⁻ on the nitrile C atom to form the anion of an imidic acid, which acquires a proton to give an imidic ion intermediate that tautomerizes to an amide, and the rest proceeds the same as for amide hydrolysis in base.

8. **Reaction of an Acid Chloride with an Alcohol (Section 18.5A)** Treating an acid chloride with an alcohol gives an ester plus HCl.

Preparation of an acid-sensitive ester is carried out using an equimolar amount of triethylamine or pyridine to neutralize the HCl.

9. **Reaction of an Acid Anhydride with an Alcohol (Section 18.5B)** Treating an acid anhydride with an alcohol gives one mole of ester and one mole of carboxylic acid.

$$\text{CH}_3\text{COCCH}_3 + \text{HOEt} \longrightarrow \text{CH}_3\text{COEt} + \text{CH}_3\text{COH}$$

10. **Reaction of an Ester with an Alcohol: Transesterification (Section 18.5C)** Transesterification requires an acid catalyst and an excess of alcohol to drive the reaction to completion.

11. **Reaction of an Acid Chloride with Ammonia or an Amine (Section 18.6A)** Reaction requires two moles of ammonia or amine, one to form the amide and one to neutralize the HCl byproduct. The mechanism involves nucleophilic addition of ammonia or the amine to the carbonyl carbon, followed by a proton transfer to give a tetrahedral addition intermediate that can directly eliminate chloride and lose a proton to give products.

12. **Reaction of an Acid Anhydride with Ammonia or an Amine (Section 18.6B)** Reaction requires two moles of ammonia or amine, one to form the amide and one to neutralize the carboxylic acid byproduct. The mechanism is analogous to that of the acid chloride reaction with ammonia or amines.

$$\text{CH}_3\text{COCCH}_3 + 2\,\text{NH}_3 \longrightarrow \text{CH}_3\text{CNH}_2 + \text{CH}_3\text{CO}^-\text{NH}_4{}^+$$

13. **Reaction of an Ester with Ammonia or an Amine (Section 18.6C)** Treating an ester with ammonia or a primary or secondary amine gives an amide. The mechanism involves nucleophilic addition of the ammonia or amine to the carbonyl carbon followed by a proton transfer to give a tetrahedral addition intermediate, that can directly eliminate alkoxide and lose a proton to the alkoxide to give products.

14. Reaction of an Acid Chloride with a Carboxylic Acid Salt (Section 18.7) Treating an acid chloride with the salt of a carboxylic acid is a valuable method for synthesizing mixed anhydrides.

$$CH_3\overset{O}{\overset{\|}{C}}Cl + Na^+\,^-O\overset{O}{\overset{\|}{C}}\!-\!\!\langle\bigcirc\rangle \longrightarrow CH_3\overset{O}{\overset{\|}{C}}O\overset{O}{\overset{\|}{C}}\!-\!\!\langle\bigcirc\rangle + Na^+Cl^-$$

15. Reaction of an Ester with a Grignard Reagent (Section 18.9A) Treating a formic ester with a Grignard reagent followed by hydrolysis gives a secondary alcohol. Treating any other ester with a Grignard reagent gives a tertiary alcohol. The mechanism involves nucleophilic attack of the Grignard reagent on the carbonyl carbon atom to give a tetrahedral addition intermediate, which collapses with the departure of alkoxide ion to give a ketone intermediate (an aldehyde in the case of formate esters), which reacts with a second mole of Grignard reagent.

16. Reaction of an Acid Chloride with a Lithium Diorganocuprate (Section 18.9C) Acid chlorides react readily with lithium diorganocuprates at −78°C to give ketones.

17. Reduction of an Ester (Section 18.10A) Reduction of an ester by lithium aluminum hydride gives two alcohols. The mechanism involves initial nucleophilic attack by a hydride ion onto the carbonyl carbon to give a tetrahedral addition intermediate, which collapses through the loss of alkoxide to give an aldehyde, which reacts with a second hydride to give the product alcohol.

Reduction by diisobutylaluminum hydride (DIBALH) at low temperature gives an aldehyde and an alcohol.

18. Reduction of an Amide (Section 18.10B) Reduction of an amide by lithium aluminum hydride gives an amine. The mechanism involves initial nucleophilic reaction of a hydride equivalent with the carbonyl carbon to give a tetrahedral addition intermediate, followed by a Lewis acid-base reaction to give an oxygen-aluminum bond that rearranges electron pairs to eject an Al-O species to give an electrophilic iminium ion, which reacts with another equivalent of hydride to give the amine product.

19. Reduction of a Nitrile (Section 18.10C) Reduction of a cyano group by lithium aluminum hydride gives a primary amino group.

Chapter 19

Enolate Anions and Enamines

Crystals of tamoxifen (Problems 19.42, 21.61, and 21.62) viewed under polarizing light. *Inset:* a model of tamoxifen.

© Michael W. Davidson/Florida State University

I n this chapter, we continue the chemistry of carbonyl compounds. In Chapters 16 through 18, we concentrated on the carbonyl group itself and on nucleophilic additions to the carbonyl carbon to form tetrahedral carbonyl addition intermediates and on products derived from protonation or collapse of these intermediates. In this chapter, we expand on the chemistry of carbonyl-containing compounds and consider the consequences of the acidity of α-hydrogens and the formation of enolate anions.

19.1 | Formation and Reactions of Enolate Anions: An Overview

On the next page is the resonance-stabilized **enolate anion** (Section 16.9A) formed by treating acetaldehyde with base. This anion is best represented as a hybrid of two contributing structures. Of these, the structure with the negative charge on the more electronegative oxygen atom makes the greater contribution to the hybrid. Note that although the majority of the negative charge is on the carbonyl oxygen, there is still a significant partial negative charge on the alpha carbon.

OWL Online homework for this chapter may be assigned in OWL for Organic Chemistry.

The reaction scheme at top:

$$CH_3-\overset{\overset{\displaystyle :\ddot{O}:}{\|}}{C}-H \;+\; NaOH \;\rightleftharpoons\; H-\overset{\overset{\displaystyle :\ddot{O}:}{\|}}{\underset{\underset{\displaystyle H}{|}}{C}}-\overset{}{C}-H \;\longleftrightarrow\; H-\overset{\overset{\displaystyle :\ddot{O}:^-}{|}}{C}=\overset{\underset{\displaystyle H}{|}}{C}-H \;+\; H_2O$$

An enolate anion

The majority of the negative charge in the hybrid is on oxygen

Enolate anions are important synthetic reagents because they react at carbon to create new carbon-carbon bonds in two types of reactions. First, they can function as nucleophiles in S_N2 reactions as shown in this general reaction.

An enolate anion   A 1° haloalkane or sulfonate

Second, they function as nucleophiles in carbonyl addition reactions. Here, we show nucleophilic acyl addition of an enolate anion to the carbonyl carbon of an aldehyde.

An enolate anion   An aldehyde   A tetrahedral carbonyl addition intermediate

Enolate anions also add in this manner to the carbonyl groups of ketones and esters.

As shown by the charge distribution on the electrostatic potential map, the majority of negative charge of an enolate anion is on the carbonyl oxygen. If reaction were to occur at the carbonyl oxygen, the product would be a vinyl ether, whereas reaction at the α-carbon leads to alkylation.

product has a C=C π bond so it is less stable

A vinyl ether

reaction at oxygen

reaction at carbon

product has a C=O π bond so it is more stable

An enolate anion

Despite this charge distribution, enolate anions react primarily at carbon for two reasons. First, there is always a counterion such as the Li^+ or Na^+ ion associated with the enolate anion. These counterions are more tightly associated with the oxygen atom than the alpha carbon. As a result, the counterion to some degree blocks the approaching electrophile, thus reducing the likelihood of a productive collision with the oxygen. In fact, enolates are thought to exist in solution as larger aggregates containing several counterions associated with several enolate oxygen atoms, and the solvent, effectively amplifying this effect.

An example of an aggregate of lithium enolates with the solvent THF.

The second reason enolates react at carbon is based on product thermodynamics. We have already seen that, other factors being equal, reactions at equilibrium will favor products with stronger bonds. If an enolate anion were to react at the alpha carbon, the product would contain a C=O π bond. If it were to react at the carbonyl oxygen, the product would contain a C=C π bond. In general, C=O bonds are stronger than C=C bonds. (Recall, for example, the relative percentages of keto and enol forms present at equilibrium for simple aldehydes and ketones, Section 16.9B.) Thus, enolate anions react primarily at the alpha carbon to form new carbon-carbon bonds.

19.2 | Aldol Reaction

Unquestionably, the most important reaction of enolate anions derived from aldehydes and ketones is their nucleophilic addition to the carbonyl group of another molecule of the same or different compound, as illustrated by the following reactions.

| Ethanal | Ethanal | 3-Hydroxybutanal |
| (Acetaldehyde) | (Acetaldehyde) | (a β-hydroxyaldehyde; formed as a racemic mixture) |

| Propanone | Propanone | 4-Hydroxy-4-methyl-2-pentanone |
| (Acetone) | (Acetone) | (a β-hydroxyketone) |

The common name of the product derived from the reaction of acetaldehyde in base is aldol because it is both an *aldehyde* and an *alcohol*. Aldol is also the generic name given to any product formed in this type of reaction. The product of an **aldol reaction** is a β-hydroxyaldehyde or a β-hydroxyketone.

A. Mechanisms

Although such reactions may be catalyzed by either acid or base, base catalysis is more common. The key step in a base-catalyzed aldol reaction is nucleophilic addition of the enolate anion of one carbonyl-containing molecule to the carbonyl group of another to form a tetrahedral carbonyl addition intermediate. This mechanism is illustrated by the aldol reaction between two molecules of acetaldehyde. Notice in this three-step mechanism that OH⁻ is a catalyst; an OH⁻ is used in Step 1, but another OH⁻ is generated in Step 3.

Mechanism   Base-Catalyzed Aldol Reaction

Step 1: Take a proton away.   Removal of an α-hydrogen by base gives a resonance-stabilized enolate anion.

$$H-\ddot{\underset{\cdot\cdot}{O}}\colon^{-} + H-CH_2-\overset{\overset{\ddot{O}\colon}{\parallel}}{C}-H \;\rightleftharpoons\; H-\ddot{\underset{\cdot\cdot}{O}}-H \;+\; \left[\colon^{-}CH_2-\overset{\overset{\ddot{O}\colon}{\parallel}}{C}-H \longleftrightarrow CH_2=\overset{\overset{\colon\ddot{O}\colon^{-}}{}}{C}-H \right]$$

pKₐ 20   pKₐ 15.7   An enolate anion
(weaker acid)   (stronger acid)

Given the relative acidities of the two acids in this equilibrium, the position of this equilibrium lies considerably to the left.

Step 2: Make a new bond between a nucleophile and an electrophile.   Nucleophilic addition of the enolate anion to the carbonyl carbon of another aldehyde (or ketone) gives a tetrahedral carbonyl addition intermediate.

$$CH_3-\overset{\overset{\ddot{O}\colon}{\parallel}}{C}-H + {}^{-}CH_2-\overset{\overset{\colon O\colon}{\parallel}}{C}-H \;\rightleftharpoons\; CH_3-\overset{\overset{\colon\ddot{O}\colon^{-}}{|}}{C}H-CH_2-\overset{\overset{\colon O\colon}{\parallel}}{C}-H$$

A tetrahedral carbonyl
addition intermediate

Step 3: Add a proton.   Reaction of the tetrahedral carbonyl addition intermediate with a proton donor gives the aldol product as a racemic mixture and generates a new base catalyst.

$$CH_3-\overset{\overset{\colon\ddot{O}\colon^{-}}{|}}{C}H-CH_2-\overset{\overset{\colon O\colon}{\parallel}}{C}-H + H-\ddot{\underset{\cdot\cdot}{O}}H \;\rightleftharpoons\; CH_3-\overset{\overset{\colon\ddot{O}H}{|}}{C}H-CH_2-\overset{\overset{\colon O\colon}{\parallel}}{C}-H + {}^{-}\colon\ddot{\underset{\cdot\cdot}{O}}H$$

(racemic)

Mechanism   Acid-Catalyzed Aldol Reaction

The mechanism of an acid-catalyzed aldol reaction involves three steps, the first two of which are preparation of the aldehyde or ketone for formation of the new carbon-carbon bond. The key step is attack of the enol of one molecule on the protonated carbonyl group of a second molecule.

Step 1: Keto-enol tautomerism.   Keto and enol forms of one molecule of the aldehyde or ketone undergo acid-catalyzed equilibration (Section 16.9B).

$$CH_3-\overset{\overset{O}{\parallel}}{C}-H \;\underset{}{\overset{HA}{\rightleftharpoons}}\; CH_2=\overset{\overset{OH}{|}}{C}-H$$

Step 2: Add a proton.   Proton transfer from the acid, HA, to the carbonyl oxygen of a second molecule of aldehyde or ketone gives an oxonium ion.

$$CH_3-\overset{\overset{\colon O\colon}{\parallel}}{C}-H + H-A \longrightarrow CH_3-\overset{\overset{{}^{+}\ddot{O}-H}{\parallel}}{C}-H + \colon A^{-}$$

Step 3: Make a new bond between a nucleophile and an electrophile. Attack by the enol of one molecule on the protonated carbonyl group of another molecule forms the new carbon-carbon bond.

$$H_3C-\overset{+}{\overset{\displaystyle H}{\underset{\displaystyle }{O}}}\overset{\displaystyle }{\underset{\displaystyle }{C}}-H \;+\; H_2C=\overset{\displaystyle \overset{H}{O}}{\underset{\displaystyle }{C}}-H \;\rightleftharpoons\; H_3C-\overset{\displaystyle \overset{H}{O}}{\underset{\displaystyle \overset{|}{H}}{C}}-CH_2-\overset{+}{\overset{\displaystyle \overset{H}{O}}{C}}-H$$

(racemic)

Step 4: Take a proton away. Proton transfer to A⁻ regenerates the acid catalyst and gives the racemic aldol product.

$$H_3C-\overset{\displaystyle \overset{H}{O}}{\underset{\displaystyle \overset{|}{H}}{C}}-CH_2-\overset{+}{\overset{\displaystyle \overset{H}{O}}{C}}-H + :A^- \;\rightleftharpoons\; H_3C-\overset{\displaystyle \overset{H}{O}}{\underset{\displaystyle \overset{|}{H}}{C}}-CH_2-\overset{\displaystyle O}{C}-H + H-A$$

(racemic) (racemic)

You might compare the mechanisms of the acid- and base-catalyzed aldol reactions. Under base catalysis, the carbon-carbon bond-forming step involves attack of an enolate anion (a nucleophile) on the uncharged carbonyl carbon (an electrophile) of a second molecule of the aldehyde or ketone. Under acid catalysis, it involves attack of the enol (a nucleophile) of one molecule on the protonated carbonyl group (an electrophile) of the second molecule.

It is quite common to create chiral products during aldol reactions, as well as in the other enolate reactions we discuss in this chapter. The products will be formed as racemic mixtures unless one of the reactants is chiral and present as a single enantiomer. In cases when two chiral centers are created in the reaction, four stereoisomers are produced as two 1:1 mixtures of enantiomers. A great deal of work has gone into learning how to carry out aldol and other enolate reactions that give predominantly a single enantiomer product, but that work is beyond the scope of this text.

Example 19.1

Draw the product of the base-catalyzed aldol reaction of each compound.

(a) Butanal **(b)** Cyclohexanone

Solution

The aldol product is formed by nucleophilic addition of the α-carbon of one molecule to the carbonyl carbon of another.

(a)

new C—C bond

(four stereoisomers; two racemic mixtures)

(b)

new C—C bond

(racemic)

Problem 19.1

Draw the product of the base-catalyzed aldol reaction of each compound.

(a) Phenylacetaldehyde **(b)** Cyclopentanone

β-Hydroxyaldehydes and β-hydroxyketones are very easily dehydrated, and often the conditions necessary to bring about an aldol reaction are sufficient to cause dehydration. Dehydration is particularly facile in the case of acid-catalyzed aldol reactions. Alternatively, dehydration can be brought about by warming the aldol product in dilute acid. The major product from dehydration of an aldol is one in which the carbon-carbon double bond is conjugated with the carbonyl group; that is, the product is an α,β-unsaturated aldehyde or ketone.

$$
\underset{\text{CH}_3\text{CHCH}_2\text{CH}}{\overset{\overset{\displaystyle\text{OH}}{|}\quad\overset{\displaystyle\text{O}}{\|}}{}} \xrightarrow[\text{acid or base}]{\text{warm in either}} \underset{\text{An }\alpha,\beta\text{-unsaturated}}{\overset{\overset{\displaystyle\text{O}}{\|}}{\text{CH}_3\overset{\beta}{\text{CH}}=\overset{\alpha}{\text{CHCH}}}} + \text{H}_2\text{O}
$$

An α,β-unsaturated
aldehyde

Aldol reactions are readily reversible, especially when catalyzed by base, and there is generally little aldol product present at equilibrium, especially for aldol reactions of ketones. Equilibrium constants for dehydration, however, are generally large so that, if reaction conditions are sufficiently vigorous to bring about dehydration, good yields of product can be obtained. The alkene generally forms as an E,Z mixture, although in some cases, such as that below, the E alkene will be substantially more stable and thus the predominant product.

Mechanism Acid-Catalyzed Dehydration of an Aldol Product

Step 1: **Keto-enol tautomerism.** The acid catalyzes interconversion of the keto and enol forms of the aldehyde.

Step 2: **Add a proton.** Proton transfer from the acid catalyst, here shown as H_3O^+, to the enol gives an oxonium ion intermediate. This step has the effect of converting the —OH, a poor leaving group, into —O^+H_2, a better leaving group.

Step 3: Break a bond to give stable molecules or ions. Loss of water from the oxonium ion intermediate gives the conjugate acid of the final product. Notice that the double bond forms to give predominantly the more stable *E* alkene in this case.

Step 4: Take a proton away. Proton transfer from the conjugate acid of the final product to water completes the reaction.

An α,β-unsaturated aldehyde

Example 19.2

Draw the product of dehydration of each aldol product in Example 19.1.

Solution

Loss of H_2O from aldol product (a) gives an α,β-unsaturated aldehyde; loss of H_2O from (b) gives an α,β-unsaturated ketone.

(a)   (b)

Problem 19.2

Draw the product of dehydration of each aldol product in Problem 19.1.

B. Crossed and Intramolecular Aldol Reactions

The reactants in the key step of an aldol reaction are an enolate anion and an enolate anion acceptor. In self-reactions, both roles are played by one kind of molecule. **Crossed aldol reactions** are also possible, as for example the crossed aldol reaction between acetone and formaldehyde. Formaldehyde cannot provide an enolate anion because it has no α-hydrogen, but it can function as a particularly good enolate anion acceptor because its carbonyl group is unhindered. Acetone forms an enolate anion, but its carbonyl group, which is bonded to two alkyl groups, is less reactive than that of formaldehyde. Consequently, the crossed aldol reaction between acetone and formaldehyde gives 4-hydroxy-2-butanone.

4-Hydroxy-2-butanone

As this example illustrates, for a crossed aldol reaction to be successful, one of the two reactants should have no α-hydrogen so that an enolate anion does not form. It also helps if the compound with no α-hydrogen has the more reactive carbonyl, for example, an aldehyde. If these requirements are not met, a complex mixture of products results. Following are examples of aldehydes that have no α-hydrogens and can be used in crossed aldol reactions.

| Formaldehyde | Benzaldehyde | Furfural | 2,2-Dimethylpropanal |

Example 19.3

Draw a structural formula for the product of the base-catalyzed crossed aldol reaction between furfural and cyclohexanone and for the product formed by its dehydration.

Solution

| Furfural | Cyclohexanone | Aldol product (racemic) |

Problem 19.3

Draw the product of the base-catalyzed crossed aldol reaction between benzaldehyde and 3-pentanone and the product formed by its dehydration.

Nitro groups can be introduced into aliphatic compounds by way of an aldol reaction between the anion of a nitroalkane and an aldehyde or a ketone. The α-hydrogens of nitroalkanes are sufficiently acidic that they are removed by bases such as aqueous NaOH and KOH. The pK_a of nitromethane, for example, is 10.2. The acidity of the α-hydrogen of a nitroalkane is caused by the stabilization of the resulting anion by delocalization of its negative charge into the nitro group.

| Nitromethane pK$_a$ 10.2 (stronger acid) | Water pK$_a$ 15.7 (weaker acid) | Anion stabilized by charge delocalization |

Following is an aldol reaction between nitromethane and cyclohexanone. Reduction of the nitro group in the aldol product thus formed is a convenient synthetic route to β-aminoalcohols.

Cyclohexanone   Nitromethane   1-(Nitromethyl)cyclohexanol   1-(Aminomethyl)cyclohexanol

When both the enolate anion and the carbonyl group to which it adds are in the same molecule, the aldol reaction results in the formation of a ring. This type of **intramolecular aldol reaction** is particularly useful for the formation of five- and six-membered rings. Because they are the most stable rings, five- and six-membered rings form much more readily than four-, seven-, or larger membered rings. Intramolecular aldol reaction of 2,7-octanedione via enolate anion α_3, for example, gives a five-membered ring. Intramolecular aldol reaction of this same compound via enolate anion α_1 would give a seven-membered ring. In the case of 2,7-octadienone, the five-membered ring forms in preference to the seven-membered ring.

2,7-Octanedione

In general, smaller rings form faster than larger rings because the reacting groups are closer together. However, the formation of three- and four-membered rings is disfavored because of the strain in these rings.

Following is another example in which either a four-membered ring (via an enolate anion at α_3) or a six-membered ring (via an enolate anion at α_1) could be formed. Because of the greater stability of six-membered rings compared to four-membered rings, the six-membered ring is formed exclusively in this intramolecular aldol reaction.

(racemic)   (four stereoisomers as   (racemic)
two racemic mixtures)

C. Retrosynthetic Analysis

The functional groups created by the aldol reaction are a β-hydroxy carbonyl or an α,β-unsaturated carbonyl. Whenever you encounter these patterns in a target molecule, you should consider using an aldol reaction for its construction. Using retrosynthetic analysis (Section 7.9), the aldol product can be dissected into the proper reactants. Keep in mind that, if the substituents R$_1$, R$_2$, and R$_3$ are too large, a reaction could fail for steric reasons. In addition, aldol reactions to give the β-hydroxy carbonyl product are generally not favorable for two ketones.

α,β-Unsaturated carbonyl
R$_1$–R$_4$ = alkyl group,
aryl group, H atom

target
product

(racemic)

reactants

β-Hydroxy carbonyl
Aldol reaction product
recognition elements

target
product

(racemic)

reactants

19.3 | Claisen and Dieckmann Condensations

A. Claisen Condensation

In Chapter 18, we described reactions of esters, all of which take place at the carbonyl carbon and involve nucleophilic acyl substitution by an addition/elimination. In this section, we examine a second type of reaction characteristic of esters, namely one that involves both formation of an enolate anion and its participation in nucleophilic acyl substitution. One of the first of these reactions discovered is the **Claisen condensation,** named after the German chemist Ludwig Claisen (1851–1930). A Claisen condensation is illustrated by the reaction of two molecules of ethyl acetate in the presence of sodium ethoxide followed by acidification to give ethyl acetoacetate.

$$2 \text{ CH}_3\text{COEt} \xrightarrow[\text{2.H}_2\text{O, HCl}]{\text{1.EtO}^-\text{Na}^+} \text{CH}_3\text{CCH}_2\text{COEt} \ + \ \text{EtOH}$$

Ethyl ethanoate
(Ethyl acetate)

Ethyl 3-oxobutanoate
(Ethyl acetoacetate)

Ethanol

As this example illustrates, the product of a Claisen condensation is a **β-ketoester**.

A β-ketoester

Like aldol reactions, Claisen condensations require a base. Aqueous bases, such as NaOH, however, cannot be used in Claisen condensations because they would bring about the hydrolysis of the ester. Rather, the bases most commonly used in Claisen condensations are nonaqueous bases, such as sodium ethoxide in ethanol and sodium methoxide in methanol.

Claisen condensation of two molecules of ethyl propanoate gives the following β-ketoester as a racemic mixture.

Ethyl propanoate   **Ethyl propanoate**   **Ethyl 2-methyl-3-oxopentanoate** (racemic)

The first steps of a Claisen condensation bear a close resemblance to the first steps of an aldol reaction (Section 19.2). The carbon-carbon bond-forming step in each reaction involves nucleophilic addition of an enolate anion to the carbonyl group of another molecule.

Mechanism   Claisen Condensation

Step 1: **Take a proton away.** Removal of an α-hydrogen by base gives a resonance-stabilized enolate anion.

pK_a 22 pK_a 15.9   Resonance-stabilized enolate anion
(weaker acid)   (stronger acid)   (stronger base)

Because the α-hydrogen of an ester is the weaker acid and ethoxide ion is the weaker base, the position of equilibrium for this step lies very much toward the left; the concentration of enolate anion is very low compared with that of ethoxide ion and ester. Thus, there is an excess of ester to react with the small amount of enolate anion that forms.

Step 2: **Make a new bond between a nucleophile and an electrophile.** Attack of the enolate anion of one ester on the carbonyl carbon of another ester gives a tetrahedral carbonyl addition intermediate.

A tetrahedral carbonyl addition intermediate

Unlike similar intermediates in aldol reactions, this intermediate (a hemiacetal anion) has an ethoxy leaving group.

Step 3: **Break a bond to give stable molecules or ions.** Collapse of the tetrahedral carbonyl addition intermediate and ejection of ethoxide ion gives a β-ketoester.

Step 4: **Take a proton away.** The position of equilibrium for Steps 1–3 lies very much on the side of starting materials. The overall condensation, however, is

---

driven to completion by the acid-base reaction between the β-ketoester (the stronger acid) and ethoxide ion (the stronger base) to give ethanol (the weaker acid) and the anion of the β-ketoester (the weaker base).

$$EtÖ:^- + CH_3-\overset{\overset{\displaystyle :O:}{\|}}{C}-\overset{\overset{\displaystyle H}{|}}{CH}-\overset{\overset{\displaystyle :O:}{\|}}{C}-ÖEt \rightleftharpoons CH_3-\overset{\overset{\displaystyle :O:}{\|}}{C}-\overset{-}{CH}-\overset{\overset{\displaystyle :O:}{\|}}{C}-ÖEt + EtOH$$

pK$_a$ 10.7
(stronger acid)

(weaker base)   pK$_a$ 15.9
(weaker acid)

Overall, the Claisen condensation involves consumption of a stronger base, in this case ethoxide, and creation of a weaker base, the resonance-stabilized enolate anion of the β-ketoester. One molecule of the original base is consumed for every two molecules of ester that react. This is in contrast to an aldol reaction, in which base is catalytic (not consumed).

Step 5: Add a proton. The chemist then opens the flask and adds mild acid to protonate the enolate anion giving the β-ketoester.

$$CH_3-\overset{\overset{\displaystyle :O:}{\|}}{C}-\overset{-}{CH}-\overset{\overset{\displaystyle :O:}{\|}}{C}ÖEt + H_3O^+ \xrightarrow{HCl, H_2O} CH_3-\overset{\overset{\displaystyle :O:}{\|}}{C}-CH_2-\overset{\overset{\displaystyle :O:}{\|}}{C}ÖEt + H_2O$$

From an analysis of this mechanism, we see that the structural feature required for a successful Claisen condensation is an ester with two α-hydrogens, one to form the initial enolate anion and the second to form the enolate anion of the resulting β-ketoester.

Example 19.4

Show the product of the Claisen condensation of ethyl butanoate in the presence of sodium ethoxide followed by acidification with aqueous HCl.

Solution

The new bond formed in a Claisen condensation is between the carbonyl group of one ester molecule and the α-carbon of another.

the new C—C bond

Ethyl 2-ethyl-3-oxohexanoate
(racemic)

Problem 19.4

Show the product of Claisen condensation of ethyl 3-methylbutanoate in the presence of sodium ethoxide followed by acidification with aqueous HCl.

B. Dieckmann Condensation

An intramolecular Claisen condensation of a dicarboxylic ester to give a five- or six-membered ring is given the special name of **Dieckmann condensation**. In the presence of one equivalent of sodium ethoxide, for example, diethyl hexanedioate (diethyl adipate) undergoes an intramolecular condensation to form a five-membered ring.

Diethyl hexanedioate
(Diethyl adipate)

Ethyl 2-oxocyclo-
pentanecarboxylate
(racemic)

The mechanism of a Dieckmann condensation is identical to the mechanism we have described for the Claisen condensation. An anion formed at the α-carbon of one ester group adds to the carbonyl of the other ester group to form a tetrahedral carbonyl addition intermediate (make a bond between a nucleophile and an electrophile). This intermediate ejects ethoxide ion to regenerate the carbonyl group (break a bond to give stable molecules or ions). Cyclization is followed by formation of the conjugate base of the β-ketoester, just as in the Claisen condensation (take a proton away). The β-ketoester is isolated after acidification with aqueous acid (add a proton).

C. Crossed Claisen Condensations

In a **crossed Claisen condensation** between two different esters, each with two α-hydrogens, a mixture of four β-ketoesters is possible; therefore, crossed Claisen condensations of this type are not synthetically useful. Such condensations are useful, however, if appreciable differences in reactivity exist between the two esters, as for example when one of the esters has no α-hydrogens and can function only as an enolate anion acceptor. Following are four examples of esters without α-hydrogens.

Ethyl formate   **Diethyl carbonate**   **Diethyl ethanedioate**   **Ethyl benzoate**
                                              (Diethyl oxalate)

Crossed Claisen condensations of this type are usually carried out by using the ester with no α-hydrogens in excess. In the following illustration, methyl benzoate is used in excess.

Methyl   **Methyl**   **Methyl 2-methyl-3-oxo-**
benzoate   **propanoate**   **3-phenylpropanoate**
                                        (racemic)

Example 19.5

Complete the equation for this crossed Claisen condensation.

Solution

(racemic)

Problem 19.5

Complete the equation for this crossed Claisen condensation.

(excess)

D. Retrosynthetic Analysis

The functional group pattern created by a Claisen, mixed Claisen, or Dieckmann condensation is a β-ketoester. Using retrosynthetic analysis, a desired β-ketoester target molecule is dissected into the corresponding ester molecules. For the Dieckmann condensation, the β-ketoester is contained within a ring, generally a 5- or 6-membered ring, which is derived from the corresponding diester.

β-Ketoester

Claisen condensation
product recognition element

target
product
(racemic)

an ester

R_1–R_3 = alkyl group,
aryl group, H atom
R_4 = alkyl group or aryl group

Cyclic β-Ketoester

Dieckmann condensation
product recognition element

target
product
(racemic)

a diester

R_1 = alkyl group or aryl group
R_2–R_8 = alkyl group, aryl group, H atom

E. Hydrolysis and Decarboxylation of β-Ketoesters

Recall from Section 18.4C that hydrolysis of an ester in aqueous sodium hydroxide (saponification) followed by acidification of the reaction mixture with aqueous HCl converts an ester to a carboxylic acid. Recall also from Section 17.9 that β-ketoacids and β-dicarboxylic acids (substituted malonic acids) readily undergo decarboxylation (lose CO_2) when heated. Both the Claisen and Dieckmann condensations yield esters of β-ketoacids. The following equations illustrate the results of a Claisen condensation followed by hydrolysis of the ester, acidification, and decarboxylation.

Claisen condensation:

(racemic)

Saponification followed by acidification:

(racemic)   (racemic)

Decarboxylation:

(racemic)

The result of this five-step sequence is reaction between two molecules of ester (one furnishing a carbonyl group and the other furnishing an enolate anion) to give a ketone and carbon dioxide. In the general reaction, both ester molecules are the same, and the product is a symmetrical ketone.

The same sequence of reactions starting with a crossed Claisen condensation gives an unsymmetrical ketone.

Example 19.6

Each set of compounds undergoes a Claisen or Dieckmann condensation followed by acidification, saponification, acidification, and thermal decarboxylation.

Draw a structural formula of the product isolated after completion of this reaction sequence.

Solution

Steps 1 and 2 bring about a crossed Claisen or Dieckmann condensation to give a β-ketoester. Steps 3 and 4 bring about hydrolysis of the β-ketoester to give a β-ketoacid, and Step 5 brings about decarboxylation to give a ketone.

Problem 19.6

Show how to convert benzoic acid to 3-methyl-1-phenyl-1-butanone (isobutyl phenyl ketone) by the following synthetic strategies, each of which uses a different type of reaction to form the new carbon-carbon bond to the carbonyl group of benzoic acid.

Benzoic acid   **3-Methyl-1-phenyl-1-butanone**

(a) A lithium diorganocopper (Gilman) reagent   **(b)** A Claisen condensation

19.4 | Claisen and Aldol Condensations in the Biological World

Carbonyl condensations are among the most widely used reactions in the biological world for the assembly of new carbon-carbon bonds. One source of carbon atoms for the synthesis of biomolecules is **acetyl-CoA**, a thioester of acetic acid and the thiol group of coenzyme A (Problem 25.35). In this section, we examine the series of reactions by which the carbon skeleton of acetic acid is converted to isopentenyl pyrophosphate, a key intermediate in the synthesis of terpenes, cholesterol, steroid hormones, and bile acids. Note that, in the discussion that follows, we will not be concerned with the mechanism by which each of these enzyme-catalyzed reactions occurs. Rather, our concern is in recognizing the types of reactions that take place.

In a Claisen condensation catalyzed by the enzyme thiolase, acetyl-CoA is converted to its enolate anion, which then adds to the carbonyl group of a second molecule of acetyl-CoA to give a tetrahedral carbonyl addition intermediate. Collapse of this intermediate by elimination of coenzyme A anion (CoAS$^-$) gives acetoacetyl-CoA. Subsequent proton transfer to coenzyme A anion gives coenzyme A. The mechanism of this reaction is identical to that of the Claisen condensation (Section 19.3A).

Acetyl-CoA   **Acetoacetyl-CoA**   **Coenzyme A**

Enzyme-catalyzed aldol reaction with a third molecule of acetyl-CoA on the ketone carbonyl of acetoacetyl-CoA gives (S)-3-hydroxy-3-methylglutaryl-CoA.

(S)-3-Hydroxy-3-methylglutaryl-CoA

Note three features of this second carbonyl-condensation reaction.

- The reaction is completely enantioselective; only the S enantiomer is formed. Condensation takes place in a chiral environment created by the enzyme, 3-hydroxy-3-methylglutaryl-CoA synthetase, which induces the formation of one enantiomer of the product to the exclusion of the other.

Coronary artery disease is the leading cause of death in the United States and other Western countries, where about one half of all deaths can be attributed to atherosclerosis. Atherosclerosis results from the buildup of fatty deposits called plaque on the inner walls of arteries. A major component of plaque is cholesterol derived from low-density-lipoproteins (LDL), which circulate in blood plasma. Because more than one half of total body cholesterol in humans is synthesized in the liver from acetyl-CoA, intensive efforts have been directed toward finding ways to inhibit this synthesis. The rate-determining step in cholesterol biosynthesis is reduction of (S)-3-hydroxy-3-methylglutaryl CoA to (R)-mevalonic acid. This reduction is catalyzed by the enzyme HMG-CoA reductase and requires two moles of NADPH per mole of HMG-CoA.

Beginning in the early 1970s, researchers at the Sankyo Company in Tokyo screened more than 8000 strains of microorganisms and in 1976 announced the isolation of mevastatin, a potent inhibitor of HMG-CoA reductase, from culture broths of the fungus *Penicillium citrinum*. The same compound was isolated by researchers at Beecham Pharmaceuticals in England from cultures of *Penicillium brevicompactum*. Soon thereafter, a second, more active compound called lovastatin was isolated at the Sankyo Company from the fungus *Monascus ruber*, and at Merck Sharpe & Dohme from *Aspergillus terreus*. Both mold metabolites are extremely effective in lowering plasma concentrations of LDL. The active form of each is the 5-hydroxycarboxylic acid formed by hydrolysis of the δ-lactone.

| (S)-3-Hydroxy-3-methyl-glutaryl-CoA (HMG-CoA) | A hemithioacetal intermediate formed by the first NADPH reduction | | (R)-Mevalonate |

- Hydrolysis of the thioester group of acetyl-CoA is coupled with the aldol reaction.
- The carboxyl group is shown as it is ionized at pH 7.4, the approximate pH of blood plasma and many cellular fluids.

Enzyme-catalyzed reduction by NADPH (a phosphorylated form of NADH) of the thioester group of 3-hydroxy-3-methylglutaryl-CoA to a primary alcohol gives mevalonic acid, here shown as its anion.

(S)-3-Hydroxy-3-methylglutaryl-CoA   (R)-Mevalonate

In this reduction, note that a change occurs in the designation of configuration from *S* to *R*, not because of any change in configuration at the chiral center, but rather because priorities 2 and 3 become reversed as a result of the reduction.

Enzyme-catalyzed transfer of a phosphate group from adenosine triphosphate (ATP) to the 3-hydroxyl group of mevalonate gives a phosphoric ester at carbon 3. Enzyme-catalyzed transfer of a pyrophosphate group from a second molecule of ATP gives a pyrophosphoric ester at carbon 5. Enzyme-catalyzed β-elimination from this molecule results in loss of CO_2 and PO_4^{3-}, both good leaving groups.

Soon thereafter, Merck developed a synthesis for simvastatin (Zocor), which came onto the market in the late 1980s and is still used worldwide for the control of plasma cholesterol levels. These drugs and several synthetic modifications now available inhibit HMG-CoA reductase by forming an enzyme-inhibitor complex that

R$_1$ = R$_2$ = H, mevastatin
R$_1$ = H, R$_2$ = CH$_3$, lovastatin (Mevacor)
R$_1$ = R$_2$ = CH$_3$, simvastatin (Zocor)

hydrolysis of the δ-lactone

The active form of each drug

prevents further catalytic action of the enzyme. It is reasoned that the 3,5-dihydroxycarboxylic acid part of each drug binds tightly to the enzyme because it mimics the hemithioacetal intermediate formed by the first reduction of HMG-CoA.

Systematic studies have shown the importance of each part of the drug for effectiveness. It has been found, for example, that the carboxylate anion (—COO$^-$) is essential, as are both the 3—OH and 5—OH groups. Insertion of a bridging unit other than —CH$_2$—CH$_2$— between carbon 5 and the two fused six-membered rings reduces potency, as does almost any modification of the six-membered rings and their pattern of substitution.

a phosphoric ester

a pyrophosphoric ester

OPO$_3^{2-}$ OP$_2$O$_6^{3-}$

β-elimination

Isopentenyl pyrophosphate   + CO$_2$ + PO$_4^{3-}$

(R)-3-Phospho-5-pyrophosphomevalonate

Isopentenyl pyrophosphate has the carbon skeleton of isoprene, the unit into which terpenes can be divided (Section 5.4). This compound is, in fact, a key intermediate in the synthesis of terpenes, as well as of cholesterol and steroid hormones. We shall return to the chemistry of isopentenyl pyrophosphate in Section 26.4B and discuss its conversion to cholesterol and terpenes.

OP$_2$O$_6^{3-}$

Isopentenyl pyrophosphate

→ cholesterol → steroid hormones
→ cholesterol → bile acids
→ terpenes

Pyrrolidine **Morpholine**

Figure 19.1
Secondary amines used in the formation of enamines.

19.5 | Enamines

Enamines are formed by the reaction of a secondary amine with an aldehyde or ketone (Section 16.8A). The secondary amines most commonly used for this purpose are pyrrolidine and morpholine (Figure 19.1).

Example 19.7

Draw structural formulas for the aminoalcohol and enamine formed in the following reactions.

(a)   (b)

Solution

(a)   An aminoalcohol   An enamine   (b)   An aminoalcohol   An enamine

Problem 19.7

Following are structural formulas for two enamines.

(a)   (b)

Draw structural formulas for the secondary amine and carbonyl compound from which each enamine is derived.

The particular value of enamines in synthetic organic chemistry is the fact that the β-carbon of an enamine is a nucleophile by virtue of the conjugation of the carbon-carbon double bond with the electron pair on nitrogen. Enamines resemble enols and enolate anions in their reactions.

The use of enamines as synthetic intermediates for the alkylation and acylation at the α-carbon of aldehydes and ketones was pioneered by Gilbert Stork of Columbia University. This use of enamines is called the Stork enamine reaction.

A. Alkylation of Enamines

Enamines readily undergo S$_N$2 reactions with methyl and primary alkyl halides, α-haloketones, and α-haloesters. Enamines are superior to enolate anions for these reactions because they are less basic and consequently give higher ratios of substitution to elimination products. In addition, they also give more alkylation on carbon than do enolate anions.

Step 1: Make a new bond between a nucleophile and an electrophile. Treatment of the enamine with one equivalent of an alkylating agent gives an iminium halide.

| The morpholine enamine of cyclohexanone | 3-Bromopropene (Allyl bromide) | An iminium bromide (racemic) |

Step 2: Hydrolysis of the iminium salt gives the alkylated ketone and regenerates morpholine as its hydrochloride salt.

(racemic) + HCl/H₂O ⟶ **2-Allylcyclohexanone** (racemic) + **Morpholinium chloride**

Example 19.8

Show how to use an enamine to bring about this synthesis.

(racemic)

Solution

Prepare an enamine by treating the ketone with either morpholine or pyrrolidine. The intermediate aminoalcohol can undergo dehydration in two directions. The direction shown here is favored because of the stabilization gained by conjugation of the carbon-carbon double bond of the enamine with the aromatic ring. Treatment of the enamine with ethyl 2-chloroacetate followed by hydrolysis of the iminium chloride in aqueous hydrochloric acid gives the product.

Problem 19.8

Write a mechanism for the hydrolysis of the following iminium chloride in aqueous HCl.

(racemic)   (racemic)

B. Acylation of Enamines

Enamines undergo acylation when treated with acid chlorides and acid anhydrides. The reaction is a nucleophilic acyl substitution as illustrated by the conversion of cyclohexanone, via its pyrrolidine enamine, to 2-acetylcyclohexanone.

| The pyrrolidine enamine of cyclohexanone | Acetyl chloride | An iminium chloride (racemic) | 2-Acetylcyclo-hexanone (racemic) |

Acylation
The process of introducing an acyl group, RCO— or ArCO—, onto an organic molecule.

Thus, we can attach an acyl group to the α-carbon of an aldehyde or ketone using its enamine as an intermediate. The process of introducing an acyl group onto an organic molecule is called **acylation**.

Example 19.9

Show how to use an enamine to bring about this synthesis.

(racemic)

Solution

Treating cyclopentanone with pyrrolidine gives an enamine. Treating the enamine with hexanoyl chloride followed by hydrolysis in aqueous HCl gives the desired β-diketone.

(racemic)

Problem 19.9

Show how to use alkylation or acylation of an enamine to convert acetophenone to the following compounds.

(a) Ph‖O‖O

(b) Ph‖O‖O

(c) Ph‖O‖O‖OEt

C. Retrosynthetic Analysis

You learned that enamines can undergo both alkylation and acylation reactions. The characteristic functional group created by enamine alkylation is a carbonyl compound with an α-carbon possessing a group that can be derived from an S_N2 reaction (methyl, primary, possibly secondary, but not tertiary halide). The characteristic functional group created by enamine acylation is a β-dicarbonyl (aldehyde or ketone). Thus, in a retrosynthetic analysis, a desired alkylation product can be derived from a carbonyl reactant with an α-hydrogen, a secondary amine (usually pyrrolidine), and the new group R bonded to a leaving group (Lv). A desired acylation product can be derived from a carbonyl reactant with an α-hydrogen, a secondary amine (again usually pyrrolidine), and the appropriate acid chloride.

Alkylated α-carbon

Enamine alkylation reaction product recognition element

target product (racemic)

R_1–R_3 = alkyl group, aryl group, H atom
R_4 = methyl, primary or possibly secondary alkyl group

β-Dicarbonyl

Enamine acylation reaction product recognition element

target product (racemic)

R_1–R_4 = alkyl group or aryl group or H atom

19.6 | Acetoacetic Ester Synthesis

Acetoacetic ester and other β-ketoesters are versatile starting materials for the formation of new carbon-carbon bonds because of:

1. The acidity of α-hydrogens (pK_a 10–11) between the two carbonyl groups.
2. The nucleophilicity of the enolate anion resulting from loss of an α-hydrogen.
3. The ability of the product to undergo decarboxylation after hydrolysis of the ester.

The **acetoacetic ester synthesis** is useful for the preparation of monosubstituted and disubstituted acetones of the following types.

CH_3CCH_2COEt → CH_3CCH_2R A monosubstituted acetone

Ethyl acetoacetate → CH_3CCHR A disubstituted acetone
(Acetoacetic ester)   R′

We have already seen the chemistry of the individual steps in this synthesis, but we have not put them together in this particular sequence. Let us illustrate the acetoacetic ester synthesis by choosing 5-hexen-2-one as a target molecule. The three carbons shown in color are provided by ethyl acetoacetate. The remaining three carbons represent the —R group of a substituted acetone.

these three carbons are from ethyl acetoacetate

the —R group of a monosubstituted acetone

5-Hexen-2-one

A. Five Sequential Reactions

1. The methylene hydrogens of ethyl acetoacetate (pK_a 10.7) are considerably more acidic than hydroxyl group of ethanol (pK_a 15.9); therefore, ethyl acetoacetate is converted completely to its anion by sodium ethoxide or other alkali metal alkoxides.

Ethyl acetoacetate	Sodium ethoxide	Sodium salt of	Ethanol
pK_a 10.7	(stronger base)	ethyl acetoacetate	pK_a 15.9
(stronger acid)		(weaker base)	(weaker acid)

2. The enolate anion of ethyl acetoacetate is a nucleophile and reacts by an S_N2 pathway with methyl and primary haloalkanes, α-haloketones, and α-haloesters. Secondary haloalkanes give lower yields, and tertiary haloalkanes undergo E2 elimination exclusively. In the following example, the anion of ethyl acetoacetate is alkylated with allyl bromide.

3-Bromopropene   (racemic)
(Allyl bromide)

3, 4. Hydrolysis of the alkylated acetoacetic ester in aqueous NaOH followed by acidification with aqueous HCl (Section 18.4C) gives a β-ketoacid.

5. Heating the β-ketoacid brings about decarboxylation (Section 17.9A) to give 5-hexen-2-one.

A disubstituted acetone can be prepared by interrupting this sequence after Step 2, treating the monosubstituted acetoacetic ester with a second equivalent of base, carrying out a second alkylation, and then proceeding with Steps 3–5.

1′. Treatment with a second equivalent of base gives a second enolate anion.

2′. Treatment of this enolate anion with a haloalkane completes the second alkylation. This haloalkane should be methyl or primary for best yields due to steric considerations.

3, 4, 5. Hydrolysis of the ester in aqueous base followed by acidification and heating gives the ketone.

B. Retrosynthetic Analysis

The sequence of five steps used in the acetoacetic ester synthesis creates a carbonyl compound with an α-carbon possessing one or two groups, generally methyl, primary, or possibly secondary alkyl groups, which are derived from S_N2 substitution. This is a convenient and versatile method for making complex ketones.

Using retrosynthetic analysis can derive a desired complex ketone from the appropriate β-ketoester and alkylating agent(s) (R-Lv). Note that other β-ketoesters besides ethyl acetoacetate can be used.

Alkylated ketone
Acetoacetic ester synthesis
product recognition elements

target
product
(racemic)

(racemic)

R_1 = alkyl group, aryl group, H atom
R_2, R_3 = methyl, primary or possibly
secondary alkyl group

Example 19.10

Show how the acetoacetic ester synthesis can be used to prepare this ketone.

4-Phenyl-2-butanone

Solution

Determine which three carbons of the product originate from ethyl acetoacetate; then establish the location on the carbon chain of the —COOH lost in decarboxylation, and finally verify the bond formed in the alkylation step. On the basis of this analysis, determine that the starting materials are ethyl acetoacetate and a benzyl halide.

these carbons from
acetoacetic ester

this bond formed
by alkylation

this carbon lost by
decarboxylation

The enolate anion
of ethyl acetoacetate

Benzyl bromide

Now combine these reagents in the following way to prepare the desired ketone.

Problem 19.10

Show how the acetoacetic ester synthesis can be used to prepare these compounds.

(a)

(b)

(c)

C. Variants

We have described what is commonly known as the acetoacetic ester synthesis and have illustrated the use of ethyl acetoacetate as the starting reagent. This same synthetic strategy is applicable to any β-ketoester, as for example those that are available by the Claisen (Section 19.3A) and Dieckmann (Section 19.3B) condensations. For example, following are structural formulas for two β-ketoesters available from Dieckmann and Claisen condensations that can be made to undergo (1) formation of an enolate anion, (2) alkylation or acylation, (3) hydrolysis followed by (4) acidification, and finally (5) decarboxylation just as we have shown for ethyl acetoacetate.

Ethyl 2-oxocyclopentanecarboxylate
(racemic)

Ethyl 2-methyl-3-oxopentanoate
(racemic)

Example 19.11

Show how to convert racemic ethyl 2-oxocyclopentanecarboxylate to racemic 2-allylcyclopentanone.

Solution

Treat this β-ketoester with one equivalent of sodium ethoxide to form an anion followed by alkylation of the anion with one equivalent of an allyl halide. Subsequent hydrolysis of the ester in aqueous base followed by acidification and thermal decarboxylation gives the desired product.

Ethyl 2-oxocyclo-
pentanecarboxylate
(racemic)

1.EtO⁻Na⁺
2. CH₂=CHCH₂Br

3.NaOH, H₂O
4.HCl, H₂O
5.heat

2-Allylcyclopentanone
(racemic)

Problem 19.11

Show how to convert racemic ethyl 2-oxocyclopentanecarboxylate to this compound.

(racemic)

19.7 | Malonic Ester Synthesis

The factors that make malonic esters and other β-diesters versatile starting materials for formation of new carbon-carbon bonds are the same as those we have already seen for the acetoacetic ester synthesis, namely:

1. The acidity of α-hydrogens (pK_a 13–14) between the two carbonyl groups.
2. The nucleophilicity of the enolate anion resulting from loss of an α-hydrogen.
3. The ability of the product to undergo decarboxylation after hydrolysis of the ester.

The **malonic ester synthesis** is useful for the preparation of monosubstituted and disubstituted acetic acids of the following types.

1. The α-hydrogens of diethyl malonate (pK_a 13.3) are more acidic than ethanol (pK_a 15.9); therefore, diethyl malonate is converted completely to its anion by sodium ethoxide or other alkali metal alkoxide.

A. Five Sequential Reactions

As with the acetoacetic ester synthesis, we have already encountered all the important chemistry of the malonic ester synthesis, although not in this particular pattern. Let us illustrate this synthesis by choosing 5-methoxypentanoic acid as a target molecule. The two carbons shown in color are provided by diethyl malonate. The remaining three carbons and the methoxy group represent the —R group of a monosubstituted acetic acid.

5-Methoxypentanoic acid

1. The α-hydrogens of diethyl malonate (pK_a 13.3) are more acidic than ethanol (pK_a 15.9); therefore, diethyl malonate is converted completely to its anion by sodium ethoxide or other alkali metal alkoxide.

2. The enolate anion of diethyl malonate is a nucleophile and reacts by an S_N2 pathway with methyl and primary haloalkanes, α-haloketones, and α-haloesters. In the following example, the anion of diethyl malonate is alkylated with 1-bromo-3-methoxypropane.

3, 4. Hydrolysis of the alkylated malonic ester in aqueous NaOH followed by acidification with aqueous HCl gives a β-dicarboxylic acid.

5. Heating the β-dicarboxylic acid slightly above its melting point brings about decarboxylation and gives 5-methoxypentanoic acid.

5-Methoxypentanoic acid

A disubstituted acetic acid can be prepared by interrupting the previous sequence after Step 2, treating the monosubstituted diethyl malonate with a second equivalent of base, carrying out a second alkylation, and then proceeding with Steps 3–5.

B. Retrosynthetic Analysis

The characteristic functional group produced by the malonic acid synthesis is a carboxylic acid in which the α-carbon has one or two methyl, primary, or possibly secondary alkyl groups. Using retrosynthetic analysis derives the desired carboxylic acid from a diester of malonic acid, and one or two appropriate methyl, primary, or possibly secondary alkyl groups with a leaving group attached, sufficient to allow an S_N2 reaction.

Alkylated carboxylic acid

Malonic ester synthesis
product recognition elements

R_1, R_2 = methyl, primary or possibly
secondary alkyl group

target
product
(racemic)

target
product
(racemic)

(racemic)

Example 19.12

Show how the malonic ester synthesis can be used to prepare 3-phenylpropanoic acid.

Solution

Determine which two carbons of the product originate from diethyl malonate, the location on the carbon chain of the —COOH lost in decarboxylation, and finally the bond formed in the alkylation step. On the basis of this analysis, determine that the starting materials are diethyl malonate and a benzyl halide.

these two carbons
from malonic ester

this bond formed
by alkylation

this carbon lost
by decarboxylation

**Benzyl
bromide**

The enolate
anion of diethyl
malonate

Now combine these reagents in the following way to get the desired product.

3-Phenylpropanoic acid

Problem 19.12

Show how the malonic ester synthesis can be used to prepare the following substituted acetic acids.

(a)   (b)   (c)

Chemical Connections

Ibuprofen: The Evolution of an Industrial Synthesis

A major consideration in any industrial synthesis is atom efficiency; it is most efficient to use only reagents whose atoms appear in the final product. An example of the evolution of syntheses with greater and greater atom efficiency is the industrial synthesis of ibuprofen.

Synthesis I

One of the first industrial syntheses of ibuprofen used the following sequence to introduce a methyl group on the carboxyl side chain of 4-isobutylphenylacetic acid. Fischer esterification of 4-isobutylphenylacetic acid followed by treatment of the ethyl ester with diethyl carbonate in the presence of sodium ethoxide in a crossed Claisen condensation (Section 19.3C) gives the anion of a substituted malonic ester B. Alkylation of this anion (Section 19.7) followed by hydrolysis gives a disubstituted malonic acid D. Decarboxylation (Section 17.9B) of D gives ibuprofen.

4-Isobutylphenylacetic acid

A

B

C

D

Ibuprofen
(racemic)

Even though this synthesis gives ibuprofen in quite good yield, it wastes carbons; only 13 of the 18 carbons in intermediate C appear in ibuprofen.

Synthesis II

An alternative route with greater atom efficiency starts with 4-isobutylacetophenone. Treating this ketone with chloroacetonitrile in the presence of sodium ethoxide gives the epoxynitrile E. Treating E with lithium perchlorate, a Lewis acid, gives the α-cyanoketone F. Hydrolysis

This synthesis, while more atom-efficient than Synthesis I, uses a cyano group, neither atom of which appears in the final product.

Synthesis III

The ultimate in atom efficiency in the synthesis of ibuprofen is achieved in the following synthesis. Catalytic reduction of the carbonyl group of 4-isobutylacetophenone gives the alcohol G. Palladium-catalyzed carbonylation of G gives ibuprofen. The one carbon

4-Isobutylacetophenone

E (racemic)

F (racemic)

Ibuprofen (racemic)

of F gives ibuprofen. The α-cyanoketone group of F is similar to an acid chloride in its chemical properties, and its hydrolysis gives a carboxylic acid and cyanide ion.

atom introduced in the synthesis appears in the final product!

4-Isobutylacetophenone

G (racemic)

Ibuprofen (racemic)

19.8 | Conjugate Addition to α,β-Unsaturated Carbonyl Compounds

Thus far we have used a variety of carbon nucleophiles to form new carbon-carbon bonds.

1. Anions of terminal alkynes (Section 7.5) and the cyanide ion.
2. Organomagnesium (Grignard) reagents, organolithium reagents, and lithium diorganocopper (Gilman) reagents.
3. Enolate anions derived from aldehydes and ketones (aldol reactions), esters (Claisen and Dieckmann condensations), β-diesters (malonic ester syntheses), and β-ketoesters (acetoacetic ester syntheses).
4. Enamines (which are synthetically equivalent to enolate anions).

These species can be used to form new carbon-carbon bonds by two synthetic strategies: (1) substitution by the carbon nucleophile in an S$_N$2 reaction and (2) addition of the carbon nucleophile to a carbonyl carbon.

Conjugate addition, as it is known, presents a third synthetic strategy: addition of a carbon nucleophile to an electrophilic carbon-carbon double or triple bond conjugated with a carbonyl or other electron-withdrawing group. In this section, we study two types of conjugate additions to electrophilic double bonds: addition of enolate anions (the Michael reaction) and addition of lithium diorganocopper (Gilman) reagents.

Conjugate addition
Addition of a nucleophile to the β-carbon of an α,β-unsaturated carbonyl compound.

A. Michael Addition of Enolate Anions

Nucleophilic addition of enolate anions to α,β-unsaturated carbonyl compounds was first reported in 1887 by the American chemist Arthur Michael. Following are two examples of **Michael reactions**. In the first example, the nucleophile adding to the conjugated system is the enolate anion of diethyl malonate. In the second example, the nucleophile is the enolate anion of ethyl acetoacetate.

| **Diethyl propanedioate (Diethyl malonate)** | **3-Buten-2-one (Methyl vinyl ketone)** | | |

| **Ethyl 3-oxobutanoate (Ethyl acetoacetate)** | **2-Cyclohexenone** | (four stereoisomers are possible; racemic) |

You will recall that nucleophiles do not add to ordinary π bonds. π bonds are generally considered as weak nucleophiles that are capable of attacking strong electrophiles (Section 6.3). What activates a carbon-carbon double bond for nucleophilic attack in a Michael reaction is the presence of the adjacent carbonyl group. One important resonance structure of the α,β-unsaturated carbonyl compounds puts positive charge at the end (in this case, the β-carbon) of the double bond, making it resemble a carbonyl group in its reactivity. Thus nucleophiles can add to this type of double bond, which we call "activated" for this reason.

Elpot of α,β-unsaturated carbonyl.

Although the major fraction of the partial positive charge (blue) of an α,β-unsaturated aldehyde or ketone is on the carbonyl carbon, there is nevertheless a significant partial positive charge on the beta carbon.

Note that aldol, Claisen, and Dieckmann condensations all give primary products with oxygens in a 1,3 relationship. The Michael reaction with enolate anions gives products with oxygens in a 1,5 relationship. These relationships are a consequence of the polarization of the reagents. In aldol, Claisen, and Dieckmann condensations, the carbonyl carbon is positive, and the α-position is negative.

In a Michael reaction, the positive polarization of the carbonyl carbon is transmitted two carbons farther by the double bond.

The Michael reaction takes place with a wide variety of α,β-unsaturated carbonyl compounds as well as with α,β-unsaturated nitriles and nitro compounds. The most commonly used types of nucleophiles in Michael reactions are summarized in Table 19.1. The bases most commonly used to generate the nucleophile are metal alkoxides, pyridine, and piperidine. It is important to realize that other nucleophiles can undergo similar additions to the beta carbon of unsaturated carbonyl compounds, for example, amines, alcohols, and water.

We can write the following general mechanism for a Michael reaction. Note that in Step 3 the base, B^-, is regenerated, in accord with the experimental observation that a Michael reaction requires only a catalytic amount of base rather than a molar equivalent, provided there are no additional acidic H atoms on the product.

Table 19.1	Combinations of Reagents for Effective Michael Reactions

These Types of α,β-Unsaturated Compounds Are Nucleophile Acceptors in Michael Reactions		These Types of Compounds Provide Effective Nucleophiles for Michael Reactions	
$CH_2{=}CHCH$ (O)	Aldehyde	$CH_3CCH_2CCH_3$ (O, O)	β-Diketone
$CH_2{=}CHCCH_3$ (O)	Ketone	CH_3CCH_2COEt (O, O)	β-Ketoester
$CH_2{=}CHCOEt$ (O)	Ester	CH_3CCH_2CN (O)	β-Ketonitrile
$CH_2{=}CHCNH_2$ (O)	Amide	$EtOCCH_2COEt$ (O, O)	β-Diester
$CH_2{=}CHC{\equiv}N$	Nitrile	$CH_3C{=}CH_2$ (enamine with pyrrolidine N)	Enamine
$CH_2{=}CHNO_2$	Nitro compound	NH_3, RNH_2, R_2NH	Amine

Mechanism Michael Reaction—Conjugate Addition of Enolate Anions

Step 1: Take a proton away. Treating H—Nu with base gives the nucleophile, Nu:$^-$.

$$Nu\!-\!H + :B^- \rightleftharpoons Nu\!:^- + H\!-\!B$$

Base

Step 2: Make a new bond between a nucleophile and an electrophile. Nucleophilic addition of Nu:$^-$ to the β-carbon of the conjugated system gives a resonance-stabilized enolate anion.

$$Nu\!:^- + -\overset{|}{C}=\overset{|}{C}-\overset{\overset{\displaystyle :\ddot{O}:}{\|}}{C}- \longrightarrow Nu\!-\!\overset{|}{C}-\overset{|}{C}=\overset{\overset{\displaystyle :\ddot{O}:^-}{|}}{C}- \longleftrightarrow Nu\!-\!\overset{|}{C}-\overset{|}{\ddot{C}}-\overset{\overset{\displaystyle :O:}{\|}}{C}-$$

Resonance-stabilized enolate anion

Step 3: Add a proton. Proton transfer from H—B gives the enol.

$$Nu\!-\!\overset{|}{C}-\overset{|}{C}=\overset{\overset{\displaystyle :\ddot{O}:^-}{|}}{C}- + H\!-\!B \longrightarrow Nu\!-\!\overset{|}{\underset{4}{C}}-\overset{|}{\underset{3}{C}}=\overset{\overset{\displaystyle \overset{1}{\ddot{O}}-H}{|}}{\underset{2}{C}}- + :B^-$$

An enol
(a product of 1,4-addition)

The enol formed in Step 3 corresponds to 1,4-addition to the conjugated system of the α,β-unsaturated carbonyl compound. It is because this intermediate is formed that the Michael reaction is classified as a 1,4- or conjugate addition.

Step 4: Keto-enol tautomerism. Conversion of the less stable enol form to the more stable keto form (Section 16.9B) gives the final product.

$$Nu\!-\!\overset{|}{C}-\overset{|}{C}=\overset{\overset{\displaystyle :\ddot{O}-H}{|}}{C}- \rightleftharpoons Nu\!-\!\overset{|}{C}-\overset{\overset{\displaystyle H}{|}}{\underset{|}{C}}-\overset{\overset{\displaystyle :O:}{\|}}{C}-$$

Less stable enol form   More stable keto form

Example 19.13

Draw a structural formula for the product formed by treating each set of reactants with sodium ethoxide in ethanol under conditions of the Michael reaction.

(a) [structure: ethyl acetoacetate] + [structure: ethyl acrylate]   (b) EtO[diethyl malonate]OEt + [cyclopent-2-enone]

Solution

(a) [product structure with COOEt branch]   (b) EtOOC[cyclopentanone with CH(COOEt) branch]

(racemic)   (racemic)

Problem 19.13

Show the product formed from each Michael product in the solution to Example 19.13 after (1) hydrolysis in aqueous NaOH, (2) acidification, and (3) thermal decarboxylation of each β-ketoacid or β-dicarboxylic acid. These reactions illustrate the usefulness of the Michael reaction for the synthesis of 1,5-dicarbonyl compounds.

As noted in Table 19.1, enamines also participate in Michael reactions as illustrated by the addition of the enamine of cyclohexanone to acrylonitrile, CH$_2$=CHCN.

Pyrrolidine
enamine of
cyclohexanone
(racemic)

A final word about addition of nucleophiles to α,β-unsaturated carbonyl compounds. The Michael reaction is an example of 1,4-addition (conjugate addition) to an α,β-unsaturated carbonyl compound. In general, resonance-stabilized enolate anions and enamines are weak bases, react slowly, and give 1,4-addition products. Organolithium and organomagnesium compounds, on the other hand, are strong bases, react rapidly, and give primarily 1,2-addition products; that is, they give products formed by addition to the carbonyl carbon.

| Phenyl-lithium | 4-Methyl-3-penten-2-one | | 4-Methyl-2-phenyl-3-penten-2-ol (racemic) |

Why do the nucleophiles listed in Table 19.1 react with conjugated carbonyl compounds by 1,4-addition rather than 1,2-addition? The answer has to do with **kinetic control** versus **thermodynamic control** of product formation. It has been shown that 1,2-addition of nucleophiles to the carbonyl carbon of α,β-unsaturated carbonyl compounds is faster than conjugate addition. If formation of the 1,2-addition product is irreversible, then that is the product observed. If, however, formation of the 1,2-addition product is reversible, then an equilibrium is established between the more rapidly formed but less stable 1,2-addition product and the more slowly formed but more stable 1,4-addition product. As mentioned at the beginning of the chapter, a carbon-oxygen double bond is stronger than a carbon-carbon double bond. Thus, under conditions of thermodynamic (equilibrium) control, the more stable 1,4-Michael addition product is formed.

Kinetic control
Experimental conditions under which the composition of the product mixture is determined by the relative rates of formation of each product.

Thermodynamic control
Experimental conditions that permit the establishment of equilibrium between two or more products of a reaction. The composition of the product mixture is determined by the relative stabilities of the products.

1,2-Addition
(less stable product)

1,4-Addition
(more stable product)

B. Retrosynthetic Analysis

The Michael addition of the nucleophiles listed in Table 19.1 to α,β-unsaturated carbonyls leads to a large variety of structures involving ketones, nitriles, and carboxylic acids, with a δ-ketone. As one example of a Michael addition retrosynthetic analysis, we examine δ-diketones. Such structures can be derived from a β-ketoester and an α,β-unsaturated carbonyl. Dissecting the product, we find that the α carbon must be derived from a β-ketoester, while the β- and γ-carbons must come from the alkene of the Michael acceptor. Note in this example that the ester group is removed by a hydrolysis/decarboxylation sequence. Alternatively, the α-carbon may be nucleophilic by virtue of coming from an enamine, in which case a hydrogen is present at this position in the starting ketone. A chemist's ability to construct complex molecules using more than a single route makes organic synthesis interesting and challenging.

target
product

δ-*Diketone*
Michael addition product
recognition element

R_1–R_6 = alkyl group, aryl group, or H atom

Example 19.14

Show how the series of reactions in Example 19.13 and Problem 19.13 (Michael reaction, hydrolysis, acidification, and thermal decarboxylation) can be used to prepare 2,6-heptanedione.

Solution

As shown in the following retrosynthetic analysis, this molecule can be constructed from the carbon skeletons of ethyl acetoacetate and methyl vinyl ketone.

these three carbons from acetoacetic ester → ⇒ this bond formed in a Michael reaction ⇒

this carbon lost by decarboxylation

Ethyl acetoacetate   **Methyl vinyl ketone**

Following are the steps in their conversion to 2,6-heptanedione.

COOEt   +   → 1. EtO⁻Na⁺ EtOH →   COOEt (racemic)   → 2. H₂O, NaOH / 3. H₂O, HCl →

COOH (racemic)   → 4. heat →   **2,6-Heptanedione**

Problem 19.14

Show how the sequence of Michael reaction, hydrolysis, acidification, and thermal decarboxylation can be used to prepare pentanedioic acid (glutaric acid).

C. Robinson Annulation

Michael reaction with an α,β-unsaturated ketone followed by an intramolecular aldol reaction has proven to be a valuable method for the synthesis of 2-cyclohexenones. An especially important example of a Michael-aldol sequence is the **Robinson annulation**, in which treatment of a cyclic ketone, β-ketoester, or β-diketone with an α,β-unsaturated ketone in the presence of a base catalyst forms a cyclohexenone ring fused to the original ring. When the following racemic β-ketoester, for example, is treated with methyl vinyl ketone in the presence of sodium ethoxide in ethanol, the Michael adduct forms and then, in the presence of sodium ethoxide, undergoes a base-catalyzed intramolecular aldol reaction followed by dehydration to give a racemic substituted cyclohexenone.

COOEt   +   → 1.NaOEt, EtOH (Michael reaction) →   COOEt (racemic)

Ethyl 2-oxocyclo-hexanecarboxylate (racemic)   **3-Buten-2-one (Methyl vinyl ketone)**

→ 2.NaOEt, EtOH (aldol reaction) →   OH ... COOEt (four stereoisomers as two racemic mixtures)   → 3. −H₂O →   COOEt (racemic)

Example 19.15

Draw structural formulas for the lettered compounds (A) and (B) in the following synthetic sequence.

$$+ \xrightarrow[\text{CH}_3\text{OH}]{\text{KOH}} (\mathbf{A})\text{C}_{11}\text{H}_{16}\text{O}_3 \xrightarrow{\text{heat}} (\mathbf{B})\text{C}_{11}\text{H}_{14}\text{O}_2$$

Solution

The product is the result of Michael addition to an α,β-unsaturated ketone followed by base-catalyzed aldol reaction and dehydration.

(A)

(B)
(racemic)

Problem 19.15

Show how to bring about the following conversion.

D. Retrosynthetic Analysis

The functional group present in the product of a Robinson annulation is a six-membered ring containing an α,β-unsaturated ketone. Often, there is a carbonyl group on a carbon adjacent to the ring, two carbons away from the β-carbon of the α,β-unsaturated ketone. Retrosynthetic analysis of the Robinson annulation is a combination of those discussed for the Michael addition and the aldol reaction. The carbon-carbon double bond in the ring is derived from the aldol reaction, while another new bond is created by the Michael reaction. The challenge is to choose the proper enolate-forming β-dicarbonyl and α,β-unsaturated ketone derivatives based on these bond connections.

Cyclic α,β-unsaturated carbonyl

Robinson annulation
product recognition element

R_1–R_7 = alkyl group, aryl group, or H atom

Bond created by
Michael addition

target
product
(racemic)

Bond created by
aldol reaction

E. Conjugate Addition of Lithium Diorganocopper Reagents

Lithium diorganocopper reagents undergo 1,4-addition to α,β-unsaturated aldehydes and ketones in a reaction that is closely related to the Michael reaction. Yields are highest with primary alkyl, allylic, vinyl, and aryl organocopper reagents.

3-Methyl-2-cyclohexenone   **3,3-Dimethylcyclohexanone**

2-Cyclohexenone   **3-Phenylcyclohexanone**
(racemic)

Grignard reagents can give primarily addition reactions to the carbonyl group of α,β-unsaturated carbonyls (1,2-addition). In fact, lithium diorganocopper reagents are unique among organometallic compounds in that they give almost exclusively 1,4-addition, which makes them very valuable reagents in synthetic organic chemistry. The mechanism of conjugate addition of lithium diorganocopper reagents is not fully understood.

Example 19.16

Propose two syntheses of 4-octanone, each involving conjugate addition of a lithium diorganocopper reagent.

Solution

A lithium diorganocopper reagent adds to the beta carbon of an α,β-unsaturated aldehyde or ketone. Therefore, locate each carbon beta to the carbonyl group in this target molecule and disconnect at those points.

Synthesis 1:

4-Octanone   **1-Hexen-3-one**   **Bromoethane**

For this synthesis, add lithium diethylcopper to 1-hexen-3-one.

1-Hexen-3-one   **4-Octanone**

Synthesis 2:

4-Octanone   **1-Hepten-3-one**   **Bromomethane**

For this synthesis, add lithium dimethylcopper to 1-hepten-3-one.

$$\text{1-Hepten-3-one} \xrightarrow[\text{2.H}_2\text{O, HCl}]{\text{1.(CH}_3)_2\text{CuLi, ether, } -78°\text{C}} \text{4-Octanone}$$

1-Hepten-3-one   **4-Octanone**

Problem 19.16

Propose two syntheses of 4-phenyl-2-pentanone, each involving conjugate addition of a lithium diorganocopper reagent.

19.9 | Crossed Enolate Reactions Using LDA

A. Acid-Base Considerations

As discussed in this chapter, enolate anions are formed when a carbonyl compound containing an α-hydrogen is treated with a base such as hydroxide or an alkoxide. We noted earlier that α-hydrogens normally are considerably less acidic than water or alcohols, so the position of equilibrium in this acid-base reaction greatly favors the reactants rather than enolate products.

$$\underset{\substack{\text{p}K_a \text{ 20} \\ \text{(weaker acid)}}}{\overset{\overset{\displaystyle O}{\|}}{\text{CH}_3\text{CCH}_3}} + \underset{\text{(weaker base)}}{\text{NaOH}} \rightleftharpoons \underset{\substack{\text{A sodium} \\ \text{enolate} \\ \text{(stronger base)}}}{\overset{\overset{\displaystyle O^-\text{Na}^+}{|}}{\text{CH}_2=\text{CCH}_3}} + \underset{\substack{\text{p}K_a \text{ 15.7} \\ \text{(stronger acid)}}}{\text{H}_2\text{O}} \qquad K_{eq} = 5 \times 10^{-5}$$

$$\underset{\substack{\text{p}K_a \text{ 23} \\ \text{(weaker acid)}}}{\overset{\overset{\displaystyle O}{\|}}{\text{CH}_3\text{COC}_2\text{H}_5}} + \underset{\text{(weaker base)}}{\text{C}_2\text{H}_5\text{O}^-\text{Na}^+} \rightleftharpoons \underset{\substack{\text{A sodium} \\ \text{enolate} \\ \text{(stronger base)}}}{\overset{\overset{\displaystyle O^-\text{Na}^+}{|}}{\text{CH}_2=\text{COC}_2\text{H}_5}} + \underset{\substack{\text{p}K_a \text{ 16} \\ \text{(stronger acid)}}}{\text{C}_2\text{H}_5\text{OH}} \qquad K_{eq} = 10^{-7}$$

If, alternatively, a second electron-withdrawing group such as a carbonyl is present, as in ethyl acetoacetate or diethyl malonate, the pK_a is shifted so that the equilibrium lies largely toward the enolate products. During the acetoacetic ester and malonic diester sequence of five steps, an ester carbonyl carbon is removed as CO_2. Therefore, the primary role of the removed ester function can be thought of as making α-hydrogens more acidic, enabling efficient deprotonation with a base such as alkoxide.

$$\underset{\substack{\text{p}K_a \text{ 10.7} \\ \text{(stronger acid)}}}{\text{EtO}^- + \text{CH}_3-\overset{\overset{\displaystyle O:}{\|}}{\text{C}}-\underset{\underset{\displaystyle H}{|}}{\text{CH}}-\overset{\overset{\displaystyle O:}{\|}}{\text{C}}-\text{OEt}} \rightleftharpoons \underset{\text{(weaker base)}}{\text{CH}_3-\overset{\overset{\displaystyle O:}{\|}}{\text{C}}-\overset{-}{\text{CH}}-\overset{\overset{\displaystyle O:}{\|}}{\text{C}}-\text{OEt}} + \underset{\substack{\text{p}K_a \text{ 15.9} \\ \text{(weaker acid)}}}{\text{EtOH}} \qquad K_{eq} = 1.6 \times 10^5$$

With a substantially stronger base, the formation of an enolate anion from an aldehyde, ketone, or ester can be driven to completion without the necessity of an additional electron-withdrawing group. A commonly used base for this purpose is lithium diisopropylamide (LDA).

$$[(\text{CH}_3)_2\text{CH}]_2\text{N}^-\text{Li}^+$$

Lithium diisopropylamide

LDA is prepared by dissolving diisopropylamine in tetrahydrofuran and treating this solution with butyllithium.

$$[(CH_3)_2CH]_2NH \ + \ CH_3(CH_2)_3Li \ \longrightarrow \ [(CH_3)_2CH]_2N^-Li^+ \ + \ CH_3(CH_2)_2CH_3 \quad K_{eq} = 10^{10}$$

Diisopropylamine   **Butyllithium**   **Lithium**   **Butane**
pK_a 40   (stronger base)   **diisopropylamide** pK_a 50
(stronger acid)   (weaker base)   (weaker acid)

Although LDA is an extremely strong base, it is a poor nucleophile because of steric crowding around the nitrogen, which prevents its addition to carbonyl groups. LDA is, therefore, ideal for generating enolate anions from carbonyl-containing compounds. Using a molar equivalent of LDA can completely convert an aldehyde, ketone, or ester to the corresponding lithium enolate.

$$CH_3\overset{\overset{\displaystyle O}{\|}}{C}OC_2H_5 \ + \ [(CH_3)_2CH]_2N^-Li^+ \ \longrightarrow \ CH_2{=}\overset{\overset{\displaystyle O^-Li^+}{|}}{C}OC_2H_5 \ + \ [(CH_3)_2CH]_2NH \quad K_{eq} = 10^{17}$$

pK_a 23   **LDA**   A lithium pK_a 40
(stronger acid)   (stronger base)   enolate   (weaker acid)
                                          (weaker base)

B. Base Stoichiometry

Let's now consider the stoichiometry of the added base needed when performing an enolate reaction. In an aldol reaction carried out with an aldehyde and hydroxide, the amount of hydroxide used has a minimal effect on the yield of the reaction. This is because using hydroxide, which is a catalyst in the aldol reaction, produces very little enolate in the equilibrium established for the first deprotonation step (about one part in a thousand). The addition of more hydroxide leads to slightly more enolate intermediate (up to a few parts per thousand), speeding up the over-all aldol reaction but not changing the reaction outcome. The key is that even with an excess of hydroxide, unreacted aldehyde is still in vast excess during the reaction. Therefore, using more than an equivalent of hydroxide would not substantially change aldol reaction product yields.

However, because LDA performs essentially a complete deprotonation with an aldehyde, ketone, or ester, the amount of LDA used is critical. The proper amount of LDA to use in an aldol reaction is 0.5 equivalents, which is enough to generate 50% enolate, which then reacts with the remaining 50% aldehyde for an efficient and rapid aldol reaction. If alternatively you used one full equivalent of LDA, all the aldehyde would be deprotonated and exist as the enolate, with no additional aldehyde left in solution to react. In this case, no aldol product would form.

C. Crossed Enolate Reactions Using LDA

Using LDA allows one to direct crossed aldol and crossed Claisen reactions. For example, consider how to perform the following reaction.

Phenylacetaldehyde   **Acetone**   **4-Hydroxy-5-phenyl-2-pentanone**
                                            (racemic)

This crossed aldol reaction between acetone and an aldehyde may be carried out successfully by treating acetone with one equivalent of LDA to convert it completely to its enolate anion. The preformed enolate is then treated with the alde-hyde followed by workup in water to give the crossed aldol condensation product.

                A lithium   **4-Hydroxy-5-phenyl-2-pentanone**
                enolate   (racemic)

Example 19.17

Show how to prepare 5-hydroxy-4-methyl-3-hexanone using a crossed aldol reaction.

Solution

First recognize that the two carbonyl-containing compounds to be joined in the aldol reaction are 3-pentanone and acetaldehyde. Treat the symmetrical ketone with LDA to form its lithium enolate. Treatment of this enolate anion with acetaldehyde followed by aqueous workup gives the desired aldol product.

Problem 19.17

Show how you might prepare the following compounds using directed aldol reactions.

(a)   (b)

This chapter introduced the use of enolates and/or enamines as nucleophiles in several reactions, including aldol reactions, Claisen condensations and Michael additions, alkylations, and acylations. We can also use LDA to generate the enolate anions and perform exactly the same reactions, as shown here for cyclohexanone and a few specific electrophiles. Similar reactions are possible for aldehydes and esters with α-hydrogens. The synthetic versatility of this approach has made LDA a very popular and important reagent in modern synthetic organic chemistry.

D. Kinetic versus Thermodynamic Enolates

For a ketone with two sets of nonequivalent α-hydrogens, the following questions arise: is formation of an enolate anion regioselective, and, if so, what factors determine the degree of regioselectivity? It has been determined experimentally that a high degree of regioselectivity often exists, and that its occurrence depends upon experimental conditions. When 2-methylcyclohexanone, for example, is added to a slight excess of LDA, the ketone is converted to its lithium enolate, which consists almost entirely of the salt of the less substituted enolate anion.

When 2-methylcyclohexanone is treated with LDA under conditions in which the ketone is in slight excess, the composition of the product is quite different; it is richer in the more substituted enolate anion.

The most important factor determining the composition of an enolate anion mixture is whether the reaction is under kinetic (rate) control or thermodynamic (equilibrium) control. In a reaction under **thermodynamic control,**

- – The reaction conditions permit the equilibration of alternative products, and
- – The composition of the product mixture is determined by the relative stabilities of the alternative products.

Equilibrium among enolate anions is established when the ketone is in slight excess, a condition under which it is possible for proton-transfer reactions to occur between an enolate and an α-hydrogen of an unreacted ketone. Thus, equilibrium is established between alternative enolate anions.

Under these conditions, the more stable enolate anion predominates. The factors that determine the relative stabilities of enolate anions are the same as those that determine the relative stabilities of alkenes; the more substituted the double bond of the enolate anion, the greater its stability. Thus, the composition of the enolate anion mixture formed under conditions of thermodynamic control reflects the relative stabilities of the individual enolate anions.

> **Thermodynamic control**
> Experimental conditions that permit the establishment of equilibrium between two or more products of a reaction. The composition of the product mixture is determined by the relative stabilities of the products.

Kinetic control
Experimental conditions under which the composition of the product mixture is determined by the relative rates of formation of each product.

In a reaction under **kinetic control**, the composition of the product mixture is determined by the relative rates of formation of each product. No equilibrium among possible alternative structures is set up. In the case of formation of enolate anions, kinetic control refers to the relative rates of removal of the alternative α-hydrogens. The less hindered α-hydrogen is removed more rapidly, and thus the major product is the less substituted enolate anion. Because a slight excess of base is used, there is no ketone to serve as a proton donor, and the less stable enolate anion cannot equilibrate with a more stable one.

Summary

Section 19.1 Formation and Reactions of Enolate Anions: An Overview

- The carbon atom immediately adjacent to a carbonyl group is referred to as the **α-carbon**, and hydrogen atoms bonded to the α-carbon are referred to as **α-hydrogens**.
  - The α position of carbonyl containing compounds such as aldehydes, ketones, and esters is relatively acidic, having pK_a values in the 21–25 range.
  - This relative acidity for a carbon-bound H atom is largely due to the stability of the resonance-stabilized anion, called an **enolate anion**, which is produced upon deprotonation.
- Enolate anions are best represented as a hybrid of two contributing structures.
  - One contributing structure places the negative charge on the α-carbon atom and contains a carbonyl π bond.
  - The other contributing structure places the negative charge on the oxygen atom and contains a $C=C$ π bond.
  - The contributing structure with the negative charge on the O atom is the major contributor to the hybrid because O is more electronegative than C, so O is better able to accommodate the negative charge.
- Despite their charge distribution, enolate anions react as carbon nucleophiles that create new carbon-carbon bonds, making enolate anions important for organic synthesis.
  - Enolate anions react with alkyl halides in S_N2 reactions.
  - Enolate anions react with aldehydes, ketones, and esters in carbonyl addition reactions.
  - In reactions with electrophiles, enolate anions react primarily at C rather than O for at least two reasons.
    - Reaction at C gives more stable products that contain a relatively strong $C=O$ π bond. Reaction at O would give products with a weaker $C=C$ π bond.
    - The enolate O atom, having most of the negative charge of the resonance hybrid, is more tightly associated with a counterion such as Na^+ or Li^+. The counterion has a shielding effect that inhibits reactions with electrophiles, an effect that is amplified by the aggregate structure of enolate anions in solution.

Section 19.2 Aldol Reaction

- In the **aldol reaction**, enolate anions derived from aldehydes or ketones react with a second molecule of aldehyde or ketone to give a carbonyl addition reaction and create a new carbon-carbon bond.
  - The mechanism of the base-catalyzed aldol reactions involves initial deprotonation of the α-hydrogen in base to create an enolate anion, which is a strong nucleophile and attacks another aldehyde or ketone molecule to give a carbonyl addition intermediate, which, in turn, reacts with water to create a β-hydroxy aldehyde or ketone product and regenerates the original base.
    - Because the base is regenerated at the end of the reaction, the aldol reaction is considered to be base catalyzed.

- The aldol reaction can also be catalyzed by acid.
  - The mechanism of the acid-catalyzed aldol reaction involves an initial acid-catalyzed keto-enol tautomerization to provide the enol form; protonation of a second molecule on the carbonyl oxygen creates an electrophilic oxonium ion that is then attacked by the nucleophilic enol, followed by loss of a proton to give the β-hydroxy aldehyde or ketone product.
    - Because the proton is regenerated at the end of the reaction, the reaction is considered to be acid-catalyzed.
  - In both the acid- and base-catalyzed aldol reactions, one or two new chiral centers are often created, leading to racemic mixtures unless a starting aldehyde, ketone, or catalyst is chiral and present as a single enantiomer.

Problems: 19.1–19.3, 19.18–19.28, 19.53–19.56, 19.58, 19.59, 19.62, 19.64, 19.70, 19.72, 19.74, 19.77

- Aldol reactions are readily reversible, especially in base.
  - Equilibrium in aldol reactions favors products in the case of aldehydes, but for ketones often little product is made.
- The β-hydroxy aldehyde or ketone products of aldol reactions are easily dehydrated and lose H_2O to give an α,β-unsaturated aldehyde or ketone.
  - Dehydration can occur under the conditions of the aldol reaction, or sometimes heating in acid is used, in which case the carbonyl tautomerizes to the enol form, the other (non-enol) OH group is protonated, H_2O departs along with a proton to give the α,β-unsaturated aldehyde or ketone.
- **Crossed aldol reactions** that give high yields of a desired product are generally not possible in the presence of two different aldehydes or ketones.
  - Crossed aldol reactions *do not* generally give high yields of a desired product using catalytic hydroxide and two different aldehydes or ketones, because a statistical mixture of products results.
  - High yields of a single desired product are possible in base if the more reactive carbonyl (usually an aldehyde) has no α-hydrogens, so that only the less reactive carbonyl species (usually a ketone) can form the enolate anion.
- **Nitro groups** can be added to organic molecules though an aldol reaction between the anion of a nitroalkane and an aldehyde or ketone carbonyl.
- **Intramolecular aldol reactions** can be used to create five or six-membered rings from dicarbonyl compounds (either aldehydes or ketones), which form in preference to smaller or larger rings that may be possible.
- To become skilled at retrosynthetic analysis, it is important to recognize that the α,β-unsaturated carbonyl and β-hydroxy carbonyl functional groups are the characteristic products of an aldol reaction.

Section 19.3 Claisen and Dieckmann Condensations

- The **Claisen condensation** involves two ester molecules reacting in base to give a β-ketoester product.
  - The Claisen condensation mechanism involves reaction of one ester molecule with base to form an enolate, which reacts as a nucleophile with another molecule of ester to give a carbonyl addition intermediate, in which the —OR group is lost to give a β-ketoester, which is deprotonated at the α position by the RO$^-$.
    - The base used in a Claisen condensation is RO$^-$, with R chosen to match the alkoxy groups on the ester starting material.
    - Depending on relative acid-base strengths, when RO$^-$ is used as the base, the position of equilibrium for the initial enolate-forming step is far to the side of the starting ester, so the small amount of enolate anion formed will have plenty of ester to react with.

Problems: 19.4–19.6, 19.29–19.38, 19.53, 19.73–19.75

  -
    - The reaction is not catalytic in base, because the deprotonated β-ketoester product (β-ketoester pK_a = 10–11) is substantially less basic than the starting RO$^-$.
    - An amount of base equal to one half equivalent compared to the amount of starting ester is the minimum amount that must be used.
  - To complete the reaction, the chemist adds dilute acid to generate the neutral β-ketoester product.

- A **Dieckmann condensation** is an intramolecular Claisen condensation of a diester.
  - Five- or six-membered ring products are favored.
  - One equivalent of base relative to the amount of starting diester is used.
- Crossed Claisen reactions can be used to give high yields of a desired product from two different esters if one ester has no α-hydrogens (cannot form an enolate anion), and is used in excess.
- For retrosynthetic analysis, the β-ketoester functional group is the characteristic product of a Claisen or Dieckmann condensation.
- The product of Claisen and Dieckmann condensation reactions can be treated with aqueous base (saponification) followed by acidification to convert the β-ketoester group into a β-ketoacid that is then heated to cause decarboxylation to give a ketone product and CO_2.
  - The general case of a Claisen condensation followed by saponification, acidification, and decarboxylation gives a symmetrical ketone product.

Section 19.4 Claisen and Aldol Condensations in the Biological World

- Biological molecules are created from simple building blocks through enzyme-catalyzed reactions that often resemble the organic transformations presented in this chapter.
  - Claisen condensations and aldol reactions are common, and **acetyl-CoA** is a common starting material for these reactions.

Section 19.5 Enamines

- **Enamines**, compounds with a $C{=}C$ π bond adjacent to a C—N bond, are formed through reaction of an aldehyde or ketone with a secondary amine, most commonly pyrrolidine or morpholine.
  - Enamines are important for synthesis because the β-carbon is a nucleophile by virtue of conjugation of the $C{=}C$ π bond with the electron pair on nitrogen.
  - Enamines resemble enols and enolate anions in their reactions, yet harsh conditions (i.e., strong acid or base) are not required.
    - Enamines can be alkylated on the β-carbon in an S_N2 reaction with methyl and primary alkyl halides.
    - Enamines can be acylated on the β-carbon by treatment with acid chlorides and acid anhydrides.
    - Following the alkylation or acylation reaction, aqueous acid is used to convert the enamine back into a carbonyl group.
- For retrosynthetic analysis, an unsymmetrical aldehyde/ketone and a β-dicarbonyl are characteristic products of enamine alkylation and acylation, respectively.

Problems: 19.7–19.9, 19.39–19.43, 19.57, 19.76, 19.80

Section 19.6 Acetoacetic Ester Synthesis

- α-Hydrogens between two carbonyl groups are especially easy to remove (pK_a values of 10–14) because the resulting anion is stabilized through delocalization with both adjacent carbonyl groups.
- The **acetoacetic ester synthesis** consists of a sequence of five synthetic steps.
  - Ethyl acetoacetate (pK_a = 10.7) is converted completely to its enolate anion using one equivalent of a base such as a sodium alkoxide.
  - The enolate anion is used as a nucleophile in an S_N2 reaction with methyl or primary alkyl halides, α-haloketones, and α-haloesters.
    - These first two steps can be repeated if a doubly alkylated product is desired.
  - The alkylated acetoacetic ester is hydrolized using HO^-.
  - Acidification gives the alkylated acetoacetic acid.
  - Heating causes decarboxylation to give the alkylated ketone product.
- The same sequence of reactions can be used with other β-ketoacids, not only ethyl acetoacetate.
  - β-Ketoesters result from Claisen condensation reactions, so the product of a Claisen reaction can be easily manipulated further using the synthetic sequence described here.

Problems: 19.10, 19.11, 19.46, 19.53, 19.73

- For retrosynthetic analysis, a methyl ketone is a characteristic product of the acetoacetic ester synthesis starting with ethyl acetoacetate, while other complex ketones can be derived from other β-ketoester starting materials.

Section 19.7 Malonic Ester Synthesis

- The **malonic ester synthesis** consists of a sequence of five synthetic steps that are analogous to those of the acetoacetic ester synthesis.
  - Diethyl malonate (pK_a = 13.3) is converted completely to its enolate anion using one equivalent of sodium ethoxide.
  - The enolate anion is used as a nucleophile in an S_N2 reaction with methyl or primary alkyl halides, α-haloketones, and α-haloesters.
    - These first two steps can be repeated if a doubly alkylated product is desired.
  - The alkylated acetoacetic ester is hydrolized using HO^-.
  - Acidification gives the alkylated acetoacetic acid.
  - Heating causes decarboxylation to give the alkylated carboxylic acid product.
- For retrosynthetic analysis, a carboxylic acid with one or two alkyl groups bonded to the α-carbon is a characteristic product of the acetoacetic ester synthesis starting with diethyl malonate.

Problems: 19.12, 19.44–19.48, 19.53, 19.61–19.64, 19.69

Section 19.8 Conjugate Addition to α,β-Unsaturated Carbonyl Compounds

- In a **Michael reaction** the β-carbon of α,β-unsaturated carbonyl species is electrophilic and able to react with certain nucleophiles, especially enolate anions.
  - One important resonance-contributing structure places positive charge at the β-carbon of α,β-unsaturated carbonyl species helping to explain reaction with nucleophiles at this position.
  - Reaction of enolate anions with α,β-unsaturated carbonyl species in Michael reactions gives products with oxygen atoms positioned in a 1,5 arrangement, which is distinct from the 1,3 positioning seen in aldol and Claisen reactions.
  - Michael reactions take place with a wide variety of α,β-unsaturated carbonyl compounds (aldehydes, ketones, esters, amides) as well as α,β-unsaturated nitro and nitrile compounds.
    - These electrophiles in Michael reactions are often referred to as Michael acceptors.
  - The mechanism of the Michael reaction involves initial formation of an enolate anion in base, attack of the enolate nucleophile onto the β-carbon of the Michael acceptor to create a new resonance-stabilized enolate anion intermediate that is protonated on oxygen to create an enol and regenerate the base, then tautomerization to the keto form completes the reaction.
    - The base is catalytic in the Michael reaction.
    - This type of addition mechanism is referred to as conjugate addition, or alternatively, 1,4-addition.
- A **Robinson annulation** is a Michael reaction with an α,β-unsaturated ketone followed by an intramolecular aldol reaction creating a cyclic product.
  - Six-membered rings are produced in high yields this way.
- Gilman reagents undergo conjugate addition with α,β-unsaturated carbonyl compounds in a reaction that is closely related to the Michael reaction.

Problems: 19.13–19.16, 19.49, 19.50, 19.53, 19.54–19.56, 19.59, 19.60, 19.64, 19.68, 19.70, 19.74, 19.80

Section 19.9 Crossed Enolate Reactions Using LDA

- **Lithium diisopropylamide (LDA)** is an extremely strong base that is not nucleophilic due to steric hindrance.
  - One molar equivalent of LDA converts aldehydes, ketones, and esters completely to their enolate anions.
  - Preformed enolate anions using LDA can be used to carry out a wide variety of crossed enolate reactions including aldol reactions, Claisen condensations, Michael additions, alkylations, and acylations.

Problems: 19.17, 19.51–
19.53, 19.65–19.67, 19.71,
19.78

– If a slight excess of carbonyl species is used, an equilibrium is set up in which the more stable of the alternative enolate anions predominates, a situation known as **thermodynamic control**.

– The most highly substituted of the possible enolate anions predominates.

– If a slight excess of LDA is used, no equilibrium is established among the alternative enolate anions and the predominant enolate is the one that forms more rapidly (the removed α-hydrogen is more accessible).

– This situation is referred to as **kinetic control**.

– The less substituted of the possible enolate anions generally predominate.

🦉WL Questions and problems for this chapter are available in OWL. Use the OWL access card that came with this textbook to access assigned questions and problems for this chapter.

Key Reactions

1. Aldol Reaction (Section 19.2) The aldol reaction involves nucleophilic addition of the enolate anion of one aldehyde or ketone to the carbonyl group of another aldehyde or ketone. The product of an aldol reaction is a β-hydroxyaldehyde or a β-hydroxyketone. An aldol reaction can be base catalyzed or acid catalyzed. If base is regenerated at the end of the reaction, it is base catalyzed, and if acid is regenerated, it is acid catalyzed. In both reactions, one or two new chiral centers are often created, leading to racemic products unless a starting aldehyde, ketone, or catalyst is chiral and present as a single enantiomer.

(a racemic mixture
of four stereoisomers)

2. Dehydration of the Product of an Aldol Reaction (Section 19.2) Dehydration of the β-hydroxyaldehyde or ketone from an aldol reaction occurs very readily under acidic or basic conditions and gives an α,β-unsaturated aldehyde or ketone.

(racemic)

3. Claisen Condensation (Section 19.3A) The product of a Claisen condensation is a β-ketoester. Condensation occurs by nucleophilic acyl substitution in which the attacking nucleophile is the enolate anion of an ester. The Claisen condensation mechanism involves reaction of one ester molecule with base to form an enolate anion, which reacts as a nucleophile with another molecule of ester to give a tetrahedral carbonyl addition intermediate, in which the RO^- group is lost to give a β-ketoester, which is deprotonated at the α position by the RO^-.

(racemic)

4. Dieckmann Condensation (Section 19.3B)
An intramolecular Claisen condensation is called a Dieckmann condensation.

(racemic)

5. Alkylation of an Enamine Followed by Hydrolysis (Section 19.5A)
Enamines are reactive nucleophiles with methyl and primary alkyl halides, α-haloketones, and α-haloesters.

(racemic)

6. Acylation of an Enamine Followed by Hydrolysis (Section 19.5B)

(racemic)   (racemic)

7. Acetoacetic Ester Synthesis (Section 19.6)
This sequence is useful for the synthesis of monosubstituted and disubstituted acetones.

1. EtO⁻Na⁺
2. CH₂=CHCH₂Br
3. NaOH, H₂O
4. HCl, H₂O
5. heat

Ethyl acetoacetate

A monosubstituted acetone

8. Malonic Ester Synthesis (Section 19.7)
This sequence is useful for the synthesis of monosubstituted and disubstituted acetic acids.

1. EtO⁻Na⁺
2. CH₂=CHCH₂Br
3. NaOH, H₂O
4. HCl, H₂O
5. heat

Diethyl malonate

A monosubstituted acetic acid

9. Michael Reaction (Section 19.8A)
A Michael reaction involves the addition of a weakly basic nucleophile to a carbon-carbon double bond made electrophilic by conjugation with the carbonyl group of an aldehyde, ketone, or ester or with a nitro or cyano group. The mechanism of the Michael reaction involves initial formation of an enolate anion

in base, attack of the enolate nucleophile at the β-carbon of the Michael acceptor to create a new resonance-stabilized enolate anion intermediate that is protonated on oxygen to create an enol and regenerate the base, then tautomerization to the keto form completes the reaction. The base is catalytic in the Michael reaction.

10. **Robinson Annulation (Section 19.8C)**   A Robinson annulation comprises a Michael reaction followed by an intramolecular aldol reaction and dehydration to form a substituted 2-cyclohexenone.

11. **Conjugate Addition of Lithium Diorganocopper Reagents (Section 19.8E)**   In a reaction closely related to the Michael reaction, lithium diorganocopper reagents undergo conjugate addition to the electrophilic double bond of α,β-unsaturated aldehydes and ketones.

Chapter 20

Dienes, Conjugated Systems, and Pericyclic Reactions

© Kevin Schafer/CORBIS

Turmeric flower. Curcumin, an orange-yellow powder isolated from the spice turmeric and responsible for much of the color of curry, has recently been found to retard tumor growth. See Chemical Connections "Curry and Cancer" on page 672. Inset: a model of curcumin.

In Chapters 5 and 6, we discussed the structure and characteristic reactions of alkenes. We limited this discussion to molecules containing isolated double bonds. In this chapter, we extend our study of molecules with π bonds to include molecules that contain two or more adjacent double bonds. Such compounds are called **conjugated**. The important feature of conjugated systems is that all the adjacent $2p$ orbitals combine with each other. The result is that the π electrons are not localized between just two carbon atoms. Instead, they are best thought of as being *delocalized* throughout the entire conjugated π orbital systems.

Conjugated
A conjugated diene or carbonyl is one in which the double bonds are separated by one single bond.

20.1 | Stability of Conjugated Dienes

Dienes are compounds that contain two carbon-carbon double bonds. Dienes can be divided into three groups: unconjugated, conjugated, and cumulated. An **unconjugated diene** is one in which the double bonds are separated by two or more single bonds. A **conjugated diene** is one in which the double bonds are separated by one single bond. A **cumulated diene** is one in which two double bonds share an *sp*-hybridized carbon. Because of the geometry of this carbon, the $2p$ orbitals of the two double bonds do not overlap in a cumulated diene and are not conjugated.

Cumulated
A cumulated diene is one in which two double bonds share an *sp*-hybridized carbon.

OWL Online homework for this chapter may be assigned in OWL for Organic Chemistry.

1,4-Pentadiene	**1,3-Pentadiene**	**1,2-Pentadiene**
(an unconjugated diene)	(a conjugated diene)	(a cumulated diene)

Example 20.1

Which of these molecules contain conjugated double bonds?

Solution

Compounds (b) and (c) contain conjugated double bonds. The double bonds in compounds (a) and (d) are unconjugated.

Problem 20.1

Which of these terpenes (Section 5.4) contains conjugated double bonds?

(a)	(b)	(c)
Geraniol	**Limonene**	An aggregating pheromone of bark beetles

Given in Table 20.1 are heats of hydrogenation for several alkenes and conjugated dienes. By using these data, we can compare the relative stabilities of conjugated and unconjugated dienes.

The simplest conjugated diene is 1,3-butadiene, but because this molecule has only four carbon atoms, it has no unconjugated constitutional isomer. However, we can estimate the effect of conjugation of two double bonds in this molecule in the following way. The heat of hydrogenation of 1-butene is -127 kJ (-30.3 kcal)/mol.

Table 20.1	Heats of Hydrogenation of Several Alkenes and Conjugated Dienes	
Name	**Structural Formula**	**$\Delta H°$** kJ(kcal)/mol
1-Butene		-127 (-30.3)
1-Pentene		-126 (-30.1)
cis-2-Butene		-120 (-28.6)
trans-2-Butene		-115 (-27.6)
1,3-Butadiene		-237 (-56.5)
trans-1,3-Pentadiene		-226 (-54.1)
1,4-Pentadiene		-254 (-60.8)

A molecule of 1,3-butadiene has two terminal double bonds, each with the same degree of substitution as the one double bond in 1-butene; therefore, we might predict that the heat of hydrogenation of 1,3-butadiene should be $2(-127 \text{ kJ/mol})$ or -254 kJ (-60.6 kcal)/mol. However, the observed heat of hydrogenation of 1,3-butadiene is -237 kJ (-56.5 kcal)/mol, a value 17 kJ (4.1 kcal)/mol less than estimated.

Both reactions are exothermic and give the same product; the more stable compound (lower in enthalpy) releases less heat upon hydrogenation. The conclusion is that conjugation of two double bonds in 1,3-butadiene gives an extra stability to the molecule of approximately 17 kJ (4.1 kcal)/mol. These energy relationships are displayed graphically in Figure 20.1.

Figure 20.1
Conjugation of double bonds in butadiene gives the molecule an additional stability of approximately 17 kJ (4.1 kcal)/mol.

Calculations of this type for other conjugated and unconjugated dienes give similar results: conjugated dienes are more stable than isomeric unconjugated dienes by approximately 14.5 to 17 kJ (3.5 to 4.1 kcal)/mol. The effects of conjugation on stability are even more general. Compounds containing conjugated double bonds, not just those in dienes, are more stable than isomeric compounds containing unconjugated double bonds. For example, 2-cyclohexenone is more stable than its isomer 3-cyclohexenone.

2-Cyclohexenone   **3-Cyclohexenone**
(more stable)   (less stable)

The additional stability of conjugated dienes relative to unconjugated dienes arises from delocalization of electron density in the conjugated diene. In two unconjugated double bonds, each pair of π electrons is localized between two carbons. In a conjugated diene, however, the four π electrons are delocalized over the set of four parallel $2p$ orbitals. As we have seen many times before, delocalization leads to increased stability.

According to the molecular orbital model, the conjugated system of a diene is described as a set of four π molecular orbitals arising from combination of four $2p$ atomic orbitals. The key idea here is that in conjugated systems, the adjacent $2p$ orbitals overlap in space, even between the $2p$ orbitals on C2 and C3 in butadiene.

Figure 20.2

Structure of 1,3-butadiene—molecular orbital model. Combination of four parallel $2p$ atomic orbitals gives two π-bonding MOs and two π-antibonding MOs. In the ground state, each π-bonding MO is filled with two spin-paired electrons. The π-antibonding MOs are unoccupied.

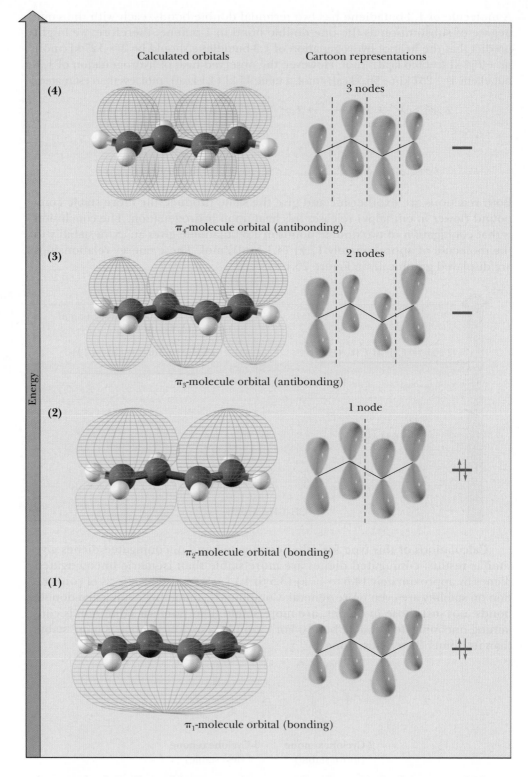

Calculated orbitals   Cartoon representations

(4)   3 nodes

π_4-molecule orbital (antibonding)

(3)   2 nodes

π_3-molecule orbital (antibonding)

(2)   1 node

π_2-molecule orbital (bonding)

(1)

π_1-molecule orbital (bonding)

As a result, they all combine to produce π molecular orbitals that cover all the atoms of the conjugated system, in this case the four carbon atoms. These MOs have zero, one, two, and three nodes, respectively, as illustrated in Figure 20.2. In the ground state, all four π electrons lie in π-bonding MOs. Because the lowest two MOs are at lower energies than that of two isolated π bonds, the net heat given off by filling these orbitals is more than would be the case for two isolated π bonds. Note that the electrons in these filled MOs are delocalized over the entire π orbital system. This π electron delocalization is the hallmark of conjugated systems and can be used to explain the spectroscopy and reactivity of conjugated molecules. Finally, it is worth pointing out that in order for maximal orbital overlap to occur, the $2p$ orbitals must be parallel, restricting the four sp^2-hybridized atoms of conjugated systems to a planar geometry.

Example 20.2

Using data from Table 20.1, estimate the extra stability that results from the conjugation of double bonds in *trans*-1,3-pentadiene.

Solution

Compare the sum of heats of hydrogenation of 1-pentene and *trans*-2-butene with the heat of hydrogenation of *trans*-1,3-pentadiene. Conjugation of double bonds in *trans*-1, 3-pentadiene imparts an added stability of approximately 15 kJ (3.6 kcal)/mol.

Problem 20.2

Estimate the stabilization gained as a result of conjugation when 1,4-pentadiene is converted to *trans*-1,3-pentadiene. Note that the answer is not as simple as comparing the heats of hydrogenation of 1,4-pentadiene and *trans*-1,3-pentadiene. Although the double bonds are moved from unconjugated to conjugated, the degree of substitution of one of the double bonds is also changed, in this case from a monosubstituted double bond to a *trans* disubstituted double bond. To answer this question, you must separate the effect that is the result of conjugation from that caused by a change in the degree of substitution.

20.2 | Electrophilic Addition to Conjugated Dienes

Conjugated dienes undergo two-step electrophilic addition reactions just like simple alkenes (Section 6.3). However, certain features are unique to the reactions of conjugated dienes.

A. 1,2-Addition and 1,4-Addition

Addition of one mole of HBr to 1,3-butadiene at $-78°C$ gives a mixture of two constitutional isomers, 3-bromo-1-butene and 1-bromo-2-butene.

$$CH_2{=}CH{-}CH{=}CH_2 + \boxed{HBr} \xrightarrow{-78°C} CH_2{=}CH{-}\overset{\overset{\displaystyle Br}{|}}{CH}{-}\overset{\overset{\displaystyle H}{|}}{CH_2} + \overset{\overset{\displaystyle Br}{|}}{CH_2}{-}CH{=}CH{-}\overset{\overset{\displaystyle H}{|}}{CH_2}$$

1,3-Butadiene	**3-Bromo-1-butene**	**1-Bromo-2-butene**
	(90%)	(10%)
	(1,2-addition)	(1,4-addition)

The designations "1,2-" and "1,4-" used here to describe additions to conjugated dienes and other systems do not refer to IUPAC nomenclature. Rather, they refer to the four-atom system of two conjugated double bonds and indicate that addition takes place at either carbons 1 and 2 or carbons 1 and 4 of the four-atom system. For example, the Michael reaction (Section 19.8A) is referred to as a 1,4-addition. The bromobutenes formed by addition of one mole of HBr to butadiene can in turn undergo addition of a second mole of HBr to give a mixture of dibromobutanes. Our concern at this point is only with the products of the first reaction.

Addition of one mole of Br_2 at $-15°C$ also gives a mixture of 1,2-addition and 1,4-addition products.

$$CH_2{=}CH{-}CH{=}CH_2 + \boxed{Br_2} \xrightarrow{-15°C} \overset{\overset{\displaystyle Br}{|}}{CH_2}{-}\overset{\overset{\displaystyle Br}{|}}{CH}{-}CH{=}CH_2 + \overset{\overset{\displaystyle Br}{|}}{CH_2}{-}CH{=}CH{-}\overset{\overset{\displaystyle Br}{|}}{CH_2}$$

1,3-Butadiene	**3,4-Dibromo-1-butene**	**1,4-Dibromo-2-butene**
	(54%, racemic)	(46%)
	(1,2-addition)	(1,4-addition)

We can account for the formation of isomeric products in the addition of HBr and Br_2 by the following mechanism.

Mechanism 1,2- and 1,4-Addition to a Conjugated Diene

***Step 1:* Make a new bond between a nucleophile (π bond) and an electrophile—add a proton.** Electrophilic addition is initiated by the reaction of a terminal carbon of one of the double bonds with HBr to give an allylic carbocation intermediate (Section 9.3B), which can best be represented as a resonance hybrid of two contributing structures. Formation of this stabilized cation is the rate-determining step.

Allylic carbocation stabilized by resonance

***Step 2:* Make a new bond between a nucleophile and an electrophile.** Reaction of bromide at one of the carbons bearing partial positive charge gives the 1,2-addition product; reaction at the other gives the 1,4-addition product.

(1,2-addition, racemic)   (1,4-addition)

Example 20.3

Addition of one mole of HBr to 2,4-hexadiene gives a mixture of 4-bromo-2-hexene and 2-bromo-3-hexene. No 5-bromo-2-hexene is formed. Account for the formation of the first two bromoalkenes and for the fact that the third bromoalkene is not formed.

Solution

2,4-Hexadiene is a conjugated diene, and you can expect products from both 1,2-addition and 1,4-addition. Reaction of the diene with HBr at C2 of the diene in Step 1, the rate-determining step, gives a resonance-stabilized 2° allylic carbocation intermediate. Reaction of this intermediate in Step 2 at one of the carbons bearing a partial positive charge gives 4-bromo-2-hexene, a 1,2-addition product; reaction at the other gives 2-bromo-3-hexene, a 1,4-addition product.

$$\text{CH}_3\text{CH}=\text{CH}-\text{CH}=\text{CHCH}_3 \quad + \quad \text{H}-\ddot{\text{Br}}: \longrightarrow$$

$$\underset{\text{H}}{\overset{\text{H}}{\text{CH}_3\text{CH}=\text{CH}-\overset{+}{\text{CH}}-\text{CHCH}_3}} \longleftrightarrow \overset{\text{H}}{\underset{\text{CH}_3\overset{+}{\text{CH}}-\text{CH}=\text{CH}-\text{CHCH}_3}{}}$$

A 2° allylic carbocation stabilized by charge delocalization

$$\downarrow :\ddot{\text{Br}}: \qquad\qquad\qquad \downarrow :\ddot{\text{Br}}:$$

$$\underset{\text{CH}_3\text{CH}=\text{CH}-\text{CH}-\text{CHCH}_3}{\overset{:\ddot{\text{Br}}:\quad\text{H}}{}} \qquad \underset{\text{CH}_3\text{CH}-\text{CH}=\text{CH}-\text{CHCH}_3}{\overset{:\ddot{\text{Br}}:\qquad\qquad\text{H}}{}}$$

4-Bromo-2-hexene   **2-Bromo-3-hexene**
(1,2-addition, racemic)   (1,4-addition, racemic)

Formation of 5-bromo-2-hexene requires reaction of the diene with HBr to give a secondary, nonallylic carbocation by protonation at C3. The activation energy for formation of this less stable 2° carbocation is considerably greater than that for formation of the resonance-stabilized 2° allylic carbocation; therefore, formation of this carbocation and the resulting 5-bromo-2-hexene does not compete effectively with formation of the observed products.

$$\text{CH}_3\text{CH}=\text{CH}-\text{CH}=\text{CHCH}_3 + \text{H}-\ddot{\text{Br}}: \longrightarrow \underset{\text{H}}{\overset{\text{H}}{\text{CH}_3\text{CH}=\text{CH}-\overset{+}{\text{CH}}-\text{CHCH}_3}} \overset{:\ddot{\text{Br}}:^-}{\longrightarrow} \underset{\text{CH}_3\text{CH}=\text{CH}-\text{CH}-\text{CHCH}_3}{\overset{\text{H}\quad:\ddot{\text{Br}}:}{}}$$

2° carbocation, but not allylic   **5-Bromo-2-hexene**
(not formed)

Problem 20.3

Predict the product(s) formed by addition of one mole of Br_2 to 2,4-hexadiene.

B. Kinetic Versus Thermodynamic Control of Electrophilic Addition

The regiochemistry of addition of HBr to conjugated dienes displays an unusual dependence on the kinetics and thermodynamics of the reaction. Before examining this dependence, let's review the regiochemistry of addition of HBr to standard alkenes.

As discussed in Section 6.3, the electrophilic addition of H—X to an alkene obeys Markovnikov's rule: hydrogen adds to the double-bonded carbon that has the greatest number of hydrogens already bonded to it (see the propene example below and Figure 6.4). The kinetic barrier for the rate-determining step in the formation of the Markovnikov product is lower because the carbocation formed is secondary and more stable than the alternative primary carbocation. The product resulting from the secondary carbocation is therefore formed faster. We noted in Section 19.9D that when the distribution of products is determined by the relative rates of formation of each, that reaction is under **kinetic (rate) control**.

With electrophilic addition to standard alkenes, such as propene, the product predicted by Markovnikov's rule is also more stable. For reactions under **thermodynamic (equilibrium) control**, the distribution of products is determined by the relative stability of each. Thus, kinetically controlled and thermodynamically controlled electrophilic additions of H—X to standard alkenes results in the same dominate product. This is the case with many reaction; the product formed faster is also the most stable product. But, it is very important to note that there is no strict relationship of this kind, and many reactions do not behave this way. As we'll see below, the addition of HBr to conjugated dienes is an example of a reaction in which kinetic and thermodynamic control gives different dominant products.

Whether a reaction is under kinetic or thermodynamic control can be manipulated by changing the experimental conditions. A common approach to switching between kinetic or thermodynamic control is by changing the temperature chosen for a reaction. In general, at lower temperatures, little to no equilibrium is established between reactants and products, and the reactions must therefore be under kinetic control. At higher temperatures, reactions become increasingly reversible, and equilibrium can be established between reactants and products, leading to thermodynamic control. The electrophilic addition of HBr to standard alkenes such as propene would give the same dominant product at all temperatures, although the extent of domination will vary somewhat with temperature. We can now return to the addition of HBr to conjugated dienes.

We saw in Section 20.2A that electrophilic addition to conjugated dienes gives a mixture of 1,2-addition and 1,4-addition products. Following are some additional experimental observations about the products of electrophilic additions to 1,3-butadiene.

1. For addition of HBr at $-78°C$ and addition of Br_2 at $-15°C$, the 1,2-addition products predominate over the 1,4-addition product. Generally at lower temperatures, the 1,2-addition products predominate over 1,4-addition products.
2. For addition of HBr and Br_2 at higher temperatures (generally 40–60°C), the 1,4-addition products predominate.
3. If the products of low temperature addition are allowed to remain in solution and then are warmed to a higher temperature, the composition of the product changes over time and becomes identical to that obtained when the reaction is carried out at higher temperature. Thus, under these higher temperature conditions, an equilibrium is established between 1,2- and 1,4-addition products in which 1,4-addition products predominate.

These experimental observations can be explained by considering kinetic versus thermodynamic products, which dominate at lower and higher temperatures, respectively. At the lower temperatures, no equilibrium is established between the 1,2- and 1,4-addition products. Because the 1,2-addition products dominate under these conditions, they must be the products formed by kinetic control, that is, the 1,2-addition is faster than 1,4-addition. Alternatively, at the higher temperatures, the 1,4-addition products dominate, and therefore we can conclude that an equilibrium is established between the 1,2- and 1,4-addition products and that the 1,4-addition products are thermodynamically more stable than 1,2-addition products.

Relationships between kinetic and thermodynamic control for electrophilic addition of HBr to 1,3-butadiene are illustrated graphically in Figure 20.3. The structure shown in the Gibbs free energy well in the center of Figure 20.3 is the resonance-stabilized allylic cation intermediate formed by proton transfer from HBr to C1 of 1,3-butadiene. The dashed lines in this intermediate show the partial double bond character between C2 and C3 and between C3 and C4 in the resonance hybrid. To the left of this intermediate is the activation energy for its

Figure 20.3
Kinetic versus thermodynamic control. A plot of Gibbs free energy versus reaction coordinate for Step 2 in the electrophilic addition of HBr to 1,3-butadiene. The resonance-stabilized allylic carbocation intermediate reacts with bromide ion by way of the transition state on the left to give the 1,2-addition product. It reacts with bromide ion by way of the alternative transition state on the right to give the 1,4-addition product.

reaction with bromide ion to form the less stable 1,2-addition product; to the right is the activation energy for its reaction with bromide ion to form the more stable 1,4-addition product. As shown in Figure 20.3, the activation energy for 1,2-addition is less than that for 1,4-addition, and the 1,2-addition product is therefore favored under kinetic control. The 1,4-addition product is the more stable and is favored when the reaction is under thermodynamic control.

To complete our discussion of electrophilic addition to conjugated dienes and of kinetic versus thermodynamic control, we need to ask the following questions.

1. Why is the 1,2-addition product (the less stable product) formed more rapidly at lower temperatures? First, we need to look at the resonance-stabilized allylic carbocation intermediate and determine which Lewis structure makes the greater contribution to the hybrid. We must consider the degree of substitution of both the positive carbon and the carbon-carbon double bond in each contributing structure.

$$CH_2\!=\!CH\!-\!\overset{+}{C}H\!-\!CH_3 \quad\longleftrightarrow\quad \overset{+}{C}H_2\!-\!CH\!=\!CH\!-\!CH_3$$

Less substituted double bond   More substituted double bond
Secondary carbocation   Primary carbocation

A secondary carbocation is more stable than a primary carbocation. If the degree of substitution of the carbon bearing the positive charge were the more important factor, the Lewis structure on the left would make the greater contribution to the hybrid. However, a more substituted double bond is more stable than a less substituted double bond (Section 6.6B). If the degree of substitution of the carbon-carbon double bond were the more important factor, the Lewis structure on the right would make the greater contribution to the hybrid.

We know from other experimental evidence that the location of the positive charge in the allylic carbocation is more important than the location of the double bond. Therefore, in the hybrid, the greater fraction of positive charge is on the secondary carbon. Reaction with bromide ion occurs more rapidly at this carbon, giving 1,2-addition, simply because it has a greater density of positive charge. The electrostatic potential map shows that the positive charge (blue) is more intense on the secondary carbon.

2. Is the 1,2-addition product also formed more rapidly at higher temperatures, even though the 1,4-addition product predominates under these conditions? The answer is yes. The factors affecting the structure of a resonance-stabilized

Electrostatic potential map of the allylic carbocation formed by protonating 1,3-butadiene.

allylic carbocation intermediate and the reaction of this intermediate with a nucleophile are not greatly affected by changes in temperature.

3. Why is the 1,4-addition product the thermodynamically more stable product? The answer to this question has to do with the relative degree of substitution of double bonds. In general, the greater the degree of substitution of a carbon-carbon double bond, the greater the stability of the compound or ion containing it. Following are pairs of 1,2- and 1,4-addition products. In each case, the more stable alkene is the 1,4-addition product.

3-Bromo-1-butene
(less stable alkene)

(E)-1-Bromo-2-butene
(more stable alkene)

3,4-Dibromo-1-butene
(less stable alkene)

(E)-1,4-Dibromo-2-butene
(more stable alkene)

However, there are cases where the 1,2-addition product is more stable and would be the product of thermodynamic control. For example, addition of bromine to 1,4-dimethyl-1,3-cyclohexadiene under conditions of thermodynamic control gives 3,4-dibromo-1,4-dimethylcyclohexene because its trisubstituted double bond is more stable than the disubstituted double bond of the 1,4-addition product.

1,4-Addition product
(less stable)

1,2-Addition product
(more stable)

4. What is the mechanism by which the thermodynamically less stable product is converted to the thermodynamically more stable product at higher temperatures? To answer this question, we must look at the relationships between kinetic energy, potential energy, and activation energy. In any chemical reaction, the reactants are constantly undergoing collisions with any other chemicals in the reaction vessel, most being collisions with the solvent. On collision, part of the kinetic energy involved in the collision is converted to potential energy that is stored in the molecules due to structural distortions (as with any collisions in the macroscopic world). If the potential energy increase involved in a structural distortion of the chemicals is equal to or greater than the activation energy for a reaction, then that reaction may occur. This observation means that in some collisions the structural distortions are so severe, that the bonds break and/or form. At higher temperatures used for electrophilic addition of HBr and Br_2 to conjugated dienes, collisions between the 1,2- and 1,4-addition products with the solvent are energetic enough to reform the resonance stabilized allylic carbocation intermediate via ionization of the C—Br bonds. When the 1,2-addition product reverts to this allylic carbocation, re-addition of bromide can give the more stable 1,4-addition product. At lower temperature, however,

the increase in potential energy of the products upon collisions is not sufficient to overcome the activation energy for C—Br bond ionization, and therefore the reactions are not reversible.

In summary, although the thermodynamically most stable product is often the most rapidly formed product, such is not always the case. Whether the thermodynamically more stable product is formed at a greater rate or not—from a reactant or a common intermediate—very much depends upon the particular reactants, the reaction mechanism, and the reaction conditions.

20.3 | UV-Visible Spectroscopy

An important property of conjugated systems is that they absorb energy in the ultraviolet-visible region of the spectrum as a result of electronic transitions (Table 12.3). In this section, we study the information this absorption gives us about the conjugation of carbon-carbon and carbon-oxygen double bonds and their substitution.

A. Introduction

The region of the electromagnetic spectrum covered by most ultraviolet spectrophotometers is from 200 to 400 nm, a region commonly referred to as the **near ultraviolet**. Wavelengths shorter than 200 nm require special instrumentation and are not used routinely. The region covered by most visible spectrophotometers runs from 400 nm (violet) to 700 nm (red), with extensions into the (near) IR region to 800 or 1000 nm available on many instruments.

Example 20.4

Calculate the energy of radiation at either end of the near-ultraviolet spectrum, that is, at 200 nm and 400 nm (review Section 12.1).

Solution

Use the relationship $E = hc/\lambda$. Be certain to express the dimension of length in consistent units.

$$E = \frac{hc}{\lambda} = 3.99 \times 10^{-13} \frac{\text{kJ} \times \text{s}}{\text{mol}} \times 3.00 \times 10^8 \frac{\text{m}}{\text{s}} \times \frac{1}{200 \times 10^{-9}\,\text{m}} = 598\ \text{kJ}\ (143\ \text{kcal})/\text{mol}$$

By a similar calculation, the energy of radiation of wavelength 400 nm is found to be 299 kJ (71.5 kcal)/mol.

Problem 20.4

Wavelengths in ultraviolet-visible spectroscopy are commonly expressed in nanometers; wavelengths in infrared spectroscopy are sometimes expressed in micrometers. Carry out the following conversions.

(a) 2.5 μm to nanometers **(b)** 200 nm to micrometers

Wavelengths and corresponding energies for near-ultraviolet and visible radiation are summarized in Table 20.2.

Table 20.2	Wavelengths and Energies of Near Ultraviolet and Visible Radiation		
Region of Spectrum	**Wavelength (nm)**	**Energy**	
		kJ/mol	**kcal/mol**
Near Ultraviolet	200–400	299–598	71.5–143
Visible	400–700	171–299	40.9–71.5

Figure 20.4

Ultraviolet spectrum of 2,5-dimethyl-2,4-hexadiene (in methanol).

Absorbance (A)

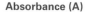

A quantitative measure of the extent to which a compound absorbs radiation of a particular wavelength. $A = \log (I_0/I)$ where I_0 is the incident radiation and I is the transmitted radiation.

Ultraviolet and visible spectral data are recorded as plots of **absorbance (A)** on the vertical axis versus wavelength on the horizontal axis.

$$\text{Absorbance } (A) = \log \frac{I_0}{I}$$

where I_0 is the intensity of radiation incident on the sample, and I is the intensity of the radiation transmitted through the sample. The quantity $(I/I_0) \times 100$ is called **percent transmittance**; many spectrophotometers read in this scale.

Typically, UV-visible spectra consist of a small number of broad absorption bands, sometimes just one. Figure 20.4 is an ultraviolet absorption spectrum of 2,5-dimethyl-2,4-hexadiene. Absorption of ultraviolet radiation by this conjugated diene begins at wavelengths below 200 nm and continues to almost 270 nm, with maximum absorption at 242 nm. This spectrum is reported as a single absorption peak using the notation λ_{max} 242 nm.

The extent of absorption of ultraviolet-visible radiation is proportional to the number of molecules capable of undergoing the observed electronic transition; therefore, ultraviolet-visible spectroscopy can be used for quantitative analysis of samples. The relationship between absorbance, concentration, and length of the sample cell (cuvette) is known as the **Beer-Lambert law**. The proportionality constant in this equation is given the name **molar absorptivity (ε)** or extinction coefficient.

Molar absorptivity (ε)

The absorbance of a 1 M solution of a compound.

$$\text{Beer-Lambert Law:} \quad A = \varepsilon \, c \, l$$

where A is the **absorbance** (unitless), ε is the molar absorptivity (in per moles per liter per centimeter, $M^{-1}cm^{-1}$), c is the concentration of solute (in moles per liter, M), and l is the length of the sample cell, or cuvette (in centimeters, cm).

The molar absorptivity is a characteristic property of a compound and is not affected by its concentration or the length of the light path. Values range from zero to 10^6 $M^{-1}cm^{-1}$. Values above 10^4 $M^{-1}cm^{-1}$ correspond to high-intensity absorptions; values below 10^4 $M^{-1}cm^{-1}$, to low-intensity absorptions. The molar absorptivity of 2,5-dimethyl-2,4-hexadiene, for example, is 13,100 $M^{-1}cm^{-1}$, a high-intensity absorption.

An interesting aspect of absorption by molecules in the visible region is that a sample will appear to our eyes as the combination of reflected wavelengths. White light is composed of all wavelengths of light in the visible region (400–740 nm), present in approximately equal intensity. Individual wavelengths of light have individual colors, as indicated on the spectrum (Figure 20.5a). Light of the given wavelength appears as the color indicated to our eyes. For example, monochromatic 400 nm light appears violet and monochromatic 700 nm light appears red.

Absorption by a substance removes the absorbed wavelengths from white light, leaving the remaining wavelengths to be reflected, the combination of which

(a)

(b)

(c)

Colored arrows are complementary

Figure 20.5

(a) Visible light color-wavelength correlation. (b) Approximate color of substance (reflected light) if a single wavelength (i.e. the wavelength listed on the numerical scale of the x-axis) is absorbed. (c) Complementary colors on a color wheel.

determines the color our eyes see. Figure 20.5b shows the approximate color that a substance would appear if a single wavelength were absorbed. For example, if a molecule absorbs strongly only at 500 nm (lighter blue light), it appears red to our eyes because the remaining reflected colors combine to appear red. Similarly, a molecule that strongly absorbs around 600 nm (orange light) appears blue, because orange is removed from the reflected light, and the remaining reflected wavelengths combine to appear blue.

The correlation between absorbed wavelength and reflection can be approximated using the concept of complementary colors illustrated by an artist's color wheel (Figure 20.5c). A molecule that absorbs light in one region of the spectrum will reflect the nonabsorbed wavelengths. A good rule of thumb is that the reflected wavelengths combine to appear more or less as the complement of the absorbed color. To a first approximation, then, a molecule that absorbs one color will appear to our eyes to be the color on the opposite side of the color wheel. The color wheel shown has approximate wavelengths of monochromatic light indicated for reference.

More complicated absorptions, with two or more strong absorptions by a single molecule, lead to a more complex interpretation of reflected color, but the concept is the same. A substance will appear to our eyes as the combination of reflected (not absorbed) wavelengths.

Example 20.5

The molar absorptivity of 2,5-dimethyl-2,4-hexadiene in methanol is 13,100 $M^{-1}cm^{-1}$. What concentration of this diene in methanol is required to give an absorbance of 1.6? Assume a light path of 1.00 cm. Calculate concentration in these units.

(a) Moles per liter   **(b)** Milligrams per milliliter

Solution

Solve the Beer-Lambert equation for concentration and substitute appropriate values for length, absorbance, and molar absorptivity.

(a) $c = \dfrac{A}{l \times \varepsilon} = \dfrac{1.6}{1.00 \text{ cm} \times 13{,}100 \text{ L} \cdot \text{mol}^{-1} \cdot \text{cm}^{-1}} = 1.22 \times 10^{-4} \text{ mol/L}$

(b) The molecular weight of 2,5-dimethyl-2,4-hexadiene is 110 g/mol. The concentration of the sample in milligrams per milliliter is

$$1.22 \times 10^{-4} \frac{\text{mol}}{\text{L}} \times \frac{110 \text{ g}}{\text{mol}} \times \frac{1 \text{ L}}{1000 \text{ ml}} \times \frac{1000 \text{ mg}}{\text{g}} = 1.34 \times 10^{-2} \text{ mg/mL}$$

Problem 20.5

The visible spectrum of β-carotene ($C_{40}H_{56}$, MW 536.89, the orange pigment in carrots) dissolved in hexane shows intense absorption maxima at 463 nm and 494 nm, both in the blue-green region. Because light of these wavelengths is absorbed by β-carotene, we perceive the color of this compound as that of the complement to blue-green, namely, red-orange.

β-Carotene

$\lambda_{\max}$ 463 nm (log ε 5.10); 494 nm (log ε 4.77)

Calculate the concentration in milligrams per milliliter of β-carotene that gives an absorbance of 1.8 at 463 nm.

B. The Origin of Transitions Between Electronic Energy Levels

Absorption of electromagnetic radiation in the ultraviolet-visible region results in promotion of electrons from a lower energy, occupied MO to a higher energy, unoccupied MO. The energy of this radiation is generally insufficient to affect electrons in the much lower energy, σ-bonding molecular orbitals. It is, however, sufficient to cause an electron in a nonbonding (lone pair) or π orbital to be promoted to an antibonding π^* orbital (called an $n \rightarrow \pi^*$ or $\pi \rightarrow \pi^*$ transition, respectively). Conjugated π systems have particularly noteworthy $\pi \rightarrow \pi^*$ transitions. Three examples of conjugated systems are

$$CH_2{=}CH{-}CH{=}CH_2 \qquad CH_2{=}CH{-}\overset{\overset{\displaystyle O}{\|}}{C}{-}CH_3$$

1,3-Butadiene **3-Buten-2-one** **Benzaldehyde**

As an example of a $\pi \rightarrow \pi^*$ transition, consider ethylene. The double bond in ethylene consists of one σ bond formed by combination of sp^2 orbitals and one π bond formed by combination of $2p$ orbitals. The relative energies of the π-bonding and π-antibonding molecular orbitals are shown schematically in Figure 20.6. The $\pi \rightarrow \pi^*$ transitions for simple, unconjugated alkenes occur below 200 nm (at 165 nm for ethylene). Because these transitions occur at extremely short wavelengths, they are not observed in conventional ultraviolet spectroscopy and, therefore, are not useful to us for determining molecular structure.

For 1,3-butadiene, the difference in energy between the highest occupied π molecular orbital and the lowest unoccupied π-antibonding molecular orbital is less than it is for ethylene with the result that a $\pi \rightarrow \pi^*$ transition for 1,3-butadiene (Figure 20.7) takes less energy (occurs at longer wavelength) than that for ethylene. This transition for 1,3-butadiene occurs at 217 nm.

Electronic excitation in molecules are accompanied by changes in vibrational or rotational energy levels. The energy levels for these excitations are considerably smaller than the energy differences between electronic excitations. These transitions are superposed on the electronic excitations, which results in a large number of absorption peaks so closely spaced that the spectrophotometer cannot resolve them. For this reason, UV-visible absorption peaks usually are much broader than IR absorption peaks.

Simple aldehydes and ketones show only weak absorption in the ultraviolet region of the spectrum owing to an n to π^* electronic transition of the carbonyl group. If, however, the carbonyl group is conjugated with one or more carbon-carbon double bonds, intense absorption ($\varepsilon = 8{,}000 - 20{,}000\ M^{-1}cm^{-1}$) occurs as a result of a π to π^* transition; as with polyenes, the position of absorption is shifted to longer wavelengths, and the molar absorptivity, ε, of the absorption maximum increases sharply. For the α,β–unsaturated ketone 3-penten-2-one, for example, $\lambda_{\max}$ is 224 nm (log ε 4.10).

Figure 20.6

A $\pi \rightarrow \pi^*$ transition in excitation of ethylene. Absorption of ultraviolet radiation causes a transition of an electron from a π-bonding MO in the ground state to a π-antibonding MO in the excited state. There is no change in electron spin.

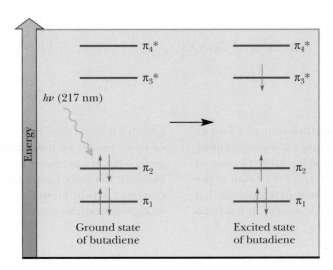

2-Pentanone
λ_{max} 180 nm (ε 900)

3-Penten-2-one
λ_{max} 224 nm (ε 12,590)

Acetophenone
λ_{max} 246 nm (ε 9,800)

The greater the extent of conjugation of unsaturated systems with the carbonyl group, the more the absorption maximum is shifted toward the visible region of the spectrum.

Like the carbonyl groups of simple aldehydes and ketones, the carboxyl group shows only weak absorption in the ultraviolet spectrum unless it is conjugated with a carbon-carbon double bond or an aromatic ring.

The important point is that conjugation decreases the energy gap between filled and unfilled π orbitals. Therefore, in general, the greater the number of double bonds in conjugation, the longer the wavelength of ultraviolet radiation absorbed. Shown in Table 20.3 are wavelengths and energies required for $\pi \rightarrow \pi*$ transitions in several conjugated alkenes.

Table 20.3	Wavelengths and Energies Required for $\pi \rightarrow \pi*$ Transitions of Ethylene and Three Conjugated Polyenes		

Name	Structural Formula	λ_{max} (nm)	Energy [kJ (kcal)/mol]
Ethylene	$CH_2{=}CH_2$	165	724 (173)
1,3-Butadiene	$CH_2{=}CHCH{=}CH_2$	217	552 (132)
(3E)-1,3,5-Hexatriene	$CH_2{=}CHCH{=}CHCH{=}CH_2$	268	448 (107)
(3E,5E)-1,3,5,7-Octatetraene	$CH_2{=}CH(CH{=}CH)_2CH{=}CH_2$	290	385 (92)

20.4 | Pericyclic Reaction Theory

To this point in this book, there have been only a few cases for which we analyzed the orbitals of reactants in order to understand the reaction mechanisms, as for example, S_N2 (Section 9.2A) and E2 (Section 9.7C). Even though we have drawn on orbital analysis infrequently, the interactions of orbitals actually dictate all chemical reactions. In fact, there is a class of reactions called *pericyclic* for which an analysis of orbitals is critical for even a rudimentary understanding of the mechanisms. **Pericyclic reactions** are those in which the reaction occurs in a single step with a

Pericyclic Reaction
A reaction that takes place in a single step without intermediates and that involves a cyclic distribution of orbitals.

Curcumin is a natural dye from the root of *Curcuma longa* L. In pure form, it is an orange-yellow crystalline powder that is isolated from the spice turmeric, one of the major ingredients of curry. Its color is a result of the highly conjugated system in curcumin (it is probable that the molecule is actually enolized as shown). It has been

known for some time that curcumin retards the growth of new cancers by inhibiting the formation of blood vessels that are necessary for the cancers to grow (angiogenesis). Recently, Korean biochemists have shown that curcumin acts by inhibiting an enzyme that is important to angiogenesis. So curry may be good for you!

Curcumin
(Natural Yellow-3, Turmeric yellow)

Although we only cover one approach to the understanding of pericyclic reactions, the various approaches developed over the years have given important different contributions to our understanding. R. B. Woodward (Harvard University), Roald Hoffmann (then at Harvard, now at Cornell University), Kenichi Fukui (Kyoto University), and Howard Zimmerman (University of Wisconsin) provided the key insights into pericyclic reaction mechanisms. Hoffman and Fukui were awarded the Nobel Prize for this work in 1981 (after the death of Woodward).

closed loop of orbitals, that is, we can draw orbitals interacting at the transition states of the reactions in a cyclic ring. Further, because the reactions occur in a single step, there are no radical or ionic intermediates. The examples given below will make this characteristic clear. Because many of these reactions involve conjugated and nonconjugated dienes, we discuss them in this chapter. One of the hallmarks of pericyclic reactions is precise control of the stereochemistry of the reactions, and the examples given below will highlight this feature.

Pericyclic reactions are routinely classified as "allowed" or "forbidden" with a particular structure for the transition state. In practice, this classification means that one geometry for the reaction either has a low energy transition state (allowed) or that a different geometry has a very high energy transition state (forbidden). To determine whether a reaction is allowed or forbidden, a handful of approaches exist. We will examine one approach: frontier molecular orbital theory.

A. Frontier Molecular Orbital Theory

Cycloaddition Reaction
A reaction in which two reactants add together in a single step to form a cyclic product.

There are five classes of pericyclic reactions, only two of which are covered in this book, cycloadditions and sigmatropic shifts. The most common, **cycloaddition** involves the reaction of a conjugated diene with an alkene, although we also now examine the reaction of an alkene with an alkene. Hence, we are examining what are commonly called 4 + 2 and 2 + 2 cycloadditions, respectively, to keep track of the number of π electrons involved in the reaction.

To analyze whether these reactions are allowed or forbidden, chemists focus on the frontier molecular orbitals of the reactants. The frontier molecular orbitals consist of the **highest occupied molecular orbitals (HOMOs)** and the **lowest unoccupied molecular orbitals (LUMOs)**. The terms "highest" and "lowest" refer to the energy of the orbitals, while the terms "occupied" and "unoccupied" refer to whether the orbital is populated with two electrons or is empty. For example, the HOMO and LUMO of butadiene would be orbitals 2 and 3 in Figure 20.2 respectively. The HOMO and LUMO of ethylene are simply the π and π^* orbitals given in Figure 1.21.

HOMO
Highest occupied molecular orbital.

LUMO
Lowest unoccupied molecular orbital.

After identifying the frontier molecular orbitals of the reactants, chemists predict a reaction geometry. The goal is to decide if the predicted geometry for a reaction is allowed or forbidden. In the analysis, we examine the HOMO of one reactant and the LUMO of the other. It does not matter which reactant is assigned the HOMO or LUMO in the analysis because the answer will be the same. What is important is how one reactant's HOMO interacts (contacts) with the other reactant's LUMO in the proposed reaction geometry. When the phasing of the orbitals that are undergoing contact matches (zero phase changes), the reaction will be allowed. In fact, if there are any even number of phase changes, the reaction is allowed. However, if there is one (or any odd number) of contacts between the orbitals in which the phasing does not match, the reaction is forbidden.

To demonstrate the frontier molecular orbital analysis, let's analyze a collision geometry for butadiene and ethylene that is called suprafacial for each reactant. A **suprafacial** interaction occurs when the same face (or side) of the π-system of an individual reactant undergoes collision (an alternative interaction is referred to as **antarafacial**, shown in Example 20.6). In Figure 20.8A, the bottoms of the *p*-orbitals on carbons 1 and 4 are undergoing reaction (hence, the same face) while the tops of the *p*-orbitals of ethylene are reacting (hence again, the same face). Therefore, each reactant is interacting in a suprafacial manner. Suprafacial collision geometries are shown for both the 4 + 2 and 2 + 2 reactions in Figure 20.8.

Suprafacial
When the same face of a π-bond performs a reaction.

Antarafacial
When the opposite faces of a π-bond perform a reaction.

Note that the 4 + 2 reaction is allowed because when the HOMO and LUMO on the reactants contact each other, the phasing matches; red on red and blue on blue. The 2 + 2 reaction is forbidden because there is one red on blue contact.

The frontier molecular orbital analysis led to the conclusion that butadiene will react with ethylene to give cyclohexene if both butadiene and ethylene collide in a suprafacial manner. However, ethylene will not react with ethylene in an analogous manner. We can extend these conclusions to the reaction of any conjugated diene with any alkene, and to any alkene with another alkene, respectively. We present below several examples and practical considerations of the allowed 4 + 2 reaction (the Diels-Alder reaction). The suprafacial approach of both reactants has important consequences on the stereochemistry of the Diels-Alder reaction.

Reaction is "allowed"   Reaction is "forbidden"

HOMO
of butadiene

LUMO
of ethylene

HOMO
of ethylene

Phasing
does not
match

LUMO
of ethylene

A suprafacial approach of the HOMO of butadiene interacts constructively with a suprafacial approach of the LUMO of ethylene.

A suprafacial approach of the HOMO of ethylene interacts destructively with a suprafacial approach of the LUMO of another ethylene.

Figure 20.8
(a) A frontier molecular orbital analysis for the 4 + 2 cycloaddition of butadiene with ethylene. The phasing is found to match between the HOMO of butadiene and the LUMO of ethylene when they collide face to face, and hence the reaction is allowed. (b) A frontier molecular orbital analysis for the 2 + 2 cycloaddition of two ethylene molecules. Because the phasing does not match in one spot when the HOMO and LUMO are combined, the reaction is termed forbidden.

Example 20.6

Let's examine an alternative collision geometry for the 4 + 2 reaction. In an antarafacial interaction, one reactant collides in such a manner that the opposite faces of the *p*-orbitals in the π-bonds interact. Show that the 4 + 2 cycloaddition reaction is forbidden when butadiene interacts in an antarafacial manner.

Solution

Shown below is the reaction of butadiene with ethylene, except that we have the ethylene collide with the butadiene at a skewed angle. Recall that molecules in a reaction flask are constantly colliding with the solvent and each other in all possible orientations, but only certain orientations lead to chemical reactions. In the antarafacial geometry for the butadiene, there is one contact between the butadiene and ethylene in which the phasing does not match. Hence, this geometry for a collision between the two reactants will not lead to product and is considered forbidden.

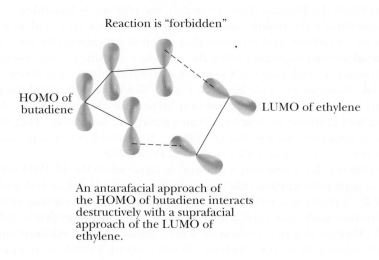

Reaction is "forbidden"

HOMO of butadiene

LUMO of ethylene

An antarafacial approach of the HOMO of butadiene interacts destructively with a suprafacial approach of the LUMO of ethylene.

Problem 20.6

The 2 + 2 cycloaddition with one suprafacial and one antarafacial interaction is allowed. Show this conclusion via a frontier molecular orbital analysis. Although the reaction is allowed, it is seldom seen. Can you think of a reason not based upon an orbital analysis of why this reaction is difficult?

20.5 | The Diels-Alder Reaction

In 1928 Otto Diels and Kurt Alder in Germany discovered a unique reaction of conjugated dienes: they undergo cycloaddition reactions with certain types of carbon-carbon double and triple bonds. For their discovery and subsequent studies of this reaction, Diels and Alder were jointly awarded the 1950 Nobel Prize for chemistry.

The compound with the double or triple bond that reacts with the diene in a Diels-Alder reaction is given the special name of **dienophile** (diene-loving), and the product of a Diels-Alder reaction is given the special name of **Diels-Alder adduct**. The designation **cycloaddition** refers to the fact that two reactants add together to give a cyclic product. Following are two examples of Diels-Alder reactions: one with a compound containing a carbon-carbon double bond, and the other containing a carbon-carbon triple bond.

Dienophile
A compound containing a double bond (consisting of one or two C, N, or O atoms) that can react with a conjugated diene to give a Diels-Alder adduct.

Diels-Alder adduct
A cyclohexene resulting from the cycloaddition reaction of a diene and a dienophile.

1,3-Butadiene　　**3-Buten-2-one**　　　**4-Cyclohexenyl methyl ketone**
(a diene)　　　　(a dienophile)　　　　　　(a Diels-Alder adduct)
　　　　　　　　　　　　　　　　　　　　　　(racemic mixture)

1,3-Butadiene　　**Diethyl 2-butynedioate**　　**Diethyl 1,4-cyclohexadiene-**
(a diene)　　　　(a dienophile)　　　　　　　**1,2-dicarboxylate**
　　　　　　　　　　　　　　　　　　　　　　(a Diels-Alder adduct)

Note that the four carbon atoms of the diene and two carbon atoms of the dienophile combine to form a six-membered ring. Note further that there are two more
σ bonds and two fewer π bonds in the product than in the reactants. This exchange
of two (weaker) π bonds for two (stronger) σ bonds is a major driving force in
Diels-Alder reactions.

We can write a Diels-Alder reaction in the following way, showing only the
carbon skeletons of the diene and dienophile. In this representation, curved arrows
are used to show that two new σ bonds are formed, three π bonds are broken, and
one new π bond is formed. It must be emphasized here that in this particular case,
the curved arrows in this diagram are not meant to show a mechanism. Rather they
are intended to show which bonds are broken, which new bonds are formed, and
how many electrons are involved (six in this case). The real mechanism is pericyclic
and is a 4 + 2 cycloaddition, as was presented in Section 20.4.

Diels-Alder reaction

The special values of the reaction discovered by Diels and Alder are that (1) it is one
of the simplest reactions that can be used to form six-membered rings, (2) it is one of
few reactions that can be used to form two new carbon-carbon bonds at the same time,
and, as we will see later in this section, (3) it is completely stereo-specific and quite
regioselective. For these reasons, the Diels-Alder reaction has proved to be enormously
valuable in synthetic organic chemistry.

Example 20.7

Draw a structural formula for the Diels-Alder adduct formed by reaction of each
diene and dienophile pair.

(a) 1,3-Butadiene and propenal
(b) 2,3-Dimethyl-1,3-butadiene and 3-buten-2-one

Solution

First draw the diene and dienophile so that each molecule is properly aligned to
form a six-membered ring. Then complete the reaction to form the six-membered
ring Diels-Alder adduct.

(a)

1,3-Butadiene   **Propenal**   (racemic mixture)
                    (Acrolein)

(b)

2,3-Dimethyl-   **3-Buten-2-one**   (racemic mixture)
1,3-butadiene   **(Methyl vinyl ketone)**

Problem 20.7

What combination of diene and dienophile undergoes Diels-Alder reaction to give each adduct?

(a)   (b)   (c)

(racemic)

Now let us look more closely at the scope and limitations, stereochemistry, and mechanism of Diels-Alder reactions.

A. Diene Must Be Able to Assume an *s-Cis* Conformation

We can illustrate the significance of conformation of the diene by reference to 1,3-butadiene. For maximum stability of a conjugated diene, overlap of the four unhybridized $2p$ orbitals making up the π system must be complete, a condition that occurs only when all four carbon atoms of the diene lie in the same plane. If the carbon skeleton of 1,3-butadiene is planar, the six atoms bonded to the skeleton of the diene are also contained in the same plane. Because of conjugation, bond rotation is actually somewhat restricted around the central single bond: if the atoms are not coplanar, conjugation is imperfect or broken completely. There are two planar conformations of 1,3-butadiene, called the ***s-trans*** **conformation** and the ***s-cis*** **conformation** where the designation *s* refers to the carbon-carbon single bond of the diene. Of these, the *s-trans* conformation is slightly lower in energy and, therefore, slightly more stable.

Although *s-trans*-1,3-butadiene is the more stable conformation, *s-cis*-1,3-butadiene is the reactive conformation in Diels-Alder reactions. In the *s-cis* conformation, carbon atoms 1 and 4 of the conjugated system are close enough to react with the carbon-carbon double or triple bond of the dienophile and to form a six-membered ring. In the *s-trans* conformation, they are too far apart for this to happen.

The energy barrier for interconversion of the *s-trans* and *s-cis* conformations for 1,3-butadiene is low, approximately 11.7 kJ (2.8 kcal)/mol; consequently, 1,3-butadiene can still be a reactive diene in Diels-Alder reactions.

s-trans conformation
(Lower in energy)

s-cis conformation
(Higher in energy)

(2*Z*,4*Z*)-2,4-Hexadiene is unreactive in Diels-Alder reactions because it is prevented by steric hindrance from assuming the required *s-cis* conformation.

s-trans conformation
(Lower in energy)

s-cis conformation
(Higher in energy)

Methyl groups would be forced closer than allowed by van der Waals radii

(2*Z*,4*Z*)-2,4-Hexadiene

Example 20.8

Which molecules can function as dienes in Diels-Alder reactions?

(a)   **(b)**   **(c)**

Solution

The dienes in both (a) and (b) are fixed in the *s-trans* conformation and, therefore, are not capable of participation in Diels-Alder reactions. The diene in (c) is fixed in the *s-cis* conformation and, therefore, has the proper orientation to participate in Diels-Alder reactions.

Problem 20.8

Which molecules can function as dienes in Diels-Alder reactions?

(a)   **(b)**   **(c)**

B. The Effect of Substituents on Rate

The simplest example of a Diels-Alder reaction is that between 1,3-butadiene and ethylene, both gases at room temperature. Although this reaction does occur, it is very slow and takes place only if the reactants are heated at 200°C under pressure.

1,3-Butadiene   **Ethylene**   **Cyclohexene**

Diels-Alder reactions are facilitated by a combination of electron-withdrawing substituents on one of the reactants and electron-releasing substituents on the other. Most commonly, the dienophile is electron-deficient and the diene is electron-rich. For example, placing a carbonyl group (electron-withdrawing because of the partial positive charge on its carbon) on the dienophile facilitates the reaction. To illustrate, 1,3-butadiene and 3-buten-2-one form a Diels-Alder adduct when heated at 140°C.

1,3-Butadiene   **3-Buten-2-one**   (racemic)

Placing electron-releasing methyl groups on the diene further facilitates reaction; 2,3-dimethyl-1,3-butadiene and 3-buten-2-one form a Diels-Alder adduct at 30°C.

2,3-Dimethyl-   **3-Buten-2-one**   (racemic)
1,3-butadiene

Several of the electron-releasing and electron-withdrawing groups most commonly encountered in Diels-Alder reactions are given in Table 20.4. Note that the ester group can be either electron-donating or -withdrawing depending on whether the oxygen or the carbonyl is attached to the double bond.

Table 20.4	Electron-Releasing and Electron-Withdrawing Groups
Electron-Releasing Groups	**Electron-Withdrawing Groups**
—CH_3	$-\overset{\overset{\displaystyle O}{\|\|}}{C}H$ (aldehyde)
—CH_2CH_3	
—$CH(CH_3)_2$	$-\overset{\overset{\displaystyle O}{\|\|}}{C}R$ (ketone)
—$C(CH_3)_3$	
—R (other alkyl groups)	$-\overset{\overset{\displaystyle O}{\|\|}}{C}OH$ (carboxyl)
—OR (ether)	$-\overset{\overset{\displaystyle O}{\|\|}}{C}OR$ (ester)
	—NO_2 (nitro)
$-O\overset{\overset{\displaystyle O}{\|\|}}{C}R$ (ester)	—C≡N (cyano)

C. Diels-Alder Reactions Can Be Used to Form Bicyclic Systems

Conjugated cyclic dienes, in which the double bonds are of necessity held in an *s-cis* conformation, are highly reactive in Diels-Alder reactions. Two particularly useful dienes for this purpose are cyclopentadiene and 1,3-cyclohexadiene. In fact, cyclopentadiene is reactive both as a diene and as a dienophile, and, on standing at room temperature, it forms a Diels-Alder self-adduct known by the common name dicyclopentadiene. When dicyclopentadiene is heated to 170°C, a reverse Diels-Alder reaction takes place, and cyclopentadiene is reformed.

Diene	Dienophile	**Dicyclopentadiene** (endo form)	From top	From side

The terms "endo" and "exo" are used for bicyclic Diels-Alder adducts to describe the orientation of substituents of the dienophile in relation to the two-carbon diene-derived bridge. **Exo** (Greek, outside) substituents are on the opposite side from the diene-derived bridge; **endo** (Greek, within) substituents are on the same side.

the double bond derived from the diene → **exo** (outside) relative to the double bond / **endo** (inside) relative to the double bond

For Diels-Alder reactions under kinetic control, the endo orientation of the dienophile is favored. Treatment of cyclopentadiene with methyl propenoate (methyl acrylate) gives the endo adduct almost exclusively. The exo adduct is not formed. Diels-Alder reactions are not always so stereoselective.

Cyclopentadiene	**Methyl propenoate**	The endo adduct (racemic)

D. The Configuration of the Dienophile Is Retained

The reaction is completely stereospecific at the dienophile. If the dienophile is a *cis* isomer, then the substituents *cis* to each other in the dienophile are *cis* in the Diels-Alder adduct. Conversely, if the dienophile is a *trans* isomer, substituents that are *trans* in the dienophile are *trans* in the adduct.

Dimethyl *cis*-2-butenedioate
(a *cis* dienophile)

**Dimethyl *cis*-4-cyclohexene-
1,2-dicarboxylate**

Dimethyl *trans*-2-butenedioate
(a *trans* dienophile)

**Dimethyl *trans*-4-cyclohexene-
1,2-dicarboxylate** (racemic)

E. The Configuration at the Diene Is Retained

The reaction is also completely stereospecific at the diene. Groups on the 1 and 4 positions of the diene retain their relative orientation.

(racemic)

A picture of the transition state will help clarify the reason for this. Bonds being formed in the transition state are shown as dashed red lines; bonds being broken are shown as dashed blue lines. The groups that are inside on the diene end up on the opposite side from the dienophile.

Transition state

Endo adduct

Side view
of
product

Front view
of
product

Figure 20.9
Mechanism of the Diels-Alder reaction. The diene and dienophile approach each other in parallel planes, one above the other, with the substituents on the dienophile endo to the diene. There is overlap of the π orbitals of each molecule and syn addition of each molecule to the other. As (1) new σ bonds form in the transition state, (2) the —CH_2— on the diene rotates upward, (3) the hydrogen atom of the dienophile becomes exo and the ester group becomes endo.

Example 20.9

Complete the following Diels-Alder reaction, showing the stereochemistry of the product.

Solution

Reaction of cyclopentadiene with this dienophile forms a disubstituted bicyclic product. The two ester groups are *cis* in the dienophile, and, given the stereoselectivity of the Diels-Alder reaction, they are *cis* and endo in the product.

Problem 20.9

What diene and dienophile might you use to prepare the following racemic Diels-Alder adduct?

F. Exploiting the Stereochemistry of the Diels-Alder Reaction

As we have mentioned repeatedly throughout the text, the synthesis of chiral products from achiral starting materials in an achiral environment invariably leads to a racemic mixture of products. Nature achieves the synthesis of single enantiomers by using enzymes, which create a chiral environment in which reaction takes place. Enzymes, in fact, show such high enantiomeric and diastereomeric selectivity that the result of an enzyme-catalyzed reaction is generally only a single one of all possible stereoisomers. Chemists have developed chiral catalysts that produce chiral products. However, these catalysts are often far less stereoselective than nature's enzyme catalysts, although great progress has been made in this field in recent years. How then do chemists achieve the synthesis of single enantiomers uncontaminated by their mirror images?

One strategy they use is resolution (Section 3.9) to separate enantiomers and recover each in pure form. The most common methods for resolution depend on (1) the different physical properties of diastereomeric salts, (2) the use of enzymes as resolving agents, and (3) chromatography on a chiral substrate. While resolution is effective in preparing pure enantiomers, half of all product prepared to the point of resolution, namely the unwanted enantiomer, is lost in the process. Thus, this strategy for the preparation of single enantiomers wastes starting materials and reagents.

We illustrate an alternative strategy, namely **asymmetric induction**, by E. J. Corey's preparation of a key intermediate in his synthesis of prostaglandins. In asymmetric induction, the reactive functional group of an achiral molecule is placed in a chiral environment by reacting it with a **chiral auxiliary**. The strategy is that the chiral auxiliary then exerts a control over the stereoselectivity of the desired reaction. The chiral auxiliary chosen by Corey was 8-phenylmenthol. This molecule has three chiral centers and can exist as a mixture of $2^3 = 8$ possible stereoisomers. It was prepared in enantiomerically pure form from naturally occurring, enantiomerically pure menthol.

Menthol
(enantiomerically pure)

8-Phenylmenthol
(a chiral auxiliary)

The initial step in Corey's prostaglandin synthesis was a Diels-Alder reaction between a substituted cyclopentadiene and the double bond of an acrylate ester. By binding the achiral acrylate reactant to enantiomerically pure 8-phenylmenthol, Corey thus placed the carbon-carbon double bond of the dienophile in a chiral environment. The result was that the diene approached the carbon-carbon double bond of the acrylate preferentially from one direction.

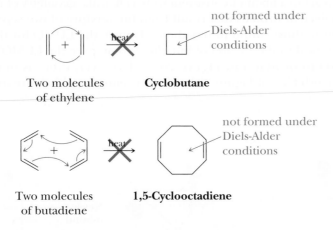

Achiral   Enantiomerically pure   97%   3%

A remarkable feature of this reaction is that it creates three chiral centers. Two of the chiral centers, namely those at the two ring junctions, are established by the Diels-Alder reaction. The third, namely the endo position of the ester group, is also established by the Diels-Alder reaction. Without the chiral auxiliary 8-phenylmenthyl group, two of the eight possible stereoisomers would be produced, namely the pair of enantiomers shown. Although both enantiomers of the bicyclic products were formed in Corey's scheme, they were formed in the ratio of 97:3, and the desired enantiomer could be separated in pure form. In subsequent steps, the 8-phenylmenthyl ester was hydrolyzed, and the pure enantiomer was converted to the so-called Corey lactone and then to enantiomerically pure prostaglandin F_{2a}.

G. A Word of Caution About Electron Pushing

Earlier we used curved arrows to show the flow of electrons that takes place in the process of bond breaking and bond forming in the Diels-Alder reaction. As discussed, these reactions involve a four-carbon diene and a two-carbon dienophile and are termed [4 + 2] cycloadditions. We can write similar electron-pushing mechanisms for the dimerization of ethylene by a [2 + 2] cycloaddition to form cyclobutane, and for the dimerization of butadiene by a [4 + 4] cycloaddition to form 1,5-cyclooctadiene.

not formed under
Diels-Alder
conditions

Two molecules   **Cyclobutane**
of ethylene

not formed under
Diels-Alder
conditions

Two molecules   **1,5-Cyclooctadiene**
of butadiene

Although [2 + 2] and [4 + 4] cycloadditions bear a formal relationship to the Diels-Alder reaction, neither, in fact, takes place under the thermal conditions required for Diels-Alder reactions (See Section 20.4) because they are forbidden as determined by the frontier molecular orbital analysis.

20.6 | Sigmatropic Shifts

The second class of pericyclic reactions that we examine is that of **sigmatropic shifts**. These reactions consist of the movement of a σ-bond across the face of one or more π-bonds. Although many examples of these reactions are known, we are only

Sigmatropic Shift
A reaction in which a σ-bond migrates across the face of one or more π-bonds.

going to analyze what is known as a [3,3]-shift. The numbering system for the nomenclature of the shift derives from assigning the number 1 to the ends of the σ-bond that is shifting and then naming the reaction to denote the number of atoms to which the σ-bond migrates. There are two common versions of this reaction, known as the Claisen and Cope rearrangements.

The Claisen Rearrangement

The Cope Rearrangement

To derive the frontier molecular orbital analysis for any [3,3]-shift, we'll use 1,5-hexatriene as the model, just as we used butadiene and ethylene as the models for the frontier molecular orbital analysis of all Diels-Alder reactions. As always, in a frontier molecular orbital analysis, we first identify a proposed geometry for the reaction. Let's propose a chairlike transition state in which the carbons on the ends of the chain react from the top of one π-bond and the bottom of the other.

The next step of a frontier molecular orbital analysis involves identifying a HOMO and a LUMO and checking to see if the HOMO and LUMO can interact with matched phasings (Figure 20.10). In this case, we assign the σ-bond that is migrating to be the HOMO; it is thus drawn as the overlap of two sp^3 hybrid orbitals (see Figure 1.18). The LUMO is assigned to be a molecular orbital that is a mixture of the two alkenes when their ends are in close proximity (Figure 20.10a) and in a trajectory to react in a manner consistent with a chairlike geometry of the transition state. The molecular orbitals that result from the mixture of two separate alkenes are analogous to those found in butadiene. Hence, the LUMO for the [3,3]-shift is phased identical to the LUMO of butadiene (compare the LUMO indicated in Figure 20.10a to orbital 3 in Figure 20.2). However, there is one important difference. In orbital 3 of Figure 20.2, the two central p-orbitals are in phase when

Figure 20.10

(a) Proper phasing of orbitals for the frontier molecular orbital analysis of a [3,3]-sigmatropic shift; note the phasing interaction between the terminal carbons when the top of one π-bond interacts with the bottom of the other in-phase.
(b) Reorientation showing how the chair conformation leads to in-phase interactions throughout the [3,3]-shift.

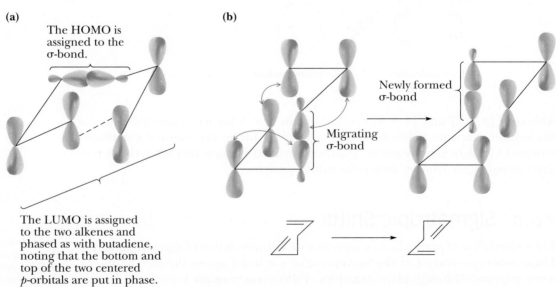

(a)

The HOMO is assigned to the σ-bond.

The LUMO is assigned to the two alkenes and phased as with butadiene, noting that the bottom and top of the two centered p-orbitals are put in phase.

(b)

Newly formed σ-bond

Migrating σ-bond

parallel. But in our analysis of the Cope reaction, the top of one of the central *p*-orbitals is placed in phase with the bottom of the other because this is the interaction geometry we are analyzing.

We can now check to see if the HOMO and LUMO phases match. In Figure 20.10b, we redraw the chair with the σ-bond vertically for clarity. The arrows show matched phasing that leads to a σ-bond and two π-bonds in the product that are all in phase and bonding. Hence, the reaction is allowed.

In summary, the frontier molecular orbital approach finds that there is an allowed geometry for reaction with the ends of the π-bonds of a 1,5-diene as in a Cope rearrangement (or analogously a Claisen rearrangement) with a chairlike transition state. Interestingly, a boatlike transition state is also allowed, although it is conformationally less stable (see Problem 20.48).

A. The Claisen Rearrangement

One example of the Claisen rearrangement transforms allyl phenyl ethers to *o*-allylphenols. Heating allyl phenyl ether, for example, the simplest member of this class of compounds, at 200–250°C results in a Claisen rearrangement to form *o*-allyphenol. In this rearrangement, an allyl group migrates from a phenolic oxygen to a carbon atom ortho to it. Carbon-14 labeling, here shown in color, has demonstrated that, during a Claisen rearrangement, carbon 3 of the allyl group becomes bonded to the ring carbon ortho to the phenolic oxygen.

Allyl phenyl ether   **2-Allylphenol**

The mechanism of a Claisen rearrangement involves a concerted redistribution of six electrons in a cyclic transition state as described above. The product of this rearrangement is a substituted cyclohexadienone, which undergoes keto-enol tautomerism to reform the aromatic ring. A new carbon-carbon bond is formed in the process.

Mechanism   The Claisen Rearrangement

Step 1: Sigmatropic shift.   Redistribution of six electrons in a cyclic transition state gives a cyclohexadienone intermediate. Dashed red lines indicate bonds being formed in the transition state, and dashed blue lines indicate bonds being broken.

Step 2: Keto-enol tautomerism.   Keto-enol tautomerism restores the aromatic character of the ring.

| **Allyl phenyl ether** | **Transition state** | **A cyclohexadienone intermediate** | **2-Allylphenol** |

Thus, we see that the transition state for the Claisen rearrangement bears a close resemblance to that for the Diels-Alder reaction. Both involve a concerted redistribution of six electrons in a cyclic transition state.

Example 20.10

Predict the product of Claisen rearrangement of *trans*-2-butenyl phenyl ether.

***trans*-2-Butenyl phenyl ether**

Solution

In the six-membered transition state for this rearrangement, carbon 3 of the allyl group becomes bonded to the ortho position of the ring.

| Transition state | A cyclohexadienone intermediate | (racemic) |

Problem 20.10

Show how to synthesize allyl phenyl ether and 2-butenyl phenyl ether from phenol and appropriate alkenyl halides.

B. The Cope Rearrangement

The Cope rearrangement of 1,5-dienes also takes place via a cyclic six-electron transition state. In this example, the product is an equilibrium mixture of isomeric dienes. The favored product is the diene on the right, which contains the more highly substituted double bonds.

Mechanism   The Cope Rearrangement

Pericyclic reaction.   Redistribution of six electrons in a cyclic transition state converts a 1,5-diene to an isomeric 1,5-diene.

| 3,3-Dimethyl-1,5-hexadiene | Transition state | 6-Methyl-1,5-heptadiene |

Example 20.11

Propose a mechanism for the following Cope rearrangement.

Solution

Redistribution of six electrons in a cyclic transition state gives the observed product.

Transition state

Problem 20.11

Propose a mechanism for the following Cope rearrangement.

320°C

C. Stereochemistry of the Cope Rearrangement

As discussed during the frontier molecular orbital theory analysis of [3,3]-sigmatropic shifts above, a chairlike transition state is allowed for these reactions. As you will show in Problem 20.48, a boatlike transition state is also allowed by frontier molecular orbital theory. However, chair conformations are more favorable than boat conformations for six-membered cyclic rings (look back at Section 2.5A). This preference influences the stereochemistry of these shifts, as we now show with an example.

Example 20.12

Upon heating the *meso* version of 3,4-dimethyl-1,5-hexadiene, three products with differing alkene stereochemistry from the Cope rearrangement are possible, but only two are found, with one being highly preferred. Show all three possible products, and predict the preference in the distribution.

Solution

The products and preference for the *cis-trans* alkenes can be explain by redrawing the reactant in chair- and boat-like conformations. These drawings reveal that the preferred product arises from a chairlike transition state.

Chair — 99.7% — *cis-trans* double bonds

Boat — 0.0% — two *cis* double bonds

Boat — 0.3% — two *trans* double bonds

Problem 20.12

Upon heating a racemic mixture of *d,l*-3,4-dimethyl-1,5-hexadiene, three products are possible, and all three are observed. The ratios are 90, 9, and nearly 1 percent. Predict which percentages correspond to which products and explain the ratio by showing the chair and boat conformations that lead to the products.

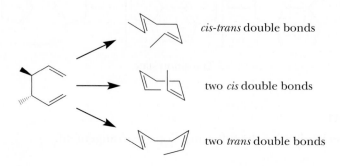

cis-trans double bonds

two *cis* double bonds

two *trans* double bonds

Problems: 20.1, 20.2, 20.6, 20.14, 20.15

Summary

Section 20.1 Stability of Conjugated Dienes

- A **conjugated diene** is one in which the double bonds are separated by only one single bond so that the $2p$ orbitals of the adjacent π bonds overlap.
  - An **unconjugated diene** is one in which the double bonds are separated by two or more single bonds.
  - A **cumulated diene** is one in which the two double bonds share an sp-hybridized carbon. In a cumulated diene, the $2p$ orbitals of the π bonds do not overlap, so they are not conjugated.
- The two conjugated double bonds in conjugated dienes are 14.5–17 kJ (3.5–4.1 kcal)/mol more stable than isomeric nonconjugated dienes, an observation that extends to all conjugated double bonds, not just in dienes.
  - The increased stability of conjugated double bonds results from delocalization of the four π electrons over the set of four parallel $2p$ orbitals.
  - According to molecular orbital theory, two conjugated double bonds are derived from four π molecular orbitals because the four parallel $2p$ orbitals overlap in space, even the $2p$ orbitals on either side of the single bond between the conjugated double bonds.
  - The lowest two π molecular orbitals have zero and one node, respectively, are bonding orbitals, and are filled with two electrons each.
  - These lowest two filled π molecular orbitals are each at an energy that is lower than isolated pi bonds, accounting for the "extra" stability of conjugated π systems.
  - The lowest filled π molecule orbital has large lobes extending over all four atoms, illustrating the delocalization of electron density in conjugated π systems.
  - In order for maximal overlap to occur, the $2p$ orbitals must be parallel, so the sp^2 atoms of the conjugated systems must be coplanar.

Section 20.2 Electrophilic Addition to Conjugated Dienes

- Conjugated dienes undergo both 1,2- and 1,4-addition reactions with electrophiles, often giving mixtures of both kinds of products.
- The ratio of 1,2-addition to 1,4-addition is temperature dependent, with 1,2-addition often predominating at lower temperature and 1,4-addition predominating at higher temperature.

– 1,2-Addition for butadiene is the predominant product formed at lower temperature under **kinetic (rate) control**, because there is usually greater positive charge at the 2 position of the allylic cation intermediate, lowering the activation barrier for reaction at this position. The lower temperature prevents equilibration between products, so relative product stability is not important.

– 1,4-Addition to butadiene is the predominant product formed at higher temperature under **thermodynamic control**, because the double bond of the 1,4-addition product is often more substituted and therefore of lower energy. The higher temperature allows equilibration of products so that product distribution depends on relative product stability.

Problems: 20.3, 20.16–20.22

– Note that the details of a conjugated diene structure will determine relative stabilities of 1,2- and 1,4-addition products, so the preceding statements concerning kinetic and thermodynamic product ratios of butadiene should be considered guidelines only and each new molecule needs to be carefully analyzed.

Section 20.3 UV-Visible Spectroscopy

• The ultraviolet region of the electromagnetic spectrum has wavelengths extending between 200 nm and 400 nm, and the visible region has wavelengths extending between 400 nm and 700 nm.

• Ultraviolet and visible spectral data are plotted as **absorbance (A)** versus wavelength, where absorbance is calculated as the log base 10 of the ratio of (I_o/I) where I_o is the intensity of light at a given wavelength irradiating a sample and I is the light transmitted through the sample.

– The quantity $(I_o/I) \times 100$ is called **percent transmittance**.

– The relationship between absorbance, concentration, and length of the sample cell (cuvette) is known as the **Beer-Lambert Law**, $A = \varepsilon c l$ where A is absorbance, ε is the molar absorptivity (also called extinction coefficient) of the molecules in the sample having the units per moles per liter per centimeter ($M^{-1}cm^{-1}$), c is concentration in moles per liter (M), and l is the length of the cuvette in centimeters (cm).

Problems: 20.4, 20.5, 20.23–20.27

– The molar absorptivity (extinction coefficient) as a function of wavelength is characteristic for a molecule and is based on the functional groups within the molecule. If the molar absorptivity is known for a given molecule, its concentration in solution can be calculated using the Beer-Lambert law.

• Absorption by molecules removes the absorbed wavelengths from white light, and a sample will appear to our eyes as the combination of reflected wavelengths.

– Wavelengths not absorbed are reflected.

– The color of combined reflected wavelengths can be roughly approximated as the complement of the absorbed color as illustrated using an artist's standard color wheel.

• Absorption of electromagnetic radiation in the ultraviolet-visible region results in promotion of electrons from a lower energy, occupied molecular orbital to a higher energy, unoccupied molecular orbital.

– The amount of energy in the ultraviolet-visible region is appropriate to excite nonbonding (lone pair) or π (bonding) electrons to π^* (antibonding) orbitals, in a process known as an n $\rightarrow \pi^*$ or $\pi \rightarrow \pi^*$ transition, respectively.

– σ bonding electrons are usually too low in energy to take part in ultraviolet or visible light absorption.

– The $\pi \rightarrow \pi^*$ transition for unconjugated alkenes is usually too high in energy (wavelength too short) to observe in the ultraviolet spectrum.

– Conjugated π systems have $\pi \rightarrow \pi^*$ transitions that can be seen in the ultraviolet or even visible absorption region, because conjugation decreases the energy difference between filled and unfilled π orbitals.

     – The greater the number of conjugated π bonds, the smaller the $\pi \rightarrow \pi^*$ energy gap, so the longer the wavelength of absorbed light.
     – Carbonyl groups can take part in conjugation along with C=C double bonds.

Section 20.4 Pericyclic Reaction Theory

- **Pericyclic reactions** occur in a single step involving a transition state that has a closed loop of orbitals.
- Although several methods exist to understand these reactions, **frontier molecular orbital theory** is the most common and easiest approach. In this approach, one follows a sequence of steps in order to predict if the reaction is allowed or forbidden.
  - First, a reaction geometry is proposed.
  - Second, the HOMO and LUMO of the reacting partners are written.
  - Third, the interaction between the HOMO and LUMO of the partners is examined in order to reveal whether an even number of phase changes (most commonly zero) or an odd number (most commonly one) exist at the points of interaction.
  - If the number of phase changes is even, the reaction is allowed, and if the number is odd, the reaction is forbidden, and is generally not observed.
- Frontier molecular orbital theory predicts that **4 + 2 cycloadditions** in which the reactants interact in a suprafacial manner is allowed but that analogous **2 + 2 cycloadditions** are forbidden.

Problems: 20.11, 20.28, 20.29

Section 20.5 The Diels-Alder Reaction

- Conjugated dienes react with certain types of molecules possessing double and triple bonds to form two new σ bonds and a ring structure in a reaction called the Diels-Alder reaction, an example of a 4 + 2 cycloaddition reaction.
  - The compound with the double or triple bond that reacts with the diene is called a **dienophile**, and the cyclic product is usually called the **Diels-Alder adduct**.
    - Three π bonds are broken and two stronger new σ bonds along with a new π bond are formed in the reaction, providing the driving force.
- The Diels-Alder reaction is facilitated by having electron-withdrawing groups such as carbonyls on one reactant (usually the dienophile) and electron-releasing groups on the other (usually the diene).
- The diene must be in the *s-cis* **conformation** to react, and dienes such as cyclopentadiene that are constrained to be in this conformation are particularly reactive.
  - When cyclic dienes are used, a bicyclic Diels-Alder adduct is produced.
  - The terms **"exo"** and **"endo"** are used with bicyclic Diels-Alder adducts. Exo substituents are on the opposite side of the newly formed ring from the diene-derived two-carbon bridge, and endo substituents are on the same side. For reactions that give kinetic products (not at equilibrium), the endo orientation of the dienophile is preferred.
- The configuration of the dienophile (i.e., E or Z) is retained in the Diels-Alder reaction, as is the relative orientation of groups on the diene, indicating a highly concerted reaction mechanism.
- The mechanism of the reaction is concerted, in that there is a single six-membered ring transition state in which the three π bonds are breaking at the same time as the two new σ bonds and one new π bond are being created.
- The Diels-Alder reaction has high stereoselectivity. One way to create enantiomerically pure target molecules is to use a chiral auxiliary, which is a chiral molecule available as a single enantiomer that is bonded to the starting material. The use of a chiral auxiliary can influence the resulting stereochemistry of a Diels-Alder reaction, producing a desired enantiomeric in excess. The chiral auxiliary is then removed.
- The arrow pushing in a cycloaddition reaction does not accurately reflect whether a particular reaction—2 + 2, 4 + 2, 4 + 4, etc.—will be allowed or forbidden. One must rely on the frontier molecular orbital approach, or other such approaches, to properly understand the mechanism of these reactions.

Problems: 20.12–20.14, 20.30–20.47

Section 20.6 Sigmatropic Shifts

- **Sigmatropic shifts** involve the migration of a σ-bond across one or more π-systems.
- One of the most common shifts is called a **[3,3]-shift** and involves the migration of a σ-bond across two flanking π-bonds.
- Frontier molecular orbital analysis shows that the reaction is allowed in a geometry that creates a chairlike transition state, although boatlike transition states can also occur.
- One example of the Claisen rearrangement transforms allyl phenol ethers to *o*-allylphenols through the redistribution of six electrons in a cyclic transition state.
- The Cope rearrangement of 1,5-dienes produces an equilibrium mixture of isomeric 1,5-dienes through the redistribution of six electrons in a cyclic transition state.
- By analyzing the possible chair- and boatlike transition states for the Cope rearrangement and by taking into account the lower energy of a chair conformation, the preferential stereochemistry of the products can be predicted.

Problems: 20.15, 20.16, 20.48–20.54

🕊WL Questions and problems for this chapter are available in OWL. Use the OWL access card that came with this textbook to access assigned questions and problems for this chapter.

Key Reactions

1. Electrophilic Addition to Conjugated Dienes (Section 20.2)   The ratio of 1,2- to 1,4-addition products depends on whether the reaction is under kinetic control or thermodynamic control. When a conjugated diene reacts with HBr, initial protonation of one of the double bonds gives a resonance-stabilized allylic cation; reaction of bromide with one of the carbons of this intermediate bearing the partial positive charge gives the 1,2-addition product, and reaction at the other gives the 1,4-addition product.

$$CH_2{=}CHCH{=}CH_2 + HBr \longrightarrow CH_3\overset{\overset{\displaystyle Br}{|}}{C}HCH{=}CH_2 + CH_3CH{=}CHCH_2Br$$

Products at $-78°C$ (kinetic control):   90%   10%
Products at $40°C$ (thermodynamic control):   15%   85%

2. The Diels-Alder Reaction: A Pericyclic Reaction (Section 20.5)   A Diels-Alder reaction takes place in a single step, without intermediates, and involves a redistribution of six π electrons in a cyclic transition state. The configuration of the diene and dienophile is preserved. Formation of the endo adduct is favored.

3. The Claisen Rearrangement: A Pericyclic Reaction (Section 20.6A)   The Claisen rearrangement transforms an allyl phenyl ether to an ortho-substituted phenol. The reaction takes place in a single step and involves the redistribution of six π electrons in a cyclic transition state.

4. The Cope Rearrangement: A Pericyclic Reaction (Section 20.6B)   The Cope rearrangement converts a 1,5-diene to give an isomeric 1,5-diene. The reaction takes place in a single step and involves the redistribution of six π electrons in a cyclic transition state.

Chapter 21

Benzene and the Concept of Aromaticity

© Douglas Brown

Peppers of the *Capsicum* family. See Chemical Connections: "Capsaicin, for Those Who Like It Hot." Inset: a model of capsaicin.

Aromatic compound
A term used initially to classify benzene and its derivatives. More accurately, it is used to classify any compound that meets the Hückel criteria for aromaticity (Section 21.2A).

OWL Online homework for this chapter may be assigned in OWL for Organic Chemistry.

Benzene, a colorless compound with a melting point of 6°C and a boiling point of 80°C, was first isolated by Michael Faraday in 1825 from the oily residue that collected in the illuminating gas lines of London. Benzene's molecular formula, C_6H_6, suggests a high degree of unsaturation. Compared with the corresponding alkane of molecular formula, C_6H_{14}, benzene's index of hydrogen deficiency is four, which can be met by an appropriate combination of rings, double bonds, and triple bonds. For example, a compound of molecular formula C_6H_6 might have four double bonds, or three double bonds and one ring, or two double bonds and two rings, or one triple bond and two rings, and so on.

Considering benzene's high degree of unsaturation, it might be expected to show many of the reactions characteristic of alkenes and alkynes. Yet, benzene is remarkably unreactive! It does not undergo the addition, oxidation, and reduction reactions characteristic of alkenes and alkynes. For example, benzene does not react with bromine, hydrogen bromide, or other reagents that usually add to carbon-carbon double and triple bonds. It is not oxidized by chromic acid under conditions that readily oxidize alkenes and alkynes. When benzene reacts, it does so by substitution, in which a hydrogen atom is replaced by another atom or group of atoms.

As noted in Chapter 5, the term **aromatic** was originally used to classify benzene and its derivatives because many of them have distinctive odors. The term aromatic, as it is now used, refers instead to the fact that these compounds are highly unsaturated and unexpectedly stable toward reagents that attack alkenes and alkynes. The term **arene** is used to describe aromatic hydrocarbons, by analogy with alkane, alkene, and alkyne. Benzene is the parent arene. Just as a group derived by removal of an H from an alkane is called an alkyl group and given the symbol R–, a group derived by removal of an H from an arene is called an **aryl group** and given the symbol Ar–.

21.1 | The Structure of Benzene

Let us put ourselves in the mid-19th century and examine the evidence on which chemists attempted to build a model for the structure of benzene. First, because the molecular formula of benzene is C_6H_6, it seemed clear that the molecule must be highly unsaturated. Yet, benzene does not show the chemical properties of alkenes, the only unsaturated hydrocarbons known at that time. Benzene does undergo chemical reactions, but its characteristic reaction is substitution rather than addition. When benzene is treated with bromine in the presence of ferric chloride, for example, only one compound with the molecular formula C_6H_5Br is formed.

$$C_6H_6 + Br_2 \xrightarrow{FeCl_3} C_6H_5Br + HBr$$

Benzene   **Bromobenzene**

Chemists concluded, therefore, that all six hydrogens of benzene must be equivalent. When bromobenzene is treated with bromine in the presence of ferric chloride as a catalyst, three isomeric dibromobenzenes are formed.

$$C_6H_5Br + Br_2 \xrightarrow{FeCl_3} C_6H_4Br_2 + HBr$$

Bromobenzene   Three isomeric
                        dibromobenzenes

For chemists in the mid-19th century, the problem was to incorporate these observations, along with the accepted tetravalence of carbon, into a structural formula for benzene. Before we examine these proposals, we should note that the problem of the structure of benzene and other aromatic hydrocarbons has occupied the efforts of chemists for over a century. Only since the 1930s has a general understanding of this problem been realized.

A. Kekulé's Model of Benzene

The first structure for benzene was proposed by August Kekulé in 1865 and consisted of a six-membered ring with one hydrogen bonded to each carbon. Although Kekulé's original structural formula provided for the equivalency of the C—H and C—C bonds, it was inadequate because all the carbon atoms were trivalent. To maintain the tetravalence of carbon, Kekulé proposed in 1872 that the ring contains three double bonds that shift back and forth so rapidly that the two forms cannot be separated. Kekulé regarded this interconversion as an equilibrium, each structure in which has become known as a **Kekulé structure** (Figure 21.1).

Now, more than 135 years after the time of Kekulé, we are apt to misunderstand what scientists in his time knew and did not know. For example, it is a given to us that covalent bonds consist of one or more pairs of shared electrons. We must remember, however, that it was not until 1897 that J. J. Thomson, professor of physics at the Cavendish Laboratory of Cambridge University, discovered the electron. Thomson was awarded the 1906 Nobel Prize in physics. That the electron played any role in chemical bonding did not become clear for another 30 years. Thus, at the time Kekulé made his proposal for the structure of benzene, the existence of electrons and their role in chemical bonding was completely unknown.

Kekulé's proposal accounted nicely for the fact that bromination of benzene gives only one bromobenzene, and that bromination of bromobenzene gives three isomeric dibromobenzenes.

Figure 21.1
Kekulé structures for benzene.

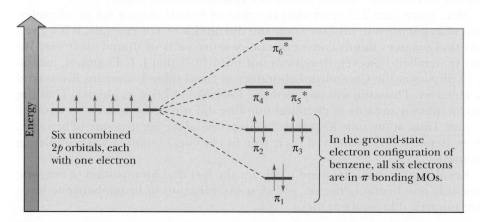

Bromobenzene

Three isomeric dibromobenzenes

Although his proposal was consistent with many experimental observations, it did not totally solve the problem and was contested for years. The major objection was that it did not account for the unusual chemical behavior of benzene. If benzene contains three double bonds, Kekulé's critics argued, why doesn't it show reactions typical of alkenes? Why, for example, doesn't benzene add three moles of bromine to form 1,2,3,4,5,6-hexabromocyclohexane? We now understand the surprising unreactivity of benzene on the basis of two complementary descriptions, the molecular orbital model and the resonance model.

B. The Molecular Orbital Model of Benzene

The carbon skeleton of benzene forms a regular hexagon with C—C—C and H—C—C bond angles of 120°. For this type of bonding, carbon uses sp^2 hybrid orbitals. Each carbon forms σ bonds to two adjacent carbons by overlap of sp^2-sp^2 hybrid orbitals, and one σ bond to hydrogen by overlap of sp^2-$1s$ orbitals. As determined experimentally, all carbon-carbon bonds are 139 pm in length, a value almost midway between the length of a single bond between sp^3 hybridized carbons (154 pm) and a double bond between sp^2 hybridized carbons (133 pm).

Each carbon also has a single unhybridized $2p$ orbital that is perpendicular to the plane of the ring and contains one electron. According to molecular orbital theory, the combination of these six parallel $2p$ atomic orbitals gives a set of six π MOs, three π-bonding MOs, and three π-antibonding MOs. Figure 21.2 shows these six molecular orbitals and their relative energies. Note that π_2 and π_3 MOs are degenerate (they have the same energy) bonding orbitals. Similarly, π^*_4 and π^*_5 are a degenerate pair of π-antibonding MOs.

In the ground-state electron configuration of benzene, the six electrons of the π system occupy the three bonding MOs (Figure 21.2). According to molecular orbital calculations, the great stability of benzene results from the fact that these three bonding MOs are much lower in energy when compared with the six uncombined $2p$ atomic orbitals.

Figure 21.2

The molecular orbital representation of the π bonding in benzene.

694   Chapter 21   Benzene and the Concept of Aromaticity

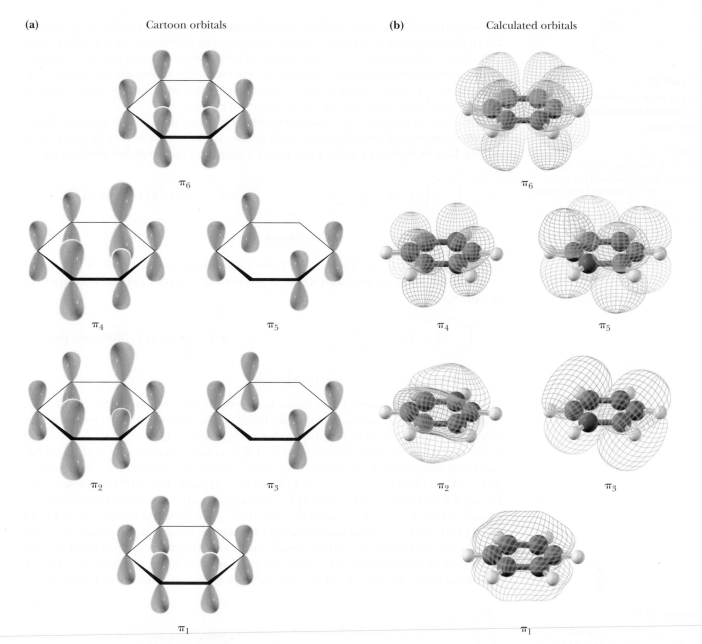

π_6

π_4 π_5

π_2 π_3

π_1

π_6

π_4 π_5

π_2 π_3

π_1

The π orbitals of benzene are shown in Figure 21.3. It is common to represent the π system of benzene as one torus (a donut-shaped region) above the plane of the ring and a second torus below it, shown in Figure 21.3 as π_1.

This picture is useful because it emphasizes the delocalization of electron density of the π system, and the equivalence among all six carbon atoms. However, this is not the whole story. The other two filled molecular orbitals (π_2 and π_3) have two nodes each, underscoring that the bond order between carbon atoms is intermediate between a double and a single bond.

C. The Resonance Model of Benzene

One of the postulates of resonance theory is that, when a molecule or ion can be represented by two or more contributing structures, it is not adequately represented by any single contributing structure. We represent benzene as a hybrid of two equivalent contributing structures, often referred to as Kekulé structures (Figure 21.4).

Each Kekulé structure makes an equal contribution to the hybrid and thus the C—C bonds are neither single nor double bonds but something intermediate.

Figure 21.3
Orbitals for the π system of benzene. (a) Cartoon representations of the six calculated orbitals that chemists routinely draw. These pictures accentuate that various combinations of parallel $2p$ orbitals lead to the π system of benzene. (b) Calculated orbitals. The three lowest in energy are occupied with electrons (see Figure 21.2). The lowest of these orbitals is the image most chemists use for the π system of benzene: a torus of electron density above and below the ring.

Figure 21.4
Benzene as a hybrid of two equivalent contributing structures.

Resonance energy
The difference in energy between a resonance hybrid and the most stable of its hypothetical contributing structures in which electrons are localized on particular atoms and in particular bonds.

We recognize that neither of these contributing structures exists (they are merely alternative ways to pair $2p$ orbitals with no reason to prefer one or the other) and that the actual structure is a superposition of both. Nevertheless, chemists continue to use a single contributing structure to represent this molecule because it is as close as we can come to an accurate structure within the limitations of classical valence bond structures and the tetravalence of carbon.

One way to estimate the **resonance energy** of benzene is to compare the heats of hydrogenation of cyclohexene and benzene. Cyclohexene is readily reduced to cyclohexane by hydrogen in the presence of a transition metal catalyst (Section 6.6A).

Benzene is reduced very slowly under these conditions to cyclohexane. It is reduced more rapidly when heated and under a pressure of several hundred atmospheres of hydrogen.

Catalytic hydrogenation of an alkene is an exothermic reaction (Section 6.6B). The heat of hydrogenation per double bond varies somewhat with the degree of substitution of the particular alkene; for cyclohexene, $\Delta H^0 = -119.7$ kJ (-28.6 kcal)/mol. If we consider benzene to be 1,3,5-cyclohexatriene, a hypothetical unsaturated compound with alternating single and double bonds, we calculate that $\Delta H^0 = 3(-119.7 \text{ kJ/mol}) = -359$ kJ (-85.8 kcal)/mol. The ΔH^0 for reduction of benzene to cyclohexane is -208 kJ (-49.8 kcal)/mol, considerably less than that calculated for 1,3,5-cyclohexatriene. The difference between these values, 151 kJ (36.0 kcal)/mol, is the **resonance energy of benzene**. Note that the product of both reductions is cyclohexane and that both reductions are exothermic. Therefore, the lower heat of hydrogenation for benzene confirms that it is more stable than 1,3,5-cyclohexatriene. These experimental results are shown graphically in Figure 21.5.

Figure 21.5
The resonance energy of benzene as determined by comparison of the heats of hydrogenation of cyclohexene, benzene, and the hypothetical compound 1,3,5-cyclohexatriene.

Several other experimental determinations of the resonance energy of benzene have been performed using different model compounds, and although these determinations differ somewhat in their results, they all agree that the resonance stabilization of benzene is large. Following are resonance energies for several other aromatic hydrocarbons.

	Benzene	Naphthalene	Anthracene	Phenanthrene
Resonance energy [kJ (kcal)/mol]	151 (36)	255 (61)	347 (83)	381 (91)

21.2 | The Concept of Aromaticity

The molecular orbital and resonance theories are powerful tools with which chemists can understand the unusual stability of benzene and its derivatives. According to resonance theory, benzene is best represented as a hybrid of two equivalent contributing structures. By analogy, cyclobutadiene and cyclooctatetraene can also be represented as hybrids of two equivalent contributing structures. Is either of these compounds aromatic?

Cyclobutadiene as a hybrid of two equivalent contributing structures

Cyclooctatetraene as a hybrid of two equivalent contributing structures

The answer for both compounds is no. Repeated attempts to isolate cyclobutadiene have all failed. It was not until 1965 that it was finally synthesized, and, even then, it could only be detected if trapped at $4°K$ ($-269°C$). Cyclobutadiene is a highly unstable compound and does not show any of the chemical and physical properties we associate with aromatic compounds. Cyclooctatetraene has chemical properties typical of alkenes. It reacts readily with halogens and halogen acids, as well as with mild oxidizing and reducing agents.

We are then faced with the broad question: "What are the fundamental principles underlying aromatic character?" In other words, what are the structural characteristics of unsaturated compounds that have a large resonance energy and do not undergo reactions typical of alkenes but rather undergo substitution reactions?

A. The Hückel Criteria for Aromaticity

The underlying criteria for aromaticity were recognized in the early 1930s by Erich Hückel, a German chemical physicist. He carried out MO energy calculations for monocyclic, planar molecules in which each atom of the ring has one $2p$ orbital available for forming sets of molecular orbitals. His calculations demonstrated that monocyclic, planar molecules with a closed loop of 2, 6, 10, 14, 18, . . . π electrons in a fully conjugated system should be aromatic. These numbers are generalized in the **($4n + 2$) π electron rule**, where n is a positive integer (0, 1, 2, 3, 4, . . .). Conversely, monocyclic, planar molecules with $4n$ π electrons (4, 8, 12, 16, 20, . . .) are especially unstable and are said to be antiaromatic. We will have more to say about antiaromaticity shortly. **Hückel's criteria for aromaticity** are summarized as follows. To be aromatic, a compound must:

1. Be cyclic.
2. Have one $2p$ orbital on each atom of the ring.

Hückel criteria for aromaticity
To be aromatic, a monocyclic compound must have one $2p$ orbital on each atom of the ring, be planar or nearly so, and have ($4n + 2$) π electrons in the cyclic arrangement of $2p$ orbitals.

Figure 21.6

Frost circles showing the number and relative energies of the π MOs for planar, fully conjugated four-, five-, and six-membered rings.

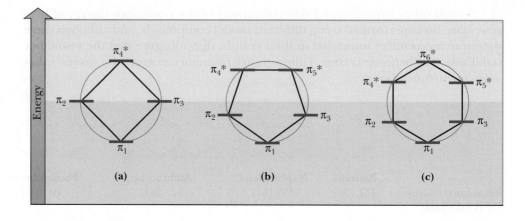

(a)   (b)   (c)

3. Be planar or nearly planar so that there is continuous or nearly continuous overlap of all $2p$ orbitals of the ring.

4. Have a closed loop of $(4n + 2)$ π electrons in the cyclic arrangement of $2p$ orbitals.

To appreciate the reasons for aromaticity and antiaromaticity, we must examine MO energy diagrams for the molecules and ions we will consider in this and the following section. The relative energies of the π MOs for planar, monocyclic, fully conjugated systems can be constructed quite easily using the **Frost circle**, or inscribed polygon method. To construct such a diagram, draw a circle and then inscribe in it a polygon of the same number of sides as the ring in question. Inscribe the polygon in such a way that one of its vertices is at the bottom of the circle. The relative energies of the MOs in the ring are then given by the points where the vertices touch the circle. Those MOs below the horizontal line through the center of the circle are bonding MOs. Those on the horizontal line are nonbonding MOs, and those above the line are antibonding MOs.

Figure 21.6 shows Frost circles describing the MOs of monocyclic, planar, and fully conjugated four-, five-, and six-membered rings. This apparently coincidental method works because it reproduces geometrically the mathematical solutions to the wave equation.

Frost circle

A graphic method for determining the relative energies of pi MOs for planar, fully conjugated, monocyclic compounds.

Example 21.1

Construct a Frost circle for a planar seven-membered ring with one $2p$ orbital on each atom of the ring, and show the relative energies of its seven π molecular orbitals. Which are bonding MOs, which are antibonding, and which are nonbonding?

Solution

Of the seven π molecular orbitals, three are bonding, and four are antibonding.

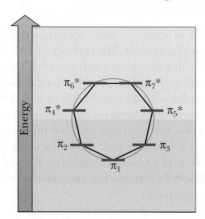

Problem 21.1

Construct a Frost circle for a planar eight-membered ring with one $2p$ orbital on each atom of the ring, and show the relative energies of its eight π molecular orbitals. Which are bonding MOs, which are antibonding, and which are nonbonding?

B. Aromatic Hydrocarbons

Cyclobutadiene, benzene, and cyclooctatetraene are the first members of a family of molecules called annulenes. An **annulene** is a cyclic hydrocarbon with a continuous alternation of single and double bonds. The name of an annulene is derived by showing the number of atoms in the ring in brackets followed by the word *annulene*. Named as annulenes, cyclobutadiene, benzene, and cyclooctatetraene are [4]annulene, [6]annulene, and [8]annulene, respectively. These compounds, however, are rarely named as annulenes.

Annulene
A cyclic hydrocarbon with a continuous alternation of single and double bonds.

Beginning in the 1960s, Franz Sondheimer and his colleagues, first at the Weizmann Institute in Israel and later at the University of London, synthesized a number of larger annulenes, primarily to test the validity of Hückel's criteria for aromaticity. They found, for example, that both [14]annulene and [18]annulene are aromatic, as predicted by Hückel. [18]Annulene has a resonance energy of approximately 418 kJ (100 kcal)/mol. Notice that, for these annulenes to achieve planarity, several of the carbon-carbon double bonds in each must have the *trans* configuration.

[14]Annulene
(aromatic)

[18]Annulene
(aromatic)

In these larger annulenes, there are two sets of equivalent hydrogens: those that point outward from the ring and those that point inward to the center of the ring. The fact is that these two sets of equivalent hydrogens have quite different ^{1}H-NMR chemical shifts.

The protons on benzene and other arenes are deshielded and appear far downfield (usually around 7–8 ppm) because of the induced ring current that occurs in aromatic molecules (Section 13.7C). The effect of induced ring current is characteristic not only of benzene and its derivatives but also of all compounds that meet the Hückel criteria for aromaticity. This concept of a circulating ring current and of an induced magnetic field predicts that hydrogen atoms on the outside of the ring should come into resonance with a downfield shift. It also predicts that a hydrogen atom in the inside of the ring should come into resonance farther upfield. Of course, no hydrogens are on the inside of the benzene ring, but with larger aromatic annulenes, as for example [18]annulene, there are both "inside" hydrogens and "outside" hydrogens. The degree of the upfield chemical shift of the inside hydrogens of [18]annulene is remarkable. They come into resonance at δ −3.00; that is, at 3.00 δ units upfield (to the right) of the TMS standard.

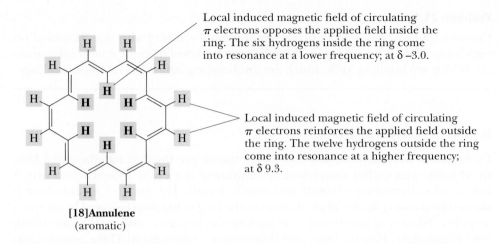

Local induced magnetic field of circulating π electrons opposes the applied field inside the ring. The six hydrogens inside the ring come into resonance at a lower frequency; at δ −3.0.

Local induced magnetic field of circulating π electrons reinforces the applied field outside the ring. The twelve hydrogens outside the ring come into resonance at a higher frequency; at δ 9.3.

[18]Annulene
(aromatic)

Example 21.2

Which hydrogens have a larger chemical shift, the six hydrogens of benzene or the eight hydrogens of cyclooctatetraene? Explain.

Solution

Benzene is an aromatic compound; its six equivalent hydrogens appear as a sharp singlet at δ 7.27. Cyclooctatetraene does not meet the Hückel criteria for aromaticity because it has $4n$ π electrons and is nonplanar. Therefore, the eight equivalent hydrogens of the cyclooctatetraene ring appear as a singlet at δ 5.8 in the region of vinylic hydrogens (δ 4.6 − δ 5.7).

Problem 21.2

Which compound gives a signal in the ^{1}H—NMR spectrum with a larger chemical shift, furan or cyclopentadiene? Explain.

According to Hückel's criteria, [10]annulene should be aromatic; it is cyclic, has one $2p$ orbital on each carbon of the ring, and has $4(2) + 2 = 10$ electrons in its π system. It has been found, however, that this molecule shows reactions typical of alkenes and is, therefore, classified as nonaromatic. The reason for its lack of aromaticity lies in the fact that the ten-membered ring is too small to accommodate the two hydrogens that point inward toward the center of the ring. Nonbonded interaction between these two hydrogens forces the ring into a nonplanar conformation in which the overlap of all ten $2p$ orbitals is no longer continuous. Therefore, because [10]annulene is not planar, it is not aromatic.

Repulsion between these two hydrogens forces the ring into a nonplanar conformation

[10]Annulene   Viewed from above   Viewed through an edge

What is remarkable is that if the two hydrogen atoms facing inward toward the center of the ring in [10]annulene are replaced by a CH_2 group, the ring is now able to assume a conformation close enough to planar that it becomes aromatic.

Bridged [10]annulene   Viewed from above   Viewed through an edge

C. Antiaromatic Hydrocarbons

According to the Hückel criteria, monocyclic, planar molecules with $4n$ π electrons (4, 8, 12, 16, 20, . . .) are especially unstable and are said to be antiaromatic. By these criteria, cyclobutadiene with 4 π electrons is antiaromatic. Using the Frost circle energy diagram from Figure 21.6, we can construct a molecular orbital energy diagram for cyclobutadiene (Figure 21.7).

In the ground-state electron configuration of cyclobutadiene, two π electrons fill the π_1-bonding MO. The third and fourth π electrons are unpaired and lie in the π_2- and π_3-nonbonding MOs. The existence of these two unpaired electrons in planar cyclobutadiene makes this molecule highly unstable and reactive compared to butadiene, a noncyclic molecule containing two conjugated double bonds. It has been found that cyclobutadiene is not planar, but slightly puckered with two shorter bonds and two longer bonds, which makes the two degenerate orbitals no longer equivalent; nevertheless, it retains some apparent diradical character.

Cyclooctatetraene shows reactions typical of alkenes and is classified as nonaromatic. X-ray studies show clearly that the most stable conformation of the molecule is a nonplanar "tub" conformation with two distinct types of carbon-carbon bonds: four longer carbon-carbon single bonds and four shorter carbon-carbon double bonds. The four single bonds are equal in length to the single bonds between sp^2 hybridized carbons (approximately 146 pm), and the four double bonds are equal in length to double bonds in alkenes (approximately 133 pm). In the tub conformation, the overlap of $2p$ orbitals on carbons forming double bonds is excellent, but almost no overlap occurs between $2p$ orbitals at the ends of carbon-carbon single bonds because these $2p$ orbitals are not parallel. Thus the π system in cyclooctatetraene is not conjugated despite having continuous sp^2 hybridized carbon atoms.

Antiaromatic compound
A monocyclic compound that is planar or nearly so, has one $2p$ orbital on each atom of the ring, and has $4n$ π electrons in the cyclic arrangement of overlapping $2p$ orbitals, where n is an integer. Antiaromatic compounds are especially unstable.

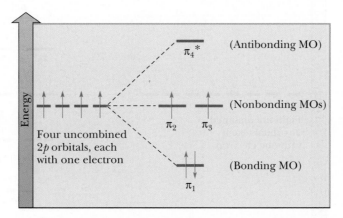

Figure 21.7
Molecular orbital energy diagram for cyclobutadiene. In the ground state, two electrons are in the low-lying π_1-bonding MO. The remaining two electrons are unpaired and occupy the degenerate π_2- and π_3-nonbonding MOs.

1,3,5,7-Cyclooctatetraene
(tub conformation with alternating
single and double bonds)

Viewed from above

Viewed through
an edge

To appreciate why planar cyclooctatetraene would be classified as antiaromatic, we need to examine the MO energy diagram for an eight-membered ring containing eight π electrons in a cyclic, fully conjugated ring. You constructed a Frost circle for this ring in answer to Problem 21.1. Note that the most stable conformation of cyclooctatetraene is not planar, but if it were planar, the Frost circle you constructed would be its MO energy diagram. The molecular orbital energy diagram for planar cyclooctatetraene is shown in Figure 21.8. In the ground state, six π electrons fill the three low-lying π_1-, π_2-, and π_3-bonding MOs. The remaining two π electrons are unpaired and lie in the degenerate π_4- and π_5-nonbonding MOs. Because of these two unpaired electrons, planar cyclooctatetraene, if it existed, would be classified as antiaromatic. Cyclooctatetraene, however, is large enough to pucker into a nonplanar conformation and become nonaromatic.

If [16]annulene were planar, it too would be antiaromatic. The size of the ring, however, is large enough that it can pucker into a nonplanar conformation in which the double bonds are no longer fully conjugated. [16]Annulene, therefore, is nonaromatic.

[16]Annulene
(nonaromatic)

Figure 21.8

Molecular orbital energy diagram for a planar conformation of cyclooctatetraene. Three pairs of electrons fill the three low-lying π-bonding molecular orbitals. Two electrons are unpaired in degenerate π-nonbonding molecular orbitals.

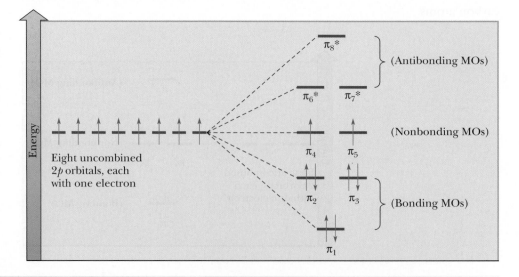

D. Heterocyclic Aromatic Compounds

Aromatic character is not limited to hydrocarbons; it is found in **heterocyclic compounds** as well. Pyridine and pyrimidine are heterocyclic analogs of benzene. In pyridine, one CH group of benzene is replaced by nitrogen, and, in pyrimidine, two CH groups are replaced by nitrogens (Figure 21.9).

Each molecule meets the Hückel criteria for aromaticity. Each is monocyclic and planar, each has one $2p$ orbital on each atom of the ring, and each has six electrons in the π system. In pyridine, nitrogen is sp^2 hybridized, and its unshared pair of electrons occupies an sp^2 orbital in the plane of the ring and is perpendicular to the $2p$ orbitals of the π system; thus, the unshared pair on the nitrogen of pyridine is not a part of the π system (Figure 21.10).

In pyrimidine, neither unshared pair of electrons of nitrogen is part of the π system. The resonance energy of pyridine is estimated to be 134 kJ (32 kcal)/mol, slightly less than that of benzene. The resonance energy of pyrimidine is estimated to be 108 kJ (26 kcal)/mol.

The five-membered ring heterocyclic compounds furan, thiophene, pyrrole, and imidazole are also aromatic.

Furan   **Thiophene**   **Pyrrole**   **Imidazole**

In these planar compounds, each heteroatom is sp^2 hybridized, and its unhybridized $2p$ orbital is part of a closed loop of five $2p$ orbitals. In furan and thiophene, one unshared pair of electrons of the heteroatom lies in the unhybridized $2p$ orbital and is a part of the π system (Figure 21.11). The other unshared pair of electrons lies in an sp^2 hybrid orbital perpendicular to the $2p$ orbitals and is not part of the π system. In pyrrole, the unshared pair of electrons on nitrogen is part of the π system. In imidazole, the unshared pair on one nitrogen is part of the aromatic sextet, and the unshared pair on the other nitrogen is not.

Nature abounds with compounds having a heterocyclic ring fused to one or more other rings. Two such compounds especially important in the biological world are indole and purine.

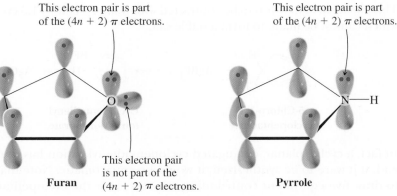

Indole   **Serotonin**   **Purine**   **Caffeine**
(a neurotransmitter)

Pyridine   **Pyrimidine**

Figure 21.9
Two heterocyclic aromatic compounds.

This sp^2 orbital is perpendicular to the six $2p$ orbitals of the π system

This electron pair is not part of the $(4n + 2)$ π electrons

Figure 21.10
Pyridine.

This electron pair is part of the $(4n + 2)$ π electrons.

This electron pair is part of the $(4n + 2)$ π electrons.

This electron pair is not part of the $(4n + 2)$ π electrons.

Furan   **Pyrrole**

Figure 21.11
Origin of the $4n + 2$ π electrons in furan and pyrrole. The estimated resonance energy of furan is 67 kJ (16 kcal)/mol, and that of pyrrole is 88 kJ (21 kcal)/mol.

Indole contains a pyrrole ring fused with a benzene ring. Compounds derived from indole include the essential amino acid L-tryptophan (Section 27.1C) and the neurotransmitter serotonin. Purine contains a six-membered pyrimidine ring fused with a five-membered imidazole ring. Caffeine is a trimethyl derivative of an oxidized purine. Compounds derived from purine and pyrimidine are building blocks of deoxyribonucleic acids (DNA) and ribonucleic acids (RNA, Chapter 28).

E. Aromatic Hydrocarbon Ions

Any neutral monocyclic unsaturated hydrocarbon with an odd number of carbons in the ring must of necessity have at least one CH_2 group in the ring and, therefore, cannot be aromatic. Examples of such hydrocarbons are cyclopropene, cyclopentadiene, and cycloheptatriene.

| Cyclopropene | Cyclopentadiene | Cycloheptatriene |

Cyclopropene has the correct number of π electrons to be aromatic, namely $4(0) + 2 = 2$, but it does not have a continuous closed loop of $2p$ orbitals. If, however, the CH_2 group becomes a CH^+ group in which the carbon atom is sp^2 hybridized and has a vacant $2p$ orbital, thus still containing only two electrons, then the overlap of orbitals is continuous, and, according to molecular orbital theory, the **cyclopropenyl cation** should be aromatic. The cyclopropenyl cation can be drawn as a resonance hybrid of three equivalent contributing structures. That we can draw three equivalent contributing structures is not the key to the aromaticity of this cation; the key is that it meets the Hückel criteria of aromaticity.

Cyclopropenyl cation
(a hybrid of three equivalent contributing structures)

As an example of the aromatic stabilization of this cation, 3-chlorocyclopropene reacts readily with antimony(V) chloride to form a stable salt.

| 3-Chloro- cyclopropene | Antimony(V) chloride (a Lewis acid) | Cyclopropenyl hexachloroantimonate |

This chemical behavior is to be contrasted with that of 5-chloro-1,3-cyclopentadiene, which cannot be made to form a stable salt.

| 5-Chloro-1,3- cyclopentadiene | | Cyclopentadienyl tetrafluoroborate |

In fact, a cyclic, planar, conjugated cyclopentadienyl cation has four π electrons, and, if it were to be synthesized, it would be antiaromatic. Note that it is possible to draw five equivalent contributing structures for the cyclopentadienyl cation.

Yet this cation is not aromatic because it has only $4n$ π electrons rather than the required $(4n + 2)$ π electrons.

To form an aromatic ion from cyclopentadiene, it is necessary to convert the CH_2 group to a CH^- group in which the carbon becomes sp^2 hybridized and has two electrons in its unhybridized $2p$ orbital. The resulting **cyclopentadienyl anion** is aromatic. Its aromatic character may also be represented by an inscribed circle with a minus sign (Figure 21.12).

Evidence of the stability of this anion is the fact that the pK_a of cyclopentadiene is approximately 16.0, which makes it the most acidic hydrocarbon known. The acidity of cyclopentadiene is comparable to that of water (pK_a 15.7) and ethanol (pK_a 15.9). Consequently, when cyclopentadiene is treated with aqueous sodium hydroxide, an equilibrium is established in which some of the hydrocarbon is converted to its aromatic anion. K_{eq} for this equilibrium is approximately 0.5.

Cyclopentadienyl anion

Figure 21.12
Cyclopentadienyl anion (aromatic).

Cyclopentadiene is converted completely to its anion by treatment with sodium amide.

Example 21.3

Construct an MO energy diagram for the cyclopentadienyl anion, and describe its ground-state electron configuration.

Solution

Refer to the Frost circle shown in Figure 21.6 for a planar, fully conjugated five-membered ring. The six π electrons occupy the π_1, π_2, and π_3 molecular orbitals, all of which are bonding MOs.

Problem 21.3

Describe the ground-state electron configuration of the cyclopentadienyl cation and radical. Assuming each species is planar, would you expect it to be aromatic or antiaromatic?

Cycloheptatriene forms an aromatic cation by conversion of its CH_2 group to a CH^+ group with this sp^2 hybridized carbon having a vacant $2p$ atomic orbital. The **cycloheptatrienyl (tropylium) cation** is planar and has six π electrons in seven $2p$ orbitals, one from each atom of the ring. It can be drawn as a resonance hybrid of seven equivalent contributing structures (Figure 21.13).

Cycloheptatrienyl cation

Figure 21.13
Cycloheptatrienyl cation (Tropylium ion) (aromatic).

Example 21.4

Construct an MO energy diagram for the cycloheptatrienyl cation, and describe its ground-state electron configuration.

Solution

Refer to the Frost circle you constructed in answer to Example 21.1. In the ground-state electron configuration of the cycloheptatrienyl cation, the six π electrons occupy the π_1, π_2, and π_3 molecular orbitals, all of which are bonding.

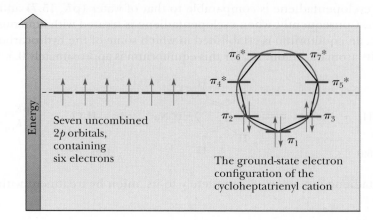

Problem 21.4

Describe the ground-state electron configuration of the cycloheptatrienyl radical and anion. Assuming that each species is planar, would you expect it to be aromatic or antiaromatic?

21.3 | Nomenclature

A. Monosubstituted Benzenes

Monosubstituted alkylbenzenes are named as derivatives of benzene, as for example ethylbenzene. The IUPAC system retains common names for several of the simpler monosubstituted alkylbenzenes. Examples are toluene (rather than methylbenzene), cumene (rather than isopropylbenzene), and styrene (rather than vinylbenzene).

Benzene   **Ethylbenzene**   **Toluene**   **Cumene**   **Styrene**

The common names phenol, aniline, benzaldehyde, benzoic acid, and anisole are also retained by the IUPAC system.

Phenol   **Aniline**   **Benzaldehyde**   **Benzoic acid**   **Anisole**

As we noted in the introduction to Chapter 5, the substituent group derived by loss of an H from benzene is a **phenyl group** abbreviated Ph–; that derived by loss of an H from the methyl group of toluene is a **benzyl group**, abbreviated Bn–.

Benzyl group, C₆H₅CH₂—
The group derived from toluene by removing a hydrogen from its methyl group.

| Benzene | Phenyl group, Ph- | Toluene | Benzyl group, Bn- |

In molecules containing other functional groups, the phenyl group and its derivatives are named as substituents.

1-Phenyl-1-pentanone   **4-(3-Methoxyphenyl)-2-butanone**   **(Z)-2-Phenyl-2-butene**

B. Disubstituted Benzenes

When two substituents occur on a benzene ring, three constitutional isomers are possible. The substituents may be located by numbering the atoms of the ring or by using the locators ortho, meta, and para. 1,2- is equivalent to **ortho** (Greek, straight or correct), 1,3- is equivalent to **meta** (Greek, in the middle, between), and 1,4- is equivalent to **para** (Greek, beyond).

When one of the two substituents on the ring imparts a special name to the compound, as for example toluene, cumene, phenol, and aniline, then the compound is named as a derivative of that parent molecule. The special substituent is assumed to occupy ring position number 1. The IUPAC system retains the common name xylene for the three isomeric dimethylbenzenes.

Ortho (o)
Refers to groups occupying 1,2-positions on a benzene ring.

Meta (m)
Refers to groups occupying 1,3-positions on a benzene ring.

Para (p)
Refers to groups occupying 1,4-positions on a benzene ring.

4-Bromotoluene
(**p**-Bromotoluene)

3-Chloroaniline
(**m**-Chloroaniline)

2-Nitrobenzoic acid
(**o**-Nitrobenzoic acid)

m-Xylene

Where neither group imparts a special name, then the two substituents are located and listed in alphabetical order before the ending *-benzene*. The carbon of the benzene ring with the substituent of lower alphabetical ranking is numbered C-1.

1-Chloro-4-ethylbenzene
(**p**-Chloroethylbenzene)

1-Bromo-2-nitrobenzene
(**o**-Bromonitrobenzene)

1,3-Dinitrobenzene
(**m**-Dinitrobenzene)

C. Polysubstituted Benzenes

When three or more substituents are present on a ring, their locations are specified by numerals. If one of the substituents imparts a special name, then the compound is named as a derivative of that parent molecule. If none of the substituents imparts a special name, the substituents are numbered to give the smallest set of numbers and are listed in alphabetical order before the ending -*benzene*. In the following examples, the first compound is a derivative of toluene, and the second is a derivative of phenol. Because there is no special name for the third compound, its three substituents are listed in alphabetical order, and the atoms of the ring are numbered using the lowest possible set of numbers.

4-Chloro-2-nitrotoluene   **2,4,6-Tribromophenol**   **2-Bromo-1-ethyl-4-nitrobenzene**

Example 21.5

Write names for these compounds.

(a)   (b)   (c)   (d) Ph ⌒⌇

Solution

(a) Ethyl 4-aminobenzoate (ethyl *p*-aminobenzoate)
(b) 3,4-Dimethoxybenzaldehyde
(c) 4-Nitrophenyl ethanoate (*p*-nitrophenyl acetate)
(d) 3-Phenylpropene (allyl benzene)

Problem 21.5

Write names for these compounds.

(a)   (b)   (c)   (d)

Polynuclear aromatic hydrocarbon (PAH)
A hydrocarbon containing two or more fused benzene rings.

Polynuclear aromatic hydrocarbons (PAHs) contain two or more benzene rings, each pair of which shares two ring carbon atoms. Naphthalene, anthracene, and phenanthrene, the most common PAHs, and substances derived from them are found in coal tar and high-boiling petroleum residues.

Naphthalene   **Anthracene**   **Phenanthrene**

A **carcinogen** is a compound that causes cancer. The first carcinogens to be identified were a group of polynuclear aromatic hydrocarbons, all of which have at least four aromatic rings. Among them is benzo[a]pyrene, one of the most carcinogenic of the aromatic hydrocarbons. It forms whenever there is incomplete combustion of organic compounds. Benzo[a]pyrene is found, for example, in cigarette smoke, automobile exhaust, and charcoal-broiled meats.

Benzo[a]pyrene causes cancer in the following way. After it is absorbed or ingested, the body converts it into a more soluble compound that can be excreted easily. By a series of enzyme-catalyzed reactions, benzo[a]pyrene is transformed into a diol (two —OH groups) epoxide (a three-membered ring, one atom of which is oxygen).

The diol epoxide attaches to DNA by reacting with one of its amino groups, thereby altering the structure of DNA and producing a cancer-causing mutation.

Benzo[a]pyrene   A diol epoxide

At one time, naphthalene was used as a moth repellent and insecticide in preserving woolens and furs, but its use has decreased owing to the introduction of chlorinated hydrocarbons, such as *p*-dichlorobenzene. In numbering PAHs, carbon atoms common to two or more rings are not numbered because they have no replaceable hydrogens.

Also found in petroleum and coal tar are lesser amounts of the following PAHs.

Pyrene   **Benzo[a]pyrene**   **Coronene**

These compounds can be found in the exhausts of gasoline-powered internal combustion engines (for example, automobile engines) and in cigarette smoke. Benzo[a]pyrene has attracted particular interest because it is a very potent carcinogen (cancer-causing substance) and mutagen.

21.4 | Phenols

A. Structure and Nomenclature

The functional group of a **phenol** is a hydroxyl group bonded directly to a benzene ring. Substituted phenols are named either as derivatives of phenol, as benzenols, or by common names.

Phenol
A compound that contains an —OH bonded to a benzene ring; a benzenol.

Image copyright ultimathule 2010. Used under license from Shutterstock.com

Thymol is a constituent of garden thyme, Thymus vulgaris.

Phenol	**3-Methylphenol** (*m*-Cresol)	**1,2-Benzenediol** (Catechol)	**1,3-Benzenediol** (Resorcinol)	**1,4-Benzenediol** (Hydroquinone)

Phenols are widely distributed in nature. Phenol itself and the isomeric cresols (*o*-, *m*-, and *p*-cresol) are found in coal tar and petroleum. Thymol and vanillin are important constituents of thyme and vanilla beans.

2-Isopropyl-5-methylphenol (**Thymol**)	**4-Hydroxy-3-methoxybenzaldehyde** (**Vanillin**)

Phenol, or carbolic acid as it was once called, is a low-melting solid that is soluble in water. In sufficiently high concentrations, it is harmful to all kinds of cells. In dilute solutions, it has some antiseptic properties and was introduced into the practice of surgery by Joseph Lister who demonstrated his technique of aseptic surgery in the surgical theater of the University of Glasgow School of Medicine in 1865. Phenol's medical use is now limited. It has been replaced by antiseptics that are both more powerful and have fewer undesirable side effects. Among these is hexylresorcinol, which is widely used in nonprescription preparations as a mild antiseptic and disinfectant. Eugenol, which can be isolated from the flower buds (cloves) of *Eugenia aromatica*, is used as a dental antiseptic and analgesic. Urushiol is the main component of the irritating oil of poison ivy.

© blickwinkel/Alamy

West Indian vanilla, Vanilla pompona.

Hexylresorcinol	**Eugenol**	**Urushiol**

B. Acidity of Phenols

Phenols and alcohols both contain a hydroxyl group. Phenols, however, are grouped as a separate class of compounds because their chemical properties are quite different from those of alcohols. One of the most important of these differences is that phenols are significantly more acidic than alcohols. The acid ionization constant of phenol is 10^6 times larger than that of ethanol.

© Cengage Learning/Charles D. Winters

Poison ivy.

$$\text{C}_6\text{H}_5\text{—OH} + \text{H}_2\text{O} \rightleftharpoons \text{C}_6\text{H}_5\text{—O}^- + \text{H}_3\text{O}^+ \qquad K_a = 1.1 \times 10^{-10} \qquad pK_a = 9.95$$

$$\text{CH}_3\text{CH}_2\text{OH} + \text{H}_2\text{O} \rightleftharpoons \text{CH}_3\text{CH}_2\text{O}^- + \text{H}_3\text{O}^+ \qquad K_a = 1.3 \times 10^{-16} \qquad pK_a = 15.9$$

Capsaicin, the pungent principle from the fruit of various species of peppers (from the genus *Capsicum* and the family Solanaceae), was isolated in 1876, and its structure was determined in 1919.

Capsaicin
(from various types of peppers)

The inflammatory properties of capsaicin are well known; as little as one drop in 5 L of water can be detected by the human tongue. We all know of the burning sensation in the mouth and sudden tearing in the eyes caused by a good dose of hot chili peppers. Capsaicin-containing extracts from these flaming foods are also used in sprays to ward off dogs or other animals that might nip at your heels while you are running or cycling.

Ironically, capsaicin is able to cause pain and relieve it as well. Currently, two capsaicin-containing creams, Mioton and Zostrix, are prescribed to treat the burning pain associated with postherpetic neuralgia, a complication of shingles. They are also prescribed for diabetics to relieve persistent foot and leg pain.

The mechanism by which capsaicin relieves these pains is not fully understood. The suggestion has been made that after application, the nerve endings in the area responsible for the transmission of pain remain temporarily numb. Capsaicin remains bound to specific receptor sites on these pain-transmitting neurons, blocking them from further action. Eventually, capsaicin is removed from these receptor sites, but in the meantime, its presence provides needed relief from pain.

Red chili peppers being dried.

© Chuck Pefley/Stone/Getty Images

Another way to compare the relative acid strengths of ethanol and phenol is to look at the hydrogen ion concentration and pH of a 0.1 M aqueous solution of each (Table 21.1). For comparison, the hydrogen ion concentration and pH of 0.1 M HCl are also included.

In aqueous solution, alcohols are neutral substances, and the hydrogen ion concentration of 0.1 M ethanol is the same as that of pure water. A 0.1 M solution of phenol is slightly acidic and has a pH of 5.4. By contrast, 0.1 M HCl, a strong acid (completely ionized in aqueous solution), has a pH of 1.0.

Table 21.1	Relative Acidities of 0.1 M Solutions of Ethanol, Phenol, and HCl		
Acid Ionization Equation		**[H⁺]**	**pH**
$CH_3CH_2OH + H_2O \rightleftharpoons CH_3CH_2O^- + H_3O^+$		1×10^{-7}	7.0
$C_6H_5OH + H_2O \rightleftharpoons C_6H_5O^- + H_3O^+$		3.3×10^{-6}	5.4
$HCl + H_2O \rightleftharpoons Cl^- + H_3O^+$		0.1	1.0

The greater acidity of phenol is a result of the greater stability of the phenoxide ion compared with an alkoxide ion. The negative charge on the phenoxide ion is delocalized by resonance. The two contributing structures on the left place the negative charge on oxygen. The three contributing structures on the right place it on the ortho and para positions of the ring. Taken together, these contributing structures delocalize the negative charge of the phenoxide ion over four atoms. There is no possibility for delocalization of charge in an alkoxide ion.

These two Kekulé structures are equivalent

These three contributing structures delocalize the negative charge onto carbon atoms of the ring

Note that although the charge-delocalization resonance model gives us a way of understanding why phenol is a stronger acid than ethanol, it does not provide us with any quantitative means of predicting just how much stronger an acid it might be. To find out how much stronger one acid is compared with another, we must determine their pK_a values experimentally and compare them.

Ring substituents, particularly halogens and nitro groups, have marked effects on the acidities of phenols by a combination of inductive and resonance effects. Both *m*-cresol and *p*-cresol are weaker acids than phenol itself; *m*-chlorophenol and *p*-chlorophenol are stronger acids than phenol.

| **Phenol** | ***m*-Cresol** | ***p*-Cresol** | ***m*-Chlorophenol** | ***p*-Chlorophenol** |
| pK_a 9.95 | pK_a 10.01 | pK_a 10.17 | pK_a 8.85 | pK_a 9.18 |

The acid-weakening effect of alkyl-substituted phenols can be understood in the following way. The sp^2-hybridized carbon of an aromatic ring is more electronegative than the sp^3-hybridized atom of an alkyl substituent. Alkyl substituents are electron releasing toward the aromatic ring. Because they are electron releasing, they destabilize phenoxide ion-contributing structures and in effect reduce the acidity of substituted phenols.

Polarization of this C—C bond by the electron-releasing inductive effect of the sp^3 carbon of the methyl group destabilizes this contributing structure

The inductive effect of the halogens is opposite to that of alkyl substituents. Because the halogens are more electronegative than carbon, they withdraw electron density from the aromatic ring and stabilize the halophenoxide ion compared to phenoxide ion itself. Fluorine, the most electronegative halogen, has the greatest acid-strengthening effect in halophenols; the effect is less for chlorophenols and still less for bromophenols.

We find the operation of both the inductive and resonance effects in the nitrophenols.

Phenol
pK_a 9.95

m-Nitrophenol
pK_a 8.28

p-Nitrophenol
pK_a 7.15

Both *m*-nitrophenol and *p*-nitrophenol are stronger acids than phenol. The acid-strengthening effect of the nitro group is greater in the para position, even though it is farther away from the —OH group. Part of the acid-strengthening property of the nitro group is the result of its electron-withdrawing inductive effect. In addition, nitro substitution in the ortho or para positions increases acidity because the negative charge of the phenoxide ion is delocalized onto an oxygen of the nitro group as shown in the contributing structure on the right.

Delocalization of negative charge onto oxygen further increases the resonance stabilization of phenoxide ion

The combined inductive and resonance acid-strengthening effects of the nitro group are such that 2,4,6-trinitrophenol (picric acid) is a stronger acid than phosphoric acid or the hydrogen sulfate ion.

2,4,6-Trinitrophenol
(Picric acid)
pK_a 0.38

H_3PO_4

Phosphoric acid
pK_a 2.1

HSO_4^-

Hydrogen sulfate ion
pK_a 1.92

Example 21.6

Arrange these compounds in order of increasing acidity: 2,4-dinitrophenol, phenol, and benzyl alcohol.

Solution

Benzyl alcohol, a primary alcohol, has a pK_a of approximately 16–18 (Section 10.3). The pK_a of phenol is 9.95. Nitro groups are electron withdrawing and increase the acidity of the phenolic —OH group. In order of increasing acidity, they are

Benzyl alcohol
pK_a 16–18

Phenol
pK_a 9.95

2,4-Dinitrophenol
pK_a 3.96

Problem 21.6

Arrange these compounds in order of increasing acidity: 2,4-dichlorophenol, phenol, cyclohexanol.

C. Acid-Base Reactions of Phenols

Phenols are weak acids and react with strong bases, such as NaOH, to form water-soluble salts.

$$\text{C}_6\text{H}_5-\text{OH} + \text{NaOH} \longrightarrow \text{C}_6\text{H}_5-\text{O}^-\text{Na}^+ + \text{H}_2\text{O}$$

Phenol	**Sodium**	**Sodium**	**Water**
pK_a 9.95	**hydroxide**	**phenoxide**	pK_a 15.7
(stronger acid)			(weaker acid)

Most phenols do not react with weaker bases such as sodium bicarbonate; they do not dissolve in aqueous sodium bicarbonate. Here you would do well to review Section 4.4.

Carbonic acid is a stronger acid than phenol, and consequently the equilibrium for the reaction of phenol and bicarbonate ion lies far to the left.

$$\text{C}_6\text{H}_5-\text{OH} + \text{NaHCO}_3 \rightleftharpoons \text{C}_6\text{H}_5-\text{O}^-\text{Na}^+ + \text{H}_2\text{CO}_3$$

Phenol	**Sodium**	**Sodium**	**Carbonic acid**
pK_a 9.95	**bicarbonate**	**phenoxide**	pK_a 6.36
(weaker acid)			(stronger acid)

Phenols do, however, form water-soluble salts with sodium carbonate, a stronger base than sodium bicarbonate.

The fact that phenols are weakly acidic whereas alcohols are neutral provides a very convenient way to separate water-insoluble phenols from water-insoluble alcohols. Suppose that we want to separate 4-methylphenol (*p*-cresol) from cyclohexanol. Each is only slightly soluble in water; therefore, they cannot be separated on the basis of their water solubility. They can be separated, however, on the basis of their differences in acidity. First, the mixture of the two is dissolved in diethyl ether or some other water-immiscible solvent. Next, the ether solution is placed in a separatory funnel and shaken with dilute aqueous NaOH. Under these conditions, 4-methylphenol reacts with NaOH and is converted to a water-soluble phenoxide salt. The upper layer in the separatory funnel is now diethyl ether (density 0.74 g/cm^3) containing only dissolved cyclohexanol. The lower aqueous layer contains the dissolved phenoxide salt. The layers are separated, and removal of the ether (bp 35°C) by distillation leaves pure cyclohexanol (bp 161°C). Acidification of the aqueous phase with 0.1 *M* HCl or other strong acid converts the phenoxide salt to 4-methylphenol, which is more soluble in ether than in water and can be extracted with ether and recovered in pure form. These experimental steps are summarized in Figure 21.14.

D. Preparation of Alkyl-Aryl Ethers

Alkyl-aryl ethers can be prepared from a phenoxide salt and an alkyl halide (the Williamson synthesis, Section 11.4A). They cannot be prepared from an aryl halide and alkoxide salt, however, because aryl halides are quite unreactive under the conditions of Williamson synthesis; they do not undergo nucleophilic displacement by either S$_N$1 or S$_N$2 mechanisms.

Alkyl-aryl ethers are often synthesized by carefully controlling solubility. Both the alkyl halide and phenol are dissolved in dichloromethane; then, the solution is mixed with an aqueous solution of sodium hydroxide. Phenol, a poor nucleophile, reacts with sodium hydroxide in the aqueous phase to form the phenoxide ion, a good nucleophile. Alkyl-aryl ethers can be synthesized by treating the sodium salt of a phenol with an alkyl halide. The following example illustrates the

Figure 21.14
Experimental steps to separate water-insoluble phenols from water-insoluble alcohols.

Williamson synthesis of alkyl-aryl ethers. The $Bu_4N^+Br^-$ is used to facilitate reaction between the polar phenoxide salt and the hydrophobic alkyl halide in the mixed solvent.

Phenol   **3-Chloropropene**
(Allyl chloride)

Phenyl 2-propenyl ether
(Allyl phenyl ether)

The synthesis of anisole illustrates the use of dimethyl sulfate as a methylating agent.

Phenol   **Dimethyl sulfate**

Methyl phenyl ether
(Anisole)

An alkyl-aryl ether, ArOR, is cleaved by hydrohalic acids, HX, to form an alkyl halide and a phenol.

2-Phenoxypropane
(Isopropyl phenyl ether)

Phenol   **2-Iodopropane**
(Isopropyl iodide)

This cleavage illustrates the fact that nucleophilic substitution is not likely to occur at an aromatic carbon, and that phenols, unlike alcohols, are not converted to aryl halides by treatment with concentrated HCl, HBr, or HI.

E. Kolbe Carboxylation: Synthesis of Salicylic Acid

Phenoxide ions react with carbon dioxide to give a carboxylic acid salt as shown by the industrial synthesis of salicylic acid, the starting material for the production of aspirin (Section 18.5B). Phenol is dissolved in aqueous NaOH, and this solution is then saturated with CO_2 under pressure to give sodium salicylate.

| **Phenol** | **Sodium phenoxide** | **Sodium salicylate** | **Salicylic acid** |

This process is referred to as high-pressure Kolbe carboxylation of sodium phenoxide. Upon acidification of the alkaline solution, salicylic acid is isolated as a solid, mp 157–159°C.

The importance of salicylic acid in industrial organic chemistry is demonstrated by the fact that over 6×10^6 kg of aspirin are synthesized in the United States each year.

Mechanism   Kolbe Carboxylation of Phenol

Step 1: **Make a new bond between a nucleophile (arene ring) and an electrophile.** The phenoxide ion reacts like an enolate anion; it is a strong nucleophile. Nucleophilic attack of the phenoxide anion on a carbonyl group of carbon dioxide gives a substituted cyclohexadienone intermediate.

| Sodium phenoxide | | Cyclohexadienone intermediate |

Step 2: **Keto-enol tautomerism.**   Keto-enol tautomerism of the cyclohexadienone intermediate gives the product salicylate anion. Note that in this case, the enol, owing to its aromatic character, is the more stable of the two tautomers.

Salicylate anion

F. Oxidation to Quinones

Because of the presence of the electron-donating —OH group on the ring, phenols are susceptible to oxidation by a variety of strong oxidizing agents. For example, oxidation of phenol itself by potassium dichromate gives 1,4-benzoquinone (*p*-quinone).

Phenol

1,4-Benzoquinone
(*p*-Quinone)

By definition, a quinone is a cyclohexadienedione. Those with carbonyl groups ortho to each other are called *o*-quinones; those with carbonyl groups para to each other are called *p*-quinones.

Quinones can also be obtained by oxidation of 1,2-benzenediol (catechol) or 1,4-benzenediol (hydroquinone).

1,2-Benzenediol
(**Catechol**)

1,2-Benzoquinone
(*o*-Quinone)

1,4-Benzenediol
(**Hydroquinone**)

1,4-Benzoquinone
(*p*-Quinone)

Perhaps the most important chemical property of quinones is that they are readily reduced to benzenediols. For example, *p*-quinone is readily reduced to hydroquinone by sodium dithionite in neutral or alkaline solution. There are many other ways to carry out this reduction. The point is that it can be done very easily, as can the corresponding oxidation of a hydroquinone.

1,4-Benzoquinone
(*p*-Quinone)

1,4-Benzenediol
(**Hydroquinone**)

There are many examples in both chemistry and biology in which the reversible oxidation/reduction of hydroquinones or quinones is important. One such example is coenzyme Q, alternatively known as ubiquinone. The name of this important biomolecule is derived from the Latin *ubique* (everywhere) + quinone.

Coenzyme Q
(oxidized form)

Coenzyme Q
(reduced form)

Coenzyme Q, a carrier of electrons in the respiratory chain, contains a long hydrocarbon chain of between 6 and 10 isoprene units that serves to anchor it firmly in the nonpolar environment of the mitochondrial inner membrane. The oxidized form of coenzyme Q is a two-electron oxidizing agent. In subsequent steps of the respiratory chain, the reduced form of coenzyme Q transfers the two electrons to another link until they are eventually delivered to a molecule of oxygen, which is in turn reduced to water.

Figure 21.15

Vitamin K_2 with seven isoprene units.

Vitamin K_2

Another quinone important in biological systems is vitamin K_2. This compound was discovered in 1935 as a result of a study of newly hatched chicks with a fatal disease in which their blood was slow to clot. It was later discovered that the delayed clotting time of blood was caused by a deficiency of prothrombin. We now know that a prothrombin deficiency is, in turn, caused by a deficiency in vitamin K_2, which is essential to the synthesis of prothrombin in the liver. The natural form of vitamin K_2 has a chain of five to eight isoprene units bonded to a 1,4-naphthoquinone ring. Figure 21.15 shows seven isoprene units in the side chain.

The natural vitamins of the K family have for the most part been replaced by synthetic preparations in food supplements. Menadione, one such synthetic material with vitamin K activity, has only hydrogen in place of the long alkyl side chain. Menadione is prepared by chromic acid oxidation of 2-methylnaphthalene under mild conditions.

2-Methylnaphthalene   **2-Methyl-1,4-naphthoquinone**
                                          (Menadione)

A commercial process that uses a quinone is black-and-white photography. Black-and-white film is coated with an emulsion containing silver bromide or silver iodide crystals, which become activated by exposure to light. The activated silver ions are reduced in the developing stage to metallic silver by hydroquinone, which at the same time is oxidized to quinone. Following is an equation showing the relationship between these species.

1,4-Benzenediol   **1,4-Benzoquinone**
(Hydroquinone)   **(*p*-Quinone)**

All silver halide not activated by light and then reduced by interaction with hydroquinone is removed in the fixing process, and the result is a black image (a negative) left by deposited metallic silver where the film has been struck by light. Other compounds are now used to reduce "light-activated" silver bromide, but the result is the same—a deposit of metallic silver in response to exposure of film to light.

21.5 | Reactions at a Benzylic Position

In this section, we study two reactions of substituted aromatic hydrocarbons that occur preferentially at the **benzylic position**.

Benzylic position
An sp^3 hybridized carbon bonded to a benzene ring.

Benzyl group

Reactions involving alkyl side chains of aromatic compounds occur preferentially at the benzylic position for two reasons. First, the benzene ring is especially resistant to reaction with many of the reagents that normally attack alkanes. Second, benzylic cations and benzylic radicals are easily formed because of resonance stabilization of these intermediates. A benzylic cation or radical is a hybrid of five contributing structures: two Kekulé structures and three that delocalize the positive charge (or the lone electron) onto carbons of the aromatic ring. Following are contributing structures for a benzylic cation. Similar contributing structures can be written for a benzylic radical and anion. Benzylic contributing structures are closely analogous to allylic structures in stabilizing cations, radicals, and anions.

The benzyl cation as a hybrid of five contributing structures

A. Oxidation

Benzene is unaffected by strong oxidizing agents, such as H_2CrO_4 and $KMnO_4$. However, when toluene is treated with these oxidizing agents under quite vigorous conditions, the side-chain methyl group is oxidized to a carboxyl group to give benzoic acid.

Toluene   **Benzoic acid**

Halogen and nitro substituents on an aromatic ring are unaffected by these oxidations. 2-Chloro-4-nitrotoluene, for example, is oxidized to 2-chloro-4-nitrobenzoic acid.

2-Chloro-4-nitrotoluene   **2-Chloro-4-nitrobenzoic acid**

Ethyl and isopropyl side chains are also oxidized to carboxyl groups. The side chain of *tert*-butylbenzene, however, is not oxidized.

tert-**Butylbenzene**

From these observations, we conclude that if a benzylic hydrogen exists, then the benzylic carbon is oxidized to a carboxyl group, and all other carbons of the side chain are removed as CO_2. If no benzylic hydrogen exists, as in the case of *tert*-butylbenzene, no oxidation of the side chain occurs.

If more than one alkyl side chain exists, each is oxidized to —COOH. Oxidation of *m*-xylene gives 1,3-benzenedicarboxylic acid, more commonly named isophthalic acid.

m-**Xylene**   **1,3-Benzenedicarboxylic acid**
     (Isophthalic acid)

Example 21.7

Draw structural formulas for the product of vigorous oxidation of 1,4-dimethylbenzene (*p*-xylene) by H_2CrO_4.

Solution

Both alkyl groups are oxidized to —COOH groups. The product is terephthalic acid, one of two monomers required for the synthesis of Dacron polyester and Mylar (Section 29.5B).

| 1,4-Dimethylbenzene | 1,4-Benzenedicarboxylic acid |
| (*p*-Xylene) | (Terephthalic acid) |

Problem 21.7

Predict the products resulting from vigorous oxidation of each compound by H_2CrO_4.

(a)   (b)

Studying these side-chain oxidations and formulating mechanisms for them is difficult. Available evidence, however, supports the formation of unstable intermediates that are either benzyl radicals or benzylic carbocations.

Naphthalene is oxidized to phthalic acid by molecular oxygen in the presence of a vanadium(V) oxide (vanadium pentoxide) catalyst.

| Naphthalene | 1,2-Benzenedicarboxylic acid |
| | (Phthalic acid) |

This conversion, which is the basis for an industrial synthesis of this aromatic dicarboxylic acid, illustrates the ease of oxidation of condensed benzene rings compared with benzene itself.

B. Halogenation

Reaction of toluene with chlorine in the presence of heat or light results in formation of chloromethylbenzene and HCl.

| Toluene | Chloromethylbenzene |
| | (Benzyl chloride) |

Bromination is easily accomplished by using *N*-bromosuccinimide (NBS) in the presence of a peroxide catalyst.

Toluene **N-Bromosuccinimide** **Bromomethylbenzene** **Succinimide**
 (NBS) **(Benzyl bromide)**

Halogenation of a larger alkyl side chain is highly regioselective, as illustrated by the halogenation of ethylbenzene. When treated with NBS, the only monobromo organic product formed is 1-bromo-1-phenylethane. This regioselectivity is dictated by the resonance stabilization of the benzylic radical intermediate. The mechanism of radical bromination at a benzylic position is identical to that for allylic bromination (Section 8.6A).

Ethylbenzene **1-Bromo-1-phenylethane**
 (racemic)

When ethylbenzene is treated with chlorine under radical reaction conditions, two products are formed in the ratio of 9:1.

Ethylbenzene **1-Chloro-1-phenylethane** **1-Chloro-2-phenylethane**
 (90%) (10%)
 (racemic)

Thus, chlorination of alkyl side chains is also regioselective but not to the same high degree as bromination. Recall that we observed this same pattern in the regioselectivities of bromination and chlorination of alkanes (Section 8.4A).

Combining the information on product distribution for bromination and chlorination of hydrocarbons, we conclude that the order of stability of radicals is

$$\text{methyl} < 1° < 2° < 3° < \text{allylic} \cong \text{benzylic}$$

This order reflects the C—H bond dissociation enthalpies (BDE) for formation of these radicals (Appendix 3).

C. Hydrogenolysis of Benzyl Ethers

Among ethers, benzylic ethers are unique in that they are cleaved under the conditions of catalytic hydrogenation as illustrated by the **hydrogenolysis** of benzyl hexyl ether. In this illustration, the benzyl group is converted to toluene, and the alkyl group is converted to an alcohol.

Hydrogenolysis
Cleavage of a single bond by H_2, most commonly accomplished by treating a compound with H_2 in the presence of a transition metal catalyst.

This bond
is cleaved.

Benzyl hexyl ether **1-Hexanol** **Toluene**

Hydrogenolysis is the cleavage of a single bond by H_2. In the hydrogenolysis of a benzylic ether, the single bond between the benzylic carbon and its attached oxygen is cleaved and replaced by a carbon-hydrogen bond.

Benzyl ethers are formed by treatment of an alcohol or phenol with benzyl chloride in the presence of a base such as triethylamine or pyridine. The particular value of benzylic ethers is that they can serve as protecting groups for the —OH groups of alcohols and phenols. Suppose, for example, we want to treat 2-allylphenol with diborane followed by hydrogen peroxide to bring about non-Markovnikov hydration of the carbon-carbon double bond. This scheme will not give the desired result because the phenolic —OH group is sufficiently acidic to react with BH_3 and destroy it. The desired product can be prepared, however, by protection of the phenolic —OH group as the benzylic ether, hydroboration/oxidation of the carbon-carbon double bond, and hydrogenolysis of the benzylic ether.

2-(2-Propenyl)phenol
(2-Allylphenol)

2-(3-Hydroxypropyl)phenol

Summary

Section 21.1 The Structure of Benzene

- Benzene and its derivatives are classified as **aromatic**.
  - Aromatic molecules display a remarkable stability that makes them unreactive toward reagents that attack other unsaturated species such as alkenes and alkynes.
  - **Aromaticity** is the term used to describe this special stability of benzene and its derivatives, and the term **arene** refers to aromatic hydrocarbons.
- August Kekulé proposed that benzene is composed of six carbon atoms in a ring, with one hydrogen atom bonded to each carbon.
  - The six carbon atoms of the ring are equivalent, and the carbon-carbon bond lengths are all intermediate between a single and double bond.
  - It is not accurate to think of benzene as having alternating single and double bonds that are static, because this would predict alternating longer and shorter carbon-carbon bonds.
- According to molecular orbital theory, benzene's structure is described as follows:
  - Each carbon atom of the ring is sp^2 hybridized.
  - Each carbon atom of the ring makes σ bonds by sp^2-sp^2 overlap with the two adjacent carbon atoms and sp^2-$1s$ overlap with a hydrogen atom.
  - Each carbon atom also has a single unhybridized $2p$ orbital containing one electron.
  - The six $2p$ orbitals overlap to form six π molecular orbitals.
    - These molecular orbitals are arranged in a 1:2:2:1 pattern with respect to energy.

Problems: 21.12–21.14

- The six π electrons fill the three π bonding molecular orbitals, all of which are at lower energy than the six isolated $2p$ orbitals, explaining why the benzene π system is so unreactive.
  - The lowest-lying filled molecular orbital has two torus-shaped lobes, one above and one below the plane of the ring, emphasizing the delocalized nature of the π system.
  - The other two filled molecular orbitals have one node apiece, underscoring that the bond order between carbon atoms is intermediate between a double and single bond.
- Benzene is best represented as a resonance hybrid composed of two resonance forms in which the locations of the double bonds are reversed. For simplicity, benzene is often represented as a single contributing structure or as a hexagon with a circle drawn on the inside.
- **Resonance energy** is the difference in energy between a resonance hybrid and the most stable hypothetical contributing structure in which electron density is localized on particular atoms and bonds.
  - The resonance energy for benzene is large, namely 36.0 kcal/mol, meaning that the π system of benzene is extremely stable, and less reactive than would be expected for an alkene under conditions like catalytic hydrogenation.

Section 21.2 The Concept of Aromaticity

- Not all cyclic hydrocarbons with alternating double bonds possess the aromatic characteristics of benzene. Erich Hückel used molecular orbital calculations to define a set of requirements for aromaticity.
  - The molecule must be cyclic.
  - There must be a $2p$ orbital on each atom of the ring (there cannot be any sp^3 hybridized atoms in the ring).
  - The molecule must be planar, or nearly so, to allow overlap of the $2p$ orbitals.
  - There must be $(4n + 2)$ π electrons in the aromatic π system, where n is a positive integer $(0, 1, 2, 3, 4, 5, \ldots)$ for a total of $2, 6, 10, 14, \ldots \pi$ electrons.
- Some hydrocarbons are considered antiaromatic, because they are much *less* stable (more reactive) than an acyclic analog with the same number of π electrons.
  - An antiaromatic hydrocarbon has $4n$, not $(4n + 2)$, π electrons.
  - Antiaromatic hydrocarbons must also be cyclic, have a $2p$ orbital on each atom of the ring, and must be planar to display full antiaromatic properties.
  - This instability of antiaromatic hydrocarbons, such as cyclobutadiene, can be explained by using molecular orbital theory.
    - For cyclobutadiene, the four $2p$ orbitals of the π system form four molecular orbitals in a 1:2:1 pattern.
    - The four π electrons fill these orbitals to give one filled bonding π molecular orbital and two half-filled degenerate nonbonding molecular orbitals.
    - The presence of the two unpaired electrons makes cyclobutadiene reactive and unstable relative to aromatic hydrocarbons.
  - To avoid antiaromaticiy, larger potentially antiaromatic structures like cyclooctatetraene adopt a non-planar geometry with alternating double and single bonds. As a result, two different carbon-carbon bond lengths are observed corresponding to single and double bonds, respectively.
- To predict the pattern of molecular orbitals found on a molecular orbital energy diagram, it is helpful to use the inscribed polygon method **(Frost circles)**.
  - The shape of the polygon being analyzed (for example a hexagon for benzene) is drawn in a ring with one vertex down, and the relative energies of the molecular orbitals are indicated by the vertices of the polygon that touch the circle.
  - A horizontal line is drawn through the center of the figure.
  - Bonding molecular orbitals are below the line, nonbonding molecular orbitals (if any) are on the line, and antibonding molecular orbitals are above the line.

Problems: 21.1–21.4, 21.15–21.18

- An **annulene** is a planar, cyclic hydrocarbon that contains continuously overlapping $2p$ orbitals, and is named by adding the number of atoms in the ring in brackets in front of the word *annulene*.
  - Cyclobutadiene and benzene are annulenes, namely [4]annulene and [6]annulene, respectively.
  - Annulenes can be much larger, such as [14]annulene and [18]annulene, which are both aromatic because they are planar and have a Hückel number of $(4n + 2)$ π electrons.
  - [10]Annulene is not aromatic because the relatively small ring cannot adopt a planar geometry.
- A **heterocyclic compound** is one that contains more than one kind of atom in a ring.
  - Certain heterocycles can be aromatic if the Hückel criteria are met.
  - Nature is filled with aromatic heterocycles such as indoles, purines, and pyrimidines.
- An important parameter to keep track of in aromatic heterocycles is whether lone pairs of electrons are part of the aromatic π system or not.
  - In pyridine (C_5H_5N), the lone pair of electrons on nitrogen is in an sp^2 hybrid orbital and is perpendicular to the six $2p$ orbitals of the aromatic 6 π electron system. This lone pair is not part of the aromatic π system, and is free to take part in interactions with other species.
  - In pyrrole (C_4H_5N), the lone pair of electrons on nitrogen is in a $2p$ orbital and is part of the π system to allow for a total of six π electrons and aromaticity. This lone pair of electrons is not as available to take part in interactions with other species.
- Because charged ring systems can satisfy the Hückel criteria for aromaticity, they are highly stabilized compared to other nonaromatic cations or anions. Examples include the cyclopropenyl cation, cycloheptatrienyl cation, and cyclopentadienyl anion.

Section 21.3 Nomenclature

- The IUPAC system retains certain common names for several of the simpler benzene derivatives including toluene, cumene, styrene, xylene, phenol, aniline, benzoic acid, and anisole.
  - These common names are used as the parent name if their characteristic functional groups are present on a benzene derivative.
- In molecules with other functional groups, the benzene ring is named as a substituent on a parent chain.
  - The C_6H_5 group is given the name **phenyl** and abbreviated Ph.
  - The $C_6H_5CH_2$— group has characteristic reactivity, so it is given the name **benzyl** and is abbreviated Bn.
- For benzene rings with two substituents, the three possible constitutional isomers are named **ortho** (1,2-substitution), **meta** (1,3-substitution), and **para** (1,4-substitution), which are abbreviated as *o*, *m*, and *p*, respectively.
  - It is also acceptable to name these species with numbers as locators (such as 1,2- or 1,3-).
  - When one of the substituents has a special name (if NH_2 is present, the molecule is an aniline, etc.), the molecule is named after that parent molecule and the key group is assigned the number 1.
  - If none of the groups impart a special name, the substituents are listed in alphabetical order followed by the word *benzene*. For example, 1-chloro-4-ethylbenzene and *p*-chloroethylbenzene are acceptable names for the same molecule.
- **Polynuclear aromatic hydrocarbons (PAHs)** contain more than one benzene ring, each pair of which shares two carbon atoms.
  - Naphthalene contains two benzene rings fused together and anthracene contains three benzene rings fused together in a linear fashion. Other common PAHs include phenanthrene, pyrene, coronene, and benzo[a]pyrene.

Problems: 21.15, 21.8–21.11, 21.19–21.31

Section 21.4 Phenols

- The characteristic feature of **phenols** is a hydroxyl group bonded to a benzene ring.
- Phenols (pK_a around 10) are more acidic than simple alcohols, because the negative charge of the phenoxide anion is highly delocalized into the aromatic ring as indicated by resonance contributing structures.
  - Substituents on the ring that lead to a further stabilization of the phenoxide anion increase acidity of a phenol, while substituents that destabilize the phenoxide anion decrease acidity of a phenol.
  - According to the inductive effect, electron-withdrawing groups (more electronegative than sp^2 hybridized carbon) such as halogens stabilize a phenoxide anion by absorbing some of the negative charge, thus making a phenol more acidic, while electron-releasing groups (less electronegative than sp^2 hybridized carbon) such as alkyl groups destabilize a phenoxide anion by pushing even more electron density into the ring, making a phenol less acidic.
  - According to the resonance effect, certain groups, such as nitro groups, make phenols more acidic (especially at the ortho and para positions) because they can stabilize phenoxide anions through additional resonance delocalization of the negative charge.

Problems: 21.6, 21.7, 21.32–21.40

- Phenols react with strong bases to create water-soluble salts, a procedure that is useful for their isolation from mixtures.
- Phenoxides can be used in a Williamson ether synthesis by reaction with alkyl halides (S_N2 process) to create alkyl-aryl ethers.
- In the Kolbe reaction, phenoxide ion reacts with CO_2 to produce salicylic acid through a mechanism in which the phenoxide ion reacts like an enolate anion attacking the electrophilic carbon of CO_2, followed by loss of a proton and keto-enol tautomerization to give the salicylate anion.
- Phenols, especially benzenediols (**hydroquinones**), are easily oxidized to **quinones**, and quinones are easily reduced back to benzenediols, forming a reversible oxidation/reduction couple that is the basis for important biological processes.

Section 21.5 Reactions at a Benzylic Position

- There are several characteristic reactions that occur at the benzylic position of aromatic hydrocarbons because benzylic cation and radical intermediates are stabilized through delocalization into the aromatic ring.
- Compounds with at least one benzylic hydrogen react with $KMnO_4$ in aqueous base or H_2CrO_4 ($Na_2Cr_2O_7$ in aqueous sulfuric acid) to produce benzoic acids. Other groups bonded to the benzylic carbon atom are removed in the process.
- Benzylic hydrogens can be replaced by bromine or chlorine in the presence of light or heat.
  - In compounds such as toluene, more than one of the benzylic hydrogens can be replaced when excess halogen in used.
  - In molecules with alkyl groups larger than methyl bonded to the benzylic carbon, the reaction is selective for replacement of the benzylic hydrogens because the benzylic radical is more stable than the other possible radicals.
  - Bromine is more selective than chlorine in these reactions, and NBS is often used as the bromination reagent.

Problems: 21.41–21.62

- Benzylic ethers are unique among ethers in that they can be cleaved using catalytic hydrogenation (H_2 and Pd) to give toluene and an alcohol, a process referred to as **hydrogenolysis**.
  - Hydrogenolysis of benzyl ethers makes them useful as an alcohol protecting group.

⬥WL **Questions and problems for this chapter are available in OWL. Use the OWL access card that came with this textbook to access assigned questions and problems for this chapter.**

1. **Acidity of Phenols (Section 21.4B)** Phenols are weak acids, pK_a approximately 10. Ring substituents may increase or decrease acidity by a combination of resonance and inductive effects.

$$\text{C}_6\text{H}_5-\text{OH} + H_2O \rightleftharpoons \text{C}_6\text{H}_5-\text{O}^- + H_3O^+ \qquad K_a = 1.1 \times 10^{-10} \qquad pK_a = 9.95$$

2. **Reaction of Phenols with Strong Bases (Section 21.4C)** Water-insoluble phenols react quantitatively with strong bases to form water-soluble salts.

$$\text{C}_6\text{H}_5-\text{OH} + NaOH \longrightarrow \text{C}_6\text{H}_5-\text{O}^-Na^+ + H_2O$$

Phenol	**Sodium hydroxide**	**Sodium phenoxide**	**Water**
pK_a 9.95			pK_a 15.7
(stronger acid)			(weaker acid)

3. **Kolbe Synthesis: Carboxylation of Phenols (Section 21.4E)** Nucleophilic addition of a phenoxide ion to carbon dioxide gives a substituted cyclohexadienone, which then undergoes keto-enol tautomerism to regenerate the aromatic ring.

4. **Oxidation of Phenols to Quinones (Section 21.4F)** Oxidation by H_2CrO_4 gives 1,2-quinones (*o*-quinones) or 1,4-quinones (*p*-quinones), depending on the structure of the particular phenol.

Phenol **1,4-Benzoquinone** (*p*-**Quinone**)

5. **Oxidation at a Benzylic Position (Section 21.5A)** A benzylic carbon bonded to at least one hydrogen is oxidized to a carboxyl group.

6. **Halogenation at a Benzylic Position (Section 21.5B)** Halogenation is regioselective for a benzylic position and occurs by a radical chain mechanism. Bromination shows a higher regioselectivity for a benzylic position than chlorination. The reaction occurs via a radical chain mechanism initiated when the X_2 is converted to two $X\cdot$ radicals, then an

X· radical abstracts the benzylic hydrogen to create a resonance-stabilized benzylic radical that reacts with another molecule of X₂ to give the halogenated product and a new X· radical that continues the chain reaction. NBS can be used as the source of Br₂.

(racemic)

7. Hydrogenolysis of Benzylic Ethers (Section 21.5C) Benzylic ethers are cleaved under the conditions of catalytic hydrogenation.

Chapter 22

Reactions of Benzene and Its Derivatives

2,6-Di-*tert*-butyl-4-methylphenol, alternatively known as butylated hydroxytoluene or BHT (see Problem 22.23) is often used as an antioxidant to retard spoilage. Inset: a model of BHT.

© Richard Levine/Alamy

By far the most characteristic reaction of aromatic compounds is substitution at a ring carbon. In this reaction, one of the ring hydrogens is replaced by another atom or group of atoms. Some groups that can be introduced directly on the ring are the halogens, the nitro ($-NO_2$) group, the sulfonic acid ($-SO_3H$) group, alkyl ($-R$) groups, and acyl ($RCO-$) groups. Each of these substitution reactions is represented in the following equations.

Halogenation:

Chlorobenzene

Nitration:

Nitrobenzene

WL Online homework for this chapter may be assigned in OWL for Organic Chemistry.

728

Sulfonation:

$$\text{C}_6\text{H}_5\text{—H} + \text{SO}_3 \xrightarrow{\text{H}_2\text{SO}_4} \text{C}_6\text{H}_5\text{—SO}_3\text{H}$$

Benzenesulfonic acid

Alkylation:

$$\text{C}_6\text{H}_5\text{—H} + \text{RX} \xrightarrow{\text{AlX}_3} \text{C}_6\text{H}_5\text{—R} + \text{HX}$$

An alkylbenzene

Acylation:

$$\text{C}_6\text{H}_5\text{—H} + \underset{\underset{\text{RCX}}{\overset{\overset{\text{O}}{\|}}{}}{} \xrightarrow{\text{AlX}_3} \text{C}_6\text{H}_5\text{—}\underset{\underset{\text{CR}}{\overset{\overset{\text{O}}{\|}}{}}{} + \text{HX}$$

An acylbenzene

We will take these reactions one at a time and examine their common mechanistic theme.

22.1 | Electrophilic Aromatic Substitution

In an **electrophilic aromatic substitution**, a hydrogen atom of an aromatic ring is replaced by an electrophile, E^+.

$$\text{C}_6\text{H}_5\text{—H} + E^+ \longrightarrow \text{C}_6\text{H}_5\text{—E} + H^+$$

We study several common types of electrophiles, how each is generated, and the mechanism by which it replaces hydrogen on an aromatic ring.

Electrophilic aromatic substitution A reaction in which there is substitution of an electrophile, E^+, for a hydrogen on an aromatic ring.

A. Chlorination and Bromination

Chlorine alone does not react with benzene, in contrast to its instantaneous addition to cyclohexene. However, in the presence of a Lewis acid catalyst, such as ferric chloride or aluminum chloride (Section 4.7), benzene reacts with chlorine to give chlorobenzene and HCl. As shown in the following mechanism, this reaction involves a series of Lewis acid/base reactions.

Mechanism Electrophilic Aromatic Substitution—Chlorination

Step 1: **Lewis acid-Lewis base reaction.** Reaction between chlorine and the Lewis acid catalyst gives a molecular complex with a positive charge on chlorine and a negative charge on iron. Redistribution of electrons in this complex generates a chloronium ion, Cl^+, a very strong electrophile, as part of an ion pair.

$$\overset{..}{\underset{..}{\text{Cl}}}\text{—}\overset{..}{\underset{..}{\text{Cl}}}\!: \; + \; \underset{\underset{\text{Cl}}{|}}{\overset{\overset{\text{Cl}}{|}}{\text{Fe}}}\text{—Cl} \; \rightleftharpoons \; \overset{..}{\underset{..}{\text{Cl}}}\text{—}\overset{+}{\underset{..}{\text{Cl}}}\text{—}\underset{\underset{\text{Cl}}{|}}{\overset{\overset{\text{Cl}}{|}}{\text{Fe}}}\text{—Cl} \; \rightleftharpoons \; \overset{..}{\underset{..}{\text{Cl}}}{}^+ \; \text{FeCl}_4^-$$

| **Chlorine** (a Lewis base) | **Ferric chloride** (a Lewis acid) | A molecular complex with a positive charge on chlorine and a negative charge on iron | An ion pair containing the chloronium ion |

Cation Intermediate

Step 2: Make a new bond between a nucleophile (arene ring) and an electrophile. Attack of the chloronium ion (a strong electrophile) on the π system (a weak nucleophile) of the aromatic ring gives a resonance-stabilized cation intermediate, here represented as a hybrid of three contributing structures. Notice that the positive charge is located primarily at the ortho and para positions of the resonance-stabilized cation intermediate. This distribution of positive charge is visible in the electrostatic potential surface of the cation intermediate, as the dark blue color at the ortho and para positions.

Step 3: Take a proton away. Proton transfer from the cation intermediate to $FeCl_4^-$ forms HCl, regenerates the Lewis acid catalyst, and gives chlorobenzene.

Treating benzene with bromine in the presence of ferric chloride or aluminum chloride gives bromobenzene and HBr. The mechanism of this reaction is the same as that for the chlorination of benzene.

We can write the following general two-step mechanism for electrophilic aromatic substitution. The first and rate-determining step is attack of the strong electrophile, E^+, by the weakly nucleophilic π electrons of the aromatic ring (electrophilic aromatic addition) to give a resonance-stabilized cation intermediate. The second and faster step, loss of H^+ from the cation intermediate, regenerates aromaticity in the ring and gives the product.

Step 1:

Step 2:

The major difference between addition of halogen to an alkene and halogen substitution on an aromatic ring centers on the fate of the cationic intermediate formed in the first step of each reaction. Recall from Section 6.3D that addition of chlorine or bromine to an alkene is a two-step process, the first and slower step of which is formation of a bridged halonium ion intermediate. This cationic intermediate then reacts with chloride or bromide ion to complete the addition. With aromatic compounds, the cationic intermediate instead loses H^+ to regenerate the aromatic ring and regain the large resonance stabilization. No such resonance stabilization is regained in the case of an alkene. The energy diagram in Figure 22.1 shows both addition and substitution reactions of benzene. Addition causes loss of the aromatic resonance energy and is disfavored except under extreme circumstances.

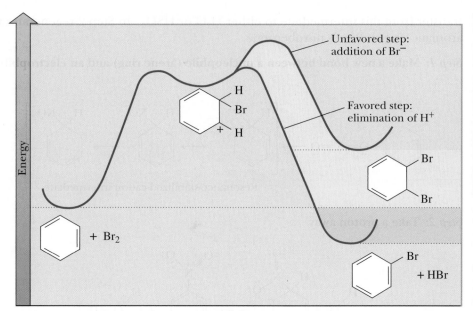

Figure 22.1
Energy diagram for the reaction of benzene with bromine. Formation of the addition product results in loss of the resonance stabilization of the aromatic ring. Formation of a substitution product regenerates the resonance-stabilized aromatic ring.

Reaction coordinate

B. Nitration and Sulfonation

The sequence of steps for nitration and sulfonation of benzene is similar to that for chlorination and bromination. For nitration, the **nitronium ion**, NO_2^+, a very strong electrophile, is generated by the reaction between nitric acid and sulfuric acid.

Mechanism   Formation of the Nitronium Ion

Step 1: **Add a proton.**   Proton transfer from sulfuric acid to the OH group of nitric acid gives the conjugate acid of nitric acid.

| Sulfuric acid | Nitric acid | Conjugate acid of nitric acid |

Step 2: **Break a bond to give stable molecules or ions.**   Loss of water from this conjugate acid gives the nitronium ion, a very strong electrophile.

Nitronium ion

Example 22.1

Write a stepwise mechanism for the nitration of benzene.

Solution

The nitronium ion (a strong electrophile) is attacked by the benzene ring (a weak nucleophile) in Step 1 to give a resonance-stabilized cation intermediate. Proton transfer from this intermediate to either H_2O or HSO_4^- in Step 2 regenerates the aromatic ring and gives nitrobenzene.

Step 1: **Make a new bond between a nucleophile (arene ring) and an electrophile**

Resonance-stabilized cation intermediate

Step 2: **Take a proton away**

An important feature of nitration is that the resulting nitro group can be reduced to a primary amino group, —NH_2, by hydrogenation in the presence of a transition metal catalyst such as nickel, palladium, or platinum under fairly mild conditions.

4-Nitrobenzoic acid **4-Aminobenzoic acid**

As illustrated by this example, neither a —COOH nor an aromatic ring is reduced under these conditions. Catalytic reduction of a nitro group has the potential disadvantage, however, that other susceptible groups such as carbon-carbon double bonds and aldehyde and ketone carbonyl groups may also be reduced.

Alternatively, a nitro group can be reduced to a primary amino group by a metal in aqueous acid. The most commonly used metal-reducing agents are iron, zinc, and tin in dilute HCl. The reductant is electrons from the metal. When reduced with a metal and hydrochloric acid, the amine is obtained as a salt, which is then treated with strong base to liberate the free amine.

2,4-Dinitrotoluene **4-Methyl-1,3-benzenediamine (2,4-Diaminotoluene)**

The reduction of a nitro group to an amino group is important because the amino group cannot be substituted directly onto an aromatic ring; it must be done by this indirect method.

Sulfonation of benzene is carried out using concentrated sulfuric acid containing dissolved sulfur trioxide (fuming sulfuric acid). The electrophile is either SO_3 or HSO_3^+, depending on experimental conditions.

In the following equation, the sulfonating agent is shown as sulfur trioxide.

Benzene   **Benzenesulfonic acid**

Problem 22.1

Write the stepwise mechanism for sulfonation of benzene by hot, concentrated sulfuric acid. In this reaction, the electrophile is SO_3 formed as shown in the following equation.

$$H_2SO_4 \rightleftharpoons SO_3 + H_2O$$

C. Friedel-Crafts Alkylation and Acylation

Alkylation of aromatic hydrocarbons was discovered in 1877 by the French chemist Charles Friedel and a visiting American chemist, James Crafts. They discovered that mixing benzene, an alkyl halide, and $AlCl_3$ results in the formation of an alkylbenzene and HX. **Friedel-Crafts alkylation**, which is among the most important methods for forming new carbon-carbon bonds to aromatic rings, is illustrated here by the reaction of benzene with 2-chloropropane in the presence of aluminum chloride.

Friedel-Crafts reaction
An electrophilic aromatic substitution in which a hydrogen of an aromatic ring is replaced by an alkyl or acyl group.

Benzene   **2-Chloropropane**   **Cumene**
       (Isopropyl chloride)   **(Isopropylbenzene)**

The mechanism for Friedel-Crafts alkylation, like that for halogenation, nitration, and sulfonation, involves the attack of the aromatic ring onto a strong electrophile, in this case a carbocation formed by reaction between the alkyl halide and the Lewis acid catalyst.

Mechanism   Friedel-Crafts Alkylation

Step 1: **Lewis acid-Lewis base reaction.** The alkyl halide (a Lewis base) and aluminum chloride (a Lewis acid) form a complex in which aluminum has a negative charge and the halogen of the alkyl halide has a positive charge. The alkyl group can also be written as a carbocation. It is unlikely, however, that a free carbocation is actually formed, especially in the case of the relatively unstable primary and secondary carbocations. Nonetheless, we often represent the reactive intermediate as a carbocation to simplify the notation of the mechanism.

A molecular complex   An ion pair
with a positive charge on   containing
chlorine and a negative   a carbocation
charge on aluminum

Step 2: Make a new bond between a nucleophile (arene ring) and an electrophile. Reaction of the carbocation (a strong electrophile) with the π electrons (a weak nucleophile) of the aromatic ring gives a resonance-stabilized cation intermediate.

The positive charge is delocalized onto
three atoms of the ring

Step 3: Take a proton away. Proton transfer regenerates the ring aromaticity.

Halogen atoms on sp^2-hybridized carbons (vinylic and aryl halides) do not react to produce electrophiles under conditions of the Friedel-Crafts alkylation because of the high activation energy required to form these carbocations.

There are three major limitations on the Friedel-Crafts alkylation. The first is the possibility for rearrangement of the alkyl group, which occurs in the following way. Friedel-Crafts alkylation involves the generation of a carbocation, and, as we have already seen in Section 6.3C, carbocations may rearrange to a more stable carbocation. Carbocation rearrangements are common in Friedel-Crafts alkylations. For example, reaction of benzene with 1-chloro-2-methylpropane (isobutyl chloride) gives only 2-methyl-2-phenylpropane (*tert*-butylbenzene).

1-Chloro-2-methylpropane
(Isobutyl chloride)

2-Methyl-2-phenylpropane
(*tert*-Butylbenzene)

In this case, the isobutyl chloride/AlCl$_3$ complex rearranges directly to the *tert*-butyl cation/AlCl$_4^-$ ion pair, which is the electrophile in this example of Friedel-Crafts alkylation.

Isobutyl chloride

Isobutyl chloride/aluminum
chloride complex

tert-Butyl cation/
AlCl$_4^-$ ion pair

In practice, alkylation with primary halides is not a useful synthetic reaction, and alkylbenzenes containing a primary alkyl group other than —CH$_2$CH$_3$ must be prepared by other means. Alkylation is useful for introducing isopropyl, *tert*-butyl, and other alkyl groups, the cations of which tend not to rearrange.

A second limitation on Friedel-Crafts alkylation is that it fails altogether on benzene rings bearing one or more strongly electron-withdrawing groups. As we shall see in the following section, substituents of the type shown in the following table have a dramatic effect on a benzene ring's reactivity toward further electrophilic aromatic substitution.

$$\text{(benzene with Y)} + RX \xrightarrow{\text{AlCl}_3} \text{No reaction}$$

When Y Equals Any of These Groups, the Benzene Ring Does Not Undergo Friedel-Crafts Alkylation

$-\overset{\overset{\displaystyle O}{\|\|}}{C}H$	$-\overset{\overset{\displaystyle O}{\|\|}}{C}R$	$-\overset{\overset{\displaystyle O}{\|\|}}{C}OH$	$-\overset{\overset{\displaystyle O}{\|\|}}{C}OR$	$-\overset{\overset{\displaystyle O}{\|\|}}{C}NH_2$
$-SO_3H$	$-C\equiv N$	$-NO_2$	$-NR_3^+$	
$-CF_3$	$-CCl_3$			

The third limitation on Friedel-Crafts alkylation is that it is hard to stop the reaction at monoalkylation because the alkylated product is more reactive than benzene itself. We will discuss reactivity in detail in Section 22.2, but, in general, alkylated benzenes are more reactive than unsubstituted compounds. This limitation can be overcome if it is feasible to use a large excess of benzene, often as both the solvent and as the reactant.

Friedel and Crafts also discovered that treating an aromatic hydrocarbon with an acyl halide (Section 18.1A) in the presence of aluminum chloride gives a ketone. Because a RCO— group is known as an acyl group, reaction of an aromatic hydrocarbon with an acyl halide is known as Friedel-Crafts acylation, as illustrated by the reaction of benzene and acetyl chloride in the presence of aluminum chloride to form acetophenone.

$$\text{Benzene} + CH_3\overset{\overset{\displaystyle O}{\|\|}}{C}Cl \xrightarrow{\text{AlCl}_3} \text{Acetophenone} + HCl$$

Benzene **Acetyl chloride** **Acetophenone**
(an acyl halide) (a ketone)

The fact that an acylbenzene is less reactive than the starting material (unreactive in most cases) overcomes the third limitation of the alkylation reaction.

The following example of electrophilic aromatic substitution involves intramolecular acylation to form a six-membered ring.

$$\text{4-Phenylbutanoyl chloride} \xrightarrow{\text{AlCl}_3} \alpha\text{-Tetralone} + HCl$$

4-Phenylbutanoyl chloride **α-Tetralone**

Mechanism   Friedel-Crafts Acylation—Generation of an Acylium Ion

Step 1: **Lewis acid-Lewis base reaction.**   Friedel-Crafts acylation begins with the donation of a pair of electrons from the halogen of the acyl halide to aluminum chloride to form a molecular complex similar to what we drew for Friedel-Crafts alkylations. In this complex, halogen has a positive formal charge and aluminum has a negative formal charge.

$$R-\overset{\overset{\displaystyle :O:}{\|\|}}{C}-\overset{..}{\underset{..}{Cl}}: + \overset{Cl}{\underset{Cl}{\overset{|}{Al}}}-Cl \rightleftharpoons R-\overset{\overset{\displaystyle :O:}{\|\|}}{C}-\overset{..}{\underset{..}{Cl}}{}^+-\overset{Cl}{\underset{Cl}{\overset{|}{Al}}}{}^--Cl$$

An acyl chloride (a Lewis base)   **Aluminum chloride** (a Lewis acid)   A molecular complex with a positive charge on chlorine and a negative charge on aluminum

Step 2: Break a bond to give stable molecules or ions. Redistribution of electrons of the carbon-chlorine bond gives an ion pair containing an **acylium ion**.

$$
\text{R}-\overset{\overset{\displaystyle :O:}{\|}}{\text{C}}-\overset{+}{\underset{\underset{\displaystyle :\overset{..}{\underset{..}{C}}l:}{|}}{\overset{..}{\underset{..}{C}}l}}-\overset{-}{\underset{|}{Al}}-\overset{..}{\underset{..}{C}}l: \longrightarrow \text{R}-\overset{+}{\text{C}}\!=\!\overset{..}{\underset{..}{O}}\quad AlCl_4^-
$$

A molecular complex with a positive charge on chlorine and a negative charge on aluminum

An ion pair containing an acylium ion

Of the two major contributing structures that can be drawn for an acylium ion, the one with complete valence shells for both carbon and oxygen makes the greater contribution to the hybrid.

Both atoms have complete valence shells

$$
\text{R}-\overset{+}{\text{C}}\!=\!\overset{..}{\underset{..}{O}}: \longleftrightarrow \text{R}-\text{C}\!\equiv\!\overset{+}{\text{O}}:
$$

The more important contributing structure

Friedel-Crafts acylation is free of a second major limitation on Friedel-Crafts alkylations: acylium ions do not undergo rearrangement. Thus, the carbon skeleton of an acyl halide is transferred unchanged to the aromatic ring.

Example 22.2

Write a structural formula for the product from Friedel-Crafts alkylation or acylation of benzene for each compound.

(a) $C_6H_5CH_2Cl$

Benzyl chloride

(b) $\overset{\overset{\displaystyle O}{\|}}{C_6H_5CCl}$

Benzoyl chloride

Solution

(a) Benzyl chloride in the presence of a Lewis acid catalyst gives the benzyl cation (an electrophile), which is then attacked by benzene (a weak nucleophile) followed by proton transfer to give diphenylmethane. In this example, the benzyl cation, although primary, cannot rearrange.

Benzyl cation

Diphenylmethane

(b) Treating benzoyl chloride with aluminum chloride gives an acylium ion (an electrophile). Reaction of this cation with the π electrons of the aromatic ring (a weak nucleophile) followed by proton transfer gives benzophenone.

Benzoyl cation   **Benzophenone**

Problem 22.2

Write a structural formula for the product from Friedel-Crafts alkylation or acylation of benzene with each compound.

(a)   **(b)**   **(c)** Ph

A special value of Friedel-Crafts acylations in synthesis is for the preparation of unrearranged alkylbenzenes, as illustrated by the preparation of isobutylbenzene.

| **2-Methyl-propanoyl chloride** | **2-Methyl-1-phenyl-1-propanone** | **Isobutylbenzene** |

Treating benzene with 2-methylpropanoyl chloride in the presence of aluminum chloride gives 2-methyl-1-phenyl-1-propanone. Wolff-Kishner or Clemmensen reduction of the carbonyl group to a methylene group (Section 16.11C) gives isobutylbenzene.

D. Other Electrophilic Aromatic Alkylations

After the discovery that Friedel-Crafts alkylations and acylations involve cationic electrophiles, it was realized that the same reactions can be accomplished by other combinations of reagents and catalysts. We study two of these reactions: generation of carbocations from alkenes and from alcohols.

As we saw in Section 6.3, treatment of an alkene with a strong acid, most commonly HX, H_2SO_4, H_3PO_4, or HF/BF_3, generates a carbocation. Cumene, an intermediate in the industrial synthesis of both acetone and phenol (Problem 16.65), is synthesized industrially by treating benzene with propene in the presence of phosphoric acid as a catalyst.

Benzene   **Propene**   **Cumene**

Alkylation with an alkene can also be carried out with a Lewis acid catalyst. Treatment of benzene with cyclohexene in the presence of aluminum chloride gives phenylcyclohexane.

Benzene   **Cyclohexene**   **Phenylcyclohexane**

Carbocations can also be generated by treatment of an alcohol with H_2SO_4, H_3PO_4, or HF (Section 10.5).

Benzene   **2-Methyl-2-propanol**   **2-Methyl-2-**
               (*tert*-Butyl alcohol)   **phenylpropane**
                                                (*tert*-Butylbenzene)

Example 22.3

Write a mechanism for the formation of isopropylbenzene (cumene) from benzene and propene in the presence of phosphoric acid.

Solution

Step 1: Make a new bond between a nucleophile (π bond) and an electrophile. Proton transfer from phosphoric acid to propene gives the isopropyl cation.

Propene   **Isopropyl cation**

Step 2: Make a new bond between a nucleophile (arene ring) and an electrophile. Reaction of the isopropyl cation with the π electrons of the benzene ring gives a resonance-stabilized carbocation intermediate.

Step 3: Take a proton away. Proton transfer to dihydrogen phosphate ion gives cumene.

Cumene

Problem 22.3

Write a mechanism for the formation of *tert*-butylbenzene from benzene and *tert*-butyl alcohol in the presence of phosphoric acid.

22.2 | Disubstitution and Polysubstitution

A. Effects of a Substituent Group on Further Substitution

In electrophilic aromatic substitution of a monosubstituted benzene, three products are possible: the new group may become oriented ortho, meta, or para to the existing group. Table 22.1 shows the orientation of nitration of a series of monosubstituted benzenes.

Based on the information in Table 22.1 and other studies like it, we can make the following generalizations about the manner in which existing groups influence further substitution reactions.

1. *Substituents affect the orientation of new groups.* Certain substituents (for example, —OCH$_3$ and —Cl) direct an incoming group preferentially to the ortho and para positions; other substituents (for example, —NO$_2$ and —COOH) direct it preferentially to the meta position. In other words, substituents on a benzene ring can be classified as **ortho-para directing** or as **meta directing**.
2. *Substituents affect the rate of further substitution.* Certain substituents cause the rate of a second substitution to be greater than that for benzene itself, whereas other substituents cause the rate of a second substitution to be lower than that for benzene. In other words, groups on a benzene ring can be classified as **activating** or **deactivating** toward further substitution.

Activating group
Any substituent on a benzene ring that causes the rate of electrophilic aromatic substitution to be greater than that for benzene.

Deactivating group
Any substituent on a benzene ring that causes the rate of electrophilic aromatic substitution to be lower than that for benzene.

These directing and activating-deactivating effects can be seen by comparing the products and rates of nitration of anisole and nitration of benzoic acid. Nitration of anisole proceeds at a rate considerably greater than that for benzene (the methoxy group is activating), and the product is a mixture of *o*-nitroanisole and *p*-nitroanisole (the methoxy group is ortho-para directing).

Anisole	*o*-Nitroanisole	*p*-Nitroanisole
	(44%)	(55%)

Quite another situation is seen in the nitration of benzoic acid. First, the reaction requires the more reactive fuming nitric acid and a higher temperature than for benzene. Because nitration of benzoic acid proceeds much more slowly than nitration of benzene itself, we say that a carboxyl group is strongly deactivating.

Table 22.1	Orientation on Nitration of Monosubstituted Benzenes				
Substituent	ortho	meta	para	ortho + para	meta
—OCH$_3$	44	—	55	99	trace
—CH$_3$	58	4	38	96	4
—Cl	70	—	30	100	trace
—Br	37	1	62	99	1
—COOH	18	80	2	20	80
—CN	19	80	1	20	80
—NO$_2$	6.4	93.2	0.3	6.7	93.2

Second, the product formed consists of approximately 80% of the meta isomer and 20% of the ortho and para isomers combined; thus we say that the carboxyl group is meta directing.

Benzoic acid   *o*-Nitro-   *m*-Nitro-   *p*-Nitro-
benzoic acid   benzoic acid   benzoic acid
(18%)   (80%)   (2%)

Listed in Table 22.2 are the directing and activating-deactivating effects for the major functional groups with which we are concerned in this text. If we compare these ortho-para and meta directors for structural similarities and differences, we can make the following generalizations.

1. Alkyl groups, phenyl groups, and substituents in which the atom bonded to the ring has an unshared pair of electrons are ortho-para directing. All other substituents are meta directing.
2. Most ortho-para directing groups are activating. The exception to this generalization is the halogens, which are weakly deactivating.

The fact that alkyl groups are weakly activating is why it is difficult to stop Friedel-Crafts alkylations at monoalkylation. When a first alkyl group is introduced onto an aromatic ring, the ring is activated toward further alkylation and, unless reaction conditions are very carefully controlled, a mixture of di-, tri-, and polyalkylation products is formed. Friedel-Crafts acylations, on the other hand, never go beyond monoacylation because an acyl group is deactivating toward further substitution.

We can illustrate the usefulness of these generalizations by considering the synthesis of two different disubstituted derivatives of benzene. Suppose we wish to

Table 22.2	Directing Effects of Substituents on Further Substitution

prepare *m*-bromonitrobenzene from benzene. This conversion can be done in two steps: nitration and bromination. If the steps are carried out in just that order, the major product is indeed *m*-bromonitrobenzene. The nitro group is a meta director and, therefore, directs bromination to a meta position.

Nitrobenzene   ***m*-Bromonitro-
                                       benzene**

If, however, we reverse the order of the steps and first form bromobenzene, we now have an ortho-para directing group on the ring, and nitration takes place preferentially at the ortho and para positions.

Bromobenzene   ***o*-Bromonitro-   ***p*-Bromonitro-
                            benzene**   benzene**

As another example of the importance of the order in electrophilic aromatic substitutions, consider the conversion of toluene to *p*-nitrobenzoic acid. The nitro group can be introduced with a nitrating mixture of nitric and sulfuric acids. The carboxyl group can be produced by oxidation of the methyl group of toluene (Section 21.5A). Nitration of toluene yields a product with the two substituents in the desired para relationship. Nitration of benzoic acid, on the other hand, yields a product with the substituents meta to each other.

***p*-Nitrobenzoic acid**

***m*-Nitrobenzoic acid**

Again, we see that the order in which the reactions are performed is critical.

Note that, in this last example, we showed nitration of toluene producing only the para isomer. Because methyl is an ortho-para directing group, both the ortho and para isomers are formed (Table 22.1). In problems of this type in which you are asked to prepare the para isomer, assume that both ortho and para isomers are formed but that there are physical methods by which they can be separated and the desired isomer obtained.

Example 22.4

Complete these electrophilic aromatic substitution reactions. Where you predict meta substitution, show only the meta product. Where you predict ortho-para substitution, show both products.

(a) [benzene with Br] + H₂SO₄ →(heat)   (b) [benzene with SO₃H] + HNO₃ →(H₂SO₄)

Solution

Bromine in (a) is ortho-para directing and weakly deactivating. The sulfonic acid group in (b) is meta directing and moderately deactivating.

(a) [benzene with Br and SO₃H ortho]   +   [benzene with Br and SO₃H para]   (b) [benzene with SO₃H and NO₂ meta]

o-Bromobenzene-sulfonic acid	*p*-Bromobenzene-sulfonic acid	*m*-Nitrobenzene-sulfonic acid

Problem 22.4

Draw structural formulas for the product of nitration of each compound. Where you predict ortho-para substitution, show both products.

(a) [benzene]—$\overset{O}{\overset{\|}{C}}$OCH₃   (b) [benzene]—$\overset{O}{\overset{\|}{O}}$CCH₃

B. Theory of Directing Effects

As we have just seen, a group on a benzene ring exerts a major effect on the pattern of further substitution. We can account for these patterns by starting with the general mechanism first presented in Section 22.1 for electrophilic aromatic substitution and carrying it a step further to consider how groups already present on the ring affect the energetics of further substitution. In this regard, we need to consider both resonance and inductive effects and the relative importance of each.

Nitration of Anisole

The rate of electrophilic aromatic substitution is limited by the slowest step in the mechanism. For the nitration of anisole, and for almost every other substitution we consider, the slow and rate-determining step is attack of the electrophile by the aromatic ring. The rate of this step depends on the stability of the transition state for this step. The more stable the transition state, the faster the rate-determining step and, thus, the overall reaction.

Shown in Figure 22.2 is the cation intermediate formed by addition of the nitronium ion meta to the methoxy group. Also shown in the figure is the cation intermediate formed by addition para to the methoxy group. Note that, in terms of electronic effects, structural formulas for the cation formed by attack ortho to the methoxy group are essentially the same as those for para attack, so, for convenience, we deal only with para attack. The cation intermediate formed by meta attack is a hybrid of contributing structures (a), (b), and (c). The cation intermediate formed by para attack is a hybrid of contributing structures (d), (e), (f), para attack

The most important
contributing structure

Figure 22.2

Nitration of anisole. Electrophilic attack meta and para to the methoxy group.

and (g). For each orientation, we can draw three contributing structures that place the positive charge on carbon atoms of the benzene ring. These three structures are the only ones that can be drawn for meta attack. However, for para attack (and for ortho attack as well), a fourth contributing structure, (f), can be drawn that involves an unshared pair of electrons on the oxygen atom of the methoxy group and places a positive charge on this oxygen. Structure (f) contributes more than structures (d), (e), or (g) because, in it, all atoms have complete octets. Because the cation formed by ortho or para attack on anisole has a greater degree of charge delocalization, and therefore a lower activation energy for its formation, nitration of anisole occurs faster in the ortho and para positions.

Nitration of Benzoic Acid

Shown in Figure 22.3 are resonance-stabilized cation intermediates formed by attack of the nitronium ion meta to the carboxyl group and then para to it. Each cation in Figure 22.3 is a hybrid of three contributing structures; no additional ones can be drawn. Now we need to compare the relative resonance stabilization

Figure 22.3

Nitration of benzoic acid. Electrophilic attack meta and para to the carboxyl group.

The most disfavored
contributing structure

of each hybrid. If we draw a Lewis structure for the carboxyl group showing the partial positive charge on the carboxyl carbon, we see that contributing structure (e) in Figure 22.3 places positive charges on adjacent atoms.

(e)

Because of the electrostatic repulsion thus generated, this structure is very unstable and makes only a negligible contribution to the hybrid.

None of the contributing structures for meta attack places positive charges on adjacent atoms. As a consequence, resonance stabilization of the cation for meta attack is greater than that for para (or ortho) attack. Stated alternatively, the activation energy for meta attack is less than that for para attack.

Comparison of the entries in Table 22.1 shows that almost all the ortho-para directing groups have an unshared pair of electrons on the atom bonded to the aromatic ring. Thus, the directing effect of these groups is primarily attributable to a resonance effect, due to the ability of the atom bonded to the ring to further delocalize the positive charge of the cation intermediate formed when electrophilic attack occurs at the ortho or para positions. Recall that, all things being equal, delocalization of a charge stabilizes a charged species.

To account for the fact that alkyl groups are also ortho-para directing, we need to consider their inductive effect on stability of the cation intermediate. In the case of alkyl groups, there is an inductive polarization of electrons from the alkyl substituent toward the cationic ring of the intermediate. This polarization amounts to a further delocalization of the positive charge, thereby stabilizing the cationic intermediate. The alkyl groups are activating because, compared to benzene alone, the cationic intermediates are lower in energy with alkyl substituents at ortho or para positions. Recall that we used the electron-releasing inductive effect (that is, hyperconjugation) of alkyl groups in Section 6.3A to account for the relative stabilities of methyl, primary, secondary, and tertiary carbocations as well.

The inductive polarization of electrons from alkyl groups (called hyperconjugation) is most effective at delocalizing the positive charge of the cation intermediate when the alkyl group is bonded directly to the ring atom carrying significant positive charge. In other words, when the alkyl group is ortho or para to the location of the incoming electrophile.

A part of the positive charge on the cation intermediate is delocalized onto methyl via hyperconjugation.

C. Theory of Activating-Deactivating Effects

We account for the activating-deactivating effects of substituent groups by much the same combination of resonance and inductive effects.

1. Any resonance effect, such as that of —NH₂, —OH, and —OR, that delocalizes the positive charge of the cation intermediate lowers the activation energy for

its formation and has an activating effect toward further electrophilic aromatic substitution.

2. Any resonance or inductive effect, such as that of $-NO_2$, $-C\equiv N$, $-C=O$, $-SO_2-$, and $-SO_3H$, that decreases electron density on the ring deactivates the ring to further substitution.

3. Any inductive effect, such as that of $-CH_3$ or another alkyl group, that releases electron density toward the cationic intermediate activates the ring toward further substitution.

4. Any inductive effect, such as that of a halogen, $-NR_3^+$, $-CCl_3$, and $-CF_3$, that decreases electron density on the ring deactivates the ring to further substitution.

The halogens represent an interesting combination of the resonance and inductive effects, the two operating in opposite directions. Recall from Table 22.2 that halogens are ortho-para directing but, unlike other ortho-para directors listed in the table, they are weakly deactivating. These observations can be accounted for in the following way.

1. The inductive effect of halogens: the halogens are relatively electronegative and have an electron-withdrawing inductive effect. Aryl halides, therefore, react more slowly in electrophilic aromatic substitution than benzene.

2. The resonance effect of halogens: when a halogen-substituted aromatic ring is attacked by an electrophile to form a cation intermediate, a halogen ortho or para to the site of electrophilic attack can help to stabilize the cation intermediate by delocalization of the positive charge through resonance involving unshared electron pairs.

Thus, the inductive and resonance effects of the halogens are counter to each other, but the former is somewhat stronger than the latter. The net effect of this opposition is that the halogens are weakly deactivating but ortho-para directing.

For cases involving electrophilic aromatic substitution of aromatic rings with two or more substituents already on the ring, the more activating group will dominate the orientation preference of the incoming group. This is because activating groups will activate the ring for more substitution according to their orientation preference, while deactivating groups only serve to deactivate the ring toward further substitution. In practice, this means that ortho-para directing groups will dominate meta directing groups when they are both on a ring.

Example 22.5

Predict the major product of each electrophilic aromatic substitution.

Solution

The key to predicting orientation of electrophilic aromatic substitution on each molecule is that ortho-para directing groups activate the ring toward further substitution, whereas meta directing groups deactivate. Therefore, where there is competition between ortho-para and meta directing groups, ortho-para directing groups win out. In these examples, the major product is that resulting from substitution ortho or para to the activating group. For (a) the next substitution is directed ortho/para to the strongly activating —OH group; the isomer with bromine between the —OH and —NO$_2$ groups is a very minor product because of steric hindrance. For (b) the incoming group is directed ortho and para to the strongly activating —OH group. For (c) the next substitution is directed ortho to the weakly activating —CH$_3$ group.

Problem 22.5

Predict the major product(s) of each electrophilic aromatic substitution.

22.3 | Nucleophilic Aromatic Substitution

Nucleophilic aromatic substitution A reaction in which a nucleophile, most commonly a halogen, on an aromatic ring is replaced by another nucleophile.

One of the important chemical characteristics of aryl halides is that they undergo relatively few reactions involving the carbon-halogen bond. Aryl halides, for example, do not undergo substitution by either of the S_N1 or S_N2 pathways that are characteristic of nucleophilic aliphatic substitutions. They do, however, undergo **nucleophilic aromatic substitution** under certain conditions but by mechanisms quite different from those for nucleophilic aliphatic substitutions. Nucleophilic aromatic substitution reactions are far less common than electrophilic aromatic substitution reactions and have only limited application for the synthesis of organic compounds. We study these reactions not only for their synthetic usefulness but also for the additional insights they give us into the unique chemical properties of aromatic compounds.

A. Nucleophilic Substitution by Way of a Benzyne Intermediate

An apparent exception to the generalization about the lack of reactivity of aryl halides to nucleophilic substitution is an early industrial process for the synthesis of phenol from chlorobenzene. When heated at 300°C under high pressure with aqueous NaOH, chlorobenzene is converted to sodium phenoxide. Neutralization of this salt with aqueous acid gives phenol.

Chlorobenzene **Sodium phenoxide**

In later technological developments, the discovery was made that chlorobenzene can be hydrolyzed to phenol by steam under pressure at 500°C. Each of these reactions appears to involve nucleophilic substitution of —OH for —Cl on the benzene ring. However, this reaction is not as simple as it might seem, as is illustrated by the reaction of substituted halobenzenes with NaOH. For example, *o*-chlorotoluene under these conditions gives a mixture of 2-methylphenol (*o*-cresol) and 3-methylphenol (*m*-cresol).

2-Methylphenol
(*o*-Cresol)

3-Methylphenol
(*m*-Cresol)

The same type of reaction can be brought about by the use of sodium amide in liquid ammonia. Under these conditions, for example, *p*-chlorotoluene gives a mixture of 4-methylaniline (*p*-toluidine) and 3-methylaniline (*m*-toluidine) in approximately equal amounts.

4-Methylaniline
(*p*-Toluidine)

3-Methylaniline
(*m*-Toluidine)

The difference in this reaction compared with other substitution reactions we have dealt with so far is that the entering group appears not only at the position occupied by the leaving group but also at a position adjacent to it.

To account for these experimental observations, it has been proposed that an elimination of HX occurs to form a **benzyne intermediate** that then undergoes nucleophilic addition to the triple bond to give the products observed.

Benzyne intermediate
A reactive intermediate formed by *β*-elimination from adjacent carbon atoms of a benzene ring and having a triple bond in the benzene ring. The second *π* bond of the benzyne triple bond is formed by weak overlap of coplanar *sp²* orbitals on adjacent carbons.

Mechanism   Nucleophilic Aromatic Substitution via a Benzyne Intermediate

Step 1: **Take a proton away and simultaneously break a bond to make stable molecules or ions.**   Dehydrohalogenation of the benzene ring gives a benzyne intermediate.

A benzyne
intermediate

Step 2: **Make a new bond between a nucleophile and an electrophile.**   Nucleophilic addition of amide ion to a carbon of the benzyne triple bond gives a carbanion intermediate. Addition to either carbon of the triple bond is possible.

A carbanion intermediate

Step 3: Add a proton. Proton transfer from ammonia to the carbanion intermediate gives one of the observed substitution products and generates a new amide ion.

3-Methylaniline
(*m*-Toluidine)

π bonding, but of reduced strength because of poor orbital overlap

A benzyne intermediate
Cartoon orbital

The bonding in a benzyne intermediate, and also the reason for its extremely reactive nature, can be pictured in the following way. According to molecular orbital theory, the benzene ring retains its planarity, π bonding, and aromatic character. The adjacent sp^2 orbitals formerly bonding to a halogen and a hydrogen now overlap to form the second π bond of the benzyne triple bond. The problem is that the atomic orbitals forming this π bond are not parallel as in acetylene and unstrained alkynes but, rather, lie at an angle of approximately 120° to the bond axis connecting them. Consequently, the overlap between these orbitals is reduced. Reduced overlap, in turn, means a weaker and more reactive π bond. Therefore, the second π bond of the benzyne intermediate undergoes addition very readily to form two new and stronger σ bonds. The relatively high energy of the benzyne intermediate is presumably why such high temperatures are required for these reactions.

B. Nucleophilic Substitution by Addition-Elimination

Aromatic halides are normally quite inert to the types of nucleophiles that readily displace halide ions from alkyl halides. However, when an aromatic compound contains strong electron-withdrawing nitro groups ortho or para (or both) to the halogen, nucleophilic aromatic substitution occurs quite readily. For example, when 1-chloro-2,4-dinitrobenzene is heated at reflux in aqueous sodium carbonate followed by treatment with aqueous acid, it is converted in nearly quantitative yield to 2,4-dinitrophenol.

1-Chloro-2,4-dinitrobenzene →(Na₂CO₃, H₂O, 100°C)→ **Sodium 2,4-dinitrophenoxide** →(HCl, H₂O)→ **2,4-Dinitrophenol**

One application of this reaction is the synthesis of 2,4-dinitrophenylhydrazine, a reagent that was once commonly used to prepare derivatives of aldehydes and ketones (Section 16.8B).

1-Chloro-2,4-dinitrobenzene + H₂NNH₂ **Hydrazine** ⟶ **2,4-Dinitrophenylhydrazine** + HCl

This type of nucleophilic aromatic substitution for halogen has been studied extensively, and it has been determined that reaction occurs in two steps: nucleophilic addition followed by elimination. For the majority of reactions of this type, addition of the nucleophile in Step 1 is the slow, rate-determining step. Elimination of halide ion in Step 2 gives the product. This reaction thus resembles reactions of carboxylic acid derivatives in that it proceeds by an addition-elimination mechanism rather than by direct substitution.

Mechanism   Nucleophilic Aromatic Substitution by Addition-Elimination

Step 1: **Make a new bond between a nucleophile and an electrophile.**   The nucleophile adds to the aromatic ring at the carbon bearing the halogen. This addition places a negative charge on the ring, which is stabilized by a resonance interaction with the nitro or other strongly electron-withdrawing groups in the ortho or para positions. Such intermediates are named Meisenheimer complexes after the German chemist who first characterized them. Note that nitro groups on both ortho and para positions participate in delocalization of the negative charge in the complex.

A Meisenheimer complex

Step 2: **Break a bond to give stable molecules and ions.**   Elimination of halide ion regenerates the aromatic ring and gives the observed product.

A Meisenheimer complex

Example 22.6

What is the state of hybridization of each ring carbon atom in the Meisenheimer complex just shown?

Solution

The carbon atom bonded to both the leaving group and entering nucleophile (—Cl and —Nu in the structure shown) is sp^3 hybridized. The other five carbons of the ring are sp^2 hybridized.

Problem 22.6

In S_N2 reactions of alkyl halides, the order of reactivity is $RI > RBr > RCl > RF$. Alkyl iodides are considerably more reactive than alkyl fluorides, often by factors as great as 10^6. All 1-halo-2,4-dinitrobenzenes, however, react at approximately the same rate in nucleophilic aromatic substitutions. Account for this difference in relative reactivities.

Section 22.1 Electrophilic Aromatic Substitution

Problems: 22.1–22.3,
22.7–22.13, 22.22–22.27,
22.32–22.61, 22.63

- Aromatic rings, such as benzene, react with very strong, usually positively charged electrophiles, in a reaction that results in substitution of a ring hydrogen.
  - The general mechanism involves attack on the electrophile by the weakly nucleophilic aromatic π cloud to form a resonance-stabilized cation intermediate on the ring that loses a proton to give a substituted arene.
  - The resonance-stabilized cation intermediate is called an **arenium ion**.
- In a halogenation reaction, aromatic rings react with Cl_2 in the presence of the Lewis acid catalyst $FeCl_3$ to give aryl chlorides. They react with Br_2 in the presence of the Lewis acid catalyst FeX_3 to give aryl bromides.
- In a sulfonation reaction, aromatic rings react with SO_3 in the presence of sulfuric acid to yield arylsulfonic acids.
- In a nitration reaction, aromatic rings react with nitric acid in the presence of sulfuric acid to yield nitroarenes, which are useful synthetic intermediates because the nitro group can be reduced to an amino group by reaction with H_2 and a transition metal catalyst, or alternatively, by using iron, zinc, or tin in HCl followed by base. There is no electrophilic aromatic substitution reaction that introduces the amino group directly onto the aromatic ring.
- In a **Friedel-Crafts alkylation**, aromatic rings react with alkyl halides in the presence of a Lewis acid like $AlCl_3$ to produce alkylbenzenes. Rearrangements and over-alkylation can be a problem.
- In a **Friedel-Crafts acylation**, aromatic rings react with acid chlorides in the presence of a Lewis acid like $AlCl_3$ to produce acylbenzenes.
  - The acylbenzene products of Friedel-Crafts acylation reactions can be reduced to the corresponding alkylbenzene using Clemmenson or Wolff-Kishner reductions, providing a convenient method of producing alkylbenzenes that cannot be made in high yield using the Friedel-Crafts alkylation due to rearrangement or over-alkylation problems.
- A variety of other electrophilic aromatic substitution reactions involve very strong electrophiles reacting with the weakly nucleophilic aromatic π cloud to form an intermediate resonance-stabilized cation on the ring that loses a proton to give the substituted arene.
  - Reactions include using an alkene in the presence of a strong acid to give a carbocation that generates an alkylbenzene, using an alkene in the presence of a Lewis acid to generate an alkylbenzene, and using an alcohol in the presence of strong acid to give a carbocation that generates an alkylbenzene.

Section 22.2 Disubstitution and Polysubstitution

- Substituent groups, other than hydrogen, on an aromatic ring influence the reaction rate and substitution pattern of electrophilic aromatic substitution reactions.
  - In particular, substituents can direct new groups meta or ortho-para, and can either speed up (activate) or slow down (deactivate) the reaction.
- Substituents can be divided into three broad classes:
  - Alkyl groups and all groups in which the atom bonded to the ring has an unshared pair of electrons are ortho-para directing and most are electron releasing; therefore they are **activating** toward electrophilic aromatic substitution compared to benzene itself.
  - Halogens are exceptions in that they are ortho-para directing, but electron-withdrawing: therefore they are weakly deactivating toward electrophilic aromatic substitution compared to benzene itself.
  - All groups with a partial positive charge on the atom attached to the ring are meta directing and electron withdrawing; therefore they are **deactivating** toward electrophilic aromatic substitution compared to benzene itself.

- Orientation and activating/deactivating effects have a large practical significance, since in synthesizing polysubstituted aromatics, the *order of addition of the substituents must be taken into account.*
  - For example, when making *m*-bromonitrobenzene from benzene, the nitro group (meta directing) must be added before the bromine atom (ortho-para directing).
  - Conversely, when making *o*-bromonitrobenzene and *p*-bromonitrobenzene from benzene, the bromine (ortho-para directing) must be added first, followed by the nitro group (meta directing).
- Substituent directing and activation/deactivation effects are the result of two types of interactions operating on the cation intermediate:
  - An **inductive effect** in which (relative to H atoms) the substituent withdraws more electron density out of (deactivating), or releases more electron density into (activating), the positively charged intermediate.
  - An always activating resonance effect in which lone pairs on atoms bonded to the ring create an extra contributing structure when they are ortho or para to the incoming electrophile. The added resonance stabilizes the cation intermediate by further distributing its positive charge over the molecule.
- These inductive and resonance effects either raise or lower the energy of the cation intermediate, which, in turn, raises (decreasing rate, deactivation) or lowers (increasing rate, activation) the rate-limiting energy barrier (activation energy) of reaction.
- These inductive and resonance effects have different levels of influence depending on their position relative to the incoming electrophile.
  - The effects of substituents that *activate* the ring are most *activating* ortho and para to the position of the incoming electrophile, because these orientations provide for the strongest cation intermediate stabilizing interactions.
  - The effects of substituents that *deactivate* the ring are less *deactivating* meta to the position of the incoming electrophile, because the meta orientation minimizes destabilization of the cation intermediate relative to an ortho or para orientation.
- The three basic types of substituents can be understood in the context of the foregoing ideas.
  - Electron-releasing groups are activating and thus always ortho-para directing.
  - Electron-withdrawing groups other than the halogens are deactivating and thus meta directing.
  - Halogens show deactivation, but an ortho-para substitution preference. The stabilizing resonance effect of the unshared electron pair directs ortho and para, and takes orientation precedence over the overall inductive deactivation due to electronegativity.
- For cases involving electrophilic aromatic substitution of aromatic rings with two or more substituents already on the ring, the more activating group will dominate the orientation preference of the incoming group.

Problems: 22.4, 22.5, 22.14–22.21, 22.32–22.61, 22.63

Section 22.3 Nucleophilic Aromatic Substitution

- Electrophilic aromatic substitution is by far the most common mechanism for reactions with aromatic rings, but in some situations aromatic rings will react with nucleophiles.
  - Aryl halides react with very strong bases ($NaNH_2$) or moderate bases (NaOH) at high temperature (300°C to 500°C) to yield products in which the halogen is replaced.
  - The base/nucleophile group ends up on the ring carbon atom that was originally bonded to the halogen, as well as positions adjacent (ortho) to it due to the benzyne intermediates.
  - Aryl halides that are activated by having ortho and/or para strongly electron withdrawing groups react with strong nucleophiles such as hydrazine to give substitution in a regioselective manner.

Problems: 22.6, 22.28, 22.29, 22.62

ⓌWL Questions and problems for this chapter are available in OWL. Use the OWL access card that came with this textbook to access assigned questions and problems for this chapter.

Key Reactions

1. **Halogenation (Section 22.1A)** The electrophile is a halonium ion formed as an ion pair by interaction of chlorine or bromine with a Lewis acid. The mechanism involves an initial reaction between Cl_2 and $FeCl_3$ to generate a molecular complex that can rearrange to give a Cl^+, $FeCl_4^-$ ion pair. The Cl^+ reacts as a very strong electrophile with the weakly nucleophilic aromatic π cloud to form a resonance-stabilized cation intermediate that loses a proton to give the aryl chloride product.

Halogenation of an aromatic ring substituted by strongly activating groups (such as —OH, —OR, and —NH_2) does not require a Lewis acid catalyst.

2. **Nitration (Section 22.1B)** The electrophile is the nitronium ion, NO_2^+, formed by interaction of nitric acid and sulfuric acid. The mechanism involves an initial protonation of nitric acid by sulfuric acid followed by loss of water to yield the nitronium ion NO_2^+. The nitronium ion reacts as a very strong electrophile with the weakly nucleophilic aromatic π cloud to form a resonance-stabilized cation intermediate that loses a proton to give the final product.

3. **Sulfonation (Section 22.1B)** The electrophile is either sulfur trioxide, SO_3, or HSO_3^+ depending on experimental conditions. The mechanism involves reaction of SO_3 as a very strong electrophile with the weakly nucleophilic aromatic π cloud to form a resonance-stabilized cation intermediate that loses a proton to give the arylsulfonic acid product.

4. **Friedel-Crafts Alkylation (Section 22.1C)** The electrophile is a carbocation formed as an ion pair by interaction of an alkyl halide with a Lewis acid. Rearrangements from a less stable carbocation to a more stable carbocation are common. The mechanism involves an initial reaction between the alkyl halide and Lewis acid $AlCl_3$ to yield an intermediate that can be thought of as a carbocation, AlX_4^- ion pair. The carbocation portion of the ion pair reacts as a very strong electrophile with the weakly nucleophilic aromatic π cloud to form a resonance-stabilized cation intermediate that loses a proton to give the final

product. Because carbocations are involved in the mechanism, rearrangements can be a problem, especially with primary or secondary alkyl halides, or any other alkyl halide that will generate a carbocation prone to rearrangement.

The reaction fails when there is one or more strongly electron-withdrawing groups on the ring. It can be hard to stop the reaction after a single alkylation, because the aromatic monoalkylation product is more reactive than the starting material.

5. **Friedel-Crafts Acylation (Section 22.1C)** The electrophile is an acyl cation (an acylium ion) formed as an ion pair by interaction of an acyl halide with a Lewis acid. The mechanism involves an initial reaction between the acid chloride and Lewis acid $AlCl_3$ to yield an intermediate that can be thought of as a resonance-stabilized acylium ion, $AlCl_4^-$ ion pair. The acylium ion portion of the ion pair reacts as a very strong electrophile with the weakly nucleophilic aromatic π cloud to form a resonance-stabilized cation intermediate that loses a proton to give the final product. Because acylium ions do not rearrange like carbocations, no rearrangements occur. The reaction fails when there is one or more strongly electron-withdrawing groups on the ring. It is easy to stop the reaction after a single acylation because the aromatic monoacylation product is less reactive than the starting material.

6. **Alkylation Using an Alkene (Section 22.1D)** The electrophile is a carbocation formed by interaction of the alkene with a Brønsted or Lewis acid.

7. **Alkylation Using an Alcohol (Section 22.1D)** The electrophile is a carbocation formed by treatment of the alcohol with a Brønsted or Lewis acid.

8. **Nucleophilic Aromatic Substitution: A Benzyne Intermediate (Section 22.3A)** The mechanism involves an initial elimination reaction between the aryl halide and strong base to give a benzyne intermediate, which undergoes addition at either sp hybridized carbon atom to give the products.

9. Nucleophilic Aromatic Substitution: Addition-Elimination (Section 22.3B)   The mechanism involves a nucleophilic attack of the ring carbon containing the halogen to give a negatively charged Meisenheimer complex, followed by loss of halogen to give the substituted product. This reaction does not occur unless there are electron-withdrawing groups ortho and/or para to the halogen, because these groups activate the ring toward nucleophilic attack.

Chapter 23

Amines

© Frank Orel/Stone/Getty

Morphine, a potent painkiller isolated from the ripe seed heads of the opium poppy, has been a lead drug for chemists in search of potent but less addicting synthetic painkillers. See Problem 23.21. Inset: a model of morphine.

Carbon, hydrogen, and oxygen are the three most common elements in organic compounds. Because of the wide distribution of amines in the biological world, nitrogen is the fourth most common element in organic compounds. The lone pair of electrons on the nitrogen of amines is a powerful electron source, so the most important chemical properties of amines are their basicity and nucleophilicity.

23.1 | Structure and Classification

Amines are derivatives of ammonia in which one or more hydrogens are replaced by alkyl or aryl groups. Amines are classified as primary, secondary, or tertiary, depending on the number of carbon atoms bonded directly to nitrogen (Section 1.3B).

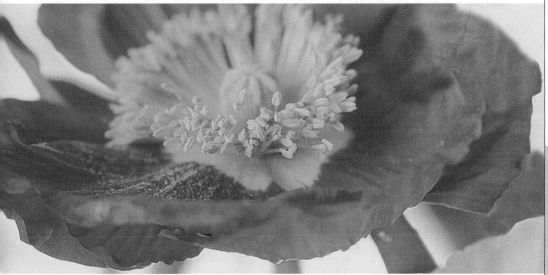

Methylamine
(a 1° amine)

Dimethylamine
(a 2° amine)

Trimethylamine
(a 3° amine)

Amines are further divided into aliphatic and aromatic amines. In an **aliphatic amine**, all carbons bonded to nitrogen are derived from alkyl groups; in an **aromatic amine**, one or more of the groups bonded to nitrogen are aryl groups.

Aliphatic amine
An amine in which nitrogen is bonded only to alkyl groups.

Aromatic amine
An amine in which nitrogen is bonded to one or more aryl groups.

OWL Online homework for this chapter may be assigned in OWL for Organic Chemistry.

Aniline
(a 1° aromatic amine)

N-Methylaniline
(a 2° aromatic amine)

Benzyldimethylamine
(a 3° aliphatic amine)

An amine in which the nitrogen atom is part of a ring is classified as a **heterocyclic amine**. When the nitrogen is part of an aromatic ring (Section 21.2D), the amine is classified as a **heterocyclic aromatic amine**. Following are structural formulas for two heterocyclic aliphatic amines and two heterocyclic aromatic amines.

Pyrrolidine   **Piperidine**
(heterocyclic aliphatic amines)

Pyrrole   **Pyridine**
(heterocyclic aromatic amines)

Example 23.1

Alkaloids are basic nitrogen-containing compounds of plant origin, many of which are physiologically active when administered to humans. Ingestion of coniine, isolated from water hemlock, can cause weakness, labored respiration, paralysis, and eventually death. Coniine is the toxic substance in the "poison hemlock" used in the death of Socrates. In small doses, nicotine is an addictive stimulant. In larger doses, it causes depression, nausea, and vomiting. In still larger doses, it is a deadly poison. Solutions of nicotine in water are used as insecticides. Cocaine is a central nervous system stimulant obtained from the leaves of the coca plant.

(a)   **(b)**   **(c)**

(S)-Coniine   **(S)-Nicotine**   **Cocaine**

Classify each amino group in these alkaloids according to type (primary, secondary, tertiary, aliphatic, aromatic, heterocyclic).

Solution

(a) A secondary aliphatic heterocyclic amine
(b) A tertiary aliphatic heterocyclic amine and a heterocyclic aromatic amine
(c) A tertiary aliphatic heterocyclic amine

Problem 23.1

Identify all carbon chiral centers in coniine, nicotine, and cocaine.

23.2 | Nomenclature

A. Systematic Names

Systematic names for aliphatic amines are derived just as they are for alcohols. The suffix -*e* of the parent alkane is dropped and is replaced by -*amine.*

2-Propanamine

(*S*)-1-Phenylethanamine

1,6-Hexanediamine

Example 23.2

Write systematic names for these amines.

(a) $CH_3(CH_2)_5NH_2$ (b) H_2N⌒⌒⌒NH_2 (c) Ph⌒⌒NH_2

Solution

(a) 1-Hexanamine   (b) 1,4-Butanediamine   (c) 2-Phenylethanamine

Problem 23.2

Write structural formulas for these amines.

(a) 2-Methyl-1-propanamine   (b) Cyclohexanamine   (c) (*R*)-2-Butanamine

IUPAC nomenclature retains the common name aniline for $C_6H_5NH_2$, the simplest aromatic amine. Its simple derivatives are named using the prefixes *o*-, *m*-, and *p*-, or numbers to locate substituents. Several derivatives of aniline have common names that are still widely used. Among these are toluidine for a methyl-substituted aniline and anisidine for a methoxyl-substituted aniline.

Aniline   4-Nitroaniline (*p*-Nitroaniline)   4-Methylaniline (*p*-Toluidine)   3-Methoxyaniline (*m*-Anisidine)

Secondary and tertiary amines are commonly named as *N*-substituted primary amines. For unsymmetrical amines, the largest group is taken as the parent amine; then the smaller group(s) bonded to nitrogen are named, and their location is indicated by the prefix *N* (indicating that they are bonded to nitrogen).

N-Methylaniline   *N*,*N*-Dimethyl-cyclopentanamine

Following are names and structural formulas for four heterocyclic aromatic amines, the common names of which have been retained in the IUPAC system.

Indole   Purine   Quinoline   Isoquinoline

Among the various functional groups discussed in this text, the —NH_2 group is one of the lowest in precedence (Table 16.1). The following compounds each contain a functional group of higher precedence than the amino group, and accordingly, the amino group is indicated by the prefix *amino.*

2-Aminoethanol   (*S*)-2-Amino-3-methyl-1-butanol   4-Aminobenzoic acid

B. Common Names

Common names for most aliphatic amines are derived by listing the alkyl groups bonded to nitrogen in alphabetical order in one word ending in the suffix -*amine,* that is, they are named as alkylamines.

CH₃NH₂

Methylamine   *tert*-**Butylamine**   **Dicyclopentylamine**   **Triethylamine**

Example 23.3

Write structural formulas for these amines.

(a) Isopropylamine   **(b)** Cyclohexylmethylamine   **(c)** Benzylamine

Solution

Problem 23.3

Write structural formulas for these amines.

(a) Isobutylamine   **(b)** Triphenylamine   **(c)** Diisopropylamine

---

When four atoms or groups of atoms are bonded to a nitrogen atom, the compound is named as a salt of the corresponding amine. The ending -*amine* (or -*aniline, pyridine,* and so on) is replaced by -*ammonium* (or *anilinium, pyridinium,* and so on), and the name of the anion is added.

Et₃NH⁺ Cl⁻

Triethylammonium   **Pyridinium acetate**
chloride

Example 23.4

Write the IUPAC name and, where possible, a common name for each compound.

(a) (C₆H₅)₂NH   **(b)**   **(c)**

(racemic)

Solution

(a) Diphenylamine

(b) *trans*-2-Aminocyclohexanol

(c) Its systematic name is (*S*)-1-phenyl-2-propanamine. Its common name is amphetamine. The dextrorotatory isomer of amphetamine (shown here) is a central nervous system stimulant and is manufactured and sold under several trade names. The salt with sulfuric acid is marketed as Dexedrine sulfate.

Problem 23.4

Write the IUPAC and, where possible, a common name for each compound.

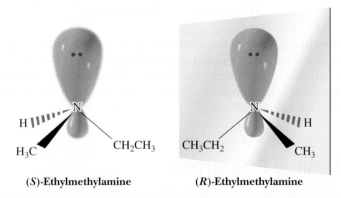

An ion containing a nitrogen atom bonded to any combination of four alkyl or aryl groups is classified as a **quaternary (4°) ammonium ion**. Compounds containing such ions have properties characteristic of salts. Cetylpyridinium chloride is used as a topical antiseptic and disinfectant.

Quaternary (4°) ammonium ion
An ion in which nitrogen is bonded to four carbons and bears a positive charge.

Me_4N^+ Cl^-

$\overset{Cl^-}{\underset{}{}}$ $\overset{+}{N}CH_2(CH_2)_{12}CH_3$

$Ph{-}CH_2\overset{+}{N}Me_3$ OH^-

Tetramethylammonium chloride

Tetradecylpyridinium chloride (Cetylpyridinium chloride)

Benzyltrimethylammonium hydroxide

23.3 | Chirality of Amines and Quaternary Ammonium Ions

The geometry of a nitrogen atom bonded to three other atoms or groups of atoms is trigonal pyramidal (Section 1.4). The sp^3-hybridized nitrogen atom is at the apex of the pyramid, and the three groups bonded to it extend downward to form the triangular base of the pyramid. If we consider the unshared pair of electrons on nitrogen as a fourth group, then the arrangement of "groups" around nitrogen is approximately tetrahedral. Because of this geometry, an amine with three different groups bonded to nitrogen is chiral and can exist as a pair of enantiomers, as illustrated by the nonsuperposable mirror images of ethylmethylamine. In assigning configuration to these enantiomers, the group of lowest priority on nitrogen is the unshared pair of electrons.

Several over-the-counter mouthwashes contain an N-alkylpyridinium chloride as an antibacterial agent.

H |||||| N
H₃C   CH₂CH₃

(S)-Ethylmethylamine

CH₃CH₂   N |||||| H
        CH₃

(R)-Ethylmethylamine

In principle, a chiral amine should be resolvable; that is, it should be separable into a pair of enantiomers. Except for special cases, however, the enantiomers cannot be resolved because they undergo rapid interconversion by a process known as pyramidal inversion. **Pyramidal inversion** is the rapid oscillation of a nitrogen atom from one side of the plane of the three atoms bonded to it to the other side of that plane.

To visualize this process, imagine the sp^3-hybridized nitrogen atom lying above the plane of the three atoms to which it is bonded. In the transition state for pyramidal inversion, the nitrogen atom and the three groups to which it is bonded become coplanar, and the molecule becomes achiral. In this planar transition state, nitrogen is sp^2 hybridized, and its lone pair of electrons lies in its unhybridized $2p$ orbital. Nitrogen then completes the inversion, becomes sp^3 hybridized again, and now lies below the plane of the three atoms to which it is bonded.

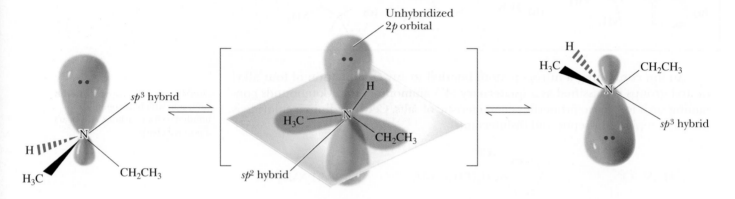

S Enantiomer   Planar transition state   R Enantiomer

As a result of pyramidal inversion, a chiral amine quite literally turns itself inside out, like an umbrella in a strong wind, and in the process becomes a racemic mixture. The activation energy for pyramidal inversion of simple amines is about 25 kJ (6 kcal)/mol. For ammonia at room temperature, the rate of nitrogen inversion is approximately 2×10^{11} s^{-1}. For simple amines, the rate is less rapid but nonetheless sufficient to make resolution impossible.

Pyramidal inversion is not possible for quaternary ammonium ions, and their salts can be resolved.

S Enantiomer   R Enantiomer

Phosphorus, in the same family as nitrogen, forms trivalent compounds called phosphines, which also have trigonal pyramidal geometry. The activation energy for pyramidal inversion of trivalent phosphorus compounds is considerably greater than it is for trivalent compounds of nitrogen, with the result that a number of chiral phosphines have been resolved.

23.4 | Physical Properties

Amines are polar compounds, and both primary and secondary amines form intermolecular hydrogen bonds (Figure 23.1).

hydrogen bonding

Figure 23.1

Intermolecular association by hydrogen bonding in primary and secondary amines. Nitrogen is approximately tetrahedral in shape with the axis of the hydrogen bond along the fourth position of the tetrahedron.

The Noanamá and Embrá peoples of the jungles of western Colombia have used poison blow darts for centuries, perhaps millennia. The poisons are obtained from the skin secretions of several brightly colored frogs of the genus *Phyllobates* (*neará* and *kokoi* in the language of the native peoples). A single frog contains enough poison for up to 20 darts. For the most poisonous species (*Phyllobates terribilis*), just rubbing a dart over the frog's back suffices to charge the dart with poison.

Scientists at the National Institutes of Health in the United States became interested in studying these poisons when it was discovered that they act on cellular ion channels, which would make them useful tools in basic research on mechanisms of ion transport. A field station was, therefore, established in western Colombia to collect the relatively common poison dart frogs. From 5000 frogs, 11 mg of two toxins, given the names batrachotoxin and batrachotoxinin A, were isolated. These names are derived from *batrachos*, the Greek word for frog. A combination of NMR spectroscopy, mass spectrometry,

© Nature Picture Library/Alamy

voltage-gated Na^+ channels in nerve and muscle cells to be blocked in the open position, which leads to a huge influx of Na^+ ions into the affected cell.

The batrachotoxin story illustrates several common themes in drug discovery. First, information

Batrachotoxin

Batrachotoxinin A

and single-crystal x-ray diffraction was used to determine the structures of these compounds.

Batrachotoxin and batrachotoxinin A are among the most lethal poisons ever discovered. It is estimated that as little as 200 μg of batrachotoxin is sufficient to induce irreversible cardiac arrest in a human being. It has been determined that they act by causing

about the kinds of biologically active compounds and their sources is often obtained from the native peoples of a region. Second, tropical rain forests are a rich source of structurally complex, biologically active substances. Third, the entire ecosystem, not just the plants, is a potential source of fascinating organic molecules.

An N—H···N hydrogen bond is weaker than an O—H···O hydrogen bond because the difference in electronegativity between nitrogen and hydrogen (3.0 − 2.1 = 0.9) is less than that between oxygen and hydrogen (3.5 − 2.1 = 1.4). The effect of intermolecular hydrogen bonding can be illustrated by comparing the boiling points of methylamine and methanol. Both are polar molecules and interact in the pure liquid by hydrogen bonding. Because hydrogen bonding is stronger in methanol than in methylamine, methanol has the higher boiling point.

	CH_3CH_3	CH_3NH_2	CH_3OH
MW (g/mol)	30.1	31.1	32.0
bp (°C)	−88.6	−6.3	65.0

All classes of amines form hydrogen bonds with water and are more soluble in water than hydrocarbons of comparable molecular weight. Most low-molecular-weight amines are completely soluble in water (Table 23.1). Higher molecular weight amines are only moderately soluble or are insoluble.

Table 23.1 | Physical Properties of Selected Amines

Name	Structural Formula	mp (°C)	bp (°C)	Solubility in Water
Ammonia	NH_3	−78	−33	Very soluble
Primary Amines				
Methylamine	CH_3NH_2	−95	−6	Very soluble
Ethylamine	$CH_3CH_2NH_2$	−81	17	Very soluble
Propylamine	$CH_3CH_2CH_2NH_2$	−83	48	Very soluble
Isopropylamine	$(CH_3)_2CHNH_2$	−95	32	Very soluble
Butylamine	$CH_3(CH_2)_3NH_2$	−49	78	Very soluble
Benzylamine	$C_6H_5CH_2NH_2$	—	185	Very soluble
Cyclohexylamine	$C_6H_{11}NH_2$	−17	135	Slightly soluble
Secondary Amines				
Dimethylamine	$(CH_3)_2NH$	−93	7	Very soluble
Diethylamine	$(CH_3CH_2)_2NH$	−48	56	Very soluble
Tertiary Amines				
Trimethylamine	$(CH_3)_3N$	−117	3	Very soluble
Triethylamine	$(CH_3CH_2)_3N$	−114	89	Slightly soluble
Aromatic Amines				
Aniline	$C_6H_5NH_2$	−6	184	Slightly soluble
Aromatic Heterocyclic Amines				
Pyridine	C_5H_5N	−42	116	Very soluble

23.5 | Basicity

Like ammonia, all amines are weak bases, and aqueous solutions of amines are basic. The following acid-base reaction between an amine and water is written using curved arrows to emphasize that, in these proton-transfer reactions, the unshared pair of electrons on nitrogen forms a new covalent bond with hydrogen and displaces hydroxide ion.

Methylamine   **Methylammonium hydroxide**

It is common to discuss the basicity of amines by reference to the acid ionization constant of the corresponding conjugate acid, as illustrated for the ionization of the methylammonium ion.

$$CH_3NH_3^+ + H_2O \rightleftharpoons CH_3NH_2 + H_3O^+$$

$$K_a = \frac{[CH_3NH_2][H_3O^+]}{[CH_3NH_3^+]} = 2.29 \times 10^{-11} \quad pK_a = 10.64$$

Table 23.2 — Acid Strengths, pK_a, of the Conjugate Acids of Selected Amines

Amine	Structure	pK_a of Conjugate Acid
Ammonia	NH_3	9.26
Primary Amines		
Methylamine	CH_3NH_2	10.64
Ethylamine	$CH_3CH_2NH_2$	10.81
Cyclohexylamine	$C_6H_{11}NH_2$	10.66
Secondary Amines		
Dimethylamine	$(CH_3)_2NH$	10.73
Diethylamine	$(CH_3CH_2)_2NH$	10.98
Tertiary Amines		
Trimethylamine	$(CH_3)_3N$	9.81
Triethylamine	$(CH_3CH_2)_3N$	10.75
Aromatic Amines		
Aniline	(phenyl)—NH_2	4.63
4-Methylaniline	CH_3—(phenyl)—NH_2	5.08
4-Chloroaniline	Cl—(phenyl)—NH_2	4.15
4-Nitroaniline	O_2N—(phenyl)—NH_2	1.0
Aromatic Heterocyclic Amines		
Pyridine	(pyridine ring)	5.25
Imidazole	(imidazole ring)	6.95

Table 23.2 gives values of pK_a for the conjugate acids of selected amines. Keep in mind that the weaker the conjugate acid (the larger its pK_a), the greater the basicity of the amine.

Example 23.5

Predict the position of equilibrium for this acid-base reaction.

$$CH_3NH_2 + CH_3COOH \rightleftharpoons CH_3NH_3^+ + CH_3COO^-$$

Solution

Use the approach we developed in Section 4.4 to predict the position of equilibrium in acid-base reactions. Equilibrium favors reaction of the stronger acid with the stronger base to give the weaker acid and weaker base.

$$CH_3NH_2 + CH_3COOH \rightleftharpoons CH_3NH_3^+ + CH_3COO^-$$

	pK_a 4.76	pK_a 10.64
	(stronger acid)	(weaker acid)

p$K_{eq} = -5.88$

$K_{eq} = 7.6 \times 10^5$

Problem 23.5

Predict the position of equilibrium for this acid-base reaction.

$$CH_3NH_3^+ + H_2O \rightleftharpoons CH_3NH_2 + H_3O^+$$

A. Aliphatic Amines

All aliphatic amines have about the same base strength, pK_a of the conjugate acid 10.0–11.0, and are slightly stronger bases than ammonia. The increase in basicity compared with ammonia can be attributed to the greater stability of an alkylammonium ion, as for example $RCH_2NH_3^+$ compared with the ammonium ion, NH_4^+. This greater stability arises from the electron-releasing effect of alkyl groups and the resulting partial delocalization of the positive charge from nitrogen onto carbon in the alkylammonium ion.

Positive charge is partially delocalized onto the alkyl group.

B. Aromatic Amines

Aromatic amines are considerably weaker bases than aliphatic amines. Compare, for example, values of pK_a for aniline and cyclohexylamine. The ionization constant for the conjugate acid of aniline is larger (the smaller the value of pK_a, the weaker the base) than that for cyclohexylamine by a factor of 10^6.

Cyclohexylamine   **Cyclohexylammonium hydroxide** $pK_a = 10.66; K_a = 2.19 \times 10^{-11}$

Aniline   **Anilinium hydroxide** $pK_a = 4.63; K_a = 2.34 \times 10^{-5}$

Aromatic amines are less basic than aliphatic amines because of a combination of two factors. First is the resonance stabilization of the free base form of aromatic amines.

Interaction of the electron pair on nitrogen with the π system of the aromatic ring

No resonance is possible with alkylamines.

For aniline and other arylamines, the resonance stabilization is the result of the interaction of the unshared pair on nitrogen with the π system of the aromatic ring. The resonance energy of benzene is approximately 151 kJ (36 kcal)/mol. For aniline, it is 163 kJ (39 kcal)/mol. Because of this resonance interaction, the electron pair on nitrogen is less available for reaction with acid. No such resonance stabilization is possible for alkylamines. Therefore, the electron pair on the nitrogen of an alkylamine is more available for reaction with an acid; alkylamines are stronger bases than arylamines.

The second factor contributing to the decreased basicity of aromatic amines is the electron-withdrawing inductive effect of the sp^2-hybridized carbons of the aromatic ring compared with the sp^3-hybridized carbons of aliphatic amines. The unshared pair of electrons on nitrogen in an aromatic amine is pulled toward the ring and, therefore, is less available for protonation to form the conjugate acid of the amine. These factors are the same two that operate to make phenoxide ion less basic than alkoxide ions (Section 21.4B).

Electron-releasing groups (for example, methyl, ethyl, and other alkyl groups) increase the basicity of aromatic amines, whereas electron-withdrawing groups (for example, nitro and carbonyl groups) decrease their basicity. The decrease in basicity on halogen substitution is the result of the electron-withdrawing inductive effect of the electronegative halogen. The decrease in basicity on nitro substitution is caused by a combination of inductive and resonance effects as can be seen by comparing the pK_a values for the conjugate acids of 3-nitroaniline and 4-nitroaniline.

3-Nitroaniline
pK_a 2.47

4-Nitroaniline
pK_a 1.0

The basicity-decreasing effect of nitro substitution in the 3-position is almost entirely the result of its inductive effect, whereas that of nitro substitution in the 4-position is attributable to both inductive and resonance effects. In the case of para substitution (as well as ortho substitution), delocalization of the lone pair on the amino nitrogen involves not only the carbons of the aromatic ring but also oxygen atoms of the nitro group.

delocalization of the nitrogen lone pair onto
the oxygen atoms of the nitro group

C. Heterocyclic Aromatic Amines

Heterocyclic aromatic amines are weaker bases than aliphatic heterocyclic amines. Compare, for example, the pK_a values for the conjugate acids of piperidine, pyridine, and imidazole.

Piperidine
pK_a 10.75

Pyridine
pK_a 5.25

Imidazole
pK_a 6.95

pK_a's of conjugate acids

We discussed the structure and bonding in pyridine and imidazole in Section 21.2D. In accounting for the relative basicities of these and other heterocyclic aromatic amines, it is important to determine first if the unshared pair of electrons on nitrogen is or is not a part of the $(4n + 2)$ π electrons giving rise to aromaticity. In the case of pyridine, the unshared pair of electrons is not a part of the aromatic sextet.

Rather, it lies in an sp^2 hybrid orbital in the plane of the ring and perpendicular to the six $2p$ orbitals containing the aromatic sextet.

This electron pair is in an sp^2 orbital so it is not part of the aromatic sextet.

Aromaticity is maintained, even when protonated.

$$\text{Pyridine} + H_2O \rightleftharpoons \text{Pyridinium ion} + HO^-$$

Pyridine　　　　**Pyridinium ion**

CONNECTIONS TO BIOLOGICAL CHEMISTRY

The Planarity of —NH₂ Groups on Aromatic Rings

An important aspect of the molecular structure of biomolecules is that amino groups bonded to an aromatic

sp^2-like and near planar

Aniline　　**Aniline** (viewed through an edge)

ring are mostly sp^2 hybridized and therefore almost planar. As described for aniline in the preceding section,

this structure is the result of resonance delocalization of the amino group lone pair into the aromatic ring. Resonance delocalization requires the nitrogen to be sp^2 hybridized with the lone pair in a $2p$ orbital to allow overlap with the aromatic π system.

The planarity of amino groups on aromatic rings has a profound influence on the properties and folding of nucleic acids. Three of the four common heterocyclic aromatic amine bases of nucleic acid bases have planar amino groups (Figure 1). Not only does the sp^2 hybridization of the amino group allow for a relatively flat overall base structure (perfect for stacking), but the geometry of the planar amino group is ideal for making specific, highly directional hydrogen bonds with the complementary base.

Two hydrogen bonds　　sp^2 hybridized and planar

Thymine (T)

Adenine (A)

sp^2 hybridized and planar　　Three hydrogen bonds

Cytosine (C)　　**Guanine (G)**

sp^2 hybridized and planar

Figure 1
The structures of the T—A and C—G base pairs showing the locations of planar —NH₂ groups bonded to the aromatic bases as well as the specific patterns of hydrogen bonds responsible for recognition between complementary strands of DNA.

The DNA duplex (Chapter 28) stores genetic information, based largely on patterns of specific hydrogen bonds. Adenine (A)-thymine (T) base pairs have two hydrogen bonds between them; guanine (G)-cytosine (C) base pairs have three hydrogen bonds (Figure 1). Moreover, the pattern of hydrogen bond donors and acceptors is different for the two base pairs. These two distinct patterns of hydrogen bonding make possible the recognition of DNA strands of the complementary sequence. In addition, the hydrogen bonds help hold two complementary DNA strands together in the familiar double helix (Figure 2). If the amino groups on the DNA bases were not sp^2 hybridized and planar, the DNA base pairs would not be flat enough to stack in the interior of the assembled helix.

Figure 2
A section of a DNA double helix illustrating the planarity of the bases, including the planarity of the —NH$_2$ groups. Base-pair hydrogen bonds are shown as green broken lines.

Proton transfer from water or other acid to pyridine does not involve the electrons of the aromatic sextet. Why, then, is pyridine a considerably weaker base than aliphatic amines? The answer is that the unshared pair of electrons on the pyridine nitrogen lies in a relatively electronegative sp^2 hybrid orbital, whereas in aliphatic amines, the unshared pair lies in an sp^3 hybrid orbital. This effect decreases markedly the basicity of the electron pair on an sp^2-hybridized nitrogen compared with that on an sp^3 hybridized nitrogen.

There are two nitrogen atoms in imidazole, each with an unshared pair of electrons. One unshared pair lies in a $2p$ orbital and is an integral part of the $(4n + 2)$ π electrons of the aromatic system. The other unshared pair lies in an sp^2 hybrid orbital and is not a part of the aromatic sextet; this pair of electrons functions as the proton acceptor.

This electron pair is an
sp2 orbital so it is not
part of the aromatic sextet.

Aromaticity is maintained,
even when protonated.

This electron pair
is in a 2*p* orbital
and is part of the
aromatic sextet.

$$\text{Imidazole} + H_2O \rightleftharpoons \text{Imidazolium ion} + HO^-$$

Imidazole

Imidazolium ion

As is the case with pyridine, the unshared pair of electrons functioning as the proton acceptor in imidazole lies in an sp^2 hybrid orbital and has markedly decreased basicity compared with an unshared pair of electrons in an sp^3 hybrid orbital. The positive charge on the imidazolium ion is delocalized on both nitrogen atoms of the ring; therefore, imidazole is a stronger base than pyridine.

Like pyridine and imidazole, pyrrole is an aromatic heterocycle, but it is not nearly as basic as pyridine or imidazole. Pyrrole's lack of base strength can be understood by noticing that the lone pair on the nitrogen atom is in a 2*p* orbital and is part of the aromatic sextet of electrons. As a consequence, a protonated pyrrole cannot maintain aromaticity because the protonated nitrogen would be sp^3 hybridized and there would only be 4 π electrons remaining (in violation of two of Hückel's aromaticity rules). A loss of aromaticity is energetically very costly and severely limits the ability of pyrrole to accept a proton despite its structural similarity to pyridine and imidazole.

Aromaticity would be
lost when protonated.

This electron pair
is in a 2*p* orbital
and is part of the
aromatic sextet.

$$\text{Pyrrole} + H_2O \rightleftharpoons + HO^-$$

Pyrrole

Energetically unfavorable
due to a loss of aromaticity

Example 23.6

Select the stronger base in each pair of amines.

(a) (A) or (B)

(b) (C) or (D)

(c) (E) or (F)

(A)   (B)   (C)   (D)   (E)   (F)

Solution

(a) Morpholine (B) is the stronger base (conjugate acid pK_a 8.2). It has a basicity comparable to that of secondary aliphatic amines. Pyridine (A), a heterocyclic aromatic amine (pK_a 5.25), is considerably less basic than aliphatic amines.

(b) Tetrahydroisoquinoline (C) has a basicity comparable to that of secondary aliphatic amines (pK_a ~ 10.8) and is the stronger base. Tetrahydroquinoline (D) has a basicity comparable to that of *N*-substituted anilines (pK_a ~ 4.4) and is the weaker base.

(c) Benzylamine (F) is the stronger base (pK_a 9.6). Its basicity is comparable to that of other aliphatic amines. The basicity of *o*-toluidine (E), an aromatic amine, is comparable to that of aniline (pK_a 4.6).

Problem 23.6

Select the stronger acid from each pair of compounds.

(a) O_2N—⬡—NH_3^+ or CH_3—⬡—NH_3^+ **(b)** ⬡NH^+ or ⬡—NH_3^+

       (A)   (B)   (C)   (D)

D. Guanidine

Guanidine, pK_a 13.6, is almost as basic as hydroxide ion. Its conjugate acid is a weaker acid than almost any other protonated amine.

$$H_2N-\underset{\underset{NH}{\|}}{C}-NH_2 + H_2O \;\rightleftharpoons\; H_2N-\underset{\underset{NH_2^+}{\|}}{C}-NH_2 + OH^- \quad pK_a = 13.6$$

     Guanidine   **Guanidinium ion**

The remarkable basicity of guanidine is attributed to the fact that the positive charge on the guanidinium ion is delocalized equally over the entire functional group as shown by these three equivalent contributing structures. This delocalization increases the stability of the guanidinium ion relative to ammonium ions.

$$H_2N-\overset{\overset{+NH_2}{\|}}{C}-NH_2 \longleftrightarrow H_2N=\overset{\overset{NH_2}{|}}{C}-NH_2 \longleftrightarrow H_2N-\overset{\overset{NH_2}{|}}{C}=NH_2$$

Three equivalent contributing structures

23.6 | Reactions with Acids

Amines, whether soluble or insoluble in water, react quantitatively with strong acids to form water-soluble salts as illustrated by the reaction of norepinephrine (noradrenaline) with aqueous HCl to form a hydrochloride salt.

(structure) **(R)-Norepinephrine** + HCl $\xrightarrow{H_2O}$ (structure) **(R)-Norepinephrine hydrochloride**

    (R)-Norepinephrine   **(R)-Norepinephrine hydrochloride**
  (only slightly soluble in water)   (a water-soluble salt)

Norepinephrine, secreted by the medulla of the adrenal gland, is a neurotransmitter. The suggestion has been made that it acts in those areas of the brain that mediate emotional behavior. Note that the common aliphatic amino group in biological molecules, such as (R)-norephinephenol and amino acids, are protonated and thus positively charged at biological pH values.

Example 23.7

Complete each acid-base reaction and name the salt formed.

(a) $Et_2NH + HCl \longrightarrow$　　**(b)** $PhCH_2NH_2 + CH_3COOH \longrightarrow$

Solution

(a) $Et_2NH_2^+Cl^-$　　　　　**(b)** $PhCH_2NH_3^+ \; CH_3COO^-$
   Diethylammonium   **Benzylammonium**
      chloride   **acetate**

Problem 23.7

Complete each acid-base reaction and name the salt formed.

(a) $Et_3N + HCl \longrightarrow$　　**(b)** ⬡$NH + CH_3COOH \longrightarrow$

Example 23.8

Following are two structural formulas for (S)-serine, one of the building blocks of proteins (Chapter 27).

$$\text{HO} \overset{\text{O}}{\underset{\text{NH}_2}{\diagup}} \text{OH} \rightleftharpoons \text{HO} \overset{\text{O}}{\underset{\text{NH}_3^+}{\diagup}} \text{O}^-$$

(A)   (B)

Is (S)-serine better represented by structural formula A or B?

Solution

Structural formula A contains both an amino group (a base) and a carboxyl group (an acid). Proton transfer from the stronger acid (—COOH) to the stronger base (—NH₂) gives an internal salt; therefore, B is the better representation for (S)-serine. Within the field of amino acid chemistry, the internal salt represented by B is called a **zwitterion** (Section 27.2).

Problem 23.8

Following are structural formulas for propanoic acid and the conjugate acids of isopropylamine and alanine, along with pK_a values for each functional group.

pK_a 10.78   O pK_a 4.78   O pK_a 2.35

$$\underset{\underset{\text{NH}_3^+}{|}}{\text{CH}_3\text{CHCH}_3} \qquad \text{CH}_3\text{CH}_2\text{COH} \qquad \underset{\underset{\text{NH}_3^+}{|}}{\text{CH}_3\text{CHCOH}} \; \text{p}K_a \; 9.87$$

Conjugate acid of   **Propanoic acid**   Conjugate acid of
isopropylamine   alanine

(a) How do you account for the fact that the —NH₃⁺ group of the conjugate acid of alanine is a stronger acid than the —NH₃⁺ group of the conjugate acid of isopropylamine?

(b) How do you account for the fact that the —COOH group of the conjugate acid of alanine is a stronger acid than the —COOH group of propanoic acid?

The basicity of amines and the solubility in water of amine salts can be used to separate water-insoluble amines from water-insoluble, nonbasic compounds. Shown in Figure 23.2 is a flow chart for the separation of aniline from acetanilide, a neutral compound.

Figure 23.2

Separation and purification of an amine and a neutral compound.

Example 23.9

Here is a flow chart for the separation of a mixture of a primary aliphatic amine (RNH$_2$, pK_a 10.8), a carboxylic acid (RCOOH, pK_a 5), and a phenol (ArOH, pK_a 10). Assume that each is insoluble in water but soluble in diethyl ether. The mixture is separated into fractions A, B, and C. Which fraction contains the amine, which the carboxylic acid, and which the phenol?

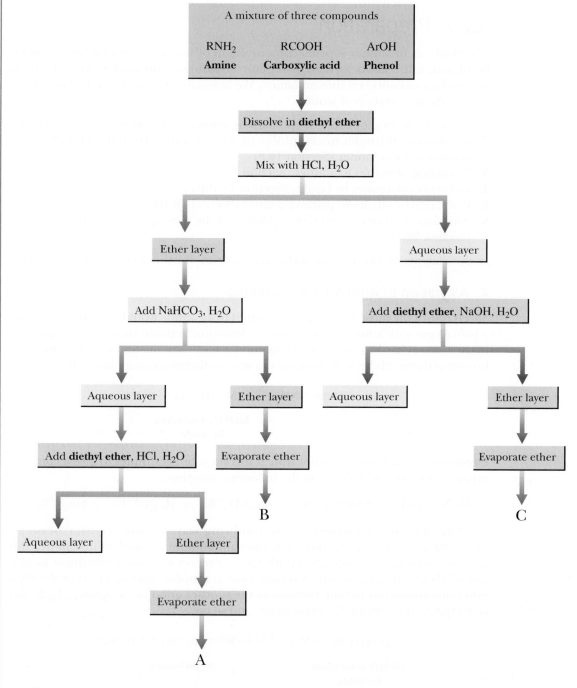

Solution

Fraction C contains RNH$_2$, fraction B contains ArOH, and fraction A contains RCOOH.

Problem 23.9

In what way(s) might the results of the separation and purification procedure outlined in Example 23.9 be different if the following conditions exist?

(a) Aqueous NaOH is used in place of aqueous $NaHCO_3$.
(b) The starting mixture contains an aromatic amine, $ArNH_2$, rather than an aliphatic amine, RNH_2.

23.7 | Preparation

The synthesis of amines is primarily a problem of how to form a carbon-nitrogen bond and, if the newly formed nitrogen-containing compound is not already an amine, how to convert it into an amine. We have already seen the following methods for the preparation of amines.

1. Nucleophilic ring opening of epoxides by ammonia and amines (Section 11.9B)
2. Addition of nitrogen nucleophiles to the carbonyl group of aldehydes and ketones to form imines (Section 16.8)
3. Reduction of imines to amines (Section 16.8)
4. Reduction of amides by $LiAlH_4$ (Section 18.10B)
5. Reduction of nitriles to primary amines (Section 18.10C)
6. Nitration of arenes followed by reduction of the nitro group to a primary amine (Section 22.1B)

In this chapter, we present two additional methods for the preparation of amines.

A. Alkylation of Ammonia and Amines

Surely one of the most direct synthetic routes to an amine would seem to be treating a haloalkane with ammonia or an amine. Reaction between these two compounds by an S_N2 mechanism gives an alkylammonium salt, as illustrated by treatment of bromomethane, MeBr, with ammonia to give methylammonium bromide.

$$MeBr + NH_3 \xrightarrow{S_N2} MeNH_3^+ Br^-$$

**Methylammonium
bromide**

Unfortunately, reaction does not stop at this stage but continues to give a complex mixture of products as shown in the following equation.

$$MeBr + NH_3 \longrightarrow MeNH_3^+Br^- + Me_2NH_2^+Br^- + Me_3NH^+Br^- + Me_4N^+Br^-$$

This mixture is formed in the following way. Proton transfer between ammonia and methylammonium ion gives ammonium ion and methylamine, also a good nucleophile, which then undergoes reaction with bromomethane to give dimethylammonium bromide. A second proton transfer reaction converts the dimethylammonium ion to dimethylamine, yet another good nucleophile, which also participates in nucleophilic substitution, and so on.

$$MeNH_3^+Br^- + NH_3 \underset{\longleftarrow}{\overset{\text{proton transfer}}{\longrightarrow}} MeNH_2 + NH_4^+Br^-$$

**Methylammonium   Methylamine
bromide**

$$MeBr + MeNH_2 \xrightarrow{S_N2} Me_2NH_2^+Br^-$$

Dimethylammonium bromide

$$Me_2NH_2^+Br^- + NH_3 \underset{\longleftarrow}{\overset{\text{proton transfer}}{\longrightarrow}} Me_2NH + NH_4^+Br^-$$

Dimethylamine

The final product from such a series of nucleophilic substitution and proton transfer reactions is a tetraalkylammonium halide. The relative proportions of the various alkylation products depend on the ratio of alkyl halide to ammonia in the reaction mixture. Whatever the starting mixture, however, the product is almost invariably a mixture of alkylated products. For this reason, alkylation of ammonia or amines is not a generally useful laboratory method for the preparation of more complex amines. However, primary amines are easily prepared because ammonia is inexpensive and can be used in large excess. Other amines can also be prepared in this way if the nucleophilic amine is inexpensive enough to be used in large excess.

B. Alkylation of Azide Ion

As we have just seen, alkylation of ammonia or amines is generally not an efficient method for the preparation of amines. One strategy for eliminating the problem of overalkylation is to use a form of nitrogen that can function as a nucleophile but that is no longer an effective nucleophile after it has formed a new carbon-nitrogen bond. One such nucleophilic form of nitrogen is the azide ion, N_3^-. Alkyl azides are easily prepared from sodium or potassium azide and a primary or secondary haloalkane by an S_N2 reaction. Azides are, in turn, reduced to primary amines by a variety of reducing agents, including lithium aluminum hydride ($LiAlH_4$).

Azide ion
(a good nucleophile)

An alkyl azide

Benzyl chloride   **Benzyl azide**   **Benzylamine**

The azide ion can also be used for stereoselective ring opening of epoxides. Reduction of the resulting β-azidoalcohol gives a β-aminoalcohol as illustrated by the conversion of cyclohexene to *trans*-2-aminocyclohexanol.

Cyclohexene   **1,2-Epoxycyclohexane**   ***trans*-2-Azidocyclohexanol**
(racemic)

***trans*-2-Aminocyclohexanol**
(racemic)

Oxidation of cyclohexene by a peroxyacid (Section 11.8C) gives an epoxide. Stereoselective nucleophilic attack by azide ion anti to the leaving oxygen of the epoxide ring (Section 11.9B) followed by reduction of the azide with lithium aluminum hydride gives racemic *trans*-2-aminocyclohexanol.

Example 23.10

Show how to convert 4-methoxybenzyl chloride to each amine.

(a)   **(b)**

Solution

(a) Two methods might be used: (1) alkylation using a large molar excess of NH_3 to reduce the extent of overalkylation or (2) nucleophilic displacement of chloride using azide ion (from NaN_3) followed by $LiAlH_4$ reduction of the azide. Of these methods, nucleophilic displacement by azide is the more convenient on a laboratory scale.

(b) Nucleophilic displacement of chloride by cyanide ion is followed by reduction of the cyano group with lithium aluminum hydride.

Problem 23.10

Show how to bring about each conversion in good yield. In addition to the given starting material, use any other reagents as necessary.

(a) MeO—⟨benzene⟩—NH₂ ⟶ MeO—⟨benzene⟩—N(CH₂CH₂OH)₂

(b)

(racemic)   (racemic)

23.8 | Reaction with Nitrous Acid

Nitrous acid, HNO_2, is an unstable compound that is prepared by adding sulfuric or hydrochloric acid to an aqueous solution of sodium nitrite, $NaNO_2$. Nitrous acid is a weak oxygen acid and ionizes according to the following equation.

$$HNO_2 + H_2O \rightleftharpoons H_3O^+ + NO_2^- \qquad pK_a = 3.37$$

Nitrous acid undergoes reaction with amines in different ways, depending on whether the amine is primary, secondary, or tertiary and whether it is aliphatic or aromatic. These reactions are all related by the facts that nitrous acid (1) participates in proton-transfer reactions and (2) is a source of the nitrosyl cation, a weak electrophile.

Mechanism   Formation of the Nitrosyl Cation

Step 1: **Add a proton.** Protonation of the OH group of nitrous acid gives an oxonium ion.

Step 2: **Break a bond to give stable molecules or ions.** Loss of water gives the nitrosyl cation, which here is represented as a hybrid of two contributing structures.

Nitrosyl cation

A. Tertiary Aliphatic Amines

When treated with nitrous acid, tertiary aliphatic amines, whether water-soluble or water-insoluble, are protonated to form water-soluble salts. No further reaction occurs beyond salt formation. This reaction is of no practical use.

B. Tertiary Aromatic Amines

Tertiary aromatic amines are bases and can also form salts with nitrous acid. An alternative pathway, however, is open to tertiary aromatic amines, namely, electrophilic aromatic substitution. The nitrosyl cation, a very weak electrophile, reacts only with aromatic rings containing strongly activating, ortho-para directing groups, such as the hydroxyl and dialkylamino groups. When treated with nitrous acid, these compounds undergo nitrosation, predominantly in the para position to give blue or green aromatic nitroso compounds.

N,N-Dimethylaniline **N,N-Dimethyl-4-nitrosoaniline**

C. Secondary Aliphatic and Aromatic Amines

Secondary amines, whether aliphatic or aromatic, undergo reaction with nitrous acid to give *N*-nitrosamines as illustrated by the reaction of piperidine with nitrous acid.

Piperidine **N-Nitrosopiperidine**

Mechanism | Reaction of a 2° Amine with the Nitrosyl Cation to Give an *N*-Nitrosamine

Step 1: **Make a new bond between a nucleophile and an electrophile.** Reaction of the 2° amine (a nucleophile) with the nitrosyl cation (an electrophile) gives an *N*-nitrosammonium ion.

Step 2: **Take a proton away.** Proton transfer to solvent gives the *N*-nitrosamine.

N-Nitrosamines are of little synthetic or commercial value. They have received considerable attention in recent years, however, because many of them are potent carcinogens. Following are structural formulas of two *N*-nitrosamines, each of which is a known carcinogen.

N-Nitrosodimethylamine
(found in cigarette smoke
and when bacon "preserved"
with sodium nitrite is fried)

N-Nitrosopyrrolidine
(formed when bacon "preserved"
with sodium nitrite is fried)

Common practice within the food industry has been to add sodium nitrite to processed meats to "retard spoilage," that is, to inhibit the growth of *Clostridium botulinum*, the bacterium responsible for botulism poisoning. Although this practice was well grounded before the days of adequate refrigeration, it is of questionable value today. Sodium nitrite is also added to prevent red meats from turning brown. If you buy some nice red hamburger in a food market and find it is gray or brown inside, you can be sure that the outside has been treated with sodium nitrite. Controversy over the use of sodium nitrite has been generated by the demonstration that nitrite ion in the presence of acid converts secondary amines to *N*-nitrosamines and that many *N*-nitrosamines are powerful carcinogens. This demonstration led in turn to pressure by consumer groups to force the Food and Drug Administration (FDA) to ban the use of nitrite additives in foods. The strength of the argument to ban nitrites was weakened with the finding that enzymes in our mouths and intestinal tracts have the ability to catalyze the reduction of nitrate to nitrite. Nitrate ion is normally found in a wide variety of foods and in drinking water. To date, there is no evidence that nitrite as a food additive poses any risk not already present through our existing dietary habits. The FDA has established the current permissible level of sodium nitrite in processed meats as 50 to 125 ppm (that is, 50–125 μg nitrite per gram of cured meat).

D. Primary Aliphatic Amines

Treatment of a primary aliphatic amine with nitrous acid results in the loss of nitrogen, N_2 and the formation of substitution, elimination, and rearrangement products as illustrated by the treatment of butylamine with nitrous acid.

OH
(25%)

Cl
(5.2%)

} Substitution

NH$_2$ $\xrightarrow[\text{0–5°C}]{\text{NaNO}_2, \text{HCl}}$

OH
(13.2%) (racemic)

} Substitution involving rearrangement

(25.9%)

} Elimination

(10.6%)

} Elimination involving rearrangement

The mechanism by which this mixture of products is formed involves formation of a **diazonium ion**. The conversion of a primary amine to a diazonium ion is called **diazotization**.

Diazonium ion
An ArN$_2^+$ or RN$_2^+$ ion.

Mechanism Reaction of a 1° Amine with Nitrous Acid

Step 1: Make a new bond between a nucleophile and an electrophile. Reaction of a 1° amine (nucleophile) with the nitrosyl cation (electrophile) is the first step of the mechanism.

$$R-\overset{..}{N}H_2 \;+\; :N\!\!\equiv\!\!\overset{+}{O}: \;\longrightarrow\; R-\overset{\overset{\displaystyle H}{|}}{\underset{\underset{\displaystyle H}{|}}{\overset{+}{N}}}-\overset{..}{N}\!\!=\!\!\overset{..}{\underset{..}{O}}$$

A 1° aliphatic
amine

Step 2: Take a proton away. Removal of a proton gives an *N*-nitrosamine.

$$R-\overset{\overset{\displaystyle H}{|}}{\underset{\underset{\displaystyle H}{|}}{\overset{+}{N}}}-\overset{..}{N}\!\!=\!\!\overset{..}{\underset{..}{O}} \;+\; H-\overset{..}{\underset{..}{O}}-H \;\longrightarrow\; R-\overset{\overset{\displaystyle H}{|}}{N}-\overset{..}{N}\!\!=\!\!\overset{..}{\underset{..}{O}} \;+\; H-\overset{\overset{\displaystyle H}{|}}{\overset{+}{O}}-H$$

An *N*-nitrosamine

Step 3: Keto-enol tautomerism. The *N*-nitrosamine undergoes keto-enol tautomerism (Section 16.9) to give a diazotic acid, which is so named because it has two (*di-*) nitrogen (*-azot-*) atoms within its structure.

$$R-\overset{\overset{\displaystyle ..}{}}{\underset{\underset{\displaystyle H}{|}}{N}}-\overset{..}{N}\!\!=\!\!\overset{..}{O} \;\underset{\text{tautomerism}}{\overset{\text{keto-enol}}{\rightleftharpoons}}\; R-\overset{..}{N}\!\!=\!\!\overset{..}{N}-\overset{..}{\underset{..}{O}}-H$$

A diazotic acid

Step 4: Add a proton. The diazotic acid is protonated.

$$R-\overset{..}{N}\!\!=\!\!\overset{..}{N}-\overset{..}{\underset{..}{O}}-H \;+\; H-\overset{..}{\underset{..}{O}}-H \;\rightleftharpoons\; R-\overset{..}{N}\!\!=\!\!\overset{..}{N}-\overset{\overset{\displaystyle H}{|}}{\overset{+}{O}}-H \;+\; H-\overset{..}{\underset{..}{O}}-H$$

Step 5: Break a bond to give stable molecules or ions. The protonated diazotic acid loses H_2O to give a diazonium ion.

$$R-\overset{..}{N}\!\!=\!\!\overset{..}{N}-\overset{\overset{\displaystyle H}{|}}{\overset{+}{O}}-H \;\longrightarrow\; R-N\!\!\equiv\!\!N: \;+\; H-\overset{..}{\underset{..}{O}}-H$$

A diazonium ion

Step 6: Break a bond to give stable molecules or ions. The diazonium ion then loses the very stable N_2 to give a carbocation.

$$R-\overset{+}{N}\!\!\equiv\!\!N: \;\longrightarrow\; R^+ \;+\; :N\!\!\equiv\!\!N:$$

A diazonium ion   A carbocation

Aliphatic diazonium ions are unstable, even at 0°C, and immediately lose nitrogen to give carbocations and nitrogen gas. The driving force for this reaction is the fact that N_2 is one of the best leaving groups because it is an extraordinarily weak base and has very strong nitrogen triple bond. It is removed from the reaction mixture as a gas as it is formed. The carbocation now has open to it the three reactions in the repertoire of aliphatic carbocations: (1) loss of a proton to give an alkene, (2) reaction with a nucleophile to give a substitution product, and (3) rearrangement to a more stable carbocation and then reaction further by (1) or (2).

Because treatment of a primary aliphatic amine with HNO_2 gives a mixture of products, it is generally not a useful reaction. An exception is the Tiffeneau-Demjanov reaction in which a cyclic β-aminoalcohol is treated with nitrous acid to give a ring-expanded ketone, with evolution of nitrogen.

A β-aminoalcohol   Cycloheptanone

We account for this molecular rearrangement as shown in the following mechanism.

Mechanism The Tiffeneau-Demjanov Reaction

Several steps: Reaction of the 1° amine with nitrous acid gives a diazonium ion. This reaction takes several steps as detailed in the mechanism box above.

A diazonium ion

Step 2: Break a bond to give stable molecules or ions and 1,2-shift. Simultaneous loss of N_2 and rearrangement by a 1,2-shift give the conjugate acid of the final product as a resonance-stabilized cation.

A resonance-stabilized cation

Step 3: Take a proton away. Proton transfer from this cation to solvent completes the reaction.

Cycloheptanone

The driving force for this molecular rearrangement is precisely what we already saw for other cation rearrangements: transformation of a less stable cation into a more stable cation. This reaction is analogous to the pinacol rearrangement (Section 10.7) with the leaving group being N_2 rather than H_2O.

Example 23.11

The following sequence of reactions gives cyclooctanone. Propose a structural formula for compound A and a mechanism for its conversion to cyclooctanone.

Solution

Catalytic hydrogenation using hydrogen over a platinum catalyst reduces the carbon-nitrogen triple bond to a single bond (Section 18.10C) and gives a β-aminoalcohol. Treatment of the β-aminoalcohol with nitrous acid results in loss of N_2 and expansion of the seven-membered ring to an eight-membered cyclic ketone.

A

Problem 23.11

How might you bring about this conversion?

E. Primary Aromatic Amines

Primary aromatic amines react with nitrous acid to form arenediazonium salts, which unlike their aliphatic counterparts, are stable at 0°C and can be kept in solution for short periods without decomposition. When an arenediazonium salt is treated with an appropriate reagent, nitrogen is lost and replaced by another atom or functional group. What makes reactions of primary aromatic amines with nitrous acid so valuable is the fact that the —NH_2 group can be replaced by the groups shown.

Aromatic amines can be converted to phenols by first forming the arenediazonium salt in aqueous sulfuric acid and then heating the solution. In this manner, 2-bromo-4-methylaniline is converted to 2-bromo-4-methylphenol.

2-Bromo-4-methylaniline　　　　　　**2-Bromo-4-methylphenol**

The intermediate in the decomposition of an arenediazonium ion in water is an aryl cation, which then undergoes reaction with water to form the phenol.

Benzenediazonium ion — An aryl cation — Phenol

Note that the aryl cation is so unstable that it can be formed only with N_2 as the leaving group. This reaction of arenediazonium salts represents the main laboratory preparation of phenols.

Example 23.12

What reagents and experimental conditions will bring about each step in the conversion of toluene to 4-hydroxybenzoic acid?

Solution

Step 1: Nitration of the aromatic ring (Section 22.1B) using HNO_3 in H_2SO_4 followed by separation of the ortho and para isomers gives 4-nitrotoluene.

Step 2: Oxidation at the benzylic carbon (Section 21.5A) using $K_2Cr_2O_7$ in H_2SO_4 gives 4-nitrobenzoic acid.

Step 3: Catalytic reduction of the nitro group (Section 22.1B) to an amino group using H_2 in the presence of Ni or other transition metal catalyst gives 4-aminobenzoic acid. Alternatively, reduction of the nitro group to a primary amine can be brought about using Zn, Sn, or Fe in aqueous HCl followed by aqueous NaOH.

Step 4: Reaction of the aromatic amine with $NaNO_2$ in aqueous H_2SO_4 followed by heating gives 4-hydroxybenzoic acid.

Problem 23.12

Show how to convert toluene to 3-hydroxybenzoic acid using the same set of reactions as in Example 23.12 but changing the order in which two or more of the steps are carried out.

The **Schiemann reaction** is the most common method for the introduction of fluorine onto an aromatic ring. It is carried out by treatment of a primary aromatic amine with sodium nitrite in aqueous HCl followed by addition of HBF_4 or $NaBF_4$. The diazonium fluoroborate salt precipitates and is collected and dried. Heating the dry salt brings about its decomposition to an aryl fluoride, nitrogen, and boron trifluoride. The Schiemann reaction is also thought to involve an aryl cation intermediate.

A diazonium fluoroborate — **Fluorobenzene**

Treatment of a primary aromatic amine with nitrous acid followed by heating with HCl/CuCl, HBr/CuBr, or KCN/CuCN results in replacement of the diazonium group by —Cl, —Br, or —CN, respectively, and is known as the **Sandmeyer reaction**. The Sandmeyer reaction fails when attempted with CuI or CuF.

HCl/CuCl
heat

2-Chlorotoluene

NaNO$_2$, H$_3$O$^+$
0–5°C

HBr/CuBr
heat

2-Bromotoluene

2-Methylaniline
(*o*-Toluidine)

KCN/CuCN
heat

2-Methylbenzonitrile

Treating an arenediazonium ion with iodide ion, generally from potassium iodide, is the best and most convenient method for introducing iodine onto an aromatic ring.

NaNO$_2$, HCl
0–5°C

KI
25°C

2-Methylaniline
(*o*-Toluidine)

2-Iodotoluene

Treating an arenediazonium ion with hypophosphorous acid, H$_3$PO$_2$, results in reduction of the diazonium group and its replacement by —H as illustrated by the conversion of aniline to 1,3,5-trichlorobenzene. Recall that —NH$_2$ is a powerful activating group (Section 22.2A). Treating aniline with chlorine requires no catalyst and gives 2,4,6-trichloroaniline. To complete the conversion, the —NH$_2$ group is removed by treatment with nitrous acid followed by hypophosphorous acid to give 1,3,5-trichlorobenzene.

Cl$_2$

NaNO$_2$, HCl
0–5°C

H$_3$PO$_2$

Aniline   **2,4,6-Trichloroaniline**   **2,4,6-Trichlorobenzene-diazonium chloride**   **1,3,5-Trichlorobenzene**

Example 23.13

Show reagents and conditions to convert toluene to 3-bromo-4-methylphenol.

(1)   (2)   (3)   (4)

NO$_2$ NO$_2$ NH$_2$ OH

Solution

Step 1: HNO$_3$ in H$_2$SO$_4$. Methyl is ortho-para directing and slightly activating.

Step 2: Treat 4-nitrotoluene with bromine in the presence of FeCl$_3$.

Step 3: Reduce the nitro group either using H_2/Ni or using Sn, Zn, or Fe in aqueous HCl followed by aqueous NaOH.

Step 4: Diazotize the amine with $NaNO_2$ in aqueous sulfuric acid followed by warming of the solution replaces $-N_2^+$ by $-OH$.

Problem 23.13

Starting with 3-nitroaniline, show how to prepare the following compounds.

(a) 3-Nitrophenol **(b)** 3-Bromoaniline **(c)** 1,3-Dihydroxybenzene (resorcinol)
(d) 3-Fluoroaniline **(e)** 3-Fluorophenol **(f)** 3-Hydroxybenzonitrile

23.9 | Hofmann Elimination

When a quaternary ammonium halide is treated with moist silver (I) oxide (a slurry of Ag_2O in H_2O), silver halide precipitates, leaving a solution of a quaternary ammonium hydroxide.

| (Cyclohexylmethyl)trimethyl-ammonium iodide | Silver oxide | (Cyclohexylmethyl)trimethyl-ammonium hydroxide |

In the mid-19th century, Augustus Hofmann discovered that when a quaternary ammonium hydroxide is heated, it decomposes to an alkene, a tertiary amine, and water. Thermal decomposition of a quaternary ammonium hydroxide to an alkene is known as the **Hofmann elimination**.

| (Cyclohexylmethyl)trimethyl-ammonium hydroxide | Methylenecyclohexane | Trimethylamine |

Hofmann elimination
When treated with a strong base, a quaternary ammonium halide undergoes β-elimination by an E2 mechanism to give the less substituted alkene as the major product.

The Hofmann elimination has most of the characteristics of an E2 reaction (Section 9.7). First, Hofmann eliminations are concerted, meaning that bond-breaking and bond-forming steps occur simultaneously or nearly so. Second, Hofmann eliminations are stereoselective for anti elimination, meaning that $-H$ and the leaving group must be anti to each other. The following mechanism illustrates the concerted nature of bond forming and bond breaking, as well as the anti arrangement of $-H$ and the trialkylamino leaving group.

Mechanism The Hofmann Elimination

Take a proton away and break a bond to give stable molecules or ions. Removal of a β-hydrogen by base, collapse of the electron pair of the C—H bond to become the π bond of the alkene, and loss of the trialkylamino group occur simultaneously. The reaction shows anti stereoselectivity.

When we studied E2 reactions of alkyl halides in Section 9.7, we saw that a β-hydrogen must be anti to the leaving group. If only one β-hydrogen meets this requirement, then the double bond is formed in that direction. If, however, two β-hydrogens meet this requirement, then elimination follows Zaitsev's rule: elimination occurs preferentially to form the more substituted double bond.

$$CH_3CH_2\overset{\overset{\displaystyle Br}{|}}{C}HCH_3 \xrightarrow[\text{E2}]{CH_3CH_2O^-Na^+} CH_3CH=CHCH_3 + CH_3CH_2CH=CH_2$$

(75%)   (25%)

Thermal decomposition of quaternary ammonium hydroxides is different because elimination occurs preferentially to form the least substituted double bond. Thermal decomposition of *sec*-butyltrimethylammonium hydroxide, for example, gives 1-butene as the major product.

$$CH_3CH_2\overset{\overset{\displaystyle HO^-}{\overset{\displaystyle +}{\overset{\displaystyle N(CH_3)_3}{|}}}}{C}HCH_3 \xrightarrow[\text{heat}]{\text{E2}} CH_3CH=CHCH_3 + CH_3CH_2CH=CH_2 + (CH_3)_3N + H_2O$$

(5%)   (95%)

Elimination reactions that give the less substituted alkene as the major product are said to follow the **Hofmann rule**.

Hofmann rule
Predicts that β-elimination will occur preferentially to give the less substituted alkene as the major product.

Example 23.14

Draw a structural formula of the major alkene formed in each β-elimination.

(a) [structure with NH₂] 1. CH₃I (excess), K₂CO₃ 2. Ag₂O, H₂O 3. heat →

(racemic)

(b) [structure with I] CH₃O⁻Na⁺ →

(racemic)

Solution

Thermal decomposition of the quaternary ammonium hydroxide in (a) follows the Hofmann rule and gives 1-octene as the major product. E2 elimination from an alkyl iodide in (b) by sodium methoxide follows Zaitsev's rule and gives *trans*-2-octene as the major product.

(a) [structure] + (CH₃)₃N + H₂O   (b) [structure] + CH₃OH

1-Octene   ***trans*-2-Octene**

Problem 23.14

The procedure of methylation of amines and thermal decomposition of quaternary ammonium hydroxides was first reported by Hofmann in 1851, but its value as a means of structure determination was not appreciated until 1881 when he published a report of its use to determine the structure of piperidine. Following are the results obtained by Hofmann.

$$C_5H_{11}N \xrightarrow[\substack{\text{1. CH}_3\text{I (excess), K}_2\text{CO}_3 \\ \text{2. Ag}_2\text{O, H}_2\text{O} \\ \text{3. heat}}]{} C_7H_{15}N \xrightarrow[\substack{\text{4. CH}_3\text{I (excess), K}_2\text{CO}_3 \\ \text{5. Ag}_2\text{O, H}_2\text{O} \\ \text{6. heat}}]{} CH_2=CHCH_2CH=CH_2$$

Piperidine   **(A)**   **1,4-Pentadiene**

(a) Show that these results are consistent with the structure of piperidine (Section 23.1).

(b) Propose two additional structural formulas (excluding stereoisomers) for $C_5H_{11}N$ that are also consistent with the results obtained by Hofmann.

In summary, both Hofmann and Zaitsev eliminations are always preferentially anti. If only one β-hydrogen is anti to the leaving group, then that one will be removed. If more than one β-hydrogen is anti, then there will be competition between Hofmann and Zaitsev elimination.

1. Eliminations involving a negatively charged leaving group, for example Cl^-, Br^-, I^-, and OTs^-, almost always follow Zaitsev's rule, unless a bulky base is used.
2. Eliminations involving a neutral leaving group, for example $N(CH_3)_3$ and $S(CH_3)_2$, almost always follow Hofmann's rule.
3. The bulkier the base, the greater the percentage of Hofmann product; compare, for example, $(CH_3)_3CO^-K^+$, which gives mostly Hofmann elimination, with $CH_3O^-Na^+$, which gives mostly Zaitsev elimination.

One of the likeliest explanations for the formation of the less stable carbon-carbon double bond is that Hofmann elimination is governed largely by steric factors, namely the bulk of the $—NR_3^+$ group. The hydroxide ion preferentially approaches and removes the least hindered α-hydrogen and gives the least substituted alkene as product. For the same reason, bulky bases, such as $(CH_3)_3CO^-K^+$, also give Hofmann elimination from alkyl halides.

23.10 | Cope Elimination

Treatment of a tertiary amine with hydrogen peroxide results in oxidation of the amine to an amine oxide.

A 3° amine   An amine oxide

When an amine oxide with at least one β-hydrogen is heated, it undergoes thermal decomposition to form an alkene and an N, N-dialkylhydroxylamine. Thermal decomposition of an amine oxide to an alkene is known as a **Cope elimination** after its discoverer, Arthur C. Cope, of the Massachusetts Institute of Technology.

Methylenecyclohexane   N,N-Dimethyl-
                            hydroxylamine

All experimental evidence indicates that the Cope elimination is syn stereoselective and concerted.

| **Mechanism** | The Cope Elimination |

Within the same molecule, take a proton away and break a bond to give stable molecules or ions. The transition state involves a planar or nearly planar arrangement of the five participating atoms and a cyclic flow of three pairs of electrons. Elimination shows syn stereoselectivity.

Transition state

an alkene

N,N-dimethyl-
hydroxylamine

If two or more syn β-hydrogens can be removed in a Cope elimination, there is little preference for one over another except when the double bond is conjugated with an aromatic ring. Therefore, as a method of preparation of alkenes, Cope eliminations are best used where only one alkene is possible.

Example 23.15

When 2-dimethylamino-3-phenylbutane is treated with hydrogen peroxide and then made to undergo a Cope elimination, the major alkene formed is 2-phenyl-2-butene.

2-Dimethylamino-3-phenylbutane   **2-Phenyl-2-butene**

(a) How many stereoisomers are possible for 2-dimethylamino-3-phenylbutane?
(b) How many stereoisomers are possible for 2-phenyl-2-butene?
(c) Suppose that the starting amine is the 2R, 3S isomer. What is the configuration of the product?

Solution

(a) There are two chiral centers in the starting amine. Four stereoisomers are possible: two pair of enantiomers.
(b) There is one carbon-carbon double bond about which stereoisomerism is possible. Two stereoisomers are possible: one E, Z pair.
(c) Following is a stereodrawing of the 2R, 3S stereoisomer showing a syn conformation of the dimethylamino group and the β-hydrogen. Cope elimination on this stereoisomer gives (E)-2-phenyl-2-butene.

2R,3S isomer   **(E)-2-Phenyl-2-butene**

Problem 23.15

In Example 23.15, you considered the product of Cope elimination from the 2R, 3S stereoisomer of 2-dimethylamino-3-phenylbutane. What is the product of a Cope elimination from the following stereoisomers? What is the product of a Hofmann elimination from each stereoisomer?

(a) 2S, 3R stereoisomer   (b) 2S, 3S stereoisomer

Summary

Section 23.1 Structure and Classification

- Amines are derivatives of ammonia that have one or more of the hydrogens replaced with alkyl and/or aryl groups.
  - Primary amines have one hydrogen of ammonia replaced by a carbon in the form of an alkyl or aryl group.
  - Secondary amines have two hydrogens of ammonia replaced by a carbon in the form of alkyl and/or aryl groups.

Problems: 23.1, 23.18–23.22

– Tertiary amines have all three hydrogens of ammonia replaced by a carbon in the form of alkyl and/or aryl groups.
– Quaternary ammonium ions have four alkyl and/or aryl groups bonded to nitrogen, resulting in a positively charged species.
– **Aliphatic amines** have alkyl groups only bonded to nitrogen, while **aromatic amines** have at least one aromatic ring bonded to the nitrogen atom.
– **Heterocyclic amines** have the nitrogen atom as part of a ring, and **heterocyclic aromatic amines** have the nitrogen atom as part of an aromatic ring.
– **Alkaloids** are basic, nitrogen-containing compounds of plant origin, many of which are physiologically active in humans.

Section 23.2 Nomenclature

Problems: 23.2, 23.4, 23.16, 23.17

- In IUPAC names, aliphatic amines are named like alcohols, except the suffix -*amine* is used and a number is used to locate the position of the amine group.
- IUPAC uses the common name aniline for simple derivatives of $C_6H_5NH_2$, although certain common names are retained for some substituted anilines such as toluidine and anisidine.
- Secondary and tertiary amines are named as *N*-substituted primary amines, and the largest group is taken as the parent amine; then the smaller groups bonded to nitrogen are named, given the prefix *N*. Examples are *N*-methylaniline and *N,N*-dimethylcyclopentanamine.
- Several common heterocycles retain their common names in IUPAC nomenclature, including pyridine, indole, purine, quinoline, and isoquinoline.
- Common names are derived by listing the alkyl groups bonded to the nitrogen atom in alphabetical order, followed by the suffix -*amine*.
- **Quaternary ammonium ions** are named by replacing the -*amine* suffix with -*ammonium* (or -*anilinium*, etc.) and the name of the anion (for example, Tetramethylammonium chloride).

Section 23.3 Chirality of Amines and Quaternary Ammonium Ions

- Secondary or tertiary amines with three different groups bonded to nitrogen are chiral, but they cannot usually be resolved because at room temperature they undergo a process called **pyramidal inversion** that rapidly interconverts the two enantiomers.
  – Phosphines are the phosphorus equivalents of amines, and because phosphines do not undergo pyramidal inversion at room temperature, chiral phosphines can be resolved.
- Quaternary ammonium salts cannot undergo pyramidal inversion, so they can be resolved.

Section 23.4 Physical Properties

Problems: 23.23, 23.24

- Amines are polar compounds, and primary or secondary amines can make intramolecular hydrogen bonds.
  – As a result of the hydrogen bonding, primary and secondary amines interact with solutes primarily through hydrogen bonding, have substantially higher melting and boiling points, and are more soluble in water than analogous hydrocarbons.
  – Because H—N hydrogen bonds are weaker than H—O hydrogen bonds, amines have lower boiling points than analogous alcohols.

Section 23.5 Basicity

- Amines are weak bases, so aqueous solutions of amines are basic. The pK_a values of the conjugate acids of aliphatic amines are in the 10 to 11 range.
  – Alkyl groups make amines slightly more basic because electron-releasing alkyl groups stabilize the alkylammoniun ion.

- Amines are protonated and positively charged near neutral pH and in biological solutions.
- Aromatic amines are considerably less basic than aliphatic amines, the pK_a values of their conjugate acids being in the 4–5 range.
  - Aromatic amines are less basic because the nitrogen lone pair takes part in resonance with the aromatic π system, a stabilizing interaction that is lost upon protonation.

Problems: 23.5–23.6, 23.25–23.29, 23.31, 23.32

  - This resonance interaction requires that the N atom of aromatic amines is sp^2 hybridized and therefore planar, a situation that is critical to nucleic acid base stacking and hydrogen bonding.
  - Electron-releasing substituents on the ring increase the basicity of anilines; electron-withdrawing groups decrease the basicity of anilines.
- The basicity of N atoms within heterocyclic aromatic amines depends on whether the N lone pair is part of the aromatic π system.
  - In pyridine, the lone pair on N is not part of the π system, so protonation does not disrupt aromaticity.
    - The pK_a of the conjugate acid of pyridine is 5.25, lower than that of an alkyl amine because the pyridine lone pair is in a relatively electronegative sp^2 orbital.
  - In imidazole, only one N atom can be protonated because the lone pair on the other N atom is part of the aromatic π system and aromaticity would be lost upon protonation.
- Guanidine groups are very basic organic groups because the protonated guanidinium ion is highly stabilized by charge delocalization.

Section 23.6 Reactions with Acids

- Amines react with strong acids to give water-soluble salts, allowing the separation of amines from water-insoluble molecules.

Problems: 23.7–23.9, 23.30

Section 23.7 Preparation

- Amines can be prepared using reactions including epoxide ring opening, addition of nitrogen nucleophiles to carbonyls followed by reduction, reduction of amides, reduction of nitriles, and nitration of arenes followed by reduction.

Problems: 23.10, 23.33, 23.34, 23.52–23.55, 23.57–23.60, 23.62–23.66, 23.71

- Alkylation of amines generally results in overalkylation.
- Primary amines can be prepared in high yield by reacting a haloalkane with the strong nucleophile (and weak base) sodium or potassium azide, followed by reduction with $LiAlH_4$.
  - Azides can also be used to ring open epoxides followed by reduction to give amino alcohols with *trans* stereoselectivity.

Section 23.8 Reaction with Nitrous Acid

- Nitrous acid, often prepared *in situ* by reacting $NaNO_2$ with acid, reacts with amines in different ways depending on whether they are primary, secondary, tertiary, or aromatic.
  - Nitrous acid participates in proton transfer reactions and is a source of the nitrosyl cation, a weak but very important electrophile.
- Tertiary aromatic amines can undergo electrophilic aromatic substitution with nitrous acid.

Problems: 23.11, 23.35–23.38

- Secondary amines react with nitrous acid to give *N*-nitrosamines.
- Primary amines react with nitrous acid to give **diazonium ion** intermediates that lose N_2 and give a variety of substitution and elimination products, so the reaction is not generally synthetically useful.
  - A synthetically useful version of the reaction is the Tiffeneau-Demjanov reaction, which gives a one-carbon ring expansion of a cyclic β-aminoalcohol and produces a cyclic ketone.

Problems: 23.12, 23.13,
23.45–23.51, 23.56, 23.61,
23.67–23.70, 23.72

- Primary aromatic amines (anilines) react with nitrous acid to give aryl diazonium ions, which are very versatile and useful synthetic intermediates in the synthesis of a variety of substituted aromatic rings. Recall that aromatic amines can be prepared through the combination of nitration followed by reduction.
  - Reaction of aryl diazonium ions with water gives phenols.
  - Reaction of aryl diazonium ions with HBF$_4$ gives aryl fluorides in a reaction known as the **Schiemann reaction**.
  - Reaction of aryl diazonium ions with HCl/CuCl, HBr/CuBr, or KCN/CuCN replaces the diazonium group with a —Cl, —Br, or —CN group, respectively, in a reaction known as the **Sandmeyer reaction**.
  - Reaction of aryl diazonium ions with KI gives aryl iodides.
  - Reaction of aryl diazonium ions with hypophosphorous acid (H$_3$PO$_2$) replaces the diazonium group with an H atom, so that the reaction sequence of reacting an aniline with nitrous acid followed by hypophosphorous acid will remove the aryl NH$_2$ group.

Section 23.9 Hofmann Elimination

Problems: 23.14, 23.39,
23.41

- Reaction of a quaternary ammonium halide with moist silver oxide to produce a quaternary ammonium hydroxide followed by heating to give an alkene is a reaction known as the **Hofmann elimination**.
  - Hofmann elimination reactions are stereoselective for anti eliminations and give predominantly the least substituted alkene, an observation that is counter to Zaitsev's rule.
    - Hofmann elimination regiochemistry is thought to derive from the steric bulk of the ammonium group that directs deprotonation by base to the least hindered site, leading to formation of the less substituted alkene.
    - Eliminations observed to produce primarily the least substituted alkene are said to be following the **Hofmann rule**.

Section 23.10 Cope Elimination

Problems: 23.15, 23.40,
23.42

- Treatment of a tertiary amine with hydrogen peroxide gives an amine oxide, which when heated gives an alkene and an N,N-dialkylhydroxyamine in a reaction known as a **Cope elimination**.
  - The Cope elimination is syn stereoselective and gives little preference for regiochemistry, unless a conjugated double bond can be created in which case the conjugated product predominates.

ᗠWL Questions and problems for this chapter are available in OWL. Use the OWL access card that came with this textbook to access assigned questions and problems for this chapter.

Key Reactions

1. **Basicity of Aliphatic Amines (Section 23.5A)** Aliphatic amines are slightly stronger bases than ammonia owing to the electron-releasing effect of alkyl groups and partial delocalization of positive charge in the alkylammonium ion.

$$CH_3NH_2 + H_2O \rightleftharpoons CH_3NH_3^+ + OH^- \qquad pK_a = 10.64$$

2. **Basicity of Aromatic Amines (Section 23.5B)** Aromatic amines are considerably weaker bases than aliphatic amines. Resonance stabilization by interaction of the unshared electron pair on nitrogen with the π system decreases its availability for protonation.

$$\text{C}_6\text{H}_5-NH_2 + H_2O \rightleftharpoons \text{C}_6\text{H}_5-NH_3^+ + OH^- \qquad pK_a = 4.63$$

3. **Basicity of Aromatic Heterocyclic Amines (Section 23.5C)** Aromatic heterocyclic amines are considerably weaker bases than aliphatic amines.

$$\text{(imidazole)} + H_2O \rightleftharpoons \text{(imidazolium)} + OH^- \qquad pK_a = 6.95$$

4. **Reaction of Amines with Strong Acids (Section 23.6)** All amines react quantitatively with strong acids to form water-soluble salts.

$$\text{—N(CH}_3)_2 + HCl \longrightarrow \text{—}\overset{+}{N}(CH_3)_2 \ H \ Cl^-$$

Insoluble in water   A water-soluble salt

5. **Alkylation of Ammonia and Amines (Section 23.7A)** This method is seldom used for preparation of pure amines because of overalkylation and the difficulty of separating products.

$$\text{—CH}_2NH_2 + CH_3I \longrightarrow \text{—CH}_2\overset{+}{N}CH_3 \ H \ I^-$$

6. **Alkylation of Azide Ion Followed by Reduction (Section 23.7B)** Azides are prepared by treatment of a primary or secondary alkyl halide of an epoxide with KN_3 and are reduced to primary amines by a variety of reducing agents including lithium aluminum hydride.

(racemic)   (racemic)

7. **Nitrosation of Tertiary Aromatic Amines (Section 23.8B)** The nitrosyl cation is a very weak electrophile and participates in electrophilic aromatic substitution only with highly activated aromatic rings.

$$(CH_3)_2N\text{—} \xrightarrow[\text{2. NaOH, H}_2O]{\text{1. NaNO}_2, \text{ HCl, 0–5°C}} (CH_3)_2N\text{—}\text{—N}=\text{O}$$

8. **Formation of N-Nitrosamines from Secondary Amines (Section 23.8C)** Reaction of the nitrosyl cation (an electrophile) with a 2° amine (a nucleophile) gives an N-nitrosamine.

$$\text{N—H} + HNO_2 \longrightarrow \text{N—N}=\text{O} + H_2O$$

9. **Reaction of Primary Aliphatic Amines with Nitrous Acid (Section 23.8D)** Treating a primary aliphatic amine with nitrous acid gives an unstable diazonium salt that loses N_2 to give a carbocation. The carbocation may (1) lose a proton to give an alkene, (2) react with a nucleophile, or (3) rearrange, followed by (1) or (2). The mechanism of diazonium ion formation involves reaction of the primary amine (a nucleophile) with the nitrosyl cation to give an N-nitrosamine that tautomerizes to the more stable diazotic acid, which is protonated on the O atom, followed by loss of water to give a diazonium ion.

$$RCH_2NH_2 \xrightarrow[\text{0–5°C}]{\text{NaNO}_2, \text{ HCl}} [RCH_2N_2^+] \longrightarrow RCH_2^+ + N_2$$

10. **Reaction of Cyclic β-Aminoalcohols with Nitrous Acid (Section 23.8D)** Treating a cyclic β-aminoalcohol with nitrous acid leads to rearrangement and a ring-expanded ketone. The mechanism involves an initial formation of a diazonium ion, followed by simultaneous loss of N_2 and rearrangement by a 1,2-shift to give a resonance-stabilized cation that loses a proton to give the product cyclic ketone.

$$\text{(cyclohexanol with } CH_2NH_2 \text{)} + HNO_2 \longrightarrow \text{(cycloheptanone)} + 2\ H_2O + N_2$$

11. **Formation of Arenediazonium Salts (Diazotization) (Section 23.8E)** Arenediazonium salts are stable in aqueous solution at 0°C for short periods.

$$\text{(benzene)}-NH_2 + HNO_2 \xrightarrow[0°C]{HCl} \text{(benzene)}-N_2^+Cl^- + H_2O$$

The mechanism is analogous to diazonium ion formation using nitrous acid and aliphatic primary amines.

12. **Conversion of a Primary Arylamine to a Phenol (Section 23.8E)** Formation of an arenediazonium salt followed by loss of nitrogen gives an aryl cation intermediate, which then reacts with water to give a phenol.

$$\text{(aniline with Br, } CH_3\text{)} \xrightarrow[\text{2. } H_2O, \text{ heat}]{\text{1. } NaNO_2, H_2SO_4, H_2O} \text{(phenol with Br, } CH_3\text{)} + N_2$$

13. **Schiemann Reaction (Section 23.8E)** Heating an arenediazonium fluoroborate is the most common synthetic method for introduction of fluorine onto an aromatic ring.

$$\text{(aniline)} \xrightarrow[\text{2. } HBF_4]{\text{1. } NaNO_2, HCl, 0°C} \text{(benzene)}-N_2^+\ BF_4^- \xrightarrow{\text{heat}} \text{(fluorobenzene)} + N_2 + BF_3$$

14. **Sandmeyer Reaction (Section 23.8E)** Treatment of an arenediazonium salt with CuCl, CuBr, or CuCN results in replacement of the diazonium group by —Cl, —Br, or —CN, respectively.

$$\text{(toluidine, } CH_3, NH_2\text{)} \xrightarrow[0-5°C]{NaNO_2, H_3O^+} \text{(toluene, } CH_3, N_2^+\text{)} \xrightarrow[\text{heat}]{KCN/CuCN} \text{(toluene, } CH_3, C\equiv N\text{)}$$

15. **Reaction of an Arenediazonium Salt with KI (Section 23.8E)** Treatment of an arenediazonium salt with KI is the most convenient method for introducing iodine onto an aromatic ring.

$$\text{(toluidine, } CH_3, NH_2\text{)} \xrightarrow[0-5°C]{NaNO_2, HCl} \text{(toluene, } CH_3, N_2^+Cl^-\text{)} \xrightarrow[25°C]{KI} \text{(toluene, } CH_3, I\text{)}$$

16. Reduction of an Arenediazonium Salt with Hypophosphorous Acid (Section 23.8E)

An —NO_2 or —NH_2 group can be used to control orientation of further substitution and then removed after it has served its purpose.

17. Hofmann Elimination (Section 23.9)

Anti stereoselective elimination of quaternary ammonium hydroxides occurs preferentially to form the least substituted carbon-carbon double bond (Hofmann's rule). The mechanism involves the simultaneous deprotonation of a β-hydrogen by base and loss of the amino group in an anti geometry.

18. Cope Elimination: Pyrolysis of a Tertiary Amine Oxide (Section 23.10)

Elimination is syn stereoselective and involves a cyclic flow of six electrons in a planar transition state.

Chapter 24

Catalytic Carbon-Carbon Bond Formation

A ruthenium-containing organometallic catalyst for alkene metathesis reactions. See Section 24.5 for the structure of this catalyst and a discussion of this reaction.

Organic chemists have learned over the course of the last one hundred twenty or so years how to synthesize amazingly complex molecules. In recent years, particularly, they have focused on compounds of medicinal interest, and a large number of current pharmaceuticals are synthesized from simpler compounds. Many of these pharmaceuticals are natural products or their analogs, and others are either simpler analogs or unrelated compounds that have been found to be active against certain organisms or diseased cells, specific cellular receptors, or specific enzyme targets. A key development that has allowed synthesis of these compounds has been the discovery of many catalytic methods of carbon-carbon bond formation. It is now possible, using a combination of new and classical methods, to carry out the synthesis of molecules with sensitive functionality and amazingly complex carbon skeletons from simple and inexpensive starting materials, often with excellent stereo- and regiocontrol. In this chapter, we make a dramatic leap from the more classical organic reactions covered in previous chapters of this book to survey several particularly useful catalytic methods of carbon-carbon bond formation, some of which represent very recent developments. Further, the catalytic methods we present are part of a thrust in industry to move to green chemistry in which less chemical waste is generated. We have room for only a few representative examples out of the wealth of carbon-carbon bond forming reactions that are now available to the modern synthetic chemist. Finally, a number of problems based on modern organic syntheses are given to illustrate the use of these reactions and their combination with other reactions.

OWL **Online homework for this chapter may be assigned in OWL for Organic Chemistry.**

24.1 | Carbon-Carbon Bond-Forming Reactions from Earlier Chapters

Let us list the methods of carbon-carbon bond formation that you have already studied as a review. All these reactions should be available to you for synthetic problems.

Nucleophilic displacement of a leaving group by a carbon nucleophile

- Gilman (organocuprate) reagents (Section 15.2C), if the leaving group is a halogen atom or toylate.
- Grignard reagents, organolithium reagents (Section 15.1C) and Gilman reagents if the leaving group is the oxygen of an epoxide.
- Alkyne (Sections 7.5 and 9.1) and cyanide (Sections 9.1 and 9.5) anions. The leaving group can be the oxygen of an epoxide (Section 11.9B).
- Enolate anion alkylations (acetoacetic ester and malonic ester synthesis; Sections 19.6 and 19.7).
- Enamine alkylations (Section 19.5).

Nucleophilic addition to a carbonyl or a carboxyl group

- Grignard reagents (Sections 16.5A and 18.9A), organolithium reagents (Sections 16.5B and 18.9B), and Gilman reagents (Section 18.8C).
- Alkyne (Section 16.5C) and cyanide (Section 16.5D) anions.
- Aldol reactions (Section 19.2).
- Claisen (Section 19.3A) and Dieckmann (Section 19.3B) condensations.
- Enamine acylations (Section 19.5B).
- Wittig reaction (Section 16.6 for C=C double bonds).

Conjugate addition to α,β-unsaturated carbonyl compounds

- Michael reaction (Section 19.8A)

Pericyclic reactions

- Diels-Alder reaction (Section 20.5)
- Claisen rearrangement (Section 20.6)
- Cope rearrangement (Section 20.6)

Carbene/carbenoid additions (Section 15.3)

Aromatic substitution

- Friedel-Crafts alkylation and acylation of aromatics (Section 22.1C)
- Reaction of cyanide with aromatic diazonium compounds (Section 23.8E)

This list already includes many different reactions, but you will find that the new reactions in this chapter are of somewhat different character.

None of the C—C bond-forming reactions summarized above are catalytic. Recall that a catalyst is a species that becomes involved in the mechanism of a reaction and lowers the barrier to that reaction, thereby accelerating the reaction rate. Catalysts do not change the thermodynamics of a reaction, but instead they alter the kinetics. Further, a catalyst is regenerated at the end of the reaction in the same form as at the start of the reaction, and hence, a catalyst is available to be used over and over—a property that is often referred to as "turn-over."

This chapter focuses on catalytic C—C bond-forming reactions. In other words, a chemical (the catalyst) is added to the reaction vessel and causes C—C bond formation in a manner that is faster than would occur in the absence of the catalyst. In fact, most of the reactions we examine in this chapter (particularly those of Section 24.5) would essentially never occur without the catalyst. The catalytic processes we examine are reasonably new compared to many of those discussed in this book, and the ones we examine all involve organometallic catalysts. Consequently, we start the chapter with a brief overview of the key reactions performed by organometallic compounds.

24.2 | Organometallic Compounds and Catalysis

We introduced organometallic compounds in Chapter 15. In the next few sections, we discuss several reactions of transition metals that are particularly useful for preparation of new carbon-carbon bonds.

A. Oxidative Addition and Reductive Elimination

Oxidative addition

Addition of a reagent to a metal center causing it to add two substituents and to increase its oxidation state by two.

Reductive elimination

Elimination of two substituents at a metal center, causing the oxidation state of the metal to decrease by two.

Ligand

A Lewis base bonded to a metal atom in a coordination compound. It may bond strongly or weakly.

Two extremely important reactions of transition metals and transition metal compounds are **oxidative addition** and its reverse **reductive elimination**. In oxidative addition, a reagent adds to a metal, causing its coordination to increase by two; reductive elimination is the opposite. These reactions are called oxidative or reductive because the formal charge of the metal changes by two during the reaction. Oxidative addition can occur with a metal coordinated with one or more **ligands** (L_n, where n is the number); it can also occur with a free metal, $M(0)$. Haloalkanes, hydrogen, halogens, and many other types of compounds can take part in these reactions. The reactivity of different substrates depends greatly on the metal.

$$ML_n + X_2 \xrightleftharpoons[\substack{\text{reductive}\\\text{elimination}}]{\substack{\text{oxidative}\\\text{addition}}} \begin{matrix} X \\ \diagdown \\ ML_n \\ \diagup \\ X \end{matrix}$$

Although there are numerous other reactions that are unique and essential to the action of organometallic catalysts, the key steps of the catalytic processes we discuss in this chapter involve oxidative additions and reductive eliminations. Thus, in an introductory chapter on catalytic C—C bond-forming reactions, we need go no deeper.

B. Key Features of the Utility of Catalytic C—C Bond Formation

The reactions and mechanisms covered in this chapter represent a growing modern trend in organic chemistry. Organometallic catalysts facilitate reactions that are otherwise impossible, are very difficult, or would require many synthetic steps in order to accomplish. The use of these catalysts causes a net decrease in the number and quantity of reagents, solvents, and purifications necessary in an overall synthetic sequence. This decrease means that the chemical waste from an industrial process can be dramatically reduced. Intentionally designing a chemical procedure or process so as to decrease waste and toxic byproducts is now a whole chemical field in and of itself that is called **green chemistry**.

It is a goal of green chemistry to use effectively each of the atoms involved in a reaction so that atoms are not "thrown away" by being incorporated into byproducts of the reactions. This concept is called **atom economy**, which describes the efficiency of a chemical process in terms of all the atoms involved. An ideal reaction would consist of the mass of the product equaling the mass of all the reactants used. In such a case, each atom of the reactants would be completely incorporated into the product, and no waste would be generated. Such reactions are rare, but the goal of achieving the highest atom economy clearly has both an environmental and economic benefit. Hence, "going green" is permeating society in many ways, with the chemical industry recognizing and embracing the value of green.

24.3 | The Heck Reaction

A. The Nature of the Reaction

In the early 1970s, Richard Heck, at the Hercules Company and later at the University of Delaware, discovered a palladium-catalyzed reaction in which the carbon group of a haloalkene or haloarene is substituted for a hydrogen on the carbon-carbon double bond (a vinylic hydrogen) of an alkene. This reaction, now known as the **Heck reaction**, is particularly valuable in synthetic organic chemistry because it is the only general method yet discovered for this type of substitution.

Substitution for a vinylic hydrogen by the Heck reaction is highly regioselective; formation of the new carbon-carbon bond most commonly occurs at the less substituted carbon of the double bond. In addition, where an E or Z configuration is possible at the new carbon-carbon double bond of the product, the Heck reaction is highly stereoselective, often giving almost exclusively the E configuration of the product.

Bromobenzene   **Methyl 2-propenoate**   **Methyl (E)-3-phenyl-2-propenoate**
                        (Methyl acrylate)   **(Methyl cinnamate)**

In addition, the Heck reaction is completely stereospecific with regard to the haloalkene; the configuration of the double bond in the haloalkene is preserved.

(Z)-3-Iodo-3-hexene   Phenylethene   ($1E,3Z$)-3-Ethyl-1-phenyl-1,3-hexadiene
                        (Styrene)

(E)-3-Iodo-3-hexene   Phenylethene   ($1E,3E$)-3-Ethyl-1-phenyl-1,3-hexadiene
                        (Styrene)

Preparation of the Catalyst

The form of the palladium catalyst most commonly added to the reaction medium is palladium(II) acetate, $Pd(OAc)_2$. This and other Pd(II) compounds are better termed precatalysts because the catalytically active form of the metal is a complex of Pd(0) formed *in situ* by reduction of Pd(II) to Pd(0).

Palladium(II)   An alkene   Oxidized
acetate   (reductant)   alkene

Reaction of Pd(0) with good ligands, L, gives the actual Heck catalyst, PdL_2. Without the ligand, Pd(0) is insoluble.

$$Pd^0 \ + \ 2\,L \ \longrightarrow \ PdL_2$$

Ligand   The Heck
                catalyst

Among the most common ligands, L, used for coordination of the Pd(0) is triphenylphosphine, $(C_6H_5)_3P$. Many other ligands can be used as well, including chiral ones such as BINAP (Section 6.7) that can lead to a significant excess of a single enantiomer in the case of chiral products.

The Haloalkane

The most common halides used in Heck reactions are aryl, heterocyclic, benzylic, and vinylic iodides and bromides, with iodides being generally most reactive. The reactivity of substrates with leaving groups on sp^2 carbons contrasts with nucleophilic substitution reactions, where such substrates are essentially unreactive. Haloalkanes in which there is an acidic beta hydrogen are rarely used because of the ease with which they undergo β-elimination under conditions of the Heck reaction to form alkenes. Triflates (trifluoromethanesulfonates, $CF_3SO_2O—$), which are easily prepared by treating an alcohol with trifluoromethanesulfonyl chloride, are also excellent substrates.

$$CF_3\overset{\displaystyle O}{\underset{\displaystyle O}{\overset{\|}{\underset{\|}{S}}}}—Cl \quad + \quad HO—R \quad \longrightarrow \quad CF_3\overset{\displaystyle O}{\underset{\displaystyle O}{\overset{\|}{\underset{\|}{S}}}}—OR \quad + \quad HCl$$

Trifluoromethanesulfonyl chloride   Alcohol   A trifluoromethanesulfonate
(a triflate)

The halide or triflate (RX) reacts with PdL_2 by oxidative addition to give a square planar Pd(II) species, which is the reaction intermediate.

$$PdL_2 \quad + \quad RX \quad \longrightarrow \quad L—\overset{\displaystyle R}{\underset{\displaystyle L}{\overset{|}{\underset{|}{Pd}}}}—X$$

The Heck
catalyst

A particular advantage of the Heck reaction is the wide range of functional groups, including alcohols, ethers, aldehydes, ketones, and esters, that may be present elsewhere in the organic halogen compound or alkene without reacting themselves or affecting the Heck reaction.

The Alkene

The reactivity of the alkene is a function of steric crowding about the carbon-carbon double bond. Ethylene and monosubstituted alkenes are the most reactive; the greater the degree of substitution on the double bond, the slower the reaction and the lower the yield of product. These steric effects also control the regiochemistry of the addition, with the alkyl group adding to the less hindered carbon of the alkene.

The Base

Commonly used bases are tertiary amines such as triethylamine, Et_3N, sodium or potassium acetate, and sodium hydrogen carbonate.

The Solvent

Polar aprotic solvents (Section 9.3D) such as *N,N*-dimethylformamide (DMF), acetonitrile, and dimethyl sulfoxide (DMSO) are commonly used. It is also possible to carry out some Heck reactions in aqueous methanol. The polar solvents are needed to dissolve the $Pd(OAc)_2$ at the beginning of the reaction.

B. Mechanism of the Reaction

The mechanism of the Heck reaction is divided into two stages: formation of the Heck catalyst and the catalytic cycle. As you study the catalytic cycle, note in particular that both Steps 2 and 4 are syn stereoselective; reaction will not proceed if these syn relationships cannot be obtained. Step 2 involves syn addition of R and PdL_2X

Stage 1: Formation of the Heck Catalyst, PdL₂

A two-electron reduction of Pd(II) to Pd(0) accompanied by its complex formation with two molecules of a ligand, L, gives the Heck catalyst, PdL_2. A common reducing agent is triethylamine or, as in the following example, the alkene itself. Because the catalyst is present only in small amounts, an insignificant amount of the alkene is lost to this reaction. In the reaction shown here, L is triphenylphosphine, $(C_6H_5)_3P$. As mentioned previously, this is actually a two-step reaction: reduction of the palladium, followed by reaction of the palladium with the ligand. We show the two steps combined here for simplicity.

$$Pd(OAc)_2 \quad + \quad \text{(alkene)} \quad + \quad 2\ Ph_3P \quad \longrightarrow \quad Pd(PPh_3)_2 \quad + \quad \text{(oxidized alkene)} \quad + \quad HOAc$$

| **Palladium(II) acetate** | An alkene (reductant) | **Triphenylphosphine** (ligand) | A Pd(0) complex (abbreviated PdL₂) | Oxidized alkene | |

Stage 2: The Catalytic Cycle

The catalytic cycle of the Heck reaction involves five steps. In Step 1, oxidative addition of the haloalkene or haloarene, RX, to PdL_2 gives a tetracoordinated Pd(II) complex containing both R and X groups bonded to Pd. Syn addition of the R and PdL_2X of this complex to the alkene gives an intermediate in which Pd is bonded to the more substituted carbon for steric reasons. Because of the long Pd—C bond, the palladium is sterically less demanding than the organic group and, therefore, ends up on the more hindered carbon. This intermediate must undergo internal rotation about the central carbon-carbon single bond in Step 3 to place H and PdL_2X syn to each other. Syn elimination of H and PdL_2X in Step 4 gives the new alkene and $HPdL_2X$. Reductive elimination in Step 5 releases the acid HX and regenerates the PdL_2 catalyst. HX is then neutralized by the added base.

The catalytic cycle of the Heck reaction

In this cycle, the alkene, haloalkane compound, and base are required in equimolar amounts; the Pd(0) species is required in only a catalytic amount. Note also the inversion of the configuration (R_2 and R_3 are originally *cis* to each other but in the product are *trans*). This inversion is a consequence of the consecutive syn addition and elimination steps. The complete mechanism for this reaction has additional intermediates (involving π complexes of the alkene with the palladium), but those shown here are the important ones for understanding the reaction and its stereochemistry.

to the double bond. Step 4 involves syn elimination of H and the Pd(II) species to generate a new double bond. These syn additions and eliminations contrast with most of the addition and elimination reactions we have seen, which prefer the anti geometry. Additions of boron hydrides (Section 6.4, hydroboration) and osmium tetroxide (Section 6.5A) or ozone (Section 6.5B) to alkenes are some examples of syn additions that you have already seen.

Example 24.1

Complete these Heck reactions.

(a)

cat. Pd(OAc)$_2$

Ph$_3$P, K$_2$CO$_3$, DMF

(b)

cat. Pd(OAc)$_2$

Ph$_3$P, K$_2$CO$_3$, DMF

Solution

In (a), 1-iodohexene has the *E* configuration, and this double bond retains its configuration in the product. Furthermore, the carbon-carbon double bond adjacent to the ester in the product now has the possibility for *cis, trans* isomerism. The Heck reaction is highly stereoselective, and this double bond has the more stable *E* configuration as well. In (b), the major product is (*E*)-1,2-diphenylethene.

(a)

Methyl (2*E*,4*E*)-2,4-nonadienoate

(b)

(*E*)-1,2-Diphenylethene
(*trans*-Stilbene)

Problem 24.1

Show how you might prepare each compound by a Heck reaction using methyl 2-propenoate as the starting alkene.

Methyl 2-propenoate
(**Methyl acrylate**)

(a)

(the *E* isomer)

(b)

(the 2*E*,4*Z* isomer)

The usual pattern in a Heck reaction of acyclic alkenes is replacement of one of the hydrogens on the double bond by an organo group. If the organopalladium group attacks the double bond so that the R group in the original RX is bonded to a carbon that lacks a hydrogen, or if the only syn hydrogen is on a neighboring carbon, the double bond shifts away from the original position. Note that the product of the following reaction contains a chiral center, but, because it is formed from achiral reagents in an achiral environment, it is formed as a racemic mixture.

Formed as a
racemic mixture

As mentioned earlier, a particularly valuable feature of the Heck reaction is that, when used with a chiral ligand, it can give chiral products in significant enantiomeric excess (ee). In the following, the chirality is provided by the chiral ligand (*R*)-BINAP (Section 6.7C).

This enantiomer
is formed in 71% ee

For this reaction to yield a chiral product, the hydrogen eliminated cannot be on the carbon that subsequently obtains the aryl substituent, because, if this were the case, the substituent would be attached to a double bond, and the product would be achiral.

Because of the chiral ligand, the activation energy for the transition state in the syn addition to the alkene (Step 2 of the catalytic cycle) is different depending on which side of the alkene the metal complex approaches (the two transition states are diastereomers). This difference in activation energy means that approach to one side of the alkene is favored and results in an excess of one enantiomer of the product. Note that this reaction is not a normal Heck reaction in that it forms a carbon-carbon bond to the more substituted carbon, and the double bond shifts. Attack at the other carbon, because of the requirement for syn elimination, cannot lead to a normal Heck product, and therefore the reaction reverses. The attack takes place at the more substituted carbon less often, but in this case there is a hydrogen that can undergo elimination.

No H here

A syn H here

Attack at more
substituted carbon

Attack at less
substituted carbon

No syn
hydrogen here

Example 24.2

Heck reaction of bromobenzene and (E)-3-hexene gives a mixture of (Z)-3-phenyl-3-hexene and (E)-4-phenyl-2-hexene in roughly equal amounts. Account for the formation of these two products.

(E)-3-Hexene + C_6H_5Br $\xrightarrow[\text{(CH}_3\text{CH}_2)_3\text{N}]{\text{Pd(OAc)}_2,\ 2\text{Ph}_3\text{P}}$

(Z)-3-Phenyl-3-hexene + **(E)-4-Phenyl-2-hexene**

Solution

Syn addition gives the product shown. After rotation, syn elimination of the H on the original double bond gives (Z)-3-phenyl-3-hexene; syn elimination on the neighboring carbon (in its most stable conformation) gives (E)-4-phenyl-2-hexene.

(E)-3-Hexene + C_6H_5Br $\xrightarrow[\text{(CH}_3\text{CH}_2)_3\text{N}]{\text{Pd(OAc)}_2,\ 2\ \text{Ph}_3\text{P}}$

Rotate

Syn elimination

Syn elimination

(Z)-3-Phenyl-3-hexene + **(E)-4-Phenyl-2-hexene** (racemic)

Problem 24.2

Give reagents and conditions for the following reaction.

24.4 | Catalytic Allylic Alkylation

Substitution mechanisms were covered in Chapter 9. It was noted that the S_N2 mechanism occurs by a single-step process wherein a leaving group (often a halogen) on an alkyl group is replaced with a nucleophile. When the alkyl group is allylic, metals can catalyze the reaction, i.e., palladium, platinum, and rhodium, among others.

Two of the most common catalysts for this reaction are $Pd(L)_4$ (frequently L is triphenylphosphine) and $PdCl_2$. Two examples of the reaction are given below. One interesting feature is the retention of stereochemistry at the carbon with the leaving group. Note that this outcome is the opposite of what would occur in an S_N2 mechanism. A particularly useful feature of the reaction is the ability to use enolates as the nucleophile, thus resulting in C—C bond formation. The enolates most commonly used are those derived by deprotonation of a hydrogen that is alpha to two electron-withdrawing groups: ketones, aldehydes, nitro, esters, sulfonates, and cyanides are some examples. The leaving groups can be halogens as with S_N2 reactions, but this catalytic reaction is particularly useful with esters as the leaving group. Acetate, written as OAc, is most commonly used (below in both examples).

Example 24.3

Write the products of the following reactions.

Solution

As stated above, the reaction occurs with retention of stereochemical configuration at the carbon with the acetate leaving group. Consequently, in order to write the products, we simply replace the acetate with the enolate, writing a C—C bond between the enolate carbon and the carbon that bears the acetate.

Problem 24.3

Write the product(s) of the following reaction. How many stereoisomers of the product are formed?

Although allylic alkylation with halogens as the leaving group occurs readily without a catalyst with enolate nucleophiles, several advantages exist in having the reaction be catalytic. There is, of course, the advantage of having the reaction occur faster and potentially under milder conditions. Also acetate is not normally a good leaving group in an S_N2 reaction. However, the real primary advantage is in reversing the stereochemical outcome of the reaction as compared to S_N2. Before examining the reason behind the stereochemistry of the reaction, let's first take a look at the mechanism involving one particularly common catalyst, $Pd(PPh_3)_4$.

A. The Mechanism of Catalytic Allylic Alkylation

The mechanism of the reaction is a combination of the simple steps we have seen before. The phosphine ligands reversibly dissociate and associate, and an oxidative addition occurs. The new step that occurs in catalytic allylic alkylation is nucleophilic attack on an allyl-ligand coordinated to a metal.

Mechanism   The Catalytic Cycle for Allylic Alkylation

The catalytic cycle of allylic alkylation has several steps, six of which are shown below. The cycle is initiated by dissociation of a ligand ($L = PPh_3$, step 1), followed in step 2 by coordination of the allylic species to make a π–complex and another ligand loss (step 3). After step 3, the Pd is in the zero oxidation state and can undergo oxidative addition of the coordinated allylic species. The oxidative addition leads to expulsion of the leaving group and replacement of the allyl-X bond with an allyl-Pd bond. This newly formed complex has two contributing structures in which either terminal carbon of the allyl group can be envisioned as having the Pd-C bond with the remaining two carbons involved in a π–complex. It is common in organometallic chemistry to represent the two contributing structures involved in an allyl complex with three carbons and an arc (see the figure on the next page). The interaction of three carbons to one metal is denoted by the prefix η^3; such a complex is called an η^3-allyl complex (η is pronounced "eta").

After the oxidative addition, the coordinated allyl group is susceptible to nucleophilic attack from solution (step 5). This attack completes the substitution, and all that is left to start the cycle again would be coordination of a phosphine ligand (L) and loss of the organic product.

B. Stereochemical and Regiochemical Issues

Several steps in the catalytic cycle are stereoselective and therefore result in the ability to control the configuration at chiral carbons. First, the oxidative addition of the allylic-LG species occurs with clean inversion of configuration (step 4 of the mechanism). The nucleophilic attack on the η^3-allyl complex occurs from solution in an analogous fashion to an S_N2 reaction, and therefore also occurs with clean inversion of stereochemistry. The net effect of two consecutive inversions of stereochemistry is overall retention of stereochemistry. The following example highlights the stereochemistry of the two steps that we are considering (Y is an electron-withdrawing group).

The reaction is also very regioselective. Nucleophilic attack occurs at the less substituted end of the of the η^3-allyl complex regardless of the initial position of the leaving group.

Pd(PPh₃)₄ / NaCH(CO₂CH₃)₂

dominant product

Pd(PPh₃)₄ / NaCH(CO₂CH₃)₂

dominant product

24.5 | Palladium-Catalyzed Cross-Coupling Reactions

Catalytic Cross-Coupling Reaction
A reaction wherein a C—C bond is formed in a catalytic fashion between alkyl, aryl, alkenyl, or alkynyl groups.

Transmetallation
Interchange of ligands between two metals or metalloids.

Arguably, the largest impact that organometallic chemistry has had on organic synthesis involves a series of reactions that are classified as **cross-coupling reactions**, many of which are catalyzed by palladium. A cross-coupling reaction is defined as one that creates a C—C bond by coupling together two alkyl, aryl, alkenyl, or alkynyl groups, as we saw with the Gilman reaction in Section 15.2. Yet, the reactions we are now examining are catalytic. There are a large number of these reactions, most of which are named after the chemists primarily associated with their creation. We will examine three in this book: the Suzuki, Stille, and Sonogashira couplings.

Cross-couplings reactions involve a transmetallation step. A **transmetallation** is a pairwise interchange of ligands between two different metals or metalloids. In the case of palladium-catalyzed cross-coupling reactions, the other metal/metalloid is commonly Zr, Sn, B, Zn, Cu, or Mg, which we designate as M in the following general example.

$$R—Pd \ + \ R'—M \ \longrightarrow \ R'—Pd \ + \ R—M$$

A. General Mechanism for Cross-Coupling Reactions

In 2010, Richard Heck, Ei-ichi Negishi, and Akira Suzuki shared the Nobel Prize in Chemistry for their work on C—C coupling reactions.

For simplicity sake, all the catalytic cross-coupling reactions can be represented by one general mechanism, even though each has subtle differences that we describe in the sections below. The catalytic cycle is so simple that it is presented here with only three steps. The differences between M, L, L', and X are what differentiate and classify a particular reaction.

Mechanism The Catalytic Cycle of Cross-Coupling

Various Pd(0) or Pd(II) species are used in the catalytic reactions. Step 1 involves oxidative addition of one of the organic species to the palladium. A transmetallation in step 2 results in the palladium having two carbon-based ligands. A reductive elimination in step 3 couples the two carbon fragments together. The relative simplicity and variability of each component involves has made this catalytic cycle a very powerful one.

B. The Suzuki Coupling

The Suzuki coupling was developed by Professor Akira Suzuki of Hokkaido University. The Suzuki coupling uses a boron compound (R'-BY_2) and an alkenyl, aryl, or alkynyl halide or triflate (RX) as the carbon sources, with a palladium salt as the catalyst. Bromides and iodides are the most commonly used halides; chlorides are less reactive. Alkyl halides can sometimes be used but are subject to elimination. A base is also required. The boron compound can be a borane (R'_3B), a borate ester ($R'B(OR)_2$), or a boric acid ($R'B(OH)_2$), where R' is alkyl, alkenyl, or aryl. The general reaction is shown in the following scheme, where X is halide or triflate and Y is alkyl, alkoxyl, or OH. A list of the types of components that can be used is given in Table 24.1. This reaction is one of the principal methods now used to prepare biaryls.

$$RX + R'\text{-}BY_2 \xrightarrow[\substack{PdL_4 \\ Base}]{} R\text{-}R' + XBY_2$$

Boranes are easily prepared from alkenes or alkynes by hydroboration (Section 6.4); borates are made from aryl or alkyl lithium compounds and trimethyl borate, among other routes.

Following are three examples of the reaction that show its versatility.

Table 24.1 | Suzuki Coupling Components Where One of the Organoboron Compounds Couples with One of the Coupling Reagents Shown

Organoboron compounds	Coupling reagents X = halide or triflate
(phenyl)—B	(phenyl)—X
RCH=CH—B	RCH=CH—X
Alkyl—B	RC≡CH—X
	Alkyl-X (Difficult)

The mechanism of the reaction starts with an oxidative addition, followed by a transmetallation in which the substituent on the borane replaces the ligand on the palladium, concluding with a reductive elimination of the palladium to form the new C—C bond. The base may serve as a new, labile ligand for the palladium, or the base may activate the borane by coordination.

Oxidative addition and ligand exchange:

$$R—X \xrightarrow{PdL_n} R—Pd—X \xrightarrow{RO^-} R—Pd—OR$$

Borane activation:

$$R_3B \xrightarrow{RO^-} R_3B^- — OR$$

Reaction:

Example 24.4

Show how the following penicillin analog can be prepared from the indicated starting material and any other necessary compounds.

Solution

Problem 24.4

Show how the following compound can be prepared from starting materials containing eight carbons or less.

C. The Stille Coupling

The second Pd(0) catalyzed cross-coupling reaction we cover is the Stille coupling, which involves the use of vinyl or aryl tin reagents (called stannanes) as the transmetallating agents. Coupling with another vinyl or aryl group leads to the creation of conjugated dienes or an alkenylarene. The coupling occurs not only with regioselectivity but also retains the stereochemistry from the reactants to the products.

Many stannane reagents are commercially available, or they can be readily synthesized via reaction between a Grignard reagent and tri-n-butyl tin chloride. The reactants that most commonly react with the transmetallated group are a vinyl triflate (C=C—OSO$_2$CF$_3$) and a vinyl iodide. Vinyl triflates are prepared from the reaction of an enolate with N-phenyl triflimide (PhNTf$_2$, Tf = OSO$_2$CF$_3$).

Example 24.5

When a conjugated diene is the desired product in a reaction, a Stille coupling is a logical choice to invoke during the synthesis. Write the correct stannane starting material and a vinyl iodide reactant that would couple with Pd(0) to give the following product.

Solution

To consider the proper starting materials for a Stille coupling, dissect the central C—C bond of the diene into two parts in a retrosynthetic fashion. One reactant should be a vinyl iodide (or triflate) while the other reactant is a stannane.

Problem 24.5

What is the product of the following reaction?

D. The Sonogashira Coupling

The last Pd(0) catalyzed cross-coupling reaction covered in this chapter is the Sonogashira coupling. It involves transmetallation of an alkynyl-Cu(I) species to Pd, followed by coupling to an aryl or vinyl iodide or triflate. The Cu(I) alkynyl complex is created in situ by the reaction of a terminal alkyne with CuI in the presence of triethylamine. The reaction is most commonly used to create diaryl alkynyl products.

Example 24.6

Two Sonogashira coupling reactions can be used to make unsymmetrical diaryl alkynes by first using trimethylsilyl acetylene. The trimethylsilyl protecting group can be removed by addition of fluoride (usually tetrabutyl ammonium fluoride, see Section 11.6). Show how such a sequence of reactions can be used to construct the following product when one reactant is phenyl iodide.

Solution

Sonogashira coupling conditions using trimethylsilyl acetylene give phenyl acetylene after deprotection using fluoride. Another such coupling using the phenyl iodide reactant shown gives the product.

Problem 24.6

What sequence of reactions will produce the following product if starting with trimethylsilyl acetylene and the appropriate two aryl iodides?

24.6 | Alkene Metathesis

Recently, a novel catalytic reaction leading to alkene metathesis has been developed. Robert Grubbs of the California Institute of Technology and Richard Schrock of the Massachusetts Institute of Technology made major contributions to this chemistry. Together their work has provided a remarkably easy and general way to generate carbon-carbon double bonds, even in complex molecules. In an **alkene metathesis** reaction, two alkenes interchange the carbons attached to their double bonds.

In 2005, Robert Grubbs, Richard Schrock, and Yves Chauvin shared the Nobel prize in chemistry for their work on metathesis reactions.

Alkene metathesis
In an alkene metathesis reaction, two alkenes interchange the carbons attached to their double bonds.

A. Stable Nucleophilic Carbenes

We discussed carbenes and carbenoids (derivatives of divalent carbon) in Section 15.3, where we saw that these compounds provide one of the best routes to three-membered rings, making two C—C bonds in the process. Certain carbenes with strongly electron-donating substituents are particularly stable. Their stability can be enhanced further by adding sterically bulky substituents that hinder self-reactions. For example, the following cyclic carbene is stable enough to isolate. In this case, the large 2,4,6-trimethylphenyl substituents protect the carbene from attack by nucleophiles or oxygen. Rather than being electron-deficient like most carbenes, these compounds are nucleophiles because of the strong electron donation by the nitrogens. Because of their nucleophilicity, they are excellent ligands (resembling phosphines) for certain transition metals.

B. Ring-Closing Alkene Metathesis Using Nucleophilic Carbene Catalysts

These stable carbenes (and others that are less stable) provide ligands for certain metals that are catalysts for the alkene metathesis reaction. As we saw at the beginning of this section, this reaction is an equilibrium. However, it can be an effective means of forming new carbon-carbon double bonds if the equilibrium can be driven in the desired direction. For example, if the reaction involves two 2,2-disubstituted alkenes of the type $R_2C\!=\!CH_2$, one of the products is ethylene. Loss of gaseous ethylene drives the reaction to the right, giving a single alkene as product.

Ethylene

A particularly useful variant of this reaction uses a starting material in which both alkenes are in the same molecule. In this case, the product is a cycloalkene, and the reaction is called ring-closing alkene metathesis. Ring sizes up to 26 and higher have been prepared by ring-closing alkene metathesis. This reaction is amazingly general and synthetically useful.

Example 24.7

Show how the following compound can be prepared from an acyclic diene.

Solution

Ring-closing alkene metathesis gives the product in one step.

(racemic)   (racemic)

Problem 24.7

Show the product of the following reaction.

A particularly useful alkene metathesis catalyst consists of ruthenium complexes with a nucleophilic carbene and another carbenoid ligand, C_6H_5CH=[M], where [M] is the metal with its ligands.

For a model of the catalyst shown here, see the opening page of this chapter.

C. Mechanism of the Metathesis Reaction

The mechanism of the alkene metathesis reaction also involves a catalytic cycle. A key step involves addition of the metallocarbenoid to the alkene to give a four-membered metallacycle. This metallacycle is unstable and can either revert to starting material or eliminate an alkene in the opposite direction to give a new alkene. Addition is not regioselective; consequently, all possible combinations of R_1 and R_2 result. In this scheme, the catalyst is R_1CH=[M].

cis or *trans*   A metallacycle

In this section, we have concentrated on the use of transition-metal nucleophilic-carbene catalysts to bring about ring-closing alkene metathesis reactions. These same types of compounds can also be used to catalyze a remarkable reaction called ring-opening alkene metathesis polymerization (ROMP). A special value of ROMP is that it can be used to prepare highly unsaturated polymers. For a discussion of ROMP techniques, see Section 29.6E.

Summary

Section 24.1 Carbon-Carbon Bond-Forming Reactions from Earlier Chapters

- The classical methods for making carbon-carbon bonds during synthesis can be grouped broadly into these categories:
  - Displacement of a leaving group by a carbon nucleophile (Gilman reagents, alkyne anions, enolate anions, and enamine alkylations)
  - Nucleophilic addition to a carbonyl or a carboxyl group, usually involving enolate nucleophiles (Grignard, alkyne anion, cyanide, aldol, Claisen, enamine, Wittig)
  - Conjugate addition to an α,β-unsaturated compound (Michael reaction)
  - Aromatic substitution (Friedel-Crafts)

Section 24.2 Organometallic Compounds and Catalysis

- Two important reactions of transition metals and their compounds are **oxidative addition** and its complement, **reductive elimination**.
  - Oxidative addition occurs when a reagent adds to a metal, causing its coordination to increase by two ligands, and reductive elimination is the reverse.
  - The terms "oxidative" and "reductive" refer to the change in formal charge on the metal that occurs during these reactions.

– Reagents such as organohalogen, hydrogen, halogens, and many others can react with metals in these ways.
• The catalytic properties of transition metals are manipulated by adding different **ligands**, which are Lewis bases that coordinate to the metal.
– Ligands can be used to modify the electronic properties, steric crowding, and even chirality around the metal in some situations.
• **Green chemistry** and **atom economy** are relatively new thrusts in the chemical industry worldwide; their goals are well complemented by organometallic catalytic reactions.

Section 24.3 The Heck Reaction

• In the **Heck reaction**, the H atom on an alkene (vinylic hydrogen) is substituted by a haloalkene or haloarene in the presence of base and a small amount of Pd catalyst.
– When there is a difference, the substitution occurs at the less substituted carbon of the alkene, and is often stereoselective for the E product.
– The configuration of the haloalkene (when appropriate) is conserved.
– A significant advantage of the Heck reaction is that alcohol, ether, aldehyde, ketone, and ester functional groups are compatible with the reaction.
– The organohalogen, alkene, and base are used in stoichiometric amounts, and the Pd catalyst is used in small amounts.
– When there is no syn hydrogen from the original alkene double bond that can eliminate in the last step of the reaction, the double bond shifts away from the original position so that a syn elimination of an H atom can take place.

Problems: 24.1, 24.2, 24.8–24.18, 24.32

Section 24.4 Catalytic Allylic Alkylation

• In catalytic allylic alkylation, a nucleophile, commonly an enolate of a doubly activated α-carbon, replaces an allylic leaving group, commonly an ester such as acetate. The reaction occurs in the presence of a catalytic amount of Pd(0).
– In contrast to S_N2 allylic alkylation, stereochemistry at the alkylated carbon is retained.
– The mechanism consists of a multistep cycle involving η^3-allyl Pd complexes.
– The reaction is regioselective with allylation occurring preferentially at the less substituted end of the η^3-allyl complex irrespective of the initial position of the leaving group.

Problems: 24.3, 24.19–24.22

Section 24.5 Palladium-Catalyzed Cross-Coupling Reactions

• Cross-coupling reactions all have very similar mechanisms. The first step is oxidative addition to Pd of an R-X species, followed by **transmetallation** of an R′ group from an R′-metal/metalloid species. Reductive elimination of R-R′ from palladium completes the cycle.
• The **Suzuki coupling** uses a boron reagent (R′-BY$_2$) with an alkenyl, aryl, or alkynyl halide (usually Br or I) or triflate with a palladium salt to give a new carbon-carbon bond.
– The boron compound can be a borane (R′$_3$B), borate ester (R′-B(OR)$_2$), or boric acid (R′-B(OH)$_2$) where R′ is an alkyl, alkenyl, or aryl group. Boranes are made using hydroboration of alkenes or alkynes. Borates are made from aryl or alkyl lithium compounds and trimethyl borate.
– The Suzuki reaction is particularly good for the construction of biaryl compounds.

Problems: 24.3, 24.23, 24.24, 24.38

Problems: 24.5, 24.25–24.30

• The **Stille coupling** uses a tin reagent (Bu$_3$Sn-R) in which the R-group is commonly a vinyl species with an alkenyl, aryl, or alkynyl halide (often iodide or triflate) with palladium to give a new carbon-carbon bond.
– The tin reagent (a stannane) is created from a Grignard reagent of the R-group and n-Bu$_3$SnCl, while the triflates are often derived from enolates.
– The Stille coupling is most commonly used to make conjugate dienes or alkenyl aryl systems.

- The **Sonogashira coupling** starts with a terminal alkyne with CuI and triethylamine to create a Cu-alkyne complex. This undergoes vinyl or aryl iodides.
  - This coupling is routinely used to create diaryl alkynes, aryl alkenyl alkynes, or dialkenyl alkynes.
  - It is common to use trimethylsilyl acetylene with two sequential Sonogashira reactions when unsymmetrical alkynes are desired as the final products.

Problems: 24.6, 24.25–24.28, 24.31

Section 24.6 Alkene Metathesis

- In an **alkene metathesis** reaction, two alkenes interchange the carbons attached to their double bonds.

Problems: 24.4, 24.33

- The catalyst is a transition metal complex such as Ru complexes of stable nucleophilic carbenes (highly sterically hindered nitrogen heterocycles).
- The metathesis reaction is usually an equilibrium process, driven to completion by using two terminal alkenes that give gaseous ethylene as a product, which bubbles out of the reaction.
  - A particularly useful version of the metathesis reaction, called ring-closing alkene metathesis, involves two terminal alkenes on the same molecule, leading to an intramolecular reaction that creates a cycloalkene product.
  - Ring-closing alkene metathesis has been used to construct very large ring sizes that are hard to make in other ways.

◕WL Questions and problems for this chapter are available in OWL. Use the OWL access card that came with this textbook to access assigned questions and problems for this chapter.

Key Reactions

1. **The Heck Reaction (Section 24.3)** In a palladium(0)-catalyzed reaction, the carbon group of a haloalkene (a vinylic halide) or haloarene is substituted for a hydrogen on a carbon-carbon double bond (a vinylic hydrogen) of an alkene. Reaction generally proceeds with a high degree of both stereoselectivity and regioselectivity. The small amount of palladium catalyst is generally introduced as a precatalyst in the form of $Pd(OAc)_2$, which is Pd(II), and is reduced in the reaction to Pd(0) by either reaction with the alkene (only a small amount of which is lost because the catalyst is used in small amounts) or a reagent such as triethylamine. The Pd(0) then reacts with two ligands, generally phosphine ligands (L), to create the active catalyst PdL_2. The phosphine ligands can be chiral, such as BINAP, so that single enantiomer products are possible for reactions that create new chiral centers.

   The catalytic cycle involves five steps, the most important of which are oxidative addition of the organohalogen to the catalyst, syn addition of the alkene, syn elimination of the new alkene product, and reductive elimination of HX (neutralized by the added base) to regenerate the catalyst.

2. **Catalytic Allylic Alkylation (Section 24.4)** Catalytic allylic alkylation commonly takes allyl acetate species and substitutes the acetate with a nucleophile. Some of the most useful nucleophiles are enolates derived from methylenes that are flanked by two electron-withdrawing groups. The mechanism of the reaction involves the oxidative addition of the allyl acetate to palladium and results in the intermediacy of η^3-allyl complexes that give inversion of configuration at the carbon with the acetate leaving group.

Attack by the nucleophile from solution gives a second inversion of configuration and, by virtue of having two S_N2-like reactions, results in overall retention of configuration. The regiochemistry in the reaction is also highly selective, with nucleophilic attack from solution being preferential at the least-substituted carbon of the η^3-allyl complex.

3. The Suzuki Coupling (Section 24.5B) The Suzuki coupling reaction is a palladium-catalyzed reaction of an organoboron compound with an organic halide or triflate. The mechanism involves transmetallation, in which the substituent on the borane replaces a ligand on palladium, followed by reductive elimination to form the new C—C bond.

4. The Stille Coupling (Section 24.5C) The Stille coupling is the palladium-catalyzed reaction of a vinyl tin reagent with an organic halide or triflate. The mechanism involves oxidative addition of the organic halide/triflate, transmetallation of the vinyl group on Sn to Pd, and reductive elimination to form the new C—C bond.

5. The Sonogashira Coupling (Section 24.5D) The Sonogashira coupling is the palladium-catalyzed reaction of a Cu(I)-alkynyl complex with a vinyl or aryl iodide. The Cu(I)-alkynyl compound is created by the reaction of a terminal alkyne with CuI in the presence of an amine base. The coupling mechanism involves oxidative addition of the organic iodide to Pd, transmetallation of the alkynyl group to Pd from Cu, and reductive elimination to form the new C—C bond.

6. Alkene Metathesis (Section 24.6) The alkene metathesis reaction is an organometallic-catalyzed reaction in which two alkenes exchange carbons of their double bonds. In a ring-closing alkene metathesis reaction, both alkenes are in the same molecule, and the product is a cycloalkene. Catalysts with Ru and Mo are often used; a nucleophilic carbene complex of Ru is particularly useful. The catalytic cycle involves reaction of the metal catalyst with the alkenes to form a four-membered ring metallacycle, which decomposes either to give starting materials or, by elimination in the opposite direction, to give a new alkene.

Chapter 25

Carbohydrates

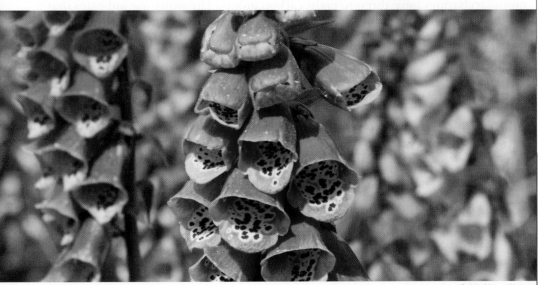

© Keith M. Law/Alamy

Foxglove *(Digitalis purpurea)*, an ornamental flowering plant, is the source of digitoxin and digitalis, medicines used in cardiology to reduce pulse rate, regularize heart rhythm, and strengthen heart beat. Inset: digitoxose, a monosaccharide obtained on hydrolysis of digitoxin. See Problem 25.15.

Carbohydrates are the most abundant organic compounds in the plant world. They act as storehouses of chemical energy (glucose, starch, glycogen); are components of supportive structures in plants (cellulose), crustacean shells (chitin), and connective tissues in animals (glucosaminoglycans); and are essential components of nucleic acids (D-ribose and 2-deoxy-D-ribose). Carbohydrates make up about three fourths of the dry weight of plants. Animals (including humans) get their carbohydrates by eating plants, but they do not store much of what they consume. Less than 1% of the body weight of animals is made up of carbohydrates.

The name *carbohydrate* means hydrate of carbon and derives from the formula $C_n(H_2O)_m$. Following are two examples of carbohydrates with molecular formulas that can be written alternatively as hydrates of carbon.

Glucose (blood sugar): $C_6H_{12}O_6$, or alternatively $C_6(H_2O)_6$
Sucrose (table sugar): $C_{12}H_{22}O_{11}$, or alternatively $C_{12}(H_2O)_{11}$

Not all carbohydrates, however, have this general formula. Some contain too few oxygen atoms to fit this formula, and some others contain too many oxygens. Some also contain nitrogen. The term carbohydrate has become so firmly rooted in chemical nomenclature that, although not completely accurate, it persists as the name for this class of compounds.

At the molecular level, most **carbohydrates** are polyhydroxyaldehydes, polyhydroxyketones, or compounds that yield either of these after hydrolysis. Therefore, the chemistry of carbohydrates is essentially the chemistry of hydroxyl groups and carbonyl groups and of the acetal bonds formed between these two functional groups.

Carbohydrate
A polyhydroxyaldehyde, a polyhydroxyketone, or a substance that gives these compounds on hydrolysis.

OWL Online homework for this chapter may be assigned in OWL for Organic Chemistry.

The fact that carbohydrates have only two types of functional groups, however, belies the complexity of their chemistry. All but the simplest carbohydrates contain multiple chiral centers. For example, glucose, the most abundant carbohydrate in the biological world, contains one aldehyde group, one primary and four secondary hydroxyl groups, and four chiral centers. Working with molecules of this complexity presents enormous challenges to organic chemists and biochemists alike.

25.1 | Monosaccharides

A. Structure and Nomenclature

Monosaccharides have the general formula $C_nH_{2n}O_n$ with one of the carbons being the carbonyl group of either an aldehyde or a ketone. The most common monosaccharides have three to eight carbon atoms. The suffix *-ose* indicates that a molecule is a carbohydrate, and the prefixes *tri-*, *tetr-*, *pent-*, and so forth indicate the number of carbon atoms in the chain. Monosaccharides containing an aldehyde group are classified as **aldoses**; those containing a ketone group are classified as **ketoses**.

Monosaccharides Classified by Number of Carbon Atoms	
Name	**Formula**
Triose	$C_3H_6O_3$
Tetrose	$C_4H_8O_4$
Pentose	$C_5H_{10}O_5$
Hexose	$C_6H_{12}O_6$
Heptose	$C_7H_{14}O_7$
Octose	$C_8H_{16}O_8$

There are only two trioses: the aldotriose glyceraldehyde and the ketotriose dihydroxyacetone.

Glyceraldehyde
(an aldotriose)

Dihydroxyacetone
(a ketotriose)

Often the designations *aldo-* and *keto-* are omitted, and these molecules are referred to simply as trioses, tetroses, and the like.

Glyceraldehyde is a common name; the IUPAC name for this monosaccharide is 2,3-dihydroxypropanal. Similarly, dihydroxyacetone is a common name; its IUPAC name is 1,3-dihydroxypropanone. The common names for these and other monosaccharides, however, are so firmly rooted in the literature of organic chemistry and biochemistry that they are used almost exclusively to refer to these compounds. Therefore, throughout our discussions of the chemistry and biochemistry of carbohydrates, we use the names most common in the literature of chemistry and biochemistry.

B. Fischer Projection Formulas

Glyceraldehyde contains a chiral center and therefore exists as a pair of enantiomers.

Monosaccharide
A carbohydrate that cannot be hydrolyzed to a simpler carbohydrate.

Aldose
A monosaccharide containing an aldehyde group.

Ketose
A monosaccharide containing a ketone group.

1,3-Dihydroxypropanone, more commonly known as dihydroxyacetone, is the active ingredient in artificial tanning agents such as Man-Tan and Magic Tan.

$$\underset{\substack{\text{CHO} \\ \text{H}\blacktriangleright\text{C}\blacktriangleleft\text{OH} \\ \text{CH}_2\text{OH}}}{}$$

(R)-Glyceraldehyde

$$\underset{\substack{\text{CHO} \\ \text{HO}\blacktriangleright\text{C}\blacktriangleleft\text{H} \\ \text{CH}_2\text{OH}}}{}$$

(S)-Glyceraldehyde

Chemists commonly use two-dimensional representations called **Fischer projections** (Section 3.4C) to show the configuration of carbohydrates. Following is an illustration of how a three-dimensional representation is converted to a Fischer projection.

$$\underset{\substack{\text{CHO} \\ \text{H}\blacktriangleright\text{C}\blacktriangleleft\text{OH} \\ \text{CH}_2\text{OH}}}{} \xrightarrow[\text{Fischer projection}]{\text{convert to a}} \quad \text{H}\!-\!\!\!\begin{array}{c}\text{CHO}\\ \rule[0.5ex]{1.2em}{0.4pt}\\ \text{CH}_2\text{OH}\end{array}\!\!\!-\text{OH}$$

(R)-Glyceraldehyde
(three-dimensional
representation)

(R)-Glyceraldehyde
(Fischer projection)

The horizontal segments of a Fischer projection represent bonds directed toward you, and the vertical segments represent bonds directed away from you. The only atom in the plane of the paper is the chiral center.

Fischer projection
A two-dimensional representation for showing the configuration of chiral centers; horizontal lines represent bonds projecting forward, and vertical lines represent bonds projecting to the rear.

C. D- and L-Monosaccharides

Even though the *R*, *S* system is widely accepted today as a standard for designating configuration, the configuration of carbohydrates as well as those of amino acids and many other compounds in biochemistry is commonly designated by the D,L system proposed by Emil Fischer in 1891. At that time, it was known that one enantiomer of glyceraldehyde has a specific rotation of +13.5; the other has a specific rotation of −13.5. Fischer proposed that these enantiomers be designated D and L (for dextro- and levorotatory) but he had no experimental way to determine which enantiomer has which specific rotation. Fischer, therefore, did the only possible thing—he made an arbitrary assignment. He assigned the dextrorotatory enantiomer an arbitrary configuration and named it D-glyceraldehyde. He named its enantiomer L-glyceraldehyde.

D-Glyceraldehyde
$[\alpha]_\text{D}^{25} = +13.5$

L-Glyceraldehyde
$[\alpha]_\text{D}^{25} = -13.5$

Fischer could have been wrong, but by a stroke of good fortune, he was correct, as proven in 1952 by a special application of X-ray crystallography.

D- and L-glyceraldehyde serve as reference points for the assignment of relative configuration to all other aldoses and ketoses. The reference point is the chiral center farthest from the carbonyl group. Because this chiral center is always the next to the last carbon on the chain, it is called the **penultimate carbon**. A **D-monosaccharide** has the same configuration at its penultimate carbon as D-glyceraldehyde (its —OH is on the right when written as a Fischer projection); an **L-monosaccharide** has the same configuration at its penultimate carbon as L-glyceraldehyde (its —OH is on the left).

Note that for monosaccharides with two or more chiral centers, the designations D or L refer only to the configuration of the highest-numbered chiral center, that is, the chiral center farthest from the aldehyde or ketone carbonyl group.

D-Monosaccharide
A monosaccharide that, when written as a Fischer projection, has the —OH on its penultimate carbon to the right.

L-Monosaccharide
A monosaccharide that, when written as a Fischer projection, has the —OH on its penultimate carbon to the left.

Also, note that the D or L designation of a given monosaccharide does not specify the sign of the specific rotation of the compound. If the sign of the rotation of plane-polarized light is to be specified in a name, it is indicated by a + (plus) or a − (minus) sign preceding the name. Thus, D-glucose, which is dextrorotatory, is designated D(+)glucose, and D-fructose, which is levorotatory, is designated D(−)fructose.

Table 25.1 shows names and Fischer projections for all D-aldotetroses, pentoses, and hexoses. Each name consists of three parts. The letter D specifies the configuration of the penultimate carbon. Prefixes such as *rib-*, *arabin-*, and *gluc-* specify the configuration of all other chiral centers in the monosaccharide. The suffix *-ose* shows that the compound is a carbohydrate.

The three most abundant hexoses in the biological world are D-glucose, D-galactose, and D-fructose. The first two are D-aldohexoses; the third is a D-2-ketohexose. Glucose, by far the most common hexose, is also known as dextrose because it is dextrorotatory. Other names for this monosaccharide are grape sugar and blood sugar. Human blood normally contains 65–110 mg of

| Table 25.1 | Configurational Relationships Among the Isometric D-Aldotetroses, D-Aldopentoses, and D-Aldohexoses |

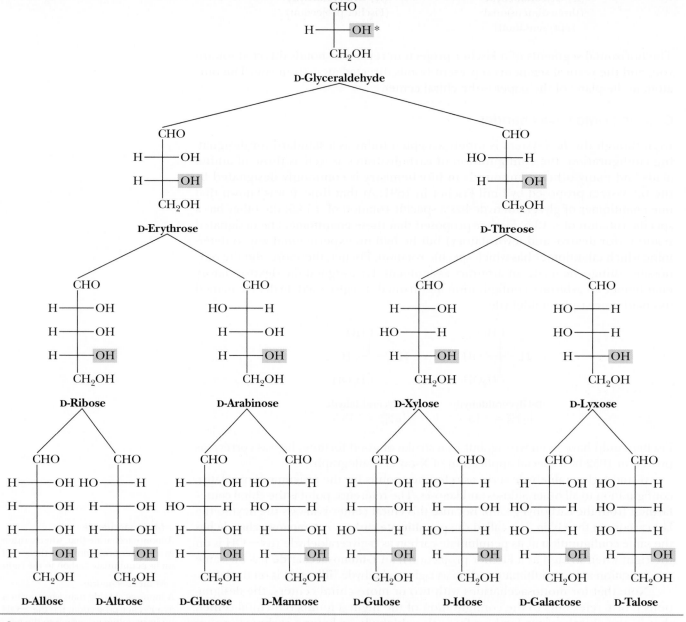

*The configuration of the reference —OH on the penultimate carbon is shown in color.

glucose/100 mL of blood. Glucose is synthesized by chlorophyll-containing plants using sunlight as a source of energy. In the process called photosynthesis, plants convert carbon dioxide from the air and water from the soil to glucose and oxygen.

$$6\ CO_2\ +\ 6\ H_2O\ +\ \text{energy} \xrightarrow[\text{chlorophyll}]{\text{sunlight}} C_6H_{12}O_6\ +\ 6\ O_2$$

Carbon dioxide   **Water**   **Glucose**   **Oxygen**

D-Fructose is found combined with D-glucose in the disaccharide sucrose (table sugar, Section 25.4A). D-Galactose is obtained with D-glucose in the disaccharide lactose (milk sugar, Section 25.4B).

D-Ribose and 2-deoxy-D-ribose, the most abundant pentoses in the biological world, are essential building blocks of nucleic acids; D-ribose in ribonucleic acids (RNA) and 2-deoxy-D-ribose in deoxyribonucleic acids (DNA).

Example 25.1

Draw Fischer projections for the four aldotetroses. Which are D-monosaccharides, which are L-monosaccharides, and which are enantiomers? Refer to Table 25.1, and write the name of each aldotetrose.

Solution

Following are Fischer projections for the four aldotetroses. The letters D- and L- refer to the configuration of the penultimate carbon, which, in the case of aldotetroses, is carbon 3. In the Fischer projection of a D-aldotetrose, the —OH on carbon 3 is on the right, and, in an L-aldotetrose, it is on the left. The erythroses are each diastereomers of each of the threoses.

D-**Erythrose**   L-**Erythrose**   D-**Threose**   L-**Threose**

Problem 25.1

Draw Fischer projections for all 2-ketopentoses. Which are D-2-ketopentoses, which are L-2-ketopentoses, and which are enantiomers?

D. Amino Sugars

Amino sugars contain an —NH₂ group in place of an —OH group. Only three amino sugars are common in nature: D-glucosamine, D-mannosamine, and D-galactosamine.

D-**Glucosamine**   D-**Mannosamine** (C-2 stereoisomer of D-glucosamine)   D-**Galactosamine** (C-4 stereoisomer of D-glucosamine)   N-**Acetyl-D-glucosamine**

N-Acetyl-D-glucosamine, a derivative of D-glucosamine, is a component of many polysaccharides, including chitin, the hard shell-like exoskeleton of lobsters, crabs, shrimp, and other shellfish. Many other amino sugars are components of naturally occurring antibiotics.

E. Physical Properties

Monosaccharides are colorless, crystalline solids, although they often crystallize with difficulty. Because hydrogen bonding is possible between their polar —OH groups and water, all monosaccharides are very soluble in water. They are only slightly soluble in ethanol and are insoluble in nonpolar solvents such as diethyl ether, chloroform, and benzene.

25.2 | The Cyclic Structure of Monosaccharides

We saw in Section 16.7B that aldehydes and ketones react with alcohols to form hemiacetals. We also saw that cyclic hemiacetals form very readily when hydroxyl and carbonyl groups are part of the same molecule and their interaction can form a five- or six-membered ring. For example, 4-hydroxypentanal forms a five-membered cyclic hemiacetal.

4-Hydroxypentanal

A cyclic hemiacetal

Note that 4-hydroxypentanal contains one chiral center and that a second chiral center is generated at carbon 1 as a result of hemiacetal formation.

Monosaccharides have hydroxyl and carbonyl groups in the same molecule. As a result, they too exist almost exclusively as five- and six-membered cyclic hemiacetals.

A. Haworth Projections

Haworth projection
A way to view furanose and pyranose forms of monosaccharides. The ring is drawn flat and most commonly viewed through its edge with the anomeric carbon on the right and the oxygen atom of the ring in the rear.

Anomeric carbon
The hemiacetal or acetal carbon of the cyclic form of a carbohydrate.

Anomers
Carbohydrates that differ in configuration only at their anomeric carbons.

A common way of representing the cyclic structure of monosaccharides is the **Haworth projection**, named after the English chemist Sir Walter N. Haworth (1937 Nobel Prize for chemistry). In a Haworth projection, a five- or six-membered cyclic hemiacetal is represented as a planar pentagon or hexagon, as the case may be, lying perpendicular to the plane of the paper. Groups bonded to the carbons of the ring then lie either above or below the plane of the ring. The new chiral center created in forming the cyclic structure is called an **anomeric carbon**. Stereoisomers that differ in configuration only at the anomeric carbon are called **anomers**. The anomeric carbon of an aldose is carbon 1; that of D-fructose, the most common ketose, is carbon 2.

Haworth projections are most commonly written with the anomeric carbon to the right and the hemiacetal oxygen to the back (Figure 25.1). In the terminology

β-D-Glucopyranose
(β-D-Glucose)

α-D-Glucopyranose
(α-D-Glucose)

Figure 25.1
Haworth projections for α-D-glucopyranose and β-D-glucopyranose.

of carbohydrate chemistry, the designation β means that the —OH on the anomeric carbon of the cyclic hemiacetal is on the same side of the ring as the terminal —CH₂OH. Conversely, the designation α means that the —OH on the anomeric carbon of the cyclic hemiacetal is on the side of the ring opposite from the terminal —CH₂OH.

A six-membered hemiacetal ring is indicated by the infix -*pyran*-, and a five-membered hemiacetal ring is indicated by the infix -*furan*-, The terms **furanose** and **pyranose** are used because monosaccharide five- and six-membered rings correspond to the heterocyclic compounds furan and pyran.

Furanose
A five-membered cyclic form of a monosaccharide.

Pyranose
A six-membered cyclic form of a monosaccharide.

Furan   Pyran

Because the α and β forms of glucose are six-membered cyclic hemiacetals, they are named α-D-glucopyranose and β-D-glucopyranose. These infixes are not always used in monosaccharide names, however. Thus, the glucopyranoses, for example, are often named simply α-D-glucose and β-D-glucose.

You would do well to remember the configuration of groups on the Haworth projections of α-D-glucopyranose and β-D-glucopyranose as reference structures. Knowing how the open-chain configuration of any other aldohexoses differs from that of D-glucose, you can then construct its Haworth projection by reference to the Haworth projection of D-glucose.

Example 25.2

Draw Haworth projections for the α and β anomers of D-galactopyranose.

Solution

One way to arrive at these projections is to use the α and β forms of D-glucopyranose as reference and to remember (or discover by looking at Table 25.1) that D-galactose differs from D-glucose only in the configuration at carbon 4. Thus, begin with the Haworth projections shown in Figure 25.1 and then invert the configuration at carbon 4.

Configuration differs from that of D-glucose at C-4.

α-D-Galactopyranose
(α-D-Galactose)

β-D-Galactopyranose
(β-D-Galactose)

α-D-Ribofuranose
(α-D-Ribose)

Problem 25.2

Mannose exists in aqueous solution as a mixture of α-D-mannopyranose and β-D-mannopyranose. Draw Haworth projections for these molecules.

β-2-Deoxy-D-ribofuranose
(β-2-Deoxy-D-ribose)

Figure 25.2
Haworth projections for some cyclic hemiacetals.

Aldopentoses also form cyclic hemiacetals. The most prevalent forms of D-ribose and other pentoses in the biological world are furanoses. Figure 25.2 shows Haworth projections for α-D-ribofuranose (α-D-ribose) and β-2-deoxy-D-ribofuranose (β-2-deoxy-D-ribose). The prefix 2-deoxy indicates the absence of oxygen at carbon 2.

The structure of L-ascorbic acid (vitamin C) resembles that of a monosaccharide. In fact, this vitamin is synthesized both biochemically by plants and some animals and commercially from D-glucose. Humans do not have the enzymes required for this synthesis and, therefore, we must obtain it in the food we eat or as a vitamin supplement. Approximately 66 million kilograms of vitamin C are synthesized every year in the United States.

L-Ascorbic acid is very easily oxidized to L-dehydroascorbic acid, a diketone. Both L-ascorbic acid and L-dehydroascorbic acid are physiologically active and are found together in most body fluids.

Ascorbic acid is one of the most important antioxidants (the H in the enolic OH is weakly bonded and easily abstracted by radicals). One of the most important roles it plays may be to replenish the lipid-soluble antioxidant α-tocopherol by transferring a hydrogen atom to the tocopherol radical, formed by reaction with radicals in the autoxidation process (see Section 8.7).

L-Ascorbic acid (Vitamin C) — oxidation/reduction — **L-Dehydroascorbic acid**

Units of D-ribose and 2-deoxy-D-ribose in nucleic acids and most other biological molecules are found almost exclusively in the β-configuration.

Other monosaccharides also form five-membered cyclic hemiacetals. Following are the five-membered cyclic hemiacetals of fructose.

α-D-Fructofuranose — **D-Fructose** — **β-D-Fructofuranose**

The β-D-fructofuranose form is found in the disaccharide sucrose (Section 25.4A).

B. Conformation Representations

A five-membered ring is so close to being planar that Haworth projections are adequate representations of furanoses. For pyranoses, however, the six-membered ring is more accurately represented as a chair conformation. Following are structural formulas for α-D-glucopyranose and β-D-glucopyranose drawn as chair conformations. Also shown is the open chain or free aldehyde form with which the cyclic hemiacetal forms are in equilibrium in aqueous solution.

β-D-Glucopyranose
(β-D-Glucose)

rotate about
C-1 to C-2 bond

α-D-Glucopyranose
(α-D-Glucose)

Notice that each group, including the anomeric —OH, on the chair conformation of β-D-glucopyranose is equatorial. Notice also that the —OH group on the anomeric carbon is axial in α-D-glucopyranose. Because of the equatorial orientation of the —OH on its anomeric carbon, β-D-glucopyranose is more stable and predominates in aqueous solution.

At this point, you should compare the relative orientations of groups on the D-glucopyranose ring in the Haworth projection and the chair conformation. The orientations of groups on carbons 1 through 5 of β-D-glucopyranose, for example, are up, down, up, down, and up in both representations.

β-D-Glucopyranose
(Haworth projection)

β-D-Glucopyranose
(chair conformation)

Example 25.3

Draw chair conformations for α-D-galactopyranose and β-D-galactopyranose. Label the anomeric carbon in each.

Solution

D-Galactose differs in configuration from D-glucose only at carbon 4. Therefore, draw the α and β forms of D-glucopyranose and then interchange the positions of the —OH and —H groups on carbon 4.

β-D-Galactopyranose
(β-D-Galactose)
$[\alpha]_D = +52.8$

D-Galactose

α-D-Galactopyranose
(α-D-Galactose)
$[\alpha]_D = +150.7$

Problem 25.3

Draw chair conformations for α-D-mannopyranose and β-D-mannopyranose. Label the anomeric carbon in each.

Mutarotation

The change in specific rotation that occurs when an α or β hemiacetal form of a carbohydrate in aqueous solution is converted to an equilibrium mixture of the two forms.

C. Mutarotation

Mutarotation is the change in specific rotation that accompanies the interconversion of α- and β-anomers in aqueous solution. As an example, a solution prepared by dissolving crystalline α-D-glucopyranose in water shows an initial rotation of +112, which gradually decreases to an equilibrium value of +52.7 as α-D-glucopyranose reaches an equilibrium with β-D-glucopyranose. A solution of β-D-glucopyranose also undergoes mutarotation, during which the specific rotation changes from an initial value of +18.7 to the same equilibrium value of +52.7. The equilibrium mixture consists of 64% β-D-glucopyranose and 36% α-D-glucopyranose. It contains only traces (0.003%) of the open-chain form. Mutarotation is common to all carbohydrates that exist in hemiacetal forms.

25.3 | Reactions of Monosaccharides

A. Formation of Glycosides (Acetals)

We saw in Section 16.7B that treatment of an aldehyde or ketone with one molecule of alcohol gives a hemiacetal and that treatment of the hemiacetal with a molecule of alcohol gives an acetal. Treatment of monosaccharides, all of which exist almost exclusively in a cyclic hemiacetal form, also gives acetals, as illustrated by the reaction of β-D-glucopyranose with methanol.

β-D-Glucopyranose
(β-D-Glucose)

Methyl β-D-glucopyranoside
(Methyl β-D-glucoside)

Methyl α-D-glucopyranoside
(Methyl α-D-glucoside)

Glycoside

A carbohydrate in which the —OH on its anomeric carbon is replaced by —OR.

Glycosidic bond

The bond from the anomeric carbon of a glycoside to an —OR group.

A cyclic acetal derived from a monosaccharide is called a **glycoside**, and the bond from the anomeric carbon to the —OR group is called a **glycosidic bond**. Mutarotation is not possible in a glycoside because an acetal is no longer in equilibrium with the open-chain carbonyl-containing compound. Glycosides are stable in water and aqueous base, but like other acetals (Section 16.7), they are hydrolyzed in aqueous acid to an alcohol and a monosaccharide.

Glycosides are named by listing the alkyl or aryl group bonded to oxygen followed by the name of the carbohydrate in which the ending -*e* is replaced by -*ide*. For example, the glycosides derived from β-D-glucopyranose are named β-D-glucopyranosides; those derived from β-D-ribofuranose are named β-D-ribofuranosides.

Example 25.4

Draw a structural formula for methyl β-D-ribofuranoside (methyl β-D-riboside). Label the anomeric carbon and the glycosidic bond.

Problem 25.4

Draw a Haworth projection and a chair conformation for methyl α-D-mannopyranoside (methyl α-D-mannoside). Label the anomeric carbon and the glycosidic bond.

Figure 25.3
Structural formulas of the five most important pyrimidine and purine bases found in DNA and RNA. The hydrogen atom shown in grey is lost in forming an *N*-glycoside.

Uracil   Cytosine   Thymine   Adenine   Guanine

Just as the anomeric carbon of a cyclic hemiacetal undergoes reaction with the —OH group of an alcohol to form a glycoside, it also undergoes reaction with the N—H group of an amine to form an *N*-glycoside. Especially important in the biological world are the *N*-glycosides formed between D-ribose and 2-deoxy-D-ribose, each as a furanose, and the heterocyclic aromatic amines uracil, cytosine, thymine, adenine, and guanine (Figure 25.3). *N*-Glycosides of these pyrimidine and purine bases are structural units of nucleic acids (Chapter 28).

Example 25.5

Draw a structural formula for cytidine, the β-*N*-glycoside formed between D-ribofuranose and cytosine.

Solution

Cytidine

Problem 25.5

Draw a structural formula for the β-*N*-glycoside formed between 2-deoxy-D-ribofuranose and adenine.

B. Reduction to Alditols

The carbonyl group of a monosaccharide can be reduced to a hydroxyl group by a variety of reducing agents, including sodium borohydride and hydrogen in the presence of a transition metal catalyst. The reduction products are known as **alditols**. Reduction of D-glucose gives D-glucitol, more commonly known as D-sorbitol. Note that D-glucose is shown here in the open-chain form. Only a small amount of this form is present in solution, but, as it is reduced, the rapid equilibrium between cyclic hemiacetal forms and the open-chain form replaces it.

Alditol
The product formed when the C=O group of a monosaccharide is reduced to a CHOH group.

The reaction shows structures:

β-D-Glucopyranose ⇌ D-Glucose $\xrightarrow{NaBH_4}$ D-Glucitol (D-Sorbitol)

Sorbitol is found in the plant world in many berries and in cherries, plums, pears, apples, seaweed, and algae. It is about 60% as sweet as sucrose (table sugar) and is used in the manufacture of candies and as a sugar substitute for diabetics. D-Sorbitol is an important food additive, usually added to prevent dehydration of foods and other materials on exposure to air because it binds water strongly.

Other alditols common in the biological world are erythritol, D-mannitol, and xylitol. Xylitol is used as a sweetening agent in "sugarless" gum, candy, and sweet cereals.

Erythritol   D-Mannitol   Xylitol

Many "sugar-free" products contain sugar alcohols, such as D-sorbitol and xylitol.

© Cengage Learning/Gregory Smolin

C. Oxidation to Aldonic Acids: Reducing Sugars

As we saw in Section 16.10A, aldehydes (RCHO) are oxidized to carboxylic acids (RCOOH) by several oxidizing agents, including oxygen, O_2. Similarly, the aldehyde group of an aldose can be oxidized, under basic conditions, to a carboxylate group. Oxidizing agents for this purpose include bromine in aqueous calcium carbonate (Br_2, $CaCO_3$, H_2O) and Tollens' solution [$Ag(NH_3)_2^+$]. Under these conditions, the cyclic form of an aldose is in equilibrium with the open-chain form, which is then oxidized by the mild oxidizing agent. D-Glucose, for example, is oxidized to D-gluconate (the anion of D-gluconic acid).

β-D-Glucopyranose (β-D-Glucose) ⇌ D-Glucose $\xrightarrow[\text{basic solution}]{\text{oxidizing agent}}$ D-Gluconate

Any carbohydrate that reacts with an oxidizing agent to form an **aldonic acid** is classified as a **reducing sugar** (it reduces the oxidizing agent).

Surprisingly, 2-ketoses are also reducing sugars. Carbon 1 (a CH$_2$OH group) of a 2-ketose is not oxidized directly. Rather, under the basic conditions of this oxidation, a 2-ketose is in equilibrium with an aldose by way of an enediol intermediate. The aldose is then oxidized by the mild oxidizing agent.

Aldonic acid
The product formed when the —CHO group of an aldose is oxidized to a —COOH group.

Reducing sugar
A carbohydrate that reacts with an oxidizing agent to form an aldonic acid. In this reaction, the carbohydrate reduces the oxidizing agent.

A 2-ketose   An enediol   An aldose   An aldonate

D. Oxidation to Uronic Acids

Enzyme-catalyzed oxidation of the primary hydroxyl group at carbon 6 of a hexose yields a uronic acid. Enzyme-catalyzed oxidation of D-glucose, for example, yields D-glucuronic acid, shown here in both its open-chain and cyclic hemiacetal forms.

D-Glucose   Fischer projection   Chair conformation

D-Glucuronic acid
(a uronic acid)

D-Glucuronic acid is widely distributed in both the plant and animal world. In humans, it is an important component of the glucosaminoglycans of connective tissues (Section 25.6). It is also used by the body to detoxify foreign hydroxyl-containing compounds, such as phenols and alcohols. In the liver, these compounds are converted to glycosides of glucuronic acid (glucuronides) and excreted in the urine. The intravenous anesthetic propofol, for example, is converted to the following glucuronide and excreted in the urine.

Propofol   A urine-soluble glucuronide

E. Oxidation by Periodic Acid

Oxidation by periodic acid, $HIO_4 \cdot 2H_2O$ or H_5IO_6, has proven useful in structure determinations of carbohydrates, particularly in determining the size of glycoside rings. Recall from Section 10.8C that periodic acid cleaves the carbon-carbon bond of a glycol in a reaction that proceeds through a cyclic periodic ester. In this reaction, iodine(VII) of periodic acid is reduced to iodine(V) of iodic acid.

| A glycol | **Periodic acid** | A cyclic periodic ester | **Iodic acid** |

Periodic acid also cleaves carbon-carbon bonds of α-hydroxyketones and α-hydroxyaldehydes by a similar mechanism. Following are abbreviated structural formulas for these functional groups and the products of their oxidative cleavage by periodic acid. As a way to help you understand how each set of products is formed, each carbonyl in a starting material is shown as a hydrated intermediate that is then oxidized. In this way, each oxidation can be viewed as being analogous to the oxidation of a glycol.

| α-Hydroxyketone | Hydrated intermediate |

| α-Hydroxyketone | Hydrated intermediate |

| α-Hydroxyaldehyde | Hydrated intermediate |

As an example of the usefulness of this reaction in carbohydrate chemistry, oxidation of methyl β-D-glucoside consumes two moles of periodic acid and produces one mole of formic acid. This stoichiometry and the formation of formic acid are possible only if —OH groups are on three adjacent carbon atoms.

The analytical procedure most often performed in a clinical chemistry laboratory is the determination of glucose in blood, urine, or other biological fluids. This is true because of the high incidence of diabetes mellitus. Approximately 15 million known diabetics live in the United States, and it is estimated that another 1 million are undiagnosed.

Diabetes mellitus is characterized by insufficient blood levels of the hormone insulin. If the blood concentration of insulin is too low, muscle and liver cells do not absorb glucose from the blood, which, in turn, leads to increased levels of blood glucose (hyperglycemia),

of this disease. In addition to being rapid, a test must also be specific for D-glucose; it must give a positive test for D-glucose but not react with any other substance normally present in biological fluids.

Blood glucose levels are now measured by an enzyme-based procedure using the enzyme glucose oxidase. This enzyme catalyzes the oxidation of β-D-glucose to D-gluconic acid.

Glucose oxidase is specific for β-D-glucose. Therefore, complete oxidation of any sample containing both β-D-glucose and α-D-glucose requires conversion of the

$$\beta\text{-D-Glucopyranose} \;(\beta\text{-D-Glucose}) + O_2 + H_2O \xrightarrow{\text{glucose oxidase}} \text{D-Gluconic acid} + H_2O_2$$

β-D-Glucopyranose
(β-D-Glucose)

D-Gluconic acid **Hydrogen peroxide**

impaired metabolism of fats and proteins, ketosis, and possible diabetic coma. A rapid test for blood glucose levels is critical for early diagnosis and effective management

α form to the β form. Fortunately, this interconversion is rapid and complete in the short time required for the test.

Molecular oxygen, O_2, is the oxidizing agent in this reaction and is reduced to hydrogen peroxide, H_2O_2. In one procedure, hydrogen peroxide formed in the glucose oxidase-catalyzed reaction is used to oxidize colorless *o*-toluidine to a colored product in a reaction catalyzed by the enzyme peroxidase. The concentration of the colored oxidation product is determined spectrophotometrically and is proportional to the concentration of glucose in the test solution.

$$\text{2-Methylaniline} (o\text{-Toluidine}) + H_2O_2 \xrightarrow{\text{peroxidase}} \text{colored product}$$

2-Methylaniline
(*o*-Toluidine)

© Cengage Learning/Charles D. Winters

A test kit for the presence of glucose in urine.

Several commercially available test kits use the glucose oxidase reaction for qualitative determination of glucose in urine.

This is evidence that methyl β-D-glucoside is indeed a pyranoside.

Methyl β-D-glucopyranoside

Methyl β-D-fructoside consumes only one mole of periodic acid and produces neither formaldehyde nor formic acid. Thus, oxidizable groups exist on adjacent carbons only at one site in the molecule. The fructoside, therefore, must be a five-membered ring (a fructofuranoside).

Methyl β-D-fructofuranoside

25.4 | Disaccharides and Oligosaccharides

Most carbohydrates in nature contain more than one monosaccharide unit. Those that contain two units are called **disaccharides**, those that contain three units are called **trisaccharides**, and so forth. The general term **oligosaccharide** is often used for carbohydrates that contain from four to ten monosaccharide units. Carbohydrates containing larger numbers of monosaccharide units are called **polysaccharides**.

In a disaccharide, two monosaccharide units are joined together by a glycosidic bond between the anomeric carbon of one unit and an —OH of the other. Three important disaccharides are sucrose, lactose, and maltose.

A. Sucrose

Sucrose (table sugar) is the most abundant disaccharide in the biological world. It is obtained principally from the juice of sugar cane and sugar beets. In sucrose, carbon 1 of α-D-glucopyranose is joined to carbon 2 of β-D-fructofuranose by an α-1,2-glycosidic bond.

Sucrose

Disaccharide
A carbohydrate containing two monosaccharide units joined by a glycosidic bond.

Oligosaccharide
A carbohydrate containing four to ten monosaccharide units, each joined to the next by a glycosidic bond.

Polysaccharide
A carbohydrate containing a large number of monosaccharide units, each joined to the next by one or more glycosidic bonds.

Note that glucose is a six-membered (pyranose) ring, whereas fructose is a five-membered (furanose) ring. Because the anomeric carbons of both the glucopyranose and fructofuranose units are involved in formation of the glycosidic bond, sucrose is a nonreducing sugar.

B. Lactose

Lactose is the principal sugar present in milk. It makes up about 5–8% of human milk and 4–6% of cow's milk. It consists of D-galactopyranose bonded by a β-1,4-glycosidic bond to carbon 4 of D-glucopyranose. Lactose is a reducing sugar.

These products help individuals with lactose intolerance meet their calcium needs.

Lactose

C. Maltose

Maltose derives its name from its presence in malt, the juice from sprouted barley and other cereal grains (from which beer is brewed). Maltose consists of two molecules of D-glucopyranose joined by an α-1,4-glycosidic bond between carbon 1 (the anomeric carbon) of one unit and carbon 4 of the other unit. Following are representations for β-maltose, so named because the —OH on the anomeric carbon of the glucose unit on the right is beta.

Maltose

Maltose is a reducing sugar because the hemiacetal group on the right unit of D-glucopyranose is in equilibrium with the free aldehyde and can be oxidized to a carboxylic acid.

D. Relative Sweetness of Some Carbohydrate and Artificial Sweeteners

Although all monosaccharides are sweet to the taste, some are sweeter than others (Table 25.2). D-Fructose tastes the sweetest, even sweeter than sucrose (table sugar, Section 25.4A). The sweet taste of honey is attributable largely to

D-fructose and D-glucose. Lactose (Section 25.4B) has almost no sweetness. It occurs in many milk products and is sometimes added to foods as a filler. Some people lack an enzyme that allows them to tolerate lactose well; they should avoid these foods.

Table 25.2 — Relative Sweetness of Some Carbohydrates and Artificial Sweetening Agents*

Carbohydrate	Sweetness Relative to Sucrose	Artificial Sweetener	Sweetness Relative to Sucrose
Fructose	1.74	Saccharin	450
Invert sugar	1.25	Acesulfame-K	200
Sucrose (table sugar)	1.00	Aspartame	160
Honey	0.97		
Glucose	0.74		
Maltose	0.33		
Galactose	0.32		
Lactose (milk sugar)	0.16		

*We have no mechanical way to measure sweetness. Such testing is done by having a group of people taste solutions of varying sweetness and ranking them in order by taste.

Example 25.6

Draw a chair conformation for the β anomer of a disaccharide in which two units of D-glucopyranose are joined by an α-1,6-glycosidic bond.

Solution

First draw a chair conformation of α-D-glucopyranose. Then connect the anomeric carbon of this monosaccharide to carbon 6 of a second D-glucopyranose unit by an α-glycosidic bond. The resulting molecule is either α or β depending on the orientation of the —OH group on the reducing end of the disaccharide. The disaccharide shown here is β.

Problem 25.6

Draw a chair conformation for the α form of a disaccharide in which two units of D-glucopyranose are joined by a β-1,3-glycosidic bond.

Membranes of animal plasma cells have large numbers of relatively small carbohydrates bound to them. In fact, the outsides of most plasma cell membranes are literally "sugar-coated." These membrane-bound carbohydrates are part of the mechanism by which cell types recognize each other and, in effect, act as

Among the first discovered and best understood of these membrane-bound carbohydrates are those of the ABO blood group system, discovered in 1900 by Karl Landsteiner (1868–1943). Whether an individual has type A, B, AB, or O blood is genetically determined and depends on the type of trisaccharide or tetrasaccharide

CHO

HO———H

H———OH

H———OH

HO———H

CH₃

An L-monosaccharide; this —OH is on the left in the Fischer projection

Carbon 6 is —CH₃ rather than —CH₂OH

L-Fucose

biochemical markers. Typically, these membrane-bound carbohydrates contain from 4 to 17 monosaccharide units consisting primarily of relatively few monosaccharides, including D-galactose, D-mannose, L-fucose, N-acetyl-D-glucosamine, and N-acetyl-D-galactosamine. L-Fucose is a 6-deoxyaldohexose.

bound to the surface of the person's red blood cells. The monosaccharides of each blood group and the type of glycosidic bond joining them are shown in the figure. The configurations of the glycosidic bonds are shown in parentheses.

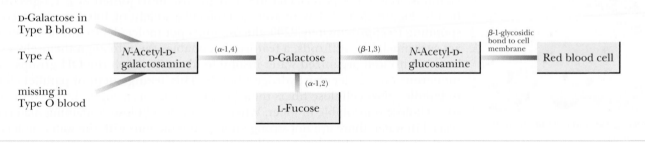

25.5 | Polysaccharides

Polysaccharides consist of large numbers of monosaccharide units bonded together by glycosidic bonds. Three important polysaccharides, all made up of glucose units, are starch, glycogen, and cellulose.

A. Starch: Amylose and Amylopectin

Starch is used for energy storage in plants. It is found in all plant seeds and tubers and is the form in which glucose is stored for later use. Starch can be separated into two principal polysaccharides: amylose and amylopectin. Although the starch from each plant is unique, most starches contain 20–25% amylose and 75–80% amylopectin.

Complete hydrolysis of both amylose and amylopectin yields only D-glucose. Amylose is composed of unbranched chains of up to 4000 D-glucose units joined by α-1,4-glycosidic bonds. Amylopectin contains chains up to 10,000 D-glucose units also joined by α-1,4-glycosidic bonds. In addition, there is considerable branching

Figure 25.4

Amylopectin is a branched polymer of approximately 10,000 D-glucose units joined by α-1,4-glycosidic bonds. Branches consist of 24–30 D-glucose units started by α-1,6-glycosidic bonds.

α-1,6-glycosidic bond

α-1,4-glycosidic bonds

Breads, grains, and pasta are sources of starches.

© Cengage Learning/Charles D. Winters

from this linear network. At branch points, new chains of 24 to 30 units are started by α-1,6-glycosidic bonds (Figure 25.4).

B. Glycogen

Glycogen is the energy-reserve carbohydrate for animals. Like amylopectin, glycogen is a branched polysaccharide of approximately 10^6 glucose units joined by α-1,4- and α-1,6-glycosidic bonds. The total amount of glycogen in the body of a well-nourished adult human is about 350 g, divided almost equally between liver and muscle.

C. Cellulose

Cellulose, the most widely distributed plant skeletal polysaccharide, constitutes almost half of the cell wall material of wood. Cotton is almost pure cellulose. Cellulose is a linear polysaccharide of D-glucose units joined by β-1,4-glycosidic bonds (Figure 25.5). It has an average molecular weight of 400,000 g/mol, corresponding to approximately 2200 glucose units per molecule. Cellulose molecules act very much like stiff rods, a feature that enables them to align themselves side by side into well-organized water-insoluble fibers in which the OH groups form numerous intermolecular hydrogen bonds. This arrangement of parallel chains in bundles gives cellulose fibers their high mechanical strength. It is also the reason cellulose is insoluble in water. When a piece of cellulose-containing material is placed in water, there are not strong enough interactions with the water molecules on the surface of the fiber to pull individual cellulose molecules away from the strongly hydrogen-bonded fiber.

Humans and other animals cannot use cellulose as food because our digestive systems do not contain β-glucosidases, enzymes that catalyze hydrolysis of β-glucosidic bonds. Instead, we have only α-glucosidases; hence, the polysaccharides we use as sources of glucose are starch and glycogen. On the other hand, many bacteria and microorganisms do contain β-glucosidases and so can digest cellulose. Termites are fortunate (much to our regret) to have such bacteria in their guts and can use wood as their principal food. Ruminants (cud-chewing animals) and horses can also digest grasses and hay because β-glucosidase-containing microorganisms are present in their alimentary systems.

Figure 25.5

Cellulose is a linear polysaccharide of up to 2200 units of D-glucose joined by β-1,4-glycosidic bonds.

β-1,4-glycosidic bonds

If you read the labels of soft drinks and other artificially sweetened food products, you will find that many of them contain high-fructose corn syrup. Looking at Table 25.3, you will see one reason for its use: fructose is more than 70% sweeter than sucrose (table sugar).

The production of high-fructose corn syrup begins with the partial hydrolysis of corn starch catalyzed by the enzyme α-amylase. This enzyme catalyzes the hydrolysis of α-glucosidic bonds and breaks corn starch into small polysaccharides called dextrins.

The enzyme glucoamylase then catalyzes the hydrolysis of the dextrins to D-glucose. Finally, enzyme-catalyzed isomerization of D-glucose gives D-fructose. Several billion pounds of high-fructose corn syrup are produced each year in this way for use by the food processing industry. Commercial high fructose corn syrup is actually 55–60% glucose and 40–45% fructose. This composition was inspired by natural honey, which has similar proportions of fructose and glucose along with some other carbohydrates in minor amounts.

$$\text{Corn starch} \xrightarrow{\alpha\text{-amylase}} \text{Dextrins} \xrightarrow{\text{Glucoamylase}} \textbf{D-Glucose} \xrightarrow[\text{isomerase}]{\text{glucose}} \textbf{D-Fructose}$$

(a polysaccharide of α-D–glucose units)   Oligosaccharides of 6–10 α-D-glucose units

D. Textile Fibers from Cellulose

Cotton is almost pure cellulose. Both rayon and acetate rayon are made from chemically modified cellulose and were the first commercially important synthetic textile fibers. In the production of rayon, cellulose-containing materials are treated with carbon disulfide, CS_2, in aqueous sodium hydroxide. In this reaction, some of the —OH groups on a cellulose fiber are converted to the sodium salts of a xanthate ester, which causes the fibers to dissolve in alkali as a viscous colloidal dispersion.

an —OH group in a cellulose fiber

$$\text{Cellulose—OH} \xrightarrow{\text{NaOH}} \text{Cellulose—}\overset{..}{\underset{..}{O}}:^-Na^+ \xrightarrow{:S=C=S:} \text{Cellulose—O—}\overset{\overset{:S}{\|}}{C}\text{—}\overset{..}{\underset{..}{S}}:^-Na^+$$

Cellulose
(insoluble in water)

Sodium salt of a xanthate ester
(a viscous colloidal suspension)

The solution of cellulose xanthate is separated from the alkali insoluble parts of wood and then forced through a spinneret, a metal disc with many tiny holes, into dilute sulfuric acid to hydrolyze the xanthate ester groups and precipitate regenerated cellulose. Regenerated cellulose extruded as a filament is called viscose rayon thread.

In the industrial synthesis of acetate rayon, cellulose is treated with acetic anhydride.

A glucose unit in a cellulose fiber

$$+ \; 3 \; CH_3\overset{O}{\overset{\|}{C}}O\overset{O}{\overset{\|}{C}}CH_3 \longrightarrow$$

Acetic anhydride

A fully acetylated glucose unit

$$+ \; 3 \; CH_3\overset{O}{\overset{\|}{C}}OH$$

Acetylated cellulose is then dissolved in a suitable solvent, precipitated, and drawn into fibers known as acetate rayon. Today, acetate rayon fibers rank fourth in production in the United States, surpassed only by Dacron polyester, nylon, and rayon.

25.6 | Glucosaminoglycans

Glucosaminoglycans are a group of linear polysaccharides composed of repeating disaccharides in which one of the monosaccharide units has either a negatively charged carboxylate group (COO^-) or negatively charged sulfate group ($-OSO_3^-$). Members of this family of carbohydrates include hyaluronic acid, heparin, chondroitin sulfate, and keratan sulfate, which are components of cartilage, tendons; and other connective tissues, and dermatan sulfate, which is a component of the extracellular matrix of the skin. A general characteristic of this class of polysaccharides is a repeating disaccharide consisting of units of a uronic acid and an aminohexose with a 1,4-glycosidic bond between the aminohexose and the uronic acid.

A. Hyaluronic Acid

Hyaluronic acid is present in connective tissue. It has a molecular weight of between 10^5 and 10^7 g/mol and contains from 3000 to 100,000 repeating units, depending on the organ in which it occurs. It is most abundant in embryonic tissues and in specialized connective tissues such as synovial fluid, the lubricant of joints in the body, and the vitreous humor of the eye, where it provides a clear, elastic gel that maintains the retina in its proper position.

The repeating disaccharide unit in hyaluronic acid is D-glucuronic acid linked by a β-1,3-glycosidic bond to N-acetyl-D-glucosamine.

The repeating unit of **hyaluronic acid**

B. Heparin

Heparin is a heterogeneous mixture of variably sulfonated polysaccharide chains, ranging in molecular weight from 6000 to 30,000 g/mol. This polysaccharide is synthesized and stored in mast cells of various tissues, particularly the liver, lungs, and gut.

Figure 25.6

A pentasaccharide unit of heparin.

Heparin has many biological functions, the best known and understood of which is its anticoagulant activity. It binds strongly to antithrombin III, a plasma protein involved in terminating the clotting process. Heparin is used medically as an anticoagulant.

The repeating monosaccharide units of heparin are *N*-acetyl-D-glucosamine, D-glucuronic acid, D-glucosamine, and L-ioduronic acid bonded by a combination of α-1,4- and β-1,4-glycosidic bonds. Figure 25.6 shows a pentasaccharide unit of heparin that binds to and inhibits the enzymatic activity of antithrombin III.

Summary

Section 25.1 Monosaccharides

- **Carbohydrates** are polyhydroxylated aldehydes or ketones or compounds that produce polyhydroxylated aldehydes or ketones upon hydrolysis.
- Carbohydrates are the most abundant organic molecules in the world. They are essential to all forms of life and perform such functions as energy storage (glucose, starch, glycogen), structural reinforcement (cellulose), and genetic information storage as components of nucleic acids (DNA, RNA).
- **Monosaccharides** usually have molecular formulas of $C_nH_{2n}O_n$ ($3 < n < 8$) and are the monomers from which larger carbohydrates are constructed.
- Monosaccharides are named by using the suffix *-ose*. The prefixes *tri-, tetra-, penta-* are used to indicate three, four, or five carbon atoms, respectively.
  - An aldehyde carbohydrate is called an **aldose** and is sometimes designated with an *aldo-* prefix.
  - A ketone carbohydrate is called a **ketose,** and is sometimes designated with a *keto-* prefix. For example, glyceraldehyde is an aldotriose, and fructose is a ketohexose.
  - The nomenclature of monosaccharides is dominated by common names. Even though IUPAC names can be derived for each different monosaccharide, the common names are much simpler and are used almost exclusively.
- Monosaccharides usually have one or more chiral centers, so stereochemistry is of major importance with monosaccharides.
  - A **Fischer projection** of a monosaccharide is used to show the structure of a monosaccharide and thus keep track of stereochemistry.
  - In a Fischer projection, the monosaccharide is usually drawn in the open chain form and the carbonyl carbon atom is placed at the top of the structure. Horizontal lines represent groups projecting above the plane of the paper, and vertical lines represent groups projecting below the plane of the paper.
  - A modified Fischer projection is the same, except the carbon atoms of chiral centers are not labeled. They are assumed to be located at the crossing points of bonds.
- The overall stereochemistry of monosaccharides is classified as D or L, based on a comparison to glyceraldehyde stereochemistry.
  - In a monosaccharide, the point of reference is the chiral center that is farthest from the carbonyl group. Because this is a carbon atom that is next to the last carbon atom in the chain (notice that the last carbon atom of the chain has two —H atoms, so it is not a chiral center), it is referred to as the **penultimate carbon** atom.
  - A monosaccharide that has the same configuration about its penultimate carbon as D-glyceraldehyde is classified as a **D-monosaccharide**. In this case the —OH group is on the right side of the carbon atom in the Fischer projection.
  - An **L-monosaccharide** has a configuration about its penultimate carbon atom that is the same as the configuration of L-glyceraldehyde, with the —OH group on the left in a Fischer projection.
  - The enantiomer of a given monosaccharide is not produced by simply changing the configuration of the penultimate carbon atom but rather by reversing the configuration of all of the chiral centers.

Problems: 25.1, 25.6–25.15

- Some sugars have an amino group (—NH₂) instead of all —OH groups, and these are called **amino sugars**. Amino sugars are much less common than normal carbohydrates, but important examples include D-glucosamine and D-galactosamine.
- Monosaccharides are all very soluble in water due to all of the —OH groups that can take part in hydrogen bonding with the water molecules.

Section 25.2 The Cyclic Structure of Monosaccharides

- The open-chain monosaccharides are in equilibrium with a cyclic hemiacetal structure.
  - The cyclic hemiacetal is greatly favored and thus is found in large excess at equilibrium.
  - When a carbohydrate forms a six-membered hemiacetal ring it is called a **pyranose**, and when a carbohydrate forms a five-membered hemiacetal ring, it is called a **furanose**.
- Cyclic monosaccharide structures are often drawn as **Haworth projections** in which the five-membered or six-membered cyclic hemiacetal is drawn as planar and perpendicular to the plane of the paper.

Problems: 25.2, 25.3, 25.16–25.23

  - The anomeric carbon is usually to the right with the hemiacetal oxygen atom in the back.
  - A more accurate "chair" conformation can be drawn for six-membered ring hemiacetals, showing which groups are axial and which are equatorial.
- Two cyclic diastereomers are possible, and these are referred to as **anomers**.
  - The two anomers are distinguished by the relative orientation of the **anomeric OH** group (the —OH group on the so-called **anomeric carbon** atom, the carbon that was a carbonyl in the open-chain form).
  - The two anomers are named as α or β.
    - The **β-anomer** is the one that has the anomeric —OH group on the same side of the ring as the —CH₂OH group in a Haworth projection.
    - The **α-anomer** has the anomeric —OH group and the terminal —CH₂OH group on opposite sides of the ring in a Haworth projection.
    - With D-glucose in the cyclic hemiacetal form, the α anomer is the one with the anomeric —OH group axial, while for the β anomer the anomeric —OH group is equatorial.
- **Mutarotation** is the change in specific rotation that accompanies the interconversion of α- and β-anomers in aqueous solution.

Section 25.3 Reactions of Monosaccharides

- A **glycoside** is an acetal derived from a monosaccharide, and the bond from the anomeric carbon to the —OR group is called the **glycosidic bond**.
  - The name of the glycoside is composed of the name of the alkyl or aryl group bonded to the acetal oxygen atom followed by the name of the monosaccharide in which the terminal -e has been replaced by -ide.
  - Glycosidic bonds can be made by reacting a saccharide with an alcohol in water, following the standard acetal formation mechanism.

Problems: 25.4, 25.5, 25.24–25.35

- An **alditol** is a polyhydroxy compound formed by reduction of the carbonyl group of a monosaccharide to a hydroxyl group. Reduction of D-glucose, for example, gives D-glucitol.
- An **aldonic acid** is a carboxylic acid formed by oxidation of the aldehyde group of an aldose. Oxidation of D-glucose, for example, gives D-gluconic acid.
- **Reducing sugars** are oxidized by mild oxidizing agents to aldonic acids.
- Enzyme-catalyzed oxidation of the primary hydroxyl group at carbon 6 of a hexose yields a **uronic acid**, examples of which are common in both the plant and animal worlds.
- **Periodic acid** cleaves the carbon-carbon bond of a glycol in a reaction that proceeds through a cyclic periodic ester. This reaction was once useful in carbohydrate structure determination.

Section 25.4 Disaccharides and Oligosaccharides

- A **disaccharide** contains two monosaccharide units joined by a glycosidic bond.
- Terms applied to carbohydrates containing larger numbers of monosaccharides are **trisaccharide**, **tetrasaccharide**, **oligosaccharide**, and **polysaccharide**.
- **Sucrose** is a disaccharide containing D-glucose joined to D-fructose by an α-1,2-glycosidic bond.
- **Lactose** is a disaccharide consisting of D-galactose joined to D-glucose by a β-1,4-glycosidic bond.
- **Maltose** is a disaccharide of two molecules of D-glucose joined by a β-1,4-glycosidic bond.

Problems: 25.6, 25.36–25.39, 25.44, 25.45

Section 25.5 Polysaccharides

- **Starch** can be separated into two fractions given the names amylose and amylopectin.
  - **Amylose** is a linear polymer of up to 4000 units of D-glucopyranose joined by α-1,4-glycosidic bonds.
  - **Amylopectin** is a highly branched polymer of D-glucopyranose joined by α-1,4-glycosidic bonds and, at branch points, by α-1,6-glycosidic bonds.
- **Glycogen**, the energy reserve carbohydrate of animals, is a highly branched polymer of D-glucopyranose joined by α-1,4-glycosidic bonds and, at branch points, by α-1,6-glycosidic bonds.
- **Cellulose**, the skeletal polysaccharide of plants, is a linear polymer of D-glucopyranose joined by β-1,4-glycosidic bonds.
  - The strength of cellulose comes from the ribbon-like structure of the individual chains that fit together perfectly through hydrogen bonding to create incredibly strong structures.
  - Rayon is made from chemically modified and regenerated cellulose. Acetate rayon is made by acetylation of cellulose.

Problems: 25.40–25.43

Section 25.6 Glucosaminoglycans

- **Hyaluronic acid** is a carboxylate containing polysaccharide that is found in connective tissue.
- **Heparin**, an anticoagulent, is a heterogeneous mixture of variably sulfonated polysaccharide chains and is found in the liver, lungs, and gut.
  - The carboxyl and sulfate groups of acidic polysaccharides are ionized as —COO^- and —SO_3^- at the pH of body fluids, which gives these polysaccharides net negative charges.

Problems: 25.46–25.49

ⓄWL Questions and problems for this chapter are available in OWL. Use the OWL access card that came with this textbook to access assigned questions and problems for this chapter.

Key Reactions

1. **Formation of Cyclic Hemiacetals (Section 25.2A)** A monosaccharide existing as a five-membered ring is a furanose; one existing as a six-membered ring is a pyranose. A pyranose is most commonly drawn as either a Haworth projection or a chair conformation.

D-Glucose

β-D-Glucopyranose
(β-D-Glucose)

2. **Mutarotation (Section 25.2C)**   Anomeric forms of a monosaccharide are in equilibrium in aqueous solution. Mutarotation is the change in specific rotation that accompanies this equilibration.

β-D-Glucopyranose
$[\alpha]_D^{25}$ +18.7

Open-chain form

α-D-Glucopyranose
$[\alpha]_D^{25}$ +112

3. **Formation of Glycosides (Section 25.3A)**   Treatment of a monosaccharide with an alcohol in the presence of an acid catalyst forms a cyclic acetal called a glycoside. The bond to the new —OR group is called a glycosidic bond, and the mechanism is the same actual formation you saw in Section 16.7.

$+ CH_3OH \xrightarrow[-H_2O]{H^+}$

4. **Formation of *N*-Glycosides (Section 25.3A)**   *N*-Glycosides formed between a monosaccharide and a heterocyclic aromatic amine are especially important in the biological world.

$(-H_2O)$

a β-*N*-glycosidic bond

anomeric carbon

5. **Reduction to Alditols (Section 25.3B)**   Reduction of the carbonyl group of an aldose or ketose to a hydroxyl group yields a polyhydroxy compound called an alditol.

$\xrightarrow{NaBH_4}$

D-Glucose

D-Glucitol
(D-Sorbitol)

6. Oxidation to an Aldonic Acid (Section 25.3C)

Oxidation of the aldehyde group of an aldose to a carboxyl group by a mild oxidizing agent gives a polyhydroxycarboxylic acid called an aldonic acid.

D-Glucose → D-Gluconic acid

7. Oxidation by Periodic Acid (Section 25.3E)

Periodic acid oxidizes and cleaves carbon-carbon bonds of glycol, α-hydroxyketone, and α-hydroxyaldehyde groups.

periodic acid cleavage at these two bonds

Methyl β-D-glucopyranoside

$2H_5IO_6$

Chapter 26
Lipids

© Nevada Wier/The Image Bank/Getty Images

Sea lions are marine mammals that require a heavy layer of fat in order to survive in cold waters. Inset: linoleic acid, a major component of unsaturated triglycerides and phospholipids (Sections 26.1 and 26.5).

Lipid
A biomolecule isolated from plant or animal sources by extraction with nonpolar organic solvents, such as diethyl ether and hexane.

Triglyceride (triacylglycerol)
An ester of glycerol with three fatty acids.

⬥WL **Online homework for this chapter may be assigned in OWL for Organic Chemistry.**

Lipids are a heterogeneous group of naturally occurring organic compounds (many related to fats and oils), classified together on the basis of their common solubility properties. Lipids are insoluble in water but soluble in non-polar aprotic organic solvents, including diethyl ether, dichloromethane, and acetone.

Lipids are divided into two main groups. First are those lipids that contain both a relatively large nonpolar hydrophobic region, most commonly aliphatic in nature, and a polar hydrophilic region. Found among this group are fatty acids, triglycerides, phospholipids, prostaglandins, and the fat-soluble vitamins. Second are those lipids that contain the tetracyclic ring system called the steroid nucleus, including cholesterol, steroid hormones, and bile acids. In this chapter, we describe the structures and biological functions of each group of lipids.

26.1 | Triglycerides

Animal fats and vegetable oils, the most abundant naturally occurring lipids, are triesters of glycerol and long-chain carboxylic acids. Fats and oils are also referred to as **triglycerides** or **triacylglycerols**. Hydrolysis of a triglyceride in aqueous base followed by acidification gives glycerol and three fatty acids.

$$
\begin{array}{c}
\text{O} \quad \text{CH}_2\text{OCR} \\
\overset{\text{O}}{\underset{||}{}} \\
\text{R}'\text{COCH} \quad \text{O} \\
\text{CH}_2\text{OCR}''
\end{array}
\quad \xrightarrow[\text{2.HCl, H}_2\text{O}]{\text{1.NaOH, H}_2\text{O}} \quad
\begin{array}{c}
\text{CH}_2\text{OH} \\
\text{HOCH} \\
\text{CH}_2\text{OH}
\end{array}
\quad + \quad
\begin{array}{c}
\text{RCOOH} \\
\text{R}'\text{COOH} \\
\text{R}''\text{COOH}
\end{array}
$$

A triglyceride   **1,2,3-Propanetriol**   Fatty acids
                        (Glycerol, glycerin)

A. Fatty Acids

More than 500 different **fatty acids** have been isolated from various cells and tissues. Given in Table 26.1 are common names and structural formulas for the most abundant of these. The number of carbons in a fatty acid and the number of carbon-carbon double bonds in its hydrocarbon chain are shown by two numbers separated by a colon. In this notation, linoleic acid, for example, is designated as an 18:2 fatty acid; its 18-carbon chain contains two carbon-carbon double bonds. Following are several characteristics of the most abundant fatty acids in higher plants and animals.

1. Nearly all fatty acids have an even number of carbon atoms, most between 12 and 20, in an unbranched chain.
2. The three most abundant fatty acids in nature are palmitic acid (16:0), stearic acid (18:0), and oleic acid (18:1).
3. In most unsaturated fatty acids, the *cis* isomer predominates; the *trans* isomer is rare.
4. Unsaturated fatty acids have lower melting points than their saturated counterparts. The greater the degree of unsaturation, the lower the melting point (see Connections to Biological Chemistry: "The Importance of *cis* Double Bonds in Fats versus Oils" in Section 5.4). Compare, for example, the melting points of linoleic acid, a **polyunsaturated fatty acid**, and stearic acid, a saturated fatty acid.

> **Fatty acid**
> A long, unbranched-chain carboxylic acid, most commonly of 12 to 20 carbons, derived from the hydrolysis of animal fats, vegetable oils, or the phospholipids of biological membranes.

> **Polyunsaturated fatty acid**
> A fatty acid with two or more carbon-carbon double bonds in its hydrocarbon chain.

Table 26.1	The Most Abundant Fatty Acids in Animal Fats, Vegetable Oils, and Biological Membranes		
Carbon Atoms/ Double Bonds*	**Structure**	**Common Name**	**Melting Point (°C)**
Saturated Fatty Acids			
12:0	$CH_3(CH_2)_{10}COOH$	**Lauric acid**	44
14:0	$CH_3(CH_2)_{12}COOH$	**Myristic acid**	58
16:0	$CH_3(CH_2)_{14}COOH$	**Palmitic acid**	63
18:0	$CH_3(CH_2)_{16}COOH$	**Stearic acid**	70
20:0	$CH_3(CH_2)_{18}COOH$	**Arachidic acid**	77
Unsaturated Fatty Acids			
16:1	$CH_3(CH_2)_5CH{=}CH(CH_2)_7COOH$	**Palmitoleic acid**	1
18:1	$CH_3(CH_2)_7CH{=}CH(CH_2)_7COOH$	**Oleic acid**	16
18:2	$CH_3(CH_2)_4(CH{=}CHCH_2)_2(CH_2)_6COOH$	**Linoleic acid**	−5
18:3	$CH_3CH_2(CH{=}CHCH_2)_3(CH_2)_6COOH$	**Linolenic acid**	−11
20:4	$CH_3(CH_2)_4(CH{=}CHCH_2)_4(CH_2)_2COOH$	**Arachidonic acid**	−49

*The first number is the number of carbons in the fatty acid; the second is the number of carbon-carbon double bonds in its hydrocarbon chain.

Example 26.1

Draw a structural formula of a triglyceride derived from one molecule each of palmitic acid, oleic acid, and stearic acid, the three most abundant fatty acids in the biological world.

Among the components of beeswax is triacontyl palmitate, $CH_3(CH_2)_{14}COO(CH_2)_{29}CH_3$, an ester of palmitic acid.

Oil

When used in the context of fats a mixture of triglycerides that is liquid at room temperature.

Fat

A mixture of triglycerides that is semisolid or is solid at room temperature.

Figure 26.1
Tristearin, a saturated triglyceride.

Solution

In this structure, palmitic acid is esterified at carbon 1 of glycerol; oleic acid, at carbon 2; and stearic acid, at carbon 3.

A triglyceride

Problem 26.1

(a) How many constitutional isomers are possible for a triglyceride containing one molecule each of palmitic acid, oleic acid, and stearic acid?

(b) Which of these constitutional isomers are chiral?

B. Physical Properties

The physical properties of a triglyceride depend on its fatty acid components. In general, the melting point of a triglyceride increases as the number of carbons in its hydrocarbon chains increases and as the number of carbon-carbon double bonds decreases. Triglycerides rich in oleic acid, linoleic acid, and other unsaturated fatty acids are generally liquids at room temperature and are called **oils**, as for example corn oil and olive oil. Olive oil, which contains mainly the mono-unsaturated oleic acid, solidifies in the refrigerator, whereas the more unsaturated corn oil will not. Triglycerides rich in palmitic, stearic, and other saturated fatty acids are generally semisolids or solids at room temperature and are called **fats**, as for example human fat and butter fat. Fats of land animals typically contain approximately 40–50% saturated fatty acids by weight (Table 26.2). Most plant oils, on the other hand, contain 20% or less saturated fatty acids and 80% or more unsaturated fatty acids. The notable exception to this generalization about plant oils are the **tropical oils** (as for example coconut and palm oils), which are considerably richer in low-molecular-weight saturated fatty acids.

The lower melting points of triglycerides rich in unsaturated fatty acids are related to differences in three-dimensional shape between the hydrocarbon chains of their unsaturated and saturated fatty acid components. Shown in Figure 26.1 is a space-filling model of tristearin, a saturated triglyceride. In this model, the hydrocarbon chains lie parallel to each other, giving the molecule an ordered, compact

	Saturated Fatty Acids			Unsaturated Fatty Acids	
Fat or Oil	**Lauric (12:0)**	**Palmitic (16:0)**	**Stearic (18:0)**	**Oleic (18:1)**	**Linoleic (18:2)**
Human fat	—	24.0	8.4	46.9	10.2
Beef fat	—	27.4	14.1	49.6	2.5
Butter fat	2.5	29.0	9.2	26.7	3.6
Coconut oil	45.4	10.5	2.3	7.5	trace
Corn oil	—	10.2	3.0	49.6	34.3
Olive oil	—	6.9	2.3	84.4	4.6
Palm oil	—	40.1	5.5	42.7	10.3
Peanut oil	—	8.3	3.1	56.0	26.0
Soybean oil	0.2	9.8	2.4	28.9	50.7

Table 26.2 Grams of Fatty Acid per 100 g of Triglyceride of Several Fats and Oils*

*Only the most abundant fatty acids are given; other fatty acids are present in lesser amounts.

shape. Because of this compact three-dimensional shape and the resulting strength of the dispersion forces between hydrocarbon chains of adjacent molecules, triglycerides rich in saturated fatty acids have melting points above room temperature.

The three-dimensional shape of an unsaturated fatty acid is quite different from that of a saturated fatty acid. Recall from Section 26.1A that unsaturated fatty acids of higher organisms are predominantly of the *cis* configuration; *trans* configurations are rare. Figure 26.2 shows a space-filling model of a **polyunsaturated triglyceride** derived from one molecule each of stearic acid, oleic acid, and linoleic acid. Each double bond in this polyunsaturated triglyceride has the *cis* configuration.

Polyunsaturated triglycerides have a less ordered structure and do not pack together as closely or as compactly as saturated triglycerides. Intramolecular and intermolecular dispersion forces are weaker, with the result that polyunsaturated triglycerides have lower melting points than their saturated counterparts.

C. Reduction of Fatty Acid Chains

For a variety of reasons, in part convenience and in part dietary preference, conversion of oils to fats has become a major industry. The process is called **hardening** of oils and involves catalytic reduction (Section 6.6A) of some or all carbon-carbon double bonds. In practice, the degree of hardening is carefully controlled to produce fats of a desired consistency. The resulting fats are sold for kitchen use (Crisco, Spry, and others). Margarine and other butter substitutes are produced by partial hydrogenation of polyunsaturated oils derived from corn, cottonseed, peanut, and soybean oils. To the hardened oils are added β-carotene (to give the final product a yellow color and make it look like butter), salt, and about 15% milk by volume to form the final emulsion. Vitamins A and D are also often added. Because the product at this stage is tasteless, acetoin and diacetyl are often added. These two compounds mimic the characteristic flavor of butter.

Polyunsaturated triglyceride
A triglyceride having several carbon-carbon double bonds in the hydrocarbon chains of its three fatty acids.

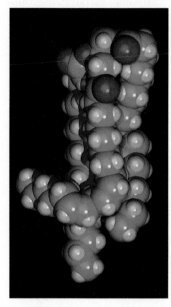

Figure 26.2
A polyunsaturated triglyceride.

$$CH_3-\underset{\underset{\displaystyle OH}{|}}{CH}-\underset{\underset{\displaystyle O}{\|}}{C}-CH_3 \qquad CH_3-\underset{\underset{\displaystyle O}{\|}}{C}-\underset{\underset{\displaystyle O}{\|}}{C}-CH_3$$

3-Hydroxy-2-butanone   **2,3-Butanedione**
(Acetoin)   (Diacetyl)

As described in the Connections to Biological Chemistry: "*Trans* Fatty Acids: What They Are and How to Avoid Them" in Section 6.6, the process of hardening oils generally involves using a hydrogenation catalyst such as Ni and a limiting amount of H_2. An unintended consequence of the hardening process is that some of the *cis* double bonds of the triglycerides are isomerized to the more stable *trans* configuration on the metal surface. Recall that the π bond is broken when the alkene adsorbs on the metal surface, so unfortunately it can isomerize if there is not enough H_2 present to react first. Research has shown that these so-called *trans* fats increase dramatically the risk of heart disease, especially atherosclerosis. The FDA has called for labeling of all products that contain *trans* fats, and many leading food producers and restaurants are currently limiting their content in the foods they sell. Of course, daily exercise is beneficial, regardless of diet.

26.2 | Soaps and Detergents

A. Structure and Preparation of Soaps

Natural **soaps** are prepared most commonly from a blend of tallow and coconut oils. In the preparation of tallow, the solid fats of cattle are melted with steam, and the tallow layer formed on the top is removed. The preparation of soaps begins by boiling these triglycerides with sodium hydroxide. The reaction that takes place is called **saponification** (Latin: *saponem*, soap). At the molecular level, saponification corresponds to base-promoted hydrolysis of the ester groups in triglycerides

Liquid vegetable oils contain mostly unsaturated fatty acids.

Soap
A sodium or potassium salt of a fatty acid.

Figure 26.3

Soap micelles. Nonpolar (hydrophobic) hydrocarbon chains are clustered in the interior of the micelle, and polar (hydrophilic) carboxylate groups are on the surface of the micelle. Soap micelles repel each other because of their negative surface charges.

(a) A soap

Polar "head"

Nonpolar "tail"

(b) Cross section of a soap micelle in water

Na⁺ ions

(Section 18.4C). The resulting soaps contain mainly the sodium salts of palmitic, stearic, and oleic acids from tallow and the sodium salts of lauric and myristic acids from coconut oil.

$$
\begin{array}{c}
\text{O} \\
\parallel \\
\text{O} \quad \text{CH}_2\text{OCR} \\
\parallel \quad | \\
\text{RCOCH} \quad \text{O} \\
| \\
\text{CH}_2\text{OCR}
\end{array}
\quad + \quad 3\ \text{NaOH}
\xrightarrow{\text{saponification}}
\begin{array}{c}
\text{CH}_2\text{OH} \\
| \\
\text{CHOH} \\
| \\
\text{CH}_2\text{OH}
\end{array}
\quad + \quad
\begin{array}{c}
\text{O} \\
\parallel \\
3\ \text{RCO}^-\text{Na}^+
\end{array}
$$

A triglyceride   **1,2,3-Propanetriol**   Sodium soaps
                                     (Glycerol; glycerin)

After hydrolysis is complete, sodium chloride is added to precipitate the soap as thick curds. The water layer is then drawn off, and glycerol is recovered by vacuum distillation. The crude soap contains sodium chloride, sodium hydroxide, and other impurities. These are removed by boiling the curd in water and reprecipitating with more sodium chloride. After several purifications, the soap can be used without further processing as an inexpensive industrial soap. Other treatments transform the crude soap into pH-controlled cosmetic soaps, medicated soaps, and the like.

Micelle
A spherical arrangement of organic molecules in water solution clustered so that their hydrophobic parts are buried inside the sphere and their hydrophilic parts are on the surface of the sphere and in contact with water.

Soap micelle with "dissolved" grease

Grease

Soap

Figure 26.4

A soap micelle with a "dissolved" oil or grease droplet.

B. How Soaps Clean

Soap owes its remarkable cleansing properties to its ability to act as an emulsifying agent. Because the long hydrocarbon chains of natural soaps are insoluble in water, they tend to cluster in such a way as to minimize their contact with surrounding water molecules. The polar carboxylate groups, on the other hand, tend to remain in contact with the surrounding water molecules. Thus, in water, soap molecules spontaneously cluster into **micelles** (Figure 26.3).

Most of the things we commonly think of as dirt (such as grease, oil, and fat stains) are nonpolar and insoluble in water. When soap and this type of dirt are mixed together, as in a washing machine, the nonpolar hydrocarbon inner parts of the soap micelles "dissolve" the nonpolar dirt molecules. In effect, new soap micelles are formed, this time with nonpolar dirt molecules in the center (Figure 26.4). In this way, nonpolar organic grease, oil, and fat are "dissolved" and washed away in the wash water.

Soaps are not without their disadvantages. Foremost among these is the fact that they form insoluble salts when used in water containing Ca(II), Mg(II), or Fe(III) ions (hard water).

$$2 \text{ CH}_3(\text{CH}_2)_{14}\text{COO}^- \text{ Na}^+ \quad + \quad \text{Ca}^{2+} \longrightarrow [\text{CH}_3(\text{CH}_2)_{14}\text{COO}^-]_2\text{Ca}^{2+} \quad + \quad 2 \text{ Na}^+$$

A sodium soap	Calcium salt of a fatty acid
(soluble in water as micelles)	(insoluble in water)

These calcium, magnesium, and iron salts of fatty acids create problems, including rings around the bathtub, films that spoil the luster of hair, and grayness and roughness that build up on textiles after repeated washings.

C. Synthetic Detergents

After the cleansing action of soaps was understood, synthetic detergents could be designed. Molecules of a good detergent must have a long hydrocarbon chain, preferably 12 to 20 carbon atoms long, and a polar group at one end of the molecule that does not form insoluble salts with the Ca(II), Mg(II), or Fe(III) ions present in hard water. Chemists recognized that these essential characteristics of a soap could be produced in a molecule containing a sulfate or sulfonate group instead of a carboxylate group. Calcium, magnesium, and iron salts of monoalkylsulfuric and sulfonic acids are much more soluble in water than comparable salts of fatty acids.

The most widely used synthetic detergents are the linear alkylbenzenesulfonates (LAS). One of the most common of these is sodium 4-dodecylbenzenesulfonate. To prepare this type of detergent, a linear alkylbenzene is treated with sulfuric acid (Section 22.1B) to form an alkylbenzenesulfonic acid. The sulfonic acid is then neutralized with NaOH; the product is mixed with builders and spray-dried to give a smooth-flowing powder. The most common builder is sodium silicate.

$$\text{CH}_3(\text{CH}_2)_{10}\text{CH}_2 \!-\!\!\!\left\langle\right\rangle\!\!\!-\xrightarrow[\text{2.NaOH}]{\text{1.H}_2\text{SO}_4} \text{CH}_3(\text{CH}_2)_{10}\text{CH}_2 \!-\!\!\!\left\langle\right\rangle\!\!\!-\text{SO}_3^-\text{Na}^+$$

Dodecylbenzene	**Sodium 4-dodecylbenzenesulfonate**
	(an anionic detergent)

Alkylbenzenesulfonate detergents were introduced in the late 1950s, and today they command close to 90% of the market once held by natural soaps.

Among the most common additives to detergent preparations are foam stabilizers, bleaches, and optical brighteners. A foam stabilizer frequently added to liquid soaps but not laundry detergents (for obvious reasons: think of a top-loading washing machine with foam spewing out the lid!) is the amide prepared from dodecanoic acid (lauric acid) and 2-aminoethanol (ethanolamine). The most common bleach is sodium perborate tetrahydrate, which decomposes at temperatures above 50°C to give hydrogen peroxide, the actual bleaching agent.

$$\overset{\displaystyle O}{\underset{\displaystyle \|}{\text{CH}_3(\text{CH}_2)_{10}\text{C}}}\text{NHCH}_2\text{CH}_2\text{OH} \qquad\qquad \text{O}\!=\!\text{B}\!-\!\text{O}\!-\!\text{O}^-\text{Na}^+\!\cdot 4\text{H}_2\text{O}$$

***N*-(2-Hydroxyethyl)dodecanamide**	**Sodium perborate tetrahydrate**
(a foam stabilizer)	(a bleach)

Also added to laundry detergents are optical brighteners, known also as optical bleaches. They are absorbed into fabrics and, after absorbing ambient light, fluoresce with a blue color, offsetting the yellow color caused by fabric aging. Quite literally, these optical brighteners produce a "whiter-than-white" appearance. You most certainly have observed the effects of optical brighteners if you have seen the surprisingly intense blue glow of "white" shirts or blouses when exposed to black light (UV radiation).

Flavin adenine dinucleotide (FAD) is a central component in the transfer of electrons in metabolic oxidations and reductions. In FAD, flavin is bonded to the five-carbon monosaccharide ribitol, which is, in turn, bonded to the terminal phosphate group of adenosine diphosphate:

flavin, the operative part of
the molecule in oxidation-
reduction reactions

riboflavin
(a vitamin of
the B complex)

ribitol

the adenosine
diphosphate (ADP)
building block
is shown in blue

Flavin adenine dinucleotide
(FAD)

FAD participates in several types of enzyme-catalyzed oxidation-reduction reactions, one of which is the oxidation of a carbon-carbon single bond in the hydrocarbon chain of a fatty acid. In the process FAD is reduced to $FADH_2$:

Oxidation of a hydrocarbon chain

$$—CH_2—CH_2— \quad + \quad FAD \quad \longrightarrow \quad —CH=CH— \quad + \quad FADH_2$$

a portion of the
hydrocarbon
chain of a fatty acid

The mechanism by which FAD oxidizes $—CH_2—CH_2—$ to $—CH=CH—$ involves the transfer of a hydride ion from the hydrocarbon chain of the fatty acid to FAD:

Mechanism Oxidation of a Fatty Acid $—CH_2—CH_2—$ to $—CH=CH—$ by FAD

The individual curved arrows in this mechanism are numbered 1–6 to help you follow the flow of electrons in the transformation.

Step 1: A basic group, Enz-B:, on the surface of the enzyme removes a hydrogen atom from the carbon adjacent to the carboxyl group.

Step 2: Electrons from this C—H sigma bond become the pi electrons of the new C—C double bond.

Step 3: A hydride ion transfers from the carbon beta to the carboxyl group to a nitrogen atom of flavin.

Step 4: The pi electrons within flavin become redistributed.

Step 5: Electrons of the C=N bond remove a hydrogen from the enzyme.

Step 6: A new basic group forms on the surface of the enzyme.

FAD

FADH₂

Note that one of the two hydrogen atoms added to FAD to produce FADH₂ comes from the hydrocarbon chain and the other comes from an acidic group on the surface of the enzyme catalyzing this oxidation. Note also that one group on the enzyme functions as a proton acceptor and that another functions as a proton donor.

26.3 | Prostaglandins

The **prostaglandins** are a family of compounds all having the 20-carbon skeleton of prostanoic acid.

Prostanoic acid

Prostaglandin
A member of the family of compounds having the 20-carbon skeleton of prostanoic acid.

The story of the discovery and structure determination of these remarkable compounds began in 1930 when gynecologists Raphael Kurzrok and Charles Lieb reported that human seminal fluid stimulates contraction of isolated uterine muscle. A few years later in Sweden, Ulf von Euler confirmed this report and noted that human seminal fluid also produces contraction of intestinal smooth muscle and lowers blood pressure when injected into the bloodstream. Von Euler proposed the name *prostaglandin* for the mysterious substance(s) responsible for these diverse effects because it was believed at the time that they were synthesized in the prostate gland. Although we now know that prostaglandin production is by no means limited to the prostate gland, the name nevertheless has stuck.

Prostaglandins are not stored as such in target tissues. Rather, they are synthesized in response to specific physiological triggers. Starting materials for the biosynthesis of prostaglandins are polyunsaturated fatty acids of 20 carbons, stored until needed as membrane phospholipid esters. In response to a physiological trigger, the ester is hydrolyzed, the fatty acid is released, and the synthesis of prostaglandins

Figure 26.5

Key intermediates in the conversion of arachidonic acid to PGE$_2$ and PGF$_{2\alpha}$. PG stands for prostaglandin. The letters E, F, G, and H are different types of prostaglandins.

is initiated. Figure 26.5 outlines the steps in the synthesis of several prostaglandins from arachidonic acid. A key step in this biosynthesis is the enzyme-catalyzed reaction of arachidonic acid with two molecules of O$_2$ to form prostaglandin G$_2$ (PGG$_2$). The antiinflammatory and anticlotting effects of aspirin and other nonsteroidal antiinflammatory drugs (NSAIDs) results from their ability to inhibit the enzyme that catalyzes this step.

Research on the involvement of prostaglandins in reproductive physiology and the inflammatory process has produced several clinically useful prostaglandin derivatives. The observations that PGF$_{2\alpha}$ stimulates contractions of uterine smooth muscle led to a synthetic derivative that is used as a therapeutic abortifacient. A problem with the use of the natural prostaglandins for this purpose is that they are very rapidly degraded within the body. In the search for less rapidly degraded prostaglandins, a number of analogs have been prepared, one of the most effective of which is carboprost. This synthetic prostaglandin is 10 to 20 times more potent than the natural PGF$_{2\alpha}$ and is only slowly degraded in the body. The comparison of these two prostaglandins illustrates how a simple change in the structure of a drug can make a significant change in its effectiveness.

PGF$_{2\alpha}$

Carboprost
(15S)-15-Methyl-PGF$_{2\alpha}$

extra methyl group at C-15

The PGEs along with several other PGs suppress gastric ulceration and appear to heal gastric ulcers. The PGE$_1$ analog, misoprostol, is currently used primarily for prevention of ulceration associated with aspirin-like NSAIDs (partly caused by their inhibition of clotting).

PGE$_1$

Misoprostol

Prostaglandins are members of an even larger family of compounds called **eicosanoids**. Eicosanoids contain 20 carbons and are derived from fatty acids. They include not only the prostaglandins but also the leukotrienes, thromboxanes, and prostacyclins. The eicosanoids are extremely widespread, and members of this family of compounds have been isolated from almost every tissue and body fluid.

Leukotriene C$_4$(LTC$_4$)
(a smooth muscle constrictor)

L-cysteine

glycine

L-glutamic acid

Thromboxane A$_2$
(a potent vasoconstrictor)

Prostacyclin
(a platelet aggregation inhibitor)

Leukotrienes are derived from arachidonic acid and are found primarily in leukocytes (white blood cells). Leukotriene C_4 (LTC_4), a typical member of this family, has three conjugated double bonds (hence the suffix -*triene*) and contains the amino acids L-cysteine, glycine, and L-glutamic acid (Chapter 27). An important physiological action of LTC_4 is constriction of smooth muscles, especially those of the lungs. The synthesis and release of LTC_4 is prompted by allergic reactions. Drugs that inhibit the synthesis of LTC_4 show promise for the treatment of the allergic reactions associated with asthma. Thromboxane A_2 is a very potent vasoconstrictor; its release triggers the irreversible phase of platelet aggregation and constriction of injured blood vessels. It is thought that aspirin and aspirin-like drugs act as mild anticoagulants because they inhibit cyclooxygenase, the enzyme that initiates the synthesis of thromboxane A_2.

26.4 | Steroids

Steroids are a group of plant and animal lipids that have the tetracyclic ring system shown in Figure 26.6. The features common to the tetracyclic ring system of most naturally occurring steroids are illustrated in Figure 26.7.

1. The fusion of rings is *trans,* and each atom or group at a ring junction is axial. Compare, for example, the orientations of —H at carbon 5 and —CH_3 at carbon 10.
2. The pattern of atoms or groups along the points of ring fusion (carbons 5 to 10 to 9 to 8 to 14 to 13) is nearly always *trans*-anti-*trans*-anti-*trans*.
3. Because of the *trans*-anti-*trans*-anti-*trans* arrangement of atoms or groups along the points of ring fusion, the tetracyclic steroid ring system is nearly flat and quite rigid.
4. Many steroids have axial methyl groups at carbons 10 and 13 of the tetracyclic ring system.

A. Structure of the Major Classes of Steroids

Cholesterol

Cholesterol is a white, water-insoluble, waxy solid found in blood plasma and in all animal tissues. This substance is an integral part of human metabolism in two ways: (1) It is an essential component of biological membranes. The body of a healthy adult contains approximately 140 g of cholesterol, about 120 g of which is present in membranes. Membranes of the central and peripheral nervous systems, for example, contain about 10% cholesterol by weight. (2) It is the compound from which sex hormones, adrenocorticoid hormones, bile acids, and vitamin D are synthesized. Thus, cholesterol is, in a sense, the parent steroid.

Steroid
A plant or animal lipid having the characteristic tetracyclic ring structure of the steroid nucleus, namely three six-membered rings and one five-membered ring.

Figure 26.6
The tetracyclic ring system characteristic of steroids.

Figure 26.7
Features common to the tetracyclic ring system of many steroids.

Methyl groups at C–10 and C–13 are axial and above the plane of the rings

Cholesterol has eight chiral centers, and a molecule with this number of chiral centers can exist as 2^8, or 256, stereoisomers (128 pairs of enantiomers). Only one of these stereoisomers is known to exist in nature: the stereoisomer with the configuration shown in Figure 26.8.

Cholesterol is insoluble in blood plasma but can be transported as a plasma-soluble complex formed with proteins called lipoproteins. **Low-density lipoproteins** (**LDL**, often called "bad cholesterol") transport cholesterol from the site of its synthesis in the liver to the various tissues and cells of the body where it is to be used. Primarily, the cholesterol associated with LDLs builds up in atherosclerotic deposits in blood vessels. **High-density lipoproteins** (**HDL**, often called "good cholesterol") transport excess and unused cholesterol from cells back to the liver for its degradation to bile acids and eventual excretion in the feces. It is thought that HDLs retard or reduce atherosclerotic deposits.

© Carolina Biological Supply Company/Phototake, NYC

Human gallstones are almost pure cholesterol; this gallstone is about 0.5 cm in diameter.

Cholesterol

Figure 26.8
Cholesterol is found in blood plasma and in all animal tissues.

Steroid Hormones

Given in Table 26.3 are representations of each major class of steroid hormones, along with the principal functions of each.

The two most important female sex hormones, or **estrogens**, are estrone and estradiol. In addition, progesterone, another type of steroid hormone, is essential for preparing the uterus for implantation of a fertilized egg. After the role of progesterone in inhibiting ovulation was understood, its potential as a possible contraceptive was realized. Progesterone itself is relatively ineffective when taken orally. As a result of a massive research program in both industrial and academic laboratories, many synthetic progesterone-mimicking steroids became available in the 1960s. When taken regularly, these drugs prevent ovulation yet allow women to maintain a normal menstrual cycle. Some of the most effective of these preparations contain a progesterone analog, such as norethindrone, combined with a smaller amount of an estrogen-like material to help prevent irregular menstrual flow during prolonged use of contraceptive pills.

Low-density lipoprotein (LDL)
Plasma particles, density 1.02–1.06 g/mL, consisting of approximately 26% proteins, 50% cholesterol, 21% phospholipids, and 4% triglycerides.

High-density lipoprotein (HDL)
Plasma particles, density 1.06–1.21 g/mL, consisting of approximately 33% proteins, 30% cholesterol, 29% phospholipids, and 8% triglycerides.

Estrogen
A steroid hormone, such as estrone and estradiol, that mediates the development of sexual characteristics in females.

"Nor" refers to the absence of a methyl group here. The methyl group is present in ethindrone.

Norethindrone
(a synthetic progesterone analog)

Androgen
A steroid hormone, such as testosterone, that mediates the development of sexual characteristics of males.

The chief function of testosterone and other **androgens** is to promote normal growth of male reproductive organs (primary sex characteristics) and development

Table 26.3	Selected Steroid Hormones

Structure	Source and Major Effects

Testosterone   **Androsterone**

Androgens (male sex hormones)—synthesized in the testes; responsible for development of male secondary sex characteristics

Progesterone   **Estrone**

Estrogens (female sex hormones)—synthesized in the ovaries; responsible for development of female secondary sex characteristics and control of the menstrual cycle

Cortisone   **Cortisol**

Glucocorticoid hormones—synthesized in the adrenal cortex; regulate metabolism of carbohydrates, decrease inflammation, and are involved in the reaction to stress

Aldosterone

A mineralocorticoid hormone—synthesized in the adrenal cortex; regulates blood pressure and volume by stimulating the kidneys to absorb Na^+, Cl^-, and HCO_3^-

Anabolic steroid
A steroid hormone, such as testosterone, that promotes tissue and muscle growth and development.

of the characteristic deep voice, pattern of body and facial hair, and musculature (secondary sex characteristics). Although testosterone produces these effects, it is not active when taken orally because it is metabolized in the liver to an inactive steroid. A number of oral **anabolic steroids** have been developed for use in rehabilitation medicine, particularly when muscle atrophy occurs during recovery from an injury. Among the synthetic anabolic steroids most widely prescribed for this purpose are methandrostenolone and stanozolol. The structural formula of methandrostenolone differs from that of testosterone by introduction of (1) a methyl group at carbon 17 and (2) an additional carbon-carbon double bond between carbons 1 and 2. In stanozolol, ring A is modified by attachment of a pyrazole ring.

Methandrostenolone **Stanozolol**

Among certain athletes, the misuse of anabolic steroids to build muscle mass and strength, particularly for sports that require explosive action, is common. The risks associated with abuse of anabolic steroids for this purpose are enormous: heightened aggressiveness, sterility, impotence, and risk of premature death from complications of diabetes, coronary artery disease, and liver cancer.

Bile Acids

Shown in Figure 26.9 is a structural formula for cholic acid, a constituent of human bile. The molecule is shown as an anion, as it is ionized in bile and intestinal fluids. **Bile acids**, or more properly, bile salts, are synthesized in the liver, stored in the gallbladder, and secreted into the intestine, where their function is to emulsify dietary fats and thereby aid in their absorption and digestion. Furthermore, bile salts are the end products of the metabolism of cholesterol and, thus, are a principal pathway for the elimination of this substance from the body. A characteristic structural feature of bile salts is a *cis* fusion of rings A and B.

Bile acid
A cholesterol-derived detergent molecule, such as cholic acid, which is secreted by the gallbladder into the intestine to assist in the absorption of dietary lipids.

B. Biosynthesis of Cholesterol

The biosynthesis of cholesterol illustrates a point we first made in our introduction to the structure of terpenes (Section 5.4). In building large molecules, one of the common patterns in the biological world is to begin with one or more smaller subunits, join them by an iterative process, and then chemically modify the completed carbon skeleton by oxidation, reduction, cross-linking, addition, elimination, or related processes to give a biomolecule with a unique identity.

The building block from which all carbon atoms of steroids are derived is the two-carbon acetyl group of acetyl-CoA (Problem 25.35). The American biochemist, Konrad Bloch, who shared the 1964 Nobel Prize for medicine or physiology with German biochemist Feodor Lynen for their discoveries concerning the biosynthesis of cholesterol and fatty acids, showed that 15 of the 27 carbon atoms of cholesterol

Figure 26.9
Cholic acid, an important constituent of human bile. Each six-membered ring is in a chair conformation.

Figure 26.10

Several key intermediates in the synthesis of cholesterol from acetyl groups of acetyl CoA. Eighteen moles of acetyl CoA are required for the synthesis of one mole of cholesterol.

The acetyl group has no chiral center

(C_2) Acetyl coenzyme A

(C_6) (R)-Mevalonate

(C_5) Isopentenyl pyrophosphate

(C_{10}) Geranyl pyrophosphate → Monoterpenes

(C_{15}) Farnesyl pyrophosphate → Sesquiterpenes, Diterpenes

(C_{30}) Squalene → Triterpenes

Cholesterol has 8 chiral centers and is one of 256 possible stereoisomers.

Cholesterol (C_{27})

are derived from the methyl group of acetyl-CoA. The remaining 12 carbon atoms are derived from the carbonyl group of acetyl-CoA (Figure 26.10).

A remarkable feature of this synthetic pathway is that the biosynthesis of cholesterol from acetyl-CoA is completely stereoselective; it is synthesized as only one of 256 possible stereoisomers. We cannot duplicate this exquisite degree of stereoselectivity in the laboratory. Cholesterol is, in turn, the key intermediate in the synthesis of most other steroids.

cholesterol
- bile acids (e.g., **cholic acid**)
- sex hormones (e.g., **testosterone and estrone**)
- mineralocorticoid hormones (e.g., **aldosterone**)
- glucocorticoid hormones (e.g., **cortisone**)

26.5 | Phospholipids

A. Structure

Phospholipid
A lipid containing glycerol esterified with two molecules of fatty acid and one molecule of phosphoric acid.

Phospholipids, or phosphoacylglycerols as they are more properly named, are the second most abundant group of naturally occurring lipids. They are found almost exclusively in plant and animal membranes, which typically consist of about 40–50% phospholipids and 50–60% proteins. The most abundant phospholipids are derived from a phosphatidic acid (Figure 26.11).

The fatty acids most common in phosphatidic acids are palmitic and stearic acids (both fully saturated) and oleic acid (one double bond in the hydrocarbon chain). Further esterification of a phosphatidic acid with a low-molecular-weight alcohol gives a phospholipid. Several of the most common alcohols forming phospholipids are given in Table 26.4. All functional groups in this table and in Figure 26.11 are

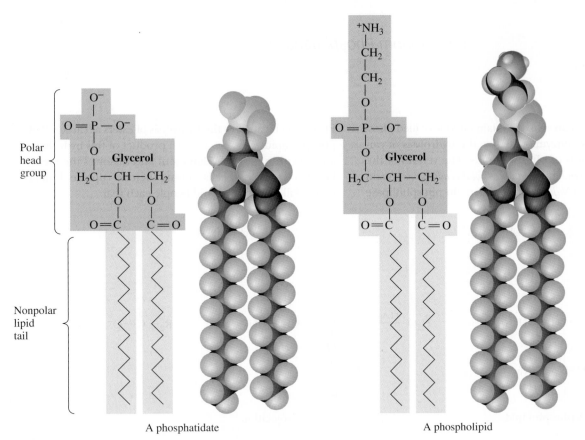

Figure 26.11
In a phosphatidic acid, glycerol is esterified with two molecules of fatty acid and one molecule of phosphoric acid. Further esterification of the phosphoric acid group with a low-molecular-weight alcohol gives a phospholipid.

shown as they are ionized at pH 7.4, the approximate pH of blood plasma and of many other biological fluids. Under these conditions, each phosphate group bears a negative charge and each amino group bears a positive charge.

B. Lipid Bilayers

Figure 26.12 shows a space-filling model of a lecithin (a phosphatidycholine). It and other phospholipids are elongated, almost rodlike molecules, with the nonpolar (hydrophobic) hydrocarbon chains lying roughly parallel to one another and the polar (hydrophilic) phosphoric ester group pointing in the opposite direction.

Lipid bilayer
A back-to-back arrangement of phospholipid monolayers, often forming a closed vesicle or membrane.

Table 26.4	Low-Molecular-Weight Alcohols Most Common to Phospholipids	

Alcohols Found in Phospholipids

Structural Formula	Name	Name of Phospholipid
$HOCH_2CH_2NH_2$	Ethanolamine	Phosphatidylethanolamine (Cephalin)
$HOCH_2CH_2\overset{+}{N}(CH_3)_3$	Choline	Phosphatidylcholine (Lecithin)
$HOCH_2CHCOO^-$ $\quad\quad\mid$ $\quad\quad NH_3^+$	Serine	Phosphatidylserine
Inositol structure	Inositol	Phosphatidylinositol

All of these products contain lecithin.

The venoms of certain snakes contain enzymes called phospholipases. These enzymes catalyze the hydrolysis of carboxylic ester bonds of phospholipids. The venom of the eastern diamondback rattlesnake (*Crotalus adamanteus*) and the Indian cobra (*Naja naja*) both contain phospholipase

PLA$_2$, which catalyzes the hydrolysis of esters at carbon 2 of phospholipids. The breakdown product of this hydrolysis, a lysolecithin, acts as a detergent and dissolves the membranes of red blood cells causing them to rupture. Indian cobras kill several thousand people each year.

$$CH_2CH_2NH_3^+$$
$$|$$
$$O$$
$$|$$
$$O{=}P{-}O^-$$
$$|$$
$$O$$
$$|$$
$$CH_2{-}CH{-}CH_2$$
$$|\quad\quad|$$
$$O\quad\quad O$$
$$|\quad\quad|$$
$$O{=}C\ \ O{=}C$$
$$|\quad\quad|$$
$$R_1\quad\quad R_2$$

PLA$_2$ catalyzes hydrolysis of this ester bond.

A phospholipid

$+\quad H_2O \xrightarrow{PLA_2}$

$$CH_2CH_2NH_3^+$$
$$|$$
$$O$$
$$|$$
$$O{=}P{-}O^-$$
$$|$$
$$O$$
$$|$$
$$CH_2{-}CH{-}CH_2$$
$$|\quad\quad|$$
$$O\quad\quad OH$$
$$|$$
$$O{=}C$$
$$|$$
$$R_1$$

A lysolecithin

$$+\quad R_2{-}\overset{\displaystyle O}{\overset{\|}{C}}{-}O^- + H^+$$

Figure 26.12
Space-filling model of a lecithin.

Fluid-mosaic model
A biological membrane that consists of a phospholipid bilayer with proteins, carbohydrates, and other lipids on the surface and embedded in the surface of the bilayer.

When placed in aqueous solution, phospholipids spontaneously form a **lipid bilayer** (Figure 26.13) in which polar head groups lie on the surface, giving the bilayer an ionic coating. Nonpolar hydrocarbon chains of fatty acids lie buried within the bilayer. This self-assembly of phospholipids into a bilayer is a spontaneous process, driven by two types of noncovalent forces: (1) hydrophobic effects, which result when nonpolar hydrocarbon chains cluster together and exclude water molecules, and (2) electrostatic interactions, which result when polar head groups interact with water and other polar molecules in the aqueous environment.

Recall from Section 26.2B that formation of soap micelles is driven by these same noncovalent forces; the polar (hydrophilic) carboxylate groups of soap molecules lie on the surface of the micelle and associate with water molecules, and the nonpolar (hydrophobic) hydrocarbon chains cluster within the micelle and thus are removed from contact with water.

The arrangement of hydrocarbon chains in the interior of a phospholipid bilayer varies from rigid to fluid, depending on the degree of unsaturation of the hydrocarbon chains themselves. Saturated hydrocarbon chains tend to lie parallel and closely packed, increasing the rigidity of the bilayer. Unsaturated hydrocarbon chains, on the other hand, have one or more *cis* double bonds, which cause "kinks" in the chains. As a result, they do not pack as closely and with as great an order as saturated chains. The disordered packing of unsaturated hydrocarbon chains leads to increased fluidity of the bilayer.

Biological membranes are made of lipid bilayers. The most satisfactory current model for the arrangement of phospholipids, proteins, and cholesterol in plant and animal membranes is the **fluid-mosaic model** proposed in 1972 by S. J. Singer and G. Nicolson (Figure 26.13). The term "mosaic" signifies that the various components in the membrane coexist side by side, as discrete units, rather than combining to form new molecules or ions. "Fluid" signifies that the same sort of fluidity exists in membranes that we have already seen for lipid bilayers. Furthermore, the protein components of membranes "float" in the bilayer and can move laterally along the plane of the membrane.

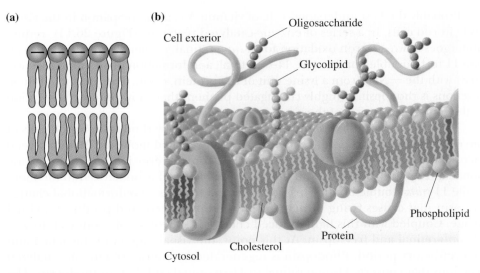

Figure 26.13

Fluid-mosaic model of a biological membrane, showing the lipid bilayer and membrane proteins oriented on the inner and outer surfaces of the membrane and penetrating the entire thickness of the membrane.

26.6 | Fat-Soluble Vitamins

Vitamins are divided into two broad classes on the basis of solubility, those that are fat-soluble (and hence classed as lipids) and those that are water-soluble. The fat-soluble vitamins include A, D, E, and K.

A. Vitamin A

Vitamin A, or retinol, occurs only in the animal world, where the best sources are cod-liver oil and other fish-liver oils, animal liver, and dairy products. Vitamin A in the form of a precursor, or provitamin, is found in the plant world in a group of tetraterpene (C_{40}) pigments called carotenes. The most common of these is β-carotene, abundant in carrots but also found in some other vegetables, particularly yellow and green ones. β-Carotene has activity as an antioxidant; one of its functions in green plants is to quench singlet oxygen, which can be produced as a byproduct of photosynthesis. β-Carotene has no vitamin A activity; however, after ingestion, it is cleaved at the central carbon-carbon double bond followed by reduction of the newly formed aldehyde to give retinol (vitamin A).

Cleavage of this
C=C gives vitamin A.

β-Carotene

enzyme-catalyzed cleavage
and reduction in the liver

CH₂OH

2

Retinol (Vitamin A)

Probably the best understood role of vitamin A is its participation in the visual cycle in rod cells. In a series of enzyme-catalyzed reactions (Figure 26.14), retinol undergoes a two-electron oxidation to all-*trans*-retinal, isomerization about the carbon 11 to 12 double bond to give 11-*cis*-retinal, and formation of an imine (Section 16.8) with the —NH₂ from a lysine unit of the protein, opsin. The product of these reactions is rhodopsin, a highly conjugated pigment that shows intense absorption in the blue-green region of the visual spectrum.

The primary event in vision is absorption of light by rhodopsin in rod cells of the retina of the eye to produce an electronically excited molecule. Within several picoseconds (1 ps = 10^{-12} s), the excess electronic energy is converted to vibrational and rotational energy, and the 11-*cis* double bond is isomerized to the more stable 11-*trans* double bond. This isomerization triggers a conformational change in opsin that causes firing of neurons in the optic nerve and produces a visual image. Coupled with this light-induced change is hydrolysis of rhodopsin to give 11-*trans*-retinal and free opsin. At this point, the visual pigment is bleached and in a refractory period. Rhodopsin is regenerated by a series of enzyme-catalyzed reactions that converts 11-*trans*-retinal to 11-*cis*-retinal and then to rhodopsin. The visual cycle is shown in abbreviated form in Figure 26.14.

B. Vitamin D

Vitamin D is the name for a group of structurally related compounds that play a major role in the regulation of calcium and phosphorus metabolism. A deficiency of vitamin D in childhood is associated with rickets, a mineral-metabolism disease that leads to bone defects that form bowlegs, knock-knees, and enlarged joints. Vitamin D₃, the most abundant form of the vitamin in the circulatory system, is produced in the skin of mammals by the action of ultraviolet radiation on 7-dehydrocholesterol (cholesterol with a double bond between carbons 7 and 8). In the liver, vitamin D₃ undergoes an enzyme-catalyzed, two-electron oxidation at carbon 25 of the side chain to form 25-hydroxyvitamin D₃; the oxidizing agent is molecular oxygen, O_2.

Figure 26.14

The primary chemical reaction of vision in rod cells is absorption of light by rhodopsin followed by isomerization of a carbon-carbon double bond from a *cis* configuration to a *trans* configuration.

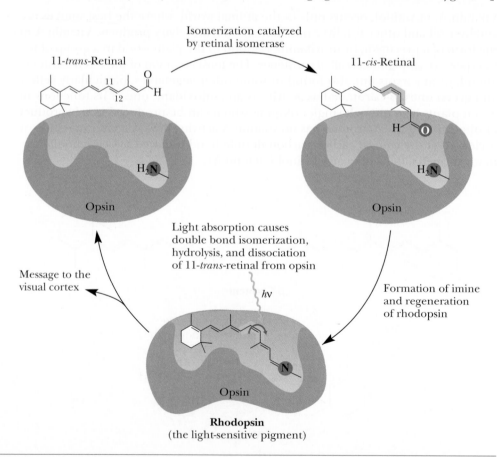

Rhodopsin
(the light-sensitive pigment)

Vitamin K is a fat-soluble vitamin, which must be obtained from the diet. A vitamin K deficiency results in slowed blood clotting, which can be a serious threat to a wounded animal or human. In the process of blood clotting, the natural vitamin, a quinone, is converted to its active hydroquinone form by reduction.

Vitamin K_1 quinone   **Vitamin K_1 hydroquinone**

In the presence of vitamin K, O_2, CO_2, and an enzyme called microsomal carboxylase, side chains of glutamate units in prothrombin, a protein essential for blood clotting, are modified by addition of carboxyl groups (from CO_2) to form γ-carboxyglutamate units. Note that this reaction is the reverse of the decarboxylation of β-dicarboxylic (malonic) acids seen in Section 17.9B. The two carboxyl groups of the chemically modified glutamate now form a tight bidentate ("two teeth") complex with Ca^{2+} during the blood-clotting process. Although there is more to be understood about blood clotting, it is at least clear that if prothrombin is not carboxylated, it does not bind calcium, and blood does not clot.

These facts have been known for many years. Until quite recently, however, the role of vitamin K_1 in this process remained a mystery. The problem was this. The anion of vitamin K_1 hydroquinone is a weak base (pK_a of approximately 9) derived from a phenol. To remove a proton from a glutamate side chain requires a very strong base derived from a conjugate acid of pK_a approximately 27. How can molecular oxygen increase the base strength of vitamin K_1 hydroquinone by 18 orders of magnitude?

Recently Paul Dowd discovered that the vitamin K_1 hydroquinone anion reacts with oxygen to give the peroxide anion intermediate 1, which is converted to compound 2. Notice that compound 2 contains a weak O—O bond in a highly strained four-membered ring. Compound 2 then rearranges to vitamin K_1 base, a strong, sterically hindered alkoxide base.

The weak O—O bond is, in this way, replaced by a stronger C—O bond in vitamin K_1 base. The extra stability provides the driving force for turning a weak phenoxide base into a strong alkoxide base, which is able to remove a proton from glutamate side chains. This makes possible the addition of CO_2 to form γ-carboxyglutamate side chains, which bind calcium ions during the clotting cascade. Thus, it is now understood why O_2, CO_2, and vitamin K_1 are essential for this phase of blood clotting. Synthetic vitamin K analogs are as effective in this process as naturally occurring vitamin K_1.

Glutamate side chain of prothrombin

Carboxylated glutamate side chain

Carboxylated glutamate side chain binding calcium ion

Vitamin K_1 hydroquinone anion (a weak base)

1

2

Vitamin K_1 base (a very strong base)

25-Hydroxyvitamin D_3 undergoes further oxidation in the kidneys, also by O_2, to form 1,25-dihydroxyvitamin D_3, the hormonally active form of the vitamin.

1. Opening of ring B by ultraviolet light
2. Enzyme-catalyzed oxidation by O_2 at C-1 and C-25

7-Dehydrocholesterol

1,25-Dihydroxyvitamin D_3

C. Vitamin E

Vitamin E was first recognized in 1922 as a dietary factor essential for normal reproduction in rats, hence its name tocopherol from the Greek: *tocos*, birth, and *pherein*, to bring about. Vitamin E is a group of compounds of similar structure, the most active of which is α-tocopherol. This vitamin occurs in fish oil, in other oils such as cottonseed and peanut oil, and in leafy green vegetables. The richest source of vitamin E is wheat germ oil.

four isoprene units, joined head-to-tail, beginning here and ending at the aromatic ring

Vitamin E (α-Tocopherol)

In the body, vitamin E functions as an antioxidant; it traps peroxy radicals of the type HOO· and ROO· formed as a result of enzyme-catalyzed oxidation by molecular oxygen of the unsaturated hydrocarbon chains in membrane phospholipids (see Chemical Connection: "Antioxidants" in Section 8.7). There is speculation that peroxy radicals play a role in the aging process and that vitamin E and other antioxidants may retard that process. Vitamin E is also necessary for the proper development and function of the membranes of red blood cells.

Summary

Section 26.1 Triglycerides

- **Lipids** are a heterogeneous class of compounds grouped together on the basis of their solubility properties; they are insoluble in water and soluble in diethyl ether, acetone, and dichloromethane.
- **Triglycerides** (**triacylglycerols**), the most abundant lipids, are triesters of glycerol and fatty acids.
- **Fatty acids** are long-chain carboxylic acids derived from the hydrolysis of animal fats, vegetable oils, and the phospholipids of biological membranes.
  - Fatty acids are named by the number of carbons in the chain and the number of carbon-carbon double bonds present (e.g., 18:2).
  - Nearly all fatty acids have an even number of carbon atoms, usually between 12 and 20.
  - The carbon-carbon double bonds in naturally occurring fatty acids are almost exclusively found in the *cis* configuration.

Problems: 26.1–26.10

- The melting point of a triglyceride increases as (1) the length of its hydrocarbon chains increases and (2) its degree of saturation increases (*cis* double bonds add kinks to the chains, limiting packing and decreasing melting points).
- Triglycerides rich in saturated fatty acids are generally solids or semisolids at room temperature and are called **fats**.
- Triglycerides rich in unsaturated fatty acids are generally liquids at room temperature and are called **oils**.
  - Oils are subjected to partial hydrogenation (limiting H_2 in the presence of a metal catalyst) in order to reduce some unsaturated fatty acids to saturated fatty acids, a process called **hardening** that converts oils into fats for the food industry.
  - An unintended consequence of hardening is that some triglyceride *cis* double bonds are isomerized to the more stable *trans* form. Triglycerides containing *trans* double bonds, referred to as ***trans* fats** in the media, increase significantly the risk of heart disease and are being limited in food production.

Section 26.2 Soaps and Detergents

- **Soaps** are sodium or potassium salts of fatty acids.
  - In water, soaps form **micelles**, which "dissolve" nonpolar organic grease and oil.
  - Natural soaps precipitate as water-insoluble salts with Mg(II), Ca(II), and Fe(III) ions in hard water because of the strong interaction between carboxylates and these ions.
  - Synthetic detergents utilize sulfonates instead of carboxylates because metal ion salts of sulfonates are much more soluble in water.
  - The most common and most widely used synthetic detergents are linear alkylbenzenesulfonates.

Problems: 26.11–26.16

Section 26.3 Prostaglandins

- **Prostaglandins** are a group of extremely biologically active compounds having the 20-carbon skeleton of prostanoic acid.
  - Prostaglandins are synthesized in response to physiological triggers, from phospholipid-bound arachidonic acid and other 20-carbon fatty acids. Aspirin and other nonsteroidal antiinflammatory drugs (NSAIDs) inhibit this conversion.
- **Eicosanoids** are a very broad class of natural lipids that contain 20 carbons, are derived from fatty acids, and include the prostaglandins as well as leukotrienes, thromboxanes, and prostacyclins, all common and important biologically active molecules.

Problems: 26.17–26.19

Section 26.4 Steroids

- **Steroids** are a group of plant and animal lipids that have a characteristic tetracyclic structure composed of three six-membered rings and one five-membered ring.
  - The steroid skeleton is rigid and relatively flat owing to a common *trans*-anti-*trans*-anti-*trans* orientation of ring fusions.
- **Cholesterol** is an integral part of animal membranes, and it is the compound from which human sex hormones, adrenocorticoid hormones, bile acids, and vitamin D are biosynthesized.
  - **Low-density lipoproteins** (**LDLs**) transport cholesterol from the site of its synthesis in the liver to tissues and cells where it is to be used.
  - **High-density lipoproteins** (**HDLs**) transport cholesterol from cells back to the liver for its degradation to bile acids and eventual excretion in the feces.
- **Oral contraceptives** contain a synthetic progestin, for example norethindrone, which prevents ovulation yet allows women to maintain an otherwise normal menstrual cycle.
- A variety of synthetic **anabolic steroids** are available for use in rehabilitation medicine where muscle tissue has weakened or deteriorated as a result of an injury.
- **Bile acids**, used to aid digestion of fats, differ from most other steroids in that they have a *cis* configuration at the junction of rings A and B.
- The carbon skeleton of cholesterol and those of all biomolecules derived from it originate with the acetyl group (a two-carbon unit) of acetyl-CoA.

Problems: 26.20–26.25

Section 26.5 Phospholipids

- **Phospholipids**, the second most abundant group of naturally occurring lipids, are derived from phosphatidic acids, compounds containing glycerol esterified with two molecules of fatty acid and a molecule of phosphoric acid.
  - Further esterification of the phosphoric acid part with a low-molecular-weight alcohol, most commonly ethanolamine, choline, serine, or inositol, gives a phospholipid.
- Phospholipids are the major component of biological membranes.
  - When placed in aqueous solution, phospholipids spontaneously form **lipid bilayers**.
  - According to the **fluid-mosaic model**, membrane phospholipids form lipid bilayers with membrane proteins associated with the bilayer as both peripheral and integral proteins.

Problems: 26.26–26.30

Section 26.6 Fat-Soluble Vitamins

Problems: 26.31–26.33

- **Vitamin A** occurs only in the animal world.
  - The carotenes of the plant world are tetraterpenes (C_{40}) and are cleaved, after ingestion, into vitamin A.
  - The best-understood role of vitamin A is its participation in the visual cycle where it is used to make the photoactive component of rhodopsin, the light-sensitive pigment in our eyes.
- **Vitamin D** is the name for a structurally related set of molecules that play a major role in the regulation of calcium and phosphorus metabolism. Vitamin D_3 is synthesized in the skin of mammals by the action of ultraviolet radiation on 7-dehydrocholesterol.
- **Vitamin E** is a group of compounds of similar structure, the most active of which is α-tocopherol. In the body, vitamin E functions as an antioxidant by trapping peroxy radicals.

ⓌWL Questions and problems for this chapter are available in OWL. Use the OWL access card that came with this textbook to access assigned questions and problems for this chapter.

Chapter 27

Amino Acids and Proteins

© Hans Strand/Stone/Getty

Spider silk is a fibrous protein that exhibits unmatched strength and toughness. Inset: alanine and glycine are major components of the fibrous protein of spider silk. See Chemical Connections: "Spider Silk."

We begin this chapter with a study of amino acids, compounds whose chemistry is built on amines (Chapter 23) and carboxylic acids (Chapter 17). We concentrate in particular on the acid-base properties of amino acids because these properties are so important in determining many of the properties of proteins, including the catalytic functions of enzymes. With this understanding of the chemistry of amino acids, we then examine the structure of proteins themselves.

27.1 | Amino Acids

A. Structure

An **amino acid** is a compound that contains both a carboxyl group and an **amino group**. Although many types of amino acids are known, the **α-amino acids** are the most significant in the biological world because they are the monomers from which proteins are constructed. We have already introduced amino acids (see Connections to Biological Chemistry: "Amino Acids" in Section 3.9). A general structural formula of an α-amino acid is shown in Figure 27.1.

Although Figure 27.1(a) is a common way of writing structural formulas for amino acids, it is not accurate because it shows an acid (—COOH) and a base (—NH₂) within the same molecule. These acidic and basic groups react with each other to form a dipolar ion or internal salt [Figure 27.1(b)]. The internal salt of an amino acid is given the special name **zwitterion**. Note that a zwitterion has no net charge; it contains one positive charge and one negative charge.

Amino acid
A compound that contains both an amino group and a carboxyl group.

α-Amino acid
An amino acid in which the amino group is on the carbon adjacent to the carboxyl group.

Zwitterion
An internal salt of an amino acid; the carboxylate anion is negatively charged, and the ammonium group is positively charged.

OWL Online homework for this chapter may be assigned in OWL for Organic Chemistry.

Figure 27.1

An α-amino acid. (a) Unionized representation and (b) the more accurate internal salt (zwitterion) form.

Because they exist as zwitterions, amino acids have many of the properties associated with salts. They are crystalline solids with high melting points and are fairly soluble in water but insoluble in nonpolar organic solvents such as ether and hydrocarbon solvents.

B. Chirality

To review, with the exception of glycine, H_2NCH_2COOH, all protein-derived amino acids have at least one chiral center and, therefore, are chiral. Figure 27.2 shows Fischer projection formulas for the enantiomers of alanine. The vast majority of carbohydrates in the biological world are of the D-series, whereas the vast majority of α-amino acids in the biological world are of the L-series. As we saw for carbohydrates, the L designation for amino acids is based on analogy to the chiral center of glyceraldehyde. Note the similarity of the L-alanine and L-glyceraldehyde chiral centers in Figure 27.2. Some students assume that the L designation is based upon optical activity and that samples of all L-amino acids must rotate the plane of plane-polarized light in the levorotatory (−) direction. This is not true. In fact, many samples of L-amino acids rotate the plane of plane-polarized light in the dextrorotatory (+) direction. The L designation of amino acids refers only to *structural* similarity to the reference compound, L-glyceraldehyde, not any experimentally measured parameter.

The alternative *R,S* convention is also used to specify the configurations of amino acids. According to this convention, L-alanine is designated (*S*)-alanine. Although the D,L method of designating the stereochemistry of amino acid chiral centers is awkward, the historical use of this system is so deeply rooted in the scientific literature that it is still the most commonly used when referring to amino acids as well as carbohydrates. We will therefore use the D,L convention throughout the remainder of this chapter.

C. Protein-Derived Amino Acids

Table 27.1 gives common names, structural formulas, and standard three-letter and one-letter abbreviations for the 20 common L-amino acids found in proteins. The amino acids in this table are divided into four categories: those with nonpolar side chains, polar but unionized side chains, acidic side chains, and basic side chains. The following structural features of these amino acids should be noted.

1. All 20 of these protein-derived amino acids are α-amino acids, meaning that the amino group is located on the carbon alpha to the carboxyl group.
2. For 19 of the 20 amino acids, the α-amino group is primary. Proline is different; its α-amino group is secondary.
3. With the exception of glycine, the α-carbon of each amino acid is a chiral center. Although not shown in this table, all 19 chiral amino acids have the same relative configuration at the α-carbon. In the D,L convention, all are L-amino acids. According to the *R,S* convention, amino acid α-carbons, with the exception of cysteine, have the *S* configuration. Because of priority rules, the presence of the sulfhydryl group on the side chain of L-cysteine gives the chiral center the *R* configuration.
4. Isoleucine and threonine contain a second chiral center. Four stereoisomers are possible for each amino acid, but only one is found in proteins.
5. The sulfhydryl group of cysteine, the imidazole group of histidine, and the phenolic hydroxyl of tyrosine are partially ionized at pH 7.0, but the ionic form is not the major form present at this pH.

Figure 27.2

The enantiomers of alanine. The vast majority of α-amino acids in the biological world have the L configuration at the α-carbon. The structure of L-glyceraldehyde is added for comparison.

Table 27.1 The Common Amino Acids Found in Proteins

Nonpolar Side Chains

Alanine (Ala, A)

Glycine (Gly, G)

Isoleucine (Ile, I)

Leucine (Leu, L)

Methionine (Met, M)

Phenylalanine (Phe, F)

Proline (Pro, P)

Tryptophan (Trp, W)

Valine (Val, V)

Polar Side Chains

Asparagine (Asn, N)

Glutamine (Gln, Q)

Serine (Ser, S)

Threonine (Thr, T)

Acidic Side Chains

Aspartic acid (Asp, D)

Glutamic acid (Glu, E)

Cysteine (Cys, C)

Tyrosine (Tyr, Y)

Basic Side Chains

Arginine (Arg, R)

Histidine (His, H)

Lysine (Lys, K)

* Each ionizable group is shown in the form present in highest concentration at pH 7.0.

Example 27.1

Of the 20 protein-derived amino acids shown in Table 27.1, how many contain: (a) an aromatic ring, (b) a side-chain hydroxyl group, (c) a phenolic —OH group, and (d) sulfur?

Solution

(a) Phenylalanine, tryptophan, tyrosine, and histidine contain aromatic rings.
(b) Serine and threonine contain side-chain hydroxyl groups.
(c) Tyrosine contains a phenolic —OH group.
(d) Methionine and cysteine contain sulfur.

Problem 27.1

Of the 20 protein-derived amino acids shown in Table 27.1, which contain (a) no chiral center, (b) two chiral centers?

D. Some Other Common L-Amino Acids

Although the vast majority of plant and animal proteins are constructed from just these 20 α-amino acids, many other amino acids are also found in nature. Ornithine and citrulline, for example, are found predominantly in the liver and are an integral part of the urea cycle, the metabolic pathway that converts ammonia to urea.

carboxamide derivative
of L-ornithine

L-Ornithine   **L-Citrulline**

Thyroxine and triiodothyronine, two of several hormones derived from the amino acid tyrosine, are found in thyroid tissue. Their principal function is to stimulate metabolism in other cells and tissues.

L-Thyroxine, T$_4$ **L-Triiodothyronine, T$_3$**

4-Aminobutanoic acid is found in high concentration (0.8 mM) in the brain but in no significant amounts in any other mammalian tissue. It is synthesized in neural tissue by decarboxylation of the α-carboxyl group of glutamic acid and is a neurotransmitter in the central nervous system of invertebrates and in humans as well.

enzyme-catalyzed
decarboxylation

Glutamic acid   **4-Aminobutanoic acid**
(γ-Aminobutyric acid, GABA)

Only L-amino acids are found in proteins, and only rarely are D-amino acids a part of the metabolism of higher organisms. Several D-amino acids, however, along with their L-enantiomers, are found in lower forms of life. D-Alanine and D-glutamic acid, for example, are structural components of the cell walls of certain bacteria. Several D-amino acids are also found in peptide antibiotics.

27.2 | Acid-Base Properties of Amino Acids

A. Acidic and Basic Groups of Amino Acids

Among the most important chemical properties of amino acids are their acid-base properties; all are weak polyprotic acids because of their —COOH and —NH$_3^+$ groups. Given in Table 27.2 are pK_a values for each ionizable group of the 20 protein-derived amino acids.

Table 27.2 || pK$_a$ Values for Ionizable Groups of Amino Acids

Amino Acid	pK$_a$ of α-COOH	pK$_a$ of α-NH$_3^+$	pK$_a$ of Side Chain	Isoelectric Point (pI)
Alanine	2.35	9.87	—	6.11
Arginine	2.01	9.04	12.48	10.76
Asparagine	2.02	8.80	—	5.41
Aspartic acid	2.10	9.82	3.86	2.98
Cysteine	2.05	10.25	8.00	5.02
Glutamic acid	2.10	9.47	4.07	3.08
Glutamine	2.17	9.13	—	5.65
Glycine	2.35	9.78	—	6.06
Histidine	1.77	9.18	6.10	7.64
Isoleucine	2.32	9.76	—	6.04
Leucine	2.33	9.74	—	6.04
Lysine	2.18	8.95	10.53	9.74
Methionine	2.28	9.21	—	5.74
Phenylalanine	2.58	9.24	—	5.91
Proline	2.00	10.60	—	6.30
Serine	2.21	9.15	—	5.68
Threonine	2.09	9.10	—	5.60
Tryptophan	2.38	9.39	—	5.88
Tyrosine	2.20	9.11	10.07	5.63
Valine	2.29	9.72	—	6.00

Acidity of α-Carboxyl Groups

The average value of pK$_a$ for an α-carboxyl group of a protonated amino acid is 2.19. Thus, the α-carboxyl group is a considerably stronger acid than acetic acid (pK$_a$ 4.76) and other low-molecular-weight aliphatic carboxylic acids. This greater acidity is accounted for by the electron-withdrawing inductive effect of the adjacent —NH$_3^+$ group. Recall that we used similar reasoning in Section 17.4A to account for the relative acidities of acetic acid and its mono-, di-, and trichloroderivatives.

The ammonium group has an electron-withdrawing inductive effect.

$$RCHCOOH + H_2O \rightleftharpoons RCHCOO^- + H_3O^+ \qquad pK_a = 2.19$$
$$\overset{|}{NH_3^+} \qquad\qquad\qquad \overset{|}{NH_3^+}$$

Acidity of Side-Chain Carboxyl Groups

Owing to the electron-withdrawing inductive effect of the α-NH$_3^+$ group, the side-chain carboxyl groups of protonated aspartic and glutamic acids are also stronger acids than acetic acid (pK$_a$ 4.76). Notice that this acid-strengthening inductive effect decreases with increasing distance of the —COOH from the —NH$_3^+$. Compare the acidities of the α-COOH of alanine (pK$_a$ 2.35), the β-COOH of aspartic acid (pK$_a$ 3.86), and the γ-COOH of glutamic acid (pK$_a$ 4.07).

Acidity of α-Ammonium Groups

The average value of pK$_a$ for an α-ammonium group, —NH$_3^+$, is 9.47 compared with an average value of 10.60 for primary aliphatic ammonium ions (Section 23.5A). Thus, the α-ammonium group of an amino acid is a slightly stronger acid than a primary aliphatic ammonium ion. Conversely, an α-amino group is a slightly weaker base than a primary aliphatic amine.

$$\underset{\underset{NH_3^+}{|}}{RCHCOO^-} + H_2O \rightleftharpoons \underset{\underset{NH_2}{|}}{RCHCOO^-} + H_3O^+ \qquad pK_a = 9.47$$

$$\underset{\underset{NH_3^+}{|}}{CH_3CHCH_3} + H_2O \rightleftharpoons \underset{\underset{NH_2}{|}}{CH_3CHCH_3} + H_3O^+ \qquad pK_a = 10.60$$

Basicity of the Guanidine Group of Arginine

The side-chain guanidine group of arginine is a considerably stronger base than an aliphatic amine. As we saw in Section 23.5D, guanidine (pK_a 13.6) is a very strong base for a neutral organic compound. The remarkable basicity of the guanidine group of arginine is attributed to the large resonance stabilization of the protonated form relative to the neutral form.

The guanidinium ion side chain of arginine is a hybrid of three contributing structures

No resonance stabilization without charge separation

$$pK_a = 12.48$$

Basicity of the Imidazole Group of Histidine

Because the imidazole group on the side chain of histidine contains six π electrons in a planar, fully conjugated ring, imidazole is classified as a heterocyclic aromatic amine (Section 21.2D). The unshared pair of electrons on one nitrogen is a part of the aromatic sextet, whereas that on the other nitrogen is not. The pair of electrons that is not part of the aromatic sextet is responsible for the basic properties of the imidazole ring. Protonation of this nitrogen produces a resonance-stabilized cation.

This lone pair is not a part of the aromatic sextet; it is the proton acceptor

$$pK_a = 6.10$$

Resonance-stabilized imidazolium cation

B. Titration of Amino Acids

Values of pK_a for the ionizable groups of amino acids are most commonly obtained by acid-base titration and by measuring the pH of the solution as a function of added base (or added acid, depending on how the titration is done). To illustrate this experimental procedure, consider a solution containing 1.00 mol of glycine to which has been added enough strong acid that both the amino and carboxyl groups are fully protonated. Next, this solution is titrated with 1.00 M NaOH; the volume of base added and the pH of the resulting solution are recorded and then plotted as shown in Figure 27.3.

The most acidic group and the one to react first with added sodium hydroxide is the carboxyl group. When exactly 0.50 mol of NaOH has been added, the carboxyl group is half neutralized. At this point, the concentration of the zwitterion

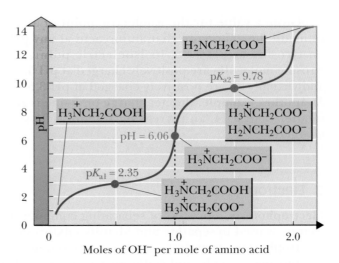

Figure 27.3
Titration of glycine with sodium hydroxide.

equals that of the positively charged ion, and the pH of 2.35 equals the pK_a of the carboxyl group (pK_{a1}).

$$\text{At pH} = \text{p}K_{a1} \quad [\overset{+}{\text{H}_3\text{NCH}_2\text{COOH}}] = [\overset{+}{\text{H}_3\text{NCH}_2\text{COO}^-}]$$

$$\text{Positive ion} \qquad \text{Zwitterion}$$

The end point of the first part of the titration is reached when 1.00 mol of NaOH has been added. At this point, the predominant species present is the zwitterion, and the observed pH of the solution is 6.06.

The next section of the curve represents titration of the —NH$_3^+$ group. When another 0.50 mole of NaOH has been added (bringing the total to 1.50 mol), half of the —NH$_3^+$ groups are neutralized and converted to —NH$_2$. At this point, the concentrations of the zwitterion and negatively charged ion are equal, and the observed pH is 9.78, the pK_a of the amino group of glycine (pK_{a2}).

$$\text{At pH} = \text{p}K_{a2} \quad [\overset{+}{\text{H}_3\text{NCH}_2\text{COO}^-}] = [\text{H}_2\text{NCH}_2\text{COO}^-]$$

$$\text{Zwitterion} \qquad \text{Negative Ion}$$

The second end point of the titration is reached when a total of 2.00 mol of NaOH have been added and glycine is converted entirely to an anion.

C. Isoelectric Point

Titration curves such as that for glycine permit us to determine pK_a values for the ionizable groups of an amino acid. They also permit us to determine another important property: isoelectric point. **Isoelectric point, pI**, for an amino acid is the pH at which the majority of molecules in solution have a net charge of zero (they are zwitterions). By examining the titration curve, you can see that the isoelectric point for glycine falls half way between the pK_a values for the carboxyl and amino groups.

Isoelectric point (pI)
The pH at which an amino acid, polypeptide, or protein has no net charge.

$$\text{pI} = \frac{1}{2}(\text{p}K_a\,\alpha\text{-COOH} + \text{p}K_a\,\alpha\text{-NH}_3^+)$$

$$= \frac{1}{2}(2.35 + 9.78) = 6.06$$

At pH 6.06, the predominant form of glycine molecules is the dipolar ion; furthermore, at this pH, the concentration of positively charged glycine molecules equals the concentration of negatively charged glycine molecules.

Given a value for the isoelectric point of an amino acid, it is possible to estimate the charge on that amino acid at any pH. For example, the charge on tyrosine at pH 5.63, its isoelectric point, is zero. A small fraction of tyrosine molecules are positively charged at pH 5.00 (0.63 unit less than its pI), and virtually all are positively charged at pH 3.63 (2.00 units less than its pI). As another example, the net charge on lysine is zero at pH 9.74. At pH values smaller than 9.74, an increasing fraction of lysine molecules are positively charged.

D. Electrophoresis

Electrophoresis
The process of separating compounds on the basis of their electric charge.

Electrophoresis, a process of separating compounds on the basis of their total charge, is used to separate and identify mixtures of amino acids and proteins. Electrophoretic separations can be carried out using paper, starch, polyacrylamide and agarose gels, and cellulose acetate as solid supports. In paper electrophoresis, a paper strip saturated with an aqueous buffer of predetermined pH serves as a bridge between two electrode vessels (Figure 27.4). Next, a sample of amino acids is applied as a spot on the paper strip. When an electrical potential is then applied to the electrode vessels, amino acids migrate toward the electrode carrying the charge opposite to their own. Molecules having a high charge density move more rapidly than those with a lower charge density. Any molecule already at its isoelectric point remains at the origin. After separation is complete, the strip is dried and sprayed with a dye to make the separated components visible.

A dye commonly used to detect amino acids is ninhydrin (1,2,3-indanetrione monohydrate). Ninhydrin reacts with α-amino acids to produce an aldehyde, carbon dioxide, and a purple-colored anion. This reaction is used very commonly in both qualitative and quantitative analysis of amino acids.

$$\underset{\substack{\text{An }\alpha\text{-amino}\\\text{acid}}}{\text{RCHCO}^-} + 2\underset{\textbf{Ninhydrin}}{\text{(indanetrione)}} \longrightarrow \underset{\text{Purple-colored anion}}{\text{(purple anion)}} + \text{RCH} + CO_2 + H_3O^+$$

Nineteen of the 20 protein-derived α-amino acids have primary amino groups and give the same purple-colored ninhydrin-derived anion. Proline, a secondary amine, gives a different, orange-colored compound.

Figure 27.4

An apparatus for electrophoresis of a mixture of amino acids. Those with a negative charge move toward the positive electrode; those with a positive charge move toward the negative electrode; those with no charge remain at the origin.

Example 27.2

The isoelectric point of tyrosine is 5.63. Toward which electrode does tyrosine migrate on paper electrophoresis at pH 7.0?

Solution

On paper electrophoresis at pH 7.0 (more basic than its isoelectric point), tyrosine has a net negative charge and migrates toward the positive electrode.

Problem 27.2

The isoelectric point of histidine is 7.64. Toward which electrode does histidine migrate on paper electrophoresis at pH 7.0?

Example 27.3

Electrophoresis of a mixture of lysine, histidine, and cysteine is carried out at pH 7.64. Describe the behavior of each amino acid under these conditions.

Solution

The isoelectric point of histidine is 7.64. At this pH, histidine has a net charge of zero and does not move from the origin. The pI of cysteine is 5.02; at pH 7.64 (more basic than its isoelectric point), cysteine has a net negative charge and moves toward the positive electrode. The pI of lysine is 9.74; at pH 7.64 (more acidic than its isoelectric point), lysine has a net positive charge and moves toward the negative electrode.

Problem 27.3

Describe the behavior of a mixture of glutamic acid, arginine, and valine on paper electrophoresis at pH 6.0.

27.3 | Polypeptides and Proteins

In 1902, Emil Fischer proposed that proteins are long chains of amino acids joined together by amide bonds between the α-carboxyl group of one amino acid and the α-amino group of another. For these amide bonds, Fischer proposed the special name **peptide bond**. Figure 27.5 shows the peptide bond formed between serine and alanine in the dipeptide serylalanine.

Peptide bond
The special name given to the amide bond formed between the α-amino group of one amino acid and the α-carboxyl group of another amino acid.

Figure 27.5
The peptide bond in serylalanine.

Dipeptide
A molecule containing two amino acid units joined by a peptide bond.

Tripeptide
A molecule containing three amino acid units, each joined to the next by a peptide bond.

Polypeptide
A macromolecule containing many amino acid units, each joined to the next by a peptide bond.

N-Terminal amino acid
The amino acid at the end of a polypeptide chain having the free —NH$_3^+$ group.

C-Terminal amino acid
The amino acid at the end of a polypeptide chain having the free —COO$^-$ group.

Proteins
Biological macromolecule, consisting of one or more polypeptides, having an overall molecular weight of greater than 5000.

Peptide is the name given to a short polymer of amino acids. Peptides are classified by the number of amino acid units in the chain. A molecule containing 2 amino acids joined by an amide bond is called a **dipeptide**. Those containing 3 to 10 amino acids are called **tripeptides**, **tetrapeptides**, **pentapeptides**, and so on. Molecules containing more than 10 but fewer than 20 amino acids are called **oligopeptides**. Those containing several dozen or more amino acids are called **polypeptides**. **Proteins** are biological macromolecules of molecular weight 5000 or greater, consisting of one or more polypeptide chains. The distinctions in this terminology are not precise.

By convention, polypeptides are written from the left, beginning with the amino acid having the free —NH$_3^+$ group and proceeding to the right toward the amino acid with the free —COO$^-$ group. The amino acid with the free —NH$_3^+$ group is called the **N-terminal amino acid** and that with the free —COO$^-$ group is called the **C-terminal amino acid**. Notice the repeating pattern in the peptide chain of N-α-carbon-carbonyl, etc.

Ser-Phe-Asp

Example 27.4

Draw a structural formula for Cys-Arg-Met-Asn. Label the *N*-terminal amino acid and the *C*-terminal amino acid. What is the net charge on this tetrapeptide at pH 6.0?

Solution

The backbone of this tetrapeptide is a repeating sequence of nitrogen-α-carbon-carbonyl. The net charge on this tetrapeptide at pH 6.0 is +1.

Problem 27.4

Draw a structural formula for Lys-Phe-Ala. Label the *N*-terminal amino acid and the *C*-terminal amino acid. What is the net charge on this tripeptide at pH 6.0?

27.4 | Primary Structure of Polypeptides and Proteins

Primary structure of proteins
The sequence of amino acids in the polypeptide chain, read from the N-terminal amino acid to the C-terminal amino acid.

The **primary (1°) structure** of a polypeptide or protein refers to the sequence of amino acids in its polypeptide chain. In this sense, primary structure is a nearly complete description of all covalent bonding in a polypeptide or protein.

In 1953, Frederick Sanger of Cambridge University, England, reported the primary structure of the two polypeptide chains of the hormone insulin. Not only was this a remarkable achievement in analytical chemistry, but it also clearly established that the molecules of a given protein all have the same amino acid composition and the same amino acid sequence.

A. Amino Acid Analysis

A method for determining the amino acid composition of a polypeptide is hydrolysis and quantitative analysis. Recall from Section 18.4D that amide bonds are very resistant to hydrolysis. Typically, samples of protein are hydrolyzed in 6 M HCl in sealed glass vials at 110°C for 24 to 72 hours. This hydrolysis can be done in a microwave oven in a shorter time. After the polypeptide is hydrolyzed, the resulting mixture of amino acids is analyzed by ion-exchange chromatography. Amino acids are detected as they emerge from the column by reaction with ninhydrin (Section 27.2D) followed by absorption spectroscopy. Current procedures for hydrolysis of polypeptides and analysis of amino acid mixtures have been refined to the point that it is possible to obtain amino acid composition from as little as 50 nanomoles (50×10^{-9} mole) of polypeptide. Figure 27.6 shows the analysis of a polypeptide hydrolysate by ion-exchange chromatography. Note that during hydrolysis, the side-chain amide groups of asparagine and glutamine are hydrolyzed, and these amino acids are detected as aspartic acid and glutamic acid. For each glutamine or asparagine hydrolyzed, an equivalent amount of ammonium chloride is formed.

Figure 27.6

Analysis of a mixture of amino acids by ion-exchange chromatography using Amberlite IR-120, a sulfonated polystyrene resin. The resin contains phenyl-$SO_3^-Na^+$ groups. The amino acid mixture is applied to the column at low pH (3.25) under which conditions the acidic amino acids (Asp, Glu) are weakly bound to the resin, and the basic amino acids (Lys, His, Arg) are tightly bound. Sodium citrate buffers at two different concentrations, and three different values of pH are used to elute the amino acids from the column. Cysteine is determined as cystine, Cys-S-S-Cys, the disulfide of cysteine.

Figure 27.7
Cleavage by cyanogen bromide, BrCN, of a peptide bond formed by the carboxyl group of methionine.

A substituted γ-lactone of the amino acid homoserine

Methyl thiocyanate

B. Sequence Analysis

It is possible to determine the order in which the amino acids are joined in a polypeptide chain. The classical sequencing strategy is to cleave the polypeptide at specific peptide bonds (using, for example, cyanogen bromide or certain proteolytic enzymes), determine the sequence of each fragment (using, for example, the Edman degradation), and then match overlapping fragments to arrive at the sequence of the polypeptide.

Cyanogen Bromide

Cyanogen bromide (BrCN) is specific for cleavage of peptide bonds formed by the carboxyl group of methionine (Figure 27.7). The products of this cleavage are a substituted γ-lactone (Section 18.1C) derived from the N-terminal portion of the polypeptide and a second fragment containing the C-terminal portion of the polypeptide.

A three-step mechanism can be written for this reaction. Cyanogen bromide cleavage depends on chemical manipulation of the leaving ability of the sulfur atom of methionine. Because CH_3S^- is the anion of a weak acid, it is a very poor leaving group, just as OH^- is a poor leaving group (Section 9.3C). Yet, just as the oxygen atom of an alcohol can be transformed into a better leaving group by converting it into an oxonium ion (by protonation), so too can the sulfur atom of methionine be transformed into a better leaving group by converting it into a sulfonium ion.

Mechanism Cleavage of a Peptide Bond at Methionine by Cyanogen Bromide

Step 1: **Make a new bond between a nucleophile and an electrophile and simultaneously break a bond to give stable molecules or ions** Reaction is initiated by nucleophilic attack of the divalent sulfur atom of methionine on the carbon of cyanogen bromide displacing bromide ion. The product of this nucleophilic displacement is a sulfonium ion.

Step 2: Make a new bond between a nucleophile and an electrophile and simultaneously break a bond to give stable molecules or ions An internal S_N2 reaction in which the oxygen of the methionine carbonyl group attacks the γ-carbon and displaces methyl thiocyanate gives a five-membered ring. Note that the oxygen of a carbonyl group is at best a weak nucleophile. This displacement is facilitated, however, because the sulfonium ion is a very good leaving group and because of the ease with which a five-membered ring is formed.

An iminolactone hydrobromide   Methyl thiocyanate

Step 3: Hydrolysis of the imino group gives a γ-lactone derived from the *N*-terminal end of the original polypeptide.

A substituted γ-lactone of the amino acid homoserine

Enzyme-Catalyzed Hydrolysis of Peptide Bonds

A group of proteolytic enzymes, among them trypsin and chymotrypsin, can be used to catalyze the hydrolysis of specific peptide bonds. Trypsin catalyzes the hydrolysis of peptide bonds formed by the carboxyl groups of arginine and lysine; chymotrypsin catalyzes the hydrolysis of peptide bonds formed by the carboxyl groups of phenylalanine, tyrosine, and tryptophan (Table 27.3).

Table 27.3	Cleavage of Specific Peptide Bonds Catalyzed by Trypsin and Chymotrypsin
Enzyme	**Catalyzes Hydrolysis of Peptide Bond Formed by Carboxyl Group of**
Trypsin	Arginine, lysine
Chymotrypsin	Phenylalanine, tyrosine, tryptophan

Example 27.5

Which of these dipeptides are hydrolyzed by trypsin? By chymotrypsin?
(a) Arg-Glu-Ser   **(b)** Phe-Gly-Lys

Solution

(a) Trypsin catalyzes the hydrolysis of peptide bonds formed by the carboxyl groups of lysine and arginine. Therefore, the peptide bond between arginine and glutamic acid is hydrolyzed in the presence of trypsin.

$$\text{Arg-Glu-Ser} + H_2O \xrightarrow{\text{trypsin}} \text{Arg} + \text{Glu-Ser}$$

Chymotrypsin catalyzes the hydrolysis of peptide bonds formed by the carboxyl groups of phenylalanine, tyrosine, and tryptophan. Because none of these three aromatic amino acids is present, tripeptide (a) is not affected by chymotrypsin.

(b) Tripeptide (b) is not affected by trypsin. Although lysine is present, its carboxyl group is at the *C*-terminal end and is not involved in peptide bond formation. Tripeptide (b) is hydrolyzed in the presence of chymotrypsin.

$$\text{Phe-Gly-Lys} + \text{H}_2\text{O} \xrightarrow{\text{chymotrypsin}} \text{Phe} + \text{Gly-Lys}$$

Problem 27.5

Which of these tripeptides are hydrolyzed by trypsin? By chymotrypsin?
(a) Tyr-Gln-Val   **(b)** Thr-Phe-Ser   **(c)** Thr-Ser-Phe

Edman Degradation

Edman degradation
A method for selectively cleaving and identifying the *N*-terminal amino acid of a polypeptide chain.

Of the various chemical methods developed for determining the amino acid sequence of a polypeptide, the one most widely used today is the **Edman degradation**, introduced in 1950 by Pehr Edman of the University of Lund, Sweden. In this procedure, a polypeptide is treated with phenyl isothiocyanate, $C_6H_5N{=}C{=}S$, and then with acid. The effect of Edman degradation is to remove the *N*-terminal amino acid selectively as a substituted phenylthiohydantoin (Figure 27.8), which is then separated and identified.

The key feature of the Edman degradation is successive $C{=}N$, $C{=}O$, and $C{=}O$ addition reactions.

Figure 27.8
Edman degradation. Treatment of a polypeptide with phenyl isothiocyanate followed by acid selectively cleaves the *N*-terminal amino acid as a substituted phenylthiohydantoin.

Mechanism   Edman Degradation—Cleavage of an *N*-Terminal Amino Acid

Step 1: **Make a new bond between a nucleophile and an electrophile**   Nucleophilic addition of the *N*-terminal amino group to the $C{=}N$ bond of phenyl isothiocyanate gives a derivative of *N*-phenylthiourea.

Step 2: **Make a new bond between a nucleophile and an electrophile followed by break a bond to give stable molecules or ions**   Heating the derivatized polypeptide in HCl at 100°C results in nucleophilic addition of sulfur to the carbonyl of the adjacent amide group to give a tetrahedral carbonyl addition intermediate, which collapses to give a thiazolinone ring derived from the *N*-terminal amino acid after the transfer of two protons to the amino group of the departing amine.

Step 3: The thiazolinone ring undergoes isomerization by ring opening followed by reclosure to give a more stable phenylthiohydantoin, which is separated and identified by chromatography.

A thiazolinone   A phenylthiohydantoin

The special value of the Edman degradation is that it cleaves the *N*-terminal amino acid from a polypeptide without affecting any other bonds in the chain. Furthermore, Edman degradation can be repeated on the shortened polypeptide, causing the next amino acid in the sequence to be cleaved and identified. In practice, it is now possible to sequence as many as the first 20 to 30 amino acids in a polypeptide by this method using only a few milligrams or less of material.

Most polypeptides in nature are longer than 20 to 30 amino acids, the practical limit to the number of amino acids that can be sequenced by repetitive Edman degradation. The special value of cleavage with cyanogen bromide, trypsin, and chymotrypsin is that a long polypeptide chain can be cleaved at specific peptide bonds into smaller polypeptide fragments, and each fragment can then be sequenced separately. The entire procedure, from the chemical cleavage step to the separation of the product, has been automated. Samples of a pure protein can be added to a machine and the first several *N*-terminal amino acids determined automatically.

Example 27.6

Deduce the amino acid sequence of a pentapeptide from the following experimental results. Note that under the column labeled "Amino Acid Composition," the amino acids are listed in alphabetical order. In no way does this listing give any information about primary structure.

Experimental Procedure	Amino Acid Composition
Pentapeptide	Arg, Glu, His, Phe, Ser
Edman degradation	Glu
Hydrolysis catalyzed by chymotrypsin	
Fragment A	Glu, His, Phe
Fragment B	Arg, Ser
Hydrolysis catalyzed by trypsin	
Fragment C	Arg, Glu, His, Phe
Fragment D	Ser

Solution

Edman degradation cleaves Glu from the pentapeptide; therefore, glutamic acid must be the *N*-terminal amino acid.

Glu-(Arg, His, Phe, Ser)

Fragment A from chymotrypsin-catalyzed hydrolysis contains Phe. Because of the specificity of chymotrypsin, Phe must be the *C*-terminal amino acid of fragment A. Fragment A also contains Glu, which we already know is the *N*-terminal amino acid.

From these observations, conclude that the first three amino acids in the chain must be Glu-His-Phe and then write the following partial sequence:

Glu-His-Phe-(Arg, Ser)

The fact that trypsin cleaves the pentapeptide means that Arg must be within the pentapeptide chain; it cannot be the *C*-terminal amino acid. Therefore, the complete sequence must be:

Glu-His-Phe-Arg-Ser

Problem 27.6

Deduce the amino acid sequence of an undecapeptide (11 amino acids) from the experimental results shown in the table.

Experimental Procedure	Amino Acid Composition
Undecapeptide	Ala, Arg, Glu, Lys$_2$, Met, Phe, Ser, Thr, Trp, Val
Edman degradation	Ala
Trypsin-catalyzed hydrolysis	
Fragment E	Ala, Glu, Arg
Fragment F	Thr, Phe, Lys
Fragment G	Lys
Fragment H	Met, Ser, Trp, Val
Chymotrypsin-catalyzed hydrolysis	
Fragment I	Ala, Arg, Glu, Phe, Thr
Fragment J	Lys$_2$, Met, Ser, Trp, Val
Treatment with cyanogen bromide	
Fragment K	Ala, Arg, Glu, Lys$_2$, Met, Phe, Thr, Val
Fragment L	Trp, Ser

Sequencing by Mass Spectrometry

As mentioned in Chapter 14, mass spectrometry is increasingly being used for the direct sequencing of proteins in very small quantities. This topic was fully addressed in Connections to Biological Chemistry: "Mass Spectra of Biological Macromolecules" in Section 14.3 and will not be further developed here. Sequencing of proteins by mass spectrometry has largely replaced automated Edman degradation in most research laboratories.

Sequencing Proteins from the Coding Nucleotide Sequence

As it has become easier to determine nucleotide sequences (see Section 28.5), it is now often easier to sequence the nucleic acid that codes for a protein than to sequence the protein itself. In some cases, this has lead to discovery of new proteins of unknown function. Often comparison of the revealed protein sequences with those of known proteins from simpler organisms discloses sequence homologies that suggest the function of the new proteins. Comparisons with yeast, whose genome was one of the first to be sequenced and for which most of the proteins coded for have known functions, have been particularly fruitful.

27.5 | Synthesis of Polypeptides

A. The Problems

The problem in peptide synthesis is to join the carboxyl group of amino acid 1 (aa$_1$) by an amide (peptide) bond to the amino group of amino acid 2 (aa$_2$).

$$\overset{+}{H_3}NCHCO^- + \overset{+}{H_3}NCHCO^- \xrightarrow{?} \overset{+}{H_3}NCHCNHCHCO^- + H_2O$$

aa$_1$ aa$_2$ aa$_1$ aa$_2$

B. The Strategy

A rational strategy for the synthesis of peptide bonds and polypeptides requires three steps.

1. Protect the α-amino group of amino acid aa$_1$ to reduce its nucleophilicity so that it does not participate in nucleophilic addition to the carboxyl group of either aa$_1$ or aa$_2$.
2. Protect the α-carboxyl group of amino acid aa$_2$ so that it is not susceptible to nucleophilic attack by the α-amino group of another molecule of aa$_2$.
3. Activate the α-carboxyl group of amino acid aa$_1$ so that it is susceptible to nucleophilic attack by the α-amino group of aa$_2$.

After dipeptide aa$_1$—aa$_2$ has been formed, the protecting group Z can be removed, and chain growth can be continued from the *N*-terminal end of the dipeptide. Alternatively, the protecting group X can be removed and chain growth can be continued from the *C*-terminal end. The range of protecting groups and activating groups is large, and experimental conditions have been found to attach and remove them as desired.

C. Amino-Protecting Groups

The most common strategy for protecting amino groups and reducing their nucleophilicity is to react them with carbonyl derivatives. The reagents most commonly used for this purpose are benzyloxycarbonyl chloride and di-*tert*-butyl dicarbonate. In the terminology adopted by the IUPAC, the benzyloxycarbonyl group is given the symbol Z "or Cbz", and the *tert*-butoxycarbonyl group is given the symbol BOC—.

Treatment of an amino group with either of these reagents forms a new functional group called a carbamate. A carbamate is an ester of carbamic acid; that is, it is an ester of the monoamide of carbonic acid.

The special advantage of the carbamate group is that it is stable to dilute base but can be removed by treatment with HBr in acetic acid or other similarly acidic conditions.

A study of the mechanism for removal of this protecting group has shown that the reaction is first order in $[H^+]$ and involves formation of a carbocation and a carbamic acid. A carbamic acid spontaneously loses carbon dioxide to form the free amine. The carbocation reacts with an available nucleophile such as halide ion to form an alkyl halide.

$$PhCH_2-O-C-NH-peptide$$

$$\rightarrow PhCH_2^+ \xrightarrow{:Br:^-} PhCH_2-Br:$$

$$\rightarrow O=C-NH-peptide \longrightarrow CO_2 + H_3N^+-peptide$$

A carbamic acid

Note that because acid-catalyzed removal of this protecting group is carried out in nonaqueous media, there is no danger of simultaneous acid-catalyzed hydrolysis of peptide (amide) bonds within the newly synthesized polypeptide. This relationship exists because water is required for hydrolysis of a peptide bond.

The benzyloxycarbonyl group can also be removed by treatment with H_2 in the presence of a transition metal catalyst (hydrogenolysis, Section 21.5C). In hydrogenolysis of a Z-protecting group, one product is toluene. The other is a carbamic acid, which undergoes spontaneous decarboxylation to give carbon dioxide and the unprotected peptide.

$$PhCH_2OCNH-peptide + H_2 \xrightarrow{Pd} PhCH_3 + CO_2 + H_2N-peptide$$

A Z-protected peptide   **Toluene**   Unprotected peptide

D. Carboxyl-Protecting Groups

Carboxyl groups are most often protected by conversion to methyl, ethyl, or benzyl esters. Methyl and ethyl esters are prepared by Fischer esterification (Section 17.7A) and are removed by hydrolysis in aqueous base (Section 18.4C) under mild conditions. Benzyl esters are conveniently removed by hydrogenolysis with H_2 over a palladium or platinum catalyst (Section 21.5C). Benzyl groups can also be removed by treatment with HBr in acetic acid.

E. Peptide-Bond Forming Reactions

The reagent most commonly used to bring about peptide bond formation is 1,3-dicyclohexylcarbodiimide (DCC). This reagent is the anhydride of a disubstituted urea, and when treated with water, it is converted to N,N'-dicyclohexylurea (DCU).

1,3-Dicyclohexylcarbodiimide
(DCC)

N,N'-**Dicyclohexylurea**
(DCU)

When an amino-protected aa_1 and a carboxyl-protected aa_2 are treated with DCC, this reagent acts as a dehydrating agent; it removes —OH from the carboxyl group and —H from the amino group to form an amide bond. More specifically, DCC activates the α-carboxyl group of aa_1 toward nucleophilic acyl substitution by converting its —OH group into a better leaving group.

Z—NHCHC—OH + H₂NCHCOCH₃ + (cyclohexyl)—N=C=N—(cyclohexyl) →(CHCl₃)

$$Z-\underset{R_1}{NHCHC}(=O)-OH \ + \ H_2N\underset{R_2}{CHCOCH_3} \ + \ \text{cyclohexyl}-N=C=N-\text{cyclohexyl} \ \xrightarrow{CHCl_3}$$

Amino-protected aa₁   Carboxyl-protected aa₂   **1,3-Dicyclohexylcarbodiimide (DCC)**

$$Z-\underset{R_1}{NHCHC}(=O)-\underset{R_2}{NHCHCOCH_3} \ + \ \text{cyclohexyl}-NH-C(=O)-NH-\text{cyclohexyl}$$

Amino and carboxyl protected dipeptide   ***N,N′*-Dicyclohexylurea (DCU)**

An abbreviated mechanism for this intermolecular dehydration is shown in Figure 27.9. An acid-base reaction in Step 1 between the carboxyl group of aa₁ and a nitrogen of DCC (Take a proton away) followed in Step 2 by addition of the carboxylate anion to the C=N double bond (Make a new bond between a nucleophile and an electrophile). The *O*-acylisourea formed is the nitrogen analog of a mixed anhydride. Nucleophilic addition of the amino group of aa₂ to the carbonyl group of the *O*-acylisourea in Step 3 (Make a new bond between a nucleophile and an electrophile) generates a tetrahedral carbonyl addition intermediate that collapses in Step 4 to give a dipeptide and DCU (Break a bond to give stable molecules or ions in a pericyclic reaction).

Figure 27.9

The role of 1,3-dicyclohexylcarbodiimide (DCC) in the formation of a peptide bond between an amino-protected amino acid (aa₁) and a carboxyl-protected amino acid (aa₂).

F. Solid-Phase Synthesis

A major problem associated with polypeptide synthesis is purification of intermediates after each protection, activation, coupling, and deprotection step. If unreacted starting materials are not removed after each step, the final product is contaminated by polypeptides missing one or more amino acids. The required purification steps are not only laborious and time consuming, but they also inevitably result in some loss of the desired product. These losses become especially severe in the synthesis of larger polypeptides.

A major advance in polypeptide synthesis came in 1962 when R. Bruce Merrifield of the Rockefeller University described a solid-phase synthesis (alternatively called polymer-supported synthesis) of the tetrapeptide, Leu-Ala-Gly-Ala, by a technique that now bears his name. Merrifield was awarded the 1984 Nobel Prize for chemistry for his work in developing the solid-phase method for peptide synthesis.

The solid support used by Merrifield was a type of polystyrene in which about 5% of the phenyl groups carry a chloromethyl (—CH₂Cl) group in their para positions (Figure 27.10). These chloromethyl groups, like all benzylic halides, are particularly reactive in nucleophilic substitution reactions.

Figure 27.10

The support used for the Merrifield solid-phase synthesis is a chloromethylated polystyrene resin.

cross-linking of polystyrene chains by copolymerization of styrene and 2% *p*-divinylbenzene

Following polymerization, about 5% of benzene rings are chloromethylated.

In the Merrifield method, the *C*-terminal amino acid is joined as a benzyl ester to the solid polymer support, and then the polypeptide chain is extended one amino acid at a time from the *N*-terminal end. The advantage of polypeptide synthesis on a solid support is that the polymer beads with the peptide chains anchored on them are completely insoluble in the solvents used in the synthesis. Furthermore, excess reagents (for example, DCC) and byproducts (for example, DCU) are removed after each step simply by washing the polymer beads. When synthesis is completed, the polypeptide is released from the polymer beads by cleavage of the benzyl ester. The steps in solid-phase synthesis of a polypeptide are summarized in Figure 27.11.

Thanks to automation, the synthesis of polypeptides is now a routine procedure in chemical research. It is common for researchers to order several peptides at a time for use in fields as diverse as medicine, biology, material science, and biomedical engineering.

A dramatic illustration of the power of the solid-phase method was the synthesis of the enzyme ribonuclease by Merrifield in 1969. The synthesis involved 369 chemical reactions and 11,931 operations, all of which were performed by an automated machine and without any intermediate isolation stages. Each of the 124 amino acids was added as an *N-tert*-butoxycarbonyl derivative and coupled using DCC. Cleavage from the resin and removal of all protective groups gave a mixture that was purified by ion-exchange chromatography. The specific activity of the synthetic enzyme was 13–24% of that of the natural enzyme. The fact that the specific activity of the synthetic enzyme was lower than that of the natural enzyme was probably attributable to the presence of polypeptide byproducts closely related to but not identical to the natural enzyme. Synthesizing ribonuclease (124 amino acids) requires forming 123 peptide bonds. If each peptide bond is formed in 99% yield, the yield of homogeneous polypeptide is $0.99^{123} = 29\%$. If each peptide bond is formed in 98% yield, the yield is 8%. Thus, even with yields as high as 99% in each peptide bond-forming step, a large portion of the synthetic polypeptides have one or more sequence defects. Many of these, nonetheless, may be fully or partially active.

27.6 | Three-Dimensional Shapes of Polypeptides and Proteins

A. Geometry of a Peptide Bond

In the late 1930s, Linus Pauling began a series of studies to determine the geometry of a peptide bond. One of his first and most important discoveries was that a peptide bond itself is planar. As shown in Figure 27.12, the four atoms of a peptide bond and the two α-carbons joined to it all lie in the same plane.

As fully discussed in Connections to Biological Chemistry: "The Unique Structure of Amide Bonds" in Section 18.2, the carbon and nitrogen of an amide are actually planar with approximately 120° bond angles about each because of resonance of the nitrogen lone pair with the carbonyl. Two configurations are possible

Figure 27.11
Steps in the Merrifield solid-phase polypeptide synthesis.

① Attach BOC-protected *C*-terminal amino acid to resin as benzyl ester. The ester protects the carboxyl group.

② Remove BOC-protecting group.

③ Couple the second BOC-protected amino acid.

④ Remove BOC-protecting group.

Repeat ③ - ④ cycle to build the polypeptide chain. Last to be attached is the *N*-terminal amino acid.

⑤ Cleave completed polypeptide from the resin by cleaving the benzyl ester.

polypeptide chain
(in β sheet)

120.0°

121.5°

118.5°

Figure 27.12

Planarity of a peptide bond. Bond angles about the carbonyl carbon and the amide nitrogen are approximately 120°.

for the atoms of a planar peptide bond. In one, the two α-carbons are *cis* to each other; in the other, they are *trans* to each other. The *trans* configuration is more favorable because the α-carbons with the bulky groups bonded to them are farther from each other than they are in the *cis* configuration. Almost all peptide bonds in naturally occurring proteins have the *trans* configuration. Proline is found *cis* most of the time, but there are some well-known examples of other *cis* peptide bonds as well. The *trans* to *cis* peptide bond conversion can have critical biological consequences in proteins and is currently an active area of research.

s-*trans* configuration   s-*cis* configuration

B. Secondary Structure

Secondary structure of proteins
The ordered arrangements (conformations) of amino acids in localized regions of a polypeptide or protein.

α-Helix
A type of secondary structure in which a section of polypeptide chain coils into a spiral, most commonly a right-handed spiral.

Amide plane   Hydrogen bonding

Figure 27.13

Hydrogen bonding between amide groups.

Secondary (2°) structure refers to ordered arrangements (conformations) of amino acids in localized regions of a polypeptide or protein molecule. The first studies of polypeptide conformations were carried out by Linus Pauling and Robert Corey beginning in 1939. They assumed that in conformations of greatest stability, all atoms in a peptide bond lie in the same plane and there is hydrogen bonding between the N—H of one peptide bond and the C=O of another, as shown in Figure 27.13.

On the basis of model building, Pauling proposed that two types of secondary structure should be particularly stable: the α-helix and the antiparallel β-pleated sheet. X-ray crystallography has validated this prediction completely.

The α-Helix

In an **α-helix** pattern shown in Figure 27.14, a polypeptide chain is coiled in a spiral. As you study this section of α-helix, note the following:

1. The helix is coiled in a clockwise, or right-handed, manner. Right-handed means that if you turn the helix clockwise, it twists away from you. In this sense, a right-handed helix is analogous to the right-handed thread of common wood or machine screw.
2. There are 3.6 amino acids per turn of the helix.
3. Each peptide bond is *trans* and planar.

hydrogen bonding

Figure 27.14

An α-helix. The peptide chain shown contains repeating units of L-alanine.

4. The N—H group of each peptide bond points roughly downward, parallel to the axis of the helix, and the C=O of each peptide bond points roughly upward, also parallel to the axis of the helix.
5. The carbonyl group of each peptide bond is hydrogen-bonded to the N—H group of the peptide bond four amino acid units away from it. Hydrogen bonds are shown as dashed lines.
6. All R— groups point outward from the helix.

Almost immediately after Pauling proposed the α-helix conformation, other researchers proved the presence of α-helix conformations in keratin, the protein of hair and wool. It soon became obvious that the α-helix is one of the fundamental folding patterns of polypeptide chains.

The β-Pleated Sheet

An antiparallel **β-pleated sheet** consists of extended polypeptide chains with neighboring chains running in opposite (antiparallel) directions. In a parallel β-pleated sheet, the polypeptide chains run in the same direction. Unlike the α-helix arrangement, N—H and C=O groups lie in the plane of the sheet and are roughly perpendicular to the long axis of the sheet. The C=O group of each peptide bond is hydrogen-bonded to the N—H group of a peptide bond of a neighboring chain (Figure 27.15).

As you study this section of β-pleated sheet, note the following.

1. The three polypeptide chains lie adjacent to each other and run in opposite (antiparallel) directions.
2. Each peptide bond is planar, and the α-carbons are *trans* to each other.
3. The C=O and N—H groups of peptide bonds from adjacent chains point at each other and are in the same plane so that hydrogen bonding is possible between adjacent polypeptide chains.

β-Pleated sheet
A type of secondary structure in which sections of polypeptide chains are aligned parallel or antiparallel to one another.

Figure 27.15
β-Pleated sheet conformation with three polypeptide chains running in opposite (antiparallel) directions. Hydrogen bonding between chains is indicated by dashed lines.

4. The R-groups on any one chain alternate, first above and then below the plane of the sheet and so on.

The β-pleated sheet conformation is stabilized by hydrogen bonding between N—H groups of one chain and C=O groups of an adjacent chain. By comparison, the α-helix is stabilized by hydrogen bonding between N—H and C=O groups within the same polypeptide chain.

C. Tertiary Structure

Tertiary (3°) structure refers to the overall folding pattern and arrangement in space of all atoms in a single polypeptide chain. No sharp dividing line exists between secondary and tertiary structures. Secondary structure refers to the spatial arrangement of amino acids close to one another on a polypeptide chain, whereas tertiary structure refers to the three-dimensional arrangement of all atoms of a polypeptide chain. Among the most important factors in maintaining 3° structure are disulfide bonds, hydrophobic interactions, hydrogen bonding, and salt linkages.

Disulfide bonds (Section 10.9G) play an important role in maintaining tertiary structure. Disulfide bonds are formed between side chains of two cysteine units by oxidation of their thiol groups (—SH) to form a disulfide bond. Treatment of a disulfide bond with a reducing agent regenerates the thiol groups.

Figure 27.16 shows the amino acid sequence of human insulin. This protein consists of two polypeptide chains: an A chain of 21 amino acids and a B chain of 30 amino acids. The A chain is bonded to the B chain by two interchain disulfide bonds. An intrachain disulfide bond also connects the cysteine units at positions 6 and 11 of the A chain.

As an example of 2° and 3° structure, let us look at the three-dimensional structure of myoglobin—a protein found in skeletal muscle and particularly abundant in diving mammals, such as seals, whales, and porpoises. Myoglobin and its structural relative, hemoglobin, are the oxygen storage and transport molecules of vertebrates. Hemoglobin binds molecular oxygen in the lungs and transports it to myoglobin in muscles. Myoglobin stores molecular oxygen until it is required for metabolic oxidation.

Myoglobin consists of a single polypeptide chain of 153 amino acids. Myoglobin also contains a single heme unit. Heme consists of one Fe^{2+} ion coordinated in a square planar array with the four nitrogen atoms of a molecule of porphyrin (Figure 27.17).

Determination of the three-dimensional structure of myoglobin represented a milestone in the study of molecular architecture. For their contribution to this research, John C. Kendrew and Max F. Perutz, both of Britain, shared the 1962 Nobel Prize for chemistry. The secondary and tertiary structures of myoglobin are shown in Figure 27.18. The single polypeptide chain is folded into a complex, almost boxlike shape.

Following are important structural features of the three-dimensional shape of myoglobin.

1. The backbone consists of eight relatively straight sections of α-helix, each separated by a bend in the polypeptide chain. The longest section of α-helix has

Three views of the protein ribonuclease A. The upper view shows atoms colored according to atom type. In the lower two views, red indicates regions of α-helix and blue, β-sheets (section 27.6). Loop regions are shown in white.

A chain

N-terminal Gly Ile Val Glu Gln Cys Cys Thr Ser Ile Cys Ser Leu Tyr Gln Leu Glu Asn Tyr Cys Asn *C*-terminal

B chain

N-terminal Phe Val Asn Gln His Leu Cys Gly Ser His Leu Val Glu Ala Leu Try Leu Val Cys Gly Glu Arg Gly Phe Phe Tyr Thr Pro Lys Ala *C*-terminal

Figure 27.16

Human insulin. The A chain of 21 amino acids and B chain of 30 amino acids are connected by interchain disulfide bonds between A7 and B7 and between A20 and B19. In addition, a single intrachain disulfide bond occurs between A6 and A11.

Figure 27.17
The structure of heme, found in myoglobin and hemoglobin.

24 amino acids, the shortest has 7. Some 75% of the amino acids are found in these eight regions of α-helix.

2. Hydrophobic side chains of phenylalanine, alanine, valine, leucine, isoleucine, and methionine are clustered in the interior of the molecule where they are shielded from contact with water. **Hydrophobic interactions** are a major factor in directing the folding of the polypeptide chain of myoglobin into this compact, three-dimensional shape.

3. The outer surface of myoglobin is coated with hydrophilic side chains, such as those of lysine, arginine, serine, glutamic acid, histidine, and glutamine, which interact with the aqueous environment by **hydrogen bonding**. The only polar side chains that point to the interior of the myoglobin molecule are those of two histidine units, which point inward toward the heme group.

4. Oppositely charged amino acid side chains close to each other in the three-dimensional structure interact by electrostatic attractions called **salt linkages**. An example of a salt linkage is the attraction of the side chains of lysine ($-NH_3^+$) and glutamic acid ($-COO^-$).

The tertiary structures of hundreds of proteins have also been determined. It is clear that proteins contain α-helix and β-pleated sheet structures but that wide variations exist in the relative amounts of each. Lysozyme, with 129 amino acids in a single polypeptide chain, has only 25% of its amino acids in α-helix regions. Cytochrome C, with 104 amino acids in a single polypeptide chain, has no α-helix structure but does contain several regions of β-pleated sheet. Yet, whatever the proportions of α-helix, β-pleated sheet, or other periodic structure, most nonpolar side chains of water-soluble proteins are directed toward the interior of the molecule, whereas polar side chains are on the surface of the molecule and in contact with the aqueous environment. Note that this arrangement of polar and nonpolar groups in water-soluble proteins very much resembles the arrangement of polar and nonpolar groups of soap molecules in micelles (Figure 26.3). It also resembles the arrangement of phospholipids in lipid bilayers (Figure 26.13).

© Brent Iverson, University of Texas

Figure 27.18
Ribbon model of myoglobin. The polypeptide chain is shown in yellow, the heme ligand in red, and the Fe atom as a white sphere.

Spider silk has some remarkable properties. Research is currently concentrated on the strong dragline silk that forms the spokes of a web of the Golden Orb Weaver (*Nephila clavipes*). This silk has three times the impact strength of Kevlar and is 30% more flexible than nylon. The commercial application of spider silk is not a novel concept. Eighteenth-century French entrepreneur Bon de Saint-Hilaire attempted to mass-produce silk in his high-density spider farms but failed because of cannibalism among his territorial arachnid workers. In contrast, native New Guineans continue to successfully collect and utilize spider silk for a wide range of applications including bags and fishing nets. Today, the only way to obtain large amounts of silk is to extract it from the abdomens of immobilized spiders, but scientific advances make the mass production and industrial application of spider silk increasingly possible.

Biologically produced dragline silk is a combination of two liquid proteins, Spidroin 1 and 2, which become oriented and solidify as they travel through a complex

© Tom Bean/Stone/Getty Images

duct system in the spider's abdomen. These proteins are composed largely of alanine and glycine, the two smallest amino acids. Although glycine comprises almost 42% of each protein, the short, 5 to 10 peptide chains of alanine, which account for 25% of each protein's composition, are more important for the properties. Nuclear magnetic resonance (NMR) techniques have vastly improved the level of understanding of spider silk's structure, which was originally determined by x-ray crystallography. NMR data of spidroins containing deuterium-tagged alanine have shown that all alanines are configured into β-pleated sheets. Furthermore, the NMR data suggest that 40% of the alanine β-sheets are highly structured while the other 60% are less oriented, forming fingers that reach out from each individual strand. These fingers are believed to join the oriented alanine β-sheets and the glycine-rich, amorphous "background" sectors of the polypeptide.

Currently, genetically modified *Escherichia coli* is used to mass-produce Spidroin 1 and 2. However, DNA redundancy initially caused synthesis problems when the spider genes were transposed into the bacteria. The *E. coli* did not transcribe some of the codons in the same way that spider cells would, forcing scientists to modify the DNA. When the proteins could be synthesized, it was necessary to develop a system to mimic the natural production of spider silk while preventing the silk from contacting the air and subsequently hardening. After the two proteins are separated from the *E. coli*, they are drawn together into methanol through separate needles. Another approach is to dissolve the silk in formic acid or to add hydrophilic amino acids, in this case histidine and arginine, to keep the artificial silk pliable. The industrial and practical applications of spider silk will not be fully known until it can be synthesized and manipulated in large quantities.

Example 27.7

With which of the following amino acid side chains can the side chain of threonine form hydrogen bonds?

(a) Valine   (b) Asparagine   (c) Phenylalanine
(d) Histidine (e) Tyrosine   (f) Alanine

Solution

The side chain of threonine contains a hydroxyl group that can participate in hydrogen bonding in two ways: its oxygen has a partial negative charge and can function as a hydrogen bond acceptor, and its hydrogen has a partial positive charge and can function as a hydrogen bond donor. Therefore, the side chain of threonine can form hydrogen bonds with the side chains of tyrosine, asparagine, and histidine.

Problem 27.7

At pH 7.4, with what amino acid side chains can the side chain of lysine form salt linkages?

D. Quaternary Structure

Most proteins of molecular weight greater than 50,000 consist of two or more non-covalently linked polypeptide chains. The arrangement of protein monomers into an aggregation is known as **quaternary (4°) structure**. A good example is hemoglobin, a protein that consists of four separate polypeptide chains: two α-chains of 141 amino acids each and two β-chains of 146 amino acids each. The quaternary structure of hemoglobin is shown in Figure 27.19.

A major factor stabilizing the aggregation of protein subunits is the **hydrophobic effect**. When separate polypeptide chains fold into compact three-dimensional shapes to expose polar side chains to the aqueous environment and shield nonpolar side chains from water, hydrophobic "patches" still appear on the surface, in contact with water. These patches can be shielded from water if two or more monomers assemble so that their hydrophobic patches are in contact. The numbers of subunits of several proteins of known quaternary structure are shown in Table 27.4. Other important factors include correctly located complementary hydrogen bonding and charged sites on different subunits. The formation of aggregates of well-defined structure based on specific structural units on the subunits is being explored in the new field of molecular recognition, also referred to as supramolecular chemistry.

Quaternary structure
The arrangement of polypeptide monomers into a noncovalently bonded aggregate.

Hydrophobic effect
The tendency of nonpolar groups to cluster to shield themselves from contact with an aqueous environment.

Table 27.4	Quaternary Structure of Selected Proteins
Protein	**Number of Subunits**
Alcohol dehydrogenase	2
Aldolase	4
Hemoglobin	4
Lactate dehydrogenase	4
Insulin	6
Glutamine synthetase	12
Tobacco mosaic virus protein disc	17

© Brent Iverson, University of Texas

Figure 27.19
Ribbon model of hemoglobin. The α-chains are shown in purple, the β-chains in yellow, the heme ligands in red, and the Fe atoms as white spheres.

Summary

Section 27.1 Amino Acids

- **Amino acids** are compounds that contain both an amino group and a carboxyl group.
- With the exception of glycine, all protein-derived amino acids are chiral.
  - In the D,L convention, all of the 20 commonly occurring amino acids are L-amino acids. Note that this designation is based on a structural analogy to L-glyceraldehyde, not on the measured optical activity of amino acid samples.
- In the *R,S* convention, 18 amino acids are (*S*)-amino acids.
  - Although cysteine has the same absolute configuration, it is an (*R*)-amino acid because of the manner in which priorities are assigned about its tetrahedral chiral center.
  - Isoleucine and threonine contain a second chiral center.
- The 20 protein-derived amino acids are commonly divided into four categories: nine with nonpolar side chains, four with polar but unionized side chains, four with acidic side chains, and three with basic side chains.

Problems: 27.1, 27.8–27.22

Section 27.2 Acid-Base Properties of Amino Acids

- At neutral pH, amino acids exist as **zwitterions** because the amino group is protonated and positively charged and the carboxyl group is deprotonated and negatively charged.
  - Because of the inductive effect of the electron-withdrawing $-NH_3^+$ group, amino acid carboxyl groups are more acidic than is acetic acid.
  - The α-amino group is slightly less basic than a primary aliphatic amine.

Problems: 27.2, 27.3, 27.23–27.39

- The **isoelectric point**, **pI**, of an amino acid, polypeptide, or protein is the pH at which it has no net charge.
- Electrophoresis is the process of separating compounds on the basis of their charge in an electric field.
  - Compounds having a high charge density move more rapidly than those with a lower charge density.
  - Any amino acid or protein in a solution with a pH that equals the pI of the compound remains at the origin.
- Ninhydrin is used to detect amino acids and proteins because it reacts with primary amino groups to create a bright purple dye.

Section 27.3 Polypeptides and Proteins

- A **peptide bond** is the special name given to the amide bond formed between α-amino acids.

Problem: 27.4

- A **polypeptide** is a biological macromolecule containing many amino acids, each joined to the next by a peptide bond. By convention, the sequence of amino acids in a polypeptide is written beginning with the **N-terminal amino acid** toward the **C-terminal amino acid**.

Section 27.4 Primary Structure of Polypeptides and Proteins

- **Primary (1°) structure** of a polypeptide is the sequence of amino acids in the polypeptide chain.
  - The procedure known as **amino acid analysis** is used to determine relative amino acid composition of a protein. All of the amide bonds are hydrolyzed in acid, followed by chromatography to separate and quantify the different amino acids present. No sequence information is determined this way.
- The primary structure of a polypeptide or protein can be determined directly by fragmenting the protein using cyanogen bromide and enzymes, followed by amino acid sequencing of the resulting fragments.

Problems: 27.41–27.50

  - Cyanogen bromide treatment cleaves proteins at the carboxyl group of methionine residues.
  - Protein-cleaving enzymes such as trypsin or chymotrypsin cleave between certain amino acids, leading to specific fragments.
  - The Edman degradation uses phenyl isothiocyanate to remove and identify the N-terminal amino acid of peptides or proteins. The Edman degradation can be repeated and automated, so that the sequence of up to 20 to 30 N-terminal amino acids can be determined.
  - Sequencing of overlapping fragments allows the reconstruction of a protein's entire sequence.
- Today, mass spectrometry of a protein or nucleic acid sequencing of its gene are the preferred methods for establishing amino acid sequence of a protein.

Section 27.5 Synthesis of Polypeptides

- **Peptide synthesis** requires that the amino group of one amino acid and the carboxyl group of the other are protected so that only a desired coupling will take place.
  - The most common amino-protecting groups are carbamates such as the Z (benzyloxycarbonyl) or BOC (*tert*-butyloxycarbonyl) groups that are removed in acid.
  - Carboxyl groups are protected in solution as esters or during solid phase synthesis by attachment to the solid support, often as esters.

- Peptide bonds are made using reagents such as carbodiimides that activate a carboxyl group for nucleophilic attack by an amino group.
- In **solid-phase synthesis**, or polymer-supported synthesis of polypeptides, the *C*-terminal amino acid is joined to a chloromethylated polystyrene resin as a benzyl ester.
  - The polypeptide chain is then extended one amino acid at a time.
  - The main advantage of solid-phase synthesis is that all of the reagent exchange and washing steps are carried out by simple filtration, and the entire operation has been automated.
  - When synthesis is completed, the polypeptide chain is released from the solid support by cleavage of the benzyl ester linkage.

Problems: 27.40, 27.51–27.53

Section 27.6 Three-Dimensional Shapes of Polypeptides and Proteins

- A peptide bond is planar; that is, the four atoms of the amide and the two α-carbons of a peptide bond lie in the same plane.
  - The planarity is due to resonance involving the amide N atom.
  - Bond angles about the amide nitrogen and the amide carbonyl carbon are about $120°$.
- **Secondary (2°) structure** refers to the ordered arrangement (conformations) of amino acids in localized regions of a polypeptide or protein. The two most important types of secondary structure are the α-helix and the β-pleated sheet.
- **Tertiary (3°) structure** refers to the overall folding pattern and arrangement in space of all atoms in a single polypeptide chain.
  - Solvation of amino acid side chains is important for protein folding, and it is observed that hydrophobic side chains tend to be located in the hydrophobic interior of a protein, while hydrophilic side chains tend to be on the surface, exposed to the aqueous environment.
- **Quaternary (4°) structure** is the arrangement of polypeptide monomers into a noncovalently bonded aggregate.
  - A major factor stabilizing the ordered assembly of proteins into specific quaternary structures is the **hydrophobic effect** in which complementary hydrophobic patches on each interacting partner contact each other, thereby providing a driving force for assembly by relieving unfavorable contacts of the hydrophobic patches with water.

Problems: 27.7, 27.54–27.57

☉WL Questions and problems for this chapter are available in OWL. Use the OWL access card that came with this textbook to access assigned questions and problems for this chapter.

Key Reactions

1. **Acidity of an α-Carboxyl Group (Section 27.2A)** An α-COOH (pK_a approximately 2.19) of a protonated amino acid is a considerably stronger acid than acetic acid (pK_a 4.76) or other low-molecular-weight aliphatic carboxylic acid owing to the electron-withdrawing inductive effect of the α-NH$_3^+$ group.

$$\underset{\underset{NH_3^+}{|}}{RCHCOOH} + H_2O \rightleftharpoons \underset{\underset{NH_3^+}{|}}{RCHCOO^-} + H_3O^+ \qquad pK_a = 2.19$$

2. **Acidity of an α-Ammonium Group (Section 27.2A)** An α-NH$_3^+$ group (pK_a approximately 9.47) is a slightly stronger acid than a primary aliphatic ammonium ion (pK_a approximately 10.76).

$$\underset{\underset{NH_3^+}{|}}{RCHCOO^-} + H_2O \rightleftharpoons \underset{\underset{NH_2}{|}}{RCHCOO^-} + H_3O^+ \qquad pK_a = 9.47$$

3. **Reaction of an α-Amino Acid with Ninhydrin (Section 27.2D)** Treatment of an α-amino acid with ninhydrin gives a purple-colored solution. Treatment of proline with ninhydrin gives an orange-colored solution.

An α-amino   **Ninhydrin**   Purple-colored anion
acid

4. **Cleavage of a Peptide Bond by Cyanogen Bromide (Section 27.4B)** Cleavage is regio-selective for a peptide bond formed by the carboxyl group of methionine. The mechanism involves reaction of the electrophilic carbon of cyanogen bromide with the nucleophilic S atom of methionine to create a sulfonium ion intermediate. The sulfonium ion intermediate reacts with the carboxyl O atom to give a cyclic structure containing an imino group that is hydrolyzed to complete cleavage of the peptide bond and give the γ-lactone product.

A substituted γ-lactone of
the amino acid homoserine

5. **Edman Degradation (Section 27.4B)** Treatment with phenyl isothiocyanate followed by acid removes the *N*-terminal amino acid as a substituted phenylthiohydantoin, which is then separated and identified. The mechanism involves reaction of the electrophilic C atom of phenyl isothiocyanate with the nucleophilic terminal amino group to give an *N*-phenylthiourea intermediate that decomposes upon heating by cyclization to give a thiazolinone intermediate as the *C*-terminal peptide bond is cleaved. The thiazolinone intermediate isomerizes to the phenylthiohydantoin product.

**Phenyl
isothiocyanate**   A phenylthiohydantoin

6. **The Benzyloxycarbonyl (Z—) Protecting Group (Section 27.5C)** The benzyloxycarbonyl protecting group is prepared by treatment of an unprotected α-NH_2 group with benzyloxycarbonyl chloride. It is removed by treatment with HBr in acetic acid or by hydrogenolysis.

**Benzyloxycarbonyl
chloride**
(Z—Cl)   **Alanine**   ***N*-Benzyloxycarbonylalanine
(Z—Ala)**

7. Peptide Bond Formation Using 1,3-Dicyclohexylcarbodiimide (Section 27.5E) This substituted carbodiimide is a dehydrating agent and is converted to a disubstituted urea. The reaction is efficient and yields are generally very high. The mechanism involves initial proton transfer from the carboxylic acid to the 1,3-dicyclohexylcarbodiimide N atom, generating an electrophilic intermediate that reacts with the carboxylate to give an *O*-acylisourea. The *O*-acylisourea intermediate is an activated ester that reacts with the amino group of another amino acid to give a tetrahedral addition intermediate that decomposes to give a new peptide bond and dicyclohexylurea.

$$
\underset{\substack{\text{Amino protected} \\ \text{aa}_1}}{Z-\underset{R_1}{NHCHC}-OH} \;+\; \underset{\substack{\text{Carboxyl protected} \\ \text{aa}_2}}{H_2N\underset{R_2}{CHCOCH_3}} \;+\; \underset{\substack{\textbf{1,3-Dicyclohexylcarbo-} \\ \textbf{diimide (DCC)}}}{\bigcirc-N{=}C{=}N-\bigcirc} \;\xrightarrow{\;CHCl_3\;}
$$

$$
\underset{\substack{\text{Amino and carboxyl} \\ \text{protected dipeptide}}}{Z-\underset{R_1}{NHCHC}-\underset{R_2}{NHCHCOCH_3}} \;+\; \underset{\substack{\textbf{\textit{N,N}'-Dicyclohexylurea} \\ \textbf{(DCU)}}}{\bigcirc-NH-\overset{O}{\overset{\|}{C}}-NH-\bigcirc}
$$

Chapter 28

Nucleic Acids

False-colored transmission electron micrograph of a plasmid DNA. If the cell wall of a bacterium such as *Escherichia coli* is partially digested and the cell is then osmotically shocked by dilution with water, its contents are extruded to the exterior. Shown here is the bacterial chromosome surrounding the cell. *Inset:* 2'-deoxyadenosine 5'-monophosphate (dAMP), a building block of DNA (Section 28.2).

© Professor Stanley N. Cohen/Stone/Photo Researchers, Inc.

The organization, maintenance, and regulation of cellular function require a tremendous amount of information, all of which must be passed on each time a cell is replicated. With very few exceptions, genetic information is stored and transmitted from one generation to the next in the form of deoxyribonucleic acids (DNA). Genes, the hereditary units of chromosomes, are long stretches of double-stranded DNA. If the DNA in a human chromosome in a single cell were uncoiled, it would be approximately 1.8 meters long!

Genetic information is expressed in two stages: transcription from DNA to ribonucleic acids (RNA) and then translation for the synthesis of proteins.

$$\text{DNA} \xrightarrow{\text{transcription}} \text{RNA} \xrightarrow{\text{translation}} \text{proteins}$$

Thus, DNA is the repository of genetic information in cells, whereas RNA serves in the transcription and translation of this information, which is then expressed through the synthesis of proteins.

In this chapter, we examine the structure of nucleosides and nucleotides and the manner in which these monomers are covalently bonded to form **nucleic acids**. Then, we examine the manner in which genetic information is encoded on molecules of DNA, the function of the three types of ribonucleic acids, and finally how the primary structure of a DNA molecule is determined.

Using chemistry that we will not discuss in the text, DNA can be synthesized on a solid phase analogous to the method described for the synthesis of oligopeptides [Section 27.5(F)]. In fact, both DNA and RNA can now be synthesized with high yield and the entire process has been automated. It is commonplace for molecular

Nucleic acid
A biopolymer containing three types of monomer units: heterocyclic aromatic amine bases derived from purine and pyrimidine, the monosaccharides D-ribose or 2-deoxy-D-ribose, and phosphoric acid.

⬚WL **Online homework for this chapter may be assigned in OWL for Organic Chemistry.**

biologists to place an order for a specific DNA or RNA on the Internet and have the sample arrive one or two days later. Such efficiency, made possible by automation and high-yielding synthetic reactions, has revolutionized molecular biology and biotechnology.

28.1 | Nucleosides and Nucleotides

Controlled hydrolysis of nucleic acids yields three components: heterocyclic aromatic amine bases, the monosaccharides D-ribose or 2-deoxy-D-ribose (Section 25.1A), and phosphate ions. The five heterocyclic aromatic amine bases most common to nucleic acids are shown in Figure 28.1. Uracil, cytosine, and thymine are referred to as pyrimidine bases after the name of the parent base; adenine and guanine are referred to as purine bases.

Figure 28.1

Names and one-letter abbreviations for the heterocyclic aromatic amine bases most common to DNA and RNA. Bases are numbered according to the patterns of the parent compounds, pyrimidine and purine.

A **nucleoside** is a compound containing D-ribose or 2-deoxy-D-ribose bonded to a heterocyclic aromatic amine base by a β-N-glycosidic bond (Section 25.3A). The monosaccharide component of DNA is 2-deoxy-D-ribose, whereas that of RNA is D-ribose. The glycosidic bond is between C-1′ (the anomeric carbon) of ribose or 2-deoxyribose and N-1 of a pyrimidine base or N-9 of a purine base. Figure 28.2 shows a structural formula for uridine, a nucleoside derived from ribose and uracil.

A **nucleotide** is a nucleoside in which a molecule of phosphoric acid is esterified with a free hydroxyl of the monosaccharide, most commonly either the 3′-hydroxyl or the 5′-hydroxyl. A nucleotide is named by giving the name of the parent nucleoside followed by the word "monophosphate." The position of the phosphoric ester is specified by the number of the carbon to which it is bonded. Figure 28.3 shows a structural formula of 5′-adenosine monophosphate, AMP. Monophosphoric esters are diprotic acids with pK_a values of approximately 1 and 6. Therefore, at pH 7,

Nucleoside

A building block of nucleic acids, consisting of D-ribose or 2-deoxy-D-ribose bonded to a heterocyclic aromatic amine base by a β-N-glycosidic bond.

Nucleotide

A nucleoside in which a molecule of phosphoric acid is esterified with an —OH of the monosaccharide, most commonly either the 3′-OH or the 5′-OH.

Figure 28.2

Uridine, a nucleoside. Atom numbers on the monosaccharide rings are primed to distinguish them from atom numbers on the heterocyclic aromatic amine bases.

Uridine

Figure 28.3

Adenosine 5′-monophosphate, AMP. The phosphate group is fully ionized at pH 7.0 giving this nucleotide a charge of -2.

AMP

ATP

Figure 28.4

Adenosine triphosphate, ATP.

the two hydrogens of a phosphoric monoester are fully ionized giving a nucleotide a charge of -2.

Nucleoside monophosphates can be further phosphorylated to form nucleoside diphosphates and nucleoside triphosphates. Shown in Figure 28.4 is a structural formula for adenosine 5′-triphosphate, ATP.

Nucleoside diphosphates and triphosphates are also polyprotic acids and are extensively ionized at pH 7.0. pK_a values of the first three ionization steps for adenosine triphosphate are less than 5.0. The value of pK_{a4} is approximately 7.0. Therefore, at pH 7.0, approximately 50% of adenosine triphosphate is present as ATP^{4-}, and 50% is present as ATP^{3-}.

As described in the Connections to Biological Chemistry: "The Planarity of —NH_2 Groups on Aromatic Rings" in Section 23.5, the N atoms of the exocyclic amine groups of the nucleic acid bases adenine, cytosine, and guanine are sp^2 hybridized and therefore have a planar geometry. Like other aromatic amines, resonance with the aromatic π cloud is responsible for this geometry because it accommodates maximum orbital overlap. The important point is that the nucleic acid bases are all essentially planar, facilitating stacking in helical structures as well as allowing for optimum hydrogen bonding to complementary bases.

Example 28.1

Draw a structural formula for each nucleotide.

(a) 2′-Deoxycytidine 5′-diphosphate   **(b)** 2′-Deoxyguanosine 3′-monophosphate

Solution

(a) Cytosine is joined by a β-N-glycosidic bond between N-1 of cytosine and C-1 of the cyclic hemiacetal form of 2-deoxy-D-ribose. The 5′-hydroxyl of the pentose is bonded to a phosphate group by an ester bond, and this phosphate is, in turn, bonded to a second phosphate group by an anhydride bond.

(b) Guanine is joined by a β-N-glycosidic bond between N-9 of guanine and C-1 of the cyclic hemiacetal form of 2-deoxy-D-ribose. The 3′-hydroxyl group of the pentose is joined to a phosphate group by an ester bond.

Problem 28.1

Draw a structural formula for each nucleotide.

(a) 2′-Deoxythymidine 5′-monophosphate
(b) 2′-Deoxythymidine 3′-monophosphate

28.2 | The Structure of DNA

In Chapter 27 we saw that the four levels of structural complexity in polypeptides and proteins are primary, secondary, tertiary, and quaternary structures. There are three levels of structural complexity in nucleic acids. Although these levels are somewhat comparable to those in polypeptides and proteins, they also differ in significant ways.

A. Primary Structure—The Covalent Backbone

Deoxyribonucleic acids consist of a backbone of alternating units of deoxyribose and phosphate in which the 3′-hydroxyl of one deoxyribose unit is joined by a phosphodiester bond to the 5′-hydroxyl of another deoxyribose unit (Figure 28.5). This pentose-phosphodiester backbone is constant throughout an entire DNA molecule. A heterocyclic aromatic amine base—adenine, guanine, thymine, or cytosine—is bonded to each deoxyribose unit by a β-N-glycosidic bond. **Primary structure** of DNA refers to the order of heterocyclic bases along the pentose-phosphodiester backbone. The sequence of bases is usually written from the 5′ end to the 3′ end.

Primary structure of nucleic acids
The sequence of bases along the pentose-phosphodiester backbone of a DNA or RNA molecule read from the 5′ end to the 3′ end.

Example 28.2

Draw a structural formula for the DNA dinucleotide TG that is phosphorylated at the 5′ end only.

Solution

Draw a structural formula for the section of DNA that contains the base sequence CTG and is phosphorylated at the 3' end only.

Figure 28.5
A tetranucleotide section of a single-stranded DNA.

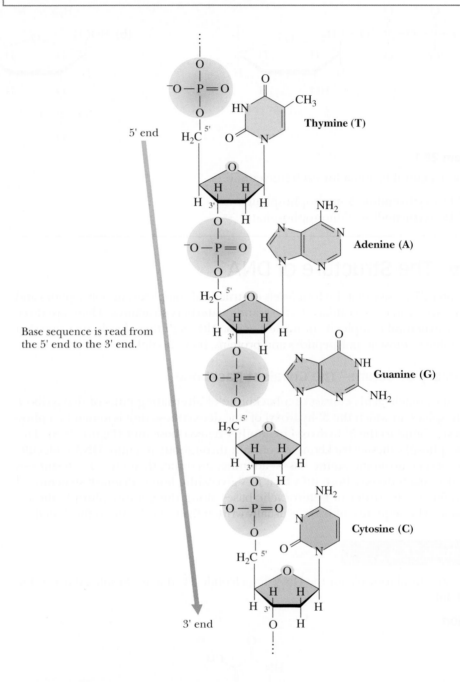

Base sequence is read from the 5' end to the 3' end.

B. Secondary Structure—The Double Helix

By the early 1950s, it was clear that DNA molecules consist of chains of alternating units of deoxyribose and phosphate joined by 3′,5′-phosphodiester bonds with a base attached to each deoxyribose unit by a β-N-glycosidic bond. In 1953, the American biologist James D. Watson and the British physicist Francis H. C. Crick proposed a double-helix model for the **secondary structure** for DNA. Watson, Crick, and Maurice Wilkins shared the 1962 Nobel Prize for physiology or medicine for "their discoveries concerning the molecular structure of nucleic acids, and its significance for information transfer in living material." Although Rosalind Franklin also played an important part in this research, she did not share in the Nobel Prize because of her death in 1958.

Secondary structure of nucleic acids
The ordered arrangement of nucleic acid strands.

The **Watson-Crick model** was based on molecular modeling and two lines of experimental observations: chemical analyses of DNA base compositions and mathematical analyses of x-ray diffraction patterns of crystals of DNA.

Base Composition

At one time, it was thought that the four principal bases occur in the same ratios and perhaps repeat in a regular pattern along the pentose-phosphodiester backbone of DNA for all species. However, more precise determinations of base composition by Erwin Chargaff revealed that bases do not occur in the same ratios (Table 28.1).

Watson-Crick model
A double-helix model for the secondary structure of a DNA molecule.

Watson and Crick with their model of DNA.

Table 28.1	Comparison in Base Composition, in Mole-Percent, of DNA from Several Organisms						
	Purines		**Pyrimidines**				**Purines/**
Organism	**A**	**G**	**C**	**T**	**A/T**	**G/C**	**Pyrimidines**
Human	30.4	19.9	19.9	30.1	1.01	1.00	1.01
Sheep	29.3	21.4	21.0	28.3	1.04	1.02	1.03
Yeast	31.7	18.3	17.4	32.6	0.97	1.05	1.00
E. coli	26.0	24.9	25.2	23.9	1.09	0.99	1.04

Researchers drew the following conclusions from this and related data. To within experimental error,

1. The mole-percent base composition in any organism is the same in all cells of the organism and is characteristic of the organism.
2. The mole-percents of adenine (a purine base) and thymine (a pyrimidine base) are equal. The mole-percents of guanine (a purine base) and cytosine (a pyrimidine base) are also equal.
3. The mole-percents of purine bases (A + G) and pyrimidine bases (C + T) are equal.

Analyses of X-Ray Diffraction Patterns

Additional information about the structure of DNA emerged when x-ray diffraction photographs taken by Rosalind Franklin and Maurice Wilkins were analyzed. These diffraction patterns revealed that, even though the base composition of DNA isolated from different organisms varies, DNA molecules themselves are remarkably uniform in thickness. They are long and fairly straight, with an outside diameter of approximately 2000 pm, and not more than a dozen atoms thick. Furthermore, the crystallographic pattern repeats every 3400 pm. Herein lay one of the chief problems to be solved. How could the molecular dimensions of DNA be so regular even though the relative percentages of the various bases differ so widely? With this accumulated information, the stage was set for the development of a hypothesis about DNA structure.

The Watson–Crick Double Helix

The heart of the Watson-Crick model is the postulate that a molecule of DNA is a complementary **double helix**. It consists of two antiparallel polynucleotide strands coiled in a right-handed manner about the same axis to form a double helix. As illustrated in the ribbon models in Figure 28.6, chirality is associated with a double helix; left-handed and right-handed double helices are related by reflection just as enantiomers are related by reflection.

Figure 28.6
A DNA double helix has a chirality associated with the helix. Right-handed and left-handed double helices of otherwise identical DNA chains are nonsuperposable mirror images.

Double helix
A type of secondary structure of DNA molecules in which two antiparallel polynucleotide strands are coiled in a right-handed manner about the same axis.

The search for antiviral drugs has been more difficult than the search for antibacterial drugs primarily because viral replication depends on the metabolic processes of the invaded cell. Thus, antiviral drugs are also likely to cause harm to the cells that harbor the virus. The challenge in developing antiviral drugs has been to understand the biochemistry of viruses and to develop drugs that target processes specific to them. Compared with the large number of antibacterial drugs that are available, there are only a handful of antiviral drugs, and they have nowhere near the effectiveness that antibiotics have on bacterial infections.

Acyclovir is one of the first of a new family of drugs for the treatment of infectious diseases caused by DNA viruses called herpesvirus. Herpes infections in humans are of two kinds: herpes simplex type 1, which gives rise to mouth and eye sores, and herpes simplex type 2, which gives rise to serious genital infections. Acyclovir is highly effective against herpesvirus-caused genital infections. The drug is activated *in vivo* by conversion of the primary —OH (which corresponds to the 5'-OH of a riboside or a deoxyriboside) to a triphosphate. Because

of its close resemblance to deoxyguanosine triphosphate, an essential precursor for DNA synthesis, acyclovir triphosphate is taken up by viral DNA polymerase to form an enzyme-substrate complex on which no 3'-OH exists for replication to continue. Thus, the enzyme-substrate complex is no longer active (it is a dead-end complex), viral replication is disrupted, and the virus is destroyed.

Perhaps the best known of the new viral antimetabolites is zidovudine (azidothymidine, AZT), an analog of deoxythymidine in which the 3'-OH has been replaced by an azido group, N_3. AZT is effective against HIV-1, a retrovirus that is the causative agent of AIDS. It is converted *in vivo* by cellular enzymes to the 5'-triphosphate, recognized as deoxythymidine 5'-triphosphate by viral RNA-dependent DNA polymerase (reverse transcriptase), and added to a growing DNA chain. There it stops chain elongation because no 3'-OH exists on which to add the next deoxynucleotide. AZT owes its effectiveness to the fact that it binds more strongly to viral reverse transcriptase than it does to human DNA polymerase.

Acyclovir
(drawn to show its structural relationship to 2-deoxyguanosine)

Zidovudine
(Azidothymidine; AZT)

© Science Source/Photo Researchers, Inc.

Rosalind Franklin (1920–1958). In 1951 she joined the Biophysical Laboratory at King's College, London, where she used x-ray diffraction methods to study DNA. She is credited with discoveries that established the density of DNA and its helical conformation. Her work was important to the model of DNA developed by Watson and Crick. She died in 1958 at age 37, and because the Nobel Prize is never awarded posthumously, she did not share in the 1962 Nobel Prize for physiology or medicine with Watson, Crick, and Wilkins. Although her relation with Watson and Crick was initially strained, Watson said, "we later came to appreciate. . . . the struggles the intelligent woman faces to be accepted by the scientific world which often regards women as mere diversions from serious thinking."

Thymine ⚌ Adenine   Cytosine ⚌ Guanine   Thymine ⚌ Adenine   Cytosine ⚌ Guanine

1110 pm   1080 pm

2 Hydrogen bonds   3 Hydrogen bonds

Figure 28.7

Base-pairing between adenine and thymine (A-T) and between guanine and cytosine (G-C). An A-T base pair is held by two hydrogen bonds, whereas a G-C base pair is held by three hydrogen bonds.

To account for the observed base ratios and uniform thickness of DNA, Watson and Crick postulated that purine and pyrimidine bases project inward toward the axis of the helix and are always paired in a very specific manner. According to scale models, the dimensions of an adenine-thymine base pair are almost identical to the dimensions of a guanine-cytosine base pair, and the length of each pair is consistent with the core thickness of a DNA strand (Figure 28.7). Thus, if the purine base in one strand is adenine, then its complement in the antiparallel strand must be thymine. Similarly, if the purine in one strand is guanine, its complement in the antiparallel strand must be cytosine. The "fits" between the TA pair and between the CG pair are remarkable and represent another important example of molecular recognition and supramolecular complex formation. There are two specific hydrogen bonds formed between each TA base pair, and three hydrogen bonds formed between each GC base pair. No other purine-pyrimidine combination of the four DNA bases can form stable hydrogen bonded pairs. The specific hydrogen bonding holds the two strands together very strongly.

A significant feature of Watson and Crick's model is that no other base-pairing is consistent with the observed thickness of a DNA molecule. A pair of pyrimidine bases is too small to account for the observed thickness, whereas a pair of purine bases is too large. Thus, according to the Watson-Crick model, the repeating units in a double-stranded DNA molecule are not single bases of differing dimensions but rather base pairs of almost identical dimensions.

To account for the periodicity observed from x-ray data, Watson and Crick postulated that base pairs are stacked one on top of the other with a distance of 340 pm between base pairs and with ten base pairs in one complete turn of the helix. There is one complete turn of the helix every 3400 pm. Shown in Figure 28.8 is a ribbon model of double-stranded **B-DNA**, the predominant form of DNA in dilute aqueous solution and thought to be the most common form in nature.

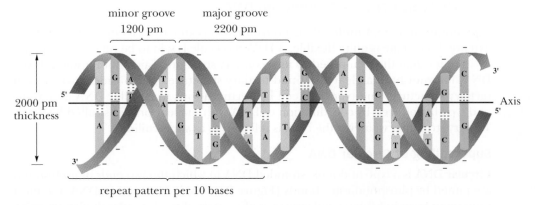

minor groove 1200 pm   major groove 2200 pm

2000 pm thickness

Axis

repeat pattern per 10 bases

Figure 28.8

Ribbon model of double-stranded B-DNA. Each ribbon shows the pentose-phosphodiester backbone of a single-stranded DNA molecule. The strands are antiparallel; one strand runs to the left from the 5′ end to the 3′ end, the other runs to the right from the 5′ end to the 3′ end. Hydrogen bonds are shown by three dotted lines between each G-C base pair and two dotted lines between each A-T base pair.

In the double helix, the bases in each base pair are not directly opposite one another across the diameter of the helix but rather are slightly displaced. This displacement and the relative orientation of the glycosidic bonds linking each base to the sugar-phosphate backbone leads to two differently sized grooves, a major groove and a minor groove (Figure 28.8). Each groove runs along the length of the cylindrical column of the double helix. The major groove is approximately 2200 pm wide; the minor groove is approximately 1200 pm wide.

Figure 28.9 shows more detail of an idealized B-DNA double helix. The major and minor grooves are clearly recognizable in this model.

Other forms of secondary structure are known that differ in the distance between stacked base pairs and in the number of base pairs per turn of the helix. One of the most common of these, **A-DNA**, also a right-handed helix, is thicker than **B-DNA** and has a repeat distance of only 2900 pm. There are ten base pairs per turn of the helix with a spacing of 290 pm between base pairs.

Figure 28.9
An idealized model of B-DNA.

Professor Stanley Cohen/SPL/Photo Researchers, Inc.

Example 28.3

One strand of a DNA molecule contains the base sequence 5'-ACTTGCCA-3'. Write its complementary base sequence.

Solution

Remember that the base sequence is always written from the 5' end of the strand to the 3' end, that A pairs with T, and that G pairs with C. In double-stranded DNA, the two strands run in opposite (antiparallel) directions so that the 5' end of one strand is associated with the 3' end of the other strand.

original strand ⟶ 5'-A-C-T-T-G-C-C-A-3'

3'-T-G-A-A-C-G-G-T-5' ⟵ complementary strand

Written from the 5' end, the complementary strand is 5'-TGGCAAGT-3'.

Problem 28.3

Write the complementary DNA base sequence for 5'-CCGTACGA-3'.

C. Tertiary Structure—Supercoiled DNA

The length of a DNA molecule is enormously greater than its diameter, and the extended molecule is quite flexible. A DNA molecule is said to be relaxed if it has no twists other than those imposed by its secondary structure. Said another way, relaxed DNA does not have a clearly defined tertiary structure. We consider two types of **tertiary structure**, one type induced by perturbations in circular DNA and a second type introduced by coordination of DNA with nuclear proteins called histones. Tertiary structure, whatever the type, is referred to as **supercoiling**.

Supercoiling of Circular DNA

Tertiary structure of nucleic acids
The three-dimensional arrangement of all atoms of a nucleic acid, commonly referred to as supercoiling.

Circular DNA
A type of double-stranded DNA in which the 5' and 3' ends of each strand are joined by phosphodiester groups.

Circular DNA is a type of double-stranded DNA in which the two ends of each strand are joined by phosphodiester bonds [Figure 28.10(a)]. This type of DNA, the most prominent form in bacteria and viruses, is also referred to as circular duplex (because it is double-stranded) DNA. One strand of circular DNA may be opened, partially unwound, and then rejoined. The unwound section introduces a strain into the molecule because the nonhelical gap is less stable than hydrogen-bonded, base-paired helical sections. The strain can be localized in the nonhelical gap. Alternatively, it may be spread uniformly over the entire circular DNA by introduction of **superhelical** twists, one twist

(a) Relaxed: circular (duplex) DNA

(b) Slightly strained: circular (duplex) DNA with four twists of the helix unwound

(c) Strained: supercoiled circular DNA

Figure 28.10
Relaxed and supercoiled DNA. (a) Circular DNA is relaxed. (b) One strand is broken, unwound by four turns, and the ends then rejoined. The strain of unwinding is localized in the nonhelical gap. (c) Supercoiling by four twists distributes the strain of unwinding uniformly over the entire molecule of circular DNA.

for each turn of a helix unwound. The circular DNA shown in Figure 28.10(b) has been unwound by four complete turns of the helix. The strain introduced by this unwinding is spread uniformly over the entire molecule by introduction of four superhelical twists [Figure 28.10(c)]. Interconversion of relaxed and supercoiled DNA is catalyzed by groups of enzymes called topoisomerases and gyrases.

Supercoiling of Linear DNA

Supercoiling of linear DNA in plants and animals takes another form and is driven by interaction between negatively charged DNA molecules and a group of positively charged proteins called **histones**. Histones are particularly rich in lysine and arginine, and at the pH of most body fluids, they have an abundance of positively charged sites along their length. The complex between negatively charged DNA and positively charged histones is called **chromatin**. Histones associate to form core particles about which double-stranded DNA then wraps. Further coiling of DNA produces the chromatin found in cell nuclei. The entire complex, with DNA wound around the histone protein cores, can be thought of as being analogous to beads on a string.

Histone
A protein, particularly rich in the basic amino acids lysine and arginine, that is found associated with DNA molecules.

28.3 | Ribonucleic Acids

Ribonucleic acids are similar to deoxyribonucleic acids in that they, too, consist of long, unbranched chains of nucleotides joined by phosphodiester groups between the 3′-hydroxyl of one pentose and the 5′-hydroxyl of the next. There are, however, three major differences in structure between RNA and DNA.

1. The pentose unit in RNA is β-D-ribose rather than β-2-deoxy-D-ribose.
2. The pyrimidine bases in RNA are uracil and cytosine rather than thymine and cytosine (Figure 28.1).
3. RNA is single-stranded rather than double-stranded.

Cells contain up to eight times as much RNA as DNA; in contrast to DNA, RNA occurs in different forms and in multiple copies of each form. RNA molecules are classified, according to their structure and function, into three major types: ribosomal RNA, transfer RNA, and messenger RNA. The molecular weight, number of nucleotides, and percent cellular abundance of these types in cells of *Escherichia coli* are summarized in Table 28.2.

Supercoiled DNA from a mitochondrion.

© medicalpicture/Alamy

Table 28.2	Types of RNA Found in Cells of *E. coli*		
Type	**Molecular-Weight Range (g/mol)**	**Number of Nucleotides**	**Percentage of Cell RNA**
mRNA	25,000–1,000,000	75–3000	2
tRNA	23,000–30,000	73–94	16
rRNA	35,000–1,100,000	120–2904	82

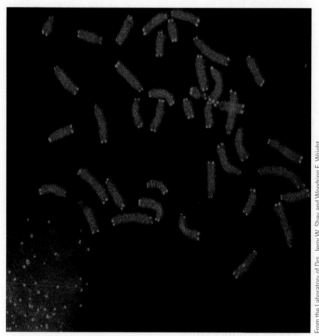

Telomeres are the repeating DNA strings (TTAGGC in vertebrates) that cap chromosomes.

From the Laboratory of Drs. Jerry W. Shay and Woodring E. Wright

In 1997 a sheep named Dolly became the first mammal to be born through cloning. By the time she was three years old, however, her genetic material was aging at the rate of the six-year-old sheep from which she was cloned.

It turned out that Dolly had shortened telomeres, and—as a consequence—her genetic structure was considerably "older" than Dolly herself.

Telomeres are the physical ends of eukaryotic chromosomes, and they surrender a bit of themselves during each cell division because the DNA-replicating machinery normally fails to copy the DNA at a chromosome's tips. This process limits the total number of divisions a cell can undergo because the telomeres are ultimately consumed. In egg and cancer cells, among others, an enzyme called telomerase rebuilds the telomere cap after each division, making such cells effectively immortal. Consequently, telomerase may be the fountain of youth because it is responsible for the extension of telomeres in most species. If telomerase activity is limited, as it is in normal cells, the telomeres will shorten, and cells will age like Dolly's. Telomeres are the repeating DNA strings (TTAGGG in vertebrates) that cap chromosomes.

Researchers are trying to alter the integrity of telomeres by controlling the expression and location of telomerase. Preserving telomere caps may rejuvenate aging organisms by enabling cells to survive additional divisions. Conversely, doing away with telomerase activity in tumors may cause cancer cells to cease their endless replications and die.

Based on a chemistry honors paper by James Stinebaugh, UCLA.

Laguna Design/Photo Library

Venkatraman Ramakrishnan, Thomas A. Steitz, and Ada E. Yonath were awarded the Nobel prize in chemistry in 2009 for their studies of the structure and function of the ribosome.

A. Ribosomal RNA

The bulk of **ribosomal RNA (rRNA)** is found in the cytoplasm in subcellular particles called ribosomes, which contain about 60% RNA and 40% protein. Ribosomes are the sites in cells at which protein synthesis takes place.

Ribosomal RNA (rRNA)
A ribonucleic acid found in ribosomes, the sites of protein synthesis.

B. Transfer RNA

Transfer RNA (tRNA) molecules have the lowest molecular weight of all nucleic acids. They consist of 73–94 nucleotides in a single chain. The function of tRNA is to carry amino acids to the sites of protein synthesis on the ribosomes. Each amino acid has at least one tRNA dedicated specifically to this purpose. Several amino acids have more than one. In the transfer process, the amino acid is joined to its specific tRNA by an ester bond between the α-carboxyl group of the amino acid and the 3′ hydroxyl group of the ribose unit at the 3′ end of the tRNA.

Transfer RNA (tRNA)
A ribonucleic acid that carries a specific amino acid to the site of protein synthesis on ribosomes.

C. Messenger RNA

Messenger RNAs (mRNA) are present in cells in relatively small amounts and are very short-lived. They are single-stranded, and their synthesis is directed by information encoded on DNA molecules. Double-stranded DNA is unwound, and a complementary strand of mRNA is synthesized along one strand of the DNA template, beginning from the 3′ end. The synthesis of mRNA from a DNA template is called transcription because genetic information contained in a sequence of bases of DNA is transcribed into a complementary sequence of bases on mRNA. The name "messenger" is derived from the function of this type of RNA, which is to carry coded genetic information from DNA to the ribosomes for the synthesis of proteins.

Messenger RNA (mRNA)
A ribonucleic acid that carries coded genetic information from DNA to the ribosomes for the synthesis of proteins.

Example 28.4

Following is a base sequence from a portion of DNA. Write the sequence of bases of the mRNA synthesized using this section of DNA as a template.

3′-A-G-C-C-A-T-G-T-G-A-C-C-5′

Solution

RNA synthesis begins at the 3′ end of the DNA template and proceeds toward the 5′ end. The complementary mRNA strand is formed using the bases C, G, A, and U. Uracil (U) is the complement of adenine (A) on the DNA template.

Reading from the 5′ end, the sequence of mRNA is 5′-UCGGUACACUGG-3′.

Problem 28.4

Here is a portion of the nucleotide sequence in phenylalanine tRNA.

3'-ACCACCUGCUCAGGCCUU-5'

Write the nucleotide sequence of its DNA complement.

28.4 | The Genetic Code

A. Triplet Nature of the Code

It was clear by the early 1950s that the sequence of bases in DNA molecules constitutes the store of genetic information and directs the synthesis of messenger RNA, which, in turn, directs the synthesis of proteins. However, the statement that "the sequence of bases in DNA directs the synthesis of proteins" presents the following problem. How can a molecule containing only four variable units (adenine, cytosine, guanine, and thymine) direct the synthesis of molecules containing up to 20 variable units (the protein-derived amino acids)? How can an alphabet of only 4 letters code for the order of letters in the 20-letter alphabet that occurs in proteins?

An obvious answer is that there is not one base but rather a combination of bases coding for each amino acid. If the code consists of nucleotide pairs, there are $4^2 = 16$ combinations; this code is more extensive, but it is still not extensive enough to code for 20 amino acids. If the code consists of nucleotides in groups of three, there are $4^3 = 64$ combinations, which is more than enough to code for the primary structure of a protein. This appears to be a very simple solution to a system that must have taken eons to evolve. Yet nature does indeed use this simple three-letter or triplet code to store genetic information. A triplet of nucleotides is called a **codon**.

Codon
A triplet of nucleotides on mRNA that directs incorporation of a specific amino acid into a polypeptide sequence.

B. Deciphering the Genetic Code

The next question is, which of the 64 triplets code for which amino acid? In 1961 Marshall Nirenberg provided a simple experimental approach to the problem, based on the observation that synthetic polynucleotides direct polypeptide synthesis in much the same manner as do natural mRNAs. Nirenberg incubated ribosomes, amino acids, tRNAs, and appropriate protein-synthesizing enzymes. With only these components, there was no polypeptide synthesis. However, when he added synthetic polyuridylic acid (poly U), a polypeptide of high molecular weight was synthesized. What was more important, the synthetic polypeptide contained only phenylalanine. With this discovery, the first element of the genetic code was deciphered: the triplet UUU codes for phenylalanine.

Similar experiments were carried out with different synthetic polyribonucleotides. It was found, for example, that polyadenylic acid (poly A) leads to the synthesis of polylysine and that polycytidylic acid (poly C) leads to the synthesis of polyproline. By 1964, all 64 codons had been deciphered (Table 28.3).

C. Properties of the Genetic Code

Several features of the genetic code are evident from a study of Table 28.3.

1. Only 61 triplets code for amino acids. The remaining three (UAA, UAG, and UGA) are signals for chain termination; they signal to the protein-synthesizing machinery of the cell that the primary sequence of the protein is complete. The three chain termination triplets are indicated in Table 28.3 by "Stop."

2. The code is degenerate, which means that several amino acids are coded for by more than one triplet. Only methionine and tryptophan are coded for by just one triplet. Leucine, serine, and arginine are coded for by six triplets, and the remaining amino acids are coded for by two, three, or four triplets.

3. For the 15 amino acids coded for by two, three, or four triplets, only the third letter of the code varies. For example, glycine is coded for by the triplets GGA, GGG, GGC, and GGU.

4. There is no ambiguity in the code, meaning that each triplet codes for only one amino acid.

Table 28.3 | The Genetic Code—mRNA Codons and the Amino Acid Each Codon Directs

First Position (5-end)	Second Position				Third Position (3-end)
	U	**C**	**A**	**G**	
U	UUU Phe	UCU Ser	UAU Tyr	UGU Cys	U
	UUC Phe	UCC Ser	UAC Tyr	UGC Cys	C
	UUA Leu	UCA Ser	UAA Stop	UGA Stop	A
	UUG Leu	UCG Ser	UAG Stop	UGG Trp	G
C	CUU Leu	CCU Pro	CAU His	CGU Arg	U
	CUC Leu	CCC Pro	CAC His	CGC Arg	C
	CUA Leu	CCA Pro	CAA Gln	CGA Arg	A
	CUG Leu	CCG Pro	CAG Gln	CGG Arg	G
A	AUU Ile	ACU Thr	AAU Asn	AGU Ser	U
	AUC Ile	ACC Thr	AAC Asn	AGC Ser	C
	AUA Ile	ACA Thr	AAA Lys	AGA Arg	A
	AUG* Met	ACG Thr	AAG Lys	AGG Arg	G
G	GUU Val	GCU Ala	GAU Asp	GGU Gly	U
	GUC Val	GCC Ala	GAC Asp	GGC Gly	C
	GUA Val	GCA Ala	GAA Glu	GGA Gly	A
	GUG Val	GCG Ala	GAG Glu	GGG Gly	G

*AUG also serves as the principal initiation codon.

Finally, we must ask one last question about the genetic code. Is the code universal, that is, is it the same for all organisms? Every bit of experimental evidence available today from the study of viruses, bacteria, and higher animals, including humans, indicates that the code is universal. Furthermore, the fact that it is the same for all these organisms means that it has been the same over billions of years of evolution.

Example 28.5

During transcription, a portion of mRNA is synthesized with the following base sequence.

$$5'\text{-AUG-GUA-CCA-CAU-UUG-UGA-}3'$$

(a) Write the nucleotide sequence of the DNA from which this portion of mRNA was synthesized.

(b) Write the primary structure of the polypeptide coded for by this section of mRNA.

Solution

(a) During transcription, mRNA is synthesized from a DNA strand, beginning from the 3′ end of the DNA template. The DNA strand must be the complement of the newly synthesized mRNA strand.

Note that the codon UGA codes for termination of the growing polypeptide chain; therefore, the sequence given in this problem codes for a pentapeptide only.

(b) The sequence of amino acids is shown in the following mRNA strand.

5'-AUG-GUA-CCA-CAU-UUG-UGA-3'
met—val—pro—his—leu—stop

Problem 28.5

The following section of DNA codes for oxytocin, a polypeptide hormone.

3'ACG-ATA-TAA-GTT-TTA-ACG-GGA-GAA-CCA-ACT-5'

(a) Write the base sequence of the mRNA synthesized from this section of DNA.
(b) Given the sequence of bases in part (a), write the primary structure of oxytocin.

28.5 | Sequencing Nucleic Acids

As recently as 1975, the task of determining the primary structure of a nucleic acid was thought to be far more difficult than determining the primary structure of a protein. Nucleic acids, it was reasoned, contain only 4 different units, whereas proteins contain 20 different units. With only 4 different units, there are fewer specific sites for selective cleavage, distinctive sequences are more difficult to recognize, and there is greater chance of ambiguity in the assignment of sequence. Two breakthroughs reversed this situation. First was the development of a type of electrophoresis called polyacrylamide gel electrophoresis, a technique so sensitive that it is possible to separate nucleic acid fragments that differ from one another in only a single nucleotide. The second breakthrough was the discovery of a class of enzymes called restriction endonucleases, isolated chiefly from bacteria.

A. Restriction Endonucleases

Restriction endonuclease
An enzyme that catalyzes hydrolysis of a particular phosphodiester bond within a DNA strand.

A **restriction endonuclease** recognizes a set pattern of four to eight nucleotides and cleaves a DNA strand by hydrolysis of the linking phosphodiester bonds at any site that contains that particular sequence. Close to 1000 restriction endonucleases have been isolated and their specificities characterized; each cleaves at specific sites, often with unique specificity. *E. coli*, for example, has a restriction endonuclease, EcoRI, that recognizes the hexanucleotide sequence, GAATTC, and cleaves it between G and A.

cleavage here
5'---G-A-A-T-T-C---3' $\xrightarrow{\text{EcoRI}}$ 5'---G + 5'-A-A-T-T-C---3'

Note that the action of restriction endonucleases is analogous to the action of trypsin (Section 27.4B), which catalyzes hydrolysis of amide bonds formed by the carboxyl groups of Lys and Arg, and of chymotrypsin, which catalyzes cleavage of amide bonds formed by the carboxyl groups of Phe, Tyr, and Trp.

Example 28.6

The following is a section of the gene coding for bovine rhodopsin along with several restriction endonucleases, their recognition sequences, and their hydrolysis sites. Which endonucleases will catalyze cleavage of this section of DNA?

5'-GCCGTCTACAACCCGGTCATCTACTATCATGATCAACAAGCAGTTCCGGAACT-3'

Enzyme	Recognition Sequence	Enzyme	Recognition Sequence
AluI	AG ↓ CT	H*pa*II	C ↓ CGG
BalI	TGG ↓ CCA	M*bo*I	↓ GATC
F*nu*DII	CG ↓ CG	NotI	GC ↓ GGCCGC
H*ea*III	GG ↓ CC	SacI	GAGCT ↓ C

Solution

Only restriction endonucleases H*pa*II and M*bo*I catalyze cleavage of this polynucleotide: H*pa*II at two sites and M*bo*I at one site.

H*pa*II   M*bo*I   H*pa*II

5′-GCCGTCTACAACCCGGTCATCTACTATCATGATCAACAAGCAGTTCCGGAACT-3′

Problem 28.6

The following is another section of the bovine rhodopsin gene. Which of the endonucleases given in Example 28.6 will catalyze cleavage of this section?

5′-ACGTCGGGTCGTCGTCCTCTCGCGGTGGTGAGTCTTCCGGCTCTTCT-3′

B. Methods for Sequencing Nucleic Acids

Any sequencing of DNA begins with site-specific cleavage of double-stranded DNA by one or more restriction endonucleases into smaller fragments called restriction fragments. Each restriction fragment is then sequenced separately, overlapping base sequences are identified, and the entire sequence of bases is then deduced.

Two methods for sequencing restriction fragments have been developed. The first of these, developed by Allan Maxam and Walter Gilbert and known as the Maxam-Gilbert method, depends on base-specific chemical cleavage. The second method, developed by Frederick Sanger and known as the chain termination or dideoxy method, depends on interruption of DNA-polymerase catalyzed synthesis. Sanger and Gilbert shared the 1980 Nobel Prize for biochemistry for their "development of chemical and biochemical analysis of DNA structure." Sanger's dideoxy method is currently more widely used, and it is on this method that we concentrate.

Sanger dideoxy method
A method developed by Frederick Sanger for sequencing DNA molecules.

C. DNA Replication *in Vitro*

To appreciate the rationale for the dideoxy method, we must first understand certain aspects of the biochemistry of DNA replication. During replication, the sequence of nucleotides in one strand is copied as a complementary strand to form the second strand of a double-stranded DNA molecule. Synthesis of the complementary strand is catalyzed by the enzyme DNA polymerase. DNA polymerase will also carry out this synthesis *in vitro* using single-stranded DNA as a template, provided that both the four deoxynucleotide triphosphate (dNTP) monomers and a primer are present. A **primer** is an oligonucleotide capable of forming a short section of double-stranded DNA (dsDNA) by base-pairing with its complement on a single-stranded DNA (ssDNA). Because a new DNA strand grows from its 5′ to 3′ end, the primer must have a free 3′-OH group to which the first nucleotide of the growing chain is added (Figure 28.11).

Single-stranded DNA

5′   T   C   A   A   C   G   A   T   C   T   G   A   OH 3′

                       G   A   C   T

dATP, dTTP, dCTP, dGTP   3′ HO —   **Primer**   5′

direction of synthesis catalyzed by DNA polymerase

Figure 28.11

DNA polymerase catalyzes the synthesis *in vitro* using single-stranded DNA as a template provided that both the four dNTP monomers and a primer are present. The primer provides a short stretch of double-stranded DNA by base-pairing with its complement on the single-stranded DNA.

D. The Chain Termination or Dideoxy Method

The key to the chain termination method is the addition to the synthesizing medium of a 2′,3′-dideoxynucleoside triphosphate (ddNTP). Because a ddNTP has no —OH group at the 3′ position, it cannot serve as an acceptor for the next nucleotide to be added to the growing polynucleotide chain. Thus, chain synthesis is terminated at any point where a ddNTP becomes incorporated, hence the designation chain termination method.

A 2′,3′-dideoxynucleoside triphosphate
(ddNTP)

In the chain termination method, a single-stranded DNA of unknown sequence is mixed with primer and divided into four separate reaction mixtures. To each reaction mixture are added all four dNTPs, one of which is labeled in the 5′ phosphoryl group with ^{32}P so that the newly synthesized fragments can be visualized by autoradiography.

$$^{32}_{15}P \rightarrow {}^{32}_{16}S + \text{Beta particle} + \text{Gamma rays}$$

Also added to each reaction mixture are DNA polymerase and one of the four ddNTPs. The ratio of dNTPs to ddNTP in each reaction mixture is adjusted so that incorporation of a ddNTP takes place infrequently. In each reaction mixture, DNA synthesis takes place; however, in a given population of molecules, synthesis is interrupted at every possible site (Figure 28.12).

When gel electrophoresis of each reaction mixture is completed, a piece of x-ray film is placed over the gel, and gamma rays released by radioactive decay of ^{32}P darken the film and create a pattern on it that is an image of the resolved oligonucleotides. The base sequence of the complement to the original single-stranded template is then read directly from bottom to top of the developed film.

A variation on this method is to use a single reaction mixture with each of the four ddNTPs labeled with a different fluorescent indicator. Each label is then detected by its characteristic spectrum following separation of the fragments by polyacrylamide gel electrophoresis or some chromatography method. First generation automated DNA sequencing machines use this fluorescent variation.

E. Sequencing the Human Genome

The sequencing of the human genome was announced in the spring of 2000 by two competing groups, the Human Genome Project, a loosely linked consortium of publicly funded groups, and a private company called Celera Genomics. Actually, this milestone didn't represent a complete sequence but rather a so-called rough draft, comprising about 85% of the entire genome. The methodology used is based on a refinement of the techniques described earlier using massively parallel separations of fragments by electrophoresis in capillary tubes. The Celera approach used some 300 of the fastest sequencing machines in parallel, each operating on many parallel DNA fragments. Supercomputers were used to assemble and compare millions of overlapping sequences. Figure 28.13 shows the apparatus used by the Celera group for this milestone achievement.

It is now much easier to obtain nucleic acid sequences than protein sequences. Sequencing a gene immediately gives the investigator the sequence of the protein

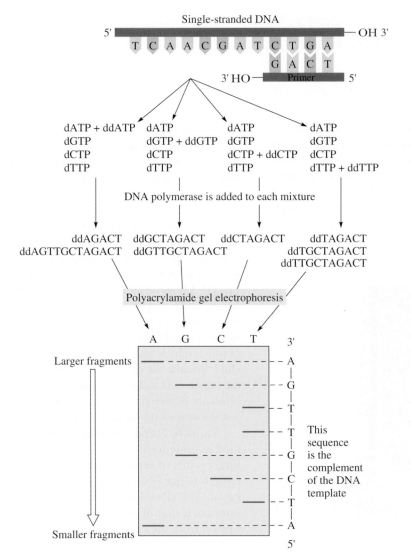

Single-stranded DNA

5' ▢▢▢▢▢▢▢▢▢▢▢▢▢▢▢▢▢▢ — OH 3'

T C A A C G A T C T G A

G A C T

3' HO — ▢▢▢▢ — 5'
Primer

dATP + ddATP	dATP	dATP	dATP
dGTP	dGTP + ddGTP	dGTP	dGTP
dCTP	dCTP	dCTP + ddCTP	dCTP
dTTP	dTTP	dTTP	dTTP + ddTTP

DNA polymerase is added to each mixture

ddAGACT	ddGCTAGACT	ddCTAGACT	ddTAGACT
ddAGTTGCTAGACT	ddGTTGCTAGACT		ddTGCTAGACT
			ddTTGCTAGACT

Polyacrylamide gel electrophoresis

A G C T 3'

Larger fragments

— A
| G
| T
| T
| G
| C
| T
— A

This sequence is the complement of the DNA template

Smaller fragments

5'

If the complement of the DNA template is 5' A–T–C–G–T–T–G–A–3'
Then the original DNA template must be 5'–T–C–A–A–C–G–A–T–3'

Figure 28.12
The chain termination or dideoxy method of DNA sequencing. The primer-DNA template is divided into four separate reaction mixtures. To each mixture is added the four dNTPs, DNA polymerase, and one of the four ddNTPs. Synthesis is interrupted at every possible site. The mixtures of oligonucleotides are separated by polyacrylamide gel electrophoresis. The base sequence of the DNA complement is read from the bottom to the top (from the 5' end to the 3' end) of the developed gel.

Figure 28.13
Photo of sequencing equipment.

© David Parker/Photo Researchers, Inc.

Each human being has a genetic makeup consisting of approximately 3 billion base pairs of nucleotides, and, except for identical twins, the base sequence of DNA in one individual is different from that of every other individual. As a result, each person has a unique DNA fingerprint. To determine a DNA fingerprint, a sample of DNA from a trace of blood, skin, or other tissue is treated with a set of restriction endonucleases, and

9 8 7 6 5 4 3 2 1

Courtesy of Dr. Lawrence Kobilinsky

A DNA "fingerprint."

the 5′ end of each restriction fragment is labeled with phosphorus-32. The resulting ^{32}P-labeled restriction fragments are then separated by polyacrylamide gel electrophoresis and visualized by placing a photographic plate over the developed gel.

In the DNA fingerprint patterns shown in the figure, lanes 1, 5, and 9 represent internal standards or control lanes. They contain the DNA fingerprint pattern of a standard virus treated with a standard set of restriction endonucleases. Lanes 2, 3, and 4 were used in a paternity suit. The DNA fingerprint of the mother in lane 4 contains five bands, which match with five of the six bands in the DNA fingerprint of the child in lane 3. The DNA fingerprint of the alleged father in lane 2 contains six bands, three of which match with bands in the DNA fingerprint of the child. Because the child inherits only half of its genes from the father, only half of the child's and father's DNA fingerprints are expected to match. In this instance, the paternity suit was won on the basis of the DNA fingerprint matching.

Lanes 6, 7, and 8 contain DNA fingerprint patterns used as evidence in a rape case. Lanes 7 and 8 are DNA fingerprints of semen obtained from the rape victim. Lane 6 is the DNA fingerprint pattern of the alleged rapist. The DNA fingerprint patterns of the semen do not match that of the alleged rapist and excluded the suspect from the case.

produced by that gene because the code is known. Because many organisms have been sequenced and the functions of many of the coded proteins are known, it is possible to take the sequence of a protein of unknown function and determine sequence homologies with the vast number of proteins of known function in a repository known as the Protein Data Bank and thereby make an educated guess about the function of a protein the new gene codes for. This technology is enabling an information revolution in chemistry, biology, and medicine.

This achievement represents the beginning of a new era of molecular medicine in which specific genetic deficiencies leading to inherited diseases will be understood on a molecular basis, and new therapies targeted at shutting down undesired genes or turning on desired ones will be developed. In addition, different individuals may respond differently to different drug therapies. In the future, medicines may be specifically tailored to the genetic makeup of individual patients.

Next Generation DNA Sequencing

Next generation sequencing methods have recently been created by marrying newer versions of the enzymatic DNA sequencing reactions with clonal amplification, microfabrication techniques, microfluidic technologies, and automated sample preparation. Using massively parallel approaches, an astounding 300 Gigabases (300 billion bases) of DNA can be sequenced in a single run, which now costs only a few thousand dollars. By exploiting special optical technologies, one of the very latest sequencing approaches involves sequencing at the single molecule level! Sequencing of entire genomes, especially microorganisms, is now routine.

Amazingly, each sequencing reaction only reads 50 to 400 bases. Hundreds of thousands to hundreds of millions of these reads are carried out simultaneously in parallel fashion on microfabricated devices. Although it sounds almost too remarkable to believe, entire genomes as large as several Gigabases in length are deduced by overlapping thousands to millions of these 50 to 400 base pair sequence reads. The key to this approach is having extensive redundancy built into the sequencing protocols so that each DNA segment of the genome is covered by multiple overlapping sequence reads. Through the use of sophisticated and highly efficient computational techniques, this overlapping enables the unambiguous assembly of each unique sequence read into a complete genome. The relatively new field of bioinformatics is largely devoted to developing new tools with which to analyze and exploit the massive amounts of sequence data that are now being generated at an exponentially accelerating pace. Indeed, managing and analyzing such large datasets is one of the key challenges to the field. It is appropriate to say that the awesome power of next generation sequencing technologies has led to a revolution in biology research and will have far-reaching positive impacts on many aspects of future medicine.

Summary

Section 28.1 Nucleosides and Nucleotides

- **Nucleic acids** are composed of three types of monomer units: heterocyclic aromatic amine bases derived from purine and pyrimidine, the monosaccharides D-ribose or 2-deoxy-D-ribose, and phosphate ions.
- A **nucleoside** is a compound containing D-ribose or 2-deoxy-D-ribose bonded to a heterocyclic aromatic amine base by a β-N-glycosidic bond.
  - For DNA the four bases are **adenine**, **cytosine**, **guanine**, and **thymine**.
  - For RNA the four bases are **adenine**, **cytosine**, **guanine**, and **uracil**.
  - Adenine and guanine are **purines**; cytosine, thymine, and uracil are **pyrimidines**.

Problems: 28.1, 28.7–28.13

- A **nucleotide** is a nucleoside in which at least one molecule of phosphoric acid is esterified with an -OH of the monosaccharide, most commonly either the 3′-OH or the 5′-OH.
  - Nucleoside mono-, di-, and triphosphates are strong polyprotic acids and are extensively ionized at pH 7.0.
  - At pH 7.0, adenosine triphosphate, for example, is a 50:50 mixture of ATP^{3-} and ATP^{4-}.
  - The N atoms of the amino groups of adenine, cytosine, and guanine, which are not in the rings, are sp^2 hybridized and planar, allowing for base stacking as well as hydrogen bonding to complementary bases.

Section 28.2 The Structure of DNA

- The **primary structure** of **deoxyribonucleic acids** (**DNA**) consists of units of 2-deoxyribose bonded by 3′,5′-phosphodiester bonds.
  - A heterocyclic aromatic amine base is attached to each deoxyribose unit by a β-N-glycosidic bond.
  - The sequence of bases is read from the 5′ end of the polynucleotide strand to the 3′ end.

- The heart of the **Watson-Crick model** is the postulate that a molecule of DNA consists of two antiparallel polynucleotide strands coiled in a right-handed manner about the same axis to form a **double helix**.
  – Purine and pyrimidine bases point inward toward the axis of the helix and are always paired G-C and A-T due to specific patterns of hydrogen bonds.
- In **B-DNA**, base pairs are stacked one on top of another with a spacing of 340 pm and ten base pairs per 3400-pm helical repeat.
- In **A-DNA**, bases are stacked with a spacing of 290 pm between base pairs and ten base pairs per 2900-pm helical repeat.
- The **tertiary structure** of DNA is commonly referred to as **supercoiling**.
  – **Circular DNA** is a type of double-stranded DNA in which the ends of each strand are joined by phosphodiester groups.
  – Opening of one strand followed by partial unwinding and rejoining the ends introduces strain in the nonhelical gap.
  – This strain can be spread over the entire molecule of circular DNA by introduction of **superhelical twists**.
- **Histones** are the DNA binding proteins found in cell nuclei.
  – They are particularly rich in lysine and arginine and, therefore, have an abundance of positive charges.
  – The association of DNA and histones produces a structure called **chromatin**.

Section 28.3 Ribonucleic Acids

- There are two important differences between the primary structure of ribonucleic acids (RNA) and DNA.
  – The monosaccharide unit in RNA is D-ribose.
  – Both RNA and DNA contain the purine bases adenine (A) and guanine (G) and the pyrimidine base cytosine (C). As the fourth base, however, RNA contains uracil (U), whereas DNA contains thymine (T).

- RNA exists in three main forms in cells.
  – Ribosomal RNA, the most abundant form, is part of the ribosome, the protein-synthesizing machine of the cell.
  – Transfer RNA carries amino acids to the sites of protein synthesis on ribosomes.
  – Messenger RNAs are copies of DNA gene sequences and carry the coded genetic information from DNA to the ribosomes for the synthesis of proteins.

Section 28.4 The Genetic Code

- The genetic code consists of nucleosides in groups of three; that is, it is a triplet code.
  – An amino acid is coded for by one triplet.
  – Only 61 triplets code for amino acids; the remaining three code for termination of polypeptide synthesis.

  – The code is degenerate, meaning that several amino acids are coded for by more than one triplet.
  – For the 15 amino acids coded for by two, three, or four triplets, only the last base of the triplet varies.
  – There is no ambiguity in the code: Each triplet codes only for a single amino acid or termination signal.
  – The genetic code appears to be universal for all different forms of life on Earth.

Section 28.5 Sequencing Nucleic Acids

- **Restriction endonucleases** recognize a set pattern of four to eight nucleotides and cleave a DNA strand by hydrolysis of the linking phosphodiester bonds at any site that contains that particular sequence.
  – Restriction endonucleases isolated from different sources recognize different sequences.

  – Restriction endonucleases are used to produce fragments from genomic DNA.

- In the **chain termination** or **dideoxy method** of DNA sequencing developed by Frederick Sanger, a primer-DNA template is divided into four separate reaction mixtures.
  - To each is added the four dNTPs, one of which is labeled with ^{32}P.
  - Also added are DNA polymerase and one of the four ddNTPs.
  - Synthesis is interrupted when a ddNTP is incorporated.
  - The mixtures of newly synthesized oligonucleotides are separated by polyacrylamide gel electrophoresis and visualized by autoradiography.
  - The base sequence of the DNA complement to the original DNA template is read from the bottom to the top (from the 5′ end to the 3′ end) of the developed photographic plate.
  - Alternatively, the four ddNTPs each contain a different fluorescent label and the sequence is read as the differently fluorescing strands elute from the gel.

⚈WL Questions and problems for this chapter are available in OWL. Use the OWL access card that came with this textbook to access assigned questions and problems for this chapter.

This chapter was originally authored by Dr. Bruce Novak, Howard J. Schaeffer Distinguished Professor of Polymer Science and Organic Chemistry, North Carolina State University.

Sea of umbrellas on a rainy day in Shanghai, China. Inset: models of adipic acid and hexamethylenediamine, the two monomers of nylon 66.

Chapter 29

Organic Polymer Chemistry

© Gavin Hellier/Stone/Getty

WL **Online homework for this chapter may be assigned in OWL for Organic Chemistry.**

The technological advancement of any society is inextricably tied to the materials available to it. Indeed, historians have used the emergence of new materials as a way of establishing a timeline to mark the development of human civilization. As part of the search to discover new materials, scientists have made increasing use of organic chemistry for the preparation of synthetic polymers. The versatility afforded by these polymers allows for the creation and fabrication of materials with ranges of properties unattainable using such materials as wood, metals, and ceramics. Deceptively simple changes in the chemical structure of a given polymer, for example, can change its mechanical properties from those of a sandwich bag to those of a bulletproof vest. Furthermore, structural changes can introduce properties never before imagined in organic polymers. For example, using well-defined organic reactions, one type of polymer can be made into an insulator (for example, the rubber sheath that surrounds electrical cords), or, if treated differently, it can be made into an electrical conductor with a conductivity nearly equal to that of metallic copper!

The years since the 1930s have seen extensive research and development in polymer chemistry, and an almost explosive growth in plastics, coatings, and rubber technology has created a worldwide multibillion-dollar industry. A few basic characteristics account for this phenomenal growth. First, the raw materials for synthetic polymers are derived mainly from petroleum. With the development of petroleum-refining processes, raw materials for the synthesis of polymers became generally cheap and plentiful. Second, within broad limits, scientists have learned how to tailor polymers to the requirements of the end use. Third, many consumer products can be fabricated more cheaply from synthetic polymers than from competing materials, such as wood, ceramics, and metals. For example, polymer technology created the water-based (latex) paints that have revolutionized the coatings industry; plastic films and foams have done the same for the packaging industry. The list could go on and on as we think of the manufactured items that are everywhere around us in our daily lives.

29.1 | The Architecture of Polymers

Polymers (Greek: *poly + meros*, many parts) are long-chain molecules synthesized by linking **monomers** (Greek: *mono + meros*, single part) through chemical reactions. The molecular weights of polymers are generally high compared with those of common organic compounds and typically range from 10,000 g/mol to more than 1,000,000 g/mol. The architectures of these macromolecules can also be quite diverse. Types of polymer architecture include linear and branched chains as well as those with comb, ladder, and star structures (Figure 29.1). Additional structural variations can be achieved by introducing covalent cross links between individual polymer chains.

Polymer
From the Greek, *poly + meros*, many parts. Any long-chain molecule synthesized by linking together many single parts called monomers.

Monomer
From the Greek, *mono + meros*, single part. The simplest nonredundant unit from which a polymer is synthesized.

Linear Branched Comb Ladder Star Crosslinked network Dendritic

Figure 29.1 Various polymer architectures.

Linear and branched polymers are often soluble in solvents, such as chloroform, benzene, toluene, DMSO, and THF. In addition, many linear and branched polymers can be melted to form highly viscous liquids. In polymer chemistry, the term **plastic** refers to any polymer that can be molded when hot and retains its shape when cooled. **Thermoplastics** are polymers that can be melted and become sufficiently fluid that they can be molded into shapes that are retained when they are cooled. **Thermosetting plastics** or thermosets can be molded when they are first prepared, but once they cool, they harden irreversibly and cannot be remelted. Because of these very different physical characteristics, thermoplastics and thermosets must be processed differently and are used in very different applications.

The single most important property of polymers at the molecular level is the size and shape of their chains. A good example of the importance of size is a comparison of paraffin wax, a natural polymer, and polyethylene, a synthetic polymer. These two distinct materials have identical repeat units, namely —CH_2—, but differ greatly in chain size. Paraffin wax has between 25 and 50 carbon atoms per chain, whereas polyethylene has between 1000 and 3000 carbon atoms per chain. Paraffin wax, such as in birthday candles, is soft and brittle but polyethylene, such as in plastic beverage bottles, is strong, flexible, and tough. These vastly different properties arise directly from the difference in size and molecular architecture of the individual polymer chains.

Plastic
A polymer that can be molded when hot and retains its shape when cooled.

Thermoplastic
A polymer that can be melted and molded into a shape that is retained when it is cooled.

Thermoset plastic
A polymer that can be molded when it is first prepared but once cooled, hardens irreversibly and cannot be remelted.

29.2 | Polymer Notation and Nomenclature

We show the structure of a polymer by placing parentheses around the **repeat unit**, which is the smallest molecular fragment that contains all the nonredundant structural features of the chain. Thus, the structure of an entire polymer chain can be reproduced by repeating the enclosed structure in both directions. A subscript n, called the **average degree of polymerization**, is placed outside the parentheses to indicate that this unit is repeated n times.

The polymers formed from symmetric monomer units, such as polyethylene, $+CH_2CH_2\}_n$, and polytetrafluoroethylene, $+CF_2CF_2\}_n$, are an exception to this notation. Although the simplest repeat units are the $-CH_2-$ and $-CF_2-$ groups, respectively, we show two methylene groups and two difluoromethylene groups because they originate from ethylene ($CH_2=CH_2$) and tetrafluoroethylene ($CF_2=CF_2$), the monomer units from which these polymers are derived.

The most common method of naming a polymer is to attach the prefix *poly-* to the name of the monomer from which the polymer is derived, as for example polyethylene and polystyrene. In the case of a more complex monomer or where the name of the monomer is more than one word, as for example the monomer vinyl chloride, parentheses are used to enclose the name of the monomer.

| Polystyrene | Styrene | Poly(vinyl chloride) (PVC) | Vinyl chloride |

is synthesized from

is synthesized from

Example 29.1

Given the following structure, determine the polymer's repeat unit, redraw the structure using the simplified parenthetical notation, and name the polymer.

Solution

The repeat unit is $-CH_2CF_2-$ and the polymer is written $+CH_2CF_2\}_n$. The repeat unit is derived from 1,1-difluoroethylene, and the polymer is named poly(1,1-difluoroethylene). This polymer is used in microphone diaphragms.

Problem 29.1

Given the following structure, determine the polymer's repeat unit, redraw the structure using the simplified parenthetical notation, and name the polymer.

29.3 | Molecular Weights of Polymers

All synthetic polymers and most naturally occurring polymers are mixtures of individual polymer molecules of variable molecular weights. When defining molecular weights in polymer chemistry, the two most common definitions are the number average and weight average molecular weights. The **number average molecular weight**, M_n, is calculated by counting the number of polymer chains of a particular molecular weight, multiplying each number by the molecular weight of its chain, summing these values, and dividing by the total number of polymer chains. The **weight average molecular weight**, M_w, is calculated by recording the total weight of each chain of a particular length, summing these weights, and dividing by the total weight of the sample. Because the larger chains in a sample weigh more than the smaller chains, the weight average molecular weight is skewed to higher values, and M_w is always greater than M_n (Figure 29.2).

Figure 29.2

The distribution of molecular weights in a given polymer sample.

Both M_n and M_w are useful values, and their ratio, M_w/M_n, called the **polydispersity index**, provides a measure of the breadth of the molecular-weight distribution. When the M_w/M_n ratio is equal to one, all the polymer molecules in a sample are the same length, and the polymer is said to be **monodisperse**. No synthetic polymers are ever monodisperse unless the individual molecules are carefully fractionated using time-consuming, rigorous separation techniques based on molecular size. On the other hand, natural polymers, such as polypeptides and DNA, that are formed using biological processes are monodisperse polymers.

29.4 | Polymer Morphology—Crystalline Versus Amorphous Materials

Polymers, like small organic molecules, tend to crystallize upon precipitation or as they are cooled from a melt. Acting to inhibit this tendency are their very large molecules, which tend to slow diffusion, and their sometimes complicated or irregular structures, which prevent efficient packing of the chains. The result is that polymers in the solid state tend to be composed of both ordered **crystalline domains** (crystallites) and disordered **amorphous domains**. The relative amounts of crystalline and amorphous domains differ from polymer to polymer and often depend upon the manner in which the material is processed.

Crystalline domain
An ordered crystalline region in the solid state of a polymer. Also called crystallites.

Amorphous domain
A disordered, noncrystalline region in the solid state of a polymer.

High degrees of crystallinity are most often found in polymers with regular, compact structures and strong intermolecular forces, such as hydrogen bonding and dipolar interactions. The temperature at which crystallites melt corresponds to the **melt transition** (T_m) of the polymer. As the degree of crystallinity of a polymer increases, its T_m increases, and it becomes more opaque owing to scattering of the light by the crystalline domains. There is also a corresponding increase in strength and stiffness with increase in crystallinity. For example, poly(6-aminohexanoic acid) has a $T_m = 223°C$. At and well above room temperature, this polymer is a hard durable material that does not undergo any appreciable change in properties even on a very hot summer afternoon. Its uses range from textile fibers to shoe heels.

Melt transition, T_m
The temperature at which crystalline regions of a polymer melt.

Amorphous domains are characterized by the absence of long-range order. Highly amorphous polymers are sometimes referred to as glassy polymers. Because they lack crystalline domains that scatter light, amorphous polymers are transparent. In addition, they are typically weak polymers, in terms both of their greater flexibility and smaller mechanical strength. On being heated, amorphous polymers are transformed from a hard glass to a soft, flexible, rubbery state. The temperature at which this transition occurs is called the **glass transition temperature** (T_g). Amorphous polystyrene, for example, has a $T_g = 100°C$. At room temperature, it is a rigid solid used for drinking cups, foamed packaging materials, disposable medical wares, tape reels, and so forth. If it is placed in boiling water, it becomes soft and rubbery.

Glass transition temperature, T_g
The temperature at which a polymer undergoes the transition from a hard glass to a rubbery state.

This relationship between mechanical properties and the degree of crystallinity can be illustrated by poly(ethylene terephthalate) (**PET**).

Poly(ethylene terephthalate) (PET)

PET can be made with a percent of crystalline domains ranging from 0% to about 55%. Completely amorphous PET is formed by cooling the melt quickly. By prolonging the cooling time, more molecular diffusion occurs, and crystallites form as the chains become more ordered. The differences in mechanical properties between these forms of PET are substantial. PET with a low degree of crystallinity is used for plastic beverage bottles, whereas fibers drawn from highly crystalline PET are used for textile fibers and tire cords.

Elastomer
A material that, when stretched or otherwise distorted, returns to its original shape when the distorting force is released.

Rubber materials must have low T_g values to behave as **elastomers (elastic polymers)**. If the temperature drops below its T_g value, then the material is converted to a rigid glassy solid, and all elastomeric properties are lost. A poor understanding of this behavior of polymers contributed to the *Challenger* spacecraft disaster in 1985. The elastomeric O-rings used to seal the solid booster rockets had a T_g value around 0°C. When the temperature dropped to an unanticipated low on the morning of the *Challenger* launch, the O-ring seals dropped below their T_g value and consequently changed from elastomers to rigid glasses, losing any sealing capabilities. The rest is tragic history. The physicist Richard Feynman sorted this out publicly in a famous televised hearing in which he put a *Challenger*-type O-ring in ice water and showed that its elasticity was lost!

29.5 | Step-Growth Polymerizations

Step-growth polymerization
A polymerization in which chain growth occurs in a stepwise manner between difunctional monomers as, for example, between adipic acid and hexamethylenediamine to form nylon 66. Also called condensation polymerization.

Condensation polymerization
A polymerization in which chain growth occurs in a stepwise manner between difunctional monomers. Also called step-growth polymerization.

Polymerizations in which chain growth occurs in a stepwise manner are called **step-growth** or **condensation polymerizations**. Step-growth polymers are formed by reaction between difunctional molecules, with each new bond created in a separate step. During polymerization, monomers react with monomers to form dimers, dimers react with dimers to form tetramers, tetramers react with monomers to form pentamers and so on. This stepwise construction of polymer chains has important consequences for both their molecular weights and molecular-weight distributions. Probability tells us that the most abundant species tend to co-condense. Thus, at the early stages of polymerization, small chains are most likely to react with monomers or other small chains to generate many low-molecular-weight oligomers rather than a small number of high-molecular-weight polymers. This tendency persists until most monomer units are used up. As a result, high-molecular-weight polymers are not produced until very late in the reaction, typically past 99% conversion of monomers to higher molecular-weight chains. Only at this point is there the probability of larger chains reacting with one another to form high-molecular-weight polymer molecules. This restriction points to an important distinction between small-molecule organic reactions and step-growth polymerizations. Although a reaction that typically yields 85% of the desired product is considered "good" in organic synthesis, the same reaction is essentially useless for step-growth polymerizations because high-molecular-weight polymers are rarely formed at such low conversions.

There are two common types of step-growth processes: (1) reaction between A-A and B-B type monomers to give (A-A-B-B)$_n$ polymers and (2) the self-condensation of A-B monomers to give (A-B)$_n$ polymers. In each case, an A functional group reacts exclusively with a B functional group, and a B functional group reacts exclusively with an A functional group. New covalent bonds in step-growth polymerizations are generally formed by polar reactions, as for example nucleophilic acyl substitution. In this section, we discuss five types of step-growth polymers: polyamides, polyesters, polycarbonates, polyurethanes, and epoxy resins.

A. Polyamides

Polyamide
A polymer in which each monomer unit is joined to the next by an amide bond, as, for example, nylon 66.

In the years following World War I, a number of chemists recognized the need for developing a basic knowledge of polymer chemistry. One of the most creative of these was Wallace M. Carothers. In the early 1930s, Carothers and his associates at E. I. DuPont de Nemours & Company began fundamental research into the reactions of aliphatic dicarboxylic acids and diols. From adipic acid and ethylene glycol, they obtained a polyester of high molecular weight that could be drawn into fibers.

Hexanedioic acid
(Adipic acid)   **1,2-Ethanediol**
(Ethylene glycol)   **Poly(ethylene adipate)**

These first polyester fibers had melt transitions (T_m) too low for use as textile fibers, and they were not investigated further. Carothers then turned his attention to the reactions of dicarboxylic acids and diamines to form polyamides and, in 1934, synthesized nylon 66, the first purely synthetic fiber. Nylon 66 is so named because it is synthesized from two different monomers, each containing six carbon atoms.

In the synthesis of nylon 66, hexanedioic acid (adipic acid) and 1,6-hexanediamine (hexamethylenediamine) are dissolved in aqueous ethanol where they react to form a one-to-one salt called nylon salt. Nylon salt is then heated in an autoclave to 250°C where the internal pressure rises to about 15 atm. Under these conditions, —COO⁻ groups from adipic acid and —NH₃⁺ groups from hexamethylenediamine react with loss of H_2O to form amide groups.

Hexanedioic acid
(Adipic acid)

1,6-Hexanediamine
(Hexamethylenediamine)

Nylon salt

heat

Nylon 66

Nylon 66 formed under these conditions has a T_m of 250−260°C and has a molecular-weight range of 10,000 to 20,000 g/mol.

In the first stage of fiber production, crude nylon 66 is melted, spun into fibers, and cooled. Next, the melt-spun fibers are **cold-drawn** (drawn at room temperature) to about four times their original length to increase their degree of crystallinity. As the fibers are drawn, individual polymer molecules become oriented in the direction of the fiber axis, and hydrogen bonds form between carbonyl oxygens of one chain and amide hydrogens of another chain (Figure 29.3). The effects of orientation of polyamide molecules on the physical properties of the fiber are dramatic; both tensile strength and stiffness are increased markedly. Cold-drawing is an important step in the production of most synthetic fibers.

Figure 29.3
The structure of cold-drawn nylon 66. Hydrogen bonds between adjacent polymer chains provide additional tensile strength and stiffness to the fibers.

The current raw material base for the production of nylon 66 is benzene, which is derived almost entirely from catalytic cracking and reforming of petroleum. Catalytic reduction of benzene to cyclohexane followed by catalyzed air oxidation gives a mixture of cyclohexanol and cyclohexanone. Oxidation of this mixture by nitric acid gives adipic acid.

| Benzene | Cyclohexane | Cyclohexanol   Cyclohexanone | Hexanedioic acid (Adipic acid) |

Adipic acid, in turn, is a starting material for the synthesis of hexamethylenediamine. Treatment of adipic acid with ammonia gives an ammonium salt, which when heated, gives adipamide. Catalytic reduction of adipamide gives hexamethylenediamine. Thus, carbon sources for the production of nylon 66 are derived entirely from petroleum, which unfortunately is not a renewable resource.

Ammonium hexanedioate (Ammonium adipate)   **Hexanediamide (Adipamide)**

1,6-Hexanediamine (Hexamethylenediamine)

Bulletproof vests have a thick layer of Kevlar.

Aramid
A polyaromatic amide; a polymer in which the monomer units are an aromatic diamine and an aromatic dicarboxylic acid.

The nylons are a family of polymers, the members of which have subtly different properties that suit them to one use or another. The two most widely used members of this family are nylon 66 and nylon 6. Nylon 6 is so named because it is synthesized from caprolactam, a six-carbon monomer. In the synthesis of nylon 6, caprolactam is partially hydrolyzed to 6-aminohexanoic acid and then heated to 250°C to bring about polymerization. Nylon 6 is fabricated into fibers, brush bristles, rope, high-impact moldings, and tire cords.

Caprolactam   **Nylon 6**

Based on extensive research into relationships between molecular structure and bulk physical properties, scientists at DuPont reasoned that a polyamide containing aromatic rings would be stiffer and stronger than either nylon 66 or nylon 6. In early 1960, DuPont introduced Kevlar, a polyaromatic amide (**aramid**) fiber synthesized from terephthalic acid and *p*-phenylenediamine.

1,4-Benzenedicarboxylic acid (Terephthalic acid)   **1,4-Benzenediamine (*p*-Phenylenediamine)**   **Kevlar**

One of the remarkable features of Kevlar is its light weight compared with other materials of similar strength. For example, a 3-in. cable woven of Kevlar has a strength equal to that of a similarly woven 3-in. steel cable. Whereas the steel cable weighs about 20 lb/ft, the Kevlar cable weighs only 4 lb/ft. Kevlar now finds use in such articles as anchor cables for offshore drilling rigs and reinforcement fibers for automobile tires. Kevlar is also woven into a fabric that is so tough that it can be used for bulletproof vests, jackets, and raincoats.

B. Polyesters

Polyester
A polymer in which each monomer unit is joined to the next by an ester bond, as, for example, poly(ethylene terephthalate).

Recall that, in the early 1930s, Carothers and his associates had concluded that polyester fibers from aliphatic dicarboxylic acids and ethylene glycol were not suitable for textile use because their melting points are too low. Winfield and Dickson at the Calico Printers Association in England further investigated polyesters in the 1940s and reasoned that a greater resistance to rotation in the polymer backbone would stiffen the polymer, raise its melting point, and thereby lead to a more acceptable polyester fiber. To create stiffness in the polymer chain, they used 1,4-benzenedicarboxylic acid (terephthalic acid). Polymerization of this aromatic dicarboxylic acid with ethylene glycol gives poly(ethylene terephthalate), abbreviated PET (also PETE).

1,4-Benzenedicarboxylic acid (Terephthalic acid) + **1,2-Ethanediol** (Ethylene glycol) $\xrightarrow{\text{heat}}$ **Poly(ethylene terephthalate)** (Dacron, Mylar) + $2n\ H_2O$

The crude polyester can be melted, extruded, and then cold-drawn to form the textile fiber Dacron polyester, outstanding features of which are its stiffness (about four times that of nylon 66), very high strength, and remarkable resistance to creasing and wrinkling. Because the early Dacron polyester fibers were harsh to the touch owing to their stiffness, they were usually blended with cotton or wool to make acceptable textile fibers. Newly developed fabrication techniques now produce less harsh Dacron polyester textile fibers. PET is also fabricated into Mylar films and recyclable plastic beverage containers.

Ethylene glycol for the synthesis of PET is obtained by air oxidation of ethylene to ethylene oxide (Section 11.8A) followed by hydrolysis to the glycol (Section 11.9A). Ethylene is, in turn, derived entirely from cracking either petroleum or ethane derived from natural gas (Section 2.9A). Terephthalic acid is obtained by oxidation of *p*-xylene, an aromatic hydrocarbon obtained along with benzene and toluene from catalytic cracking and reforming of naphtha and other petroleum fractions (Section 2.9B).

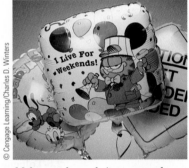

Mylar can be made into extremely strong films. Because the film has very tiny pores, it is used for balloons that can be inflated with helium; the helium atoms diffuse only slowly through the pores of the film.

$$CH_2{=}CH_2 \xrightarrow[\text{catalyst}]{O_2} CH_2{-}CH_2 \xrightarrow{H^+,\ H_2O} HOCH_2CH_2OH$$

Ethylene → **Oxirane** (Ethylene oxide) → **1,2-Ethanediol** (Ethylene glycol)

$$H_3C{-}\bigcirc{-}CH_3 \xrightarrow[\text{catalyst}]{O_2} HOC{-}\bigcirc{-}COH$$

p-Xylene → **Terephthalic acid**

C. Polycarbonates

Polycarbonate
A polyester in which the carboxyl groups are derived from carbonic acid.

Polycarbonates, the most familiar of which is Lexan, are a class of commercially important engineering polyesters. In the production of Lexan, the disodium salt of bisphenol A (Problem 22.22) reacts with phosgene to form the polymer.

The chemical equation at top:

n Na$^+$O—[aromatic ring]—C(CH$_3$)$_2$—[aromatic ring]—O$^-$Na$^+$ + n Cl—C(=O)—Cl → ⟮[aromatic ring]—C(CH$_3$)$_2$—[aromatic ring]—O—C(=O)—O⟯$_n$ + $2n$ NaCl

Disodium salt of bisphenol A   **Phosgene**   **Lexan** (a polycarbonate)

Lexan is a tough, transparent polymer with high-impact and tensile strengths, and it retains its properties over a wide temperature range. It has found significant use in sporting equipment, such as bicycle, football, motorcycle, and snowmobile helmets as well as hockey and baseball catchers' face masks. In addition it is used to make light, impact-resistant housings for household appliances and automobile and aircraft equipment and to manufacture safety glass and unbreakable windows.

A polycarbonate hockey mask.

© Cengage Learning/Charles D. Winters

D. Polyurethanes

Polyurethane
A polymer containing the —NHCO$_2$ groups as a repeating unit.

A urethane, or carbamate, is an ester of carbamic acid, H$_2$NCOOH. Carbamates are most commonly prepared by treatment of an isocyanate with an alcohol.

$$RN=C=O + R'OH \longrightarrow RNHCOR'$$

     An isocyanate   A carbamate

Polyurethanes consist of flexible polyester or polyether units (blocks) alternating with rigid urethane units (blocks). The rigid urethane blocks are derived from a diisocyanate, commonly a mixture of 2,4- and 2,6-toluene diisocyanate. The more flexible blocks are derived from low-molecular-weight (MW 1000–4000) polyesters or polyethers with —OH groups at each end of the polymer chain. Polyurethane fibers are fairly soft and elastic and have found use as Spandex and Lycra, the "stretch" fabrics used in bathing suits, leotards, and undergarments.

n O=C=N—[toluene ring with CH$_3$]—N=C=O + n HO—polymer—OH ⟶ ⟮C(=O)NH—[toluene ring with CH$_3$]—NHC(=O)O—polymer—O⟯$_n$

2,6-Toluene diisocyanate   Low-molecular-weight polyester or polyether   A polyurethane

Polyurethane foams for upholstery and insulating materials are made by adding small amounts of water during polymerization. Water reacts with isocyanate groups to produce gaseous carbon dioxide, which then acts as the foaming agent.

$$RN=C=O + H_2O \longrightarrow \left[RNH—C(=O)—OH \right] \longrightarrow RNH_2 + CO_2$$

    An isocyanate   A carbamic acid (unstable)

E. Epoxy Resins

Epoxy resins are materials prepared by a polymerization in which one monomer contains at least two epoxy groups. Within this range, there are a large number of polymeric materials possible, and epoxy resins are produced in forms ranging from low-viscosity liquids to high-melting solids. The most widely used epoxide monomer is the diepoxide prepared by treatment of one mole of bisphenol A (Problem 22.22) with two moles of epichlorohydrin (Section 11.10). To prepare the following epoxy resin, the diepoxide monomer is treated with 1,2-ethanediamine (ethylene diamine). This component is usually called the "catalyst" in the two-component formulations that can be bought in any hardware store; it is also the component with the pungent smell. It is not a catalyst but a reagent.

A diepoxide   A diamine

An epoxy resin

Epoxy resins are widely used as adhesives and insulating surface coatings. They have good electrical insulating properties, which leads to their use for encapsulating electrical components ranging from integrated circuit boards to switch coils and insulators for power transmission systems. They are also used as composites with other materials, such as glass fiber, paper, metal foils, and other synthetic fibers to create structural components for jet aircraft, rocket motor casings, and so on.

Example 29.2

Write a mechanism for the acid-catalyzed polymerization of 1,4-benzenediisocyanate and ethylene glycol. To simplify your mechanism, consider only the reaction of one —NCO group with one —OH group.

1,4-Benzenediisocyanate   **Ethylene glycol**   A polyurethane

Solution

A mechanism is shown in three steps. Proton transfer in Step 1 from the acid, HA, to nitrogen followed by addition of ROH to the carbonyl carbon in Step 2 gives an oxonium ion. Proton transfer in Step 3 from the oxonium ion to A$^-$ gives the carbamate ester.

Problem 29.2

Write the repeating unit of the polymer formed from the following reaction and propose a mechanism for its formation.

A diepoxide   A diamine

Medical science has advanced very rapidly in the last few decades. Some procedures considered routine today, such as organ transplantation and the use of lasers in surgery, were unimaginable 60 years ago. As the technological capabilities of medicine have grown, the demand for synthetic materials that can be used inside the body has increased as well. Polymers have many of the characteristics of an ideal biomaterial: they are lightweight and strong, are inert or biodegradable

Whereas most medical uses of polymeric materials require biostability, applications that use the biodegradable nature of some macromolecules have been developed. An example is the use of poly(glycolic acid) and glycolic acid/lactic acid copolymers as absorbable sutures, which go under the trade name Lactomer.

Traditional suture materials such as catgut must be removed by a health care specialist after they have served their purpose. Stitches of Lactomer, however,

Glycolic acid

Lactic acid

copolymerization

$-nH_2O$

Lactomer
A copolymer of
poly(glycolic acid)-
poly(lactic acid)

depending on their chemical structure, and have physical properties (softness, rigidity, elasticity) that are easily tailored to match those of natural tissues. Carbon-carbon backbone polymers are degradation resistant and are used widely in permanent organ and tissue replacements.

are hydrolyzed slowly over a period of approximately two weeks, and by the time, the torn tissues have fully healed, the stitches have fully degraded, and no suture removal is necessary. Glycolic and lactic acids formed during hydrolysis of the stitches are metabolized and excreted by existing biochemical pathways.

F. Thermosetting Polymers

Thermosetting polymers are composed of long chains that are cross linked by covalent bonds. In effect, a thermosetting polymer is one giant molecule. The first thermosetting polymer was produced by Leo Baekeland (1863–1944) in 1907 by reacting phenol with formaldehyde to form the following three-dimensional structure. The product, known as Bakelite, is a good electrical insulator.

Phenol   **Formaldehyde**

Bakelite

In the preparation of a thermoset, one of the monomers must be trifunctional. In the case of Bakelite, the trifunctional monomer is phenol. Alkyl thermosets are polyesters of an organic diacid, HOOC—R—COOH, and a trialcohol such as glycerol. Urea-formaldehyde thermosets are polyamides in which one molecule of urea, H_2N—CO—NH_2, can condense with up to four molecules of formaldehyde.

Manufacture of thermosets begins with a fluid mixture of the two monomers. The fluid is first shaped and then polymerized, either by heating or by being mixed with an initiator. The product of the polymerization is a network of covalently bonded atoms that is a solid, even at high temperatures. When heated to high temperatures, thermoset polymers char and decompose, but they do not melt.

29.6 | Chain-Growth Polymerizations

From the perspective of the chemical industry, the single most important reaction of alkenes is **chain-growth polymerization**, a type of polymerization in which monomer units are joined together without loss of atoms. An example is the formation of polyethylene from ethylene.

$$n\ CH_2{=}CH_2 \xrightarrow{\text{catalyst}} {+}CH_2CH_2{+}_n$$

Ethylene　　　　　**Polyethylene**

Chain-growth polymerization
A polymerization that involves sequential addition reactions, either to unsaturated monomers or to monomers possessing other reactive functional groups.

The mechanism of this type of polymerization differs greatly from the mechanism of step-growth polymerizations. In the latter, all monomers plus the polymer endgroups possess equally reactive functional groups, allowing for all possible combinations of reactions to occur, including monomer with monomer, dimer with dimer, and so forth. In contrast, chain-growth polymerizations involve endgroups possessing reactive intermediates that react with a monomer only. The reactive intermediates used in chain-growth polymerizations include radicals, carbanions, carbocations, and organometallic complexes.

The number of monomers that undergo chain-growth polymerizations is large and includes such compounds as alkenes, alkynes, allenes, isocyanates, and cyclic compounds, such as lactones, lactams, ethers, and epoxides. We concentrate on the chain-growth polymerizations of ethylene and substituted ethylenes and show how these compounds can be polymerized by radical, cation, anion, and organometallic-mediated mechanisms.

An alkene

Table 29.1 lists several important polymers derived from ethylene and substituted ethylenes along with their common names and most important uses.

A. Radical Chain-Growth Polymerizations

Among the initiators used for radical chain-growth polymerizations are diacyl peroxides, such as dibenzoyl peroxide, which decompose as shown upon heating. In the first step, homolytic cleavage of the weak O—O peroxide bond yields two acyloxy radicals. Each acyloxy radical then decomposes to form an aryl radical and CO_2.

Dibenzoyl peroxide　　　**Benzoyloxy radical**　　　**Phenyl radical**

The low thermal conductivity of polystyrene makes it a good insulating material.

© Cengage Learning/Charles D. Winters

Table 29.1	Polymers Derived from Ethylene and Substituted Ethylenes	
Monomer Formula	**Common Name**	**Polymer Name(s) and Common Uses**
$CH_2{=}CH_2$	Ethylene	**Polyethylene**, **Polythene**; break-resistant containers and packaging materials
$CH_2{=}CHCH_3$	Propylene	**Polypropylene**, **Herculon**; textile and carpet fibers
$CH_2{=}CHCl$	Vinyl chloride	**Poly(vinyl chloride)**, **PVC**; construction tubing
$CH_2{=}CCl_2$	1,1-Dichloroethylene	**Poly(1,1-dichloroethylene)**, **Saran Wrap**; food packaging
$CH_2{=}CHCN$	Acrylonitrile	**Polyacrylonitrile**, **Orlon**; acrylics and acrylates
$CF_2{=}CF_2$	Tetrafluoroethylene	**Poly(tetrafluoroethylene)**, **PTFE**; **Teflon**, nonstick coatings
$CH_2{=}CHC_6H_5$	Styrene	**Polystyrene**, **Styrofoam**; insulating materials
$CH_2{=}CHCOOCH_2CH_3$	Ethyl acrylate	**Poly(ethyl acrylate)**; latex paints
$CH_2{=}CCOOCH_3$ $\quad\mid$ $\quad CH_3$	Methyl methacrylate	**Poly(methyl methacrylate)**, **Lucite**, **Plexiglass**; glass substitutes

Another common class of initiators used in radical polymerizations is azo compounds, such as azoisobutyronitrile (AIBN), which decompose upon heating or by the absorption of UV light to produce alkyl radicals and nitrogen gas.

Azoisobutyronitrile (AIBN)   Alkyl radicals

The chain initiation, propagation, and termination steps for radical polymerization of a substituted ethylene monomer are shown for the monomer $RCH{=}CH_2$. Dissociation of the initiator produces a radical that reacts with the double bond of a monomer. Once initiated, the chains continue to propagate through successive additions of monomers.

Mechanism	Radical Polymerization of a Substituted Ethylene

Step 1: Initiation: radicals form from nonradical compounds.

$$In{-}In \xrightarrow{\text{heat or light}} 2\ In\cdot$$

Step 2: Propagation: reaction of a radical and a molecule gives a new radical.

Step 3: Chain termination: radicals are destroyed.

In radical reactions, the chain termination involves combination of radicals to produce a nonradical molecule or molecules. One common termination step is **radical coupling** to form a new carbon-carbon bond linking two growing polymer chains. This type of termination step is a diffusion-controlled process that occurs without an activation energy barrier. Another common termination process is **disproportionation**, which involves the abstraction of a hydrogen atom from the beta position to the propagating radical of one chain by the radical endgroup of another chain. This process results in two dead chains, one terminated in an alkyl group and the other in an alkenyl group.

Radical reactions with double bonds almost always give the more stable (more substituted) radical. Because additions are biased in this fashion, the polymerizations of vinyl monomers tend to yield polymers with head-to-tail linkages. Vinyl polymers made by radical processes generally have no more than 1–2% head-to-head linkages.

Because organic radicals are highly reactive species, it is not surprising that radical polymerizations are often complicated by unwanted side reactions. A frequently observed side reaction is hydrogen abstraction by the radical endgroup from a growing polymer chain, a solvent molecule, or another monomer. These side reactions are called **chain-transfer reactions** because the activity of the endgroup is "transferred" from one chain to another.

Chain transfer is illustrated by radical polymerization of ethylene. Polyethylene formed by radical polymerization exhibits a number of butyl branches on the polymer main chain. These four-carbon branches are generated in a "back-biting" chain-transfer reaction in which the radical endgroup abstracts a hydrogen from the fourth carbon back (the fifth carbon in the chain). Abstraction of this hydrogen is particularly facile because the transition state associated with the process can adopt a conformation like that of a chair cyclohexane. Continued polymerization of monomer from this new radical center leads to branches four carbons long.

A six-membered transition
state leading to
1,5-hydrogen abstraction

As a result of these various abstraction reactions, polymers synthesized by radical processes can have highly branched structures. The number of butyl branches depends on the relative stability of the propagating-radical endgroup and varies depending on the polymer. Polyethylene chains propagate through highly reactive primary radicals, which tend to be susceptible to 1,5-hydrogen abstraction reactions; these polymers typically have 15 to 30 branches per 500 monomer units. In contrast, polystyrene chains propagate through substituted benzyl radicals, which are stabilized by delocalization of the unpaired electron into the aromatic ring. These stabilized

Disproportionation
A termination process that involves the abstraction of a hydrogen atom from the beta position of the propagating radical of one chain by the radical endgroup of another chain.

Chain-transfer reaction
The transfer of reactivity of an endgroup from one chain to another during a polymerization.

The influence of chemical structure on the properties of an organic compound is clearly seen in the electrical conducting properties of certain organic polymers. Most organic polymers are insulators. For example, polytetrafluoroethylene with the repeating unit —CF_2CF_2— or poly(vinyl chloride) with the repeating unit —CH_2CHCl— have conductivities of 10^{-18} S/cm. On the other end of the scale, the conductivity of copper is almost 10^6 S/cm.

Can organic polymers approach the conductivity of copper? Research carried out over the last 20 years shows that the answer is yes. When acetylene is passed through a solution containing certain transition metal catalysts, it can be polymerized to a shiny film of polyacetylene.

Polyacetylene

By itself, polyacetylene is not a conductor. However, by a process called doping, which involves introducing small amounts of electron-donating or electron-accepting compounds, it is possible to produce a polyacetylene that shows a conductivity of 1.5×10^5 S/cm.

The purpose of the doping agent is either to remove electrons from the π system (p-doping) or add electrons to the π system (n-doping). A p-doped polyacetylene can be represented as a conjugated polyalkene chain containing positively charged carbons at several points along the chain.

A p-doped polyacetylene

We can think of the positive charge as a defect that can move to the left or to the right along the polymer chain, thus giving rise to conductivity.

In crude polyacetylene, the polymer chains are jumbled, pointing in all directions. However, by stretching the film, the chains can be made to line up in a more ordered fashion. The conductivity of doped and oriented polyacetylene chains is greater along the direction of the chain than it is perpendicular to the chain. This result suggests that it is much easier for electrons to travel along a chain than to hop from one chain to the next.

Applications for conducting organic polymers are beginning to be developed. A rechargeable battery with electrodes of p-doped and n-doped polyacetylene already has been produced. Given the atomic weight of carbon, organic polymer batteries should be lighter than nickel-cadmium or lead-acid batteries. Weight is an important consideration for battery powered electric cars. In addition, many metals used in today's batteries (mercury, nickel, lead) are toxic. If research leads to practical organic batteries, waste disposal problems could be considerably lessened.

Figure 29.4
Fabrication of a LDPE film.

radicals are less likely to undergo hydrogen abstraction reactions. Polystyrene typically exhibits only one branch per 4000 to 10,000 monomer units.

The first commercial process for ethylene polymerization used peroxide catalysts at temperatures of 500°C and pressures of 1000 atm and produced a soft, tough polymer known as low-density polyethylene (LDPE). At the molecular level, chains of LDPE are highly branched owing to chain-transfer reactions. Because this extensive chain branching prevents polyethylene chains from packing efficiently, LDPE is largely amorphous and transparent, with only a small amount of crystallites of a size too small to scatter light. LDPE has a density between 0.91 and 0.94 g/cm³ and a melt transition temperature (T_m) of about 108°C. Because its T_m is only slightly above 100°C, it cannot be used for products that will be exposed to boiling water.

Approximately 65% of all low-density polyethylene is used for the manufacture of films. Fabrication of LDPE films is done by a blow-molding technique illustrated in Figure 29.4. A tube of molten LDPE along with a jet of compressed air is forced through an opening and blown into a giant, thin-walled bubble. The film is then cooled and taken up onto a roller. This double-walled film can be slit down the side to give LDPE film or it can be sealed at points along its length to make LDPE bags. LDPE film is inexpensive, which makes it ideal for trash bags and for packaging for such consumer items as baked goods, vegetables, and other produce.

B. Ziegler-Natta Chain-Growth Polymerizations

An alternative method for polymerization of alkenes, which does not involve radicals, was developed by Karl Ziegler of Germany and Giulio Natta of Italy in the 1950s. For their pioneering work, they were awarded the 1963 Nobel Prize in chemistry. The early Ziegler-Natta catalysts were highly active, heterogeneous catalysts composed of a $MgCl_2$ support, a Group 4B transition metal halide, such as $TiCl_4$, and an alkylaluminum compound, such as $Al(CH_2CH_3)_2Cl$. These catalysts bring about polymerization of ethylene and propylene at 1 to 4 atm and at temperatures as low as 60°C. Polymerizations under these conditions do not involve radicals.

$$n CH_2{=}CH_2 \xrightarrow[\text{MgCl}_2]{\text{TiCl}_4/\text{Al(CH}_2\text{CH}_3)_2\text{Cl}} \left(\!\!\bigvee\!\!\right)_n$$

Ethylene   **Polyethylene**

The active catalyst in a Ziegler-Natta polymerization is thought to be an alkyltitanium compound, which is formed by alkylation of the titanium halide by $Al(CH_2CH_3)_2Cl$ on the surface of a $MgCl_2/TiCl_4$ particle. Once formed, this species repeatedly inserts ethylene into the titanium-carbon bond to yield polyethylene.

Mechanism Ziegler-Natta Catalysis of Ethylene Polymerization

Step 1: A titanium-ethyl bond forms.

Diethylaluminum chloride

Step 2: Ethylene inserts into the titanium-carbon bond.

Over 2.5×10^{11} kg of polyethylene are produced worldwide every year using optimized Ziegler-Natta catalysts, and large-scale reactors can yield up to 1.25×10^5 kg of polyethylene per hour. Production of polymer at this scale is partly attributable to the mild conditions required for a Ziegler-Natta polymerization and the fact that the polymer obtained has substantially different physical and mechanical properties from that obtained by radical polymerization. Polyethylene from Ziegler-Natta systems, termed high-density polyethylene (HDPE), has a higher density (0.96 g/cm^3) and T_m ($133°C$) than low-density polyethylene, is three to ten times stronger, and is opaque rather than transparent. This added strength and opacity is the result of a much lower degree of chain branching and the resulting higher degree of crystallinity of HDPE compared with LDPE.

Approximately 45% of all HDPE used in the United States is blow molded. In blow molding, a short length of HDPE tubing is placed in an open die [Figure 29.5(a)], and the die is closed, sealing the bottom of the tube. Compressed air is then forced into the hot polyethylene/die assembly, and the tubing is literally blown up to take the

Polyethylene films are produced by extruding the molten plastic through a ring-like gap and inflating the film into a balloon.

© James Holmes/Zeckor/Photo Researchers, Inc.

Figure 29.5

Blow molding of a HDPE
container.

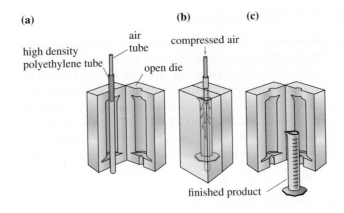

finished product

shape of the mold [Figure 29.5(b)]. After cooling, the die is opened [Figure 29.5(c)], and there is the container!

Even greater improvements in properties of HDPE can be realized through special processing techniques. In the melt state, HDPE chains adopt random coiled conformations similar to those of cooked spaghetti. Engineers have developed special extrusion techniques that force the individual polymer chains of HDPE to uncoil and adopt an extended linear conformation. These extended chains then align with one another to form highly crystalline materials. HDPE processed in this fashion is stiffer than steel and has approximately four times the tensile strength of steel! Because the density of polyethylene (≈ 1.0 g/cm^3) is considerably less than that of steel (8.0 g/cm^3), these comparisons of strength and stiffness are even more favorable if they are made on a weight basis.

In recent years, there have been several important advances made in catalysts used in Ziegler-Natta-type polymerizations. One of the most important has been the discovery of soluble complexes that catalyze the polymerization of ethylene and propylene at extraordinary rates. Because these new homogeneous catalysts are substantially different in structure from the early Ziegler-Natta systems, these polymerizations are referred to as **coordination polymerizations**. Catalysts for coordination polymerizations are frequently formed by allowing bis-(cyclopentadienyl) dimethylzirconium, [Cp$_2$Zr(CH$_3$)$_2$], to react with methaluminoxane (MAO). MAO is a complex mixture of methylaluminum oxide oligomers, [—(CH$_3$)AlO—)$_n$], formed by allowing trimethylaluminum to react with small amounts of water. It is thought that MAO activates the zirconium by abstracting a methyl anion to form a zirconium cation that is the active polymerization catalyst.

Mechanism Homogeneous Catalysis for Ziegler-Natta Coordination Polymerization

Step 1: The zirconium catalyst is activated.

Bis(cyclopentadienyl)-dimethylzirconium   **MAO**   A zirconium cation (the active form of the catalyst)

Step 2: Ethylene inserts into the zirconium-carbon bond.

Some of these coordination-polymerization catalysts polymerize up to 20,000 ethylene monomer units per second, a rate reached only by enzyme-catalyzed biological reactions. Another important characteristic of these catalysts is that they show high reactivity toward 1-alkenes, allowing the formation of copolymers, such as that of ethylene and 1-hexene.

Ethylene   **1-Hexene**   **Ethylene-(1-hexene) copolymer**

Copolymers of this type with these moderate length branches (C_4, C_6, and so on) are called linear low-density polyethylene, or LLDPE. These are useful materials because they have many of the properties of LDPE made from radical reactions but are formed at the substantially milder conditions associated with Ziegler-Natta polymerizations.

C. Stereochemistry and Polymers

Thus far, we have written the formula of a substituted ethylene polymer in the following manner and have not been concerned with the configuration of each chiral center along the chain.

Nevertheless, the relative configurations of these chiral centers are important in determining the properties of a polymer. Polymers with identical configurations at all chiral centers along the chain are called **isotactic** polymers. Those with alternating configurations are called **syndiotactic** polymers, and those with completely random configurations are called **atactic** polymers (Figure 29.6).

In general, the more stereoregular the chiral centers are, that is, the more highly isotactic or highly syndiotactic the polymer is, the more crystalline it is. A random placement of the substituents, such as in atactic materials, results in a polymer that cannot pack well and is usually highly amorphous. Atactic polystyrene, for

Isotactic polymer
A polymer with identical configurations (either all *R* or all *S*) at all chiral centers along its chain, as, for example, isotactic polypropylene.

Syndiotactic polymer
A polymer with alternating *R* and *S* configurations at the chiral centers along its chain, as, for example, syndiotactic polypropylene.

Atactic polymer
A polymer with completely random configurations at the chiral centers along its chain, as, for example, atactic polypropylene.

Isotactic polymer
(identical configurations)

Syndiotactic polymer
(alternating configurations)

Figure 29.6
Relative configurations of chiral centers in polymers with different tacticities.

Atactic polymer
(random configurations)

Anionic polymerizations are most common for monomers substituted with electron-withdrawing groups.

Styrene

Alkyl methacrylates COOR

Alkyl acrylates COOR

Acrylonitrile CN

Alkyl cyanoacrylates CN COOR

Cationic polymerizations are most common for monomers substituted with electron-donating groups.

Styrene

Isobutylene

Vinyl ethers OR

Vinyl thioethers SR

example, is an amorphous glass, whereas isotactic polystyrene is a crystalline fiber-forming polymer with a high melt transition. The control over the relative configuration, or tacticity, along a polymer backbone is, therefore, an area of considerable interest in modern polymer synthesis.

D. Ionic Chain-Growth Polymerizations

Chain-growth polymers can also be synthesized using reactions that rely on either anionic or cationic species in the propagation steps. The choice of ionic procedure depends greatly on the electronic nature of the monomers to be polymerized. Vinyl monomers with electron-withdrawing groups, which stabilize carbanions, are used in anionic polymerizations, whereas vinyl monomers with electron-donating groups, which stabilize cations, are used in cationic polymerizations (Table 29.2).

Styrene is conspicuous among the monomers given in Table 29.2 because it can be polymerized using either anionic or cationic techniques as well as radical techniques. This characteristic particular to styrene is attributable to the fact that the phenyl group can stabilize cationic, anionic, and radical benzylic intermediates.

Resonance stabilization of a benzylic cation endgroup

Resonance stabilization of a benzylic anion endgroup

Anionic Polymerizations

Anionic polymerizations can be initiated by addition of a nucleophile to an activated alkene. The most common nucleophiles used for this purpose are metal alkyls, such as methyl- or *sec*-butyllithium. The newly formed carbanion then acts as a nucleophile and adds to another monomer unit, and the propagation continues.

Step 1: Make a new bond between a nucleophile and an electrophile Polymerization is initiated by addition of a nucleophile, shown here as a carbanion derived from an organolithium compound, to an activated carbon-carbon double bond to give a carbanion.

Step 2: Make a new bond between a nucleophile and an electrophile This carbanion adds to the activated double bond of a second alkene molecule to give a dimer.

Step 3: Make a new bond between a nucleophile and an electrophile Chain growth continues to give a polymer

A dimer   A polymer

An alternative method for the initiation of anionic polymerizations involves a one-electron reduction of the monomer by lithium or sodium to form a radical anion. The radical anion thus formed is either further reduced to form a dianion or dimerizes to form a dimer dianion.

Step 1: A one-electron reduction of the diene by lithium metal gives a radical anion.

Step 2: One-electron reduction of this radical anion gives a monomer dianion.

Step 3: Alternatively, radical coupling gives a dimer dianion.

Butadiene

A radical anion   A monomer dianion

radical coupling

A dimer dianion

In either case, a single initiator can now propagate chains from both ends, by virtue of its two active endgroup carbanions. These reactions are heterogeneous and involve transfer of the electron from the surface of the metal. To improve the efficiency of this process, soluble reducing agents such as sodium naphthalide are used. Sodium undergoes electron-transfer reactions with extended aromatic compounds, such as naphthalene, to form soluble radical anions.

Naphthalene   **Sodium naphthalide**
(a radical anion)

The naphthalide radical anion is a powerful reducing agent. For example, styrene undergoes a one-electron reduction to form the styryl radical anion, which couples to form a dianion. The latter then propagates polymerization at both ends, growing chains in two directions simultaneously.

Styrene   Styryl radical anion

A distyryl dianion

Polystyrene

The propagation characteristics of anionic polymerizations are similar to those of radical polymerizations, but with the important difference that many of the chain-transfer and chain-termination reactions that plague radical processes are absent. Furthermore, because the propagating chain ends carry the same charge, bimolecular coupling and disproportionation reactions are also averted. An interesting set of circumstances arises when chain-transfer and chain-termination steps are no longer significant. Under these conditions, polymer chains are initiated and continue to grow until either all the monomer is consumed or some external agent is added to terminate the chains. Polymerizations of this type are called **living polymerizations** because they will restart if more monomer is added after it is initially consumed.

The absence of chain-transfer and chain-termination steps in living polymerizations has far-reaching consequences. One of the most visible of these is in the area of molecular-weight control. The molecular weight of a polymer originating from living polymerizations is determined directly by the monomer-to-initiator ratio. It is, therefore, relatively easy to obtain polymers of a well-defined size simply by controlling the stoichiometry of the reagents. In contrast, the average sizes of polymer chains formed from nonliving, chain-growth processes (radical, Ziegler-Natta, and so on) vary from system to system and are determined by the ratio of the rate of propagation to the rate of termination. In most cases, precise control over the molecular weight of the product obtained in nonliving systems is not possible because it is very difficult to change one of the rates involved without affecting the other.

After consumption of the monomer under living, anionic conditions, electrophilic terminating agents can be added to functionalize the chain ends. Examples of terminating reagents include CO_2 and ethylene oxide, which, after protonation, form carboxylic acid and alcohol-terminated chains, respectively.

Living polymer
A polymer chain that continues to grow without chain-termination steps until either all of the monomer is consumed or some external agent is added to terminate the chain. The polymer chains will continue to grow if more monomer is added.

Used in everything from model planes to passenger planes, Superglue is one of the best-known modern glues. Indeed, anyone who has unwittingly glued his or her fingers together while using the product can acknowledge its remarkable (and insidious) adhesive properties. The curing process that facilitates these properties is a chain polymerization reaction. The ingredient that gives Superglue its adhesive ability is methyl cyanoacrylate. This compound is just one member of a larger family of cyanoacrylates with the following general structure.

$$H_2C=C\begin{array}{c} C\equiv N \\ C=O \\ O \\ R \end{array}$$

Contrary to popular understanding, Superglue does not "air dry." In fact, cyanoacrylates cure (convert from liquid to solid) in the presence of weak nucleophiles such as water. Under normal conditions, a thin layer of water is present on almost all surfaces. This accounts for many unintended adhesions involving appendages and/or expensive tools! The curing process involves an

produces a new anion, which is also a weak nucleophile and adds to another monomer, ultimately creating the powerfully adhesive polymer chains of cured Superglue.

Superglue has a rich history. It begins during World War II when cyanoacrylates were discovered in an attempt to produce optically clear gun sights. Tests with Superglue's main ingredient failed for obvious reasons: the compound stuck to all instruments. Approximately ten years later, researchers at Eastman Kodak rediscovered cyanoacrylates, and Kodak first marketed them in 1958. It is said that researchers discovered Superglue's properties when they attempted to take the refractive index of the monomer. A refractometer, the instrument used for this purpose, has two prisms that come together on the liquid whose refractive index is to be determined. The prisms of the instrument were permanently glued together, and a new product was born!

Superglue has been improved since that time with stabilizers and other supplementary ingredients that reinforce its adhesive strength. In recent years, that same adhesive strength has captured attention in new fields. Medical-grade Superglues such as 2-octylcyanoacrylate and 2-butylcyanoacrylate are now commonly used as sutures in laceration repair. They have also proven effective in skin, bone, and cartilage grafts. Dentists

anionic chain polymerization reaction, which occurs as follows.

In the chain-initiating step, a weak nucleophile donates an electron pair to a cyanoacrylate monomer. The CH_2 group is highly electropositive as a result of the electron withdrawal by the cyano and ester groups, which makes this Michael reaction very easy. The addition

use cyanoacrylates in dental cements and fillings, and paleontologists use them to reorganize fragile fossil specimens. Superglue has many fascinating applications, all facilitated by the anionic chain polymerization reaction.

Based on a chemistry honors paper by Carrie Brubaker, UCLA.

Telechelic polymer
A polymer in which its growing chains are terminated by formation of new functional groups at both ends of its chains. These new functional groups are introduced by adding reagents, such as CO_2 or ethylene oxide, to the growing chains.

In a similar fashion, polymer chains with functional groups at both ends, called **telechelic polymers**, can be prepared by addition of these same reagents (CO_2, ethylene oxide, and so on) to solutions of chains with two active ends initiated by sodium naphthalide.

Example 29.3

Show how to prepare polybutadiene that is terminated at both ends with carboxylate groups.

Solution

Form a growing chain with two active endgroups by treatment of butadiene with two moles of lithium metal to form a dianion followed by addition of monomer units and formation of a living polymer. Cap the active endgroups with a carboxylate group by treatment of the living polymer with carbon dioxide.

Problem 29.3

Show how to prepare polybutadiene that is terminated at both ends with primary alcohol groups.

Cationic Polymerizations

Only alkenes with electron-donating substituents, such as alkyl, aryl, ether, thioether, and amino groups, undergo useful cationic polymerizations. The two most common methods of generating cationic initiators are (1) the reaction of a strong protic acid with an organic monomer and (2) the abstraction of a halide from the organic initiator by a Lewis acid. Cationic chain-growth polymerizations are generally effective only for monomers yielding relatively stable carbocations, that is monomers that form either 3° carbocations or cations stabilized by electron-donating groups, such as ether, thioether, or amino groups.

Initiation by protonation of an alkene requires the use of a strong acid with a nonnucleophilic anion to avoid 1,2-addition across the alkene double bond. Suitable acids with nonnucleophilic anions include HF/AsF_5 and HF/BF_3. In the following general equation, initiation is by proton transfer from $H^+BF_4^-$ to the alkene to form a tertiary carbocation, which then continues the cationic chain growth polymerization.

Mechanism Initiation of Cationic Polymerization of an Alkene by $HF \cdot BF_3$

Step 1: **Make a new bond between a π bond and an electrophile—add a proton**
Proton transfer from the $HF \cdot BF_3$ complex to the alkene gives a carbocation.

Step 2: **Make a new bond between a π bond and an electrophile** Propagation continues the polymerization.

The second common method for generating carbocations involves the reaction between an alkyl halide and a Lewis acid, such as BF_3, $SnCl_4$, $AlCl_3$, $Al(CH_3)_2Cl$, and $ZnCl_2$. When a trace of water is present, the mechanism of initiation using some Lewis acids is thought to involve protonation of the alkene.

2-Methylpropene
(Isobutylene)

In the absence of water, the Lewis acid removes a halide ion from the alkyl halide to form the initiating carbocation.

Mechanism Initiation of Cationic Polymerization of an Alkene by a Lewis Acid

Step 1: **Lewis acid–Lewis base reaction followed by break a bond to give stable molecules or ions** Reaction of the chloroalkane (a Lewis base) with tin(IV) chloride (a Lewis acid) gives a carbocation from which polymerization then proceeds.

2-Chloro-2-phenylpropane

The polymerization of alkenes then propagates by the electrophilic attack of the carbocation on the double bond of the alkene monomer. The regiochemistry of the addition is determined by the formation of the more stable (the more highly substituted) carbocation.

Example 29.4

Write a mechanism for the polymerization of 2-methylpropene (isobutylene) initiated by treatment of 2-chloro-2-phenylpropane with $SnCl_4$. Label the initiation, propagation, and termination steps.

Solution

Lewis acid-Lewis base reaction followed by break a bond to give stable molecules or ions Chain initiation: cations form from nonionic materials.

2-Chloro-2-phenylpropane

Make a new bond between a π bond and an electrophile Chain propagation: a cation and a molecule react to give a new cation.

2-Methylpropene

Chain termination: destruction of cations.

Problem 29.4

Write a mechanism for the polymerization of methyl vinyl ether initiated by 2-chloro-2-phenylpropane and SnCl$_4$. Label the initiation, propagation, and termination steps.

E. Ring-Opening Metathesis Polymerizations

During early investigations into the polymerizations of cycloalkenes by transition metal catalysts such as those used in Ziegler-Natta polymerizations, polymers of unexpected structures that contained the same number of double bonds as originally present in the monomers were formed. This process is illustrated by the polymerization of norbornene.

Norbornene

1,2-Addition polymer

ROMP polymer

If reaction had proceeded in the same manner as Ziegler-Natta polymerization of ethylene and substituted ethylenes (Section 29.6B), a 1,2-addition polymer would have been formed. What is formed, however, is an unsaturated polymer in which the number of double bonds in the polymer is the same as that in the monomers polymerized. This process is called **ring-opening metathesis polymerization**, or ROMP, after the olefin metathesis involving reaction of acyclic alkenes and nucleophilic carbene catalysts described in Section 24.5.

The fact that ROMP polymers are unsaturated requires that this polymerization proceed by a mechanism substantially different from that involved in polymerization of ethylene and substituted ethylenes by the same catalyst mixtures. Following lengthy and detailed studies, chemists discovered that ROMP involves the same metallacyclobutane species as in ring-closing alkene metathesis reactions (Section 24.5B). The intermediate metallacyclobutane derivative undergoes a ring-opening reaction to give a new substituted carbene. Repetition of these steps leads to the formation of the unsaturated polymer as illustrated here by ROMP of cyclopentene.

A metal carbene

ROMP polymer from cyclopentene

All steps in ROMP are reversible, and the reaction is driven in the forward direction by the release of ring strain that accompanies the opening of the ring. The reactivity of the following cycloalkenes toward ROMP decreases in this order.

| | 125(29.8) | 113(27) | 24.7(5.9) | 5.9(1.4) |

Ring strain
[kJ (kcal)/mol]

ROMP reactions are unique in that all the unsaturation present in the monomers is conserved in the polymeric product. This feature makes ROMP techniques especially attractive for the preparation of highly unsaturated, fully conjugated materials. One example is the direct preparation of polyacetylene by the ROMP technique through one of the double bonds of cyclooctatetraene. For further discussion of polyacetylene, see Chemical Connections: "Organic Polymers That Conduct Electricity" in this chapter.

Cyclooctatetraene   **Polyacetylene**

An important polymer in electrooptical applications is poly(phenylene vinylene) (PPV), which has alternating phenyl and vinyl groups. One of the routes to this polymer starts with a substituted bicyclooctadiene which is polymerized using ROMP techniques to form a soluble, processable polymer. Heating the processed polymer results in elimination of two equivalents of acetic acid, which aromatizes the six-membered ring and completes the conjugation.

A substituted
bicyclooctadiene

Poly(phenylene vinylene)

Polymers, in the form of plastics, are materials upon which our society is incredibly dependent. Durable and lightweight, plastics are probably the most versatile synthetic materials in existence; in fact, their current production in the United States exceeds that of steel. Plastics have come under criticism, however, for their role in the trash crisis. They comprise 21% of the volume and 8% of the weight of solid wastes, most of which is derived from disposable packaging and wrapping. Of the 2.5×10^7 kg of thermoplastic materials produced in 1993 in America, less than 2% was recycled.

Why aren't more plastics being recycled? The durability and chemical inertness of most plastics make them ideally suited for reuse. The answer to this question has more to do with economics and consumer habits than with technological obstacles. Because curbside pickup and centralized drop-off stations for recyclables are just now becoming common, the amount of used material available for reprocessing has traditionally been small. This limitation, combined with the need for an additional sorting and separation step, rendered the use of recycled plastics in manufacturing expensive compared with virgin materials. Until recently, consumers perceived products made from "used" materials as being inferior to new ones, so the market for recycled products has not been large. However, the increase in environmental concerns over the last few years has resulted in a greater demand for recycled products. As manufacturers adapt to satisfy this new market, plastic recycling will eventually catch up with the recycling of other materials, such as glass and aluminum.

Six types of plastics are commonly used for packaging applications. In 1988 manufacturers adopted recycling code numbers developed by the Society of the Plastics Industry. Because the plastics recycling industry still is not fully developed, only polyethylene terephthalate (PET) and high-density polyethylene (HDPE) are currently being recycled in large quantities, although outlets for the other plastics are being

These students are wearing jackets made from recycled PET soda bottles.

developed. Low-density polyethylene (LDPE), which accounts for about 40% of plastic trash, has been slow in finding acceptance with recyclers. Facilities for the reprocessing of poly(vinyl chloride) (PVC), polypropylene (PP), and polystyrene (PS) exist but are still rare.

The process for the recycling of most plastics is simple, with separation of the desired plastics from other contaminants the most labor-intensive step. PET soft drink bottles, for example, usually have a paper label, adhesive, and an aluminum cap that must be removed before the PET can be reused. The recycling process begins with hand or machine sorting, after which the bottles are shredded into small chips. An air cyclone removes paper and other lightweight materials, and any remaining labels and adhesives are eliminated with a detergent wash. The PET chips are then dried, and aluminum, the

Recycling Code	Polymer	Common Uses	Uses of Recycled Polymer
1 PET	**Poly(ethylene terephthalate)**	Soft drink bottles, household chemical bottles, films, textile fibers	Soft drink bottles, household chemical bottles, films, textile fibers
2 HDPE	**High-density polyethylene**	Milk and water jugs, grocery bags, bottles	Bottles, molded containers
3 PVC	**Poly(vinyl chloride)**	Shampoo bottles, pipes, shower curtains, vinyl siding, wire insulation, floor tiles, credit cards	Plastic floor mats
4 LDPE	**Low-density polyethylene**	Shrink wrap, trash and grocery bags, sandwich bags, squeeze bottles	Trash bags and grocery bags
5 PP	**Polypropylene**	Plastic lids, clothing fibers, bottle caps, toys, diaper linings	Mixed plastic components
6 PS	**Polystyrene**	Styrofoam cups, egg cartons, disposable utensils, packaging materials, appliances	Molded items such as cafeteria trays, rulers, Frisbees, trash cans, videocassettes
7	**All other plastics and mixed plastics**	Various	Plastic lumber, playground equipment, road reflectors

final contaminant, is removed electrostatically. The PET produced by this method is 99.9% free of contaminants and sells for about half the price of the virgin material. Unfortunately, plastics with similar densities cannot be separated with this technology, nor can plastics composed of several polymers be broken down into pure components. However, recycled mixed plastics can be molded into plastic lumber that is strong, durable, and graffiti-resistant.

An alternative to this process, which uses only physical methods of purification, is chemical recycling. Eastman Kodak salvages large amounts of its PET film scrap by a transesterification reaction. The scrap is treated with methanol in the presence of an acid catalyst to give ethylene glycol and dimethyl terephthalate.

These monomers are purified by distillation or recrystallization and used as feed stocks for the production of more PET film.

Poly(ethylene terephthalate)   **Ethylene glycol**   **Dimethyl terephthalate**
(PET)

Section 29.1 The Architecture of Polymers

- **Polymers** are long-chain molecules synthesized by linking **monomers** through chemical reactions. Types of polymer architecture include linear and branched chains, as well as those with comb, ladder, and star structures.
- The term **plastic** refers to any polymer that can be molded when hot and retains its shape when cooled.
  - **Thermoplastics** are polymers that can be melted and become sufficiently fluid that they can be molded into shapes that are retained when they are cooled.
  - **Thermosetting plastics** or **thermosets** can be molded when they are first prepared, but once they cool, they harden irreversibly owing to extensive covalent cross linking between chains.
- The properties of polymers are determined by the size and shape of their chains.

Section 29.2 Polymer Notation and Nomenclature

Problems: 29.1, 29.5, 29.6

- The structures of polymers are shown by drawing parentheses around the **repeat unit**, which is the smallest molecular fragment that contains all the nonredundant structural features of the chain.
  - The **average degree of polymerization** (the average number of repeat units per chain) is denoted by a subscript just outside the parentheses.
- Polymers are named by attaching the prefix *poly-* to the name of the monomer from which the polymer is derived.

Section 29.3 Molecular Weights of Polymers

- The **number average molecular weight**, M_n, is calculated by counting the number of polymer chains of a particular molecular weight, multiplying each number by the molecular weight of its chain, summing these values, and dividing by the total number of polymer chains.
- The **weight average molecular weight**, M_w, is calculated by recording the total weight of each chain of a particular length, summing these weights, and dividing by the total weight of the sample.
- The **polydispersity index** of a polymer is the ratio M_w / M_n.
  - If all the polymers are the same length, the polydispersity index is 1 and the sample is referred to as **monodisperse**. Synthetic polymers are never monodisperse unless they are purified.

Section 29.4 Polymer Morphology—Crystalline Versus Amorphous Materials

Problems: 29.2, 29.3, 29.14–29.26

- Polymers in the solid state tend to be composed of both ordered **crystalline domains** (crystallites) and disordered **amorphous domains**.
- The **melt transition**, T_m, of the polymer is the temperature at which the crystallites melt. As the degree of crystallinity increases, so does T_m.
- Amorphous domains have no long-range order and give rise to transparent soft materials that are referred to as glassy polymers.
  - On being heated, amorphous polymers are transformed from a hard glass to a soft, flexible, rubbery state at a temperature referred to as the **glass transition temperature**, T_g.
  - Rubber materials must have low T_g values to behave as **elastomers**, which are materials that return to their original shape upon distortion.

Section 29.5 Step-Growth Polymerizations

- Polymers in which chain growth occurs in a stepwise manner are called **step-growth** or **condensation polymers**.
  - Step-growth polymers are formed by reaction between difunctional molecules, with each new bond created in a separate step.

- At the initial stages of polymerization, monomers react with each other until they are used up.
- High-molecular-weight polymers appear only near the end of the reaction.

- **Nylons** are polyamides formed from a diacid and a diamine, or alternatively from an amino acid, and have use as fibers.
- **Polyesters** are derived from diacids and diols and have use as textile fibers such as Dacron.
- **Polycarbonates** such as Lexan are tough, transparent polymers with high tensile strength that are used for products ranging from sporting equipment to unbreakable windows.
- **Polyurethanes** consist of flexible polyester or polyether units alternating with rigid urethane (carbamate) blocks. Polyurethanes are used as flexible fibers such as Lycra or Spandex, as well as foaming materials.
- Epoxy resins are materials prepared by polymerization in which one monomer contains at least two epoxide groups. Epoxy resins are used as adhesives and insulating surface coatings.

Problems: 29.2, 29.7–29.23

Section 29.6 Chain-Growth Polymerizations

- **Chain-growth polymerization** is a type of polymerization in which monomer units are joined together without loss of atoms.
    - Once initiated, chain-growth polymerizations involve endgroups possessing reactive intermediates that react with a new monomer.
    - Because chain-growth polymers grow from the ends only and monomers do not react with themselves, the growth of polymer chains occurs linearly throughout the polymerization, unlike step-growth polymerizations.
    - Reactive intermediates used in chain-growth polymerizations include radicals, carbanions, carbocations, and organometallic complexes.
    - Monomers used for chain-growth polymerization include alkenes, alkynes, allenes, isocyanates, and cyclic compounds such as lactones, lactams, ethers, and epoxides.
- Low-density polyethylene (LDPE) can be made by radical polymerization of ethylene using peroxides as radical initiators. LDPE is soft and transparent and is made into films using a blow-molding technique.
- Polyethylene can also be made using metal catalysts such as the **Ziegler-Natta catalysts**, which can produce **high-density polyethylene (HDPE)**, which is stronger and more opaque than LDPE.
    - HDPE is used for making products such as dishes and water bottles.
    - Even better metal catalysts such as zirconium complexes have now been developed for polyethylene production.

Problems: 29.3, 29.4, 29.24–29.38

- Polymers can have chiral centers, and polymers with an identical configuration at all chiral centers are called **isotactic**, those with alternating configurations are called **syndiotactic**, and those with completely random configurations are called **atactic**.
    - The more stereoregular a polymer, the more crystalline it is.
- Anionic polymerizations of alkenes, which do not suffer from the termination problems of radical polymerizations, can be initiated by nucleophiles or one-electron reduction.
- Due to a lack of competing processes, **living polymerizations** are polymerizations that, once initiated, continue to grow at the ends of the chains until all the monomer is consumed, and adding more monomer initiates new reaction of the growing chains.
    - Living polymerizations provide good control over the size distribution of polymer chains.
- Bicyclic alkene monomers can be used in a polymerization called **ring-opening metathesis polymerization (ROMP)** using the same chemistry as alkene metathesis.

⟳WL **Questions and problems for this chapter are available in OWL. Use the OWL access card that came with this textbook to access assigned questions and problems for this chapter.**

1. Step-Growth Polymerization of a Dicarboxylic Acid and a Diamine Gives a Polyamide (Section 29.5A)

$$n\ \text{HOC}-\text{M}-\text{COH} + n\ \text{H}_2\text{N}-\text{M}'-\text{NH}_2 \xrightarrow{\text{heat}} \left(\!\!\begin{array}{c}\text{M}\\\end{array}\!\!-\overset{O}{\underset{}{\text{C}}}-\underset{H}{N}-M'-\underset{H}{N}\right)_n + 2n\ \text{H}_2\text{O}$$

2. Step-Growth Polymerization of a Dicarboxylic Acid and a Diol Gives a Polyester (Section 29.5B)

$$n\ \text{HOC}-\text{M}-\text{COH} + n\ \text{HO}-\text{M}'-\text{OH} \xrightarrow[\text{catalyst}]{\text{acid}} \left(\text{M}-\text{C}-\text{O}-\text{M}'-\text{O}\right)_n + 2n\ \text{H}_2\text{O}$$

3. Step-Growth Polymerization of a Diacyl Chloride and a Diol Gives a Polycarbonate (Section 29.5C)

$$n\ \text{Cl}-\overset{O}{\underset{}{\text{C}}}-\text{Cl} + n\ \text{HO}-\text{M}-\text{OH} \longrightarrow \left(\text{O}-\overset{O}{\underset{}{\text{C}}}-\text{O}-\text{M}\right)_n + 2n\ \text{HCl}$$

4. Step-Growth Polymerization of a Diisocyanate and a Diol Gives a Polyurethane (Section 29.5D)

$$n\ \text{OCN}-\text{M}-\text{NCO} + n\ \text{HO}-\text{M}'-\text{OH} \longrightarrow \left(\overset{O}{\underset{}{\text{C}}}-\underset{H}{N}-M-\underset{H}{N}-\overset{O}{\underset{}{\text{C}}}-\text{O}-M'-\text{O}\right)_n$$

5. Step-Growth Polymerization of a Diepoxide and a Diamine Gives an Epoxy Resin (Section 29.5E)

$$n\ \text{O}\!\!\triangleleft\!\!-\text{M}-\!\!\triangleright\!\!\text{O} + n\ \text{H}_2\text{N}-\text{M}'-\text{NH}_2 \longrightarrow \left(\underset{H}{N}-\underset{OH}{CH}-M-\underset{OH}{CH}-\underset{H}{N}-M'\right)_n$$

6. Radical Chain-Growth Polymerization of Substituted Ethylenes (Section 29.6A)

$$n \overset{\text{COOMe}}{\diagup\!\!\diagdown} \xrightarrow{\text{AIBN}} \left(\overset{\text{COOMe}}{\diagup\!\!\diagdown}\right)_n$$

7. Titanium-Mediated (Ziegler-Natta) Chain-Growth Polymerization of Ethylene and Substituted Ethylenes (Section 29.6B)

$$n \diagup\!\!\diagdown \xrightarrow[\text{MgCl}_2]{\text{TiCl}_4/\text{Al}(\text{C}_2\text{H}_5)_2\text{Cl}} \left(\!\!\diagup\!\!\diagdown\!\!\right)_n$$

8. Anionic Chain-Growth Polymerization of Substituted Ethylenes (Section 29.6D)

$$n \overset{\text{C}_6\text{H}_5}{\diagup\!\!\diagdown} \xrightarrow[(\text{Na}^+\text{Nap}^-)]{\begin{array}{c}\text{Sodium}\\\text{naphthalide}\end{array}} \left(\overset{\text{C}_6\text{H}_5}{\diagup\!\!\diagdown}\right)_n$$

9. Cationic Chain-Growth Polymerization of Substituted Ethylenes (Section 29.6D)

$$n \diagup\!\!\diagdown \xrightarrow{\text{BF}_3/\text{H}_2\text{O}} \left(\diagup\!\!\diagdown\right)_n$$

10. Ring-Opening Metathesis Polymerization (ROMP) (Section 29.6E)

$$n \bigpentagon \xrightarrow[\text{carbene catalyst}]{\text{Metal, nucleophilic}} \left(\diagup\!\!\diagdown\!\!\diagdown\!\!\diagup\right)_n$$

Appendix 1 Thermodynamics and the Equilibrium Constant

For the equilibrium $A \rightleftharpoons B$ $\quad K_{eq} = \dfrac{[B]}{[A]}$

$$\Delta G^0 = -RT \ln K_{eq}$$

R = molar gas constant = 8.3145 J (1.987 cal) $\cdot$ mol^{-1} $\cdot$ K^{-1}

T = in kelvin (K)

$$\%B = \frac{B}{A + B} \times 100$$

K_{eq}	ΔG^0 kJ/mol	ΔG^0 kcal/mol	$\ln K_{eq}$	$\log K_{eq}$	% B in Mixture
1	0.00	0.00	0.00	0.00	50.00
2	−1.72	−0.41	0.69	0.30	66.67
5	−3.97	−0.95	1.61	0.70	83.33
10	−5.69	−1.36	2.30	1.00	90.91
20	−7.41	−1.77	3.00	1.30	95.24
100	−11.4	−2.73	4.61	2.00	99.01
1,000	−17.1	−4.09	6.91	3.00	99.90
10,000	−22.8	−5.46	9.21	4.00	99.99

Class and Example	Typical pK_a	Class and Example	Typical pK_a
Sulfonic acid	0–1	**β-Ketoester**	11
$CH_3{-}C{-}CH{-}COCH_2CH_3$			
Carboxylic acid	3–5	**Water**	15.7
$CH_3CO{-}H$		$HO{-}H$	
Arylammonium ion	4–5	**Alcohol**	15–19
		$CH_3CH_2O{-}H$	
		Amide	15–19
		$CH_3CN{-}H$	
Imide	8–9	**Cyclopentadiene**	16
Thiol	8–12	**α-Hydrogen of an aldehyde or ketone**	18–20
$CH_3CH_2S{-}H$		$CH_3CCH_2{-}H$	
Phenol	9–10	**α-Hydrogen of an ester**	23–25
		$CH_3CH_2OCCH_2{-}H$	
Ammonium ion	9.24	**Alkyne**	25
$NH_3{-}H^+$		$HC{\equiv}C{-}H$	
β-Diketone	10	**Ammonia**	38
$CH_3{-}C{-}CH{-}CCH_3$		$NH_2{-}H$	
Nitroalkane	10	**Amine**	40
$H{-}CH_2NO_2$		$[(CH_3)_2CH]_2N{-}H$	
Alkylammonium ion	10–12	**Alkene**	44
$(CH_3CH_2)_3\overset{+}{N}{-}H$		$CH_2{=}CH{-}H$	
		Alkane	51
		$CH_3CH_2{-}H$	

Appendix 3 Bond Dissociation Enthalpies

Bond dissociation enthalpy (BDE) is defined as the amount of energy required to break a bond homolytically into two radicals in the gas phase at 25°C.

$$A\!-\!B \rightarrow A^{\cdot} + B^{\cdot} \qquad \Delta H^0 \text{ [kJ (kcal)/mol]}$$

Bond	ΔH^0	Bond	ΔH^0	Bond	ΔH^0
H—H bonds		**C—C multiple bonds**		**C—Br bonds**	
H—H	435 (104)	$CH_2\!=\!CH_2$	727 (174)	$CH_3\!-\!Br$	301 (72)
D—D	444 (106)	$HC\!\equiv\!CH$	966 (231)	$C_2H_5\!-\!Br$	301 (72)
				$(CH_3)_2CH\!-\!Br$	309 (74)
X—X bonds		**C—H bonds**		$(CH_3)_3C\!-\!Br$	305 (73)
F—F	159 (38)	$CH_3\!-\!H$	439 (105)	$CH_2\!=\!CHCH_2\!-\!Br$	247 (59)
Cl—Cl	247 (59)	$C_2H_5\!-\!H$	422 (101)	$C_6H_5\!-\!Br$	351 (84)
Br—Br	192 (46)	$(CH_3)_2CH\!-\!H$	414 (99)	$C_6H_5CH_2\!-\!Br$	263 (63)
I—I	151 (36)	$(CH_3)_3C\!-\!H$	405 (97)		
		$CH_2\!=\!CH\!-\!H$	464 (111)	**C—I bonds**	
H—X bonds		$CH_2\!=\!CHCH_2\!-\!H$	372 (89)	$CH_3\!-\!I$	242 (58)
H—F	568 (136)	$C_6H_5\!-\!H$	472 (113)	$C_2H_5\!-\!I$	238 (57)
H—Cl	431 (103)	$C_6H_5CH_2\!-\!H$	376 (90)	$(CH_3)_2CH\!-\!I$	238 (57)
H—Br	368 (88)	$HC\!\equiv\!C\!-\!H$	556 (133)	$(CH_3)_3C\!-\!I$	234 (56)
H—I	297 (71)			$CH_2\!=\!CHCH_2\!-\!I$	192 (46)
		C—F bonds		$C_6H_5\!-\!I$	280 (67)
O—H bonds		$CH_3\!-\!F$	481 (115)	$C_6H_5CH_2\!-\!I$	213 (51)
HO—H	497 (119)	$C_2H_5\!-\!F$	472 (113)		
$CH_3O\!-\!H$	439 (105)	$(CH_3)_2CH\!-\!F$	464 (111)	**C—N single bonds**	
$C_6H_5O\!-\!H$	376 (90)	$C_6H_5\!-\!F$	531 (127)	$CH_3\!-\!NH_2$	355 (85)
				$C_6H_5\!-\!NH_2$	435 (104)
O—O bonds		**C—Cl bonds**			
HO—OH	213 (51)	$CH_3\!-\!Cl$	351 (84)	**C—O single bonds**	
$CH_3O\!-\!OCH_3$	159 (38)	$C_2H_5\!-\!Cl$	355 (85)	$CH_3\!-\!OH$	385 (92)
$(CH_3)_3CO\!-\!OC(CH_3)_3$	159 (38)	$(CH_3)_2CH\!-\!Cl$	355 (85)	$C_6H_5\!-\!OH$	468 (112)
		$(CH_3)_3\,C\!-\!Cl$	355 (85)		
C—C single bonds		$CH_2\!=\!CHCH_2\!-\!Cl$	288 (69)		
$CH_3\!-\!CH_3$	376 (90)	$C_6H_5\!-\!Cl$	405 (97)		
$C_2H_5\!-\!CH_3$	372 (89)	$C_6H_5CH_2\!-\!Cl$	309 (74)		
$CH_2\!=\!CH\!-\!CH_3$	422 (101)				
$CH_2\!=\!CHCH_2\!-\!CH_3$	322 (77)				
$C_6H_5\!-\!CH_3$	435 (104)				
$C_6H_5CH_2\!-\!CH_3$	326 (78)				

Appendix 4 Characteristic ¹H-NMR Chemical Shifts

Type of Hydrogen (R = alkyl, Ar = aryl)	Chemical Shift (δ)*	Type of Hydrogen (R = alkyl, Ar = aryl)	Chemical Shift (δ)*
$(CH_3)_4Si$	0 (by definition)	RCH_2OH	3.4–4.0
R_2NH	0.5–5.0	RCH_2Br	3.4–3.6
ROH	0.5–6.0	RCH_2Cl	3.6–3.8
RCH_3	0.8–1.0	$RCOCH_3$ (C=O)	3.7–3.9
RCH_2R	1.2–1.4		
R_3CH	1.4–1.7	$RCOCH_2R$ (C=O)	4.1–4.7
$R_2C=CRCHR_2$	1.6–2.6	RCH_2F	4.4–4.5
$RC\equiv CH$	2.0–3.0	$ArOH$	4.5-4.7
$RCCH_3$ (C=O)	2.1–2.3	$R_2C=CH_2$	4.6–5.0
		$R_2C=CHR$	5.0–5.7
$RCCH_2R$ (C=O)	2.2–2.6	ArH	6.5–8.5
$ArCH_3$	2.2–2.5	RCH (C=O)	9.5–10.1
$ArCH_2R$	2.3–2.8		
RCH_2I	3.1–3.3	$RCOH$ (C=O)	10–13
RCH_2OR	3.3–4.0		

*Values are relative to tetramethylsilane. Other atoms within the molecule may cause the signal to appear outside these ranges.

Type of Carbon	Chemical Shift (δ)	Type of Carbon	Chemical Shift (δ)
$(CH_3)_4Si$	0 (by definition)	⟨benzene⟩$\overset{}{C}$—R	110–160
RCH_2I	0–40		
RCH_3	10–40	$R\overset{O}{\overset{\|}{C}}OR$	160–180
RCH_2R	15–55		
R_3CH	20–60	$R\overset{O}{\overset{\|}{C}}NR_2$	165–180
RCH_2Br	25–65		
RCH_2Cl	35–80	$R\overset{O}{\overset{\|}{C}}OH$	165–185
R_3COH	40–80		
R_3COR	40–80		
$RC{\equiv}CR$	65–85	$R\overset{O}{\overset{\|}{C}}H, R\overset{O}{\overset{\|}{C}}R$	180–215
$R_2C{=}CR_2$	100–150		

Appendix 6 Characteristic Infrared Absorption Frequencies

Bonding		Frequency (cm^{-1})	Intensity*	Type of Vibration (Stretching unless noted)
C—H	alkane	2850–3000	m	
	—CH$_3$	1375 and 1450	w–m	bending
	—CH$_2$—	1450–1475	m	bending
	alkene	3000–3100	w–m	
		650–1000	s	out-of-plane bending
	alkyne	3300	m-s	
	arene	3030	w-m	
		690–900	s	out-of-plane bending
	aldehyde	2720	w	
C=C	alkene	1600–1680	w–m	
	arene	1450 and 1600	m	
C≡C	alkyne	2100–2250	w	
C—O	alcohol, ether,	1000–1100 (sp^3 C—O)	s	
	ester, carboxylic acid	1200–1250 (sp^2 C—O)	s	
	anhydride	900–1300	s	
C=O	amide	1630–1680	s	
	carboxylic acid	1700–1725	s	
	ketone	1630–1820	s	
	aldehyde	1630–1820	s	
	ester	1735–1800	s	
	anhydride	1740–1760 and 1800–1850	s	
	acid chloride	1800	s	
O—H	alcohol, phenol free	3600–3650	w	
	hydrogen bonded	3200–3500	m	
	carboxylic acid	2500–3300	s	
N—H	amine and amide	3100–3550	m–s	
C≡N	nitrile	2200–2250	m	

*m = medium, s = strong, w = weak

Appendix 7 Electrostatic Potential Maps

The term *electronic structure* refers to the distribution of electron density in a molecule. According to the laws of quantum mechanics, electrons have no definite locations. Instead, they collectively produce a negatively charged region around a nucleus, measured by electron density in units of $e/\text{Å}^3$ (electrons per cubic angstrom). For an atom, density is high near the nucleus and vanishingly small far from the nucleus. Electrostatic potential maps are now easily computed for small molecules using desktop computers and various software packages. Electrostatic potential maps (elpots) in this text were produced using MacSpartan (Wavefunction, Inc.).

Electrostatic potential (elpot) maps provide a way to visualize the distribution of electron density in a molecule. Electrostatic potential is defined as the potential energy that a positively charged particle would experience in a molecule's presence. The electrostatic potential is made up of two parts.

1. The repulsive component (positive potential, repulsion) exerted by the positively charged nuclei
2. The attractive component (negative potential, attraction) exerted by the negatively charged electron cloud

An electrotatic potential map of ethylene (top and side views).

Thus, the electrostatic potential contains information about the entire electron distribution.

Electrostatic potential maps are color-coded. By convention, the most negative potential is red, and the most positive potential is blue. Intermediate potentials are coded accordingly (orange-yellow-green). While any surface might be chosen to display an electrostatic potential map, the most common is $0.002\ e/\text{Å}^3$. Nearly all of a molecule's electron density lies within this surface, which corresponds almost exactly to how closely another molecule can approach without running into severe steric repulsive forces; that is, the surface corresponds almost exactly to the van der Waals surface of the molecule.

An electrostatic potential map for ethylene shows areas of high electron density to which electrophiles will be attracted (red) over the π orbitals. There are four blue patches, one over each hydrogen; these regions are relatively electron-poor.

The methyl carbocation, CH_3^+, provides an even more dramatic visualization of an electrostatic potential. The entire ion is blue in color, corresponding to the net positive charge. The central atom is the deepest blue, corresponding to the location of the largest fraction of the positive charge.

An electrostatic potential map of the methyl cation (CH_3^+).

Appendix 8 Summary of Stereochemical Terms

Absolute Configuration The actual configuration of groups about a tetrahedral chiral center; absolute configuration is specified by the *R*,*S* system.

Atropisomers Enantiomers that lack a chiral center and differ because of hindered rotation about a carbon-carbon single bond.

Center of symmetry A point so situated that identical components of an object are located on opposite sides and equidistant from that point along any axis passing through it.

Chemioselective reaction A reaction in which one functional group in a molecule containing two or more functional groups reacts selectively with a reagent.

Chiral From the Greek, *cheir*, hand; an object that is not superposable on its mirror image; an object that has handedness.

Chiral center A tetrahedral atom, most commonly carbon, that is bonded to four different groups. In molecules containing one chiral center, the exchange of two groups makes an enantiomer. In molecules containing two or more chiral centers, the exchange of two groups on at least one (but not all) of the chiral centers gives a diastereomer.

Cis,trans **isomers** Stereoisomers that have the same connectivity of their atoms but a different configurational arrangement of their atoms in space because of the presence of either a ring or a double bond. *Cis,trans* isomers are diastereomers; that is, they are stereoisomers that are not mirror images.

Configuration The arrangement of atoms or groups of atoms bonded to a stereocenter. Configuration in alkenes is designated by the E,Z system or the *cis,trans* system; configuration in molecules containing chiral centers is designated by the *R*,*S* system.

Diastereomers Stereoisomers that are not mirror images of each other; refers to relationships among two or more objects.

Diastereoselective reaction A reaction that produces one diastereomer in preference to all others.

Enantiomeric excess (ee) The difference between the percentages of two enantiomers in a mixture. For example, if a sample contains 98% of one enantiomer and 2% of the other, the enantiomeric excess (ee) is 98% − 2% = 96%.

Enantiomers   Stereoisomers that are nonsuperposable mirror images; refers to a relationship between pairs of objects.

Enantioselective reaction   A reaction that produces one enantiomer in preference to the other. Catalytic reduction of the following alkene in the presence of an (R)-BINAP-Ru catalyst gives (S)-naproxen in greater than 98% enantiomeric excess ($> 99\% : < 1\%$).

(S)-Naproxen
(ee > 98%)

Fischer projection   A two-dimensional projection of a chiral center in a molecule; groups on the right and left of the chiral center project toward the reader, whereas those above and below the chiral center project away from the reader. The only atom in the plane on which the projection is drawn is the chiral center itself.

(R)-Glyceraldehyde
(three-dimensional
representation)

(R)-Glyceraldehyde
(Fischer projection)

Meso compound   An achiral compound possessing two or more chiral centers that also has chiral isomers. Examples of meso compounds are *cis*-1,2-cyclopentanediol and meso-2,3-butanediol. A meso compound has either a plane or a center of symmetry. Both of these examples as drawn have an internal plane of symmetry.

cis-1,2-Cyclopentanediol
(a meso compound)

(2R,3S)-2,3-Butanediol
(a meso compound)

Optical activity   The ability of a compound to rotate the plane of polarized light.

Optical purity   The specific rotation of a mixture of enantiomers divided by the specific rotation of the enantiomerically pure substance (expressed as a percent). Optical purity is numerically equal to enantiomeric excess, but experimentally determined.

Plane of symmetry   An imaginary (mirror) plane passing through an object dividing it so that one half is the mirror image of the other half.

Racemic mixture   A mixture of equal amounts of two enantiomers.

Regioselective reaction   An addition or substitution reaction in which one of two or more possible products is formed in preference to all constitutional isomers that might be formed. Addition of HBr to 1-methylcyclohexene gives 1-bromo-1-methylcyclohexane to the virtual exclusion of 1-bromo-2-methylcyclohexane.

$$\text{cyclohexene} + \text{HBr} \longrightarrow \text{1-bromo-1-methylcyclohexane}$$

Hydroboration-oxidation of a cycloalkene is both regioselective and stereoselective, but it is not enantioselective (both enantiomers are produced in equal amounts).

$$\xrightarrow[\text{2. H}_2\text{O}_2/\text{NaOH}]{\text{1. BH}_3}$$

(a racemic mixture)

Specific rotation   The observed rotation of the plane of polarized light when a sample is placed in a tube 1.0 dm in length and at a concentration expressed in g/mL (density) for a pure liquid and at a concentration of 1 g/mL for a solution. Specific rotation is in deg·mL/dm/g and is usually given without units.

Stereocenter   An atom, most commonly carbon, about which exchange of two groups produces a stereoisomer.

Stereocenters but not chiral centers

$$\text{CH}_3\text{CH}_2\text{CH} = \text{CHCH}_3$$

2-Pentene

interchange of two groups at either stereocenter

trans-2-Pentene   +   *cis*-2-Pentene

(a pair of diastereomers)

A stereocenter and also a chiral center

OH
|
$$\text{CH}_3\text{CHCH}_2\text{CH}_3$$

2-Butanol

interchange of two groups at either stereocenter

(*R*)-2-Butanol   +   (*S*)-2-Butanol

(a pair of enantiomers)

Stereoisomers   Compounds that have the same molecular formula, the same connectivity of their atoms, but a different orientation of their atoms in space. The term "stereoisomer" includes *cis,trans* isomers in cycloalkanes and alkenes as well as enantiomers, diastereomers, meso compounds, and atropisomers. Conformational isomers are also stereoisomers, whether they are isolable or not.

Stereoselective reaction   A reaction in which one stereoisomer is formed in preference to all others. A stereoselective reaction may be enantioselective or diastereoselective. For example, catalytic reduction of the following alkene in the presence of an (*R*)-BINAP-Ru catalyst is enantioselective; it gives (*S*)-naproxen in greater than 98% enantiomeric excess.

$$\text{(S)-Naproxen} \quad (\text{ee} > 98\%)$$

Stereospecific reaction   A reaction in which the stereochemistry of the product is dependent on the stereochemistry of the starting material. For example, oxidation of 2-butene by osmium tetroxide is stereospecific: oxidation of *cis*-2-butene gives meso-2,3-butanediol, whereas oxidation of *trans*-2-butene gives a racemic mixture of the enantiomers of 2,3-butanediol. (The term "regiospecific" is used analogously.)

cis-2-Butene → Meso-2,3-Butanediol

trans-2-Butene → (2S,3S)-2,3-Butanediol + (2R,3R)-2,3-Butanediol

Appendix 9 Summary of the Rules of Nomenclature

A9.1 Alkanes

The parent or root name of an alkane is that of the longest chain of carbon atoms in the compound.

Following are prefixes used in the IUPAC system to show the presence of 1 to 20 carbon atoms in an unbranched chain.

Prefix	Number of Carbon Atoms	Prefix	Number of Carbon Atoms
meth-	1	undec-	11
eth-	2	dodec-	12
prop-	3	tridec-	13
but-	4	tetradec-	14
pent-	5	pentadec-	15
hex-	6	hexadec-	16
hept-	7	heptadec-	17
oct-	8	octadec-	18
non-	9	nonadec-	19
dec-	10	eicos-	20

A substituent derived from an alkane by removal of a hydrogen atom is called an **alkyl group** and is commonly represented by the symbol R—.

Following are names for alkyl groups with one to five carbon atoms. Common names and their abbreviations are given in parentheses.

Name	Condensed Structural Formula	Name	Condensed Structural Formula
Methyl (Me)	$-CH_3$		
Ethyl (Et)	$-CH_2CH_3$	1,1-Dimethylethyl (*tert*-butyl, *t*-Bu)	$-CCH_3$ with CH_3 above and CH_3 below
Propyl (Pr)	$-CH_2CH_2CH_3$		
1-Methylethyl (isopropyl, iPr)	$-CHCH_3$ with CH_3	Pentyl	$-CH_2CH_2CH_2CH_2CH_3$
		3-Methylbutyl (isopentyl)	$-CH_2CH_2CHCH_3$ with CH_3
Butyl (Bu)	$-CH_2CH_2CH_2CH_3$		
2-Methylpropyl (isobutyl, iBu)	$-CH_2CHCH_3$ with CH_3	2-Methylbutyl	$-CH_2CHCH_2CH_3$ with CH_3
1-Methylpropyl (*sec*-butyl, *s*-Bu)	$-CHCH_2CH_3$ with CH_3	2,2-Dimethylpropyl (neopentyl)	$-CH_2CCH_3$ with CH_3 above and CH_3 below

The rules of the IUPAC system for naming alkanes are:

1. The name for the alkane with an unbranched chain of carbon atoms consists of a prefix showing the number of carbon atoms in the parent chain and the ending -*ane*.

2. For branched-chain alkanes, take the longest chain of carbon atoms as the parent chain and its name becomes the root name.

3. Give each substituent on the parent chain a name and a number. The number shows the carbon atom of the parent chain to which the substituent is bonded. Use a hyphen to connect the number to the name.

$$CH_3$$
$$|$$
$$CH_3CHCH_3$$

2-Methylpropane

4. If there is one substituent, number the parent chain from the end that gives it the lower number.

$$CH_3$$
$$|$$
$$CH_3CH_2CH_2CHCH_3$$

2-Methylpentane   (not 4-Methylpentane)

5. If there are two or more identical substituents, number the parent chain from the end that gives the lower number to the substituent encountered first. The number of times the substituent occurs is indicated by a prefix *di-, tri-, tetra-, penta-, hexa-*, and so on. Use a comma to separate position numbers.

2,4-Dimethylhexane   (not 3,5-dimethylhexane)

6. If there are two or more different substituents, list them in alphabetical order, and number the chain from the end that gives the lower number to the substituent encountered first. If there are different substituents in equivalent positions on opposite ends of the parent chain, give the substituent of lower alphabetical order the lower number.

3-Ethyl-5-methylheptane   (not 3-methyl-5-ethylheptane)

7. The prefixes *di-, tri-, tetra-,* and so on are not included in alphabetizing. Alphabetize the names of the substituents first, and then insert these prefixes. In the following example, the alphabetizing parts are *ethyl* and *methyl*, not *ethyl* and *dimethyl*.

4-Ethyl-2,2-dimethylhexane
(not 2,2-dimethyl-4-ethylhexane)

8. Where there are two or more parent chains of identical length, choose the parent chain with the greater number of substituents.

3-Ethyl-2-methylhexane   (not 3-Isopropylhexane)

A9.2 Alkenes and Alkynes

1. For both alkenes and alkynes, number the longest chain that contains the multiple bond in the direction that gives the carbon atoms of the functional group the lowest possible numbers.
2. Indicate the location of the multiple bond by the number of its first carbon.
3. Name the branched chain alkene or alkyne similar to alkanes.
4. Number the carbon atoms, locate and name substituent groups, locate the multiple bond, and name the main chain.
5. For alkenes, the ending *-ene* is used, and for alkynes, the ending *-yne* is used.

Examples of alkenes;

1-Hexene   **4-Methyl-1-hexene**   **2-Ethyl-4-methyl-1-pentene**

Examples of alkynes:

3-Methyl-1-butyne   **6,6-Dimethyl-3-heptyne**   **1,6-Heptadiyne**

6. For a molecule containing both a carbon-carbon double bond and a carbon-carbon triple bond, the compound is named as an alkenyne; the infix *-en-* shows the presence of the double bond and the infix *-yn-* shows the presence of the triple bond. In this case, the parent chain is numbered in the direction that gives the carbon atoms of the double bond the lowest set of numbers.

2-Hexen-4-yne
or alternatively **Hex-2-en-4-yne**

A9.3 Alcohols

1. The parent or root name is that of the longest chain of carbon atoms that contains the -OH group.
2. Number the parent chain from the end that gives the carbon bearing the -OH group the smaller number.

3. To show that the compound is an alcohol, change the suffix of the parent alkane from *e* to *-ol.*
4. The location of the -OH group takes precedence over alkyl groups and halogen atoms in numbering the parent chain.
5. For cyclic alcohols, numbering begins with the carbon bearing the -OH group.
6. In complex alcohols, the number for the hydroxyl group may be placed between the infix and the suffix. For example, 4-methyl-2-hexanol and 4-methylhexan-2-ol are both acceptable names for the following compound.

4-Methyl-2-hexanol
or
4-methylhexan-2-ol

A9.4 Aldehydes and Ketones

1. The parent or root name for an aldehyde is the name of the longest chain of carbon atoms that contains the functional group. Show the presence of the aldehyde group by changing the suffix of the parent alkane from *-e* to *-al.* Show the presence of a ketone group by changing the suffix of the parent alkane from *-e* to *-one.* Thus, aldehydes are named as alkan*als*, and ketones are named as alkan*ones*.
2. Because the carbonyl group of an aldehyde can only be on the end of the parent chain, and because numbering must begin with this carbon as number 1, the position of the aldehyde group is unambiguous and there is no need to use a number to locate it.
3. For unsaturated aldehydes or ketones, the presence of the carbon-carbon double or triple bond is indicated by the infix *-en-* or *-yn-*, respectively. The location of the carbonyl group determines the numbering pattern.
4. In complex aldehydes and ketones, the number for the carbonyl group may be placed between the infix and the suffix as illustrated in the following examples.

3-Methyl-2-pentenal
or
3-methylpent-2-enal

5-Methyl-3 hexanone
or
5-methylhexan-3-one

A9.5

1. Names for compounds that contain two or more functional groups indicated by a suffix, the IUPAC has established an **order of precedence of functions**. Following is the order of precedence for the functional group we concentrate on in this text.

Functional Group	Suffix if higher priority	Prefix if lower priority	Example when the functional group has lower priority	
Carboxyl	-oic acid	——		
Aldehyde	-al	oxo-	3-Oxopropanoic acid	
Ketone	-one	oxo-	3-Oxobutanoic acid	
Alcohol	-ol	hydroxy-	4-Hydroxybutanoic acid	
Amino	-amine	amino-	3-Aminobutanoic acid	
Sulfhydryl	-thiol	mercapto	2-Mercaptoethanol	

Increasing precedence

A9.6 Carboxylic Acids

1. The IUPAC name of a carboxylic acid is derived from that of the longest chain of carbon atoms (the root name) that contains the carboxyl group by dropping the suffix -*e* from the name of the parent alkane and adding the suffix -*oic* followed by the word *acid*.
2. The chain is numbered beginning with the carbon of the carboxyl group. Because the carboxyl group is understood to be on carbon-1, there is no need to give it a number.
3. In the IUPAC system, a carboxyl group takes precedence over most other functional groups, including hydroxyl, amino groups, and the carbonyl groups of aldehydes and ketones.

(R)-5-Hydroxyoctanoic acid   **5-Oxohexanoic acid**

A9.7 Carboxylic Esters

1. The IUPAC name of an ester is derived from the name of the corresponding carboxylic acid. The alkyl or aryl group bonded to oxygen is named first, followed by the name of the acid in which the suffix -*ic acid* is replaced by the suffix -*ate*.

$CH_3COCH_2CH_3$

**Ethyl ethanoate
(Ethyl acetate)**   **Methyl 3-oxobutanoate**

A9.8 Carboxylic Amides

1. The IUPAC name of an amide is derived from the name of the parent carboxylic acid by dropping the suffix *-oic acid* and adding *-amide*.
2. If the nitrogen atom of the amide is bonded to an alkyl or aryl group, the group is named, and its location on nitrogen is indicated by the prefix *N-*

Butanamide   ***N*-Methylbutanamide**

A9.9 Carboxylic Anhydrides

1. Anhydrides are named by replacing the word *acid* in the name of the parent carboxylic acid with the word *anhydride*.

$$CH_3\overset{O}{\overset{\|}{C}}O\overset{O}{\overset{\|}{C}}CH_3$$

Acetic anhydride   **Benzoic anhydride**

Example A9.1

Following is the structural formula of mevalonic acid, a key intermediate in the biosynthesis of cholesterol. Write the IUPAC name of this compound.

Mevalonic acid

Step 1: Identify all functional groups and, of these, select the one of highest precedence. This group must be in the chain of carbon atoms selected as the parent alkane, and its location on the parent chain determines the numbering of the parent chain.

Mevalonic acid contains three functional groups, two hydroxyl groups, and one carboxyl group. The carboxyl group has the highest precedence, and it is therefore, carbon 1 of the parent chain.

Accordingly, this molecule is named as a substituted pentanoic acid.

Step 2: Name and locate all other substituents on the parent chain.

The two hydroxyl groups are on carbons 3 and 5, and therefore are named and located as 3,5-dihydroxy- and the methyl group is named and located as -3-methyl-.

Step 3: Specify the configuration of all chiral centers or carbon-carbon double bonds about which *cis-trans* isomerism is possible. In this example, there is one chiral center at carbon 3 and it has the *R* configuration.

Step 1

pentanoic acid

Step 2

3,5-dihydroxy-3-methylpentanoic acid

Step 3

(3R)-3,5-Dihydroxy-3-methylpentanoic acid

Example A9.2

Following is the structural formula of the terpene farnesol. Write the IUPAC name of this compound.

Farnesol

Step 1: There is a single hydroxyl group, which must be located in the longest carbon chain, which is this example is 12 carbon atoms. Therefore, the compound is derived from dodecan-1-ol.

Step 2: There are methyl substituents on carbons 3, 7, and 11 of the parent chain, indicated by 3, 7, 11 trimethyl-

Step 3: There are three carbon-carbon double bonds indicated by the infix 2,6,10-tridecen-.

Step 4: Of the three double bonds in farnesol, those at carbons 2 and 6 can show *cis/trans* isomerism, which may be designated in common name as *trans-trans*-farnesol, but in the IUPAC system must be designated by the *E,Z* system as (2E,6E)-

Step 1

dodecan-1-ol

Step 2

3,7,11-trimethyldodecan-1-ol

Step 3

3,7,11-trimethyl-2,6,10-dodecatrien-1-ol

Step 4

(2R 6E)-3,7,11-trimethyl-2,6,10-dodecatrien-1-ol

Following is the structural formula of the Queen substance, a substance secreted by the mandibular gland of queen honeybees. It inhibits the development of ovaries in worker bees, prevents queen cells formation, and attracts male bees (drones) to virgin queens for the purpose of mating. Write the IUPAC name of this compound.

Queen substance

Step 1: There are two oxygen functional groups in this molecule. A carboxyl group and a ketone group. Of these, the carboxyl group has the higher precedence, and determines the numbering pattern of the longest chain that contains both of these functional groups.

Step 2: The presence and location of the ketone group is indicated by the prefix 9-oxo-, and the presence and location of the carbon-carbon double bond is indicated by the infix -2-ene.

Step 3: The carbon-carbon double bond has the E configuration, which could be indicated by ($2E$), but because there is only one double bond, and we know from Step 2 that it is between carbons 2 and 3, it is not necessary to specify ($2E$). It is sufficient to place (E) first in the name.

Step 1

decanoic acid

Step 2

9-oxodec-2-enoic acid

Step 3

(E)-9-Oxodec-2-enoic acid

Appendix 10 Common Mistakes in Arrow Pushing

Throughout this book arrow pushing is used to indicate the flow of electrons in the various organic reaction mechanisms that are discussed. A few simple rules for properly performing arrow pushing were introduced in Section 6.2. In this Appendix we examine some of the most common mistakes that students make when first learning arrow-pushing methods and tell you how to avoid them. The mistakes given below are the ones seen most often by the authors during their cumulative dozens of years of experience in teaching Introductory Organic Chemistry.

A10.1 Backwards Arrows

Reversing the direction of one or more arrows during a chemical step is the most common mistake made by students when writing organic reaction mechanisms. Backwards arrow pushing usually derives from a student thinking about the movement of atoms, not the movement of electrons. Hence, to avoid this mistake it is important to remember that arrows depict how electrons move, not where atoms move, within or between chemical structures. Further, one can avoid this mistake by remembering that every arrow must start at an electron source (a bond or lone pair) and terminate at an electron sink (an atom that can accept a new bond or lone pair).

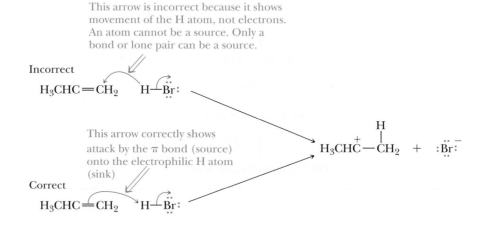

A10.2 Not Enough Arrows

A second common mistake in writing arrow-pushing schemes is to not use enough arrows. This usually results from not keeping track of all lone pairs, bonds made, or bonds broken in a mechanism step. In other words, if you analyze exactly the new position of electrons resulting from each arrow, missing arrows will become evident. In the following example we compare two arrow-pushing scenarios, one of which is missing an arrow. In the incorrect scheme there is no arrow that indicates breaking of the C-H bond of the reactant and formation of the π-bond in the alkene product. Note that when an arrow is missing, the result is commonly too many bonds and/or lone pairs on one atom (see the next section on hypervalency) and not enough bonds or lone pairs on another.

Incorrect

Correct

This arrow is missing in the above scenario.

A10.3 Hypervalency

Another frequent mistake when writing arrow-pushing schemes is to expand the valency of an atom to more electrons than an atom can accommodate, a situation referred to as hypervalency. An overarching principle of organic chemistry is that carbon has eight electrons in its valence shell when present in stable organic molecules (the Octet Rule, Section 1.2). Analogously, many of the other most common elements in organic molecules, such as nitrogen, oxygen, and chlorine, also obey the Octet Rule. There are three common ways in which students incorrectly draw hypervalent atoms: 1) Too many bonds to an atom, 2) Forgetting the presence of hydrogens, and 3) Forgetting the presence of lone pairs.

In the following case an arrow is used to depict a potential resonance contributing structure of nitromethane. However, the result is a nitrogen atoms with 10 electrons in its valence shell because there are too many bonds to N. Such mistakes can be avoided by remembering to draw all bonds and lone pairs on an atom so that the total number of electrons in each atom's valence shell is apparent.

This arrow alone is incorrect because it leads to a hypervalent N atom

There are five bonds on this nitrogen.

Incorrect

This second arrow prevents the problem

Now there are four bonds on this nitrogen.

Correct

Another common way students mistakenly end up with a hypervalent atom is to forget the presence of hydrogens that are not explicitly written.

When using stick diagrams to write organic chemical structures not all the hydrogens are drawn, and hence it is common to forget them during an arrow pushing exercise. The following example shows two proposed resonance contributing structures of an amide anion. The arrow drawn on the molecule to the left is incorrect because it depicts the formation of a new bond to a carbon that already has four bonds. When both bonds to hydrogen are drawn explicitly as on the structure farthest to the right, it is clear there are now five bonds around the indicated carbon atom. Notice also that the negative charge was lost upon drawing the contributing structures on the right, providing another clear signal that something was wrong because overall charge is always conserved when arrows are drawn correctly.

Another common way to make a hypervalency mistake is by forgetting to count all lone pairs of electrons. The following example shows a negatively charged nucleophile incorrectly adding to the formal positive charge on an alkylated ketone. This may look correct because atoms with positive and negative charges are being directly combined, but when counting bonds and lone pairs of electrons, it is found that the oxygen ends up with 10 electrons overall. Hence, this is a mistake.

A10.4 Mixed Media Errors

Acids and bases are catalysts, reactants, products, and intermediates in many organic chemistry transformations. When writing mechanisms for reactions involving acids and bases, there are three general rules that will help guide you in depicting the correct mechanism.

- Do not show the creation of a strong acid for the mechanism of a reaction that is performed in strongly basic media.
- Do not show the creation of a strong base for the mechanism of a reaction that is performed in strongly acidic media.
- In strongly acidic media, all the intermediates and products will be either neutral or positively charged, while in strongly basic media, all the products and intermediates will be neutral or negatively charged.

The reason for these rules is that significant amounts of strong acids and bases cannot co-exist simultaneously in the same medium because they would rapidly undergo a proton transfer reaction before anything else would happen in the solution.

An example of a mixed media error is given below. The first equation shows a strong base being created although the reaction is performed under acidic conditions (see conditions over the first equilibrium arrows). Not shown are the three steps that lead to the intermediate. A mistake is made in the arrow pushing because a strong base (methoxide) is generated as the leaving group even though the reaction is run in strong acid. In the correct mechanism, the next step would be protonation of the ether oxygen atom followed by loss of methanol (not shown) to give a carboxylic acid product.

Incorrect

three steps

Incorrect creation of a strong base in acidic media

Correct

three steps

This leaves as methanol in the next step of the mechanism

A10.5 Failing to Conserve Charge

Overall charge must be conserved in all mechanism steps. Failure to conserve overall charge could be caused by some of the preceding errors (hypervalency, failure to draw arrows, mixed media errors), but we mention it by itself because it is always helpful to check that your arrow pushing is consistent by confirming that overall charge conservation is obeyed. In the example shown below, an arrow is missing leading to a neutral intermediate even though the overall charge on the left side of the equation was minus one. Notice there are five bonds to carbon on the intermediate (hypervalency), providing another obvious indication that something was incorrect in the mechanism step as drawn.

This side of the
equation has an
overall negative
charge due to the
hydroxide anion

As drawn there is no
negative charge on this
intermediate and this C
atom has 5 bonds.

Incorrect

$$H_3C-\overset{\displaystyle :O:}{\underset{\displaystyle :\ddot{O}-H}{\overset{\displaystyle \|}{C}}}-OCH_3 \longrightarrow H_3C-\overset{\displaystyle :O:}{\underset{\displaystyle :OH}{\overset{\displaystyle \|}{C}}}-OCH_3$$

This C atom has only four
bonds and there is an
overall negative charge on
both sides of the equation

Correct

$$H_3C-\overset{\displaystyle :O:}{\underset{\displaystyle :\ddot{O}-H}{\overset{\displaystyle \|}{C}}}-OCH_3 \longrightarrow H_3C-\overset{\displaystyle :\ddot{O}:^-}{\underset{\displaystyle :\ddot{O}H}{\overset{\displaystyle |}{C}}}-OCH_3$$

Appendix 11 Organic Chemistry Road Maps

An organic chemistry road map is a graphical representation of the many organic reactions in the context of the important functional groups. The functional groups of an organic chemistry road map are analogous to cities on a real road map, and the reactions are the roads between those cities. Arrows are used to represent routes that are known between functional groups, and the reagents required to bring about each reaction are written next to the corresponding arrow. For the road maps that follow, the arrows and reagents are color coded to denote the chapter in which the reaction is first described.

Multi-step synthesis questions are often the most challenging for organic chemistry students even though synthesis is at the core of organic chemistry as a discipline. The problem comes down to keeping track of the different reactions encountered throughout the course in such a way that they can be recalled in the context of transforming simpler molecules into more complex molecules.

The power of the organic chemistry road map is that it visualizes the reactions introduced in different chapters in a context that emphasizes how these reactions can be used in specific sequences to interconvert key functional groups in multi-step synthesis problems. Often it is not possible to change one functional group into another with a single reaction. The road map helps you deduce a pathway that is possible when several different reactions are required. For example, you will notice that you cannot create an alkyne directly from an alkane. However, by looking at the road map for Chapters 6–11, you observe that it is possible to convert an alkane into a haloalkane (Br_2 and hν), followed by an E2 elimination (strong base) to give an alkene. The alkene can then be reacted with X_2 to give a vicinal dihaloalkane, which is then reacted with $NaNH_2$ in NH_3 to give the alkyne. Of course, you always need to keep track of both regiochemistry (i.e., Markovnikov addition to an alkene, replacement of an H atom at the most substituted carbon, etc.) and stereochemistry (syn vs. anti, inversion of a chiral center, etc.) in order to predict accurately the products of a reaction sequence.

In order to avoid having too much information on a single page, several different road maps have been created. In particular, there is a road map for Chapters 6–11, Chapters 15–18, Chapter 19, and Chapters 20–23. Although these road maps are intended to be a useful reference, you will benefit from making and using your own road maps. See the end of chapter problems throughout this book. The authors' students have been making and using road maps for almost two decades now, and these road maps are universally credited with making organic chemistry lecture courses a much richer learning experience.

Reactions that make carbon-carbon bonds are particularly useful for organic synthesis because it allows the construction of larger molecules from smaller fragments. All of the many carbon-carbon bond-forming reactions are indicated on the following road maps as reagents with solid backgrounds. For the two reactions involving the cleavage of carbon-carbon bonds, the reagents are circled.

Roadmap for Reactions from Chapters 6–11

Carbon-carbon bond forming reactions are indicated by reagents written with solid backgrounds. Reactions that cleave carbon-carbon bonds are indicated by reagents that are circled.

Chapter 6 is in red
Chapter 7 is in blue
Chapter 8 is in light green
Chapter 9 is in gold
Chapter 10 is in magenta
Chapter 11 is in black

Legend on the next page.

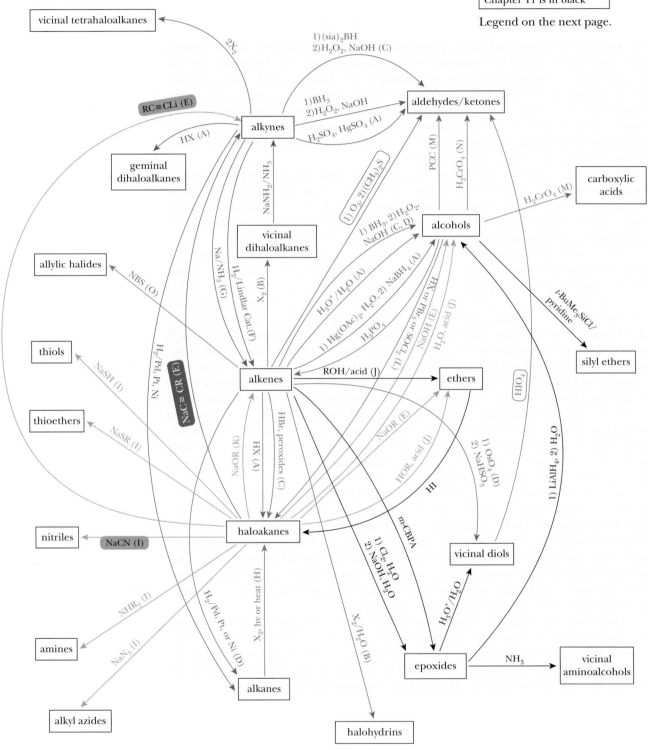

(A) Regiochemistry: Markovnikov addition to a π bond

(B) Stereochemistry: anti-addition

(C) Regiochemistry: non-Markovnikov addition to a π bond

(D) Stereochemistry: syn-addition

(E) Works well for methyl and $1°$ haloalkanes

(F) Stereochemistry: gives *cis*-alkenes as products

(G) Stereochemistry: gives *trans*-alkenes as products

(H) Reactivity of C-H bonds follows $3° > 2° > 1°$

(I) Works for methyl, $1°$, and $2°$ haloalkanes

(J) Works for $2°$ and $3°$ haloalkanes, may see rearrangements

(K) Works for all haloalkanes except methyl, although a bulky (non-nucleophilic) base must be used for $1°$ haloalkanes. Regiochemistry: follows Zaitzev's rule so the more substituted alkene predominates. Stereochemistry: requirement for the X and H to be eliminated with anti-periplanar geometry.

(L) PBr_3 and SOX_2 works for methyl, $1°$, and $2°$ Haloalkanes. HX can give rearrangements.

(M) For $1°$ alcohols.

(N) For $2°$ alcohols.

(O) Regiochemistry: the product with the more substituted alkene predominates.

Roadmap for Reactions from Chapters 15–18

Carbon-carbon bond forming reactions are indicated by reagents written in solid backgrounds with white lettering.

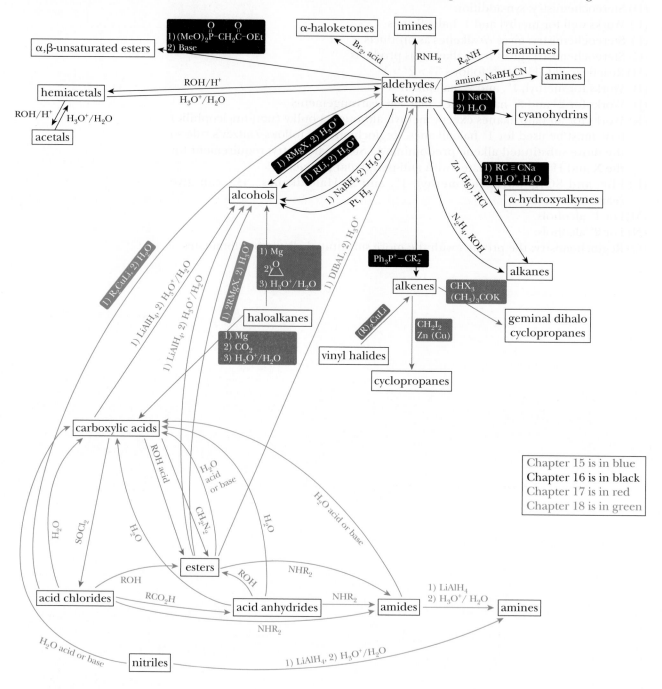

Chapter 15 is in blue
Chapter 16 is in black
Chapter 17 is in red
Chapter 18 is in green

Roadmap for Reactions from Chapter 19

Carbon-carbon bond forming reactions are indicated by reagents written with solid backgrounds and white lettering.

Roadmap for Reactions from Chapters 20–23

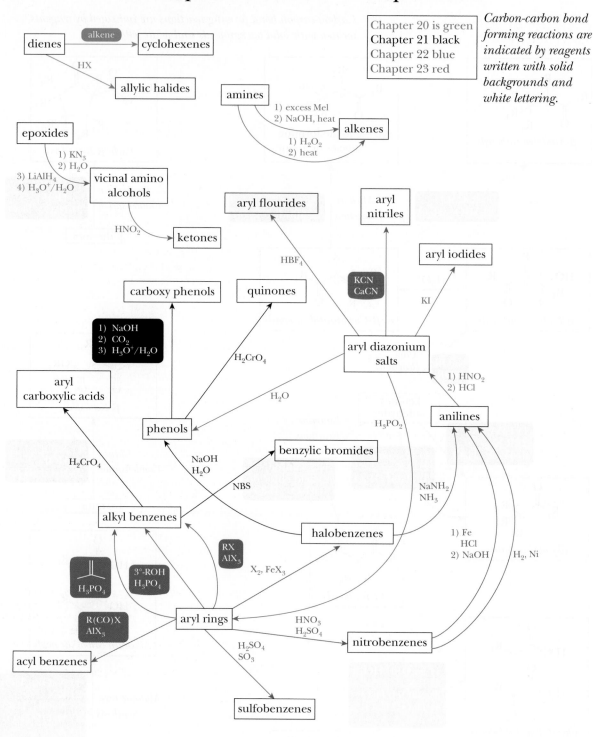

Chapter 20 is green
Chapter 21 black
Chapter 22 blue
Chapter 23 red

Carbon-carbon bond forming reactions are indicated by reagents written with solid backgrounds and white lettering.

dienes →(alkene)→ cyclohexenes

dienes →(HX)→ allylic halides

amines →(1) excess MeI 2) NaOH, heat)→ alkenes

amines →(1) H₂O₂ 2) heat)→ alkenes

epoxides →(1) KN₃ 2) H₂O 3) LiAlH₄ 4) H₃O⁺/H₂O)→ vicinal amino alcohols

vicinal amino alcohols →(HNO₂)→ ketones

aryl flourides

aryl nitriles

aryl iodides

carboxy phenols

quinones

1) NaOH
2) CO₂
3) H₃O⁺/H₂O

KCN
CaCN

aryl carboxylic acids

aryl diazonium salts

aryl diazonium salts →(HBF₄)→ aryl flourides

aryl diazonium salts →(KI)→ aryl iodides

aryl diazonium salts →(1) HNO₂ 2) HCl)← anilines

phenols →(H₂CrO₄)→ quinones

phenols ←(H₂O)— aryl diazonium salts

aryl diazonium salts →(H₃PO₂)→

phenols

benzylic bromides

aryl carboxylic acids ←(H₂CrO₄)—

alkyl benzenes →(NaOH H₂O)→ phenols

alkyl benzenes →(NBS)→ benzylic bromides

alkyl benzenes →(X₂, FeX₃)→ halobenzenes

RX
AlX₃

3°-ROH
H₃PO₄

H₃PO₄

R(CO)X
AlX₃

halobenzenes →(NaNH₂ NH₃)→ anilines

nitrobenzenes →(1) Fe HCl 2) NaOH)→ anilines

nitrobenzenes →(H₂, Ni)→ anilines

aryl rings →(HNO₃ H₂SO₄)→ nitrobenzenes

aryl rings →(H₂SO₄ SO₃)→ sulfobenzenes

aryl rings → acyl benzenes

aryl rings → alkyl benzenes

sulfobenzenes

Glossary

1,2-Shift (Section 6.3C) A type of rearrangement in which an atom or group of atoms with its bonding electrons moves from one atom to an adjacent electron-deficient atom.

Absolute configuration (Section 3.3) Which of the two possible isomers an enantiomer is (i.e., whether it is the right- or left-handed isomer).

Absorbance (A) (Section 20.3A) A quantitative measure of the extent to which a compound absorbs radiation of a particular wavelength. $A = \log (I_0/I)$ where I_0 is the incident radiation and I is the transmitted radiation.

Acetal (Section 16.7B) A molecule containing two —OR or —OAr groups bonded to the same carbon.

Aceto group (Section 17.2B) A CH_3CO— group; also called an acetyl group.

Achiral (Section 3.1) An object that lacks chirality; an object that has no handedness.

Activating group (Section 22.2A) Any substituent on a benzene ring that causes the rate of electrophilic aromatic substitution to be greater than that for benzene.

Activation energy (Section 4.5B) The difference in Gibbs free energy between reactants and a transition state.

Acylation (Section 19.5B) The process of introducing an acyl group, RCO— or ArCO—, onto an organic molecule.

Acyl group (Section 18.1A) An RCO— or ArCO— group.

Acylium ion (Section 22.1C) A resonance-stabilized cation with the structure $[RC{\equiv}O]^+$ or $[ArC{\equiv}O]^+$. The positive charge is delocalized over both the carbonyl carbon and the carbonyl oxygen.

Addition reaction (Section 6.1) A reaction in which two atoms or groups of atoms react with a double bond, forming a compound with the two new groups bonded to the carbons of the original double bond.

Alcohol (Section 1.3A) A compound containing an —OH (hydroxyl) group bonded to a carbon atom.

Aldehyde (Section 1.3C) A compound containing a —CHO group.

Alditol (Section 25.3B) The product formed when the C=O group of a monosaccharide is reduced to a CHOH group.

Aldonic acid (Section 25.3C) The product formed when the —CHO group of an aldose is oxidized to a —COOH group.

Aldose (Section 25.1A) A monosaccharide containing an aldehyde group.

Aliphatic amine (Section 23.1) An amine in which nitrogen is bonded only to alkyl groups.

Alkaloid (Section 23.1) A basic nitrogen-containing compound of plant origin, many of which are physiologically active when administered to humans.

Alkene metathesis (Section 24.6) A reaction in which two alkenes interchange the carbons attached to their double bonds.

Alkoxy group (Section 11.2) An —OR group where R is an alkyl group.

Alkyl group (Section 2.3A) A group derived by removing a hydrogen from an alkane; given the symbol R—.

Alkylation reaction (Section 7.5A) Any reaction in which a new carbon-carbon bond to an alkyl group is formed.

Alkyne (Section 7.1) An unsaturated hydrocarbon that contains one or more carbon-carbon triple bonds.

Allene (Section 7.5B) The compound $CH_2{=}C{=}CH_2$. Any compound that contains adjacent carbon-carbon double bonds; that is, any molecule that contains a C=C=C functional group.

Allyl (Section 5.2B) A —$CH_2CH{=}CH_2$ group.

Allylic (Section 9.3B) Next to a carbon-carbon double bond.

Allylic carbocation (Section 9.3B) A carbocation in which an allylic carbon bears the positive charge.

Allylic carbon (Section 8.6) A carbon adjacent to a carbon-carbon double bond.

Allylic substitution (Section 8.6) Any reaction in which an atom or group of atoms is substituted for another atom or group of atoms at an allylic carbon.

Amino acid (Section 27.1A) A compound that contains both an amino group and a carboxyl group.

α-Amino acid (Section 27.1A) An amino acid in which the amino group is on the carbon adjacent to the carboxyl group.

Amino group (Section 1.3B) A compound containing an sp^3-hybridized nitrogen atom bonded to one, two, or three carbon atoms.

Amorphous domain (Section 29.4) A disordered, noncrystalline region in the solid state of a polymer.

Anabolic steroid (Section 26.4A) A steroid hormone, such as testosterone, that promotes tissue and muscle growth and development.

Androgen (Section 26.4A) A steroid hormone, such as testosterone, that mediates the development of sexual characteristics of males.

Angle strain (Section 2.5A) The strain that arises when a bond angle is either compressed or expanded compared to its optimal value.

Anion (Section 1.2A) An atom or group of atoms bearing a negative charge.

Annulene (Section 21.2B) A cyclic hydrocarbon with a continuous alternation of single and double bonds.

Anomeric carbon (Section 25.2A) The hemiacetal or acetal carbon of the cyclic form of a carbohydrate.

Anomers (Section 25.2A) Carbohydrates that differ in configuration only at their anomeric carbons.

Antiaromatic compound (Section 21.2C) A monocyclic compound that is planar or nearly so, has one $2p$ orbital on each atom of the ring, and has $4n\ \pi$ electrons in the cyclic arrangement of overlapping $2p$ orbitals, where n is an integer. Antiaromatic compounds are especially unstable.

Antibonding molecular orbital (Section 1.7A) A molecular orbital in which electrons have a higher energy than they would in isolated atomic orbitals.

Anti conformation (Section 2.5A) A conformation about a single bond in which two groups on adjacent carbons lie at a dihedral angle of 180°.

Anti stereoselectivity (Section 6.3D) The addition of atoms or groups of atoms to opposite faces of a carbon-carbon double bond.

Aprotic acid (Section 4.7) An acid that is not a proton donor; an acid that is an electron pair acceptor in a Lewis acid-base reaction.

Aprotic solvent (Section 9.3D) A solvent that cannot serve as a hydrogen-bond donor; nowhere in the molecule is there a hydrogen bonded to an atom of high electronegativity. Common aprotic solvents are dichloromethane, diethyl ether, and dimethyl sulfoxide.

Aramid (Section 29.5A) A polyaromatic amide; a polymer in which the monomer units are an aromatic diamine and an aromatic dicarboxylic acid.

Arene (Introduction, Chapter 5) A term used to classify benzene and its derivatives.

Aromatic amine (Section 23.1) An amine in which nitrogen is bonded to one or more aryl groups.

Aromatic compound (Introduction, Chapter 21) A term used initially to classify benzene and its derivatives. More accurately, it is used to classify any compound that meets the Hückel criteria for aromaticity (Section 21.2C).

Aryl group (Ar—) (Introduction, Chapter 5) A group derived from an arene by removal of an H.

Atactic polymer (Section 29.6C) A polymer with completely random configurations at the chiral centers along its chain, as, for example, atactic polypropylene.

Atropisomers (Section 3.2) Enantiomers that lack a chiral center and differ because of hindered rotation.

Aufbau principle (Section 1.1A) Orbitals fill in order of increasing energy, from lowest to highest.

Autoxidation (Section 8.7) Air oxidation of materials such as unsaturated fatty acids.

Axial bond (Section 2.5B) A bond to a chair conformation of cyclohexane that extends from the ring parallel to the imaginary axis through the center of the ring; a bond that lies roughly perpendicular to the equator of the ring.

Azeotrope (Section 16.7B) A liquid mixture of constant composition with a boiling point that is different from that of any of its components.

Base peak (Section 14.1) The peak caused by the most abundant ion in a mass spectrum; the most intense peak. It is assigned an arbitrary intensity of 100.

Basicity (Section 9.3E) An equilibrium property measured by the position of equilibrium in an acid-base reaction, as, for example, the acid-base reaction between ammonia and water.

Benzyl group ($C_6H_5CH_2$—) (Section 21.3A) The group derived from toluene by removing a hydrogen from its methyl group.

Benzylic position (Section 21.5) An sp^3-hybridized carbon bonded to a benzene ring.

Benzyne intermediate (Section 22.3A) A reactive intermediate formed by β-elimination from adjacent carbon atoms of a benzene ring and having a triple bond in the benzene ring. The second π bond of the benzyne triple bond is formed by the weak overlap of coplanar $2p$ orbitals on adjacent carbons.

Betaine (Section 16.6) A neutral molecule with nonadjacent positive and negative charges. An example of a betaine is the intermediate formed by addition of a Wittig reagent to an aldehyde or ketone.

Bicycloalkane (Section 2.4B) An alkane containing two rings that share two carbons.

Bile acid (Section 26.4A) A cholesterol-derived detergent molecule, such as cholic acid, which is secreted by the gallbladder into the intestine to assist in the absorption of dietary lipids.

Bimolecular reaction (Section 9.2) A reaction in which two species are involved in the rate-determining step.

Boat conformation (Section 2.5B) A nonplanar conformation of a cyclohexane ring in which carbons 1 and 4 of the ring are bent toward each other.

Bond dipole moment (μ) (Section 1.2B) A measure of the polarity of a covalent bond. The product of the charge on either atom of a polar bond times the distance between the atoms.

Bond dissociation enthalpy (Section 1.2B) The amount of energy required to break a bond into two radicals in the gas phase at $25°C$, A—B $\longrightarrow$ A• + •B

Bonding electrons (Section 1.2C) Valence electrons involved in forming a covalent bond (i.e., shared electrons).

Bonding molecular orbital (Section 1.7A) A molecular orbital in which electrons have a lower energy than they would in isolated atomic orbitals.

Bond length (Section 1.2B) The distance between atoms in a covalent bond in picometers (pm; 1 pm $= 10^{-12}$ m) or Å (1Å $= 10^{-10}$ m).

Brønsted-Lowry acid (Section 4.2) A proton donor.

Brønsted-Lowry base (Section 4.2) A proton acceptor.

Carbanion (Section 15.1A) An ion in which carbon has an unshared pair of electrons and bears a negative charge.

Carbene (Section 15.3) A neutral molecule that contains a carbon atom surrounded by only six valence electrons (R_2C:).

Carbenoid (Section 15.3C) A compound that delivers the elements of a carbene without actually producing a free carbene.

Carbocation (Section 4.2C) A species in which a carbon atom has only six electrons in its valence shell and bears a positive charge.

Carbohydrate (Introduction, Chapter 25) A polyhydroxyaldehyde, a polyhydroxyketone, or a substance that gives these compounds on hydrolysis.

α-Carbon (Section 16.9A) A carbon atom adjacent to a carbonyl group.

Carbonyl group (Section 1.3C) A C=O group.

Carboxyl group (Section 1.3D) A —COOH group.

Carboxylic acid (Section 1.3D) A compound containing a carboxyl, —COOH, group.

Carboxylic ester (Section 1.3D) A derivative of a carboxylic acid in which H of the carboxyl group is replaced by a carbon.

Cation (Section 1.2A) An atom or group of atoms bearing a positive charge.

Center of symmetry (Section 3.1) A point so situated that identical components of an object are located on opposite sides and equidistant from that point along any axis passing through it.

Chain-growth polymerization (Section 29.6) A polymerization that involves sequential addition reactions, either to unsaturated monomers or to monomers possessing other reactive functional groups.

Chain initiation (Section 8.5B) A step in a chain reaction characterized by the formation of reactive intermediates (radicals, anions, or cations) from nonradical or noncharged molecules.

Chain length (Section 8.5B) The number of times the cycle of chain propagation steps repeats in a chain reaction.

Chain propagation (Section 8.5B) A step in a chain reaction characterized by the reaction of a reactive intermediate and a molecule to give a new reactive intermediate and a new molecule.

Chain termination (Section 8.5B) A step in a chain reaction that involves destruction of reactive intermediates.

Chain-transfer reaction (Section 29.6A) The transfer of reactivity of an endgroup from one chain to another during a polymerization.

Chair conformation (Section 2.5B) The most stable nonplanar conformation of a cyclohexane ring; all bond angles are approximately $109.5°$, and all bonds on adjacent carbons are staggered.

Chemical shift (δ) (Section 13.3) The shift in parts per million of an NMR signal relative to the signal of TMS.

Chiral (Section 3.1) From the Greek, *cheir* meaning hand; an object that is not superposable on its mirror image; an object that has handedness.

Chiral center (Section 3.2) A tetrahedral atom, most commonly carbon, that is bonded to four different groups; also called a chirality center.

Chlorofluorocarbons (CFCs, Freons) (Section 8.5C) Compounds with one or two carbons, chlorine, and fluorine, formerly used as refrigerants.

Chromatography (Section 3.9C) A separation method involving passing a vapor or solution mixture through a column packed with a material with different affinities for different components of the mixture.

Circular DNA (Section 28.2C) A type of double-stranded DNA in which the 5′ and 3′ ends of each strand are joined by phosphodiester groups.

Cis (Section 2.6A) A prefix meaning on the same side.

Cis, trans isomers (Sections 2.6A and 5.1C) Stereoisomers that have the same connectivity but a different arrangement of their atoms in space as a result of the presence of either a ring or a carbon-carbon double bond.

Clemmensen reduction (Section 16.11E) Reduction of the C=O group of an aldehyde or ketone to a CH_2 group using $Zn(Hg)$ and HCl.

Codon (Section 28.4A) A triplet of nucleotides on mRNA that directs incorporation of a specific amino acid into a polypeptide sequence.

Condensation polymerization (Section 29.5) A polymerization in which chain growth occurs in a stepwise manner between difunctional monomers. Also called step-growth polymerization.

Configuration (Section 2.6A) Refers to the arrangement of atoms about a stereocenter.

Configurational isomers (Section 3.2) Isomers that differ by the configuration of substituents on an atom. Refers to the arrangement of atoms about a stereocenter.

Conformation (Section 2.5A) Any three-dimensional arrangement of atoms in a molecule that results from rotation about a single bond.

Conjugate acid (Section 4.2A) The species formed when a base accepts a proton from an acid.

Conjugate addition (Section 19.8) Addition of a nucleophile to the β-carbon of an α,β-unsaturated carbonyl compound. (Section 20.2A) Addition to carbons 1 and 4 of a conjugated diene.

Conjugate base (Section 4.2A) The species formed when an acid transfers a proton to a base.

Conjugated (Introduction, Chapter 20) A conjugated diene or carbonyl is one in which the double bonds are separated by one single bond.

Conjugation (Section 12.4H) A situation in which two multiple bonds are separated by a single bond. Alternatively, a series of overlapping 2p orbitals. 1,3-butadiene, for example, is a conjugated diene, and 3-butene-2-one is a conjugated enone.

Constitutional isomers (Section 2.2) Compounds with the same molecular formula but a different connectivity of their atoms.

Contributing structures (Section 1.8A) Representations of a molecule or ion that differ only in the distribution of valence electrons.

Correlation tables (Section 12.3D) Tables of data on absorption patterns of functional groups.

Coupling constant (J) (Section 13.9) The separation on an NMR spectrum (in hertz) between adjacent peaks in a multiplet and a quantitative measure of the influence of the spin-spin coupling with adjacent nuclei.

Covalent bond (Section 1.2A) A chemical bond formed between two atoms by sharing one or more pairs of electrons.

Crown ether (Section 11.11) A cyclic polyether derived from ethylene glycol and substituted ethylene glycols.

Crystalline domain (Section 29.4) An ordered crystalline region in the solid state of a polymer. Also called a crystallite.

Cumulated (Section 20.1) A cumulated diene is one in which two double bonds share an sp-hybridized carbon.

Curved arrow (Section 1.8A) A symbol used to show the redistribution of valence electrons in resonance contributing structures or reactions, symbolizing movement of two electrons.

Cyanohydrin (Section 16.5D) A molecule containing an —OH group and a —CN group bonded to the same carbon.

Cycloaddition reaction (Section 20.4A) A reaction in which two reactants add together in a single step to form a cyclic product. The best known of these is the Diels-Alder reaction.

Cycloalkane (Section 2.4) A saturated hydrocarbon that contains carbons joined to form a ring.

Deactivating group (Section 22.2A) Any substituent on a benzene ring that causes the rate of electrophilic aromatic substitution to be lower than that for benzene.

Decarboxylation (Section 17.9) Loss of CO_2 from a carboxyl group.

Dehydration (Section 10.6) Elimination of water.

Dehydrohalogenation (Sections 7.5B and 9.5) Removal of —H and —X from adjacent carbons; a type of β-elimination.

Delocalization (Section 1.1) The spreading of electron density over a larger volume of space.

Deshielding (Section 13.3) The term used to express the concept of less shielding in NMR.

Dextrorotatory (Section 3.7B) Refers to a substance that rotates the plane of polarized light to the right.

Diamagnetic current in NMR (Section 13.3) The circulation of electron density in a molecule in an applied magnetic field.

Diastereomers (Section 3.2) Stereoisomers that are not mirror images of each other; refers to relationships among two or more objects.

Diastereotopic groups (Section 13.10) Atoms or groups on an atom that are bonded to an atom that is bonded to two nonidentical groups, one of which contains a chiral center. When one of the atoms or groups is replaced by another group, a new chiral center is created and a set of diastereomers results. The hydrogens of the CH_2 group of 2-butanol, for example, are diastereotopic. Diastereotopic groups have different chemical shifts under all conditions.

Diaxial interactions (Section 2.5B) Refers to the steric strain arising from interaction between an axial substituent and an axial hydrogen (or other group) on the same side of a chair conformation of a cyclohexane ring.

Diazonium ion (Section 23.8D) An ArN_2^+ or RN_2^+ ion.

Dielectric constant (Section 9.3D) A measure of a solvent's ability to insulate opposite charges from one another.

Diels-Alder adduct (Section 20.5) A cyclohexene resulting from the cycloaddition reaction of a diene and a dienophile.

Dienophile (Section 20.5) A compound containing a double bond (consisting of one or two C, N, or O atoms) that can react with a conjugated diene to give a Diels-Alder adduct.

Dihedral angle (Section 2.5A) The angle created by two intersecting planes.

Diol (Section 10.1B) A compound containing two hydroxyl groups.

Dipeptide (Section 27.3) A molecule containing two amino acid units joined by a peptide bond.

Dipole-dipole interaction (Section 10.2) The attraction between the positive end of one dipole and the negative end of another.

Disaccharide (Section 25.4) A carbohydrate containing two monosaccharide units joined by a glycosidic bond.

Dispersion forces (Section 2.7B) Very weak intermolecular forces of attraction resulting from the interaction between temporary induced dipoles.

Disproportionation (Section 29.6A) A termination process that involves the abstraction of a hydrogen atom from the beta position of the propagating radical of one chain by the radical endgroup of another chain.

Disulfide (Section 11.12A) A molecule containing an —S—S— group.

Double-headed arrow (Section 1.8A) A symbol used to show that structures on either side of it are resonance-contributing structures.

Double helix (Section 28.2B) A type of secondary structure of DNA molecules in which two antiparallel polynucleotide strands are coiled in a right-handed manner about the same axis.

Downfield (Section 13.4) A signal of an NMR spectrum that is shifted toward the left (larger chemical shift) on the chart paper.

E (Section 5.2C) From the German, *entgegen*, opposite. Specifies that groups of higher priority on the carbons of a double bond are on opposite sides.

E,Z system (Section 5.2C) A system to specify the configuration of groups about a carbon-carbon double bond.

E1 (Section 9.6A) A unimolecular β-elimination reaction.

E2 (Section 9.6B) A bimolecular β-elimination reaction.

Eclipsed conformation (Section 2.5A) A conformation about a carbon-carbon single bond in which the atoms or groups on one carbon are as close as possible to the atoms or groups on an adjacent carbon.

Edman degradation (Section 27.4B) A method for selectively cleaving and identifying the *N*-terminal amino acid of a polypeptide chain.

Elastomer (Section 29.4) A material that, when stretched or otherwise distorted, returns to its original shape when the distorting force is released.

Electromagnetic radiation (Section 12.1) Light and other forms of radiant energy.

Electronegativity (Section 1.2B) A measure of the force of an atom's attraction for electrons.

Electron affinity (Section 1.2B) Energy added or released when an electron is added to an atom or molecule.

Electrophile (Section 6.2E and Introduction to Chapter 9) From the Greek meaning electron loving. Any species that can accept a pair of electrons to form a new covalent bond; alternatively, a Lewis acid.

Electrophilic aromatic substitution (Section 22.1) A reaction in which there is substitution of an electrophile, E^+, for a hydrogen on an aromatic ring.

Electrophoresis (Section 27.2D) The process of separating compounds on the basis of their electric charge.

β-Elimination (Introduction, Chapter 9) A reaction in which a molecule, such as HCl, HBr, HI, or HOH, is split out or eliminated from adjacent carbons.

Enamine (Section 16.8A) An unsaturated compound derived by the reaction of an aldehyde or ketone and a secondary amine followed by loss of H_2O; R_2C=CR—NR_2.

Enantiomeric excess (ee) (Section 3.7D) The difference between the percentage of two enantiomers in a mixture.

Enantiomers (Section 3.2) Stereoisomers that are nonsuperposable mirror images of each other; refers to a relationship between pairs of objects.

Enantioselective reaction (Section 6.7C) A reaction that produces one enantiomer in preference to the other.

Enantiotopic groups (Section 13.10) Atoms or groups on an atom that give a chiral center when one of the groups is replaced by another group. A pair of enantiomers results. The hydrogens of the CH_2 group of ethanol, for example, are enantiotopic. Replacing one of them by deuterium gives (*R*)-1-deuteroethanol; replacing the other gives (*S*)-1-deuteroethanol. Enantiotopic groups have identical chemical shifts in achiral environments but different chemical shifts in chiral environments.

Endergonic reaction (Section 4.5B) A reaction in which the Gibbs free energy of the products is higher than that of the reactants. The position of equilibrium for an endergonic reaction favors starting materials.

Endothermic reaction (Section 4.5B) A reaction in which the enthalpy of the products is higher than the enthalpy of the reactants; a reaction in which heat is absorbed.

Energy (Section 1.1B) The ability to do work.

Energy-level diagrams (Section 1.1A) A graph showing the changes in energy that occur during a chemical reaction; energy is plotted on the vertical axis, and reaction progress is plotted on the horizontal axis. Also called a reaction coordinate diagram.

Enol (Section 7.7A) A compound containing a hydroxyl group bonded to a doubly bonded carbon atom.

Enolate anion (Section 16.9A) An anion derived by loss of a hydrogen from a carbon alpha to a carbonyl group; the anion of an enol.

Enthalpy change, ΔH^0 (Section 4.5B) The difference in total bond strengths and solvation between various points under comparison on a reaction coordinate diagram.

Entropy (S) (Section 4.5B) Measures chaos versus order and chaos is favorable.

Epoxide (Section 11.7) A cyclic ether in which oxygen is one atom of a three-membered ring.

Equatorial bond (Section 2.5B) A bond to a chair conformation of cyclohexane that extends from the ring roughly perpendicular to the imaginary axis through the center of the ring; a bond that lies roughly along the equator of a cyclohexane ring.

Equivalent hydrogens (Section 13.5) Hydrogens that have the same chemical environment.

Ester (Section 1.3E) A derivative of a carboxylic acid in which H of the carboxyl group is replaced by a carbon.

Estrogen (Section 26.4A) A steroid hormone, such as estrone and estradiol, that mediates the development of sexual characteristics in females.

Ether (Section 11.1) A compound containing an oxygen atom bonded to two carbon atoms.

Excited state (Section 1.1B) A state of a system at higher energy than the ground state.

Exergonic reaction (Section 4.5B) A reaction in which the Gibbs free energy of the products is lower than that of the reactants. The position of equilibrium for an exergonic reaction favors products.

Exothermic reaction (Section 4.5B) A reaction in which the enthalpy of the products is lower than that of the reactants; a reaction in which heat is released.

Fat (Section 26.1B) A mixture of triglycerides that is semisolid or solid at room temperature.

Fatty acid (Section 26.1A) A long, unbranched-chain carboxylic acid, most commonly of 12 to 20 carbons, derived from the hydrolysis of animal fats, vegetable oils, or the phospholipids of biological membranes.

Fingerprint region (Section 12.3D) Vibrations in the region 1500 to 400 cm^{-1} of an IR spectrum are complex and difficult to analyze but are characteristic for different molecules.

First ionization potential (Section 1.1B) The energy needed to remove the most loosely held electron from an atom or molecule.

Fischer esterification (Section 17.7A) The process of forming an ester by refluxing a carboxylic acid and an alcohol in the presence of an acid catalyst, commonly H_2SO_4, $ArSO_3H$, or HCl.

Fischer projection (Section 3.4C and Section 25.1B) A two-dimensional representation of a molecule; in these projections, groups on the right and left are by convention in front, while those at the top and bottom are to the rear.

Fishhook arrow (Section 8.3D) A barbed curved arrow used to show the change in position of a single electron.

Fluid-mosaic model (Section 26.5B) A biological membrane that consists of a phospholipid bilayer with proteins, carbohydrates, and other lipids on the surface and embedded in the bilayer.

Formal charge (Section 1.2D) The charge on an atom in a polyatomic ion or molecule.

Fourier transform NMR (FT-NMR) (Section 13.4) The modern NMR method that is based on a constant magnetic field, a short pulse of electromagnetic radiation, and a mathematical Fourier transform to produce the spectrum.

Frequency (Section 12.1) The number of full cycles of a wave that pass a given point in a second, and reported in hertz (Hz), which has the units s^{-1}.

Friedel-Crafts reaction (Section 22.1C) An electrophilic aromatic substitution in which a hydrogen of an aromatic ring is replaced by an alkyl or acyl group.

Frost circle (Section 21.2A) A graphic method for determining the relative energies of π MOs for planar, fully conjugated, monocyclic compounds.

Functional group (Section 1.3) An atom or group of atoms within a molecule that shows a characteristic set of physical and chemical properties.

Furanose (Section 25.2A) A five-membered cyclic form of a monosaccharide.

Gauche conformation (Section 2.5A) A conformation about a single bond of an alkane in which two groups on adjacent carbons lie at a dihedral angle of 60°.

Geminal coupling (Section 13.9D) Spin-spin coupling that occurs between nonequivalent H atoms bonded to the same C atom. The H atoms are generally nonequivalent owing to restricted bond rotation in the molecule.

Gibbs free energy change ($\Delta G°$) (Section 4.5B) The energy that dictates the position of chemical equilibria and rates of chemical reactions. A thermodynamic function of enthalpy, entropy, and temperature, given by the equation $\Delta G° = \Delta H° - T\Delta S°$. If $\Delta G° < 0$, the position of equilibria for the reaction favors products. If $\Delta G° > 0$, the position of equilibria favors reactants.

Glass transition temperature (T_G) (Section 29.4) The temperature at which a polymer undergoes the transition from a hard glass to a rubbery state.

Glycol (Section 6.5A) A compound with hydroxyl (—OH) groups on adjacent carbons.

Glycoside (Section 25.3A) A carbohydrate in which the —OH on its anomeric carbon is replaced by —OR.

Glycosidic bond (Section 25.3A) The bond from the anomeric carbon of a glycoside to an —OR group.

Ground state (Section 1.1A) The lowest energy state of a system.

Ground-state electron configuration (Section 1.1A) The lowest-energy electron configuration for an atom or molecule.

Haloalkane (alkyl halide) (Section 8.1) A compound containing a halogen atom covalently bonded to an sp^3-hybridized carbon atom. Given the symbol R—X.

Haloalkene (vinylic halide) (Section 8.1) A compound containing a halogen atom bonded to one of the carbons of a carbon-carbon double bond.

Haloarene (aryl halide) (Section 8.1) A compound containing a halogen atom bonded to a benzene ring. Given the symbol Ar—X.

Haloform (Section 8.2B) A compound of the type CHX_3 where X is a halogen.

Halohydrin (Section 6.3E) A compound containing a halogen atom and a hydroxyl group on adjacent carbons; those containing Br and OH are bromohydrins, and those containing Cl and OH are chlorohydrins.

Hammond's postulate (Section 8.5D) The structure of the transition state for an exothermic step looks more like the reactants of that step than the products. Conversely, the structure of the transition state for an endothermic step looks more like the products of that step than the reactants.

Haworth projection (Section 25.2A) A way to view furanose and pyranose forms of monosaccharides. The ring is drawn flat and most commonly viewed through its edge with the anomeric carbon on the right and the oxygen atom of the ring to the rear.

Heat of combustion (ΔH^0) (Section 2.8A) Standard heat of combustion is the heat released when one mole of a substance in its standard state (gas, liquid, solid) is oxidized completely to carbon dioxide and water.

Heat of reaction (ΔH^0) (Section 4.5B) The difference in enthalpy between reactants and products. If the enthalpy of products is lower than that of the reactants, heat is released and the reaction is exothermic. If the enthalpy of the products is higher than that of the reactants, energy is absorbed, and the reaction is endothermic.

α-Helix (Section 27.6B) A type of secondary structure in which a section of polypeptide chain coils into a spiral, most commonly a right-handed spiral.

Hemiacetal (Section 16.7B) A molecule containing an —OH and an —OR or —OAr group bonded to the same carbon.

Hertz (Hz) (Section 12.1) The unit in which frequency is measured: s^{-1} (read "per second").

Heterocycle (Section 11.2) A cyclic compound whose ring contains more than one kind of atom. Oxirane (ethylene oxide), for example, is a heterocycle whose ring contains two carbon atoms and one oxygen atom.

Heterocyclic amine (Section 23.1) An amine in which nitrogen is one of the atoms of a ring.

Heterocyclic aromatic amine (Section 23.1) An amine in which nitrogen is one of the atoms of an aromatic ring.

Heterolytic bond cleavage (Section 8.3D) Cleavage of a bond so that one fragment retains both electrons and the other retains none.

High-density lipoprotein (HDL) (Section 26.4A) Plasma particles, density 1.06–1.21 g/mL, consisting of approximately 33% proteins, 30% cholesterol, 29% phospholipids, and 8% triglycerides.

High-resolution mass spectrometry (Section 14.2A) Instrumentation that is capable of separating ions that differ in mass by as little as 0.0001 amu.

Histone (Section 28.2C) A protein, particularly rich in the basic amino acids lysine and arginine, that is found associated with DNA molecules.

Hofmann elimination (Section 23.9) When treated with a strong base, a quaternary ammonium halide undergoes β-elimination by an E2 mechanism to give the less-substituted alkene as the major product.

Hofmann rule (Section 23.9) Any β-elimination that occurs preferentially to give the less substituted alkene as the major product.

Homolytic bond cleavage (Section 8.3D) Cleavage of a bond so that each fragment retains one electron; formation of radicals.

Homotopic groups (Section 13.10) Atoms or groups on an atom that give an achiral molecule when one of the groups is replaced by another group. The hydrogens of the CH_2 group of propane, for example, are homotopic. Replacing either one of them with deuterium gives 2-deuteropropane, which is achiral. Homotopic groups have identical chemical shifts under all conditions.

Hückel criteria for aromaticity (Section 21.2A) To be aromatic, a monocyclic compound must have one $2p$ orbital on each atom of the ring, be planar or nearly so, and have $(4n + 2)$ π electrons in the cyclic arrangement of $2p$ orbitals.

Hund's rule (Section 1.1A) When orbitals of equal energy are available but there are not enough electrons to fill all of them completely, one electron is put in each before a second electron is added to any.

Hybridization (Section 1.7B) The combination of atomic orbitals of different types.

Hybrid orbital (Section 1.7B) An orbital formed by the combination of two or more atomic orbitals.

Hydration (Section 6.3B) The addition of water.

Hydride ion (Section 16.11A) A hydrogen atom with two electrons in its valence shell; $H:^-$.

Hydroboration-oxidation (Section 6.4) A method for converting an alkene to an alcohol. The alkene is treated with borane (BH_3) to give a trialkylborane, which is then oxidized with alkaline hydrogen peroxide to give an alcohol.

α-Hydrogen (Section 16.9A) A hydrogen on a carbon alpha to a carbonyl group.

Hydrogen bonding (Section 10.2) The attractive interaction between a hydrogen atom bonded to an atom of high electronegativity (most commonly O or N) and a lone pair of electrons on another atom of high electronegativity (again, most commonly O or N).

Hydrogenolysis (Section 21.5C) Cleavage of a single bond by H_2, most commonly accomplished by treating a compound with H_2 in the presence of a transition metal catalyst.

Hydroperoxide (Section 11.5B) A compound containing an —OOH group.

Hydrophilic (Section 17.3) From the Greek, meaning water-loving.

Hydrophobic (Section 17.3) From the Greek, meaning water-fearing.

Hydrophobic effect (Section 27.6D) The tendency of nonpolar groups to cluster so as to shield them from contact with an aqueous environment.

Hydroxyl group (Section 1.3A) An —OH group.

Hyperconjugation (Section 6.3A) Interaction of electrons in a σ-bonding orbital with the vacant $2p$ orbital of an adjacent positively charged carbon.

Imide (Section 18.1D) A functional group in which two acyl groups, RCO— or ArCO—, are bonded to a nitrogen atom.

Imine (Section 16.8A) A compound containing a carbon-nitrogen double bond, $R_2C=NR'$; also called a Schiff base.

Index of hydrogen deficiency (Section 5.1A) The sum of the number of rings and π bonds in a molecule.

Inductive effect (Sections 4.6D and 6.3A) The polarization of the electron density of a covalent bond caused by the electronegativity of a nearby atom.

Infrared active (Section 12.3B) Any molecular vibration that leads to a substantial change in dipole moment and is observed in an IR spectrum.

Infrared (IR) spectroscopy (Section 12.3A) A spectroscopic technique in which a compound is irradiated with infrared radiation, absorption of which causes covalent bonds to change from a lower vibration state to a higher one. Infrared spectroscopy is particularly valuable for determining the kinds of functional groups present in a molecule.

Ionization potential (IP) (Section 14.1) The minimum energy required to remove an electron from an atom or molecule to a distance where there is no electrostatic interaction between the resulting ion and electron.

Isoelectric point (pI) (Section 27.2C) The pH at which an amino acid, polypeptide, or protein has no net charge.

Isomers (Section 1.2C) Different compounds with the same molecular formula.

Isotactic polymer (Section 29.6C) A polymer with identical configurations (either all R or all S) at all chiral centers along its chain, as, for example, isotactic polypropylene.

Keto-enol tautomerism (Section 7.7A) A type of isomerism involving keto (from ketone) and enol tautomers.

Ketone (Section 1.3C) A compound containing a carbonyl group bonded to two carbons.

Ketose (Section 25.1A) A monosaccharide containing a ketone group.

Kinetic control (Section 19.8A) Experimental conditions under which the composition of the product mixture is determined by the relative rates of formation of each product.

Lactam (Section 18.1D) A cyclic amide.

Lactone (Section 18.1C) A cyclic ester.

Leaving group (Introduction, Chapter 9) The group that is displaced in a substitution reaction or the Lewis base that is lost in an elimination reaction.

Levorotatory (Section 3.7B) Refers to a substance that rotates the plane of polarized light to the left.

Lewis acid (Section 4.7) Any molecule or ion that can form a new covalent bond by accepting a pair of electrons.

Lewis base (Section 4.7) Any molecule or ion that can form a new covalent bond by donating a pair of electrons.

Lewis dot structure (Section 1.1C) The symbol of an element surrounded by a number of dots equal to the number of electrons in the valence shell of the atom.

Ligand (Section 24.2) A Lewis base bonded to a metal atom in a coordination compound. It may bond strongly or weakly.

Lindlar catalyst (Section 7.8A) Finely powdered palladium metal deposited on solid calcium carbonate that has been specially modified with lead salts. Its particular use is as a catalyst for the reduction of an alkyne to a *cis* alkene.

Line-angle formula (Section 2.1) An abbreviated way to draw structural formulas in which vertices and line endings represent carbons.

Lipid (Introduction, Chapter 26) A biomolecule isolated from plant or animal sources by extraction with nonpolar organic solvents, such as diethyl ether and hexane.

Lipid bilayer (Section 26.5B) A back-to-back arrangement of phospholipid monolayers, often forming a closed vesicle or membrane.

Living polymer (Section 29.6D) A polymer chain that continues to grow without chain-termination steps until either all of the monomer is consumed or some external agent is added to terminate the chain. The polymer chains will continue to grow if more monomer is added.

Low-density lipoprotein (LDL) (Section 26.4A) Plasma particles, density 1.02–1.06 g/mL, consisting of approximately 26% proteins, 50% cholesterol, 21% phospholipids, and 4% triglycerides.

Low-resolution mass spectrometry (Section 14.2A) Instrumentation that is capable of separating only ions that differ in mass by 1 or more amu.

Markovnikov's rule (Section 6.3A) In the addition of HX, H_2O, or ROH to an alkene, hydrogen adds to the carbon of the double bond having the greater number of hydrogens.

Mass spectrometry (Introduction Chapter 14) An analytical technique for measuring the mass-to-charge ratio (m/z) of ions.

Mass spectrum (Section 14.1) A plot of the relative abundance of ions versus their mass-to-charge ratio.

Melt transition (T_m) (Section 29.4) The temperature at which crystalline regions of a polymer melt.

Mercaptan (Section 10.9B) A common name for a thiol; that is, any compound that contains an -SH (sulfhydryl) group.

Meso compound (Section 3.4B) An achiral compound possessing two or more chiral centers that also has chiral isomers.

Messenger RNA (mRNA) (Section 28.3C) A ribonucleic acid that carries coded genetic information from DNA to the ribosomes for the synthesis of proteins.

Meta (m) (Section 21.3B) Refers to groups occupying 1,3-positions on a benzene ring.

Methylene (Section 5.2B) A —CH_2— group.

Micelle (Section 26.2B) A spherical arrangement of organic molecules in water solution clustered so that their hydrophobic parts are buried inside the sphere and their hydrophilic parts are on the surface of the sphere and in contact with water.

Molar absorptivity (ε) (Section 20.3A) The absorbance of a 1 M solution of a compound.

Molecular dipole moment (μ) (Section 1.5) The vector sum of individual bond dipoles.

Molecular ion (M^+) (Section 14.1) The radical cation formed by removal of a single electron from a parent molecule in a mass spectrometer.

Molecular orbital (MO) theory (Section 1.7A) A theory of chemical bonding in which electrons in molecules occupy molecular orbitals that extend over the entire molecule and are formed by the combination of the atomic orbitals that make up the molecule.

Molecular spectroscopy (Section 12.2) The study of which frequencies of radiation are absorbed or emitted by a particular substance and the correlation of these frequencies with details of molecular structure.

Monomer (Section 29.1) From the Greek, *mono + meros,* meaning single part. The simplest nonredundant unit from which a polymer is synthesized.

Monosaccharide (Section 25.1A) A carbohydrate that cannot be hydrolyzed to a simpler carbohydrate.

D-Monosaccharide (Section 25.1C) A monosaccharide that, when written as a Fischer projection, has the —OH on its penultimate carbon to the right.

L-Monosaccharide (Section 25.1C) A monosaccharide that, when written as a Fischer projection, has the —OH on its penultimate carbon to the left.

Mutarotation (Section 25.2C) The change in specific rotation that occurs when an α or β hemiacetal form of a carbohydrate in aqueous solution is converted to an equilibrium mixture of the two forms.

($n + 1$) *rule* (Section 13.8) If a hydrogen has n hydrogens nonequivalent to it but equivalent among themselves on the same or adjacent atom(s), its ^{1}H—NMR signal is split into ($n + 1$) peaks.

Newman projection (Section 2.5A) A way to view a molecule by looking along a carbon-carbon single bond.

Nitrile (Section 18.1E) A compound containing a —C≡N (cyano) group bonded to a carbon atom.

Nitrogen rule (Section 14.3) A rule stating that the molecular ion of a compound with an odd number of nitrogen atoms has an odd m/z ratio; if zero or an even number of nitrogen atoms, the molecular ion has an even m/z ratio.

Node (Section 1.6A) A point in space where the value of a wave function is zero.

Nonbonding electrons (Section 1.2C) Valence electrons not involved in forming covalent bonds. Also called unshared pairs or lone pairs.

Nonpolar covalent bond (Section 1.2B) A covalent bond between atoms whose difference in electronegativity is less than approximately 0.5.

Nuclear magnetic resonance (NMR) spectroscopy (Introduction, Chapter 13) A spectroscopic technique that gives information about the number and types of atoms in a molecule, for example, hydrogens (^{1}H—NMR) and carbons (^{13}C—NMR).

Nucleic acid (Introduction, Chapter 28) A biopolymer containing three types of monomer units: heterocyclic aromatic amine bases derived from purine and pyrimidine, the monosaccharides D-ribose or 2-deoxy-D-ribose, and phosphoric acid.

Nucleophile (Section 6.2E and Introduction to Chapter 9) From the Greek meaning nucleus-loving. Any species that can donate a pair of electrons to form a new covalent bond; alternatively, a Lewis base.

Nucleophilic acyl substitution (Section 18.3B) A reaction in which a nucleophile bonded to the carbon of an acyl group is replaced by another nucleophile.

Nucleophilic aromatic substitution (Section 22.3) A reaction in which a nucleophile, most commonly a halogen, on an aromatic ring is replaced by another nucleophile.

Nucleophilic substitution (Introduction, Chapter 9) Any reaction in which one nucleophile is substituted for another at a tetravalent carbon atom.

Nucleophilicity (Section 9.3E) A kinetic property measured by the rate at which a nucleophile causes nucleophilic substitution on a reference compound under a standardized set of experimental conditions.

Nucleoside (Section 28.1) A building block of nucleic acids, consisting of D-ribose or 2-deoxy-D-ribose bonded to a heterocyclic aromatic amine base by a β-N-glycosidic bond.

Nucleotide (Section 28.1) A nucleoside in which a molecule of phosphoric acid is esterified with an —OH of the monosaccharide, most commonly either the 3′—OH or the 5′—OH.

Observed rotation (Section 3.7B) the number of degrees through which a compound rotates the plane of polarized light.

Octet rule (Section 1.2) Group 1A–7A elements react to achieve an outer shell of eight valence electrons.

Oil (Section 26.1B) When used in the context of fats and oils, a mixture of triglycerides that is liquid at room temperature.

Oligosaccharide (Section 25.4) A carbohydrate containing four to ten monosaccharide units, each joined to the next by a glycosidic bond.

Optical purity (Section 3.7D) The specific rotation of a mixture of enantiomers divided by the specific rotation of the enantiomerically pure substance (expressed as a percent). Optical purity is numerically equal to enantiomeric excess, but experimentally determined.

Optically active (Section 3.7) Refers to a compound that rotates the plane of plane-polarized light.

Orbital (Section 1.1) A region of space that can hold two electrons.

Order of precedence of functions (Section 16.2B) A ranking of functional groups in order of priority for the purposes of IUPAC nomenclature.

Organic synthesis (Section 7.9) A series of reactions by which a set of organic starting materials is converted to a more complicated structure.

Organometallic compound (Introduction, Chapter 15) A compound that contains a carbon-metal bond.

Ortho (*o*) (Section 21.3B) Refers to groups occupying 1,2-positions on a benzene ring.

Orthogonal (Section 1.1) Having no net overlap.

Oxidation (Section 6.5) The loss of electrons. Alternatively, either the loss of hydrogens, the gain of oxygens, or both.

Oxidative addition (Section 24.2) Addition of a reagent to a metal center causing it to add two substituents and to increase its oxidation state by two.

Oxonium ion (Section 6.3B) An ion in which oxygen bears a positive charge.

Oxymercuration-reduction (Section 6.3F) A method for converting an alkene to an alcohol. The alkene is treated with mercury(II) acetate followed by reduction with sodium borohydride.

Para (*p*) (Section 21.3B) Refers to groups occupying 1,4-positions on a benzene ring.

Part per million (ppm) (Section 13.3) Units used on NMR spectra to record chemical shift relative to the TMS standard.

Pauli exclusion principle (Section 1.1A) No more than two electrons may be present in an orbital. If two electrons are present, their spins must be paired.

Peptide bond (Section 27.3) The special name given to the amide bond formed between the α-amino group of one amino acid and the α-carboxyl group of another amino acid.

Pericyclic reaction (Section 20.4) A reaction that takes place in a single step, without intermediates, and involves a cyclic redistribution of bonding electrons.

Phasing (Section 1.6A) The sign of the wave function at particular coordinates in space, either plus or minus. Phasing is often represented by colors, such as red or blue.

Phenol (Section 21.4A) A compound that contains an —OH bonded to a benzene ring; a benzenol.

Phenyl group (Introduction, Chapter 5) A group derived by removing an H from benzene; abbreviated C_6H_5— or Ph—.

Phospholipid (Section 26.5A) A lipid containing glycerol esterified with two molecules of fatty acid and one molecule of phosphoric acid.

Photolysis (Section 15.3A) Cleavage by light.

Photons (Section 12.1) An alternative way to describe electromagnetic radiation as a stream of particles.

Pi (π) bond (Section 1.7D) A covalent bond formed by the overlap of parallel $2p$ orbitals.

Pi (π) molecular orbital (Section 1.7D) A molecular orbital formed by overlapping parallel $2p$ orbitals on adjacent atoms; its electron density lies above and below the line connecting the atoms.

Plane of symmetry (Section 3.1) An imaginary plane passing through an object dividing it so that one half is the mirror image of the other half.

Plane-polarized light (Section 3.7A) Light oscillating in only parallel planes.

Plastic (Section 29.1) A polymer that can be molded when hot and retains its shape when cooled.

β-Pleated sheet (Section 27.6B) A type of polypeptide secondary structure in which sections of polypeptide chains are aligned parallel or antiparallel to one another.

Polar covalent bond (Section 1.2B) A covalent bond between atoms whose difference in electronegativity is between approximately 0.5 and 1.9.

Polarimeter (Section 3.7B) An instrument for measuring the ability of a compound to rotate the plane of plane-polarized light.

Polarizability (Section 8.3B) A measure of the ease of distortion of the distribution of electron density about an atom or group in response to interaction with other molecules or ions. Fluorine, which has a high electronegativity and holds its electrons tightly, has a very low polarizability. Iodine, which has a lower electronegativity and holds its electrons less tightly, has a very high polarizability.

Polyamide (Section 29.5A) A polymer in which each monomer unit is joined to the next by an amide bond, as, for example, nylon 66.

Polycarbonate (Section 29.5C) A polyester in which the carboxyl groups are derived from carbonic acid.

Polyester (Section 29.5B) A polymer in which each monomer unit is joined to the next by an ester bond, as, for example, poly(ethylene terephthalate).

Polymer (Section 29.1) From the Greek, *poly + meros*, meaning many parts. Any long-chain molecule synthesized by linking together many single parts called monomers.

Polynuclear aromatic hydrocarbon (PAH) (Section 21.3C) A hydrocarbon containing two or more fused benzene rings.

Polypeptide (Section 27.3) A macromolecule containing many amino acid units, each joined to the next by a peptide bond.

Polysaccharide (Section 25.4) A carbohydrate containing a large number of monosaccharide units, each joined to the next by one or more glycosidic bonds.

Polyunsaturated fatty acid (Section 26.1A) A fatty acid with two or more carbon-carbon double bonds in its hydrocarbon chain.

Polyunsaturated triglyceride (Section 26.1B) A triglyceride having several carbon-carbon double bonds in the hydrocarbon chains of its three fatty acids.

Polyurethane (Section 29.5D) A polymer containing the —NHCO$_2$— group as a repeating unit.

Potential energy (Section 1.1B) The energy that can be released if given an opportunity.

Primary (1°) amine (Section 1.3B) An amine in which nitrogen is bonded to one carbon and two hydrogens.

Primary structure of nucleic acids (Section 28.2A) The sequence of bases along the pentose-phosphodiester backbone of a DNA or RNA molecule read from the 5' end to the 3' end.

Primary structure of proteins (Section 27.4) The sequence of amino acids in the polypeptide chain, read from the *N*-terminal amino acid to the *C*-terminal amino acid.

Principle of microscopic reversibility (Section 10.6) This principle states that the sequence of transition states and reactive intermediates in the mechanism of any reversible reaction must be the same, but in reverse order, for the reverse reaction as for the forward reaction.

Prostaglandin (Section 26.3) A member of the family of compounds having the 20-carbon skeleton of prostanoic acid.

Protecting group (Section 11.6) Reversibly creating an unreactive group for the purpose of preventing a functional group from potentially reacting to give an unwanted product or products.

Protic acid (Section 4.7) An acid that is a proton donor in an acid-base reaction.

Protic solvent (Section 9.3D) A solvent that is a hydrogen-bond donor. Common protic solvents are water, low-molecular-weight alcohols, and low-molecular weight carboxylic acids.

Pyranose (Section 25.2A) A six-membered cyclic form of a monosaccharide.

Quantum mechanics (Section 1.6A) The branch of science that studies the interaction of matter and radiation.

Quantized (Section 1.1) Having specific values for energy and momentum.

Quaternary (4°) ammonium ion (Section 23.2B) An ion in which nitrogen is bonded to four carbons and bears a positive charge.

Quaternary structure (Section 27.6D) The arrangement of polypeptide monomers into a noncovalently bonded aggregate.

R (Section 3.3) From the Latin, *rectus*, straight, correct; used in the *R,S* convention to show that the order of priority of groups on a chiral center is clockwise.

R,S System (Section 3.3) A set of rules for specifying absolute configuration about a chiral center; also called the Cahn-Ingold-Prelog system.

Racemic mixture (Section 3.7C) A mixture of equal amounts of two enantiomers.

Radical (Section 8.3D) Any chemical species that contains one or more unpaired electrons.

Radical cation (Section 14.1) A species formed when a neutral molecule loses one electron; it contains both an odd number of electrons and a positive charge.

Radical inhibitor (Section 8.7) A compound such as a phenol that selectively reacts with radicals to remove them from a chain reaction and terminate the chain.

Raman spectroscopy (Section 12.3B) A vibrational molecular spectroscopy that is complementary to infrared (IR) spectroscopy in that infrared inactive vibrations are seen in Raman spectroscopy.

Rate-determining step (Section 6.2A) The step in a multistep reaction sequence that crosses the highest energy barrier.

Reaction coordinate diagram (Section 4.5B) A graph showing the energy changes that occur during a chemical reaction; energy is plotted on the vertical axis and reaction progress is plotted on the horizontal axis.

Reaction mechanism (Section 4.5) A step-by-step description of how a chemical reaction occurs.

Reactive intermediate (Section 6.2A) A high-energy species formed between two successive reaction steps, that lies in an energy minimum between the two transition states.

Rearrangement (Section 6.3C) A change in connectivity of the atoms in a product compared with the connectivity of the same atoms in the starting material.

Reducing sugar (Section 25.3C) A carbohydrate that reacts with an oxidizing agent to form an aldonic acid. In this reaction, the carbohydrate reduces the oxidizing agent.

Reduction (Section 6.5) The gain of electrons. Alternatively, either the gain of hydrogen, loss of oxygen, or both.

Reductive amination (Section 16.11D) A method for preparing substituted amines by treating an aldehyde or ketone with an amine in the presence of a reducing agent.

Reductive elimination (Section 24.2) Elimination of two substituents at a metal center, causing the oxidation state of the metal to decrease by two.

Regioselective reaction (Section 6.3A) An addition or substitution reaction in which one of two or more possible products is formed in preference to all others that might be formed.

Resolution (Sections 3.9 and 14.2A) Separation of a racemic mixture into its enantiomers; in mass spectrometry, a measure of how well a mass spectrometer separates ions of different mass.

Resonance (Section 1.8A) A theory that many molecules and ions are best described as a hybrid of several Lewis structures.

Resonance energy (Section 21.1C) The difference in energy between a resonance hybrid and the most stable of its hypothetical contributing structures in which electrons are localized on particular atoms and in particular bonds.

Resonance hybrid (Section 1.8A) A molecule, ion, or radical described as a composite of a number of contributing structures.

Resonance in NMR spectroscopy (Section 13.3) The absorption of electromagnetic radiation by a precessing nucleus and the resulting "flip" of its nuclear spin from the lower energy state to the higher energy state.

Restriction endonuclease (Section 28.5A) An enzyme that catalyzes the hydrolysis of a particular phosphodiester bond within a DNA strand.

Retrosynthesis (Section 7.9A) A process of reasoning backwards from a target molecule to a suitable set of starting materials.

Ribosomal RNA (rRNA) (Section 28.3A) A ribonucleic acid found in ribosomes, the sites of protein synthesis.

Ring current (Section 13.7C) An applied magnetic field causes the π electrons of an aromatic ring to circulate, giving rise to the so-called ring current and an associated magnetic field that opposes the applied field in the middle of the ring but reinforces the applied field on the outside of the ring.

S (Section 3.3) From the Latin, *sinister*, left; used in the *R,S* convention to show that the order of priority of groups on a chiral center is counterclockwise.

Sanger dideoxy method (Section 28.5B) A method developed by Frederick Sanger for sequencing DNA molecules.

Saponification (Section 18.4C) Hydrolysis of an ester in aqueous NaOH or KOH to an alcohol and the sodium or potassium salts of carboxylic acids.

Schiff base (Section 16.8A) An alternative name for an imine.

Secondary (2°) amine (Section 1.3B) An amine in which nitrogen is bonded to two carbons and one hydrogen.

Secondary structure of nucleic acids (Section 28.2B) The ordered arrangement of nucleic acid strands.

Secondary structure of proteins (Section 27.6B) The ordered arrangements (conformations) of amino acids in localized regions of a polypeptide or protein.

Shell (Section 1.1) A region of space around a nucleus that can be occupied by electrons, corresponding to a principal quantum number.

Shielding in NMR (Section 13.3) Also called diamagnetic shielding; the term refers to the reduction in magnetic field strength experienced by a nucleus underneath electron density induced to circulate when the molecule is placed in a strong magnetic field.

Sigma (σ) molecular orbital (Section 1.7A) A molecular orbital in which electron density is concentrated between two nuclei, along the axis joining them, and is cylindrically symmetrical.

Signal (Section 13.3) A recording in an NMR spectrum of a nuclear magnetic resonance.

Signal splitting in NMR (Section 13.8) Spin-spin coupling with adjacent nuclei split NMR signals depending on the extent of coupling and the number of adjacent equivalent nuclei.

S_N1 reaction (Section 9.2B) A unimolecular nucleophilic substitution reaction.

S_N2 reaction (Section 9.2A) A bimolecular nucleophilic substitution reaction.

Soap (Section 26.2A) A sodium or potassium salt of a fatty acid.

Solvolysis (Section 9.2B) A nucleophilic substitution in which the solvent is also the nucleophile.

sp Hybrid orbital (Section 1.7B) A hybrid atomic orbital formed by the combination of one s atomic orbital and one 2p atomic orbital.

sp^2 Hybrid orbital (Section 1.7B) A hybrid atomic orbital formed by the combination of one s atomic orbital and two 2p atomic orbitals.

sp^3 Hybrid orbital (Section 1.7 B) A hybrid atomic orbital formed by the combination of one s atomic orbital and three 2p atomic orbitals.

Specific rotation (Section 3.7B) The observed rotation of the plane of polarized light when a sample is placed in a tube 1.0 dm in length and at a concentration of 1 g/mL for a solution. For a pure liquid, concentration is expressed in g/mL (density).

Spin-spin coupling (Section 13.9) An interaction in which nuclear spins of adjacent atoms influence each other and lead to the spitting of NMR signals.

Staggered conformation (Section 2.5A) A conformation about a carbon-carbon single bond in which the atoms or groups on one carbon are as far apart as possible from atoms or groups on an adjacent carbon.

Step-growth polymerization (Section 29.5) A polymerization in which chain growth occurs in a stepwise manner between difunctional monomers as, for example, between adipic acid and hexamethylenediamine to form nylon 66. Also called condensation polymerization.

Stereocenter (Sections 2.6A and 3.2) An atom, most commonly carbon, about which exchange of two groups produces a stereoisomer. Chiral centers are one type of stereocenter.

Stereochemistry (Introduction, Chapter 3) The study of three-dimensional arrangements of atoms in molecules.

Stereoisomers (Sections 2.6 and 3.2) Isomers that have the same molecular formula and the same connectivity of their atoms but a different orientation of their atoms in space.

Stereoselective reaction (Section 6.3D) A reaction in which one stereoisomer is formed in preference to all others.

A stereoselective reaction may be enantioselective or diastereoselective, as the case may be.

Stereospecific reaction (Section 6.7A) A special type of stereoselective reaction in which the stereochemistry of the product is dependent on the stereochemistry of the starting material.

Steric hindrance (Section 9.3B) The ability of groups, because of their size, to hinder access to a reaction site within a molecule.

Steric strain (Section 2.5A) The strain that arises when nonbonded atoms separated by four or more bonds are forced closer to each other than their atomic (contact) radii would allow. Steric strain is also called non-bonded interaction strain, or van der Waals strain.

Steroid (Section 26.4) A plant or animal lipid having the characteristic tetracyclic ring structure of the steroid nucleus, namely three six-membered rings and one five-membered ring.

Substitution (Section 8.4) A reaction in which an atom or group of atoms in a compound is replaced by another atom or group of atoms.

Sulfide (Section 11.12) The sulfur analog of an ether; a molecule containing a sulfur atom bonded to two carbon atoms. Sulfides are also called thioethers.

Syndiotactic polymer (Section 29.6C) A polymer with alternating R and S configurations at the chiral centers along its chain, as, for example, syndiotactic polypropylene.

Syn stereoselective (Section 6.4) The addition of atoms or groups of atoms to the same face of a carbon-carbon double bond.

Tautomers (Section 7.7A) Constitutional isomers in equilibrium with each other that differ in the location of a hydrogen atom and a double bond relative to a heteroatom, most commonly O, N, or S.

Telechelic polymer (Section 29.6D) A polymer in which its growing chains are terminated by formation of new functional groups at both ends of its chains. These new functional groups are introduced by adding reagents, such as CO_2 or ethylene oxide, to the growing chains.

C-Terminal amino acid (Section 27.3) The amino acid at the end of a polypeptide chain having the free —COOH group.

N-Terminal amino acid (Section 27.3) The amino acid at the end of a polypeptide chain having the free —NH_2 group.

Terpene (Section 5.4) A compound whose carbon skeleton can be divided into two or more units identical with the carbon skeleton of isoprene.

Tertiary (3°) amine (Section 1.3B) An amine in which nitrogen is bonded to three carbons.

Tertiary structure of nucleic acids (Section 28.2C) The three-dimensional arrangement of all atoms of a nucleic acid, commonly referred to as supercoiling.

Tertiary structure of proteins (Section 27.6C) The three-dimensional arrangement in space of all atoms in a single polypeptide chain.

Tesla (T) (Section 13.2) The SI unit for magnetic field strength.

Thermochemistry (Section 4.5) The study of the energy of chemical structures.

Thermodynamic control (Section 19.8A) Experimental conditions that permit the establishment of equilibrium between two or more products of a reaction. The composition of the product mixture is determined by the relative stabilities of the products.

Thermolysis (Section 15.3A) Cleavage by heating.

Thermoplastic (Section 29.1) A polymer that can be melted and molded into a shape that is retained when it is cooled.

Thermoset plastic (Section 29.1) A polymer that can be molded when it is first prepared, but once cooled, hardens irreversibly and cannot be remelted.

Thiol (Section 10.9A) A compound containing an —SH (sulfhydryl) group bonded to an sp^3-hybridized carbon.

Tollens' reagent (Section 16.10A) A solution prepared by dissolving Ag_2O in aqueous ammonia; used for selective oxidation of an aldehyde to a carboxylic acid.

Torsional strain (Section 2.5A) Strain that arises when nonbonded atoms separated by three bonds are forced from a staggered conformation to an eclipsed conformation. Torsional strain is also called eclipsed-interaction strain.

Trans (Section 2.6A) A prefix meaning across from.

Transesterification (Section 18.5C) Exchange of the —OR or —OAr group of an ester for another —OR or —OAr group.

Transfer RNA (tRNA) (Section 28.3B) A ribonucleic acid that carries a specific amino acid to the site of protein synthesis on ribosomes.

Transition state (Section 4.5A) The highest energy point on a reaction coordinate diagram. The chemical structure at this point is commonly called an activated complex.

Triglyceride (triacylglycerol) (Section 26.1) An ester of glycerol with three fatty acids.

Triol (Section 10.1B) A compound containing three hydroxyl groups.

Tripeptide (Section 27.3) A molecule containing three amino acid units, each joined to the next by a peptide bond.

Twist-boat conformation (Section 2.5B) A nonplanar conformation of a cyclohexane ring that is twisted from and slightly more stable than a boat conformation.

Unimolecular reaction (Section 9.2B) A reaction in which only one species is involved in the rate-determining step.

Unsaturated hydrocarbon (Introduction, Chapter 5) A hydrocarbon containing one or more carbon-carbon double or triple bonds. The three classes of unsaturated hydrocarbons are alkenes, alkynes, and arenes.

Upfield (Section 13.4) A signal of an NMR spectrum that is shifted toward the right (smaller chemical shift) on the chart paper.

Valence Bond Theory (Section 1.7B) A model of bonding that places electron pairs between adjacent atoms to create bonds.

Valence electrons (Section 1.1C) Electrons in the valence (outermost) shell of an atom.

Valence shell (Section 1.1C) The outermost occupied electron shell of an atom.

Valence-shell electron-pair repulsion (VSEPR) (Section 1.4) A method for predicting bond angles based on the idea that electron pairs repel each other and keep as far apart as possible.

van der Waals forces (Section 8.3B) A group of intermolecular attractive forces including dipole-dipole, dipole-induced dipole, and induced dipole-induced dipole (dispersion) forces.

van der Waals radius (Section 8.3B) The minimum distance of approach to an atom that does not cause nonbonded interaction strain.

Vibrational infrared region (Section 12.3A) The portion of the infrared region that extends from 4000 to 400 cm^{-1}.

Vicinal coupling (Section 13.9) A common type of spin-spin coupling involving the H atoms on two C atoms that are bonded to each other.

Vinyl group (Section 5.2B) a —CH=CH_2 group.

Vinylic carbocation (Section 7.6B) A carbocation in which the positive charge is on one of the carbons of a carbon-carbon double bond.

Watson-Crick model (Section 28.2B) A double-helix model for the secondary structure of a DNA molecule.

Wave function (Section 1.6A) A solution to a set of equations that defines the energy of an electron in an atom and the region of space it may occupy.

Wavelength (λ) (Section 12.1) The distance between consecutive peaks on a wave.

Wavenumbers ($\overline{\nu}$) (Section 12.3A) The frequency of electromagnetic radiation expressed as the number of waves per centimeter, with units cm^{-1} (read: reciprocal centimeters).

Williamson ether synthesis (Section 11.4A) A general method for the synthesis of dialkyl ethers by an S_N2 reaction between a haloalkane and an alkoxide ion.

Wolff-Kishner reduction (Section 16.11E) Reduction of the C=O group of an aldehyde or ketone to a CH_2 group using hydrazine and a base.

Ylide (Section 16.6) A neutral molecule with positive and negative charges on adjacent atoms.

Z (Section 5.2C) From the German, *zusammen,* meaning opposite. Specifies that groups of higher priority on the carbons of a double bond are on the same side.

Zaitsev's rule (Section 9.5) A rule stating that the major product of a β-elimination reaction is the most stable alkene; that is, it is the alkene with the greatest number of substituents on the carbon-carbon double bond.

Zwitterion (Section 27.1A) An internal salt of an amino acid; the carboxylate is negatively charged, and the ammonium group is positively charged.

Index

Key: **Boldface** indicates a glossary entry; *italics* indicate a figure; *t* indicates a table.

of carboxylic acids, 544–548
of α-hydrogens, 516–518
of imides, 568–570
of phenols, 810–813, 811*t*
of side-chain carboxyl groups, 869
of sulfonamides, 568–570
of thiols, 363
Acids, 132–158. *See also* Nucleic acids
aprotic, **155**
Arrhenius, 132–133
Brønsted-Lowry, **133**–139
conjugate, **133**
fatty, 172–173
Lewis, **153**–155, 182–183
molecular structure and, 172–173
protic, **155**
reactions of amines with, 769–772
relative strengths of, 139–142, 140*t*
trans fatty, 216
Acrolein, *486, 676*
Acrylic acid, *539*
Acrylonitrile, *498, 930t, 936t*
Activated complex, 145
Activating groups, **739**
Activating-deactivating effects, theory of, 744–746
Acyclic molecules, with two or more chiral centers, 108–114
Acyclovir, *902*
Acyl (RCO—) groups, **562**
Acylation, **628**
of benzene, *729,* 733–737
of enamines, 628–629
Friedel-Crafts, 735–736
Acylbenzene, *729*
Acylium ions, 735–736
Adamantane, 88, *88*
Addition reactions, 176*t*
addition-elimination, 748–749
anti, **242**
bromination, 176*t*
bromo(halo)hydrin, 176*t*
of carbon nucleophiles, 492–498
conjugate, **638**–646
cyclo-, **672,** 674
diol formation (oxidation), 176*t*
electrophilic, **186**–204
of Grignard reagents, 492–494
halogenation (*See* Halogenation reactions)
halohydrin formation, **200**–202
hydration, 176*t,* 193–195
hydroboration, 176*t,* **205**–208
hydrochlorination, 176*t*
hydrohalogenation, 231
of nitrogen nucleophiles, 510–516
nucleophilic acyl, **490,** 570–571
of oxygen nucleophiles, 502–510
oxymercuration, 176*t,* 202–205
radical, 274–277
syn, **241**
thermodynamics of, 178
Adenine (A), *766, 767, 825, 897, 900*
3'-Adenine monophosphate, *37*
Adenosine diphosphate (ADP), 358, 848

Adenosine triphosphate (ATP), 624, *898*
-adien-, 168
Adipamide, *294*
Adipic acid, *521, 539, 540, 922, 923, 924*
synthesis of hexamethylenediamine, 924, *924*
Adipoyl dichloride, *562*
A-DNA, **904**
ADP. *See* Adenosine diphosphate (ADP)
Aging, telomeres and, 906
AIBN. *See* Azoisobutyronitrile (AIBN)
-al suffix, 63
Alanine, *867, 869t, 873, 881, 894*
D-Alanine, *866*
L-Alanine, *514, 540, 866*
Alanine transaminase, *514*
Alcohol dehydrogenase, 891*t*
Alcohol group, order of precedence, 488*t*
Alcohols, **16**–18, 330–367
acidity of, 337–338, 337*t*
addition of, to aldehydes and ketones, 502–508
addition of, to alkenes, 374
basicity of, 337–338
conversion of, to haloalkanes, 339–443
conversion of, to sulfonates, 343–345
dehydration of, acid-catalyzed, 345–350, 384–385
hydroboration-oxidation and formation of, 205
infrared spectra of, 407, 407*t*
mass spectra of, 462–463
NMR spectra of, 444
nomenclature of, 331–333
oxidation of, 352–360
physical properties of, 333–337, 334*t,* 542*t*
pinacol rearrangement in, 350–352
primary (1°), **17,** *18,* **332**
reactions of, with active metals, 338
reactions of, with carboxylic acid derivatives, 581–591
secondary (2°), **17,** *18,* **332**
structure of, 331
tertiary (3°), **17,** **332**
unsaturated, *333*
Aldehyde group, order of precedence, 488*t*
Aldehydes, **19**–20, 485–537
addition of carbon nucleophiles to, 492–498
addition of nitrogen nucleophiles to, 510–516
addition of oxygen nucleophiles to, 502–510
bonding in, 485–486
formation of imine from, 511–512
hydration of alkynes to, 236
infrared spectra of, 409–410, 412*t*
keto-enol tautomerism in, 516–519
mass spectra of, 463–464
NMR spectra of, 445
nomenclature of, 486–489
oxidation of, 520–521
physical properties of, 489–490, 490*t,* 542*t*
reactions of, 586–588
reduction of, 521–527

structure of, 485–486
Wittig reaction and, **498**–501
Alder, Kurt, 674
Aldimine, *515*
Alditols, **825**–826
D-Aldohexoses, 818, 818*t*
Aldol condensations, in biological world, 623–625
intramolecular, **643**
mechanisms, 744–748
retrosynthetic analysis, 644
Aldolase, 891*t*
Aldonic acids, 826–**827**
Aldopentoses, 821
D-Aldopentoses, 818*t*
Aldoses, **816**
Aldosterone, *854*
D-Aldotetroses, 818*t*
Aliphatic amines, 764
primary, 776–779
secondary, 775–776
tertiary, 775
Aliphatic carboxylic acids, 540, 541*t*
Aliphatic diazonium ions, 777
Aliphatic hydrocarbons, **55,** *56*
Alkaloids, **756**
Alkanes, 55–98
acidity of, 153*t*
boiling points of, 88, 89*t*
bond lengths and bond strengths in, 51
conformations of, 67–72
constitutional isomerism in, 57–59
cyclo- (*see* Cycloalkanes)
density of, 89*t,* 90
halogenation of, 259–262
heats of combustion of, 92–93
infrared spectra of, 404–405, 405*t*
mass spectra of, 459–461
melting points of, 89*t,* 90
molecular formula, 57
NMR spectra of, 443
nomenclature of, 59–64
oxidation of, 91–92
physical properties of, 88–91, 89*t*
polymerized by anionic and cationic chain-gravity mechanisms, 936*t*
reactions of, 91–93
sources/importance of, 93–96
structure of, 55–57
Alkenes, 159–174
acidity of, 153*t*
addition of alcohols to, 374
addition of hydrogen halides to, 186–193
addition of water to, 193–195
addition reactions of, 175–176, 176*t*
alkynes from, 230–232
anionic polymerization of, 937
bond lengths and bond strengths in, 51
bridgehead, 767
carbocation rearrangement in addition of hydrogen chloride to, 196
cationic polymerization of, 940, 941
cis, trans isomerism in, 163, 165, *210*
designating configuration in, 165–166
electrophilic additions to, **186**–205
epoxidation of, 382

Nitrogen rule, **459**
Nitromethane, *615, 616*
1-(Nitromethyl)cyclohexanol, *616*
Nitronium ion, **731**
m-Nitrophenol, *713*
p-Nitrophenol, *713*
N-Nitrosamines, 775–776
N-Nitrosodimethylamine, *776*
N-Nitrosopiperidine, *775*
N-Nitrosopyrrolidine, *776*
Nitrosyl cation
　formation of, *774*
　reaction of, with secondary amine, 775
4-Nitrotoluene, 981
Nitrous acid, reactions of, with amines, 774–782
NMR spectroscopy. *See* Nuclear magnetic resonance (NMR) spectroscopy
Nodal plane, **27**, *29, 31*
Node, **27**
NOE. *See* Nuclear Overhauser enhancement (NOE)
Nomenclature
　of acid anhydrides, 562–563
　of acid halides, 562
　of aldehydes, 485–486
　of alkanes, 59–64
　of alkenes, 163–196
　of alkyl groups, 61*t*
　of alkynes, 267–268
　of amides, 356–366
　of amines, 756–759
　of benzene, 706–707
　of carboxylic acids, 539–542, 541*t*
　of chiral centers, 102–105
　of cycloalkanes, 64–65
　of epoxides, 379–380
　of esters, 563–565
　of ethers, 369–370
　of imides, 566
　IUPAC system, 59, 63–64
　of ketones, 485–486
　of monosaccharides, 816
　of nitriles, 567–568
　of phenols, 709–710
　of polymers, 919–920
　R,S system of, 105–108
　of sulfides, 391
　of thiols, 360–361
Nonadecane, 57*t*
Nonane, 57*t*
　infrared spectrum of, *398*
　physical properties of, 89*t*
Nonbonding electrons, **12**
Non-Markovnikov addition, radical-initiated, 275–276
Non-Markovnikov hydration, 205, **206**
Nonpolar covalent bonds, **10,** 10*t*
Nonpolar molecules, 24–26
Nonpolar side chains, 867*t*
Nonpolar solvent, **296**
Nonsteroidal antiinflammatory drugs (NSAIDs), 543, 850
Norbornane, *66*
Norbornene, *167, 942*
Norepinephrine, *769*

Norepinephrine hydrochloride, *769*
Norethindrone, *853*
Notation, polymer, 919–920
Noyori, Ryoji, 383
NSAIDs. *See* Nonsteroidal antiinflammatory drugs (NSAIDs)
N-terminal amino acids, **874,** 878–879
　cleavage of, 878–879
Nuclear magnetic "resonance," **418,** 418–420
Nuclear magnetic resonance (NMR) spectrometer, *420,* 420–421
Nuclear magnetic resonance (NMR) spectroscopy, **415**–448
　¹³C-, 415, 419, **441**–442
　chemical shifts and, **424**–428
　equivalent hydrogens and, 421–424
　Fourier transform, 420
　interpretation of, 443–448
　(*n* + 1) rule and, 428–430
　nuclear magnetic resonance and, 418–420
　nuclear spin states and, 415–416
　orientation of nuclear spins in an applied magnetic field, 416–417
　signal areas and, 423–424
　signal splitting and, 428–430
　solving spectral problems, 446–448
Nuclear Overhauser enhancement (NOE), 445
Nuclear spin. *See* Spin (nuclear)
Nuclear spin quantum number, **416,** 416*t*
Nucleic acids, 869–915
　DNA, 896–897, 899–905
　genetic code and, 908–910
　nucleosides and nucleotides in, 897–899
　RNA, 896–897, 905–908, 905*t*
　sequencing of, 910–915
Nucleophiles, **155, 182**–183, 186, 188, **281**–281
　carbon, 492–498
　moderate, **300**
　nitrogen, 606–612
　oxygen, 502–510
　strong, **300**
　structure of, 298–301, 319–320
　weak, **300**
Nucleophilic acyl additions, **490,** 570–571
Nucleophilic carbenes, 809
　ring-closing alkene metathesis using, 810–811
Nucleophilic ring opening, 386–387
Nucleophilic substitution reactions, **281**–282
　acyl, **562,** *628*
　by addition-elimination, 748–749
　aliphatic, 284–287
　analysis of, 303–305
　aromatic, **692**–693
　β-elimination *vs.*, 316–320
　in haloalkanes, 282–284, 283*t*
　in halohydrins, 380–381
　ring opening, 386–387
Nucleophilicity, **298**–301, 299*t*
　effect of shape on, 301
　solvation effects on, 300–301

Nucleosides, **897**–899
Nucleotides, **897**–899
　coding sequence, 880
Nucleus, atomic, 1, *2*
Number average molecular weight (*M*ₙ), **920**
Nylon 6, 924, *924*
Nylon 6, 922
Nylon 66, 521, 738, *918, 922,* 922–924

O

Observed rotation, **120**
Octanal, *238*
Octanamide, *599*
1-Octanamine, *599*
Octane
　boiling point of, *126*
　formulas for, 57*t*
　heat of combustion of, 92, 92*t*
　mass spectrum of, *454*
　octane rating of, 95
　physical properties of, 89*t*
Octane rating, 95
2,7-Octanedione, *616*
Octanoic acid, 541*t*
(*R*)-2-Octanol, *589*
(*S*)-2-Octanol, *343, 344*
2-Octanone, *464*
4-Octanone, *645, 646*
(3*E*,5*E*)-1,3,5,7-Octatetraene, 671*t*
6-Octen-1-amine, *602*
1-Octene, *271, 783*
trans-2-Octene, *783*
trans-4-Octene, *242*
6-Octenenitrile, *602*
Octet rule, **6**
　exceptions to, 16
Octyl acetate, 553
2-Octyl acetate, *345*
2-Octyl *p*-toluenesulfonate, *589*
(*R*)-2-Octyl tosylate, *589*
(*S*)-2-Octyl tosylate, *152, 344, 345*
2-Octylcyanoacrylate, 939
1-Octyne, *237, 238*
　infrared spectrum of, 406, *406*
　physical properties of, 230*t*
4-Octyne, *242*
Odor
　of carboxylic acids, 544
　thiols and, 361
-oic acid suffix, 63
Oils, **844,** 844*t, 845*
　cis double bonds in, 172–173
　essential, 170–172, *171*
　trans fatty acids, 216
　tropical, **844**
-ol suffix, 63
Oleic acid, *172, 173,* 843*t,* 844*t*
Oligopeptides, **874**
Oligosaccharides, **830**–833
1,4-addition, 661–667
1,2-addition, 661–667
-one suffix, 63
1,2-shift, 195, 223, **349,**
Opium poppy, 755
OPSIN, *512*
Optical activity, 119–121